Physicochemical Properties of Ionic Liquid Mixtures

Suojiang Zhang • Qing Zhou
Xingmei Lu • Yuting Song
Xinxin Wang
Editors

Physicochemical Properties of Ionic Liquid Mixtures

Springer

Editors
Suojiang Zhang
Beijing Key Laboratory of Ionic Liquids
Clean Process, Institute of Process
Engineering, Chinese Academy of Sciences
Beijing, China People's Republic

College of Chemistry and Chemical
Engineering, University of Chinese
Academy of Sciences
Beijing, China People's Republic

Xingmei Lu
Beijing Key Laboratory of Ionic Liquids
Clean Process, Institute of Process
Engineering, Chinese Academy of Sciences
Beijing, China People's Republic

College of Chemistry and Chemical
Engineering, University of Chinese
Academy of Sciences
Beijing, China People's Republic

Xinxin Wang
Beijing Key Laboratory of Ionic Liquids
Clean Process, Institute of Process
Engineering, Chinese Academy of Sciences
Beijing, China People's Republic

Qing Zhou
Beijing Key Laboratory of Ionic Liquids
Clean Process, Institute of Process
Engineering, Chinese Academy of Sciences
Beijing, China People's Republic

College of Chemistry and Chemical
Engineering, University of Chinese
Academy of Sciences
Beijing, China People's Republic

Yuting Song
Beijing Key Laboratory of Ionic Liquids
Clean Process, Institute of Process
Engineering, Chinese Academy of Sciences
Beijing, China People's Republic

ISBN 978-94-017-7571-7 ISBN 978-94-017-7573-1 (eBook)
DOI 10.1007/978-94-017-7573-1

Library of Congress Control Number: 2016954242

Printed on acid-free paper

This Springer imprint is published by Springer Nature
The registered company is Springer Science+Business Media B.V.
The registered company address is: Van Godewijckstraat 30, 3311 GX Dordrecht, The Netherlands

Preface

In the past two decades over 70,000 ionic liquid publications have been emerging, embracing a large number of physical chemical properties data. However, these hundreds of thousands of data are dispersed in more than hundreds of journals, reports, books and patent documents, which cause great difficulties for scientific researches and industrial applications. Therefore, it is imperative and necessary to collect those highly scattered data for providing comprehensive and systemic information. Professor Suojiang Zhang with his group is dedicating to such an important work for about 15 years, and published a handbook of pure ionic liquids entitled *Ionic Liquids: Physicochemical Properties* in 2009. In the present book, he attempts to present physicochemical properties data in particular for mixed systems, which are becoming more and more important with growing of the variety of functional ionic liquid systems and their applications.

This handbook featured not only abundant data but also measurement methods, fitting and prediction models. The work will be of huge value, on the one hand, it provides the possibility to extract common structure–property relationship information from various types of ionic liquids; on the other hand, it will provide indispensable information for validating the industrial practices based on ionic liquids. I hope that this work could play a fundamental role in stimulating and promoting more innovations in basic theories and industrial applications of ionic liquids, and finally contributing to the green dream of chemistry and chemical engineering.

Feb. 3, 2015 Mingyuan He
Beijing, China

Contents

Part I

Ionic Liquids Database

Ionic Liquids Database

1

1.1 Introduction

The nature of ionic liquids has been 'hot' with many attracted to study these remarkable substances which challenge our classical theory of the liquid state and provide opportunity for reform of traditional product engineering. In the past decade, as the research frontiers and focuses in the field of international green chemistry, its fundamental and applied research have made rapid progress, demonstrating their important scientific value and the huge potential application. Especially, ionic liquids have been designed as greener solvents to replace conventional volatile solvents and are showing increasingly promising perspectives in the diverse fields of synthesis, catalysis/biocatalysis, materials science, electrochemistry, and separation technology both at the laboratory and the industrial scale. For any kind of novel and promising substances, its basic and applied research exist side by side and play a part together; ionic liquids are no exception. With the persistent expansion and deepening of the application scope and level, the demand for the fundamental knowledge becomes more and more urgent, for instance, the properties and behavior law of correlative ionic liquids systems/mixtures in real engineering processes. Therefore, the reports on the properties of ionic liquids mixtures dramatically increased in just a few years, their total data points reached hundreds of thousands of properties data, including volume properties, transport properties, electrochemical properties, phase equilibrium, and so on. The above researches are very meaningful, which not only can obtain structure-activity relationship information to construct a new system of scientific theories but also have a certain role in promoting and facilitating for the related industrial development.

However, these hundreds of thousands of properties data have been dispersed in more than hundred kinds of international journals, reports, and patents. Therefore, it is necessary to collect, analyze data of all the properties of ionic liquids systems, and summarize laws, in order to adapt to the requirement of development and provide useful references for scientific workers. Derived from this original intention, we established the

S. Zhang et al., *Physicochemical Properties of Ionic Liquid Mixtures*,
DOI 10.1007/978-94-017-7573-1_1

original ionic liquids physicochemical properties database in 2004, updated annually, expanding from the pure ionic liquids to ionic liquids mixtures, and based on this database, we have made a series researches of structure-activity relationship.

A comprehensive database on physicochemical properties of pure ionic liquids and its mixtures, which has been collected from more than 1000 dependable literature sources in the period from 1982 to 2009, is available. There are more than 19,000 data points on the 29 kinds of physicochemical properties for 2311 available pure ILs, from which 907 kinds of cations and 229 kinds of anions were extracted. And there are more than 65,050 data points on the 39 kinds of physicochemical properties for 1388 mixture systems. These systems include 1047 binary, 336 ternary, 3 quaternary, and 2 quinary mixed systems. Almost all kinds of important physicochemical properties of ionic liquids mixtures were collected in this database. It covered 39 kinds of different physicochemical properties, such as crystallization temperature, density, viscosity, surface tension, conductivity, sound of speed, refractive index, phase equilibrium, isentropic compressibility, isothermal compressibility, expansibility, vapor pressure, thermodynamic functions of solution, etc.

In 2009, on the basis of database the physicochemical properties handbook of pure ionic liquids, entitled *Ionic Liquids: Physicochemical Properties*, was published by Elsevier. Herein, we planned a handbook of ionic liquid mixture. The handbook is divided into two parts. The first part mainly introduces the properties of ionic liquid mixtures, such as volumetric properties, transport properties, phase equilibria, etc., which includes the methods of measurements, fitting, and prediction. The second part contains nine chapters which are about physical and chemical properties data of ionic liquids mixtures; according to the classification of cations structure, the chapters respectively contain imidazolium, triazolium, pyrrolidinium, piperidinium, pyridinium, isoquinolinium, sulfonium, ammonium, and phosphonium. We hope that this handbook could provide useful references for scientific workers, promote the basic theoretical innovation of ionic liquids, and help push on the industrial application process of ionic liquids.

1.2 Properties of Ionic Liquid Mixtures

1.2.1 Volumetric Properties

The volumetric properties refer to be closely related with volume change of substance, such as density, molar volume, partial molar volume, excess molar volume, compressibility, expansibility, etc. The multicomponent mixtures containing ionic liquids are very complex systems, because they depend not only on molecular-molecular interaction but also on the ion–ion and ion–molecular interactions. Meanwhile, one of the factors affecting structural effects also has to be considered that arise from interstitial accommodation due to differences in molar volume and free volume between solution components. The knowledge of volumetric properties may help to test molecular simulation theories or models related to ionic liquids.

Furthermore, they will provide information and available data for the engineering design and optimization. So, the volumetric properties are basic and important thermodynamic properties.

The density is one of the volumetric properties, and major properties are calculated based on them; therefore, the reports and investigations about density are many more, including mixtures of organic solvents and ionic liquids, gases and ionic liquids, solid and ionic liquids, etc. In a general way, the measurement densities of ionic liquids mixtures depend on temperature, pressure, and component. Densities of ionic liquid mixtures at different temperatures for the whole range of composition can be correlated by using the linear equation or second-order polynomial.

$$\rho = \sum_{i=0}^{n} A_i T^i \tag{1.1}$$

where ρ is the density of mixtures, A is coefficients of equation, T is the temperature, $n = 1$ or 2.

Usually, the densities are fitted with composition of mixture containing ionic liquid distinctly nonlinear; this phenomenon is different from general organic solvent mixtures. Geppert-Rybczyńska et al. [1] applied the Connors and Wright equation [2] that is used for correlating concentration dependencies of surface tensions to describe the change of density with composition, and the correlation is satisfactory.

$$\rho = \rho_1 - \left[1 + \frac{bx_1}{(1 - ax_1)}\right](1 - x_1)(\rho_1 - \rho_2) \tag{1.2}$$

where ρ_1 and ρ_2 are densities of pure compounds, x_i is the mole fraction of component i of the mixture, a and b are fitting constants.

The molar volume (V_m) and the excess molar volume (V_m^E) can be calculated from above measuring densities data of mixtures. The molar volume is defined as follows:

$$V_m = \frac{x_1 M_1^o + x_2 M_2^o}{\rho} \tag{1.3}$$

where M_i^o is the molar mass of the pure component i, x_i is the mole fraction of component i of the mixture.

The excess molar volume can be calculated according to the following equation:

$$V_m^E = V_m - \sum_{i=1}^{2} x_i V_{m,i} \tag{1.4}$$

where $V_{m,i}$ is the molar volume of component i of the mixture. The values of V_m^E can be fitted by Redlich–Kister type polynomial equation: [3–5]

$$V_{\mathrm{m}}^{\mathrm{E}} = x(1-x)\sum_{i=0}^{n} A_i(2x-1)^i \tag{1.5}$$

where x is the mole fraction of the one component of mixture, A_i are the adjustable parameters, and n is the order of polynomial equation, normally, $n = 4$ or 5.

In most cases, the V^{E} are all negative and asymmetric over the entire composition range, only V^{E} of a few ionic liquid mixtures are all positive or both negative and positive. The value of V^{E} is affected by the intermolecular interaction forces between the components of the mixtures and the differences in size and shape of molecules.

Further, the partial molar volumes $\overline{V}_1$ and $\overline{V}_2$ can be evaluated using the following equations:

$$\overline{V}_1 = V_{\mathrm{m}}^{\mathrm{E}} + V_{\mathrm{m},1} + (1-x)\left(\partial V_{\mathrm{m}}^{\mathrm{E}}/\partial x\right)_{p.T} \tag{1.6}$$

$$\overline{V}_2 = V_{\mathrm{m}}^{\mathrm{E}} + V_{\mathrm{m},2} + x\left(\partial V_{\mathrm{m}}^{\mathrm{E}}/\partial x\right)_{p.T} \tag{1.7}$$

Differentiation of Eq. 1.5 with respect to x_1 and combination of the results of differentiation with Eqs. 1.6 and 1.7 leads to the following equations for the partial molar volumes:

$$\overline{V}_1 = V_{\mathrm{m},1} + (1-x)^2\sum_{i=0}^{n} A_i(1-2x)^i + 2x(1-x)^2\sum_{i=0}^{n} A_i(t)(1-2x)^{i-1} \tag{1.8}$$

$$\overline{V}_2 = V_{\mathrm{m},2} + x^2\sum_{i=0}^{n} A_i(1-2x)^i + 2x^2(1-x)\sum_{i=0}^{n} A_i(t)(1-2x)^{i-1} \tag{1.9}$$

The partial molar volume at infinite dilution provides information about solute–solvent interactions that are independent of the composition. It is obtained from the following equations:

$$\overline{V}_1{}^{\infty} = V_{\mathrm{m},1} + \sum_{i=0}^{n} A_i \tag{1.10}$$

$$\overline{V}_2{}^{\infty} = V_{\mathrm{m},2} + \sum_{i=0}^{n} A_i(-1)^i \tag{1.11}$$

So, from Eqs. 1.10 and 1.11, the partial excess molar volumes at infinite dilution can be calculated:

$$\overline{V_1{}^{\mathrm{E}\infty}} = \sum_{i=0}^{n} A_i \tag{1.12}$$

$$\overline{V_2{}^{\mathrm{E}\infty}} = \sum_{i=0}^{n} A_i(-1)^i \tag{1.13}$$

Matkowska D. et al. [6] take into account the effect of both composition and temperature on the density, and the densities of mixtures can be expressed by the Eq. 1.14. So, the excess molar volumes are calculated from Eq. 1.15.

$$\rho(T,x_1) = \frac{M_1x_1 + M_2x_2}{\dfrac{M_1x_1}{\rho_1^0(T)} + \dfrac{M_2x_2}{\rho_2^0(T)} + V^{\mathrm{E}}(T,x_1)} \tag{1.14}$$

$$V^E = M_1x_1\left(\frac{1}{\rho} - \frac{1}{\rho_1}\right) + M_2x_2\left(\frac{1}{\rho} - \frac{1}{\rho_2}\right) \tag{1.15}$$

where M_1 and M_2 are molecular weights of the pure components and ρ, ρ_1, and ρ_2 are densities of the mixture and the pure components, respectively.

Because densities of pure compounds considering pressure were correlated by the Tait equation in the form [7]

$$\rho(T,p) = \frac{\rho_0(T,p_0)}{1 - A\ln\dfrac{B(T)+p}{B(T)+p_0}} \tag{1.16}$$

where the reference pressure $p_0 = 0.1$ MPa, the $\rho(T)$ and $B(T)$ have the following functions of temperature

$$\rho_0(T,p_0) = \rho_{00} + \rho_{01}(T/K) + \rho_{02}(T/K)^2 + \rho_{03}(T/K)^3 \tag{1.17}$$

$$B(T) = B_0 + B_1(T/K) + B_2(T/K)^2 \tag{1.18}$$

So, the densities of the mixture were correlated by the following twelve-parameter equation in which the excess volume was formally represented by a sum of van Laar terms in which two such terms turned out to be suitable.

$$\rho(T,p,x_1) = \frac{M_1x_1 + M_2x_2}{\dfrac{M_1x_1}{\rho_1(T,p)} + \dfrac{M_2x_2}{\rho_2(T,p)} + x_1x_2\left(\dfrac{a_0(T,p)}{b_0(T,p)x_1 + x_2} + \dfrac{a_1(T,p)}{b_1(T,p)x_1 + x_2}\right)} \tag{1.19}$$

Correspondingly, the excess molar volumes are reproduced by the equation

$$V^{\mathrm{E}}(\mathrm{T},p,x_1) = x_1x_2\left(\frac{a_0(T,p)}{b_0(T,p)x_1 + x_2} + \frac{a_1(T,p)}{b_1(T,p)x_1 + x_2}\right) \tag{1.20}$$

where the $a_i(T, p)$ and $b_i(T, p)$ coefficients depend linearly on T and p.

The apparent molar volumes, $V_{\phi 1}$ and $V_{\phi 2}$ can be calculated with the equations:

$$V_{\phi 1} = \left[V_m - (1-x)V_{m,2}\right]/x \tag{1.21}$$

$$V_{\phi 2} = \left[V_m - xV_{m,1}\right]/(1-x) \tag{1.22}$$

Combining Eq. 1.4, $V_{\phi 1}$ and $V_{\phi 2}$ can be expressed as:

$$V_{\phi 1} = V_{\mathrm{m},1} - V_{\mathrm{m}}^{\mathrm{E}}/x \tag{1.23}$$

$$V_{\phi 2} = V_{\mathrm{m},2} - V_{\mathrm{m}}^{\mathrm{E}}/(1-x) \tag{1.24}$$

The isobaric thermal expansivity (α_p) can be calculated on the basis of the volumetric data:

$$\alpha_p = \frac{1}{V_{\mathrm{m}}}\left(\frac{\partial V_{\mathrm{m}}}{\partial T}\right)_p = -\frac{1}{\rho}\left(\frac{\partial \rho}{\partial T}\right)_p \tag{1.25}$$

The excess thermal expansion coefficient α_p^E is defined as follows:

$$\alpha_p^{\mathrm{E}} = \alpha_p - \left(\phi_1\alpha_{p,1} + \phi_2\alpha_{p,2}\right) \tag{1.26}$$

where $\alpha_{p,i}$ are the thermal expansion coefficients for pure components i, ϕ_i are the volume fractions for pure components i, defined as:

$$\phi_i = \frac{x_i V_{\mathrm{m},i}}{\sum_{i=1}^{2} x_i V_{\mathrm{m},i}} \tag{1.27}$$

The limiting apparent expansivity ϕ_E^0 could be estimated using the following relation:

$$\phi_E^0 = \partial V_\phi^0/\partial T \tag{1.28}$$

where V_ϕ^0 is the limiting values of apparent molar volume (equal to the partial molar volume at infinite dilution).

Isentropic compressibility k_S is defined as

$$k_S = -\frac{1}{V_{\mathrm{m}}}\left(\frac{\partial V_{\mathrm{m}}}{\partial p}\right)_S = \frac{1}{\rho}\left(\frac{\partial \rho}{\partial p}\right)_S \tag{1.29}$$

and is also given by the Newton–Laplace equation

$$k_S = \frac{1}{\rho u^2} = \frac{V_{\mathrm{m}}}{M_{\mathrm{m}}u^2} \tag{1.30}$$

where M_{m} and u are the molar mass and speed of sound of the mixture, respectively. Molar isentropic compression $k_{S,m}$ is defined from the isentropic compressibility as

$$k_{S,m} = -\left(\frac{\partial V_m}{\partial p}\right)_S = V_m k_S = \frac{V_m^2}{M_m u^2} \tag{1.31}$$

Apparent molar adiabatic compressibility k_ϕ can be calculated from the following equation:

$$k_\phi = \frac{M k_S}{\rho} - \frac{1000(k_{S,0}\rho - k_S \rho_0)}{m \rho \rho_0} \tag{1.32}$$

where M is the molar mass of the solute, m is the molality, ρ_0 and $k_{S,0}$ are the densities and isentropic compressibility of solvent.

Industrial applications often require specific data to support the design process, but it makes harder to obtain accurate experimental data at some harsh conditions in the applications; at this time, the models predict seem essential based on a large amounts of existing experimental data. Therefore, many researchers focus on the study of models that could predict the volumetric properties of ionic liquid mixtures.

The models that predict the densities of mixtures are mostly semi-theoretical equation of state (EOS). Hosseini et al. [8] used a simple equation to predict the volumetric properties of binary mixtures including an ionic liquid, the equation is proposed by Spencer and Danner [9], and the mixing rule is developed by Li, Spencer, and Danner [10]. The corresponding equations are as follows:

$$\rho_{mix.} = \left(\sum_i \frac{P_{Ci}}{R_{xi} T_{Ci}}\right) Z_{RA\ mix.}^{-\left(1+\tau_{mix.}^{2/7}\right)} \tag{1.33}$$

where:

$$Z_{RA\ mix.} = \sum_i x_i Z_{RAi} \tag{1.34}$$

and

$$\tau_{mix.} = 1 - (T/T_{C\ mix.}) \tag{1.35}$$

the mixing rule for the reduced temperature is expressed as:

$$T_{C\ mix.} = \sum_i x_i T_{Ci} \tag{1.36}$$

Yousefi [11] applied Tao and Mason equation of state (TM EOS) [12] based on statistical–mechanical perturbation theory to binary mixtures of ionic liquids. The equations are expressed as follows:

$$\frac{P}{\rho KT} = 1 + (B - a)\rho + \frac{\alpha\rho}{1 - \lambda b\rho} + A_1(\alpha - B)b\rho^2 \frac{\left(e\dfrac{kT_c}{T} - A_2\right)}{1 + 1.8(b\rho)^4} \tag{1.37}$$

where

$$A_1 = 0.143 \tag{1.38}$$

$$A_2 = 1.64 + 2.65\left[e^{(k-1.093)} - 1\right] \tag{1.39}$$

$$k = 1.093 + 0.26\left[(\omega + 0.002)^{\frac{1}{2}} + 4.50(\omega + 0.002)\right] \tag{1.40}$$

where ω is the Pitzer acentric factor, λ is an adjustable parameter, ρ is the number density, T_c is the critical temperature, kT has usual meaning, B is the second virial coefficient, α is the scaling parameter, and b is the effective van der Waals covolume.

Hosseini et al. [13–15] developed a perturbed hard-sphere equation of state (PHS EOS) for predicting the densities of several ionic liquids and their mixtures. The general frame of the PHS EOS for ionic liquids mixtures has the form:

$$\frac{P}{\rho kT} = \frac{1 + \eta + \eta^2 - \eta^3}{(1 - \eta)^3} - \frac{\rho}{kT}\sum_i^m\sum_j^m x_i x_j a(T)_{ij} \tag{1.41}$$

where x_i and x_j are the mole fractions of the ith and jth components, respectively. η is the packing fraction of mixtures of hard sphere [16]. This parameter is defined by the following expression:

$$\eta = \frac{\rho}{4}\sum_i^m x_i b(T)_i \tag{1.42}$$

As for binary mixtures, the hard-sphere covolumes, $b(T)_{ij}$ are additive according to following expression:

$$b(T)_{ij} = 1/8\left[b(T)_i^{1/3} + b(T)_j^{1/3}\right]^3 \tag{1.43}$$

The attractive forces between two hard-sphere species of mixtures including i and j components can be written as follows:

$$a(T)_{ij} = \frac{2\pi}{3(\rho^*)_{ij}} kT^*(F_a)_{ij} \tag{1.44}$$

Hosseini et al. [17] applied the following combining rules to T_{ref}, ρ^* and the universal function F_a:

$$(T_{ref})_{ij} = \sqrt{T_{ref,i} T_{ref,j}} \tag{1.45}$$

$$\left(\rho_{ij}^*\right)^{-1/3} = \frac{1}{2}\left[\left(\rho_i^*\right)^{-\frac{1}{3}} + \left(\rho_j^*\right)^{-1/3}\right] \tag{1.46}$$

$$(F_a)_{ij} = \sqrt{(F_a)_{ii}(F_a)_{jj}} \tag{1.47}$$

In case of V_m^E, the Prigogine-Flory-Paterson (PFP) [18] and the Flory-Benson-Treszczanowicz (FBT) [19] models are common use prediction models. They are very useful in predicting and correlating the thermodynamic properties of binary mixtures of polar and nonpolar molecules.

1.2.2 Transport Properties

The transport properties of ionic liquids with molecular solvents that are commonly used for process analyses in industrial applications are the viscosity, the diffusion coefficient, the surface tension, the speed of sound, etc. Overall, there is still a lack of data on their transport properties because of the wide variety of ionic liquids.

1.2.2.1 Viscosity

Generally, ionic liquids are much more viscous than conventional organic solvents, which to some extend limit their range of industrial applications. However, the viscosity of ionic liquids is different from the density, in that it shows a remarkable influence of impurity: the presence of a trace of impurities can result in dramatic decreases in viscosity [20]. Therefore, the viscosity of ionic liquid mixtures is much smaller than pure ionic liquid.

The common measurements of the viscosity of ionic liquids are rolling sphere [21–23], falling body [24–28], and moving piston [29, 30]. Among them, the most commonly used method is falling body.

The viscosity of ionic liquids mixtures is related with temperature, the structure of ionic liquid, the size of solvent, the interaction between solute and solvent, and so on. Generally, the values of viscosity decrease with increasing temperature, which can be a result of the disruption of intramolecular hydrogen bonding between the ionic liquid molecules. As for the influence of ionic liquid structures and the solvent, the viscosities increase with increasing number of carbon atoms in the linear alkyl group in the cations of ionic liquid[31, 32]. Domańska et al. [33–35] found that the values of viscosities increase as the alkyl chain of alcohol solutions increases at the same temperature (see Fig. 1.1). The same result can be seen in ref [36]. Gómez et al. [37] measured the viscosities of water + 1-hexyl-3-methylimidazolium chloride and water + 1-octyl-3-methylimidazolium chloride; they found that the modification of viscosity deviations is related to interactions of the type water–ion and ion–ion.

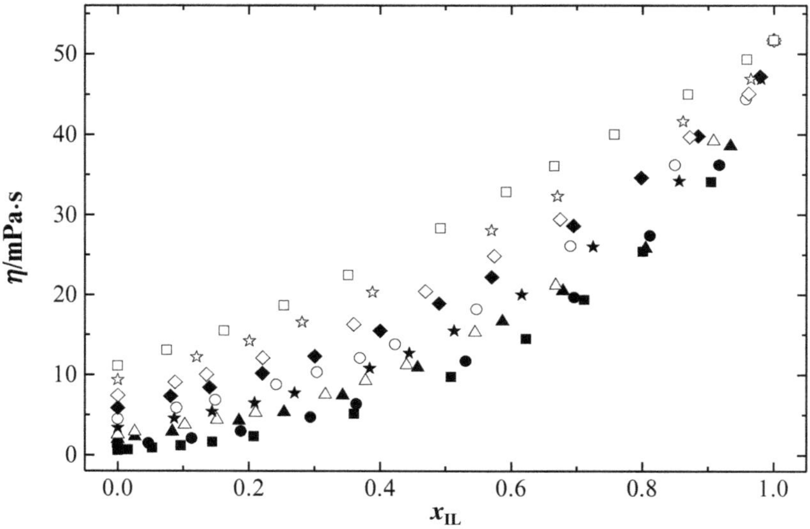

Fig. 1.1 The viscosities of ionic liquid alcohol solutions as a function of mole fractions of 1-butyl-3-methylimida-zolium thiocyanate ([BMIM][SCN], x_1) at $T = 298.15$ K. ■: [BMIM][SCN] + methanol, ●: [BMIM][SCN] + ethanol, ▲: [BMIM][SCN] + 1-propanol, △: [BMIM][SCN] + 1-butanol, ★: [BMIM][SCN] + 1-pentanol, ○: [BMIM][SCN] + 1-hexanol, ◆: [BMIM][SCN] + 1-heptanol, ◇: [BMIM][SCN] + 1-octanol, ☆: [BMIM][SCN] +1-nonanol, □: [BMIM][SCN] + 1-decanol

The relationship between viscosities of ionic liquid mixtures and temperature can be correlated by Vogel-Tammann-Fulcher (VTF) equation [38]:

$$\ln\eta = A + \frac{B}{T - T_0} \tag{1.48}$$

where η is the viscosities of mixtures, T is thermodynamic temperature, A, B, and T_0 are parameters.

In order to study solvent–solute interaction, the viscosity deviations ($\Delta\eta$) were calculated by the following relation:

$$\Delta\eta = \eta - \sum_{i=1}^{n} x_i\eta_i \tag{1.49}$$

where η and η_i are the viscosity of mixtures and pure components i, respectively.

As for binary mixtures, the values of $\Delta\eta$ can be fitted by Redlich-Kister-type polynomial equation: [3–5]

$$\Delta\eta = x_1x_2\sum_{i=0}^{n} A_i(1 - 2x_2)^i \tag{1.50}$$

where x_1 and x_2 are the mole fraction of the components 1 and 2, A_i are the adjustable parameters, and n is the order of polynomial equation. The optimum number of adjustable parameters was ascertained from an examination of the standard deviation (S):

$$S = \left[\frac{\sum\left(\Delta\eta_{\text{exp}} - \Delta\eta_{\text{cal}}\right)^2}{n}\right]^{1/2} \tag{1.51}$$

where n is the total number of experimental points, $\Delta\eta_{\text{exp}}$ is the experimental value, and $\Delta\eta_{\text{cal}}$ is the calculated value from Eq. 1.49.

So far, the related estimation methods of predicting the viscosities of ionic liquid mixtures are very scarce. Lashkarblooki et al. [39] used multilayer perceptron artificial neural network model to predict the viscosities of ternary mixtures containing ionic liquids. Their results indicated that this model is able to predict viscosity of ionic liquids ternary systems with a considerably high accuracy. Hu et al. [40] made the ionic liquid mixtures as electrolyte solutions and used a series of equations for prediction of the viscosity of mixed electrolyte solutions in predicting corresponding properties of mixed ionic liquid solutions. The results indicate that the method can provide good predictions for the multicomponent ionic liquid solutions.

1.2.2.2 Diffusion Coefficient

The diffusion coefficient (D) is a physical constant dependent on molecule size and other properties of the diffusing substance as well as on temperature and pressure. It is an important parameter to describe suitably diffusion mass transport phenomena [41, 42], and the diffusion coefficient of ionic liquids in some solvents can affect their use in separation processes [43] and catalytic reactions [44–46].

There are various methods available to measure diffusion coefficients of mixtures, such as impedance spectroscopy [47], polarization measurements,[47, 48] chronoamperometry [49, 50], diaphragm cell method [51], optical method [52], Taylor dispersion method [53, 54], and dynamic light scattering (DLS).[55–58] Among this methods, the Taylor dispersion method has been shown to be relatively fast, but it can be applied only in the regime of infinite dilution. DLS method is a suitable technique that is used in measuring mutual diffusivity coefficients of ionic liquids mixtures over a range of concentration [59, 60].

The diffusion coefficients of ionic liquid in solvents can be fitted to Arrhenius plots of the form:

$$D_{21} = D_0 exp\left(-\frac{E_a}{RT}\right) \tag{1.52}$$

where D_0 is a preexponential factor, E_a is the activation energy of the solute for diffusion. The subscript 1 and 2 represent ionic liquid and solvents, respectively.

The relationship between the diffusion coefficients of ionic liquid and the viscosities of solvents can be expressed by Stokes–Einstein equation:

$$D_{21} = \frac{k_B T}{C\pi\eta_1 r_2} \tag{1.53}$$

where k_B is the Boltzmann constant, T is the temperature, the constant C is dependent on the boundary condition of the surface flow (four for the slip, six for the stick boundary condition), η_1 is the viscosity of the solvent, r_2 is the effective hydrodynamic radius of the solute.

The effect factors of self-diffusion coefficients and mutual diffusion coefficients are distinguished. So we discussed them separately.

The self-diffusion coefficients can be mainly influenced by size of ion, shape of ion, and magnitude of interaction between ions. For example, the self-diffusion coefficients of the ions in the 1-alkyl-3-methylimidazolium bis(trifluoromethylsulfonyl)amide decrease with increasing alkyl chain length[61]. Tokuda et al. [62] found that the sequence of self-diffusion coefficients of the ions in ionic liquids is 1-butyl-3-methylimidazolium bis(trifluoromethylsulfonyl)amide ([BMIM][TFSA]) > N-butylpyridinium bis(trifluoromethylsulfonyl)amide ([bpy][TFSA]) > N-butyl-N,N,N-trimethylammonium bis(trifluoromethylsulfonyl)amide ([btma][TFSA]) > N-butyl-N-methyl-pyrrolidinium bis(trifluoromethylsulfonyl)amide ([bmpro][TFSA]), which can be seen that the shape of the cations could influence the self-diffusion coefficients of the ions. Tsuzuki et al. [63] calculated the stabilization energies of [EMIM][BF_4], [EMIM][CF_3SO_3], and [EMIM][TFSA] and found that the self-diffusion coefficients of the ions in [EMIM][TFSA] are larger than those in the other two ionic liquids, although the size of $TFSA^-$ is larger than other anions. This is because the self-diffusion coefficients depend on the magnitude of the interaction between the anion and cation.

The mutual diffusion coefficients can be mainly influenced by the temperature, the size of solute, the interaction between solute and solvent, and the viscosity of solute. The mutual diffusion coefficients increase with increasing temperature. Sarraute et al. [64] found that diffusion coefficients decrease, in water, acetonitrile, and methanol, with increasing alkyl side chain in the cation for the 1-alkyl-3-methylimidazolium bis(trifluoromethanesulfonyl)imide ([C_nMIM][TFSA]) ionic liquids. They also found that for the ionic liquids based on the [BMIM] cation, the diffusivities sequence in water are [BMIM][TFSA] < [BMIM][BF_4] ~ [BMIM][PF_6] < [BMIM]Cl. This is because as for different size of anions, larger ionic liquids diffuse more slowly than the smaller ones. Rausch et al. [60] found that for the ionic liquids-ethanol systems, the mutual diffusion coefficients are distinctly lower than those obtained for the mixtures with acetone. This can be explained for: The formation of hydrogen bonds between ethanol and the ions retard the mutual diffusion process. They also found that the larger the viscosities of pure ionic liquids, the lower the mutual diffusivities in the solvent. This is because the increasing viscosity reduced the mobility of the molecules.

The diffusion coefficients of organic molecules can be estimated by the Wilke–Chang equation [65]:

$$D^0_{\mathrm{AB}} = \frac{a(\phi M_{\mathrm{B}})^{1/2} T}{\eta_{\mathrm{B}} V_A^{0.6}} \tag{1.54}$$

where D^0_{AB} is the diffusion coefficient of solute A (ionic liquid) in solvent B, a is a constant, ϕ is the association parameter for the solvent, M_{B}, η_{B} are molecular weight and viscosity of solvent, respectively, V_{A} is molar volume of solute.

The diffusion of ionic liquids in water could also be regarded as the diffusion of electrolyte, and the diffusion coefficients of electrolytes are related to infinite dilution conductance by the Nernst–Haskell equation [66]:

$$D_{AB}^{0} = \frac{RT}{F^2} \frac{|z_+| + |z_-|}{|z_+ z_-|} \frac{\Lambda_+^0 \Lambda_-^0}{\Lambda_+^0 + \Lambda_-^0} \tag{1.55}$$

where D_{AB}^0 is the diffusion coefficient of ionic liquid in water at infinite dilution, R is the gas constant, T is the absolute temperature, F is the Faraday's constant, z_+ and z_- are the charge numbers of cation and anion molecules, respectively, Λ_+^0 and Λ_-^0 are the infinite dilution conductances of the cation and anion molecules, respectively.

The self-diffusion coefficients of the ions at infinite dilution could also be evaluated using the Nernst expression [66]:

$$D_{ion}^{0} = \frac{RT\Lambda_{ion}^0}{|z_{ion}|F^2} \tag{1.56}$$

where D_{ion}^0 is the self-diffusion coefficient of ions at infinite dilution, R is the gas constant, T is the absolute temperature, Λ_{ion}^0 is the infinite dilution molar conductance of the ion, and z_{ion} is the charge number of the ion.

1.2.2.3 Surface Tension

Surface tension refers to resistance properties of liquid substances on some force, and it is one of the most important properties of ionic liquids. The surface tension data is an important factor for engineers considered in chemical process and reactor engineering, biochemical engineering, electrical engineering, etc. [67].

The surface tension values of ionic liquids are mainly dependent upon the structure of cations and anions that are present at the surface of ionic liquids [68]. Wang et al. [69] measured surface tension of ionic liquids based on 1-C_nH_{2n+1}-3-methylimidazolium ([$C_n$mim]) with two kinds of fluorinated anions (BF_4^- and PF_6^-) ($n = 2-7$; $4-7$, respectively) by using a Wilhelmy plate apparatus. It can be seen that the surface tension of the same ionic liquid decreases linearly with a rise in temperature. The dependence could be described by the empirical Eötvös equation [70]:

$$\sigma = \frac{k(T_c - T)}{V_M^{2/3}} \tag{1.57}$$

where k is an empirical constant, T_C is the critical temperature. V_M is the molecular volume of ionic liquids, respectively.

It also can be seen that the surface tension of the ionic liquids in the same anions decreases with the increase in alkyl chain length at a given temperature (see Table 1.1). This conclusion may be explained by Langmuir's principle of independent surface action [71]. As for the influence of anions, Xu et al. [68] found that the

Table 1.1 Values of the surface tension of the ionic liquids

Ionic liquids	$\sigma/(\mathrm{mJ \cdot m^{-2}})$ at 298.15 K
$[C_2mim][BF_4]$	50.4
$[C_3mim][BF_4]$	48.6
$[C_4mim][BF_4]$	43.9
$[C_5mim][BF_4]$	40.8
$[C_6mim][BF_4]$	37.0
$[C_7mim][BF_4]$	36.1
$[C_4mim][PF_6]$	48.0
$[C_5mim][PF_6]$	43.0
$[C_6mim][PF_6]$	38.6
$[C_7mim][PF_6]$	35.7

surface tension of carboxylic acids decreases in the series $HCOOH > CH_3COOH > CH_3CH_2COOH > CH_3(CH_2)_2COOH$.

There are some correlation methods for the prediction of the surface tension of ionic liquids. The common methods include parachors, corresponding state theory (CST), and group contribution methods.

The parachors method is based on an empirical equation (Eq. 1.58) proposed by MacLeod [72]:

$$\sigma^{\frac{1}{4}} = K\rho \tag{1.58}$$

where K is a temperature independent constant characteristic of the compound.

Sugden [25] proposed a modification and defined parachor (P_{ch}).

$$P_{ch} = KM_{\mathrm{w}} = M_{\mathrm{w}}\sigma^{\frac{1}{4}}/\rho \tag{1.59}$$

where M_{w} is the molecular weight.

Deetlefs et al. [73] were the first to employ the parachor for prediction of the surface tension of ionic liquids, and the parachors were calculated using group contribution method by Knotts et al. [74]. Then, Gardas and Coutinho [75] proposed another correlation for the surface tension of ionic liquids based on the molecular volume of the ion pair:

$$\sigma = {}^{d}\big/{}_{V^{2/3}} \tag{1.60}$$

where V is the molecular volume in Å^3, and $d = 2147.761 \pm 18.277(\mathrm{mN \cdot m^{-1}})$ Å^2.

Although these two models could provide a good estimation of the surface tension of ionic liquids, they can be used just for imidazolium-based ionic liquids.

Recently, Mousazadeh and Faramarzi [76] proposed the corresponding states theory (CST) correlation for the surface tension of ionic liquids. They used the melting and boiling points of ionic liquids instead of critical properties:

$$\sigma = 0.819\left(\frac{T_b - T}{T_b - T_m}\right)\sigma_m + 0.500\left(\frac{T}{T_b}\right)\sigma_m \tag{1.61}$$

where σ_m is the surface tension at melting temperature T_m, and T_b is the boiling temperature.

The drawback of the model is its lack of predictive power although it revealed a small average absolute relative deviation (AARD). On the other hand, many ionic liquids do not have a melting point, and that the boiling temperatures of the ionic liquids are as elusive as their critical temperatures, which limit its application to a certain extent.

Gharagheizi et al. [77] used the group contribution method to estimate surface tension of ionic liquids. Their method includes 12 substructures related to chemical structure of anion, 7 substructures related to the chemical structure of cation and temperature (see Eq. 1.62).

$$\sigma = \sum_{i=1}^{N_a} N_{ai}\sigma_{ai} + \sum_{i=1}^{N_c} N_{ci}\sigma_{ci} + A + BT \tag{1.62}$$

where N_{ai}, N_{ci}, σ_{ai}, σ_{ci}, A, B, N_a, N_c are the number of occurrence of ith substructure of anions and cations, the contribution of the ith substructure of anions and cations, the intercept of Eq. 1.62, the temperature coefficient, and the number of total substructures of anions and cations.

This method shows an AARD of 3.6 % from experimental data and can be used to predict the surface tension of various ionic liquids.

Wu et al. [78] proposed a corresponding-states group contribution method to estimate the surface tension of ionic liquids and developed a new equation based on the method:

$$\sigma = \sum_{i=0}^{3} a_i \left(\sum_{j=1}^{k} n_j \Delta\sigma_{0,j}\right)^i (1 - T_r)^{11/9} \tag{1.63}$$

where n_j is the number of groups of type j, and k is the total number of different groups in the molecular. The parameters a_i and $\Delta\sigma_{0,j}$ were estimated by minimizing the objective function O.F. in Eq. 1.64:

$$\text{O.F.} = \sum_{i=1}^{N_P} \left[\left(\sigma_{calc} - \sigma_{exp}\right)^2\right]_i \tag{1.64}$$

where N_P represents the number of data points.

This method offers a simple approach for estimating surface tension of new ionic liquids and has a good accuracy.

So far, experimental data for surface tension of ionic liquids mixtures is very scarce. Domańska et al. [79] measured the surface tension of binary mixtures of several ionic liquids combined with different alcohols at different temperatures. It is found that surface tension values of mixtures increase with ionic liquids mole fraction increase. But as for the same ionic liquid, the increase of surface tension in different solvents presents different trends. For example, surface tension values of ([MMIM][CH_3SO_4] + methanol) mixtures showed a liner increase with the ionic liquid mole fraction increase, which indicates that the methanol molecules have strongly joined on the surface of ionic liquids. However, surface tension values of ([MMIM][CH_3SO_4] + ethanol/1-butanol) mixtures occurred at a breakpoint, and the values increase slowly before the breakpoint, whereas it increase sharply after breakpoint. It has also been found that the values of breakpoint are different for different ionic liquids mixtures. This may be caused by anionic and cationic hydrophobic nature or the hydrophobic nature of alcohols.

The surface tension of the ionic liquids aqueous solutions shows a different tendency compared to (ionic liquids + alcohols) mixtures. Rilo et al. [80] measured the surface tension of 1-alkyl-3-methylimidazolium tetrafluoroborate ([C_nMIM][BF_4], $n = 2,4,6$) aqueous solutions. Since surface tension values of these ionic liquids are smaller than surface tension values of water, the surface tension decreases with the increasing of the ionic liquids mole fraction. Moreover, it was found that the surface tension decreases when the length of the alkyl chain increases over the whole composition. Jiang et al. [81] measured the surface tension of 1-butyl-3-methyl-imidazolium L-lactate + water mixtures. It was observed that at low mole fraction of ionic liquids (x_{IL}), the surface tension decreases sharply, and then decreases slowly while $x_{IL} > 0.1$. This indicates 1-butyl-3-methyl-imidazolium L-lactate acts as a surface active agent in aqueous solution.

The tendency of temperature effects on the surface tension of ionic liquid mixtures is the same as the pure ionic liquid. The surface tension measured at a given composition decreases with increasing temperature. The reason is that the surface tension drop caused by the molecular solvent becomes less significant at higher temperatures [82]. However, the temperature effect on the surface tension of ionic liquid mixtures is quite diverse, although the general trend of surface tension with temperatures is consistent [83]. (For example, for ([C_8mim][PF_6] + water) mixtures, the dependency of the surface tension with temperature is almost equal to that of the pure ionic liquid while for the [C_6mim][TFSA] + 1-octene) mixtures, the surface tension dependence on temperature is less significant than the pure ionic liquid).

Surface tension deviations of binary mixtures are obtained by Eq. 1.65:

$$\delta\sigma = \sigma - \sum_{i=1}^{2} x_i \sigma_i \tag{1.65}$$

where σ and σ_i are the surface tension of mixture and component i, respectively; x_i is the mole fraction of component i.

1.2.2.4 Speed of Sound

The speed of sound (u) is a useful property, since it can derive many thermophysical properties including isentropic and isothermal compressibilities, isobaric thermal expansion coefficient, thermal pressure coefficient, the Joule–Thomson coefficient, isobaric and isochoric heat capacities, ratio of isobaric and isochoric heat capacities, and the reduced bulk modulus along with other properties [84].

The isentropic compressibility (K_s) was evaluated for the investigated systems from Laplace–Newton's equation

$$K_{\mathrm{s}} = \frac{1}{\rho u^2} \tag{1.66}$$

where u and ρ are speed of sound and density of the solution, respectively.

The isothermal compressibility (K_T) could be calculated from the speed of sound [85]:

$$K_{\mathrm{T}} = \frac{1}{\rho}\left(\frac{1}{U^2} + \frac{T a_{\mathrm{P}}^2 M}{C_{\mathrm{P,m}}^0}\right) \tag{1.67}$$

where ρ is the density ($\mathrm{kg \cdot m^{-3}}$), M is the molar mass ($\mathrm{kg \cdot mol^{-1}}$), $a_{\mathrm{P}}$is the thermal expansion coefficient ($\mathrm{K^{-1}}$), $C_{\mathrm{P,m}}^0$ is isobaric heat capacity ($\mathrm{J \cdot K^{-1} \cdot mol^{-1}}$).

We can also obtain p-ρ-T data from the experimental speed of sound since it can be measured accurately at some pressures and it is a key quantity in the equation of state formulation. Dzida et al. [86] calculated the densities of 1-ethyl-3-methylimidazolium bis(trifluoromethylsulfonyl)imide ([EMIm][TFSA]) under high pressures from speed of sound. Equations are as follows:

$$\left(\frac{\partial \rho}{\partial T}\right)_T = \frac{1}{u^2} + \frac{T\alpha^2}{C_{\mathrm{P}}} \tag{1.68}$$

$$\Delta\rho = \int_{P_1}^{P_2} \left(\frac{1}{u^2} + \frac{T\alpha^2}{C_{\mathrm{P}}}\right) dP \approx \int_{P_1}^{P_2} \frac{1}{u^2} dP + \frac{T\alpha^2}{C_{\mathrm{P}}} \Delta P \tag{1.69}$$

where $\alpha = -\rho^{-1}(\partial\rho/\partial T)_P$ is the thermal expansion coefficient, C_{P} is the specific isobaric heat capacity. To calculate the density at P_2, ρ, α, and C_{P} at P_1 should be known, in addition to the speed of sound as a function of pressure.

The calculated data is in a good agreement with the experimental data reported in the literature.

The temperature dependence of speed of sound was correlated by means of a polynomial type equation:

$$u = \sum_{i=0}^{2} b_i T^i \tag{1.70}$$

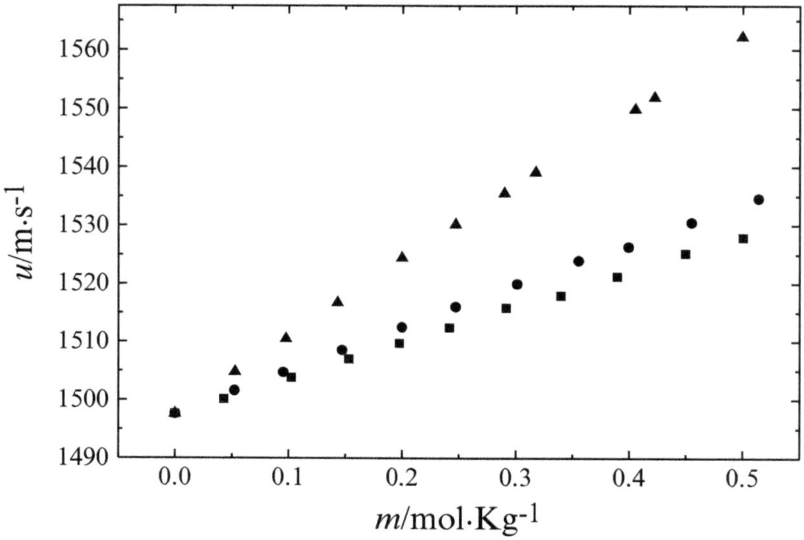

Fig. 1.2 The speed of sound as a function of molality of ionic liquids (m) at $T = 298.15$ K. ■: [Im]Cl; ●: [Mim]Cl; ▲: [Bmim]Cl

The speed of sound, u in m/s units can be estimated using the theoretical Auerbach's relation [87]:

$$u = \left(\frac{\sigma}{6.33 \times 10^{-10} \times \rho}\right)^{2/3} \tag{1.71}$$

where σ and ρ are the surface tension in $N \cdot m^{-1}$ units and density in $kg \cdot m^{-3}$ units, respectively. Gardas et al. [84] used this empirical relation to estimate the speeds of sound of imidazolium-based ionic liquids and found that this method had a better accuracy.

As for ionic liquids mixtures, their speed of sound could be influenced by temperatures, the structures of ionic liquid, the interaction between solute and solvent, and so on. Generally, the values of speed of sound decrease as temperature increases. Gardas et al. [88] measured the speed of sound of aqueous solutions of imidazolium chloride ([Im]Cl), 1-methyl imidazolium chloride ([Mim]Cl), and 1-butyl-3-methyl imidazolium chloride ([Bmim]Cl) in the concentration range of 0.05 $mol \cdot Kg^{-1}$ to 0.5 $mol \cdot Kg^{-1}$ at 298.15 K. From Fig. 1.2, we can see that the speed of sound increases with increasing of alkyl substituents on the imidazolium cation. The sequence is [Bmim]Cl > [Mim]Cl > [Im]Cl. Govinda et al. [89] measured the speed of sound of trimethylammonium acetate $[(CH_3)_3NH][CH_3COO]$ (TMAA), trimethylammonium hydrogen sulfate $[(CH_3)_3NH][HSO_4]$ (TMAS), and trimethylammonium dihydrogen phosphate $[(CH_3)_3NH][H_2PO_4]$ (TMAP) in dimethylsulfoxide (DMSO) over the whole concentration range at a wide range of

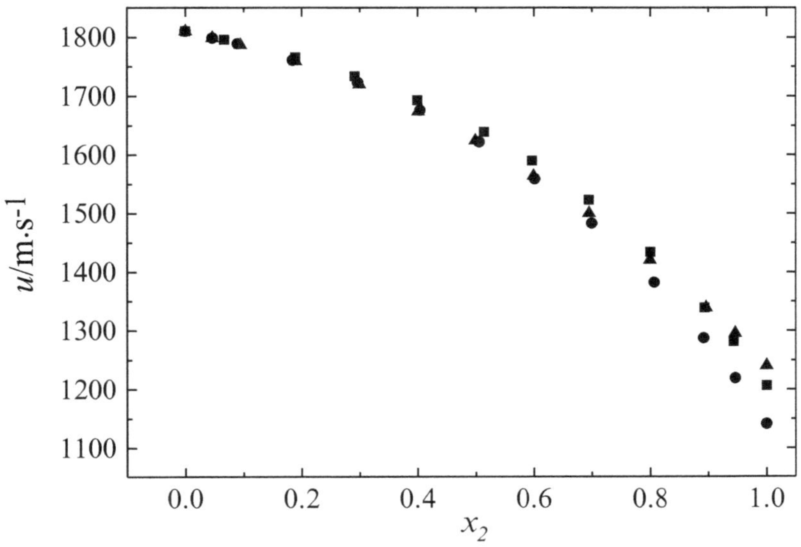

Fig. 1.3 The speed of sound as a function of mole fraction of ionic liquids (x_1) at $T = 298.15$ K. ■: 1-Propanol (1) + [BMpyr][dca] (2); ●: 2-Propanol (1) + [BMpyr][dca] (2); ▲: 1-Butanol (1) + [BMpyr][dca] (2)

temperatures. They found that the larger the anion size of ionic liquid, the higher the value of speed of sound. The sequence is TMAP > TMAS > TMAA. González et al. [90] found that the speed of sound of 1-butyl-1-methylpyrrolidinium dicyanamide ([BMpyr][dca]) mixtures is related to the interaction between [BMpyr][dca] and alcohol solvents (see Fig. 1.3).

1.2.3 Refractive Index

Refractive index (n_D) is a dimensionless number that describes how light, or any other radiation, propagates through that medium. It is one of the most important physical properties of transparent solid, liquid, and gas, and it can be determined accurately and conveniently. As a standard of purity of liquid, it is more reliable than the boiling point. In general, it is influenced by concentration, temperature, wavelength, and pressure. Now it is usually used in designing of laser systems, monitoring of environment pollution, and measuring material homogeneity of solidified transparent materials [91].

To better understand the nature of ionic liquids and expand their applications, detailed knowledge on the physical properties of ionic liquids and ionic liquids mixtures, such as refractive index, is required [92]. So, currently, much refractive index data of ionic liquids and ionic liquids mixtures is determined to understand their thermodynamic behavior.

Seki et al. [93] studied the relationship between refractive index of ionic liquid and the molecular polarizability of ion pair by measuring 17 types of room temperature ionic liquids at various temperatures. The result is that the refractive index of ionic liquids increase as the polarizability increases and shows a good linear tendency. The ionic liquids can be designed with high refractive indexes by using the above conclusion.

Generally, as for ionic liquids mixtures, it can be seen that the refractive index of the mixtures increases with the increase of the mole fraction of ionic liquids, and it decreases with the increase of temperature [94]. Iglesias-Otero et al. [95] found that the refractive index of [Bmim][BF_4] systems data can be fit well with the expression:

$$n_D = \frac{C_1 + C_2 x}{1 + C_3 x} \tag{1.72}$$

where x is the molar fraction of [Bmim][BF_4], C_i are parameters.

Shekaari et al. [96] measured the refractive index of 1-Pentyl-3-methylimidazolium Chloride ([PnMIm]Cl) aqueous solutions and found that the refractive index of the mixtures can be fitted with a polynomial expansion of the form:

$$n_D - n_D^* = \sum_{i=1}^{3} A_i m^i \tag{1.73}$$

where m is the molality of the ionic liquid, n_D^* is the refractive index of pure water, and A_i are fitting parameters.

To date, the estimation methods for the refractive index of room temperature ionic liquids are very few. Rilo et al. [97] proposed a very simple model (Eq. 1.74) to reproduce the refractive index of an ionic liquid mixture.

$$n_D = \left(\varnothing_{IL} n_{IL}^2 + \varnothing_S n_S^2\right)^{1/2} \tag{1.74}$$

where $\varnothing_{IL}$ and $\varnothing_S$ are the volume fraction of ionic liquid and solvent, respectively. And that they can be calculated by Eqs. 1.75 and 1.76:

$$\varnothing_{IL} = \frac{x_{IL} M_{IL} \rho_S}{x_{IL} M_{IL} \rho_S + x_S M_S \rho_{IL}} \tag{1.75}$$

$$\varnothing_S = \frac{x_S M_S \rho_{IL}}{x_{IL} M_{IL} \rho_S + x_S M_S \rho_{IL}} \tag{1.76}$$

where x_{IL} and x_S are the molar fraction of ionic liquid and solvent, M_{IL} and M_S are the molar mass of ionic liquid and solvent, and ρ_{IL} and ρ_S are the density of ionic liquid and solvent, respectively.

The model is very convenient, and as long as we know the real density for all compositions, the refractive index could be estimated accurately.

Kurnia et al. [98] predicted the refractive index of ionic liquids mixtures by using the following expressions proposed by Lorentz–Lorenz (Eq. 1.77), Dale–Gladstone (Eq. 1.78), and Eykman (Eq. 1.79) [99–101]:

$$\frac{n^2_{\mathrm{Dm,\,calc.}} - 1}{n^2_{\mathrm{Dm,\,calc.}} + 2} = \sum_{i=1}^{N}\left[\varphi_i\left(\frac{n^2_{\mathrm{Di}} - 1}{n^2_{\mathrm{Di}} + 2}\right)\right] \tag{1.77}$$

$$n_{\mathrm{Dm,\,calc.}} - 1 = \sum_{i=1}^{N}\left[\varphi_1\left(n_{\mathrm{Di}} - 1\right)\right] \tag{1.78}$$

$$\frac{n^2_{\mathrm{Dm,\,calc.}} - 1}{n^2_{\mathrm{Dm,\,calc.}} + 0.4} = \sum_{i=1}^{N}\left[\varphi_i\left(\frac{n^2_i - 1}{n^2_i + 0.4}\right)\right] \tag{1.79}$$

where $n_{\mathrm{Dm,calc.}}$ is the refractive index of the mixture, and $n_{\mathrm{D}i}$and φ_i are the refractive index and the volumetric fraction of components i, respectively.

This method shows a small deviation when compared with the experimental refractive index data for all the four binary mixtures.

There are different ways of defining the deviations in the refractive index (Δn_{D}). The earlier one is defined as the following equation [102] like any other deviation of a physical property of a liquid mixture. And it is the most commonly used equation.

$$\Delta n_{\mathrm{D}} = n_{\mathrm{D}} - \left(x_1 n_{\mathrm{D,\,1}} + x_2 n_{\mathrm{D,\,2}}\right) \tag{1.80}$$

where x_i is the mole fraction, and the subscripts 1 and 2 represent pure component 1 and component 2.

This is a convenient means of reporting high-quality data but is not physically significant, so some new expressions were proposed.

Brocos et al. [103] defined the deviations in the refractive index as the following equation:

$$\Delta n_{\mathrm{D}} = n_{\mathrm{D}} - \left(\varnothing_1 n_{\mathrm{D,\,1}} + \varnothing_2 n_{\mathrm{D,\,2}}\right) \tag{1.81}$$

where $\varnothing$ is the volume fraction, and the subscripts 1 and 2 represent pure component 1 and component 2.

Iglesias-Otero et al. [104] proposed a more rigorous definition of the deviations in the refractive index, and the equation is as follows:

$$\Delta n_{\mathrm{D}} = n_{\mathrm{D}} - \left[\frac{\left(n_{\mathrm{D,\,1}}\right)^2\left(n_{\mathrm{D,\,2}}\right)^2 + 2\varnothing_1\left(n_{\mathrm{D,\,1}}\right)^2 + 2\varnothing_2\left(n_{\mathrm{D,\,2}}\right)^2}{2 + \varnothing_1\left(n_{\mathrm{D,\,2}}\right)^2 + \varnothing_2\left(n_{\mathrm{D,\,1}}\right)^2}\right]^{1/2} \tag{1.82}$$

where $\varnothing$ is the volume fraction, and the subscripts 1 and 2 represent pure component 1 and component 2.

Reis et al. [105] defined the deviations in the refractive index as the following equation:

$$\Delta n_D = n_D - \left[\varnothing_1 (n_{D,1})^2 + \varnothing_2 (n_{D,2})^2\right]^{1/2} \tag{1.83}$$

where $\varnothing$ is the volume fraction, and the subscripts 1 and 2 represent pure component 1 and component 2.

The values of Δn_D that calculated by any above equations could be correlated with the ionic liquid mole fraction (x_1) or volume fraction ($\varnothing_1$) by the Redlich-Kister [106] equation.

The values of Δn_D are influenced by the structures of ionic liquids and solvents. Vercher et al. [94] found that that Δn_D decreases with increasing the length of alkyl chains of imidazolium ionic liquids. The deviations in the refractive index follow the sequence [emim][triflate] (1-ethyl-3-methylimidazolium trifluoromethanesulfonate) > [bmim][triflate] (3-butyl-1-methylimidazolium trifluoromethanesulfonate) > [beim][triflate] (3-butyl-1-ethylimidazolium trifluoromethanesulfonate). This is caused by the volume expansion of the substituents. González et al. [107] found that the refractive index deviation decreases as the length of the chain of alcohol increases. They also found that the refractive index deviation of the OH group that is in the secondary position on the chain of alcohol is higher than its OH group that is in the last position on the chain of alcohol. Namely, the sequence is methanol > 2-propanol > 1-propanol. The same behavior for mixtures of ionic liquids with alkanols can be seen from reports by Arce et al. [108] and Vercher et al. [102] .

Iglesias-Otero et al. [104] found that the positive values of the deviations in the refractive index corresponded to negative excess volume values, and the peaks appeared almost at the same molar fraction for ionic liquids mixtures. The reason is that if the value of excess volume is positive, there is more free volume in the solution. So, as a result, its refractive index will be lower than in an ideal solution.

The molar refraction R of ionic liquids mixtures is very important, and it is a measure of the ability of molecular orbitals to be deformed under an electric field [103]. The molar refraction R is defined as the following equation [109]:

$$R = V_m (n_D^2 - 1)/(n_D^2 + 2) \tag{1.84}$$

where V_m is molar volume.

1.2.4 Vapor–Liquid Equilibrium (VLE)

Vapor–liquid equilibrium (VLE) behavior data are very important for the design and development of the gas absorption and separation, distillation process, etc. Ionic liquids are usually used as absorbents or entrainers since they have the advantage of high separation factors, easy operation, and no entrainment in the top product of the column for the distillation [110].

1.2.4.1 Solubility of Gas in Ionic Liquid Systems

As the vapor pressure of ionic liquids is very low, it is usually neglected, and the gaseous phase is assumed to completely consist of pure-gas component. Consequently, the VLE condition is only applied to that gas component. Its results can be correlated by the extended Henry's law:

$$k_{\mathrm{H,G}}(T,p)a_{\mathrm{G}}(T,m_{\mathrm{G}}) = f_{\mathrm{G}}(T,p) \tag{1.85}$$

where $k_{\mathrm{H,G}}(T, p)$is the Henry's constant of gas G in ionic liquid at temperature $T$and pressure p (based on the molality scale). $a_{\mathrm{G}}(T, m_{\mathrm{G}})$ is the activity of the gas in the ionic liquid, and the influence of pressure on that activity is neglected. $f_G(T, p)$ is the fugacity of the gas in the vapor phase.

The influence of pressure on Henry's law constant is expressed as:

$$k_{\mathrm{H,G}}(T,p) = k^{(0)}_{\mathrm{H,G}}(T)\exp\left(\frac{V^{(\infty)}_{m,G}p}{RT}\right) \tag{1.86}$$

where $k^{(0)}_{\mathrm{H,G}}(T)$ is the Henry's constant at zero pressure, $V^{(\infty)}_{m,G}$ is the partial molar volume of the gas at infinite dilution in the ionic liquid, and R is the universal gas constant.

The activity of the respective gas in the ionic liquid (on the molality scale) is

$$a_G = \frac{m_G}{m^o}\gamma_G \tag{1.87}$$

where m^o=1 mol · kg^{-1}. The activity coefficient a_G is calculated by employing the virial expansion for the excess Gibbs energy (also on the molality scale) according to Pitzer [111, 112]:

$$\ln\gamma_{\mathrm{G}} = 2\frac{m_G}{m^o}\beta^{(0)}_{\mathrm{G,G}} + 3\left(\frac{m_G}{m^o}\right)^2\mu_{\mathrm{G,G,G}} \tag{1.88}$$

Parameters $\beta^{(0)}_{\mathrm{G,G}}$ and $\mu_{\mathrm{G,G,G}}$ describe binary and ternary interactions, respectively, between gas molecules in the solvent.

The fugacity of the pure gas f_{G} at equilibrium temperature and pressure is the product of the total pressure p and the fugacity coefficient $\Phi_G(T, p)$

$$f_{\mathrm{G}}(T,p) = \mathrm{p}\Phi_{\mathrm{G}}(T,p) \tag{1.89}$$

The Henry's constant of the gas in ionic liquid at zero pressure $k^{(0)}_{\mathrm{H,G}}(\mathrm{T})$ can be extrapolated from the experimental results.

$$k^{(0)}_{\mathrm{H,G}}(\mathrm{T}) = \lim_{p\to 0}\left[\frac{f_{\mathrm{G}}(T,p)}{m_{\mathrm{G}}/m^o}\right] \tag{1.90}$$

Table 1.2 Henry's law constants for CO_2 in several ionic liquids

Henry's law constants (atm) 303.15 K		Henry's law constants (bar) 298.15 K	
$[N_{4,1,1,1}][Tf_2N]^a$	60	$[bmim][PF_6]^c$	53.4
$[N_{6,1,1,1}][Tf_2N]^a$	43	$[bmmim][PF_6]^c$	61.8
$[N_{10,1,1,1}][Tf_2N]^a$	44	$[bmim][BF_4]^c$	56.5
$[N_{4,1,1,3}][Tf_2N]^a$	63	$[bmmim][BF_4]^c$	61
$[N_{6,1,1,3}][Tf_2N]^a$	51	$[emim][Tf_2N]^c$	35.6
$[N_{10,1,1,3}][Tf_2N]^a$	42	$[emmim][Tf_2N]^c$	39.6
$[N_{1,4,4,4}][Tf_2N]^a$	50	$[Bmim][Tf_2N]^d$	33
$[N_{1,8,8,8}][Tf_2N]^a$	28	$[Hmim][Tf_2N]^d$	31.6
$[N_{6,2,2,2}][Tf_2N]^a$	58	$[Hmim][eFAP]^d$	25.2
$[P_{14,6,6,6}]Cl^b$	35	$[Hmim][pFAP]^d$	21.6
$[P_{14,6,6,6}][DCA]^b$	29.3	$[p_5mim][bFAP]^d$	20.2
$[P_{14,6,6,6}][Tf_2N]^b$	33	$[C_6H_4F_9mim][Tf_2N]^d$	28.4
$[P_{2,4,4,4}][DEP]^b$	69	$[C_8H_4F_{13}mim][Tf_2N]^d$	27.3
$[P_{14,4,4,4}][DBS]^b$	30	$[Et_3NBH_2mim][Tf_2N]^d$	33.1
		$[Hmpy][Tf_2N]^d$	32.8

[a]From ref. [120]
[b]From ref. [121]
[c]From ref. [113]
[d]From ref. [122]

Up to now, there has been research about the gases including CO_2, SO_2, NH_3, N_2, O_2, etc. Among the gases, CO_2 has been investigated more. In the case of the physical absorption, CO_2 fills the available free molar volume of the ionic liquid and interacts preferentially with the anion rather than the cation. [113, 114] Generally, the CO_2 solubility increases with increasing alkyl side chain of cations and decreases with increasing temperatures [115–117].

In order to compare the CO_2 solubility in different ionic liquids, the Henry's law constants of solution were collected in Table 1.2. Henry's law constants for the phosphonium-based room temperature ionic liquids are, generally, less than the values for ammonium-based room temperature ionic liquids, and fluoroalkyl chains are frequently added to substrates to increase their solubility in supercritical CO_2. Recently, many amino-functionalized ionic liquids were synthesized and used in the CO_2 absorption. It was found that the results of the supported absorption of CO_2 by the amino-functionalized ionic liquids are very good [118, 119].

1.2.4.2 Vapor–Liquid Phase Equilibrium

As for the vapor–liquid equilibrium formed by liquid mixture in a closed system, there is a single variable that remains independent when the liquid compositions are given. For example, this can be temperature or pressure.

The VLE data are generally measured at constant temperature or constant pressure. For isothermal measurements, there are generally two measuring techniques: One is headspace measurement in which the vapor composition is analyzed [123, 124], and the other is static measurement in which the total pressure is obtained

[125]. For isobaric measurements, the method of boiling point analysis is the most commonly used method [126–128].

For the binary mixtures, experimental data can be fitted using the NRTL model. The bubble-point pressure of a binary mixture may be estimated using the classical Eq. 1.91:

$$P = \frac{P_1^{sat} x_1 \gamma_1}{C_1} + \frac{P_2^{sat} x_2 \gamma_2}{C_2} \tag{1.91}$$

with

$$C_i = \frac{\varphi_i^{vap}(T, P, y_1, y_2)}{\varphi_i^*(T, P_i^{sat})} \exp\left(\frac{1}{RT}\int_P^{P_i^{sat}} v_{liq}^* dP\right) \quad i = 1, 2 \tag{1.92}$$

$$F_P = \exp\left(\frac{1}{RT}\int_P^{P_i^{sat}} v_{liq}^* dP\right) \quad i = 1, 2 \tag{1.93}$$

where P_i^{sat} is the vapor pressure of the pure component i at system temperature, x_i represents mole fraction of component i in the liquid phase, φ_i^{vap} is the fugacity coefficient of component i in the gas phase, φ_i^* is the fugacity coefficient of pure component i at the saturation, F_P is the Poynting correction.

The separation factors (α_{ij}) for each system were calculated using the following relation:

$$\alpha_{ij} = \frac{K_i}{K_j} = \frac{y_i/x_i}{y_j/x_j} \tag{1.94}$$

where K_i is the ratio of mole fraction of component i in the vapor (y_i) to that in the liquid phase (x_i).

The relative volatility a_{ij} can be obtained from the following equation:

$$a_{12} = \frac{\gamma_1 P_1^o}{\gamma_2 P_2^o} \tag{1.95}$$

where γ_i is the activity coefficient of component i and P_i^o is the vapor pressure of solvent i at equilibrium temperature which was calculated by using the Antoine coefficients obtained with the same recirculating still [129].

Thus far, there are several thermodynamic models that are used to correlate the VLE, such as activity coefficient models [130–132], equations of state [133, 134], and lattice theory [135–138].

Activity coefficient models mainly include equation of Wilson, equation of NRTL, equation of UNIQUAC, and equation of UNIFAC, in which equation of NRTL is most widely used. These equations are listed in Table 1.3.

Table 1.3 Several activity coefficient models

Name	Description
Wilson	$\ln\gamma_i = 1 - \ln\left(\sum_j \Lambda_{ij}x_j\right) + \sum_k \frac{\Lambda_{ki}x_k}{\sum_j \Lambda_{kj}x_j}$ $\Lambda_{ij} = \frac{V_j}{V_i}\exp\left(-\frac{g_{ij}-g_{ii}}{RT}\right)$
NRTL	$\ln\gamma_i = \frac{\sum_j \tau_{ji}G_{ji}x_j}{\sum_k \tau_{ki}G_{ki}x_k} + \sum_j \frac{G_{ij}x_j}{G_{ki}x_k}\left(\tau_{ij} - \frac{\sum_l \tau_{lj}G_{lj}x_l}{\sum_k G_{kj}x_k}\right)$ $\tau_{ji} = \frac{g_{ji}-g_{ii}}{RT}$, $G_{ji} = \exp(-\alpha_{ji})\,\tau_{ji}$, $\alpha_{ij} = \alpha_{ji}$
UNIQUAC	$\ln\gamma_i = ln\frac{\phi_i}{x_i} + \left(\frac{Z}{2}\right)q_i ln\frac{\theta_i}{\phi_i} + l_i - \frac{\phi_i}{x_i}\sum_j x_j l_j - q_i\ln\left(\sum_j \theta_j\tau_{ji}\right) + q_i - q_i\sum_j \frac{\theta_j\tau_{ij}}{\sum_k \theta_k\tau_{kj}}$ $l_i = \frac{Z}{2}(r_i - q_i) - (r_i - 1)$, $\theta_i = \frac{q_i x_i}{\sum_j q_j x_j}$, $\phi_i = \frac{r_i x_i}{\sum_j r_j x_j}$, $\tau_{ji} = \exp\left(-\frac{u_{ji}-u_{ii}}{RT}\right)$
UNIFAC	$\ln\gamma_i = \ln\gamma_i^c + \ln\gamma_i^R$ $\ln\gamma_i^c = \ln\frac{\phi_i}{x_i} + \left(\frac{Z}{2}\right)q_i ln\frac{\theta_i}{\phi_i} + l_i - \frac{\phi_i}{x_i}\sum_j x_j l_j$ $\ln\gamma_i^R = \sum v_k^i\left(\ln\Gamma_k - \ln\Gamma_k^i\right)$ $\ln\Gamma_k = Q_K\left[1 - \ln\left(\sum_m \theta_m\Psi_{mk}\right) - \sum_m \frac{\theta_m\Psi_{mk}}{\sum_n \theta_n\Psi_{nm}}\right]$ $\ln\Gamma_k^i = Q_K\left[1 - \ln\left(\sum_m \theta_m^i\Psi_{mk}\right) - \sum_m \frac{\theta_m^i\Psi_{mk}}{\sum_n \theta_n^i\Psi_{nm}}\right]$ $\theta_i = \frac{q_i x_i}{\sum_j q_j x_j}$, $\phi_i = \frac{r_i x_i}{\sum_j r_j x_j}$, $q_i = \sum v_k^i Q_k$, $r_i = \sum v_k^i R_k$ $\theta_m = \frac{q_m x_m}{\sum_n q_n x_n}$, $\Psi_{nm} = \exp\left(-\frac{a_{nm}}{T}\right)$, $x_m = \frac{\sum_j v_m^i x_j}{\sum_j\sum_j v_k^i x_i}$

The calculation of phase behavior using an equation of state (EOS) has been explored relatively less. The generic vdW (van der Waals) type of cubic EOS is written in the following form:

$$P = \frac{RT}{V-b} - \frac{a(T)}{V^2} \tag{1.96}$$

$$a(T) = 0.421875\frac{R^2T_c^2}{P_c}\alpha(T) \tag{1.97}$$

$$b = 0125\frac{RT_c}{P_c} \tag{1.98}$$

For the present EOS form, the fugacity coefficient ϕ_i of ith species is reduced to a simple form as follows:

$$\ln\phi_i = \ln\frac{RT}{P(V-b)} + \frac{b_i'}{V-b} - \frac{a_i' + a}{VRT} \tag{1.99}$$

where $a_i' \equiv \left(\frac{\partial n_a}{\partial n_i}\right)_{n_{j\neq i}}$ and $b_i' \equiv \left(\frac{\partial n_b}{\partial n_i}\right)_{n_{j\neq i}}$: n = total mole number and n_i = mole number of ith species (or $x_i = n_i/n$)

Since ionic liquids are usually composed of cations and anions with long alkyl chains or a long chainlike structure and have negligible vapor pressures, several authors tried to model ionic liquid by the use of lattice theory. Hu et al. [134, 139] extended and applied the model to the description of thermodynamic properties of the [$C_n$mim][BF_4] and [PF_6] families and the solubility of CO_2. The resulting lattice fluid equation was expressed as:

$$\begin{aligned}\tilde{p} = \tilde{T}\left\{-\ln(1-\tilde{\rho}) + \frac{Z}{2}\ln\left[\frac{2}{Z}\left(\frac{1}{r}-1\right)\tilde{\rho}+1\right]\right\} - \frac{Z}{2}\tilde{\rho} \\ -\frac{Z}{4}\frac{1}{\tilde{T}}\left(3\tilde{\rho}^4 - 4\tilde{\rho}^3 + \tilde{\rho}^2\right) - \frac{Z}{12}\frac{1}{\tilde{T}^2}\left(10\tilde{\rho}^6 - 24\tilde{\rho}^5 + 21\tilde{\rho}^4 - 8\tilde{\rho}^3 + \tilde{\rho}^2\right) \\ +\frac{r-1+\lambda}{r}\tilde{T}\tilde{\rho}^2\left[\frac{[1+D(1-\tilde{\rho})]^2 - 1}{[1+D(1-\tilde{\rho})][1+D\tilde{\rho}(1-\tilde{\rho})]}\right]\end{aligned} \tag{1.100}$$

with

$$D = \exp\left(\frac{1}{\tilde{T}}\right) - 1 \tag{1.101}$$

where z is the coordination number, $\tilde{T}$ is the reduced temperature, $\tilde{p}$ is the reduced pressure, and $\tilde{\rho}$ is the reduced density.

By far there are numerous approaches for predicting VLE of ionic liquids mixtures, such as the COSMO-RS (Conductor-like screening model for real solvents) model [140–142], the quantitative structure-property relationship (QSPR) method [143], the regular solution model [144–148], and the UNIFAC (Universal quasichemical functional group activity coefficients) model [149–154].

The COSMO-RS model is an efficient method for the prediction of thermophysical data, such as the phase behavior of solvent-ionic liquid mixtures. It is based on quantum chemistry and uses only atom-specific parameters [155].

For the mixtures, the σ-profile of a solvent S, $p_S(\sigma)$ is the result of adding the individual $p^{X_i}(\sigma)$ weighed by their mole fractions, x_i, as expressed in Eq. 1.102:

$$p_S(\sigma) = \sum_{i\in S} x_i p^{X_i}(\sigma) \tag{1.102}$$

The statistical thermodynamics is expedient to consider a normalized ensemble and since the integral of $p^{X_i}(\sigma)$ over the entire σ-range is the total surface area A^{X_i} of a compound X_i, the normalized σ-profile, $p'_S(\sigma)$, of the overall system is defined as follows:

$$p'_S(\sigma) = \frac{\mathrm{p}_S(\sigma)}{A_S} = \frac{\mathrm{p}_S(\sigma)}{\sum_{i \in S} x_i A^{X_i}} \tag{1.103}$$

The interaction energies of the contact segments are due to specific contributions from misfit (Eq. 1.104), hydrogen bonding interactions (Eq. 1.105), and van der Waals (Eq. 1.106) interactions:

$$E_{misfit}(\sigma, \sigma') = a_{eff} \frac{\alpha'}{2} \left(\sigma + \sigma^2\right)^2 \tag{1.104}$$

$$E_{HB} = a_{eff} c_{HB} \min\left(0; \min(0; \sigma_{donor} + \sigma_{HB}\right) \times \max\left(0; \sigma_{acceptor} - \sigma_{HB}\right)\right) \tag{1.105}$$

$$E_{vdW} = a_{eff}\left(\tau_{vdW} + \tau'_{vdW}\right) \tag{1.106}$$

where α' is the coefficient for electrostatic misfit interactions, a_{eff} is the effective contact area between two surface segments, c_{HB} is the coefficient for hydrogen bond strength, σ_{HB} is the threshold for hydrogen bonding and τ_{vdW} and τ'_{vdW} are element-specific van der Waals coefficients.

In a word, COSMO-RS is very useful in designing new ionic liquids for specific applications before extensive experimental measurements.

1.2.4.3 Activity Coefficients at Infinite Dilution

Activity coefficients at infinite dilution (γ^∞) describe the behavior of a single solute molecule completely surrounded by solvent, and it indicates a maximum nonideality [156]. As for binary mixtures, γ^∞ can be represented as follows:

$$\gamma_1^\infty = \lim_{x_1 \to 0} \gamma_1 \tag{1.107}$$

$$\gamma_2^\infty = \lim_{x_2 \to 0} \gamma_2 \tag{1.108}$$

Activity coefficients at infinite dilution provide the information of interaction between solute and solvent and have important theoretical and application value [156–158]. So, it is not only the importance parameter of evaluating the solvent separation performance but also the gist of characterizing of the strength of interactions between the solute and solvent. Now it is widely used in the design of units operation, such as distillation, absorption, and extraction.

There are a number of different methods available for measurement of γ^∞, and the commonly used methods are: extrapolation method, differential ebulliometry, differential static cell, gas stripping, and gas–liquid chromatography [158, 159]. Now gas chromatography is the most widely used method since it is simple and has a reliable precision. Since ionic liquids have some attractive properties, such as negligible vapor pressure and high thermal stability, they can be used as stationary phase to measure activity coefficients at infinite dilution for organic solutes by gas–liquid chromatography [160]. Until 2011, the values of activity coefficient at infinite dilution of organic compounds in 58 available ionic liquids have been published in many valuable papers [161]. Moreover, the number of these papers presents the trend of growth in recent years.

As for the method of gas–liquid chromatography, the γ^{∞} for organic solutes and water in ionic liquids are calculated from the following equation developed by Everett [162] and Cruickshank et al. [163].

$$\ln\gamma_{1,3}^{\infty} = \ln\left(\frac{n_3RT}{V_NP_1^0}\right) - \frac{B_{11} - v_1^0}{RT}P_1^0 + \frac{2B_{12} - v_1^{\infty}}{RT}JP_0 \tag{1.109}$$

In Eq. 1.109, the subscript 1 refers to the solute, the subscript 2 refers to carrier gas, the subscript 3 refers to the solvent, $\gamma_{1,3}^{\infty}$ is the activity coefficient of solute at infinite dilution in the stationary phase, n_3 is the number of moles of solvent on the column packing, R is the gas constant, T is the column temperature, V_N is the net retention volume of the solute, P_1^0 is the saturated vapor pressure of the solute at temperature T, B_{11} is the second virial coefficient of pure solute, v_1^0 is the molar volume of the solute, v_1^{∞} is the partial molar volume of the solute at infinite dilution in the solvent, B_{12} is the mixed second virial coefficient of the solute and carrier gas and is calculated using the Tsonopolous equation [10], and JP_0 is the mean column pressure.

The net retention volume of the solute V_N is given by

$$V_N = JU_0(t_r - t_G)\frac{T_{col}}{T_f}\left[1 - \frac{P_{0w}}{P_o}\right] \tag{1.110}$$

where t_r is the retention time (min), t_G is the dead time (min), J is the pressure correction term, U_0 is the flow rate (ml·min^{-1}), T_{col} is the column temperature (K), T_f is the flow meter temperature (K), P_{0w} is saturation pressure of water at T_f, P_o is the pressure at the column outlet.

The pressure correction term J is given by Grant [164].

$$J = \frac{3(P_i/P_o)^2 - 1}{2(P_i/P_o)^3 - 1} \tag{1.111}$$

where P_i and P_o are the inlet and outlet pressures of the GC column, respectively.

So far, the common prediction models of activity coefficients at infinite dilution for organic solutes in ionic liquids are UNIFAC model and COMOS-RS models.

UNIFAC is one of the most successful group contribution models, and it was originally used to predict vapor–liquid equilibria (VLE) [165]. The advantage of UNIFAC model is that the number of functional groups is much smaller than the number of chemical compounds. In theory, as long as the value of functional groups and the group interaction parameters are available, some data of thermodynamic properties of these structural groups can be calculated. Gmehling and his team proposed modified UNIFAC (Do) [166–170]. They mainly fitted a large number of reliable experimental data, such as vapor–liquid equilibria, h^E, and γ^∞ to obtain group interaction parameters. The predicted results using this method proved to be very reliable. Since the group contribution methods are one of the important thermodynamic tools, more accurate modified models are carried out.

COSMO-RS is a dielectric continuum solvation model [171–173]. In this model, molecules are divided into surface of the same size segments, and it is assumed that any segment can get in contact with every other segment [165]. Some basic parameters fitted with a few experimental data was used when performing COSMO-RS calculations, so this model is usually used in calculating substances whose experimental data are missing [174]. This model originally was used to predict vapor pressures and partition coefficients, and till now, it has been used to predict the activity coefficients at infinite dilution of ionic liquids in organic compounds [175, 176]. In all, COSMO-RS is a convenient tool and has a better accuracy.

The temperature dependence of γ^∞ can be correlated according to the following equation:

$$\ln\gamma^\infty = a + \frac{b}{T} \tag{1.112}$$

According to Gibbs-Helmholtz equation (Eq. 1.113), the partial molar excess enthalpy at infinite dilution ($H^{E,\infty}$) can be obtained from the slope of lines fitted by Eq. 1.112.

$$\frac{\partial \ln\gamma^\infty}{\partial(1/T)} = \frac{H^{E,\infty}}{R} \tag{1.113}$$

The selectivity values at infinite dilution $\left(S_{ij}^\infty\right)$ are used to select an appropriate solvent in the process of separation, and it can be determined from the γ^∞. The equation [177] is as follows:

$$S_{ij}^\infty = \frac{\gamma_{ik}^\infty}{\gamma_{jk}^\infty} \tag{1.114}$$

The solubility parameter of solvents (δ_s) is a physical quantity that has good predictability. However, the data about solubility parameter of ionic liquids is very little. The activity coefficients at infinite dilution of different solutes in a given solvent can be used to estimate solubility parameter of the solvents. According to the Flory–Huggins equation and the Scatchard–Hildebrand equation, the equations for calculating solubility parameter of ionic liquids are as follows [178, 179]:

$$ln\gamma_i^\infty = ln\gamma_i^{\infty\mathrm{comb}} + ln\gamma_i^{\infty\mathrm{res}} \tag{1.115}$$
$$ln\gamma_i^{\infty\mathrm{comb}} = \ln(r_i/r_s)^{2/3} + 1 - (r_i/r_s)^{2/3} \tag{1.116}$$

$$ln\gamma_i^{\infty res} = (v_i/RT)(\delta_i - \delta_s)^2 \tag{1.117}$$

$$Y = -\frac{ln\gamma_i^{\infty res}}{v_i} + \frac{\delta_i^2}{RT} = \frac{2\delta_s}{RT}\delta_i - \frac{\delta_s^2}{RT} \tag{1.118}$$

where $ln\gamma_i^{\infty comb}$ is the combinatorial term, $ln\gamma_i^{\infty res}$ is the residual term, r_i and r_s are the van der Waals volumes of solute and solvent, respectively, v_i is the solute molar volume, δ_i and δ_s are the solubility parameters of solute and solvent, respectively. v_i, δ_i and r_i were obtained from the literature [180]. The van der Waals volumes of ionic liquids were calculated by group contribution method [181]. From the Eq. 1.118, we can see that there is a linear relation between a residual function (Y_i) and the solute solubility parameter δ_i for a given solvent and temperature T. Wang et al. [161] have shown that the linear correlation has a high accuracy. The solubility parameter of ionic liquids can be calculated from the value of the solvent solubility parameter δ_s which can be obtained from the slope of this line.

1.2.5 Liquid–Liquid Equilibrium (LLE)

The determination of liquid–liquid equilibrium (LLE) data is a common procedure for the evaluation of the ability of a solvent. Therefore, LLE data is very important for the design and development of separation processes, and its use is fundamental to wide industrial applications in chemical engineering [182].

Various models have been used for correlating experimental data of liquid–liquid equilibrium of ionic liquid systems, such as NRTL, UNIQUAC, the modified Flory–Huggins equation, and a lattice model based on polymer-solution models.

The NRTL model [183] is the most commonly used model for the correlation of these LLE data. According to this model, the nonideal liquid phase activity coefficient (γ of component i) is given by the following equation.

$$\ln\gamma_i = \frac{\sum_{j=1}^{c}\tau_{ji}G_{ji}x_j}{\sum_{k=1}^{c}G_{ki}x_k} + \sum_{j=1}^{c}\left[\frac{G_{ij}x_j}{\sum_{k=1}^{c}G_{kj}x_k}\left(\tau_{ij} - \frac{\sum_{i=1}^{c}\tau_{ij}G_{ij}x_i}{\sum_{k=1}^{c}G_{kj}x_k}\right)\right] \tag{1.119}$$

where

$$G_{ji} = \exp(-\alpha_{ji}\tau_{ji}) \tag{1.120}$$

$$\tau_{ji} = \frac{g_{ji} - g_{ii}}{RT} \tag{1.121}$$

where x is the mole fraction, τ represents interaction parameter, $g_{ji} - g_{ii}$ are the fitting parameters, the nonrandomness parameter $\alpha_{ji} = \alpha_{ij} = \alpha$ was optimized.

The UNIQUAC model [184] is less commonly used than NRTL due to the lack of pure component volume and area parameters for ionic liquids. It gives the following equations for the nonideal activity coefficient for component *i*.

$$\ln\gamma_i = \ln\left(\frac{\Phi_i}{x_i}\right) + \frac{Z}{2}q_i\ln\left(\frac{\theta_i}{\Phi_i}\right) + l_i - \frac{\Phi_i}{x_i}\sum_{j=1}^{c} x_j l_j$$
$$+ q_i\left(1 - ln\sum_{j=1}^{c}\theta_j\tau_{ji} - \sum_{j=1}^{c}\frac{\theta_j\tau_{ij}}{\sum_{k=1}^{c}\theta_k\tau_{kj}}\right) \quad (1.122)$$

where

$$\tau_{ji} = \frac{g_{ji} - g_{ii}}{RT} = \frac{A_{ji}}{T} \quad (1.123)$$

$$q_T = \sum_k q_k x_k \quad (1.124)$$

$$\theta_i = \frac{q_i x_i}{q_T} \quad (1.125)$$

$$\Phi_i = \frac{r_i x_i}{r_T} \quad (1.126)$$

$$r_T = \sum_k r_k x_k \quad (1.127)$$

$$l_i = \frac{Z}{2}(r_k - q_k) + 1 - r_k \quad (1.128)$$

where x is the mole fraction, θ and Φ are area fraction and segment fraction, respectively.

The Flory–Huggins model modification [185] is applied successfully in ionic liquids mixtures by Rebelo et al. [186]. The model expresses the dimensionless excess Gibbs energy as follows:

$$\frac{G^E}{RT} = (rx_1 + x_2)\chi(T)\varphi_1\varphi_2 \quad (1.129)$$

where x_i are mole fractions, $\chi(T)$ is the interaction parameter, the temperature dependence of which is given by the Eq. 1.130, and φ_i are the segment fractions related to the mole fractions by the Eq. 1.131

$$\chi(T) = d_0 + \frac{d_1}{T} - d_2 lnT \quad (1.130)$$

$$\varphi_1 = \frac{rx_1}{rx_1 + x_2}, \varphi_2 = \frac{rx_2}{rx_1 + x_2} \quad (1.131)$$

where r is the number of segments occupied by component 1, subscripts 1 and 2 referring to the larger and smaller molecule, respectively. Usually, r does not differ greatly from the ratio of molar volumes of the components ($r \approx V_1/V_2$).

In the lattice model [187] of that molecular thermodynamic model, the Gibbs energy of mixing is expressed as:

$$\begin{aligned}\frac{G^M}{RT} &= \frac{\varphi_1}{r_1} ln\varphi_1 + \frac{\varphi_2}{r_2} ln\varphi_2 + \frac{Z}{2}\left(\frac{\varphi_1 q_1}{r_1} ln\frac{\theta_1}{\varphi_1} + \frac{\varphi_2 q_2}{r_2} ln\frac{\theta_2}{\varphi_2}\right) \\ &+ \frac{Z\varphi_1\varphi_2}{2(1-\varphi_1\varphi_2)}\left[\frac{\varepsilon}{kT} - \ln(1+\varphi_1\varphi_2 C)\right] - \frac{b_1\varphi_1}{r_1}\ln\left(\frac{1-\varphi_2 C}{1+\varphi_1\varphi_2 C}\right) \\ &- \frac{b_2\varphi_2}{r_2}\ln\left(\frac{1-\varphi_1 C}{1+\varphi_1\varphi_2 C}\right)\end{aligned} \quad (1.132)$$

with

$$C = \exp\left(\frac{\varepsilon}{kT}\right) - 1 \quad (1.133)$$

where ε is the interchange energy, Z is the coordination number, φ_i is the volume (or segment) fraction, r_i is the number of segments of molecule, b_i is the number of chemical bonds in the molecule:

$$b_i = r_i - 1 \quad (1.134)$$

q_i is the structural parameter (zq_i is the surface of molecule):

$$zq_i = zr_i - 2b_i \quad (1.135)$$

θ_i is the surface fraction, and it can be defined as:

$$\theta_i = \frac{q_i N_i}{q_1 N_1 + q_2 N_2} \quad (1.136)$$

N_i is the number of lattice sites occupied by the molecule. Temperature dependence of the interchange energy is expressed as:

$$\frac{\varepsilon}{kT} = Aexp(-BT) \quad (1.137)$$

$$B = 2\alpha_T \quad (1.138)$$

where α_T is the coefficient of thermal expansion.

The solute distribution ratio (β) and the selectivity (S) are important parameters in liquid–liquid extraction. They can be defined as follows:

$$\beta = \frac{x_i^{\mathrm{II}}}{x_i^{\mathrm{I}}} \tag{1.139}$$

$$S = \frac{x_2^{\mathrm{II}} x_1^{\mathrm{I}}}{x_2^{\mathrm{I}} x_1^{\mathrm{II}}} \tag{1.140}$$

where x_i^{I} and x_i^{II} are mole fractions of compound i ($i = 1$ or 2) in phase I and II, respectively. They can be influenced by the temperature and the rule is that the β and S decrease as temperature increases.

The LLE of ionic liquid mixtures can be influenced by temperatures, ions, the size of solvent, etc. Generally, a temperature increase leads to an increased miscibility with a contraction of the immiscibility region. Yang et al. [188] found that a longer alkyl chain length in ionic liquid molecules enhances the solubility of ionic liquid in alcohols. Arce et al. [189, 190] measured the LLE data of 1-alkyl-3-methyl-imidazolium bis[(trifluoromethyl)sulfonyl]amide ionic liquids ([$C_n$mim][NTf_2], $n = 2,4,8,10,12$) mixtures, and their conclusion shows that the selectivity increases as the cation alkyl chain length decreases. Ferreira et al. [191] investigated the influence of the size of ionic liquid cation core on the solubility of mixtures with the same anion ([NTf_2]). The results indicate that the solubility increases in the order ammonium < imidazolium < pyridinium < phosphonium. The corresponding selectivity decreases in the order ammonium > imidazolium > pyridinium > phosphonium. They also studied the influence of the size of ionic liquid anion core on the solubility and selectivity with the same cation ([C_2mim]). The solubility order is [DCA] $\approx$ [SCN] $\approx$ [$MeSO_4$], and the corresponding selectivity order is [DCA] > [SCN] > [$MeSO_4$]. As for the effect of size of solvents on the phase behavior, Ferreira et al. [192] found that the mutual solubilities between hydrocarbons (with a constant carbon number) and ionic liquids follow the increasing order aromatics > cyclic aliphatic hydrocarbons > *n*-alkanes in another investigation. Domańska et al. [193] reported that the solubility of *n*-alkane decreases with an increase of the alkyl chain length of the *n*-alkane from C_3 to C_{10} in 1,3-dimethylimidazolium methylsulfate ([mmim][CH_3SO_4]) and 1-butyl-3-methylimidazolium methylsulfate ([bmim][CH_3SO_4]).

In order to expand the application of ionic liquids mixtures, we need to use the models to predict the phase behavior in real processes. So far, the commonly used models are COSMO-RS model [172, 176, 194], lattice-fluid model [195, 196], analytical solution of groups (ASOG) model [197–200], etc.

Successful applications of COSMO-RS to the predictions of the thermodynamic properties of ionic liquids have been reported. For example, Banerjee et al. [201] used the COSMO-RS model to predicted LLE for 32 ternary systems with different ionic liquids.

Xu et al. [135] developed a new multicomponent lattice-fluid model to predict LLE of ternary systems containing ionic liquids. The new model has only one adjustable binary parameter k_{ij}, which can be determined by experimental thermodynamic data of binary systems. This model can well predict LLE of ternary systems containing ionic liquids by a comparison of model prediction with the experimental data.

ASOG model mainly predicts the activity coefficients by using a group contribution model. Robles [202] used the ASOG model to predict LLE data of 28 systems. The results of the prediction were satisfactory, and it can be concluded that the ASOG model is able to predict the LLE of binary and ternary systems including these ionic liquids.

1.2.6 Solid–Liquid Equilibrium (SLE)

Solid–liquid equilibrium (SLE) are an important phase equilibrium property for ionic liquid mixtures since nowadays many ionic liquids exist as solids, such as pyrrolidinium or pyridinium-based cations.[203] The data of SLE is usually used in the design of separation processes, especially cooling, evaporation, and antisolvent crystallization.

The study of the SLE is less mature than vapor–liquid equilibrium (VLE) and liquid–liquid equilibrium (LLE). The models used in SLE are from VLE or LLE and then do a few small modifications. So far, the most commonly used methods are activity coefficient equation, equation of state, λh equation, and so on.

Activity coefficient equations mainly include Wilson, NRTL, UNIQUAC, UNIFAC, etc. These equations are listed in Table 1.3.

The existing equation of state can be divided into three categories: (1) Cubic equations of state, such as van der Waals equation, RK equation, SRK equation, and PR equation; (2) Multi-parameter equations of state, such as BWR equation, MH equation; and (3) Theory equations of state, such as virial equation. These equations are mainly used in the study of VLE. They expressed directly the solubility of solute as a function of fugacity and fugacity coefficient, which avoid using of the activity coefficient of solid solute. When these equations are used in the study of the SLE, they could have a good precision in both solid and liquid phase, which limits their applications in the study of SLE.

λh equation is proposed specially for the study of SLE. Its form is very simple (Eq. 1.141)

$$\ln\left[1+\frac{\lambda(1-x_i)}{x_i}\right]=\lambda h\left(\frac{1}{T}-\frac{1}{T_{mi}}\right) \tag{1.141}$$

where λ and h are equation parameters, x_i is the mole fraction of the solute, T is temperature, and T_{mi} is the melting point of solute. It is convenient to use in the SLE, since it only needs to know the melting point of pure substance. Nevertheless, it has a great error when used in multicomponent system [204].

So far, there are numerous studies on SLE of ionic liquid mixtures. A general condition for SLE between pure solid and saturated solution of 1 in solvent is defined as [205]:

$$\ln a_1 = ln x_1 \gamma_1 = -\frac{\Delta_{fus} H_1^o}{RT}\left(1-\frac{T}{T_{fus,1}^o}\right) - \frac{\Delta_{fus} c_{p,1}^o}{R}\left(1-\frac{T_{fus,1}^o}{T} + ln\frac{T_{fus,1}^o}{T}\right) \quad (1.142)$$

where $T_{fus,1}^o$, $\Delta_{fus}H_1^o$, and $\Delta_{fus}c_{p,1}^o$ are basic thermal properties of pure solid at $p = 0.1$ MPa, namely, fusion temperature, fusion enthalpy, and isobaric heat capacity change due to fusion. R is the gas constant and x_1 is the solid solubility (in mole fraction) at temperature T.

If the solid–solid phase transitions were observed in ionic liquids mixtures, and the change of heat capacity at the melting temperature was not known, a general thermodynamic equation relating temperature, T_{SLE} and the mole fraction of the ionic liquid, x_1 in the solvent could be fitted as follows [205]:

$$\ln x_1 \gamma_1 (x_1, T, a) = -\frac{\Delta_{fus} H_1}{R}\left(\frac{1}{T} - \frac{1}{T_{fus,1}}\right) \quad (1.143)$$

where $\Delta_{fus}H$ and $T_{fus,1}$ are melting enthalpy and temperature of ionic liquid. At temperature below solid–solid phase transition, the equation must include the transition term as follows [132]:

$$\ln x_1 \gamma_1 (x_1, T, a) = -\frac{\Delta_{fus} H_1}{R}\left(\frac{1}{T} - \frac{1}{T_{fus,1}}\right) - \frac{\Delta_{tr} H_1}{R}\left(\frac{1}{T} - \frac{1}{T_{tr,1}}\right) \quad (1.144)$$

where $\Delta_{tr}H$ and $T_{tr,1}$ are the enthalpy and the temperature of the solid–solid phase transition, respectively.

We can use the Wilson [206], UNIQUAC [184], and the NRTL [183] equations to fit the solute activity coefficients, γ_1 to describe the Gibbs excess free energy of mixing (G^E). These equations have two adjustable parameters P_1 and P_2 (and the parameter α which is fixed, additionally for the NRTL equation), and they can be defined as follows:

$$\mathrm{F}(P_1, P_2) = \sum_{i=1}^{n} \left[T_{exp,i} - T_{calc,i}(x_i, P_1, P_2)\right]^2 \quad (1.145)$$

where n is the number of experimental points. $T_{exp,i}$ and $T_{calc,i}$ are the experimental and calculated equilibrium temperatures, respectively.

As a measure of the reliability of the correlations, the root-mean-square deviation of temperature σ_T can be obtained from the following equation:

$$\sigma_T = \left[\sum_{1}^{n} \frac{\left(T_{exp,i} - T_{calc,i}\right)^2}{n - 2} \right]^{1/2} \tag{1.146}$$

Parameters r_i and q_i of the UNIQUAC ASM model were calculated with following relationships:

$$r_i = 0.029281V_m \tag{1.147}$$

$$q_i = \frac{(Z-2)r_i}{2} + \frac{2(1-l_i)}{2} \tag{1.148}$$

Where $V_m$is the molar volume of pure compound i at 298.15 K. Z is the coordination number and was assumed to be 10; l_i is the bulk factor.

There are a numerous factors influencing the phase behavior of ionic liquids with other liquids. In general, the higher the melting point of the ionic liquid, the smaller the solubility of ionic liquid in the solution. It also can be proven that the smaller solubility of ionic liquid is corresponded to higher melting enthalpy. Dománska et al. [207] found that the solubionic liquidity of ionic liquid decreases as alkyl chain length of the alkane (solvent) increases in all systems with imidazolium and ammonium salts. And as for the same size of alcohols, different structures could influence the solubility of ionic liquid. Dománska et al. [208–211] measured the solid–liquid equilibrium of 1-butyl-, decyl-, or dodecyl-3-methylimidazolium chloride [C_4, C_{10}, C_{12}mim]Cl in alcohols, and the results indicate that the solubility of ionic liquid in alcohols decreases with an increase of the carbon chain of an alcohol from C_2 to C_8, with the exception of the solubility of [C_4mim]Cl in 1-butanol. From the experimental data, solubility of [C_4mim]Cl increases in the order 1-butanol > 2-butanol > tert-butyl alcohol [208], that is the best solubility in 1-butanol. This is because of the best packing effect in the solution for the same number of carbon atoms in the solvent and butyl substituent at the imidazolium ring [212]. The size of anions could also influence the solubility of ionic liquid, and it was shown that the size of the two-liquid phase area order is [bmim][PF_6] > [bmim][CH_3SO_4] > [bmim]Cl [213]. Dománska et al. [207] summarized the results of the solubility of ionic liquids in typical solvents from different publications; the conclusion is that the miscibility gap in the liquid phase increases in the order alcohol < aromatic hydrocarbon < cycloalkane < alkane.

Recently, researchers predict the SLE phenomena of ionic liquid mixtures via quantum chemical calculations. The commonly used model is COSMO-RS. The model uses molecular structure as the only information and is based on the principles of pair-wise interacting segments [214, 215]. Verma et al. [203] used the model to predict the solubility of four ionic liquids mixtures. And the deviations are within satisfactory limits by comparing with the literature values.

1.3 The Usage of Handbook

In order to facilitate the readers to access data, this section will introduce the general rule and search method of this manual.

The information of cations and anions of ionic liquids in mixtures are listed in Table 1.4, which include ionic name and ID number in the database. The ionic ID number consists of five digits, the first two digits indicate the ionic types, and the following three digits indicate the order of ion appearance. The ionic liquids, compounds, mixtures, properties, and symbols are listed in Appendix. The lists of ionic liquids and mixtures are arranged in alphabetical order of ionic liquids names. The list of compounds is arranged in the appearance order and number of elements C, H, and O. The list of properties includes all properties of the involved in the second part. At the same time, all properties were classified. In addition, the abbreviation, molecular formulae, molar weight, and structure are presented in the second part of this book.

Readers can obtain ID number of cation and anion from Table 1.4. The new number combined by the ID number of cation and anion is just of ionic liquid number. And then in the second part, readers can retrieve the properties of corresponding ionic liquid mixtures through the new number. Of course, this has the premise that there is combined ionic liquid in this book, the reader is expected to get the desired results. Alternatively, the reader can also find the corresponding ID number by ionic liquid name directly in the Appendix ionic liquids list. If a reader wants to understand the properties of someone ionic liquid mixture, first confirm whether there is target mixture in the Appendix mixture list, and if so, directly search based on the ID number in the second part.

For example, for the ionic liquid 1-butyl-3-methylimidazolium tetrafluoroborate, we can easily find ID number of cation and anion from Table 1.4; 1-butyl-3-methylimidazolium ID is "02011", tetrafluoroborate ID is "02001", so 1-butyl-3-methylimidazolium tetrafluoroborate ID is "02011-02001". This ID can also be found in the ionic liquids list in the Appendix. There are 48 binary mixtures and 23 ternary mixtures containing 1-butyl-3-methylimidazolium tetrafluoroborate in mixtures list. We can directly refer to the properties of the above 71 mixtures by the ID 02011-02001 in ▶ Chap 2.

Table 1.4 Information of cations and anions

ID	Name of cations	ID	Name of anions
01001	1-Methylimidazolium	01001	Chloride
02001	1,3-Dimethylimidazolium	01002	Bromide
02003	1-Ethyl-3-methylimidazolium	01003	Iodine
02007	1-Ethyl-3-ethylimidazolium	01004	Perchlorate
02008	1-Propyl-3-methylimidazolium	02001	Tetrafluoroborate
02011	1-Butyl-3-methylimidazolium	03001	Bis((trifluoromethyl)sulfonyl)imide
02016	1-Amyl-3-methylimidazolium	03003	Bis((perfluoroethane)sulfonyl)imide
02021	1-Hexyl-3-methylimidazolium	03004	Dicyanoamide
02024	1-Heptyl-3-methylimidazolium	03009	Nitrate
02029	1-Octyl-3-methylimidazolium	04002	Methanesulfonate
02039	1-Decyl-3-methylimidazolium	04003	Trifluoromethanesulfonate
02046	1-Dodecyl-3-methylimidazolium	04007	Tosylate
02059	1-Octadecyl-3-methylimidazolium	04008	Octylsulfate
02085	3-Methyl-1-vinylimidazolium	04017	Methylsulfate
02118	1,3-Hexyloxymethylimidazolium	04018	Ethylsulfate
02125	1-(2-Hydroxyethyl)-3-methylimidazolium	04019	Diethyleneglycolmonomethylethersulphate
02199	1-Methyl-3-allylimidazolium	04021	Hydrogen sulfate
02236	1-Methyl-3-ethylimidazolium	04028	1,1,2-Trifluoro-2-(perfluoroethoxy) ethanesulfonate
02237	1-Hexyloxymethyl-3-methylimidazolium	04029	1,1,2,3,3,3-Hexafluoropropanesulfonate
02238	1-Isobutenyl-3-methylimidazolium	04030	1,1,2,2-Tetrafluoroethanesulfonate
02239	1-Hydroxy-1-propyl-3-methylimidazolium	04031	2-(1,2,2,2-Tetrafluoroethoxy)-1,1,2,2-tetrafluoroethanesulfonate
02242	*n*-Acryloyloxypropyl-*N*-methylimidazolium	04032	Ethoxyethylsulfate
03007	1, 2-Dimethyl-3-propylimidazolium	04033	1,1,2-Trifluoro-2-(trifluoromethoxy) ethanesulfonate
03009	1-Propyl-2,3-dimethylimidazolium	04036	Ethylenglykolmonomethylethersulfate
03010	1-Butyl-2,3-dimethylimidazolium	05001	Hexafluorophosphate
03017	1-Decyl-2,3-dimethylimidazolium	05011	Dimethylphosphate

(*continued*)

Table 1.4 (continued)

ID	Name of cations	ID	Name of anions
07068	1-Ethyl-3-butylbenzotriazolium	05012	Diethylphosphate
08003	*N*-methyl-*N*-propyl-pyrrolidinium	05013	Dibutylphosphate
08004	*N*-methyl-*N*-butyl-pyrrolidinium	06001	Trifluoroacetate
08005	*N*-methyl-*N*-hexylpyrrolidinium	06003	Acetate
08014	1-Methyl-1-octylpyrrodinium	06005	Formate
08017	Pyrrolidinium	06007	Lactate(2-hydroxypropanoate)
09001	*N*-methyl-*N*-propylpiperidinium	06010	Alanine
09012	1-Methyl-1-octylpiperidinium	06011	Valinate
11001	*N*-butyl pyridinium	06034	Glycolate(2-hydroxyethanoate)
11003	1-Octyl pyridinium	06042	Pricrate
11009	4-Methyl-*N*-butylpyridinium	07001	Tri(trifluoromethylsulfonyl)methide
11010	1-Dodecyl-3-methylpyridinium	07002	Tricyanomethane
11012	1-Tetradecyl-3-methylpyridinium	15003	Thiocyanate
11025	Ethyl pyridinium		
11038	1-Ethyl-3-methylpyridinium		
11041	1-Butyl-3-methylpyridinium		
11046	*N* -Hexyl-3 -methylpyridinium		
11069	1-Butyl-2-methylpyridinium		
11074	3-Methyl-1-propylpyridinium		
11075	Pyridinium		
11076	1-Decyl-3-methylpyridinium		
11077	1,4-Dimethylpyridinium		
11078	2,4,6-Collidine		
12006	*N*-tetradecyl-isoquinolinium		
14002	Triethylsulphonium		
15013	Trimethyl-butylammonium		

15032	Tetrabutylammonium		
15039	Tetraheptylammonium		
15042	Tetradecyl(trihexyl)phosphonium		
15067	*N*, *N*, *N*-trioctyl-*N*-methylammonium		
15084	Ethylammonium		
15086	2-Hydroxyethylammonium		
15094	Choline or 2-hydroxyethyltrimethylammonium		
15096	Didecyldimethylammonium		
15099	Butyl(2-hydroxyethyl)dimethylammonium		
15132	Propyl(2-hydroxyethyl)dimethylammonium		
15142	Diethanolammonium		
15143	Triethanolammonium		
15163	Hexyl(2-hydroxyethyl)dimethylammonium		
15164	Ethyl(2-hydroxyethyl)dimethylammonium		
15165	Diisopropyl-ethylammonium		
15167	*N*,*N*-dimethylethanolammonium		
15168	Tris(2-hydroxyethyl)methylammonium		
15170	(Benzyl)dimethyldodecylammonium		
15171	(Benzyl)dimethyltetradecylammonium		
15172	Acetylcholinium		
17004	Trihexyl-tetradecylphosphonium		
17005	Tetra(*n*-butyl)phosphonium		
17017	Tributyl(tetradecyl)phosphonium		
17040	Tri-*iso*-butyl(methyl)phosphonium		

Acknowledgement This research was supported financially by National Natural Scientific Fund of China (No. 21210006, No. 21336002, No. 21376242, No. 21476234), and Natural Science Foundation of Beijing of China (No. 2132055).

References

1. Geppert-Rybczynska M, Heintz A, Lehmann JK, Golus A (2010) Volumetric properties of binary mixtures containing ionic liquids and some aprotic solvents. J Chem Eng Data 55 (9):4114–4120
2. Connors KA, Wright JL (1989) Dependence of surface-tension on composition of binary aqueous organic solutions. Anal Chem 61(3):194–198
3. Ren R, Zuo Y, Zhou Q, Zhang HL, Zhang SJ (2011) Density, excess molar volume and conductivity of binary mixtures of the ionic liquid 1,2-dimethyl-3-hexylimidazolium bis (trifluoromethylsulfonyl)imide and dimethyl carbonate. J Chem Eng Data 56(1):27–30
4. Rodriguez H, Brennecke JF (2006) Temperature and composition dependence of the density and viscosity of binary mixtures of water plus ionic liquid. J Chem Eng Data 51(6):2145–2155
5. Zhou Q, Wang LS, Chen HP (2006) Densities and viscosities of 1-butyl-3-methylimidazolium tetrafluoroborate + H2O binary mixtures from (303.15 to 353.15) K. J Chem Eng Data 51 (3):905–908
6. Matkowska D, Hofman T (2013) Volumetric properties of the ionic liquids: [C6mim] [MeSO4], [C6mim][EtSO4], [C4mim][EtSO4] and their mixtures with methanol or ethanol. J Mol Liq 177:301–305
7. Matkowska D, Hofman T (2013) Volumetric properties of the {[Cmim][MeSO] + (1 -) MeOH} system at temperatures from (283.15 to 333.15) K and pressures from (0.1 to 35) MPa. J Solut Chem 42(5):979–990
8. Hosseini SM, Papari MM, Fadaei-Nobandegani F, Moghadasi J (2013) A simple equation for predicting the volumetric properties of mixtures involving ionic liquids. J Solut Chem 42 (9):1854–1862
9. Spencer CF, Danner RP (1972) Improved equation for prediction of saturated liquid density. J Chem Eng Data 17(2):236–241
10. P BE, P JM, OC JP (2001) The properties of gases and liquids, 5th edn. McGraw-Hill, New York
11. Yousefi F (2012) Correlation of volumetric properties of binary mixtures of some ionic liquids with alcohols using equation of state. Ionics 18(8):769–775
12. Tao FM, Mason EA (1994) Statistical-mechanical equation of state for nonpolar fluids: prediction of phase boundaries. J Chem Phys 100(12):9075–9087
13. Hosseini SM (2010) A perturbed hard-sphere equation of state for phosphonium-, pyridinium-, and pyrrolidinium-based ionic liquids. Ionics 16(6):571–575
14. Hosseini SM, Moghadasi J, Papari MM (2010) A perturbed hard-sphere equation of state extended to imidazolium-based ionic liquids. Ionics 16(8):757–761
15. Hosseini SM, Moghadasi J, Papari MM, Nobandegani FF (2011) Modeling the volumetric properties of mixtures involving ionic liquids using perturbed hard-sphere equation of state. J Mol Liq 160(2):67–71
16. Boublik T (1970) Hard-sphere equation of state. J Chem Phys 53(1):471
17. Hosseini SM, Moghadasi J, Papari MM, Nobandegani FF (2012) Modeling the volumetric properties of ionic liquids using modified perturbed hard-sphere equation of state: application to pure and binary mixtures. Ind Eng Chem Res 51(2):758–766
18. Letcher TM, Baxter RC (1989) Application of the Prigogine-Flory-Patterson theory part I. Mixtures ofn-alkanes with bicyclic compounds, benzene, cyclohexane andn-hexane. J Solut Chem 18(1):65–80

19. Treszczanowicz AJ, Benson GC (1985) Excess volumes of alkanol + alkane binary systems in terms of an association model with a Flory contribution term. Fluid Phase Equilib 23 (2–3):117–135
20. Zhou Q, Lu XM, Zhang SJ, Guo LL (2014) Physicochemical properties of ionic liquids. In: Ionic liquids further UnCOILed: critical expert overviews. Wiley, Hoboken
21. Tomida D, Kumagai A, Qiao K, Yokoyama C (2006) Viscosity of bmim PF6 and bmim BF4 at high pressure. Int J Thermophys 27(1):39–47
22. Tomida D, Kumagai A, Kenmochi S, Qiao K, Yokoyama C (2007) Viscosity of 1-hexyl-3-methylimidazolium hexafluorophosphate and 1-octyl-3-methylimidazolium hexafluorophosphate at high pressure. J Chem Eng Data 52(2):577–579
23. Tomida D, Kumagai A, Qiao K, Yokoyama C (2007) Viscosity of 1-butyl-3-methylimidazolium hexafluorophosphate plus CO2 mixture. J Chem Eng Data 52(5):1638–1640
24. Harris KR, Woolf LA, Kanakubo M (2005) Temperature and pressure dependence of the viscosity of the ionic liquid 1-butyl-3-methylimidazolium hexaftuorophosphate. J Chem Eng Data 50(5):1777–1782
25. Harris KR, Kanakubo M, Woolf LA (2006) Temperature and pressure dependence of the viscosity of the ionic liquids 1-methyl-3-octylimidazolium hexafluorophosphate and 1-methyl-3-octylimidazolium tetrafluoroborate. J Chem Eng Data 51(3):1161–1167
26. Harris KR, Kanakubo M, Woolf LA (2007) Temperature and pressure dependence of the viscosity of the ionic liquids 1-hexyl-3-methylimidazolium hexafluorophosphate and 1-butyl-3-methylimidazolium bis(trifluoromethylsulfonyl)imide. J Chem Eng Data 52(3):1080–1085
27. Harris KR, Kanakubo M, Woolf LA (2007) Temperature and pressure dependence of the viscosity of the ionic liquid 1-butyl-3-methylimidazolium tetrafluoroborate: viscosity and density relationships in ionic liquids. J Chem Eng Data 52(6):2425–2430
28. Harris KR, Woolf LA, Kanakubo M, Ruther T (2011) Transport properties of N-butyl-N-methylpyrrolidinium bis(trifluoromethylsulfonyl)amide. J Chem Eng Data 56(12):4672–4685
29. Aparicio S, Alcalde R, Garcia B, Leal JM (2009) High-pressure study of the methylsulfate and tosylate imidazolium ionic liquids. J Chem Eng Data 113(16):5593–5606
30. Ahosseini A, Scurto AM (2008) Viscosity of imidazolium-based ionic liquids at elevated pressures: cation and anion effects. Int J Thermophys 29(4):1222–1243
31. Zhu AL, Wang JJ, Han LJ, Fan MH (2009) Measurements and correlation of viscosities and conductivities for the mixtures of imidazolium ionic liquids with molecular solutes. Chem Eng J 147(1):27–35
32. Tomida D, Kenmochi S, Qiao K, Bao QX, Yokoyama C (2011) Viscosity of ionic liquid mixtures of 1-alkyl-3-methylimidazolium hexafluorophosphate + CO_2. Fluid Phase Equilib 307(2):185–189
33. Domanska U, Laskowska M (2009) Temperature and composition dependence of the density and viscosity of binary mixtures of {1-butyl-3-methylimidazolium thiocyanate + 1-alcohols}. J Chem Eng Data 54(7):2113–2119
34. Domanska U, Krolikowska M (2010) Density and viscosity of binary mixtures of {1-butyl-3-methylimidazolium thiocyanate + 1-heptanol, 1-octanol, 1-nonanol, or 1-decanol}. J Chem Eng Data 55(9):2994–3004
35. Domanska U, Laskowska M (2009) Effect of temperature and composition on the density and viscosity of binary mixtures of ionic liquid with alcohols. J Solut Chem 38(6):779–799
36. Shao DB, Lu XX, Fang WJ, Guo YS, Xu L (2012) Densities and viscosities for binary mixtures of the ionic liquid N-ethyl piperazinium propionate with n-alcohols at several temperatures. J Chem Eng Data 57(3):937–942
37. Gomez E, Gonzalez B, Dominguez A, Tojo E, Tojo J (2006) Dynamic viscosities of a series of 1-alkyl-3-methylimidazolium chloride ionic liquids and their binary mixtures with water at several temperatures. J Chem Eng Data 51(2):696–701
38. Vila J, Gines P, Pico JM, Franjo C, Jimenez E, Varela LM, Cabeza O (2006) Temperature dependence of the electrical conductivity in EMIM-based ionic liquids – evidence of Vogel-Tamman-Fulcher behavior. Fluid Phase Equilibr 242(2):141–146

39. Lashkarblooki M, Hezave AZ, Al-Ajmi AM, Ayatollahi S (2012) Viscosity prediction of ternary mixtures containing ILs using multi-layer perceptron artificial neural network. Fluid Phase Equilib 326:15–20
40. Peng XM, Hu YF, Chu HD, Miao C, Li ZY, Wang ZX (2012) Prediction of density and viscosity of ternary systems C(2)q Br + C(4)q Br + H2O, C(2)q Br + C(6)q Br + H2O, and C(4)q Br + C (6)q Br + H2O at different temperatures using their binary subsystems data. Pet Sci 9(2):242–250
41. Bard AJ, Faulkner LR (2001) Electrochemical methods, fundamentals and applications, vol chapter 5 and 9. Wiley, New York
42. Baur JE (2007) Handbook of electrochemistry, vol chapter 19. Elsevier, Amsterdam
43. Scovazzo P, Kieft J, Finan DA, Koval C, DuBois D, Noble R (2004) Gas separations using non-hexafluorophosphate PF6 (-) anion supported ionic liquid membranes. J Membr Sci 238 (1–2):57–63
44. Haumann M, Dentler K, Joni J, Riisager A, Wasserscheid P (2007) Continuous gas-phase hydroformylation of 1-butene using supported ionic liquid phase (SILP) catalysts. Adv Synth Catal 349(3):425–431
45. Chauvin Y, Mußmann L, Olivier H (1995) Flüssige 1,3-Dialkylimidazoliumsalze als Lösungsmittel für die Katalyse in Zweiphasensystemen: durch Rhodiumkomplexe katalysierte Hydrierung, Isomerisierung und Hydroformylierung von Alkenen. Angew Chem 107 (23–24):2941–2943
46. Steines S, Wasserscheid P, Driessen-Holscher B (2000) An ionic liquid as catalyst medium for stereoselective hydrogenations of sorbic acid with ruthenium complexes. J Fur Praktische Chemie-Chemiker-Zeitung 342(4):348–354
47. Papageorgiou N, Athanassov Y, Armand M, Bonhote P, Pettersson H, Azam A, Gratzel M (1996) The performance and stability of ambient temperature molten salts for solar cell applications. J Electrochem Soc 143(10):3099–3108
48. Hauch A, Georg A (2001) Diffusion in the electrolyte and charge-transfer reaction at the platinum electrode in dye-sensitized solar cells. Electrochim Acta 46(22):3457–3466
49. Baur JE, Wightman RM (1991) Diffusion coefficients determined with microelectrodes. J Electroanal Chem 305(1):73–81
50. Evans RG, Klymenko OV, Price PD, Davies SG, Hardacre C, Compton RG (2005) A comparative electrochemical study of diffusion in room temperature ionic liquid solvents versus acetonitrile. Chemphyschem 6(3):526–533
51. Stokes RH (1950) An improved diaphragm-cell for diffusion studies, and some tests of the method. J Am Chem Soc 72(2):763–767
52. Gosting LJ (1950) A study of the diffusion of potassium chloride in water at 25°C with the Gouy interference method. J Am Chem Soc 72(10):4418–4422
53. Taylor G (1953) Dispersion of soluble matter in solvent flowing slowly through a tube. In: Proceedings of the Royal Society of London series a-mathematical and physical sciences 219 (1137):186–203
54. Taylor G (1954) Conditions under which dispersion of a solute in a stream of solvent can be used to measure molecular diffusion. In: Proceedings of the Royal Society of London series a-mathematical and physical sciences 225(1163):473–477
55. Gulari E, Brown RJ, Pings CJ (1973) Measurement of mutual diffusion coefficients and thermal diffusivities by quasi-elastic light scattering. Aiche J 19(6):1196–1204
56. Czworniak KJ, Andersen HC, Pecora R (1975) Light-scattering measurement and theoretical interpretation of mutual diffusion-coefficients in binary liquid mixtures. Chem Phys 11(3):451–473
57. Krahn W, Schweiger G, Lucas K (1983) Light scattering measurements of mutual diffusion coefficients in binary liquid mixtures. J Phys Chem 87(22):4515–4519
58. Siddiqi MA, Krahn W, Lucas K (1987) Mutual diffusion coefficients in some binary liquid mixtures. J Chem Eng Data 32(1):48–50
59. Rausch MH, Lehmann J, Leipertz A, Froba AP (2011) Mutual diffusion in binary mixtures of ionic liquids and molecular liquids by dynamic light scattering (DLS). Phys Chem Chem Phys 13(20):9525–9533

60. Rausch MH, Hopf L, Heller A, Leipertz A, Froba AP (2013) Binary diffusion coefficients for mixtures of ionic liquids EMIM N(CN)(2), EMIM NTf2, and HMIM NTf2 with acetone and ethanol by Dynamic Light Scattering (DLS). J Phys Chem B 117(8):2429–2437
61. Tokuda H, Hayamizu K, Ishii K, Susan M, Watanabe M (2005) Physicochemical properties and structures of room temperature ionic liquids. 2. Variation of alkyl chain length in imidazolium cation. J Phys Chem B 109(13):6103–6110
62. Tokuda H, Ishii K, Susan M, Tsuzuki S, Hayamizu K, Watanabe M (2006) Physicochemical properties and structures of room-temperature ionic liquids. 3. Variation of cationic structures. J Phys Chem B 110(6):2833–2839. doi: 10.1021/jpO53396f
63. Tsuzuki S, Shinoda W, Saito H, Mikami M, Tokuda H, Watanabe M (2009) Molecular dynamics simulations of ionic liquids: cation and anion dependence of self-diffusion coefficients of ions. J Phys Chem B 113(31):10641–10649
64. Sarraute S, Gomes MFC, Padua AAH (2009) Diffusion coefficients of 1-alkyl-3-methylimidazolium ionic liquids in water, methanol, and acetonitrile at infinite dilution. J Chem Eng Data 54(9):2389–2394
65. Wilke CR, Chang P (1955) Correlation of diffusion coefficients in dilute solutions. Aiche J 1 (2):264–270
66. Robinson RA, Stokes RH (1965) Electrolyte solutions. Butterworths, London
67. Rooney D, Jacquemin J, Gardas R (2009) Thermophysical properties of ionic liquids. Topics in current chemistry. Springer-Verlag, Berlin/Heidelberg
68. Xu AR, Wang JJ, Zhang YJ, Chen QT (2012) Effect of alkyl chain length in anions on thermodynamic and surface properties of 1-butyl-3-methylimidazolium carboxylate ionic liquids. Ind Eng Chem Res 51(8):3458–3465
69. Wang Y, Zhao FY, Liu YM, Hu YQ (2007) Study on surface tension of a series of 1-alkyl-3-methylimidazolium room temperature ionic liquids. Acta Chim Sin 65(15):1443–1448
70. Adamson AW, Gast AP (1997) Physical chemistry of surfaces. Wiley, New York
71. Law G, Watson PR (2001) Surface tension measurements of N-alkylimidazolium ionic liquids. Langmuir 17(20):6138–6141
72. Macleod DB (1923) On a relation between surface tension and density. Trans Faraday Soc 19 (1):38–41
73. Deetlefs M, Seddon KR, Shara M (2006) Predicting physical properties of ionic liquids. Phys Chem Chem Phys 8(5):642–649
74. Knotts TA, Wilding WV, Oscarson JL, Rowley RL (2001) Use of the DIPPR database for development of QSPR correlations: surface tension. J Chem Eng Data 46(5):1007–1012
75. Gardas RL, Coutinho JAP (2008) Applying a QSPR correlation to the prediction of surface tensions of ionic liquids. Fluid Phase Equilib 265(1–2):57–65
76. Mousazadeh MH, Faramarzi E (2011) Corresponding states theory for the prediction of surface tension of ionic liquids. Ionics 17(3):217–222
77. Gharagheizi F, Ilani-Kashkouli P, Mohammadi AH (2012) Group contribution model for estimation of surface tension of ionic liquids. Chem Eng Sci 78:204–208
78. Wu KJ, Zhao CX, He CH (2012) A simple corresponding-states group-contribution method for estimating surface tension of ionic liquids. Fluid Phase Equilib 328:42–48
79. Domanska U, Pobudkowska A, Rogalski M (2008) Surface tension of binary mixtures of imidazolium and ammonium based ionic liquids with alcohols, or water: cation, anion effect. J Colloid Interface Sci 322(1):342–350
80. Rilo E, Pico J, Garcia-Garabal S, Varela LM, Cabeza O (2009) Density and surface tension in binary mixtures of CnMIM-BF4 ionic liquids with water and ethanol. Fluid Phase Equilib 285(1–2):83–89
81. Jiang HC, Zhao Y, Wang JY, Zhao FY, Liu RJ, Hu YQ (2013) Density and surface tension of pure ionic liquid 1-butyl-3-methyl-imidazolium L-lactate and its binary mixture with alcohol and water. J Chem Thermodyn 64:1–13
82. Ahosseini A, Sensenich B, Weatherley LR, Scurto AM (2010) Phase equilibrium, volumetric, and interfacial properties of the ionic liquid, 1-hexyl-3-methylimidazolium bis(trifluoromethylsulfonyl)amide and 1-octene. J Chem Eng Data 55(4):1611–1617

83. Tariq M, Freire MG, Saramago B, Coutinho JAP, Lopes JNC, Rebelo LPN (2012) Surface tension of ionic liquids and ionic liquid solutions. Chem Soc Rev 41(2):829–868
84. Gardas RL, Coutinho JAP (2008) Estimation of speed of sound of ionic liquids using surface tensions and densities: a volume based approach. Fluid Phase Equilib 267(2):188–192
85. Paulechka YU, Zaitsau DH, Kabo GJ (2004) On the difference between isobaric and isochoric heat capacities of liquid cyclohexyl esters. J Mol Liq 115(2–3):105–111
86. Dzida M, Chorazewski M, Geppert-Rybczynska M, Zorebski E, Zorebski M, Zarska M, Czech B (2013) Speed of sound and adiabatic compressibility of 1-ethyl-3-methylimidazolium bis (trifluoromethylsulfonyl)imide under pressures up to 100 MPa. J Chem Eng Data 58 (6):1571–1576
87. Auerbach R (1948) Surface tension and sound velocity. Ezperientia 4:473–474
88. Gardas RL, Dagade DH, Terdale SS, Coutinho JAP, Patil KJ (2008) Acoustic and volumetric properties of aqueous solutions of imidazolium based ionic liquids at 298.15 K. J Chem Thermodyn 40(4):695–701
89. Govinda V, Reddy PM, Attri P, Venkatesu P, Venkateswarlu P (2013) Influence of anion on thermophysical properties of ionic liquids with polar solvent. J Chem Thermodyn 58:269–278
90. Gonzalez EJ, Gonzalez B, Macedo EA (2013) Thermophysical properties of the pure ionic liquid 1-butyl-1-methylpyrrolidinium dicyanamide and its binary mixtures with alcohols. J Chem Eng Data 58(6):1440–1448
91. Singh S (2002) Refractive index measurement and its applications. Phys Scr 65(2):167–180
92. Huddleston JG, Visser AE, Reichert WM, Willauer HD, Broker GA, Rogers RD (2001) Characterization and comparison of hydrophilic and hydrophobic room temperature ionic liquids incorporating the imidazolium cation. Green Chem 3(4):156–164
93. Seki S, Tsuzuki S, Hayamizu K, Umebayashi Y, Serizawa N, Takei K, Miyashiro H (2012) Comprehensive refractive index property for room-temperature ionic liquids. J Chem Eng Data 57(8):2211–2216
94. Vercher E, Llopis FJ, Gonzalez-Alfaro V, Miguel PJ, Martinez-Andreu A (2011) Refractive indices and deviations in refractive indices of trifluoromethanesulfonate-based ionic liquids in water. J Chem Eng Data 56(12):4499–4504
95. Iglesias-Otero MA, Troncoso J, Carballo E, Romani L (2007) Density and refractive index for binary systems of the ionic liquid Bmim BF4 with methanol, 1,3-dichloropropane, and dimethyl carbonate. J Solut Chem 36(10):1219–1230
96. Shekaari H, Mousavi SS, Mansoori Y (2009) Thermophysical properties of ionic liquid, 1-pentyl-3-methylimidazolium chloride in water at different temperatures. Int J Thermophys 30(2):499–514
97. Rilo E, Dominguez-Perez M, Vila J, Segade L, Garcia M, Varela LM, Cabeza O (2012) Easy prediction of the refractive index for binary mixtures of ionic liquids with water or ethanol. J Chem Thermodyn 47:219–222
98. Kurnia KA, Taib MM, Mutalib MIA, Murugesan T (2011) Densities, refractive indices and excess molar volumes for binary mixtures of protic ionic liquids with methanol at T = 293.15 to 313.15 K. J Mol Liq 159(3):211–219
99. Soriano AN, Doma BT, Li MH (2009) Measurements of the density and refractive index for 1-n-butyl-3-methylimidazolium-based ionic liquids. J Chem Thermodyn 41(3):301–307
100. Soriano AN, Doma BT, Li MH (2010) Density and refractive index measurements of 1-ethyl-3-methylimidazolium-based ionic liquids. J Taiwan Inst Chem Eng 41(1):115–121
101. Cota I, Gonzalez-Olmos R, Iglesias M, Medina F (2007) New short aliphatic chain ionic liquids: synthesis, physical properties, and catalytic activity in aldol condensations. J Phys Chem B 111(43):12468–12477
102. Vercher E, Llopis FJ, Gonzalez-Alfaro V, Martinez-Andreu A (2010) Refractive indices and deviations in refractive indices for binary mixtures of 1-ethyl-3-methylimidazolium trifluoromethanesulfonate with methanol, ethanol, I-propanol, and 2-propanol at several temperatures. J Chem Eng Data 55(3):1430–1433

103. Brocos P, Pineiro A, Bravo R, Amigo A (2003) Refractive indices, molar volumes and molar refractions of binary liquid mixtures: concepts and correlations. Phys Chem Chem Phys 5 (3):550–557
104. Iglesias-Otero MA, Troncoso J, Carballo E, Romani L (2008) Density and refractive index in mixtures of ionic liquids and organic solvents: correlations and predictions. J Chem Thermodyn 40(6):949–956
105. Reis JCR, Lampreia IMS, Santos AFS, Moita M, Douheret G (2010) Refractive index of liquid mixtures: theory and experiment. Chemphyschem 11(17):3722–3733
106. Redlich O, Kister AT (1948) Algebraic representation of thermodynamic properties and the classification of solutions. Ind Eng Chem 40(2):345–348
107. Gonzalez EJ, Gonzalez B, Calvar N, Dominguez A (2007) Physical properties of binary mixtures of the ionic liquid 1-ethyl-3-methylimidazolium ethyl sulfate with several alcohols at T = (298.15, 313.15, and 328.15) K and atmospheric pressure. J Chem Eng Data 52 (5):1641–1648
108. Arce A, Rodil E, Soto A (2006) Physical and excess properties for binary mixtures of 1-methyl-3-octylimidazolium tetrafluoroborate, Omim BF4, ionic liquid with different alcohols. J Solut Chem 35(1):63–78
109. McNaught AD, Wilkinson A (1997) IUPAC compendium of chemical terminology. Blackwell Science, Oxford
110. Lei ZG, Chen BH, Ding ZW (2005) Special distillation processes. Elsevier, Amsterdam
111. Pitzer KS (1973) Thermodynamics of electrolytes. 1. Theoretical basis and general equations. J Phys Chem 77(2):268–277
112. KS P (1991) Ion interaction approach: theory and data correlation. In: Activity coefficients in electrolyte solutions. CRC Press, Boca Raton
113. Cadena C, Anthony JL, Shah JK, Morrow TI, Brennecke JF, Maginn EJ (2004) Why is CO_2 so soluble in imidazolium-based ionic liquids? J Am Chem Soc 126(16):5300–5308
114. Shi W, Maginn EJ (2008) Molecular simulation and regular solution theory modeling of pure and mixed gas absorption in the ionic liquid 1-n-hexyl-3-methylimidazolium bis(trifluoromethylsulfonyl)amide (hmim Tf2N). J Phys Chem B 112(51):16710–16720
115. Shi W, Maginn EJ (2008) Atomistic simulation of the absorption of carbon dioxide and water in the ionic liquid 1-n-hexyl-3-methylimidazolium bis(trifluoromethylsulfonyl)imide ([hmim] [Tf2N]. J Phys Chem B 112(7):2045–2055
116. Aki S, Mellein BR, Saurer EM, Brennecke JF (2004) High-pressure phase behavior of carbon dioxide with imidazolium-based ionic liquids. J Phys Chem B 108(52):20355–20365
117. Shiflett MB, Yokozeki A (2007) Solubility of CO_2 in room temperature ionic liquid hmim Tf2N. J Phys Chem B 111(8):2070–2074
118. Zhang YQ, Zhang SJ, Lu XM, Zhou Q, Fan W, Zhang XP (2009) Dual amino-functionalised phosphonium ionic liquids for CO_2 capture. Chem-a Eur J 15(12):3003–3011
119. Zhang JM, Zhang SJ, Dong K, Zhang YQ, Shen YQ, Lv XM (2006) Supported absorption of CO_2 by tetrabutylphosphonium amino acid ionic liquids. Chem-a Eur J 12(15):4021–4026
120. Condemarin R, Scovazzo P (2009) Gas permeabilities, solubilities, diffusivities, and diffusivity correlations for ammonium-based room temperature ionic liquids with comparison to imidazolium and phosphonium RTIL data. Chem Eng J 147(1):51–57
121. Ferguson L, Scovazzo P (2007) Solubility, diffusivity, and permeability of gases in phosphonium-based room temperature ionic liquids: data and correlations. Ind Eng Chem Res 46(4):1369–1374
122. Anderson JL, Dixon JK, Brennecke JF (2007) Solubility of CO_2, CH4, C2H6, C2H4, O-2, and N-2 in 1-hexyl-3-methylpyridinium bis(trifluoromethylsulfonyl)imide: comparison to other ionic liquids. Acc Chem Res 40(11):1208–1216
123. Jork C, Kristen C, Pieraccini D, Stark A, Chiappe C, Beste YA, Arlt W (2005) Tailor-made ionic liquids. J Chem Thermodyn 37(6):537–558
124. Lei ZG, Arlt W, Wasserscheid P (2006) Separation of 1-hexene and n-hexane with ionic liquids. Fluid Phase Equilib 241(1–2):290–299

125. Doker M, Gmehling J (2005) Measurement and prediction of vapor-liquid equilibria of ternary systems containing ionic liquids. Fluid Phase Equilib 227(2):255–266
126. Zhao J, Jiang XC, Li CX, Wang ZH (2006) Vapor pressure measurement for binary and ternary systems containing a phosphoric ionic liquid. Fluid Phase Equilib 247(1–2):190–198
127. Jiang XC, Wang JF, Li CX, Wang LM, Wang ZH (2007) Vapour pressure measurement for binary and ternary systems containing water methanol ethanol and an ionic liquid 1-ethyl-3-ethylimidazolium diethylphosphate. J Chem Thermodyn 39(6):841–846
128. Wang JF, Li CX, Wang ZH, Li ZJ, Jiang YB (2007) Vapor pressure measurement for water, methanol, ethanol, and their binary mixtures in the presence of an ionic liquid 1-ethyl-3-methylimidazolium dimethylphosphate. Fluid Phase Equilib 255(2):186–192
129. Orchilles AV, Miguel PJ, Vercher E, Martinez-Andreu A (2008) Isobaric vapor-liquid equilibria for 1-propanol plus water plus 1-ethyl-3-methylimidazolium trifluoromethanesulfonate at 100 kPa. J Chem Eng Data 53(10):2426–2431
130. Domanska U, Paduszynski K (2008) Phase equilibria study in binary systems (tetra-n-butylphosphonium tosylate ionic liquid + 1-alcohol, or benzene, or n-alkylbenzene). J Phys Chem B 112(35):11054–11059
131. Domanska U, Krolikowski M, Paduszynski K (2009) Phase equilibria study of the binary systems (N-butyl-3-methylpyridinium tosylate ionic liquid plus an alcohol). J Chem Thermodyn 41(8):932–938
132. Domanska U, Krolikowska M, Paduszynski K (2011) Physico-chemical properties and phase behaviour of piperidinium-based ionic liquids. Fluid Phase Equilib 303(1):1–9
133. Wang TF, Peng CJ, Liu HL, Hu Y, Jiang JW (2007) Equation of state for the vapor-liquid equilibria of binary systems containing imidazolium-based ionic liquids. Ind Eng Chem Res 46(12):4323–4329
134. Xu XC, Peng CJ, Liu HL, Hu Y (2009) Modeling pVT properties and phase equilibria for systems containing ionic liquids using a new lattice-fluid equation of state. Ind Eng Chem Res 48(24):11189–11201
135. Xu XC, Peng CJ, Liu HL, Hu Y (2011) A lattice-fluid model for multi-component ionic-liquid systems. Fluid Phase Equilib 302(1–2):260–268
136. Tsioptsias C, Tsivintzelis I, Panayiotou C (2010) Equation-of-state modeling of mixtures with ionic liquids. Phys Chem Chem Phys 12(18):4843–4851
137. Paduszynski K, Chiyen J, Ramjugernath D, Letcher TM, Domanska U (2011) Liquid-liquid phase equilibrium of (piperidinium-based ionic liquid plus an alcohol) binary systems and modelling with NRHB and PCP-SAFT. Fluid Phase Equilib 305(1):43–52
138. Paduszynski K, Domanska U (2011) Solubility of aliphatic hydrocarbons in piperidinium ionic liquids: measurements and modeling in terms of perturbed-chain statistical associating fluid theory and nonrandom hydrogen-bonding theory. J Phys Chem B 115(43):12537–12548
139. Xu X, Liu H, Peng C, Hu Y (2008) A new molecular-thermodynamic model based on lattice fluid theory: application to pure fluids and their mixtures. Fluid Phase Equilib 265 (1–2):112–121
140. Lei ZG, Chen BH, Li CY (2007) COSMO-RS modeling on the extraction of stimulant drugs from urine sample by the double actions of supercritical carbon dioxide and ionic liquid. Chem Eng Sci 62(15):3940–3950
141. Diedenhofen M, Eckert F, Klamt A (2003) Prediction of infinite dilution activity coefficients of organic compounds in ionic liquids using COSMO-RS. J Chem Eng Data 48(3):475–479
142. Kato R, Gmehling J (2005) Systems with ionic liquids: measurement of VLE and gamma (infinity) data and prediction of their thermodynamic behavior using original UNIFAC, mod. UNIFAC(Do) and COSMO-RS(O1). J Chem Thermodyn 37(6):603–619
143. Eike DM, Brennecke JF, Maginn EJ (2004) Predicting infinite-dilution activity coefficients of organic solutes in ionic liquids. Ind Eng Chem Res 43(4):1039–1048
144. Carlisle TK, Bara JE, Gabriel CJ, Noble RD, Gin DL (2008) Interpretation of CO_2 solubility and selectivity in nitrile-functionalized room-temperature ionic liquids using a group contribution approach. Ind Eng Chem Res 47(18):7005–7012

145. Camper D, Becker C, Koval C, Noble R (2005) Low pressure hydrocarbon solubility in room temperature ionic liquids containing imidazolium rings interpreted using regular solution theory. Ind Eng Chem Res 44(6):1928–1933
146. Finotello A, Bara JE, Camper D, Noble RD (2008) Room-temperature ionic liquids: temperature dependence of gas solubility selectivity. Ind Eng Chem Res 47(10):3453–3459
147. Kilaru PK, Condemarin PA, Scovazzo P (2008) Correlations of low-pressure carbon dioxide and hydrocarbon solubilities in imidazolium-, phosphonium-, and ammonium-based room-temperature ionic liquids. Part 1. Using surface tension. Ind Eng Chem Res 47(3):900–909
148. Kilaru PK, Scovazzo P (2008) Correlations of low-pressure carbon dioxide and hydrocarbon solubilities in imidazolium-, phosphonium-, and ammonium-based room-temperatuire ionic liquids. Part 2. Using activation energy of viscosity. Ind Eng Chem Res 47(3):910–919
149. Skjoldjorgensen S, Kolbe B, Gmehling J, Rasmussen P (1979) Vapor-liquid equilibria by UNIFAC group contribution. Ind Eng Chem Process Des Dev 18(4):714–722
150. Gmehling J, Rasmussen P, Fredenslund A (1982) Vapor-liquid equilibria by UNIFAC group contribution. 2. Revision and extension. Ind Eng Chem Process Des Dev 21(1):118–127
151. Macedo EA, Weidlich U, Gmehling J, Rasmussen P (1983) Vapor-liquid equilibria by UNIFAC group contribution. 3. Revision and extension. Ind Eng Chem Process Des Dev 22 (4):676–678
152. Tiegs D, Gmehling J, Rasmussen P, Fredenslund A (1987) Vapor-liquid equilibria by UNIFAC group contribution. 4. Revision and extension. Ind Eng Chem Res 26(1):159–161
153. Hansen HK, Rasmussen P, Fredenslund A, Schiller M, Gmehling J (1991) Vapor-liquid equilibria by UNIFAC group contribution. 5. Revision and extension. Ind Eng Chem Res 30 (10):2352–2355
154. Wittig R, Lohmann J, Gmehling J (2003) Vapor-liquid equilibria by UNIFAC group contribution. 6. Revision and extension. Ind Eng Chem Res 42(1):183–188
155. Yang L, Sandler SI, Peng CJ, Liu HL, Hu Y (2010) Prediction of the phase behavior of ionic liquid solutions. Ind Eng Chem Res 49(24):12596–12604
156. Kojima K, Zhang SJ, Hiaki T (1997) Measuring methods of infinite dilution activity coefficients and a database for systems including water. Fluid Phase Equilib 131(1–2):145–179
157. Sandler SI (1996) Infinite dilution activity coefficients in chemical, environmental and biochemical engineering. Fluid Phase Equilib 116(1–2):343–353
158. Eckert CA, Sherman SR (1996) Measurement and prediction of limiting activity coefficients. Fluid Phase Equilib 116(1–2):333–342
159. Krummen M, Wasserscheid P, Gmehling J (2002) Measurement of activity coefficients at infinite dilution in ionic liquids using the dilutor technique. J Chem Eng Data 47 (6):1411–1417
160. Heintz A, Kulikov DV, Verevkin SP (2001) Thermodynamic properties of mixtures containing ionic liquids. 1. Activity coefficients at infinite dilution of alkanes, alkenes, and alkylbenzenes in 4-methyl-n-butylpyridinium tetrafluoroborate using gas-liquid chromatography. J Chem Eng Data 46(6):1526–1529
161. Wang LS, Wang XX, Li Y, Jiang K, Shao XZ, Du CJ (2013) Ionic liquids: solubility parameters and selectivities for organic solutes. Aiche J 59(8):3034–3041
162. Everett DH (1965) Effect of gas imperfection on G.L.C. measurements: a refined method for determining activity coefficients and second virial coefficients. Trans Faraday Soc 61:1637–1645
163. Cruickshank AJB, Windsor ML, Young CL (1966) The use of gas-liquid chromatography to determine activity coefficients and second virial coefficients of mixtures. I. Theory and verification of method of data analysis. In: Proceedings of the Royal Society of London series a mathematical and physical sciences 295(1442):259–270
164. Dw G (1971) Gas-liquid chromatography. Van Nostrand Reinhold, London
165. Xue ZM, Mu TC, Gmehling J (2012) Comparison of the a priori COSMO-RS models and group contribution methods: original UNIFAC, modified UNIFAC(Do), and modified UNIFAC(Do) consortium. Ind Eng Chem Res 51(36):11809–11817

166. Weidlich U, Gmehling J (1987) A modified UNIFAC model. 1. Prediction of VLE, hE, and $\gamma\infty$. Ind Eng Chem Res 26(7):1372–1381
167. Gmehling J, Li JD, Schiller M (1993) A modified UNIFAC model. 2. Present parameter matrix and results for different thermodynamic properties. Ind Eng Chem Res 32(1):178–193
168. Gmehling J, Lohmann J, Jakob A, Li JD, Joh R (1998) A modified UNIFAC (Dortmund) model. 3. Revision and extension. Ind Eng Chem Res 37(12):4876–4882
169. Gmehling J, Wittig R, Lohmann J, Joh R (2002) A modified UNIFAC (Dortmund) model. 4. Revision and extension. Ind Eng Chem Res 41(6):1678–1688
170. Jakob A, Grensemann H, Lohmann J, Gmehling J (2006) Further development of modified UNIFAC (Dortmund): revision and extension 5. Ind Eng Chem Res 45(23):7924–7933
171. Klamt A, Jonas V, Burger T, Lohrenz JCW (1998) Refinement and parametrization of COSMO-RS. J Phys Chem A 102(26):5074–5085
172. Klamt A, Eckert F (2000) COSMO-RS: a novel and efficient method for the a priori prediction of thermophysical data of liquids. Fluid Phase Equilib 172(1):43–72
173. Klamt A (1995) Conductor-like screening model for real solvents a new approach to the quantitative calculation of solvation phenomena. J Phys Chem 99(7):2224–2235
174. Gutierrez JP, Meindersma GW, de Haan AB (2012) COSMO-RS-based ionic-liquid selection for extractive distillation processes. Ind Eng Chem Res 51(35):11518–11529
175. Anantharaj R, Banerjee T (2011) COSMO-RS based predictions for the desulphurization of diesel oil using ionic liquids: effect of cation and anion combination. Fuel Process Technol 92 (1):39–52
176. Diedenhofen M, Klamt A (2010) COSMO-RS as a tool for property prediction of IL mixtures-a review. Fluid Phase Equilib 294(1–2):31–38
177. Tiegs D, Gmehling J, Medina A, Soares M, Bastos J, Alessi P, Kikic I (1986) DECHEMA Chemistry Data. vol Series IX. DECHEMA, Frankfurt/Main
178. Foco GM, Bottini SB, Quezada N, de la Fuente JC, Peters CJ (2006) Activity coefficients at infinite dilution in 1-alkyl-3-methylimidazolium tetrafluoroborate ionic liquids. J Chem Eng Data 51(3):1088–1091
179. Kikic I, Alessi P, Rasmussen P, Fredenslund A (1980) On the combinatorial part of the UNIFAC and UNIQUAC models. Can J Chem Eng 58(2):253–258
180. Cl Y (1999) Chemical properties handbook. McGraw-Hill Book Co., New York
181. Bondi A (1968) Physical properties of molecular crystals, liquids and glasses. Wiley, New York
182. de los Rios AP, Fernandez FJH, Gomez D, Rubio M, Villora G (2012) (Liquid plus liquid) Equilibrium for ternary systems containing of an ionic liquid, n-hexane and an organic compound involved in the kinetic resolution of rac-2-pentanol. Sep Sci Technol 47(2): 300–311
183. Renon H, Prausnit JM (1968) Local compositions in thermodynamic excess functions for liquid mixtures. Aiche J 14(1):135–144
184. Abrams DS, Prausnitz JM (1975) Statistical thermodynamics of liquid mixtures: a new expression for the excess Gibbs energy of partly or completely miscible systems. Aiche J 21(1):116–128
185. De Sousa HC, Rebelo LPN (2000) A continuous polydisperse thermodynamic algorithm for a modified Flory-Huggins model: the (polystyrene plus nitroethane) example. J Polym Sci B-Polym Phys 38(4):632–651
186. Rebelo LPN, Najdanovic-Visak V, Visak ZP, da Ponte MN, Szydlowski J, Cerdeirina CA, Troncoso J, Romani L, Esperanca J, Guedes HJR, de Sousa HC (2004) A detailed thermodynamic analysis of C(4)mim BF4 plus water as a case study to model ionic liquid aqueous solutions. Green Chem 6(8):369–381
187. Qin Y, Prausnitz JM (2005) Solubilities of solutes in ionic liquids from a simple perturbed-hard-sphere theory. Zeitschrift für Physikalische Chemie 219:1223–1241
188. Yang JY, Peng CJ, Liu HL, Hu Y (2006) Calculation of vapor-liquid and liquid-liquid phase equilibria for systems containing ionic liquids using a lattice model. Ind Eng Chem Res 45(20):6811–6817

189. Arce A, Earle MJ, Katdare SP, Rodriguez H, Seddon KR (2008) Application of mutually immiscible ionic liquids to the separation of aromatic and aliphatic hydrocarbons by liquid extraction: a preliminary approach. Phys Chem Chem Phys 10(18):2538–2542
190. Arce A, Earle MJ, Rodriguez H, Seddon KR (2007) Separation of benzene and hexane by solvent extraction with 1-alkyl-3-methylimidazolium bis{(trifluoromethyl)sulfonyl}amide ionic liquids: effect of the alkyl-substituent length. J Phys Chem B 111(18):4732–4736
191. Ferreira AR, Freire MG, Ribeiro JC, Lopes FM, Crespo JG, Coutinho JAP (2012) Overview of the liquid-liquid equilibria of ternary systems composed of ionic liquid and aromatic and aliphatic hydrocarbons, and their modeling by COSMO-RS. Ind Eng Chem Res 51(8):3483–3507
192. Ferreira AR, Freire MG, Ribeiro JC, Lopes FM, Crespo JG, Coutinho JAP (2011) An overview of the liquid-liquid equilibria of (ionic liquid plus hydrocarbon) binary systems and their modeling by the conductor-like screening model for real solvents. Ind Eng Chem Res 50(9):5279–5294
193. Domanska U, Pobudkowska A, Eckert F (2006) Liquid-liquid equilibria in the binary systems (1,3-dimethylimidazolium, or 1-butyl-3-methylimidazolium methylsulfate plus hydrocarbons). Green Chem 8(3):268–276
194. Klamt A (2005) Summary, limitations, and perspectives. In: COSMO-RS. Elsevier, Amsterdam
195. Yang JY, Yan QL, Liu HL, Hu Y (2006) A molecular thermodynamic model for compressible lattice polymers. Polymer 47(14):5187–5195
196. Yang JY, Peng CJ, Liu HL, Hu Y, Jiang JW (2006) A generic molecular thermodynamic model for linear and branched polymer solutions in a lattice. Fluid Phase Equilib 244(2):188–192
197. Kojima K, Tochigi K (1979) Prediction of vapor–liquid equilibria by the ASOG method. Elsevier, Tokyo
198. Tochigi K, Tiegs D, Gmehling J, Kojima K (1990) Determination of new ASOG parameters. J Chem Eng Jpn 23(4):453–463
199. Tochigi K, Yoshida K, Kurihara K, Ochi K, Murata J, Yasumoto M, Sako T (2001) Prediction of vapor-liquid equilibrium for systems containing hydrofluoroethers using ASOG group contribution method. Fluid Phase Equilib 183:173–182
200. Tochigi K, Yoshida K, Kurihara K, Ochi K, Murata J, Urata S, Otake K (2002) Determination of ASOG parameters for selecting azeotropic mixtures containing hydrofluoroethers. Fluid Phase Equilib 194:653–662
201. Banerjee T, Verma KK, Khanna A (2008) Liquid-liquid equilibrium for ionic liquid systems using COSMO-RS: effect of cation and anion dissociation. Aiche J 54(7):1874–1885
202. Robles PA, Graber TA, Aznar M (2009) Prediction by the ASOG method of liquid-liquid equilibrium for binary and ternary systems containing 1-alkyl-3-methylimidazolium hexafluorophosphate. Fluid Phase Equilib 287(1):43–49
203. Verma NR, Gopal G, Anantharaj R, Banerjee T (2012) (Solid plus liquid) equilibria predictions of ionic liquid containing systems using COSMO-RS. J Chem Thermodyn 48:246–253
204. Domanska U (1987) Solid-liquid phase relations of some normal long-chain fatty acids in selected organic one- and two-component solvents. Ind Eng Chem Res 26(6):1153–1162
205. Prausnitz JM, Lichtenthaler RN, de Azevedo EG (1999) Molecular thermodynamics of fluid-phase equilibria, 3rd edn. Prentice Hall, Englewood Cliffs
206. Wilson GM (1964) Vapor-liquid equilibrium. XI. A new expression for the excess free energy of mixing. J Am Chem Soc 86(2):127–130
207. Domanska U, Zolek-Tryznowska Z, Krolikowski M (2007) Thermodynamic phase behavior of ionic liquids. J Chem Eng Data 52(5):1872–1880
208. Domanska U, Bogel-Lukasik E (2004) Solid-liquid equilibria for systems containing 1-butyl-3-methylimidazolium chloride. Fluid Phase Equilib 218(1):123–129
209. Domanska U, Bogel-Lukasik E, Bogel-Lukasik R (2003) Solubility of 1-dodecyl-3-methylimidazolium chloride in alcohols (C-2-C-12). J Phys Chem B 107(8):1858–1863
210. Domanska U, Bogel-Lukasik E, Bogel-Lukasik R (2003) 1-octanol/water partition coefficients of 1-alkyl-3-methylimidazolium chloride. Chem-a Eur J 9(13):3033–3041

211. Domanska U, Bogel-Lukasik E (2003) Measurements and correlation of the (solid plus liquid) equilibria of 1-decyl-3-methylimidazolium chloride plus alcohols (C-2-C-12). Ind Eng Chem Res 42(26):6986–6992
212. Domanska U (2005) Solubilities and thermophysical properties of ionic liquids. Pure Appl Chem 77(3):543–557
213. Domanska U, Mazurowska L (2004) Solubility of 1,3-dialkylimidazolium chloride or hexafluorophosphate or methylsulfonate in organic solvents: effect of the anions on solubility. Fluid Phase Equilib 221(1–2):73–82
214. Eckert F, Klamt A (2002) Fast solvent screening via quantum chemistry: COSMO-RS approach. Aiche J 48(2):369–385
215. Lin ST, Sandler SI (2002) A priori phase equilibrium prediction from a segment contribution solvation model. Ind Eng Chem Res 41(5):899–913

Part II

Properties of Ionic Liquid Mixtures

Mixtures Containing Imidazolium Ionic Liquids

2

2.1 Mixtures Containing 1-Alkyl Imidazolium Ionic Liquids

2.1.1 00001-01001: Imidazolium chloride

Abbreviation:	[Im][Cl], [Im]Cl
Molecular Formula:	$C_3H_5N_2Cl$
Molar Mass:	104.537
Structure:	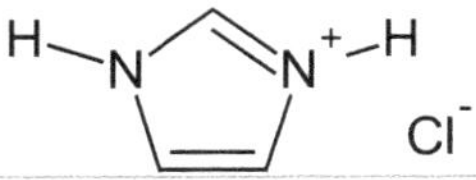

2.1.1.1 Binary Mixtures

Mixture 1: H_2O (1) + [Im]Cl (2)

(1) Volumetric Properties, Speed of Sound and Compressibility Properties

Ref. [1]

m_2/ (mol · kg^{-1})	ρ/ (kg · m^{-3})	u/ (m · s^{-1})	$10^{-3} \cdot \varphi_V$/ (mm^3 · mol^{-1})	$10^{-3} \cdot \overline{V}_2$/ (mm^3 · mol^{-1})	$10^{-3} \cdot \overline{V}_1$/ (mm^3 · mol^{-1})	$10^{11} \cdot \beta_S$/ Pa^{-1}	φ_{K_S}/(mm^3 · MPa^{-1} · mol^{-1})
$T = 298.15$ K							
0.00000	997.05	1497.6	104.40[a]	–	18.068	44.72	−13.8[a]
0.04296	998.24	1500.1	104.70	104.69	18.068	44.52	−13.0 ± 4.2
0.10274	999.89	1503.8	104.52	104.35	18.068	44.23	−14.3 ± 1.7
0.15300	1001.26	1507.0	104.38	104.05	18.069	43.98	−14.9 ± 1.2
0.19764	1002.47	1509.7	104.26	103.77	18.070	43.77	−14.5 ± 0.9
0.24165	1003.72	1512.4	104.13	103.48	18.071	43.56	−14.9 ± 0.7
0.29145	1004.65	1515.8	104.03	103.20	18.072	43.32	−14.3 ± 0.6
0.33998	1006.45	1517.9	103.84	102.83	18.074	43.12	−14.1 ± 0.5
0.38926	1007.62	1521.2	103.72	102.53	18.076	42.89	−14.3 ± 0.4
0.44952	1008.96	1525.2	103.59	102.16	18.080	42.61	−14.2 ± 0.4
0.50034	1010.31	1527.9	103.45	101.82	18.083	42.40	−13.8 ± 0.3

[a]Extrapolated values to infinite dilution

S. Zhang et al., *Physicochemical Properties of Ionic Liquid Mixtures*,
DOI 10.1007/978-94-017-7573-1_2

2.1.2 01001-01001: 1-Methylimidazolium chloride

Abbreviation:	$[C_1Im]Cl$, [Mim]Cl
Molecular Formula:	$C_4H_7N_2Cl$
Molar Mass:	118.56
Structure:	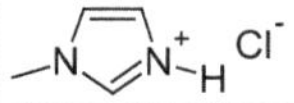

2.1.2.1 Binary Mixtures

Mixture 1: H_2O (1) + [Mim]Cl (2)

(1) Volumetric Properties, Speed of Sound and Compressibility Properties

Ref. [1]

m_2/ $(mol \cdot kg^{-1})$	ρ/ $(kg \cdot m^{-3})$	u/ $(m \cdot s^{-1})$	$10^{-3} \cdot \varphi_V$/ $(mm^3 \cdot mol^{-1})$	$10^{-3} \cdot \overline{V}_2$/ $(mm^3 \cdot mol^{-1})$	$10^{-3} \cdot \overline{V}_1$/ $(mm^3 \cdot mol^{-1})$	$10^{11} \cdot \beta_S$/ Pa^{-1}	$\varphi_{K_S}/(mm^3 \cdot MPa^{-1} \cdot mol^{-1})$
T = 298.15 K							
0.00000	997.05	1497.6	118.46	–	18.068	44.72	−14.9
0.05234	998.29	1501.6	118.74	118.72	18.068	44.43	−14.1 ± 3.4
0.09548	999.30	1504.7	118.62	118.48	18.068	44.20	−12.7 ± 1.9
0.14703	1000.49	1508.5	118.48	118.17	18.069	43.92	−12.6 ± 1.2
0.19981	1001.69	1512.5	118.34	117.85	18.070	43.64	−12.7 ± 0.9
0.24717	1002.77	1516.0	118.21	117.56	18.071	43.39	−12.6 ± 0.7
0.30106	1003.97	1519.9	118.07	117.23	18.073	43.12	−12.3 ± 0.6
0.35550	1005.18	1523.9	117.93	116.88	18.075	42.84	−12.3 ± 0.5
0.39940	1006.13	1526.3	117.81	116.61	18.077	42.66	−11.0 ± 0.4
0.45507	1007.35	1530.5	117.67	116.25	18.080	42.38	−11.2 ± 0.4
0.51388	1008.62	1534.6	117.52	115.88	18.083	42.10	−11.1 ± 0.3

2.2 Mixtures Containing 1,3-Dialkyl Imidazolium Ionic Liquids

2.2.1 02001-01001: 1,3-Dimethylimidazolium chloride

Abbreviation:	$[C_1MIm]Cl$, [mmim]Cl, [MeMeIm]Cl, [Mmim]Cl
Molecular Formula:	$C_5H_9N_2Cl$
Molar Mass:	132.59
Structure:	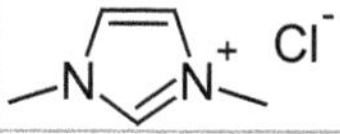

2.2.1.1 Binary Mixtures

Mixture 1: H_2O (1) + [Mmim]Cl (2)

(1) Volumetric Properties

Ref. [2]

ρ (g · cm^{-3})						
w_1	293.15 K	298.15 K	303.15 K	308.15 K	313.15 K	318.15 K
0.20045	1.15778	1.15479	1.15215	1.14931	1.14646	1.14359
0.30148	1.13529	1.13240	1.12993	1.12715	1.12433	1.12185
0.40195	1.11177	1.10903	1.10669	1.10422	1.10107	1.09834
0.50008	1.08904	1.08714	1.08455	1.08268	1.08038	1.07858
0.59922	1.06756	1.06577	1.06334	1.06163	1.05944	1.05708
0.70034	1.04752	1.04581	1.04370	1.04180	1.03952	1.03712
0.80030	1.03053	1.02798	1.02615	1.02453	1.02266	1.02105
0.89800	1.01407	1.01220	1.01057	1.00886	1.00720	1.00565
$\overline{V}_2$ (cm^3 · mol^{-1})						
m (mol · kg^{-1})	293.15 K	298.15 K	303.15 K	308.15 K	313.15 K	318.15 K
0.8567	112.3179	113.0365	113.4205	113.4209	113.6063	113.8360
1.8820	111.7937	111.9877	112.3891	112.6143	113.0391	113.4199
3.2271	110.9296	110.8396	111.3061	111.7070	112.1349	112.5123
5.0443	109.5601	109.9167	110.3339	110.2289	110.3028	110.4667
7.5395	109.2368	109.5254	109.7433	110.2493	110.6012	110.6206
11.2216	109.0288	109.7791	109.8417	110.3115	110.5250	111.1940
17.4746	109.8408	109.6940	110.1756	110.3987	110.9918	110.7470
30.0832	108.6946	111.2950	110.4672	111.2710	110.3477	113.3409
$\alpha \times 10^4$ (K^{-1})						
m (mol · kg^{-1})	293.15 K	298.15 K	303.15 K	308.15 K	313.15 K	318.15 K
0.0000	2.06	2.53	2.98	3.40	3.81	4.20
0.8567	4.81	4.84	4.87	4.89	4.92	4.94
1.8820	4.87	4.84	4.81	4.78	4.75	4.72
3.2271	4.23	4.46	4.68	4.90	5.11	5.31
5.0443	4.17	4.06	3.96	3.86	3.76	3.67
7.5395	3.63	3.77	3.91	4.04	4.17	4.30
11.2216	3.28	3.60	3.90	4.19	4.47	4.74
17.4746	4.45	4.10	3.78	3.46	3.16	2.88
30.0832	3.56	3.45	3.35	3.25	3.15	3.06
$^{\varphi}E \times 10^{-6}$ (m^3 · mol^{-1} · K^{-1})						
m (mol · kg^{-1})	293.15 K	298.15 K	303.15 K	308.15 K	313.15 K	318.15 K
0.8567	0.376	0.325	0.277	0.231	0.186	0.144
1.8820	0.204	0.178	0.152	0.128	0.104	0.082
3.2271	0.114	0.110	0.106	0.102	0.098	0.095
5.0443	0.088	0.076	0.064	0.052	0.041	0.031
7.5395	0.061	0.058	0.056	0.054	0.051	0.049
11.2216	0.047	0.049	0.052	0.054	0.056	0.058
17.4746	0.063	0.054	0.046	0.039	0.031	0.024
30.0832	0.044	0.041	0.038	0.036	0.033	0.030

2.2.2 02001-03001: 1,3-Dimethylimidazolium bis((trifluoromethyl)sulfonyl)imide

Abbreviation:	$[C_1MIm][NTf_2]$, $[C_1MIm][TFSI]$, $[C_1MIm][Tf_2N]$, $[MMIM][(CF_3SO_2)_2N]$, [MMIM][BTI]
Molecular Formula:	$C_7H_9F_6N_3O_4S_2$
Molar Mass:	377.28
Structure:	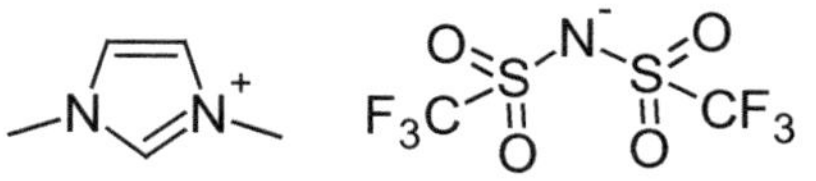

2.2.2.1 Binary Mixtures

Mixture 1: $CH_2(CH_2)_4CH_2$ (1) + $[MMIM][(CF_3SO_2)_2N]$ (2)
(1) Phase Equilibrium

Ref. [3]

x_1	P (kPa)	x_1	P (kPa)	x_1	P (kPa)	x_1	P (kPa)
353.15 K							
0.0073	9.87	0.0795	95.50	0.5656	99.25	0.9583	99.19
0.0136	18.44	0.1028	99.11	0.6441	99.08	0.9784	99.19
0.0216	28.98	0.1331	99.27	0.6461	99.24	0.9900	99.18
0.0285	38.10	0.1749	99.32	0.7176	99.22	0.9960	99.17
0.0363	48.01	0.2328	99.27	0.7273	99.09	0.9983	99.15
0.0426	55.79	0.3181	99.27	0.8055	99.12	0.9995	99.13
0.0484	62.78	0.3984	99.26	0.8746	99.16	1.0000	99.10
0.0621	78.41	0.4808	99.26	0.9251	99.18		

(2) Partial Molar Excess Enthalpies at Infinite Dilution

Ref. [4]

Solute (1)	$H_{12}^{E,\infty}$/kJ · mol^{-1}
Cyclohexane	6.517

Mixture 2: C_6H_6 (1) + $[MMIM][(CF_3SO_2)_2N]$ (2)
(1) Phase Equilibrium

Ref. [3]

x_1	P (kPa)	x_1	P (kPa)	x_1	P (kPa)	x_1	P (kPa)
353.15 K							
0.0086	1.36	0.2484	35.26	0.7038	98.98	0.9706	101.11
0.0173	2.58	0.3000	42.61	0.7383	101.13	0.9804	101.11
0.0266	3.87	0.3537	50.32	0.7690	101.15	0.9879	101.11
0.0363	5.23	0.4132	58.97	0.7697	101.07	0.9929	101.11

(*continued*)

x_1	P (kPa)	x_1	P (kPa)	x_1	P (kPa)	x_1	P (kPa)
0.0457	6.56	0.4698	67.24	0.8294	101.09	0.9961	101.10
0.0751	10.74	0.5246	75.30	0.8763	101.08	0.9977	101.10
0.1104	15.68	0.5753	82.65	0.9114	101.09	0.9988	101.09
0.1518	21.53	0.6220	89.15	0.9377	101.08	1.0000	101.10
0.1981	28.08	0.6649	94.68	0.9574	101.11		

(2) Partial Molar Excess Enthalpies at Infinite Dilution

Ref. [4]

Solute (1)	$H_{12}^{E,\infty}$/kJ · mol^{-1}
Benzene	−0.601

Mixture 3: C_5H_{12} (1) + [MMIM][BTI] (2)
(1) Partial Molar Excess Enthalpies at Infinite Dilution

Ref. [4]

Solute (1)	$H_{12}^{E,\infty}$/kJ · mol^{-1}
n-Pentane	5.694

Mixture 4: C_6H_{14} (1) + [MMIM][BTI] (2)
(1) Partial Molar Excess Enthalpies at Infinite Dilution

Ref. [4]

Solute (1)	$H_{12}^{E,\infty}$/kJ · mol^{-1}
n-Hexane	6.669

Mixture 5: C_7H_{16} (1) + [MMIM][BTI] (2)
(1) Partial Molar Excess Enthalpies at Infinite Dilution

Ref. [4]

Solute (1)	$H_{12}^{E,\infty}$/kJ · mol^{-1}
n-Heptane	7.795

Mixture 6: 1-C_5H_{10} (1) + [MMIM][BTI] (2)
(1) Partial Molar Excess Enthalpies at Infinite Dilution

Ref. [4]

Solute (1)	$H_{12}^{E,\infty}$/kJ · mol^{-1}
1-Pentene	2.898

Mixture 7: 1-C_6H_{12} (1) + [MMIM][BTI] (2)
(1) Partial Molar Excess Enthalpies at Infinite Dilution

Ref. [4]

Solute (1)	$H_{12}^{E,\infty}$/kJ · mol^{-1}
1-Hexene	3.812

Mixture 8: 1-C_7H_{14} (1) + [MMIM][BTI] (2)
(1) Partial Molar Excess Enthalpies at Infinite Dilution

Ref. [4]

Solute (1)	$H_{12}^{E,\infty}$/kJ · mol^{-1}
1-Heptene	5.326

Mixture 9: $CH_2(CH_2)_3CH_2$ (1) + [MMIM][BTI] (2)
(1) Partial Molar Excess Enthalpies at Infinite Dilution

Ref. [4]

Solute (1)	$H_{12}^{E,\infty}$/kJ · mol^{-1}
Cyclopentane	5.402

Mixture 10: CH = $CHCH_2CH_2CH_2$ (1) + [MMIM][BTI] (2)
(1) Partial Molar Excess Enthalpies at Infinite Dilution

Ref. [4]

Solute (1)	$H_{12}^{E,\infty}$/kJ · mol^{-1}
Cyclopentene	2.228

Mixture 11: CH = $CHCH_2CH_2CH_2CH_2$ (1) + [MMIM][BTI] (2)
(1) Partial Molar Excess Enthalpies at Infinite Dilution

Ref. [4]

Solute (1)	$H_{12}^{E,\infty}$/kJ · mol^{-1}
Cyclohexene	4.100

Mixture 12: $C_6H_5CH_3$ (1) + [MMIM][BTI] (2)
(1) Partial Molar Excess Enthalpies at Infinite Dilution

Ref. [4]

Solute (1)	$H_{12}^{E,\infty}$/kJ · mol^{-1}
Toluene	−0.747

Mixture 13: H_2O (1) + [MMIM][BTI] (2)
(1) Partial Molar Excess Enthalpies at Infinite Dilution

Ref. [4]

Solute (1)	$H_{12}^{E,\infty}$/kJ · mol^{-1}
Water	9.977

2.2.3 02001-04017: 1,3-Dimethylimidazolium methylsulfate

Abbreviation:	[Mmim][$MeSO_4$], [MMIM][$MeSO_4$], MMimMSO$_4$
Molecular Formula:	$C_6H_{12}N_2O_4S$
Molar Mass:	208.24
Structure:	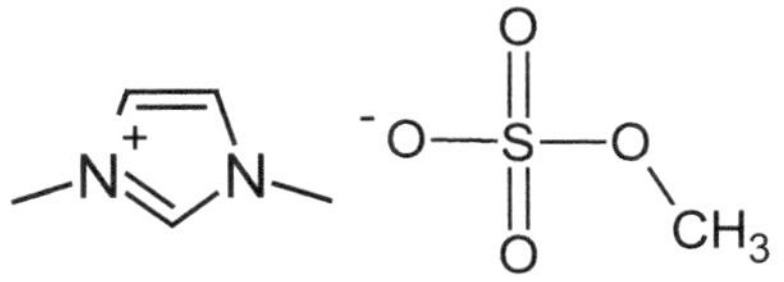

2.2.3.1 Binary Mixtures

Mixture 1: H_2O (1) + [Mmim][$MeSO_4$] (2)

(1) Osmotic Coefficient, Activity and Vapor Pressure

Ref. [5]

m_2/(mol · kg^{-1})	Φ_1	a_1	p/kPa	m_2/(mol · kg^{-1})	Φ_1	a_1	p/kPa
313.15 K				333.15 K			
0.0323	0.994	0.9988	7.369	0.0323	0.998	0.9988	19.901
0.0477	0.992	0.9983	7.365	0.0477	0.978	0.9983	19.891
0.0958	0.987	0.9966	7.352	0.0958	0.964	0.9967	19.858
0.1871	0.966	0.9935	7.329	0.1871	0.947	0.9936	19.797
0.3147	0.939	0.9894	7.299	0.3147	0.922	0.9896	19.716
0.4227	0.928	0.9860	7.274	0.4227	0.910	0.9862	19.649
0.5263	0.919	0.9827	7.249	0.5263	0.896	0.9832	19.587
0.7346	0.904	0.9764	7.202	0.7346	0.881	0.9770	19.463
0.9792	0.896	0.9689	7.147	0.9792	0.870	0.9698	19.319
1.2694	0.883	0.9604	7.085	1.2694	0.856	0.9617	19.156
1.6569	0.885	0.9486	6.997	1.6569	0.848	0.9506	18.936
2.0878	0.895	0.9349	6.896	2.0878	0.851	0.9381	18.684
2.4860	0.899	0.9227	6.805	2.4860	0.852	0.9265	18.453
2.8974	0.907	0.9096	6.709	2.8974	0.860	0.9141	18.205
3.2928	0.911	0.8975	6.620	3.2928	0.864	0.9026	17.974

(2) Volumetric Properties, Transport Properties, Refractive Index and Thermodynamic Properties

Ref. [6]

x_1	ρ/(g · cm^{-3})	n_D	η/(mPa · s)	V^E/(cm^3 · mol^{-1})	Δn_D	$\Delta\eta$/ (mPa · s)	ΔG^{*E}/ (J · mol^{-1})
298.15 K							
0.0000	1.32912	1.48296	77.7	0.000	0.0000	0.000	0.0
0.0497	1.32699	1.48187	68.1	0.016	0.0064	−5.788	378.8
0.1475	1.32281	1.47986	53.6	−0.019	0.0191	−12.858	1152.8

(continued)

x_1	ρ/(g · cm^{-3})	n_D	η/(mPa · s)	V^E/(cm^3 · mol^{-1})	Δn_D	$\Delta\eta$/ (mPa · s)	ΔG^{*E}/ (J · mol^{-1})
0.2128	1.31954	1.47824	44.93	−0.044	0.0273	−16.467	1618.9
0.3086	1.31375	1.47536	35.23	−0.076	0.0388	−18.800	2316.7
0.3869	1.30771	1.47252	27.57	−0.087	0.0478	−20.439	2747.1
0.4947	1.29697	1.46739	19.55	−0.103	0.0589	−20.177	3276.8
0.5959	1.28277	1.46071	13.88	−0.109	0.0674	−18.065	3660.9
0.6947	1.26217	1.45105	9.599	−0.098	0.0726	−14.753	3863.3
0.7966	1.22682	1.43483	5.928	−0.035	0.0717	−10.599	3684.0
0.8962	1.16206	1.40572	2.925	0.042	0.0576	−5.945	2697.0
0.9485	1.10041	1.37848	1.745	0.070	0.0382	−3.104	1653.8
1.0000	0.99720	1.33251	0.890	0.000	0.0000	0.000	0.0
313.15 K							
0.0000	1.31856		41.68	0.000		0.000	0.0
0.0497	1.31644		37.51	0.019		−2.128	426.7
0.1475	1.31231		30.44	−0.012		−5.188	1243.4
0.2128	1.30903		25.86	−0.030		−7.088	1713.4
0.3086	1.30327		20.76	−0.056		−8.257	2429.9
0.3869	1.29725		16.93	−0.062		−8.871	2926.2
0.4947	1.28656		12.41	−0.070		−8.970	3481.9
0.5959	1.27329		9.041	−0.067		−8.188	3870.1
0.6947	1.25191		6.368	−0.050		−6.807	4050.7
0.7966	1.21682		4.027	0.017		−4.972	3841.6
0.8962	1.15298		2.069	0.086		−2.843	2828.8
0.9485	1.09264		1.261	0.099		−1.504	1745.8
1.0000	0.99221		0.653	0.000		0.000	0.0
328.15 K							
0.0000	1.30813		25.21	0.000		0.000	0.0
0.0497	1.30609		23.95	0.013		−0.031	561.8
0.1475	1.30196		19.85	−0.011		−0.178	1410.4
0.2128	1.29864		17.11	−0.021		−2.839	1900.2
0.3086	1.29281		13.79	−0.036		−3.792	2598.9
0.3869	1.28689		10.50	−0.044		−5.154	2877.9
0.4947	1.27622		8.521	−0.047		−4.468	3667.1
0.5959	1.26204		6.336	−0.036		−4.152	4063.6
0.6947	1.24905		4.518	0.019		−3.528	4222.9
0.7966	1.20664		2.917	0.058		−2.614	3991.9
0.8962	1.14343		1.548	0.121		−1.521	2953.3
0.9485	1.08403		0.990	0.122		−0.786	1913.4
1.0000	0.98569		0.504	0.000		0.000	0.0

Mixture 2: C_2H_5OH (1) + [Mmim][$MeSO_4$] (2)
(1) Volumetric Properties, Speed of Sound, Transport Properties, Refractive Index, Compressibility Properties, Thermodynamic Properties and Surface Tension

Ref. [6]

x_1	$\rho/(g \cdot cm^{-3})$	n_D	$\eta/(mPa \cdot s)$	$V^E/(cm^3 \cdot mol^{-1})$	Δn_D	$\Delta\eta/(mPa \cdot s)$	$\Delta G^{*E}/(J \cdot mol^{-1})$
298.15 K							
0.0000	1.32912	1.48296	77.7	0.000	0.0000	0.000	0.0
0.0669	1.31647	1.48028	62.1	−0.179	0.0055	−10.548	204.9
0.1228	1.30519	1.47773	51.6	−0.348	0.0098	−16.726	380.2
0.2212	1.28246	1.47336	36.97	−0.589	0.0176	−23.816	659.9
0.2944	1.26318	1.46936	29.56	−0.760	0.0225	−25.617	919.7
0.3984	1.23122	1.46252	20.26	−0.966	0.0285	−26.946	1126.5
0.5000	1.19280	1.45421	12.83	−1.062	0.0326	−26.580	1095.7
0.5926	1.15085	1.44554	8.784	−1.148	0.0353	−23.532	1139.6
0.6921	1.09430	1.43248	5.706	−1.104	0.0345	−18.984	1106.1
0.7970	1.01810	1.41494	3.513	−0.965	0.0298	−13.132	967.0
0.8957	0.92316	1.39329	2.096	−0.652	0.0203	−6.980	654.2
0.9450	0.86426	1.37874	1.602	−0.420	0.0118	−3.696	457.3
1.0000	0.78546	1.36023	1.082	0.000	0.0000	0.000	0.0
313.15 K							
0.0000	1.31856		41.68	0.000		0.000	0.0
0.0669	1.30589		34.42	−0.204		−4.522	240.4
0.1228	1.29459		29.38	−0.393		−7.278	440.2
0.2212	1.27177		22.80	−0.663		−9.841	849.0
0.2944	1.25246		17.97	−0.860		−11.675	1016.5
0.3984	1.22040		12.87	−1.093		−12.532	1249.8
0.5000	1.18188		8.823	−1.216		−12.427	1327.7
0.5926	1.13974		6.258	−1.316		−11.210	1380.4
0.6921	1.08298		4.309	−1.280		−9.096	1404.3
0.7970	1.00641		2.649	−1.133		−6.470	1157.4
0.8957	0.91098		1.603	−0.787		−3.482	778.2
0.9450	0.85176		1.225	−0.521		−1.848	526.9
1.0000	0.77200		0.827	0.000		0.000	0.0
328.15 K							
0.0000	1.30813		25.21	0.000		0.000	0.0
0.0669	1.29553		21.43	−0.240		−2.139	283.1
0.1228	1.28420		18.72	−0.451		−3.476	516.5
0.2212	1.26138		15.07	−0.762		−4.705	977.0
0.2944	1.24197		12.28	−0.977		−5.700	1191.1
0.3984	1.20957		9.141	−1.220		−6.281	1470.6
0.5000	1.17112		6.461	−1.387		−6.465	1564.5
0.5926	1.12874		4.727	−1.496		−5.923	1640.9
0.6921	1.07175		3.246	−1.472		−4.961	1590.7
0.7970	0.99472		2.077	−1.310		−3.552	1370.9
0.8957	0.89870		1.265	−0.921		−1.938	926.3
0.9450	0.83906		0.961	−0.613		−1.030	618.1
1.0000	0.75855		0.641	0.000		0.000	0.0

Ref. [7]

x_2	$\rho/(g \cdot cm^{-3})$	$V^E_m/(cm^3 \cdot mol^{-1})$	n_D	Δn_D	$u/(m \cdot s^{-1})$	$\kappa_s/T \cdot Pa^{-1}$	$\Delta\kappa_s/T \cdot Pa^{-1}$
$T = 293.15$ K							
0	0.7893	0	1.36138	0	1160	941	0
0.0738	0.8915	−0.503	1.38810	0.0177	1241	728	−160
0.1356	0.9597	−0.743	1.40426	0.0263	1305	612	−232
0.2275	1.0406	−0.955	1.42272	0.0335	1396	493	−285
0.3317	1.1116	−1.068	1.43794	0.0359	1490	405	−299
0.4290	1.1639	−1.095	1.44915	0.0352	1568	350	−284
0.5395	1.2111	−1.014	1.45944	0.0320	1642	306	−249
0.6331	1.2436	−0.865	1.46650	0.0275	1693	281	−207
0.7308	1.2720	−0.669	1.47200	0.0211	1737	260	−158
0.8142	1.2930	−0.487	1.47624	0.0151	1769	247	−111
0.9081	1.3133	−0.243	1.48029	0.0076	1800	235	−56
0.9512	1.3217	−0.119	1.48211	0.0042	1812	230	−30
1	1.3309	0	1.48392	0	1826	225	0
$T = 298.15$ K							
0	0.7851	0	1.35929	0	1143	974	0
0.0738	0.8874	−0.530	1.38734	0.0189	1226	750	−170
0.1356	0.9557	−0.780	1.40398	0.0279	1291	628	−245
0.2275	1.0367	−1.000	1.42173	0.0344	1383	504	−300
0.3317	1.1078	−1.117	1.43648	0.0363	1477	414	−314
0.4290	1.1602	−1.142	1.44761	0.0354	1555	356	−298
0.5395	1.2075	−1.056	1.45819	0.0323	1629	312	−260
0.6331	1.2399	−0.902	1.46510	0.0277	1681	286	−217
0.7308	1.2684	−0.697	1.47068	0.0212	1725	265	−165
0.8142	1.2893	−0.507	1.47489	0.0151	1757	251	−116
0.9081	1.3097	−0.255	1.47895	0.0076	1787	239	−59
0.9512	1.3181	−0.126	1.48079	0.0041	1800	234	−31
1	1.3272	0	1.48270	0	1813	229	0
$T = 303.15$ K							
0	0.7807	0	1.35725	0	1127	1009	0
0.0738	0.8833	−0.558	1.38689	0.0205	1211	772	−179
0.1356	0.9517	−0.819	1.40322	0.0292	1277	644	−259
0.2275	1.0328	−1.048	1.42109	0.0356	1370	516	−316
0.3317	1.1040	−1.167	1.43560	0.0372	1465	422	−329
0.4290	1.1565	−1.191	1.44677	0.0363	1542	363	−312
0.5395	1.2038	−1.099	1.45699	0.0328	1617	318	−272
0.6331	1.2363	−0.940	1.46385	0.0281	1668	291	−227
0.7308	1.2648	−0.726	1.46940	0.0215	1713	270	−172
0.8142	1.2857	−0.526	1.47353	0.0153	1745	256	−121
0.9081	1.3061	−0.266	1.47763	0.0077	1775	243	−61
0.9512	1.3145	−0.132	1.47949	0.0043	1788	238	−33
1	1.3237	0	1.48129	0	1801	233	0

Ref. [8]

x_2	ρ (g · cm^{-3})	σ (mN · m^{-1})
298.15 K		
1.0000	1.3290[a]	56.46
0.9703	1.3261	54.07
0.8911	1.3127	46.70
0.8004	1.2931	38.85
0.7179	1.2716	31.04
0.6433	1.2491	24.10
0.4688	1.1825	22.28
0.3258	1.1068	22.22
0.1372	0.9592	22.70
0.0000	0.7848[b]	21.55[c]

[a]Volumetric Properties is 1.328 g · cm^{-3}, or 1.32725 g · cm^{-3} at 298.15 K. The surface tension is 59.8 mN · m^{-1} at 298.15 K
[b]Volumetric Properties is 0.7855 g · cm^{-3} at 298.15 K
[c]Surface tension is 21.82 mN · m^{-1} at 298.15 K

(2) Osmotic Coefficient, Activity and Vapor Pressure

Ref. [9]

m/(mol · kg^{-1})	Φ_1	a_1	p/kPa
T = 323.15 K			
0.0289	0.815	0.998	29.412
0.0374	0.794	0.997	29.396
0.0789	0.719	0.995	29.322
0.0991	0.687	0.994	29.291
0.1941	0.626	0.989	29.145
0.3285	0.560	0.983	28.976
0.4326	0.530	0.979	28.854
0.5312	0.503	0.976	28.752
0.6925	0.479	0.970	28.580
0.9868	0.449	0.960	28.286
1.2835	0.424	0.951	28.021
1.6987	0.399	0.939	27.674
2.0777	0.382	0.929	27.375
2.4892	0.363	0.920	27.097
2.9367	0.345	0.911	26.821
3.3145	0.326	0.905	26.653
3.7093	0.325	0.895	26.344
4.0160	0.331	0.885	26.042

Mixture 3: CH_3OH (1) + [Mmim][$MeSO_4$] (2)
(1) Volumetric Properties and Surface Tension

Ref. [10]

T/K	p/MPa	ρ/kg · m^{-3}								
		$x_2 = 0.0595$	$x_2 = 0.1256$	$x_2 = 0.2255$	$x_2 = 0.3487$	$x_2 = 0.4358$	$x_2 = 0.5999$	$x_2 = 0.6780$	$x_2 = 0.7985$	$x_2 = 0.8788$
313.15	0.1	897.40	995.68	1099.73	1178.03	1214.89	1259.56	1272.43	1293.34	1306.67
	2.5	899.34	997.08	1103.45	1178.78	1215.88	1260.50	1273.27	1293.66	1307.60
	5.0	901.48	998.93	1105.04	1180.09	1217.10	1261.63	1274.40	1294.78	1308.84
	10.0	905.53	1002.32	1107.82	1182.55	1219.41	1263.81	1276.57	1296.86	1310.84
	15.0	909.28	1005.47	1110.58	1184.88	1221.54	1265.86	1278.54	1298.79	1312.82
	20.0	912.87	1008.72	1112.94	1187.36	1223.84	1267.94	1280.59	1300.72	1314.73
	25.0	916.36	1011.65	1115.61	1189.55	1225.91	1269.90	1282.63	1302.59	1316.65
318.15	0.1	892.75	991.27	1098.10	1173.63	1211.01	1256.25	1268.88	1289.56	1302.95
	2.5	894.76	992.74	1099.36	1175.27	1212.06	1256.87	1269.52	1289.80	1303.55
	5.0	896.93	994.61	1101.05	1176.67	1213.30	1258.23	1270.74	1290.94	1304.76
	10.0	901.12	997.98	1103.84	1179.19	1215.63	1260.12	1272.88	1293.07	1306.85
	15.0	905.02	1001.35	1106.40	1181.55	1217.87	1262.34	1274.94	1295.12	1308.87
	20.0	908.73	1004.57	1109.09	1183.86	1220.12	1264.21	1276.82	1296.98	1310.76
	25.0	912.33	1007.78	1111.62	1186.11	1222.20	1266.38	1278.86	1298.82	1312.67
323.15	0.1	888.46	987.09	1094.44	1170.54	1207.30	1253.26	1265.81	1286.17	1298.59
	2.5	890.23	989.02	1095.16	1171.75	1208.50	1253.51	1266.27	1286.42	1298.95
	5.0	892.75	990.75	1097.24	1173.09	1209.85	1254.68	1267.47	1287.58	1301.07
	10.0	896.85	994.31	1100.14	1175.63	1211.92	1256.64	1269.70	1289.70	1302.92
	15.0	900.92	997.64	1102.68	1177.98	1214.35	1258.78	1271.77	1291.70	1305.22
	20.0	904.77	1000.97	1105.59	1180.36	1216.74	1260.99	1273.77	1293.67	1307.19
	25.0	908.35	1004.08	1107.96	1182.75	1218.87	1262.92	1275.80	1295.54	1309.14
328.15	0.1	883.92	982.93	1090.95	1166.83	1203.70	1249.89	1261.59	1282.63	1295.17

	2.5	886.03	984.76	1091.73	1167.85	1204.86	1250.01	1262.64	1282.83	1295.60
	5.0	888.29	986.63	1093.17	1169.21	1206.08	1251.05	1263.86	1284.04	1297.04
	10.0	892.59	990.09	1096.09	1171.74	1208.51	1253.24	1266.15	1286.21	1299.34
	15.0	896.65	993.64	1098.86	1174.17	1210.72	1255.37	1268.18	1288.18	1301.53
	20.0	900.42	996.92	1101.57	1176.56	1212.91	1257.49	1270.32	1290.28	1303.85
	25.0	904.13	1000.12	1104.13	1178.72	1215.10	1259.44	1272.25	1292.24	1305.86
333.15	0.1	879.42	978.71	1087.29	1163.08	1200.14	1244.84	1258.96	1279.02	1291.97
	2.5	881.45	980.35	1087.93	1163.89	1201.18	1245.50	1259.28	1279.21	1292.00
	5.0	883.90	982.47	1089.62	1165.49	1202.46	1246.55	1260.49	1280.50	1293.62
	10.0	888.26	986.20	1092.34	1168.15	1204.81	1248.07	1262.75	1282.63	1295.87
	15.0	892.46	989.32	1095.34	1170.47	1207.22	1250.67	1264.91	1284.64	1297.80
	20.0	896.41	992.83	1098.06	1173.01	1209.38	1252.91	1267.00	1286.71	1300.22
	25.0	900.18	996.24	1100.60	1175.34	1211.69	1255.04	1269.05	1288.75	1302.69

		$V^E/cm^3 \cdot mol^{-1}$								
T/K	p/MPa	$x_2 = 0.0595$	$x_2 = 0.1256$	$x_2 = 0.2255$	$x_2 = 0.3487$	$x_2 = 0.4358$	$x_2 = 0.5999$	$x_2 = 0.6780$	$x_2 = 0.7985$	$x_2 = 0.8788$
313.15	0.1	−1.00	−1.64	−2.37	−2.57	−2.45	−1.71	−1.06	−0.55	−0.39
	2.5	−0.97	−1.60	−2.49	−2.54	−2.45	−1.74	−1.09	−0.55	−0.48
	5.0	−0.95	−1.57	−2.45	−2.50	−2.42	−1.72	−1.09	−0.56	−0.52
	10.0	−0.90	−1.49	−2.34	−2.39	−2.30	−1.62	−0.99	−0.47	−0.42
	15.0	−0.85	−1.42	−2.27	−2.30	−2.22	−1.56	−0.94	−0.43	−0.40
	20.0	−0.81	−1.36	−2.17	−2.24	−2.16	−1.50	−0.89	−0.39	−0.37
	25.0	−0.77	−1.30	−2.11	−2.17	−2.09	−1.45	−0.86	−0.36	−0.36
318.15	0.1	−1.03	−1.69	−2.58	−2.61	−2.51	−1.80	−1.12	−0.57	−0.41
	2.5	−1.00	−1.64	−2.54	−2.61	−2.49	−1.77	−1.10	−0.52	−0.42
	5.0	−0.97	−1.60	−2.49	−2.56	−2.43	−1.74	−1.07	−0.49	−0.40
	10.0	−0.92	−1.52	−2.39	−2.46	−2.34	−1.63	−1.00	−0.44	−0.36
	15.0	−0.88	−1.45	−2.30	−2.38	−2.27	−1.59	−0.96	−0.43	−0.36
	20.0	−0.83	−1.39	−2.22	−2.30	−2.19	−1.51	−0.90	−0.38	−0.33

(*continued*)

		$V^E/cm^3 \cdot mol^{-1}$								
T/K	p/MPa	$x_2 = 0.0595$	$x_2 = 0.1256$	$x_2 = 0.2255$	$x_2 = 0.3487$	$x_2 = 0.4358$	$x_2 = 0.5999$	$x_2 = 0.6780$	$x_2 = 0.7985$	$x_2 = 0.8788$
	25.0	−0.79	−1.34	−2.14	−2.23	−2.13	−1.49	−0.87	−0.36	−0.32
323.15	0.1	−1.07	−1.75	−2.68	−2.74	−2.60	−1.94	−1.25	−0.66	−0.40
	2.5	−1.02	−1.71	−2.58	−2.69	−2.56	−1.83	−1.17	−0.57	−0.32
	5.0	−1.00	−1.65	−2.55	−2.62	−2.49	−1.77	−1.11	−0.51	−0.38
	10.0	−0.94	−1.57	−2.45	−2.52	−2.38	−1.67	−1.06	−0.48	−0.33
	15.0	−0.90	−1.50	−2.34	−2.42	−2.30	−1.61	−1.00	−0.43	−0.33
	20.0	−0.85	−1.44	−2.27	−2.35	−2.25	−1.57	−0.96	−0.42	−0.33
	25.0	−0.81	−1.38	−2.18	−2.28	−2.18	−1.51	−0.93	−0.38	−0.32
328.15	0.1	−1.10	−1.80	−2.77	−2.80	−2.65	−1.99	−1.20	−0.66	−0.40
	2.5	−1.07	−1.76	−2.68	−2.75	−2.62	−1.89	−1.20	−0.59	−0.37
	5.0	−1.03	−1.71	−2.61	−2.67	−2.55	−1.81	−1.14	−0.54	−0.34
	10.0	−0.98	−1.61	−2.49	−2.56	−2.45	−1.73	−1.09	−0.50	−0.33
	15.0	−0.92	−1.54	−2.40	−2.47	−2.35	−1.66	−1.02	−0.45	−0.32
	20.0	−0.87	−1.47	−2.31	−2.38	−2.27	−1.60	−0.99	−0.43	−0.34
	25.0	−0.83	−1.41	−2.22	−2.29	−2.20	−1.54	−0.93	−0.40	−0.33
333.15	0.1	−1.14	−1.85	−2.84	−2.85	−2.71	−1.88	−1.30	−0.64	−0.41
	2.5	−1.09	−1.79	−2.74	−2.77	−2.66	−1.82	−1.22	−0.56	−0.32
	5.0	−1.07	−1.75	−2.69	−2.73	−2.60	−1.76	−1.19	−0.55	−0.36
	10.0	−1.00	−1.67	−2.56	−2.62	−2.49	−1.62	−1.13	−0.51	−0.34
	15.0	−0.95	−1.56	−2.46	−2.52	−2.41	−1.60	−1.08	−0.47	−0.31
	20.0	−0.90	−1.50	−2.37	−2.43	−2.32	−1.54	−1.03	−0.44	−0.33
	25.0	−0.85	−1.44	−2.28	−2.35	−2.25	−1.49	−0.98	−0.41	−0.37

Ref. [8]

x_2	ρ (g · cm^{-3})	σ (mN · m^{-1})
298.15 K		
1.0000	1.3290	56.46[a]
0.8695	1.3122	51.08
0.7410	1.2881	46.90
0.6146	1.2605	42.20
0.4786	1.2241	36.75
0.3612	1.1795	32.51
0.2647	1.1251	29.07
0.1465	1.0229	25.77
0.0000	0.7853	22.09[b]

[a]Surface tension is 59.80 mN · m^{-1}
[b]Surface tension is 22.51 mN · m^{-1}

Mixture 4: $(CH_3O)_2C = O$ (1) + [Mmim][MeSO$_4$] (2)
(1) Phase Equilibrium

Ref. [11]

x^{I}_2	x^{II}_2	T (K)
0.0004	0.5498	278.15
0.0004	0.5498	283.15
0.0004	0.5498	288.15
0.0005	0.5498	293.15
0.0005	0.5493	298.15
0.0006	0.5493	303.15
0.0006	0.5488	308.15
0.0007	0.5488	313.15
0.0007	0.5482	318.15
0.0008	0.5482	323.15
0.0008	0.5471	328.15
0.0008	0.5471	333.15
0.0008	0.5471	338.15
0.0008	0.5471	343.15

Mixture 5: $(CH_3CH_2O)_2C = O$ (1) + [Mmim][MeSO$_4$] (2)
(1) Phase Equilibrium

Ref. [11]

x^{I}_2	x^{II}_2	T (K)
0.0001	0.9101	278.15
0.0001	0.9085	283.15
0.0001	0.9069	288.15
0.0001	0.9045	293.15
0.0001	0.9019	298.15

(continued)

x^{I}_{2}	x^{II}_{2}	T (K)
0.0001	0.8993	303.15
0.0001	0.8974	308.15
0.0001	0.8945	313.15
0.0001	0.8915	318.15
0.0001	0.8883	323.15
0.0002	0.886	328.15
0.0002	0.8823	333.15
0.0002	0.8797	338.15
0.0002	0.8754	343.15

Mixture 6: CH_3COCH_3 (1) + [Mmim][$MeSO_4$] (2)
(1) Phase Equilibrium

Ref. [11]

x^{I}_{2}	x^{II}_{2}	T (K)
0.0300	0.3337	278.15
0.0284	0.3389	283.15
0.0280	0.3404	288.15
0.0278	0.3452	293.15
0.0276	0.3463	298.15
0.0270	0.3514	303.15
0.0263	0.3546	308.15
0.0257	0.3611	313.15
0.0251	0.3668	318.15

Mixture 7: 2-$CH_3COC_2H_5$ (1) + [Mmim][$MeSO_4$] (2)
(1) Phase Equilibrium

Ref. [11]

x^{I}_{2}	x^{II}_{2}	T (K)
0.0039	0.7069	278.15
0.0040	0.7063	283.15
0.0041	0.7048	288.15
0.0041	0.7036	293.15
0.0043	0.7024	298.15
0.0044	0.7012	303.15
0.0044	0.6997	308.15
0.0045	0.6988	313.15
0.0047	0.6979	318.15
0.0047	0.6961	323.15
0.0047	0.6955	328.15
0.0047	0.6949	333.15

(2) Volumetric Properties, Refractive Index, Speed of Sound and Compressibility Properties

Ref. [12]

x_2	ρ/ g · cm^{-3}	V_m^E/ cm^3 · mol^{-1}	n_D	Δn_D	u/ m · s^{-1}	κ_S/ T · Pa^{-1}	$\Delta\kappa_S$/ T · Pa^{-1}
T = 293.15 K							
0	0.8056	0	1.37879	0	1212	845	0
0.0006	0.8065	−0.037	1.37993	0.0011	1214	841	−4
0.0012	0.8073	−0.072	1.38040	0.0015	1215	839	−5
0.0029	0.8095	−0.140	1.38130	0.0022	1217	834	−9
0.7074	1.2440	−1.498	1.47023	0.0171	1703	277	−129
0.7983	1.2738	−1.033	1.47479	0.0121	1750	256	−94
0.9284	1.3106	−0.225	1.48045	0.0040	1803	235	−35
0.9743	1.3235	−0.053	1.48218	0.0010	1819	228	−13
1	1.3309	0	1.48392	0	1826	225	0
T = 298.15 K							
0	0.7997	0	1.37618	0	1192	880	0
0.0006	0.8007	−0.052	1.37720	0.0010	1194	876	−3
0.0012	0.8015	−0.075	1.37762	0.0013	1195	874	−5
0.0029	0.8039	−0.170	1.37841	0.0019	1196	869	−8
0.7074	1.2403	−1.587	1.46925	0.0177	1689	283	−137
0.7983	1.2701	−1.099	1.47366	0.0124	1737	261	−100
0.9284	1.3071	−0.255	1.47914	0.0041	1791	239	−37
0.9743	1.3199	−0.060	1.48086	0.0009	1806	232	−14
1	1.3272	0	1.48270	0	1813	229	0
T = 303.15 K							
0	0.7949	0	1.37355	0	1163	931	0
0.0006	0.7960	−0.069	1.37467	0.0011	1164	927	−3
0.0012	0.7967	−0.082	1.37513	0.0015	1165	925	−5
0.0029	0.7991	−0.172	1.37600	0.0021	1168	917	−11
0.7074	1.2365	−1.639	1.46782	0.0181	1676	288	−149
0.7983	1.2664	−1.138	1.47229	0.0127	1724	266	−108
0.9284	1.3035	−0.273	1.47781	0.0042	1778	243	−40
0.9743	1.3163	−0.072	1.47945	0.0009	1794	236	−15
1	1.3237	0	1.48129	0	1801	233	0

Mixture 8: 2-$CH_3COC_3H_7$ (1) + [Mmim][$MeSO_4$] (2)
(1) Phase Equilibrium

Ref. [11]

x^I_2	x^{II}_2	T (K)
0.0009	0.8606	278.15
0.0009	0.8598	283.15
0.0009	0.8570	288.15
0.0009	0.8563	293.15
0.0009	0.8543	298.15
0.0009	0.8522	303.15
0.0009	0.8505	308.15

(*continued*)

x^{I}_2	x^{II}_2	T (K)
0.0010	0.8495	313.15
0.0010	0.8478	318.15
0.0010	0.8471	323.15
0.0010	0.8455	328.15
0.0010	0.8435	333.15
0.0010	0.8476	338.15
0.0010	0.8387	343.15

Ref. [13]

x_2	T/K	x_2	T/K	x_2	T/K	x_2	T/K
0.5846	368.81	0.6922	361.96	0.7819	344.07	0.8909	311.04
0.6498	365.90	0.7213	356.61	0.8118	330.60	0.9465	305.40
0.6636	364.63	0.7457	352.02	0.8455	320.92		

Mixture 9: CH_3COOCH_3 (1) + [Mmim][$MeSO_4$] (2)
(1) Phase Equilibrium

Ref. [11]

x^{I}_2	x^{II}_2	T (K)
0.0003	0.6401	278.15
0.0003	0.6401	283.15
0.0003	0.6405	288.15
0.0003	0.6409	293.15
0.0003	0.6409	298.15
0.0003	0.6425	303.15
0.0003	0.6436	308.15
0.0003	0.6444	313.15
0.0003	0.6476	318.15

Mixture 10: $CH_3COOCH_2CH_3$ (1) + [Mmim][$MeSO_4$] (2)
(1) Phase Equilibrium

Ref. [11]

x^{I}_2	x^{II}_2	T (K)
0.0000	0.8419	278.15
0.0000	0.8408	283.15
0.0000	0.8402	288.15
0.0000	0.8397	293.15
0.0000	0.8385	298.15
0.0000	0.8373	303.15
0.0000	0.8373	308.15
0.0000	0.8356	313.15
0.0000	0.8344	318.15
0.0000	0.8325	323.15

(*continued*)

x^{I}_2	x^{II}_2	T (K)
0.0000	0.8325	328.15
0.0000	0.8325	333.15
0.0000	0.8325	338.15

(2) Volumetric Properties, Refractive Index, Speed of Sound and Compressibility Properties

Ref. [12]

x_2	ρ/ $g \cdot cm^{-3}$	V_m^E/ $cm^3 \cdot mol^{-1}$	n_D	Δn_D	u/ $m \cdot s^{-1}$	κ_s/ $T \cdot Pa^{-1}$	$\Delta\kappa_s$/ $T \cdot Pa^{-1}$
T = 293.15 K							
0	0.9005	0	1.37241	0	1162	822	0
0.8549	1.2966	−0.804	1.47633	0.0086	1765	247	−65
0.8746	1.3011	−0.640	1.47752	0.0076	1775	244	−56
0.9072	1.3083	−0.381	1.47884	0.0053	1792	238	−43
0.9294	1.3133	−0.224	1.48008	0.0040	1801	235	−33
0.9491	1.3177	−0.093	1.48086	0.0026	1808	232	−24
1	1.3309	0	1.48392	0	1826	225	0
T = 298.15 K							
0	0.8944	0	1.36977	0	1141	859	0
0.8549	1.2930	−0.850	1.47553	0.0092	1752	252	−69
0.8746	1.2975	−0.697	1.47640	0.0079	1765	247	−61
0.9072	1.3047	−0.415	1.47750	0.0053	1778	243	−45
0.9294	1.3098	−0.261	1.47876	0.0040	1789	239	−35
0.9491	1.3142	−0.129	1.47955	0.0026	1795	236	−25
1	1.3272	0	1.48270	0	1813	229	0
T = 303.15 K							
0	0.8882	0	1.36712	0	1111	912	0
0.8549	1.2893	−0.896	1.47440	0.0097	1739	256	−75
0.8746	1.2938	−0.730	1.47522	0.0082	1746	254	−64
0.9072	1.3010	−0.440	1.47618	0.0055	1764	247	−49
0.9294	1.3062	−0.290	1.47752	0.0043	1777	243	−38
0.9491	1.3107	−0.154	1.47822	0.0027	1782	240	−27
1	1.3237	0	1.48129	0	1801	233	0

Mixture 11: 2-CH_3CHCH_3OH (1) + [Mmim][$MeSO_4$] (2)
(1) Volumetric Properties, Refractive Index, Speed of Sound and Compressibility Properties

Ref. [12]

x_2	ρ/ $g \cdot cm^{-3}$	V_m^E/ $cm^3 \cdot mol^{-1}$	n_D	Δn_D	u/ $m \cdot s^{-1}$	κ_s/ $T \cdot Pa^{-1}$	$\Delta\kappa_s$/ $T \cdot Pa^{-1}$
T = 293.15 K							
0	0.7850	0	1.37707	0	1156	953	0
0.0490	0.8397	−0.258	1.38865	0.0063	1177	859	−58
0.0981	0.8888	−0.433	1.39897	0.0114	1205	775	−106

(continued)

x_2	ρ/ g · cm^{-3}	V_m^E/ cm^3 · mol^{-1}	n_D	Δn_D	u/ m · s^{-1}	κ_s/ T · Pa^{-1}	$\Delta\kappa_s$/ T · Pa^{-1}
0.1987	0.9757	−0.670	1.41695	0.0186	1269	636	−172
0.2961	1.0458	−0.810	1.43085	0.0221	1338	534	−204
0.4055	1.1120	−0.892	1.44372	0.0233	1431	439	−219
0.5043	1.1624	−0.886	1.45335	0.0224	1525	370	−216
0.5987	1.2040	−0.819	1.46137	0.0203	1611	320	−197
0.6937	1.2399	−0.641	1.46768	0.0165	1684	284	−164
0.7992	1.2742	−0.340	1.47404	0.0116	1743	258	−113
0.9042	1.3048	−0.093	1.47881	0.0051	1791	239	−56
0.9502	1.3175	−0.031	1.48087	0.0023	1807	232	−29
1	1.3309	0	1.48392	0	1826	225	0
T = 298.15 K							
0	0.7810	0	1.37496	0	1139	986	0
0.0490	0.8359	−0.287	1.38617	0.0059	1167	878	−72
0.0981	0.8848	−0.449	1.39708	0.0116	1198	788	−124
0.1987	0.9718	−0.701	1.41521	0.0188	1271	637	−199
0.2961	1.0420	−0.850	1.42948	0.0226	1352	525	−237
0.4055	1.1085	−0.958	1.44245	0.0238	1444	433	−247
0.5043	1.1590	−0.956	1.45208	0.0228	1534	367	−238
0.5987	1.2003	−0.852	1.46001	0.0205	1605	324	−209
0.6937	1.2363	−0.675	1.46641	0.0167	1670	290	−171
0.7992	1.2707	−0.375	1.47250	0.0114	1729	263	−118
0.9042	1.3014	−0.129	1.47743	0.0050	1777	243	−58
0.9502	1.3140	−0.052	1.47960	0.0023	1797	236	−31
1	1.3272	0	1.48270	0	1813	229	0
T = 303.15 K							
0	0.7766	0	1.37278	0	1121	1025	0
0.0490	0.8316	−0.306	1.38391	0.0058	1166	884	−102
0.0981	0.8808	−0.501	1.39509	0.0117	1207	779	−168
0.1987	0.9679	−0.764	1.41455	0.0202	1292	619	−249
0.2961	1.0383	−0.928	1.42882	0.0239	1370	513	−277
0.4055	1.1047	−1.017	1.44156	0.0248	1459	425	−278
0.5043	1.1550	−0.984	1.45073	0.0232	1534	368	−258
0.5987	1.1967	−0.909	1.45836	0.0206	1602	326	−225
0.6937	1.2327	−0.720	1.46482	0.0168	1662	294	−182
0.7992	1.2674	−0.437	1.47103	0.0115	1720	267	−125
0.9042	1.2979	−0.155	1.47603	0.0051	1768	246	−62
0.9502	1.3104	−0.058	1.47820	0.0023	1785	239	−33
1	1.3237	0	1.48129	0	1801	233	0

(2) Osmotic Coefficient, Activity and Vapor Pressure

Ref. [9]

m/(mol · kg^{-1})	Φ_1	a_1	p/kPa
T = 323.15 K			
0.0241	0.914	0.997	23.527
0.0557	0.846	0.994	23.456
0.2089	0.683	0.983	23.183
0.3053	0.614	0.978	23.057
0.3982	0.588	0.972	22.926
0.6947	0.509	0.958	22.595
0.9721	0.461	0.948	22.336
1.2737	0.442	0.935	22.023
1.6596	0.416	0.920	21.685
2.0575	0.392	0.908	21.381
2.4473	0.364	0.898	21.160
2.8266	0.343	0.890	20.957
3.3134	0.329	0.877	20.656
3.7120	0.323	0.866	20.382
4.1118	0.298	0.863	20.315

Mixture 12: 1-$C_6H_{13}OH$ (1) + [Mmim][$MeSO_4$] (2)
(1) Phase Equilibrium

Ref. [13]

x_2	T/K	x_2	T/K	x_2	T/K	x_2	T/K
0.2855	290.35	0.4502	296.00	0.5552	298.76	0.8453	305.05
0.3275	292.19	0.4769	296.80	0.6161	299.89	0.9105	306.84
0.3845	294.53	0.4963	297.24	0.6677	300.76	0.9551	307.98
0.4212	295.36	0.5140	297.73	0.7673	303.46	1.0000	308.90

Mixture 13: 1-$C_7H_{15}OH$ (1) + [Mmim][$MeSO_4$] (2)
(1) Phase Equilibrium

Ref. [13]

x_2	T/K	x_2	T/K	x_2	T/K	x_2	T/K
0.2414	289.03	0.4497	297.28	0.6348	301.67	0.8967	307.60
0.2731	291.20	0.5021	298.82	0.7247	303.64	0.9599	308.40
0.3371	293.53	0.5630	300.20	0.7997	305.12	1.0000	308.90
0.3783	295.47						

Mixture 14: 1-$C_8H_{17}OH$ (1) + [Mmim][$MeSO_4$] (2)
(1) Phase Equilibrium

Ref. [13]

x_2	T/K	x_2	T/K	x_2	T/K	x_2	T/K
0.2780	322.33	0.4425	342.97	0.5792	345.83	0.7872	333.47
0.2960	325.69	0.4439	343.22	0.5827	345.47	0.8072	332.04

(continued)

x_2	T/K	x_2	T/K	x_2	T/K	x_2	T/K
0.3140	329.33	0.4465	342.97	0.6276	342.59	0.8469	330.49
0.3296	332.45	0.4537	343.20	0.6680	341.56	0.8675	328.15
0.3555	337.03	0.4704	344.18	0.6948	340.35	0.8949	325.37
0.3873	339.47	0.4716	344.62	0.7440	335.96	0.9233	323.08
0.4159	341.56	0.4858	344.82	0.7550	335.47	0.9491	319.26
0.4409	342.87	0.5047	345.85				

Mixture 15: 1-$C_9H_{19}OH$ (1) + [Mmim][$MeSO_4$] (2)
(1) Phase Equilibrium

Ref. [13]

x_2	T/K	x_2	T/K	x_2	T/K	x_2	T/K
0.3755	351.23	0.5120	354.50	0.6546	352.38	0.8319	349.63
0.4158	352.83	0.5369	353.76	0.7042	351.71	0.8745	347.57
0.4355	354.26	0.5799	353.48	0.7628	350.57	0.9216	341.35
0.4586	354.49	0.6150	353.82	0.7975	348.92	0.9519	327.80
0.4770	355.42						

Mixture 16: 1-$C_{10}H_{21}OH$ (1) + [Mmim][$MeSO_4$] (2)
(1) Phase Equilibrium

Ref. [13]

x_2	T/K	x_2	T/K	x_2	T/K	x_2	T/K
0.1589	311.36	0.3411	332.76	0.5981	355.85	0.8256	353.98
0.2042	314.24	0.3710	338.10	0.6836	356.55	0.8592	352.75
0.2354	319.14	0.4079	343.28	0.7181	355.45	0.8964	351.37
0.2599	322.25	0.4500	349.18	0.7559	354.65	0.9304	349.19
0.2828	325.63	0.5019	353.45	0.7787	354.65	0.9679	346.65
0.3023	329.25	0.5288	332.76	0.7994	353.33		

Mixture 17: $CH_3CH_2CH_2OCH_2CH_2CH_3$ (1) + [Mmim][$MeSO_4$] (2)
(1) Phase Equilibrium

Ref. [13]

x_2	T/K	x_2	T/K	x_2	T/K	x_2	T/K
0.5698	359.36	0.6680	347.74	0.7653	335.80	0.8507	324.01
0.5865	356.97	0.6720	348.06	0.7820	333.88	0.8726	321.07
0.6000	355.48	0.6969	345.13	0.7977	331.80	0.8968	317.68
0.6081	354.39	0.7167	342.66	0.8228	328.86	0.9187	315.67
0.6209	353.28	0.7394	340.49	0.8381	325.97	0.9417	310.92
0.6409	352.20	0.7509	338.02	0.8454	324.62	0.9631	304.48
0.6520	350.02						

Mixture 18: $CH_3(CH_2)_3O(CH_2)_3CH_3$ (1) + [Mmim][$MeSO_4$] (2)
(1) Phase Equilibrium

Ref. [13]

x_2	T/K	x_2	T/K	x_2	T/K	x_2	T/K
0.7662	407.12	0.8435	368.38	0.8851	358.40	0.9362	349.17
0.7795	401.51	0.8571	365.51	0.9056	354.69	0.9614	344.77
0.7888	394.64	0.8725	361.36	0.9286	350.16	0.9800	341.67
0.8115	380.97						

Mixture 19: $(CH_3)_3COCH_3$ (1) + [Mmim][$MeSO_4$] (2)
(1) Phase Equilibrium

Ref. [13]

x_2	T/K	x_2	T/K	x_2	T/K	x_2	T/K
0.6872	324.13	0.7759	320.38	0.8407	315.68	0.9334	301.88
0.7155	322.95	0.7927	319.76	0.8654	312.43	0.9404	299.52
0.7423	321.87	0.8094	318.56	0.8774	309.71	0.9720	296.71
0.7560	321.33	0.8130	318.17	0.9247	304.12		

Mixture 20: $CH_3OC(CH_3)_2C_2H_5$ (1) + [Mmim][$MeSO_4$] (2)
(1) Phase Equilibrium

Ref. [13]

x_2	T/K	x_2	T/K	x_2	T/K	x_2	T/K
0.9654	302.77	0.8717	331.16	0.8078	341.62	0.7134	350.66
0.9421	316.82	0.8650	333.13	0.7696	344.82	0.6913	353.17
0.9232	321.23	0.8444	336.71	0.7440	347.98	0.6521	356.76
0.8988	325.57						

Mixture 21: 3-$C_2H_5COC_2H_5$ (1) + [Mmim][$MeSO_4$] (2)
(1) Phase Equilibrium

Ref. [13]

x_2	T/K	x_2	T/K	x_2	T/K	x_2	T/K
0.6186	367.51	0.7350	355.80	0.7910	339.51	0.8526	322.09
0.6556	364.49	0.7615	348.34	0.8066	334.83	0.8871	312.85
0.6888	361.77	0.7789	343.36	0.8295	327.53	0.9402	306.96
0.7128	359.18						

Mixture 22: 2-$CH_3COC_4H_9$ (1) + [Mmim][$MeSO_4$] (2)
(1) Phase Equilibrium

Ref. [13]

x_2	T/K	x_2	T/K	x_2	T/K	x_2	T/K
0.6758	394.01	0.7828	374.57	0.8386	359.60	0.8859	338.91
0.7020	389.37	0.8079	367.71	0.8569	354.59	0.9146	325.93
0.7386	381.51	0.8336	360.97	0.8717	346.17	0.9554	296.99

Mixture 23: 4-$C_3H_7COC_3H_7$ (1) + [Mmim][$MeSO_4$] (2)
(1) Phase Equilibrium

Ref. [13]

x_2	T/K	x_2	T/K	x_2	T/K	x_2	T/K
0.7473	405.46	0.8208	385.38	0.8799	367.62	0.9408	338.76
0.7748	399.01	0.8413	380.40	0.8977	358.97	0.9746	322.38
0.7915	393.39	0.8646	373.28	0.9153	349.64		

Mixture 24: $(CH_2)_4CO$ (1) + [Mmim][$MeSO_4$] (2)
(1) Phase Equilibrium

Ref. [13]

x_2	T/K	x_2	T/K	x_2	T/K	x_2	T/K
0.3944	396.80	0.4722	376.39	0.5164	364.98	0.5685	349.63
0.4337	387.43	0.4925	371.32	0.5397	358.17	0.6031	336.33
0.4522	382.27						

Mixture 25: C_5H_{12} (1) + [Mmim][$MeSO_4$] (2)
(1) Phase Equilibrium

Ref. [14]

x_2	T/K	x_2	T/K
0.7814	306.32	0.8210	298.08
0.7953	304.48	0.8398	293.63
0.8093	300.77	0.8619	288.21

Mixture 26: C_6H_{14} (1) + [Mmim][$MeSO_4$] (2)
(1) Phase Equilibrium

Ref. [14]

x_2	T/K	x_2	T/K
0.6977	341.39	0.8742	321.78
0.7303	338.44	0.8866	319.43
0.7527	336.67	0.9185	313.36
0.7791	333.56	0.9322	308.98

(continued)

x_2	T/K	x_2	T/K
0.8084	331.14	0.9582	301.52
0.8329	327.89	0.9731	295.87
0.8632	324.33		

Mixture 27: C_7H_{16} (1) + [Mmim][$MeSO_4$] (2)
(1) Phase Equilibrium

Ref. [14]

x_2	T/K	x_2	T/K
0.7020	367.52	0.8340	346.12
0.7202	364.60	0.8753	337.86
0.7421	361.67	0.8982	331.79
0.7623	358.86	0.9293	322.66
0.7854	355.71	0.9666	313.07
0.8062	353.38	0.6812	371.00
0.8204	349.57		

Mixture 28: C_8H_{18} (1) + [Mmim][$MeSO_4$] (2)
(1) Phase Equilibrium

Ref. [14]

x_2	T/K	x_2	T/K
0.5462	396.20	0.8319	354.46
0.5942	388.79	0.8611	347.21
0.6273	384.29	0.8852	341.90
0.6829	376.82	0.9033	337.11
0.7016	374.04	0.9260	330.28
0.7239	370.81	0.9351	327.80
0.7612	366.39	0.9472	324.20
0.7988	360.03	0.9595	319.32
0.8086	357.85		

Mixture 29: $C_{10}H_{22}$ (1) + [Mmim][$MeSO_4$] (2)
(1) Phase Equilibrium

Ref. [14]

x_2	T/K	x_2	T/K
0.5381	442.81	0.8257	377.83
0.5922	430.60	0.8763	363.72
0.6340	419.84	0.9176	354.00
0.6597	411.65	0.9344	350.23
0.6889	406.21	0.9585	328.50

(continued)

x_2	T/K	x_2	T/K
0.7194	399.80	0.9681	322.66
0.7621	390.63	0.9809	318.18
0.8057	381.71	1.0000	308.30

Mixture 30: $CH_2(CH_2)_4CH_2$ (1) + [Mmim][$MeSO_4$] (2)
(1) Phase Equilibrium

Ref. [14]

x_2	T/K	x_2	T/K
0.7404	348.12	0.9026	328.60
0.7857	343.98	0.9320	322.47
0.8503	337.39	0.9531	316.90
0.8787	333.17		

Mixture 31: $CH_2(CH_2)_5CH_2$ (1) + [Mmim][$MeSO_4$] (2)
(1) Phase Equilibrium

Ref. [14]

x_2	T/K	x_2	T/K
0.6145	362.50	0.8223	349.98
0.6851	360.89	0.8613	343.30
0.7518	356.03	0.8820	337.21
0.7914	353.37	0.9571	313.98

Mixture 32: C_6H_6 (1) + [Mmim][$MeSO_4$] (2)
(1) Phase Equilibrium

Ref. [14]

x_2	T/K	x_2	T/K
0.5186	354.99	0.6284	318.48
0.5280	349.40	0.6422	312.60
0.5365	348.02	0.6481	313.61
0.5400	344.40	0.6687	309.79
0.5637	339.20	0.6707	308.57
0.5834	333.76	0.6875	305.98
0.5937	328.94	0.7043	303.99
0.5956	324.92	0.7074	302.51
0.6054	321.39	0.7192	301.53
0.6179	319.76	0.7354	300.29
0.6253	316.90	0.7437	301.92

Mixture 33: $C_6H_5CH_3$ (1) + [Mmim][$MeSO_4$] (2)
(1) Phase Equilibrium

Ref. [14]

x_2	T/K	x_2	T/K
0.6319	383.03	0.7718	318.92
0.6423	371.60	0.8042	313.26
0.6680	350.59	0.8547	310.03
0.6926	339.52	0.8931	309.59
0.7222	331.37	0.9224	308.12
0.7405	323.28	1.0000	305.65

Mixture 34: $C_6H_5C_2H_5$ (1) + [Mmim][$MeSO_4$] (2)
(1) Phase Equilibrium

Ref. [14]

x_2	T/K	x_2	T/K
0.8222	408.78	0.8653	362.02
0.8280	403.74	0.8723	353.22
0.8311	399.92	0.8862	341.50
0.8394	391.97	0.9029	326.99
0.8428	387.87	0.9207	301.88
0.8560	371.42		

Mixture 35: $C_6H_5C_3H_7$ (1) + [Mmim][$MeSO_4$] (2)
(1) Phase Equilibrium

Ref. [14]

x_2	T/K	x_2	T/K
0.8570	428.26	0.9087	351.52
0.8610	416.84	0.9099	348.37
0.8875	388.86	0.9125	341.10
0.8963	368.48	0.9261	312.17

Mixture 36: o-$C_6H_4(CH_3)_2$ (1) + [Mmim][$MeSO_4$] (2)
(1) Phase Equilibrium

Ref. [14]

x_2	T/K	x_2	T/K
0.6231	410.68	0.7793	339.61
0.6516	393.06	0.8042	328.98
0.6896	374.67	0.8351	322.40
0.7204	362.93	0.8684	315.80
0.7394	355.96	0.9035	312.92
0.7538	352.72	0.9527	311.42

Mixture 37: m-$C_6H_4(CH_3)_2$ (1) + [Mmim][$MeSO_4$] (2)
(1) Phase Equilibrium

Ref. [14]

x_2	T/K	x_2	T/K
0.6238	415.97	0.7794	345.63
0.6416	402.19	0.8036	333.28
0.6579	394.46	0.8246	324.17
0.6900	380.92	0.8462	319.96
0.7134	370.98	0.8635	315.57
0.7383	362.93	0.8882	315.08
0.7733	348.56	0.9436	306.34

Mixture 38: p-$C_6H_4(CH_3)_2$ (1) + [Mmim][$MeSO_4$] (2)
(1) Phase Equilibrium

Ref. [14]

x_2	T/K	x_2	T/K
0.6297	411.50	0.7881	340.79
0.6432	404.18	0.8087	328.98
0.6743	388.35	0.8304	322.59
0.6926	378.37	0.8638	317.31
0.7088	372.96	0.8934	314.31
0.7311	365.87	0.9226	310.57
0.7507	355.93	0.9602	309.56
0.7673	349.17		

Mixture 39: 1-C_3H_7OH (1) + $MMimMSO_4$ (2)
(1) Osmotic Coefficient, Activity and Vapor Pressure

Ref. [9]

$m/(mol \cdot kg^{-1})$	Φ_1	a_1	p/kPa
T = 323.15 K			
0.0512	0.848	0.995	12.039
0.0790	0.772	0.993	12.013
0.0997	0.727	0.991	11.996
0.2023	0.670	0.984	11.904
0.2902	0.590	0.980	11.852
0.3952	0.544	0.974	11.790
0.4880	0.526	0.970	11.730
0.6881	0.492	0.960	11.613
0.9815	0.454	0.948	11.463
1.2994	0.432	0.935	11.303
1.6910	0.396	0.923	11.154
2.0974	0.375	0.910	10.998
2.5454	0.360	0.896	10.825

(*continued*)

m/(mol · kg^{-1})	Φ_1	a_1	p/kPa
2.9285	0.343	0.886	10.709
3.3226	0.334	0.875	10.574
3.7100	0.322	0.866	10.467

Mixture 40: 1-C_4H_9OH (1) + [MMIM][CH_3SO_4] (2)
(1) Volumetric Properties and Surface Tension

Ref. [8]

x_2	ρ (g · cm^{-3})	σ (mN · m^{-1})
298.15 K		
1.0000	1.3290	56.46
0.9772	1.3256	52.72
0.9661	1.3229	50.01
0.9442	1.3173	45.34
0.9164	1.3098	39.52
0.8811	1.2994	32.62
0.8565	1.2916	26.74
0.7877	1.2681	22.97
0.7172	1.2414	22.72
0.5951	1.1895	22.55
0.5343	1.1611	23.08
0.4429	1.1151	23.47
03696	1.0750	22.90
0.2407	0.9962	23.33
0.1212	0.9112	23.70
0.0000	0.8056	23.67

2.2.3.2 Ternary Mixtures

Mixture 1: C_2H_5OH (1) + H_2O (2) + [Mmim][$MeSO_4$] (3)
(1) Volumetric Properties, Transport Properties, Refractive Index and Thermodynamic Properties

Ref. [6]

x_1	x_2	ρ/(g · cm^{-3})	n_D	η/(mPa · s)	V^E/(cm^3 · mol^{-1})	Δn_D	$\Delta\eta$/ (mPa · s)	ΔG^{*E}/ (J · mol^{-1})
298.15 K								
0.0567	0.9117	1.04795	1.36927	1.857	−0.304	0.030	−1.470	1859.7
0.1168	0.8536	1.02304	1.37265	2.306	−0.595	0.032	−0.878	2407.1
0.1877	0.7851	0.99517	1.37432	2.547	−0.862	0.033	−0.468	2657.9
0.0674	0.8724	1.08892	1.38837	2.493	−0.347	0.045	−3.039	2489.4
0.1409	0.8036	1.05623	1.38872	2.793	−0.684	0.044	−2.389	2763.9
0.2203	0.7294	1.02107	1.38736	2.903	−0.908	0.041	−1.900	2851.2
0.3053	0.6498	0.98536	1.38487	2.790	−1.039	0.037	−1.608	2738.1
0.3959	0.5651	0.95067	1.38179	2.620	−1.101	0.032	−1.345	2558.6
0.6047	0.3697	0.88294	1.37434	2.050	−1.030	0.021	−0.918	1858.9
0.0395	0.9469	1.01567	1.35275	1.476	−0.180	0.017	−0.465	1312.9
0.1015	0.8858	0.99373	1.35907	2.008	−0.520	0.022	0.122	2105.4

(*continued*)

x_1	x_2	$\rho/(g \cdot cm^{-3})$	n_D	$\eta/(mPa \cdot s)$	$V^E/(cm^3 \cdot mol^{-1})$	Δn_D	$\Delta\eta/$ (mPa · s)	$\Delta G^{*E}/$ (J · mol^{-1})
0.1629	0.8252	0.97221	1.36307	2.345	−0.772	0.024	0.514	2510.6
0.2319	0.7572	0.94804	1.36540	2.430	−0.939	0.025	0.661	2610.6
0.3106	0.6797	0.92259	1.36652	2.401	−1.041	0.024	0.702	2578.5
0.5094	0.4837	0.87025	1.36644	2.023	−1.064	0.019	0.502	2070.7
0.0859	0.8132	1.12902	1.40706	3.449	−0.445	0.057	−5.215	3051.7
0.1736	0.7351	1.08808	1.40412	3.548	−0.774	0.053	−4.388	3112.3
0.2646	0.6542	1.04602	1.39976	3.378	−0.973	0.048	−3.804	2983.2
0.3603	0.5691	1.00379	1.39462	3.080	−1.076	0.041	−3.308	2745.2
0.4561	0.4839	0.96451	1.38933	2.789	−1.113	0.035	−2.805	2485.4
0.5579	0.3933	0.92582	1.38377	2.433	−1.080	0.028	−2.316	2129.1
0.1180	0.7046	1.17378	1.42814	4.992	−0.527	0.066	−9.553	3321.7
0.2333	0.6125	1.12169	1.42101	5.161	−0.862	0.059	−7.626	3451.5
0.3421	0.5256	1.07136	1.41318	4.150	−1.024	0.051	−6.976	2962.8
0.4400	0.4474	1.02633	1.40571	3.565	−1.080	0.044	−6.065	2636.3
0.5500	0.3595	0.97677	1.39703	2.981	−1.067	0.036	−4.970	2250.7
0.7386	0.2088	0.89473	1.38184	2.056	−0.860	0.021	−3.016	1432.7
0.8266	0.1385	0.85786	1.37463	1.706	−0.684	0.014	−2.023	1020.4
0.0987	0.5532	1.24598	1.45376	10.820	−0.494	0.066	−16.846	3484.7
0.1993	0.4914	1.20769	1.44719	8.770	−0.761	0.063	−15.931	3203.4
0.2974	0.4312	1.16748	1.43994	7.164	−0.956	0.058	−14.643	2937.2
0.3982	0.3694	1.12289	1.43140	5.798	−1.075	0.053	−13.037	2656.4
0.4936	0.3108	1.07751	1.42247	4.661	−1.107	0.047	−11.360	2347.6
0.5968	0.2475	1.02579	1.41190	3.677	−1.108	0.039	−9.299	2012.1
0.6995	0.1844	0.97002	1.40033	2.827	−0.980	0.031	−7.120	1614.6
0.0984	0.3754	1.27805	1.46636	19.420	−0.422	0.052	−21.927	2761.5
0.1977	0.3341	1.24733	1.46062	14.810	−0.695	0.052	−22.098	2590.4
0.2977	0.2925	1.21246	1.45368	11.290	−0.908	0.051	−21.157	2416.7
0.4003	0.2497	1.17178	1.44555	8.498	−1.048	0.049	−19.370	2220.5
0.4999	0.2083	1.12715	1.43624	6.438	−1.128	0.046	−16.980	2021.6
0.6032	0.1652	1.07411	1.42503	4.767	−1.108	0.041	−14.036	1780.1
0.7142	0.1190	1.00889	1.41078	3.381	−1.008	0.033	−10.464	1463.9
0.1018	0.1512	1.29771	1.47397	33.760	−0.157	0.026	−24.559	1194.9
0.2074	0.1334	1.27113	1.46879	24.620	−0.454	0.031	−26.975	1328.6
0.3097	0.1162	1.24115	1.46277	17.870	−0.707	0.035	−27.199	1412.8
0.4116	0.0990	1.20604	1.45547	12.800	−0.9057	0.038	−25.774	1448.7

Mixture 2: 2-$CH_3COC_2H_5$ (1) + 2-CH_3CHCH_3OH (2) + [MMIM][$MeSO_4$] (3)
(1) Distribution Ratio and Selectivity

Ref. [15]

Organic-rich phase		Ionic-liquid-rich phase			
x_1^I	x_2^I	x_1^{II}	x_2^{II}	β	S
298.15 K					
0.936	0.058	0.290	0.025	0.42	1.36
0.878	0.112	0.294	0.058	0.51	1.53
0.835	0.153	0.314	0.084	0.55	1.46

(*continued*)

Organic-rich phase		Ionic-liquid-rich phase			
x_1^{I}	x_2^{I}	x_1^{II}	x_2^{II}	β	S
0.810	0.172	0.337	0.110	0.64	1.53
0.778	0.189	0.365	0.145	0.77	1.63
0.730	0.207	0.424	0.174	0.84	1.45
0.687	0.220	0.494	0.204	0.93	1.29

Mixture 3: $C_6H_5CH_3$ (1) + C_7H_{16} (2) + [MMIM][$MeSO_4$] (3)
(1) Selectivity

Ref. [16]

n-Heptane -rich phase			[mim]CH_3SO_4-rich phase			
x_1	x_2	x_3	x_1	x_2	x_3	Experimental selectivity
0.1 MPa						
313.2 K						
0.0000	1.0000	0.0000	0.0028	0.0000	0.9972	–
0.0608	0.9325	0.0067	0.0052	0.0013	0.9936	61.35
0.1075	0.8902	0.0023	0.0083	0.0011	0.9895	62.48
0.1788	0.8183	0.0028	0.0158	0.0011	0.9826	65.74
0.2270	0.7725	0.0005	0.0194	0.0011	0.9786	60.02
0.3193	0.6768	0.0039	0.0268	0.0010	0.9721	56.81
0.3432	0.6568	0.0000	0.0285	0.0010	0.9700	54.54
0.3960	0.6007	0.0034	0.0316	0.0009	0.9675	53.26
0.4694	0.5278	0.0028	0.0365	0.0010	0.9625	41.04
0.5157	0.4843	0.0000	0.0427	0.0008	0.9565	50.12
0.5574	0.4426	0.0000	0.0478	0.0010	0.9512	37.96
0.6101	0.3899	0.0000	0.0501	0.0009	0.9490	35.58
0.6558	0.3442	0.0000	0.0560	0.0007	0.9433	41.99
0.7254	0.2746	0.0000	0.0591	0.0010	0.9400	22.37
0.7253	0.2747	0.0000	0.0628	0.0007	0.9365	33.98
0.8049	0.1951	0.0000	0.0629	0.0007	0.9360	21.78
0.8386	0.1614	0.0000	0.0721	0.0008	0.9271	17.35
0.8943	0.1057	0.0000	0.0740	0.0007	0.9253	12.49
0.9431	0.0569	0.0000	0.0830	0.0006	0.9164	8.35
0.9614	0.0386	0.0000	0.0790	0.0005	0.9207	6.34
1.0000	0.0000	0.0000	0.0971	0.0000	0.9029	–
348.2 K						
0.0000	0.9947	0.0053	0.0000	0.0012	0.9988	–
0.0595	0.9036	0.0370	0.0039	0.0013	0.9948	46.36
0.1242	0.8571	0.0188	0.0101	0.0012	0.9886	56.28
0.1725	0.8072	0.0203	0.0126	0.0014	0.9860	42.85
0.2326	0.7476	0.0198	0.0184	0.0011	0.9805	54.11
0.2727	0.7133	0.0139	0.0219	0.0013	0.9768	41.44
0.3349	0.6496	0.0155	0.0266	0.0012	0.9722	40.34
0.3885	0.5972	0.0143	0.0278	0.0009	0.9713	47.96

(*continued*)

n-Heptane -rich phase			$[mim]CH_3SO_4$-rich phase			
x_1	x_2	x_3	x_1	x_2	x_3	Experimental selectivity
0.4367	0.5473	0.0160	0.0346	0.0011	0.9643	39.46
0.4806	0.4862	0.0332	0.0404	0.0010	0.9583	42.43
0.5356	0.4519	0.0126	0.0434	0.0008	0.9558	46.03
0.5901	0.3889	0.0210	0.0423	0.0009	0.9562	33.51
0.6651	0.3349	0.0000	0.0499	0.0006	0.9495	41.95
0.6977	0.2965	0.0058	0.0616	0.0007	0.9377	41.09
0.7452	0.2540	0.0008	0.0622	0.0006	0.9372	34.10
0.7853	0.2143	0.0004	0.0659	0.0005	0.9336	34.77
0.8285	0.1715	0.0000	0.0716	0.0007	0.9277	21.49
0.8760	0.1240	0.0000	0.0768	0.0005	0.9227	21.58
0.9242	0.0758	0.0000	0.0787	0.0004	0.9209	15.10
0.9611	0.0389	0.0000	0.0826	0.0003	0.9171	12.03
1.0000	0.0000	0.0000	0.1047	0.0000	0.8953	–

(2) Solubility

Ref. [16]

Temperature					
Toluene in IL, maximum		*n*-Heptane in IL, average		IL in *n*-Heptane, average	
313.2 K	348.2 K	313.2 K	348.2 K	313.2 K	348.2 K
0.10	0.10	0.0009	0.0009	0.0012	0.0121

Mixture 4: $CH_3COOCH_2CH_3$ (1) + 2-CH_3CHCH_3OH (2) + $[MMIM][MeSO_4]$ (3)
(1) Distribution Ratio and Selectivity

Ref. [17]

Organic-rich phase		Ionic liquid-rich phase		β	S
x^{I}_{1}	x^{I}_{2}	x^{II}_{1}	x^{II}_{2}		
298.15 K					
0.981	0.019	0.183	0.033	1.74	9.31
0.958	0.041	0.188	0.055	1.32	6.74
0.933	0.067	0.193	0.081	1.22	5.87
0.911	0.089	0.198	0.121	1.36	6.27
0.890	0.110	0.206	0.171	1.56	6.73
0.872	0.127	0.232	0.240	1.89	7.09
0.860	0.139	0.276	0.288	2.07	6.46
0.839	0.159	0.324	0.327	2.05	5.31
0.819	0.175	0.378	0.346	1.97	4.28
0.781	0.203	0.419	0.347	1.71	3.19
0.715	0.255	0.507	0.336	1.32	1.86

Mixture 5: 2-$CH_3COC_2H_5$ (1) + CH_3CH_2OH (2) + [MMIM][$MeSO_4$] (3)
(1) Distribution Ratio and Selectivity

Ref. [11]

x^I_1	x^I_2	x^{II}_1	x^{II}_2	β	S
298.15 K					
0.975	0.020	0.303	0.030	1.5	5
0.970	0.024	0.308	0.047	1.9	6
0.963	0.030	0.316	0.065	2.2	7
0.931	0.059	0.334	0.093	1.6	4
0.871	0.107	0.359	0.124	1.2	3
0.843	0.126	0.390	0.157	1.3	3
0.803	0.144	0.438	0.184	1.3	2

2.2.4 02001-05011: 1-Methyl-3-methylimidazolium dimethylphosphate

Abbreviation:	[MMIM][DMP], [MMIM][$(CH_3)_2PO_4$]
Molecular Formula:	$C_7H_{15}N_2O_4P$
Molar Mass:	222.18
Structure:	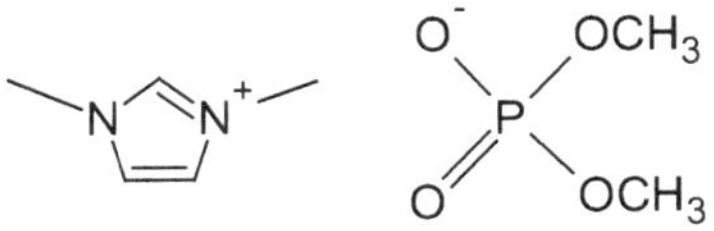

2.2.4.1 Binary Mixtures

Mixture 1: H_2O (1) + [MMIM][$(CH_3)_2PO_4$] (2)
(1) Phase Equilibrium

Ref. [18]

x_1	P (kPa)	x_1	P (kPa)	x_1	P (kPa)	x_1	P (kPa)
353.15 K							
0.037	0.09	0.738	11.56	0.927	37.59	0.993	47.24
0.078	0.14	0.804	17.78	0.929	37.99	0.996	47.48
0.113	0.20	0.851	24.19	0.947	41.07	0.998	47.75
0.182	0.33	0.885	29.84	0.962	43.45	0.999	47.77
0.273	0.59	0.886	30.72	0.974	45.09	0.999	47.78
0.432	1.57	0.909	34.41	0.983	46.16	0.999	47.81
0.555	3.45	0.91	34.28	0.989	46.71	1.000	47.82
0.656	6.74						

(2) Activity Coefficient at Infinite Dilution

Ref. [18]

Solute	γ^{∞}_1
353.15 K	
H_2O	0.028

(3) Vapor Pressure

Ref. [19]

T (K)	P^{exp} (kPa)	P^{NRTL} (kPa)	γ_1^{exp}	γ_1^{NRTL}
IL-content: 20 wt.%				
312.88	6.937	6.935	0.976	0.976
323.27	11.912	11.874	0.982	0.979
332.77	18.856	18.769	0.985	0.981
339.80	25.879	25.846	0.984	0.982
344.80	32.164	32.157	0.983	0.983
349.26	38.845	38.853	0.984	0.984
353.12	45.622	45.546	0.986	0.985
356.71	52.463	52.633	0.982	0.985
361.13	62.359	62.610	0.982	0.986
363.80	69.195	69.382	0.984	0.986
IL-content: 30 wt.%				
318.12	8.885	8.687	0.963	0.942
328.81	15.102	14.846	0.965	0.948
337.05	22.129	21.859	0.964	0.953
343.52	29.303	29.169	0.960	0.956
348.14	35.641	35.580	0.959	0.958
353.37	44.284	44.233	0.961	0.960
357.21	51.818	51.670	0.964	0.961
360.71	59.427	59.307	0.965	0.963
364.29	68.230	68.102	0.966	0.964
367.55	77.254	77.042	0.968	0.965
IL-content: 40 wt.%				
318.93	8.581	8.337	0.909	0.883
329.31	14.466	14.090	0.919	0.895
337.01	20.558	20.296	0.914	0.903
343.82	28.022	27.591	0.923	0.909
349.30	35.375	34.966	0.924	0.913
354.30	43.461	43.096	0.925	0.917
358.99	52.462	52.097	0.927	0.921
363.08	61.780	61.203	0.932	0.924
366.66	70.827	70.219	0.934	0.926
369.65	79.824	78.572	0.943	0.928

(continued)

T (K)	P^{exp} (kPa)	P^{NRTL} (kPa)	γ_1^{exp}	γ_1^{NRTL}
IL-content: 60 wt.%				
330.81	11.458	11.092	0.721	0.698
339.68	17.258	17.100	0.725	0.718
346.47	23.367	23.386	0.731	0.732
352.86	30.734	30.992	0.738	0.744
359.54	39.993	41.091	0.736	0.756
364.65	48.840	50.564	0.739	0.765
369.76	59.571	61.796	0.745	0.773
373.01	67.881	69.961	0.755	0.778

Mixture 2: CH_3COCH_3 (1) + [MMIM][$(CH_3)_2PO_4$] (2)
(1) Phase Equilibrium

Ref. [18]

x_1	P (kPa)	x_1	P (kPa)	x_1	P (kPa)	x_1	P (kPa)
353.15 K							
0.046	10.11	0.535	169.57	0.823	215.99	0.993	216.74
0.08	19.13	0.602	190.77	0.868	215.85	0.995	215.45
0.127	28.27	0.634	205.92	0.896	215.74	0.996	216.84
0.164	37.47	0.634	205.82	0.917	215.65	0.997	215.47
0.193	46.11	0.634	205.73	0.933	215.56	0.998	216.90
0.308	80.35	0.684	212.47	0.946	215.48	0.998	215.43
0.388	112.72	0.733	214.98	0.969	216.68	0.999	216.93
0.462	142.77	0.78	215.99	0.992	215.47	1.000	216.82

(2) Activity Coefficient at Infinite Dilution

Ref. [18]

Solute	γ^{∞}_1
353.15 K	
CH_3COCH_3	0.980

Mixture 3: CH_3OH (1) + [MMIM][$(CH_3)_2PO_4$] (2)
(1) Phase Equilibrium

Ref. [18]

x_1	P (kPa)	x_1	P (kPa)	x_1	P (kPa)	x_1	P (kPa)
353.15 K							
0.107	0.66	0.658	19.13	0.883	120.13	0.985	178.13
0.172	1.27	0.715	29.97	0.883	120.13	0.992	180.57
0.243	1.92	0.762	45.75	0.89	129.65	0.996	181.46
0.305	2.89	0.803	65.18	0.92	147.35	0.997	181.96
0.375	3.92	0.836	85.44	0.944	160.66	0.998	182.14
0.494	7.26	0.862	104.28	0.962	169.12	0.999	182.38
0.58	12.08	0.883	120.14	0.975	174.63	1.000	182.61

(2) Activity Coefficient at Infinite Dilution

Ref. [18]

Solute	$\gamma^{\infty}{}_{1}$
353.15 K	
CH_3OH	0.029

(3) Vapor Pressure

Ref. [19]

T (K)	P^{exp} (kPa)	P^{NRTL} (kPa)	$\gamma_1{}^{exp}$	$\gamma_1{}^{NRTL}$
IL-content: 20 wt.%				
285.96	8.170	8.211	0.971	0.976
296.36	14.517	14.551	0.975	0.977
303.39	20.818	20.893	0.975	0.978
308.34	26.505	26.646	0.974	0.979
312.23	32.041	32.075	0.978	0.979
316.62	39.400	39.292	0.982	0.980
320.86	47.657	47.533	0.983	0.980
324.19	55.019	54.983	0.981	0.980
328.57	66.577	66.240	0.986	0.981
331.53	75.022	74.882	0.983	0.981
IL-content: 30 wt.%				
286.31	7.811	7.837	0.933	0.936
296.67	13.789	13.864	0.934	0.939
304.40	20.437	20.633	0.933	0.942
309.85	26.760	26.940	0.937	0.943
314.86	33.917	34.107	0.939	0.944
318.74	40.509	40.725	0.941	0.946
322.58	48.272	48.297	0.946	0.946
326.19	56.196	56.459	0.943	0.947
329.59	64.664	65.201	0.940	0.948
332.65	73.462	73.987	0.942	0.949
IL-content: 40 wt.%				
288.33	7.856	7.858	0.864	0.864
298.78	14.002	13.909	0.877	0.871
306.38	20.494	20.491	0.876	0.876
311.98	26.898	26.904	0.879	0.879
317.14	34.379	34.224	0.886	0.882
321.32	41.651	41.357	0.890	0.884
325.13	49.091	48.887	0.890	0.886
328.73	57.357	57.049	0.893	0.888
332.45	67.008	66.635	0.895	0.890
335.52	76.224	75.531	0.899	0.891

(continued)

T (K)	P^{exp} (kPa)	P^{NRTL} (kPa)	γ_1^{exp}	γ_1^{NRTL}
IL-content: 60 wt.%				
307.57	13.522	13.416	0.605	0.601
315.38	19.910	19.746	0.617	0.612
322.36	27.059	27.403	0.613	0.621
328.38	35.341	35.903	0.619	0.629
332.53	42.658	42.974	0.630	0.634
336.72	51.127	51.266	0.638	0.640
340.40	59.954	59.612	0.648	0.644
343.71	69.025	68.056	0.657	0.648

Mixture 4: C_2H_5OH (1) + [MMIM][$(CH_3)_2PO_4$] (2)
(1) Phase Equilibrium

Ref. [18]

x_1	P (kPa)	x_1	P (kPa)	x_1	P (kPa)	x_1	P (kPa)
353.15 K							
0.034	0.53	0.618	22.96	0.804	58.53	0.965	103.75
0.066	0.98	0.685	31.98	0.818	63.37	0.979	106.09
0.097	1.44	0.709	35.94	0.818	63.37	0.99	107.66
0.135	2.07	0.709	35.93	0.818	63.37	0.994	108.24
0.172	2.70	0.709	35.91	0.849	72.14	0.997	108.58
0.272	4.82	0.739	42.46	0.889	84.68	0.998	108.74
0.367	7.55	0.757	46.05	0.922	94.16	0.999	108.90
0.455	11.07	0.783	53.30	0.947	100.09	1.000	109.08
0.539	15.97						

(2) Activity Coefficient at Infinite Dilution

Ref. [18]

Solute	γ^{∞}_1
353.15 K	
C_2H_5OH	0.115

(3) Vapor Pressure

Ref. [19]

T (K)	P^{exp} (kPa)	P^{NRTL} (kPa)	γ_1^{exp}	γ_1^{NRTL}
IL-content: 10 wt.%				
298.62	7.853	7.927	0.994	1.004
310.06	14.936	14.964	1.003	1.005
317.00	21.274	21.419	0.999	1.005
322.08	27.498	27.524	1.005	1.006
326.31	33.481	33.690	0.999	1.006
329.87	39.587	39.739	1.002	1.006
333.11	45.710	46.033	0.999	1.006

(continued)

T (K)	P^{exp} (kPa)	P^{NRTL} (kPa)	γ_1^{exp}	γ_1^{NRTL}
335.88	51.656	52.065	0.998	1.006
338.54	57.869	58.444	0.996	1.006
340.69	63.471	64.094	0.996	1.006
IL-content: 20 wt.%				
308.82	13.326	13.282	0.983	0.980
316.09	19.602	19.493	0.991	0.985
321.94	26.336	26.139	0.996	0.989
326.89	32.968	33.193	0.985	0.991
331.12	40.304	40.410	0.991	0.993
334.26	46.507	46.605	0.993	0.995
337.18	52.704	53.049	0.989	0.996
339.67	58.641	59.124	0.989	0.997
342.22	65.778	65.966	0.995	0.998
344.56	71.823	72.784	0.985	0.999
IL-content: 30 wt.%				
310.53	13.643	13.091	0.951	0.913
317.47	19.381	18.946	0.946	0.924
323.25	25.887	25.389	0.951	0.933
328.20	32.568	32.310	0.947	0.940
332.47	39.673	39.499	0.950	0.945
336.10	46.701	46.633	0.951	0.950
339.48	54.186	54.194	0.953	0.954
342.24	61.247	61.143	0.958	0.957
344.33	67.366	66.867	0.966	0.959
346.53	74.249	73.376	0.972	0.961
IL-content: 60 wt.%				
320.94	12.265	11.365	0.607	0.563
329.38	18.915	17.799	0.628	0.591
336.10	25.462	24.955	0.624	0.612
342.62	33.690	34.082	0.625	0.632
347.91	42.006	43.444	0.626	0.648
352.30	50.123	52.785	0.627	0.661
356.43	59.711	63.048	0.637	0.672
359.57	68.622	71.880	0.649	0.680

Mixture 5: $OCH_2CH_2CH_2CH_2$ (1) + [MMIM][$(CH_3)_2PO_4$] (2)
(1) Phase Equilibrium

Ref. [18]

x_1	P (kPa)	x_1	P (kPa)	x_1	P (kPa)	x_1	P (kPa)
353.15 K							
0.021	9.67	0.512	157.75	0.745	157.85	0.951	157.83
0.051	21.20	0.585	157.78	0.75	157.83	0.97	157.84
0.077	32.28	0.634	157.88	0.75	157.82	0.984	157.85

(continued)

x_1	P (kPa)	x_1	P (kPa)	x_1	P (kPa)	x_1	P (kPa)
0.103	41.64	0.634	157.87	0.75	157.81	0.992	157.84
0.139	56.74	0.634	157.87	0.8	157.84	0.995	157.84
0.23	97.54	0.651	157.80	0.85	157.83	0.997	157.83
0.291	124.04	0.689	157.85	0.893	157.83	0.998	157.84
0.359	150.34	0.705	157.82	0.926	157.82	1.000	157.86
0.435	157.67						

(2) Activity Coefficient at Infinite Dilution

Ref. [18]

Solute	$\gamma^{\infty}{}_1$
353.15 K	
$OCH_2CH_2CH_2CH_2$	3.21

2.2.4.2 Ternary Mixtures

Mixture 1: C_2H_5OH (1) + H_2O (2) + [MMIM][DMP] (3)

(1) Vapor Pressure

Ref. [19]

T (K)	P^{exp} (kPa)	P^{NRTL} (kPa)	RD (P)%
$x_1 = 0.3819, x_2 = 0.5864$			
302.91	9.348	8.840	−5.43
313.32	16.234	15.564	−4.13
320.29	23.163	22.187	−4.21
326.42	30.999	29.864	−3.66
331.60	39.397	38.005	−3.53
335.95	47.952	46.213	−3.63
339.33	55.723	53.570	−3.86
342.93	65.018	62.467	−3.92
$x_1 = 0.5202, x_2 = 0.4438$			
303.72	10.023	9.585	−4.37
312.82	16.327	15.723	−3.70
319.75	23.183	22.404	−3.36
325.46	30.484	29.618	−2.84
330.03	37.782	36.711	−2.83
334.11	45.460	44.213	−2.74
338.30	54.862	53.226	−2.98
341.99	64.044	62.400	−2.57
$x_1 = 0.7018, x_2 = 0.2566$			
299.90	8.296	7.873	−5.09
311.11	15.183	14.698	−3.20
318.54	22.242	21.601	−2.88
324.76	29.963	29.353	−2.04

(continued)

T (K)	P^{exp} (kPa)	P^{NRTL} (kPa)	RD (P)%
329.85	38.133	37.332	−2.10
334.55	47.039	46.243	−1.69
338.64	56.373	55.415	−1.70
342.29	65.914	64.854	−1.61
$x_1 = 0.8938$, $x_2 = 0.0587$			
302.37	9.605	9.201	−4.20
312.03	16.101	15.650	−2.80
318.74	22.562	22.136	−1.89
324.97	30.307	30.068	−0.79
330.06	38.472	38.221	−0.65
334.45	46.956	46.696	−0.55
338.34	55.690	55.460	−0.41
341.93	64.891	64.751	−0.22
ARD			2.80

$$\mathrm{RD}(P)\% = \left(\mathrm{p}^{\mathrm{NRTL}} - \mathrm{P}^{\mathrm{exp}}\right)/\mathrm{p}^{\mathrm{exp}}.100;$$

$$\mathrm{ARD} = \frac{1}{n}\sum_{i=1}^{n}\left|\frac{p_i^{NRTL} - p_i^{\mathrm{exp}}}{p_i^{\mathrm{exp}}}\right| \times 100$$

Ref. [20]

T/K	p/kPa	y_1	T/K	p/kPa	y_1
$x_1 = 0.1003$, $x_2 = 0.8130$					
307.20	8.06	0.4924	339.19	37.90	0.4655
318.63	14.51	0.4828	343.94	46.65	0.4616
326.57	21.36	0.4761	348.51	56.40	0.4579
333.16	28.94	0.4706	352.80	66.70	0.4544
$x_1 = 0.1866$, $x_2 = 0.7166$					
309.85	10.99	0.5842	338.50	43.13	0.5676
319.82	18.16	0.5785	343.31	52.62	0.5648
327.22	25.82	0.5742	347.34	62.17	0.5624
333.51	34.51	0.5705	350.62	71.03	0.5605
$x_1 = 0.3798$, $x_2 = 0.5008$					
305.83	9.05	0.7027	337.18	42.23	0.6926
317.41	16.67	0.6991	341.56	51.02	0.6912
325.31	24.46	0.6965	345.51	60.03	0.6899
331.75	33.14	0.6944	349.04	69.23	0.6887
$x_1 = 0.5266$, $x_2 = 0.3369$					
308.14	9.99	0.7888	337.51	41.82	0.7812
318.95	17.53	0.7860	342.04	50.99	0.7800
326.85	25.70	0.7839	345.72	59.36	0.7790
332.42	33.29	0.7825	349.87	69.72	0.7780

(*continued*)

T/K	p/kPa	y_1	T/K	p/kPa	y_1
$x_1 = 0.6906, x_2 = 0.1537$					
309.25	10.06	0.8984	338.41	41.60	0.8933
319.67	17.29	0.8965	342.92	50.55	0.8926
327.38	24.99	0.8952	346.99	59.97	0.8919
333.37	33.16	0.8942	350.44	69.00	0.8914
$x_1 = 0.7912, x_2 = 0.0413$					
309.79	10.01	0.9720	338.99	41.33	0.9702
320.17	17.14	0.9713	343.08	49.33	0.9701
328.40	25.42	0.9709	346.64	57.68	0.9699
334.45	33.57	0.9705	350.12	66.57	0.9697

(2) Phase Equilibrium

Ref. [21]

wt.% of IL	T (K)	x_1	y_1	γ_1	γ_2	α_{12}
101.32 kPa						
About 10 %						
11.526	362.53	0.965	0.968	0.680	1.374	1.124
11.379	362.59	0.903	0.919	0.687	1.270	1.211
11.858	362.59	0.826	0.868	0.710	1.147	1.381
12.100	362.59	0.742	0.823	0.748	1.037	1.615
12.115	362.59	0.664	0.786	0.797	0.962	1.855
12.260	363.01	0.593	0.752	0.841	0.904	2.089
12.350	363.42	0.500	0.718	0.938	0.823	2.556
12.255	363.88	0.412	0.694	1.079	0.745	3.239
12.261	364.44	0.333	0.669	1.258	0.695	4.046
12.539	365.15	0.261	0.643	1.506	0.658	5.110
12.215	366.28	0.191	0.618	1.892	0.616	6.832
12.151	368.02	0.126	0.586	2.564	0.578	9.840
11.862	371.91	0.069	0.492	3.440	0.578	13.083
About 20 %						
23.301	363.58	0.961	0.966	0.679	1.300	1.172
23.630	363.64	0.896	0.923	0.693	1.114	1.381
23.869	363.63	0.829 C	0.878	0.711	1.074	1.475
23.954	363.64	0.715	0.835	0.781	0.865	2.017
24.051	363.67	0.662	0.806	0.812	0.856	2.115
24.012	363.78	0.569	0.773	0.899	0.780	2.573
24.278	364.03	0.475	0.742	1.021	0.719	3.176
24.928	364.54	0.382	0.711	1.194	0.669	3.991
24.954	365.15	0.290	0.689	1.485	0.610	5.417
24.339	365.97	0.214	0.651	1.834	0.598	6.821
23.829	367.15	0.167	0.619	2.138	0.588	8.088
23.447	369.24	0.109	0.569	2.799	0.574	10.787
23.214	373.54	0.068	0.512	3.507	0.531	14.502

Mixture 2: C_2H_5OH (1) + CH_3OH (2) + [MMIM][DMP] (3)
(1) Vapor Pressure

Ref. [20]

T/K	*p*/kPa	y_1	*T*/K	*p*/kPa	y_1
$x_1 = 0.1284, x_2 = 0.7384$					
305.19	14.70	0.2704	327.47	42.58	0.2663
312.11	20.94	0.2692	331.63	50.62	0.2655
317.69	27.49	0.2682	335.72	59.83	0.2647
322.89	34.73	0.2672	339.17	68.87	0.2640
$x_1 = 0.3186, x_2 = 0.5378$					
297.92	8.32	0.5172	327.40	36.84	0.5210
308.71	14.80	0.5192	332.00	45.30	0.5211
315.93	21.47	0.5201	336.59	55.05	0.5212
322.10	28.84	0.5207	340.45	65.18	0.5212
$x_1 = 0.4688, x_2 = 0.3792$					
301.76	8.97	0.6809	329.48	35.94	0.6853
311.38	15.18	0.6829	333.97	43.76	0.6856
318.52	21.53	0.6840	338.36	53.09	0.6859
324.54	28.63	0.6848	342.74	63.67	0.6862
$x_1 = 0.6375, x_2 = 0.2012$					
304.83	9.17	0.8446	333.48	37.24	0.8468
314.30	15.08	0.8457	338.17	45.80	0.8471
321.69	21.72	0.8463	342.48	54.79	0.8471
327.97	29.09	0.8466	346.55	65.06	0.8473
$x_1 = 0.7837, x_2 = 0.0470$					
304.94	7.90	0.9668	334.72	35.68	0.9672
315.32	14.05	0.9671	339.42	43.77	0.9672
322.78	20.54	0.9672	343.5	52.29	0.9672
329.24	28.01	0.9672	348.02	63.16	0.9672

2.2.5 02003-01001: 1-Ethyl-3-methylimidazolium chloride

Abbreviation:	[C_2MIm]Cl, [emim]Cl, [EMIM]Cl, EtMeImCl, [C_2mim]Cl
Molecular Formula:	$C_6H_{11}N_2Cl$
Molar Mass:	146.62
Structure:	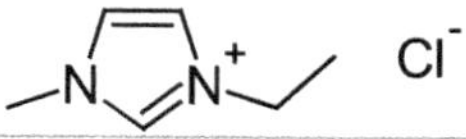

2.2.5.1 Binary Mixtures

Mixture 1: $(NH_2)_2CO$ (1) + EtMeImCl (2)
(1) Volumetric Properties, Transport Properties and Conductivity

Ref. [22]

T/K	ρ/g · cm^{-3}	η/cP	k/mS · cm^{-1}	T/K	ρ/g · cm^{-3}	η/cP	k/mS · cm^{-1}
x_1 = 25.0 mol%				x_1 = 75.0 mol%			
343	1.136	97.69	11.20	343	1.196	47.31	13.17
348	—	—	13.58	348	—	—	15.83
353	1.128	59.32	16.24	353	1.190	28.68	18.80
358	—	—	20.63	358	—	—	21.68
363	1.121	31.94	23.77	363	1.182	18.33	24.70

Mixture 2: $HO[CH_2CH_2O]_nOH$ (PEG-1500, n ≈ 34) (1) + $[C_2mim]Cl$ (2)
(1) Phase Equilibrium

Ref. [23]

T/K	x_{IL}, up	x_{IL}, low
333.15	0.789	0.986
353.15	0.769	0.993
373.15	0.741	0.993
393.15	0.709	0.996
413.15	0.682	0.997

Ref. [23]

System IL + PEG	Temperature range	w_{IL}, PEG-rich phase	w_{IL}, IL-rich phase
$[C_2mim]Cl$ + PEG-1500	333.15–413.15 K	0.17–0.27	0.88–0.97

Mixture 3: $HO[CH_2CH_2O]_nOH$ (PEG-2000, n ≈ 45) (1) + $[C_2mim]Cl$ (2)
(1) Phase Equilibrium

Ref. [23]

T/K	x_{IL}, up	x_{IL}, low
333.15	0.687	0.995
353.15	0.666	0.996
373.15	0.636	0.998
393.15	0.582	0.999
413.15	0.535	0.999

Ref. [23]

System IL + PEG	Temperature range	w_{IL}, PEG-rich phase	w_{IL}, IL-rich phase
$[C_2mim]Cl$ + PEG-2000	333.15–413.15 K	0.08–0.14	0.94–0.99

Mixture 4: $HO[CH_2CH_2O]_nOH$ (PEG-3400, n ≈ 77) (1) + $[C_2mim]Cl$ (2)
(1) Phase Equilibrium

Ref. [23]

T/K	x_{IL}, up	x_{IL}, low
333.15	0.593	0.999
353.15	0.575	0.999
373.15	0.530	0.998
393.15	0.451	0.998
413.15	0.374	0.999

Ref. [23]

System IL + PEG	Temperature range	w_{IL}, PEG-rich phase	w_{IL}, IL-rich phase
$[C_2mim]Cl$ + PEG-3400	333.15–413.15 K	0.02–0.06	0.95–0.98

2.2.5.2 Ternary Mixtures

Mixture 1: H_2O (1) + C_2H_5OH (2) + [emim]Cl (3)
(1) Phase Equilibrium

Ref. [24]

x'_2 [a]	w_3 [a]	y_2	T/K	γ_1	γ_2	α_{21} [b]
p = 100 kPa						
0.9500	0.5906	0.9848	373.52	0.43	0.65	3.40
0.9500	0.5001	0.9803	366.38	0.65	0.76	2.62
0.9501	0.3997	0.9748	360.25	0.97	0.87	2.04
0.9501	0.3000	0.9696	356.26	1.28	0.94	1.67
0.9501	0.2001	0.9634	353.68	1.63	0.98	1.38
0.9501	0.1000	0.9561	352.06	2.01	1.00	1.14

[a] Compositions reported are of equilibrium liquid
[b] $\alpha_{21} = (y_2/x'_2)/(y_1/x'_1)$

Mixture 2: H_2O (1) + $(CH_3)_3COH$ (2) + [emim]Cl (3)
(1) Phase Equilibrium

Ref. [25]

x'_2 [a]	w_3 [a]	y_2	T/K	γ_1	γ_2	$\alpha_{2,1}$ [b]
p = 100 kPa						
0.9501	0.5846	0.9928	375.65	0.22	0.83	7.22
0.9500	0.5500	0.9915	372.95	0.27	0.86	6.11
0.9500	0.5001	0.9903	369.57	0.32	0.91	5.37
0.9500	0.4501	0.9874	366.27	0.45	0.96	4.13
0.9500	0.4000	0.9855	364.10	0.53	0.98	3.58
0.9501	0.3001	0.9789	360.45	0.81	1.02	2.43
0.9505	0.2001	0.9684	357.90	1.25	1.03	1.60
0.9502	0.1000	0.9507	356.00	1.97	1.02	1.01
0.9502	0.0503	0.9415	355.35	2.33	1.01	0.84

[a] Compositions reported are of equilibrium liquid
[b] $\alpha_{2,1} = (y_2/x'_2)/(y_1/x'_1)$

2.2.6 02003-01002: 1-Ethyl-3-methylimidazolium bromide

Abbreviation:	[C_2MIm]Br, [emim]Br, [EMIM]Br
Molecular Formula:	$C_6H_{11}N_2Br$
Molar Mass:	191.07
Structure:	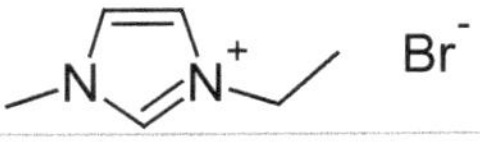

2.2.6.1 Binary Mixtures

Mixture 1: H_2O (1) + [EMIM]Br (2)

(1) Volumetric Properties

Ref. [26]

m_2/ mol · kg^{-1}	ρ/ (g · cm^{-3})	$10^{-3} \cdot \Phi_V$/ mm^3 · mol^{-1}	$10^{-3} \cdot \overline{V}_2$/ mm^3 · mol^{-1}	$10^{-3} \cdot \overline{V}_1$/ mm^3 · mol^{-1}	$10^4\alpha$/ K^{-1}	$10^6\Phi_E$/ m^3 · mol^{-1} · K^{-1}
T = 293.15 K						
0.00000	0.99821			18.047	2.125	
0.04154	1.00046	190.93	190.62	18.047	2.099	0.040
0.08422	1.00273	190.50	189.77	18.048	2.096	0.040
0.13082	1.00533	190.00	188.80	18.050	2.155	0.041
0.17084	1.00750	189.59	187.97	18.052	2.177	0.041
0.21279	1.00970	189.18	187.11	18.055	2.147	0.041
0.25638	1.01202	188.75	186.20	18.059	2.268	0.043
0.30323	1.01442	188.30	185.25	18.064	2.257	0.043
0.35442	1.01709	187.81	184.20	18.070	2.395	0.045
0.39772	1.01932	187.40	183.31	18.076	2.453	0.046
0.44346	1.02174	186.95	182.37	18.084	2.188	0.041
T = 298.15 K						
0.00000	0.99705			18.068	2.567	
0.04154	0.99928	191.15	190.86	18.068	2.548	0.049
0.08422	1.00153	190.72	190.00	18.069	2.570	0.049
0.13082	1.00410	190.24	189.03	18.071	2.603	0.050
0.17084	1.00624	189.83	188.21	18.073	2.622	0.050
0.21279	1.00846	189.41	187.34	18.076	2.618	0.050
0.25638	1.01073	188.99	186.44	18.080	2.688	0.051
0.30323	1.01316	188.54	185.48	18.085	2.705	0.051
0.35442	1.01576	188.05	184.44	18.091	2.795	0.053
0.39772	1.01795	187.65	183.56	18.097	2.808	0.053
0.44346	1.02055	187.17	182.57	18.105	2.701	0.051
T = 303.15 K						
0.00000	0.99565			18.093	3.010	
0.04154	0.99789	191.42	191.13	18.094	2.998	0.057
0.08422	1.00014	190.99	190.28	18.094	3.045	0.058
0.13082	1.00270	190.50	189.31	18.096	3.054	0.058
0.17084	1.00484	190.10	188.48	18.098	3.069	0.058
0.21279	1.00706	189.68	187.62	18.101	3.090	0.059
0.25638	1.00928	189.26	186.73	18.105	3.110	0.059

(*continued*)

m_2/ mol · kg^{-1}	ρ/ (g · cm^{-3})	$10^{-3} \cdot \Phi_V$/ mm^3 · mol^{-1}	$10^{-3} \cdot \overline{V}_2$/ mm^3 · mol^{-1}	$10^{-3} \cdot \overline{V}_1$/ mm^3 · mol^{-1}	$10^4 \alpha$/ K^{-1}	$10^{6\Phi}{}_E$/ m^3 · mol^{-1} · K^{-1}
0.30323	1.01166	188.81	185.77	18.110	3.154	0.060
0.35442	1.01424	188.34	184.73	18.116	3.197	0.060
0.39772	1.01643	187.93	183.85	18.123	3.164	0.060
0.44346	1.01891	187.47	182.89	18.130	3.217	0.060
T = 308.15 K						
0.00000	0.99404			18.123	3.455	
0.04154	0.99634	191.72	191.43	18.123	3.450	0.066
0.08422	0.99853	191.30	190.59	18.124	3.523	0.067
0.13082	1.00109	190.81	189.62	18.126	3.506	0.067
0.17084	1.00323	190.40	188.79	18.128	3.518	0.067
0.21279	1.00539	189.99	187.93	18.131	3.565	0.068
0.25638	1.00765	189.57	187.03	18.134	3.533	0.067
0.30323	1.00999	189.13	186.08	18.139	3.605	0.068
0.35442	1.01253	188.65	185.04	18.146	3.600	0.068
0.39772	1.01480	188.23	184.14	18.152	3.522	0.066
0.44346	1.01729	187.77	183.17	18.159	3.734	0.070
T = 313.15 K						
0.00000	0.99222			18.156	3.903	
0.04154	0.99445	192.08	191.80	18.156	3.905	0.075
0.08422	0.99662	191.67	190.96	18.157	4.003	0.077
0.13082	0.99918	191.17	189.99	18.159	3.961	0.076
0.17084	1.00130	190.77	189.16	18.161	3.969	0.076
0.21279	1.00346	190.36	188.29	18.164	4.042	0.077
0.25638	1.00571	189.93	187.39	18.168	3.959	0.075
0.30323	1.00803	189.50	186.44	18.173	4.059	0.077
0.35442	1.01059	189.02	185.39	18.179	4.006	0.076
0.39772	1.01286	188.59	184.49	18.185	3.882	0.073
0.44346	1.01518	188.16	183.54	18.193	4.255	0.080

(2) Speed of Sound, Compressibility Properties and Internal Pressure

Ref. [26]

m_2/ mol · kg^{-1}	u/ m · s^{-1}	$10^{11}\beta_S$/ Pa^{-1}	$10^{11}\beta_T$/ Pa^{-1}	ϕ_{K_S}/mm^3 · MPa^{-1} · mol^{-1}	ϕ_{K_T}/ mm^3 · MPa^{-1} · mol^{-1}	$10^{-5}P_i$/ Pa
T = 298.15 K						
0.00000	1497.6	44.72	45.19			1716
0.04154	1500.6	44.44	44.90	−6.1 ± 4.3	−7.4 ± 4.9	1714
0.08422	1503.7	44.16	44.63	−6.1 ± 2.1	−5.6 ± 2.4	1740
0.13082	1506.9	43.86	44.34	−6.1 ± 1.4	−4.7 ± 1.6	1774
0.17084	1509.6	43.61	44.10	−5.8 ± 1.0	−4.2 ± 1.2	1796
0.21279	1512.5	43.34	43.83	−5.8 ± 0.8	−4.5 ± 1.0	1804
0.25638	1515.7	43.07	43.58	−5.9 ± 0.7	−3.7 ± 0.8	1864
0.30323	1518.7	42.79	43.31	−5.5 ± 0.6	−3.4 ± 0.7	1887
0.35442	1522.1	42.49	43.04	−5.2 ± 0.5	−2.3 ± 0.6	1962
0.39772	1525.2	42.23	42.78	−5.4 ± 0.4	−2.6 ± 0.5	1983
0.44346	1528.1	41.96	42.47	−5.6 ± 0.4	−4.1 ± 0.5	1921

(3) Volumetric Properties and Compressibility Properties

Ref. [26]

T/K	$10^{-3}\cdot\phi_V{}^0$ /mm$^3\cdot$mol^{-1}	$10^{-3}B_V$ /mm$^6\cdot$mol^{-2}	$10^6\phi_E{}^0$ /m$^3\cdot$mol$^{-1}\cdot$K^{-1}	S_E	$\phi_{K_S}{}^0$ /mm$^3\cdot$MPa$^{-1}\cdot$mol^{-1}	$10^{-3}S_{K_S}$ /mm$^{9/2}\cdot$MPa$^{-1}\cdot$mol$^{-3/2}$	$\phi_{K_T}{}^0$ /mm$^3\cdot$MPa$^{-1}\cdot$mol^{-1}	$10^{-3}S_{K_T}$ /mm$^{9/2}\cdot$MPa$^{-1}\cdot$mol$^{-3/2}$
293.15	190.9 ± 1.2	−11.68	0.038 ± 0.002	0.0095				
298.15	191.1 ± 1.2	−11.77	0.047 ± 0.002	0.0078	−6.6 ± 6.9	1.8	−8.2 ± 7.6	8.5
303.15	191.4 ± 1.2	−11.80	0.056 ± 0.002	0.0060				
308.15	191.7 ± 1.2	−11.92	0.066 ± 0.002	0.0043				
313.15	192.0 ± 1.2	−12.02	0.075 ± 0.002	0.0025				

(4) Osmotic Properties, Activity and Thermodynamic Properties

Ref. [26]

m_2/mol · kg^{-1}	Φ	a_1	γ_1	$\gamma_\pm$	ΔG^E/J · mol^{-1}
$T = 298.15$ K					
0.01059	0.96083	0.99963	1.00001	0.89345	−0.14
0.02296	0.94545	0.99922	1.00004	0.85154	−0.44
0.04640	0.93082	0.99845	1.00011	0.80317	−1.25
0.06833	0.91986	0.99774	1.00019	0.77236	−2.18
0.08885	0.91258	0.99708	1.00027	0.74987	−3.20
0.13660	0.90033	0.99558	1.00048	0.71067	−5.93
0.18577	0.89094	0.99405	1.00071	0.68115	−9.17
0.23525	0.88315	0.99254	1.00096	0.65780	−12.74
0.28420	0.87646	0.99107	1.00121	0.63878	−16.57
0.33893	0.86979	0.98943	1.00152	0.62081	−21.09
0.39109	0.86403	0.98790	1.00182	0.60603	−25.61
0.44224	0.85881	0.98641	1.00213	0.59320	−30.24

2.2.7 02003-02001: 1-Ethyl-3-methylimidazolium tetrafluoroborate

Abbreviation:	$[C_2MIm][BF_4]$, $[emim][BF_4]$, $[EMIM][BF_4]$, $[C_2mim][BF_4]$, $EMIBF_4$
Molecular Formula:	$C_6H_{11}BF_4N_2$
Molar Mass:	197.97
Structure:	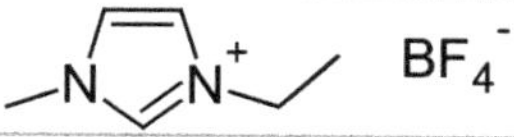

2.2.7.1 Binary Mixtures

Mixture 1: $[C_2mim][BF_4]$ (1) + $[C_6mim][BF_4]$ (2)

(1) Volumetric Properties and Transport Properties

Ref. [27]

η/mPa · s									
x_2	T = 298.15 K	T = 299.15 K	T = 301.15 K	T = 302.15 K	T = 303.15 K	T = 304.15 K	T = 305.15 K	T = 307.15 K	T = 308.15 K
0	35.80	34.52	32.15	31.06	30.06	29.01	28.13	26.38	25.54
0.0978	43.30	41.62	38.54	37.08	35.77	34.46	33.23	30.96	29.94
0.1484	48.10	46.22	42.79	41.13	39.58	38.13	36.71	34.20	33.00
0.2070	53.76	51.53	47.42	45.56	43.73	42.09	40.44	37.44	36.05
0.2941	64.27	61.64	56.52	54.22	51.99	49.86	47.93	44.34	42.66
0.3994	78.33	74.73	68.10	64.94	62.29	59.61	57.12	52.51	50.38
0.4474	86.69	82.48	75.03	71.45	68.36	65.27	62.36	57.17	54.84
0.5062	95.97	91.20	82.69	78.62	75.11	71.65	68.44	62.60	59.86
0.5540	104.47	99.34	89.87	85.60	81.57	77.71	74.29	67.76	64.76
0.5939	111.78	106.04	95.89	91.32	87.00	82.76	78.94	71.92	68.76
0.6855	128.27	121.94	109.74	104.12	99.07	94.21	89.62	81.37	77.56
0.8197	158.11	148.61	132.86	125.87	119.57	113.40	107.98	97.52	93.01
0.8804	171.73	162.17	145.51	137.43	130.23	123.68	117.65	106.04	100.77
0.9455	184.13	173.86	155.28	147.11	139.30	132.43	125.12	113.07	107.30
1	211.21	198.06	177.15	167.21	160.87	148.44	141.25	127.53	120.69

Ref. [28]

ρ/g · cm^{-3}									
x_2	T/K = 298.15	T/K = 299.15	T/K = 301.15	T/K = 302.15	T/K = 303.15	T/K = 304.15	T/K = 305.15	T/K = 307.15	T/K = 308.15
0	1.2796	1.2788	1.2772	1.2765	1.2757	1.2749	1.2742	1.2727	1.2719
0.0978	1.2616	1.2608	1.2593	1.2586	1.2579	1.2571	1.2564	1.2549	1.2542
0.1484	1.2533	1.2525	1.2510	1.2511	1.2495	1.2488	1.2480	1.2465	1.2458
0.2070	1.2437	1.2430	1.2415	1.2407	1.2400	1.2392	1.2385	1.2370	1.2363
0.2941	1.2303	1.2296	1.2281	1.2274	1.2266	1.2259	1.2252	1.2237	1.2230
0.3994	1.2151	1.2144	1.2129	1.2122	1.2114	1.2107	1.2100	1.2085	1.2078
0.4474	1.2082	1.2075	1.2061	1.2054	1.2047	1.2040	1.2033	1.2018	1.2011
0.5062	1.2006	1.1999	1.1984	1.1977	1.1970	1.1963	1.1956	1.1942	1.1934
0.5540	1.1944	1.1937	1.1923	1.1916	1.1909	1.1902	1.1895	1.1881	1.1874
0.5939	1.1895	1.1888	1.1874	1.1867	1.1860	1.1853	1.1846	1.1832	1.1825
0.6855	1.1784	1.1776	1.1763	1.1756	1.1749	1.1742	1.1735	1.1721	1.1714
0.8197	1.1636	1.1629	1.1614	1.1607	1.1600	1.1593	1.1586	1.1572	1.1565
0.8804	1.1572	1.1566	1.1552	1.1545	1.1538	1.1531	1.1524	1.1510	1.1503
0.9455	1.1508	1.1501	1.1487	1.1481	1.1473	1.1466	1.1459	1.1445	1.1438
1	1.1458	1.1451	1.1437	1.1430	1.1423	1.1415	1.1408	1.1395	1.1388

Ref. [28]

x_1	α_p/k K^{-1}	x_1	α_p/k K^{-1}
303.15 K			
0	0.5996	0.5540	0.5897
0.0978	0.5865	0.5939	0.5920
0.1484	0.5978	0.6855	0.5921
0.2070	0.5996	0.8197	0.6125
0.2941	0.6004	0.8804	0.6051
0.3994	0.6013	0.9455	0.6153
0.4474	0.5904	1	0.6161
0.5062	0.5967	–	–

(2) Excess Molar Enthalpy

Ref. [28]

x_1	H^E/J · mol^{-1}	x_1	H^E/J · mol^{-1}
303.15 K			
0.0352	37	0.4106	310
0.1856	162	0.4433	287
0.2339	214	0.6006	242
0.2862	252	0.8121	159
0.3459	288	0.9348	28
0.3504	295	–	–

Mixture 2: $[C_2mim][BF_4]$ (1) + $[C_2mim][Tf_2N]$ (2)

(1) Volumetric Properties

Ref. [29]

x_1	ρ (g/cm^3)	x_1	ρ (g/cm^3)
Room-Temperature			
0	1.50	0.90	1.35
0.25	1.52	0.95	1.30
0.50	1.48	1	1.28
0.75	1.42		

Mixture 3: $CF_3CHFCHFCF_2CF_3$ *threo*-H (1) + $[emim][BF_4]$ (2)

(1) Phase Equilibrium

Ref. [30]

T/K	x'_1/mol%	x_1/mol%	V'^{a}/ $cm^3 \cdot mol^{-1}$	V^{a}/ $cm^3 \cdot mol^{-1}$	$V^{ex\prime\, b}$/ $cm^3 \cdot mol^{-1}$	$V^{ex\, b}$/ $cm^3 \cdot mol^{-1}$
283.6 ± 0.2	94.6 ± 0.7	9.7 ± 0.6	152.9 ± 0.8	150.9 ± 0.9	−1.8 ± 0.8	−2.6 ± 0.9
293.6 ± 0.2	97.0 ± 0.6	8.4 ± 0.7	156.1 ± 0.6	152.6 ± 0.9	−1.3 ± 0.6	−1.9 ± 0.9
303.6 ± 0.2	98.0 ± 0.7	7.3 ± 1.0	159.4 ± 0.6	153.5 ± 0.6	−0.9 ± 0.6	−1.9 ± 0.6

(*continued*)

T/K	x'_1/mol%	x_1/mol%	V' [a]/ $cm^3 \cdot mol^{-1}$	V[a]/ $cm^3 \cdot mol^{-1}$	$V^{ex\prime}$ [b]/ $cm^3 \cdot mol^{-1}$	V^{ex} [b]/ $cm^3 \cdot mol^{-1}$
313.7 ± 0.2	98.3 ± 0.8	6.8 ± 0.8	162.4 ± 0.6	155.1 ± 0.6	−0.9 ± 0.6	−1.4 ± 0.6
325.4 ± 0.2	98.1 ± 1.1	6.1 ± 0.8	165.9 ± 0.5	155.7 ± 0.6	−1.3 ± 0.5	−1.8 ± 0.6
333.5 ± 0.2	98.0 ± 1.1	5.7 ± 1.0	168.4 ± 0.5	156.2 ± 0.8	−1.6 ± 0.5	−2.1 ± 0.8

[a]Observed molar volume
[b]Volumetric Properties

(2) Cloud Point

Ref. [30]

T/K	x_1/mol %	T/K	x_1/mol %
250 ± 3	69.8 ± 0.5	268 ± 1	90.1 ± 0.5
260 ± 2	80.0 ± 0.5		

Mixture 4: $CF_3CHFCHFCF_2CF_3$ *threo*-D (1) + [emim][BF_4] (2)
(1) Phase Equilibrium

Ref. [30]

T/K	x'_1/mol%	x_1/mol%	V' [a]/ $cm^3 \cdot mol^{-1}$	V[a]/ $cm^3 \cdot mol^{-1}$	$V^{ex\prime}$ [b]/ $cm^3 \cdot mol^{-1}$	V^{ex} [b]/ $cm^3 \cdot mol^{-1}$
283.7 ± 0.2	94.4 ± 0.4	10.5 ± 0.5	153.5 ± 0.8	150.3 ± 1.0	−1.8 ± 0.8	−3.3 ± 1.0
294.0 ± 0.2	96.8 ± 0.4	8.9 ± 0.7	156.1 ± 0.6	152.5 ± 1.0	−1.9 ± 0.6	−2.1 ± 1.0
303.8 ± 0.2	97.6 ± 0.7	8.0 ± 0.9	159.2 ± 0.6	153.4 ± 0.8	−1.7 ± 0.6	−2.2 ± 0.8
313.4 ± 0.2	97.9 ± 0.4	7.2 ± 1.0	162.6 ± 0.6	154.8 ± 0.8	−1.2 ± 0.6	−1.7 ± 0.8
323.5 ± 0.2	98.5 ± 0.6	6.2 ± 1.2	165.5 ± 1.0	156.0 ± 0.6	−1.8 ± 1.0	−1.4 ± 0.6
333.9 ± 0.2	97.8 ± 0.6	5.7 ± 1.4	168.5 ± 1.0	156.4 ± 0.7	−2.3 ± 1.0	−2.1 ± 0.7

[a]Observed molar volume
[b]Volumetric Properties

(1) Cloud Point

Ref. [30]

T/K	x_1/mol %	T/K	x_1/mol %
252 ± 3	69.6 ± 0.5	273 ± 1	89.8 ± 0.5
261 ± 2	80.1 ± 0.5		

Mixture 5: $CF_3CHFCHFCF_2CF_3$ *erythro*-H (1) + [emim][BF_4] (2)
(1) Phase Equilibrium

Ref. [30]

T/K	x'_1/mol%	x_1/mol%	V' [a]/ $cm^3 \cdot mol^{-1}$	V[a]/ $cm^3 \cdot mol^{-1}$	$V^{ex\prime}$ [b]/ $cm^3 \cdot mol^{-1}$	V^{ex} [b]/ $cm^3 \cdot mol^{-1}$
283.6 ± 0.2	98.4 ± 1.1	2.9 ± 1.2	157.0 ± 0.7	151.6 ± 0.9	−0.4 ± 0.7	−1.8 ± 0.9
293.6 ± 0.2	98.5 ± 1.3	2.4 ± 1.3	159.8 ± 0.6	152.4 ± 0.8	−0.3 ± 0.6	−1.9 ± 0.8
303.5 ± 0.2	98.6 ± 1.3	2.3 ± 1.5	162.9 ± 0.7	153.0 ± 0.9	−0.2 ± 0.7	−2.2 ± 0.9
313.7 ± 0.2	98.8 ± 1.2	1.8 ± 1.4	165.8 ± 0.6	154.3 ± 0.7	−0.4 ± 0.6	−1.9 ± 0.7

(*continued*)

T/K	x'_1/mol%	x_1/mol%	V' [a]/ $cm^3 \cdot mol^{-1}$	V[a]/ $cm^3 \cdot mol^{-1}$	$V^{ex\prime}$ [b]/ $cm^3 \cdot mol^{-1}$	V^{ex} [b]/ $cm^3 \cdot mol^{-1}$
325.4 ± 0.2	98.5 ± 1.2	2.5 ± 1.3	169.7 ± 0.6	154.8 ± 0.8	−0.4 ± 0.6	−2.6 ± 0.8
333.5 ± 0.2	98.5 ± 1.4	2.5 ± 1.5	172.2 ± 0.7	155.8 ± 0.4	−0.8 ± 0.7	−2.2 ± 0.4

[a]Observed molar volume
[b]Volumetric Properties

Mixture 6: $CF_3CHFCHFCF_2CF_3$ *erythro*-D (1) + [emim][BF_4] (2)
(1) Phase Equilibrium

Ref. [30]

T/K	x'_1/mol%	x_1/mol%	V' [a]/ $cm^3 \cdot mol^{-1}$	V[a]/ $cm^3 \cdot mol^{-1}$	$V^{ex\prime}$ [b]/ $cm^3 \cdot mol^{-1}$	V^{ex} [b]/ $cm^3 \cdot mol^{-1}$
283.9 ± 0.2	99.5 ± 0.5	2.0 ± 1.1	157.0 ± 0.7	152.8 ± 0.9	−0.5 ± 0.7	−0.7 ± 0.9
294.1 ± 0.2	99.0 ± 1.0	1.7 ± 1.5	159.7 ± 0.7	153.7 ± 0.8	−0.6 ± 0.7	−0.7 ± 0.8
303.9 ± 0.2	98.6 ± 1.2	1.5 ± 1.1	162.3 ± 0.6	154.8 ± 0.8	−0.9 ± 0.6	−0.4 ± 0.8
313.5 ± 0.2	98.7 ± 1.2	1.1 ± 1.0	165.2 ± 0.7	155.6 ± 0.8	−0.9 ± 0.7	−0.4 ± 0.8
323.4 ± 0.2	98.6 ± 1.1	2.5 ± 1.1	168.5 ± 0.6	156.1 ± 0.8	−0.9 ± 0.6	−1.0 ± 0.8
334.0 ± 0.2	99.1 ± 0.9	2.0 ± 1.1	172.9 ± 0.5	155.9 ± 0.7	−0.4 ± 0.5	−2.1 ± 0.7

[a]Observed molar volume
[b]Volumetric Properties

Mixture 7: $CF_3CHFCHFCF_2CF_3$ [*threo*-H (87.12 ± 0.6 %) + *erythro*-H (11.89 ± 0.6 %)] (1) + [emim][BF_4] (2)
(1) Phase Equilibrium

Ref. [30]

T/K	x'_1/mol%	x_1/mol%	V' [a]/ $cm^3 \cdot mol^{-1}$	V[a]/ $cm^3 \cdot mol^{-1}$	$V^{ex\prime}$ [b]/ $cm^3 \cdot mol^{-1}$	V^{ex} [b]/ $cm^3 \cdot mol^{-1}$
283.6 ± 0.2	97.7 ± 0.9	7.8 ± 0.8	154.4 ± 0.7	151.2 ± 0.7	−0.5 ± 0.7	−2.4 ± 0.7
293.6 ± 0.2	98.4 ± 0.9	6.4 ± 0.9	157.2 ± 0.6	152.6 ± 0.9	−0.4 ± 0.6	−1.9 ± 0.9
293.6 ± 0.2	99.0 ± 1.0	5.9 ± 0.9	160.5 ± 0.5	153.7 ± 0.8	−0.1 ± 0.5	−1.7 ± 0.8
313.7 ± 0.2	99.1 ± 1.0	5.4 ± 1.1	163.5 ± 0.6	154.7 ± 0.8	−0.2 ± 0.6	−1.7 ± 0.8
325.4 ± 0.2	98.8 ± 1.0	5.0 ± 1.1	167.0 ± 0.5	155.4 ± 0.8	−0.5 ± 0.5	−2.2 ± 0.8
333.5 ± 0.2	98.8 ± 1.0	5.0 ± 1.1	169.7 ± 0.6	156.4 ± 0.8	−0.6 ± 0.6	−2.0 ± 0.8

[a]Observed molar volume
[b]Volumetric Properties

Mixture 8: 1,3-HOC_3H_6OH (1) + [emim][BF_4] (2)
(1) Phase Equilibrium

Ref. [31]

T/K	$100x'_1$	$100x_1$	V'/ $cm^3 \cdot mol^{-1}$	V/ $cm^3 \cdot mol^{-1}$	$V^{E\prime}$/ $cm^3 \cdot mol^{-1}$	V^E/ $cm^3 \cdot mol^{-1}$
286.6 ± 0.2	23.8 ± 0.5	97.0 ± 0.2	129.5 ± 0.4	72.5 ± 0.2	−4.7 ± 0.4	−1.9 ± 0.2
287.8 ± 0.2	24.5 ± 0.5	97.0 ± 0.2	129.2 ± 0.4	72.7 ± 0.2	−4.5 ± 0.4	−1.8 ± 0.2
288.2 ± 0.2	24.9 ± 0.4	96.8 ± 0.2	128.9 ± 0.4	72.9 ± 0.2	−4.4 ± 0.4	−1.7 ± 0.2

(*continued*)

T/K	$100x'_1$	$100x_1$	V'/ $cm^3 \cdot mol^{-1}$	V/ $cm^3 \cdot mol^{-1}$	$V^{E\prime}$/ $cm^3 \cdot mol^{-1}$	V^E/ $cm^3 \cdot mol^{-1}$
291.9 ± 0.2	30.9 ± 0.4	95.8 ± 0.2	123.9 ± 0.3	74.7 ± 0.2	−4.9 ± 0.3	−1.0 ± 0.2
298.3 ± 0.2	35.8 ± 0.4	94.5 ± 0.2	120.8 ± 0.3	76.0 ± 1.2	−4.4 ± 0.3	−0.9 ± 1.2
303.3 ± 0.2	43.5 ± 0.3	92.2 ± 0.1	114.6 ± 0.2	77.5 ± 0.1	−4.6 ± 0.2	−1.6 ± 0.1

V', observed molar volume lower phase; V, observed molar volume upper phase; $V^{E\prime}$, Volumetric Properties lower phase; V^E, Volumetric Properties upper phase

Mixture 9: H_2O (1) + [emim][BF_4] (2)
(1) Boiling Point and Activity Coefficient

Ref. [32]

w_2	T/K	γ_1	w_2	T/K	γ_1
100 kPa					
0.7826	380.64	1.008	0.4001	374.61	0.993
0.7475	379.35	1.008	0.3276	374.22	0.992
0.6978	377.91	1.010	0.2005	373.68	0.990
0.6043	376.29	1.005	0.1004	373.26	0.993
0.5029	375.30	0.998			

(2) Volumetric Properties, Transport Properties, Surface Tension and Conductivity

Ref. [33]

ρ/kg · m^{-3}							
x_1	293.15 K	298.15 K	303.15 K	308.15 K	313.15 K	318.15 K	323.15 K
0	1340.7	1336.7	1332.9	1329.6	1325.9	1322.2	1319.5
0.0995	1336.1	1332.1	1328.3	1324.9	1321.2	1317.6	1314.8
0.2062	1330.0	1326.0	1322.2	1318.9	1315.2	1311.6	1308.8
0.3047	1322.9	1318.9	1315.2	1311.9	1308.1	1304.5	1301.6
0.4034	1313.4	1309.5	1305.8	1302.5	1298.7	1295.0	1292.0
0.5034	1299.8	1296.0	1292.3	1288.9	1285.3	1281.6	1278.6
0.5998	1284.2	1280.6	1276.9	1273.5	1269.7	1266.0	1263.1
0.7005	1262.0	1258.4	1254.8	1251.4	1247.5	1243.9	1240.7
0.8011	1225.9	1222.3	1218.7	1215.4	1211.8	1208.2	1205.0
0.9027	1157.4	1154.2	1151.1	1148.0	1145.0	1141.8	1138.5
1	998.2	997.0	995.7	994.0	992.2	990.1	988.0
η/mPa · s							
x_1	293.15 K	298.15 K	303.15 K	308.15 K	313.15 K	318.15 K	323.15 K
0	45.290	37.190	31.362	25.557	21.537	18.838	15.903
0.0995	34.431	29.025	24.800	20.533	17.964	15.927	13.743
0.2062	25.098	21.008	17.897	14.584	12.678	11.157	9.797
0.3047	19.252	15.965	13.577	10.931	9.634	8.324	7.470
0.4034	13.760	11.448	9.674	8.187	7.165	6.357	5.692
0.5034	10.365	8.597	7.384	6.409	5.465	4.979	4.522
0.5998	7.892	6.975	5.492	4.707	4.159	3.630	3.215
0.7005	5.797	4.822	3.896	3.531	3.133	2.584	2.512

(continued)

η/mPa · s							
x_1	293.15 K	298.15 K	303.15 K	308.15 K	313.15 K	318.15 K	323.15 K
0.8011	4.127	3.499	2.962	2.555	2.267	2.044	1.904
0.9027	3.035	2.584	2.189	1.898	1.689	1.530	1.439
1	1.002	0.890	0.798	0.719	0.653	0.599	0.547
V^E/cm^3 · mol^{-1}							
x_1	293.15 K	298.15 K	303.15 K	308.15 K	313.15 K	318.15 K	323.15 K
0	0	0	0	0	0	0	0
0.0995	0.0081	0.0137	0.0156	0.0193	0.0210	0.0224	0.0263
0.2062	0.0159	0.0246	0.0308	0.0353	0.0383	0.0407	0.0471
0.3047	0.0348	0.0445	0.0507	0.0605	0.0711	0.0804	0.0933
0.4034	0.0861	0.0963	0.1046	0.1159	0.1350	0.1456	0.1743
0.5034	0.1999	0.2109	0.2304	0.2472	0.2573	0.2708	0.2982
0.5998	0.1902	0.1964	0.2176	0.2387	0.2624	0.2814	0.2984
0.7005	0.1182	0.1323	0.1510	0.1772	0.2059	0.2239	0.2528
0.8011	0.0666	0.0875	0.1166	0.1392	0.1590	0.1774	0.2045
0.9027	0.0351	0.0560	0.0767	0.0968	0.1079	0.1199	0.1461
1	0	0	0	0	0	0	0
$\Delta\eta$/mPa · s							
x_1	293.15 K	298.15 K	303.15 K	308.15 K	313.15 K	318.15 K	323.15 K
0	0	0	0	0	0	0	0
0.0995	−6.45	−4.46	−3.25	−2.38	−1.77	−1.10	−0.63
0.2062	−11.06	−8.62	−6.92	−5.69	−4.79	−3.92	−2.94
0.3047	−12.54	−10.10	−8.26	−6.92	−5.75	−4.96	−3.75
0.4034	−13.66	−11.04	−9.18	−7.23	−6.13	−5.12	−4.01
0.5034	−12.63	−10.27	−8.44	−6.55	−5.71	−4.68	−3.65
0.5998	−10.83	−8.40	−7.42	−5.87	−4.97	−4.27	−3.48
0.7005	−8.47	−6.91	−5.97	−4.57	−3.87	−3.48	−2.63
0.8011	−5.68	−4.59	−3.86	−3.06	−2.60	−2.18	−1.70
0.9027	−2.28	−1.83	−1.55	−1.22	−1.03	−0.84	−0.60
1	0	0	0	0	0	0	0

Ref. [34]

x_2	c (mol/L)	ρ (g/cm^3)	x_2	c (mol/L)	ρ (g/cm^3)
298.15 K					
1.0000	6.463	1.2799	0.1020	3.117	1.1117
0.7774	6.228	1.2655	0.0709	2.508	1.0886
0.6910	6.106	1.2585	0.0483	1.933	1.0682
0.5923	5.933	1.2486	0.0288	1.301	1.0464
0.4987	5.721	1.2367	0.0132	0.661	1.0234
0.4551	5.601	1.2300	0.0123	0.621	1.0218
0.3118	5.048	1.2003	0.0077	0.400	1.0132
0.2191	4.466	1.1710	0.0040	0.215	1.0055
0.2141	4.425	1.1691	0.0000	0.000	0.9970
0.1418	3.699	1.1358			

Ref. [34]

x_2	c (mol/L)	σ (mN/m)	$\delta\sigma$ (mN/m)	x_2	c (mol/L)	σ (mN/m)	$\delta\sigma$ (mN/m)
298.15 K							
1.0000	6.458	53.04	0.00	0.1020	3.117	55.24	−14.24
0.9015	6.373	53.25	−1.59	0.0709	2.508	55.92	−14.13
0.7534	6.196	53.51	−4.04	0.0483	1.933	57.31	−13.15
0.6018	5.967	53.73	−6.60	0.0288	1.301	58.15	−12.67
0.5074	5.743	53.88	−8.17	0.0132	0.661	61.91	−9.19
0.4011	5.458	54.01	−9.99	0.0123	0.621	62.87	−8.25
0.2894	4.975	54.15	−11.89	0.0077	0.400	65.00	−6.21
0.2141	4.425	53.80	−13.63	0.0040	0.215	67.38	−3.89
0.1418	3.699	54.43	−14.32	0.0000	0.000	71.35	0.00

Ref. [35]

x_2	$10^{-3}c_2$/mol · m^{-3}	ρ/kg · m^{-3}	κ/S · cm^{-1}	$10^4\Lambda$/S · m^2 · mol^{-1}
298.15 K				
0.004746	0.2534	1007.69	1.874	73.9
0.01001	0.5144	1018.31	3.26	63.4
0.02226	1.052	1040.22	5.35	50.9
0.03637	1.572	1061.36	6.81	43.3
0.05733	2.200	1087.31	8.08	36.7
0.08102	2.761	1110.93	8.84	32.0
0.1150	3.376	1136.45	9.24	27.4
0.1740	4.122	1168.60	9.08	22.0
0.2564	4.790	1198.61	8.13	16.98
0.4190	5.532	1233.28	5.99	10.82
1	6.484	1283.70	1.553	2.40

(3) Phase Equilibrium

Ref. [36]

T = 373.15 K		T = 403.15 K	
x_1	P/kPa	x_1	P/kPa
0.050	2.5	0.073	10.1
0.204	11.6	0.106	15.3
0.288	19.0	0.128	20.1
0.375	24.7	0.149	24.8
0.402	29.2	0.188	30.6
0.483	39.3	0.271	45.0
0.535	45.3	0.329	59.7
0.582	49.9	0.371	70.0
0.633	56.4	0.401	80.2

(*continued*)

T = 373.15 K		T = 403.15 K	
x_1	P/kPa	x_1	P/kPa
0.663	60.4		
0.719	67.0		
0.778	75.6		

Mixture 10: 2-CH_3CHCH_3OH (1) + [emim][BF_4] (2)
(1) Boiling Point and Activity Coefficient

Ref. [32]

w_2	T/K	γ_1	w_2	T/K	γ_1
100 kPa					
0.8068	357.68	2.033	0.5218	355.53	1.299
0.7397	356.76	1.731	0.4180	355.48	1.191
0.6590	355.90	1.526	0.3260	355.45	1.123
0.5979	355.62	1.412			

Mixture 11: 1-C_3H_7OH (1) + [emim][BF_4] (2)
(1) Boiling Point and Activity Coefficient

Ref. [32]

w_2	T/K	γ_1	w_2	T/K	γ_1
100 kPa					
0.8009	374.77	1.848	0.5663	370.66	1.353
0.6898	371.87	1.551	0.4422	370.11	1.228
0.6276	371.20	1.435			

Mixture 12: $LiBF_4$ (1) + [emim][BF_4] (2)
(1) Glass Transition Temperature, Crystalline Temperature, and Melting Point

Ref. [37]

Concentrations of $LiBF_4$	T_g (K)	T_c (K)[a]	T_m (K)
Neat	187	225 (c)	288
		224 (h)	
0.25 M	188	236 (h)	282
0.50 M	188	251 (h)	274
0.75 M	192		
1.00 M	191		
1.25 M	193		
1.50 M	194		

[a]The crystalline temperatures were obtained during the cooling (c) and heating (h) scans

Mixture 13: $[CH_2\text{-}CH_2\text{-}O]_n$ (poly(ethyleneoxide)) (1) + $EMIBF_4$ (2)
(1) Conductivity

Ref. [38]

Composition (wt.%)		
Polymer	Ionic liquid	σ (mS/cm) at 298.15 K
72.1 % PEO	27.9 % $EMIBF_4$	0.5
67.2 % PEO	32.8 % $EMIBF_4$	0.6

Mixture 14: $[CH_2\text{-}CHOH]_n$ (poly(vinylalcohol)) (1) + $EMIBF_4$ (2)
(1) Conductivity

Ref. [38]

Composition (wt.%)		
Polymer	Ionic liquid	σ (mS/cm) at 298.15 K
54.2 % PVA	45.8 % $EMIBF_4$	0.005
39.8 % PVA	60.2 % $EMIBF_4$	1.6

Mixture 15: $[CH_2\text{-}CHCN]_n$ (poly(acrylonitrile)) (1) + $EMIBF_4$ (2)
(1) Conductivity

Ref. [38]

Composition (wt.%)		
Polymer	Ionic liquid	σ (mS/cm) at 298.15 K
37.1 % PAN	62.9 % $EMIBF_4$	1.1
26.6 % PAN	73.4 % $EMIBF_4$	6.6

Mixture 16: C_2H_5OH (1) + $[EMIM][BF_4]$ (2)
(1) Volumetric Properties and Surface Tension

Ref. [34]

x_2	c (mol/L)	ρ (g/cm^3)		x_2	c (mol/L)	ρ (g/cm^3)	
298.15 K							
1.0000	6.463	1.2799		0.7056	5.641	1.2255	
0.9054	6.224	1.2626		0.6087	5.228	1.1903	
0.8072	5.940	1.2416		0.0000	0.000	0.7858	
x_2	c (mol/L)	σ (mN/m)	$\delta\sigma$ (mN/m)	x_2	c (mol/L)	σ (mN/m)	$\delta\sigma$ (mN/m)
298.15 K							
1.0000	6.463	53.04	0.00	0.7056	5.641	37.37	−6.37
0.9054	6.224	47.41	−2.48	0.6087	5.228	33.58	−7.18
0.8072	5.940	42.77	−4.10	0.0000	0.000	52.81	0.00

Mixture 17: CH_3CN (1) + [emim][BF_4] (2)
(1) Volumetric Properties and Conductivity

Ref. [35]

x_2	$10^{-3}c_2$/mol · m^{-3}	ρ/kg · m^{-3}	κ/S · cm^{-1}	$10^4\Lambda$/S m^2 · mol^{-1}
298.15 K				
0.01082	0.2014	795.48	1.90	94.4
0.01646	0.3036	804.68	2.55	84.0
0.02252	0.4114	814.30	3.13	76.1
0.04929	0.8621	853.38	4.89	56.7
0.08059	1.342	893.87	6.03	44.9
0.1218	1.903	939.88	6.72	35.3
0.1736	2.513	988.57	6.92	27.6
0.2371	3.146	1038.16	6.65	21.2
0.3264	3.867	1093.25	5.90	15.26
0.4565	4.672	1153.25	4.64	9.94
0.6520	5.528	1215.55	3.10	5.61
0.8336	6.092	1255.87	2.11	3.46
1	6.484	1283.70	1.553	2.40
0.008411	0.1571	791.26	1.585	100.9
0.01528	0.2823	802.59	2.46	87.1
0.02253	0.4113	813.87	3.14	76.4
0.04942	0.8649	854.17	4.94	57.2
0.1007	1.623	916.26	6.50	40.0
0.1433	2.165	960.07	6.94	32.0
0.2043	2.829	1012.50	6.91	24.4
0.2778	3.490	1063.36	6.41	18.36
0.3840	4.245	1120.11	5.41	12.75
0.4888	4.828	1163.04	4.44	9.21
0.6478	5.500	1211.72	3.20	5.81
0.7590	5.867	1237.95	2.54	4.33
0.8902	6.221	1263.01	1.95	3.13
1	6.468	1280.38	1.553	2.40

2.2.7.2 Ternary Mixtures

Mixture 1: H_2O (1) + C_2H_5OH (2) + [emim][BF_4] (3)
(1) Phase Equilibrium

Ref. [24]

x'_2[a]	w_3[a]	y_2	T/K	γ_1	γ_2	α_{21}[b]
p = 100 kPa						
0.9499	0.9001	0.9883	365.61	0.92	1.82	4.46
0.9501	0.8001	0.9848	356.84	1.05	1.58	3.42
0.9501	0.7000	0.9808	354.00	1.20	1.40	2.68

(continued)

x'_2 [a]	w_3 [a]	y_2	T/K	γ_1	γ_2	α_{21} [b]
0.9501	0.5999	0.9765	353.00	1.34	1.27	2.18
0.9501	0.5000	0.9720	352.60	1.48	1.18	1.83
0.9501	0.4000	0.9680	352.50	1.60	1.11	1.59
0.9501	0.3000	0.9630	352.13	1.79	1.07	1.37
0.9500	0.2001	0.9585	351.79	1.96	1.04	1.21
0.9500	0.0999	0.9539	351.47	2.14	1.01	1.09

[a]Compositions reported are of equilibrium liquid
[b]$\alpha_{2,1} = (y_2/x'_2)/(y_1/x'_1)$

Ref. [36]

x_2	y_2	x_2	y_2	x_2	y_2
$T = 337.15$ K					
$x_3 = 0.1$		$x_3 = 0.3$		$x_3 = 0.5$	
0.602	0.743	0.299	0.661	0.298	0.685
0.754	0.819	0.402	0.725	0.396	0.747
0.837	0.885	0.601	0.827	0.450	0.786
0.849	0.895	0.746	0.879	0.603	0.869
0.913	0.938	0.850	0.926	0.743	0.916
0.945	0.955	0.901	0.950	0.849	0.949
		0.945	0.973	0.881	0.962
				0.901	0.968
				0.946	0.988
				0.968	0.989

Mixture 2: H_2O (1) + 2-CH_3CHCH_3OH (2) + [emim][BF_4] (3)
(1) Phase Equilibrium

Ref. [32]

x'_2 [a,b]	y_2	T/K	x_1 [a,c]	x_2 [a,c]	γ_1	γ_2	x'_2 [a,b]	y_2	T/K	x_1 [a,c]	x_2 [a,c]	γ_1	γ_2
100 kPa													
w_3 = 0.300													
0.9879	0.9869	355.37	0.0108	0.8750	2.36	1.11	0.5941	0.7105	354.01	0.3714	0.5435	1.59	1.36
0.9574	0.9559	355.08	0.0378	0.8501	2.28	1.12	0.4955	0.6673	354.11	0.4654	0.4570	1.45	1.51
0.9439	0.9430	355.05	0.0499	0.8390	2.23	1.12	0.3921	0.6397	354.43	0.5656	0.3649	1.28	1.79
0.8869	0.8914	354.75	0.1010	0.7920	2.13	1.13	0.2893	0.6219	354.82	0.6671	0.2715	1.12	2.30
0.8235	0.8435	354.44	0.1584	0.7392	1.98	1.16	0.1840	0.5917	355.89	0.7729	0.1742	1.00	3.27
0.7583	0.7966	354.18	0.2182	0.6843	1.89	1.20	0.1163	0.5730	357.37	0.8420	0.1108	0.90	4.69
0.6776	0.7498	354.04	0.2929	0.6156	1.74	1.26							
w_3 = 0.400													
0.9980	0.9981	355.46	0.0016	0.8302	2.20	1.18	0.6409	0.7550	354.25	0.3119	0.5566	1.59	1.39
0.9800	0.9811	355.35	0.0167	0.8169	2.18	1.18	0.5961	0.7286	354.24	0.3527	0.5206	1.55	1.44
0.9505	0.9549	355.17	0.0414	0.7952	2.12	1.19	0.5564	0.7096	354.26	0.3893	0.4882	1.50	1.49
0.9006	0.9125	354.92	0.0837	0.7579	2.06	1.20	0.4642	0.6781	354.43	0.4756	0.4121	1.36	1.68
0.8498	0.8767	354.72	0.1272	0.7195	1.92	1.23	0.3706	0.6424	354.79	0.5654	0.3329	1.25	1.94
0.8004	0.8433	354.53	0.1700	0.6817	1.84	1.26	0.2711	0.6179	355.48	0.6632	0.2466	1.11	2.45
0.7467	0.8108	354.40	0.2171	0.6402	1.75	1.29	0.1689	0.5971	356.94	0.7663	0.1557	0.95	3.54
0.6893	0.7794	354.30	0.2682	0.5951	1.66	1.34	0.1116	0.5786	358.80	0.8252	0.1037	0.86	4.79
w_3 = 0.600													
0.9905	0.9942	355.60	0.0066	0.6820	1.68	1.42	0.5689	0.7606	355.23	0.3271	0.4317	1.42	1.74
0.9677	0.9806	355.54	0.0224	0.6697	1.66	1.43	0.4632	0.7187	355.54	0.4180	0.3607	1.29	1.94
0.9126	0.9476	355.44	0.0612	0.6394	1.65	1.45	0.3688	0.6787	356.08	0.5033	0.2941	1.20	2.20
0.8502	0.9103	355.33	0.1064	0.6040	1.63	1.48	0.2457	0.6370	357.54	0.6209	0.2022	1.04	2.84

(continued)

x'_2 [a,b]	y_2	T/K	x_1 [a,c]	x_2 [a,c]	γ_1	γ_2	x'_2 [a,b]	y_2	T/K	x_1 [a,c]	x_2 [a,c]	γ_1	γ_2
0.7647	0.8670	355.21	0.1705	0.5540	1.52	1.55	0.1596	0.6100	359.75	0.7077	0.1344	0.90	3.75
0.6737	0.8192	355.16	0.2415	0.4985	1.46	1.63	0.1235	0.5751	361.29	0.7453	0.1050	0.87	4.27
$w_3 = 0.800$													
0.9806	0.9946	358.12	0.0088	0.4462	1.06	1.96	0.6119	0.8555	359.97	0.2060	0.3248	1.13	2.16
0.9483	0.9846	358.36	0.0238	0.4369	1.11	1.97	0.5013	0.7963	360.62	0.2786	0.2800	1.15	2.27
0.9015	0.9697	358.52	0.0462	0.4231	1.12	1.99	0.4025	0.7460	361.48	0.3502	0.2359	1.10	2.45
0.8439	0.9500	358.72	0.0750	0.4054	1.13	2.02	0.2998	0.6857	363.03	0.4325	0.1852	1.04	2.70
0.7734	0.9216	358.94	0.1121	0.3825	1.17	2.06	0.1952	0.5885	365.73	0.5259	0.1275	1.01	3.05
0.6992	0.8928	359.61	0.1536	0.3570	1.14	2.08	0.1433	0.5461	367.87	0.5764	0.0964	0.94	3.46

[a]Compositions reported are of equilibria liquid
[b]Mole fraction on IL-free basis
[c]Mole fraction on IL-containing basis

Mixture 3: H_2O (1) + 1-C_3H_7OH (2) + [emim][BF_4] (3)
(1) Phase Equilibrium

Ref. [32]

x'_2 [a,b]	y_2	T/K	x_1 [a,c]	x_2 [a,c]	γ_1	γ_2	x'_2 [a,b]	y_2	T/K	x_1 [a,c]	x_2 [a,c]	γ_1	γ_2
100 kPa													
$w_3 = 0.600$													
0.9756	0.9719	370.95	0.0169	0.6740	1.78	1.38	0.4414	0.5465	365.14	0.4374	0.3455	1.37	1.90
0.9028	0.8909	369.98	0.0682	0.6339	1.77	1.40	0.3428	0.4970	364.72	0.5275	0.2752	1.28	2.20
0.8428	0.8292	369.18	0.1118	0.5998	1.74	1.42	0.2419	0.4487	364.84	0.6246	0.1993	1.18	2.73
0.7353	0.7417	367.79	0.1931	0.5363	1.60	1.49	0.1537	0.4165	365.79	0.7138	0.1297	1.06	3.75
0.6217	0.6580	366.56	0.2834	0.4658	1.51	1.60	0.1218	0.4107	366.55	0.7471	0.1036	0.99	4.50
0.5375	0.5984	365.86	0.3536	0.4109	1.46	1.70							
$w_3 = 0.700$													
0.9925	0.9940	372.09	0.0044	0.5823	1.40	1.57	0.4630	0.5937	367.49	0.3724	0.3211	1.32	2.02
0.9207	0.9368	371.44	0.0475	0.5517	1.40	1.60	0.3687	0.5301	367.09	0.4524	0.2642	1.28	2.23
0.8597	0.8859	370.82	0.0856	0.5246	1.43	1.63	0.2513	0.4627	367.19	0.5599	0.1879	1.18	2.72
0.7643	0.8135	369.88	0.1481	0.4802	1.40	1.69	0.1606	0.4204	368.24	0.6495	0.1243	1.05	3.59
0.6663	0.7423	369.01	0.2163	0.4319	1.37	1.77	0.1238	0.4113	369.30	0.6877	0.0972	0.97	4.32
0.5747	0.6632	368.25	0.2840	0.3838	1.40	1.84							

[a] Compositions reported are of equilibria liquid
[b] Mole fraction on IL-free basis
[c] Mole fraction on IL-containing basis

Mixture 4: H_2O (1) + $OCH_2CH_2CH_2CH_2$ (2) + [EMIM][BF_4] (3)
(1) Phase Equilibrium

Ref. [36]

x_2	y_2	x_2	y_2	x_2	y_2
T = 337.15 K					
x_3 = 0.1		x_3 = 0.3		x_3 = 0.5	
0.428	0.827	0.255	0.827	0.254	0.870
0.623	0.855	0.429	0.887	0.428	0.919
0.792	0.894	0.623	0.902	0.624	0.947
0.849	0.913	0.793	0.944	0.795	0.973
0.949	0.961	0.849	0.956	0.850	0.976
		0.949	0.985	0.949	0.994

Ref. [36]

Tie lines								
Upper phase			Lower phase			Cloud Points		
x_2	x_1	x_3	x_2	x_1	x_3	x_2	x_1	x_3
T = 337.15 K								
0.919	0.079	0.002	0.298	0.399	0.303	0.301	0.350	0.349
0.794	0.199	0.007	0.304	0.567	0.129	0.322	0.255	0.423
0.945	0.053	0.002	0.306	0.316	0.379	0.337	0.169	0.494
0.856	0.139	0.004	0.296	0.530	0.174	0.344	0.080	0.576
0.967	0.031	0.001	0.334	0.117	0.549	0.299	0.525	0.176
0.956	0.042	0.002	0.328	0.180	0.492	0.337	0.623	0.040
						0.388	0.580	0.032

Mixture 5: $[CH_2\text{-}CHCN]_n$ (poly(acrylonitrile) [PAN]) (1) + $CH_2CH_2CH_2CH_2SO_2$ ((sulpholane) [TMS]) (2) + $EMIBF_4$ (3)
(1) Conductivity

Ref. [38]

Composition (wt.%)			
Polymer	Ionic liquid	TMS	σ (mS/cm), 298.15 K
19.3 % PAN	18.4 % $EMIBF_4$	62.3 %	1.6
20.7 % PAN	31.9 % $EMIBF_4$	47.4 %	6.7
12.9 % PAN	48.3 % $EMIBF_4$	38.8 %	15.0
11.9 % PAN	57.5 % $EMIBF_4$	30.6 %	14.0

Mixture 6: $CH_3COOCH_2CH_3$ (1) + C_2H_5OH (2) + [EMIM][BF_4] (3)
(1) Phase Equilibrium

Ref. [39]

$100x_3$	T/K	x_1'	y_1	γ_1	γ_2	α_{12}
P = 101.32 kPa						
7.198	350.25	0.102	0.191	2.023	1.020	2.079
7.176	348.60	0.195	0.315	1.841	1.029	1.898
7.178	347.35	0.309	0.431	1.657	1.047	1.694
7.232	346.55	0.415	0.510	1.500	1.101	1.467
7.176	346.30	0.510	0.570	1.375	1.165	1.274
7.157	346.36	0.607	0.641	1.296	1.210	1.156
7.235	346.44	0.653	0.674	1.264	1.241	1.099
7.174	346.50	0.696	0.714	1.253	1.239	1.090
7.144	346.94	0.732	0.744	1.223	1.235	1.064
7.099	347.80	0.764	0.771	1.180	1.211	1.040
7.190	349.24	0.893	0.881	1.101	1.311	0.887
7.180	350.15	0.950	0.941	1.073	1.341	0.839
20.110	353.45	0.092	0.180	1.215	1.046	2.166
20.135	351.05	0.192	0.324	2.064	1.066	2.017
20.244	349.73	0.291	0.444	1.950	1.055	1.946
20.262	348.35	0.410	0.545	1.778	1.097	1.724
20.260	347.65	0.517	0.631	1.671	1.118	1.598
20.125	347.27	0.619	0.704	1.574	1.152	1.464
19.938	347.55	0.650	0.725	1.526	1.149	1.420
20.577	348.25	0.681	0.749	1.482	1.128	1.398
20.126	349.50	0.717	0.775	1.390	1.077	1.360
20.012	353.00	0.903	0.915	1.162	1.032	1.156
20.094	354.46	0.948	0.953	1.102	1.007	1.112
29.862	357.54	0.098	0.207	2.396	0.991	2.403
29.748	354.84	0.213	0.369	2.134	1.001	2.161
29.497	353.10	0.326	0.495	1.970	0.997	2.027
28.982	350.75	0.412	0.577	1.945	1.043	1.947
28.516	349.95	0.530	0.673	1.798	1.034	1.825
28.198	349.66	0.613	0.737	1.711	1.018	1.769
28.193	350.15	0.628	0.747	1.666	0.999	1.749
28.065	351.75	0.705	0.800	1.506	0.933	1.674
28.003	354.54	0.774	0.844	1.323	0.850	1.580
27.600	355.55	0.872	0.911	1.221	0.819	1.503
27.510	356.15	0.910	0.937	1.180	0.805	1.471

Mixture 7: I_2 (1) + [EMIM][BF_4] (2) + [MPIM]I (3)
(1) Volumetric Properties, Transport Properties and Conductivity

Ref. [40]

x_3	$\tilde{m}^a$/mol · kg^{-1}	298.15 K	303.15 K		313.15 K	323.15 K	333.15 K
ρ/kg · dm^{-3}, 0.05 mol L^{-1} I_2 in Mixture of [EMIM][BF_4] and [MPIM]I							
0.10	0.0409	1.3243	1.3204		1.3128	1.3050	1.2975
0.20	0.0376	1.3499	1.3460		1.3384	1.3305	1.3228
0.30	0.0392	1.3769	1.3730		1.3654	1.3576	1.3498
0.40	0.0363	1.4023	1.3983		1.3904	1.3825	1.3746
0.50	0.0329	1.4273	1.4232		1.4153	1.4073	1.3993
0.60	0.0325	1.4557	1.4505		1.4425	1.4344	1.4264
0.80	0.0336	1.5043	1.5002		1.4922	1.4839	1.4757
1.0	0.0315	1.5553	1.5511		1.5430	1.5347	1.5264
x_3	293.15 K	298.15 K	303.15 K	313.15 K	323.15 K	333.15 K	353.15 K
η/mPa · s, 0.05 mol L^{-1} I_2 in Mixture of [EMIM][BF_4] and [MPIM]I							
0.10	58	47	39	27	20	16	10
0.20	84	67	54	37	27	20	12
0.40	162	123	96	61	41	29	17
0.50	215	160	123	76	50	35	19
0.80	670	466	333	182	108	69	33
1.0	1535	1023	697	350	192.8	113.9	49.2
x_3	278.151 K	288.152 K		298.151 K	303.152 K	313.150 K	323.150 K
κ/mS · cm^{-1}, 0.05 mol L^{-1} I_2 in Mixture of [EMIM][BF_4] and [MPIM]I							
0.10	5.245	–	12.43	14.81	20.29	26.70	
0.20	3.720	–	9.600	11.62	16.38	22.05	
0.30	2.610	4.620	7.449	9.186	13.37	18.49	
0.40	1.782	–	5.595	7.034	10.55	14.95	
0.50	1.139	–	4.035	5.194	8.130	11.92	
0.60	0.710	–	2.847	3.757	6.134	9.324	
0.80	–	0.737	1.523	2.099	3.713	6.035	
1.0	–	–	0.656	0.956	1.867	3.306	

[a] $\tilde{m} = n_{solute}/m_{solution}$

(2) Diffusion Coefficient

Ref. [40]

x_3	298.15 K	303.15 K	313.15 K	323.15 K	333.15 K
$D \cdot 10^7$/cm^2 s^{-1}, 0.05 mol · L^{-1} I_2 in Mixture of [EMIM][BF_4] and [MPIM]I					
Triiodide diffusion coefficients were determined by steady-state cyclic voltammetry at ultramicroelectrodes					
0.10	3.6	4.4	6.5	8.9	11.6
0.20	3.2	3.9	5.7	8.6	11.6
0.30	3.1	4.0	5.6	8.4	11.9
0.40	2.9	3.8	5.8	8.9	11.7

(*continued*)

x_3	298.15 K	303.15 K	313.15 K	323.15 K	333.15 K
0.50	2.9	3.7	5.8	8.5	11.8
0.60	2.5	3.3	4.6	7.2	10.6
0.80	1.8	2.5	4.2	7.1	10.4
1.0	1.2	1.7	3.0	4.9	7.4
Triiodide diffusion coefficients were determined by polarization measurements at thin layer cells					
0.10	3.7	4.5	6.7	9.8	13.3
0.20	3.7	4.6	6.9	9.8	13.3
0.30	3.5	4.4	6.8	10.1	13.8
0.40	3.3	4.2	6.5	10.0	14.0
0.50	3.0	3.9	6.2	9.8	13.9
0.60	2.9	3.9	6.3	9.9	14.3
0.80	2.0	2.7	4.7	7.9	12.2
1.0	1.1	1.5	2.8	5.0	8.0

2.2.8 02003-03001: 1-Ethyl-3-methylimidazolium bis((trifluoromethyl)sulfonyl)imide

Abbreviation:	$[C_2MIm][NTf_2]$, $[emim][NTf_2]$, $[EMIM][NTf_2]$, $[emim][Tf_2N]$, [EMIM][BTI], $[EMIM][(CF_3SO_2)_2N]$, $[C_2mim][Tf_2N]$, $[C_1C_2Im][NTf_2]$
Molecular Formula:	$C_8H_{11}F_6N_3O_4S_2$
Molar Mass:	391.32
Structure:	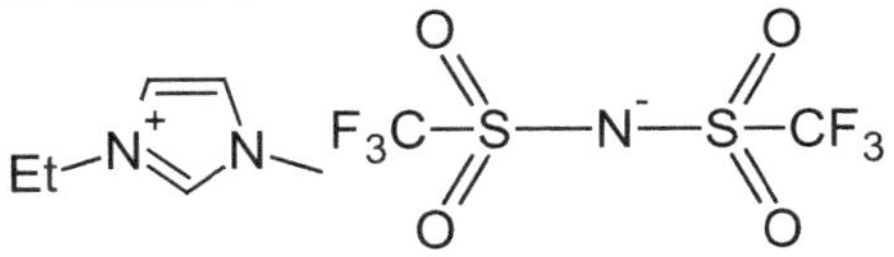

2.2.8.1 Binary Mixtures

Mixture 1: C_6H_5F (1) + $[emim][Tf_2N]$ (2)

(1) Phase Equilibrium

Ref. [41]

T/K	$100x_1'$	$100x_1$	V' ($cm^3 \cdot mol^{-1}$)	V ($cm^3 \cdot mol^{-1}$)	$V^{E\prime}$ ($cm^3 \cdot mol^{-1}$)	V^E ($cm^3 \cdot mol^{-1}$)
282.5 ± 0.2	87.5 ± 0.4	99.2 ± 0.8	112.5 ± 2.8	93.1 ± 1.0	−0.3 ± 2.8	−0.7 ± 1.0
297.9 ± 0.2	87.5 ± 0.3	99.1 ± 0.9	113.7 ± 2.0	94.5 ± 1.0	−0.9 ± 2.0	−1.1 ± 1.0
315.5 ± 0.2	87.4 ± 0.3	99.0 ± 1.0	115.0 ± 2.0	96.1 ± 1.0	−1.7 ± 2.0	−1.5 ± 1.0
333.2 ± 0.2	87.2 ± 0.3	98.9 ± 0.9	116.9 ± 2.6	98.4 ± 1.1	−2.2 ± 2.6	−1.3 ± 1.1
353.1 ± 0.2	86.9 ± 0.3	98.9 ± 0.8	119.7 ± 2.6	101.0 ± 1.2	−2.4 ± 2.6	−0.9 ± 1.2
373.0 ± 0.2	86.5 ± 0.3	98.9 ± 0.9	123.1 ± 2.8	103.5 ± 1.2	−2.2 ± 2.8	−0.8 ± 1.2

Mixture 2: 1,3-$C_6H_4F_2$ (1) + [emim][Tf_2N] (2)
(1) Phase Equilibrium

Ref. [41]

T/K	$100x_1'$	$100x_1$	V' ($cm^3 \cdot mol^{-1}$)	V ($cm^3 \cdot mol^{-1}$)	$V^{E\prime}$ ($cm^3 \cdot mol^{-1}$)	V^E ($cm^3 \cdot mol^{-1}$)
283.3 ± 0.2	75.7 ± 0.4	100.0 − 0.3	134.0 ± 3.6	98.2 ± 3.3	−2.8 ± 3.6	−0.5 ± 3.3
294.4 ± 0.2	75.5 ± 0.4	100.0 − 0.3	136.3 ± 3.6	99.3 ± 3.4	−2.2 ± 3.6	−0.7 ± 3.4
316.1 ± 0.2	75.3 ± 0.4	100.0 − 0.3	140.4 ± 3.5	102.0 ± 3.4	−1.3 ± 3.5	−0.5 ± 3.4
333.3 ± 0.2	74.8 ± 0.3	100.0 − 0.3	142.9 ± 3.4	104.0 ± 3.5	−1.9 ± 3.4	−0.6 ± 3.5
353.1 ± 0.2	74.2 ± 0.3	100.0 − 0.3	146.0 ± 3.3	106.5 ± 3.5	−2.6 ± 3.3	−0.7 ± 3.5
373.0 ± 0.2	73.6 ± 0.3	100.0 − 0.3	149.2 ± 3.3	109.0 ± 3.8	−3.3 ± 3.3	−0.9 ± 3.8

Mixture 3: 1,4-$C_6H_4F_2$ (1) + [emim][Tf_2N] (2)
(1) Phase Equilibrium

Ref. [41]

T/K	$100x_1'$	$100x_1$	V' ($cm^3 \cdot mol^{-1}$)	V ($cm^3 \cdot mol^{-1}$)	$V^{E\prime}$ ($cm^3 \cdot mol^{-1}$)	V^E ($cm^3 \cdot mol^{-1}$)
283.4 ± 0.2	81.6 ± 0.5	100.0 − 0.2	123.9 ± 2.6	97.6 ± 3.2	−3.3 ± 2.6	−0.7 ± 3.2
294.4 ± 0.2	81.4 ± 0.5	100.0 − 0.2	125.5 ± 3.0	98.8 ± 3.2	−3.2 ± 3.0	−0.6 ± 3.2
316.2 ± 0.2	81.0 ± 0.4	100.0 − 0.2	128.8 ± 3.6	101.0 ± 3.2	−3.1 ± 3.6	−0.6 ± 3.2
343.2 ± 0.2	80.7 ± 0.4	99.9 ± 0.1	133.7 ± 3.6	103.8 ± 3.0	−2.0 ± 3.6	−0.9 ± 3.0
353.1 ± 0.2	80.4 ± 0.4	99.9 ± 0.1	135.4 ± 3.6	105.0 ± 3.0	−2.0 ± 3.6	−0.8 ± 3.0
373.0 ± 0.2	79.9 ± 0.4	99.9 ± 0.1	138.7 ± 3.6	106.9 ± 3.0	−2.1 ± 3.6	−1.2 ± 3.0

Mixture 4: 1,2,4-$C_6H_3F_3$ (1) + [emim][Tf_2N] (2)
(1) Phase Equilibrium

Ref. [41]

T/K	$100x_1'$	$100x_1$	V' ($cm^3 \cdot mol^{-1}$)	V ($cm^3 \cdot mol^{-1}$)	$V^{E\prime}$ ($cm^3 \cdot mol^{-1}$)	V^E ($cm^3 \cdot mol^{-1}$)
362.4 ± 0.2	89.3 ± 0.8	97.7 ± 1.0	124.2 ± 2.6	113.5 ± 3.0	−3.5 ± 2.6	−0.8 ± 3.0
367.4 ± 0.2	89.2 ± 0.8	97.7 ± 1.0	124.7 ± 2.8	114.0 ± 3.1	−3.6 ± 2.8	−0.8 ± 3.1
372.4 ± 0.2	89.1 ± 0.7	97.7 ± 0.9	125.2 ± 2.9	114.5 ± 3.1	−3.8 ± 2.9	−0.7 ± 3.1

Mixture 5: 1,3,5-$C_6H_3F_3$ (1) + [emim][Tf_2N] (2)
(1) Phase Equilibrium

Ref. [41]

T/K	$100x_1'$	$100x_1$	V' ($cm^3 \cdot mol^{-1}$)	V ($cm^3 \cdot mol^{-1}$)	$V^{E\prime}$ ($cm^3 \cdot mol^{-1}$)	V^E ($cm^3 \cdot mol^{-1}$)
283.5 ± 0.2	49.7 ± 2.3	100.0 – 0.8	175.4 ± 6.0	101.3 ± 1.5	−3.9 ± 6.0	−1.2 ± 1.5
298.2 ± 0.2	49.8 ± 2.2	100.0 – 0.7	176.2 ± 6.5	103.5 ± 1.5	−4.9 ± 6.5	−0.8 ± 1.5
316.6 ± 0.2	48.7 ± 2.3	100.0 – 0.6	179.0 ± 6.6	105.7 ± 1.6	−6.9 ± 6.6	−1.0 ± 1.6
333.2 ± 0.2	48.2 ± 2.3	100.0 – 0.7	181.5 ± 6.5	108.0 ± 1.8	−8.8 ± 6.5	−1.0 ± 1.8
353.1 ± 0.2	47.8 ± 2.3	100.0 – 0.8	183.0 ± 6.1	110.8 ± 1.8	−10.2 ± 6.1	−1.0 ± 1.8
373.0 ± 0.2	48.2 ± 2.2	100.0 – 0.8	184.6 ± 6.0	114.4 ± 1.8	−11.4 ± 6.0	−0.4 ± 1.8

Mixture 6: 1,2,3,5-$C_6H_2F_4$ (1) + [emim][Tf_2N] (2)
(1) Phase Equilibrium

Ref. [41]

T/K	$100x_1'$	$100x_1$	V' ($cm^3 \cdot mol^{-1}$)	V ($cm^3 \cdot mol^{-1}$)	$V^{E\prime}$ ($cm^3 \cdot mol^{-1}$)	V^E ($cm^3 \cdot mol^{-1}$)
283.5 ± 0.2	65.9 ± 0.5	100.0 – 0.3	152.3 ± 4.8	105.3 ± 1.5	−4.2 ± 4.8	0.0 ± 1.5
298.3 ± 0.2	65.2 ± 0.7	100.0 – 0.3	156.2 ± 5.1	106.9 ± 1.7	−3.5 ± 5.1	−0.3 ± 1.7
316.7 ± 0.2	64.7 ± 0.6	100.0 – 0.3	157.9 ± 5.1	109.6 ± 1.8	−5.3 ± 5.1	−0.1 ± 1.8
333.2 ± 0.2	63.8 ± 0.7	100.0 – 0.3	160.5 ± 5.1	111.9 ± 2.0	−6.6 ± 5.1	−0.2 ± 2.0
353.1 ± 0.2	63.1 ± 0.7	100.0 – 0.3	164.0 ± 5.4	114.9 ± 2.5	−7.4 ± 5.4	−0.1 ± 2.5
373.0 ± 0.2	62.3 ± 0.7	100.0 – 0.3	168.8 ± 5.4	118.0 ± 2.5	−7.2 ± 5.4	−0.1 ± 2.5

Mixture 7: 1,2,4,5-$C_6H_2F_4$ (1) + [emim][Tf_2N] (2)
(1) Phase Equilibrium

Ref. [41]

T/K	$100x_1'$	$100x_1$	V' ($cm^3 \cdot mol^{-1}$)	V ($cm^3 \cdot mol^{-1}$)	$V^{E\prime}$ ($cm^3 \cdot mol^{-1}$)	V^E ($cm^3 \cdot mol^{-1}$)
283.3 ± 0.2	73.6 ± 0.5	99.9 ± 0.1	142.2 ± 1.5	103.0 ± 1.0	−1.8 ± 1.5	−1.2 ± 1.0
298.2 ± 0.2	73.1 ± 0.5	99.9 ± 0.1	144.9 ± 1.7	104.7 ± 1.1	−1.8 ± 1.7	−1.3 ± 1.0
315.5 ± 0.2	72.5 ± 0.6	100.0 – 0.2	147.3 ± 2.7	107.4 ± 1.0	−2.7 ± 2.7	−0.6 ± 1.0
333.2 ± 0.2	71.8 ± 0.6	100.0 – 0.2	150.8 ± 2.6	109.5 ± 1.3	−2.9 ± 2.6	−0.9 ± 1.3
353.1 ± 0.2	71.0 ± 0.6	100.0 – 0.2	155.1 ± 2.5	112.4 ± 1.4	−2.9 ± 2.5	−0.7 ± 1.4
373.0 ± 0.2	70.2 ± 0.6	100.0 – 0.2	159.5 ± 2.4	115.7 ± 1.5	−2.8 ± 2.4	−0.3 ± 1.5

Mixture 8: C_6HF_5 (1) + [emim][Tf_2N] (2)
(1) Phase Equilibrium

Ref. [41]

T/K	$100x_1'$	$100x_1$	V' ($cm^3 \cdot mol^{-1}$)	V ($cm^3 \cdot mol^{-1}$)	$V^{E\prime}$ ($cm^3 \cdot mol^{-1}$)	V^E ($cm^3 \cdot mol^{-1}$)
283.4 ± 0.2	71.5 ± 0.8	99.7 ± 0.3	148.4 ± 2.8	109.3 ± 2.0	−2.6 ± 2.8	−0.5 ± 2.0
298.2 ± 0.2	70.4 ± 0.6	99.8 ± 0.2	152.3 ± 2.9	110.9 ± 2.1	−2.5 ± 2.9	−0.8 ± 2.1
316.6 ± 0.2	69.6 ± 0.4	99.7 ± 0.3	155.2 ± 2.9	113.9 ± 2.1	−3.7 ± 2.9	−0.6 ± 2.1
333.2 ± 0.2	68.1 ± 0.5	100.0 – 0.4	160.5 ± 3.0	115.5 ± 2.2	−3.1 ± 3.0	−1.1 ± 2.2
353.1 ± 0.2	66.9 ± 0.5	100.0 – 0.4	165.2 ± 3.0	118.7 ± 2.3	−3.5 ± 3.0	−1.0 ± 2.3
373.0 ± 0.2	65.7 ± 0.6	100.0 – 0.4	169.2 ± 3.1	122.8 ± 2.4	−4.7 ± 3.1	−0.2 ± 2.4

Mixture 9: C_6H_6 (1) + [emim][Tf_2N] (2)

(1) Phase Equilibrium

Ref. [42]

T (K)	x_1^{L1a} (mol %)	x_1^{L2a} (mol %)	$\overline{V}_{L1}$ [b](cm^3 · mol^{-1})	$\overline{V}_{L2}$ [b](cm^3 · mol^{-1})	$\overline{V}_{L1}^{exc}$ (cm^3 · mol^{-1})	$\overline{V}_{L2}^{exc}$ (cm^3 · mol^{-1})
282.5 ± 0.2	77.5 ± 0.4	100.0 − 0.2	123.4 ± 1.0	87.5 ± 0.7	−2.1 ± 1.0	−0.1 ± 0.7
298.0 ± 0.2	77.2 ± 0.4	100.0 − 0.2	125.6 ± 1.0	89.2 ± 0.7	−2.1 ± 1.0	−0.1 ± 0.7
317.1 ± 0.2	77.0 ± 0.4	100.0 − 0.2	128.3 ± 1.0	91.4 ± 0.7	−2.4 ± 1.0	−0.1 ± 0.7

[a]Denote one liquid phase and coexisting another liquid phase, respectively
[b]Observed molar volume
[c]Volumetric Properties

(2) Phase Equilibrium

Ref. [3]

x_1	P (kPa)	x_1	P (kPa)	x_1	P (kPa)	x_1	P (kPa)
353.15 K							
0.0096	1.47	0.3247	42.21	0.7622	100.96	0.9875	100.99
0.0226	3.07	0.3808	50.03	0.7959	101.05	0.9920	100.99
0.0338	4.45	0.4413	58.63	0.8220	100.99	0.9951	100.99
0.0447	5.79	0.4989	67.09	0.8301	101.05	0.9972	100.99
0.0553	7.12	0.5576	75.80	0.8709	101.00	0.9982	100.99
0.0876	11.04	0.6150	84.20	0.9099	100.99	0.9991	100.99
0.1241	15.62	0.6669	91.34	0.9428	101.00	0.9996	100.98
0.1684	21.36	0.7038	95.80	0.9614	100.99	1.0000	100.98
0.2167	27.63	0.7126	96.82	0.9726	101.00		
0.2695	34.66	0.7563	100.71	0.9815	101.00		

(3) Excess Enthalpy

Ref. [3]

x_1	H^E (J · mol^{-1})	x_1	H^E (J · mol^{-1})	x_1	H^E (J · mol^{-1})	x_1	H^E (J · mol^{-1})
323.15 K, 1.342 MPa							
0.0354	−36.0	0.1323	−151	0.4201	−537	0.7798	−572
0.0692	−72.0	0.2435	−291	0.6348	−730	0.9530	−119

(4) Partial Molar Excess Enthalpies at Infinite Dilution

Ref. [4]

Solute (1)	$H_{12}^{E,\infty}$/kJ · mol^{-1}
Benzene	−0.659

Mixture 10: C_6F_6 (1) + [emim][Tf_2N] (2)

(1) Phase Equilibrium

Ref. [42]

T (K)	x_1^{L1a} (mol %)	x_1^{L2a} (mol %)	$\overline{V}_{L1}{}^{b}$ ($cm^3 \cdot mol^{-1}$)	$\overline{V}_{L2}{}^{b}$ ($cm^3 \cdot mol^{-1}$)	$\overline{V}^{exc}_{L1}$ ($cm^3 \cdot mol^{-1}$)	$\overline{V}^{exc}_{L2}$ ($cm^3 \cdot mol^{-1}$)
282.5 ± 0.2	100.0 – 0.3	66.6 ± 1.4	113.4 ± 1.5	158.4 ± 2.5	0.0 ± 1.5	−2.3 ± 2.5
297.9 ± 0.2	100.0 – 0.3	66.2 ± 1.8	115.5 ± 1.4	160.4 ± 2.4	−0.1 ± 1.4	−3.3 ± 2.4
317.2 ± 0.2	100.0 – 0.5	64.4 ± 1.5	119.9 ± 1.5	165.8 ± 2.6	0.0 ± 1.5	−3.4 ± 2.6

[a]*Denote* one liquid phase and coexisting another liquid phase, respectively
[b]Observed molar volume
[c]Volumetric Properties

Mixture 11: CF_3CF_2H (1) + [emim][Tf_2N] (2)

(1) Solubility

Ref. [43]

T/K	P/MPa	$100x_1$	T/K	P/MPa	$100x_1$
283.1	0.0100	0.8	323.1	0.0102	0.4
283.1	0.1000	8.5	323.1	0.0998	3.0
283.1	0.1997	16.8	323.1	0.2000	5.9
283.1	0.2977	25.0	323.1	0.2998	8.8
283.1	0.3997	33.1	323.1	0.3996	11.7
283.1	0.4999	41.3	323.1	0.5002	15.2
283.1	0.5997	49.6	323.1	0.5997	17.6
283.1	0.6999	58.3	323.1	0.7999	22.8
283.1	0.7996	68.1	323.1	1.0001	28.5
298.1	0.0102	0.6	348.1	0.0102	0.4
298.1	0.0998	5.5	348.1	0.1000	1.8
298.1	0.1997	10.9	348.1	0.1997	3.4
298.1	0.2997	16.1	348.1	0.2998	5.0
298.2	0.3996	21.3	348.1	0.3998	7.4
298.2	0.4999	26.7	348.1	0.5001	9.7
298.2	0.5997	31.7	348.1	0.5996	10.5
298.1	0.8001	42.2	348.2	0.7998	14.0
298.1	0.9998	52.1	348.1	0.9997	17.1

(2) Phase Equilibrium

Ref. [43]

T/K	$100x_1'$	$100x_1$	V'/ $cm^3 \cdot mol^{-1}$	V/ $cm^3 \cdot mol^{-1}$	$V^{E\prime}$/ $cm^3 \cdot mol^{-1}$	V^E/ $cm^3 \cdot mol^{-1}$
282.7 ± 0.2	81.7 ± 1.0	99.9 ± 0.1	118.2 ± 1.6	94.2 ± 1.6	−5.4 ± 1.6	−0.1 ± 1.6
293.4 ± 0.2	76.5 ± 0.8	100.0 – 0.2	128.9 ± 1.4	98.6 ± 1.5	−7.0 ± 1.4	0.1 ± 1.5
303.7 ± 0.2	71.9 ± 0.6	100.0 – 0.2	139.5 ± 1.4	103.9 ± 1.5	−7.9 ± 1.4	0.0 ± 1.5
313.1 ± 0.2	68.2 ± 0.7	100.0 – 0.2	147.5 ± 1.3	110.3 ± 1.4	−10.6 ± 1.3	0.1 ± 1.4

Mixture 12: CF_3OCF_2H (1) + [emim][Tf_2N] (2)
(1) Solubility

Ref. [43]

T/K	P/MPa	$100x_1$	T/K	P/MPa	$100x_1$
283.1	0.0101	1.0	323.1	0.0102	0.4
283.1	0.0498	4.7	323.1	0.0497	1.6
283.1	0.1001	9.5	323.1	0.0999	3.1
283.1	0.1997	19.4	323.1	0.2000	6.5
283.1	0.2997	30.1	323.1	0.2999	9.1
283.1	0.4002	42.3	323.1	0.3997	12.5
283.1	0.4997	56.5	323.1	0.4999	15.7
298.1	0.0103	0.6	323.1	0.5998	18.7
298.1	0.0500	3.0	348.1	0.0102	0.2
298.1	0.0998	6.0	348.1	0.0500	0.9
298.1	0.1997	12.1	348.2	0.1002	1.7
298.1	0.3000	18.5	348.1	0.2001	4.0
298.1	0.3997	24.7	348.2	0.3001	6.0
298.1	0.4999	31.4	348.1	0.4000	7.4
298.1	0.6000	38.3			

(2) Phase Equilibrium

Ref. [43]

T/K	$100x_1'$	$100x_1$	V'/ $cm^3 \cdot mol^{-1}$	V/ $cm^3 \cdot mol^{-1}$	$V^{E\prime}$/ $cm^3 \cdot mol^{-1}$	V^E/ $cm^3 \cdot mol^{-1}$
282.9 ± 0.2	66.4 ± 1.5	99.5 ± 0.5	150.9 ± 5.6	101.6 ± 1.5	−2.0 ± 5.6	−0.3 ± 1.5
293.6 ± 0.2	62.2 ± 1.6	99.4 ± 0.5	159.8 ± 5.7	105.3 ± 1.8	−2.7 ± 5.7	−0.6 ± 1.8
303.4 ± 0.2	58.5 ± 1.7	99.4 ± 0.5	168.8 ± 6.0	109.4 ± 1.8	−2.4 ± 6.0	−0.6 ± 1.8
313.5 ± 0.2	55.0 ± 1.8	99.3 ± 0.5	176.5 ± 6.3	114.4 ± 1.9	−3.5 ± 6.3	−0.8 ± 1.9

Mixture 13: CF_3CH_3 (1) + [emim][Tf_2N] (2)
(1) Phase Equilibrium

Ref. [43]

T/K	$100x_1'$	$100x_1$	V'/ $cm^3 \cdot mol^{-1}$	V/ $cm^3 \cdot mol^{-1}$	$V^{E\prime}$/ $cm^3 \cdot mol^{-1}$	V^E/ $cm^3 \cdot mol^{-1}$
293.0 ± 0.2	51.9 ± 2.4	100 – 0.4	164.5 ± 7.0	88.0 ± 3.0	−4.9 ± 7.0	−0.5 ± 3.0
303.3 ± 0.2	50.9 ± 2.1	100 – 0.4	167.8 ± 6.0	92.5 ± 2.6	−6.2 ± 6.0	−0.1 ± 2.6
313.2 ± 0.2	50.2 ± 2.0	99.7 ± 0.3	171.1 ± 5.5	97.9 ± 2.5	−7.7 ± 5.5	−0.3 ± 2.5

Mixture 14: CF_3OCH_3 (1) + [emim][Tf_2N] (2)
(1) Phase Equilibrium

Ref. [43]

T/K	$100x_1'$	$100x_1$	V'/ $cm^3 \cdot mol^{-1}$	V/ $cm^3 \cdot mol^{-1}$	$V^{E\prime}$/ $cm^3 \cdot mol^{-1}$	V^E/ $cm^3 \cdot mol^{-1}$
283.2 ± 0.2	59.2 ± 1.5	99.9 ± 0.1	147.8 ± 7.0	87.6 ± 5.0	−8.4 ± 7.0	−0.4 ± 5.0
293.4 ± 0.2	57.6 ± 1.1	99.9 ± 0.1	152.5 ± 6.0	90.3 ± 4.5	−7.0 ± 6.0	−0.5 ± 4.5
303.6 ± 0.2	56.7 ± 1.0	99.9 ± 0.1	158.8 ± 5.0	92.9 ± 4.0	−6.1 ± 5.0	−0.3 ± 4.0
313.7 ± 0.2	56.1 ± 1.0	99.9 ± 0.1	162.5 ± 5.0	96.1 ± 3.0	−6.0 ± 5.0	−0.3 ± 3.0

Mixture 15: [C_2mim][BF_4] (1) + [C_2mim][Tf_2N] (2)
(1) Volumetric Properties

Ref. [29]
The mixture has appareed, its properties data is seen mixture 2 in IL 02003-02001.
Mixture 16: H_2O (1) + [C_2mim][Tf_2N] (2)
(1) Volumetric Properties

Ref. [44]

x_1	ρ/g · cm^{-3}	x_1	ρ/g · cm^{-3}
298.19 K		323.17 K	
0	1.5193	0	1.4947
0.0209	1.5183	0.0353	1.4931
0.0499	1.5169	0.0581	1.4920
0.0631	1.5162	0.0834	1.4906
0.1205	1.5130	0.1164	1.4887
0.1608	1.5107	0.1611	1.4862
0.2032	1.5081	0.2221	1.4824
0.3025	1.5013		
x_1	$V^E_m/cm^3 \cdot mol^{-1}$	x_1	$V^E_m/cm^3 \cdot mol^{-1}$
298.21 K		323.21 K	
0	0	0	0
0.0209	0.033	0.0353	0.057
0.0499	0.076	0.0581	0.098
0.0631	0.095	0.0834	0.145
0.1205	0.193	0.1164	0.221
0.1608	0.236	0.1611	0.278
0.2032	0.280	0.2221	0.349
0.3025	0.313		

(2) Solubility

Ref. [44]

T/K	x_1	T/K	x_1	T/K	x_1	T/K	x_1
285.15	0.2750	303.69	0.3214	338.38	0.4380	345.11	0.9980
292.62	0.2970	315.62	0.3572	345.75	0.4855	325.22	0.9985
298.25	0.3031	323.17	0.3921	353.18	0.5160	302.43	0.9991
299.76	0.3040	329.44	0.4060	358.09	0.5245	298.15	0.9992
303.22	0.3137	332.75	0.4150	361.17	0.5330	292.64	0.9995

Ref. [45]

T/K	$x_1 \pm \sigma^a$	$10^4\ (x_2 \pm \sigma^a)$
288.15	0.2755 ± 0.0005	7.66 ± 0.01
293.15	0.2869 ± 0.0008	8.03 ± 0.02
298.15	0.2982 ± 0.0009	8.38 ± 0.05
303.15	0.3119 ± 0.0007	8.75 ± 0.01
308.15	0.3280 ± 0.0004	9.33 ± 0.08
313.15	0.3446 ± 0.0002	9.85 ± 0.05
318.15	0.3603 ± 0.0004	10.5 ± 0.1

[a]Standard deviation

(3) Thermodynamic Properties

Ref. [45]

$(\Delta_{sol}H_m{}^0 \pm \sigma^a)$/ kJ · mol^{-1}	$(\Delta_{sol}G_m{}^0 \pm \sigma^a)$/ kJ · mol^{-1}	$(\Delta_{sol}S_m{}^0 \pm \sigma^a)$/ J · K^{-1} · mol^{-1}	$(\Delta_{svt}H_m{}^0 \pm \sigma^a)$/ kJ · mol^{-1}	$(\Delta_{svt}Gm^0 \pm \sigma^a)$/ kJ · mol^{-1}	$(\Delta_{svt}S_m{}^0 \pm \sigma^a)$/ J · K^{-1} · mol^{-1}	$(\Delta_{svt}H_m{}^* \pm \sigma^a)$/ kJ · mol^{-1}	$(\Delta_{svt}G_m{}^* \pm \sigma^a)$/ kJ · mol^{-1}	$(\Delta_{svt}S_m{}^* \pm \sigma^a)$/ J · K^{-1} · mol^{-1}
298.15 K								
7.1 ± 1.5	17.562 ± 0.016	−35.1 ± 5.1	−128.9 ± 6.2	−62.228 ± 0.016	−224 ± 21	−126.8 ± 6.2	−80.136 ± 0.016	−156 ± 21

[a]Standard deviation

(4) Phase Equilibrium

Ref. [18]

x_1	P (kPa)	x_1	P (kPa)	x_1	P (kPa)	x_1	P (kPa)
353.15 K							
0.029	3.19	0.439	40.72	0.843	47.36	0.945	47.39
0.060	6.45	0.520	45.50	0.870	47.36	0.967	47.38
0.089	9.54	0.596	47.33	0.892	47.37	0.980	47.38
0.145	15.30	0.660	47.35	0.910	47.37	0.989	47.37
0.193	20.15	0.716	47.35	0.924	47.38	0.994	47.38
0.270	27.38	0.768	47.36	0.935	47.38	1.000	47.37
0.351	34.27	0.810	47.36				

Ref. [46]

x_1	P (mbar)	x_1	P (mbar)	x_1	P (mbar)	x_1	P (mbar)
353.15 K							
0.0289	31.93	0.3513	342.89	0.7166	473.95	0.8919	474.09
0.0598	64.53	0.4388	407.48	0.7682	474.03	0.9095	474.13
0.0890	95.43	0.5202	455.29	0.8097	474.08	0.9238	474.15
0.1448	152.96	0.5972	473.67	0.8434	474.12	0.9354	474.14
0.1933	201.52	0.6609	473.82	0.8703	474.10	0.9450	474.22
0.2700	273.86						

(5) Activity Coefficient at Infinite Dilution and Partial Molar Excess Enthalpies at Infinite Dilution

Ref. [18]

Solute (1)	$\gamma^{\infty}{}_1$
353.15 K	
Water	2.32

Ref. [4]

Solute (1)	$H_{12}{}^{E,\infty}$/kJ · mol^{-1}
Water	9.661

Mixture 17: $CF_3CF_2CF_2OCH_3$ (1) + [emim][Tf_2N] (2)
(1) Phase Equilibrium

Ref. [47]

T/K	$100x_1'$	$100x_1$	V'/ cm^3 · mol^{-1}	V/ cm^3 · mol^{-1}	$V^{E\prime}$/ cm^3 · mol^{-1}	V^E/ cm^3 · mol^{-1}
283.0 ± 0.2	17.3 ± 1.3	99.7 ± 0.3	233.7 ± 1.1	138.6 ± 0.6	−1.3 ± 1.1	−0.3 ± 0.6
294.9 ± 0.2	18.0 ± 0.9	99.9 ± 0.1	234.8 ± 1.1	141.5 ± 0.3	−1.8 ± 1.1	−0.4 ± 0.3
303.4 ± 0.2	18.3 ± 1.0	99.8 ± 0.2	236.2 ± 0.4	144.0 ± 0.6	−1.7 ± 0.4	−0.5 ± 0.6

(continued)

T/K	$100x_1'$	$100x_1$	V'/ $cm^3 \cdot mol^{-1}$	V/ $cm^3 \cdot mol^{-1}$	$V^{E\prime}$/ $cm^3 \cdot mol^{-1}$	V^E/ $cm^3 \cdot mol^{-1}$
313.7 ± 0.2	19.0 ± 0.7	99.7 ± 0.3	237.2 ± 0.5	146.8 ± 0.6	−1.9 ± 0.5	−1.1 ± 0.6
323.7 ± 0.2	19.1 ± 0.7	99.7 ± 0.3	238.9 ± 0.4	150.0 ± 0.6	−2.1 ± 0.4	−1.0 ± 0.6
333.0 ± 0.2	19.3 ± 0.6	99.7 ± 0.3	239.8 ± 0.6	153.0 ± 0.6	−3.0 ± 0.6	−1.1 ± 0.6

V', observed molar volume lower phase; V, observed molar volume upper phase; $V^{E\prime}$, Volumetric Properties lower phase; V^E, Volumetric Properties upper phase

Mixture 18: $CF_3CF_2CF_2CF_2OCH_3$ (1) + [emim][Tf_2N] (2)
(1) Phase Equilibrium

Ref. [47]

T/K	$100x_1'$	$100x_1$	V'/ $cm^3 \cdot mol^{-1}$	V/ $cm^3 \cdot mol^{-1}$	$V^{E\prime}$/ $cm^3 \cdot mol^{-1}$	V^E/ $cm^3 \cdot mol^{-1}$
283.0 ± 0.2	10.4 ± 1.8	99.8 ± 0.2	244.2 ± 3.0	161.8 ± 2.0	−1.2 ± 3.0	0.3 ± 2.0
297.1 ± 0.2	10.4 ± 1.8	99.6 ± 0.4	246.7 ± 2.9	165.5 ± 1.7	−1.3 ± 2.9	0.0 ± 1.7
303.4 ± 0.2	10.8 ± 1.8	99.6 ± 0.4	247.0 ± 2.8	167.0 ± 1.7	−1.8 ± 2.8	−0.3 ± 1.7
313.7 ± 0.2	11.2 ± 1.2	99.5 ± 0.4	248.5 ± 2.8	169.8 ± 1.0	−1.9 ± 2.8	−0.5 ± 1.0
323.7 ± 0.2	11.5 ± 0.9	99.5 ± 0.4	250.2 ± 2.8	172.8 ± 1.0	−1.9 ± 2.8	−0.6 ± 1.0
333.1 ± 0.2	11.9 ± 0.9	99.6 ± 0.4	250.9 ± 2.2	176.2 ± 1.0	−2.2 ± 2.2	−0.4 ± 1.0

V', observed molar volume lower phase; V, observed molar volume upper phase; VE′, Volumetric Properties lower phase; VE, Volumetric Properties upper phase

Mixture 19: $CF_3CF_2CF_2CF_2OCH_2CH_3$ (1) + [emim][Tf_2N] (2)
(1) Phase Equilibrium

Ref. [47]

T/K	$100x_1'$	$100x_1$	V'/ $cm^3 \cdot mol^{-1}$	V/ $cm^3 \cdot mol^{-1}$	$V^{E\prime}$/ $cm^3 \cdot mol^{-1}$	V^E/ $cm^3 \cdot mol^{-1}$
283.1 ± 0.2	6.9 ± 1.0	99.5 ± 0.5	247.6 ± 1.0	181.1 ± 0.8	−2.5 ± 1.0	−0.4 ± 0.8
297.2 ± 0.2	7.3 ± 0.9	100.0 -0.3	250.3 ± 1.3	184.6 ± 0.9	−2.1 ± 1.3	−0.5 ± 0.9
303.4 ± 0.2	7.4 ± 0.9	99.6 ± 0.4	251.2 ± 1.1	186.7 ± 0.8	−2.2 ± 1.1	−0.6 ± 0.8
313.7 ± 0.2	7.5 ± 0.9	100.0 -0.3	253.0 ± 1.0	189.6 ± 0.7	−2.3 ± 1.0	−0.5 ± 0.7
323.8 ± 0.2	8.0 ± 0.8	99.7 ± 0.3	254.4 ± 0.9	193.0 ± 0.7	−2.4 ± 0.9	−0.6 ± 0.7
333.1 ± 0.2	8.1 ± 0.8	100.0 -0.3	256.0 ± 1.0	196.2 ± 0.7	−2.6 ± 1.0	−0.4 ± 0.7

V', observed molar volume lower phase; V, observed molar volume upper phase; VE′, Volumetric Properties lower phase; VE, Volumetric Properties upper phase

Mixture 20: $CF_3CF_2CF(OCH_3)CF(CF_3)_2$ (1) + [emim][Tf_2N] (2)
(1) Phase Equilibrium

Ref. [47]

T/K	$100x_1'$	$100x_1$	V'/ $cm^3 \cdot mol^{-1}$	V/ $cm^3 \cdot mol^{-1}$	$V^{E\prime}$/ $cm^3 \cdot mol^{-1}$	V^E/ $cm^3 \cdot mol^{-1}$
283.1 ± 0.2	99.9 ± 0.1	2.8 ± 0.7	206.6 ± 1.0	254.0 ± 1.5	−0.2 ± 1.0	−1.0 ± 1.5
296.6 ± 0.2	100.0–0.2	2.9 ± 0.6	210.7 ± 1.2	255.1 ± 1.3	−0.1 ± 1.2	−1.1 ± 1.3
303.5 ± 0.2	100.0–0.2	2.9 ± 0.7	212.7 ± 1.1	257.1 ± 1.1	−0.2 ± 1.1	−0.8 ± 1.1

(continued)

T/K	$100x_1'$	$100x_1$	V'/ $cm^3 \cdot mol^{-1}$	V/ $cm^3 \cdot mol^{-1}$	$V^{E\prime}$/ $cm^3 \cdot mol^{-1}$	V^E/ $cm^3 \cdot mol^{-1}$
313.1 ± 0.2	99.9 ± 0.1	3.8 ± 0.8	215.5 ± 1.0	258.3 ± 0.9	−0.5 ± 1.0	−0.8 ± 0.9
322.8 ± 0.2	99.9 ± 0.1	4.8 ± 0.8	219.0 ± 1.0	258.5 ± 1.3	−0.1 ± 1.0	−1.6 ± 1.3
333.3 ± 0.2	99.9 ± 0.1	3.6 ± 0.7	222.0 ± 0.9	261.7 ± 1.1	−0.5 ± 0.9	−1.4 ± 1.1

V', observed molar volume lower phase; V, observed molar volume upper phase; $V^{E\prime}$, Volumetric Properties lower phase; V^E, Volumetric Properties upper phase

Mixture 21: $CF_3CF_2CF_2CF(OCH_2CH_3)CF(CF_3)_2$ (1) + [emim][Tf_2N] (2)
(1) Phase Equilibrium

Ref. [47]

T/K	$100x_1'$	$100x_1$	V'/ $cm^3 \cdot mol^{-1}$	V/ $cm^3 \cdot mol^{-1}$	$V^{E\prime}$/ $cm^3 \cdot mol^{-1}$	V^E/ $cm^3 \cdot mol^{-1}$
283.3 ± 0.2	99.5 ± 0.5	1.3 ± 0.8	251.3 ± 2.2	255.1 ± 2.0	−0.1 ± 2.2	−0.2 ± 2.0
297.3 ± 0.2	99.5 ± 0.5	2.1 ± 1.0	255.4 ± 2.0	257.4 ± 2.0	−0.3 ± 2.0	−0.2 ± 2.0
303.5 ± 0.2	99.8 ± 0.2	1.9 ± 0.8	257.2 ± 1.6	258.9 ± 1.8	−0.6 ± 1.6	−0.4 ± 1.8
313.5 ± 0.2	99.7 ± 0.3	1.7 ± 1.1	260.6 ± 1.8	260.2 ± 1.8	−0.7 ± 1.8	−0.4 ± 1.8
323.8 ± 0.2	99.9 ± 0.1	1.4 ± 0.8	263.9 ± 1.7	262.2 ± 1.9	−0.8 ± 1.7	−0.2 ± 1.9
333.1 ± 0.2	99.5 ± 0.5	4.3 ± 0.7	267.3 ± 2.0	264.2 ± 1.3	−0.8 ± 2.0	−0.1 ± 1.3

V', observed molar volume lower phase; V, observed molar volume upper phase; $V^{E\prime}$, Volumetric Properties lower phase; V^E, Volumetric Properties upper phase

Mixture 22: CO_2 (1) + [emim][Tf_2N] (2)
(1) Bubble Point

Ref. [48]

x_1	T(K)	P(MPa)	T (K)	P (MPa)	T (K)	P (MPa)
0.123	312.17	0.626	361.51	1.273	410.91	2.103
0.123	322.00	0.742	371.39	1.428	420.81	2.247
0.123	331.87	0.852	381.26	1.583	430.71	2.412
0.123	341.79	1.002	391.12	1.748	440.59	2.572
0.123	351.66	1.137	401.02	1.923	450.49	2.743
0.212	312.10	1.212	361.51	2.493	410.89	4.053
0.212	321.95	1.442	371.35	2.793	420.80	4.383
0.212	331.84	1.697	381.24	3.088	430.71	4.723
0.212	341.74	1.933	391.11	3.423	440.63	5.078
0.212	351.64	2.198	401.00	3.733	450.48	5.418
0.303	312.14	1.902	361.54	3.952	410.85	6.538
0.303	322.00	2.267	371.36	4.437	420.80	7.078
0.303	331.85	2.622	381.25	4.937	430.69	7.588
0.303	341.76	3.052	391.09	5.463	440.58	8.123
0.303	351.64	3.497	400.97	6.008	450.49	8.648
0.392	312.14	2.738	361.55	5.908	410.85	9.863
0.392	322.00	3.278	371.43	6.653	420.76	10.683
0.392	331.88	3.848	381.22	7.418	430.68	11.523

(continued)

x_1	T(K)	P(MPa)	T (K)	P (MPa)	T (K)	P (MPa)
0.392	341.77	4.518	391.08	8.208	440.55	12.338
0.392	351.65	5.188	400.96	9.028	450.46	13.153
0.479	312.13	3.786	351.69	7.321	391.11	11.852
0.479	322.00	4.541	361.55	8.386	401.00	13.062
0.479	331.89	5.401	371.40	9.501	410.90	14.307
0.479	341.77	6.321	381.26	10.681		
0.519	312.11	4.399	341.74	7.474	371.34	11.409
0.519	321.98	5.309	351.65	8.715	381.23	12.854
0.519	331.86	6.334	361.52	10.035	391.11	14.329
0.570	312.16	5.244	341.78	9.180	370.40	14.180
0.570	322.00	6.394	351.67	10.790		
0.570	331.89	7.699	361.52	12.550		
0.593	312.14	5.789	331.88	8.634		
0.593	317.04	6.418	336.88	9.480	351.63	12.305
0.593	322.01	7.103	341.76	10.375	356.57	13.340
0.593	326.93	7.839	346.69	11.320	361.51	14.402

Ref. [48]

	Pressure (MPa)							
T (K)	$x_1 = 0.123$	$x_1 = 0.212$	$x_1 = 0.303$	$x_1 = 0.392$	$x_1 = 0.479$	$x_1 = 0.519$	$x_1 = 0.570$	$x_1 = 0.593$
313.15	0.639	1.239	1.944	2.790	3.861	4.490	5.354	5.916
323.15	0.752	1.469	2.295	3.340	4.637	5.421	6.529	7.268
333.15	0.876	1.716	2.684	3.944	5.506	6.478	7.883	8.848
343.15	1.010	1.978	3.107	4.597	6.457	7.647	9.397	10.637
353.15	1.153	2.254	3.558	5.292	7.482	8.911	11.053	12.617
363.15	1.303	2.543	4.036	6.023	8.569	10.258	12.832	14.770
373.15	1.460	2.842	4.534	6.787	9.708	11.671	14.714	
383.15	1.621	3.153	5.049	7.577	10.890	13.136		
393.15	1.787	3.472	5.577	8.387	12.103	14.639		
403.15	1.954	3.799	6.113	9.212	13.338			
413.15	2.122	4.132	6.654	10.047	14.584			
423.15	2.291	4.471	7.195	10.886				
433.15	2.457	4.815	7.733	11.724				
443.15	2.621	5.161	8.263	12.555				
453.15	2.780	5.509	8.780	13.374				

Ref. [49]

x_1	T/K	p/MPa	x_1	T/K	p/MPa	x_1	T/K	p/MPa	x_1	T/K	p/MPa	x_1	T/K	p/MPa
0.221	293.65	0.620	0.327	293.24	1.130	0.418	293.50	1.688	0.503	293.50	2.398	0.560	292.41	2.879
0.221	298.52	0.725	0.327	298.50	1.295	0.418	298.55	1.945	0.503	298.29	2.706	0.560	298.24	3.351
0.221	303.25	0.837	0.327	303.41	1.462	0.418	303.48	2.185	0.503	303.41	3.076	0.560	303.11	3.778
0.221	313.59	1.083	0.327	313.32	1.868	0.418	313.32	2.745	0.503	313.24	3.845	0.560	313.22	4.780
0.221	323.39	1.345	0.327	323.13	2.292	0.418	323.10	3.365	0.503	323.27	4.725	0.560	323.07	5.883
0.221	333.15	1.645	0.327	333.17	2.758	0.418	333.23	4.054	0.503	333.16	5.708	0.560	333.16	7.173
0.221	343.25	1.957	0.327	343.27	3.287	0.418	343.30	4.836	0.503	343.11	6.769	0.560	343.19	8.569
0.221	353.10	2.276	0.327	353.16	3.850	0.418	353.02	5.565	0.503	353.18	7.912	0.560	353.14	10.115
0.221	363.23	2.680	0.327	363.19	4.408	0.418	363.17	6.436	0.503	363.35	9.128	0.560	363.34	11.821
0.606	292.60	3.387	0.650	292.16	3.825	0.700	292.82	4.736	0.750	293.44	16.076			
0.606	298.29	3.942	0.650	297.70	4.485	0.700	298.15	5.520	0.750	298.09	19.048			
0.606	303.46	4.506	0.650	303.54	5.230	0.700	303.30	6.535	0.750	303.26	22.130			
0.606	313.30	5.686	0.650	313.35	6.715	0.700	313.24	10.180	0.750	313.19	28.187			
0.606	323.11	7.103	0.650	323.10	8.550	0.700	323.09	14.450	0.750	323.14	33.787			
0.606	332.88	8.695	0.650	333.16	10.905	0.700	333.04	18.885	0.750	333.08	38.898			
0.606	343.13	10.649	0.650	343.08	13.565	0.700	343.15	23.085	0.750	343.29	43.565			
0.606	353.07	12.650	0.650	353.05	16.435	0.700	353.22	27.066	0.750	353.08	47.850			
0.606	363.55	14.996	0.650	363.22	19.450	0.700	363.22	30.858						

(2) Thermodynamic Properties of Solution

Ref. [49]

T/K	$\Delta_{sol}H$/(J · mol^{-1})	$T\Delta_{sol}S$/(J · mol^{-1})
298.15	−14279.3	−14162.1
303.15	−14137.8	−14096.5
313.15	−13868.3	−13872.5
323.15	−13615.5	−13669.2
333.15	−13377.8	−13425.9
343.15	−13154.0	−13177.0
353.15	−12942.9	−12890.0
363.15	−12743.4	−12601.3

Mixture 23: C_6H_{14} (1) + [EMIM][BTI] (2)
(1) Excess Molar Enthalpy

Ref. [50]

x_1	H^E (J · mol^{-1})	x_1	H^E (J · mol^{-1})	x_1	H^E (J · mol^{-1})	x_1	H^E (J · mol^{-1})
353.15 K							
0.0239	124.4	0.0923	329.4	0.2162	288.0	0.6588	120.7
0.0472	223.8	0.1354	316.4	0.3916	218.0	0.8528	52.2
0.0700	276.3						

(2) Partial Molar Excess Enthalpies at Infinite Dilution

Ref. [4]

Solute (1)	$H_{12}^{E,\infty}$/kJ · mol^{-1}
n-Hexane	5.744

(3) Phase Equilibrium

Ref. [3]

x_1	P (kPa)	x_1	P (kPa)	x_1	P (kPa)	x_1	P (kPa)
353.15 K							
0.0079	18.68	0.1380	142.22	0.6169	142.37	0.9597	142.31
0.0155	35.79	0.1757	142.31	0.6711	142.38	0.9726	142.32
0.0222	50.62	0.2197	142.32	0.6842	142.26	0.9814	142.32
0.0284	64.74	0.2676	142.30	0.7210	142.37	0.9880	142.32
0.0348	79.11	0.3229	142.29	0.7572	142.27	0.9927	142.32
0.0409	91.86	0.3783	142.34	0.7652	142.38	0.9957	142.32
0.0470	104.59	0.4388	142.36	0.8198	142.27	0.9973	142.31
0.0600	129.83	0.5014	142.36	0.8719	142.28	0.9986	142.31
0.0796	141.99	0.5620	142.38	0.9173	142.30	0.9993	142.30
0.1054	142.17	0.6166	142.25	0.9437	142.30	1.0000	142.30

Mixture 24: C_2H_5OH (1) + [EMIM][BTI] (2)
(1) Excess Molar Enthalpy

Ref. [50]

x_1	H^E ($J \cdot mol^{-1}$)	x_1	H^E ($J \cdot mol^{-1}$)	x_1	H^E ($J \cdot mol^{-1}$)	x_1	H^E ($J \cdot mol^{-1}$)
323.15 K							
0.0529	380.3	0.1885	1243	0.5246	2451	0.8436	1608
0.1017	683.6	0.3291	1937	0.7259	2196	0.9687	439.0

(2) Partial Molar Excess Enthalpies at Infinite Dilution

Ref. [4]

Solute (1)	${H_{12}}^{E,\infty}/kJ \cdot mol^{-1}$
Ethanol	7.268

(3) Phase Equilibrium

Ref. [4]

x_1	P/kPa	x_1	P/kPa	x_1	P/kPa	x_1	P/kPa
353.11 K							
0.033	5.30	0.495	66.30	0.856	97.43	0.968	105.28
0.097	14.54	0.577	74.83	0.860	96.88	0.980	106.14
0.153	22.72	0.654	81.84	0.860	96.87	0.990	107.00
0.203	29.72	0.719	87.19	0.873	98.50	0.997	107.61
0.244	35.42	0.771	91.03	0.902	101.07	0.998	107.74
0.311	44.41	0.813	93.88	0.930	102.78	0.999	107.80
0.406	56.24	0.841	96.53	0.952	104.19	1.000	107.92

Mixture 25: 1-C_3H_7OH (1) + [EMIM][NTf_2] (2)
(1) Excess Molar Enthalpy

Ref. [50]

x_1	H^E ($J \cdot mol^{-1}$)	x_1	H^E ($J \cdot mol^{-1}$)	x_1	H^E ($J \cdot mol^{-1}$)	x_1	H^E ($J \cdot mol^{-1}$)
323.15 K							
0.0418	321.0	0.1536	1131	0.4629	2619	0.8082	2040
0.0812	625.7	0.2770	1889	0.6741	2608	0.9602	632.3

(2) Cloud Point

Ref. [51]

w_2	T/K	w_2	T/K
0.1070	279.3	0.4963	295.3
0.1622	286.5	0.5724	295.1
0.1983	289.4	0.6446	294.2

(*continued*)

w_2	T/K	w_2	T/K
0.3157	294.0	0.7210	291.4
0.4319	295.2	0.7686	287.8

wH_2O_1 = (205 ± 10) ppm; wH_2O_2 = (100 ± 25) ppm; precision of temperature ±0.5 K.

Ref. [51]

w_2	T/K	w_2	T/K
$w_{H2O,2}$ = (35 ± 10) ppm			
0.1731	287.3	0.5183	295.3
0.3219	294.2	0.6004	294.9
$w_{H2O,2}$ = (235 ± 25) ppm			
0.5889	294.9		

$w_{H2O,1}$ = (205 ± 10) ppm; precision of temperature ± 0.5 K.

Ref. [51]

$10^{-3}w_{H2O}$/ppm	T/K	$10^{-3}w_{H2O}$/ppm	T/K
w_2 = 0.7210			
0.18	291.4	2.92	288.3
1.06	290.3	6.27	284.6
1.97	289.3	7.94	282.8

Ref. [52]

$100w_2$	T/K	$100w_2$	T/K	$100w_2$	T/K	$100w_2$	T/K
12.78	281.0	37.71	293.3	60.49	293.0	76.16	286.6
16.36	285.4	38.93	293.8	64.38	292.7	77.71	285.1
20.46	288.5	44.85	294.1	68.64	291.5	81.51	280.9
27.10	291.8	53.38	294.1	73.31	289.0		

(3) Activity Coefficient at Infinite Dilution, Partial Molar Excess Enthalpy at Infinite Dilution and Critical Temperatures of Phase Separation

Ref. [4]

Solute (1)	$H_{12}^{E,\infty}$/kJ · mol^{-1}
1-Propanol	6.926

Ref. [53]

	γ_1^{∞}				T_c	$\gamma_1^{s,\infty\,a}$	$H_1^{E,\infty}$(J/mol)
	313 K	323 K	333 K	343 K			
1-Propanol	2.1	1.9	1.8	1.6	294	2.0	+7946

[a] $\gamma_1^{s,\infty a}$ is the estimated value of the activity coefficient, expressed at infinite dilution, above which phase splitting occurs, as dictated by eq $\ln(\gamma_2^{\infty}) \geq -\ln r + \left(1 - \frac{1}{r}\right) + \frac{1}{2}\left(1 + \frac{1}{r^{1/2}}\right)^2$

Mixture 26: CH_3CN (1) + $[C_2mim][NTf_2]$ (2)
(1) Volumetric Properties

Ref. [54]

x_1	ρ/ (kg · m^{-3})	V^E/(10^{-6} m^3 · mol^{-1})	x_1	ρ/ (kg · m^{-3})	V^E/(10^{-6} m^3 · mol^{-1})
T = 303.02 K			T = 322.41 K		
0.0000	1513.7	0.000	0.0000	1494.2	0.000
0.1024	1498.2	−0.242	0.1196	1476.3	−0.432
0.2554	1467.9	−0.444	0.2243	1456.5	−0.691
0.4878	1397.3	−0.597	0.4481	1394.2	−0.961
0.5514	1369.2	−0.578	0.5104	1369.7	−0.985
0.7615	1224.6	−0.453	0.5850	1333.9	−0.952
0.9441	939.1	−0.139	0.8367	1116.5	−0.572
1.0000	771.28	0.000	0.9354	937.95	−0.230
			1.0000	750.00	0.000

(2) Vapor Pressure

Ref. [54]

x_1	p_{solv}^{sat} (10^2 Pa)	x_1	p_{solv}^{sat} (10^2 Pa)	x_1	p_{solv}^{sat} (10^2 Pa)	x_1	p_{solv}^{sat} (10^2 Pa)	x_1	p_{solv}^{sat} (10^2 Pa)
T = 293.39 K		T = 303.48 K		T = 313.48 K		T = 323.46 K		T = 333.95 K	
0.0000	0	0.0000	0	0.000	0	0.0000	0	0.0000	0
0.4603	29.6	0.4600	45.4	0.4596	69.4	0.4590	103.7	0.4582	149.3
0.5338	40.4	0.5335	61.2	0.5330	90.8	0.5324	127.7	0.6673	258.7
0.6690	49.0	0.6688	78.4	0.6684	117.8	0.6680	172.3	0.7783	339.7
0.7792	67.2	0.7791	104.0	0.7789	157.9	0.7786	234.6	1.0000	505.6
1.0000	94.7	1.0000	150.5	1.0000	230.6	1.0000	342.8		

Mixture 27: CH_2F_2 (1) + [emim][BMeI] (2)
(1) Solubility and Diffusivity

Ref. [55]

T/K	P/MPa	100 x_1	$10^{11}D$/m^2 · s^{-1}	T/K	P/MPa	100 x_1	$10^{11}D$/m^2 · s^{-1}
283.15	0.0102	1.4	4.1	323.15	0.1003	5.5	15
283.15	0.1001	13.6	5.1	323.15	0.2497	13.3	18
283.15	0.2503	30.6	8.2	323.15	0.4001	20.0	20
283.15	0.3993	44.8	13	323.15	0.5495	26.4	
283.15	0.5493	57.0	17	323.15	0.7003	32.0	23
283.15	0.7004	67.2	19	323.15	0.8504	36.8	30
283.15	0.8494	78.6		323.15	1.0004	41.7	
298.15	0.0096	1.0	5.9	348.05	0.0103	0.2	
298.15	0.0998	9.5	9.1	348.05	0.0998	3.4	30
298.15	0.2502	21.8	11	348.05	0.2497	8.4	32
298.15	0.3994	32.5	14	348.05	0.3995	13.1	43

(*continued*)

T/K	P/MPa	100 x_1	$10^{11}D/\mathrm{m}^2 \cdot \mathrm{s}^{-1}$	T/K	P/MPa	100 x_1	$10^{11}D/\mathrm{m}^2 \cdot \mathrm{s}^{-1}$
298.15	0.5494	41.7	18	348.05	0.5503	17.5	41
298.15	0.6995	49.9	20	348.05	0.6992	21.6	
298.15	0.8495	57.0	23	348.05	0.8504	25.4	48
298.15	0.9997	64.3		348.05	1.0005	28.8	
323.15	0.0102	0.5	15				

Mixture 28: $CH_3OOC(CH_2)_4COOCH_3$ (1) + [EMIM][NTf$_2$] (2)
(1) Phase Equilibrium

Ref. [56]

T/K	x_1	p_1^{exptl}/Pa	p_1^{NRTL}/Pa	γ_1^{NRTL}
298.65	0.000	0.00	0.00	0.577
	0.044	0.05	0.11	0.608
	0.138	0.36	0.39	0.677
	0.224	0.74	0.69	0.742
	0.311	1.13	1.04	0.809
	0.457	1.50	1.73	0.912
	0.548	2.30	2.19	0.965
	0.690	2.97	2.91	1.018
	0.811	3.43	3.45	1.026
	0.907	3.81	3.81	1.012
	1.000	4.23	4.15	1.000
305.55	0.000	0.00	0.00	0.562
	0.044	0.09	0.19	0.585
	0.137	0.60	0.65	0.637
	0.223	1.18	1.13	0.686
	0.308	1.84	1.67	0.736
	0.454	2.66	2.75	0.821
	0.543	3.43	3.48	0.869
	0.690	4.58	4.76	0.936
	0.811	5.72	5.83	0.976
	0.907	6.93	6.64	0.994
	1.000	6.56	7.37	1.000
308.55	0.000	0.00	0.00	0.607
	0.044	0.20	0.32	0.627
	0.136	1.08	1.06	0.670
	0.220	1.78	1.81	0.710
	0.303	2.84	2.64	0.751
	0.450	3.92	4.29	0.824
	0.538	5.37	5.40	0.867
	0.689	7.10	7.43	0.932
	0.810	9.68	9.11	0.972
	0.906	10.3	10.41	0.993
	1.000	10.	11.58	1.000

(continued)

T/K	x_1	p_1^{exptl}/Pa	p_1^{NRTL}/Pa	γ_1^{NRTL}
313.55	0.000	0.00	0.00	0.735
	0.043	0.38	0.55	0.751
	0.135	1.75	1.79	0.785
	0.216	3.07	2.98	0.815
	0.296	4.81	4.22	0.844
	0.445	5.83	6.74	0.895
	0.532	7.93	8.31	0.922
	0.688	11.09	11.21	0.963
	0.809	14.57	13.49	0.986
	0.905	15.12	15.26	0.996
	1.000	15.08	16.93	1.000
318.65	0.000	0.00	0.00	0.743
	0.043	0.69	0.84	0.761
	0.133	2.71	2.75	0.799
	0.211	4.30	4.53	0.831
	0.282	7.19	6.26	0.859
	0.437	9.21	10.36	0.916
	0.524	13.32	12.77	0.942
	0.684	16.97	17.34	0.979
	0.807	22.00	20.78	0.995
	0.904	22.64	23.37	0.999
	1.000	22.38	25.87	1.000

Mixture 29: $C_6H_5COOC_2H_5$ (1) + [EMIM][NTf_2] (2)
(1) Phase Equilibrium

Ref. [56]

T/K	x_1	p_1^{exptl}/Pa	p_1^{NRTL}/Pa	γ_1^{NRTL}
298.65	0.000	0.00	0.00	3.675
	0.112	7.34	7.65	2.712
	0.245	15.16	12.72	2.069
	0.339	12.98	15.12	1.776
	0.508	17.34	18.19	1.426
	0.588	20.23	19.32	1.308
	0.720	21.67	20.92	1.157
	0.852	22.66	22.50	1.052
	0.952	22.85	24.06	1.006
	0.979	25.60	24.62	1.001
	1.000	24.84	25.11	1.000
305.55	0.000	0.00	0.00	2.969
	0.102	8.85	8.92	2.415
	0.202	14.71	15.07	2.055
	0.212	16.26	15.59	2.026
	0.310	20.17	20.08	1.784

(*continued*)

T/K	x_1	p_1^{exptl}/Pa	p_1^{NRTL}/Pa	γ_1^{NRTL}
	0.484	25.08	26.14	1.488
	0.581	29.23	28.76	1.364
	0.716	32.25	31.58	1.216
	0.850	33.21	33.55	1.087
	0.952	34.32	35.02	1.014
	0.979	35.57	35.64	1.003
	1.000	37.07	36.30	1.000
308.55	0.000	0.00	0.00	2.877
	0.091	11.18	11.06	2.401
	0.173	18.58	18.35	2.104
	0.182	18.82	19.11	2.074
	0.271	24.52	25.28	1.845
	0.452	35.22	34.92	1.530
	0.573	40.95	39.89	1.380
	0.711	44.63	44.27	1.233
	0.848	46.45	47.11	1.100
	0.951	48.72	48.84	1.017
	0.958	45.15	49.01	1.013
	1.000	54.62	50.50	1.000
313.55	0.000	0.00	0.00	2.789
	0.080	14.92	15.21	2.414
	0.135	25.77	23.35	2.208
	0.151	23.62	25.46	2.154
	0.230	34.32	34.57	1.917
	0.417	49.36	49.52	1.515
	0.563	58.55	57.42	1.301
	0.706	62.06	63.55	1.149
	0.846	70.65	69.40	1.047
	0.950	74.86	74.85	1.006
	0.958	73.68	75.32	1.004
	1.000	79.48	78.33	1.000
318.65	0.000	0.00	0.00	3.237
	0.069	21.43	21.62	2.740
	0.183	46.99	45.88	2.197
	0.374	69.80	71.40	1.675
	0.552	87.27	87.18	1.384
	0.700	97.55	96.55	1.209
	0.844	103.56	103.57	1.076
	0.949	105.16	109.44	1.011
	0.957	113.83	110.00	1.008
	1.000	114.31	114.04	1.000

Mixture 30: $C_6H_5CH_2NH_2$ (1) + [EMIM][NTf_2] (2)
(1) Phase Equilibrium

Ref. [56]

T/K	x_1	p_1^{exptl}/Pa	p_1^{NRTL}/Pa	γ_1^{NRTL}
298.65	0.000	0.00	0.00	1.687
	0.124	18.04	17.91	1.527
	0.193	27.34	26.64	1.455
	0.300	37.51	38.78	1.365
	0.362	44.45	45.33	1.322
	0.440	52.88	53.26	1.277
	0.570	66.60	65.81	1.218
	0.656	73.16	73.89	1.189
	0.741	83.80	81.90	1.166
	0.876	94.67	94.01	1.133
	1.000	94.34	94.77	1.000
305.65	0.000	0.00	0.00	2.765
	0.109	31.78	32.08	2.213
	0.185	42.49	48.10	1.953
	0.288	66.83	65.33	1.704
	0.347	75.91	73.75	1.600
	0.429	80.37	84.78	1.484
	0.561	101.84	101.14	1.357
	0.640	111.93	110.81	1.301
	0.736	121.83	122.13	1.248
	0.873	135.21	136.33	1.174
	1.000	133.64	132.99	1.000
308.65	0.000	0.00	0.00	4.233
	0.080	48.33	48.42	3.231
	0.172	70.19	81.16	2.540
	0.271	104.12	104.96	2.076
	0.326	116.74	115.40	1.900
	0.415	131.83	130.57	1.690
	0.548	152.37	151.77	1.486
	0.619	163.85	162.82	1.412
	0.728	174.76	179.49	1.323
	0.871	196.29	197.05	1.215
	1.000	187.91	186.25	1.000
313.65	0.000	0.00	0.00	7.980
	0.042	66.41	66.41	6.043
	0.153	107.84	141.44	3.561
	0.249	169.11	170.07	2.637
	0.296	177.76	180.80	2.356
	0.395	215.45	200.94	1.966
	0.532	225.75	227.78	1.653

(*continued*)

T/K	x_1	p_1^{exptl}/Pa	p_1^{NRTL}/Pa	γ_1^{NRTL}
	0.591	232.86	238.97	1.561
	0.718	248.75	260.42	1.401
	0.868	277.43	271.76	1.208
	1.000	262.37	259.10	1.000
318.65	0.000	0.00	0.00	6.624
	0.132	177.43	179.15	3.736
	0.219	224.85	229.32	2.892
	0.262	260.81	247.57	2.609
	0.368	283.34	283.91	2.134
	0.511	329.00	324.80	1.756
	0.552	326.21	335.40	1.679
	0.704	356.80	368.27	1.445
	0.865	386.27	378.28	1.208
	1.000	363.86	362.08	1.000

Mixture 31: [EMIM][BTI] (1) + [BMIM][BTI] (2)
(1) Volumetric Properties

Ref. [4]

T/K	ρ/g · cm^{-3}	T/K	ρ/g · cm^{-3}
(1):(2) = 50:50 weight percent			
293.15	1.483	328.15	1.449
298.15	1.479	333.15	1.445
303.15	1.474	338.15	1.440
308.15	1.469	343.15	1.435
313.15	1.464	348.15	1.430
318.15	1.459	353.15	1.425
323.15	1.454	358.15	1.421

Mixture 32: CH_3OH (1) + [EMIM][BTI] (2)
(1) Phase Equilibrium

Ref. [4]

x_1	P/kPa	x_1	P/kPa	x_1	P/kPa	x_1	P/kPa
353.10 K							
0.103	18.61	0.475	91.55	0.772	144.45	0.903	164.53
0.191	36.05	0.565	108.35	0.819	151.87	0.903	164.52
0.264	49.98	0.643	122.64	0.855	157.34	0.903	164.52
0.323	61.46	0.712	134.62	0.882	161.45	1.000	179.95
0.373	71.29						

(2) Partial Molar Excess Enthalpies at Infinite Dilution

Ref. [4]

Solute (1)	$H_{12}^{E,\infty}$/kJ · mol^{-1}
Methanol	6.290

Mixture 33: $CH_2(CH_2)_4CHOH$ (1) + $[C_2MIm][NTf_2]$ (2)
(1) Cloud Point

Ref. [51]

w_2	T/K	w_2	T/K
First run		Second run	
0.1274	313.5	0.0711	306.1
0.1796	316.4	0.0877	308.9
0.2352	318.0	0.2654	318.4
0.3395	319.2	0.4727	318.9
0.4201	319.2	0.5718	317.6
0.5324	318.4	0.8033	300.6
0.6428	315.2		
0.7048	311.5		
0.7499	307.5		

Mixture 34: CH_3COCH_3 (1) + $[EMIM][(CF_3SO_2)_2N]$ (2)
(1) Phase Equilibrium

Ref. [18]

x_1	P (kPa)	x_1	P (kPa)	x_1	P (kPa)	x_1	P (kPa)
353.15 K							
0.013	1.14	0.339	39.81	0.726	127.22	0.938	196.92
0.038	3.48	0.402	50.26	0.763	138.75	0.967	205.80
0.054	4.83	0.465	61.86	0.795	149.16	0.984	210.74
0.068	6.07	0.525	74.28	0.822	158.37	0.993	213.25
0.095	8.51	0.584	87.87	0.846	166.29	0.997	214.40
0.152	14.88	0.637	101.46	0.865	173.07	0.999	214.94
0.208	21.66	0.685	114.73	0.897	183.74	1.000	215.16
0.273	30.11						

Ref. [46]

x_1	P (mbar)	x_1	P (mbar)	x_1	P (mbar)	x_1	P (mbar)
353.15 K							
0.0133	11.39	0.3393	398.11	0.7264	1272.2	0.9376	1969.2
0.0383	34.85	0.4022	502.63	0.7631	1387.5	0.9671	2058.0
0.0539	48.25	0.4647	618.62	0.7948	1491.6	0.9844	2107.4
0.0683	60.73	0.5249	742.81	0.8222	1583.7	0.9933	2132.5
0.0949	85.09	0.5839	878.70	0.8455	1662.9	0.9972	2144.0
0.1521	148.84	0.6369	1014.6	0.8653	1730.7	0.9990	2149.4
0.2081	216.60	0.6846	1147.3	0.8967	1837.5	1.0000	2151.6
0.2735	301.15						

(2) Activity Coefficient at Infinite Dilution and Partial Molar Excess Enthalpies at Infinite Dilution

Ref. [18]

Solute (1)	$\gamma^{\infty}{}_1$
353.15 K	
Acetone	0.438

Ref. [4]

Solute (1)	$H_{12}{}^{E,\infty}$/kJ · mol^{-1}
Acetone	−1.243

Mixture 35: 2-CH_3CHCH_3OH (1) + [EMIM][BTI] (2)
(1) Phase Equilibrium

Ref. [18]

x_1	P (kPa)	x_1	P (kPa)	x_1	P (kPa)	x_1	P (kPa)
353.15 K							
0.012	1.73	0.295	38.61	0.740	82.08	0.960	90.64
0.024	3.21	0.354	45.94	0.752	82.82	0.975	91.14
0.036	4.84	0.412	52.86	0.782	84.38	0.986	91.52
0.049	6.50	0.475	59.81	0.802	85.29	0.992	91.86
0.061	8.08	0.533	65.84	0.809	85.62	0.996	92.05
0.095	12.54	0.587	70.90	0.859	87.57	0.998	92.16
0.135	17.88	0.636	74.96	0.904	89.04	0.999	92.21
0.185	24.43	0.679	78.22	0.936	89.96	1.000	92.26
0.239	31.42	0.718	80.81				

Ref. [46]

x_1	P (mbar)	x_1	P (mbar)	x_1	P (mbar)	x_1	P (mbar)
353.15 K							
0.0185	22.09	0.2725	338.24	0.7122	798.57	0.8941	885.34
0.0405	46.50	0.3299	410.41	0.7429	820.79	0.9361	897.46
0.0598	69.77	0.3896	484.49	0.7501	822.25	0.9663	906.22
0.0802	93.71	0.4497	555.18	0.7831	840.64	0.9840	912.35
0.0978	116.18	0.5086	620.02	0.7860	843.50	0.9931	916.47
0.1179	138.14	0.5658	678.22	0.8119	854.80	0.9968	918.46
0.1356	159.93	0.6213	728.75	0.8364	865.40	0.9988	920.06
0.1746	211.66	0.6694	767.83	0.8411	867.26	1.0000	921.34
0.2196	269.55	0.6909	788.24	0.8577	873.78		

(2) Activity Coefficient at Infinite Dilution and Partial Molar Excess Enthalpies at Infinite Dilution

Ref. [18]

Solute	$\gamma^{\infty}{}_1$
353.15 K	
2-Propanol	1.42

Ref. [4]

Solute (1)	$H_{12}{}^{E,\infty}$/kJ · mol^{-1}
2-Propanol	8.332

Mixture 36: $OCH_2CH_2CH_2CH_2$ (1) + [EMIM][BTI] (2)
(1) Phase Equilibrium

Ref. [18]

x_1	P (kPa)	x_1	P (kPa)	x_1	P (kPa)	x_1	P (kPa)
353.15 K							
0.013	1.94	0.356	52.68	0.776	129.09	0.974	154.16
0.026	3.69	0.411	62.01	0.778	129.13	0.983	154.64
0.039	5.45	0.471	72.37	0.813	135.47	0.988	154.97
0.054	7.43	0.527	82.50	0.834	138.70	0.992	155.23
0.074	10.06	0.585	93.14	0.844	140.56	0.995	155.47
0.110	15.11	0.640	103.66	0.880	145.35	0.997	155.64
0.149	20.67	0.691	113.37	0.916	149.59	0.999	155.76
0.198	27.90	0.722	118.90	0.947	152.31	0.999	155.83
0.248	35.53	0.735	121.58	0.964	153.53	1.000	155.92
0.301	43.78						

(2) Activity Coefficient at Infinite Dilution

Ref. [18]

Solute	$\gamma^{\infty}{}_1$
353.15 K	
Tetrahydrofuran	0.800

Mixture 37: $[C_2mim][NTf_2]$ (1) + $[C_{10}mim][NTf_2]$ (2)
(1) Volumetric Properties

Ref. [57]

	V (pure component) or V^E/cm^3 · mol^{-1}	
x_1	T = 298.15 K	T = 333.15 K
1	257.873 (257.9)	263.969 (264.0)
0.798	0.15_9	0.14_2
0.603	0.24_9	0.22_2

(continued)

	V (pure component) or V^E/cm^3 · mol^{-1}	
0.499	0.24_5	0.23_4
0.400	0.25_0	0.22_8
0.205	0.16_8	0.16_5
0	404.061 (404.0)	413.775 (413.5)

Mixture 38: $[C_2mim][NTf_2]$ (1) + $[C_8mim][NTf_2]$ (2)
(1) Volumetric Properties

Ref. [57]

	V (pure component) or V^E/cm^3 · mol^{-1}	
x_1	T = 298.15 K	T = 333.15 K
1	257.985 (257.9)	264.064 (264.0)
0.750	0.11_5	0.11_8
0.500	0.12_1	0.12_4
0.250	0.12_8	0.12_5
0	362.539 (361.8)	371.182 (370.5)

Mixture 39: $C_6H_5OCH_3$ (1) + $[EMIM][NTf_2]$ (2)
(1) Phase Equilibrium

Ref. [58]

T/K	x_1	p_1^{exptl}/Pa	p_1^{NRTL}/Pa	γ_1^{NRTL}	$\ln\gamma_1^{NRTL}$	T/K	x_1	p_1^{exptl}/Pa	p_1^{NRTL}/Pa	γ_1^{NRTL}	$\ln\gamma_1^{NRTL}$
298.6	0.000	0.00	0.00	1.591	0.464	298.6	0.679	431.62	403.51	1.188	0.172
	0.200	155.51	152.46	1.521	0.419		0.796	430.15	436.23	1.096	0.091
	0.377	261.75	267.95	1.420	0.351		0.920	496.64	469.12	1.020	0.019
	0.569	345.76	363.70	1.278	0.245		1.000	500.11	500.11	1.000	0.000
303.6	0.000	0.00	0.00	1.812	0.594	303.6	0.661	580.89	606.63	1.364	0.311
	0.174	192.98	204.81	1.748	0.558		0.783	640.17	641.01	1.216	0.195
	0.232	285.00	268.59	1.719	0.542		0.917	670.31	650.40	1.053	0.052
	0.493	529.33	509.11	1.535	0.429		1.000	673.07	673.07	1.000	0.000
308.6	0.000	0.00	0.00	2.140	0.761	308.6	0.764	875.72	847.19	1.243	0.217
	0.146	247.23	259.38	1.986	0.686		0.914	913.59	860.04	1.054	0.053
	0.185	341.20	320.78	1.945	0.665		1.000	892.11	892.11	1.000	0.000
	0.642	753.94	805.51	1.406	0.341						
313.6	0.000	0.00	0.00	2.724	1.002	313.6	0.621	1035.40	1088.07	1.486	0.396
	0.118	312.06	335.48	2.409	0.879		0.740	1159.89	1138.35	1.306	0.267
	0.143	414.76	395.21	2.351	0.855		0.911	1156.25	1143.64	1.065	0.063
	0.370	892.65	826.05	1.895	0.639		1.000	1178.31	1178.31	1.000	0.000
318.6	0.000	0.00	0.00	3.455	1.240	318.6	0.598	1402.91	1449.89	1.572	0.452
	0.088	395.31	400.39	2.952	1.083		0.713	1607.42	1526.47	1.387	0.327
	0.108	465.41	474.49	2.859	1.050		0.907	1546.61	1523.16	1.088	0.084
	0.303	1095.63	1015.30	2.174	0.776		1.000	1543.31	1543.31	1.000	0.000
323.6	0.000	0.00	0.00	4.648	1.536	323.6	0.572	1738.92	1783.24	1.555	0.441
	0.059	492.44	464.67	3.960	1.376		0.686	1925.27	1839.40	1.338	0.291
	0.076	558.03	573.82	3.792	1.333		0.904	1950.05	1898.54	1.048	0.047
	0.239	1180.83	1269.14	2.644	0.972		1.000	2004.25	2004.25	1.000	0.000

Mixture 40: $CH_3C_6H_4OH$ ((Hydroxymethyl)-benzene) (1) + [EMIM][NTf_2] (2)
(1) Phase Equilibrium

Ref. [58]

T/K	x_1	p_1^{exptl}/Pa	p_1^{NRTL}/Pa	γ_1^{NRTL}	$\ln\gamma_1^{NRTL}$	T/K	x_1	p_1^{exptl}/Pa	p_1^{NRTL}/Pa	γ_1^{NRTL}	$\ln\gamma_1^{NRTL}$
298.6	0.000	0.00	0.00	1.937	0.661	298.6	0.549	8.19	7.93	0.982	−0.018
	0.049	1.14	1.14	1.586	0.461		0.670	10.42	9.72	0.986	−0.014
	0.154	2.87	2.76	1.217	0.196		0.851	11.66	12.48	0.996	−0.004
	0.270	3.89	4.21	1.060	0.058		0.963	13.90	14.17	1.000	0.000
	0.395	5.83	5.80	0.998	−0.002		1.000	13.02	14.72	1.000	0.000
	0.482	7.16	6.99	0.984	−0.016						
303.6	0.000	0.00	0.00	2.378	0.866	303.6	0.539	12.35	11.56	0.956	−0.045
	0.048	1.88	1.87	1.734	0.550		0.668	15.53	14.58	0.973	−0.028
	0.151	3.99	4.03	1.190	0.174		0.850	17.64	18.95	0.994	−0.006
	0.265	5.95	5.97	1.007	0.007		0.963	20.92	21.58	1.000	0.000
	0.391	7.94	8.36	0.954	−0.047		1.000	19.90	22.42	1.000	0.000
	0.478	10.53	10.19	0.951	−0.051						
308.6	0.000	0.00	0.00	1.974	0.680	308.6	0.527	18.62	17.85	0.954	−0.047
	0.046	2.57	2.53	1.547	0.436		0.550	20.43	18.65	0.957	−0.044
	0.144	5.94	5.85	1.148	0.138		0.666	21.93	22.99	0.973	−0.027
	0.183	6.62	6.97	1.074	0.071		0.849	29.06	29.94	0.994	−0.006
	0.259	8.70	9.11	0.993	−0.007		0.963	31.65	34.14	1.000	0.000
	0.473	17.38	15.94	0.949	−0.052		1.000	30.25	35.47	1.000	0.000
313.6	0.000	0.00	0.00	2.392	0.872	313.6	0.514	27.38	24.85	0.967	−0.033
	0.045	4.07	4.05	1.820	0.599		0.663	32.80	32.40	0.978	−0.022
	0.134	8.67	8.70	1.299	0.261		0.848	42.30	42.11	0.995	−0.005
	0.252	13.06	13.35	1.059	0.057		0.963	45.19	48.05	1.000	0.000
	0.382	18.55	18.69	0.981	−0.020		1.000	45.36	49.93	1.000	0.000

	0.467	21.99	22.55	0.968	−0.032						
318.6	0.000	0.00	0.00	2.393	0.872	318.6	0.498	40.20	36.13	0.998	−0.002
	0.042	5.82	5.84	1.927	0.656		0.540	42.25	39.02	0.993	−0.007
	0.123	13.13	12.89	1.445	0.368		0.660	48.14	47.49	0.990	−0.010
	0.245	19.94	20.47	1.148	0.138		0.846	59.51	61.31	0.997	−0.003
	0.376	27.73	28.28	1.034	0.033		0.963	67.53	69.96	1.000	0.000
	0.459	30.21	33.52	1.005	0.005		1.000	67.17	72.67	1.000	0.000
323.6	0.000	0.00	0.00	2.445	0.894	323.6	0.535	60.97	57.60	1.047	0.046
	0.039	8.71	8.35	2.105	0.744		0.655	68.57	68.59	1.018	0.018
	0.112	18.05	19.37	1.687	0.523		0.844	83.95	86.93	1.002	0.002
	0.371	43.54	43.57	1.142	0.133		0.962	96.10	98.98	1.000	0.000
	0.448	48.51	50.05	1.087	0.083		1.000	98.27	102.84	1.000	0.000
	0.480	56.58	52.83	1.070	0.067						

Mixture 41: 1,2-$HOCH_2CH_2OH$ (1) + [EMIM][NTf_2] (2)
(1) Phase Equilibrium

Ref. [58]

T/K	x_1	p_1^{exptl}/Pa	p_1^{NRTL}/Pa	γ_1^{NRTL}	$\ln\gamma_1^{NRTL}$	T/K	x_1	p_1^{exptl}/Pa	p_1^{NRTL}/Pa	γ_1^{NRTL}	$\ln\gamma_1^{NRTL}$
339.6	0.000	0.00	0.00	2.672	0.983	339.6	0.629	246.26	255.90	1.535	0.428
	0.133	87.93	89.17	2.534	0.930		0.713	249.64	256.56	1.357	0.305
	0.294	176.04	176.87	2.268	0.819		0.860	251.50	252.30	1.107	0.101
	0.436	239.12	227.81	1.971	0.679		0.951	261.38	256.14	1.016	0.015
	0.545	253.85	249.09	1.723	0.544		1.000	273.38	265.22	1.000	0.000
343.6	0.000	0.00	0.00	2.857	1.050	343.6	0.596	332.24	329.87	1.541	0.433
	0.128	115.35	120.23	2.619	0.963		0.612	320.06	331.20	1.507	0.410
	0.283	244.43	230.92	2.276	0.822		0.848	331.62	335.44	1.102	0.097
	0.413	304.44	291.27	1.964	0.675		0.948	353.03	345.03	1.014	0.014
	0.522	306.08	319.86	1.707	0.535		1.000	357.00	358.95	1.000	0.000
348.6	0.000	0.00	0.00	2.851	1.048	348.6	0.491	455.91	470.77	1.914	0.649
	0.122	154.14	164.79	2.701	0.994		0.552	460.36	487.26	1.761	0.566
	0.268	342.76	326.92	2.437	0.891		0.832	452.64	481.89	1.155	0.144
	0.367	439.70	407.71	2.216	0.796		0.944	512.00	483.19	1.021	0.021
	0.384	434.78	418.76	2.176	0.777		1.000	494.08	501.13	1.000	0.000
353.6	0.000	0.00	0.00	3.315	1.199	353.6	0.494	637.20	656.76	1.985	0.686

	0.115	218.76	236.00	3.063	1.119		0.813	604.61	648.42	1.192	0.175
	0.251	494.27	454.79	2.703	0.994		0.940	681.03	644.33	1.025	0.025
	0.348	605.80	563.64	2.423	0.885		1.000	677.55	669.13	1.000	0.000
	0.455	618.30	640.17	2.101	0.742						
358.6	0.000	0.00	0.00	3.752	1.322	358.6	0.423	845.16	879.21	2.417	0.882
	0.108	302.27	322.12	3.467	1.243		0.790	806.38	869.36	1.279	0.246
	0.235	637.36	619.89	3.072	1.122		0.936	872.85	832.00	1.034	0.033
	0.305	817.24	742.94	2.833	1.041		1.000	921.01	860.09	1.000	0.000
	0.416	880.08	873.70	2.441	0.892						
363.6	0.000	0.00	0.00	3.792	1.333	363.6	0.376	1167.75	1208.72	2.662	0.979
	0.101	380.66	433.76	3.547	1.266		0.762	1169.93	1263.69	1.374	0.318
	0.219	960.47	846.76	3.198	1.163		0.931	1221.86	1170.98	1.042	0.042
	0.261	1065.44	965.14	3.061	1.119		1.000	1241.41	1206.35	1.000	0.000
	0.345	1125.28	1154.06	2.774	1.020						

Mixture 42: 1,4-$HOCH_2CH_2CH_2CH_2OH$ (1) + [EMIM][NTf_2] (2)
(1) Phase Equilibrium

Ref. [58]

T/K	x_1	p_1^{exptl}/Pa	p_1^{NRTL}/Pa	γ_1^{NRTL}	$\ln\gamma_1^{NRTL}$	T/K	x_1	p_1^{exptl}/Pa	p_1^{NRTL}/Pa	γ_1^{NRTL}	$\ln\gamma_1^{NRTL}$
339.6	0.000	0.00	0.00	7.296	1.987	339.6	0.790	34.65	32.72	1.060	0.058
	0.059	11.91	11.72	5.066	1.623		0.855	34.91	34.36	1.029	0.028
	0.152	18.18	19.44	3.280	1.188		0.923	36.29	36.35	1.008	0.008
	0.184	21.86	20.86	2.902	1.065		0.965	36.59	37.75	1.002	0.002
	0.333	24.29	24.60	1.889	0.636		0.969	38.92	37.90	1.001	0.001
	0.362	26.00	25.10	1.775	0.574		0.980	36.30	38.30	1.001	0.001
	0.398	24.90	25.69	1.651	0.501		1.000	35.38	39.06	1.000	0.000
343.6	0.000	0.00	0.00	7.971	2.076	343.6	0.788	48.29	44.42	1.064	0.062
	0.053	15.95	15.75	5.594	1.722		0.853	45.19	46.56	1.031	0.031
	0.136	24.36	26.28	3.649	1.294		0.922	49.38	49.26	1.009	0.009
	0.172	30.36	28.59	3.147	1.146		0.964	51.14	51.16	1.002	0.002
	0.307	35.16	33.36	2.050	0.718		0.968	52.59	51.35	1.002	0.002
	0.346	35.19	34.25	1.867	0.624		0.980	49.01	51.92	1.001	0.001
	0.387	31.82	35.11	1.712	0.538		1.000	47.75	52.97	1.000	0.000
348.6	0.000	0.00	0.00	9.441	2.245	348.6	0.785	68.72	66.24	1.074	0.071
	0.043	23.86	23.10	6.772	1.913		0.849	68.95	69.18	1.037	0.036
	0.110	35.79	39.09	4.504	1.505		0.920	72.24	73.04	1.011	0.011
	0.156	45.27	44.24	3.608	1.283		0.963	76.24	75.84	1.002	0.002
	0.202	48.51	47.41	2.991	1.096		0.967	77.60	76.13	1.002	0.002
	0.271	52.70	50.44	2.371	0.863		0.980	74.11	77.02	1.001	0.001
	0.330	50.13	52.31	2.016	0.701		1.000	69.45	78.56	1.000	0.000
353.6	0.000	0.00	0.00	8.719	2.165	353.6	0.720	114.45	105.28	1.365	0.311
	0.087	51.12	49.58	5.349	1.677		0.782	94.66	105.65	1.262	0.233

	0.143	64.18	64.32	4.204	1.436		0.846	100.10	105.22	1.162	0.150
	0.181	67.54	71.04	3.662	1.298		0.918	104.99	104.41	1.062	0.060
	0.238	77.65	78.42	3.073	1.123		0.962	108.81	104.76	1.017	0.017
	0.299	92.41	84.36	2.630	0.967		0.967	111.22	104.90	1.013	0.013
	0.313	75.45	85.49	2.551	0.937		0.979	105.20	105.46	1.006	0.006
	0.529	105.12	99.25	1.750	0.560		1.000	99.95	107.10	1.000	0.000
	0.556	98.98	100.46	1.688	0.523						
358.6	0.000	0.00	0.00	12.668	2.539	358.6	0.709	158.13	148.97	1.367	0.313
	0.067	73.89	75.28	7.279	1.985		0.778	141.19	149.76	1.253	0.225
	0.095	97.54	88.85	6.086	1.806		0.841	144.85	149.49	1.158	0.147
	0.132	91.60	100.43	4.970	1.603		0.916	147.66	148.92	1.059	0.058
	0.137	103.23	101.70	4.842	1.577		0.961	155.86	149.90	1.016	0.016
	0.209	105.66	114.09	3.560	1.270		0.966	149.53	150.16	1.012	0.012
	0.267	128.40	120.60	2.941	1.079		0.979	149.18	151.10	1.005	0.005
	0.510	147.77	139.98	1.786	0.580		1.000	142.39	153.57	1.000	0.000
	0.540	138.95	141.83	1.710	0.537						

Mixture 43: $CH_2(CH_2)_4CH_2$ (1) + [EMIM][$(CF_3SO_2)_2N$] (2)
(1) Phase Equilibrium

Ref. [3]

x_1	P (kPa)	x_1	P (kPa)	x_1	P (kPa)	x_1	P (kPa)
353.15 K							
0.0073	7.52	0.0475	45.85	0.2071	99.37	0.5761	99.32
0.0141	14.40	0.0542	51.69	0.2646	99.38	0.6265	99.31
0.0206	20.78	0.0704	64.98	0.3275	99.38	0.6748	99.29
0.0270	26.88	0.0963	84.53	0.3875	99.37	0.7184	99.28
0.0339	33.42	0.1224	97.20	0.4566	99.35	0.7633	99.26
0.0409	39.90	0.1609	99.22	0.5191	99.33		

(2) Excess Enthalpy

Ref. [3]

x_1	H^E ($J \cdot mol^{-1}$)	x_1	H^E ($J \cdot mol^{-1}$)	x_1	H^E ($J \cdot mol^{-1}$)	x_1	H^E ($J \cdot mol^{-1}$)
323.15 K, 1.307 MPa							
0.0293	161	0.1114	412	0.3733	375	0.7444	168
0.0576	284	0.2093	452	0.5884	253	0.9434	44.0

(3) Partial Molar Excess Enthalpies at Infinite Dilution

Ref. [4]

Solute (1)	$H_{12}^{E,\infty}/kJ \cdot mol^{-1}$
Cyclohexane	6.025

Mixture 44: CH = $CHCH_2CH_2CH_2CH_2$ (1) + [EMIM][$(CF_3SO_2)_2N$] (2)
(1) Phase Equilibrium

Ref. [3]

x_1	P (kPa)	x_1	P (kPa)	x_1	P (kPa)	x_1	P (kPa)
353.15 K							
0.0077	4.64	0.1701	83.36	0.6758	92.66	0.9689	92.69
0.0158	9.20	0.2124	92.56	0.7220	92.71	0.9790	92.69
0.0235	13.44	0.2614	92.69	0.7377	92.66	0.9858	92.68
0.0309	17.52	0.3141	92.70	0.7672	92.71	0.9910	92.68
0.0391	21.89	0.3741	92.70	0.8021	92.66	0.9945	92.68
0.0473	26.24	0.4335	92.70	0.8063	92.72	0.9968	92.68
0.0556	30.51	0.4966	92.70	0.8555	92.67	0.9981	92.67
0.0792	42.19	0.5600	92.71	0.8986	92.67	0.9991	92.66
0.1044	54.56	0.6195	92.71	0.9354	92.68	0.9997	92.66
0.1360	69.23	0.6719	92.71	0.9563	92.68	1.0000	92.65

(2) Excess Enthalpy

Ref. [3]

x_1	H^E $(J \cdot mol^{-1})$	x_1	H^E $(J \cdot mol^{-1})$	x_1	H^E $(J \cdot mol^{-1})$	x_1	H^E $(J \cdot mol^{-1})$
323.15 K, 1.307 MPa							
0.0612	225	0.4588	354	0.8841	69.0	0.9797	12.0
0.1180	400	0.7177	170	0.9468	32.0	0.9900	7.00
0.2665	493						

(3) Partial Molar Excess Enthalpies at Infinite Dilution

Ref. [4]

Solute (1)	$H_{12}^{E,\infty}/kJ \cdot mol^{-1}$
Cyclohexene	3.948

Mixture 45: 1-C_6H_{12}(1) + [EMIM][$(CF_3SO_2)_2N$] (2)
(1) Excess Enthalpy

Ref. [3]

x_1	H^E $(J \cdot mol^{-1})$	x_1	H^E $(J \cdot mol^{-1})$	x_1	H^E $(J \cdot mol^{-1})$	x_1	H^E $(J \cdot mol^{-1})$
323.15 K, 1.342 MPa							
0.0254	73.0	0.0978	262	0.3399	266	0.7157	124
0.0502	149	0.1862	304	0.5527	186	0.9351	31.0

Mixture 46: $CH_3(CH_2)_7CHO$ (1) + [EMIM][NTf_2] (2)
(1) Phase Equilibrium

Ref. [59]

T (K)	x_1	p_1^{exp} (Pa)	p_1^{NRTL} (Pa)	γ_1^{NRTL}	$\ln \gamma_1^{NRTL}$
298.6	0.000	0.000	0.000	4.020	1.391
	0.079	12.50	12.58	3.109	1.134
	0.149	20.20	19.88	2.598	0.955
	0.239	26.13	26.64	2.166	0.773
	0.314	31.44	30.99	1.923	0.654
	0.434	36.63	36.76	1.650	0.501
	0.563	41.01	41.82	1.445	0.368
	0.673	47.14	45.19	1.306	0.267
	0.718	45.25	46.30	1.254	0.226
	0.879	48.59	48.85	1.082	0.079
	0.938	49.77	49.58	1.029	0.028
	1.000	50.05	51.39	1.000	0.000
303.6	0.000	0.00	0.00	5.040	1.617
	0.071	19.18	20.13	3.831	1.343
	0.130	33.15	30.71	3.174	1.155
	0.153	35.67	33.85	2.978	1.091

(*continued*)

T (K)	x_1	p_1^{exp} (Pa)	p_1^{NRTL} (Pa)	γ_1^{NRTL}	$\ln \gamma_1^{NRTL}$
	0.221	37.66	41.42	2.513	0.921
	0.412	55.06	54.53	1.779	0.576
	0.552	60.39	60.71	1.476	0.389
	0.665	66.17	64.34	1.298	0.261
	0.712	64.86	65.54	1.236	0.212
	0.875	67.76	69.20	1.062	0.060
	0.937	72.10	71.09	1.019	0.019
	1.000	72.35	74.48	1.000	0.000
308.6	0.000	0.00	0.00	6.826	1.921
	0.053	27.19	26.70	4.747	1.557
	0.060	27.45	29.06	4.555	1.516
	0.104	42.51	40.24	3.646	1.294
	0.137	48.40	46.12	3.181	1.157
	0.204	51.29	54.96	2.547	0.935
	0.280	59.13	62.57	2.113	0.748
	0.388	76.79	71.59	1.745	0.557
	0.541	84.60	82.43	1.441	0.366
	0.657	86.35	89.19	1.283	0.249
	0.705	90.70	91.53	1.227	0.205
	0.871	98.96	98.11	1.065	0.063
	0.935	100.84	100.94	1.021	0.020
	1.000	104.11	105.79	1.000	0.000
313.6	0.000	0.00	0.00	8.073	2.089
	0.042	33.23	34.58	5.649	1.732
	0.050	39.87	38.71	5.343	1.676
	0.120	65.29	62.19	3.571	1.273
	0.184	72.90	74.04	2.767	1.018
	0.258	78.76	84.16	2.246	0.809
	0.363	101.52	96.59	1.830	0.604
	0.525	114.10	113.59	1.487	0.397
	0.647	123.11	124.22	1.320	0.278
	0.697	125.91	127.79	1.261	0.232
	0.865	139.59	136.36	1.084	0.080
	0.933	138.00	139.38	1.027	0.027
	1.000	148.08	145.39	1.000	0.000
318.6	0.000	0.00	0.00	9.667	2.269
	0.031	44.19	46.28	7.126	1.964
	0.038	56.23	53.10	6.723	1.906
	0.160	104.42	106.90	3.235	1.174
	0.235	117.00	120.07	2.474	0.906
	0.337	141.67	133.89	1.921	0.653
	0.510	154.01	153.86	1.460	0.379
	0.688	168.03	171.36	1.204	0.186
	0.860	182.76	187.02	1.052	0.051

(continued)

T (K)	x_1	p_1 ^{exp} (Pa)	p_1 ^{NRTL} (Pa)	γ_1^{NRTL}	ln γ_1^{NRTL}
	0.931	199.96	195.29	1.014	0.014
	1.000	208.30	206.72	1.000	0.000
323.6	0.000	0.00	0.00	13.585	2.609
	0.022	58.91	61.71	9.639	2.266
	0.028	76.64	72.76	8.855	2.181
	0.132	137.82	136.78	3.597	1.280
	0.212	147.10	154.55	2.529	0.928
	0.310	183.02	174.59	1.952	0.669
	0.494	210.39	211.49	1.482	0.393
	0.679	240.56	242.17	1.235	0.211
	0.929	274.00	273.69	1.020	0.020
	1.000	289.93	288.73	1.000	0.000

Mixture 47: 4-$CH_3C_6H_4CHO$ (1) + [EMIM][NTf_2] (2)
(1) Phase Equilibrium

Ref. [59]

T (K)	x_1	p_1 ^{exp} (Pa)	p_1 ^{NRTL} (Pa)	γ_1^{NRTL}	ln γ_1^{NRTL}
298.6	0.000	0.00	0.00	1.510	0.412
	0.075	5.49	4.90	1.422	0.352
	0.085	5.10	5.53	1.411	0.345
	0.404	19.43	21.43	1.158	0.146
	0.587	30.29	28.84	1.073	0.070
	0.784	35.66	36.63	1.019	0.019
	0.900	39.68	41.40	1.004	0.004
	0.972	47.14	44.54	1.000	0.000
	0.984	46.53	45.09	1.000	0.000
	1.000	43.94	45.82	1.000	0.000
303.6	0.000	0.00	0.00	0.425	1.530
	0.072	7.71	6.64	0.366	1.442
	0.081	7.32	7.36	0.359	1.432
	0.125	9.98	11.00	0.326	1.385
	0.423	27.30	31.05	0.142	1.152
	0.458	30.88	33.06	0.125	1.133
	0.571	40.18	39.37	0.078	1.081
	0.673	49.75	44.92	0.045	1.046
	0.779	51.05	50.69	0.021	1.021
	0.897	56.34	57.46	0.004	1.005
	0.984	65.41	62.73	0.000	1.000
	1.000	62.90	63.76	0.000	1.000
308.6	0.000	0.00	0.00	1.464	0.381
	0.069	10.59	9.26	1.391	0.330
	0.076	9.83	10.07	1.385	0.326
	0.119	14.01	15.38	1.344	0.296
	0.442	42.18	47.80	1.126	0.119

(continued)

T (K)	x_1	p_1^{exp} (Pa)	p_1^{NRTL} (Pa)	γ_1^{NRTL}	$\ln \gamma_1^{NRTL}$
	0.554	59.85	57.43	1.079	0.076
	0.659	72.23	66.18	1.045	0.044
	0.772	71.29	75.61	1.020	0.020
	0.970	96.02	93.22	1.000	0.000
	0.983	95.22	94.44	1.000	0.000
	1.000	88.00	96.04	1.000	0.000
313.6	0.000	0.00	0.00	1.530	0.425
	0.065	14.89	12.86	1.450	0.371
	0.070	13.48	13.72	1.445	0.368
	0.114	19.14	21.61	1.396	0.334
	0.424	61.96	66.26	1.152	0.141
	0.537	78.31	79.93	1.095	0.091
	0.643	99.48	92.24	1.056	0.054
	0.765	103.33	106.39	1.024	0.023
	0.969	130.52	131.73	1.000	0.000
	0.983	139.18	133.52	1.000	0.000
	1.000	123.71	135.83	1.000	0.000
318.6	0.000	0.000	0.00	1.768	0.570
	0.062	19.68	18.87	1.651	0.501
	0.063	18.83	18.94	1.650	0.501
	0.401	83.15	90.19	1.227	0.205
	0.518	100.20	108.45	1.142	0.132
	0.624	140.40	124.09	1.084	0.080
	0.757	142.09	143.60	1.034	0.034
	0.885	153.98	163.56	1.008	0.008
	0.968	190.93	177.68	1.001	0.001
	1.000	172.43	183.37	1.000	0.000
323.6	0.000	0.00	0.00	1.853	0.617
	0.055	22.94	23.42	1.735	0.551
	0.059	25.97	25.14	1.726	0.546
	0.378	107.19	118.36	1.269	0.239
	0.496	152.12	143.11	1.169	0.156
	0.598	175.77	162.84	1.105	0.100
	0.748	192.56	191.86	1.040	0.039
	0.880	215.75	218.89	1.009	0.009
	0.967	229.43	238.69	1.001	0.001
	1.000	236.37	246.59	1.000	0.000

Mixture 48: 2-$CH_3COC_7H_{14}$ (1) + [EMIM][NTf_2] (2)
(1) Phase Equilibrium

Ref. [59]

T (K)	x_1	p_1^{exp} (Pa)	p_1^{NRTL} (Pa)	γ_1^{NRTL}	$\ln \gamma_1^{NRTL}$
298.6	0.000	0.00	0.00	3.638	1.291
	0.056	11.26	12.02	3.159	1.150
	0.098	20.05	18.93	2.861	1.051

(continued)

T (K)	x_1	$p_1{}^{exp}$ (Pa)	$p_1{}^{NRTL}$ (Pa)	$\gamma_1{}^{NRTL}$	$\ln \gamma_1{}^{NRTL}$
	0.276	35.57	36.75	1.969	0.677
	0.412	43.99	43.59	1.563	0.447
	0.443	49.40	44.76	1.493	0.401
	0.518	48.05	47.35	1.349	0.299
	0.539	45.50	48.01	1.316	0.275
	0.700	53.74	53.22	1.123	0.116
	0.727	55.49	54.18	1.101	0.096
	0.837	60.88	58.65	1.035	0.034
	0.916	57.34	62.56	1.009	0.009
	0.950	63.78	64.54	1.003	0.003
	1.000	62.11	67.70	1.000	0.000
303.6	0.000	0.00	0.00	4.406	1.483
	0.054	18.25	18.88	3.771	1.327
	0.094	30.11	29.53	3.376	1.217
	0.274	56.21	55.58	2.186	0.782
	0.407	66.90	63.67	1.684	0.521
	0.438	62.85	64.97	1.598	0.468
	0.515	67.98	67.79	1.417	0.349
	0.535	68.59	68.45	1.379	0.321
	0.697	73.37	74.15	1.146	0.136
	0.836	84.43	80.78	1.041	0.040
	0.915	86.36	85.88	1.011	0.011
	0.950	88.05	88.54	1.004	0.004
	0.986	86.98	91.61	1.000	0.000
	1.000	89.80	92.86	1.000	0.000
308.6	0.000	0.00	0.00	4.727	1.553
	0.051	27.85	27.34	4.053	1.399
	0.090	42.10	43.37	3.615	1.285
	0.268	85.29	81.78	2.296	0.831
	0.402	99.36	92.94	1.743	0.556
	0.433	88.11	94.65	1.648	0.499
	0.512	91.11	98.33	1.448	0.370
	0.530	100.34	99.10	1.409	0.343
	0.693	104.89	106.45	1.157	0.146
	0.835	121.30	115.54	1.043	0.042
	0.914	122.30	122.70	1.011	0.011
	0.950	127.91	126.46	1.004	0.004
	0.986	130.32	130.87	1.000	0.000
	1.000	127.99	132.66	1.000	0.000
313.6	0.000	0.00	0.00	5.599	1.723
	0.048	45.20	42.74	4.758	1.560
	0.087	61.13	68.02	4.199	1.435
	0.262	125.96	124.23	2.554	0.938
	0.285	132.74	127.53	2.413	0.881
	0.397	137.09	137.76	1.872	0.627
	0.417	137.11	138.91	1.797	0.586

(continued)

T (K)	x_1	p_1^{exp} (Pa)	p_1^{NRTL} (Pa)	γ_1^{NRTL}	$\ln \gamma_1^{NRTL}$
	0.429	137.09	139.54	1.754	0.562
	0.446	149.24	140.38	1.697	0.529
	0.525	142.15	143.67	1.474	0.388
	0.690	156.36	151.04	1.180	0.166
	0.702	154.26	151.78	1.165	0.153
	0.730	151.85	153.53	1.134	0.126
	0.834	174.78	162.21	1.049	0.048
	0.914	167.32	171.72	1.013	0.013
	0.916	174.54	172.03	1.012	0.012
	0.949	179.86	176.86	1.004	0.004
	0.951	167.89	177.08	1.004	0.004
	0.986	174.62	183.00	1.000	0.000
	0.986	176.31	183.05	1.000	0.000
	1.000	178.80	185.51	1.000	0.000

Mixture 49: $C_6H_5C_2H_4COCH_3$ (1) + [EMIM][NTf$_2$] (2)
(1) Phase Equilibrium

Ref. [59]

T (K)	x_1	p_1^{exp} (Pa)	p_1^{NRTL} (Pa)	γ_1^{NRTL}	$\ln \gamma_1^{NRTL}$
298.6	0.000	0.000	0.000	2.942	1.079
	0.158	2.49	2.27	2.150	0.765
	0.276	2.95	3.26	1.760	0.565
	0.443	3.90	4.15	1.398	0.335
	0.571	4.83	4.67	1.219	0.198
	0.669	4.97	5.04	1.125	0.118
	0.795	5.56	5.57	1.046	0.045
	0.963	6.80	6.46	1.001	0.001
	1.000	6.42	6.70	1.000	0.000
303.6	0.000	0.00	0.00	3.007	1.101
	0.154	3.78	3.32	2.20	0.788
	0.272	4.30	4.80	1.79	0.583
	0.441	5.36	6.09	1.41	0.345
	0.568	7.03	6.84	1.23	0.205
	0.666	7.16	7.38	1.13	0.123
	0.793	8.50	8.15	1.05	0.047
	0.963	10.10	9.45	1.00	0.001
	1.000	9.78	9.80	1.00	0.000
308.6	0.000	0.00	0.00	3.649	1.294
	0.148	6.41	5.23	2.562	0.941
	0.267	6.14	7.40	2.005	0.695
	0.438	7.95	9.11	1.506	0.409
	0.564	9.64	9.98	1.278	0.246
	0.663	10.56	10.62	1.158	0.147
	0.791	12.04	11.57	1.058	0.057

(*continued*)

T (K)	x_1	$p_1{}^{exp}$ (Pa)	$p_1{}^{NRTL}$ (Pa)	$\gamma_1{}^{NRTL}$	$\ln \gamma_1{}^{NRTL}$
	0.963	15.01	13.34	1.002	0.002
	1.000	14.68	13.82	1.000	0.000
313.6	0.000	0.00	0.00	3.172	1.154
	0.137	7.85	6.83	2.363	0.860
	0.259	9.78	10.32	1.884	0.633
	0.434	11.46	13.26	1.448	0.370
	0.560	13.52	14.78	1.251	0.224
	0.659	15.71	15.92	1.144	0.134
	0.788	19.03	17.52	1.053	0.052
	0.963	22.60	20.36	1.002	0.002
	1.000	21.67	21.12	1.000	0.000
318.65	0.000	0.00	0.00	2.816	1.035
	0.125	10.20	8.70	2.210	0.793
	0.252	12.06	14.19	1.785	0.580
	0.430	17.86	18.99	1.401	0.337
	0.554	19.96	21.48	1.229	0.206
	0.654	23.14	23.36	1.132	0.124
	0.784	28.51	25.96	1.050	0.048
	0.962	32.08	30.41	1.001	0.001
	1.000	31.53	31.56	1.000	0.000
323.65	0.000	0.00	0.00	3.250	1.179
	0.113	13.08	12.28	2.529	0.928
	0.244	19.12	20.58	1.963	0.674
	0.424	27.24	26.98	1.479	0.391
	0.547	30.15	29.99	1.274	0.242
	0.648	29.38	32.29	1.157	0.146
	0.779	35.32	35.53	1.059	0.057
	0.962	45.66	41.47	1.002	0.002
	1.000	45.23	43.05	1.000	0.000

Mixture 50: 1-C_4H_9OH (1) + [EMIM][NTf_2] (2)
(1) Cloud Point

Ref. [51]

w_2	T/K	w_2	T/K
0.4247	320.9	0.3608	320.5

$w_{H2O,\ 1}$ = (771 ± 10) ppm; wH_2O, 2 (70 ± 10) ppm, precision of temperature ±0.5 K.

Ref. [51]

$10^{-3}w_{H2O}$/ppm	T/K	$10^{-3}w_{H2O}$/ppm	T/K
w_2 = 0.3608			
0.32	320.5	4.58	316.9
2.44	318.8	6.46	315.4
3.56	317.8		

Ref. [52]

$100w_2$	T/K	$100w_2$	T/K	$100w_2$	T/K	$100w_2$	T/K
5.67	293.4	27.01	319.1	47.99	320.7	69.75	316.3
10.56	307.3	32.56	319.9	54.34	320.6	71.21	315.8
13.98	311.6	37.02	320.3	59.87	319.8	80.58	306.3
21.39	317.0	45.05	320.8	64.02	318.5	89.05	285.9

(2) Activity Coefficient at Infinite Dilution, Partial Molar Excess Enthalpy at Infinite Dilution and Critical Temperatures of Phase Separation

Ref. [53]

	γ_1^{∞}				T_c	$\gamma_1^{s,\infty}$	$H_1^{E,\infty}$(J/mol)
	313 K	323 K	333 K	343 K			
1-Butanol	3.0	2.7	2.4	2.2	321	2.5	+9307

Mixture 51: 1-$C_5H_{11}OH$ (1) + [EMIM][NTf_2] (2)
(1) Cloud Point

Ref. [51]

w_2	T/K	w_2	T/K
0.1054	327.9	0.4499	340.4
0.1738	334.8	0.5087	340.1
0.2385	337.9	0.5405	340.2
0.2639	338.7	0.6295	338.9
0.3159	339.7	0.7222	334.9
0.3390	340.0	0.7380	333.9
0.3848	340.2	0.8462	318.9
0.4354	340.4		

Ref. [51]

$10^{-3}w_{H2O}$/ppm	T/K	$10^{-3}w_{H2O}$/ppm	T/K
$w_2 = 0.7222$		$w_2 = 0.3159$	
0.16	334.9	0.09	339.7
1.06	334.0	1.53	338.6
2.04	333.1	4.06	336.5
3.00	332.2	5.64	335.2
5.05	330.2		
7.06	328.2		

Ref. [52]

$100w_2$	T/K	$100w_2$	T/K	$100w_2$	T/K	$100w_2$	T/K
13.14	330.9	28.15	338.8	41.50	339.8	66.96	336.8
21.11	336.6	36.23	339.7	55.16	339.5		

(2) Activity Coefficient at Infinite Dilution, Partial Molar Excess Enthalpy at Infinite Dilution and Critical Temperatures of Phase Separation

Ref. [53]

	γ_1^{∞}						
	313 K	323 K	333 K	343 K	T_c	$\gamma_1^{s,\infty a}$	$H_1^{E,\infty}$(J/mol)
1-Pentanol	4.3	3.8	3.3	3.0	340	3.0	+11217

[a]$\gamma_1^{s,\infty a}$ is the estimated value of the activity coefficient, expressed at infinite dilution, above which phase splitting occurs, as dictated by eq $\ln(\gamma_2^{\infty}) \geq -\ln r + \left(1-\frac{1}{r}\right) + \frac{1}{2}\left(1 + \frac{1}{r^{1/2}}\right)^2$

Mixture 52: C_5H_{12} (1) + [EMIM][BTI] (2)
(1) Partial Molar Excess Enthalpy at Infinite Dilution

Ref. [4]

Solute (1)	$H_{12}^{E,\infty}$/kJ · mol^{-1}
n-Pentane	4.847

Mixture 53: C_7H_{16} (1) + [EMIM][BTI] (2)
(1) Partial Molar Excess Enthalpy at Infinite Dilution

Ref. [4]

Solute (1)	$H_{12}^{E,\infty}$/kJ · mol^{-1}
n-Heptane	6.763

Mixture 54: 1-C_5H_{10} (1) + [EMIM][BTI] (2)
(1) Partial Molar Excess Enthalpy at Infinite Dilution

Ref. [4]

Solute (1)	$H_{12}^{E,\infty}$/kJ · mol^{-1}
1-Pentene	2.108

Mixture 55: 1-C_6H_{12} (1) + [EMIM][BTI] (2)
(1) Partial Molar Excess Enthalpy at Infinite Dilution

Ref. [4]

Solute (1)	$H_{12}^{E,\infty}$/kJ · mol^{-1}
1-Hexene	2.968

Mixture 56: 1-C_7H_{14} (1) + [EMIM][BTI] (2)
(1) Partial Molar Excess Enthalpy at Infinite Dilution

Ref. [4]

Solute (1)	$H_{12}^{E,\infty}$/kJ · mol^{-1}
1-Heptene	4.193

Mixture 57: $CH_2(CH_2)_3CH_2$ (1) + [EMIM][BTI] (2)
(1) Partial Molar Excess Enthalpy at Infinite Dilution

Ref. [4]

Solute (1)	$H_{12}^{E,\infty}$/kJ · mol^{-1}
Cyclopentane	4.226

Mixture 58: CH = $CHCH_2CH_2CH_2$ (1) + [EMIM][BTI] (2)
(1) Partial Molar Excess Enthalpy at Infinite Dilution

Ref. [4]

Solute (1)	$H_{12}^{E,\infty}$/kJ · mol^{-1}
Cyclopentene	2.122

Mixture 59: $C_6H_5CH_3$ (1) + [EMIM][BTI] (2)
(1) Partial Molar Excess Enthalpy at Infinite Dilution

Ref. [4]

Solute (1)	$H_{12}^{E,\infty}$/kJ · mol^{-1}
Toluene	0.130

Mixture 60: 2-$CH_3COC_2H_5$ (1) + [EMIM][BTI] (2)
(1) Partial Molar Excess Enthalpy at Infinite Dilution

Ref. [4]

Solute (1)	$H_{12}^{E,\infty}$/kJ · mol^{-1}
2-Butanone	−1.964

Mixture 61: 2-$CH_3COC_3H_7$ (1) + [EMIM][BTI] (2)
(1) Partial Molar Excess Enthalpy at Infinite Dilution

Ref. [4]

Solute (1)	$H_{12}^{E,\infty}$/kJ · mol^{-1}
2-Pentanone	−2.139

Mixture 62: CH_3OH (1) + [EMIM][BTI] (2)
(1) Partial Molar Excess Enthalpy at Infinite Dilution

Ref. [4]

Solute (1)	$H_{12}^{E,\infty}$/kJ · mol^{-1}
Methanol	6.290

Mixture 63: $[C_2mim][NTf_2]$ (1) + $[P_{666,14}][NTf_2]$ (2)
(1) Phase Equilibrium

Ref. [60]

	Upper phase		Lower phase	
T/K	x_1	x_2	x_1	x_2
298.15	0.300	0.700	0.980	0.020
318.15	0.357	0.643	0.932	0.068
338.15	0.414	0.586	0.920	0.080
348.15	0.426	0.574	0.892	0.108
358.15	0.499	0.501	0.857	0.143
368.15	0.530	0.470	0.825	0.175
378.15	0.592	0.408	0.820	0.180
388.15	0.691	0.309	0.784	0.216

Mixture 64: C_6H_5Cl (1) + $[emim][Tf_2N]$ (2)
(1) Phase Equilibrium

Ref. [61]

T/K	$100x_1'$	$100x_1$	$V'/cm^3 \cdot mol^{-1}$	$V/cm^3 \cdot mol^{-1}$	$V^{E\prime}/cm^3 \cdot mol^{-1}$	$V^E/cm^3 \cdot mol^{-1}$
293.3 ± 0.2	70.6 ± 0.3	99.7 ± 0.3	145.6 ± 1.4	102.1 ± 0.3	−1.7 ± 1.4	−0.2 ± 0.3
313.2 ± 0.2	71.1 ± 0.3	99.7 ± 0.3	147.3 ± 1.7	103.7 ± 0.4	−1.7 ± 1.7	−0.4 ± 0.4
332.7 ± 0.2	71.8 ± 0.4	99.7 ± 0.3	148.3 ± 1.9	105.9 ± 0.4	−2.0 ± 1.9	−0.4 ± 0.4
353.1 ± 0.2	72.9 ± 0.4	99.7 ± 0.3	146.9 ± 1.8	108.9 ± 0.6	−4.3 ± 1.8	−0.1 ± 0.6
372.9 ± 0.2	73.6 ± 0.6	99.6 ± 0.3	147.5 ± 2.2	111.4 ± 0.9	−5.3 ± 2.2	−0.1 ± 0.9

Mixture 65: C_6H_5Br (1) + $[emim][Tf_2N]$ (2)
(1) Phase Equilibrium

Ref. [61]

T/K	$100x_1'$	$100x_1$	$V'/cm^3 \cdot mol^{-1}$	$V/cm^3 \cdot mol^{-1}$	$V^{E\prime}/cm^3 \cdot mol^{-1}$	$V^E/cm^3 \cdot mol^{-1}$
293.0 ± 0.2	59.4 ± 0.6	100.0 − 0.2	164.5 ± 0.7	104.7 ± 0.4	−2.1 ± 0.7	−0.3 ± 0.4
313.2 ± 0.2	60.7 ± 0.4	100.0 − 0.2	164.9 ± 0.7	106.7 ± 0.4	−2.2 ± 0.7	−0.2 ± 0.4
332.3 ± 0.2	62.1 ± 0.6	100.0 − 0.2	165.4 ± 0.9	108.6 ± 0.3	−2.0 ± 0.9	−0.2 ± 0.3
353.1 ± 0.2	63.4 ± 0.6	100.0 − 0.2	166.0 ± 0.8	110.5 ± 0.6	−2.2 ± 0.8	−0.4 ± 0.6
372.9 ± 0.2	64.7 ± 0.3	100.0 − 0.2	166.4 ± 1.0	112.3 ± 0.6	−3.4 ± 1.0	−0.8 ± 0.6

Mixture 66: C_6H_5I (1) + $[emim][Tf_2N]$ (2)
(1) Phase Equilibrium

Ref. [61]

T/K	$100x_1'$	$100x_1$	$V'/cm^3 \cdot mol^{-1}$	$V/cm^3 \cdot mol^{-1}$	$V^{E\prime}/cm^3 \cdot mol^{-1}$	$V^E/cm^3 \cdot mol^{-1}$
293.0 ± 0.2	99.8 ± 0.2	33.8 ± 0.4	111.2 ± 0.3	206.9 ± 1.0	−0.4 ± 0.3	−0.6 ± 1.0
313.2 ± 0.2	99.9 ± 0.1	36.1 ± 0.4	113.1 ± 0.3	205.6 ± 1.0	−0.1 ± 0.3	−1.5 ± 1.0
332.3 ± 0.2	100.0 − 0.2	38.7 ± 0.4	115.1 ± 0.3	203.8 ± 1.0	0 ± 0.3	−2.2 ± 1.0
353.1 ± 0.2	99.8 ± 0.2	41.5 ± 0.3	116.1 ± 0.5	202.8 ± 2.0	−0.9 ± 0.5	−2.0 ± 2.0
372.9 ± 0.2	99.9 ± 0.1	44.3 ± 0.4	118.2 ± 0.3	199.9 ± 1.0	−0.8 ± 0.3	−3.6 ± 1.0

2.2.8.2 Ternary Mixtures

Mixture 1: CH_3CN (1) + CO_2 (2) + [C_2mim][NTf_2] (3)

(1) Henry's Law Constant

Ref. [54]

T (K)	p_{eq} (10^2 Pa)	K_H (10^5 Pa)	x_2 (10^{-2})	Dev (%)
$x_1 = 0.0000$				
303.63	479.2	38.08	2.614	−0.3
303.87	464.1	38.61	2.578	+0.6
313.87	495.9	45.85	2.172	−0.1
313.92	492.9	46.22	2.155	+0.6
313.95	500.6	45.43	2.192	−1.2
323.96	519.6	54.15	1.840	−0.1
323.98	520.4	54.64	1.823	+0.7
324.05	433.0	53.92	1.848	−0.7
324.12	521.6	54.47	1.829	+0.2
334.07	542.2	62.93	1.584	−0.5
334.10	546.1	63.43	1.571	+0.3
334.16	448.4	63.90	1.560	+0.9
343.99	563.7	72.51	1.375	−0.3
344.23	571.0	72.90	1.368	−0.1
$x_1 = 0.4523$				
293.76	534.8	42.00	2.368	+1.1
303.71	579.0	49.81	1.998	−2.9
313.72	635.4	62.57	1.591	+1.4
324.01	696.7	73.79	1.350	+1.1
334.07	765.6	83.60	1.192	−0.9
$x_1 = 0.5261$				
293.53	500.0	46.49	2.140	−0.2
303.62	548.7	58.83	1.692	+0.7
313.59	601.9	70.73	1.408	−0.7
323.85	662.7	85.54	1.165	+0.2
$x_1 = 0.6610$				
293.76	526.9	49.36	2.015	−0.4
303.15	587.5	61.99	1.605	+1.1
313.27	652.9	73.11	1.362	−0.9
334.38	841.4	96.66	1.031	+0.2
$x_1 = 0.7730$				
292.66	493.7	64.27	1.548	−1.2
303.03	556.7	76.98	1.293	+2.7
312.79	627.4	83.79	1.188	−0.4
322.86	721.1	91.09	1.094	−2.5
333.37	847.8	104.15	0.9568	+1.4

Mixture 2: CH_3COCH_3 (1) + 2-CH_3CHCH_3OH (2) + [EMIM][$(CF_3SO_2)_2N$] (3)
(1) Phase Equilibrium

Ref. [46]

x_1	x_2	P (mbar)	x_1	x_2	P (mbar)
353.15 K					
20.59 mol.% acetone			60.65 mol.% acetone		
0.0074	0.0393	62.29	0.0188	0.0123	40.45
0.0109	0.0578	92.90	0.0359	0.0234	81.09
0.0181	0.0942	150.18	0.0544	0.0354	119.86
0.0219	0.1123	176.99	0.0727	0.0473	163.10
0.0268	0.1370	222.42	0.0894	0.0581	198.11
0.0383	0.1899	306.85	0.1351	0.0875	293.22
0.0512	0.2464	398.04	0.1804	0.1164	396.32
0.0658	0.3071	497.49	0.2281	0.1469	511.51
0.0808	0.3661	595.83	0.2766	0.1776	638.00
0.0958	0.4218	689.04	0.3198	0.2048	757.70
0.1102	0.4731	773.16	0.3597	0.2298	876.72
0.1236	0.5185	846.42	0.3955	0.2520	989.57
0.1361	0.5593	910.67	0.4294	0.2730	1102.2
0.1471	0.5943	964.45	0.4581	0.2906	1202.5
0.1572	0.6252	1012.0	0.4825	0.3055	1290.5
0.1656	0.6507	1051.4	0.5028	0.3179	1366.0
0.1724	0.6711	1083.5	0.5198	0.3282	1430.6
0.1783	0.6883	1111.5	0.5340	0.3367	1485.5
0.1832	0.7027	1135.8	0.5458	0.3438	1531.9
0.1874	0.7147	1157.3	0.5556	0.3497	1570.6
0.1909	0.7248	1175.7	0.5639	0.3546	1603.5
0.1932	0.7312	1187.9	0.5692	0.3578	1624.2
39.95 mol.% acetone			0.5735	0.3604	1641.5
0.0145	0.0241	50.67	78.78 mol.% acetone		
0.0279	0.0460	95.50	0.0368	0.0091	50.49
0.0422	0.0692	142.44	0.0687	0.0170	95.01
0.0517	0.0845	175.65	0.1000	0.0248	142.08
0.0609	0.0993	207.75	0.1299	0.0322	188.49
0.0864	0.1395	292.85	0.1551	0.0385	229.17
0.1134	0.1815	387.79	0.2138	0.0532	328.68
0.1438	0.2280	495.32	0.2726	0.0681	435.85
0.1730	0.2717	602.60	0.3324	0.0833	558.35
0.2017	0.3138	708.45	0.3888	0.0977	683.95
0.2287	0.3524	809.87	0.4411	0.1113	813.66
0.2533	0.3871	903.57	0.4870	0.1233	937.45
0.2769	0.4197	994.38	0.5288	0.1343	1058.9
0.2971	0.4471	1073.4	0.5689	0.1449	1183.9
0.3143	0.4701	1141.7	0.6030	0.1541	1297.2

(continued)

x_1	x_2	P (mbar)	x_1	x_2	P (mbar)
0.3287	0.4892	1200.1	0.6322	0.1620	1398.6
0.3407	0.5048	1250.1	0.6566	0.1687	1486.1
0.3508	0.5179	1293.2	0.6770	0.1743	1561.3
0.3592	0.5287	1330.3	0.6941	0.1791	1624.2
0.3662	0.5377	1361.9	0.7084	0.1831	1676.8
0.3721	0.5452	1389.5	0.7203	0.1864	1720.6
0.3760	0.5500	1407.2	0.7304	0.1893	1757.2
0.3791	0.5539	1422.1	0.7368	0.1911	1781.9

Mixture 3: 2-CH_3CHCH_3OH (1) + H_2O (2) + [EMIM][$(CF_3SO_2)_2N$] (3)
(1) Phase Equilibrium

Ref. [46]

x_1	x_2	P (mbar)	x_1	x_2	P (mbar)
353.15 K					
19.92 mol.% 2-propanol			59.75 mol.% 2-propanol		
0.0069	0.0369	45.80	0.0212	0.0147	43.54
0.0132	0.0701	88.82	0.0380	0.0263	77.53
0.0211	0.1096	137.62	0.0593	0.0410	121.38
0.0286	0.1465	183.52	0.0774	0.0534	158.43
0.0403	0.2022	255.65	0.0975	0.0672	200.26
0.0570	0.2778	350.15	0.1475	0.1015	303.92
0.0757	0.3564	444.41	0.1990	0.1365	407.53
0.0959	0.4369	530.71	0.2529	0.1731	513.55
0.1141	0.5051	593.68	0.3023	0.2064	604.63
0.1301	0.5631	640.75	0.3474	0.2366	681.91
0.1434	0.6102	675.09	0.3857	0.2622	741.71
0.1542	0.6476	702.23	0.4182	0.2838	788.70
0.1634	0.6795	727.53	0.4487	0.3041	829.15
0.1705	0.7036	748.54	0.4737	0.3207	859.86
0.1759	0.7220	767.36	0.4947	0.3346	883.87
0.1801	0.7363	784.27	0.5117	0.3458	902.50
0.1835	0.7477	799.43	0.5256	0.3550	917.28
0.1862	0.7566	813.56	0.5370	0.3625	929.11
0.1884	0.7638	826.21	0.5464	0.3687	938.84
0.1901	0.7696	837.48	0.5540	0.3737	946.85
0.1915	0.7743	847.74	0.5604	0.3779	953.62
0.1926	0.7779	856.42	0.5635	0.3800	956.94
0.1934	0.7807	863.43	0.5656	0.3822	959.84
41.20 mol.% 2-propanol			80.04 mol.% 2-propanol		

(*continued*)

x_1	x_2	P (mbar)	x_1	x_2	P (mbar)
0.0175	0.0281	61.41	0.0434	0.0099	65.67
0.0300	0.0480	105.10	0.0714	0.0163	108.95
0.0467	0.0742	161.64	0.0987	0.0226	151.84
0.0622	0.0981	212.45	0.1282	0.0295	198.01
0.0876	0.1370	292.08	0.1517	0.0350	234.77
0.1236	0.1906	394.08	0.2088	0.0485	322.76
0.1593	0.2425	486.42	0.2672	0.0625	413.25
0.1966	0.2956	570.39	0.3307	0.0780	509.89
0.2304	0.3428	633.76	0.3922	0.0933	596.10
0.2605	0.3843	684.06	0.4491	0.1076	673.09
0.2863	0.4195	723.69	0.4977	0.1201	733.33
0.3079	0.4487	755.01	0.5415	0.1314	782.50
0.3263	0.4735	781.32	0.5846	0.1427	826.15
0.3413	0.4936	803.04	0.6186	0.1516	856.96
0.3536	0.5099	822.49	0.6475	0.1592	880.84
0.3634	0.5230	839.59	0.6715	0.1656	899.12
0.3714	0.5337	855.24	0.6914	0.1709	913.43
0.3779	0.5424	869.36	0.7079	0.1753	924.74
0.3833	0.5494	882.37	0.7217	0.1790	934.05
0.3877	0.5552	894.44	0.7330	0.1820	941.64
0.3913	0.5600	904.51	0.7425	0.1846	947.94
0.3941	0.5637	913.55	0.7486	0.1862	952.12
0.3960	0.5668	921.07	0.7528	0.1883	955.58

Mixture 4: C_7H_{16} (1) + $C_6H_5CH_3$ (2) + $[C_2mim][NTf_2]$ (3)
(1) Distribution Ratio and Selectivity

Ref. [62]

Upper phase			Lower phase				
x_1	x_2	x_3	x_1	x_2	x_3	β	S
298.15 K							
1.000	0.000	0.000	0.023	0.000	0.977	–	–
0.936	0.064	0.000	0.028	0.053	0.919	0.83	27.7
0.793	0.207	0.000	0.025	0.166	0.809	0.80	25.4
0.744	0.256	0.000	0.027	0.204	0.769	0.80	22.0
0.582	0.418	0.000	0.025	0.301	0.674	0.72	16.8
0.459	0.541	0.000	0.023	0.375	0.602	0.69	13.8
0.345	0.655	0.000	0.023	0.435	0.542	0.66	10.0
0.169	0.831	0.000	0.017	0.545	0.438	0.66	6.5
0.068	0.932	0.000	0.008	0.606	0.386	0.65	5.5
0.000	1.000	0.000	0.000	0.654	0.346	0.65	–

Estimated uncertainty: ±0.011 mole fraction

Mixture 5: C_8H_{18} (1) + $C_6H_5C_2H_5$ (2) + $[C_2mim][NTf_2]$ (3)

(1) Distribution Ratio and Selectivity

Ref. [62]

Upper phase			Lower phase				
x_1	x_2	x_3	x_1	x_2	x_3	β	S
298.15 K							
1.000	0.000	0.000	0.015	0.000	0.985	–	–
0.954	0.046	0.000	0.015	0.022	0.963	0.48	30.4
0.816	0.184	0.000	0.016	0.090	0.894	0.49	24.9
0.716	0.282	0.002	0.021	0.141	0.838	0.50	17.0
0.558	0.442	0.000	0.019	0.209	0.772	0.47	13.9
0.471	0.529	0.000	0.022	0.252	0.726	0.48	10.2
0.383	0.617	0.000	0.026	0.297	0.677	0.48	7.1
0.286	0.714	0.000	0.022	0.350	0.628	0.49	6.4
0.202	0.797	0.001	0.017	0.397	0.586	0.50	5.9
0.113	0.887	0.000	0.010	0.436	0.554	0.49	5.6
0.050	0.950	0.000	0.007	0.492	0.501	0.52	3.7
0.000	1.000	0.000	0.000	0.510	0.490	0.51	–

Estimated uncertainty: ± 0.014 mole fraction

Mixture 6: C_6H_{14} (1) + [EMIM][BTI] (2) + [BMIM][BTI] (3) (w_2:w_3 = 50:50 weight percent)

(1) Activity Coefficient at Infinite Dilution

Ref. [4]

	$\gamma^{\infty}{}_1$ (gas–liquid chromatography)			
Solute (1)	303.15 K	313.15 K	323.15 K	333.15 K
n-Hexane	18.8	18.3	17.1	16.7

Mixture 7: C_7H_{16} (1) + [EMIM][BTI] (2) + [BMIM][BTI] (3) (w_2:w_3 = 50:50 weight percent)

(1) Activity Coefficient at Infinite Dilution

Ref. [4]

	$\gamma^{\infty}{}_1$ (gas–liquid chromatography)			
Solute (1)	303.15 K	313.15 K	323.15 K	333.15 K
n-Heptane	28.3	27.1	25.6	24.7

Mixture 8: C_8H_{18} (1) + [EMIM][BTI] (2) + [BMIM][BTI] (3) (w_2:w_3 = 50:50 weight percent)
(1) Activity Coefficient at Infinite Dilution

Ref. [4]

	γ^{∞}_1 (gas–liquid chromatography)			
Solute (1)	303.15 K	313.15 K	323.15 K	333.15 K
n-Octane	43.0	40.3	36.2	35.5

Mixture 9: 1-C_5H_{10} (1) + [EMIM][BTI] (2) + [BMIM][BTI] (3) (w_2:w_3 = 50:50 weight percent)
(1) Activity Coefficient at Infinite Dilution

Ref. [4]

	γ^{∞}_1 (gas–liquid chromatography)			
Solute (1)	303.15 K	313.15 K	323.15 K	333.15 K
1-Pentene	6.43	6.29	6.15	6.11

Mixture 10: 2-$CH_3COC_2H_5$ (1) + [EMIM][BTI] (2) + [BMIM][BTI] (3) (w_2:w_3 = 50:50 weight percent)
(1) Activity Coefficient at Infinite Dilution

Ref. [4]

	γ^{∞}_1 (gas–liquid chromatography)			
Solute (1)	303.15 K	313.15 K	323.15 K	333.15 K
2-Butanone	0.526	0.548	0.562	0.584

Mixture 11: 2-$CH_3COC_3H_7$ (1) + [EMIM][BTI] (2) + [BMIM][BTI] (3) (w_2:w_3 = 50:50 weight percent)
(1) Activity Coefficient at Infinite Dilution

Ref. [4]

	γ^{∞}_1 (gas–liquid chromatography)			
Solute (1)	303.15 K	313.15 K	323.15 K	333.15 K
2-Pentanone	0.706	0.738	0.759	0.781

Mixture 12: H_2O (1) + [EMIM][BTI] (2) + [BMIM][BTI] (3) (w_2:w_3 = 50:50 weight percent)
(1) Activity Coefficient at Infinite Dilution

Ref. [4]

	γ^{∞}_1 (gas–liquid chromatography)			
Solute (1)	303.15 K	313.15 K	323.15 K	333.15 K
Water	3.25	2.98	2.77	2.59

Mixture 13: $[C_2mim][NTf_2]$ (1) + $[C_4mim][NTf_2]$ (2) + $[P_{666,14}][NTf_2]$ (3)
(1) Phase Equilibrium

Ref. [60]

Top phase			Bottom phase		
x_1	x_2	x_3	x_1	x_2	x_3
298.15 K					
0.300	0.000	0.700	0.980	0.000	0.020
0.304	0.044	0.652	0.872	0.075	0.053
0.274	0.120	0.606	0.701	0.211	0.088
0.254	0.203	0.543	0.561	0.319	0.120
0.232	0.292	0.476	0.430	0.402	0.168
0.206	0.389	0.405	0.323	0.451	0.226

2.2.9 02003-03003: 1-Ethyl-3-methylimidazolium bis((perfluoroethane)sulfonyl)imide

Abbreviation:	$[C_2MIm][BETI]$, [emim][BETI], [EMIM][BETI], [emim][BEI]
Molecular Formula:	$C_{10}H_{11}F_{10}N_3O_4S_2$
Molar Mass:	491.33
Structure:	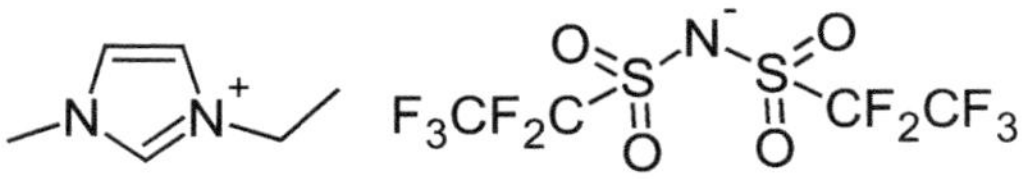

2.2.9.1 Binary Mixtures

Mixture 1: CH_3OH (1) + [EMIM][BETI] (2)
(1) Volumetric Properties

Ref. [63]

x_1	ρ $(g \cdot cm^{-3})$	V^E_m $(cm^3 \cdot mol^{-1})$	ρ $(g \cdot cm^{-3})$	V^E_m $(cm^3 \cdot mol^{-1})$	ρ $(g \cdot cm^{-3})$	V^E_m $(cm^3 \cdot mol^{-1})$
	$T = 298.15$ K		$T = 303.15$ K		$T = 313.15$ K	
0.0000	1.5932	0.00	1.5876	0.00	1.5767	0.00
0.0584	1.5862	0.06	1.5805	0.03	1.5695	0.06
0.1116	1.5794	0.08	1.5737	0.06	1.5626	0.10
0.1686	1.5713	0.11	1.5656	0.09	1.5545	0.13
0.2956	1.5497	0.14	1.544	0.12	1.5328	0.16
0.5104	1.4946	0.09	1.4889	0.07	1.4777	0.09
0.6083	1.4553	0.04	1.4496	0.02	1.4383	0.04
0.7060	1.3989	−0.01	1.3932	−0.02	1.3820	−0.03
0.8237	1.2859	−0.04	1.2802	−0.05	1.2693	−0.08
0.9033	1.1482	−0.04	1.1426	−0.05	1.1320	−0.08
1.0000	0.7866	0.00	0.7819	0.00	0.7726	0.00

Mixture 2: CH_3COCH_3 (1) + [EMIM][BETI] (2)
(1) Volumetric Properties

Ref. [63]

x_1	ρ (g · cm^{-3})	V^E_m (cm^3 · mol^{-1})	ρ (g · cm^{-3})	V^E_m (cm^3 · mol^{-1})	ρ (g · cm^{-3})	V^E_m (cm^3 · mol^{-1})
	T = 298.15 K		T = 303.15 K		T = 313.15 K	
0.0000	1.5932	0.00	1.5873	0.00	1.5767	0.00
0.0807	1.5768	−0.08	1.5711	−0.10	1.5602	−0.11
0.1417	1.5631	−0.15	1.5574	−0.18	1.5464	−0.19
0.2521	1.5346	−0.33	1.5288	−0.35	1.5177	−0.39
0.2791	1.5266	−0.36	1.5208	−0.39	1.5096	−0.42
0.4150	1.4792	−0.53	1.4733	−0.56	1.4622	−0.66
0.5195	1.4318	−0.68	1.4258	−0.72	1.4148	−0.86
0.5676	1.4055	−0.76	1.3995	−0.79	1.3884	−0.94
0.7092	1.3031	−0.92	1.2972	−0.98	1.2857	−1.12
0.7550	1.2582	−0.92	1.2522	−0.98	1.2408	−1.13
0.8137	1.1884	−0.89	1.1825	−0.95	1.1708	−1.08
0.8927	1.0620	−0.72	1.0559	−0.75	1.0443	−0.87
0.9502	0.9350	−0.50	0.9291	−0.53	0.9176	−0.62
1.000	0.7841	0.00	0.7784	0.00	0.7669	0.00

Mixture 3: 1,1,1,2-$C_2H_2F_4$ (1) + [emim][BEI] (2)
(1) Solubility and Diffusivity

Ref. [64]

T/K	P/MPa	$100x_1$	10^{11} D/m^2 · s^{-1}	T/K	P/MPa	$100x_1$	10^{11} D/m^2 · s^{-1}
283.10	0.0103	3.0	1.9	323.15	0.0105	1.1	5.8
283.15	0.0504	13.7	–	323.15	0.0505	4.6	6.7
283.10	0.1003	25.8	3.0	323.15	0.1004	8.9	8.2
283.15	0.1502	36.8	3.7	323.15	0.1505	13.0	7.7
283.15	0.2001	47.3	4.9	323.15	0.2004	16.9	8.4
283.10	0.2501	57.5	6.6	323.15	0.2503	20.7	8.8
283.10	0.3004	67.3	8.7	323.15	0.3003	24.3	9.4
283.10	0.3505	79.5	-	323.15	0.3504	27.8	–
298.05	0.0105	1.9	2.6	348.10	0.0104	0.6	13
298.10	0.0505	8.6	3.8	348.10	0.0506	2.7	14
298.10	0.1005	16.4	4.2	348.10	0.1004	5.3	14
298.10	0.1504	23.6	4.8	348.10	0.1504	7.9	13
298.10	0.2004	30.5	5.7	348.10	0.2004	10.3	15
298.10	0.2503	37.0	6.7	348.05	0.2496	12.6	14
298.10	0.3004	43.2	7.9	348.10	0.3005	14.9	17
298.05	0.3505	49.1	–	348.05	0.3503	17.1	–

Mixture 4: CH_2F_2 (1) + [emim][BEI] (2)
(1) Solubility and Diffusivity

Ref. [55]

T/K	P/MPa	100 x_1	$10^{11}D$/m^2 · s^{-1}	T/K	P/MPa	100 x_1	$10^{11}D$/m^2 · s^{-1}
283.15	0.0101	1.4	3.8	323.15	0.0997	5.9	13
283.15	0.1000	14.5	4.8	323.15	0.2497	14.6	18
283.15	0.2495	32.5	7.4	323.15	0.3996	22.1	19
283.15	0.3995	47.0	12	323.15	0.5495	28.8	22
283.15	0.5496	59.3	15	323.15	0.6995	34.8	23
283.15	0.6994	70.3	17	323.15	0.8504	40.3	27
283.15	0.8505	80.2	17	323.15	0.9993	45.4	
298.15	0.0096	1.0	7.5	348.05	0.0101	0.1	
298.15	0.0997	10.4	7.9	348.05	0.1000	3.8	26
298.15	0.2496	23.8	11	348.05	0.2501	9.5	33
298.15	0.3996	34.9	13	348.05	0.3992	14.7	
298.15	0.5493	44.5	16	348.05	0.5496	19.6	35
298.15	0.6993	52.9	18	348.05	0.6996	24.1	34
298.15	0.8503	60.3	21	348.05	0.8504	28.3	35
298.15	1.0005	67.2		348.05	0.9994	32.1	
323.15	0.0100	0.4					

2.2.9.2 Ternary Mixtures

Mixture 1: CH_3OH (1) + CH_3COCH_3 (2) + [EMIM][BETI] (3)
(1) Volumetric Properties

Ref. [63]

x_3	x_1	ρ (g · cm^{-3})	V^E_m (cm^3 · mol^{-1})	ρ (g · cm^{-3})	V^E_m (cm^3 · mol^{-1})	ρ (g · cm^{-3})	V^E_m (cm^3 · mol^{-1})
		T = 298.15 K		T = 303.15 K		T = 313.15 K	
0.0211	0.7657	0.8894	−0.33	0.8841	−0.34	0.8735	−0.35
0.0245	0.6613	0.8975	−0.40	0.8922	−0.42	0.8813	−0.44
0.0272	0.2130	0.8825	−0.45	0.8770	−0.49	0.8657	−0.55
0.0549	0.4019	0.9781	−0.54	0.9724	−0.57	0.9610	−0.62
0.0795	0.5957	1.0640	−0.46	1.0584	−0.49	1.0471	−0.53
0.1454	0.5467	1.1941	−0.45	1.1884	−0.48	1.1767	−0.51
0.1999	0.6292	1.2871	−0.25	1.2812	−0.26	1.2699	−0.30
0.2294	0.1188	1.2533	−0.69	1.2475	−0.75	1.2356	−0.83
0.2818	0.4481	1.3501	−0.36	1.3441	−0.37	1.3328	−0.43
0.3221	0.2227	1.3548	−0.53	1.3490	−0.58	1.3373	−0.64
0.4927	0.2348	1.4633	−0.31	1.4576	−0.34	1.4459	−0.34
0.5038	0.0711	1.4502	−0.47	1.4445	−0.52	1.4328	−0.55
0.5472	0.2722	1.4923	−0.18	1.4867	−0.21	1.4751	−0.20
0.6055	0.3074	1.5192	−0.07	1.5135	−0.07	1.5022	−0.07

2.2.10 02003-03004: 1-Ethyl-3-methylimidazolium dicyanoamide

Abbreviation:	[C_2MIm][dca], [emim][dca], [EMIM][dca], [emim][$N(CN)_2$]
Molecular Formula:	$C_8H_{11}N_5$
Molar Mass:	177.21
Structure:	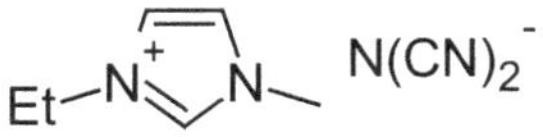

2.2.10.1 Ternary Mixtures

Mixture 1: H_2O (1) + C_2H_5OH (2) + [emim][$N(CN)_2$] (3)

(1) Phase Equilibrium

Ref. [24]

x'_2 [a]	w_3 [a]	y_2	T/K	γ_1	γ_2	$\alpha_{2,1}$ [b]
p = 100 kPa						
0.9500	0.7998	0.9857	374.41	0.54	0.88	3.62
0.9501	0.7001	0.9820	365.92	0.74	0.94	2.86
0.9501	0.6000	0.9778	361.09	0.95	0.97	2.31
0.9501	0.5000	0.9729	357.65	1.20	0.99	1.88
0.9501	0.3999	0.9694	355.44	1.38	1.00	1.66
0.9501	0.3000	0.9628	353.83	1.70	1.01	1.36
0.9500	0.1999	0.9576	352.58	1.95	1.01	1.19
0.9500	0.1000	0.9535	351.64	2.14	1.01	1.08

[a]Compositions reported are of equilibrium liquid

[b]$\alpha_{2,1} = (y_2/x'_2)/(y_1/x'_1)$

2.2.11 02003-04003: 1-Ethyl-3-methylimidazolium trifluoromethanesulfonate

Abbreviation:	[C_2MIm][TfO], [emim][TfO], [EMIM][TfO], [emim][triflate], [emim][OTf]
Molecular Formula:	$C_7H_{11}F_3N_2O_3S$
Molar Mass:	260.24
Structure:	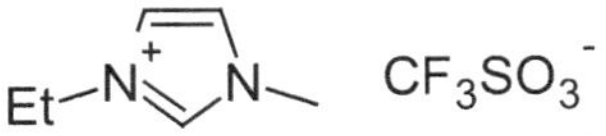

2.2.11.1 Binary Mixtures

Mixture 1: C_2H_5OH (1) + [emim][triflate] (2)

(1) Excess Molar Enthalpy

Ref. [65]

x_2	H^E (J · mol^{-1})	x_2	H^E (J · mol^{-1})	x_2	H^E (J · mol^{-1})
T = 303.15 K					
0.1271	1218	0.5006	1772	0.7501	1086
0.2489	1719	0.6176	1519	0.869	599
0.3752	1879				

(2) Phase Equilibrium

Ref. [66]

x_2	T/K	x_2	T/K	x_2	T/K
100 kPa					
0.0000	351.21	0.1192	353.39	0.2390	356.30
0.0133	351.51	0.1326	353.70	0.2562	356.83
0.0234	351.77	0.1529	354.17	0.2668	357.25
0.0242	351.70	0.1627	354.36	0.2885	357.90
0.0362	351.84	0.1875	354.98	0.2969	358.25
0.0393	352.00	0.1913	355.02	0.2999	358.39
0.0500	352.11	0.1985	355.19	0.3062	358.60
0.0540	352.23	0.2018	355.27	0.3113	358.74
0.0634	352.48	0.2100	355.54	0.3221	359.28
0.0683	352.44	0.2142	355.62	0.3235	359.25
0.0770	352.67	0.2191	355.83	0.3330	359.64
0.0865	352.84	0.2275	356.09	0.3331	359.75
0.1036	353.13	0.2282	356.04	0.3661	361.19
0.1172	353.34	0.2377	356.39	0.3766	361.75

(3) Volumetric Properties, Speed of Sound and Compressibility Properties

Ref. [67]

x_2	ρ $kg \cdot m^{-3}$	V^E_m $cm^3 \cdot mol^{-1}$	u $m \cdot s^{-1}$	u^E $m \cdot s^{-1}$	κ_S TPa^{-1}	κ_S^E TPa^{-1}	x_2	ρ $kg \cdot m^{-3}$	V^E_m $cm^3 \cdot mol^{-1}$	u $m \cdot s^{-1}$	u^E $m \cdot s^{-1}$	κ_S TPa^{-1}	κ_S^E TPa^{-1}
$T = 278.15$ K													
0.0499	802.14	−0.263	1212.52	29.57	847.96	−39.86	0.5932	1264.18	−0.367	1394.83	63.43	406.58	−37.99
0.1000	892.80	−0.378	1232.96	47.60	736.79	−57.02	0.6928	1299.43	−0.300	1416.18	49.87	383.72	−27.50
0.1503	965.56	−0.439	1252.69	60.44	659.99	−65.20	0.8020	1331.10	−0.201	1435.87	32.80	364.38	−16.69
0.1999	1025.77	−0.474	1273.50	69.91	601.11	−68.69	0.8517	1359.80	−0.158	1454.20	24.82	347.76	−12.22
0.3009	1075.78	−0.502	1294.55	79.65	554.68	−66.57	0.9017	1371.25	−0.127	1461.89	16.39	341.24	−7.86
0.4001	1156.45	−0.463	1333.90	79.84	485.99	−58.42	0.9492	1382.03	−0.077	1468.98	8.83	335.31	−4.13
0.5015	1216.18	−0.416	1366.81	73.29	440.13	−47.90							
$T = 288.15$ K													
0.0499	793.65	−0.288	1177.69	31.78	908.46	−46.89	0.5932	1255.54	−0.400	1370.61	68.83	423.98	−43.95
0.1000	884.25	−0.415	1201.13	51.46	783.87	−67.19	0.6928	1290.81	−0.325	1392.43	54.31	399.57	−31.84
0.1503	956.95	−0.482	1223.08	65.53	698.56	−76.76	0.8020	1322.51	−0.216	1412.31	35.96	379.09	−19.39
0.1999	1017.12	−0.521	1245.56	75.89	633.72	−80.72	0.8517	1351.24	−0.169	1430.73	27.19	361.54	−14.17
0.3009	1067.12	−0.550	1267.89	86.17	582.94	−77.60	0.9017	1362.70	−0.135	1438.29	18.16	354.74	−9.20
0.4001	1147.78	−0.508	1308.55	86.24	508.82	−67.75	0.9492	1373.50	−0.081	1445.46	9.79	348.46	−4.83
0.5015	1207.52	−0.455	1342.17	79.17	459.72	−55.36							
$T = 298.15$ K													
0.0499	785.10	−0.316	1142.88	34.80	975.15	−56.23	0.5932	1246.96	−0.439	1346.92	74.44	442.04	−50.72
0.1000	875.66	−0.456	1170.11	56.05	834.09	−79.84	0.6928	1282.26	−0.355	1369.18	58.84	416.01	−36.68
0.1503	948.32	−0.531	1194.17	71.24	739.46	−90.74	0.8020	1313.99	−0.236	1389.28	39.06	394.30	−22.33
0.1999	1008.47	−0.573	1218.21	82.33	668.18	−94.93	0.8517	1342.74	−0.184	1407.84	29.56	375.75	−16.31
0.3009	1058.46	−0.606	1241.66	93.20	612.80	−90.54	0.9017	1354.22	−0.146	1415.44	19.80	368.58	−10.60
0.4001	1139.14	−0.560	1283.72	93.04	532.70	−78.55	0.9492	1365.03	−0.087	1422.67	10.58	361.95	−5.50

(continued)

x_2	ρ kg · m^{-3}	V^E_m cm^3 · mol^{-1}	u m · s^{-1}	u^E m · s^{-1}	κ_S TPa^{-1}	κ_S^E TPa^{-1}	x_2	ρ kg · m^{-3}	V^E_m cm^3 · mol^{-1}	u m · s^{-1}	u^E m · s^{-1}	κ_S TPa^{-1}	κ_S^E TPa^{-1}
0.5015	1198.91	−0.501	1317.99	85.41	480.16	−63.94							
T = 308.15 K													
0.0499	776.44	−0.348	1109.16	37.19	1046.89	−66.01	0.5932	1238.43	−0.486	1323.63	80.22	460.89	−58.36
0.1000	867.01	−0.503	1139.42	60.24	888.40	−93.79	0.6928	1273.76	−0.392	1346.39	63.58	433.08	−42.18
0.1503	939.65	−0.587	1165.70	76.71	783.18	−106.37	0.8020	1305.52	−0.261	1366.72	42.35	410.07	−25.67
0.1999	999.81	−0.634	1191.36	88.37	704.69	−110.62	0.8517	1334.31	−0.203	1385.42	32.12	390.46	−18.77
0.3009	1049.81	−0.669	1215.68	100.08	644.54	−104.99	0.9017	1345.80	−0.159	1393.07	21.66	382.89	−12.25
0.4001	1130.53	−0.621	1259.25	99.78	557.82	−90.61	0.9492	1356.63	−0.094	1400.42	11.55	375.86	−6.34
0.5015	1190.34	−0.555	1294.16	91.72	501.59	−73.57							
T = 318.15 K													
0.0499	767.63	−0.384	1075.80	39.73	1125.60	−77.60	0.5932	1229.94	−0.540	1300.74	86.31	480.54	−67.12
0.1000	858.25	−0.558	1109.14	64.53	947.14	−110.03	0.6928	1265.32	−0.436	1323.94	68.67	450.88	−48.51
0.1503	930.92	−0.652	1137.53	82.24	830.16	−124.41	0.8020	1297.12	−0.290	1344.61	45.90	426.41	−29.52
0.1999	991.10	−0.704	1164.71	94.47	743.78	−128.66	0.8517	1325.93	−0.225	1363.47	34.84	405.68	−21.56
0.3009	1041.14	−0.743	1189.87	107.22	678.41	−121.66	0.9017	1337.44	−0.173	1371.13	23.68	397.71	−14.15
0.4001	1121.93	−0.691	1235.07	106.89	584.32	−104.55	0.9492	1348.27	−0.101	1378.68	12.58	390.21	−7.28
0.5015	1181.81	−0.619	1270.70	98.43	524.04	−84.69							
T = 328.15 K													
0.0499	758.62	−0.428	1042.79	42.34	1212.22	−91.26	0.5932	1221.51	−0.605	1278.31	92.86	500.99	−77.28
0.1000	849.37	−0.622	1079.19	68.99	1010.90	−129.14	0.6928	1256.93	−0.484	1301.77	74.44	469.48	−56.03
0.1503	922.10	−0.728	1109.70	87.98	880.67	−145.54	0.8020	1288.75	−0.325	1323.04	50.08	443.29	−34.19
0.1999	982.34	−0.787	1138.38	100.82	785.53	−149.71	0.8517	1317.62	−0.252	1342.09	38.22	421.35	−25.05
0.3009	1032.44	−0.830	1164.35	114.74	714.45	−141.09	0.9017	1329.14	−0.191	1349.85	26.27	412.91	−16.59
0.4001	1113.33	−0.774	1211.21	114.54	612.26	−120.87	0.9492	1339.98	−0.112	1357.58	13.99	404.92	−8.55
0.5015	1173.30	−0.693	1247.65	105.78	547.53	−97.76							

Mixture 2: $CHCl_3$ (1) + [emim][triflate] (2)
(1) Phase Equilibrium

Ref. [68]

x_2	T/K	x_2	T/K	x_2	T/K
0.0000	333.96	0.2810	334.98	0.3881	339.12
0.1554	334.00	0.2952	335.51	0.4014	339.70
0.1978	333.98	0.3106	335.83	0.4204	340.80
0.2185	333.96	0.3284	336.48	0.4289	341.88
0.2342	334.07	0.3491	337.41	0.4524	342.99
0.2523	334.36	0.3585	337.79	0.4666	344.44
0.2706	334.68	0.3724	338.37	0.4709	344.54

Ref. [68]

	[emim][triflate]-rich phase		Chloroform-rich phase	
T/K	x_1	x_2	x_1	x_2
293.15	0.80519	0.19481	0.99810	0.00190
313.15	0.78685	0.21315	0.99818	0.00182
323.15	0.77563	0.22437	0.99822	0.00178
333.15	0.76840	0.23160	0.99825	0.00175

Mixture 3: 1-C_3H_7OH (1) + [emim][triflate] (2)
(1) Phase Equilibrium

Ref. [69]

x_2	T/K	x_2	T/K	x_2	T/K
100 kPa					
0.0000	369.86	0.0778	371.09	0.2016	373.39
0.0081	370.05	0.0822	371.22	0.2164	373.69
0.0106	370.13	0.0948	371.37	0.2238	373.83
0.0143	370.15	0.0989	371.48	0.2390	374.18
0.0186	370.29	0.1128	371.69	0.2554	374.64
0.0230	370.30	0.1139	371.76	0.2705	374.95
0.0279	370.43	0.1275	371.95	0.2777	375.25
0.0345	370.46	0.1325	372.02	0.2859	375.41
0.0397	370.56	0.1421	372.25	0.2969	375.63
0.0477	370.65	0.1557	372.48	0.3027	375.73
0.0521	370.77	0.1634	372.66	0.3210	376.39
0.0627	370.90	0.1780	372.96	0.3354	376.63
0.0660	371.02	0.1782	372.85	0.3455	377.04

(2) Volumetric Properties, Speed of Sound and Compressibility Properties

Ref. [67]

x_2	ρ kg · m^{-3}	V^E_m cm^3 · mol^{-1}	u m · s^{-1}	u^E m · s^{-1}	κ_S T Pa^{-1}	κ_S^E T Pa^{-1}	x_2	ρ kg · m^{-3}	V^E_m cm^3 · mol^{-1}	u m · s^{-1}	u^E m · s^{-1}	κ_S T Pa^{-1}	κ_S^E T Pa^{-1}
$T = 278.15$ K													
0.0501	885.81	−0.149	1279.79	14.42	689.26	−17.12	0.6007	1280.35	−0.207	1412.25	45.31	391.60	−27.01
0.1001	946.17	−0.228	1286.31	23.45	638.76	−25.73	0.7000	1316.97	−0.166	1432.27	36.88	370.15	−20.25
0.1504	998.62	−0.237	1296.00	30.72	596.20	−30.94	0.7996	1348.61	−0.122	1450.34	26.12	352.51	−13.32
0.2004	1044.76	−0.255	1308.30	37.22	559.20	−34.80	0.8503	1363.09	−0.095	1459.01	20.07	344.64	−9.88
0.3002	1122.37	−0.279	1335.85	46.63	499.29	−38.17	0.8905	1373.86	−0.070	1465.75	15.13	338.79	−7.25
0.3993	1184.64	−0.278	1363.10	50.70	454.32	−36.93	0.9510	1389.10	−0.040	1475.06	6.96	330.86	−3.22
0.5010	1237.17	−0.245	1389.17	50.02	418.85	−32.73							
$T = 288.15$ K													
0.0501	877.77	−0.166	1247.29	16.46	732.30	−21.29	0.6007	1271.85	−0.235	1387.95	49.96	408.15	−31.78
0.1001	938.00	−0.254	1255.93	26.94	675.87	−32.05	0.7000	1308.46	−0.187	1408.42	40.87	385.28	−23.87
0.1504	990.36	−0.267	1267.29	35.31	628.71	−38.50	0.7996	1340.11	−0.137	1426.71	29.11	366.60	−15.75
0.2004	1036.43	−0.289	1280.69	42.40	588.27	−42.85	0.8503	1354.58	−0.106	1435.49	22.51	358.26	−11.74
0.3002	1113.94	−0.315	1309.80	52.44	523.27	−46.22	0.8905	1365.36	−0.078	1441.91	16.72	352.27	−8.47
0.3993	1176.17	−0.313	1337.94	56.49	474.96	−44.16	0.9510	1380.60	−0.043	1451.22	7.70	343.93	−3.76
0.5010	1228.67	−0.276	1364.56	55.40	437.10	−38.78							
$T = 298.15$ K													
0.0501	869.67	−0.186	1215.57	18.46	778.19	−26.04	0.6007	1263.40	−0.269	1364.32	54.62	425.23	−37.08
0.1001	929.80	−0.283	1226.27	30.39	715.22	−39.28	0.7000	1300.00	−0.214	1385.16	44.72	400.92	−27.79
0.1504	982.09	−0.302	1239.17	39.78	663.11	−47.04	0.7996	1331.65	−0.155	1403.79	31.99	381.07	−18.35
0.2004	1028.09	−0.328	1253.67	47.50	618.87	−51.95	0.8503	1346.12	−0.120	1412.66	24.77	372.25	−13.68
0.3002	1105.54	−0.357	1284.29	58.18	548.40	−55.25	0.8905	1356.91	−0.088	1418.96	18.27	366.02	−9.80
0.3993	1167.74	−0.356	1313.39	62.30	496.44	−52.29	0.9510	1372.15	−0.048	1428.32	8.38	357.23	−4.32
0.5010	1220.23	−0.315	1340.54	60.74	456.04	−45.52							
$T = 308.15$ K													
0.0501	861.50	−0.210	1184.31	20.48	827.59	−31.60	0.6007	1255.01	−0.311	1341.14	59.66	443.00	−43.26

0.1001	921.55	−0.319	1197.09	33.96	757.23	−47.81	0.7000	1291.61	−0.247	1362.34	48.90	417.15	−32.35
0.1504	973.78	−0.345	1211.40	44.30	699.78	−56.94	0.7996	1323.26	−0.179	1381.34	35.17	396.05	−21.41
0.2004	1019.75	−0.375	1227.00	52.71	651.36	−62.51	0.8503	1337.74	−0.138	1390.29	27.27	386.74	−15.96
0.3002	1097.15	−0.409	1259.12	64.10	574.91	−65.71	0.8905	1348.52	−0.102	1396.49	20.04	380.25	−11.38
0.3993	1159.35	−0.408	1289.20	68.36	518.97	−61.70	0.9510	1363.76	−0.055	1405.94	9.25	370.96	−5.04
0.5010	1211.84	−0.364	1316.90	66.39	475.83	−53.32							
T = 318.15 K													
0.0501	853.22	−0.239	1153.43	22.59	880.96	−38.21	0.6007	1246.67	−0.363	1318.36	65.03	461.51	−50.42
0.1001	913.23	−0.363	1168.11	37.50	802.52	−57.66	0.7000	1283.29	−0.290	1339.97	53.40	434.00	−37.65
0.1504	965.42	−0.397	1183.85	48.87	739.08	−68.44	0.7996	1314.94	−0.209	1359.35	38.59	411.56	−24.95
0.2004	1011.37	−0.433	1200.55	58.01	686.01	−74.75	0.8503	1329.41	−0.161	1368.23	29.81	401.81	−18.49
0.3002	1088.78	−0.473	1234.23	70.22	602.93	−77.85	0.8905	1340.19	−0.118	1374.46	21.92	394.97	−13.18
0.3993	1150.97	−0.472	1265.31	74.69	542.68	−72.60	0.9510	1355.43	−0.062	1384.01	10.13	385.16	−5.83
0.5010	1203.49	−0.424	1293.63	72.38	496.52	−62.37							
T = 328.15 K													
0.0501	844.80	−0.274	1122.65	24.72	939.20	−46.02	0.6007	1238.39	−0.428	1295.93	71.00	480.82	−58.93
0.1001	904.80	−0.417	1139.17	41.08	851.67	−69.23	0.7000	1275.01	−0.340	1317.93	58.43	451.55	−43.94
0.1504	956.99	−0.462	1156.49	53.69	781.28	−82.14	0.7996	1306.67	−0.245	1337.64	42.40	427.72	−29.12
0.2004	1002.94	−0.505	1174.29	63.62	723.06	−89.29	0.8503	1321.14	−0.188	1346.57	32.78	417.44	−21.56
0.3002	1080.39	−0.552	1209.53	76.74	632.68	−92.20	0.8905	1331.92	−0.138	1352.89	24.23	410.20	−15.42
0.3993	1142.62	−0.550	1241.71	81.54	567.62	−85.52	0.9510	1347.16	−0.072	1362.50	11.30	399.86	−6.87
0.5010	1195.18	−0.498	1270.71	78.97	518.17	−73.14							
T = 338.15 K													
0.0501	836.19	−0.319	1092.05	27.11	1002.79	−55.71	0.6007	1230.15	−0.504	1273.89	77.58	500.93	−69.03
0.1001	896.24	−0.484	1110.24	44.79	905.19	−83.03	0.7000	1266.78	−0.401	1296.24	63.94	469.82	−51.34
0.1504	948.48	−0.543	1129.26	58.81	826.77	−98.59	0.7996	1298.46	−0.288	1316.46	46.71	444.38	−34.11
0.2004	994.46	−0.593	1148.19	69.59	762.75	−106.69	0.8503	1312.91	−0.218	1325.31	36.03	433.64	−25.14
0.3002	1072.00	−0.650	1185.00	83.68	664.31	−109.24	0.8905	1323.72	−0.163	1331.71	26.72	425.98	−18.03
0.3993	1134.29	−0.646	1218.28	88.86	593.99	−100.76	0.9510	1338.96	−0.083	1341.44	12.57	415.04	−8.07
0.5010	1186.91	−0.588	1248.07	86.12	540.88	−85.87							

Mixture 4: H_2O (1) + [emim][triflate] (2)
(1) Volumetric Properties, Speed of Sound, Compressibility Properties and Transport Properties

Ref. [67]

x_2	ρ kg · m^{-3}	V^E_m cm^3 · mol^{-1}	u m · s^{-1}	u^E m · s^{-1}	κ_S T Pa^{-1}	κ_S^E T Pa^{-1}	x_2	ρ kg · m^{-3}	V^E_m cm^3 · mol^{-1}	u m · s^{-1}	u^E m · s^{-1}	κ_S T Pa^{-1}	κ_S^E T Pa^{-1}
T = 278.15 K													
0.0178	1065.40	−0.048	1510.49	99.62	411.39	−61.23	0.4976	1364.71	0.006	1510.68	59.52	321.08	−26.86
0.0502	1146.23	−0.116	1562.95	163.00	357.14	−89.96	0.5495	1370.64	0.030	1505.27	49.50	321.99	−22.17
0.1004	1218.92	−0.133	1575.03	175.06	330.71	−89.48	0.6039	1375.99	0.049	1500.77	40.63	322.67	−18.07
0.1501	1262.90	−0.149	1569.54	163.39	321.43	−80.42	0.6476	1379.90	0.049	1497.41	34.06	323.20	−15.09
0.2002	1292.15	−0.138	1559.57	145.74	318.18	−70.01	0.6996	1383.92	0.057	1493.96	27.11	323.75	−11.93
0.2507	1312.86	−0.106	1548.97	127.43	317.46	−60.14	0.7510	1387.36	0.071	1491.05	21.02	324.21	−9.18
0.2980	1327.87	−0.093	1539.76	111.43	317.64	−52.00	0.8496	1392.97	0.090	1486.62	11.20	324.83	−4.76
0.3498	1340.33	−0.055	1530.48	95.34	318.52	−43.99	0.8942	1395.40	0.070	1484.81	7.21	325.06	−3.04
0.4000	1350.21	−0.035	1522.78	81.67	319.39	−37.38	0.9498	1398.09	0.048	1482.92	2.80	325.26	−1.14
0.4498	1358.15	−0.007	1516.22	69.74	320.28	−31.66							
T = 288.15 K													
0.0178	1062.57	−0.028	1528.98	84.01	402.57	−48.77	0.4976	1356.19	0.073	1488.54	49.16	332.78	−22.86
0.0502	1140.95	−0.066	1563.67	137.61	358.46	−73.59	0.5495	1362.12	0.091	1483.02	40.88	333.80	−18.91
0.1004	1211.96	−0.061	1564.93	147.69	336.92	−74.60	0.6039	1367.47	0.103	1478.28	33.51	334.63	−15.40
0.1501	1255.19	−0.067	1554.13	137.17	329.85	−67.57	0.6476	1371.38	0.098	1474.81	28.10	335.25	−12.88
0.2002	1284.06	−0.052	1541.34	121.84	327.81	−59.08	0.6996	1375.40	0.099	1471.28	22.42	335.88	−10.23
0.2507	1304.57	−0.021	1529.19	106.19	327.80	−50.88	0.7510	1378.85	0.105	1468.21	17.41	336.44	−7.87
0.2980	1319.46	−0.009	1519.17	92.71	328.39	−44.12	0.8496	1384.47	0.112	1463.58	9.45	337.20	−4.16
0.3498	1331.85	0.026	1509.22	79.07	329.64	−37.33	0.8942	1386.90	0.084	1461.71	6.23	337.47	−2.73
0.4000	1341.71	0.041	1501.12	67.60	330.76	−31.75	0.9498	1389.61	0.055	1459.69	2.65	337.74	−1.13
0.4498	1349.63	0.064	1494.29	57.66	331.83	−26.92							

T = 298.15 K													
0.0178	1058.88	−0.011	1541.84	71.81	397.26	−40.00	0.4976	1347.68	0.131	1466.70	41.16	344.93	−19.74
0.0502	1135.19	−0.024	1561.14	117.44	361.45	−61.58	0.5495	1353.62	0.144	1461.07	34.19	346.07	−16.32
0.1004	1204.72	0.001	1552.87	125.88	344.23	−63.40	0.6039	1358.98	0.150	1456.21	28.00	347.01	−13.29
0.1501	1247.30	0.004	1537.62	116.45	339.10	−57.82	0.6476	1362.90	0.139	1452.67	23.45	347.70	−11.11
0.2002	1275.85	0.023	1522.51	103.08	338.13	−50.72	0.6996	1366.93	0.134	1449.06	18.71	348.40	−8.82
0.2507	1296.20	0.055	1509.12	89.64	338.75	−43.79	0.7510	1370.38	0.134	1445.90	14.51	349.04	−6.78
0.2980	1311.01	0.066	1498.42	78.14	339.72	−38.05	0.8496	1376.01	0.131	1441.05	7.84	349.96	−3.56
0.3498	1323.37	0.097	1488.01	66.48	341.28	−32.20	0.8942	1378.45	0.097	1439.22	5.27	350.23	−2.38
0.4000	1333.21	0.109	1479.63	56.75	342.61	−27.41	0.9498	1381.16	0.061	1437.12	2.30	350.57	−1.00
0.4498	1341.13	0.127	1472.59	48.34	343.85	−23.24							
T = 308.15 K													
0.0178	1054.44	0.003	1549.85	61.99	394.82	−33.52	0.4976	1339.19	0.184	1445.00	34.99	357.62	−17.30
0.0502	1128.98	0.013	1555.73	101.01	365.97	−52.39	0.5495	1345.14	0.192	1439.33	29.07	358.85	−14.31
0.1004	1197.22	0.056	1539.08	108.08	352.62	−54.63	0.6039	1350.52	0.192	1434.39	23.81	359.88	−11.66
0.1501	1239.24	0.068	1520.07	99.68	349.23	−50.11	0.6476	1354.45	0.177	1430.80	19.94	360.64	−9.76
0.2002	1267.53	0.091	1503.06	88.01	349.21	−44.10	0.6996	1358.50	0.166	1427.15	15.94	361.41	−7.76
0.2507	1287.75	0.125	1488.67	76.41	350.41	−38.15	0.7510	1361.96	0.161	1423.95	12.39	362.11	−5.98
0.2980	1302.52	0.134	1477.44	66.56	351.72	−33.22	0.8496	1367.61	0.147	1418.93	6.71	363.17	−3.13
0.3498	1314.85	0.163	1466.72	56.57	353.53	−28.14	0.8942	1370.05	0.109	1417.16	4.65	363.43	−2.16
0.4000	1324.69	0.171	1458.13	48.25	355.05	−23.97	0.9498	1372.76	0.067	1415.00	2.14	363.82	−0.97
0.4498	1332.62	0.185	1450.96	41.09	356.44	−20.34							
T = 318.15 K													
0.0178	1049.35	0.016	1553.49	53.79	394.88	−28.51	0.4976	1330.70	0.235	1423.37	29.85	370.92	−15.19
0.0502	1122.35	0.046	1547.70	87.23	371.96	−45.04	0.5495	1336.69	0.237	1417.70	24.81	372.22	−12.57
0.1004	1189.47	0.108	1523.69	93.04	362.12	−47.39	0.6039	1342.09	0.232	1412.73	20.31	373.34	−10.24
0.1501	1231.02	0.129	1501.54	85.57	360.30	−43.68	0.6476	1346.04	0.212	1409.15	17.02	374.13	−8.58
0.2002	1259.09	0.156	1482.97	75.37	361.14	−38.52	0.6996	1350.10	0.196	1405.49	13.62	374.95	−6.83

(*continued*)

x_2	ρ kg · m^{-3}	V^E_m cm^3 · mol^{-1}	u m · s^{-1}	u^E m · s^{-1}	κ_S T Pa^{-1}	κ_S^E T Pa^{-1}	x_2	ρ kg · m^{-3}	V^E_m cm^3 · mol^{-1}	u m · s^{-1}	u^E m · s^{-1}	κ_S T Pa^{-1}	κ_S^E T Pa^{-1}
0.2507	1279.22	0.191	1467.81	65.35	362.84	−33.37	0.7510	1353.58	0.186	1402.28	10.60	375.70	−5.26
0.2980	1293.96	0.199	1456.11	56.80	364.49	−29.07	0.8496	1359.26	0.161	1397.05	5.60	376.94	−2.67
0.3498	1306.30	0.226	1445.27	48.30	366.49	−24.66	0.8942	1361.70	0.120	1395.46	4.08	377.12	−1.95
0.4000	1316.15	0.230	1436.58	41.18	368.16	−21.02	0.9498	1364.42	0.073	1393.27	1.95	377.56	−0.91
0.4498	1324.11	0.241	1429.34	35.05	369.66	−17.84							
T = 328.15 K													
0.0178	1043.69	0.029	1553.11	46.84	397.22	−24.51	0.4976	1322.22	0.284	1401.73	25.59	384.92	−13.37
0.0502	1115.22	0.080	1537.13	75.43	379.50	−38.93	0.5495	1328.25	0.280	1396.18	21.36	386.22	−11.11
0.1004	1181.47	0.157	1506.73	80.10	372.83	−41.22	0.6039	1333.68	0.271	1391.20	17.49	387.41	−9.05
0.1501	1222.62	0.187	1482.00	73.50	372.40	−38.15	0.6476	1337.66	0.246	1387.65	14.68	388.24	−7.60
0.2002	1250.52	0.219	1462.19	64.58	374.02	−33.68	0.6996	1341.75	0.225	1384.04	11.81	389.07	−6.09
0.2507	1270.60	0.255	1446.49	55.97	376.15	−29.21	0.7510	1345.25	0.210	1380.83	9.22	389.87	−4.70
0.2980	1285.34	0.262	1434.60	48.69	378.02	−25.51	0.8496	1350.97	0.175	1375.47	4.80	391.25	−2.33
0.3498	1297.70	0.287	1423.60	41.33	380.23	−21.62	0.8942	1353.41	0.130	1374.04	3.72	391.36	−1.83
0.4000	1307.60	0.288	1414.91	35.26	382.00	−18.46	0.9498	1356.14	0.078	1371.85	1.90	391.82	−0.92
0.4498	1315.58	0.294	1407.67	30.02	383.60	−15.68							
T = 338.15 K													
0.0178	1037.50	0.042	1548.94	40.72	401.74	−21.16	0.4976	1313.75	0.332	1380.03	21.66	399.68	−11.55
0.0502	1107.94	0.109	1524.17	65.01	388.52	−33.70	0.5495	1319.83	0.323	1374.73	18.23	400.91	−9.69
0.1004	1173.24	0.206	1488.23	68.53	384.83	−35.64	0.6039	1325.30	0.309	1369.70	14.82	402.19	−7.82
0.1501	1214.07	0.244	1461.46	62.72	385.64	−33.07	0.6476	1329.31	0.279	1366.21	12.43	403.03	−6.56
0.2002	1241.85	0.281	1440.72	54.93	387.95	−29.17	0.6996	1333.43	0.253	1362.68	10.02	403.87	−5.27
0.2507	1261.90	0.319	1424.67	47.55	390.43	−25.30	0.7510	1336.96	0.233	1359.53	7.83	404.67	−4.06
0.2980	1276.67	0.325	1412.93	41.60	392.36	−22.25	0.8496	1342.74	0.187	1354.21	4.01	406.10	−1.96
0.3498	1289.07	0.348	1401.67	35.04	394.85	−18.69	0.8942	1345.17	0.141	1352.86	3.22	406.18	−1.61
0.4000	1299.02	0.345	1393.04	29.87	396.70	−15.96	0.9498	1347.92	0.083	1350.68	1.67	406.66	−0.82
0.4498	1307.06	0.348	1385.89	25.42	398.33	−13.55							

Ref. [70]

	T/K							
x_1	278.15	288.15	298.15	308.15	318.15	328.15	338.15	348.15
ρ/g · cm^{-3}								
0.0000	1.40052	1.39204	1.38360	1.37522	1.36690	1.35863	1.35043	1.34230
0.2295	1.38830	1.37981	1.37136	1.36295	1.35459	1.34627	1.33801	1.32977
0.4310	1.37238	1.36387	1.35539	1.34693	1.33849	1.33007	1.32168	1.31329
0.6157	1.34734	1.33887	1.33037	1.32186	1.31333	1.30478	1.29619	1.28758
0.7825	1.29991	1.29174	1.28347	1.27510	1.26662	1.25804	1.24935	1.24055
0.8608	1.25476	1.24716	1.23938	1.23142	1.22327	1.21494	1.20643	1.19774
0.9353	1.17190	1.16597	1.15965	1.15294	1.14588	1.13849	1.13076	1.12273
0.9712	1.09749	1.09364	1.08908	1.08390	1.07818	1.07195	1.06525	1.05813
0.9924	1.03035	1.02864	1.02588	1.02224	1.01786	1.01283	1.00721	1.00105
1.0000	0.99999	0.99913	0.99707	0.99406	0.99023	0.98571	0.98056	0.97484
V^E/cm^3 · mol^{-1}								
0.0000	0.000	0.000	0.000	0.000	0.000	0.000	0.000	0.000
0.2295	0.104	0.135	0.162	0.187	0.210	0.232	0.252	0.276
0.4310	0.062	0.120	0.170	0.216	0.260	0.302	0.343	0.385
0.6157	−0.040	0.037	0.106	0.170	0.230	0.287	0.346	0.404
0.7825	−0.123	−0.037	0.039	0.108	0.174	0.237	0.300	0.363
0.8608	−0.142	−0.061	0.008	0.071	0.131	0.188	0.243	0.299
0.9353	−0.124	−0.065	−0.016	0.027	0.067	0.105	0.142	0.179
0.9712	−0.075	−0.043	−0.016	0.007	0.028	0.049	0.069	0.088
0.9924	−0.018	−0.010	−0.003	0.002	0.008	0.013	0.018	0.024
1.0000	0.000	0.000	0.000	0.000	0.000	0.000	0.000	0.000
η/mPa · s								
0.0000	96	61	41	29	21	16	13	11
0.2286	47	32	23	17	13	10	9	7
0.4324	28	20	15	12	10	8	7	6
0.6159	18	13	10	8	7	6	5	5
0.7830	10	7	6	5	4	4	4	3
0.9352	5	4	3	3	3	3	2	2
$\Delta\eta$/mPa · s								
0.0000	0	0	0	0	0	0	0	0
0.2286	−27	−15	−9	−6	−4	−2	−1	−1
0.4324	−27	−15	−9	−5	−3	−1	−1	0
0.6159	−20	−11	−7	−4	−2	−1	0	1
0.7830	−12	−7	−4	−2	−1	0	1	1
0.9352	−3	−1	0	0	1	1	1	1

(2) Phase Equilibrium

Ref. [69]

x_2	T/K	x_2	T/K	x_2	T/K
100 kPa					
0.0000	372.81	0.1091	375.60	0.2482	380.90
0.0029	372.95	0.1190	375.87	0.2635	381.75
0.0069	373.13	0.1298	376.21	0.2836	382.56
0.0143	373.37	0.1413	376.56	0.3010	383.81
0.0251	373.67	0.1546	376.99	0.3199	384.89
0.0367	373.94	0.1699	377.55	0.3359	385.70
0.0489	374.19	0.1868	378.25	0.3561	387.14
0.0603	374.44	0.2016	378.84	0.3741	388.41
0.0725	374.71	0.2167	379.51	0.3971	389.81
0.0874	375.05	0.2327	380.15	0.4145	391.10
0.0988	375.34				

(3) Excess Molar Enthalpy

Ref. [71]

x_2	$H^E/(J \cdot mol^{-1})$	x_2	$H^E/(J \cdot mol^{-1})$	x_2	$H^E/(J \cdot mol^{-1})$
303.15 K					
0.1012	487	0.4031	1071	0.6997	763
0.1943	794	0.4921	1059	0.7805	601
0.2959	991	0.5929	964	0.8440	448

Mixture 5: $CH_3COOCH_2CH_3$ (1) + [emim][triflate] (2)
(1) Phase Equilibrium

Ref. [66]

x_2	T/K	x_2	T/K	x_2	T/K
100 kPa					
0.0000	349.84	0.2255	350.19	0.3525	351.20
0.0187	349.98	0.2478	350.26	0.3690	351.51
0.0468	350.02	0.2547	350.29	0.3833	351.85
0.0769	349.99	0.2695	350.34	0.4054	352.31
0.1133	350.04	0.2890	350.55	0.4090	352.36
0.1464	350.05	0.3106	350.73	0.4133	352.38
0.1748	350.07	0.3149	350.84	0.4155	352.50
0.2005	350.07	0.3349	351.00		

Mixture 6: CH_3OH (1) + [emim][triflate] (2)

(1) Volumetric Properties, Speed of Sound and Compressibility Properties

Ref. [67]

x_2	ρ kg · m^{-3}	V^E_m cm^3 · mol^{-1}	u m · s^{-1}	u^E m · s^{-1}	κ_S T Pa^{-1}	κ_S^E T Pa^{-1}	x_2	ρ kg · m^{-3}	V^E_m cm^3 · mol^{-1}	u m · s^{-1}	u^E m · s^{-1}	κ_S T Pa^{-1}	κ_S^E T Pa^{-1}
T = 278.15 K													
0.0505	930.38	−0.336	1216.34	52.35	726.49	−72.50	0.6012	1330.42	−0.378	1432.72	63.87	366.18	−36.16
0.1005	1018.16	−0.473	1253.28	78.68	625.30	−92.81	0.6994	1353.20	−0.298	1447.85	47.31	352.53	−25.01
0.1502	1084.09	−0.548	1284.95	93.21	558.68	−96.63	0.7965	1371.53	−0.216	1460.58	31.32	341.78	−15.64
0.1999	1135.25	−0.574	1312.52	100.89	511.32	−93.73	0.8443	1379.25	−0.161	1466.36	23.84	337.19	−11.58
0.3005	1210.76	−0.573	1356.49	102.43	448.86	−79.94	0.8984	1387.51	−0.131	1472.01	15.08	332.62	−7.18
0.4007	1262.73	−0.514	1388.69	93.46	410.66	−63.88	0.9489	1394.45	−0.092	1477.51	7.74	328.50	−3.64
0.4926	1298.36	−0.467	1411.35	80.86	386.67	−50.26							
T = 288.15 K													
0.0505	921.13	−0.367	1185.98	55.43	771.84	−84.12	0.6012	1321.68	−0.406	1409.11	68.97	381.05	−41.57
0.1005	1008.98	−0.516	1225.28	83.79	660.16	−107.68	0.6994	1344.52	−0.318	1424.43	51.44	366.56	−28.86
0.1502	1074.98	−0.598	1258.61	99.70	587.24	−112.09	0.7965	1362.92	−0.228	1437.19	34.31	355.22	−18.13
0.1999	1126.21	−0.626	1287.07	107.96	536.02	−108.45	0.8443	1370.67	−0.170	1442.97	26.24	350.39	−13.48
0.3005	1201.81	−0.622	1332.02	109.74	468.97	−92.13	0.8984	1378.95	−0.137	1448.67	16.87	345.55	−8.47
0.4007	1253.87	−0.557	1364.73	100.38	428.21	−73.52	0.9489	1385.93	−0.096	1453.96	8.70	341.31	−4.30
0.4926	1289.56	−0.504	1387.66	87.11	402.71	−57.81							
T = 298.15 K													
0.0505	911.87	−0.401	1156.48	58.62	819.96	−97.55	0.6012	1313.03	−0.439	1386.04	73.93	396.44	−47.44
0.1005	999.81	−0.564	1197.97	88.90	696.93	−124.56	0.6994	1335.93	−0.343	1401.54	55.34	381.07	−32.96
0.1502	1065.90	−0.653	1232.80	106.06	617.30	−129.43	0.7965	1354.37	−0.246	1414.37	37.01	369.09	−20.71
0.1999	1117.20	−0.683	1262.10	114.89	561.93	−124.87	0.8443	1362.15	−0.183	1420.23	28.38	363.96	−15.42

(continued)

x_2	ρ kg · m^{-3}	V^E_m cm^3 · mol^{-1}	u m · s^{-1}	u^E m · s^{-1}	κ_S T Pa^{-1}	κ_S^E T Pa^{-1}	x_2	ρ kg · m^{-3}	V^E_m cm^3 · mol^{-1}	u m · s^{-1}	u^E m · s^{-1}	κ_S T Pa^{-1}	κ_S^E T Pa^{-1}
0.3005	1192.91	−0.677	1307.97	116.88	490.00	−105.63	0.8984	1370.46	−0.146	1425.92	18.28	358.87	−9.69
0.4007	1245.07	−0.607	1341.25	107.18	446.47	−84.15	0.9489	1377.46	−0.100	1431.14	9.36	354.45	−4.88
0.4926	1280.83	−0.547	1364.43	93.18	419.38	−66.07							
T = 308.15 K													
0.0505	902.59	−0.439	1127.52	61.78	871.49	−112.84	0.6012	1304.44	−0.479	1363.41	79.15	412.40	−54.12
0.1005	990.64	−0.617	1171.25	94.11	735.85	−143.87	0.6994	1327.40	−0.373	1379.13	59.53	396.09	−37.66
0.1502	1056.83	−0.713	1207.40	112.42	649.07	−149.07	0.7965	1345.89	−0.266	1391.99	39.91	383.46	−23.65
0.1999	1108.22	−0.746	1237.54	121.88	589.19	−143.48	0.8443	1353.69	−0.199	1398.03	30.80	377.96	−17.69
0.3005	1184.06	−0.739	1284.33	124.17	512.00	−120.92	0.8984	1362.03	−0.157	1403.71	19.92	372.61	−11.15
0.4007	1236.32	−0.662	1318.17	114.18	465.51	−96.18	0.9489	1369.05	−0.106	1408.80	10.15	368.03	−5.58
0.4926	1272.16	−0.595	1341.66	99.55	436.69	−75.47							
T = 318.15 K													
0.0505	893.26	−0.482	1099.19	65.21	926.56	−130.90	0.6012	1295.91	−0.524	1341.24	84.71	428.96	−61.75
0.1005	981.42	−0.674	1144.83	99.33	777.43	−165.96	0.6994	1318.93	−0.408	1357.34	64.14	411.53	−43.10
0.1502	1047.74	−0.778	1182.47	119.01	682.60	−171.71	0.7965	1337.48	−0.290	1370.14	43.06	398.28	−27.04
0.1999	1099.25	−0.817	1213.45	129.15	617.82	−164.88	0.8443	1345.30	−0.217	1376.24	33.29	392.46	−20.23
0.3005	1175.24	−0.807	1261.19	131.85	534.95	−138.49	.8984	53.66	−0.169	1381.97	21.64	386.81	−12.79
0.4007	1227.62	−0.725	1295.51	121.52	485.35	−109.93	0.9489	1360.70	−0.112	1386.98	11.00	382.03	−6.37
0.4926	1263.55	−0.651	1319.40	106.33	454.63	−86.25							

(2) Phase Equilibrium

Ref. [72]

x_2	T/K	x_2	T/K	x_2	T/K	x_2	T/K
100 kPa							
0.0030	337.58	0.0903	340.21	0.1866	343.69	0.2921	348.45
0.0061	337.68	0.1048	340.68	0.1980	344.13	0.3065	349.35
0.0117	337.87	0.1126	340.90	0.2110	344.67	0.3246	350.35
0.0192	338.10	0.1247	341.37	0.2228	345.19	0.3442	351.55
0.0281	338.36	0.1352	341.69	0.2334	345.67	0.3656	352.75
0.0373	338.63	0.1447	342.05	0.2455	346.21	0.3912	354.55
0.0480	338.94	0.1523	342.30	0.2564	346.70	0.4235	356.55
0.0563	339.19	0.1612	342.70	0.2687	347.32	0.4504	358.35
0.0649	339.42	0.1669	342.88	0.2782	347.78	0.4743	360.15
0.0774	339.81	0.1798	343.30				

Mixture 7: CH_3COOCH_3 (1) + [emim][triflate] (2)
(1) Phase Equilibrium

Ref. [73]

x_2	T/K	x_2	T/K	x_2	T/K	x_2	T/K
100 kPa							
0.0000	329.72	0.1577	330.61	0.2681	332.36	0.3634	335.04
0.0193	329.89	0.1779	330.84	0.2815	332.72	0.3767	335.55
0.0374	329.94	0.1959	331.07	0.2982	333.11	0.3933	336.25
0.0591	330.00	0.2107	331.30	0.3082	333.38	0.4096	336.85
0.0846	330.11	0.2289	331.58	0.3204	333.79	0.4258	337.35
0.1105	330.23	0.2397	331.78	0.3344	334.09	0.4307	337.55
0.1346	330.40	0.2555	332.14	0.3474	334.55	0.4318	337.55

Mixture 8: CH_3COCH_3 (1) + [emim][triflate] (2)
(1) Phase Equilibrium

Ref. [72]

x_2	T/K	x_2	T/K	x_2	T/K	x_2	T/K
100 kPa							
0.0157	329.45	0.1721	333.25	0.2866	337.45	0.4022	343.45
0.0263	329.69	0.1867	333.75	0.2945	337.95	0.4105	343.55
0.0395	330.01	0.2010	334.15	0.3095	338.45	0.4125	344.05
0.0544	330.34	0.2124	334.65	0.3236	339.15	0.4282	344.95
0.0700	330.69	0.2240	334.95	0.3435	340.25	0.4349	345.45
0.0832	331.01	0.2359	335.35	0.3532	340.60	0.4399	345.95
0.1004	331.38	0.2481	335.85	0.3636	341.21	0.4513	346.45
0.1150	331.73	0.2596	336.30	0.3686	341.45	0.4609	347.25
0.1282	332.05	0.2702	336.65	0.3791	342.05	0.4722	347.95
0.1448	332.55	0.2769	337.05	0.3908	342.77	0.4876	348.95
0.1585	332.85						

Mixture 9: CH_3NO_2 (1) + [emim][triflate] (2)
(1) Excess Molar Enthalpy

Ref. [74]

x_2	$H^E/(J \cdot mol^{-1})$	x_2	$H^E/(J \cdot mol^{-1})$	x_2	$H^E/(J \cdot mol^{-1})$
303.15 K					
0.1234	−126	0.5026	−201	0.7438	−125
0.2533	−191	0.6253	−167	0.8668	−58
0.3774	−196				

2.2.11.2 Ternary Mixtures

Mixture 1: $CHCl_3$ (1) + C_2H_5OH (2) + [emim][triflate] (3)
(1) Phase Equilibrium

Ref. [68]

x_3	x_1'	y_1	T/K
100 kPa			
0.0525	0.000	0.000	352.28
0.0532	0.031	0.091	350.53
0.0535	0.069	0.188	348.52
0.0537	0.114	0.284	346.33
0.0541	0.157	0.372	344.53
0.0545	0.204	0.449	342.46
0.0550	0.257	0.516	340.47
0.0555	0.316	0.578	338.64
0.0557	0.380	0.633	336.97
0.0558	0.446	0.685	335.58
0.0561	0.514	0.725	334.47
0.0559	0.587	0.758	333.59
0.0554	0.642	0.784	333.00
0.0548	0.719	0.818	332.53
0.0561	0.821	0.864	332.28
0.0628	0.906	0.907	332.41
0.0626	0.943	0.929	332.72
0.0627	0.972	0.953	333.14
0.0611	0.989	0.974	333.39
0.0600	1.000	1.000	333.96
0.1170	0.000	0.000	353.50
0.1183	0.030	0.078	351.78
0.1195	0.069	0.175	349.92
0.1210	0.113	0.271	347.89
0.1217	0.164	0.365	345.68
0.1247	0.217	0.447	343.62
0.1259	0.272	0.522	341.67
0.1272	0.331	0.585	339.88
0.1287	0.393	0.641	338.15

(*continued*)

x_3	x_1'	y_1	T/K
0.1292	0.461	0.696	336.59
0.1290	0.525	0.738	335.31
0.1285	0.601	0.781	334.16
0.1286	0.669	0.813	333.40
0.1273	0.744	0.848	332.83
0.1276	0.847	0.893	332.57
0.1296	0.916	0.926	332.84
0.1314	0.978	0.969	333.38
0.1300	1.000	1.000	333.96
0.1920	0.000	0.000	355.12
0.1948	0.029	0.066	353.44
0.1970	0.067	0.150	351.91
0.1998	0.103	0.229	350.36
0.2026	0.148	0.321	348.53
0.2050	0.200	0.409	346.61
0.2082	0.256	0.483	344.84
0.2111	0.318	0.551	342.92
0.2136	0.380	0.616	341.30
0.2149	0.444	0.674	339.69
0.2150	0.509	0.722	338.26
0.2136	0.574	0.766	336.95
0.2118	0.629	0.802	335.94
0.2116	0.699	0.837	334.97
0.2122	0.804	0.889	334.03
0.2117	0.885	0.932	333.52
0.2081	0.951	0.968	333.43
0.2185	1.000	1.000	333.96
0.3062	0.000	0.000	358.60
0.3064	0.033	0.076	356.81
0.3078	0.060	0.139	355.77
0.3088	0.099	0.214	354.33
0.3093	0.140	0.286	352.74
0.3114	0.192	0.373	350.95
0.3155	0.264	0.474	349.09
0.3171	0.335	0.557	347.16
0.3204	0.412	0.628	345.22
0.3209	0.485	0.691	343.38
0.3217	0.560	0.749	341.81
0.3234	0.629	0.795	340.65
0.3268	0.707	0.840	339.48
0.3286	0.804	0.891	338.29
0.3305	0.876	0.928	337.39
0.3288	0.935	0.961	336.78
0.3260	0.973	0.984	336.56
0.3228	1.000	1.000	336.26

Mixture 2: 1-C_3H_7OH (1) + H_2O (2) + [emim][triflate] (3)
(1) Phase Equilibrium

Ref. [69]

x_3	x_1'	y_1	T/K	x_3	x_1'	y_1	T/K	x_3	x_1'	y_1	T/K
100 kPa				100 kPa				100 kPa			
0.0550	0.000	0.000	374.28	0.0996	0.323	0.439	363.67	0.1882	0.761	0.740	368.91
0.0543	0.009	0.059	372.38	0.0992	0.403	0.473	363.45	0.1863	0.829	0.799	369.75
0.0538	0.030	0.189	369.25	0.0985	0.483	0.511	363.53	0.1883	0.883	0.848	370.68
0.0533	0.079	0.301	365.30	0.0983	0.571	0.563	363.93	0.1877	0.923	0.893	371.54
0.0550	0.157	0.373	363.42	0.0983	0.658	0.616	364.58	0.1886	0.955	0.932	372.28
0.0559	0.238	0.410	362.60	0.0981	0.737	0.678	365.51	0.1782	1.000	1.000	372.85
0.0558	0.334	0.445	362.14	0.0974	0.798	0.728	366.47	0.3010	0.000	0.000	383.81
0.0566	0.434	0.481	362.08	0.0987	0.851	0.778	367.53	0.3164	0.013	0.036	383.33
0.0576	0.523	0.516	362.32	0.1003	0.897	0.841	368.62	0.3165	0.029	0.081	382.06
0.0583	0.604	0.554	362.71	0.1030	0.935	0.882	369.76	0.3120	0.066	0.166	379.78
0.0592	0.680	0.603	363.40	0.1041	0.966	0.936	370.51	0.3158	0.126	0.270	377.54
0.0597	0.752	0.663	364.30	0.0989	1.000	1.000	371.48	0.3142	0.184	0.344	375.79
0.0594	0.820	0.724	365.37	0.2069	0.000	0.000	378.88	0.3143	0.248	0.405	374.19
0.0581	0.879	0.784	366.51	0.2010	0.023	0.081	376.61	0.3136	0.326	0.465	373.08
0.0587	0.928	0.846	367.86	0.1980	0.054	0.158	375.05	0.3113	0.403	0.522	372.46
0.0590	0.961	0.901	369.02	0.1956	0.093	0.232	372.30	0.3061	0.471	0.577	372.10
0.0589	0.980	0.945	369.82	0.1925	0.146	0.304	370.18	0.3029	0.548	0.633	372.14
0.0627	1.000	1.000	370.90	0.1910	0.211	0.371	368.51	0.3000	0.628	0.690	372.25
0.1015	0.000	0.000	375.27	0.1909	0.284	0.427	367.48	0.2988	0.705	0.741	372.62
0.1004	0.024	0.107	372.67	0.1904	0.361	0.478	367.02	0.2940	0.768	0.794	372.96
0.1024	0.058	0.198	370.54	0.1908	0.450	0.527	366.77	0.2951	0.837	0.848	373.74
0.1009	0.106	0.279	367.41	0.1903	0.542	0.585	366.97	0.2945	0.894	0.896	374.29
0.0990	0.164	0.346	365.32	0.1897	0.624	0.640	367.56	0.2955	0.956	0.951	374.97
0.0991	0.240	0.401	364.23	0.1887	0.699	0.694	368.11	0.2969	1.000	1.000	375.63

Mixture 3: $CH_3COOCH_2CH_3$ (1) + C_2H_5OH (2) + [emim][triflate] (3)
(1) Phase Equilibrium

Ref. [66]

x_3	x_1'	y_1	T/K	x_3	x_1'	y_1	T/K
100 kPa							
0.054	0.000	0.000	352.29	0.202	0.218	0.343	350.23
0.054	0.020	0.045	351.45	0.204	0.231	0.353	350.08
0.054	0.045	0.096	350.68	0.204	0.289	0.417	349.39
0.054	0.096	0.178	349.28	0.203	0.299	0.433	349.25
0.055	0.152	0.257	348.13	0.205	0.362	0.484	348.70
0.055	0.216	0.333	347.17	0.205	0.379	0.499	348.63
0.055	0.285	0.392	346.46	0.208	0.430	0.554	348.27
0.056	0.354	0.450	345.97	0.208	0.505	0.612	348.00
0.056	0.425	0.499	345.63	0.210	0.589	0.667	347.86
0.056	0.497	0.546	345.49	0.217	0.594	0.682	347.97

(continued)

x_3	x_1'	y_1	T/K	x_3	x_1'	y_1	T/K
0.057	0.566	0.592	345.50	0.210	0.660	0.720	347.84
0.057	0.633	0.636	345.67	0.215	0.666	0.728	347.87
0.057	0.710	0.695	345.97	0.213	0.727	0.770	348.00
0.057	0.779	0.750	346.48	0.209	0.730	0.774	347.95
0.058	0.849	0.813	347.23	0.210	0.783	0.815	348.18
0.058	0.905	0.872	348.06	0.206	0.801	0.826	348.24
0.058	0.948	0.925	348.85	0.206	0.845	0.861	348.48
0.047	1.000	1.000	350.02	0.199	0.873	0.881	348.69
0.096	0.000	0.000	353.07	0.199	0.900	0.907	348.93
0.097	0.021	0.045	352.09	0.194	0.924	0.927	349.12
0.098	0.043	0.087	351.44	0.194	0.941	0.943	349.33
0.100	0.092	0.177	350.24	0.192	0.962	0.963	349.55
0.101	0.156	0.268	348.89	0.192	0.971	0.970	349.68
0.104	0.226	0.346	347.82	0.201	1.000	1.000	350.07
0.105	0.298	0.416	347.04	0.297	0.000	0.000	358.25
0.105	0.375	0.479	346.53	0.296	0.018	0.039	357.68
0.105	0.457	0.538	346.20	0.296	0.042	0.088	357.01
0.104	0.535	0.601	346.07	0.299	0.091	0.173	355.79
0.104	0.608	0.651	346.17	0.301	0.147	0.257	354.66
0.104	0.681	0.696	346.36	0.303	0.209	0.339	353.55
0.103	0.740	0.743	346.70	0.305	0.277	0.417	352.56
0.103	0.801	0.790	347.14	0.304	0.352	0.491	351.68
0.104	0.869	0.850	347.85	0.308	0.425	0.558	351.05
0.105	0.919	0.903	348.58	0.304	0.501	0.620	350.49
0.105	0.957	0.947	349.25	0.304	0.571	0.676	350.14
0.113	1.000	1.000	350.04	0.307	0.626	0.725	350.03
0.188	0.000	0.000	355.14	0.312	0.711	0.777	349.95
0.194	0.022	0.048	354.41	0.308	0.785	0.833	349.83
0.195	0.045	0.090	353.78	0.300	0.857	0.886	349.91
0.196	0.100	0.184	352.33	0.299	0.914	0.929	350.13
0.198	0.103	0.191	352.18	0.293	0.957	0.965	350.31
0.199	0.161	0.271	351.12	0.289	1.000	1.000	350.55
0.202	0.168	0.278	351.01				

Mixture 4: CH_3COOCH_3 (1) + CH_3OH (2) + [emim][triflate] (3)
(1) Phase Equilibrium
Ref. [73]

x_3	x_1'	y_1	T/K	x_3	x_1'	y_1	T/K	x_3	x_1'	y_1	T/K	x_3	x_1'	y_1	T/K
100 kPa															
0.051	0.000	0.000	338.65	0.113	0.437	0.586	329.42	0.172	0.000	0.000	343.03	0.250	0.507	0.667	333.40
0.052	0.021	0.067	337.45	0.115	0.494	0.629	328.93	0.173	0.020	0.057	341.88	0.251	0.572	0.712	332.85
0.057	0.038	0.113	336.35	0.112	0.560	0.670	328.58	0.177	0.045	0.116	340.81	0.248	0.640	0.760	332.30
0.052	0.055	0.155	335.55	0.114	0.619	0.709	328.45	0.177	0.077	0.195	339.37	0.249	0.700	0.801	332.10
0.057	0.092	0.234	334.39	0.112	0.679	0.748	328.38	0.178	0.119	0.266	337.83	0.246	0.756	0.834	331.80
0.052	0.103	0.253	333.96	0.113	0.738	0.786	328.44	0.179	0.179	0.355	336.05	0.243	0.810	0.872	331.66
0.053	0.166	0.346	332.01	0.110	0.805	0.834	328.63	0.180	0.245	0.435	334.30	0.239	0.866	0.907	331.53
0.059	0.313	0.477	329.14	0.106	0.871	0.885	328.97	0.180	0.312	0.494	333.21	0.240	0.916	0.942	331.62
0.059	0.396	0.548	328.30	0.106	0.939	0.942	329.52	0.182	0.378	0.556	332.19	0.240	0.962	0.973	331.73
0.059	0.454	0.592	327.85	0.111	1.000	1.000	330.23	0.182	0.439	0.605	331.48	0.240	1.000	1.000	331.78
0.059	0.535	0.633	327.52	0.141	0.000	0.000	341.79	0.180	0.507	0.651	330.90	0.289	0.000	0.000	348.35
0.059	0.605	0.674	327.32	0.141	0.017	0.050	340.83	0.181	0.572	0.697	330.42	0.295	0.018	0.050	347.57
0.058	0.672	0.713	327.31	0.142	0.039	0.107	339.70	0.180	0.636	0.739	330.13	0.297	0.043	0.110	346.36
0.057	0.733	0.752	327.43	0.142	0.070	0.175	338.31	0.179	0.695	0.779	329.95	0.302	0.074	0.183	345.02
0.056	0.786	0.788	327.62	0.143	0.108	0.248	336.87	0.176	0.751	0.815	329.87	0.298	0.119	0.254	343.53
0.056	0.838	0.829	327.93	0.143	0.160	0.328	335.21	0.174	0.807	0.853	329.85	0.303	0.176	0.351	341.87

0.055	0.894	0.879	328.43	0.144	0.221	0.401	333.66	0.170	0.863	0.892	329.95	0.304	0.234	0.422	340.40
0.055	0.941	0.928	329.02	0.145	0.287	0.466	332.30	0.173	0.915	0.931	330.23	0.307	0.290	0.484	339.22
0.055	0.979	0.973	329.62	0.146	0.358	0.533	331.30	0.174	0.963	0.967	330.52	0.309	0.352	0.540	338.11
0.055	1.000	1.000	330.01	0.146	0.437	0.595	330.41	0.178	1.000	1.000	330.84	0.313	0.408	0.593	337.56
0.107	0.000	0.000	340.66	0.147	0.503	0.641	329.83	0.234	0.000	0.000	345.50	0.311	0.472	0.645	336.64
0.108	0.017	0.053	339.63	0.147	0.569	0.686	329.47	0.233	0.020	0.056	344.23	0.317	0.532	0.697	335.90
0.109	0.037	0.106	338.57	0.148	0.629	0.726	329.27	0.235	0.045	0.116	343.34	0.315	0.598	0.741	335.44
0.109	0.067	0.171	337.31	0.147	0.688	0.764	329.23	0.236	0.079	0.186	342.00	0.311	0.659	0.781	334.87
0.109	0.102	0.241	335.89	0.147	0.741	0.798	329.19	0.238	0.123	0.264	340.40	0.308	0.721	0.825	334.23
0.110	0.152	0.320	334.22	0.147	0.794	0.837	329.29	0.243	0.178	0.355	338.80	0.306	0.782	0.865	333.83
0.111	0.207	0.392	332.80	0.144	0.847	0.873	329.42	0.243	0.247	0.430	337.22	0.300	0.847	0.905	333.32
0.112	0.261	0.449	331.69	0.144	0.903	0.918	329.73	0.249	0.307	0.493	336.25	0.296	0.903	0.939	333.17
0.112	0.322	0.491	330.66	0.146	0.956	0.960	330.15	0.243	0.382	0.561	335.00	0.298	0.957	0.972	333.13
0.113	0.377	0.545	329.95	0.146	1.000	1.000	330.51	0.248	0.442	0.614	334.10	0.298	1.000	1.000	333.11

Mixture 5: CH_3COCH_3 (1) + CH_3OH (2) + [emim][triflate] (3)
(1) Phase Equilibrium

Ref. [72]

x_3	x'_1	y_1	T/K	x_3	x'_1	y_1	T/K
100 kPa							
0.056	0.000	0.000	339.19	0.168	0.345	0.457	336.43
0.057	0.034	0.069	338.09	0.168	0.398	0.513	335.83
0.058	0.074	0.137	337.11	0.167	0.459	0.564	335.29
0.059	0.120	0.206	336.09	0.166	0.522	0.623	334.65
0.059	0.174	0.276	335.03	0.163	0.581	0.670	334.14
0.060	0.250	0.367	333.75	0.162	0.636	0.718	333.72
0.060	0.316	0.430	332.83	0.163	0.696	0.764	333.35
0.060	0.388	0.492	332.02	0.158	0.746	0.804	333.11
0.060	0.455	0.550	331.40	0.165	0.829	0.865	333.25
0.059	0.530	0.613	330.87	0.160	0.889	0.911	332.93
0.059	0.596	0.662	330.49	0.169	0.927	0.943	333.13
0.059	0.675	0.723	330.17	0.168	0.979	0.984	333.05
0.058	0.749	0.780	329.97	0.160	1.000	1.000	332.93
0.057	0.815	0.833	329.89	0.256	0.000	0.000	346.70
0.057	0.875	0.884	329.90	0.247	0.032	0.059	345.45
0.056	0.932	0.934	330.01	0.254	0.116	0.187	343.97
0.055	0.980	0.981	330.18	0.259	0.163	0.250	343.40
0.054	1.000	1.000	330.30	0.262	0.212	0.308	342.80
0.125	0.000	0.000	341.37	0.258	0.314	0.423	340.92
0.123	0.016	0.035	340.60	0.259	0.364	0.472	340.29
0.121	0.082	0.147	338.90	0.254	0.418	0.522	339.77
0.123	0.122	0.204	338.13	0.256	0.466	0.572	339.33
0.123	0.163	0.263	337.39	0.258	0.516	0.620	338.70
0.123	0.209	0.326	336.60	0.258	0.585	0.663	338.25
0.124	0.259	0.380	335.80	0.255	0.689	0.751	337.30
0.125	0.306	0.428	335.18	0.253	0.738	0.801	336.86
0.122	0.358	0.479	334.47	0.250	0.790	0.841	336.66
0.121	0.399	0.522	333.93	0.249	0.839	0.880	336.30
0.120	0.464	0.570	333.37	0.248	0.892	0.921	336.02
0.120	0.520	0.619	332.89	0.247	0.940	0.956	335.78
0.119	0.578	0.668	332.48	0.248	1.000	1.000	335.85
0.118	0.632	0.710	332.18	0.355	0.000	0.000	352.05
0.119	0.689	0.757	331.94	0.360	0.048	0.083	351.20
0.117	0.758	0.803	331.73	0.365	0.092	0.151	350.66
0.116	0.817	0.850	331.64	0.368	0.138	0.212	350.04
0.117	0.873	0.895	331.62	0.370	0.189	0.277	349.33
0.114	0.924	0.936	331.56	0.367	0.258	0.357	348.13
0.116	0.979	0.983	331.66	0.367	0.323	0.431	347.14
0.115	1.000	1.000	331.73	0.367	0.395	0.496	346.22

(*continued*)

x_3	x'_1	y_1	T/K	x_3	x'_1	y_1	T/K
0.167	0.000	0.000	342.88	0.365	0.469	0.570	345.35
0.163	0.021	0.046	342.01	0.363	0.536	0.642	344.52
0.167	0.057	0.104	341.34	0.362	0.610	0.703	343.71
0.169	0.085	0.147	340.78	0.361	0.686	0.770	342.72
0.168	0.123	0.201	340.05	0.356	0.756	0.825	342.02
0.168	0.160	0.250	339.30	0.355	0.831	0.886	341.46
0.171	0.201	0.302	338.75	0.351	0.900	0.937	340.85
0.172	0.248	0.356	338.11	0.358	1.000	1.000	340.62
0.166	0.298	0.412	336.96				

2.2.12 02003-04007: 1-Ethyl-3-methylimidazolium tosilate

Abbreviation:	[emim][TOS]
Molecular Formula:	$C_{13}H_{18}N_2O_3S$
Molar Mass:	282.37
Structure:	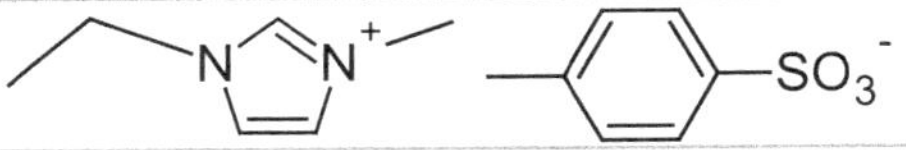

2.2.12.1 Ternary Mixtures

Mixture 1: C_6H_{14} (1) + CH_3CH_2COOH (2) + [emim][TOS] (3)
(1) Selectivity

Ref. [75]

Hexane-rich phase		RTIL-rich phase		
x_3'	x_2'	x_3''	x_2''	Selectivity (S)
0.0000	0.5221	0.0850	0.8050	7
0.0000	0.2083	0.1700	0.7300	28
0.0000	0.0573	0.3450	0.5800	127
0.0000	0.0212	0.5450	0.4000	336

Mixture 2: C_6H_{14} (1) + $CH_3CH_2COOCH = CH_2$ (Vinyl propionate) (2) + [emim] [TOS] (3)
(1) Selectivity

Ref. [75]

Hexane-rich phase		RTIL-rich phase		
x_3'	x_2'	x_3''	x_2''	Selectivity (S)
0.0008	0.8213	0.1176	0.8824	–
0.0007	0.6272	0.2533	0.7467	–
0.0005	0.4271	0.4315	0.5685	–
0.0002	0.2213	0.6762	0.3238	–
0.0000	0.1129	0.8266	0.1734	–

Mixture 3: C_6H_{14} (1) + $CH_3CH(Ph)COOCH_2CH_3$ (*rac*-1-Phenylethyl propionate) (2) + [emim][TOS] (3)
(1) Selectivity

Ref. [75]

Hexane-rich phase		RTIL-rich phase		
x_3'	x_2'	x_3''	x_2''	Selectivity (*S*)
0.0016	0.7761	0.4312	0.5687	1629
0.0015	0.5627	0.6510	0.3485	540
0.0012	0.3403	0.7033	0.2966	5739
0.0011	0.1622	0.8637	0.1363	–
0.0000	0.0873	0.9627	0.0373	–

Mixture 4: C_6H_{14} (1) + $CH_3CH(Ph)OH$ (*rac*-1-Phenylethanol) (2) + [emim][TOS] (3)
(1) Selectivity

Ref. [75]

Hexane-rich phase		RTIL-rich phase		
x_3'	x_2'	x_3''	x_2''	Selectivity (*S*)
0.0003	0.0491	0.0950	0.6500	79
0.0002	0.0246	0.2200	0.6450	189
0.0001	0.0060	0.4450	0.4800	1060
0.0000	0.0038	0.6450	0.3100	1806

2.2.13 02003-04018: 1-Ethyl-3-methylimidazolium ethylsulfate

Abbreviation:	[Emim][$EtSO_4$], [C_2mim][$EtSO_4$], EMISE, EMimESO_4, [emim][$C_2H_5SO_4$], [C_1C_2Im][$EtSO_4$], [EMIM][ES]
Molecular Formula:	$C_8H_{16}N_2O_4S$
Molar Mass:	236.29
Structure:	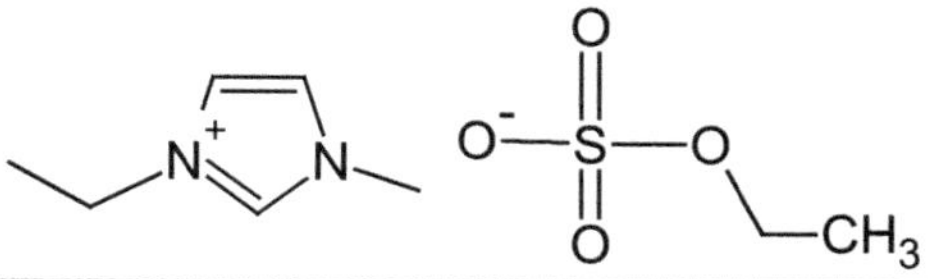

2.2.13.1 Binary Mixtures

Mixture 1: C_2H_5OH (1) + [C_2mim][$EtSO_4$] (2)
(1) Volumetric Properties, Transport Properties, Refractive Index, Speed of Sound and Compressibility Properties

Ref. [76]

x_1	ρ/g · cm^{-3}	$10^3\eta$/Pa · s	n_D	u/m · s^{-1}	k_s/T · Pa^{-1}	V^E/cm^3 · mol^{-1}	$10^3\Delta\eta$/Pa · s	Δn_D	Δk_s/T · Pa^{-1}
T = 298.15 K									
0.0000	1.23763	97.58	1.47940	1679.0	286.61	0.000	0.000	0.000	0.00
0.0623	1.22890	79.12	1.47717	1669.8	291.85	−0.049	−12.44	0.005	−37.30
0.2300	1.20126	44.93	1.47031	1626.6	314.62	−0.219	−30.45	0.018	−128.95
0.4113	1.16168	22.965	1.46017	1564.0	351.48	−0.464	−34.93	0.030	−215.77
0.5393	1.12363	13.580	1.45018	1506.5	392.16	−0.595	−31.96	0.035	−262.43
0.6788	1.06670	7.224	1.43532	1428.8	459.23	−0.672	−24.85	0.037	−290.59
0.7662	1.01790	4.598	1.42247	1367.2	525.59	−0.629	−19.04	0.034	−283.87
0.8303	0.97300	3.178	1.41052	1315.6	593.79	−0.574	−14.28	0.030	−259.37
0.9554	0.84860	1.573	1.37687	1195.9	823.98	−0.267	−3.811	0.011	−114.57
0.9640	0.83740	1.427	1.37373	1186.2	848.67	−0.218	−3.133	0.009	−95.71
0.9863	0.80610	1.271	1.36514	1160.7	920.86	−0.081	−1.133	0.003	−38.76
1.0000	0.78546	1.082	1.36023	1146.3	968.90	0.000	0.000	0.000	−0.07
T = 313.15 K									
0.0000	1.22751	50.01		1641.6	302.29	0.000	0.000		0.00
0.0623	1.21887	42.18		1633.6	307.41	−0.085	−4.766		−43.65
0.2300	1.19117	26.189		1590.1	332.05	−0.308	−12.51		−150.18
0.4113	1.15132	14.716		1527.7	372.14	−0.587	−15.07		−251.88
0.5393	1.11300	8.577		1459.1	422.03	−0.735	−14.91		−302.13
0.6788	1.05570	5.208		1381.0	496.66	−0.825	−11.41		−336.67
0.7662	1.00660	3.413		1318.9	571.09	−0.782	−8.910		−330.62
0.8303	0.9613	2.398		1266.2	648.81	−0.710	−6.776		−303.00
0.9554	0.8362	1.200		1143.3	914.83	−0.361	−1.819		−134.87
0.9640	0.8247	1.095		1132.7	945.01	−0.287	−1.504		−111.37
0.9863	0.7934	0.971		1106.2	1030.08	−0.144	−0.529		−43.77
1.0000	0.7720	0.827		1092.8	1084.57	0.000	0.000		0.00

(continued)

x_1	ρ/g · cm^{-3}	$10^3\eta$/Pa · s	n_D	u/m · s^{-1}	k_s/T · Pa^{-1}	V^E/cm^3 · mol^{-1}	$10^3\Delta\eta$/Pa · s	Δn_D	Δk_s/T · Pa^{-1}
T = 328.15 K									
0.0000	1.21747	29.176		1604.5	319.04	0.000	0.000		0.00
0.0623	1.20877	25.355		1594.8	325.25	−0.101	−2.042		−49.43
0.2300	1.18101	16.839		1551.0	351.96	−0.385	−5.774		−172.34
0.4113	1.14094	10.140		1488.2	395.75	−0.707	−7.301		−290.31
0.5393	1.10230	6.204		1431.1	442.96	−0.871	−7.584		−357.34
0.6788	1.04470	3.934		1350.7	524.68	−0.984	−5.872		−400.16
0.7662	0.99520	2.642		1286.7	606.93	−0.934	−4.669		−395.91
0.8303	0.94960	1.874		1232.2	693.52	−0.854	−3.610		−366.47
0.9554	0.8235	0.941		1102.6	998.85	−0.437	−0.972		−172.81
0.9640	0.8120	0.862		1091.4	1033.95	−0.361	−0.808		−145.34
0.9863	0.7803	0.759		1063.1	1134.05	−0.180	−0.272		−65.18
1.0000	0.7586	0.641		1043.2	1211.45	0.000	0.000		0.00

(2) Excess Molar Enthalpy

Ref. [65]

x_2	H^E (J · mol^{-1})	x_2	H^E (J · mol^{-1})	x_2	H^E (J · mol^{-1})
T = 303.15 K					
0.0486	522	0.3917	1243	0.7963	494
0.1198	1086	0.4937	1153	0.8774	306
0.1961	1263	0.6087	922		
0.2952	1317	0.7017	715		

(3) Phase Equilibrium

Ref. [77]

T/K	x_1	γ_1
101.3 kPa		
351.44	1.000	1.000
351.99	0.980	0.999
352.49	0.960	1.000
353.15	0.936	0.999
353.77	0.912	1.001
354.67	0.889	0.992
355.65	0.863	0.984
356.89	0.832	0.973
358.92	0.789	0.950
361.25	0.742	0.926
363.32	0.708	0.899
366.92	0.649	0.861
369.60	0.618	0.823
373.05	0.572	0.789
386.38	0.443	0.656
402.88	0.338	0.521

(4) Activity Coefficient at Infinite Dilution and Partial Molar Excess Enthalpy at Infinite Dilution

Ref. [78]

	$\gamma^{\infty}{}_1$					
Solvent	303.15 K	313.15 K	323.15 K	333.15 K	343.15 K	353.15 K
[Emim][EtSO$_4$]	–	–	0.57	0.60	0.65	0.70

Ref. [79]

Solute i	Temp range/K	γ_i^{∞} (298 K)	$H_i^{E,\infty}$/kJ · mol^{-1}
ethanol	302.4–354.3	0.89	4.5

(5) Osmotic Coefficient, Activity and Vapor Pressure

Ref. [80]

m/(mol · kg^{-1})	Φ_1	a_1	p/kPa
T = 323.15 K			
0.0565	0.744	0.996	29.361
0.0734	0.729	0.995	29.329
0.1084	0.701	0.993	29.267
0.1932	0.623	0.989	29.145
0.3175	0.584	0.983	28.969
0.3923	0.553	0.980	28.883
0.4921	0.543	0.976	28.746
0.6773	0.538	0.967	28.485
0.9888	0.504	0.955	28.130
1.2900	0.481	0.944	27.809
1.6972	0.450	0.932	27.438
2.0633	0.440	0.920	27.070
2.5861	0.411	0.907	26.678
2.9298	0.391	0.900	26.474
3.3224	0.384	0.889	26.155
3.7188	0.367	0.882	25.933
4.1154	0.359	0.873	25.660

(6) Vapor Pressure and Activity Coefficient

Ref. [81]

T/K	P^{exp}/kPa	P^{NRTL}/kPa	γ_1^{exp}	$\gamma_1^{exp} - \gamma_1^{NRTL}$
x_1 = 0.9788				
299.82	8.514	8.477	1.0044	0.0043
309.12	14.234	14.186	1.0035	0.0034
315.58	19.901	19.861	1.0022	0.0021
321.28	26.388	26.369	1.0008	0.0007
326.63	34.054	34.057	1.0000	−0.0001
332.21	44.000	44.007	0.9999	−0.0002
337.70	56.084	56.102	0.9998	−0.0003
342.13	67.782	67.819	0.9995	−0.0006

(*continued*)

T/K	P^{exp}/kPa	P^{NRTL}/kPa	γ_1^{exp}	$\gamma_1^{exp} - \gamma_1^{NRTL}$
$x_1 = 0.9229$				
298.99	7.742	7.626	1.0162	0.0152
306.41	11.681	11.563	1.0111	0.0101
313.87	17.287	17.171	1.0078	0.0068
320.94	24.548	24.475	1.0040	0.0030
326.82	32.469	32.426	1.0023	0.0013
332.60	42.304	42.274	1.0017	0.0007
337.59	52.691	52.706	1.0007	−0.0003
342.96	66.174	66.255	0.9989	−0.0021
$x_1 = 0.8369$				
300.55	7.718	7.576	1.0215	0.0188
311.00	13.526	13.434	1.0097	0.0069
317.30	18.685	18.572	1.0090	0.0061
322.92	24.612	24.486	1.0081	0.0051
328.62	32.084	32.038	1.0045	0.0015
333.58	40.123	40.130	1.0029	−0.0001
339.52	51.997	52.054	1.0020	−0.0011
345.43	66.579	66.712	1.0011	−0.0020
$x_1 = 0.6874$				
305.35	8.385	8.143	1.0327	0.0298
314.52	13.612	13.260	1.0302	0.0266
321.14	18.711	18.465	1.0175	0.0134
326.74	24.248	24.144	1.0088	0.0043
331.63	30.278	30.259	1.0054	0.0006
335.42	35.663	35.840	1.0000	−0.0050
338.76	41.105	41.470	0.9963	−0.0089
345.75	54.978	55.657	0.9933	−0.0122

Mixture 2: H_2O (1) + EMISE (2)
(1) Volumetric Properties, Transport Properties, Refractive Index, Speed of Sound and Compressibility Properties

Ref. [44]

x_1	ρ/g · cm^{-3}										
	293.15 K	298.15 K	303.15 K	308.15 K	313.15 K	318.15 K	323.15 K	328.15 K	333.15 K	338.15 K	343.15 K
0	1.2424	1.2394	1.2363	1.2333	1.2302	1.2272	1.2241	1.2211	1.2181	1.2151	1.2120
0.0514	1.2414	1.2383	1.2352	1.2322	1.2291	1.2261	1.2230	1.2200	1.2169	1.2138	1.2107
0.1300	1.2398	1.2368	1.2336	1.2306	1.2274	1.2244	1.2213	1.2182	1.2152	1.2121	1.2089
0.1783	1.2388	1.2357	1.2326	1.2295	1.2263	1.2233	1.2201	1.2169	1.2139	1.2109	1.2078
0.2168	1.2378	1.2347	1.2315	1.2285	1.2254	1.2223	1.2191	1.2160	1.2129	1.2099	1.2067
0.2642	1.2364	1.2333	1.2302	1.2271	1.2239	1.2208	1.2176	1.2146	1.2115	1.2084	1.2053
0.3054	1.2350	1.2320	1.2288	1.2258	1.2226	1.2195	1.2163	1.2132	1.2101	1.2070	1.2039
0.3568	1.2330	1.2300	1.2269	1.2239	1.2207	1.2176	1.2144	1.2112	1.2082	1.2051	1.2019
0.4008	1.2312	1.2280	1.2250	1.2219	1.2187	1.2156	1.2124	1.2093	1.2062	1.2031	1.2000
0.4613	1.2281	1.2252	1.2220	1.2189	1.2156	1.2124	1.2093	1.2061	1.2031	1.2000	1.1968
0.5002	1.2259	1.2229	1.2197	1.2166	1.2134	1.2101	1.2069	1.2038	1.2007	1.1976	1.1943
0.5584	1.2218	1.2188	1.2156	1.2124	1.2092	1.2060	1.2028	1.1996	1.1964	1.1932	1.1899
0.6124	1.2171	1.2139	1.2108	1.2076	1.2044	1.2011	1.1978	1.1946	1.1914	1.1881	1.1847
0.6698	1.2107	1.2074	1.2041	1.2009	1.1976	1.1943	1.1910	1.1877	1.1843	1.1809	1.1774
0.7125	1.2043	1.2008	1.1976	1.1943	1.1909	1.1876	1.1843	1.1810	1.1775	1.1739	1.1704
0.7635	1.1942	1.1904	1.1871	1.1840	1.1806	1.1772	1.1737	1.1702	1.1668	1.1635	1.1599
0.8006	1.1842	1.1805	1.1772	1.1736	1.1703	1.1670	1.1635	1.1600	1.1566	1.1532	1.1495
0.8497	1.1654	1.1621	1.1588	1.1555	1.1523	1.1489	1.1452	1.1416	1.1383	1.1349	1.1313
0.8963	1.1393	1.1356	1.1328	1.1290	1.1260	1.1230	1.1196	1.1164	1.1129	1.1093	1.1060
1	0.9982	0.9970	0.9956	0.9940	0.9922	0.9901	0.9880	0.9857	0.9832	0.9805	0.9778
	V^E_m/cm^3 · mol^{-1}										
0.0514	−0.036	−0.020	−0.019	−0.018	−0.017	−0.016	−0.015	−0.015	0.001	0.016	0.016

0.1300	−0.110	−0.107	−0.090	−0.087	−0.071	−0.070	−0.068	−0.053	−0.052	−0.038	−0.023
0.1783	−0.171	−0.153	−0.149	−0.132	−0.116	−0.114	−0.098	−0.070	−0.069	−0.069	−0.068
0.2168	−0.204	−0.186	−0.168	−0.164	−0.160	−0.145	−0.129	−0.115	−0.101	−0.101	−0.087
0.2642	−0.239	−0.221	−0.214	−0.197	−0.181	−0.166	−0.150	−0.148	−0.135	−0.122	−0.121
0.3054	−0.265	−0.257	−0.238	−0.232	−0.216	−0.201	−0.185	−0.171	−0.158	−0.147	−0.146
0.3568	−0.294	−0.284	−0.275	−0.268	−0.251	−0.237	−0.221	−0.197	−0.195	−0.184	−0.172
0.4008	−0.332	−0.302	−0.302	−0.284	−0.266	−0.252	−0.236	−0.223	−0.211	−0.200	−0.199
0.4613	−0.365	−0.362	−0.342	−0.323	−0.297	−0.273	−0.267	−0.244	−0.242	−0.232	−0.221
0.5002	−0.397	−0.384	−0.363	−0.344	−0.326	−0.295	−0.279	−0.266	−0.255	−0.246	−0.226
0.5584	−0.428	−0.413	−0.391	−0.365	−0.346	−0.324	−0.309	−0.288	−0.270	−0.253	−0.235
0.6124	−0.455	−0.425	−0.409	−0.383	−0.364	−0.336	−0.313	−0.293	−0.276	−0.253	−0.229
0.6698	−0.477	−0.441	−0.411	−0.385	−0.359	−0.332	−0.310	−0.285	−0.256	−0.229	−0.200
0.7125	−0.471	−0.423	−0.399	−0.367	−0.336	−0.310	−0.288	−0.264	−0.231	−0.195	−0.168
0.7635	−0.446	−0.385	−0.355	−0.334	−0.304	−0.274	−0.242	−0.209	−0.184	−0.167	−0.137
0.8006	−0.406	−0.352	−0.321	−0.277	−0.252	−0.228	−0.197	−0.166	−0.142	−0.122	−0.090
0.8497	−0.311	−0.275	−0.244	−0.213	−0.192	−0.165	−0.128	−0.095	−0.077	−0.058	−0.034
0.8963	−0.219	−0.170	−0.153	−0.107	−0.091	−0.077	−0.053	−0.036	−0.012	0.011	0.024

Ref. [70]

	T/K							
x_1	278.15	288.15	298.15	308.15	318.15	328.15	338.15	348.15
ρ/g · cm^{-3}								
0.0000	1.25112	1.24413	1.23737	1.23061	1.22390	1.21723	1.21060	1.20401
0.2112	1.24659	1.23979	1.23299	1.22621	1.21945	1.21274	1.20605	1.19940
0.4119	1.23956	1.23270	1.22585	1.21902	1.21221	1.20541	1.19863	1.19191
0.5939	1.22774	1.22080	1.21387	1.20692	1.19997	1.19301	1.18603	1.17905
0.7669	1.20183	1.19485	1.18781	1.18072	1.17356	1.16634	1.15905	1.15168
0.8487	1.17563	1.16883	1.16192	1.15490	1.14777	1.14052	1.13314	1.12570
0.9292	1.12063	1.11502	1.10914	1.10300	1.09659	1.08992	1.08300	1.07583
0.9683	1.06686	1.06325	1.05904	1.05427	1.04901	1.04329	1.03715	1.03061
0.9916	1.02113	1.01946	1.01678	1.01326	1.00903	1.00416	0.99872	0.99276
1.0000	0.99999	0.99913	0.99707	0.99406	0.99023	0.98571	0.98056	0.97484
V^E/cm^3 · mol^{-1}								
0.0000	0.000	0.000	0.000	0.000	0.000	0.000	0.000	0.000
0.2112	−0.211	−0.215	−0.195	−0.181	−0.166	−0.156	−0.144	−0.136
0.4119	−0.399	−0.372	−0.334	−0.304	−0.278	−0.253	−0.231	−0.215
0.5939	−0.524	−0.470	−0.415	−0.367	−0.325	−0.285	−0.248	−0.213
0.7669	−0.514	−0.439	−0.367	−0.306	−0.249	−0.196	−0.146	−0.097
0.8487	−0.450	−0.371	−0.300	−0.238	−0.182	−0.130	−0.081	−0.037
0.9292	−0.246	−0.185	−0.134	−0.091	−0.054	−0.019	0.012	0.043
0.9683	−0.058	−0.028	−0.003	0.017	0.034	0.050	0.065	0.079
0.9916	−0.013	−0.005	0.001	0.005	0.009	0.013	0.017	0.020
1.0000	0.000	0.000	0.000	0.000	0.000	0.000	0.000	0.000
η/mPa · s								
0.0000	308	166	95	62	41	30	21	17
0.2245	150	86	54	36	26	19	15	12
0.4084	83	49	33	23	17	13	11	9
0.5976	45	27	19	14	11	9	7	6
0.7675	22	14	10	8	6	6	5	4
0.9292	6	5	4	4	3	3	3	2
$\Delta\eta$/mPa · s								
0.0000	0	0	0	0	0	0	0	0
0.2245	−89	−43	−20	−12	−6	−4	−2	−1
0.4084	−100	−49	−24	−14	−7	−5	−2	−1
0.5976	−80	−40	−20	−12	−6	−4	−2	−1
0.7675	−51	−25	−13	−7	−3	−2	−1	0
0.9292	−17	−8	−3	−2	0	0	1	1

Ref. [76]

x_1	ρ/g · cm^{-3}	$10^3\eta$/Pa · s	n_D	u/m · s^{-1}	k_s/T · Pa^{-1}	V^E/cm^3 · mol^{-1}	$10^3\Delta\eta$/Pa · s	Δn_D	Δk_s/T · Pa^{-1}
T = 298.15 K									
0.0000	1.23763	97.58	1.47940	1679.1	286.61	0.000	0.000	0.000	0.00
0.0715	1.23614	82.01	1.47827	1687.0	284.24	−0.037	−8.655	0.009	−13.88
0.2089	1.23272	57.84	1.47600	1696.7	281.78	−0.121	−19.54	0.027	−38.46
0.3227	1.2290	42.89	1.47344	1707.2	279.19	−0.188	−23.49	0.041	−59.38
0.5041	1.2200	24.739	1.46726	1729.0	274.19	−0.294	−24.10	0.062	−93.58
0.5975	1.2129	17.516	1.46243	1741.6	271.82	−0.352	−22.29	0.071	−111.00
0.7441	1.1922	10.576	1.44916	1773.0	266.83	−0.340	−15.05	0.079	−139.59
0.8540	1.1592	5.848	1.42825	1792.6	268.44	−0.273	−9.162	0.074	−155.66
0.9506	1.0844	2.316	1.38383	1709.2	315.66	−0.051	−3.349	0.044	−124.00
1.0000	0.9971	0.890	1.33251	1496.9	447.62	0.000	0.000	0.000	0.00
T = 313.15 K									
0.0000	1.22751	50.01		1641.6	302.29	0.000	0.000		0.00
0.0715	1.22597	43.36		1640.1	303.22	−0.023	−3.122		−8.25
0.2089	1.22253	32.093		1660.8	296.56	−0.095	−7.607		−32.56
0.3227	1.21887	24.756		1671.0	293.82	−0.166	−9.325		−49.92
0.5041	1.20982	15.112		1692.6	288.51	−0.251	−10.02		−78.53
0.5975	1.20243	10.982		1703.9	286.44	−0.281	−9.536		−92.61
0.7441	1.18155	6.715		1734.5	281.32	−0.249	−6.567		−116.56
0.8540	1.14619	3.818		1749.2	285.15	−0.089	−4.042		−126.84
0.9506	1.07610	1.608		1693.1	324.19	−0.008	−1.482		−100.22
1.0000	0.9922	0.653		1529.6	430.75	0.000	0.000		0.00

(continued)

x_1	ρ/g · cm^{-3}	$10^3\eta$/Pa · s	n_D	u/m · s^{-1}	k_s/T · Pa^{-1}	V^E/cm^3 · mol^{-1}	$10^3\Delta\eta$/Pa · s	Δn_D	Δk_s/T · Pa^{-1}
T = 328.15 K									
0.0000	1.21747	29.176		1604.5	319.04	0.000	0.000		0.00
0.0715	1.21587	25.800		1617.1	314.51	−0.011	−1.326		−12.09
0.2089	1.21240	19.817		1622.1	313.46	−0.072	−3.370		−27.67
0.3227	1.20882	15.800		1631.7	310.70	−0.148	−4.123		−42.47
0.5041	1.19954	10.024		1652.9	305.15	−0.204	−4.699		−67.21
0.5975	1.19202	7.450		1665.9	302.27	−0.223	−4.594		−79.97
0.7441	1.1708	4.622		1695.3	297.20	−0.168	−3.218		−100.55
0.8540	1.1353	2.596		1704.3	303.25	−0.006	−2.095		−106.12
0.9506	1.0669	1.189		1668.1	336.83	0.030	−0.731		−82.76
1.0000	0.9857	0.504		1545.4	424.82	0.000	0.000		0.00

(2) Phase Equilibrium

Ref. [77]

T/K	x_1	γ_1
101.3 kPa		
373.15	1.000	1.000
373.54	0.995	0.991
373.83	0.987	0.989
374.16	0.980	0.984
374.50	0.973	0.980
374.86	0.967	0.973
375.34	0.958	0.966
376.02	0.945	0.956
376.85	0.932	0.942
378.02	0.915	0.921
379.36	0.893	0.901
381.00	0.875	0.869
382.72	0.855	0.839
385.42	0.820	0.800
389.07	0.787	0.740
392.90	0.747	0.689
397.52	0.705	0.631
400.29	0.677	0.604
402.33	0.658	0.584
403.64	0.648	0.571
405.32	0.634	0.555
410.58	0.580	0.520

Ref. [79]

T/K	x_1	p_1^{exp}/Pa	p_1^{NRTL}/Pa	γ_1^{NRTL}
302.9	0	0	0	0.23
	0.278	345	345	0.30
	0.424	639	638	0.37
	0.572	1070	1094	0.47
	0.613	1199	1265	0.50
	0.620	1439	1294	0.51
	0.832	2711	2676	0.79
	0.853	2600	2868	0.82
	0.959	3834	3840	0.98
	1	4095	4095	1.00
312.9	0	0	0	0.26
	0.480	1533	1524	0.45
	0.501	1555	1638	0.46
	0.54	1893	1877	0.49
	0.551	2054	1947	0.50

(continued)

T/K	x_1	p_1^{exp}/Pa	p_1^{NRTL}/Pa	γ_1^{NRTL}
	0.810	4702	4519	0.79
	0.874	5045	5444	0.88
	0.888	5317	5656	0.90
	0.953	6907	6591	0.98
	1	7072	7072	1.00
322.9	0	0	0	0.20
	0.284	1036	1071	0.31
	0.302	1128	1173	0.32
	0.440	2434	2214	0.41
	0.463	2561	2433	0.43
	0.518	3237	3049	0.49
	0.542	3115	3356	0.51
	0.647	4832	5013	0.64
	0.669	5074	5424	0.67
	0.783	8315	7988	0.84
	0.858	9884	9878	0.95
	0.941	11626	11519	1.01
	1	12141	12141	1.00

(3) Thermal Conductivity

Ref. [82]

w_1	λ/W · m^{-1} · K^{-1}	w_1	λ/W · m^{-1} · K^{-1}
293 K			
0	0.181	0.20	0.232
0.01	0.181	0.50	0.352
0.05	0.190	0.70	0.445
0.10	0.201	1.00	0.607

(4) Osmotic Coefficient, Activity Properties and Vapor Pressure

Ref. [83]

	Φ			
m_2 (mol · kg^{-1})	T = 298.15 K	T = 308.15 K	T = 318.15 K	T = 328.15 K
0.0086	0.973	0.981	1.019	1.018
0.0185	0.957	0.943	0.972	0.975
0.0284	0.937	0.921	0.934	0.955
0.0376	0.921	0.917	0.93	0.936
0.0570	0.871	0.862	0.868	0.886
0.0749	0.852	0.862	0.859	0.858
0.0938	0.830	0.838	0.845	0.850
0.1269	0.815	0.8170	0.831	0.842
0.1704	0.782	0.797	0.813	0.815
0.2087	0.774	0.779	0.797	0.815

(*continued*)

	Φ			
m_2 (mol · kg^{-1})	T = 298.15 K	T = 308.15 K	T = 318.15 K	T = 328.15 K
0.2648	0.754	0.764	0.797	0.800
0.3276	0.751	0.766	0.793	0.796
0.4235	0.736	0.755	0.776	0.787
0.5106	0.724	0.745	0.772	0.781
0.6979	0.712	0.727	0.742	0.755
0.8919	0.706	0.721	0.730	0.747
1.1198	0.697	0.711	0.717	0.729
1.6061	0.696	0.707	0.715	0.719
1.8889	0.694	0.707	0.717	0.726
2.7627	0.733	0.747	0.759	0.766

Ref. [84]

m/(mol · kg^{-1})	Φ_1	α_1	p/kPa	m/(mol · kg^{-1})	Φ_1	α_1	p/kPa
T = 313.15 K				T = 333.15 K			
0.1847	0.970	0.9936	7.330	0.1847	0.936	0.9938	19.796
0.3033	0.949	0.9897	7.301	0.3033	0.918	0.9900	19.718
0.3896	0.941	0.9869	7.280	0.3896	0.896	0.9875	19.662
0.5094	0.923	0.9832	7.253	0.5094	0.865	0.9842	19.588
0.7105	0.895	0.9773	7.210	0.7105	0.847	0.9785	19.470
1.0472	0.861	0.9680	7.141	1.0472	0.812	0.9698	19.284
1.2943	0.842	0.9615	7.093	1.2943	0.801	0.9634	19.153
1.6782	0.835	0.9508	7.013	1.6782	0.790	0.9534	18.938
2.1131	0.831	0.9387	6.924	2.1131	0.784	0.9421	18.697
2.4905	0.829	0.9283	6.847	2.4905	0.788	0.9317	18.488
2.9466	0.844	0.9143	6.743	2.9466	0.797	0.9188	18.208
3.2369	0.852	0.9055	6.678	3.2369	0.809	0.9100	18.031
3.7222	0.885	0.8881	6.550	3.7222	0.831	0.8945	17.683
4.1131	0.921	0.8724	6.434	4.1131	0.855	0.8810	17.369

Ref. [81]

T/K	P^{exp}/kPa	P^{NRTL}/kPa	γ_1^{exp}	$\gamma_1^{exp} - \gamma_1^{NRTL}$
x_1 = 0.9915				
316.93	8.843	8.891	0.9937	−0.0054
328.27	15.566	15.648	0.9939	−0.0053
335.25	21.593	21.682	0.9952	−0.0041
344.29	32.228	32.358	0.9953	−0.0040
348.27	38.203	38.309	0.9966	−0.0027
352.99	46.426	46.523	0.9973	−0.0021
358.06	56.897	56.979	0.9980	−0.0014
363.71	70.899	70.884	0.9996	0.0002
x_1 = 0.9683				
318.06	9.018	9.110	0.9788	−0.0100

(continued)

T/K	P^{exp}/kPa	P^{NRTL}/kPa	γ_1^{exp}	$\gamma_1^{exp} - \gamma_1^{NRTL}$
325.10	12.858	12.977	0.9804	−0.0091
332.58	18.414	18.558	0.9825	−0.0077
339.37	25.062	25.265	0.9828	−0.0079
345.48	32.753	32.980	0.9844	−0.0068
351.71	42.577	42.800	0.9865	−0.0051
356.54	51.819	52.020	0.9881	−0.0038
364.69	71.142	71.353	0.9895	−0.0029
$x_1 = 0.9291$				
321.47	9.988	10.023	0.9502	−0.0034
327.87	13.708	13.763	0.9521	−0.0039
334.76	18.953	19.054	0.9533	−0.0050
343.69	28.279	28.412	0.9566	−0.0045
349.08	35.550	35.758	0.9570	−0.0056
354.16	43.748	44.102	0.9563	−0.0077
359.21	53.509	53.955	0.9573	−0.0080
366.60	71.023	71.668	0.9583	−0.0087
$x_1 = 0.8489$				
325.55	10.030	9.991	0.8532	0.0033
334.18	15.148	15.181	0.8565	−0.0018
340.27	20.028	20.100	0.8607	−0.0031
346.15	25.941	26.075	0.8644	−0.0044
352.47	33.899	34.109	0.8685	−0.0054
357.64	41.547	42.144	0.8653	−0.0125
364.54	54.343	55.310	0.8673	−0.0154
369.29	65.276	66.268	0.8727	−0.0133

(5) Excess Molar Enthalpy

Ref. [71]

x_2	H^E/(J · mol^{-1})	x_2	H^E/(J · mol^{-1})	x_2	H^E/(J · mol^{-1})
303.15 K					
0.0504	−482	0.2921	−968	0.5727	−655
0.0997	−767	0.3935	−853	0.6687	−520
0.1468	−915	0.4806	−794	0.8212	−285
0.2004	−986				

Mixture 3: C_6H_{14} (1) + [EMIM][EtSO$_4$] (2)

(1) Phase Equilibrium

Ref. [85]

x_2	$T^{LLE\ a}$/K	x_2	$T^{LLE\ a}$/K	x_2	$T^{LLE\ a}$/K
0.8552	301.24	0.7325	321.82	0.6772	339.18
0.7678	313.09	0.7262	324.55		

[a]The results are for example (311.75 ± 0.05) K

Mixture 4: C_8H_{18} (1) + [EMIM][EtSO$_4$] (2)
(1) Phase Equilibrium

Ref. [85]

x_2	$T^{LLE\ a}$/K	x_2	$T^{LLE\ a}$/K	x_2	$T^{LLE\ a}$/K
0.9871	317.15	0.9463	340.23	0.8986	358.15
0.9770	326.15	0.9352	345.10	0.8842	360.68
0.9684	331.15	0.9253	346.65	0.8546	365.04
0.9633	334.75	0.9184	351.45	0.8184	368.97

[a]The results are for example (311.75 ± 0.05) K

(2) Activity Coefficient at Infinite Dilution and Partial Molar Excess Enthalpy at Infinite Dilution

Ref. [79]

Solute i	Temp range/K	γ_i^∞ (298 K)	$H_i^{E,\infty}$/kJ · mol^{-1}
1-octane	301.9–333.5	243.6	16.4

Mixture 5: C_6H_6 (1) + [EMIM][EtSO$_4$] (2)
(1) Phase Equilibrium

Ref. [85]

x_2	$T^{LLE\ a}$/K	x_2	$T^{LLE\ a}$/K	x_2	$T^{LLE\ a}$/K
0.4883	288	0.3915	311.11	0.2648	336.51
0.4418	294.60	0.3576	314.12	0.2445	346.31
0.4205	302.56	0.3277	319.64		
0.4048	309.09	0.2893	328.22		

[a]The results are for example (311.75 ± 0.05) K

Ref. [3]

x_1	P (kPa)	x_1	P (kPa)	x_1	P (kPa)	x_1	P (kPa)
353.15 K							
0.0046	0.28	0.1590	6.84	0.6448	15.98	0.8930	15.94
0.0108	0.57	0.2387	9.82	0.6664	15.98	0.9473	15.95
0.0183	0.92	0.3330	12.88	0.7178	15.97	0.9778	15.96
0.0330	1.58	0.4319	15.41	0.7313	15.97	0.9893	15.97
0.0548	2.54	0.5270	15.98	0.8130	15.95	1.0000	15.96
0.0955	4.25	0.6055	15.98				

(2) Activity Coefficient at Infinite Dilution and Partial Molar Excess Enthalpy at Infinite Dilution

Ref. [79]

Solute i	Temp range/K	γ_i^∞ (298 K)	$H_i^{E,\infty}$/kJ · mol^{-1}
Benzene	301.8–364.7	3.1	2.6

Mixture 6: $C_6H_5CH_3$ (1) + [EMIM][$EtSO_4$] (2)
(1) Phase Equilibrium

Ref. [85]

x_2	$T^{LLE\ a}$/K	x_2	$T^{LLE\ a}$/K	x_2	$T^{LLE\ a}$/K
0.7246	311.75	0.7080	324.78	0.6823	333.44
0.7153	319.55	0.7012	325.78	0.6351	359.15

[a]The results are for example (311.75 ± 0.05) K

Ref. [86]

Organic phase	IL-rich phase	
x_2'	x_2''	T/K
0.0061	0.7588	293.15
0.0046	0.7608	303.15
0.0058	0.7561	313.15
0.0046	0.7594	323.15

(2) Activity Coefficient at Infinite Dilution and Partial Molar Excess Enthalpy at Infinite Dilution

Ref. [79]

Solute i	Temp range/K	γ_i^∞ (298 K)	$H_i^{E,\infty}$/kJ · mol^{-1}
Toluene	312.3–364.7	6.2	3.0

Mixture 7: $C_6H_5C_2H_5$ (1) + [EMIM][$EtSO_4$] (2)
(1) Phase Equilibrium

Ref. [85]

x_2	$T^{LLE\ a}$/K	x_2	$T^{LLE\ a}$/K	x_2	$T^{LLE\ a}$/K
0.8890	297.12	0.8598	324.87	0.8206	342.75
0.8806	304.87	0.8487	334.14	0.7500	375.15
0.8697	313.20	0.8411	336.66		
0.8666	322.60	0.8318	340.52		

[a]The results are for example (311.75 ± 0.05) K

(2) Activity Coefficient at Infinite Dilution and Partial Molar Excess Enthalpy at Infinite Dilution

Ref. [79]

Solute i	Temp range/K	γ_i^∞ (298 K)	$H_i^{E,\infty}$/kJ · mol^{-1}
Ethyl benzene	354.3–385.7	13.5	6.0

Mixture 8: 2-$CH_3COC_3H_7$ (1) + [EMIM][$EtSO_4$] (2)
(1) Phase Equilibrium

Ref. [85]

x_2	$T^{LLE\ a}$/K	x_2	$T^{LLE\ a}$/K	x_2	$T^{LLE\ a}$/K
0.6122	281.90	0.5881	311.26	0.5441	338.48
0.6064	293.59	0.5819	316.97	0.5277	349.27
0.6012	297.78	0.5751	321.66	0.5089	358.95
0.5977	302.91	0.5549	332.57		

[a]The results are for example (311.75 ± 0.05) K

Mixture 9: 3-$C_2H_5COC_2H_5$ (1) + [EMIM][$EtSO_4$] (2)
(1) Phase Equilibrium

Ref. [85]

x_2	$T^{LLE\ a}$/K	x_2	$T^{LLE\ a}$/K	x_2	$T^{LLE\ a}$/K
0.6305	288.15	0.5679	336.15	0.5351	350.50
0.6244	299.00	0.5836	327.65	0.5256	354.01
0.6180	307.60	0.5760	331.37	0.5024	360.23
0.6107	314.89	0.5599	340.66	0.4858	365.15
0.6054	319.05	0.5528	343.08		

[a]The results are for example (311.75 ± 0.05) K

Mixture 10: 2-$CH_3COC_4H_9$ (1) + [EMIM][$EtSO_4$] (2)
(1) Phase Equilibrium

Ref. [85]

x_2	$T^{LLE\ a}$/K	x_2	$T^{LLE\ a}$/K	x_2	$T^{LLE\ a}$/K
0.7818	273.62	0.7347	334.08	0.6773	367.68
0.7666	306.66	0.7212	342.30	0.6646	371.97
0.7627	314.63	0.7141	349.05	0.6487	379.93
0.7545	320.05	0.7032	354.77	0.6273	383.71
0.7492	323.11	0.6957	360.55	0.5988	393.45
0.7430	328.02	0.6874	363.41	0.5653	399.55

[a]The results are for example (311.75 ± 0.05) K

Mixture 11: 4-$C_3H_7COC_3H_7$ (1) + [EMIM][$EtSO_4$] (2)
(1) Phase Equilibrium

Ref. [85]

x_2	$T^{LLE\ a}$/K	x_2	$T^{LLE\ a}$/K	x_2	$T^{LLE\ a}$/K
0.9783	302.15	0.9423	341.65	0.8731	375.15
0.9674	306.70	0.9327	347.20	0.8500	383.15
0.9605	312.75	0.9201	353.10	0.8234	392.78
0.9508	325.36	0.9082	361.47	0.7961	399.49
0.9471	331.20	0.8960	367.71	0.7491	407.38

[a]The results are for example (311.75 ± 0.05) K

Mixture 12: $CH_3(CH_2)_3O(CH_2)_3CH_3$ (1) + [EMIM][$EtSO_4$] (2)
(1) Phase Equilibrium

Ref. [85]

x_2	$T^{LLE\ a}$/K	x_2	$T^{LLE\ a}$/K	x_2	$T^{LLE\ a}$/K
0.9931	316.10	0.9556	344.12	0.9067	408.95
0.9865	321.13	0.9480	353.09	0.0040	338.97
0.9824	322.77	0.9427	371.90	0.0029	336.24
0.9740	327.10	0.9300	385.58		
0.9660	334.32	0.9199	398.66		

[a]The results are for example (311.75 ± 0.05) K

Mixture 13: $(CH_3)_3COCH_3$ (1) + [EMIM][$EtSO_4$] (2)
(1) Phase Equilibrium

Ref. [85]

x_2	$T^{LLE\ a}$/K	x_2	$T^{LLE\ a}$/K	x_2	$T^{LLE\ a}$/K
0.8966	308.15	0.8740	325.47	0.8352	345.08
0.8795	320.15	0.8566	335.33	0.7930	368.06

[a]The results are for example (311.75 ± 0.05) K

Mixture 14: $(CH_3)_3COC_2H_5$ (1) + [EMIM][$EtSO_4$] (2)
(1) Phase Equilibrium

Ref. [85]

x_2	$T^{LLE\ a}$/K	x_2	$T^{LLE\ a}$/K	x_2	$T^{LLE\ a}$/K
0.9701	305.69	0.9109	331.86	0.8485	357.78
0.9451	316.01	0.8916	342.06	0.8095	371.85
0.9341	324.01	0.8731	349.34		

[a]The results are for example (311.75 ± 0.05) K

Mixture 15: $OCH_2CH_2CH_2CH_2$ (1) + [EMIM][$EtSO_4$] (2)
(1) Phase Equilibrium

Ref. [85]

x_2	$T^{LLE\ a}$/K	x_2	$T^{LLE\ a}$/K	x_2	$T^{LLE\ a}$/K
0.5730	313.75	0.5498	341.60	0.4211	406.23
0.5679	317.78	0.5403	346.00	0.3905	419.29
0.5626	330.53	0.5278	347.50	0.0014	>433.15
0.5587	335.04	0.4953	358.00		
0.5548	338.40	0.4562	385.56		

[a]The results are for example (311.75 ± 0.05) K

Mixture 16: CH_3SOCH_3 (1) + [EMIM][$EtSO_4$] (2)
(1) Phase Equilibrium

Ref. [85]

x_2	$T^{SLE\ a}$/K	x_2	$T^{SLE\ a}$/K	x_2	$T^{SLE\ a}$/K
0.4596	250.52	0.3057	266.57	0.0940	283.84
0.4481	252.20	0.2689	269.98	0.0771	285.06
0.4283	254.30	0.2428	271.95	0.0570	286.89
0.4077	255.85	0.2220	273.50	0.0384	288.14
0.3821	258.95	0.1931	275.87	0.0225	289.48
0.3589	261.49	0.1680	277.93	0.0107	290.57
0.3477	262.92	0.1447	279.96	0.0025	291.40
0.3298	264.86	0.1188	281.89	0.0000	291.66

[a]The results are for example (311.75 ± 0.05) K

Mixture 17: NH_3 (1) + [Emim][$EtOSO_3$] (2)
(1) Solubility

Ref. [87]

T (K)	P (MPa)	$100x_1$ (mol %)	T (K)	P (MPa)	$100x_1$ (mol %)
282.7	0.287	53.6 ± 0.9	347.5	1.051	42.4 ± 4.4
282.7	0.427	70.7 ± 0.6	347.5	1.819	61.3 ± 3.8
282.7	0.517	80.5 ± 0.3	347.5	2.500	74.4 ± 2.6
282.7	0.544	83.9 ± 0.2	347.5	2.790	79.0 ± 2.1
282.7	0.586	87.5 ± 0.1	347.5	3.091	84.4 ± 1.3
–	–	–	–	–	–
297.6	0.418	51.8 ± 1.4	372.3	2.461	56.2 ± 6.2
297.6	0.651	69.4 ± 0.9	372.3	3.593	69.7 ± 5.1
297.6	0.802	79.8 ± 0.5	372.3	4.118	74.7 ± 4.5
297.6	0.855	83.3 ± 0.4	372.3	4.777	81.2 ± 3.2
297.6	0.916	87.1 ± 0.2	–	–	–
–	–	–	–	–	–
322.3	0.706	47.7 ± 2.6	298.1	0.421	51.8 ± 1.4
322.3	1.166	66.1 ± 1.9	298.1	0.653	69.4 ± 0.9
322.3	1.510	77.8 ± 1.2	298.1	0.812	79.8 ± 0.5
322.3	1.641	81.8 ± 0.9	298.1	0.869	83.3 ± 0.4
322.3	1.771	86.2 ± 0.5	298.1	0.933	87.1 ± 0.2
–	–	–	–	–	–

Mixture 18: CH_3OH (1) + EMISE (2)

(1) Volumetric Properties, Transport Properties, Refractive Index, Speed of Sound and Compressibility Properties

Ref. [88]

	$\rho/(kg \cdot m^{-3})$ at p/MPa								
x_2	0.1	2.5	5.0	10.0	15.0	20.0	25.0	30.0	35.0
$T = 283.15$ K									
0.04781	895.14	896.98	898.83	902.39	905.85	909.18	912.41	915.51	918.59
0.10257	968.87	970.47	972.09	975.24	978.32	981.31	984.20	987.02	989.77
0.19771	1052.50	1053.84	1055.23	1057.94	1060.59	1063.21	1065.75	1068.22	1070.64
0.24323	1081.02	1082.31	1083.64	1086.24	1088.80	1091.31	1093.76	1096.15	1098.50
0.30257	1111.91	1113.14	1114.41	1116.86	1119.26	1121.66	1123.99	1126.28	1128.48
0.38873	1145.18	1146.33	1147.51	1149.84	1152.14	1154.40	1156.62	1158.82	1160.95
0.48505	1173.12	1174.24	1175.35	1177.57	1179.75	1181.92	1184.04	1186.14	1188.17
0.58896	1195.65	1196.74	1197.79	1199.93	1202.04	1204.13	1206.18	1208.21	1210.17
0.72888	1218.63	1219.71	1220.74	1222.79	1224.82	1226.84	1228.82	1230.77	1232.67
0.85211	1233.50	1234.47	1235.48	1237.48	1239.47	1241.45	1243.37	1245.28	1247.16
0.93975	1242.49	1243.52	1244.49	1246.47	1248.43	1250.39	1252.30	1254.18	1256.03
$T = 293.15$ K									
0.04781	886.34	888.27	890.20	893.93	897.52	900.98	904.34	907.55	910.71
0.10257	960.49	962.13	963.82	967.09	970.28	973.36	976.36	979.27	982.13
0.19771	1044.52	1045.92	1047.38	1050.18	1052.91	1055.60	1058.22	1060.75	1063.26
0.24323	1073.23	1074.55	1075.93	1078.61	1081.24	1083.82	1086.34	1088.79	1091.21
0.30257	1104.25	1105.51	1106.81	1109.35	1111.82	1114.27	1116.67	1119.01	1121.32
0.38873	1137.80	1138.99	1140.22	1142.61	1144.95	1147.27	1149.55	1151.75	1153.96
0.48505	1165.78	1166.91	1168.06	1170.35	1172.59	1174.79	1176.97	1179.09	1181.21
0.58896	1188.53	1189.55	1190.65	1192.84	1194.99	1197.11	1199.20	1201.26	1203.30

0.72888	1211.60	1212.63	1213.69	1215.79	1217.88	1219.92	1221.94	1223.92	1225.88
0.85211	1226.42	1227.39	1228.45	1230.49	1232.52	1234.53	1236.51	1238.43	1240.37
0.93975	1235.42	1236.43	1237.46	1239.51	1241.50	1243.48	1245.42	1247.34	1249.28
$T = 298.15$ K									
0.04781	881.90	883.82	885.81	889.63	893.33	896.87	900.30	903.63	906.81
0.10257	956.23	957.85	959.60	962.97	966.22	969.37	972.43	975.41	978.30
0.19771	1040.54	1041.91	1043.39	1046.26	1049.05	1051.78	1054.44	1057.05	1059.58
0.24323	1069.25	1070.58	1071.99	1074.73	1077.43	1080.03	1082.59	1085.11	1087.54
0.30257	1100.31	1101.59	1102.92	1105.51	1108.07	1110.55	1112.96	1115.37	1117.69
0.38873	1134.01	1135.21	1136.46	1138.92	1141.32	1143.68	1145.98	1148.26	1150.46
0.48505	1162.03	1163.15	1164.36	1166.69	1168.99	1171.24	1173.44	1175.62	1177.74
0.58896	1184.83	1185.86	1187.01	1189.25	1191.45	1193.61	1195.76	1197.87	1199.91
0.72888	1207.95	1208.92	1210.02	1212.17	1214.29	1216.37	1218.42	1220.46	1222.43
0.85211	1222.76	1223.17	1224.62	1226.79	1228.88	1230.93	1232.93	1234.93	1236.86
0.93975	1231.87	1232.94	1234.01	1236.07	1238.12	1240.12	1242.11	1244.08	1245.99
$T = 303.15$ K									
0.04781	877.37	879.39	881.43	885.44	890.01	892.79	896.28	899.65	902.86
0.10257	951.86	953.61	955.38	958.89	962.19	965.40	968.54	971.57	974.42
0.19771	1036.37	1037.84	1039.36	1042.38	1045.21	1047.97	1050.68	1053.34	1055.86
0.24323	1065.20	1066.61	1068.04	1070.92	1073.64	1076.28	1078.90	1081.44	1083.87
0.30257	1096.34	1097.65	1099.00	1101.75	1104.31	1106.82	1109.29	1111.72	1114.04
0.38873	1130.14	1131.40	1132.67	1135.24	1137.67	1140.05	1142.39	1144.70	1146.89
0.48505	1158.24	1159.40	1160.61	1163.06	1165.38	1167.67	1169.90	1172.11	1174.20
0.58896	1181.07	1182.15	1183.33	1185.69	1187.91	1190.10	1192.25	1194.38	1196.40
0.72888	1204.24	1205.29	1206.40	1208.67	1210.81	1212.92	1214.99	1217.04	1219.00
0.85211	1219.11	1220.12	1221.21	1223.41	1225.51	1227.56	1229.58	1231.59	1233.49

(*continued*)

x_2	ρ/(kg · m^{-3}) at p/MPa								
	0.1	2.5	5.0	10.0	15.0	20.0	25.0	30.0	35.0
0.93975	1228.28	1229.37	1230.44	1232.62	1234.67	1236.70	1238.71	1240.68	1242.56
T = 313.15 K									
0.04781	868.71	870.87	873.00	877.12	881.03	884.83	888.46	891.58	896.52
0.10257	943.67	945.44	947.28	950.87	954.31	957.65	960.89	964.01	967.07
0.19771	1028.79	1030.21	1031.77	1034.78	1037.72	1040.60	1043.41	1046.12	1048.80
0.24323	1057.64	1059.08	1060.59	1063.49	1066.28	1069.03	1071.73	1074.34	1076.92
0.30257	1088.94	1090.22	1091.62	1094.33	1097.00	1099.60	1102.15	1104.62	1107.08
0.38873	1122.96	1124.22	1125.52	1128.06	1130.57	1133.05	1135.46	1137.80	1140.13
0.48505	1151.19	1152.35	1153.60	1156.03	1158.42	1160.76	1163.07	1165.30	1167.55
0.58896	1174.15	1175.21	1176.41	1178.75	1181.04	1183.30	1185.51	1187.68	1189.83
0.72888	1197.48	1198.52	1199.65	1201.89	1204.09	1206.26	1208.39	1210.48	1212.56
0.85211	1212.41	1213.23	1214.31	1216.48	1218.62	1220.72	1222.81	1224.83	1226.88
0.93975	1221.68	1222.70	1223.82	1225.96	1228.06	1230.16	1232.21	1234.23	1236.23
T = 323.15 K									
0.04781	859.88	862.14	864.43	868.74	872.87	876.82	880.61	884.24	887.77
0.10257	935.30	937.17	939.08	942.80	946.41	949.87	953.21	956.44	959.59
0.19771	1020.88	1022.39	1024.01	1027.14	1030.20	1033.16	1036.04	1038.84	1041.58
0.24323	1049.97	1051.46	1052.99	1055.97	1058.89	1061.72	1064.46	1067.16	1069.79
0.30257	1081.45	1082.65	1084.11	1086.92	1089.69	1092.38	1094.99	1097.52	1100.04
0.38873	1115.61	1116.83	1118.22	1120.88	1123.48	1126.02	1128.50	1130.90	1133.28
0.48505	1143.95	1145.15	1146.43	1148.93	1151.40	1153.83	1156.17	1158.47	1160.75
0.58896	1167.00	1168.12	1169.35	1171.75	1174.11	1176.44	1178.70	1180.91	1183.11
0.72888	1190.46	1191.53	1192.70	1194.99	1197.27	1199.49	1201.68	1203.82	1205.91
0.85211	1205.38	1206.30	1207.44	1209.67	1211.88	1214.04	1216.15	1218.23	1220.28
0.93975	1214.87	1215.90	1217.06	1219.28	1221.46	1223.61	1225.70	1227.76	1229.79

$T = 333.15$ K									
0.04781	850.98	853.34	855.70	860.18	864.49	868.61	872.54	876.32	879.95
0.10257	926.85	928.81	930.83	934.71	938.46	942.07	945.55	948.89	952.14
0.19771	1012.95	1014.54	1016.23	1019.48	1022.66	1025.73	1028.71	1031.60	1034.42
0.24323	1042.28	1043.81	1045.40	1048.48	1051.51	1054.43	1057.29	1060.06	1062.76
0.30257	1073.72	1075.11	1076.58	1079.47	1082.32	1085.07	1087.76	1090.37	1092.96
0.38873	1108.15	1109.48	1110.88	1113.59	1116.27	1118.87	1121.41	1123.90	1126.35
0.48505	1136.70	1137.95	1139.28	1141.86	1144.40	1146.86	1149.29	1151.66	1153.97
0.58896	1159.94	1161.11	1162.35	1164.82	1167.24	1169.62	1171.94	1174.22	1176.46
0.72888	1183.49	1184.59	1185.79	1188.13	1190.47	1192.77	1194.98	1197.18	1199.35
0.85211	1198.51	1199.19	1200.35	1202.62	1204.87	1207.10	1209.26	1211.40	1213.49
0.93975	1207.93	1209.04	1210.20	1212.45	1214.69	1216.87	1219.03	1221.13	1223.21
	$V^E/(cm^3 \cdot mol^{-1})$ at p/MPa								
$T = 283.15$ K									
0.04781	−0.47	−0.45	−0.44	−0.41	−0.39	−0.37	−0.35	−0.33	−0.32
0.10257	−0.66	−0.64	−0.62	−0.58	−0.55	−0.52	−0.49	−0.46	−0.44
0.19771	−0.77	−0.74	−0.71	−0.66	−0.62	−0.58	−0.55	−0.52	−0.49
0.24323	−0.78	−0.76	−0.73	−0.69	−0.64	−0.61	−0.57	−0.54	−0.51
0.30257	−0.85	−0.83	−0.80	−0.75	−0.71	−0.67	−0.64	−0.61	−0.57
0.38873	−0.82	−0.79	−0.77	−0.72	−0.68	−0.65	−0.62	−0.59	−0.56
0.48505	−0.76	−0.74	−0.72	−0.68	−0.64	−0.61	−0.58	−0.56	−0.53
0.58896	−0.65	−0.63	−0.61	−0.58	−0.55	−0.52	−0.50	−0.48	−0.46
0.72888	−0.51	−0.51	−0.50	−0.47	−0.45	−0.43	−0.42	−0.40	−0.38
0.85211	−0.32	−0.31	−0.30	−0.29	−0.28	−0.27	−0.26	−0.25	−0.24
0.93975	−0.22	−0.23	−0.22	−0.21	−0.21	−0.20	−0.20	−0.20	−0.20
$T = 293.15$ K									
0.04781	−0.50	−0.49	−0.47	−0.44	−0.42	−0.39	−0.37	−0.35	−0.33
0.10257	−0.72	−0.69	−0.67	−0.63	−0.59	−0.55	−0.52	−0.49	−0.46

(continued)

	$V^E/(cm^3 \cdot mol^{-1})$ at p/MPa								
0.19771	−0.83	−0.80	−0.77	−0.72	−0.67	−0.63	−0.59	−0.55	−0.52
0.24323	−0.85	−0.82	−0.79	−0.74	−0.70	−0.65	−0.61	−0.58	−0.54
0.30257	−0.92	−0.89	−0.86	−0.81	−0.76	−0.72	−0.68	−0.65	−0.61
0.38873	−0.89	−0.86	−0.84	−0.79	−0.74	−0.70	−0.67	−0.63	−0.60
0.48505	−0.82	−0.80	−0.77	−0.73	−0.69	−0.65	−0.62	−0.59	−0.56
0.58896	−0.71	−0.69	−0.66	−0.63	−0.59	−0.56	−0.54	−0.51	−0.49
0.72888	−0.57	−0.55	−0.53	−0.51	−0.49	−0.47	−0.45	−0.43	−0.41
0.85211	−0.34	−0.33	−0.32	−0.31	−0.29	−0.28	−0.27	−0.26	−0.25
0.93975	−0.23	−0.23	−0.22	−0.22	−0.22	−0.21	−0.21	−0.20	−0.20
$T = 298.15$ K									
0.04781	−0.52	−0.51	−0.49	−0.46	−0.43	−0.40	−0.38	−0.36	−0.34
0.10257	−0.75	−0.72	−0.69	−0.65	−0.61	−0.57	−0.53	−0.51	−0.48
0.19771	−0.87	−0.83	−0.80	−0.75	−0.70	−0.65	−0.61	−0.57	−0.54
0.24323	−0.89	−0.85	−0.82	−0.77	−0.72	−0.67	−0.63	−0.59	−0.56
0.30257	−0.95	−0.92	−0.89	−0.84	−0.79	−0.74	−0.70	−0.66	−0.63
0.38873	−0.93	−0.90	−0.87	−0.82	−0.77	−0.73	−0.69	−0.65	−0.62
0.48505	−0.85	−0.82	−0.80	−0.75	−0.71	−0.67	−0.64	−0.61	−0.58
0.58896	−0.73	−0.71	−0.69	−0.65	−0.61	−0.58	−0.55	−0.53	−0.51
0.72888	−0.58	−0.55	−0.54	−0.51	−0.49	−0.47	−0.45	−0.43	−0.41
0.85211	−0.34	−0.25	−0.29	−0.29	−0.28	−0.27	−0.25	−0.25	−0.24
0.93975	−0.23	−0.24	−0.24	−0.23	−0.23	−0.22	−0.22	−0.21	−0.21
$T = 303.15$ K									
0.04781	−0.53	−0.51	−0.50	−0.46	−0.48	−0.41	−0.39	−0.37	−0.35
0.10257	−0.76	−0.74	−0.71	−0.66	−0.62	−0.58	−0.55	−0.52	−0.49
0.19771	−0.89	−0.86	−0.83	−0.77	−0.71	−0.67	−0.63	−0.59	−0.55

0.24323	−0.91	−0.88	−0.85	−0.79	−0.74	−0.69	−0.65	−0.61	−0.58
0.30257	−0.98	−0.95	−0.91	−0.86	−0.80	−0.76	−0.72	−0.68	−0.64
0.38873	−0.95	−0.92	−0.89	−0.84	−0.79	−0.75	−0.71	−0.67	−0.64
0.48505	−0.88	−0.84	−0.82	−0.77	−0.73	−0.69	−0.66	−0.62	−0.59
0.58896	−0.76	−0.72	−0.71	−0.66	−0.63	−0.60	−0.57	−0.54	−0.52
0.72888	−0.59	−0.57	−0.55	−0.53	−0.50	−0.48	−0.46	−0.44	−0.42
0.85211	−0.35	−0.33	−0.32	−0.31	−0.30	−0.28	−0.27	−0.26	−0.25
0.93975	−0.25	−0.24	−0.24	−0.24	−0.23	−0.23	−0.22	−0.22	−0.21
T = 313.15 K									
0.04781	−0.59	−0.57	−0.55	−0.51	−0.48	−0.45	−0.42	−0.38	−0.44
0.10257	−0.84	−0.81	−0.78	−0.72	−0.67	−0.63	−0.59	−0.56	−0.52
0.19771	−0.99	−0.94	−0.91	−0.84	−0.78	−0.72	−0.68	−0.64	−0.60
0.24323	−1.00	−0.96	−0.93	−0.86	−0.80	−0.75	−0.70	−0.66	−0.62
0.30257	−1.07	−1.02	−0.99	−0.92	−0.86	−0.81	−0.76	−0.72	−0.68
0.38873	−1.04	−1.00	−0.97	−0.90	−0.85	−0.80	−0.76	−0.71	−0.68
0.48505	−0.95	−0.91	−0.88	−0.83	−0.78	−0.74	−0.70	−0.66	−0.63
0.58896	−0.82	−0.78	−0.76	−0.71	−0.67	−0.64	−0.60	−0.57	−0.55
0.72888	−0.64	−0.61	−0.60	−0.57	−0.54	−0.51	−0.49	−0.47	−0.45
0.85211	−0.38	−0.33	−0.32	−0.30	−0.29	−0.27	−0.26	−0.25	−0.24
0.93975	−0.27	−0.26	−0.26	−0.25	−0.25	−0.24	−0.24	−0.23	−0.23
T = 323.15 K									
0.04781	−0.65	−0.62	−0.60	−0.56	−0.52	−0.49	−0.46	−0.43	−0.41
0.10257	−0.92	−0.88	−0.85	−0.78	−0.73	−0.68	−0.63	−0.60	−0.56
0.19771	−1.08	−1.03	−0.99	−0.91	−0.84	−0.78	−0.73	−0.68	−0.64
0.24323	−1.10	−1.06	−1.01	−0.94	−0.87	−0.81	−0.75	−0.71	−0.67
0.30257	−1.16	−1.11	−1.07	−0.99	−0.93	−0.87	−0.81	−0.76	−0.72
0.38873	−1.13	−1.08	−1.04	−0.98	−0.91	−0.86	−0.81	−0.76	−0.73
0.48505	−1.02	−0.99	−0.95	−0.89	−0.84	−0.79	−0.74	−0.70	−0.67

(*continued*)

	$V^E/(cm^3 \cdot mol^{-1})$ at p/MPa								
0.58896	−0.88	−0.84	−0.81	−0.76	−0.72	−0.68	−0.64	−0.60	−0.58
0.72888	−0.68	−0.66	−0.64	−0.60	−0.57	−0.54	−0.51	−0.49	−0.47
0.85211	−0.39	−0.36	−0.34	−0.33	−0.31	−0.29	−0.27	−0.25	−0.25
0.93975	−0.29	−0.29	−0.28	−0.28	−0.27	−0.27	−0.25	−0.25	−0.25
T = 333.15 K									
0.04781	−0.71	−0.68	−0.66	−0.60	−0.56	−0.53	−0.49	−0.46	−0.43
0.10257	−1.01	−0.96	−0.92	−0.85	−0.79	−0.73	−0.69	−0.64	−0.60
0.19771	−1.18	−1.12	−1.08	−0.99	−0.91	−0.85	−0.79	−0.74	−0.69
0.24323	−1.20	−1.15	−1.10	−1.02	−0.94	−0.88	−0.82	−0.76	−0.72
0.30257	−1.25	−1.19	−1.15	−1.06	−0.99	−0.92	−0.86	−0.81	−0.76
0.38873	−1.21	−1.16	−1.12	−1.04	−0.97	−0.91	−0.86	−0.81	−0.76
0.48505	−1.10	−1.06	−1.02	−0.95	−0.89	−0.84	−0.79	−0.75	−0.71
0.58896	−0.94	−0.91	−0.87	−0.82	−0.77	−0.72	−0.68	−0.65	−0.61
0.72888	−0.73	−0.69	−0.67	−0.63	−0.60	−0.57	−0.54	−0.52	−0.50
0.85211	−0.41	−0.33	−0.32	−0.30	−0.28	−0.26	−0.25	−0.24	−0.22
0.93975	−0.29	−0.28	−0.28	−0.27	−0.26	−0.26	−0.25	−0.25	−0.25

Ref. [89]

x_1	ρ/g · cm^{-3}	u/m · s^{-1}	η/mPa · s	n_D	V^E/cm^3 · mol^{-1}	$K_{S,m}$/m^3 · TPa^{-1} · mol^{-1}	$K_{S,m}{}^E$/m^3 · TPa^{-1} · mol^{-1}	Δn_D	$\Delta\eta$/mPa · s
T = 298.15 K									
0.0000	1.2376	1679	97.58	1.47940	0.000	0.05472	0.00000	0.0000	0.000
0.1346	1.2242	1663	60.93	1.47491	−0.148	0.05038	−0.00271	0.0160	−23.58
0.2595	1.2091	1640	39.32	1.47006	−0.368	0.04661	−0.00496	0.0302	−33.08
0.3267	1.1994	1624	30.59	1.46675	−0.486	0.04469	−0.00607	0.0372	−35.29
0.4741	1.1719	1583	16.38	1.45742	−0.725	0.04052	−0.00845	0.0503	−35.20
0.5701	1.1468	1547	10.12	1.44913	−0.802	0.03808	−0.00973	0.0567	−32.14
0.6664	1.1139	1500	6.450	1.43808	−0.910	0.03588	−0.01075	0.0603	−26.466
0.7757	1.0590	1423	3.232	1.42043	−0.912	0.03428	−0.01103	0.0593	−19.083
0.8294	1.0211	1371	2.351	1.40780	−0.859	0.03413	−0.01053	0.0549	−14.753
0.9230	0.9249	1251	1.639	1.37377	−0.662	0.03568	−0.00784	0.0352	−6.377
0.9626	0.8637	1187	1.263	1.35299	−0.418	0.03775	−0.00529	0.0204	−2.914
0.9917	0.8060	1123	1.039	1.33319	−0.132	0.04118	−0.00151	0.0051	−0.314
1.0000	0.7864	1103	0.546	1.32687	0.000	0.04259	0.00000	0.0000	0.000
T = 313.15 K									
0.0000	1.2275	1641	50.01		0.000	0.05824	0.00000		0.000
0.1346	1.2141	1627	33.83		−0.188	0.05351	−0.00338		−9.51
0.2595	1.1990	1603	23.40		−0.443	0.04962	−0.00603		−13.75
0.3267	1.1891	1587	18.94		−0.566	0.04761	−0.00736		−14.88
0.4741	1.1613	1546	11.02		−0.818	0.04326	−0.01024		−15.49
0.5701	1.1359	1509	7.173		−0.898	0.04079	−0.01176		−14.580
0.6664	1.1029	1461	4.776		−1.031	0.03858	−0.01301		−12.206
0.7757	1.0475	1382	2.508		−1.036	0.03715	−0.01335		−9.055
0.8294	1.0092	1330	1.860		−0.969	0.03713	−0.01284		−7.047
0.9230	0.9122	1207	0.901		−0.750	0.03941	−0.00963		−3.362

(continued)

x_1	ρ/g · cm^{-3}	u/m · s^{-1}	η/mPa · s	n_D	V^E/cm^3 · mol^{-1}	$K_{S,m}$/m^3 · TPa^{-1} · mol^{-1}	$K_{S,m}{}^E$/m^3 · TPa^{-1} · mol^{-1}	Δn_D	$\Delta\eta$/mPa · s
0.9626	0.8504	1137	0.660		−0.475	0.04244	−0.00620		−1.640
0.9917	0.7922	1076	0.516		−0.156	0.04643	−0.00192		−0.342
1.0000	0.7723	1055	0.446		0.000	0.04827	0.00000		0.000
T = 328.15 K									
0.0000	1.2175	1605	29.18		0.000	0.06188	0.00000		0.000
0.1346	1.2040	1588	20.92		−0.230	0.05712	−0.00385		−4.38
0.2595	1.1886	1564	15.25		−0.486	0.05304	−0.00709		−6.45
0.3267	1.1788	1549	12.65		−0.647	0.05092	−0.00875		−7.12
0.4741	1.1507	1506	7.887		−0.922	0.04644	−0.01225		−7.634
0.5701	1.1251	1469	5.387		−1.020	0.04387	−0.01416		−7.368
0.6664	1.0918	1420	3.661		−1.166	0.04168	−0.01571		−6.319
0.7757	1.0359	1340	2.020		−1.175	0.04041	−0.01624		−4.811
0.8294	0.9973	1287	1.501		−1.107	0.04060	−0.01568		−3.783
0.9230	0.8993	1162	0.666		−0.853	0.04374	−0.01191		−1.921
0.9626	0.8369	1090	0.551		−0.546	0.04768	−0.00770		−0.896
0.9917	0.7777	1029	0.430		−0.162	0.05268	−0.00251		−0.178
1.0000	0.7578	1006	0.369		0.000	0.05513	0.00000		0.000

Ref. [90]

x_2	ρ/g · cm^{-3}	V^E/cm^3 · mol^{-1}	x_2	ρ/g · cm^{-3}	V^E/cm^3 · mol^{-1}
298.15 K					
0.0338	0.8583	−0.432	0.4449	1.1510	−1.153
0.0353	0.8601	−0.398	0.4707	1.1579	−1.138
0.1229	0.9795	−0.931	0.5619	1.1787	−1.073
0.1234	0.9798	−0.922	0.5665	1.1794	−1.046
0.2522	1.0757	−1.144	0.7601	1.2086	−0.616
0.3582	1.1232	−1.171			
303.15 K					
0.0338	0.8537	−0.411	0.3582	1.1195	−1.183
0.0353	0.8556	−0.382	0.4707	1.1542	−1.138
0.1229	0.9753	−0.929	0.5619	1.1747	−1.032
0.1234	0.9756	−0.921	0.5665	1.1755	−1.014
0.2026	1.0420	−1.130	0.6105	1.1830	−0.878
0.2522	1.0719	−1.157			
313.15 K					
0.0338	0.8451	−0.438	0.4453	1.1406	−1.255
0.0353	0.8472	−0.419	0.4917	1.1529	−1.238
0.1234	0.9672	−0.955	0.6083	1.1771	−1.098
0.1805	1.0163	−1.021	0.8753	1.2096	−0.177
0.2771	1.0761	−1.142			
0.3582	1.1117	−1.209			

(2) Activity Coefficient at Infinite Dilution and Partial Molar Excess Enthalpy at Infinite Dilution

Ref. [79]

Solute *i*	Temp range/K	γ_i^∞ (298 K)	$H_i^{E,\infty}$/kJ · mol^{-1}
Methanol	312.4–354.2	0.43	2.7

(3) Vapor Pressure and Activity Coefficient

Ref. [81]

T/K	P^{exp}/kPa	P^{NRTL}/kPa	γ_1^{exp}	γ_1^{exp}–γ_1^{NRTL}
x_1 = 0.9852				
291.66	11.684	11.787	0.9897	−0.0087
299.14	17.400	17.541	0.9905	−0.0080
304.28	22.581	22.753	0.9911	−0.0075
308.71	28.062	28.252	0.9920	−0.0066
313.71	35.592	35.784	0.9934	−0.0053
318.45	44.233	44.415	0.9947	−0.0040
322.57	53.123	53.302	0.9957	−0.0031
327.74	66.303	66.487	0.9961	−0.0027
x_1 = 0.9451				

(*continued*)

T/K	P^{exp}/kPa	P^{NRTL}/kPa	γ_1^{exp}	$\gamma_1^{exp}-\gamma_1^{NRTL}$
292.00	11.213	11.315	0.9717	−0.0089
295.26	13.371	13.492	0.9724	−0.0087
302.87	19.786	20.011	0.9713	−0.0111
307.75	25.181	25.476	0.9718	−0.0113
314.46	34.639	35.019	0.9733	−0.0107
319.77	44.019	44.554	0.9729	−0.0118
325.29	56.146	56.711	0.9756	−0.0098
329.17	66.186	66.832	0.9763	−0.0095
x_1 = 0.8806				
294.24	11.210	11.254	0.9240	−0.0037
300.06	15.229	15.326	0.9250	−0.0059
304.34	18.911	19.057	0.9260	−0.0071
310.83	25.922	26.194	0.9266	−0.0097
317.71	35.615	36.124	0.9262	−0.0132
321.94	43.163	43.678	0.9302	−0.0111
327.98	55.934	56.719	0.9306	−0.0131
331.77	65.500	66.471	0.9313	−0.0138
x_1 = 0.7597				
298.28	10.617	10.437	0.8195	0.0139
301.76	12.716	12.538	0.8212	0.0115
307.97	17.313	17.195	0.8224	0.0057
313.20	22.240	22.186	0.8244	0.0021
318.56	28.533	28.527	0.8280	0.0002
323.07	34.955	34.998	0.8312	−0.0010
327.83	43.000	43.114	0.8344	−0.0022
333.16	53.789	54.036	0.8375	−0.0039

Mixture 19: 1-C_3H_7OH (1) + [Emim][$EtSO_4$] (2)
(1) Volumetric Properties, Transport Properties, Refractive Index, Speed of Sound and Compressibility Properties

Ref. [89]

x_1	ρ/ $g \cdot cm^{-3}$	u/ $m \cdot s^{-1}$	η/ $mPa \cdot s$	n_D	V^E/ $cm^3 \cdot mol^{-1}$	$K_{S,m}$/ $m^3 \cdot TPa^{-1} \cdot mol^{-1}$	$K_{S,m}{}^E$/ $m^3 \cdot TPa^{-1} \cdot mol^{-1}$	Δn_D	$\Delta\eta$/ $mPa \cdot s$
T = 298.15 K									
0.0000	1.2376	1679	97.58	1.47940	0.000	0.05472	0.00000	0.0000	0.000
0.1619	1.2081	1637	60.85	1.47317	−0.209	0.05319	−0.00314	0.0094	−21.25
0.2104	1.1979	1620	52.11	1.47102	−0.263	0.05290	−0.00391	0.0119	−25.36
0.3360	1.1674	1574	34.41	1.46449	−0.334	0.05245	−0.00560	0.0174	−31.06
0.5159	1.1116	1492	18.01	1.45240	−0.406	0.05286	−0.00698	0.0227	−30.27
0.7298	1.0155	1382	7.296	1.43139	−0.380	0.05468	−0.00727	0.0223	−20.540
0.8046	0.9700	1337	5.491	1.42139	−0.334	0.05620	−0.00649	0.0195	−15.196
0.8894	0.9077	1284	3.577	1.40743	−0.287	0.05859	−0.00494	0.0137	−9.008
0.9401	0.8625	1249	2.776	1.39739	−0.180	0.06088	−0.00315	0.0085	−4.964
0.9727	0.8298	1227	2.419	1.39003	−0.097	0.06261	−0.00175	0.0043	−2.209
1.0000	0.7996	1207	2.017	1.38310	0.000	0.06463	0.00000	0.0000	0.000
T = 313.15 K									
0.0000	1.2275	1642	50.01		0.000	0.05816	0.00000		0.000
0.1619	1.1979	1600	33.73		−0.250	0.05656	−0.00395		−8.41
0.2104	1.1876	1584	29.56		−0.306	0.05637	−0.00485		−10.23
0.3360	1.1571	1537	20.72		−0.418	0.05599	−0.00705		−12.97
0.5159	1.1008	1456	11.73		−0.484	0.05660	−0.00905		−13.22
0.7298	1.0044	1340	5.125		−0.483	0.05946	−0.00929		−9.430
0.8046	0.9586	1294	3.918		−0.419	0.06143	−0.00840		−7.002
0.8894	0.8959	1238	2.580		−0.342	0.06470	−0.00636		−4.222
0.9401	0.8506	1201	1.997		−0.223	0.06770	−0.00409		−2.342

(continued)

x_1	ρ/ g · cm^{-3}	u/ m · s^{-1}	η/ mPa · s	n_D	V^E/ cm^3 · mol^{-1}	$K_{S,m}$/ m^3 · TPa^{-1} · mol^{-1}	$K_{S,m}{}^E$/ m^3 · TPa^{-1} · mol^{-1}	Δn_D	$\Delta\eta$/ mPa · s
0.9727	0.8178	1178	1.729		−0.125	0.06994	−0.00233		−1.028
1.0000	0.7874	1155	1.430		0.000	0.07266	0.00000		0.000
T = 328.15 K									
0.0000	1.2175	1605	29.18		0.000	0.06188	0.00000		0.000
0.1619	1.1876	1565	20.84		−0.271	0.06015	−0.00499		−3.78
0.2104	1.1773	1548	18.59		−0.348	0.05998	−0.00613		−4.66
0.3360	1.1467	1501	13.61		−0.497	0.05978	−0.00886		−6.11
0.5159	1.0902	1419	8.159		−0.606	0.06075	−0.01150		−6.502
0.7298	0.9932	1300	3.774		−0.606	0.06461	−0.01195		−4.869
0.8046	0.9472	1252	2.918		−0.540	0.06722	−0.01085		−3.620
0.8894	0.8840	1193	1.923		−0.425	0.07156	−0.00821		−2.229
0.9401	0.8385	1154	1.482		−0.286	0.07546	−0.00533		−1.244
0.9727	0.8055	1130	1.272		−0.163	0.07849	−0.00296		−0.538
1.0000	0.7748	1105	1.041		0.000	0.08199	0.00000		0.000

Ref. [90]

x_2	ρ/g · cm^{-3}	V^E/cm^3 · mol^{-1}	x_2	ρ/g · cm^{-3}	V^E/cm^3 · mol^{-1}
298.15 K					
0.0648	0.8683	−0.214	0.5114	1.1213	−0.569
0.1142	0.9111	−0.375	0.5814	1.1436	−0.519
0.1579	0.9447	−0.435	0.7276	1.1821	−0.347
0.2832	1.0237	−0.593	0.8789	1.2143	−0.148
0.4697	1.1067	−0.599	0.9519	1.2274	−0.029
303.15 K					
0.0648	0.8626	−0.173	0.5114	1.1181	−0.628
0.1142	0.9066	−0.336	0.5814	1.1405	−0.603
0.1579	0.9404	−0.406	0.7276	1.1794	−0.451
0.2832	1.0194	−0.558	0.8789	1.2111	−0.202
0.4697	1.1032	−0.640	0.9519	1.2242	−0.078
313.15 K					
0.0648	0.8561	−0.235	0.5114	1.1115	−0.677
0.1142	0.8998	−0.374	0.5814	1.1338	−0.614
0.1579	0.9335	−0.435	0.7276	1.1728	−0.452
0.2832	1.0129	−0.625	0.8789	1.2046	−0.192
0.4697	1.0967	−0.691	0.9519	1.2179	−0.086

(2) Activity Coefficient at Infinite Dilution and Partial Molar Excess Enthalpy at Infinite Dilution

Ref. [79]

Solute i	Temp range/K	γ_i^∞ (298 K)	$H_i^{E,\infty}$/kJ · mol^{-1}
1-propanol	322.8–375.4	1.5	6.5

(3) Osmotic Coefficient, Activity and Vapor Pressure

Ref. [80]

$m/(\mathrm{mol \cdot kg^{-1}})$	Φ_1	a_1	p/kPa
$T = 323.15$ K			
0.0464	0.914	0.995	12.040
0.0854	0.860	0.991	11.995
0.0991	0.826	0.990	11.982
0.2144	0.770	0.980	11.862
0.3907	0.702	0.968	11.705
0.4826	0.668	0.962	11.637
0.6749	0.644	0.949	11.479
0.9772	0.596	0.932	11.274
1.2548	0.583	0.916	11.072
1.6504	0.551	0.896	10.834
2.0823	0.552	0.871	10.522
2.5319	0.541	0.848	10.244
2.9028	0.531	0.831	10.035
3.3062	0.524	0.812	9.805

Mixture 20: 2-CH_3CHCH_3OH (1) + [Emim][$EtSO_4$] (2)
(1) Volumetric Properties, Transport Properties, Refractive Index, Speed of Sound and Compressibility Properties

Ref. [89]

x_1	ρ/ g · cm^{-3}	u/m · s^{-1}	η/ mPa · s	n_D	V^E/ cm^3 · mol^{-1}	$K_{S,m}$/ m^3 · TPa^{-1} · mol^{-1}	$K_{S,m}{}^E$/ m^3 · TPa^{-1} · mol^{-1}	Δn_D	$\Delta\eta$/ mPa · s
T = 298.15 K									
0.0000	1.2376	1679	97.58	1.47940	0.000	0.05472	0.00000	0.0000	0.000
0.1228	1.2143	1644	69.71	1.47434	−0.163	0.05386	−0.00349	0.0078	−16.15
0.2066	1.1964	1611	53.67	1.47057	−0.309	0.05381	−0.00534	0.0127	−24.18
0.3090	1.1709	1570	40.10	1.46502	−0.413	0.05381	−0.00753	0.0179	−27.97
0.3990	1.1449	1529	29.79	1.45926	−0.480	0.05417	−0.00910	0.0215	−29.68
0.4977	1.1118	1475	21.14	1.45177	−0.558	0.05526	−0.01013	0.0244	−28.91
0.6341	1.0542	1399	12.40	1.43881	−0.512	0.05727	−0.01104	0.0256	−24.62
0.7223	1.0085	1342	8.462	1.42818	−0.512	0.05952	−0.01067	0.0242	−20.145
0.8551	0.9202	1251	4.475	1.40784	−0.436	0.06462	−0.00842	0.0178	−11.446
0.9706	0.8141	1164	2.499	1.38306	−0.149	0.07270	−0.00281	0.0050	−2.387
1.0000	0.7807	1138	2.082	1.37495	0.000	0.07614	0.00000	0.0000	0.000
T = 313.15 K									
0.0000	1.2275	1643	50.01		0.000	0.05809	0.00000		0.000
0.1228	1.2044	1608	37.79		−0.247	0.05723	−0.00436		−6.25
0.2066	1.1864	1576	30.34		−0.415	0.05718	−0.00681		−9.61
0.3090	1.1609	1534	23.56		−0.561	0.05734	−0.00956		−11.42
0.3990	1.1347	1491	18.23		−0.637	0.05799	−0.01148		−12.37
0.4977	1.1013	1439	13.41		−0.716	0.05917	−0.01311		−12.38
0.6341	1.0435	1360	8.274		−0.696	0.06194	−0.01423		−10.882
0.7223	0.9974	1302	5.753		−0.680	0.06465	−0.01403		−9.116
0.8551	0.9086	1207	3.081		−0.582	0.07120	−0.01127		−5.324

(continued)

x_1	ρ/ $g \cdot cm^{-3}$	u/$m \cdot s^{-1}$	η/ $mPa \cdot s$	n_D	V^E/ $cm^3 \cdot mol^{-1}$	$K_{S,m}$/ $m^3 \cdot TPa^{-1} \cdot mol^{-1}$	$K_{S,m}{}^E$/ $m^3 \cdot TPa^{-1} \cdot mol^{-1}$	Δn_D	$\Delta\eta$/ $mPa \cdot s$
0.9706	0.8017	1113	1.658		−0.211	0.08199	−0.00377		−1.125
1.0000	0.7678	1085	1.354		0.000	0.08660	0.00000		0.000
$T = 328.15$ K									
0.0000	1.2175	1605	29.18		0.000	0.06188	0.00000		0.000
0.1228	1.1945	1569	22.97		−0.334	0.06111	−0.00539		−2.74
0.2066	1.1763	1537	18.93		−0.521	0.06115	−0.00850		−4.41
0.3090	1.1507	1494	15.20		−0.711	0.06153	−0.01197		−5.25
0.3990	1.1243	1451	12.10		−0.810	0.06237	−0.01451		−5.80
0.4977	1.0907	1399	9.184		−0.918	0.06382	−0.01677		−5.929
0.6341	1.0326	1318	5.878		−0.929	0.06725	−0.01847		−5.380
0.7223	0.9862	1259	4.150		−0.920	0.07072	−0.01832		−4.617
0.8551	0.8966	1161	2.225		−0.784	0.07902	−0.01501		−2.789
0.9706	0.7885	1063	1.147		−0.295	0.09292	−0.00545		−0.602
1.0000	0.7539	1031	0.919		0.000	0.09948	0.00000		0.000

Ref. [90]

x_2	ρ/g · cm^{-3}	V^E/cm^3 · mol^{-1}	x_2	ρ/g · cm^{-3}	V^E/cm^3 · mol^{-1}
298.15 K					
0.0665	0.8522	−0.192	0.5203	1.1172	−0.388
0.1114	0.8926	−0.286	0.5882	1.1378	−0.340
0.2553	0.9954	−0.438	0.7084	1.1744	−0.246
0.3374	1.0403	−0.452	0.8825	1.2147	−0.082
0.4452	1.0889	−0.440	0.9545	1.2289	−0.026
303.15 K					
0.0665	0.8483	−0.211	0.5203	1.1146	−0.539
0.1114	0.8890	−0.323	0.5882	1.1371	−0.468
0.2553	0.9924	−0.555	0.7084	1.1714	−0.334
0.3374	1.0378	−0.622	0.8825	1.2115	−0.125
0.4452	1.0864	−0.606	0.9545	1.2258	−0.046
313.15 K					
0.0665	0.8396	−0.220	0.5203	1.1078	−0.617
0.1114	0.8810	−0.392	0.5882	1.1302	−0.509
0.2553	0.9851	−0.664	0.7084	1.1642	−0.330
0.3374	1.0306	−0.717	0.8825	1.2051	−0.129
0.4452	1.0794	−0.687	0.9545	1.2193	−0.032

(2) Activity Coefficient at Infinite Dilution and Partial Molar Excess Enthalpy at Infinite Dilution

Ref. [79]

Solute *i*	Temp range/K	γ_i^∞ (298 K)	$H_i^{E,\infty}$/kJ · mol^{-1}
iso-propanol	302.5–343.9	1.5	5.4

(3) Osmotic Coefficient, Activity and Vapor Pressure

Ref. [80]

m/(mol · kg^{-1})	Φ_1	a_1	p/kPa
T = 323.15 K			
0.0601	0.957	0.993	23.426
0.0757	0.952	0.991	23.385
0.1148	0.905	0.988	23.294
0.2001	0.800	0.981	23.135
0.3055	0.734	0.973	22.954
0.4052	0.686	0.967	22.804
0.4942	0.664	0.961	22.665
0.7072	0.610	0.949	22.382
0.9702	0.575	0.935	22.040
1.2750	0.553	0.919	21.648
1.6603	0.536	0.899	21.163
2.0610	0.518	0.880	20.710

(continued)

$m/(\text{mol}\cdot\text{kg}^{-1})$	Φ_1	a_1	p/kPa
2.4737	0.504	0.861	20.266
2.8311	0.495	0.845	19.887
3.3196	0.482	0.825	19.413
3.7067	0.478	0.808	19.013

Mixture 21: $CH_2(CH_2)_4CH_2$ (1) + [EMIM][$C_2H_5OSO_3$] (2)
(1) Phase Equilibrium

Ref. [3]

x_1	P (kPa)	x_1	P (kPa)	x_1	P (kPa)	x_1	P (kPa)
303.15 K							
0.0024	3.12	0.0574	16.27	0.3398	16.25	0.7776	16.23
0.0049	5.97	0.0808	16.27	0.4030	16.25	0.8781	16.23
0.0074	8.73	0.1098	16.27	0.4824	16.24	0.9329	16.25
0.0099	11.45	0.1451	16.26	0.5801	16.24	0.9660	16.28
0.0159	16.11	0.1860	16.26	0.6005	16.25	0.9829	16.29
0.0229	16.22	0.2314	16.25	0.6708	16.24	0.9992	16.29
0.0299	16.25	0.2848	16.25	0.6828	16.24	1.0000	16.25
0.0413	16.27						

(2) Activity Coefficient at Infinite Dilution and Partial Molar Excess Enthalpy at Infinite Dilution

Ref. [79]

Solute i	Temp range/K	γ_i^{∞} (298 K)	$H_i^{E,\infty}/\text{kJ}\cdot\text{mol}^{-1}$
cyclohexane	301.6–364.7	45.3	17.3

Mixture 22: H_2 (1) + [C_2C_1Im][C_2SO_4] (2)
(1) Henry's Law Constant

Ref. [91]

T/K	$p/10^2$ Pa	$K_H/10^5$ Pa	$x_1/10^{-4}$	Dev%
283.16	779.45	2492	4.015	−0.3
292.93	804.85	2143	4.669	+0.1
303.39	499.66	2056	4.866	+0.6
303.39	832.19	2062	4.853	+0.3
303.42	921.92	2071	4.831	−0.1
316.29	959.48	2280	4.389	−0.4
323.29	980.04	2544	3.933	−1.1
324.27	531.49	2516	3.976	+1.7
333.12	1008.87	3071	3.258	−0.4
333.22	545.39	3107	3.220	−1.3
343.13	937.32	3924	2.550	+0.1
343.13	1038.24	3894	2.569	+0.9

(2) Thermodynamic Properties

Ref. [91]

T/K	$\Delta_{solv}G$/kJ · mol^{-1}	$\Delta_{solv}H$/kJ · mol^{-1}	$\Delta_{solv}S$/J · mol^{-1} · K^{-1}
283	18.40	+14.4	−14.1
293	18.69	+6.43	−41.8
303	19.23	−1.04	−66.9
313	20.02	−8.03	−89.6
323	21.02	−14.6	−110
333	22.21	−20.7	−129
343	23.59	−26.6	−146

Mixture 23: CO_2 (1) + $[C_2C_1Im][C_2SO_4]$ (2)
(1) Henry's Law Constant

Ref. [91]

T/K	$p/10^2$ Pa	$K_H/10^5$ Pa	$x_1/10^{-4}$	Dev%
283.32	380.10	62.13	160.0	−0.1
293.48	402.72	77.22	128.8	+0.2
303.38	836.77	94.28	105.6	+0.1
303.39	777.62	94.52	105.3	−0.1
313.28	869.49	113.2	88.00	+0.4
313.28	443.24	114.1	87.30	−0.4
314.68	813.40	117.0	85.15	−0.4
323.18	901.28	134.2	74.22	+0.6
323.24	839.67	135.7	73.41	−0.4
333.12	932.51	158.1	63.02	+0.5
333.14	869.20	159.0	62.68	−0.0
333.16	479.90	159.5	62.48	−0.3
343.08	898.46	186.0	53.58	−0.4
343.12	963.32	184.7	53.96	+0.3

(2) Thermodynamic Properties

Ref. [91]

T/K	$\Delta_{solv}G$/kJ · mol^{-1}	$\Delta_{solv}H$/kJ · mol^{-1}	$\Delta_{solv}S$/J · mol^{-1} · K^{-1}
283	9.698	−15.0	−87.3
293	10.57	−14.9	−87.0
303	11.44	−14.8	−86.7
313	12.30	−14.8	−86.5
323	13.17	−14.7	−86.2
333	14.03	−14.6	−86.0
343	14.89	−14.6	−85.9

Mixture 24: C_9H_{20} (1) + [EMIM][$EtSO_4$] (2)
(1) Activity Coefficient at Infinite Dilution and Partial Molar Excess Enthalpy at Infinite Dilution

Ref. [79]

Solute *i*	Temp range/K	γ_i^{∞} (298 K)	$H_i^{E,\infty}$/kJ · mol^{-1}
1-Nonane	301.8–333.5	392.4	15.3

Mixture 25: $C_{10}H_{22}$ (1) + [EMIM][$EtSO_4$] (2)
(1) Activity Coefficient at Infinite Dilution and Partial Molar Excess Enthalpy at Infinite Dilution

Ref. [79]

Solute *i*	Temp range/K	γ_i^{∞} (298 K)	$H_i^{E,\infty}$/kJ · mol^{-1}
1-Decane	301.8–333.5	538.4	11.0

Mixture 26: $C_{11}H_{24}$ (1) + [EMIM][$EtSO_4$] (2)
(1) Activity Coefficient at Infinite Dilution and Partial Molar Excess Enthalpy at Infinite Dilution

Ref. [79]

Solute *i*	Temp range/K	γ_i^{∞} (298 K)	$H_i^{E,\infty}$/kJ · mol^{-1}
1-Undecane	301.8–354.3	700.9	16.5

Mixture 27: $C_{12}H_{26}$ (1) + [EMIM][$EtSO_4$] (2)
(1) Activity Coefficient at Infinite Dilution and Partial Molar Excess Enthalpy at Infinite Dilution

Ref. [79]

Solute *i*	Temp range/K	γ_i^{∞} (298 K)	$H_i^{E,\infty}$/kJ · mol^{-1}
1-Dodecane	323.0–364.8	878.5	14.0

Mixture 28: 1-C_6H_{12} (1) + [EMIM][$EtSO_4$] (2)
(1) Activity Coefficient at Infinite Dilution and Partial Molar Excess Enthalpy at Infinite Dilution

Ref. [79]

Solute *i*	Temp range/K	γ_i^{∞} (298 K)	$H_i^{E,\infty}$/kJ · mol^{-1}
1-Hexene	301.9–344.0	67.9	−1.4

Mixture 29: 1-C_7H_{14} (1) + [EMIM][$EtSO_4$] (2)
(1) Activity Coefficient at Infinite Dilution and Partial Molar Excess Enthalpy at Infinite Dilution

Ref. [79]

Solute *i*	Temp range/K	γ_i^{∞} (298 K)	$H_i^{E,\infty}$/kJ · mol^{-1}
1-Heptene	301.9–344.0	106.0	5.1

Mixture 30: 1-C_8H_{16} (1) + [EMIM][EtSO$_4$] (2)

(1) Activity Coefficient at Infinite Dilution and Partial Molar Excess Enthalpy at Infinite Dilution

Ref. [79]

Solute i	Temp range/K	γ_i^∞ (298 K)	$H_i^{E,\infty}$/kJ · mol^{-1}
1-Octene	302.1–344.0	156.2	5.6

Mixture 31: 1-C_9H_{18} (1) + [EMIM][EtSO$_4$] (2)

(1) Activity Coefficient at Infinite Dilution and Partial Molar Excess Enthalpy at Infinite Dilution

Ref. [79]

Solute i	Temp range/K	γ_i^∞ (298 K)	$H_i^{E,\infty}$/kJ · mol^{-1}
1-Nonene	301.9–343.9	226.4	12.6

Mixture 32: 1-$C_{10}H_{20}$ (1) + [EMIM][EtSO$_4$] (2)

(1) Activity Coefficient at Infinite Dilution and Partial Molar Excess Enthalpy at Infinite Dilution

Ref. [79]

Solute i	Temp range/K	γ_i^∞ (298 K)	$H_i^{E,\infty}$/kJ · mol^{-1}
1-Decene	301.9–322.9	353.6	11.3

Mixture 33: 1-$C_{11}H_{22}$ (1) + [EMIM][EtSO$_4$] (2)

(1) Activity Coefficient at Infinite Dilution and Partial Molar Excess Enthalpy at Infinite Dilution

Ref. [79]

Solute i	Temp range/K	γ_i^∞ (298 K)	$H_i^{E,\infty}$/kJ · mol^{-1}
1-Undecene	301.9–354.3	480.9	9.4

Mixture 34: 1-$C_{12}H_{24}$ (1) + [EMIM][EtSO$_4$] (2)

(1) Activity Coefficient at Infinite Dilution and Partial Molar Excess Enthalpy at Infinite Dilution

Ref. [79]

Solute i	Temp range/K	γ_i^∞ (298 K)	$H_i^{E,\infty}$/kJ · mol^{-1}
1-Dodecene	312.5–354.3	606.2	8.7

Mixture 35: $CH = CHCH_2CH_2CH_2CH_2$ (1) + [EMIM][EtSO$_4$] (2)

(1) Activity Coefficient at Infinite Dilution and Partial Molar Excess Enthalpy at Infinite Dilution

Ref. [79]

Solute i	Temp range/K	γ_i^∞ (298 K)	$H_i^{E,\infty}$/kJ · mol^{-1}
Cyclohexene	301.6–343.8	21.8	10.9

Mixture 36: $C_6H_5C_3H_7$ (1) + $[EMIM][EtSO_4]$ (2)
(1) Activity Coefficient at Infinite Dilution and Partial Molar Excess Enthalpy at Infinite Dilution

Ref. [79]

Solute *i*	Temp range/K	γ_i^{∞} (298 K)	$H_i^{E,\infty}$/kJ · mol^{-1}
Propyl benzene	343.8–385.7	23.4	6.2

Mixture 37: $C_6H_5C_4H_9$ (1) + $[EMIM][EtSO_4]$ (2)
(1) Activity Coefficient at Infinite Dilution and Partial Molar Excess Enthalpy at Infinite Dilution

Ref. [79]

Solute *i*	Temp range/K	γ_i^{∞} (298 K)	$H_i^{E,\infty}$/kJ · mol^{-1}
Butyl benzene	343.7–396.1	42.4	7.4

Mixture 38: $C_6H_5C_5H_{11}$ (1) + $[EMIM][EtSO_4]$ (2)
(1) Activity Coefficient at Infinite Dilution and Partial Molar Excess Enthalpy at Infinite Dilution

Ref. [79]

Solute *i*	Temp range/K	γ_i^{∞} (298 K)	$H_i^{E,\infty}$/kJ · mol^{-1}
Pentyl benzene	354.3–396.1	73.6	8.2

Mixture 39: 1-C_4H_9OH (1) + $[EMIM][EtSO_4]$ (2)
(1) Activity Coefficient at Infinite Dilution and Partial Molar Excess Enthalpy at Infinite Dilution

Ref. [79]

Solute *i*	Temp range/K	γ_i^{∞} (298 K)	$H_i^{E,\infty}$/kJ · mol^{-1}
1-Butanol	323.0–365.0	2.0	5.9

Mixture 40: 1-$C_5H_{11}OH$ (1) + $[EMIM][EtSO_4]$ (2)
(1) Activity Coefficient at Infinite Dilution and Partial Molar Excess Enthalpy at Infinite Dilution

Ref. [79]

Solute *i*	Temp range/K	γ_i^{∞} (298 K)	$H_i^{E,\infty}$/kJ · mol^{-1}
1-Pentanol	333.4–364.7	3.3	6.3

Mixture 41: 1-$C_6H_{13}OH$ (1) + $[EMIM][EtSO_4]$ (2)
(1) Activity Coefficient at Infinite Dilution and Partial Molar Excess Enthalpy at Infinite Dilution

Ref. [79]

Solute *i*	Temp range/K	γ_i^{∞} (298 K)	$H_i^{E,\infty}$/kJ · mol^{-1}
1-Hexanol	343.8–375.2	5.3	7.6

Mixture 42: $(CH_3)_2CHCH_2OH$ (1) + [EMIM][$EtSO_4$] (2)
(1) Activity Coefficient at Infinite Dilution and Partial Molar Excess Enthalpy at Infinite Dilution

Ref. [79]

Solute i	Temp range/K	γ_i^{∞} (298 K)	$H_i^{E,\infty}$/kJ · mol^{-1}
iso-butanol	322.7–386.0	1.9	5.4

Mixture 43: $CH_3CH_2CH(OH)CH_3$ (1) + [EMIM][$EtSO_4$] (2)
(1) Activity Coefficient at Infinite Dilution and Partial Molar Excess Enthalpy at Infinite Dilution

Ref. [79]

Solute i	Temp range/K	γ_i^{∞} (298 K)	$H_i^{E,\infty}$/kJ · mol^{-1}
sec-butanol	312.5–364.9	2.1	5.7

Mixture 44: $(CH_3)_3COH$ (1) + [EMIM][$EtSO_4$] (2)
(1) Activity Coefficient at Infinite Dilution and Partial Molar Excess Enthalpy at Infinite Dilution

Ref. [79]

Solute i	Temp range/K	γ_i^{∞} (298 K)	$H_i^{E,\infty}$/kJ · mol^{-1}
tert-butanol	302.0–354.4	2.1	5.1

Mixture 45: $CH_3CH_2C(CH_3)_2OH$ (1) + [EMIM][$EtSO_4$] (2)
(1) Activity Coefficient at Infinite Dilution and Partial Molar Excess Enthalpy at Infinite Dilution

Ref. [79]

Solute i	Temp range/K	γ_i^{∞} (298 K)	$H_i^{E,\infty}$/kJ · mol^{-1}
tert-pentanol	302.0–354.4	2.6	3.4

Mixture 46: CH_3CH_2CHO (1) + [EMIM][$EtSO_4$] (2)
(1) Activity Coefficient at Infinite Dilution and Partial Molar Excess Enthalpy at Infinite Dilution

Ref. [79]

Solute i	Temp range/K	γ_i^{∞} (298 K)	$H_i^{E,\infty}$/kJ · mol^{-1}
Propanal	333.5–375.2	2.2	2.7

Mixture 47: $CH_3(CH_2)_2CHO$ (1) + [EMIM][$EtSO_4$] (2)
(1) Activity Coefficient at Infinite Dilution and Partial Molar Excess Enthalpy at Infinite Dilution

Ref. [79]

Solute i	Temp range/K	γ_i^{∞} (298 K)	$H_i^{E,\infty}$/kJ · mol^{-1}
Butanal	322.8–385.9	3.6	1.6

Mixture 48: $CH_3(CH_2)_3CHO$ (1) + [EMIM][$EtSO_4$] (2)
(1) Activity Coefficient at Infinite Dilution and Partial Molar Excess Enthalpy at Infinite Dilution

Ref. [79]

Solute *i*	Temp range/K	γ_i^∞ (298 K)	$H_i^{E,\infty}$/kJ · mol^{-1}
Pentanal	333.5–375.2	5.8	2.6

Mixture 49: $CH_3(CH_2)_4CHO$ (1) + [EMIM][$EtSO_4$] (2)
(1) Activity Coefficient at Infinite Dilution and Partial Molar Excess Enthalpy at Infinite Dilution

Ref. [79]

Solute *i*	Temp range/K	γ_i^∞ (298 K)	$H_i^{E,\infty}$/kJ · mol^{-1}
Hexanal	312.2–333.4	11.6	1.01

Mixture 50: $CH_3(CH_2)_5CHO$ (1) + [EMIM][$EtSO_4$] (2)
(1) Activity Coefficient at Infinite Dilution and Partial Molar Excess Enthalpy at Infinite Dilution

Ref. [79]

Solute *i*	Temp range/K	γ_i^∞ (298 K)	$H_i^{E,\infty}$/kJ · mol^{-1}
Heptanal	303.3–354.2	16.7	5.89

Mixture 51: $CH_3(CH_2)_6CHO$ (1) + [EMIM][$EtSO_4$] (2)
(1) Activity Coefficient at Infinite Dilution and Partial Molar Excess Enthalpy at Infinite Dilution

Ref. [79]

Solute *i*	Temp range/K	γ_i^∞ (298 K)	$H_i^{E,\infty}$/kJ · mol^{-1}
Octanal	303.3–343.8	30.2	2.1

Mixture 52: $CH_3OOCCH_2CH_3$ (1) + [Emim][$EtSO_4$] (2)
(1) Activity Coefficient at Infinite Dilution and Partial Molar Excess Enthalpy at Infinite Dilution

Ref. [79]

Solute *i*	Temp range/K	γ_i^∞ (298 K)	$H_i^{E,\infty}$/kJ · mol^{-1}
Methyl propanoate	322.9–375.3	5.7	5.3

Mixture 53: $CH_3OOCCH_2CH_2CH_3$ (1) + [EMIM][$EtSO_4$] (2)
(1) Activity Coefficient at Infinite Dilution and Partial Molar Excess Enthalpy at Infinite Dilution

Ref. [79]

Solute *i*	Temp range/K	γ_i^∞ (298 K)	$H_i^{E,\infty}$/kJ · mol^{-1}
Methyl butanoate	343.9–385.9	9.8	5.4

Mixture 54: $CH_3OOCCH_2CH_2CH_2CH_3$ (1) + [EMIM][EtSO$_4$] (2)
(1) Activity Coefficient at Infinite Dilution and Partial Molar Excess Enthalpy at Infinite Dilution

Ref. [79]

Solute i	Temp range/K	γ_i^∞ (298 K)	$H_i^{E,\infty}$/kJ · mol^{-1}
Methyl pentanoate	333.6–385.9	16.8	8.2

Mixture 55: $CH_3OOCCH_2CH_2CH_2CH_2CH_3$ (1) + [EMIM][EtSO$_4$] (2)
(1) Activity Coefficient at Infinite Dilution and Partial Molar Excess Enthalpy at Infinite Dilution

Ref. [79]

Solute i	Temp range/K	γ_i^∞(298 K)	$H_i^{E,\infty}$/kJ · mol^{-1}
Methyl hexanoate	301.7–343.9	20.7	4.8

Mixture 56: $CH_3OOCCH_2CH_2CH_2CH_2CH_2CH_3$ (1) + [EMIM][EtSO$_4$] (2)
(1) Activity Coefficient at Infinite Dilution and Partial Molar Excess Enthalpy at Infinite Dilution

Ref. [79]

Solute i	Temp range/K	γ_i^∞ (298 K)	$H_i^{E,\infty}$/kJ · mol^{-1}
Methyl heptanoate	301.8–333.4	30.5	4.0

Mixture 57: CH_3COCH_3 (1) + [EMIM][EtSO$_4$] (2)
(1) Activity Coefficient at Infinite Dilution and Partial Molar Excess Enthalpy at Infinite Dilution

Ref. [79]

Solute i	Temp range/K	γ_i^∞ (298 K)	$H_i^{E,\infty}$/kJ · mol^{-1}
Acetone	323.1–386.0	1.7	1.4

Mixture 58: CH_3CN (1) + [EMIM][EtSO$_4$] (2)
(1) Activity Coefficient at Infinite Dilution and Partial Molar Excess Enthalpy at Infinite Dilution

Ref. [79]

Solute i	Temp range/K	γ_i^∞ (298 K)	$H_i^{E,\infty}$/kJ · mol^{-1}
Acetonitrile	312.2–354.3	0.83	1.10

Mixture 59: 1,4-$CH_2CH_2OCH_2CH_2O$ (1) + [EMIM][EtSO$_4$] (2)
(1) Activity Coefficient at Infinite Dilution and Partial Molar Excess Enthalpy at Infinite Dilution

Ref. [79]

Solute i	Temp range/K	γ_i^∞ (298 K)	$H_i^{E,\infty}$/kJ · mol^{-1}
1,4-dioxane	312.4–354.2	2.1	0.4

Mixture 60: $CHCl_3$ (1) + [EMIM][$EtSO_4$] (2)
(1) Activity Coefficient at Infinite Dilution and Partial Molar Excess Enthalpy at Infinite Dilution

Ref. [79]

Solute *i*	Temp range/K	γ_i^∞ (298 K)	$H_i^{E,\infty}$/kJ · mol^{-1}
Trichloromethane	301.6–354.2	0.44	−7.0

Mixture 61: $NH(CH_2)_5C = O$ (1) + [EMIM][$EtSO_4$] (2)
(1) Activity Coefficient at Infinite Dilution and Partial Molar Excess Enthalpy at Infinite Dilution

Ref. [79]

Solute *i*	Temp range/K	γ_i^∞ (298 K)	$H_i^{E,\infty}$/kJ · mol^{-1}
ε-caprolactam	412.3–427.1	5.5	

Mixture 62: $CH_2(CH_2)_3CH_2C = NOH$ (1) + [EMIM][$EtSO_4$] (2)
(1) Activity Coefficient at Infinite Dilution and Partial Molar Excess Enthalpy at Infinite Dilution

Ref. [79]

Solute *i*	Temp range/K	γ_i^∞ (298 K)	$H_i^{E,\infty}$/kJ · mol^{-1}
Cyclohexanone oxime	417.2–427.5	0.49	

Mixture 63: C_7H_{16} (1) + [Emim][$C_2H_5SO_4$] (2)
(1) Phase Equilibrium

Ref. [86]

Organic phase	IL-rich phase	
x_2'	x_2''	*T*/K
0.0038	0.9968	293.15
0.0042	0.9961	303.15
0.0040	0.9954	313.15
0.0047	0.9929	323.15
0.0095	0.9942	333.15

Mixture 64: $CH_3CH(CH_2)_4CH_2$ (1) + [Emim][$C_2H_5SO_4$] (2)
(1) Phase Equilibrium

Ref. [86]

Organic phase	IL-rich phase	
x_2'	x_2''	*T*/K
0.0145	0.9792	293.15
0.0095	0.9777	303.15
0.0094	0.9742	313.15
0.0071	0.9690	323.15
0.0063	0.9600	333.15

Mixture 65: CH_2Cl_2 (1) + [emim][$EtSO_4$] (2)
(1) Volumetric Properties and Conductivity

Ref. [92]

w_2	x_2	c_2/mol · L^{-1}	ρ/kg · m^{-3}	κ/S · m^{-1}	$\Lambda/10^{-4}$ S · m^2 · mol^{-1}
298.15 K					
0	0	0	1316.98		
0.02479	0.009055	0.1382	1317.42	0.06340	4.587
0.05011	0.01861	0.2793	1317.25	0.1366	4.889
0.09979	0.03832	0.5560	1316.44	0.3103	5.580
0.1997	0.08232	1.110	1313.25	0.6636	5.978
0.3002	0.1336	1.662	1307.67	0.9352	5.628
0.3995	0.1930	2.195	1298.30	1.085	4.942
0.5007	0.2650	2.737	1291.58	1.112	4.062
0.6031	0.3532	3.270	1281.25	1.027	3.140
0.6936	0.4487	3.734	1272.16	0.8935	2.393
0.7936	0.5802	4.235	1261.09	0.7280	1.719
0.8973	0.7584	4.748	1250.30	0.5558	1.171
0.9694	0.9194	5.097	1242.23	0.4362	0.8559
1	1	5.237	1242.3	0.3820	0.7295

Mixture 66: CH_3NO_2 (1) + [emim][$EtSO_4$] (2)
(1) Excess Molar Enthalpy

Ref. [74]

x_2	H^E/(J · mol^{-1})	x_2	H^E/(J · mol^{-1})	x_2	H^E/(J · mol^{-1})
303.15 K					
0.1001	−288	0.4001	−644	0.6953	−451
0.1977	−485	0.5071	−620	0.8026	−308
0.3075	−594	0.6026	−547	0.8668	−209

2.2.13.2 Ternary Mixtures

Mixture 1: C_6H_{14} (1) + SCH = CHCH = CH (thiophene) (2) + [C_2mim][$EtSO_4$] (3)
(1) Phase Equilibrium

Ref. [93]

IL-rich phase			Hydrocarbon-rich phase		
x^{II}_3	x^{II}_2	x^{II}_1	x^I_3	x^I_2	x^I_1
T = 298.15 K					
0.993	0.000	0.007	0.000	0.000	1.000
0.860	0.131	0.009	0.000	0.082	0.918
0.733	0.257	0.010	0.000	0.188	0.812
0.649	0.343	0.008	0.010	0.381	0.609
0.488	0.503	0.009	0.014	0.641	0.345
0.465	0.524	0.011	0.005	0.732	0.263

(continued)

IL-rich phase			Hydrocarbon-rich phase		
x^{II}_3	x^{II}_2	x^{II}_1	x^{I}_3	x^{I}_2	x^{I}_1
0.448	0.543	0.009	0.005	0.807	0.188
0.352	0.641	0.007	0.004	0.901	0.095
0.293	0.703	0.004	0.000	0.953	0.047
0.288	0.712	0.000	0.000	1.000	0.000

Mixture 2: C_7H_{16} (1) + SCH = CHCH = CH (thiophene) (2) + [C_2mim][$EtSO_4$] (3)
(1) Phase Equilibrium

Ref. [93]

IL-rich phase			Hydrocarbon-rich phase		
x^{II}_3	x^{II}_2	x^{II}_1	x^{I}_3	x^{I}_2	x^{I}_1
$T = 298.15$ K					
1.000	0.000	0.000	0.000	0.000	1.000
0.845	0.155	0.000	0.006	0.092	0.902
0.672	0.319	0.009	0.006	0.311	0.683
0.567	0.422	0.011	0.001	0.494	0.505
0.484	0.511	0.005	0.000	0.628	0.372
0.422	0.566	0.012	0.000	0.870	0.130
0.369	0.623	0.008	0.000	0.937	0.063
0.337	0.663	0.000	0.000	0.967	0.033
0.288	0.712	0.000	0.000	1.000	0.000

Mixture 3: $C_{12}H_{26}$ (1) + SCH = CHCH = CH (thiophene) (2) + [C_2mim][$EtSO_4$] (3)
(1) Phase Equilibrium

Ref. [93]

IL-rich phase			Hydrocarbon-rich phase		
x^{II}_3	x^{II}_2	x^{II}_1	x^{I}_3	x^{I}_2	x^{I}_1
$T = 298.15$ K					
1.000	0.000	0.000	0.000	0.000	1.000
0.959	0.041	0.000	0.000	0.083	0.917
0.906	0.094	0.000	0.000	0.190	0.810
0.830	0.170	0.000	0.000	0.303	0.697
0.774	0.226	0.000	0.000	0.428	0.572
0.703	0.297	0.000	0.000	0.558	0.442
0.647	0.353	0.000	0.000	0.672	0.328
0.623	0.377	0.000	0.000	0.783	0.217
0.621	0.379	0.000	0.000	0.838	0.162
0.582	0.418	0.000	0.000	0.906	0.094
0.581	0.419	0.000	0.000	0.963	0.037
0.564	0.436	0.000	0.000	0.938	0.062
0.555	0.445	0.000	0.000	0.985	0.015
0.288	0.712	0.000	0.000	1.000	0.000

Mixture 4: $C_{16}H_{34}$ (1) + SCH = CHCH = CH (thiophene) (2) + $[C_2mim][EtSO_4]$ (3)
(1) Phase Equilibrium

Ref. [93]

IL-rich phase			Hydrocarbon-rich phase		
x^{II}_3	x^{II}_2	x^{II}_1	x^{I}_3	x^{I}_2	x^{I}_1
T = 298.15 K					
1.000	0.000	0.000	0.000	0.000	1.000
0.912	0.088	0.000	0.000	0.130	0.870
0.880	0.120	0.000	0.000	0.290	0.710
0.837	0.163	0.000	0.000	0.436	0.564
0.810	0.190	0.000	0.000	0.576	0.424
0.738	0.262	0.000	0.000	0.665	0.335
0.727	0.273	0.000	0.000	0.716	0.284
0.612	0.388	0.000	0.000	0.879	0.121
0.579	0.421	0.000	0.000	0.977	0.023
0.545	0.455	0.000	0.000	0.919	0.081
0.436	0.564	0.000	0.000	0.955	0.045
0.288	0.712	0.000	0.000	1.000	0.000

Mixture 5: $CH_3C(CH_3)_2CH_2CH(CH_3)CH_3$ (1) + SCH=CHCH=CH (thiophene) (2) + $[C_2mim][EtSO_4]$ (3)
(1) Phase Equilibrium

Ref. [94]

IL-rich phase			Hydrocarbon-rich phase		
x_1^{II}	x_2^{II}	x_3^{II}	x_1^{I}	x_2^{I}	x_3^{I}
298.15 K					
0.000	0.000	1.000	1.000	0.000	0.000
0.000	0.093	0.907	0.903	0.097	0.000
0.000	0.212	0.788	0.784	0.213	0.003
0.000	0.317	0.683	0.572	0.424	0.004
0.000	0.385	0.615	0.441	0.550	0.009
0.000	0.431	0.569	0.341	0.659	0.000
0.000	0.445	0.555	0.147	0.853	0.000
0.000	0.499	0.501	0.097	0.903	0.000
0.000	0.557	0.443	0.057	0.943	0.000
0.000	0.712	0.288	0.000	1.000	0.000

Mixture 6: $C_6H_5CH_3$ (1) + SCH = CHCH = CH (thiophene) (2) + [C_2mim][$EtSO_4$] (3)
(1) Phase Equilibrium

Ref. [94]

IL-rich phase			Hydrocarbon-rich phase		
x_1^{II}	x_2^{II}	x_3^{II}	x_1^{I}	x_2^{I}	x_3^{I}
298.15 K					
0.268	0.000	0.732	1.000	0.000	0.000
0.258	0.047	0.695	0.936	0.064	0.000
0.244	0.106	0.650	0.863	0.137	0.000
0.225	0.174	0.601	0.752	0.248	0.000
0.194	0.254	0.552	0.646	0.354	0.000
0.169	0.328	0.503	0.533	0.467	0.000
0.151	0.383	0.466	0.458	0.542	0.000
0.130	0.428	0.442	0.382	0.618	0.000
0.115	0.472	0.413	0.331	0.669	0.000
0.078	0.552	0.370	0.212	0.788	0.000
0.057	0.602	0.341	0.135	0.865	0.000
0.036	0.637	0.327	0.085	0.915	0.000
0.000	0.712	0.288	0.000	1.000	0.000

Mixture 7: C_2H_5OH (1) + H_2O (2) + EMISE (3)
(1) Phase Equilibrium

Ref. [77]

T/K	x_1	x_2	y_1	γ_1	γ_2
357.34	0.346	0.545	0.690	2.300	3.329
358.58	0.232	0.690	0.604	3.267	2.503
359.34	0.179	0.755	0.581	4.129	2.218
360.40	0.137	0.810	0.544	5.167	1.985
361.50	0.104	0.851	0.507	6.544	1.812
362.69	0.082	0.880	0.464	7.901	1.674
359.01	0.152	0.811	0.534	4.905	2.092
357.20	0.210	0.756	0.574	3.801	2.412
356.23	0.262	0.706	0.610	3.166	2.685
355.20	0.347	0.623	0.629	2.489	3.167
354.48	0.430	0.542	0.645	2.065	3.748
353.93	0.538	0.435	0.668	1.687	4.776
355.12	0.347	0.635	0.623	2.500	3.118
356.54	0.229	0.755	0.586	3.581	2.477
357.40	0.175	0.813	0.561	4.527	2.226
358.40	0.137	0.852	0.550	5.588	2.040
362.41	0.062	0.922	0.444	10.639	1.615
364.45	0.062	0.906	0.390	9.818	1.521
360.81	0.102	0.866	0.498	6.820	1.828
363.79	0.089	0.847	0.465	7.057	1.667

(*continued*)

T/K	x_1	x_2	y_1	γ_1	γ_2
366.52	0.074	0.840	0.420	7.692	1.518
370.40	0.053	0.836	0.351	9.416	1.320
370.46	0.119	0.754	0.379	4.130	1.462
375.60	0.060	0.755	0.370	6.917	1.214
354.18	0.520	0.445	0.672	1.729	4.621
356.00	0.408	0.513	0.689	2.051	3.727
357.32	0.374	0.520	0.695	2.129	3.487
359.24	0.347	0.521	0.696	2.136	3.231
361.72	0.301	0.530	0.693	2.245	2.884
365.11	0.268	0.518	0.689	2.226	2.593
353.65	0.638	0.336	0.673	1.438	6.254
353.79	0.606	0.355	0.694	1.506	5.893
354.60	0.518	0.420	0.707	1.707	4.813
355.81	0.456	0.453	0.717	1.851	4.257
357.32	0.415	0.464	0.643	1.917	3.912
356.35	0.495	0.394	0.743	1.670	4.788
355.71	0.546	0.347	0.761	1.551	5.572
357.33	0.510	0.352	0.762	1.562	5.158
359.48	0.446	0.375	0.777	1.645	4.447
364.48	0.387	0.366	0.787	1.575	3.760
360.05	0.531	0.264	0.835	1.352	6.165

(2) Volumetric Properties, Speed of Sound, Refractive Index, Compressibility Properties, Transport Properties and Thermodynamic Properties

Ref. [95]

x_1	x_2	ρ/ g · cm^{-3}	u/ m · s^{-1}	n_D	$K^E_{S,m}$/ m^3 · TPa^{-1} · mol^{-1}	$10^3\eta$/ Pa · s
298.15 K						
0.0898	0.1724	1.21812	1668.3	1.47286	0.04397	43.71
0.3528	0.1226	1.16262	1576.8	1.45898	0.04238	18.65
0.4657	0.1012	1.13017	1526.6	1.45066	0.04220	12.87
0.6548	0.0654	1.05615	1421.4	1.43155	0.04324	6.067
0.8545	0.0276	0.93319	1277.3	1.39925	0.04767	2.451
0.0906	0.3653	1.20700	1679.5	1.46675	0.03390	26.40
0.1867	0.3267	1.18516	1640.8	1.46136	0.03424	19.71
0.6042	0.1590	1.04741	1424.7	1.42761	0.03891	5.533
0.9667	0.0134	0.81749	1172.7	1.36802	0.05384	1.317
0.2030	0.4762	1.15574	1632.9	1.44988	0.02632	11.23
0.3097	0.4125	1.12171	1570.5	1.44198	0.02814	8.578
0.4044	0.3559	1.08842	1514.0	1.43426	0.03008	6.856
0.6484	0.2101	0.98706	1364.7	1.41004	0.03697	3.500
0.2130	0.6721	1.07337	1608.2	1.41571	0.01647	4.695
0.3172	0.5831	1.03243	1522.0	1.40876	0.01971	4.026
0.7178	0.2410	0.88202	1275.2	1.38045	0.03727	2.045

(*continued*)

x_1	x_2	ρ/ g · cm^{-3}	u/ m · s^{-1}	n_D	$K^E_{S,m}$/ m^3 · TPa^{-1} · mol^{-1}	$10^3\eta$/ Pa · s
0.9072	0.0792	0.81704	1187.9	1.36663	0.04929	1.343
0.0829	0.8208	1.10117	1722.3	1.41144	0.01150	4.237
0.1776	0.7360	1.06192	1631.9	1.40690	0.01394	3.981
0.3540	0.5781	0.99203	1479.1	1.39703	0.01986	3.217
0.4582	0.4849	0.95275	1404.8	1.39078	0.02417	2.967
0.5636	0.3905	0.91567	1342.2	1.38464	0.02903	2.443
0.6675	0.2976	0.88130	1288.8	1.37857	0.03438	2.007
0.7723	0.2038	0.85023	1242.6	1.37282	0.04022	1.728
0.0723	0.8613	1.07912	1719.8	1.39785	0.01003	3.311
0.1532	0.7862	1.04425	1646.3	1.39590	0.01203	3.265
0.2354	0.7098	1.01132	1565.2	1.39287	0.01460	3.148
0.3235	0.6281	0.97697	1487.2	1.38917	0.01784	3.109
0.4195	0.5389	0.94271	1416.8	1.38499	0.02179	2.799
0.5210	0.4447	0.90982	1355.4	1.38047	0.02638	2.385
0.6339	0.3399	0.87660	1297.9	1.37558	0.03207	1.976
0.8653	0.1250	0.81680	1199.3	1.36553	0.04628	1.406
0.0603	0.8923	1.05992	1706.3	1.38604	0.00919	2.787
0.1307	0.8254	1.03207	1658.6	1.38628	0.01067	2.848
0.2036	0.7562	1.00216	1584.5	1.38518	0.01289	3.061
0.2884	0.6757	0.97039	1506.7	1.38307	0.01588	2.836
0.3804	0.5884	0.93821	1434.4	1.38030	0.01960	2.550
0.4793	0.4944	0.90735	1371.3	1.37712	0.02403	2.444
0.7086	0.2767	0.84706	1260.1	1.36966	0.03608	1.737
0.0540	0.9172	1.03613	1677.3	1.37164	0.00855	2.239
0.1162	0.8569	1.01270	1652.9	1.37430	0.00969	2.603
0.1798	0.7952	0.98790	1593.2	1.37537	0.01151	2.666
0.2581	0.7192	0.95950	1518.5	1.37524	0.01423	2.751
0.3439	0.6361	0.93060	1446.8	1.37423	0.01767	2.653
0.4403	0.5426	0.90210	1382.8	1.37261	0.02191	2.299
0.5486	0.4376	0.87400	1323.7	1.37050	0.02721	2.109
0.0462	0.9365	1.01820	1638.4	1.36004	0.00830	1.767
0.1018	0.8819	0.99850	1647.0	1.36450	0.00903	2.162
0.1618	0.8230	0.97720	1605.4	1.36735	0.01051	2.558
0.2358	0.7503	0.95100	1531.1	1.36892	0.01305	2.537
0.3196	0.6680	0.92390	1458.3	1.36932	0.01635	2.391
0.4138	0.5756	0.89710	1392.7	1.36892	0.02046	2.347
0.6523	0.3414	0.84330	1273.3	1.36615	0.03269	1.735
0.0425	0.9493	1.00020	1602.0	1.34953	0.00818	1.473
0.0885	0.9037	0.98520	1631.8	1.35476	0.00859	2.015
0.1473	0.8454	0.96600	1607.7	1.35959	0.00984	2.302
0.2156	0.7777	0.94350	1543.4	1.36274	0.01204	2.443
0.2972	0.6968	0.91790	1469.5	1.36457	0.01520	2.342
0.3862	0.6086	0.89380	1404.8	1.36534	0.01902	2.334
0.4911	0.5046	0.86910	1343.9	1.36541	0.02399	2.027

Ref. [95]

x_1	x_2	V^E/ $cm^3 \cdot mol^{-1}$	Δn_D	$K^E_{S,m}$/ $m^3 \cdot TPa^{-1} \cdot mol^{-1}$	$10^3 \Delta\eta$/ $Pa \cdot s$	ΔG^{*E}/ $J \cdot mol^{-1}$
298.15 K						
0.0898	0.1724	−0.179	0.02948	−0.00291	−28.54	1674.2
0.3528	0.1226	−0.566	0.03963	−0.00737	−33.04	1909.1
0.4657	0.1012	−0.684	0.04162	−0.00879	−29.99	1964.8
0.6548	0.0654	−0.731	0.03979	−0.00981	−22.005	1677.4
0.8545	0.0276	−0.553	0.02573	−0.00757	−10.008	987.35
0.0906	0.3653	−0.370	0.05181	−0.00398	−27.12	3175.2
0.1867	0.3267	−0.518	0.05220	−0.00564	−28.27	2992.8
0.6042	0.1590	−0.792	0.04357	−0.00967	−18.369	2102.5
0.9667	0.0134	−0.212	0.00579	−0.00230	−1.687	329.00
0.2030	0.4762	−0.658	0.06463	−0.00663	−20.71	3704.9
0.3097	0.4125	−0.792	0.06008	−0.00799	−19.233	3323.7
0.4044	0.3559	−0.869	0.05533	−0.00890	−17.287	3021.3
0.6484	0.2101	−0.871	0.03877	−0.00932	−11.197	1996.4
0.2130	0.6721	−0.857	0.06042	−0.00736	−7.347	3687.4
0.3172	0.5831	−1.028	0.05281	−0.00848	−6.565	3293.9
0.7178	0.2410	−0.869	0.02199	−0.00773	−2.967	1587.2
0.9072	0.0792	−0.410	0.00698	−0.00365	−1.031	536.01
0.0829	0.8208	−0.519	0.06249	−0.00512	−5.980	3713.6
0.1776	0.7360	−0.797	0.05678	−0.00684	−5.292	3530.3
0.3540	0.5781	−1.066	0.04475	−0.00864	−4.299	2947.5
0.4582	0.4849	−1.064	0.03721	−0.00891	−3.511	2715.3
0.5636	0.3905	−0.998	0.02978	−0.00867	−2.985	2197.7
0.6675	0.2976	−0.856	0.02243	−0.00787	−2.387	1672.1
0.7723	0.2038	−0.734	0.01539	−0.00662	−1.623	1255.8
0.0723	0.8613	−0.475	0.05358	−0.00468	−4.016	3265.0
0.1532	0.7862	−0.725	0.05023	−0.00636	−3.518	3208.8
0.2354	0.7098	−0.925	0.04579	−0.00752	−3.082	3090.9
0.3235	0.6281	−1.018	0.04058	−0.00828	−2.527	3029.7
0.4195	0.5389	−1.061	0.03474	−0.00869	−2.191	2728.7
0.5210	0.4447	−1.046	0.02848	−0.00870	−1.922	2281.4
0.6339	0.3399	−0.961	0.02165	−0.00813	−1.571	1749.1
0.8653	0.1250	−0.500	0.00762	−0.00444	−0.583	749.40
0.0603	0.8923	−0.355	0.04489	−0.00405	−2.698	2911.4
0.1307	0.8254	−0.670	0.04371	−0.00584	−2.308	2950.0
0.2036	0.7562	−0.850	0.04112	−0.00700	−1.753	3115.0
0.2884	0.6757	−1.000	0.03729	−0.00794	−1.581	2899.2
0.3804	0.5884	−1.061	0.03265	−0.00848	−1.436	2599.7
0.4793	0.4944	−1.068	0.02746	−0.00865	−1.078	2443.5
0.7086	0.2767	−0.842	0.01535	−0.00723	−0.711	1437.1
0.0540	0.9172	−0.321	0.03339	−0.00352	−1.449	2395.0
0.1162	0.8569	−0.626	0.03461	−0.00532	−0.913	2771.4
0.1798	0.7952	−0.822	0.03420	−0.00651	−0.675	2830.0
0.2581	0.7192	−0.987	0.03226	−0.00750	−0.375	2897.0

(*continued*)

x_1	x_2	V^E/ $cm^3 \cdot mol^{-1}$	Δn_D	$K^E_{S,m}$/ $m^3 \cdot TPa^{-1} \cdot mol^{-1}$	$10^3\Delta\eta$/ Pa · s	ΔG^{*E}/ $J \cdot mol^{-1}$
0.3439	0.6361	−1.066	0.02925	−0.00811	−0.236	2783.5
0.4403	0.5426	−1.096	0.02539	−0.00843	−0.324	2384.0
0.5486	0.4376	−1.054	0.02077	−0.00826	−0.217	2100.0
0.0462	0.9365	−0.258	0.02370	−0.00285	−0.807	1802.1
0.1018	0.8819	−0.551	0.02677	−0.00478	−0.325	2318.4
0.1618	0.8230	−0.785	0.02812	−0.00617	0.166	2746.9
0.2358	0.7503	−0.952	0.02784	−0.00718	0.260	2730.1
0.3196	0.6680	−1.053	0.02613	−0.00789	0.245	2570.5
0.4138	0.5756	−1.090	0.02338	−0.00829	0.349	2490.5
0.6523	0.3414	−0.936	0.01463	−0.00749	0.110	1569.2
0.0425	0.9493	−0.219	0.01464	−0.00236	−0.215	1332.6
0.0885	0.9037	−0.475	0.01865	−0.00418	0.356	2134.8
0.1473	0.8454	−0.733	0.02192	−0.00576	0.679	2487.7
0.2156	0.7777	−0.922	0.02327	−0.00687	0.864	2651.2
0.2972	0.6968	−1.030	0.02294	−0.00765	0.814	2548.4
0.3862	0.6086	−1.091	0.02136	−0.00813	0.864	2517.8
0.4911	0.5046	−1.078	0.01865	−0.00823	0.622	2113.1

(3) Vapor Pressure and Activity Coefficient

Ref. [81]

T/K	P^{exp}/kPa	P^{NRTL}/kPa	γ_1^{NRTL}	γ_2^{NRTL}
$x_2 = 0.8307, x_1 = 0.0813$				
308.27	8.360	8.775	0.9461	3.8544
317.67	13.528	14.223	0.9511	3.7387
324.54	18.813	19.829	0.9544	3.6590
330.77	24.984	26.436	0.9572	3.5900
336.67	32.638	34.322	0.9596	3.5276
341.52	40.227	42.212	0.9614	3.4781
346.21	48.919	51.247	0.9631	3.4319
350.66	58.751	61.259	0.9646	3.3894
$x_2 = 0.7205, x_1 = 0.1880$				
302.37	7.507	7.775	0.9907	2.5973
313.71	13.598	14.239	0.9969	2.5371
320.20	18.672	19.674	0.9999	2.5044
326.48	25.225	26.526	1.0025	2.4739
332.47	33.053	34.856	1.0048	2.4458
337.45	41.146	43.370	1.0065	2.4231
342.36	50.690	53.386	1.0081	2.4014
346.84	61.230	64.185	1.0094	2.3821
$x_2 = 0.5626, x_1 = 0.3302$				
300.88	7.384	7.653	1.0508	1.7976
311.17	12.914	13.413	1.0592	1.7754
318.83	19.147	19.799	1.0645	1.7595
324.22	24.722	25.721	1.0678	1.7488

(continued)

T/K	P^{exp}/kPa	P^{NRTL}/kPa	γ_1^{NRTL}	γ_2^{NRTL}
330.37	33.063	34.244	1.0712	1.7368
335.07	40.996	42.253	1.0736	1.7279
339.84	50.456	51.949	1.0758	1.7191
344.39	61.139	62.885	1.0777	1.7109
x_2 = 0.3394, x_1 = 0.5312				
301.49	8.144	8.121	1.1207	1.3157
310.37	12.989	13.198	1.1346	1.3084
317.47	18.752	19.024	1.1445	1.3028
323.20	24.832	25.215	1.1517	1.2984
329.06	32.683	33.221	1.1585	1.2941
333.93	40.838	41.427	1.1637	1.2906
338.66	50.208	50.987	1.1683	1.2873
343.38	61.278	62.277	1.1727	1.2840

Mixture 8: 1-C_3H_7OH (1) + H_2O (2) + EMISE (3)
(1) Volumetric Properties, Transport Properties, Refractive Index and Thermodynamic Properties

Ref. [96]

x_1	x_2	ρ/g · cm^{-3}	n_D	η/mPa · s	V^E/cm^3 · mol^{-1}	Δn_D	$\Delta\eta$/mPa · s	ΔG^{*E}/J · mol^{-1}
298.15 K								
0.1280	0.6147	1.1486	1.44560	10.128	−0.580	0.069	−15.785	4563.0
0.2412	0.5349	1.1022	1.43789	8.497	−0.645	0.060	−14.315	4058.6
0.3498	0.4584	1.0575	1.43022	7.028	−0.643	0.052	−12.809	3518.8
0.4649	0.3772	1.0108	1.42206	5.826	−0.629	0.043	−10.855	2983.9
0.5725	0.3013	0.9678	1.41456	4.938	−0.599	0.035	−8.795	2511.3
0.6642	0.2367	0.9314	1.40804	3.908	−0.543	0.027	−7.312	1856.0
0.0826	0.7820	1.1180	1.42504	5.717	−0.530	0.068	−8.353	4490.2
0.1695	0.7080	1.0745	1.42032	5.451	−0.628	0.061	−7.478	4229.2
0.2588	0.6318	1.0332	1.41545	5.009	−0.665	0.054	−6.747	3858.4
0.3494	0.5546	0.9953	1.41067	4.634	−0.683	0.046	−5.931	3496.9
0.4500	0.4688	0.9573	1.40580	4.260	−0.682	0.039	−4.983	3093.6
0.6559	0.2933	0.8899	1.39678	3.103	−0.576	0.024	−3.435	1852.6
0.0662	0.8408	1.0970	1.41152	4.244	−0.454	0.062	−5.714	4054.3
0.1383	0.7759	1.0585	1.40610	4.326	−0.581	0.054	−5.019	3996.6
0.2141	0.7076	1.0216	1.40625	4.247	−0.642	0.051	−4.453	3821.3
0.2999	0.6304	0.9848	1.40315	4.063	−0.678	0.045	−3.907	3546.3
0.3963	0.5436	0.9487	1.39982	3.873	−0.683	0.038	−3.277	3226.9
0.5062	0.4446	0.9132	1.39642	3.283	−0.658	0.031	−2.933	2542.7
0.7206	0.2516	0.8569	1.39049	3.041	−0.517	0.017	−1.352	1834.3
0.0559	0.8764	1.0792	1.40006	3.394	−0.395	0.055	−4.104	3612.1
0.1139	0.8226	1.0471	1.39969	3.913	−0.524	0.052	−3.248	3924.7
0.1845	0.7571	1.0117	1.39839	3.683	−0.609	0.048	−3.069	3653.7
0.2669	0.6805	0.9758	1.39664	3.675	−0.661	0.043	−2.598	3505.4
0.3557	0.5981	0.9428	1.39485	3.481	−0.680	0.038	−2.277	3183.5

(*continued*)

x_1	x_2	$\rho/g \cdot cm^{-3}$	n_D	$\eta/mPa \cdot s$	$V^E/cm^3 \cdot mol^{-1}$	Δn_D	$\Delta\eta/mPa \cdot s$	$\Delta G^{*E}/J \cdot mol^{-1}$
0.5658	0.4030	0.8820	1.39098	2.887	−0.626	0.025	−1.650	2190.4
0.8287	0.1590	0.8281	1.38633	2.391	−0.398	0.010	−0.620	966.9
0.0474	0.9071	1.0577	1.38662	2.604	−0.330	0.045	−2.737	2986.7
0.1023	0.8549	1.0278	1.38842	3.042	−0.496	0.044	−2.108	3357.6
0.1670	0.7932	0.9949	1.38901	3.216	−0.579	0.042	−1.708	3432.0
0.2397	0.7240	0.9640	1.38906	3.406	−0.655	0.039	−1.264	3477.2
0.3161	0.6512	0.9356	1.38884	3.146	−0.667	0.036	−1.258	3121.7
0.4188	0.5534	0.9044	1.38834	3.184	−0.670	0.031	−0.862	2934.1
0.0422	0.9290	1.0361	1.37367	2.255	−0.283	0.035	−1.469	2627.4
0.0901	0.8825	1.0108	1.37757	2.506	−0.457	0.036	−1.133	2889.0
0.1467	0.8276	0.9815	1.37995	2.968	−0.525	0.036	−0.570	3307.8
0.2082	0.7680	0.9555	1.38151	2.933	−0.605	0.035	−0.495	3201.3
0.2888	0.6898	0.9268	1.38288	3.140	−0.663	0.033	−0.144	3257.7
0.4746	0.5095	0.8770	1.38445	2.712	−0.646	0.026	−0.241	2452.0
0.0384	0.9460	1.0140	1.36098	1.865	−0.241	0.024	−0.569	2106.6
0.0813	0.9039	0.9930	1.36670	2.179	−0.424	0.028	−0.236	2527.9
0.1286	0.8573	0.9697	1.37051	2.642	−0.515	0.029	0.248	3038.4
0.1909	0.7961	0.9435	1.37393	2.701	−0.592	0.030	0.333	3047.6
0.2591	0.7289	0.9189	1.37668	2.963	−0.615	0.029	0.625	3214.0
0.3408	0.6485	0.8958	1.37895	2.936	−0.650	0.028	0.634	3043.6
0.0375	0.9551	0.9973	1.35204	1.651	−0.220	0.017	0.004	1757.5
0.1198	0.8735	0.9590	1.36371	2.308	−0.493	0.024	0.629	2683.8
0.2472	0.7471	0.9110	1.37225	2.852	−0.599	0.026	1.125	3160.4
0.0961	0.1388	1.2153	1.47318	50.905	−0.242	0.023	−24.069	1598.4
0.1971	0.1233	1.1923	1.46844	38.682	−0.310	0.026	−28.139	1712.5
0.0981	0.2695	1.2078	1.46965	34.718	−0.354	0.039	−27.433	2660.1
0.2061	0.2372	1.1798	1.46404	26.738	−0.447	0.039	−28.206	2596.3
0.3040	0.2080	1.1513	1.45823	20.220	−0.499	0.039	−28.197	2415.1
0.3994	0.1795	1.1201	1.45178	15.510	−0.511	0.037	−26.550	2250.9
0.5003	0.1493	1.0830	1.44410	11.477	−0.503	0.035	−23.856	2013.2
0.6001	0.1195	1.0414	1.43535	8.367	−0.479	0.031	−20.313	1718.9
0.7013	0.0893	0.9937	1.42523	6.086	−0.472	0.026	−15.847	1414.7
0.8010	0.0595	0.9388	1.41352	4.291	−0.379	0.020	−10.995	1007.3
0.9011	0.0296	0.8753	1.39981	3.013	−0.300	0.012	−5.601	575.9
0.0962	0.4557	1.1924	1.46221	20.230	−0.483	0.059	−24.087	3990.4
0.2001	0.4034	1.1592	1.45593	16.021	−0.559	0.055	−23.433	3625.5
0.2998	0.3531	1.1247	1.44926	12.695	−0.581	0.051	−22.094	3250.2
0.4954	0.2544	1.0500	1.43456	8.001	−0.579	0.040	−17.631	2494.0
0.5960	0.2037	1.0076	1.42611	6.313	−0.558	0.034	−14.613	2102.7
0.6982	0.1522	0.9612	1.41674	4.821	−0.507	0.027	−11.319	1624.6
0.8022	0.0997	0.9101	1.40633	3.660	−0.407	0.019	−7.615	1132.9
0.8979	0.0515	0.8593	1.39585	2.770	−0.284	0.010	−4.026	609.6

Mixture 9: 2-CH_3CHCH_3OH (1) + H_2O (2) + EMISE (3)
(1) Volumetric Properties, Transport Properties, Refractive Index and Thermodynamic Properties

Ref. [96]

x_1	x_2	ρ/ g · cm^{-3}	n_D	η/ mPa · s	V^E/ cm^3 · mol^{-1}	Δn_D	$\Delta\eta$/ mPa · s	ΔG^{*E}/ J · mol^{-1}
298.15 K								
0.0949	0.1426	1.2147	1.47289	50.594	−0.324	0.024	−24.138	1616.8
0.2002	0.1260	1.1896	1.46762	38.893	−0.472	0.028	−27.380	1778.5
0.2964	0.1109	1.1633	1.46203	29.818	−0.563	0.030	−28.735	1847.3
0.4843	0.0813	1.1005	1.44825	16.603	−0.633	0.031	−26.872	1771.8
0.6918	0.0486	1.0069	1.42737	8.040	−0.615	0.027	−18.775	1418.4
0.7943	0.0324	0.9467	1.41360	5.437	−0.524	0.022	−13.152	1127.1
0.0907	0.2737	1.2086	1.46971	36.644	−0.424	0.040	−25.814	2781.8
0.1902	0.2437	1.1818	1.46423	28.198	−0.556	0.041	−27.653	2649.3
0.2845	0.2154	1.1535	1.45842	22.538	−0.644	0.040	−27.050	2584.6
0.4839	0.1553	1.0821	1.44315	12.918	−0.699	0.037	−23.428	2202.4
0.5849	0.1249	1.0388	1.43349	9.527	−0.670	0.034	−20.117	1935.2
0.6881	0.0939	0.9887	1.42243	6.871	−0.643	0.029	−15.922	1609.0
0.8921	0.0325	0.8651	1.39461	3.401	−0.405	0.013	−5.845	777.4
0.0834	0.4648	1.1950	1.46251	21.619	−0.540	0.060	−23.056	4165.0
0.1860	0.4128	1.1609	1.45589	17.075	−0.679	0.057	−22.830	3773.2
0.3755	0.3167	1.0907	1.44179	11.400	−0.791	0.048	−19.700	3134.3
0.4723	0.2676	1.0511	1.43361	9.141	−0.788	0.043	−17.461	2767.3
0.5715	0.2173	1.0082	1.42468	7.252	−0.795	0.037	−14.741	2374.3
0.6749	0.1649	0.9596	1.41446	5.595	−0.702	0.030	−11.596	1915.7
0.8849	0.0584	0.8500	1.39062	3.105	−0.431	0.012	−4.327	810.2
0.1126	0.6808	1.1365	1.43874	8.626	−0.713	0.071	−12.372	4772.9
0.2236	0.5957	1.0853	1.43083	7.597	−0.860	0.062	−11.035	4314.2
0.3310	0.5133	1.0372	1.42265	6.616	−0.916	0.053	−9.726	3829.7
0.4354	0.4332	0.9926	1.41490	5.704	−0.925	0.045	−8.413	3320.2
0.5330	0.3583	0.9527	1.40784	4.891	−0.898	0.037	−7.146	2801.8
0.6305	0.2835	0.9144	1.40090	4.158	−0.826	0.029	−5.800	2259.7
0.0850	0.7786	1.1168	1.42522	6.055	−0.656	0.069	−8.132	4619.5
0.1722	0.7044	1.0719	1.41990	5.791	−0.840	0.062	−7.244	4356.9
0.2632	0.6269	1.0279	1.41416	5.431	−0.931	0.054	−6.398	4031.5
0.3595	0.5450	0.9855	1.40819	4.964	−0.966	0.046	−5.592	3620.4
0.4571	0.4619	0.9467	1.40248	4.409	−0.976	0.039	−4.856	3120.9
0.6619	0.2877	0.8755	1.39153	3.380	−0.809	0.024	−3.175	2008.5
0.7683	0.1972	0.8432	1.38623	2.878	−0.647	0.016	−2.270	1358.4
0.8792	0.1028	0.8124	1.38090	2.445	−0.421	0.008	−1.235	687.6
0.0688	0.8355	1.0977	1.41265	4.554	−0.583	0.063	−5.668	4212.8
0.1434	0.7686	1.0570	1.40970	4.720	−0.798	0.058	−4.850	4187.6
0.2231	0.6971	1.0165	1.40584	4.649	−0.910	0.052	−4.225	4006.8
0.3084	0.6205	0.9781	1.40180	4.418	−0.976	0.046	−3.709	3706.9
0.4012	0.5373	0.9411	1.39752	4.076	−0.983	0.039	−3.240	3299.3

(continued)

x_1	x_2	ρ/ g · cm^{-3}	n_D	η/ mPa · s	V^E/ cm^3 · mol^{-1}	Δn_D	$\Delta\eta$/ mPa · s	ΔG^{*E}/ J · mol^{-1}
0.5022	0.4466	0.9059	1.39324	3.690	−0.952	0.032	−2.744	2807.7
0.0579	0.8750	1.0776	1.39993	3.573	−0.504	0.055	−3.877	3743.5
0.1209	0.8165	1.0427	1.39946	3.937	−0.744	0.053	−3.154	3916.4
0.1919	0.7505	1.0060	1.39761	4.048	−0.895	0.048	−2.639	3877.6
0.2695	0.6784	0.9699	1.39515	4.008	−0.957	0.044	−2.236	3712.1
0.3575	0.5968	0.9349	1.39241	3.843	−0.991	0.038	−1.900	3418.0
0.5641	0.4049	0.8705	1.38635	3.160	−0.915	0.025	−1.406	2393.6
0.0505	0.9031	1.0575	1.38754	2.901	−0.439	0.046	−2.532	3267.7
0.1066	0.8498	1.0270	1.38930	3.332	−0.706	0.046	−1.903	3580.2
0.1684	0.7910	0.9943	1.38931	3.601	−0.849	0.044	−1.416	3711.3
0.2400	0.7229	0.9612	1.38851	3.690	−0.944	0.040	−1.074	3663.6
0.3223	0.6446	0.9289	1.38722	3.614	−1.005	0.036	−0.859	3449.6
0.4151	0.5563	0.8979	1.38560	3.380	−0.998	0.031	−0.766	3062.4
0.0434	0.9286	1.0337	1.37318	2.306	−0.359	0.035	−1.339	2682.8
0.0925	0.8810	1.0087	1.37750	2.885	−0.642	0.037	−0.680	3258.3
0.1486	0.8265	0.9798	1.37965	3.214	−0.809	0.037	−0.260	3502.2
0.2130	0.7640	0.9502	1.38069	3.393	−0.915	0.036	0.025	3569.3
0.2891	0.6901	0.9206	1.38109	3.412	−0.981	0.033	0.169	3459.1
0.4888	0.4963	0.8632	1.38037	2.979	−0.967	0.025	0.061	2633.7
0.0389	0.9432	1.0179	1.36330	1.977	−0.305	0.027	−0.691	2263.9
0.0834	0.8995	0.9967	1.36925	2.578	−0.596	0.031	−0.063	2970.9
0.1346	0.8492	0.9712	1.37292	2.994	−0.777	0.032	0.385	3349.6
0.1967	0.7884	0.9431	1.37527	3.248	−0.898	0.032	0.677	3511.8
0.3555	0.6325	0.8884	1.37773	3.179	−1.016	0.028	0.704	3183.7
0.5934	0.3990	0.8359	1.37794	2.643	−0.927	0.019	0.314	2057.9
0.0353	0.9565	0.9994	1.35229	1.663	−0.256	0.017	−0.069	1776.4
0.0754	0.9166	0.9823	1.35984	2.321	−0.546	0.023	0.575	2686.5
0.1225	0.8699	0.9602	1.36520	2.780	−0.742	0.026	1.017	3169.5

Mixture 10: $C_6H_5CH_3$ (1) + C_7H_{16} (2) + [emim][$C_2H_5SO_4$] (3)

(1) Selectivity

Ref. [16]

n-Heptane-rich phase			[emim][$C_2H_5SO_4$]-rich phase			Experimental Selectivity
x_1	x_2	x_3	x_1	x_2	x_3	
0.1 MPa						
313.2 K						
0.0000	0.9934	0.0066	0.0000	0.0051	0.9949	–
0.0590	0.9320	0.0090	0.0140	0.0040	0.9820	55.29
0.1070	0.8770	0.0160	0.0240	0.0039	0.9721	50.44
0.1590	0.8190	0.0220	0.0360	0.0038	0.9602	48.80
0.2120	0.7770	0.0110	0.0470	0.0037	0.9493	46.56
0.2600	0.7270	0.0130	0.0550	0.0036	0.9414	42.72
0.3131	0.6651	0.0218	0.0695	0.0036	0.9269	41.14
0.3634	0.6046	0.0144	0.0853	0.0039	0.9108	37.46

(*continued*)

n-Heptane-rich phase			[emim][$C_2H_5SO_4$]-rich phase			Experimental Selectivity
x_1	x_2	x_3	x_1	x_2	x_3	
0.4139	0.5643	0.0139	0.1029	0.0038	0.8933	37.63
0.4705	0.5110	0.0185	0.0964	0.0031	0.9005	33.60
0.5174	0.4688	0.0137	0.1079	0.0036	0.8885	27.04
0.5542	0.4290	0.0168	0.1165	0.0038	0.8797	23.66
0.6446	0.3489	0.0065	0.1238	0.0030	0.8732	22.40
0.6664	0.3065	0.0114	0.1429	0.0029	0.8542	23.54
0.7396	0.2593	0.0011	0.1492	0.0025	0.8483	21.10
0.7646	0.2284	0.0070	0.1645	0.0024	0.8331	20.22
0.8179	0.1769	0.0099	0.1692	0.0022	0.8286	16.28
0.8646	0.1320	0.0034	0.1849	0.0018	0.8133	15.79
0.9189	0.0954	0.0089	0.1724	0.0013	0.8262	10.04
0.9694	0.0518	0.0084	0.2034	0.0008	0.7958	5.61
0.9675	0.0000	0.0325	0.1790	0.0000	0.8210	–
348.2 K						
0.0000	0.9913	0.0087	0.0000	0.0052	0.9948	–
0.0601	0.9364	0.0035	0.0124	0.0054	0.9822	35.55
0.1059	0.8675	0.0266	0.0229	0.0058	0.9713	32.14
0.1395	0.8231	0.0375	0.0333	0.0056	0.9611	34.80
0.2096	0.7547	0.0357	0.0465	0.0059	0.9476	28.36
0.2456	0.7129	0.0415	0.0536	0.0054	0.9410	28.69
0.3102	0.6694	0.0204	0.0682	0.0055	0.9263	26.83
0.3504	0.6204	0.0292	0.0721	0.0055	0.9224	23.18
0.3829	0.5842	0.0328	0.0826	0.0052	0.9122	24.20
0.4482	0.5229	0.0289	0.0943	0.0041	0.9016	26.65
0.5033	0.4693	0.0274	0.1101	0.0046	0.8853	22.20
0.5391	0.4335	0.0274	0.1220	0.0042	0.8738	23.62
0.5876	0.3850	0.0274	0.1190	0.0039	0.8770	19.91
0.6401	0.3386	0.0213	0.1335	0.0034	0.8632	21.03
0.6766	0.2980	0.0254	0.1421	0.0038	0.8541	16.34
0.7399	0.2416	0.0185	0.1543	0.0028	0.8430	18.11
0.7922	0.1874	0.0203	0.1612	0.0026	0.8362	14.50
0.8150	0.1531	0.0320	0.1775	0.0022	0.8203	15.12
0.8888	0.0850	0.0262	0.1826	0.0015	0.8159	12.02
0.9417	0.0440	0.0143	0.1844	0.0011	0.8145	7.94
0.9627	0.0000	0.0373	0.1808	0.0000	0.8192	–

(2) Solubility

Ref. [16]

Temperature					
Toluene in IL, maximum		n-Heptane in IL, average		IL in n-Heptane, average	
313.2 K	348.2 K	313.2 K	348.2 K	313.2 K	348.2 K
0.18	0.18	0.0030	0.0040	0.0084	0.0248

Mixture 11: $(CH_3)_3COC_2H_5$ (1) + C_2H_5OH (2) + [Emim][$EtSO_4$] (3)
(1) Volumetric Properties, Transport Properties, Refractive Index, Speed of Sound and Compressibility Properties

Ref. [97]

x_1	x_2	ρ/g · cm^{-3}	η/mPa · s	n_D	u/m · s^{-1}	k_s/TPa^{-1}	V^E/cm^{-3} · mol^{-1}	$\Delta\eta$/mPa · s	ΔR/cm^{-3} · mol^{-1}	Δk_s/TPa^{-1}
T = 298.15 K										
0.0000	0.9522	0.85234	1.589	1.37769	1200	814	−0.266	−4.234	0.020	−62
0.0527	0.9020	0.84187	1.428	1.37823	1183	848	−0.412	−4.107	0.021	−71
0.1407	0.8182	0.82609	1.208	1.37835	1163	895	−0.595	−3.848	0.009	−84
0.2556	0.7089	0.80827	0.974	1.37854	1141	950	−0.749	−3.456	0.016	−94
0.3132	0.6539	0.80017	0.878	1.37889	1116	1003	−0.779	−3.237	0.043	−70
0.0000	0.9062	0.90503	2.138	1.39213	1248	710	−0.440	−8.255	0.038	−90
0.0720	0.8410	0.88524	1.860	1.39070	1216	764	−0.628	−7.812	0.028	−98
0.1589	0.7622	0.86365	1.568	1.38882	1181	830	−0.787	−7.233	0.016	−97
0.2368	0.6916	0.84630	1.352	1.38711	1154	888	−0.883	−6.669	0.005	−91
0.0000	0.8666	0.94332	2.709	1.40265	1286	641	−0.553	−11.61	0.055	−102
0.0770	0.7999	0.91873	2.401	1.39974	1244	703	−0.768	−10.85	0.034	−110
0.1736	0.7162	0.88991	1.937	1.39636	1198	784	−0.873	−9.963	0.039	−107
0.0000	0.8060	0.99200	3.860	1.41553	1339	562	−0.649	−16.49	0.062	−109
0.0564	0.7605	0.97135	3.565	1.41298	1302	607	−0.832	−15.66	0.065	−118
0.1236	0.7064	0.94764	3.102	1.40981	1260	665	−0.970	−14.78	0.072	−120
0.0000	0.7508	1.02842	5.078	1.42508	1382	509	−0.692	−20.75	0.066	−107
0.0338	0.7254	1.01518	4.792	1.42310	1357	535	−0.816	−20.17	0.061	−114
0.0848	0.6872	0.99572	4.386	1.42047	1320	576	−0.972	−19.29	0.072	−121
0.0000	0.7174	1.04766	5.975	1.43017	1406	483	−0.707	−23.17	0.070	−104
0.0350	0.6923	1.03375	5.651	1.42838	1379	508	−0.844	−22.49	0.083	−112
0.0845	0.6567	1.01434	5.135	1.42536	1341	548	−1.005	−21.58	0.081	−120
0.0000	0.5951	1.10450	10.65	1.44516	1467	421	−0.702	−30.64	0.084	−79

0.0501	0.5653	1.08483	9.768	1.44179	1441	444	−0.983	−29.47	0.065	−103
0.0000	0.5103	1.13481	15.40	1.45298	1525	379	−0.664	−34.31	0.082	−73
0.0465	0.4866	1.11699	14.20	1.44977	1488	405	−0.970	−33.22	0.057	−90
0.0000	0.4147	1.16259	22.88	1.46005	1566	351	−0.577	−36.33	0.076	−57
0.0475	0.3950	1.14589	21.09	1.45757	1532	372	−1.033	−35.32	0.072	−78
0.0000	0.3256	1.18410	32.21	1.46557	1598	331	−0.491	−35.85	0.072	−43
0.0290	0.3162	1.17390	30.74	1.46442	1580	341	−0.737	−35.36	0.108	−57
0.0000	0.2039	1.20805	49.77	1.47144	1636	309	−0.320	−30.38	0.048	−26
0.0324	0.1973	1.19784	46.30	1.46984	1617	319	−0.695	−31.26	0.040	−42
0.0000	0.1271	1.22067	65.47	1.47457	1657	298	−0.183	−22.32	0.039	−16
0.0174	0.1248	1.21569	63.01	1.47383	1648	303	−0.434	−23.26	0.031	−25
0.0000	1.0000	0.78522	1.077	1.35923	1144	973	0.000	0.000	0.000	0
0.0000	0.9504	0.85474	1.628	1.37833	1202	810	−0.281	−4.379	0.019	−63
0.0000	0.9026	0.90892	2.189	1.39300	1253	700	−0.462	−8.564	0.031	−94
0.0000	0.7866	1.00544	4.312	1.41900	1352	544	−0.662	−17.96	0.062	−107
0.0000	0.7676	1.01799	4.678	1.42246	1369	524	−0.681	−19.48	0.070	−107
0.0000	0.7120	1.05050	6.132	1.43097	1407	481	−0.702	−23.55	0.074	−102
0.0000	0.5926	1.10555	10.83	1.44549	1480	413	−0.704	−30.71	0.087	−85
0.0000	0.5724	1.11325	11.74	1.44727	1490	405	−0.696	−31.81	0.074	−81
0.0000	0.5223	1.13096	14.78	1.45198	1516	384	−0.676	−33.75	0.081	−74
0.0000	0.4604	1.15007	18.98	1.45680	1545	364	−0.629	−35.69	0.073	−64
0.0000	0.4336	1.15767	21.39	1.45888	1558	356	−0.612	−35.94	0.080	−60
0.0000	0.3598	1.17632	28.75	1.46366	1586	338	−0.532	−35.91	0.079	−48
0.0000	0.3088	1.18770	34.69	1.46642	1603	328	−0.466	−35.04	0.066	−41
0.0000	0.2187	1.20533	47.69	1.47092	1631	312	−0.332	−30.99	0.064	−28
0.0000	0.1281	1.22061	66.09	1.47476	1655	299	−0.199	−21.59	0.056	−16
0.0000	0.0000	1.23882	100.4	1.47889	1683	285	0.000	0.000	0.000	0

(2) Distribution Ratio and Selectivity

Ref. [98]

Organic phase			Ionic liquid phase				
x_1	x_2	x_3	x_1	x_2	x_3	β	S
298.1 K							
1.000	0.000	0.000	0.031	0.000	0.969	–	–
0.991	0.009	0.000	0.028	0.064	0.908	7.11	251.7
0.976	0.024	0.000	0.046	0.141	0.813	5.88	124.7
0.960	0.040	0.000	0.052	0.224	0.724	5.60	103.4
0.916	0.084	0.000	0.059	0.358	0.583	4.26	66.2
0.875	0.125	0.000	0.060	0.458	0.482	3.66	53.4
0.794	0.206	0.000	0.078	0.540	0.382	2.62	26.7
0.709	0.289	0.002	0.103	0.602	0.295	2.08	14.3
0.651	0.347	0.002	0.133	0.632	0.235	1.82	8.9
0.546	0.441	0.013	0.180	0.646	0.174	1.46	4.4
0.481	0.493	0.026	0.223	0.636	0.141	1.29	2.8

(3) Boiling Point

Ref. [98]

T/K	x_1	x_2	y_1	y_2
101.3 kPa				
349.06	0.0369	0.7198	0.4067	0.5933
350.49	0.0288	0.7382	0.3444	0.6556
351.32	0.0197	0.7772	0.2731	0.7269
346.93	0.0480	0.7522	0.4143	0.5857
343.47	0.0707	0.7321	0.5246	0.4754
343.83	0.0876	0.7723	0.4706	0.5294
342.27	0.1213	0.7410	0.5298	0.4702
341.08	0.1582	0.7080	0.5777	0.4223
340.49	0.1930	0.6758	0.6052	0.3948
340.29	0.2395	0.6536	0.6116	0.3884
340.20	0.2828	0.6306	0.6052	0.3948
340.96	0.2326	0.6914	0.5546	0.4454
341.97	0.1783	0.7463	0.5026	0.4974
343.01	0.1416	0.7884	0.4537	0.5463
344.27	0.1081	0.8242	0.3986	0.6014
345.49	0.0848	0.8546	0.3387	0.6613
346.75	0.0702	0.8761	0.2791	0.7209
351.04	0.0100	0.9816	0.0316	0.9684
349.86	0.0251	0.9550	0.1046	0.8954
348.58	0.0423	0.9374	0.1666	0.8334
347.11	0.0687	0.9108	0.2374	0.7626
346.24	0.0801	0.8754	0.2924	0.7076

(*continued*)

T/K	x_1	x_2	y_1	y_2
344.58	0.1104	0.8450	0.3633	0.6367
342.86	0.1510	0.7859	0.4587	0.5413
341.92	0.1884	0.7503	0.4996	0.5004
340.61	0.2979	0.6503	0.5673	0.4327
340.27	0.3482	0.6041	0.5929	0.4071
340.29	0.4083	0.5690	0.5859	0.4141
340.04	0.4886	0.4929	0.6128	0.3872
340.87	0.3420	0.6403	0.5381	0.4619
341.81	0.2560	0.7278	0.4821	0.5179
343.17	0.1792	0.8064	0.4139	0.5861
344.50	0.1332	0.8540	0.3511	0.6489
354.13	0.0093	0.8570	0.0577	0.9423
353.02	0.0131	0.8526	0.1037	0.8963
353.69	0.0077	0.8759	0.0490	0.9510
353.82	0.0063	0.8878	0.0243	0.9757
351.73	0.0146	0.8838	0.1194	0.8806
348.68	0.0430	0.8536	0.2498	0.7502
347.08	0.0560	0.8412	0.3154	0.6846
349.59	0.0304	0.8258	0.2671	0.7329
351.47	0.0188	0.8466	0.1762	0.8238
353.30	0.0105	0.8127	0.1638	0.8362
364.54	0.0011	0.6777	0.0154	0.9846
360.97	0.0036	0.6859	0.1341	0.8659
355.05	0.0200	0.6879	0.3103	0.6897
346.55	0.0527	0.6751	0.5220	0.4780
340.15	0.1095	0.6589	0.6683	0.3317
342.77	0.0813	0.7262	0.5602	0.4398
346.50	0.0517	0.7131	0.4738	0.5262
343.45	0.0637	0.7130	0.5605	0.4395
340.15	0.1230	0.6554	0.6634	0.3366
369.49	0.0102	0.5609	0.1414	0.8586
365.26	0.0300	0.5798	0.1989	0.8011
366.42	0.0020	0.6061	0.1290	0.8710
357.08	0.0589	0.6079	0.2704	0.7296
352.42	0.0280	0.6613	0.4041	0.5959
348.33	0.0715	0.6154	0.5097	0.4903
398.81	0.0275	0.2601	0.0855	0.9145
386.01	0.0227	0.3591	0.1206	0.8794
382.30	0.0144	0.3825	0.1121	0.8879
377.79	0.0320	0.4038	0.1626	0.8374
372.55	0.0119	0.4927	0.1489	0.8511

Mixture 12: p-$H_2C = C(CH_3)C_6H_4CH_3$ (limonene) (1) + $(CH_3)_2C = CHCH_2CH_2C(OH)(CH_3)CH = CH_2$ (linalool) (2) + [emim][$EtSO_4$] (3)
(1) Distribution Ratio and Selectivity

Ref. [99]

	Ionic liquid-rich phase		Limonene-rich phase			
T (K)	x_1	x_2	x_1	x_2	$\beta_{linalool}$	S
298.15	0.0087	0.0000	1.0000	0.0000	–	–
	0.0100	0.0278	0.9574	0.0426	0.65	62.67
	0.0140	0.0607	0.9162	0.0838	0.72	47.40
	0.0196	0.1049	0.8594	0.1402	0.75	32.85
	0.0217	0.1239	0.8324	0.1670	0.74	28.40
	0.0276	0.1557	0.8073	0.1922	0.81	23.71
	0.0426	0.1990	0.7778	0.2210	0.90	16.46
	0.0754	0.2727	0.7403	0.2577	1.06	10.39
	0.1370	0.3551	0.6976	0.2986	1.19	6.06
	0.2018	0.4019	0.6580	0.3356	1.20	3.91
	0.3183	0.4641	0.5558	0.4259	1.09	1.90
	0.3735	0.4745	0.5158	0.4533	1.05	1.45
318.15	0.0006	0.0000	1.0000	0.0000	–	–
	0.0116	0.0275	0.9483	0.0517	0.53	43.64
	0.0174	0.0830	0.8602	0.1395	0.59	29.45
	0.0289	0.1564	0.7715	0.2285	0.68	18.24
	0.0496	0.2234	0.7227	0.2756	0.81	11.81
	0.0885	0.3031	0.6772	0.3195	0.95	7.26
	0.1453	0.3745	0.6331	0.3601	1.04	4.53
	0.2042	0.4252	0.5852	0.4040	1.05	3.02
	0.3231	0.4992	0.4673	0.4843	1.03	1.49

Mixture 13: $C_2H_5OC(CH_3)_2CH_2CH_3$ (1) + C_2H_5OH (2) + [emim][$EtSO_4$] (3)
(1) Phase Equilibrium

Ref. [100]

Organic phase		Ionic liquid phase	
x_1	x_2	x_1	x_2
T = 298.15 K			
1.0000	0.0000	0.0159	0.0000
0.9916	0.0084	0.0145	0.0806
0.9831	0.0166	0.0160	0.1407
0.9776	0.0222	0.0173	0.1895
0.9670	0.0327	0.0160	0.2366
0.9537	0.0461	0.0188	0.3038
0.9190	0.0808	0.0190	0.4057
0.8961	0.1035	0.0327	0.4725
0.8304	0.1691	0.0477	0.5603

(*continued*)

Organic phase		Ionic liquid phase	
x_1	x_2	x_1	x_2
0.7964	0.2031	0.0543	0.5922
0.7018	0.2965	0.0798	0.6476
0.4817	0.4961	0.1994	0.6703

Mixture 14: C_6H_{14} (1) + C_6H_6 (2) + [emim][$C_2H_5SO_4$] (3)
(1) Distribution Ratio and Separation Factors

Ref. [101]

Hexane-rich phase		IL-rich phase			
x_1^{I}	x_2^{I}	x_1^{II}	x_2^{II}	D_2	$\alpha_{2,1}$
T = 298.2 K					
0.9451	0.0549	0.0069	0.0381	0.694	95.1
0.8800	0.1200	0.0069	0.0825	0.688	87.7
0.8370	0.1630	0.0068	0.1117	0.685	84.3
0.7732	0.2268	0.0066	0.1515	0.668	78.3
0.7207	0.2793	0.0064	0.1730	0.619	69.8
0.6686	0.3314	0.0064	0.1994	0.602	62.9
0.5864	0.4136	0.0063	0.2390	0.578	53.8
0.5346	0.4654	0.0063	0.2652	0.570	48.4
T = 313.2 K					
0.9353	0.0647	0.0083	0.0381	0.589	66.4
0.8731	0.1269	0.0081	0.0743	0.586	63.1
0.8274	0.1726	0.0080	0.1006	0.583	60.3
0.7479	0.2521	0.0080	0.1405	0.557	52.1
0.7008	0.2992	0.0077	0.1641	0.548	49.9
0.6376	0.3624	0.0075	0.1876	0.518	44.0
0.5935	0.4065	0.0073	0.2050	0.504	41.0
0.5306	0.4694	0.0072	0.2292	0.488	36.0
T = 328.2 K					
0.9331	0.0669	0.0092	0.0324	0.484	49.1
0.8649	0.1351	0.0090	0.0661	0.489	47.0
0.8200	0.1800	0.0088	0.0857	0.476	44.4
0.7564	0.2436	0.0087	0.1164	0.478	41.5
0.6967	0.3033	0.0085	0.1391	0.459	37.6
0.6397	0.3603	0.0082	0.1612	0.447	34.9
0.5711	0.4289	0.0082	0.1891	0.441	30.7
0.5256	0.4744	0.0080	0.2048	0.432	28.4

Mixture 15: H_2O (1) + CH_3OH (2) + [EMIM][ES] (3)
(1) Vapor Pressure and Activity Coefficient

Ref. [81]

T/K	P^{exp}/kPa	P^{NRTL}/kPa	γ_1^{NRTL}	γ_2^{NRTL}
$x_1 = 0.6300, x_2 = 0.2835$				
299.40	8.180	7.751	0.9157	1.1299
307.97	12.866	12.083	0.9220	1.1272
315.30	18.311	17.290	0.9269	1.1249
321.89	24.984	23.507	0.9309	1.1228
328.23	33.107	31.203	0.9345	1.1208
333.43	41.341	38.996	0.9373	1.1192
337.97	50.099	47.094	0.9395	1.1178
342.74	60.923	57.093	0.9418	1.1164
$x_1 = 0.4915, x_2 = 0.4147$				
295.22	8.246	7.798	0.8970	1.0642
303.17	12.660	11.868	0.9046	1.0640
310.33	18.184	16.973	0.9108	1.0636
316.49	24.482	22.771	0.9158	1.0633
322.31	31.822	29.708	0.9201	1.0629
327.68	40.138	37.629	0.9239	1.0625
332.48	49.206	46.143	0.9270	1.0621
337.18	59.786	56.006	0.9300	1.0617
$x_1 = 0.2754, x_2 = 0.6196$				
291.87	8.282	8.227	0.8816	0.9870
300.22	13.124	12.878	0.8918	0.9896
307.23	18.662	18.364	0.8996	0.9914
313.81	25.593	25.215	0.9063	0.9930
320.06	34.116	33.626	0.9123	0.9943
324.96	42.423	41.775	0.9167	0.9953
330.31	53.212	52.506	0.9212	0.9962
334.59	63.468	62.702	0.9246	0.9970

Mixture 16: C_2H_5OH (1) + CH_3OH (2) + [EMIM][ES] (3)
(1) Vapor Pressure and Activity Coefficient

Ref. [81]

T/K	P^{exp}/kPa	P^{NRTL}/kPa	γ_1^{NRTL}	γ_2^{NRTL}
$x_1 = 0.1295, x_2 = 0.7444$				
296.00	11.432	11.130	1.0973	0.9004
299.03	13.348	13.085	1.0976	0.9025
304.38	17.682	17.269	1.0981	0.9060
310.21	23.475	23.075	1.0985	0.9095
315.81	30.513	30.129	1.0987	0.9128
320.50	37.631	37.379	1.0989	0.9153
325.47	46.790	46.627	1.0990	0.9179
330.38	57.630	57.571	1.0990	0.9203

(*continued*)

T/K	P^{exp}/kPa	P^{NRTL}/kPa	γ_1^{NRTL}	γ_2^{NRTL}
$x_1 = 0.2745, x_2 = 0.5918$				
298.19	11.427	11.028	1.0997	0.8612
302.22	14.159	13.669	1.0996	0.8646
307.70	18.706	18.115	1.0993	0.8691
313.01	24.211	23.549	1.0989	0.8732
318.14	30.732	30.065	1.0985	0.8769
322.78	37.857	37.207	1.0981	0.8801
327.85	47.173	46.598	1.0977	0.8835
332.64	57.673	57.220	1.0972	0.8865
$x_1 = 0.4380, x_2 = 04197$				
300.50	11.416	10.766	1.0911	0.8048
301.83	11.973	11.563	1.0909	0.8063
310.18	18.303	17.853	1.0897	0.8152
316.16	24.419	23.993	1.0887	0.8212
321.83	31.746	31.368	1.0879	0.8266
326.86	39.651	39.455	1.0871	0.8311
331.63	48.761	48.677	1.0863	0.8352
335.28	56.765	56.908	1.0857	0.8383
$x_1 = 0.6238, x_2 = 0.2242$				
303.51	11.403	10.697	1.0633	0.7252
307.04	13.626	12.923	1.0628	0.7303
313.28	18.604	17.842	1.0618	0.7390
319.87	25.492	24.714	1.0608	0.7476
325.65	33.173	32.427	1.0599	0.7549
331.45	42.889	42.174	1.0590	0.7618
335.96	52.006	51.343	1.0583	0.7670
339.87	61.187	60.615	1.0578	0.7714

2.2.14 02003-04019: 1-Ethyl-3-methylimidazolium diethyleneglycolmonomethylethersulphate

Abbreviation:	$[C_2MIm][CH_3(OCH_2CH_2)_2OSO_3]$, $[emim][CH_3(OCH_2CH_2)_2OSO_3]$, $[EMIM][CH_3(OCH_2CH_2)_2OSO_3]$
Molecular Formula:	$C_{11}H_{22}N_2O_6S$
Molar Mass:	310.37
Structure:	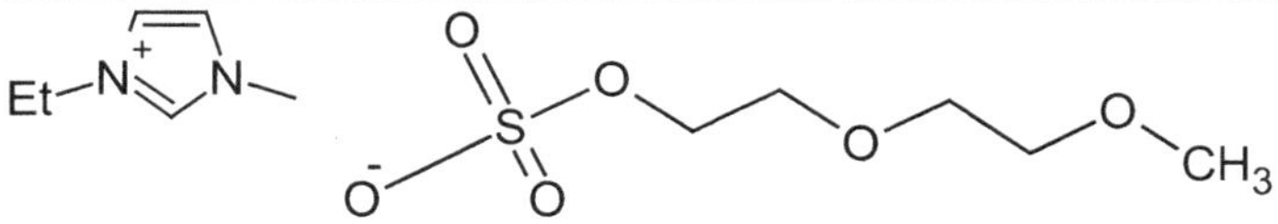

2.2.14.1 Binary Mixtures

Mixture 1: CH_3OH (1) + [EMIM][$CH_3(OCH_2CH_2)_2OSO_3$] (2)

(1) Volumetric Properties, Speed of Sound and Compressibility Properties

Ref. [63]

x_1	ρ (g · cm^{-3})	u (m · s^{-1})	ks (T · Pa^{-1})	Δks (T · Pa^{-1})	Δu (m · s^{-1})
T = 298.15 K					
0.1791	1.2220	1677.6	291	−124	79
0.3560	1.2025	1647.0	307	−244	155
0.4548	1.1874	1620.0	321	−305	188
0.5356	1.1716	1597.2	335	−353	213
0.6298	1.1473	1563.6	357	−403	237
0.7412	1.1050	1499.4	403	−442	239
0.7987	1.0722	1453.2	442	−447	228
0.8527	1.0314	1387.5	504	−427	195
0.9011	0.9807	1324.8	581	−386	161
0.9298	0.9408	1270.8	658	−331	125
0.9620	0.8831	1201.2	785	−229	74
0.9829	0.8356	1158.0	892	−138	44

Ref. [102]

x_1	ρ (g · cm^{-3})	V^E_m (cm^3 · mol^{-1})	x_1	ρ (g · cm^{-3})	V^E_m (cm^3 · mol^{-1})
T = 298.15 K					
0.038	1.2337	−0.010	0.747	1.1016	−0.933
0.103	1.2288	−0.096	0.804	1.0687	−0.920
0.189	1.2213	−0.212	0.857	1.0277	−0.879
0.238	1.2167	−0.311	0.899	0.9833	−0.805
0.373	1.2009	−0.569	0.931	0.9386	−0.654
0.439	1.1907	−0.654	0.963	0.8819	−0.451
0.530	1.1733	−0.757	0.983	0.8337	−0.251
0.641	1.1440	−0.866			
T = 303.15 K					
0.038	1.2299	−0.032	0.747	1.0977	−0.987
0.103	1.2250	−0.123	0.804	1.0646	−0.961
0.189	1.2174	−0.226	0.857	1.0235	−0.915
0.238	1.2128	−0.329	0.899	0.9790	−0.836
0.373	1.1971	−0.613	0.931	0.9342	−0.680
0.439	1.1869	−0.701	0.963	0.8774	−0.470
0.530	1.1696	−0.821	0.983	0.8292	−0.266
0.641	1.1401	−0.913			
T = 313.15 K					
0.038	1.2229	−0.039	0.747	1.0898	−1.042
0.103	1.2179	−0.123	0.804	1.0568	−1.039
0.189	1.2104	−0.261	0.857	1.0155	−0.990

(*continued*)

x_1	ρ (g · cm^{-3})	V^E_m (cm^3 · mol^{-1})	x_1	ρ (g · cm^{-3})	V^E_m (cm^3 · mol^{-1})
0.238	1.2058	−0.374	0.899	0.9708	−0.906
0.373	1.1900	−0.669	0.931	0.9258	−0.740
0.439	1.1796	−0.743	0.963	0.8687	−0.513
0.530	1.1622	−0.871	0.983	0.8202	−0.290
0.641	1.1327	−0.989			

Mixture 2: H_2O (1) + [EMIM][$CH_3(OCH_2CH_2)_2OSO_3$] (2)
(1) Volumetric Properties

Ref. [102]

x_1	ρ (g · cm^{-3})	V^E_m (cm^3 · mol^{-1})	x_1	ρ (g · cm^{-3})	V^E_m (cm^3 · mol^{-1})
T = 298.15 K					
0.111	1.2344	−0.006	0.733	1.2028	−0.390
0.170	1.2332	−0.029	0.787	1.1930	−0.390
0.259	1.2312	−0.089	0.840	1.1786	−0.367
0.351	1.2286	−0.149	0.906	1.1463	−0.278
0.410	1.2266	−0.190	0.916	1.1388	−0.254
0.485	1.2234	−0.235	0.950	1.1037	−0.145
0.564	1.2191	−0.298	0.978	1.0560	−0.051
0.650	1.2125	−0.346			
T = 303.15 K					
0.111	1.2306	−0.020	0.733	1.1990	−0.367
0.170	1.2294	−0.040	0.787	1.1891	−0.359
0.259	1.2275	−0.110	0.840	1.1748	−0.338
0.351	1.2249	−0.162	0.906	1.1428	−0.254
0.410	1.2229	−0.199	0.916	1.1354	−0.232
0.485	1.2198	−0.250	0.950	1.1008	−0.131
0.564	1.2154	−0.295	0.978	1.0537	−0.043
0.650	1.2088	−0.336			
T = 313.15 K					
0.111	1.2237	−0.032	0.733	1.1919	−0.319
0.170	1.2226	−0.065	0.787	1.1820	−0.308
0.259	1.2206	−0.111	0.840	1.1677	−0.285
0.351	1.2180	−0.156	0.906	1.1360	−0.207
0.410	1.2160	−0.189	0.916	1.1287	−0.187
0.485	1.2128	−0.223	0.950	1.0947	−0.096
0.564	1.2083	−0.254	0.978	1.0490	−0.032
0.650	1.2017	−0.292			

2.2.14.2 Ternary Mixtures

Mixture 1: H_2O (1) + CH_3OH (2) + [EMIM][$CH_3(OCH_2CH_2)_2OSO_3$] (3)
(1) Volumetric Properties

Ref. [103]

			$V^E_m/(cm^3 \cdot mol^{-1})$	
x_3	x_1	$\rho/(g \cdot cm^{-3})$	Experimental	Graph
$T = 298.15$ K				
0.0019	0.7078	0.9334	−0.871	−1.131
0.0027	0.6147	0.9152	−0.961	−1.506
0.0033	0.5048	0.8933	−0.998	−0.998
0.0040	0.4016	0.8739	−0.977	−1.816
0.0050	0.3102	0.8583	−0.869	−1.649
0.0053	0.2273	0.8434	−0.738	−1.253
0.0064	0.1101	0.8250	−0.247	−0.247
0.0871	0.6008	1.0753	−0.909	−0.890
0.1928	0.5887	1.1523	−0.719	−0.436
0.3087	0.4066	1.1722	−0.660	−0.660
0.3403	0.5555	1.1948	−0.423	−0.450
0.4820	0.4648	1.2156	−0.194	−0.078
0.5212	0.2834	1.2058	−0.320	−0.320
$T = 303.15$ K				
0.0019	0.7078	0.9300	−0.748	−0.405
0.0027	0.6147	0.9115	−0.854	−0.649
0.0033	0.5048	0.8894	−0.888	−0.888
0.0040	0.4016	0.8698	−0.828	−0.978
0.0050	0.3102	0.8541	−0.702	−0.917
0.0053	0.2273	0.8390	−0.552	−0.746
0.0064	0.1101	0.8205	−0.247	−0.247
0.0871	0.6008	1.0713	−0.742	−0.240
0.1928	0.5887	1.1484	−0.490	−0.023
0.3087	0.4066	1.1683	−0.255	−0.255
0.3403	0.5555	1.1947	−0.119	0.315
0.4820	0.4648	1.2119	−0.207	0.341
0.5212	0.2834	1.2021	−0.121	−0.121
$T = 313.15$ K				
0.0019	0.7078	0.9234	−0.364	−4.443
0.0027	0.6147	0.9042	−0.531	−2.579
0.0033	0.5048	0.8816	−0.181	−0.181
0.0040	0.4016	0.8617	−0.630	−1.306
0.0050	0.3102	0.8457	−0.174	−1.795
0.0053	0.2273	0.8305	−0.640	−1.203
0.0064	0.1101	0.8118	−0.574	−0.574
0.0871	0.6008	1.0638	−0.427	−5.638
0.1928	0.5887	1.1411	−0.059	−6.863

(continued)

			$V^E_m/(cm^3 \cdot mol^{-1})$	
x_3	x_1	$\rho/(g \cdot cm^{-3})$	Experimental	Graph
0.3087	0.4066	1.1611	−0.340	−0.340
0.3403	0.5555	1.1876	−0.476	−7.699
0.4820	0.4648	1.2049	0965	7.078
0.5212	0.2834	1.1951	0.972	0.972

2.2.15 02003-04021: 1-Ethyl-3-methylimidazolium hydrogen sulfate

Abbreviation:	[EMIM][HSO_4]
Molecular Formula:	$C_6H_{12}N_2O_4S$
Molar Mass:	208.24
Structure:	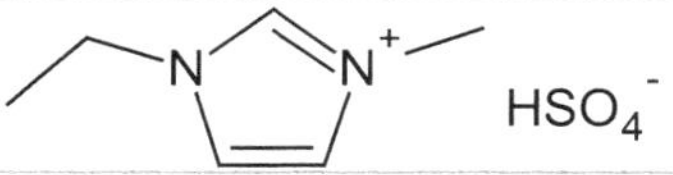

2.2.15.1 Ternary Mixtures

Mixture 1: $CH_3COOCH_2CH_3$ (1) + C_2H_5OH (2) + [EMIM][HSO_4] (3)

(1) Volumetric Properties

Ref. [104]

Ester-rich phase			[EMIM][HSO_4]-rich phase		
w_1	w_2	$\rho/(g \cdot cm^{-3})$	w_1	w_2	$\rho/(g \cdot cm^{-3})$
313 K					
1.000	0.000	0.87665	0.017	0.000	1.340
0.893	0.107	0.86289	0.039	0.103	1.24938
0.785	0.211	0.85221	0.075	0.192	1.16494
0.734	0.256	0.84918	0.119	0.243	1.10720
0.687	0.294	0.84850	0.171	0.286	1.05419
0.628	0.332	0.85086	0.277	0.337	0.97867
0.606	0.342	0.85390			
0.537	0.368	0.86703			
0.455	0.375	0.89146			

Ref. [104]

Ester-rich phase			[EMIM][HSO_4]-rich phase		
w_1	w_2	$\rho/(g \cdot cm^{-3})$	w_1	w_2	$\rho/(g \cdot cm^{-3})$
313 K					
0.771	0.223	0.85161	0.058	0.159	1.19753
0.883	0.116	0.86161	0.033	0.085	1.26657
0.508	0.375	0.87379	0.187	0.295	1.04157
0.584	0.351	0.85784	0.134	0.257	1.09020

Mixture 2: $CH_3COOCH_2CH_3$ (1) + CH_3COOH (2) + [EMIM][HSO_4] (3)
(1) Volumetric Properties

Ref. [104]

Ester-rich phase			[EMIM][HSO_4]-rich phase		
w_1	w_2	$\rho/(g \cdot cm^{-3})$	w_1	w_2	$\rho/(g \cdot cm^{-3})$
313 K					
1.000	0.000	0.87665	0.017	0.000	1.34
0.907	0.092	0.89034	0.036	0.096	1.29657
0.791	0.202	0.90833	0.057	0.165	1.26191
0.694	0.284	0.92453	0.129	0.285	1.18684
0.642	0.325	0.93569	0.178	0.333	1.14225
0.571	0.368	0.95597	0.257	0.364	1.09574
0.482	0.383	0.99049	0.314	0.380	1.06409
0.424	0.390	1.01263			

Ref. [104]

Ester-rich phase			[EMIM][HSO_4]-rich phase		
w_1	w_2	$\rho/(g \cdot cm^{-3})$	w_1	w_2	$\rho/(g \cdot cm^{-3})$
313 K					
0.926	0.073	0.88751	0.044	0.122	1.28410
0.838	0.158	0.90070	0.080	0.217	1.23322
0.767	0.223	0.91182	0.120	0.278	1.19174
0.667	0.307	0.93047	0.197	0.340	1.13241
0.610	0.347	0.94386	0.249	0.363	1.09927
0.578	0.364	0.95294	0.287	0.373	1.07879

Mixture 3: 1-C_3H_7OH (1) + $CH_3COOC_3H_7$ (2) + [EMIM][HSO_4] (3)
(1) Phase Equilibrium

Ref. [105]

Ester-rich phase			[EMIM][HSO_4]-rich phase		
w_1	w_2	w_3	w_1	w_2	w_3
313 K					
0.085	0.914	0.002	0.024	0.012	0.963
0.117	0.881	0.002	0.038	0.013	0.949
0.182	0.816	0.002	0.054	0.011	0.934
0.262	0.736	0.002	0.074	0.016	0.910
0.367	0.627	0.007	0.137	0.029	0.834
0.440	0.541	0.019	0.179	0.039	0.782
0.520	0.429	0.051	0.233	0.059	0.708

Ref. [105]

Ester-rich phase			$[EMIM][HSO_4]$-rich phase		
w_1	w_2	w_3	w_1	w_2	w_3
313 K					
0.000	1.000	0.000	0.010	0.003	0.987
0.020	0.980	0.000	0.015	0.015	0.969
0.080	0.920	0.000	0.033	0.012	0.954
0.150	0.850	0.000	0.053	0.012	0.935
0.202	0.796	0.002	0.076	0.015	0.909
0.289	0.708	0.003	0.117	0.022	0.862
0.330	0.665	0.005	0.135	0.028	0.837
0.405	0.585	0.010	0.179	0.039	0.782
0.427	0.557	0.015	0.181	0.041	0.778
0.447	0.531	0.022	0.200	0.049	0.751
0.473	0.503	0.023	0.221	0.055	0.724
0.507	0.457	0.036	0.245	0.065	0.689
0.513	0.304	0.183	0.259	0.071	0.670
0.514	0.425	0.061	0.288	0.084	0.628
0.527	0.385	0.088	0.364	0.136	0.500
0.528	0.345	0.127	0.456	0.226	0.318

Mixture 4: CH_3COOH (1) + $CH_3COOC_3H_7$ (2) + $[EMIM][HSO_4]$ (3)
(1) Phase Equilibrium

Ref. [105]

Ester-rich phase			$[EMIM][HSO_4]$-rich phase		
w_1	w_2	w_3	w_1	w_2	w_3
313 K					
0.046	0.954	0.000	0.145	0.026	0.829
0.141	0.857	0.002	0.231	0.046	0.723
0.235	0.760	0.004	0.308	0.074	0.618
0.320	0.669	0.010	0.382	0.124	0.494

Ref. [105]

Ester-rich phase			$[EMIM][HSO_4]$-rich phase		
w_1	w_2	w_3	w_1	w_2	w_3
313 K					
0.000	1.000	0.000	0.013	0.010	0.977
0.010	0.990	0.000	0.078	0.017	0.906
0.080	0.920	0.000	0.149	0.028	0.822
0.140	0.859	0.001	0.187	0.037	0.776
0.175	0.822	0.002	0.241	0.053	0.705
0.230	0.766	0.004	0.246	0.047	0.707
0.271	0.722	0.007	0.287	0.071	0.642
0.336	0.652	0.012	0.327	0.092	0.581

(continued)

Ester-rich phase			[EMIM][HSO$_4$]-rich phase		
w_1	w_2	w_3	w_1	w_2	w_3
0.390	0.590	0.020	0.352	0.108	0.541
0.414	0.559	0.028	0.357	0.104	0.538
0.462	0.467	0.071	0.391	0.140	0.469
			0.447	0.213	0.339
			0.468	0.273	0.258
			0.475	0.376	0.149

Mixture 5: 1-C_4H_9OH (1) + $CH_3COOC_4H_9$ (2) + [EMIM][HSO$_4$] (3)
(1) Phase Equilibrium

Ref. [105]

Ester-rich phase			[EMIM][HSO$_4$]-rich phase		
w_1	w_2	w_3	w_1	w_2	w_3
313 K					
0.168	0.828	0.004	0.021	0.006	0.974
0.334	0.661	0.005	0.037	0.005	0.958
0.564	0.424	0.015	0.101	0.010	0.889
0.656	0.314	0.030	0.125	0.013	0.863
0.704	0.201	0.095	0.189	0.022	0.789

Ref. [105]

Ester-rich phase			[EMIM][HSO$_4$]-rich phase		
w_1	w_2	w_3	w_1	w_2	w_3
313 K					
0.000	0.999	0.001	0.000	0.002	0.998
0.074	0.925	0.001	0.014	0.006	0.979
0.150	0.850	0.000	0.036	0.005	0.959
0.250	0.750	0.001	0.041	0.006	0.954
0.298	0.701	0.001	0.050	0.007	0.944
0.349	0.649	0.002	0.076	0.006	0.917
0.400	0.599	0.002	0.091	0.008	0.901
0.449	0.544	0.007	0.097	0.009	0.894
0.544	0.448	0.008	0.138	0.015	0.847
0.546	0.441	0.013	0.154	0.017	0.830
0.637	0.334	0.029	0.180	0.020	0.800
0.645	0.317	0.038	0.226	0.028	0.745
0.682	0.280	0.037	0.341	0.047	0.612
0.705	0.212	0.084	0.433	0.066	0.500
			0.540	0.090	0.370
			0.590	0.104	0.306
			0.641	0.122	0.237
			0.691	0.149	0.159
			0.702	0.185	0.114

Mixture 6: CH_3COOH (1) + $CH_3COOC_4H_9$ (2) + [EMIM][HSO_4] (3)
(1) Phase Equilibrium

Ref. [105]

Ester-rich phase			[EMIM][HSO_4]-rich phase		
w_1	w_2	w_3	w_1	w_2	w_3
313 K					
0.026	0.974	0.001	0.064	0.006	0.928
0.058	0.941	0.001	0.132	0.011	0.853
0.145	0.851	0.001	0.234	0.028	0.736
0.328	0.668	0.007	0.362	0.064	0.579
0.450	0.534	0.015	0.471	0.125	0.400

Ref. [105]

Ester-rich phase			[EMIM][HSO_4]-rich phase		
w_1	w_2	w_3	w_1	w_2	w_3
313 K					
0.010	0.989	0.001	0.010	0.004	0.986
0.100	0.899	0.001	0.040	0.006	0.954
0.219	0.778	0.003	0.119	0.011	0.871
0.354	0.638	0.008	0.153	0.015	0.833
0.468	0.512	0.020	0.190	0.019	0.792
0.496	0.438	0.067	0.297	0.044	0.659
0.524	0.421	0.055	0.353	0.061	0.586
			0.391	0.071	0.539
			0.440	0.107	0.453
			0.478	0.131	0.392
			0.500	0.160	0.341
			0.535	0.227	0.238
			0.549	0.331	0.120

2.2.16 02003-04030: 1-Ethyl-3-methylimidazolium 1,1,2,2-tetrafluoroethanesulfonate

Abbreviation:	[emim][TFES]
Molecular Formula:	$C_8H_{12}N_2O_3SF_4$
Molar Mass:	292.25
Structure:	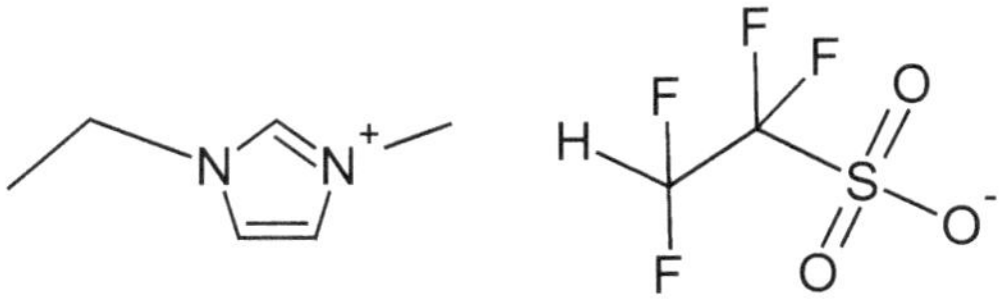

2.2.16.1 Binary Mixtures

Mixture 1: CH_2F_2 (1) + [emim][TFES] (2)
(1) Solubility and Diffusivity

Ref. [55]

T/K	*P*/MPa	100 x_1	$10^{11}D$/ $m^2 \cdot s^{-1}$	*T*/K	*P*/MPa	100 x_1	$10^{11}D$/ $m^2 \cdot s^{-1}$
298.15	0.0099	0.6		298.05	0.5496	27.9	5.9
298.05	0.0991	5.4		298.05	0.7022	34.8	10
298.05	0.2484	13.3	2.9	298.05	0.8483	41.3	10
298.05	0.3995	20.9	4.9	298.05	1.0016	47.7	12

2.2.17 02003-05001: 1-Ethyl-3-methylimidazolium hexafluorophosphate

Abbreviation:	[C_2MIm][PF_6], [emim][PF_6], [EMIM][PF_6]
Molecular Formula:	$C_6H_{11}F_6N_2P$
Molar Mass:	256.13
Structure:	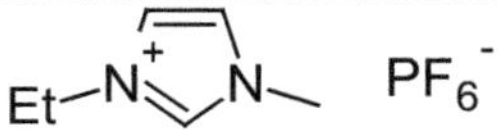

2.2.17.1 Binary Mixtures

Mixture 1: IM-C_6H_5 (2-phenylimidazole) (1) + [emim][PF_6] (2)
(1) Activity Coefficient and Phase Equilibrium

Ref. [106]

x_2	*T*/K	γ_1	x_2	*T*/K	γ_1
0.0834	415.65	1.032	0.6036	393.45	1.759
0.1061	416.65	1.072	0.6369	387.15	1.750
0.1704	411.85	1.085	0.7385	380.15	2.168
0.2579	408.75	1.164	0.8118	374.45	2.183
0.3334	405.95	1.247	0.8384	371.55	2.774
0.4453	399.25	1.309	0.9692	328.55	3.105
0.4869	397.25	1.435			

Mixture 2: $C_{15}H_{12}N_2$ (4,5-diphenylimidazole) (1) + [emim][PF_6] (2)
(1) Activity Coefficient and Phase Equilibrium

Ref. [106]

x_2	T/K	γ_1	x_2	T/K	γ_1
0.1017	497.05	0.985	0.7004	455.65	1.449
0.1650	491.45	0.969	0.7547	449.75	1.581
0.2499	485.35	0.976	0.8033	443.65	1.752
0.2997	485.15	1.041	0.8602	428.55	1.808
0.3636	477.55	1.009	0.9019	417.95	2.047
0.4711	470.95	1.083	0.9518	389.15	2.092
0.5773	466.05	1.243			

Mixture 3: $C_{21}H_{16}N_2$ (2,4,5-triphenylimidazole) (1) + [emim][PF_6] (2)
(1) Phase Equilibrium

Ref. [106]

x_2	T/K	x_2	T/K
0.0929	526.65	0.5339	512.75
0.2163	526.15	0.6324	503.75
0.3064	517.35	0.7691	493.65
0.3217	518.35	0.8824	459.75
0.4316	518.15	0.9638	409.25
0.4920	518.65		

Mixture 4: CO_2 (1) + [emim][PF_6] (2)
(1) Bubble Point

Ref. [107]

x_1	T (K)	P (MPa)	T (K)	P (MPa)	T (K)	P (MPa)
0.104	332.83	1.49	342.77	1.66	352.71	1.89
	337.63	1.56	347.84	1.78		
0.243	323.38	3.24	343.42	4.50	362.68	5.91
	328.24	3.53	348.12	4.81	366.03	6.28
	333.15	3.84	352.88	5.12		
	338.14	4.17	357.92	5.52		
0.356	322.80	5.66	337.80	7.32	352.67	9.40
	327.91	6.14	342.80	7.96		
	333.00	6.70	347.74	8.66		
0.449	313.12	7.16	327.96	10.18	342.88	14.84
	317.81	8.30	332.90	11.62	347.71	16.48
	322.98	9.26	337.86	13.10	352.67	18.04
0.508	308.14	11.70	327.98	22.28	347.74	31.66
	313.01	14.28	332.93	24.68	352.66	34.02
	318.05	17.08	337.77	26.90		
	323.02	19.64	342.74	29.42		

(*continued*)

x_1	T (K)	P (MPa)	T (K)	P (MPa)	T (K)	P (MPa)
0.549	313.04	30.64	327.97	39.10	342.80	47.04
	317.93	33.46	332.83	41.74	347.78	49.76
	323.02	36.40	337.82	44.44	352.67	52.30
0.581	313.11	56.50	327.99	65.20	342.82	73.70
	318.11	59.40	332.72	68.00	347.73	76.30
	323.07	62.10	337.75	71.10	352.48	79.10
0.599	313.08	61.92	327.99	72.74	342.63	84.00
	318.07	65.56	332.76	76.48	347.70	87.82
	323.04	69.20	337.74	80.16	352.59	91.72
0.619	313.04	79.30	323.02	86.30	332.88	93.50
	317.99	82.90	327.74	89.70	337.94	97.10

2.2.18 02003-05011: 1-Ethyl-3-methylimidazolium dimethylphosphate

Abbreviation:	[EMIM][DMP]
Molecular Formula:	$C_8H_{17}N_2O_4P$
Molar Mass:	236.21
Structure:	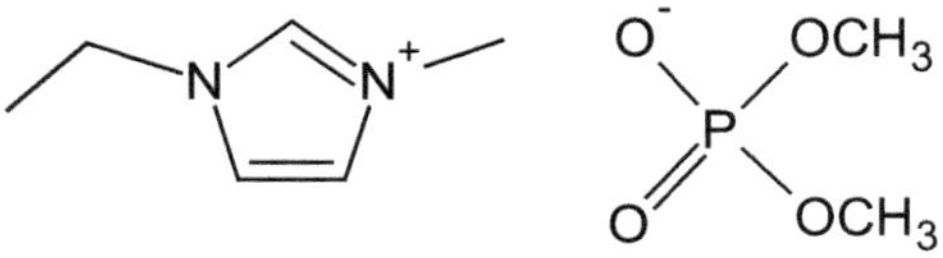

2.2.18.1 Binary Mixtures

Mixture 1: H_2O (1) + [EMIM][DMP] (2)

(1) Vapor Pressure

Ref. [108]

T/K	P^{exp}/kPa	P^{NRTL}/kPa	γ_i^{NRTL}	γ_i^{NRTL}
$x_1 = 0.9915$				
323.76	12.609	12.546	1.0026	0.9976
331.49	18.256	18.197	1.0010	0.9978
337.42	23.924	23.884	0.9996	0.9979
341.67	28.972	28.844	1.0024	0.9980
346.27	35.221	35.158	0.9998	0.9981
349.73	40.683	40.649	0.9990	0.9981
354.06	48.559	48.527	0.9989	0.9982
357.98	56.785	56.740	0.9990	0.9983
$x_1 = 0.9683$				
329.14	15.383	15.487	0.9651	0.9716

(*continued*)

T/K	P^{exp}/kPa	P^{NRTL}/kPa	γ_i^{NRTL}	γ_i^{NRTL}
338.86	24.089	24.286	0.9662	0.9740
342.95	28.840	29.100	0.9662	0.9749
345.87	32.706	33.007	0.9667	0.9756
349.44	38.006	38.373	0.9670	0.9763
353.59	45.046	45.522	0.9669	0.9771
356.92	51.484	52.054	0.9670	0.9777
360.27	58.781	59.395	0.9682	0.9783
x_1 = 0.9291				
331.82	15.082	15.387	0.8689	0.8864
337.61	19.715	20.172	0.8714	0.8916
342.89	24.888	25.584	0.8716	0.8960
347.54	30.483	31.340	0.8750	0.8996
351.65	36.194	37.294	0.8760	0.9027
356.18	43.556	44.924	0.8783	0.9059
360.46	51.763	53.323	0.8822	0.9088
363.71	58.804	60.545	0.8847	0.9109
x_1 = 0.8489				
338.84	15.216	14.739	0.6968	0.6750
343.48	18.825	18.301	0.7033	0.6837
347.73	22.775	22.180	0.7100	0.6914
351.54	26.726	26.235	0.7112	0.6981
356.93	33.494	33.016	0.7175	0.7072
363.48	43.556	43.201	0.7237	0.7178
367.26	50.463	50.207	0.7273	0.7236

$$\mathrm{ARD}(P) = 0.014 = \frac{\sum_{i=1}^{n} \left|p^{NRTL} - p^{\exp}\right| / p^{\exp}}{\mathrm{n}};$$

$$\mathrm{RMSD} = 0.018 = \sqrt{\frac{\sum_{i=1}^{n} \left(p^{cal}/p^{\exp} - 1\right)^2}{n}}.$$

Mixture 2: CH_3OH (1) + [EMIM][DMP] (2)

(1) Vapor Pressure

Ref. [108]

T/K	P^{exp}/kPa	P^{NRTL}/kPa	γ_i^{NRTL}	γ_i^{NRTL}
x_1 = 0.9851				
299.20	17.460	17.474	0.9910	0.9918
303.04	21.243	21.250	0.9919	0.9922
306.80	25.597	25.595	0.9926	0.9926
310.45	30.556	30.520	0.9941	0.9929
313.84	35.819	35.795	0.9939	0.9932
317.74	42.854	42.797	0.9948	0.9935
321.12	49.870	49.763	0.9959	0.9938

(*continued*)

T/K	P^{exp}/kPa	P^{NRTL}/kPa	γ_i^{NRTL}	γ_i^{NRTL}
324.50	57.859	57.653	0.9976	0.9940
x_1 = 0.9451				
300.94	17.001	17.013	0.9198	0.9205
303.56	19.448	19.462	0.9221	0.9228
306.15	22.184	22.175	0.9253	0.9249
309.90	26.684	26.662	0.9287	0.9279
315.14	34.161	34.212	0.9304	0.9318
319.87	42.447	42.495	0.9340	0.9350
323.49	49.844	49.939	0.9356	0.9374
326.99	58.025	58.137	0.9377	0.9395
x_1 = 0.8806				
304.51	15.969	15.905	0.7750	0.7719
309.18	20.307	20.192	0.7851	0.7806
313.42	25.144	24.898	0.7959	0.7881
317.39	30.280	30.123	0.7990	0.7948
321.86	37.098	37.080	0.8024	0.8020
326.01	44.681	44.719	0.8078	0.8084
329.69	52.477	52.545	0.8128	0.8139
333.06	60.419	60.679	0.8151	0.8187
x_1 = 0.7597				
314.99	16.761	16.776	0.5718	0.5723
318.98	20.289	20.412	0.5777	0.5813
324.41	26.098	26.423	0.5858	0.5931
327.59	30.280	30.601	0.5935	0.5998
332.42	37.434	37.996	0.6008	0.6098
336.61	44.663	45.592	0.6056	0.6182
340.40	52.360	53.496	0.6123	0.6256
343.86	60.419	61.707	0.6190	0.6321

$$\mathrm{ARD}(P) = 0.0050 = \frac{\sum_{i=1}^{n} \left| p^{NRTL} - p^{\exp} \right| / p^{\exp}}{\mathrm{n}};$$

$$\mathrm{RMSD} = 0.0081 = \sqrt{\frac{\sum_{i=1}^{n} \left(p^{cal} / p^{\exp} - 1 \right)^2}{n}}.$$

Mixture 3: C_2H_5OH (1) + [EMIM][DMP] (2)
(1) Vapor Pressure

Ref. [108]

T/K	P^{exp}/kPa	P^{NRTL}/kPa	γ_i^{NRTL}	γ_i^{NRTL}
x_1 = 0.9787				
309.95	14.808	14.801	0.9991	0.9986
314.00	18.301	18.291	0.9992	0.9987

(*continued*)

T/K	P^{exp}/kPa	P^{NRTL}/kPa	γ_i^{NRTL}	γ_i^{NRTL}
320.12	24.953	24.881	1.0017	0.9988
323.38	29.189	29.144	1.0004	0.9988
328.35	36.872	36.859	0.9993	0.9989
330.72	41.146	41.100	1.0001	0.9990
334.42	48.576	48.535	0.9999	0.9990
339.27	60.073	59.980	1.0006	0.9991
$x_1 = 0.9229$				
310.50	14.165	14.127	0.9845	0.9819
316.95	19.761	19.727	0.9850	0.9833
320.67	23.785	23.751	0.9855	0.9841
325.84	30.469	30.469	0.9851	0.9851
330.66	38.080	38.133	0.9846	0.9859
333.90	44.095	44.163	0.9850	0.9865
337.47	51.571	51.689	0.9848	0.9870
340.98	60.003	60.148	0.9852	0.9876
$x_1 = 0.8368$				
313.46	14.187	14.138	0.9319	0.9287
319.68	19.458	19.453	0.9337	0.9334
325.34	25.652	25.680	0.9364	0.9374
329.31	30.948	30.988	0.9388	0.9400
333.55	37.546	37.667	0.9396	0.9427
337.28	44.529	44.504	0.9454	0.9449
340.29	50.755	50.747	0.9468	0.9466
343.58	58.368	58.415	0.9476	0.9484
$x_1 = 0.6881$				
320.08	14.187	14.176	0.8118	0.8112
323.71	17.108	17.032	0.8206	0.8170
332.30	25.652	25.775	0.8258	0.8298
334.95	29.013	29.148	0.8297	0.8335
340.75	37.546	37.831	0.8351	0.8414
344.88	45.188	45.259	0.8454	0.8467
347.49	50.755	50.559	0.8533	0.8500
350.67	58.368	57.714	0.8635	0.8539

$$\mathrm{ARD}(P) = 0.0023 = \frac{\sum_{i=1}^{n} \left| p^{NRTL} - p^{\exp} \right| / p^{\exp}}{\mathrm{n}};$$

$$\mathrm{RMSD} = 0.0032 = \sqrt{\frac{\sum_{i=1}^{n} \left(p^{cal}/p^{\exp} - 1 \right)^2}{n}}.$$

2.2.18.2 Ternary Mixtures

Mixture 1: H_2O (1) + CH_3OH (2) + [EMIM][DMP] (3)

(1) Vapor Pressure

Ref. [108]

T/K	P^{exp}/kPa	P^{NRTL}/kPa	γ_1^{NRTL}	γ_2^{NRTL}
$x_1 = 0.6960$, $x_2 = 0.2685$				
310.90	15.628	14.315	0.9434	1.1721
317.91	21.984	20.052	0.9470	1.1694
321.90	26.594	24.132	0.9489	1.1678
326.37	32.585	29.508	0.9509	1.1660
329.48	37.406	33.817	0.9522	1.1648
333.48	44.506	40.143	0.9538	1.1632
336.79	51.100	46.095	0.9551	1.1618
340.04	58.425	52.663	0.9563	1.1606
$x_1 = 0.5196$, $x_2 = 0.4380$				
308.83	17.519	16.693	0.9161	1.1002
312.59	21.001	20.038	0.9191	1.0999
317.11	26.274	24.813	0.9226	1.0995
320.53	30.774	29.026	0.9251	1.0992
324.92	37.529	35.316	0.9281	1.0987
328.56	43.882	41.389	0.9305	1.0983
331.92	50.661	47.735	0.9326	1.0978
336.47	61.268	57.627	0.9354	1.0972
$x_1 = 0.1569$, $x_2 = 0.7920$				
302.15	16.221	16.530	0.8672	0.9708
307.60	21.319	21.744	0.8752	0.9739
313.64	28.527	29.085	0.8835	0.9769
318.17	35.166	35.906	0.8893	0.9790
321.81	41.387	42.299	0.8937	0.9806
325.54	48.635	49.835	0.8980	0.9821
328.78	55.795	57.269	0.9016	0.9834
331.88	63.700	65.225	0.9049	0.9845

$$\mathrm{ARD} = 0.057 = \frac{\sum_{i=1}^{n} \left| p^{NRTL} - p^{\exp} \right| / p^{\exp}}{\mathrm{n}};$$

$$\mathrm{RMSD} = 0.064 = \sqrt{\frac{\sum_{i=1}^{n} \left(p^{cal}/p^{\exp} - 1 \right)^2}{n}}.$$

Mixture 2: H_2O (1) + C_2H_5OH (2) + [EMIM][DMP] (3)

(1) Vapor Pressure

Ref. [108]

T/K	P^{exp}/kPa	P^{NRTL}/kPa	γ_1^{NRTL}	γ_2^{NRTL}
$x_1 = 0.8787, x_2 = 0.0859$				
317.84	15.234	15.475	0.9811	3.7733
323.20	19.936	20.084	0.9826	3.7159
328.15	25.172	25.325	0.9839	3.6644
333.29	31.397	31.954	0.9852	3.6125
337.69	38.182	38.737	0.9862	3.5693
341.56	45.076	45.673	0.9870	3.5323
345.30	52.965	53.323	0.9878	3.4973
349.11	62.117	62.203	0.9885	3.4624
$x_1 = 0.7601, x_2 = 0.1985$				
315.36	15.930	16.455	1.0205	2.5226
320.71	20.822	21.459	1.0224	2.4983
324.70	25.264	25.973	1.0237	2.4806
328.62	30.386	31.168	1.0249	2.4635
333.15	37.291	38.262	1.0261	2.4441
337.18	44.559	45.655	1.0271	2.4273
340.87	52.241	53.448	1.0279	2.4121
344.34	60.504	61.745	1.0287	2.3981
$x_1 = 0.5993, x_2 = 0.3518$				
316.63	18.392	19.037	1.0834	1.7673
319.92	21.652	22.520	1.1077	1.7507
324.54	27.061	28.144	1.1100	1.7417
329.30	33.784	35.138	1.1120	1.7328
333.10	40.192	41.750	1.1135	1.7258
336.87	47.425	49.296	1.1148	1.7190
340.27	54.897	57.048	1.1159	1.7129
343.81	63.675	66.199	1.1169	1.7068
$x_1 = 0.2062, x_2 = 0.7263$				
316.09	19.076	19.299	1.0761	1.1502
318.88	21.888	22.203	1.0842	1.1492
322.27	25.687	26.228	1.0937	1.1479
325.40	29.809	30.471	1.1021	1.1467
330.02	36.960	37.804	1.1141	1.1450
334.39	44.991	46.039	1.1248	1.1434
337.83	52.303	53.557	1.1329	1.1422
340.89	59.676	61.104	1.1398	1.1411

$$\text{ARD} = 0.024 = \frac{\sum_{i=1}^{n} \left|p^{NRTL} - p^{\exp}\right| / p^{\exp}}{\text{n}};$$

$$\text{RMSD} = 0.027 = \sqrt{\frac{\sum_{i=1}^{n} \left(p^{cal}/p^{\exp} - 1\right)^2}{n}}.$$

Mixture 3: C_2H_5OH (1) + CH_3OH (2) + [EMIM][DMP] (3)
(1) Vapor Pressure

Ref. [108]

T/K	P^{exp}/kPa	P^{NRTL}/kPa	γ_1^{NRTL}	γ_2^{NRTL}
$x_1 = 0.6831, x_2 = 0.2455$				
309.93	16.265	16.117	1.0367	0.7237
312.42	18.516	18.336	1.0370	0.7279
316.61	22.840	22.638	1.0376	0.7347
320.34	27.380	27.166	1.0380	0.7406
324.33	33.033	32.854	1.0384	0.7467
328.62	40.075	40.045	1.0387	0.7530
332.60	47.781	47.861	1.0390	0.7586
336.84	57.340	58.131	1.0392	0.7644
$x_1 = 0.4767, x_2 = 0.4569$				
308.08	17.277	17.267	1.0488	0.8185
311.55	20.569	20.606	1.0495	0.8230
315.85	25.394	25.487	1.0503	0.8285
320.48	31.624	31.806	1.0510	0.8340
325.16	39.143	39.502	1.0517	0.8393
329.39	47.135	47.747	1.0522	0.8439
333.10	55.259	56.134	1.0526	0.8477
$x_1 = 0.2974, x_2 = 0.6405$				
303.95	16.079	16.067	1.0471	0.8687
305.80	17.729	17.679	1.0477	0.8708
309.20	21.053	21.009	1.0487	0.8745
313.82	26.381	26.378	1.0500	0.8793
318.70	33.153	33.266	1.0512	0.8840
323.60	41.204	41.662	1.0522	0.8885
327.77	49.410	50.153	1.0530	0.8921
331.23	57.252	58.255	1.0536	0.8949
$x_1 = 0.0717, x_2 = 0.8720$				
303.51	18.507	18.560	1.0366	9.1539
305.22	20.237	20.241	1.0374	9.1688
309.17	24.599	24.640	1.0391	9.2016
313.75	30.657	30.726	1.0410	9.2374
318.50	37.970	38.336	1.0427	9.2721
323.25	46.729	47.470	1.0443	9.3046
327.31	55.658	56.686	1.0456	9.3306
331.07	65.190	66.501	1.0467	9.3535

$$\mathrm{ARD} = 0.0076 = \frac{\sum_{i=1}^{n} \left|p^{NRTL} - p^{\exp}\right| / p^{\exp}}{\mathrm{n}};$$

$$\mathrm{RMSD} = 0.0094 = \sqrt{\frac{\sum_{i=1}^{n} \left(p^{cal}/p^{\exp} - 1\right)^2}{n}}.$$

2.2.19 02003-05012: 1-Ethyl-3-methylimidazolium diethylphosphate

Abbreviation:	[EMIM][DEP]
Molecular Formula:	$C_{10}H_{21}N_2O_4P$
Molar Mass:	264.26
Structure:	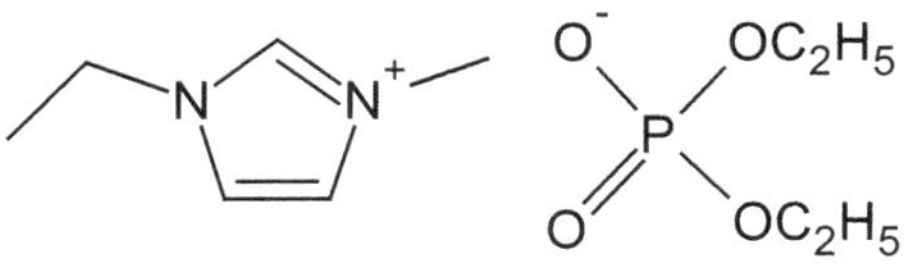

2.2.19.1 Binary Mixtures

Mixture 1: H_2O (1) + [EMIM][DEP] (2)
(1) Vapor Pressure

Ref. [19]

T (K)	P^{exp} (kPa)	P^{NRTL} (kPa)	γ_1^{exp}	γ_1^{NRTL}
IL-content: 25 wt.%				
327.26	14.424	14.532	0.980	0.988
335.12	20.885	21.054	0.982	0.990
341.15	27.385	27.621	0.983	0.991
346.07	33.933	34.187	0.985	0.993
350.31	40.541	40.875	0.985	0.993
354.16	47.516	47.856	0.987	0.994
357.58	54.432	54.870	0.987	0.995
360.99	62.188	62.725	0.987	0.995
364.58	71.275	71.979	0.986	0.996
367.71	80.167	80.935	0.987	0.996
IL-content: 35 wt.%				
324.01	11.872	11.862	0.959	0.958
333.16	18.472	18.505	0.963	0.965
340.33	25.531	25.702	0.963	0.970
346.16	32.846	33.203	0.963	0.973
351.38	40.961	41.397	0.966	0.976
355.78	49.009	49.565	0.967	0.978
359.65	57.164	57.837	0.968	0.980
363.02	65.237	65.951	0.971	0.981
366.71	74.821	75.913	0.968	0.983
369.8	83.919	85.170	0.969	0.984
IL-content: 45 wt.%				
329.92	15.449	14.759	0.955	0.912
337.27	21.781	20.952	0.959	0.922
343.22	28.272	27.465	0.957	0.930

(continued)

T (K)	P^{exp} (kPa)	P^{NRTL} (kPa)	γ_1^{exp}	γ_1^{NRTL}
348.24	35.080	34.203	0.959	0.935
352.84	42.573	41.574	0.963	0.940
357.17	50.787	49.676	0.966	0.944
360.8	58.448	57.440	0.964	0.948
364.39	66.549	66.102	0.957	0.951
367.58	74.960	74.694	0.957	0.953
370.32	82.985	82.786	0.958	0.956
IL-content: 70 wt.%				
334.93	13.038	12.041	0.701	0.647
344.10	19.428	18.874	0.695	0.675
351.72	26.507	26.828	0.689	0.697
357.82	33.996	35.092	0.692	0.714
363.23	42.086	44.118	0.694	0.728
367.80	50.099	53.190	0.696	0.739
371.93	59.534	62.702	0.712	0.749
373.26	68.727	66.037	0.783	0.752

Mixture 2: CH_3OH (1) + [EMIM][DEP] (2)
(1) Vapor Pressure

Ref. [19]

T (K)	P^{exp} (kPa)	P^{NRTL} (kPa)	γ_1^{exp}	γ_1^{NRTL}
IL-content: 20 wt.%				
285.31	7.943	7.949	0.974	0.975
295.87	14.284	14.282	0.979	0.979
303.26	20.793	20.932	0.975	0.981
309.56	28.456	28.543	0.980	0.983
313.96	35.059	35.137	0.982	0.984
317.9	42.097	42.095	0.985	0.985
321.84	50.529	50.185	0.993	0.986
324.99	57.598	57.566	0.987	0.987
328.19	66.241	65.976	0.991	0.987
331.35	75.677	75.246	0.993	0.988
IL-content: 30 wt.%				
290.14	9.790	9.790	0.934	0.934
298.78	15.607	15.673	0.938	0.942
305.32	21.895	21.932	0.946	0.947
310.83	28.660	28.738	0.949	0.951
315.62	35.992	36.040	0.953	0.955
320.27	44.642	44.552	0.959	0.958
324.22	52.990	53.067	0.958	0.960
327.57	61.405	61.333	0.963	0.962
331.13	70.710	71.255	0.956	0.964
334.14	79.563	80.672	0.952	0.965

(*continued*)

T (K)	P^{exp} (kPa)	P^{NRTL} (kPa)	γ_1^{exp}	γ_1^{NRTL}
IL-content: 40 wt.%				
289.63	8.591	8.445	0.867	0.852
300.08	15.075	15.089	0.871	0.872
306.63	21.568	21.197	0.898	0.882
312.19	28.251	27.917	0.902	0.891
317.18	35.812	35.412	0.908	0.898
321.43	43.889	43.051	0.921	0.903
325.33	51.908	51.268	0.919	0.908
328.43	59.472	58.680	0.924	0.912
331.74	69.050	67.571	0.935	0.915
334.81	78.027	76.783	0.934	0.919
IL-content: 60 wt.%				
305.46	12.559	12.199	0.606	0.588
313.41	18.999	18.759	0.626	0.618
319.78	26.200	25.974	0.646	0.641
325.55	34.314	34.400	0.659	0.660
330.03	42.187	42.440	0.671	0.675
334.44	50.686	51.829	0.673	0.688
338.48	59.839	61.909	0.677	0.700
342.19	69.403	72.529	0.680	0.711

Mixture 3: C_2H_5OH (1) + [EMIM][DEP] (2)
(1) Vapor Pressure

Ref. [19]

T (K)	P^{exp} (kPa)	P^{NRTL} (kPa)	γ_1^{exp}	γ_1^{NRTL}
IL-content: 15 wt.%				
302.30	9.591	9.545	0.992	0.987
310.38	14.939	14.845	0.994	0.988
317.14	21.191	21.056	0.995	0.989
322.91	28.148	27.984	0.995	0.990
328.29	36.217	36.109	0.993	0.990
332.87	44.660	44.504	0.994	0.991
336.64	52.892	52.629	0.996	0.991
340.18	61.569	61.329	0.995	0.991
343.31	70.297	70.013	0.996	0.992
346.36	79.864	79.459	0.997	0.992
IL-content: 25 wt.%				
302.96	9.466	9.353	0.968	0.957
312	15.457	15.315	0.970	0.961
318.86	21.995	21.782	0.973	0.964
324.57	28.924	28.815	0.970	0.966
329.36	36.297	36.105	0.973	0.968
333.57	44.010	43.776	0.974	0.969
337.78	52.895	52.740	0.973	0.970

(continued)

T (K)	P^{exp} (kPa)	P^{NRTL} (kPa)	γ_1^{exp}	γ_1^{NRTL}
341.20	61.437	61.152	0.976	0.971
344.65	70.994	70.703	0.976	0.972
347.71	80.477	80.231	0.976	0.973
IL-content: 35 wt.%				
305.78	10.146	9.972	0.918	0.902
314.42	16.153	15.914	0.924	0.910
321.23	22.779	22.521	0.927	0.917
326.87	29.84	29.625	0.928	0.921
331.86	37.569	37.424	0.929	0.925
336.25	45.597	45.656	0.927	0.928
340.14	54.106	54.189	0.930	0.931
343.58	62.697	62.793	0.932	0.933
347.00	72.250	72.489	0.932	0.935
349.82	81.091	81.353	0.934	0.937
IL-content: 60 wt.%				
314.56	9.068	9.144	0.594	0.599
322.22	14.137	13.899	0.633	0.622
329.88	20.754	20.637	0.647	0.644
336.31	28.255	28.254	0.661	0.661
341.65	36.306	36.276	0.675	0.674
346.71	45.266	45.585	0.682	0.687
352.11	56.799	57.640	0.689	0.699
356.00	66.717	67.905	0.696	0.708

2.2.19.2 Ternary Mixtures

Mixture 1: C_2H_5OH (1) + CH_3OH (2) + [EMIM][DEP] (3)

(1) Vapor Pressure

Ref. [20]

T/K	p/kPa	y_1	T/K	p/kPa	y_1
x_1 = 0.1312, x_2 = 0.7544					
294.55	8.70	0.2894	322.19	35.03	0.2833
303.70	14.29	0.2875	327.00	43.20	0.2822
311.08	20.73	0.2859	331.29	51.96	0.2811
316.55	27.08	0.2846	335.87	63.20	0.2800
x_1 = 0.3260, x_2 = 0.5503					
297.77	8.68	0.5366	326.05	36.43	0.5392
307.50	14.85	0.5381	330.80	45.00	0.5393
314.76	21.38	0.5388	334.88	53.64	0.5392
321.25	29.14	0.5391	339.01	64.04	0.5392
x_1 = 0.4805, x_2 = 0.3886					
302.37	9.68	0.6993	329.41	37.26	0.7022
310.98	15.29	0.7006	333.66	45.06	0.7023
317.92	21.63	0.7014	337.71	53.78	0.7024
323.89	28.91	0.7018	342.16	64.46	0.7025

(continued)

T/K	p/kPa	y_1	T/K	p/kPa	y_1
$x_1 = 0.6543, x_2 = 0.2065$					
303.19	8.66	0.8579	331.59	36.22	0.8587
313.08	14.84	0.8583	336.42	45.01	0.8586
320.36	21.32	0.8586	340.52	53.89	0.8587
326.25	28.29	0.8586	344.60	64.15	0.8585
$x_1 = 0.8054, x_2 = 0.0483$					
307.24	9.39	0.9707	334.42	36.41	0.9705
316.46	15.39	0.9707	339.04	44.92	0.9704
323.61	21.99	0.9707	343.55	54.83	0.9704
329.34	28.82	0.9705	347.61	64.91	0.9703

Mixture 2: C_2H_5OH (1) + H_2O (2) + [EMIM][DEP] (3)
(2) Phase Equilibrium

Ref. [21]

wt.% of IL	T (K)	x_1	y_1	γ_1	γ_2	α_{12}
101.32 kPa						
About 10 %						
10.891	362.23	0.960	0.962	0.682	1.474	1.039
11.450	362.27	0.901	0.910	0.687	1.397	1.104
11.708	362.27	0.824	0.855	0.706	1.259	1.258
11.777	362.27	0.743	0.809	0.740	1.130	1.470
11.888	362.52	0.665	0.776	0.785	1.007	1.749
12.017	362.83	0.589	0.741	0.837	0.938	1.999
12.362	363.39	0.499	0.721	0.940	0.812	2.592
12.155	363.96	0.403	0.687	1.087	0.745	3.262
11.943	364.62	0.323	0.665	1.279	0.686	4.168
11.984	365.44	0.252	0.649	1.549	0.631	5.476
11.728	366.57	0.174	0.625	2.079	0.584	7.919
11.607	368.21	0.124	0.581	2.552	0.578	9.781
11.273	371.60	0.081	0.505	3.031	0.576	11.593
About 20 %						
21.562	363.21	0.900	0.927	0.694	1.099	1.412
22.096	363.31	0.823	0.881	0.717	1.009	1.591
22.120	363.31	0.783	0.851	0.728	1.028	1.586
22.862	363.51	0.704	0.808	0.763	0.962	1.775
22.978	363.66	0.612	0.778	0.838	0.844	2.221
22.832	364.01	0.527	0.740	0.911	0.797	2.555
23.372	364.42	0.481	0.711	0.946	0.795	2.659
23.113	364.93	0.386	0.674	1.092	0.742	3.285
22.838	365.88	0.287	0.647	1.359	0.665	4.551
23.223	367.16	0.206	0.598	1.667	0.647	5.723
23.103	368.65	0.152	0.563	2.016	0.622	7.181
23.106	371.79	0.092	0.485	2.569	0.610	9.264

2.2.20 02003-06001: 1-Ethyl-3-methylimidazolium trifluoroacetate

Abbreviation:	$[C_2MIm][CF_3CO_2]$, $[emim][CF_3CO_2]$, $[EMIM][CF_3CO_2]$, [emim][TFA]
Molecular Formula:	$C_8H_{11}F_3N_2O_2$
Molar Mass:	224.18
Structure:	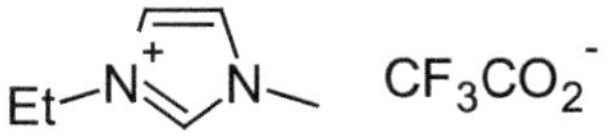

2.2.20.1 Binary Mixtures

Mixture 1: H_2O (1) + [emim][TFA] (2)

(1) Volumetric Properties and Transport Properties

Ref. [70]

	T/K							
x_1	278.15	288.15	298.15	308.15	318.15	328.15	338.15	348.15
ρ/g · cm^{-3}								
0.0000	1.30689	1.29881	1.2908	1.28285	1.27495	1.26712	1.25935	1.25163
0.2050	1.29984	1.29182	1.28385	1.27592	1.26804	1.26021	1.25242	1.24466
0.3983	1.29127	1.28325	1.27526	1.26729	1.25934	1.25141	1.24349	1.23559
0.5808	1.278	1.26993	1.26185	1.25377	1.24567	1.23755	1.22941	1.22123
0.7573	1.24838	1.24021	1.23195	1.22366	1.2153	1.20684	1.19832	1.18998
0.8420	1.21687	1.20886	1.20076	1.19257	1.18427	1.17586	1.16733	1.15869
0.9256	1.15147	1.14467	1.13765	1.1304	1.12293	1.1152	1.10728	1.09905
0.9667	1.08688	1.08251	1.07755	1.07206	1.06608	1.05966	1.05283	1.04556
0.9912	1.02668	1.02493	1.02215	1.0185	1.01411	1.00907	1.00287	0.99723
1.0000	0.99999	0.99913	0.99707	0.99406	0.99023	0.98571	0.98056	0.97484
V^E/cm^3 · mol^{-1}								
0.0000	0.000	0.000	0.000	0.000	0.000	0.000	0.000	0.000
0.2050	−0.112	−0.095	−0.080	−0.066	−0.055	−0.045	−0.034	−0.024
0.3983	−0.370	−0.331	−0.296	−0.265	−0.237	−0.209	−0.181	−0.154
0.5808	−0.651	−0.588	−0.532	−0.482	−0.435	−0.388	−0.343	−0.297
0.7573	−0.764	−0.675	−0.595	−0.524	−0.457	−0.392	−0.328	−0.278
0.8420	−0.698	−0.604	−0.523	−0.449	−0.382	−0.319	−0.257	−0.199
0.9256	−0.471	−0.392	−0.327	−0.270	−0.219	−0.171	−0.127	−0.083
0.9667	−0.236	−0.194	−0.160	−0.130	−0.104	−0.079	−0.056	−0.032
0.9912	−0.050	−0.042	−0.035	−0.030	−0.024	−0.019	−0.002	−0.007
1.0000	0.000	0.000	0.000	0.000	0.000	0.000	0.000	0.000
η/mPa · s								
0.0000	78	48	32	23	17	13	10	9
0.2042	56	35	24	17	13	10	8	7
0.3952	40	26	18	14	10	8	7	6

(*continued*)

	T/K							
x_1	278.15	288.15	298.15	308.15	318.15	328.15	338.15	348.15
0.5808	25	17	12	9	8	6	6	5
0.7565	16	11	8	6	5	5	4	4
0.9256	6	5	4	3	3	3	2	2
$\Delta\eta$/mPa · s								
0.0000	0	0	0	0	0	0	0	0
0.2042	−7	−3	−2	−1	0	0	0	0
0.3952	−8	−4	−2	−1	0	0	1	1
0.5808	−8	−4	−2	−1	0	1	1	1
0.7565	−4	−2	−1	0	1	1	1	1
0.9256	−1	0	1	1	1	1	1	1

Mixture 2: [emim][TFA] (1) + [emim][Ac] (2)
(1) Volumetric Properties

Ref. [109]

T/K	$100x_1$	ρ/g · cm^{-3}	T/K	$100x_1$	ρ/g · cm^{-3}
298.1	0.00	1.102	348.1	0.00	1.073
298.1	24.64	1.152	348.1	24.64	1.121
298.1	50.02	1.204	348.1	50.02	1.172
298.1	74.68	1.252	348.1	74.68	1.217
298.1	100.00	1.301	348.1	100.00	1.264
323.1	0.00	1.086	373.1	0.00	1.059
323.1	24.64	1.136	373.1	24.64	1.107
323.1	50.02	1.189	373.1	50.02	1.157
323.1	74.68	1.235	373.1	74.68	1.202
323.1	100.00	1.283	373.1	100.00	1.250

Mixture 3: CO_2 (1) + [emim][TFA] (2)
(1) Solubility

Ref. [109]

T/K	P/MPa	$100x_1$	T/K	P/MPa	$100x_1$	T/K	P/MPa	$100x_1$
298.1	0.0100	0.1	323.1	0.0100	0.4	348.1	0.0101	0.6
298.1	0.0498	0.9	323.1	0.0499	0.9	348.1	0.0501	1.0
298.1	0.0998	1.8	323.1	0.1000	1.5	348.1	0.1002	1.3
298.1	0.3999	6.8	323.1	0.4000	4.4	348.1	0.3997	3.2
298.1	0.6997	11.5	323.1	0.7000	7.2	348.2	0.6996	5.0
298.1	0.9999	16.0	323.1	1.0001	10.4	348.1	0.9998	7.0
298.1	1.2997	20.1	323.1	1.2997	13.4	348.1	1.2996	9.5
298.1	1.4998	22.6	323.1	1.4998	15.6	348.1	1.4997	10.8
298.1	1.9996	28.2	323.1	1.9996	20.0	348.2	1.9996	14.4

2.2.20.2 Ternary Mixtures

Mixture 1: CO_2 (1) + [emim][TFA] (2) + [emim][OAc] (3)
(1) Solubility

Ref. [109]

T/K	*P*/MPa	$100x_1$	*T*/K	*P*/MPa	$100x_1$	*T*/K	*P*/MPa	$100x_1$
298.1	0.0101	10.6	323.1	0.0100	6.3	348.1	0.0102	4.5
298.1	0.0501	14.6	323.1	0.0500	10.5	348.1	0.0501	8.1
298.1	0.1001	16.4	323.1	0.1001	12.4	348.1	0.1002	9.9
298.1	0.3997	21.5	323.1	0.4000	17.1	348.1	0.4000	14.1
298.2	0.6996	25.0	323.1	0.6996	19.9	348.1	0.7000	16.0
298.1	0.9997	28.2	323.1	0.9996	22.3	348.2	0.9998	17.6
298.1	1.2996	31.0	323.1	1.2998	24.6	348.0	1.2997	18.9
298.1	1.4995	32.7	323.1	1.4998	25.8	348.1	1.4998	19.6
298.1	1.9995	37.1	323.1	1.9996	29.1	348.1	1.9995	22.3

Ionic liquid mixture with a fixed mole ratio of 49.98 ([emim][Ac])/50.02 ([emim][TFA]).

2.2.21 02003-06003: 1-Ethyl-3-methylimidazolium acetate

Abbreviation:	[C_2MIm][OAc], [emim][OAc], [EMIM][OAc]
Molecular Formula:	$C_8H_{14}N_2O_2$
Molar Mass:	170.21
Structure:	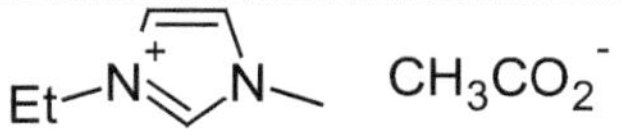

2.2.21.1 Binary Mixtures

Mixture 1: NH_3 (1) + [emim][OAc] (2)
(1) Solubility

Ref. [87]

T (K)	*P* (MPa)	$100x_1$ (mol%)	*T* (K)	*P* (MPa)	$100x_1$ (mol%)
282.5	0.321	62.4 ± 1.2	324.5	1.568	79.5 ± 1.6
282.5	0.435	74.9 ± 0.8	324.5	1.633	81.4 ± 1.6
282.5	0.488	80.2 ± 0.4	324.5	1.774	85.2 ± 1.2
282.5	0.525	83.4 ± 0.4	348.5	1.098	47.3 ± 6.8
282.5	0.535	84.7 ± 0.4	348.5	1.710	62.0 ± 6.0
282.5	0.550	87.7 ± 0.4	348.5	2.134	69.4 ± 5.2
298.3	0.470	59.9 ± 2.0	348.5	2.423	75.1 ± 4.0
298.3	0.667	73.0 ± 1.2	348.5	2.569	77.3 ± 3.6
298.3	0.765	78.8 ± 0.8	348.5	2.891	81.9 ± 2.8

(*continued*)

T (K)	P (MPa)	$100x_1$ (mol%)	T (K)	P (MPa)	$100x_1$ (mol%)
298.3	0.820	82.5 ± 0.8	298.2	0.463	60.1 ± 2.0
298.3	0.850	83.9 ± 0.8	298.2	0.662	73.1 ± 1.2
298.3	0.898	87.1 ± 0.4	298.2	0.759	78.9 ± 0.8
324.5	0.792	53.8 ± 4.0	298.2	0.818	82.5 ± 0.8
324.5	1.178	68.3 ± 3.2	298.2	0.845	83.9 ± 0.8
324.5	1.420	75.0 ± 2.4	298.2	0.896	87.1 ± 0.4

Mixture 2: [emim][TFA] (1) + [emim][Ac] (2)
(1) Volumetric Properties

Ref. [109]
These physicochemical properties data are listed in a mixture 2 of IL 02003-06001.
Mixture 3: CO_2 (1) + [emim][OAc] (2)
(1) Solubility

Ref. [109]

T/K	P/MPa	$100x_1$	T/K	P/MPa	$100x_1$	T/K	P/MPa	$100x_1$
298.1	0.0100	18.9	323.1	0.0100	13.8	348.1	0.0101	9.4
298.1	0.0499	24.6	323.1	0.0499	20.3	348.1	0.0501	15.7
298.1	0.1000	26.7	323.1	0.1000	23.0	348.1	0.1002	18.6
298.1	0.3996	31.3	323.1	0.4000	28.5	348.1	0.3997	24.1
298.1	0.6995	34.0	323.1	0.7000	31.4	348.2	0.6996	26.4
298.1	0.9996	36.2	323.1	1.0001	33.6	348.1	0.9998	28.0
298.1	1.2998	38.4	323.1	1.2997	35.0	348.1	1.2996	29.4
298.1	1.4997	39.8	323.1	1.4998	36.1	348.2	1.4997	29.8
298.1	1.9998	42.8	323.1	1.9996	39.0	348.2	1.9996	32.1

2.2.21.2 Ternary Mixtures

Mixture 1: H_2O (1) + C_2H_5OH (2) + [emim][OAc] (3)
(1) Phase Equilibrium

Ref. [24]

x'_2 [a]	w_3 [a]	y_2	T/K	γ_1	γ_2	α_{21} [b]
p = 100 kPa						
0.9500	0.6003	0.9809	380.12	0.41	0.50	2.69
0.9500	0.5002	0.9771	369.43	0.66	0.65	2.24
0.9500	0.4001	0.9729	361.94	0.96	0.79	1.88
0.9500	0.3000	0.9687	356.85	1.28	0.91	1.62
0.9500	0.2000	0.9636	353.76	1.60	0.97	1.39
0.9500	0.0999	0.9567	351.99	1.98	1.00	1.16

[a]Compositions reported are of equilibrium liquid
[b]$\alpha_{2,1} = (y_2/x'_2)/(y_1/x'_1)$

Mixture 2: H_2O (1) + $(CH_3)_3COH$ (2) + [emim][OAc] (3)
(1) Phase Equilibrium
Ref. [25]

x'_2 [a]	w_3 [a]	y_2	T/K	γ_1	γ_2	α_{21} [b]
p = 100 kPa						
0.9501	0.6026	0.9900	385.15	0.21	0.59	5.19
0.9500	0.5499	0.9879	379.12	0.29	0.66	4.31
0.9500	0.5000	0.9863	374.28	0.37	0.73	3.80
0.9500	0.4501	0.9842	369.99	0.47	0.80	3.28
0.9500	0.4001	0.9817	366.38	0.59	0.87	2.83
0.9500	0.3001	0.9758	361.13	0.88	0.96	2.12
0.9503	0.2000	0.9680	357.90	1.24	1.01	1.58
0.9501	0.1000	0.9537	355.96	1.83	1.02	1.08
0.9500	0.0499	0.9366	355.09	2.53	1.01	0.78

[a]Compositions reported are of equilibrium liquid
[b]$\alpha_{2,1} = (y_2/x'_2)/(y_1/x'_1)$

Mixture 3: CO_2 (1) + [emim][TFA] (2) + [emim][OAc] (3)
(1) Solubility

Ref. [109]
These physicochemical properties data are listed in a mixture 1 of IL 02003-06001.

2.2.22 02003-15003: 1-Ethyl-3-methyl imidazolium thiocyanate

Abbreviation:	[EMIM][SCN], [emim][SCN]
Molecular Formula:	$C_7H_{11}N_3S$
Molar Mass:	169.25
Structure:	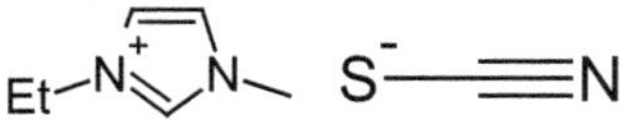

2.2.22.1 Binary Mixtures

Mixture 1: NH_3 (1) + [emim][SCN] (2)
(1) Solubility

Ref. [87]

T (K)	P (MPa)	$100x_1$ (mol %)	T (K)	P (MPa)	$100x_1$ (mol %)
283.2	0.244	45.1 ± 0.7	348.0	0.840	37.8 ± 2.7
283.2	0.364	65.2 ± 0.5	348.0	1.553	58.1 ± 2.6
283.2	0.447	73.1 ± 0.4	348.0	2.045	67.3 ± 2.2
283.2	0.502	78.6 ± 0.2	348.0	2.419	74.1 ± 1.7
283.2	0.547	81.9 ± 0.2	348.0	2.711	78.4 ± 1.4
283.2	0.590	87.6 ± 0.1	348.0	3.174	85.8 ± 0.8

(*continued*)

T (K)	P (MPa)	$100x_1$ (mol %)	T (K)	P (MPa)	$100x_1$ (mol %)
298.1	0.307	44.4 ± 0.9	372.8	1.149	34.0 ± 4.4
298.1	0.536	64.2 ± 0.7	372.8	2.144	54.2 ± 4.1
298.1	0.672	72.3 ± 0.5	372.8	2.958	63.3 ± 3.5
298.1	0.747	78.1 ± 0.4	372.8	3.576	70.8 ± 3.2
298.1	0.815	81.5 ± 0.3	372.8	4.120	75.4 ± 2.8
298.1	0.911	87.4 ± 0.1	372.8	5.007	83.9 ± 1.7
322.6	0.535	41.6 ± 1.6	298.1	0.314	44.3 ± 0.9
322.6	0.961	61.8 ± 1.4	298.1	0.540	64.2 ± 0.7
322.6	1.241	70.4 ± 1.1	298.1	0.666	72.4 ± 0.5
322.6	1.420	76.6 ± 0.8	298.1	0.772	78.0 ± 0.4
322.6	1.562	80.4 ± 0.6	298.1	0.831	81.5 ± 0.3
322.6	1.777	86.9 ± 0.3	298.1	0.930	87.4 ± 0.1

Mixture 2: SCH=CHCH=CH (thiophene) (1) + [EMIM][SCN] (2)
(1) Phase Equilibrium

Ref. [110]

x_2	T^{LLE}/K	x_2	T^{LLE}/K
0.3795	355.3	0.3147	301.5
0.3750	351.1	0.3094	296.8
0.3708	347.8	0.3031	291.1
0.3665	344.2	0.2967	287.1
0.3629	341.6	0.2942	283.6
0.3601	338.0	0.2881	279.1
0.3525	334.3	0.2878	278.2
0.3479	330.2	0.2812	271.6
0.3429	326.0	0.2757	265.8
0.3376	321.2	0.2663	257.8
0.3323	317.6	0.2585	251.6
0.3246	309.9	0.2462	242.9
0.3210	306.7	0.2388	236.6
0.3171	303.9		

2.2.23 02007-05012: 1-Ethyl-3-ethylimidazolium diethylphosphate

Abbreviation:	[EEIM][DEP], [DEIM][DEP]
Molecular Formula:	$C_{11}H_{23}N_2O_4P$
Molar Mass:	278.28
Structure:	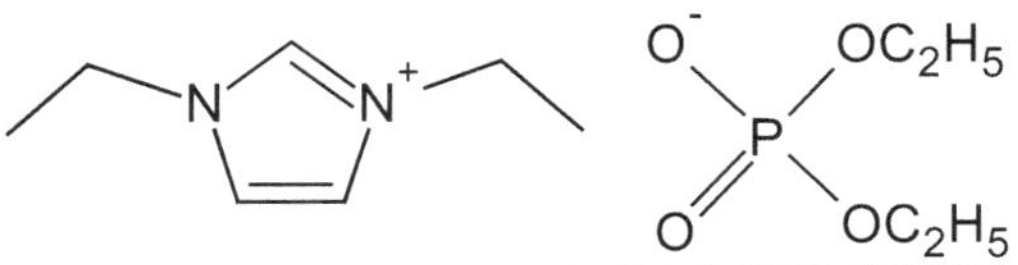

2.2.23.1 Binary Mixtures

Mixture 1: H_2O (1) + [EEIM][DEP] (2)

(1) Vapor Pressure

Ref. [111]

T/K	P^{exp}/kPa	P^{NRTL}/kPa	γ_1^{exp}	γ_1^{NRTL}
$x_1 = 0.9929$				
317.63	9.313	9.230	1.0078	0.9989
323.32	12.417	12.310	1.0077	0.9990
330.04	17.104	17.044	1.0026	0.9991
337.05	23.601	23.556	1.0001	0.9991
343.15	30.859	30.842	0.9997	0.9992
348.44	38.586	38.632	0.9980	0.9992
353.97	48.377	48.476	0.9972	0.9992
358.10	57.075	57.141	0.9981	0.9993
$x_1 = 0.9734$				
318.64	9.352	9.413	0.9801	0.9865
328.63	15.336	15.447	0.9806	0.9877
333.95	19.706	19.831	0.9821	0.9883
338.47	24.119	24.357	0.9791	0.9887
344.41	31.311	31.612	0.9799	0.9893
349.69	38.853	39.518	0.9731	0.9898
354.52	47.442	48.144	0.9757	0.9902
358.11	54.915	55.562	0.9789	0.9905
$x_1 = 0.9392$				
322.37	9.999	10.467	0.8986	0.9417
328.73	13.708	14.328	0.9039	0.9448
334.78	18.237	19.056	0.9068	0.9475
340.58	23.740	24.793	0.9096	0.9499
346.28	30.475	31.791	0.9127	0.9522
350.76	36.793	38.402	0.9138	0.9538
355.06	43.825	45.803	0.9140	0.9553
359.05	51.389	53.889	0.9151	0.9597
$x_1 = 0.7348$				
335.67	9.576	9.461	0.5844	0.5773
343.03	13.588	13.439	0.5978	0.5913
348.57	17.498	17.298	0.6083	0.6014
354.81	22.917	22.737	0.6172	0.6123
360.40	29.158	28.782	0.6298	0.6217
366.55	37.269	36.949	0.6372	0.6317
371.31	44.949	44.547	0.6449	0.6392

$$\mathrm{ARD(P)} = 0.018 = \frac{\sum_{i=1}^{n} \left|p^{NRTL} - p^{\exp}\right| / p^{\exp}}{\mathrm{n}};$$

$$\mathrm{RMSD} = 0.024 = \sqrt{\frac{\sum_{i=1}^{n} \left(p^{cal}/p^{\exp} - 1\right)^2}{n}}$$

Mixture 2: CH_3OH (1) + [EEIM][DEP] (2)
(1) Vapor Pressure

Ref. [111]

T/K	P^{exp}/kPa	P^{NRTL}/kPa	γ_1^{exp}	γ_1^{NRTL}
$x_1 = 0.9874$				
299.20	17.386	17.531	0.9845	0.9927
302.90	21.011	21.169	0.9856	0.9931
306.70	25.347	25.551	0.9855	0.9934
310.58	30.529	30.811	0.9846	0.9937
314.13	36.039	36.386	0.9845	0.9940
317.92	42.869	43.277	0.9848	0.9942
321.32	49.884	50.354	0.9851	0.9944
324.75	57.891	58.437	0.9854	0.9946
$x_1 = 0.9530$				
300.90	17.011	17.166	0.9148	0.9232
303.56	19.485	19.680	0.9162	0.9254
307.06	23.240	23.448	0.9199	0.9282
311.51	28.841	29.124	0.9224	0.9315
315.46	34.748	35.089	0.9252	0.9342
319.41	41.695	42.066	0.9286	0.9368
323.44	49.950	50.342	0.9320	0.9394
326.85	57.898	58.381	0.9336	0.9414
$x_1 = 0.8968$				
304.41	15.977	15.894	0.7652	0.7613
309.04	20.320	20.158	0.7766	0.7704
312.75	24.418	24.241	0.7831	0.7774
316.39	29.051	28.916	0.7876	0.7839
320.25	34.773	34.670	0.7930	0.7906
324.51	42.231	42.132	0.7995	0.7977
327.72	48.525	48.612	0.8013	0.8027
331.26	56.515	56.688	0.8057	0.8082
$x_1 = 0.6052$				
330.07	14.336	14.311	0.3180	0.3175
335.85	18.745	18.732	0.3289	0.3287
341.15	23.771	23.767	0.3389	0.3388
346.66	29.985	30.169	0.3471	0.3492
351.65	37.048	37.167	0.3574	0.3585

(continued)

T/K	P^{exp}/kPa	P^{NRTL}/kPa	γ_1^{exp}	γ_1^{NRTL}
356.22	44.985	44.720	0.3691	0.3669
361.20	54.429	54.407	0.3761	0.3759
366.19	65.452	65.779	0.3829	0.3848

$$\mathrm{ARD(P)} = 0.0063 = \frac{\sum_{i=1}^{n} \left|p^{NRTL} - p^{\exp}\right| / p^{\exp}}{\mathrm{n}};$$

$$\mathrm{RMSD} = 0.0071 = \sqrt{\frac{\sum_{i=1}^{n} \left(p^{cal}/p^{\exp} - 1\right)^2}{n}}.$$

Mixture 3: C_2H_5OH (1) + [EEIM][DEP] (2)
(1) Vapor Pressure

Ref. [111]

T/K	P^{exp}/kPa	P^{NRTL}/kPa	γ_1^{exp}	γ_1^{NRTL}
$x_1 = 0.9817$				
309.96	14.808	14.838	0.9955	0.9975
314.00	18.301	18.327	0.9961	0.9976
319.06	23.625	23.654	0.9964	0.9976
323.19	28.909	28.942	0.9965	0.9977
327.26	35.060	35.097	0.9966	0.9977
330.78	41.264	41.282	0.9973	0.9977
334.37	48.506	48.502	0.9978	0.9978
337.91	56.716	56.666	0.9987	0.9978
$x_1 = 0.9338$				
309.80	13.490	13.585	0.9617	0.9685
313.92	16.741	16.857	0.9623	0.9690
318.49	21.170	21.262	0.9653	0.9696
321.24	24.276	24.354	0.9668	0.9699
325.40	29.674	29.749	0.9679	0.9704
329.03	35.216	35.260	0.9696	0.9708
333.54	43.290	43.289	0.9713	0.9713
338.72	54.395	54.357	0.9725	0.9718
$x_1 = 0.8580$				
310.59	11.648	11.637	0.8669	0.8661
316.24	15.641	15.632	0.8693	0.8688
318.57	17.646	17.596	0.8724	0.8700
323.93	22.978	22.918	0.8747	0.8724
329.23	29.635	29.465	0.8799	0.8748
333.85	36.689	36.410	0.8835	0.8768
339.15	46.313	46.022	0.8845	0.8790
342.93	54.503	54.123	0.8867	0.8805

(continued)

T/K	P^{exp}/kPa	P^{NRTL}/kPa	γ_1^{exp}	γ_1^{NRTL}
$x_1 = 0.5151$				
344.56	9.139	9.175	0.2382	0.2391
349.59	11.648	11.684	0.2442	0.2450
355.84	15.641	15.686	0.2516	0.2523
359.82	18.686	18.709	0.2565	0.2568
365.11	23.205	23.222	0.2622	0.2624
370.80	29.875	29.536	0.2719	0.2688

$$\text{ARD(P)} = 0.0030 = \frac{\sum_{i=1}^{n} \left|p^{NRTL} - p^{\exp}\right| / p^{\exp}}{\text{n}};$$

$$\text{RMSD} = 0.0041 = \sqrt{\frac{\sum_{i=1}^{n} \left(p^{cal}/p^{\exp} - 1\right)^2}{n}}.$$

2.2.23.2 Ternary Mixtures

Mixture 1: H_2O (1) + CH_3OH (2) + [EEIM][DEP] (3)

(1) Vapor Pressure

Ref. [111]

T/K	P^{exp}/kPa	P^{NRTL}/kPa	γ_1^{NRTL}	γ_2^{NRTL}
$x_1 = 0.8508, x_2 = 0.1196$				
319.03	16.216	14.376	0.9973	1.0622
321.86	18.498	16.477	0.9975	1.0635
325.97	22.407	19.994	0.9977	1.0653
331.12	28.258	25.288	0.9980	1.0673
333.80	31.795	28.484	0.9982	1.0682
336.92	36.326	32.632	0.9983	1.0693
344.29	49.295	44.504	0.9987	1.0715
347.46	55.865	50.613	0.9988	1.0723
$x_1 = 0.7036, x_2 = 0.2639$				
312.90	16.315	14.579	1.0177	1.0149
315.25	18.794	16.360	1.0177	1.0162
317.88	21.302	18.566	1.0176	1.0177
321.85	25.687	22.382	1.0176	1.0198
326.63	31.881	27.836	1.0175	1.0221
331.21	38.869	34.097	1.0174	1.0242
335.99	47.656	41.838	1.0173	1.0262
340.23	56.413	49.911	1.0171	1.0278
$x_1 = 0.5228, x_2 = 0.4009$				
309.03	17.437	15.768	1.0524	0.9742
312.89	21.098	19.064	1.0520	0.9767
315.76	24.248	21.888	1.0517	0.9785
319.76	29.422	26.404	1.0512	0.9808

(*continued*)

T/K	p^{exp}/kPa	p^{NRTL}/kPa	γ_1^{NRTL}	γ_2^{NRTL}
323.72	35.377	31.629	1.0507	0.9830
328.18	43.197	38.529	1.0502	0.9853
331.26	49.418	44.024	1.0498	0.9868
334.76	57.313	51.029	1.0494	0.9884
$x_1 = 0.1577, x_2 = 0.7985$				
307.13	21.296	20.797	1.1463	0.9341
309.27	23.689	23.115	1.1458	0.9357
312.25	27.356	26.708	1.1450	0.9378
316.21	33.057	32.188	1.1440	0.9404
319.98	39.246	38.279	1.1429	0.9427
324.03	46.915	45.863	1.1418	0.9451
327.26	53.915	52.786	1.1409	0.9470
330.29	61.177	60.056	1.1400	0.9486

$$\mathrm{ARD(P)} = 0.088 = \frac{\sum_{i=1}^{n} \left|p^{NRTL} - p^{\exp}\right|/p^{\exp}}{\mathrm{n}};$$

$$\mathrm{RMSD} = 0.096 = \sqrt{\frac{\sum_{i=1}^{n} \left(p^{cal}/p^{\exp} - 1\right)^2}{n}}.$$

Mixture 2: H_2O (1) + C_2H_5OH (2) + [EEIM][DEP] (3)
(1) Vapor Pressure

Ref. [111]

T/K	p^{exp}/kPa	p^{NRTL}/kPa	γ_1^{NRTL}	γ_2^{NRTL}
$x_1 = 0.8830, x_2 = 0.0864$				
318.19	14.564	14.999	1.0243	3.1714
322.00	17.518	18.082	1.0243	3.1402
326.20	21.573	22.082	1.0243	3.1066
331.03	26.931	27.589	1.0243	3.0690
335.94	33.593	34.334	1.0243	3.0320
341.98	43.663	44.493	1.0242	2.9879
345.69	50.996	51.910	1.0241	2.9617
349.53	59.606	60.651	1.0240	2.9352
$x_1 = 0.7650, x_2 = 0.1996$				
317.10	16.644	16.678	1.1202	2.0404
320.63	19.846	19.876	1.1194	2.0316
324.60	24.038	24.064	1.1185	2.0219
328.64	29.042	29.087	1.1176	2.0121
333.33	36.163	35.999	1.1165	2.0009
337.06	42.591	42.432	1.1156	1.9921
340.80	50.190	49.840	1.1147	1.9835
343.73	56.853	56.383	1.1140	1.9767

(*continued*)

T/K	P^{exp}/kPa	P^{NRTL}/kPa	γ_1^{NRTL}	γ_2^{NRTL}
$x_1 = 0.6038\ x_2 = 0.3544$				
316.89	18.226	17.955	1.3024	1.4202
321.59	22.792	22.675	1.3003	1.4171
323.17	24.589	24.475	1.2996	1.4160
325.70	27.811	27.634	1.2984	1.4144
329.80	33.722	33.464	1.2965	1.4117
333.27	39.564	39.169	1.2948	1.4095
336.55	45.788	45.317	1.2932	1.4074
341.03	56.173	55.003	1.2910	1.4047
$x_1 = 0.2193,\ x_2 = 0.7233$				
316.49	19.076	19.188	1.8177	1.0211
319.28	21.888	22.064	1.8192	1.0209
321.07	23.916	24.112	1.8200	1.0207
324.24	27.909	28.125	1.8212	1.0205
328.47	34.097	34.343	1.8225	1.0201
332.31	40.629	40.959	1.8232	1.0198
336.77	49.568	49.968	1.8237	1.0195
340.23	57.584	58.063	1.8237	1.0193

$$\text{ARD(P)} = 0.011 = \frac{\sum_{i=1}^{n} \left|p^{NRTL} - p^{\exp}\right| / p^{\exp}}{\text{n}};$$

$$\text{RMSD} = 0.014 = \sqrt{\frac{\sum_{i=1}^{n} \left(p^{cal}/p^{\exp} - 1\right)^2}{n}}.$$

Mixture 3: C_2H_5OH (1) + CH_3OH (2) + [EEIM][DEP] (3)
(1) Vapor Pressure

Ref. [111]

T/K	P^{exp}/kPa	P^{NRTL}/kPa	γ_1^{NRTL}	γ_2^{NRTL}
$x_1 = 0.6906,\ x_2 = 0.2482$				
309.63	16.265	16.313	0.9882	0.8265
313.28	19.559	19.674	0.9877	0.8344
317.59	24.189	24.396	0.9871	0.8434
321.83	29.629	29.933	0.9865	0.8519
326.41	36.675	37.088	0.9859	0.8605
330.81	44.732	45.268	0.9854	0.8685
333.61	50.576	51.227	0.9851	0.8733
336.58	57.492	58.258	0.9848	0.8783
$x_1 = 0.4816,\ x_2 = 0.4615$				
308.12	17.495	17.847	0.9748	0.8852
311.07	20.305	20.742	0.9745	0.8899
314.95	24.571	25.135	0.9740	0.8958
318.99	29.839	30.534	0.9735	0.9016

(*continued*)

T/K	P^{exp}/kPa	P^{NRTL}/kPa	γ_1^{NRTL}	γ_2^{NRTL}
324.00	37.620	38.550	0.9729	0.9085
327.58	44.184	45.328	0.9726	0.9131
330.69	50.560	51.993	0.9722	0.9170
$x_1 = 0.3000, x_2 = 0.6467$				
306.10	18.527	18.461	0.9422	0.9155
309.67	22.007	22.101	0.9424	0.9197
313.06	25.955	26.108	0.9426	0.9235
316.01	29.856	30.086	0.9428	0.9267
319.73	35.468	35.820	0.9430	0.9305
323.47	41.970	42.491	0.9432	0.9341
327.63	50.343	51.119	0.9435	0.9379
330.81	57.627	58.651	0.9436	0.9407
$x_1 = 0.0682, x_2 = 0.8834$				
303.01	18.507	18.396	0.8751	0.9270
307.02	22.574	22.529	0.8772	0.9304
310.03	26.132	26.120	0.8787	0.9328
313.31	30.532	30.581	0.8803	0.9353
316.82	35.919	36.045	0.8820	0.9379
321.04	43.427	43.698	0.8840	0.9408
324.82	51.215	51.665	0.8857	0.9433
328.01	58.696	59.308	0.8871	0.9453

$$\mathrm{ARD(P)} = 0.012 = \frac{\sum_{i=1}^{n} \left|p^{NRTL} - p^{\exp}\right| / p^{\exp}}{\mathrm{n}};$$

$$\mathrm{RMSD} = 0.013 = \sqrt{\frac{\sum_{i=1}^{n} \left(p^{cal}/p^{\exp} - 1\right)^2}{n}}$$

2.2.24 02008-01002: 1-Propyl-3-methylimidazolium bromide

Abbreviation:	$[C_3MIm]Br$, [PMIm]Br
Molecular Formula:	$C_7H_{13}N_2Br$
Molar Mass:	205.09
Structure:	

2.2.24.1 Binary Mixtures

Mixture 1: H_2O (1) + [PMIm]Br (2)

(1) Osmotic Coefficient

Ref. [112]

	Φ			
m_2/(mol · kg^{-1})	298.15 K	308.15 K	318.15 K	328.15 K
0.0071	0.942	0.928	0.961	0.947
0.0141	0.917	0.925	0.951	0.950
0.0243	0.903	0.902	0.925	0.924
0.0338	0.894	0.904	0.920	0.926
0.0462	0.890	0.892	0.910	0.910
0.0667	0.865	0.875	0.898	0.896
0.0898	0.855	0.864	0.874	0.875
0.1081	0.839	0.851	0.867	0.864
0.1529	0.820	0.835	0.857	0.851
0.1940	0.807	0.816	0.836	0.842
0.2403	0.798	0.802	0.812	0.818
0.3018	0.783	0.792	0.805	0.813
0.3677	0.762	0.772	0.793	0.794
0.4338	0.755	0.761	0.776	0.787
0.5359	0.742	0.749	0.753	0.760
0.8595	0.706	0.720	0.728	0.735
1.0292	0.702	0.714	0.717	0.724
1.6039	0.674	0.696	0.702	0.704
2.1041	0.676	0.690	0.694	0.701
2.7049	0.671	0.683	0.701	0.709

Mixture 2: CH_3CN (1) + [C_3mim]Br (2)
(1) Volumetric Properties and Speed of Sound

Ref. [113]

	$T = 288.15$ K		$T = 293.15$ K		$T = 298.15$ K		$T = 303.15$ K		$T = 308.15$ K	
m_2 (mol · kg^{-1})	ρ (g · cm^{-3})	u (m · s^{-1})	ρ (g · cm^{-3})	u (m · s^{-1})	ρ (g · cm^{-3})	u (m · s^{-1})	ρ (g · cm^{-3})	u (m · s^{-1})	ρ (g · cm^{-3})	u (m · s^{-1})
0.0000	0.787376	1319.14	0.782007	1299.01	0.776609	1278.77	0.771183	1258.51	0.765725	1238.40
0.0204	0.788991	1320.25	0.783623	1300.10	0.778237	1279.97	0.772813	1259.67	0.767370	1239.62
0.0299	0.789726	1320.51	0.784364	1300.31	0.778979	1280.17	0.773562	1260.03	0.768119	1240.08
0.0404	0.790532	1321.07	0.785181	1301.12	0.779798	1281.00	0.774387	1260.87	0.768949	1240.79
0.0526	0.791470	1321.50	0.786115	1301.34	0.780735	1281.20	0.775331	1261.14	0.769896	1241.22
0.0611	0.792123	1321.96	0.786772	1301.98	0.781397	1281.87	0.775996	1261.79	0.770564	1241.73
0.0796	0.793533	1322.59	0.788186	1302.46	0.782814	1282.36	0.777417	1262.34	0.771996	1242.45
0.1027	0.795242	1323.52	0.789902	1303.43	0.784538	1283.37	0.779149	1263.39	0.773734	1243.54
0.1256	0.796939	1324.42	0.791606	1304.34	0.786248	1284.33	0.780864	1264.39	0.775460	1244.56
0.1496	0.798688	1325.32	0.793361	1305.29	0.788011	1285.31	0.782637	1265.38	0.777238	1245.61
0.1753	0.800578	1326.40	0.795262	1306.55	0.789919	1286.59	0.784553	1266.67	0.779160	1246.79
0.2069	0.802812	1327.35	0.797500	1307.38	0.792166	1287.47	0.786808	1267.64	0.781426	1247.96
0.2527	0.806071	1329.07	0.800772	1309.30	0.795452	1289.44	0.790107	1269.61	0.784738	1249.83
0.3139	0.810335	1330.95	0.805049	1311.13	0.799743	1291.33	0.794412	1271.62	0.789059	1252.07
[C_3mim][Br]	1.323470	1699.30	1.319844	1685.03	1.316223	1672.04	1.312600	1659.90	1.308975	1648.33

Mixture 3: CH_3SOCH_3 (1) + $[C_3mim]Br$ (2)
(1) Volumetric Properties and Speed of Sound

Ref. [113]

	T = 288.15 K		T = 293.15 K		T = 298.15 K		T = 303.15 K		T = 308.15 K	
m_2 (mol · kg^{-1})	ρ (g · cm^{-3})	u (m · s^{-1})	ρ (g · cm^{-3})	u (m · s^{-1})	ρ (g · cm^{-3})	u (m · s^{-1})	ρ (g · cm^{-3})	u (m · s^{-1})	ρ (g · cm^{-3})	u (m · s^{-1})
0.0000	1.105336	1519.12	1.100306	1502.19	1.095278	1485.21	1.090253	1468.32	1.085231	1451.47
0.0380	1.107073	1520.94	1.102058	1504.17	1.097047	1487.28	1.092039	1470.41	1.087033	1453.62
0.0501	1.107623	1521.50	1.102611	1504.70	1.097603	1487.78	1.092600	1470.93	1.087600	1454.16
0.0594	1.108045	1521.87	1.103040	1505.18	1.098036	1488.28	1.093034	1471.43	1.088033	1454.71
0.0788	1.108910	1523.00	1.103907	1506.00	1.098912	1489.16	1.093917	1472.33	1.088929	1455.72
0.1011	1.109879	1523.78	1.104885	1507.04	1.099893	1490.21	1.094907	1473.38	1.089926	1456.68
0.1261	1.110967	1524.81	1.105976	1507.80	1.100993	1490.97	1.096017	1474.22	1.091044	1457.64
0.1490	1.111952	1525.76	1.106976	1509.09	1.102003	1492.31	1.097033	1475.55	1.092065	1458.86
0.1799	1.113266	1527.05	1.108298	1510.38	1.103331	1493.60	1.098374	1476.86	1.093417	1460.27
0.2152	1.114736	1528.50	1.109775	1511.62	1.104820	1494.86	1.099870	1478.20	1.094928	1461.69
0.2578	1.116492	1530.18	1.111547	1513.54	1.106603	1496.88	1.101667	1480.18	1.096735	1463.63
0.3109	1.118621	1532.12	1.113688	1515.32	1.108760	1498.68	1.103839	1482.06	1.098928	1465.69
$[C_3mim][Br]$	1.323470	1699.30	1.319844	1685.03	1.316223	1672.04	1.312600	1659.90	1.308975	1648.33

Mixture 4: $HCON(CH_3)_2$ (1) + [C_3mim]Br (2)
(1) Volumetric Properties and Speed of Sound

Ref. [113]

	$T = 288.15$ K		$T = 293.15$ K		$T = 298.15$ K		$T = 303.15$ K		$T = 308.15$ K	
m_2 (mol · kg^{-1})	ρ (g · cm^{-3})	u (m · s^{-1})	ρ (g · cm^{-3})	u (m · s^{-1})	ρ (g · cm^{-3})	u (m · s^{-1})	ρ (g · cm^{-3})	u (m · s^{-1})	ρ (g · cm^{-3})	u (m · s^{-1})
0.0000	0.953442	1496.60	0.948685	1477.27	0.943917	1457.81	0.939142	1438.38	0.934357	1419.04
0.0295	0.955471	1498.44	0.950721	1479.23	0.945962	1459.82	0.941197	1440.48	0.936424	1421.19
0.0358	0.955896	1498.74	0.951148	1479.51	0.946392	1460.11	0.941628	1440.79	0.936859	1421.49
0.0497	0.956834	1499.53	0.952089	1480.17	0.947339	1460.80	0.942580	1441.44	0.937812	1422.18
0.0591	0.957413	1499.80	0.952669	1480.29	0.947920	1460.90	0.943162	1441.61	0.938398	1422.45
0.0801	0.958776	1500.84	0.954039	1481.59	0.949295	1462.19	0.944542	1442.92	0.939782	1423.69
0.1018	0.960154	1501.65	0.955416	1482.21	0.950675	1462.88	0.945932	1443.63	0.941175	1424.51
0.1244	0.961605	1502.70	0.956878	1483.53	0.952144	1464.24	0.947403	1444.99	0.942652	1425.82
0.1525	0.963324	1503.73	0.958602	1484.40	0.953872	1465.10	0.949136	1445.92	0.944395	1426.91
0.1759	0.964790	1504.80	0.960068	1485.44	0.955349	1466.20	0.950617	1447.05	0.945882	1428.11
0.2058	0.966570	1505.78	0.961858	1486.49	0.957139	1467.29	0.952415	1448.16	0.947684	1429.18
0.2585	0.969758	1507.84	0.965060	1488.80	0.960356	1469.65	0.955641	1450.55	0.950923	1431.56
0.3035	0.972333	1509.26	0.967639	1490.09	0.962940	1470.97	0.958234	1451.96	0.953523	1433.10
[C3mim][Br]	1.323470	1699.30	1.319844	1685.03	1.316223	1672.04	1.312600	1659.90	1.308975	1648.33

2.2.25 02008-01003: 1-Propyl-3-methylimidazolium iodine

Abbreviation:	[C_3MIm]I, [MPIM]I
Molecular Formula:	$C_7H_{13}IN_2$
Molar Mass:	252.1
Structure:	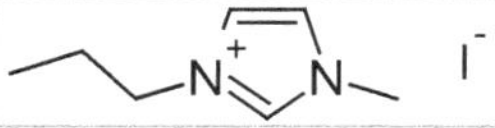

2.2.25.1 Ternary Mixtures

Mixture 1: I_2 (1) + [EMIM][BF_4] (2) + [MPIM]I (3)

(1) Volumetric Properties, Transport Properties and Electrical Conductivity

Ref. [40]

These physicochemical properties data are listed in a mixture 7 of IL 02003–02001.

(2) Transport Properties

Ref. [40]

These physicochemical properties data are listed in a mixture 7 of IL 02003-02001.

2.2.26 02008-03001: 1-Propyl-3-methylimidazolium bis((trifluoromethyl)sulfonyl)imide

Abbreviation:	[C_3MIm][NTf_2], [C_3MIm][Tf_2N], [C_3MIm][TFSI], [C_3mim][Tf_2N]
Molecular Formula:	$C_9H_{13}F_6N_3O_4S_2$
Molar Mass:	405.34
Structure:	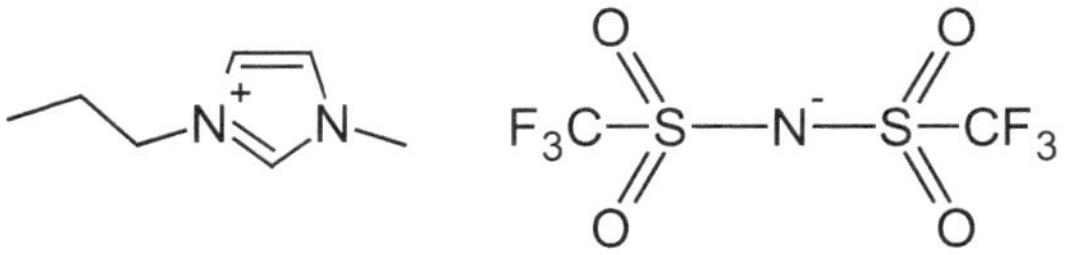

2.2.26.1 Binary Mixtures

Mixture 1: H_2O (1) + [C_3mim][Tf_2N] (2)

(1) Solubility

Ref. [45]

T/K	$x_1 \pm \sigma^a$	$10^4 (x_2 \pm \sigma^a)$
288.15	0.2502 ± 0.0004	5.15 ± 0.05
293.15	0.2581 ± 0.0004	5.27 ± 0.01
298.15	0.2715 ± 0.0007	5.40 ± 0.01
303.15	0.2861 ± 0.0009	5.68 ± 0.01
308.15	0.3013 ± 0.0007	6.04 ± 0.03
313.15	0.3172 ± 0.0005	6.49 ± 0.05
318.15	0.3324 ± 0.0007	7.01 ± 0.05

[a]Standard deviation

(2) Thermodynamic Properties

Ref. [45]

$(\Delta_{sol}H_m{}^0 \pm \sigma^a)$/ $kJ \cdot mol^{-1}$	$(\Delta_{sol}G_m{}^0 \pm \sigma^a)$/ $kJ \cdot mol^{-1}$	$(\Delta_{sol}S_m{}^0 \pm \sigma^a)$/ $J \cdot K^{-1} \cdot mol^{-1}$	$(\Delta_{svt}H_m{}^0 \pm \sigma^a)$/ $kJ \cdot mol^{-1}$	$(\Delta_{svt}Gm^0 \pm \sigma^a)$/ $kJ \cdot mol^{-1}$	$(\Delta_{svt}S_m{}^0 \pm \sigma^a)$/ $J \cdot K^{-1} \cdot mol^{-1}$	$(\Delta_{svt}H_m{}^* \pm \sigma^a)$/ $kJ \cdot mol^{-1}$	$(\Delta_{svt}G_m{}^* \pm \sigma^a)$/ $kJ \cdot mol^{-1}$	$(\Delta_{svt}S_m{}^* \pm \sigma^a)$/ $J \cdot K^{-1} \cdot mol^{-1}$
298.15 K								
5.9 ± 1.5	18.652 ± 0.001	−42.6 ± 5.0	−141.1 ± 6.2	−61.423 ± 0.001	−267 ± 21	−138.9 ± 6.2	−79.331 ± 0.001	−200 ± 21

[a]Standard deviation

2.2.27 02011-01001: 1-Butyl-3-methylimidazolium chloride

Abbreviation:	[C_4MIm]Cl, [BMIM]Cl, [bmim]Cl, [Bmim]Cl
Molecular Formula:	$C_8H_{15}N_2Cl$
Molar Mass:	174.67
Structure:	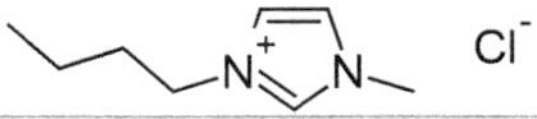

2.2.27.1 Binary Mixtures

Mixture 1: H_2O (1) + [Bmim]Cl (2)

(1) Volumetric Properties, Transport Properties, Refractive Index, Speed of Sound and Compressibility Properties

Ref. [1]

m_2/(mol · kg^{-1})	ρ/(kg · m^{-3})	u/(m · s^{-1})	$10^{-3} \cdot \varphi_V$/ (mm^3 · mol^{-1})	$10^{-3} \cdot \overline{V}_2$ /(mm^3 · mol^{-1})	$10^{-3} \cdot \overline{V}_1$/ (mm^3 · mol^{-1})	$10^{11} \cdot \beta_S$/ Pa^{-1}	φ_{K_S}/(mm^3 · MPa^{-1} · mol^{-1})
T = 298.15 K							
0.00000	997.05	1497.6	174.76[a]	–	18.068	44.72	−15.7[a]
0.05281	997.72	1504.8	175.06	175.05	18.068	44.26	−15.4 ± 3.4
0.09783	998.29	1510.5	174.96	174.85	18.068	43.91	−12.2 ± 1.8
0.14328	998.85	1516.7	174.86	174.63	18.069	43.52	−13.2 ± 1.2
0.19981	999.56	1524.4	174.73	174.33	18.069	43.05	−13.9 ± 0.9
0.24711	1000.15	1530.2	174.63	174.08	18.070	42.70	−12.7 ± 0.7
0.28987	1000.69	1535.5	174.54	173.85	18.072	42.39	−12.1 ± 0.6
0.31733	1001.04	1539.1	174.48	173.70	18.072	42.17	−12.3 ± 0.5
0.40515	1002.14	1549.9	174.29	173.22	18.076	41.54	−11.5 ± 0.4
0.42202	1002.35	1551.9	174.25	173.13	18.076	41.42	−11.4 ± 0.4
0.49947	1003.31	1562.2	174.08	172.70	18.080	40.84	−11.9 ± 0.3

[a]Extrapolated values to infinite dilution

Ref. [114]

m (mol · kg^{-1})	ρ/(g · cm^{-3})	u/(m · s^{-1})	κ_s/(T · Pa^{-1})
298.15 K			
447.62	0.1000	1.33251	1496.88
272.67	0.2213	1.52469	1838.29
275.58	0.3080	1.51959	1828.21
274.66	0.3852	1.51728	1831.31
270.73	0.5017	1.51027	1845.03
266.95	0.5976	1.50214	1859.61
264.18	0.6987	1.48926	1872.84
263.99	0.8012	1.46760	1881.46
276.86	0.8967	1.43021	1853.73
315.54	0.9504	1.39243	1754.68
447.62	1.0000	1.33251	1496.88

Ref. [115]

x_1	ρ/g · cm^{-3}	m	V_Φ	$10^3\eta$/Pa · s	x_1	ρ/g · cm^{-3}	m	V_Φ	$10^3\eta$/Pa · s
T = 298.15 K					T = 328.15 K				
0.5666	1.0810	42.46	159.77	123.5	0.5666	1.0634	42.46	162.53	30.04
0.6070	1.0799	35.94	159.62	72.16	0.6070	1.0620	35.94	162.46	19.75
0.6933	1.0765	24.56	159.26	35.66	0.6933	1.0581	24.56	162.27	11.32
0.7698	1.0711	16.60	158.91	18.08	0.7699	1.0521	16.60	162.18	6.474
0.8279	1.0638	11.53	158.76	10.19	0.8279	1.0445	11.53	162.29	4.019
0.8674	1.0571	8.489	158.54	6.628	0.8674	1.0373	8.489	162.46	2.795
0.9194	1.0423	4.868	158.65	3.543	0.9194	1.0233	4.868	163.05	1.635
0.9625	1.0223	2.163	159.42	1.885	0.9625	1.0061	2.163	164.11	0.955
0.9885	1.0052	0.648	161.21	1.175	0.9885	0.9893	0.648	170.85	0.753
T = 313.15 K					T = 343.15 K				
0.5666	1.0720	42.46	161.19	56.63	0.5666	1.0547	42.46	163.87	17.69
0.6070	1.0708	35.94	161.08	35.36	0.6070	1.0531	35.94	163.85	12.20
0.6933	1.0672	24.56	160.81	18.94	0.6933	1.0488	24.56	163.74	7.386
0.7698	1.0615	16.60	160.60	10.11	0.7699	1.0424	16.60	163.76	4.419
0.8279	1.0541	11.53	160.59	6.125	0.8279	1.0344	11.53	164.02	2.858
0.8674	1.0471	8.489	160.60	4.123	0.8674	1.0271	8.489	164.29	2.043
0.9194	1.0328	4.868	161.01	2.311	0.9194	1.0131	4.868	165.10	1.228
0.9625	1.0140	2.163	162.26	1.297	0.9625	0.9967	2.163	166.28	0.741
0.9885	0.9991	0.648	164.11	0.955	0.9885	0.9796	0.648	175.35	0.652

Ref. [116]

	n_D			
m_2/mol · kg^{-1}	T = 298.15 K	T = 308.15 K	T = 318.15 K	T = 328.15 K
0.0105	1.3330	1.3317	1.3292	1.3273
0.0230	1.3334	1.3320	1.3295	1.3275
0.0289	1.3335	1.3323	1.3296	1.3277
0.0398	1.3340	1.3326	1.3300	1.3280
0.0621	1.3343	1.3330	1.3305	1.3288
0.0803	1.3349	1.3336	1.3310	1.3290
0.0986	1.3354	1.3341	1.3312	1.3297
0.1413	1.3366	1.3351	1.3325	1.3309
0.1816	1.3377	1.3363	1.3338	1.3319
0.2223	1.3388	1.3373	1.3346	1.3331
0.2764	1.3404	1.3389	1.3362	1.3343
0.3386	1.3419	1.3403	1.3377	1.3358
0.3710	1.3428	1.3416	1.3387	1.3369
0.4863	1.3457	1.3441	1.3415	1.3398
0.5990	1.3484	1.3469	1.3442	1.3422
0.7152	1.3512	1.3497	1.3467	1.3449
0.9754	1.3571	1.3562	1.3528	1.3508
1.4915	1.3679	1.3661	1.3631	1.3611
1.9999	1.3774	1.3754	1.3723	1.3703
2.9617	1.3923	1.3904	1.3872	1.3849

(2) Osmotic Coefficient, Activity Properties, Vapor Pressure and Excess Gibbs Free-Energy

Ref. [5]

m/(mol · kg^{-1})	Φ_1	a_1	p/kPa	m/(mol · kg^{-1})	Φ_1	a_1	p/kPa
T = 313.15 K				T = 333.15 K			
0.0566	0.993	0.9980	7.362	0.0566	0.948	0.9981	19.886
0.3127	0.956	0.9893	7.298	0.3127	0.902	0.9899	19.722
0.6885	0.936	0.9770	7.207	0.6885	0.862	0.9789	19.501
1.0050	0.917	0.9673	7.136	1.0050	0.846	0.9698	19.320
1.2691	0.905	0.9595	7.078	1.2691	0.824	0.9630	19.183
1.4271	0.901	0.9547	7.043	1.4271	0.821	0.9587	19.097
1.7982	0.898	0.9435	6.959	1.7982	0.809	0.9489	18.902
1.9970	0.894	0.9377	6.916	1.9970	0.805	0.9438	18.798
2.2914	0.893	0.9289	6.852	2.2914	0.801	0.9360	18.642
2.6218	0.889	0.9194	6.782	2.6218	0.802	0.9271	18.464
2.9465	0.887	0.9101	6.712	2.9465	0.801	0.9185	18.293
3.1500	0.887	0.9042	6.669	3.1500	0.800	0.9132	18.187
3.5307	0.889	0.8931	6.586	3.5307	0.800	0.9033	17.988
4.0884	0.894	0.8766	6.465	4.0884	0.806	0.8880	17.682

Ref. [26]

m_2/mol · kg^{-1}	Φ	a_1	γ_1	$\gamma_\pm$	ΔG^E/J · mol^{-1}
T = 298.15 K					
0.00971	0.96757	0.99966	1.00001	0.89790	−0.13
0.02372	0.94340	0.99919	1.00005	0.85059	−0.45
0.04853	0.92702	0.99838	1.00013	0.80176	−1.29
0.07628	0.91697	0.99748	1.00022	0.76606	−2.51
0.10388	0.90993	0.99660	1.00033	0.73980	−3.94
0.15469	0.90032	0.99499	1.00054	0.70419	−6.97
0.20410	0.89309	0.99345	1.00076	0.67855	−10.30
0.26034	0.88626	0.99172	1.00102	0.65566	−14.42
0.31333	0.88068	0.99011	1.00128	0.63808	−18.57
0.36829	0.87551	0.98845	1.00157	0.62265	−23.10
0.42655	0.87054	0.98671	1.00187	0.60855	−28.13
0.48376	0.86605	0.98502	1.00219	0.59638	−33.25
0.54126	0.86185	0.98333	1.00251	0.58541	−38.59

Ref. [116]

	Φ			
m_2/mol · kg^{-1}	T = 298.15 K	T = 308.15 K	T = 318.15 K	T = 328.15 K
0.0105	1.004	0.948	0.958	0.978
0.0230	0.958	0.929	0.932	0.939
0.0289	0.932	0.924	0.935	0.938
0.0398	0.892	0.911	0.916	0.921
0.0621	0.855	0.866	0.881	0.906
0.0803	0.833	0.861	0.869	0.873
0.0986	0.827	0.864	0.865	0.867
0.1413	0.814	0.844	0.857	0.862
0.1816	0.789	0.817	0.823	0.832
0.2223	0.779	0.794	0.807	0.822
0.2764	0.772	0.790	0.797	0.810
0.3386	0.768	0.779	0.798	0.803
0.3710	0.764	0.775	0.784	0.790
0.4863	0.735	0.747	0.762	0.775
0.5990	0.718	0.738	0.750	0.759
0.7152	0.716	0.731	0.748	0.755
0.9754	0.699	0.709	0.723	0.733
1.4915	0.686	0.692	0.713	0.719
1.9999	0.686	0.694	0.706	0.721
2.9617	0.711	0.721	0.729	0.744

(3) Thermodynamic Properties

Ref. [26]

m_2/ $mol \cdot kg^{-1}$	$(\Delta H_{mix}$ or $H^E)^a$/ $J \cdot mol^{-1}$	ΔH_{mix} or H^E/ $J \cdot mol^{-1}$	Φ_L/ $J \cdot mol^{-1}$	$(T\Delta S)^E$/ $J \cdot mol^{-1} \cdot K^{-1}$	ΔG_{mix}/ $J \cdot mol^{-1}$	$T\Delta S_{mix}$/ $J \cdot mol^{-1}$	$(S_1 - S_1{}^0)^E$/ $J \cdot mol^{-1} \cdot K^{-1}$	$(S_2 - S_2{}^0)^E$/ $J \cdot mol^{-1}\ K^{-1}$
0.00971	7283	1.3	−8747	1.40	−15.66	16.93	0.00137	17.87
0.02372	5362	2.3	−10668	2.74	−34.57	36.86	0.00189	15.31
0.04853	4734	4.1	−11296	5.42	−64.82	68.95	−0.00070	19.15
0.07628	4819	6.6	−11211	9.12	−96.10	102.71	−0.00319	21.56
0.10388	4793	8.9	−11237	12.88	−125.55	134.49	0.00190	18.60

[a]Values are reported for per mole of solute

(4) Phase Equilibrium

Ref. [117]

T/K	x_1	γ_1	T/K	x_1	γ_1
101.3 kPa					
373.15	1.0000	1.000	376.01	0.9433	0.958
373.30	0.9991	0.996	376.37	0.9379	0.952
373.48	0.9954	0.993	376.84	0.9330	0.941
373.59	0.9915	0.993	377.67	0.9199	0.927
373.74	0.9865	0.993	379.05	0.9078	0.896
373.93	0.9827	0.990	381.09	0.8896	0.852
374.09	0.9801	0.987	383.67	0.8697	0.799
374.19	0.9781	0.985	386.78	0.8431	0.744
374.36	0.9746	0.983	388.22	0.8295	0.721
374.54	0.9703	0.981	391.19	0.8118	0.669
374.77	0.9660	0.977	394.00	0.7879	0.631
374.99	0.9638	0.972	405.82	0.7191	0.482
375.22	0.9601	0.968	407.30	0.7104	0.467
375.43	0.9548	0.966	421.97	0.6318	0.348
375.63	0.9506	0.963	441.11	0.5403	0.249

Mixture 2: C_2H_5OH (1) + [BMIM]Cl (2)

(1) Volumetric Properties, Speed of Sound, Refractive Index and Isentropic Compressibility

Ref. [114]

x_1	m (mol · kg^{-1})	ρ/(g · cm^{-3})	n_D	u/(m · s^{-1})	κ_s/(T · Pa^{-1})	$V\text{ø}$ (cm^3 · mol^{-1})
298.15 K						
0.1874	1.45	1.06263	1.52564	1744.81	309.12	153.81
0.2532	2.46	1.05369	1.51878	1711.83	323.87	155.47
0.4815	3.88	1.01424	1.49102	1598.19	386.01	156.69
0.6209	8.37	0.97970	1.46890	1514.96	444.74	158.52
0.7217	13.25	0.94687	1.44986	1443.00	507.20	159.23
0.8485	23.38	0.89165	1.41754	1332.23	631.90	159.93
0.8982	64.01	0.86340	1.40177	1279.11	707.90	160.72
0.9373	94.11	0.83750	1.38718	1232.51	786.02	160.86
1.0000		0.78545	1.36023	1146.30	968.91	

(2) Phase Equilibrium

Ref. [117]

T/K	x_1	γ_1	T/K	x_1	γ_1
101.3 kPa					
351.44	1.0000	1.000	366.16	0.7951	0.722
351.97	0.9826	0.997	370.21	0.7635	0.652
352.99	0.9562	0.984	377.00	0.7119	0.554
354.18	0.9276	0.968	386.30	0.6442	0.452
356.04	0.8926	0.937	398.67	0.5581	0.357
358.91	0.8581	0.874	412.61	0.4816	0.278
361.61	0.8298	0.817			

(3) Osmotic Coefficient, Activity and Vapor Pressure

Ref. [84]

m/(mol · kg^{-1})	Φ_1	α_1	p/kPa	m/(mol · kg^{-1})	Φ_1	α_1	p/kPa
T = 313.15 K				T = 333.15 K			
0.1278	1.020	0.9881	17.661	0.1278	0.782	0.9908	46.471
0.3749	0.887	0.9698	17.330	0.3749	0.709	0.9758	45.749
0.6198	0.802	0.9553	17.066	0.6198	0.666	0.9627	45.121
0.7747	0.783	0.9456	16.892	0.7747	0.647	0.9549	44.747
1.0960	0.742	0.9278	16.569	1.0960	0.630	0.9384	43.953
1.4368	0.717	0.9094	16.237	1.4368	0.613	0.9221	43.174
1.9039	0.685	0.8868	15.827	1.9039	0.589	0.9018	42.201
2.3185	0.680	0.8649	15.431	2.3185	0.576	0.8843	41.365
2.6876	0.657	0.8498	15.159	2.6876	0.566	0.8693	40.648
3.0181	0.653	0.8340	14.874	3.0181	0.553	0.8575	40.084
3.3481	0.660	0.8157	14.543	3.3481	0.550	0.8440	39.443
3.5208	0.660	0.8073	14.392	3.5208	0.548	0.8371	39.112

Mixture 3: $HO[CH_2CH_2O]_nOH$ (PEG-2000, n ≈ 45) (1) + $[C_4mim]Cl$ (2)
(1) Phase Equilibrium

Ref. [23]

T/K	x_{IL}, up	x_{IL}, low
333.15	0.988	0.876
353.15	0.986	0.873
373.15	0.990	0.861
393.15	0.989	0.849
413.15	0.994	0.842

Ref. [23]

System IL + PEG	Temperature range	w_{IL}, PEG-rich phase	w_{IL}, IL-rich phase
$[C_2mim]Cl$ + PEG-2000	333.15–413.15 K	0.32–0.38	0.86–0.93

Mixture 4: $HO[CH_2CH_2O]_nOH$ (PEG-3400, n ≈ 77) (1) + $[C_4mim]Cl$ (2)
(1) Phase Equilibrium

Ref. [23]

T/K	x_{IL}, up	x_{IL}, low
333.15	0.989	0.789
353.15	0.997	0.783
373.15	0.998	0.766
393.15	0.999	0.711
413.15	0.993	0.648

Ref. [23]

System IL + PEG	Temperature range	w_{IL}, PEG-rich phase	w_{IL}, IL-rich phase
$[C_2mim]Cl$ + PEG-3400	333.15–413.15 K	0.09–0.16	0.83–0.99

2.2.27.2 Ternary Mixtures

Mixture 1: C_2H_5OH (1) + H_2O (2) + [bmim]Cl (3)
(1) Volumetric Properties, Speed of Sound, Refractive Index and Isentropic Compressibility

Ref. [114]

w_1	w_2	$\rho/(g \cdot cm^{-3})$	n_D	$u/(m \cdot s^{-1})$	$\kappa_s/(T\ Pa^{-1})$	$V_ø\ (cm^3 \cdot mol^{-1})$
298.15 K						
0.0143	0.0940	1.07643	1.51094	1812.06	282.92	160.30
0.0296	0.0926	1.07076	1.50857	1792.03	290.81	160.53
0.0655	0.0892	1.05853	1.50144	1749.54	308.64	160.73
0.1101	0.0849	1.04178	1.49430	1696.19	333.64	161.06
0.1537	0.0807	1.02683	1.48523	1651.19	357.20	161.03
0.2207	0.0743	1.00435	1.47380	1588.88	394.40	160.91
0.2918	0.0676	0.98121	1.46186	1530.62	435.01	160.75
0.3985	0.0574	0.94820	1.44538	1454.91	498.23	160.31
0.5351	0.0444	0.90843	1.42481	1371.52	585.20	159.28
0.7718	0.0217	0.84458	1.39115	1249.31	758.61	155.58
0.8754	0.0119	0.81744	1.37656	1200.36	849.02	153.47
0.0312	0.1298	1.06888	1.50097	1803.70	287.57	160.17
0.0700	0.1246	1.05489	1.49421	1754.83	307.84	160.67
0.1128	0.1189	1.03996	1.48603	1704.52	330.96	160.94
0.1698	0.1113	1.02063	1.47651	1644.47	362.31	161.03
0.2340	0.1026	0.99959	1.46616	1585.83	397.80	160.93

(continued)

w_1	w_2	$\rho/(g \cdot cm^{-3})$	n_D	$u/(m \cdot s^{-1})$	$\kappa_s/(T\ Pa^{-1})$	$V_ø\ (cm^3 \cdot mol^{-1})$
0.3118	0.0922	0.97483	1.45610	1522.71	442.42	160.74
0.4001	0.0804	0.94786	1.44105	1459.76	495.10	160.41
0.5406	0.0616	0.90701	1.42138	1372.69	585.12	159.60
0.7280	0.0365	0.85600	1.39530	1272.95	720.94	157.26
0.0337	0.1875	1.06462	1.48809	1815.28	285.05	159.79
0.0726	0.1799	1.05136	1.48178	1766.55	304.79	160.33
0.1221	0.1703	1.03477	1.47422	1708.92	330.91	160.78
0.1759	0.1599	1.01693	1.46674	1652.28	360.20	161.03
0.2454	0.1464	0.99475	1.45760	1587.83	398.73	161.03
0.3249	0.1310	0.96911	1.44594	1522.89	444.93	161.19
0.4312	0.1103	0.93852	1.43133	1447.95	508.22	160.43
0.5518	0.0870	0.90428	1.41513	1373.56	586.14	159.71
0.7421	0.0500	0.85270	1.39170	1270.93	726.04	157.56
0.0369	0.2812	1.05595	1.46745	1823.92	284.67	159.48
0.0811	0.2683	1.04254	1.46284	1772.33	305.37	160.03
0.1351	0.2526	1.02543	1.45620	1711.99	332.73	160.73
0.1933	0.2356	1.00784	1.44983	1654.29	362.56	161.04
0.2673	0.2139	0.98520	1.44120	1585.12	403.97	161.29
0.3530	0.1889	0.96029	1.43110	1517.10	452.45	160.99
0.4510	0.1603	0.93227	1.42019	1447.65	511.84	160.53
0.5871	0.1206	0.89459	1.40516	1363.49	601.28	159.79
0.7632	0.0692	0.84778	1.38602	1267.75	733.92	157.78
0.0989	0.3605	1.02954	1.43546	1764.34	312.03	160.00
0.1673	0.3331	1.00991	1.43027	1697.10	343.80	160.88
0.4963	0.2015	0.91916	1.40304	1433.59	529.37	160.31
0.5987	0.1605	0.89191	1.39463	1370.28	597.12	159.37
0.6949	0.1220	0.86668	1.38665	1315.53	666.71	158.31
0.0971	0.5417	1.01537	1.40286	1752.68	320.60	159.39
0.1848	0.4891	0.99502	1.40081	1690.22	351.79	160.31
0.2933	0.4240	0.96846	1.39733	1603.66	401.51	161.22
0.3975	0.3615	0.94244	1.39315	1524.43	456.59	161.18
0.4973	0.3016	0.91715	1.38861	1454.12	515.66	160.39
0.5946	0.2432	0.89232	1.38398	1391.09	579.12	159.20
0.6950	0.1830	0.86687	1.37881	1330.99	651.18	157.65
0.7949	0.1230	0.84145	1.37328	1274.07	732.12	155.66
0.9003	0.0598	0.81404	1.36688	1212.63	835.40	148.35
0.0932	0.6347	1.00820	1.38693	1728.11	332.13	159.18
0.1824	0.5723	0.99014	1.38725	1688.74	354.14	159.64
0.2942	0.4941	0.96525	1.38626	1611.02	399.17	160.38
0.3958	0.4229	0.94056	1.38428	1533.87	451.90	161.25
0.4883	0.3582	0.91765	1.38189	1465.67	507.29	161.15
0.5944	0.2839	0.89136	1.37861	1397.85	574.15	159.76
0.0951	0.8144	0.99148	1.35629	1635.26	377.18	159.03
0.1860	0.7326	0.97685	1.36080	1653.11	374.60	158.01

(continued)

w_1	w_2	$\rho/(\mathrm{g \cdot cm^{-3}})$	n_D	$u/(\mathrm{m \cdot s^{-1}})$	$\kappa_s/(\mathrm{T\,Pa^{-1}})$	$V_ø\ (\mathrm{cm^3 \cdot mol^{-1}})$
0.2928	0.6364	0.95748	1.36455	1614.59	400.63	157.42
0.3922	0.5470	0.93666	1.36668	1549.57	444.63	158.92
0.4958	0.4538	0.91306	1.36781	1477.17	501.93	160.58
0.5971	0.3626	0.88929	1.36786	1410.01	565.60	158.58
0.6961	0.2735	0.86536	1.36730	1347.63	636.30	155.78
0.1009	0.7193	0.99850	1.37205	1688.96	351.08	159.24
0.1966	0.6427	0.98128	1.37448	1670.28	365.28	159.07
0.2997	0.5603	0.95990	1.37568	1609.15	402.33	159.99
0.3980	0.4816	0.93758	1.37570	1538.65	450.52	161.12
0.4980	0.4016	0.91377	1.37475	1467.97	507.84	161.42
0.8013	0.1590	0.83912	1.36826	1278.31	729.30	158.60

(2) Phase Equilibrium

Ref. [117]

T/K	x_1	x_2	y_1	γ_1	γ_2
101.3 kPa					
398.50	0.2036	0.4278	0.7666	0.983	1.004
393.63	0.2625	0.3849	0.8138	0.882	1.301
382.86	0.2019	0.5424	0.6578	1.608	1.315
376.07	0.1002	0.7018	0.4912	4.060	1.284
374.33	0.0886	0.7450	0.4553	4.866	1.287
373.11	0.1133	0.7163	0.4998	3.969	1.398
367.16	0.1866	0.6479	0.5984	2.969	1.921
362.62	0.2986	0.5601	0.6863	2.187	2.636
360.42	0.3667	0.5036	0.7210	1.932	3.191
359.25	0.4170	0.4608	0.7416	1.775	3.650
357.44	0.4932	0.4000	0.7695	1.607	4.514
356.04	0.5795	0.3298	0.7939	1.443	5.787
372.77	0.0381	0.8394	0.3314	11.952	1.208
367.91	0.0863	0.7980	0.4423	6.253	1.517
364.56	0.1323	0.7578	0.5373	4.600	1.810
361.19	0.2048	0.6960	0.5956	3.362	2.241
359.04	0.2759	0.6321	0.6435	2.704	2.682
357.41	0.3482	0.5681	0.6766	2.279	3.182
356.17	0.4205	0.5040	0.6998	1.978	3.767
354.66	0.5073	0.4264	0.7374	1.738	4.730
352.92	0.5723	0.3697	0.4466	1.649	5.853
353.43	0.4021	0.5575	0.6732	2.300	3.802
355.74	0.2841	0.6853	0.6103	2.977	2.818
355.44	0.2259	0.7479	0.5788	3.788	2.614
357.82	0.1666	0.8124	0.5348	4.690	2.189
357.85	0.1339	0.8479	0.5254	5.829	2.095
359.45	0.1032	0.8825	0.5014	7.120	1.891

(*continued*)

T/K	x_1	x_2	y_1	γ_1	γ_2
352.47	0.7319	0.2106	0.8540	1.312	10.463
354.25	0.6867	0.2175	0.8591	1.305	9.429
358.62	0.6354	0.2348	0.8639	1.193	7.341
362.54	0.5723	0.2592	0.8650	1.145	5.714
369.00	0.5084	0.2749	0.8635	1.021	4.228
379.75	0.3854	0.3304	0.8533	0.933	2.400
390.27	0.3482	0.3327	0.8386	0.738	1.677
353.55	0.7974	0.1399	0.9000	1.155	15.073
357.42	0.7141	0.1644	0.9010	1.111	10.992
361.12	0.6167	0.2197	0.9022	1.119	7.117
363.28	0.0969	0.8328	0.4597	6.578	1.729
360.47	0.1491	0.7837	0.5464	4.744	2.047
357.47	0.2588	0.6816	0.6128	3.059	2.646
358.34	0.1911	0.7616	0.5635	4.008	2.288
359.26	0.1461	0.8146	0.5313	5.065	2.064
360.54	0.1083	0.8589	0.4858	6.513	1.862
362.06	0.0831	0.8910	0.4444	8.022	1.693
363.91	0.0599	0.9212	0.4002	10.397	1.526
365.65	0.0458	0.9407	0.3442	12.778	1.399
367.57	0.0323	0.9580	0.2799	16.887	1.279

Ref. [21]

wt.% of IL	T (K)	x_1	y_1	γ_1	γ_2	α_{12}
101.32 kPa						
About 20 %						
20.229	353.91	0.958	0.967	0.977	1.770	1.257
20.134	353.71	0.911	0.932	0.996	1.695	1.341
20.580	353.71	0.846	0.890	1.022	1.596	1.462
21.012	353.81	0.754	0.847	1.086	1.377	1.799
21.364	353.91	0.693	0.817	1.134	1.310	1.976
21.389	354.11	0.658	0.799	1.157	1.277	2.065
21.766	354.32	0.581	0.762	1.238	1.221	2.311
23.108	354.66	0.507	0.731	1.345	1.156	2.650
22.245	355.04	0.433	0.710	1.500	1.062	3.214
22.299	355.52	0.379	0.693	1.636	1.008	3.690
22.571	356.03	0.335	0.675	1.765	0.975	4.111
22.449	356.84	0.260	0.649	2.117	0.975	5.273
21.707	358.26	0.179	0.599	2.680	0.975	6.868
21.519	359.79	0.132	0.570	3.254	0.843	8.706
21.090	364.49	0.074	0.466	3.998	0.843	10.941

(continued)

wt.% of IL	T (K)	x_1	y_1	γ_1	γ_2	α_{12}
About 30 %						
29.596	356.52	0.925	0.939	0.925	1.679	1.250
30.882	356.42	0.864	0.906	0.961	1.434	1.522
30.769	356.37	0.779	0.867	1.016	1.244	1.853
31.173	356.35	0.681	0.823	1.098	1.244	2.177
31.466	356.42	0.625	0.799	1.156	1.097	2.391
31.828	356.63	0.544	0.762	1.251	1.057	2.685
32.105	356.93	0.474	0.730	1.354	1.024	2.996
32.370	357.33	0.396	0.702	1.527	0.966	3.582
32.476	357.84	0.331	0.667	1.696	0.951	4.038
32.222	358.55	0.274	0.645	1.919	0.903	4.803
33.172	360.28	0.194	0.591	2.330	0.872	6.018
31.990	361.91	0.139	0.542	2.786	0.854	7.322
31.126	364.46	0.090	0.465	3.344	0.852	8.767

Ref. [24]

x'_1 [a]	w_3 [a]	y_1	T/K	γ_2	γ_1	α_{12} [b]
p = 100 kPa						
0.9500	0.5881	0.9769	371.16	0.67	0.67	2.22
0.9500	0.4999	0.9728	364.43	0.93	0.77	1.87
0.9500	0.3999	0.9683	358.90	1.25	0.88	1.60
0.9500	0.3009	0.9640	355.37	1.55	0.95	1.40
0.9500	0.2000	0.9595	353.21	1.82	0.99	1.24
0.9500	0.1000	0.9538	351.89	2.12	1.00	1.08

[a] Compositions reported are of equilibrium liquid
[b] $\alpha_{2,1} = (y_2/x'_2)/(y_1/x'_1)$

Ref. [36]

x_1	y_1	x_1	y_1	x_1	y_1
T = 337.15 K					
x_3 = 0.1		x_3 = 0.3		x_3 = 0.5	
0.724	0.839	0.380	0.757	0.374	0.840
0.759	0.851	0.504	0.827	0.504	0.903
0.857	0.905	0.620	0.874	0.592	0.927
0.890	0.924	0.732	0.911	0.724	0.957
0.954	0.970	0.857	0.954	0.857	0.988
0.977	0.985	0.876	0.958	0.902	0.994
		0.954	0.991		

Mixture 2: CH_3COOH (1) + $HCONH_2$ (2) + [bmim]Cl (3)
(1) Phase Equilibrium

Ref. [118]

x_2	T/K	x_2	T/K	x_2	T/K	x_2	T/K
x_1:x_3 = 9:1							
0.0000	281.09	0.3003	255.56	0.3491	258.52	0.4981	249.56
0.1055	275.74	0.3003	261.77	0.4003	254.81	0.5501	242.41
0.1968	269.58	0.3491	255.36	0.4459	253.41		

Mixture 3: CH_3CH_2COOH (1) + $HCONH_2$ (2) + [bmim]Cl (3)
(1) Phase Equilibrium

Ref. [118]

x_2	T/K	x_2	T/K	x_2	T/K	x_2	T/K
x_1:x_3 = 8.9108:1							
0.4480	244.46	0.5564	255.78	0.6245	243.37	0.7523	255.03
0.4480	251.27	0.5751	243.44	0.6245	255.23	0.8520	263.57
0.5008	244.53	0.5751	255.94	0.7042	250.52	1.0000	275.54
0.5378	244.52	0.6000	255.64	0.7167	251.82		

Mixture 4: H_2O (1) + $OCH_2CH_2CH_2CH_2$ (2) + [BMIM]Cl (3)
(1) Phase Equilibrium

Ref. [36]

x_2	y_2	x_2	y_2	x_2	y_2
T = 337.15 K					
x_3 = 0.1		x_3 = 0.3		x_3 = 0.5	
0.501	0.878	0.127	0.905	0.127	0.950
0.622	0.908	0.254	0.939	0.255	0.975
0.699	0.926	0.501	0.971	0.358	0.985
0.792	0.948	0.622	0.983	0.428	0.988
0.949	0.994	0.699	0.988	0.500	0.991
		0.792	0.993	0.623	0.993
		0.950	0.997	0.699	0.995
				0.793	0.996

Mixture 5: H_2O (1) + $(CH_3)_3COH$ (2) + [bmim]Cl (3)
(1) Phase Equilibrium

Ref. [25]

x'_2 [a]	w_3 [a]	y_2	T/K	γ_1	γ_2	$\alpha_{2,1}$ [b]
p = 100 kPa						
0.9501	0.5907	0.9838	375.59	0.47	0.78	3.19
0.9500	0.5499	0.9823	372.16	0.54	0.83	2.92

(continued)

x'_2[a]	w_3[a]	y_2	T/K	γ_1	γ_2	$\alpha_{2,1}$[b]
0.9500	0.4999	0.9794	368.05	0.69	0.90	2.51
0.9500	0.4500	0.9775	365.13	0.79	0.94	2.29
0.9499	0.3999	0.9744	362.90	0.94	0.97	2.00
0.9499	0.2999	0.9683	359.83	1.21	1.00	1.61
0.9500	0.2000	0.9576	357.34	1.67	1.02	1.19
0.9500	0.1001	0.9444	355.58	2.22	1.02	0.90
0.9499	0.0503	0.9347	354.85	2.63	1.02	0.75

[a]Compositions reported are of equilibrium liquid
[b]$\alpha_{21} = (y_2/x'_2)/(y_1/x'_1)$

2.2.28 02011-01002: 1-Butyl-3-methylimidazolium bromide

Abbreviation:	$[C_4MIm]Br$, [BMIM]Br, [bmim]Br, BMIMBr, $[C_4mim]Br$
Molecular Formula:	$C_8H_{15}N_2Br$
Molar Mass:	219.12
Structure:	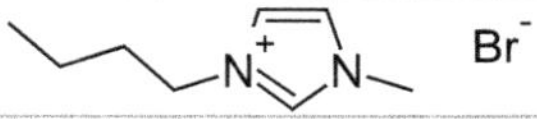

2.2.28.1 Binary Mixtures

Mixture 1: H_2O (1) + [BMIM]Br (2)

(1) Osmotic Properties, Speed of Sound, Volumetric Properties

Ref. [119]

m_2	Φ_1	m_2	Φ_1	m_2	Φ_1
0.0220	0.9440	0.2702	0.8094	0.7354	0.7260
0.0461	0.9262	0.3522	0.7852	0.9713	0.7125
0.0726	0.8939	0.3994	0.7772	1.2718	0.7006
0.1241	0.8549	0.5204	0.7551	2.0237	0.6903

Ref. [120]

Molality, M_2	u/m · s^{-1}	ρ/kg · m^{-3}	$10^6 \cdot V_\varphi$/m^3 · mol^{-1}	u/m · s^{-1}	ρ/kg · m^{-3}	$10^6 \cdot V_\varphi$/m^3 · mol^{-1}	u/m · s^{-1}	ρ/kg · m^{-3}	$10^6 \cdot V_\varphi$/m^3 · mol^{-1}
	T = 298.15 K			T = 303.15 K			T = 308.15 K		
0.0692	1504.01	1000.439	169.87	1515.86	999.001	170.60	1526.01	997.350	171.31
0.1376	1511.34	1003.696	170.00	1522.65	1002.198	170.90	1532.33	1000.488	171.80
0.2697	1523.10	1009.849	169.83	1533.26	1008.276	170.67	1541.87	1006.541	171.33
0.6117	1552.17	1024.930	169.18	1561.74	1023.151	170.02	1568.72	1021.173	170.86
1.2796	1600.31	1050.917	168.32	1603.46	1048.782	169.16	1605.58	1046.556	169.91
2.133	1645.93	1078.259	167.80	1644.90	1075.775	168.61	1644.12	1073.138	169.42
3.115	1682.71	1103.800	167.37	1678.29	1100.977	168.18	1673.34	1098.052	168.96
5.1874	1724.21	1143.515	166.85	1715.98	1140.260	167.61	1707.44	1136.943	168.35
7.7923	1741.22	1176.049	166.73	1731.38	1172.571	167.42	1721.27	1169.045	168.11
	T = 313.15 K			T = 318.15 K					
0.0692	1534.64	995.472	172.43	1541.87	993.421	173.38			
0.1376	1540.37	998.581	172.70	1546.89	996.491	173.60			
0.2697	1549.09	1004.595	172.06	1554.97	1002.472	172.77			
0.6117	1573.85	1019.011	171.70	1576.98	1016.678	172.54			
1.2796	1606.73	1044.144	170.68	1606.94	1041.589	171.44			
2.133	1640.63	1070.425	170.18	1637.45	1067.580	170.93			
3.115	1667.89	1095.037	169.72	1661.96	1091.934	170.47			
5.1874	1698.62	1133.565	169.07	1689.56	1130.133	169.78			
7.7923	1710.99	1165.494	168.78	1700.49	1161.900	169.44			

Ref. [120]

	$E^0{}_\varphi$				
Solvent	$T = 298.15$ K	$T = 303.15$ K	$T = 308.15$ K	$T = 313.15$ K	$T = 318.15$ K
H_2O	0.1432	0.1592	0.1752	0.1912	0.2072

Ref. [121]

$m_2/(\text{mol} \cdot \text{kg}^{-1})$	$\rho/(\text{kg} \cdot \text{dm}^{-3})$	φ: Apparent molar volume ($\text{cm}^3 \cdot \text{mol}^{-1}$)	$V^\infty = \varphi - 1.86\sqrt{m}$ ($\text{cm}^3 \cdot \text{mol}^{-1}$)
0.0013	0.997112	169.8	169.7
0.0019	0.997141	167.9	167.8
0.0023	0.997167	169.9	169.9
0.0033	0.997209	169.3	169.2
0.0039	0.997240	169.4	169.3
0.0113	0.997609	169.4	169.2
0.0115	0.997619	169.3	169.1
0.0146	0.997782	169.4	169.1
0.0247	0.998282	169.1	168.8
0.0248	0.998284	169.4	169.1
0.0274	0.998406	169.6	169.3
0.0423	0.999147	169.5	169.1
0.0794	1.000970	169.5	168.9
0.1010	1.002020	169.3	168.8
0.1037	1.002150	169.4	168.8
0.1058	1.002252	169.4	168.8
0.1786	1.005737	169.3	168.6
0.3067	1.011721	169.1	168.1

Ref. [121]

Behavior of salt		Behavior of water	
m_{2p} ($\text{mol} \cdot \text{kg}^{-1}$)	P_2 ($\text{cm}^3 \cdot \text{mol}^{-1}$)	m_{1p} ($\text{mol} \cdot \text{kg}^{-1}$)	P_1 ($\text{cm}^3 \cdot \text{mol}^{-1}$)
0.0187	166.8	0.6335	18.1
0.0258	167.0	0.3702	18.2
0.0441	167.1	0.2293	18.3

Ref. [121]

Molality of the salt ($\text{mol} \cdot \text{kg}^{-1}$)	X_2: Molar fraction of the salt	ρ ($\text{kg} \cdot \text{dm}^{-3}$)	V'_2 ($\text{cm}^3 \cdot \text{mol}^{-1}$)
0	0	0.997047	
1.5064	0.0264	1.059335	166.6
3.7028	0.0625	1.118315	165.3

Ref. [121]

Molality of the salt ($mol \cdot kg^{-1}$)	$\varphi_2/(cm^3 \cdot mol^{-1})$	$V'_2/(cm^3 \cdot mol^{-1})$	$V'_1/(cm^3 \cdot mol^{-1})$
0.0384	169.5	169.4	18.1
0.1307	169.4	169.2	18.1
0.1954	169.2	168.9	18.1
0.3730	169.0	168.5	18.1
0.4600	168.9	168.2	18.1
0.5648	168.8	168.0	18.1
0.8917	168.3	167.2	18.1
1.0189	168.1	166.9	18.1
1.2428	167.9	166.6	18.1
1.5100	167.8	166.3	18.1
1.7680	167.3	165.9	18.1
1.9176	167.2	165.7	18.1
3.6456	166.5	165.5	18.1
4.4408	166.2	165.5	18.1
5.7000	166.1	165.2	18.1

(2) Vapor Pressure

Ref. [122]

X_2 = 20 mol %		X_2 = 40 mol %		X_2 = 60 mol %		X_2 = 80 mol %	
T/K	P/kPa	T/K	P/kPa	T/K	P/kPa	T/K	P/kPa
304.8	3.5	320.6	4.1	355.6	8.8	379.4	11.3
320.6	7.9	332.8	7.2	361.0	11.2	387.4	14.5
326.8	10.8	341.0	10.4	369.2	15.3	392.0	17.4
333.4	14.0	353.8	16.5	376.0	19.2	396.8	19.9
338.4	17.7	359.5	21.4	381.6	24.1	402.4	24.1
343.4	21.1	365.0	25.8	386.8	29.4	407.6	27.5
346.6	24.3	370.8	31.2	395.0	36.4	413.0	32.4
351.6	29.5	375.6	37.4	400.6	42.5	418.5	37.3
356.6	36.4	380.2	43.6	406.8	50.6	422.6	41.3
361.0	43.2	384.6	49.3	411.8	58.4	426.4	47.5
365.8	51.9	387.4	56.6	416.2	68.2	430.8	51.8
368.6	58.1	392.2	64.5	420.2	75.7	434.2	58.1
375.4	72.9	395.8	71.5	425.2	87.6	437.4	62.9
379.4	82.6	400.8	84.1	431.2	103.0	441.6	70.1
383.2	95.8	407.2	103.1			446.4	77.3
385.4	103.2					450.8	83.7
						453.6	91.9
						456.2	96.5
						457.4	101.2

Mixture 2: CH_3OH (1) + BMIMBr (2)
(1) Speed of Sound and Volumetric Properties

Ref. [120]

Molality, M_2	$u/m \cdot s^{-1}$	$\rho/kg \cdot m^{-3}$	$10^6 \cdot V_\varphi/m^3 \cdot mol^{-1}$	$u/m \cdot s^{-1}$	$\rho/kg \cdot m^{-3}$	$10^6 \cdot V_\varphi/m^3 \cdot mol^{-1}$	$u/m \cdot s^{-1}$	$\rho/kg \cdot m^{-3}$	$10^6 \cdot V_\varphi/m^3 \cdot mol^{-1}$
	$T = 298.15$ K			$T = 303.15$ K			$T = 308.15$ K		
0.0473	1106.75	790.393	145.79	1090.45	785.688	145.24	1074.26	780.958	144.88
0.0767	1108.81	792.789	145.63	1092.58	788.091	145.17	1076.48	783.371	144.76
0.1497	1114.13	798.175	150.66	1097.98	793.495	150.34	1081.94	788.794	150.01
0.2612	1121.70	806.303	152.39	1105.65	801.654	152.10	1089.70	796.988	151.77
0.3434	1127.07	812.207	152.73	1111.09	807.572	152.49	1095.24	802.921	152.22
0.4912	1136.37	822.389	153.58	1120.60	817.794	153.35	1104.85	813.177	153.12
0.7273	1151.05	837.955	154.22	1135.41	833.402	154.05	1119.88	828.837	153.85
0.9756	1164.13	852.578	156.06	1148.69	848.250	155.63	1133.40	843.727	155.49
1.4045	1186.17	876.422	157.17	1171.02	871.970	157.13	1156.01	867.512	157.06
2.4462	1233.30	924.879	159.17	1218.77	920.559	159.22	1204.37	916.245	159.24
3.8956	1287.39	976.589	160.86	1273.49	972.412	160.98	1259.72	968.238	161.08
	$T = 313.15$ K			$T = 318.15$ K					
0.0473	1058.20	776.204	143.90	1042.3	771.426	142.81			
0.0767	1060.50	778.627	143.96	1044.65	773.860	143.06			
0.1497	1065.99	784.073	149.44	1050.18	779.323	148.87			
0.2612	1073.92	792.294	151.35	1058.21	787.582	150.85			
0.3434	1079.44	798.249	151.83	1063.8	793.553	151.42			
0.4912	1089.25	808.542	152.79	1073.75	803.883	152.43			
0.7273	1104.52	824.253	153.59	1089.21	819.660	153.27			
0.9756	1118.22	839.188	155.29	1103.08	834.634	155.07			
1.4045	1141.04	863.046	156.94	1126.26	858.563	156.79			
2.4462	1190.11	911.924	159.23	1175.9	907.596	159.19			
3.8956	1246.05	964.059	161.17	1232.52	959.889	161.23			

Ref. [120]

Solvent	E^0_φ $T = 298.15$ K	$T = 303.15$ K	$T = 308.15$ K	$T = 313.15$ K	$T = 318.15$ K
CH_3OH	−0.0537	−0.1105	−0.1674	−0.2243	−0.2811

Mixture 3: C_2H_5OH (1) + BMIMBr (2)
(1) Speed of Sound and Volumetric Properties

Ref. [120]

Molality, M_2	u/ $m \cdot s^{-1}$	ρ/ $kg \cdot m^{-3}$	$10^6 \cdot V_\varphi$/ $m^3 \cdot mol^{-1}$	u/ $m \cdot s^{-1}$	ρ/ $kg \cdot m^{-3}$	$10^6 \cdot V_\varphi$/ $m^3 \cdot mol^{-1}$	u/ $m \cdot s^{-1}$	ρ/ $kg \cdot m^{-3}$	$10^6 \cdot V_\varphi$/ $m^3 \cdot mol^{-1}$
	$T = 298.15$ K			$T = 303.15$ K			$T = 308.15$ K		
0.0271	1145.15	787.221	148.59	1128.33	782.917	148.20	1111.57	778.581	147.68
0.0587	1147.27	789.649	151.09	1130.45	785.356	150.68	1113.75	781.029	150.26
0.0846	1148.57	791.567	152.83	1131.91	787.288	152.34	1115.26	782.969	151.95
0.1257	1150.80	794.595	154.00	1134.00	790.319	153.69	1117.37	786.010	153.35
0.1587	1152.66	796.910	155.48	1136.10	792.647	155.15	1119.39	788.343	154.88
0.2261	1155.78	801.670	156.51	1139.74	797.404	156.37	1123.15	793.117	156.13
0.3162	1160.72	807.896	157.31	1144.33	803.664	157.10	1127.98	799.393	156.91
	$T = 313.15$ K			$T = 318.15$ K					
0.0271	1094.92	774.206	146.88	1078.34	769.798	146.25			
0.0587	1097.14	776.665	149.65	1080.60	772.267	149.13			
0.0846	1098.63	778.613	151.42	1082.11	774.213	151.15			
0.1257	1101.00	781.663	152.94	1084.46	777.288	152.47			
0.1587	1103.00	784.006	154.49	1086.55	779.628	154.21			
0.2261	1106.82	788.795	155.81	1090.46	784.413	155.70			
0.3162	1111.65	795.091	156.64	1095.42	790.753	156.39			

Ref. [120]

Solvent	E^0_φ $T = 298.15$ K	$T = 303.15$ K	$T = 308.15$ K	$T = 313.15$ K	$T = 318.15$ K
C_2H_5OH	−0.0770	−0.1300	−0.1830	−0.2350	−0.2880

Mixture 4: CF_3CH_2OH (1) + [BMIm]Br (2)
(1) Vapor Pressure

Ref. [123]

T (K)	P (kPa)	T (K)	P (kPa)	T (K)	P (kPa)	T (K)	P (kPa)	T (K)	P (kPa)	T (K)	P (kPa)
$X_2 = 0$ %		$X_2 = 40$ %		$X_2 = 50$ %		$X_2 = 60$ %		$X_2 = 80$ %		$X_2 = 90$ %	
293.2	7.2	308.4	4.9	329.6	7.4	358.2	14.1	397.6	15.3	409.4	11.1
297.9	9.5	338.0	19.7	353.8	21.0	377.6	30.6	416.0	28.1	425.6	19.2
312.2	20.9	353.0	36.6	369.2	38.1	401.2	63.7	430.0	43.4	448.4	36.8
323.0	36.0	364.0	55.8	382.6	61.4	411.8	87.0	442.0	64.1	456.4	47.0
328.5	46.6	374.0	79.6	393.4	85.1	416.6	101.0	456.2	88.3	465.0	56.8
343.2	88.3	380.8	100.1	398.8	99.9			459.6	99.9	469.4	67.1

2.2.28.2 Ternary Mixtures

Mixture 1: H_2O (1) + K_2HPO_4 (2) + $[C_4mim]Br$ (3)
(1) Phase Equilibrium

Ref. [124]

100 w_3	100 w_2	100 w_3	100 w_2
298.15 K			
61.26	0.92	10.42	22.98
53.86	1.53	9.38	23.96
48.60	2.23	8.52	24.86
38.46	5.10	7.76	25.68
35.59	6.37	7.06	26.46
32.54	7.67		
29.20	9.47		
25.79	11.47		
22.59	13.49		
19.40	15.81		
17.97	16.87		
16.12	18.26		
14.52	19.52		
12.70	20.96		
11.29	22.17		

Ref. [124]

100 w_3	100 w_2	100 w_3	100 w_2	100 w_3	100 w_2
Total composition		Top phase		Bottom phase	
298.15 K					
34.09	18.01	58.20	1.02	1.69	40.65
25.08	21.29	51.98	1.72	1.68	37.49
27.03	17.38	46.95	2.58	1.44	36.03
23.64	18.90	45.96	2.70	1.69	34.34
21.85	17.86	40.30	4.64	2.11	32.82

Mixture 2: H_2O (1) + K_3PO_4 (2) + $[C_4mim]Br$ (3)
(1) Phase Equilibrium

Ref. [124]

100 w_3	100 w_2	100 w_3	100 w_2
298.15 K			
50.40	1.84	12.99	18.43
46.93	2.38	11.17	19.77
44.33	3.09	9.49	21.11
41.08	4.00	9.46	21.14
39.37	4.53	8.35	22.14
34.41	6.37	7.87	22.48

(*continued*)

100 w_3	100 w_2	100 w_3	100 w_2
33.50	6.77	7.33	23.12
29.84	8.50	6.83	23.60
25.44	10.72	6.69	23.65
23.63	11.81	6.14	24.39
21.41	13.02	5.63	24.89
19.35	14.34	4.89	25.81
17.31	15.55	4.10	26.87
16.36	16.24	3.12	28.12
14.23	17.59	1.46	31.93

Ref. [124]

100 w_3	100 w_2	100 w_3	100 w_2	100 w_3	100 w_2
Total composition		Top phase		Bottom phase	
298.15 K					
34.00	17.91	61.47	0.60	0.27	39.19
31.28	15.63	53.50	1.22	1.83	33.33
29.83	14.97	50.35	1.73	1.46	32.26
27.08	14.83	45.91	2.48	3.06	29.24
26.08	13.01	39.90	4.23	5.51	24.96

Mixture 3: C_2H_5OH (1) + H_2O (2) + [BMIM]Br (3)
(1) Phase Equilibrium

Ref. [21]

wt.% of IL	T (K)	x_1	y_1	γ_1	γ_2	α_{12}
101.32 kPa						
About 10 %						
9.658	351.91	0.950	0.953	1.012	2.174	1.066
9.738	351.81	0.891	0.903	1.026	2.063	1.139
10.189	351.91	0.814	0.842	1.043	1.960	1.218
10.465	352.12	0.734	0.802	1.093	1.697	1.474
10.072	352.42	0.658	0.767	1.148	1.537	1.709
10.884	352.82	0.581	0.731	1.220	1.426	1.956
10.486	353.33	0.494	0.702	1.348	1.277	2.410
10.461	353.84	0.421	0.677	1.496	1.181	2.889
10.129	354.45	0.352	0.656	1.686	1.098	3.500
10.092	355.32	0.278	0.632	1.994	1.014	4.470
10.224	356.29	0.214	0.613	2.420	0.943	5.826
9.868	357.85	0.155	0.582	2.978	0.889	7.582
9.736	360.23	0.106	0.544	3.715	0.836	10.011
9.468	363.14	0.077	0.474	4.007	0.834	10.759
About 20 %						
18.435	352.92	0.962	0.964	1.001	2.187	1.045
18.621	352.85	0.903	0.922	1.022	1.845	1.264
18.871	352.93	0.833	0.873	1.044	1.730	1.379

(continued)

wt.% of IL	T (K)	x_1	y_1	γ_1	γ_2	α_{12}
19.127	353.04	0.741	0.820	1.095	1.568	1.595
19.376	353.24	0.674	0.793	1.152	1.425	1.848
19.773	353.53	0.595	0.762	1.240	1.295	2.185
19.617	353.93	0.518	0.734	1.346	1.194	2.571
19.924	354.46	0.442	0.707	1.486	1.110	3.050
19.804	354.96	0.378	0.687	1.654	1.039	3.621
19.918	355.73	0.308	0.660	1.886	0.984	4.355
19.688	356.85	0.242	0.636	2.207	0.918	5.452
19.617	358.48	0.169	0.596	2.791	0.868	7.268
18.853	360.99	0.112	0.530	3.404	0.854	8.973

Mixture 4: H_2O (1) + $C_6H_5K_3O_7 \cdot H_2O$ (Tri-potassium Citrate) (2) + [C_4mim]Br (3)
(1) Phase Equilibrium

Ref. [125]

T/K = 278.15				T/K = 298.15		T/K = 308.15		T/K = 318.15	
100 w_3	100 w_2	100 w_3	100 w_2	100 w_3	100 w_2	100 w_3	100 w_2	100 w_3	100 w_2
62.13	2.22	17.08	27.92	49.98	6.48	48.01	8.10	52.71	6.17
57.56	3.15	16.36	28.63	47.38	7.84	46.00	8.96	44.37	10.50
54.70	4.20	15.68	29.22	41.44	11.20	41.29	11.71	41.65	12.02
52.72	4.90	14.97	29.91	37.15	13.76	39.01	13.24	39.56	13.32
50.68	5.65	14.42	30.45	33.99	15.97	36.95	14.63	37.11	14.91
49.10	6.37	13.99	30.90	29.14	19.61	34.20	16.52	35.51	16.14
47.65	7.01	13.36	31.50	27.12	20.96	31.87	18.09	33.97	17.08
46.25	7.66	12.80	32.14	25.05	22.55	29.33	20.11	31.77	18.72
44.95	8.22	12.10	32.96	23.25	24.02	27.59	21.27	30.22	19.87
43.88	8.85	11.27	33.89	21.47	25.56	26.61	22.09	28.58	21.22
42.90	9.41	10.46	34.94	19.80	26.88	25.17	23.06	26.31	22.86
40.90	10.42	9.64	35.97	18.69	27.76	24.04	23.96	22.88	25.67
39.41	11.43	8.81	37.08	17.75	28.69	23.36	24.60	21.08	27.14
37.82	12.37	7.76	38.76	16.66	29.68	22.03	25.62	19.91	28.04
35.78	13.67	6.35	41.44	15.69	30.56	20.85	26.55	18.93	28.91
34.04	14.94			15.03	31.22	19.55	27.68	17.87	29.93
32.50	16.05			14.56	31.59	18.31	28.61	16.94	30.72
31.09	17.10			13.98	32.18	17.32	29.49	15.42	32.23
29.73	17.98			13.44	32.73	16.03	30.91	14.29	33.39
28.79	18.68			12.85	33.38	15.10	31.76	13.52	34.24
27.35	19.76			12.14	34.31	14.09	32.75	12.41	35.43
25.72	21.04			11.61	34.95	13.02	33.92	11.45	36.65
24.22	22.10			11.06	35.47	12.12	34.90	10.44	37.95
22.94	23.11			10.30	36.48	11.34	35.88	9.60	39.18
21.99	23.81			9.72	37.46	10.48	36.93	8.81	40.34
20.95	24.67			9.35	37.92	9.53	38.23	6.70	44.09
19.59	25.80			8.63	38.98	8.50	39.90		
18.80	26.48			8.01	40.03	7.58	41.51		
17.91	27.26			7.05	41.80				

Ref. [125]

Total composition		Top phase: ionic liquid-rich phase			Bottom phase: salt-rich phase				
100 w_3	100 w_2	100 w_3	100 w_2	100 w_1[a]	100 w_3[a]	100 w_2	100 w_1	TLL (Tie-Line Lengths)	A
T = 278.15 K									
31.12	20.19	51.93 ± 0.08 51.49[b]	5.01 ± 0.02		9.03 ± 0.17	36.90 ± 0.13		53.46	−1.35
39.52	16.06	55.92 ± 0.18 57.92[b]	3.46 ± 0.11		5.90 ± 0.27	42.17 ± 0.24		63.25	−1.29
39.65	20.36	63.89 ± 0.13	1.50 ± 0.06		2.63 ± 0.27	49.04 ± 0.24		77.54	−1.29
38.25	23.64	67.07 ± 0.37	1.06 ± 0.29		1.84 ± 0.30	52.06 ± 0.27		82.80	−1.28
37.70	27.02	70.78 ± 0.16 70.09[b]	0.55 ± 0.09	28.66	0.77 ± 0.22	55.95 ± 0.19	42.83	89.28	−1.26
38.30	26.71	71.15 ± 0.31	0.69 ± 0.23		0.64 ± 0.32	57.26 ± 0.29		90.40	−1.25
37.48	30.36	73.83 ± 0.28 74.01[b]	0.37 ± 0.19	24.89	0.37 ± 0.33	60.24 ± 0.30	38.41	94.77	−1.23
37.68	30.01	75.01 ± 0.16 75.56[b]	0.29 ± 0.06	23.93	0.48 ± 0.25	59.61 ± 0.21	39.27	95.25	−1.26
T = 298.15 K									
35.61	18.76	54.40 ± 0.19	4.75 ± 0.12		7.38 ± 0.28	40.06 ± 0.25		58.80	−1.33
35.59	22.22	62.43 ± 0.10	2.57 ± 0.02		4.27 ± 0.29	46.31 ± 0.27		72.77	−1.33
34.76	25.17	66.52 ± 0.09	1.65 ± 0.01		2.50 ± 0.33	50.78 ± 0.31		80.70	−1.30
38.46	27.82	73.30 ± 0.11	0.71 ± 0.02		1.14 ± 0.37	57.95 ± 0.35		92.10	−1.26
38.72	30.71	76.15 ± 0.11	0.42 ± 0.01		0.49 ± 0.45	63.03 ± 0.41		98.21	−1.21
T = 318.15 K									
33.98	21.07	56.15 ± 0.06	4.74 ± 0.03		7.06 ± 0.37	43.02 ± 0.34		62.25	−1.28
34.76	20.21	56.23 ± 0.07 56.64[b]	4.50 ± 0.03	38.28	6.63 ± 0.35	43.03 ± 0.33	51.42	62.81	−1.29
43.68	17.54	62.09 ± 0.07	2.36 ± 0.02		4.09 ± 0.47	50.90 ± 0.45		75.64	−1.19
43.58	20.97	67.98 ± 0.08	1.33 ± 0.01		2.82 ± 0.53	54.21 ± 0.50		83.92	−1.23
43.61	24.40	72.37 ± 0.09	0.82 ± 0.01		2.33 ± 0.56	57.59 ± 0.53		90.15	−1.23
43.70	27.85	76.48 ± 0.10	0.46 ± 0.01		1.42 ± 0.60	62.95 ± 0.57		97.67	−1.20

[a]Were determined by the Karl Fischer method
[b]Were determined by nitrogen analysis using an element analyzer instrument

2.2.29 02011-01003: 1-Butyl-3-methylimidazolium iodine

Abbreviation:	$[C_4MIm]I$, [BMIM]I, [bmim]I, [BMIm]I
Molecular Formula:	$C_8H_{15}N_2I$
Molar Mass:	266.12
Structure:	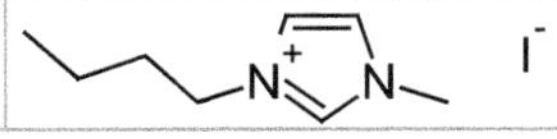

2.2.29.1 Binary Mixtures

Mixture 1: H_2O (1) + [BMIm]I (2)

(1) Refractive Index and Osmotic Coefficient

Ref. [116]

	n_D			
m_2/mol · kg^{-1}	T = 298.15 K	T = 308.15 K	T = 318.15 K	T = 328.15 K
0.0133	1.3331	1.3321	1.3305	1.3285
0.0293	1.3336	1.3325	1.3310	1.3289
0.0442	1.3343	1.3330	1.3316	1.3293
0.0505	1.3346	1.3334	1.3317	1.3298
0.0786	1.3356	1.3344	1.3328	1.3310
0.0995	1.3365	1.3353	1.3336	1.3315
0.1307	1.3376	1.3363	1.3347	1.3329
0.1782	1.3395	1.3381	1.3364	1.3347
0.2160	1.3410	1.3397	1.3380	1.3359
0.3257	1.3449	1.3435	1.3418	1.3404
0.4065	1.3478	1.3467	1.3447	1.3433
0.5184	1.3519	1.3503	1.3487	1.3464
0.6452	1.3560	1.3548	1.3529	1.3507
0.7509	1.3595	1.3580	1.3561	1.3541
0.8786	1.3635	1.3620	1.3600	1.3578
1.1538	1.3714	1.3701	1.3680	1.3657
1.8374	1.3893	1.3876	1.3853	1.3831
2.5889	1.4054	1.4035	1.4012	1.3986
	Φ			
0.0133	0.833	0.856	0.899	0.910
0.0293	0.821	0.843	0.888	0.889
0.0442	0.792	0.821	0.880	0.885
0.0505	0.784	0.814	0.869	0.878
0.0786	0.766	0.788	0.845	0.861
0.0995	0.761	0.780	0.819	0.833
0.1307	0.751	0.765	0.782	0.807
0.1782	0.711	0.734	0.758	0.778
0.2160	0.697	0.714	0.740	0.768

(continued)

	Φ			
0.3257	0.644	0.662	0.706	0.715
0.4065	0.613	0.645	0.682	0.690
0.5184	0.596	0.626	0.644	0.662
0.6452	0.564	0.581	0.610	0.619
0.7509	0.554	0.566	0.574	0.586
0.8786	0.528	0.553	0.558	0.569
1.1538	0.470	0.501	0.521	0.528
1.8374	0.405	0.413	0.429	0.450
2.5889	0.362	0.380	0.387	0.408

2.2.30 02011-01004: 1-Butyl-3-methylimidazolium perchlorate

Abbreviation:	$[C_4MIm][ClO_4]$, $[BMIM][ClO_4]$, $[bmim][ClO_4]$, $[Bmim][ClO_4]$
Molecular Formula:	$C_8H_{15}ClN_2O_4$
Molar Mass:	238.67
Structure:	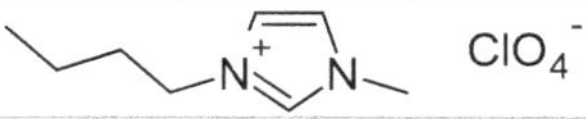

2.2.30.1 Binary Mixtures

Mixture 1: C_2H_5OH (1) + $[Bmim][ClO_4]$ (2)

(1) Volumetric Properties, Transport Properties and Refractive Index

Ref. [126]

x_1	T/K	ρ/ $g \cdot cm^{-3}$	η/ mPa · s	n_D	V^E/ $cm^3 \cdot mol^{-1}$	$\Delta\eta$/ mPa · s	Δn_D
0.100	283.15	1.2493	255.52	1.4750	−0.154	−164.35	0.0100
0.100	288.15	1.2457	188.52	1.4738	−0.159	−109.32	0.0101
0.100	293.15	1.2421	142.20	1.4725	−0.182	−75.31	0.0100
0.100	298.15	1.2386	109.99	1.4710	−0.198	−52.35	0.0100
0.100	303.15	1.2351	86.43	1.4697	−0.211	−37.72	0.0101
0.100	308.15	1.2317	69.74	1.4685	−0.222	−26.90	0.0105
0.100	313.15	1.2282	56.98	1.4670	−0.230	−20.27	0.0102
0.100	318.15	1.2247	46.98	1.4655	−0.245	−15.66	0.0105
0.100	323.15	1.2213	39.45	1.4645	−0.261	−12.11	0.0108
0.100	328.15	1.2178	33.69	1.4630	−0.274	−9.37	0.0107
0.100	333.15	1.2144	28.86	1.4615	−0.284	−7.10	0.0107
0.100	338.15	1.2110	24.94	1.4603	−0.294	−5.83	0.0108
0.100	343.15	1.2075	21.69	1.4590	−0.304	−4.90	0.0109
0.165	283.15	1.2394	181.83	1.4725	−0.320	−207.81	0.0148

(*continued*)

x_1	T/K	ρ/ $g \cdot cm^{-3}$	η/ $mPa \cdot s$	n_D	V^E/ $cm^3 \cdot mol^{-1}$	$\Delta\eta$/ $mPa \cdot s$	Δn_D
0.165	288.15	1.2357	136.86	1.4715	−0.323	−139.56	0.0152
0.165	293.15	1.2319	105.92	1.4700	−0.317	−95.97	0.0149
0.165	298.15	1.2280	83.35	1.4687	−0.285	−67.33	0.0152
0.165	303.15	1.2241	66.48	1.4673	−0.259	−48.78	0.0152
0.165	308.15	1.2203	54.25	1.4660	−0.219	−35.47	0.0156
0.165	313.15	1.2162	44.76	1.4648	−0.146	−26.97	0.0156
0.165	318.15	1.2112	37.20	1.4637	0.046	−20.96	0.0164
0.165	323.15	1.2072	31.56	1.4620	0.094	−16.33	0.0160
0.165	328.15	1.2010	27.02	1.4608	0.474	−12.98	0.0162
0.165	333.15	1.1987	23.50	1.4595	0.292	−9.91	0.0165
0.165	338.15	1.1954	20.45	1.4580	0.270	−8.14	0.0163
0.165	343.15	1.1903	18.12	1.4565	0.496	−6.59	0.0162
0.291	283.15	1.2166	87.42	1.4680	−0.616	−243.64	0.0246
0.291	288.15	1.2128	68.04	1.4667	−0.628	−166.87	0.0247
0.291	293.15	1.2089	54.90	1.4655	−0.619	−116.70	0.0249
0.291	298.15	1.2050	44.84	1.4640	−0.586	−83.26	0.0250
0.291	303.15	1.1998	36.75	1.4627	−0.421	−61.26	0.0252
0.291	308.15	1.1958	30.63	1.4615	−0.385	−45.68	0.0257
0.291	313.15	1.1931	26.13	1.4600	−0.503	−34.90	0.0256
0.291	318.15	1.1889	22.14	1.4587	−0.450	−27.36	0.0262
0.291	323.15	1.1847	19.51	1.4570	−0.385	−21.25	0.0259
0.291	328.15	1.1801	17.16	1.4560	−0.275	−16.90	0.0264
0.291	333.15	1.1752	15.43	1.4545	−0.109	−13.03	0.0265
0.291	338.15	1.1693	13.79	1.4532	0.190	−10.57	0.0267
0.291	343.15	1.1594		1.4518	1.024		0.0268
0.401	283.15	1.1920	50.26	1.4625	−0.868	−229.65	0.0315
0.401	288.15	1.1883	40.89	1.4612	−0.901	−157.77	0.0317
0.401	293.15	1.1846	33.49	1.4595	−0.928	−111.67	0.0315
0.401	298.15	1.1809	27.92	1.4580	−0.933	−80.47	0.0317
0.401	303.15	1.1772	23.38	1.4568	−0.977	−59.57	0.0321
0.401	308.15	1.1736	20.06	1.4555	−0.999	−44.55	0.0325
0.401	313.15	1.1699	17.38	1.4540	−1.015	−34.30	0.0325
0.479	283.15	1.1736	35.48	1.4590	−1.313	−208.16	0.0368
0.479	288.15	1.1698	29.17	1.4575	−1.347	−143.79	0.0369
0.479	293.15	1.1661	24.41	1.4563	−1.378	−101.99	0.0372
0.479	298.15	1.1622	20.67	1.4550	−1.371	−73.75	0.0377
0.479	303.15	1.1583	17.56	1.4535	−1.403	−54.72	0.0378
0.592	293.15	1.1296	12.98		−1.719	−86.27	
0.592	298.15	1.1258	11.13		−1.730	−63.04	
0.592	303.15	1.1220	9.60		−1.796	−47.22	
0.592	308.15	1.1175	8.48		−1.756	−35.81	
0.592	313.15	1.1148	7.43		−1.886	−28.03	
0.637	293.15	1.1104	11.66		−1.692	−76.77	

(continued)

x_1	T/K	ρ/ g · cm^{-3}	η/ mPa · s	n_D	V^E/ cm^3 · mol^{-1}	$\Delta\eta$/ mPa · s	Δn_D
0.637	298.15	1.1046	10.06		−1.511	−56.05	
0.637	303.15	1.1027	8.71		−1.767	−41.95	
0.637	308.15	1.1006	7.52		−1.967	−31.98	
0.637	313.15	1.0967	6.66		−1.990	−24.99	
0.790	293.15	1.0226	5.38		−1.412	−46.27	
0.790	298.15	1.0187	4.62		−1.412	−34.07	
0.790	303.15	1.0148	4.03		−1.501	−25.69	
0.790	308.15	1.011	3.74		−1.539	−19.49	
0.790	313.15	1.007	3.46		−1.558	−15.19	
0.860	293.15	0.9605	4.96		−0.772	−29.87	
0.860	298.15	0.9567	4.22		−0.766	−21.92	
0.860	303.15	0.9527	3.59		−0.852	−16.54	
0.860	308.15	0.9487	3.16		−0.884	−12.62	
0.860	313.15	0.9446	2.80		−0.894	−9.90	

2.2.31 02011-02001: 1-Butyl-3-methylimidazolium tetrafluoroborate

Abbreviation:	[C_4MIm][BF_4], [BMIM][BF_4], [bmim][BF_4],[BMIm][BF_4], [Bmim][BF_4], [C_4mim][BF_4], BMIMBF_4, bmimBF_4, [C_1C_4Im][BF_4]
Molecular Formula:	$C_8H_{15}BF_4N_2$
Molar Mass:	226.03
Structure:	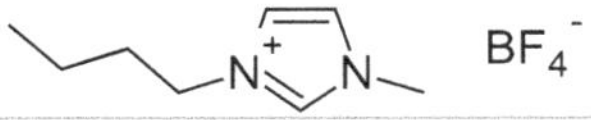

2.2.31.1 Binary Mixtures

Mixture 1: CH_3OH (1) + [BMIM][BF_4] (2)

(1) Volumetric Properties, Speed of Sound, Compressibility Properties and Conductivity

Ref. [127]

p/MPa	ρ/(kg · m^{-3})	$\varphi_{V,2}$/(cm^3 · mol^{-1})	p/MPa	ρ/(kg · m^{-3})	$\varphi_{V,2}$/(cm^3 · mol^{-1})	p/MPa	ρ/(kg · m^{-3})	$\varphi_{V,2}$/(cm^3 · mol^{-1})
x_2 = 0.0			x_2 = 0.0087			x_2 = 0.0433		
T = 298.15 K								
0.235	785.98		0.38	804.76	175.94	0.40	867.22	176.91
5.000	790.69		4.38	808.51	176.32	5.27	871.22	177.30
9.999	795.40		10.91	814.38	176.32	10.26	875.17	177.57
15.013	799.88		15.33	818.18	176.04	15.19	878.92	177.77
19.987	804.10		19.94	821.99	175.75	20.02	882.44	177.92
25.008	808.17		25.77	826.59	175.39	25.42	886.20	178.10
29.997	812.04		30.52	830.16	175.29	30.08	889.29	178.26
34.991	815.76		35.74	833.90	175.33	35.42	892.67	178.47
39.981	819.34		39.88	836.72	175.56	39.56	895.16	178.57
T = 323.15 K								
0.27	762.14		0.24	781.00	177.29	0.44	844.94	177.40
4.99	767.53		5.44	786.60	177.83	5.43	849.82	177.71
9.984	772.87		11.02	792.32	177.54	10.16	854.23	177.87
14.982	777.91		15.44	796.64	177.05	15.25	858.75	177.99
19.989	782.69		20.64	801.47	176.51	20.34	863.04	178.07
24.986	787.21		25.66	805.89	176.08	25.63	867.24	178.18
29.995	791.53		30.56	809.96	176.02	29.94	870.48	178.29
34.987	795.64		35.45	813.80	176.32	35.21	874.21	178.52
39.993	799.60		39.41	816.73	176.80	39.89	877.31	178.69
T = 348.15 K								
0.366	737.29		0.64	757.20	175.52	0.71	821.46	178.25
4.987	743.46		6.02	763.72	177.59	5.37	826.71	178.75
9.986	749.65		10.96	769.43	178.08	10.67	832.40	179.04

14.991	755.42		15.12	774.02	177.97	15.15	836.97	179.14
20.004	760.84		20.44	779.62	177.40	20.11	841.77	179.19
25.02	765.94		25.32	784.47	176.89	25.18	846.40	179.23
30.013	770.76		30.15	789.02	176.46	30.16	850.67	179.32
34.989	775.32		35.12	793.42	176.53	35.74	855.14	179.50
40.000	779.70		39.64	797.19	176.40	40.01	858.32	179.74
$T = 373.15$ K								
0.534	710.55		0.68	731.31	169.60	0.62	797.86	175.33
4.983	717.67		5.41	738.04	173.60	5.50	804.09	176.72
9.99	724.99		10.39	744.76	175.27	10.78	810.45	177.64
14.998	731.69		15.45	751.20	175.70	15.39	815.67	178.19
19.985	737.88		20.58	757.35	175.31	20.21	820.81	178.63
24.994	743.68		25.58	762.96	175.01	24.99	825.57	179.04
29.997	749.13		30.78	768.40	174.97	30.55	830.70	179.55
35.001	754.27		35.78	773.25	175.66	34.97	834.46	180.05
39.996	759.14		39.51	776.62	176.62	39.91	838.34	180.64
$T = 398.15$ K								
0.934	689.30		0.87	702.45	159.82	1.25	772.27	171.29
4.984	689.00		5.01	709.59	167.04	5.06	778.08	173.39
9.998	697.91		10.03	717.77	170.95	10.41	785.75	175.25
14.986	705.88		14.95	725.27	171.94	15.07	791.97	176.30
19.987	713.16		20.06	732.52	171.86	20.31	798.45	177.19
24.992	719.87		24.98	738.99	171.44	25.08	803.88	177.90
29.991	726.11		30.12	745.21	171.56	30.20	809.19	178.75
35.005	731.99		35.31	750.93	172.78	34.96	813.67	179.67
40.004	737.47		39.97	755.59	175.02	39.94	817.88	180.75
$x_2 = 0.1302$			$x_2 = 0.2626$			$x_2 = 0.4988$		

(*continued*)

p/MPa	ρ/(kg · m^{-3})	$\varphi_{V,2}$/(cm^3 · mol^{-1})	p/MPa	ρ/(kg · m^{-3})	$\varphi_{V,2}$/(cm^3 · mol^{-1})	p/MPa	ρ/(kg · m^{-3})	$\varphi_{V,2}$/(cm^3 · mol^{-1})
T = 298.15 K								
0.3	975.12	179.14	0.24	1065.61	182.14	0.25	1144.91	184.61
5.89	979.15	179.13	5.21	1068.88	181.93	5.31	1147.71	184.31
10.84	982.54	179.10	10.08	1071.94	181.72	10.05	1150.25	184.03
15.68	985.71	179.06	15.38	1075.12	181.49	15.62	1153.13	183.72
20.92	988.97	179.02	20.75	1078.17	181.28	20.14	1155.39	183.47
25.12	991.46	178.99	25.09	1080.52	181.12	25.41	1157.92	183.18
31.07	994.78	178.98	30.12	1083.10	180.95	30.41	1160.24	182.92
35.56	997.14	178.99	35.07	1085.49	180.80	35.74	1162.61	182.66
40.12	999.40	179.00	39.96	1087.71	180.67	39.78	1164.34	182.47
T – 323.15 K								
0.26	952.36	181.44	0.67	1046.20	184.16	0.78	1126.41	187.07
6.31	957.29	181.49	5.32	1049.53	183.99	5.31	1128.87	186.85
10.14	960.32	181.46	10.08	1052.82	183.80	10.09	1131.48	186.59
15.12	964.14	181.38	15.06	1056.14	183.58	15.42	1134.40	186.28
20.11	967.85	181.26	20.43	1059.59	183.34	20.74	1137.33	185.95
25.12	971.44	181.12	25.81	1062.91	183.10	25.03	1139.71	185.68
30.44	975.12	180.95	30.09	1065.44	182.92	30.01	1142.47	185.35
35.62	978.57	180.78	35.24	1068.37	182.70	35.24	1145.39	184.99
39.62	981.14	180.62	39.97	1070.95	182.51	39.97	1148.04	184.65
T = 348.15 K								
0.34	931.46	182.39	0.86	1026.35	186.09	0.61	1107.55	189.55
5.71	936.68	182.46	5.06	1029.49	186.03	5.41	1110.43	189.31
10.82	941.38	182.43	10.41	1033.58	185.85	10.06	1113.45	189.00
15.14	945.16	182.38	15.09	1037.07	185.65	15.12	1116.66	188.65
20.37	949.49	182.29	20.42	1040.92	185.40	20.07	1119.71	188.31

25.27	953.30	182.23	25.31	1044.36	185.14	25.31	1122.86	187.95
30.56	957.16	182.17	30.21	1047.70	184.88	30.09	1125.66	187.63
35.42	960.46	182.14	35.08	1050.92	184.62	35.26	1128.60	187.28
39.72	963.19	182.16	39.92	1054.02	184.35	39.92	1131.18	186.97
$T = 373.15$ K								
0.84	910.67	182.48	0.96	1006.13	187.69	0.88	1088.22	192.06
5.45	915.31	183.01	5.32	1009.96	187.69	5.47	1091.64	191.77
10.28	919.98	183.36	10.07	1014.10	187.57	10.08	1095.00	191.44
15.17	924.53	183.57	15.23	1018.44	187.38	15.43	1098.81	191.04
20.42	929.19	183.68	20.09	1022.37	187.16	20.61	1102.41	190.65
25.33	933.36	183.72	25.34	1026.45	186.90	25.34	1105.63	190.28
29.81	936.98	183.73	30.08	1029.97	186.67	30.08	1108.78	189.90
35.56	941.40	183.72	35.62	1033.90	186.39	35.64	1112.39	189.47
40.05	944.67	183.71	39.94	1036.83	186.17	39.97	1115.12	189.13
$T = 398.15$ K								
2.31	891.44	180.78	1.02	985.77	188.55	0.99	1069.40	194.23
5.12	894.51	181.52	5.06	989.10	188.98	5.08	1072.60	194.05
10.03	899.70	182.47	10.12	994.10	189.01	10.47	1077.23	193.64
15.02	904.75	183.12	15.45	999.14	188.91	15.21	1081.15	193.25
20.10	909.67	183.56	20.06	1003.34	188.76	20.03	1085.00	192.84
25.06	914.25	183.85	25.62	1008.18	188.52	25.32	1089.05	192.39
29.97	918.57	184.05	30.74	1012.44	188.26	30.09	1092.57	191.98
35.06	922.81	184.20	35.62	1016.31	188.01	35.04	1096.06	191.58
39.74	926.51	184.31	39.98	1019.62	187.79	39.96	1099.38	191.18
$x_2 = 0.7501$			$x_2 = 0.9102$			$x_2 = 1.0$		
$T = 298.15$ K								
0.31	1186.96	185.85	0.21	1203.93	186.35	0.28	1211.19	
5.08	1189.20	185.55	5.04	1206.16	186.02	5.64	1213.63	

(*continued*)

p/MPa	ρ/(kg · m^{-3})	$\varphi_{V,2}$/(cm^3 · mol^{-1})	p/MPa	ρ/(kg · m^{-3})	$\varphi_{V,2}$/(cm^3 · mol^{-1})	p/MPa	ρ/(kg · m^{-3})	$\varphi_{V,2}$/(cm^3 · mol^{-1})
10.03	1191.50	185.24	10.03	1208.41	185.69	10.62	1215.85	
15.42	1193.96	184.91	15.31	1210.74	185.35	15.34	1217.77	
20.89	1196.42	184.58	20.07	1212.80	185.05	20.64	1220.29	
25.68	1198.54	184.29	25.41	1215.05	184.72	25.27	1222.18	
30.08	1200.46	184.03	30.09	1216.97	184.44	30.05	1223.93	
35.29	1202.69	183.73	39.93	1220.88	183.87	35.41	1226.07	
39.91	1204.65	183.46	39.91	1220.87	183.87	39.61	1227.81	
T = 323.15 K								
0.24	1168.14	188.64	0.23	1185.20	189.23	0.73	1192.89	
5.07	1170.70	188.29	5.04	1187.61	188.87	5.64	1195.30	
10.34	1173.44	187.92	10.03	1190.07	188.50	10.61	1197.71	
15.43	1176.04	187.56	15.47	1192.72	188.10	15.44	1200.01	
20.07	1178.36	187.249	20.09	1194.92	187.77	20.46	1202.37	
25.09	1180.84	186.89	25.31	1197.38	187.40	25.36	1204.64	
30.41	1183.41	186.54	30.45	1199.76	187.04	30.74	1207.09	
35.24	1185.69	186.22	35.86	1202.22	186.67	35.65	1209.29	
39.89	1187.86	185.91	39.86	1204.01	186.40	39.85	1211.14	
T = 348.15 K								
0.87	1150.03	191.37	0.42	1166.95	192.12	0.21	1174.44	
5.07	1152.55	191.03	5.31	1169.75	191.69	5.96	1177.75	
10.06	1155.48	190.62	10.07	1172.42	191.27	10.48	1180.30	
15.87	1158.80	190.16	15.42	1175.35	190.82	15.43	1182.72	
20.08	1161.16	189.82	20.14	1177.87	190.43	20.87	1185.63	
25.62	1164.19	189.39	25.62	1180.73	189.99	25.85	1188.16	

30.07	1166.56	189.06	30.14	1183.03	189.63	30.11	1190.19	
35.01	1169.13	188.69	35.74	1185.80	189.21	35.11	1192.88	
39.99	1171.65	188.33	39.98	1187.84	188.90	39.61	1194.77	
T = 373.15 K								
0.84	1131.82	194.14	0.62	1149.53	194.94	0.47	1157.24	
5.01	1134.65	193.75	5.41	1152.54	194.46	5.35	1160.25	
10.32	1138.17	193.26	10.03	1155.38	194.01	10.08	1163.09	
15.09	1141.24	192.82	15.72	1158.79	193.47	15.71	1166.39	
20.42	1144.57	192.35	20.61	1161.64	193.02	20.97	1169.38	
25.08	1147.41	191.94	25.83	1164.62	192.55	25.82	1172.07	
30.41	1150.55	191.48	30.14	1167.01	192.17	30.12	1174.39	
35.21	1153.29	191.09	35.04	1169.67	191.75	35.53	1177.22	
39.98	1155.94	190.70	39.94	1172.26	191.35	39.82	1179.41	
T = 398.15 K								
0.98	1114.32	196.76	1.03	1132.67	197.71	0.87	1141.05	
5.31	1117.54	196.33	5.41	1135.70	197.23	6.67	1144.53	
10.41	1121.25	195.83	10.25	1138.98	196.70	10.49	1147.07	
15.96	1125.17	195.28	15.36	1142.35	196.16	15.73	1150.32	
20.34	1128.18	194.85	20.47	1145.63	195.63	20.98	1153.78	
25.72	1131.78	194.33	25.01	1148.46	195.17	25.14	1156.30	
30.08	1134.62	193.91	30.89	1152.03	194.60	30.89	1159.52	
35.21	1137.87	193.43	35.28	1154.61	194.18	35.81	1162.40	
39.94	1140.77	193.00	39.92	1157.27	193.75	39.56	1164.38	

Ref. [127]

x_2	T = 298.15 K V^E_m/(cm^3 · mol^{-1})	T = 323.15 K V^E_m/(cm^3 · mol^{-1})	T = 398.15 K V^E_m/(cm^3 · mol^{-1})
p = 0.1 MPa			
0.0	0.0	0.0	
0.0087	−0.1163	−0.1385	
0.0433	−0.4421	−0.5558	
0.1302	−0.9934	−1.0819	
0.2626	−1.1929	−1.4341	
0.4988	−1.0154	−1.2360	
0.7501	−0.5776	−0.6823	
0.9102	−0.2454	−0.2782	
1.0	0.0	0.0	
p = 10 MPa			
0.0	0.0	0.0	0.0
0.0087	−0.0875	−0.1751	−0.2259
0.0433	−0.3653	−0.7345	−0.9468
0.1302	−0.8937	−1.4391	−1.9032
0.2626	−1.1131	−1.7860	−2.1213
0.4988	−0.9559	−1.4518	−1.7116
0.7501	−0.5264	−0.7936	−0.9270
0.9102	−0.2031	−0.3019	−0.3464
1.0	0.0	0.0	0.0
P = 30 MPa			
0.0	0.0	0.0	0.0
0.0087	−0.0803	−0.1505	−0.2019
0.0433	−0.2747	−0.5583	−0.7021
0.1302	−0.8313	−1.1355	−1.4243
0.2626	−0.9781	−1.5250	−1.7679
0.4988	−0.8656	−1.2819	−1.5147
0.7501	−0.4814	−0.7186	−0.8256
0.9102	−0.1323	−0.2673	−0.3075
1.0	0.0	0.0	0.0

Ref. [127]

	T = 298.15 K		T = 323.15 K		T = 398.15 K	
x_2	$\overline{V}_1$/ (cm^3 · mol^{-1})	$\overline{V}_2$/ (cm^3 · mol^{-1})	$\overline{V}_1$/ (cm^3 · mol^{-1})	$\overline{V}_2$/ (cm^3 · mol^{-1})	$\overline{V}_1$/ (cm^3 · mol^{-1})	$\overline{V}_2$/ (cm^3 · mol^{-1})
p = 0.1 MPa						
0.0	40.77	180.79	42.05	182.75		
0.0087	40.56	180.80	41.80	182.75		
0.0433	40.14	180.81	41.31	182.77		
0.1302	39.15	180.90	40.16	182.88		
0.2626	37.82	181.23	38.61	183.25		

(*continued*)

x_2	$T = 298.15$ K		$T = 323.15$ K		$T = 398.15$ K	
	$\overline{V}_1$/ (cm^3 · mol^{-1})	$\overline{V}_2$/ (cm^3 · mol^{-1})	$\overline{V}_1$/ (cm^3 · mol^{-1})	$\overline{V}_2$/ (cm^3 · mol^{-1})	$\overline{V}_1$/ (cm^3 · mol^{-1})	$\overline{V}_2$/ (cm^3 · mol^{-1})
0.4988	36.00	182.35	36.49	184.56		
0.7501	34.82	184.30	35.13	186.83		
0.9102	34.49	185.96	34.73	188.76		
1.0	34.44	186.62	34.67	189.55		
$p = 10$ MPa						
0.0	40.28	180.60	41.44	182.75	45.91	187.03
0.0087	40.24	180.66	41.40	182.82	45.77	187.09
0.0433	40.12	180.95	41.21	183.08	45.35	187.45
0.1302	39.88	181.72	41.05	184.06	44.94	188.97
0.2626	39.77	183.14	40.82	185.55	44.89	191.88
0.4988	39.10	185.21	40.11	187.93	43.67	195.92
0.7501	37.23	186.26	37.99	189.09	39.98	197.85
0.9102	35.32	186.21	35.86	189.05	36.24	197.68
1.0	34.01	185.95	34.40	188.77	33.67	197.11
$p = 30$ MPa						
0.0	39.47	180.01	40.49	181.99	44.16	186.74
0.0087	39.43	180.05	40.44	182.04	44.03	186.77
0.0433	39.38	180.35	40.31	182.31	43.78	187.15
0.1302	39.17	181.02	40.16	183.16	43.51	188.45
0.2626	39.02	182.21	39.94	184.46	43.32	190.66
0.4988	38.42	184.00	39.30	186.54	42.26	193.89
0.7501	36.80	184.93	37.44	187.58	39.34	195.52
0.9102	35.21	184.95	35.58	187.56	36.37	195.45
1.0	34.03	184.68	34.29	187.31	34.31	195.02

Ref. [128]

m_2 (mol · kg^{-1})	$10^3\rho$ (kg · m^{-3})	u (m · s^{-1})	10^6V_φ (m^3 · mol^{-1})	$10^{12}\kappa_s$ (Pa^{-1})	$10^{14}\kappa_\varphi$ (m^3 · mol^{-1} · Pa^{-1})
$T = 298.15$ K					
0.0203	0.788033	1104.41	167.40	1040.39	−15.41
0.0390	0.789368	1105.90	169.06	1035.83	−14.45
0.0564	0.790567	1107.17	170.81	1031.89	−13.35
0.0724	0.791666	1108.20	171.57	1028.54	−12.36
0.1104	0.794278	1110.84	172.19	1020.29	−11.62
0.1461	0.796670	1113.12	172.95	1013.07	−10.82
0.2322	0.802285	1118.83	174.23	995.73	−9.97
0.2495	0.803376	1119.29	174.47	993.56	−9.20
0.3051	0.806868	1122.34	175.10	983.89	−8.50
0.4201	0.813915	1130.32	175.88	961.65	−8.50
0.5784	0.823136	1143.71	176.84	928.75	−9.27
0.6438	0.826790	1146.43	177.21	920.26	−8.45

Ref. [129]

x_2	ρ/cm^3 · mol^{-1}						
	T = 293.15 K	T = 298.15 K	T = 303.15 K	T = 308.15 K	T = 313.15 K	T = 318.15 K	T = 323.15 K
0	0.7913	0.7865	0.7818	0.7771	0.7723	0.7675	0.7626
0.0486	0.8772	0.8727	0.8682	0.8637	0.8592	0.8546	0.8501
0.0988	0.9397	0.9354	0.9310	0.9267	0.9223	0.9180	0.9136
0.1505	0.9873	0.9830	0.9788	0.9745	0.9703	0.9661	0.9618
0.1987	1.0220	1.0178	1.0137	1.0095	1.0054	1.0012	0.9971
0.2541	1.0534	1.0494	1.0453	1.0412	1.0372	1.0331	1.0291
0.3235	1.0842	1.0802	1.0762	1.0722	1.0683	1.0643	1.0604
0.3721	1.1015	1.0976	1.0936	1.0897	1.0858	1.0819	1.0780
0.4517	1.1245	1.1206	1.1168	1.1129	1.1091	1.1053	1.1015
0.5740	1.1518	1.1481	1.1443	1.1406	1.1368	1.1331	1.1294
0.6729	1.1682	1.1645	1.1608	1.1571	1.1534	1.1497	1.1461
0.7342	1.1770	1.1733	1.1697	1.1660	1.1624	1.1587	1.1551
0.7847	1.1835	1.1798	1.1762	1.1725	1.1689	1.1653	1.1617
0.8423	1.1902	1.1865	1.1829	1.1793	1.1757	1.1721	1.1686
0.9038	1.1964	1.1927	1.1892	1.1856	1.1820	1.1784	1.1749
1	1.2049	1.2013	1.1978	1.1942	1.1907	1.1872	1.1837

Ref. [130]

x_2	$T = 298.15$ K		$T = 303.15$ K		$T = 308.15$ K		$T = 313.15$ K		$T = 318.15$ K	
	ρ $(g \cdot m^{-3})$	u $(m \cdot s^{-1})$	ρ $(g \cdot cm^{-3})$	u $(m \cdot s^{-1})$	ρ $(g \cdot cm^{-3})$	u $(m \cdot s^{-1})$	ρ $(g \cdot cm^{-3})$	u $(m \cdot s^{-1})$	ρ $(g \cdot cm^{-3})$	u $(m \cdot s^{-1})$
0	0.786531	1102.68	0.781808	1086.37	0.777066	1070.11	0.772283	1053.96	0.767474	1037.91
0.0086	0.805151	1121.25	0.800482	1105.27	0.795796	1089.35	0.791081	1073.51	0.786347	1057.8
0.0473	0.871093	1177.15	0.866582	1162.4	0.862063	1147.75	0.857529	1133.11	0.852979	1118.56
0.0989	0.935734	1238.28	0.931386	1224.64	0.927032	1210.99	0.922668	1197.51	0.918304	1184.02
0.2027	1.020748	1334.41	1.016602	1321.8	1.012461	1309.16	1.008326	1296.54	1.004194	1284.03
0.3516	1.090689	1423.63	1.086808	1411.33	1.082621	1399.13	1.079104	1378	1.075215	1374.95
0.4958	1.131967	1476.98	1.12815	1465.01	1.124345	1453.07	1.120992	1441.82	1.117111	1430.06
0.6961	1.168328	1524.1	1.164632	1512.19	1.160962	1500.43	1.157297	1488.78	1.153661	1477.25
0.7982	1.181541	1541.21	1.177909	1529.49	1.174283	1517.86	1.170671	1506.29	1.167073	1494.83
1	1.20142	1576.05	1.197884	1555.47	1.194346	1543.92	1.190815	1532.52	1.187294	1521.19

Ref. [35]

x_2	$10^{-3}c_2$/mol · m^{-3}	ρ/kg · m^{-3}	κ/S · cm^{-1}	$10^4\Lambda$/S m^2 · mol^{-1}
298.15 K				
0.007409	0.1776	802.48	0.867	48.8
0.01543	0.3603	818.28	1.451	40.2
0.02435	0.5526	834.29	1.97	35.6
0.03355	0.7396	849.76	2.40	32.4
0.05717	1.173	885.02	3.30	28.1
0.08589	1.624	921.13	3.81	23.4
0.1242	2.127	961.01	4.18	19.7
0.1755	2.662	1002.48	4.26	15.99
0.2480	3.237	1046.31	3.92	12.11
0.3624	3.877	1094.85	3.09	7.98
0.5583	4.556	1145.28	1.760	3.86

Mixture 2: C_2H_5OH (1) + [BMIM][BF_4] (2)

(1) Volumetric Properties, Compressibility Properties, Refractive Index, Surface Tension and Transport Properties

Ref. [131]

x_2	ρ (g · cm^{-3})	n_D	x_2	ρ (g · cm^{-3})	n_D	x_2	ρ (g · cm^{-3})	n_D
T = 298.15 K								
0.0517	0.85045	1.3703	0.2518	1.04290	1.3946	0.6244	1.15656	1.4137
0.0918	0.89163	1.3767	0.3228	1.08197	1.3998	0.7090	1.17774	1.4161
0.1995	0.97613	1.3898	0.4219	1.11711	1.4058	0.8295	1.19274	1.4188
0.2495	1.00741	1.3943	0.5356	1.13894	1.4105			

Ref. [132]

	T/K					T/K			
x_2	293.15	303.15	313.15	323.15	x_2	293.15	303.15	313.15	323.15
ρ (g · cm^{-3})									
0	0.78938	0.78078	0.77204	0.76314	0.4219	1.08581	1.07814	1.07051	1.06291
0.0517	0.85459	0.84627	0.83783	0.82927	0.5356	1.12088	1.11335	1.10588	1.09846
0.0918	0.89567	0.88754	0.87927	0.87090	0.6244	1.14266	1.13523	1.12786	1.12054
0.1479	0.94311	0.93512	0.92702	0.91884	0.7090	1.16024	1.15289	1.14559	1.13837
0.1995	0.98007	0.97215	0.96415	0.95611	0.8295	1.18136	1.17412	1.16693	1.15980
0.2518	1.01133	1.00346	0.99556	0.98764	0.9327	1.19633	1.18916	1.18204	1.17497
0.3228	1.04679	1.03901	1.03122	1.02346	1	1.20493	1.19783	1.19074	1.18372

Ref. [133]

P (MPa)	ρ (kg · m^{-3})	φ_V (cm^3 · mol^{-1})	P (MPa)	ρ (kg · m^{-3})	φ_V (cm^3 · mol^{-1})
x_2 = 0.0701			x_2 = 0.3147		
T = 298.15 K					
0.390	875.91	178.03	0.340	1044.49	184.76
5.500	880.04	177.96	5.630	1048.05	184.45

(*continued*)

P (MPa)	ρ (kg · m^{-3})	φ_V (cm^3 · mol^{-1})	P (MPa)	ρ (kg · m^{-3})	φ_V (cm^3 · mol^{-1})
11.37	884.58	177.81	10.91	1051.47	184.14
15.13	887.37	177.71	15.28	1054.20	183.88
21.07	891.60	177.53	20.54	1057.36	183.58
25.57	894.66	177.41	25.47	1060.21	183.31
30.62	897.94	177.31	31.12	1063.34	183.01
35.86	901.17	177.25	35.80	1065.82	182.78
39.34	903.23	177.23	39.77	1067.84	182.58
T = 323.15 K					
0.290	856.52	177.35	0.340	1026.83	186.49
5.410	861.09	177.61	5.010	1030.15	186.28
10.73	865.61	177.77	10.11	1033.63	186.04
15.43	869.40	177.88	15.41	1037.07	185.80
20.33	873.16	177.98	20.75	1040.38	185.56
25.38	876.83	178.09	26.29	1043.62	185.32
30.33	880.21	178.25	30.64	1046.04	185.15
34.74	883.06	178.44	35.26	1048.48	184.97
40.01	886.25	178.64	40.12	1050.91	184.81
T = 348.15 K					
0.520	835.75	176.18	0.230	1007.50	188.33
5.340	840.67	176.77	5.740	1011.89	188.13
11.71	846.75	177.38	11.42	1016.16	187.91
15.84	850.45	177.69	15.74	1019.22	187.74
20.55	854.43	178.05	21.63	1023.15	187.52
25.38	858.25	178.43	26.12	1025.95	187.36
30.22	861.80	178.88	30.55	1028.55	187.23
35.17	865.16	179.42	35.63	1031.33	187.09
40.03	868.18	180.02	40.01	1033.56	187.00
T = 373.15 K					
0.510	812.28	174.28	0.810	987.83	189.87
5.790	818.18	176.07	5.820	992.08	189.89
11.04	823.77	177.31	10.95	996.25	189.82
15.13	827.92	178.04	15.73	999.95	189.73
21.22	833.79	178.86	19.86	1003.02	189.61
25.13	837.37	179.27	25.99	1007.35	189.43
30.26	841.81	179.76	30.67	1010.47	189.29
35.52	846.10	180.23	35.22	1013.34	189.15
39.84	849.40	180.57	39.82	1016.10	189.02
T = 398.15 K					
1.010	788.85	167.92	0.960	966.26	191.06
5.750	795.42	170.47	6.350	971.33	191.39
10.15	801.14	172.20	10.93	975.46	191.52
15.11	807.16	173.70	15.62	979.51	191.56
20.74	813.42	175.13	20.15	983.26	191.54

(continued)

P (MPa)	ρ (kg · m^{-3})	φ_V (cm^3 · mol^{-1})	P (MPa)	ρ (kg · m^{-3})	φ_V (cm^3 · mol^{-1})
25.13	817.89	176.16	25.17	987.21	191.47
30.85	823.16	177.55	30.41	991.12	191.36
35.86	827.28	178.86	35.16	994.46	191.25
39.99	830.32	179.98	39.85	997.59	191.14
x_2 = 0.5384			x_2 = 0.7452		
T = 298.15 K					
0.210	1124.67	185.82	0.180	1171.71	186.30
5.040	1127.45	185.50	4.960	1174.16	185.97
10.26	1130.36	185.17	10.06	1176.71	185.63
15.32	1133.10	184.86	15.21	1179.21	185.30
20.09	1135.59	184.57	20.36	1181.64	184.98
26.07	1138.61	184.21	25.41	1183.95	184.67
30.07	1140.56	183.98	30.09	1186.03	184.39
35.41	1143.07	183.69	35.48	1188.36	184.08
39.98	1145.14	183.45	39.92	1190.22	183.82
T = 323.15 K					
0.850	1106.94	188.20	0.420	1153.47	188.99
4.890	1109.43	187.94	5.060	1156.05	188.65
9.860	1112.40	187.62	10.74	1159.11	188.25
14.87	1115.29	187.31	15.08	1161.39	187.94
20.16	1118.22	186.98	20.31	1164.06	187.58
25.84	1121.23	186.64	25.12	1166.45	187.25
30.26	1123.49	186.39	30.08	1168.84	186.93
35.42	1126.01	186.10	35.07	1171.16	186.62
40.02	1128.17	185.85	40.01	1173.40	186.31
T = 348.15 K					
0.450	1088.13	190.66	0.380	1134.96	191.75
5.010	1091.32	190.34	5.360	1138.08	191.34
10.06	1094.71	189.99	10.09	1140.95	190.95
15.21	1098.01	189.64	15.42	1144.07	190.53
20.06	1100.98	189.32	20.08	1146.71	190.17
25.43	1104.11	188.98	25.41	1149.62	189.77
30.67	1107.00	188.66	30.07	1152.07	189.43
35.62	1109.59	188.38	35.14	1154.63	189.08
39.56	1111.54	188.17	39.94	1156.96	188.76
T = 373.15 K					
0.750	1069.65	192.92	0.520	1117.10	194.36
5.060	1072.77	192.68	5.090	1120.20	193.97
10.07	1076.30	192.39	10.61	1123.82	193.50
15.34	1079.90	192.05	15.84	1127.11	193.06
20.61	1083.37	191.71	20.09	1129.70	192.72
25.93	1086.76	191.36	25.46	1132.84	192.29
30.01	1089.27	191.09	30.07	1135.43	191.94

(continued)

P (MPa)	ρ (kg · m^{-3})	φ_V (cm^3 · mol^{-1})	P (MPa)	ρ (kg · m^{-3})	φ_V (cm^3 · mol^{-1})
35.03	1092.26	190.76	35.18	1138.19	191.56
39.94	1095.08	190.45	39.92	1140.63	191.22
T = 398.15 K					
0.920	1049.06	195.35	0.940	1098.62	197.05
5.060	1052.44	195.19	5.120	1101.74	196.68
10.42	1056.68	194.91	10.07	1105.35	196.24
15.03	1060.22	194.63	15.08	1108.88	195.78
19.97	1063.89	194.30	20.64	1112.68	195.28
24.98	1067.49	193.95	25.37	1115.80	194.85
30.03	1070.98	193.60	30.28	1118.94	194.42
35.07	1074.35	193.24	35.14	1121.93	194.01
39.94	1077.48	192.90	39.92	1124.78	193.61
x_2 = 0.9152					
T = 298.15 K			T = 323.15 K		
0.120	1199.70	186.53	0.510	1181.35	189.36
5.140	1202.06	186.18	4.870	1183.60	189.02
10.03	1204.31	185.85	10.08	1186.23	188.62
15.62	1206.82	185.48	15.21	1188.77	188.24
20.48	1208.95	185.17	20.47	1191.31	187.86
26.14	1211.36	184.82	25.01	1193.45	187.54
29.87	1212.91	184.59	30.07	1195.79	187.19
35.74	1215.30	184.25	34.96	1198.00	186.86
39.98	1216.97	184.01	40.00	1200.22	186.53
T = 348.15 K			T = 373.15 K		
0.460	1163.17	192.22	0.780	1145.53	195.06
5.040	1165.81	191.82	5.140	1148.26	194.64
11.21	1169.27	191.29	10.57	1151.58	194.12
15.08	1171.38	190.96	15.08	1154.27	193.70
20.74	1174.39	190.50	20.54	1157.45	193.20
25.14	1176.67	190.15	24.67	1159.79	192.83
29.98	1179.10	189.78	29.76	1162.61	192.39
35.07	1181.59	189.40	36.74	1166.35	191.82
39.74	1183.80	189.06	39.87	1167.99	191.55
T = 398.15 K					
0.860	1128.45	197.84			
5.840	1131.83	197.31			
9.630	1134.36	196.92			
16.47	1138.81	196.21			
20.04	1141.07	195.85			
25.12	1144.22	195.35			
30.45	1147.45	194.84			
35.74	1150.56	194.34			
39.86	1152.93	193.96			

Ref. [34]

x_2	c (mol/L)	ρ (g/cm^3)	x_2	c (mol/L)	ρ (g/cm^3)
298.15 K					
1.0000	5.313	1.2012	0.3999	3.643	1.0755
0.9059	5.149	1.1888	0.3284	3.268	1.0468
0.8040	4.946	1.1737	0.2881	3.024	1.0279
0.7107	4.727	1.1574	0.2553	2.804	1.0107
0.6082	4.441	1.1359	0.2002	2.383	0.9774
0.5024	4.080	1.1086	0.0986	1.386	0.8972
0.4202	3.738	1.0827			

Ref. [34]

x_2	c (mol/L)	σ (mN/m)	$\delta\sigma$ (mN/m)	x_2	c (mol/L)	σ (mN/m)	$\delta\sigma$ (mN/m)
298.15 K							
1.0000	5.313	45.33	0.00	0.4202	3.738	27.20	−4.61
0.9059	5.149	41.54	−1.61	0.3999	3.643	26.55	−4.78
0.8040	4.946	37.99	−2.78	0.3284	3.268	25.53	−4.14
0.7107	4.727	34.90	−3.70	0.2553	2.804	24.65	−3.31
0.6082	4.441	31.32	−4.88	0.2002	2.383	23.28	−3.39
0.5024	4.080	28.87	−4.86	0.0986	1.386	22.83	−1.47

Ref. [134]

	x_2									
T/K	0.0000	0.2000	0.3000	0.3999	0.4998	0.6000	0.6996	0.8001	0.8998	1.0000
ρ/g · cm^{-3}										
p = 0.20 MPa										
313.2	0.7695	0.9648	1.0223	1.0713	1.0976	1.1209	1.1418	1.1576	1.1731	1.1852
353.2	0.7335	0.9341	0.9922	1.0411	1.0693	1.0931	1.1248	1.1309	1.1460	1.1587
p = 0.60 MPa										
313.2	0.7697	0.9654	1.0225	1.0715	1.0979	1.1214	1.1422	1.1578	1.1735	1.1856
353.2	0.7342	0.9346	0.9926	1.0418	1.0700	1.0934	1.1252	1.1315	1.1463	1.1590
393.2	0.6906	0.9028	0.9633	1.0130	1.0416	1.0661	1.0970	1.1053	1.1207	1.1337
p = 1.00 MPa										
313.2	0.7703	0.9658	1.0229	1.0716	1.0980	1.1221	1.1429	1.1585	1.1736	1.1860
353.2	0.7348	0.9350	0.9929	1.0422	1.0704	1.0940	1.1258	1.1318	1.1465	1.1595
393.2	0.6915	0.9032	0.9639	1.0135	1.0420	1.0668	1.0977	1.1055	1.1210	1.1341
p = 1.40 MPa										
313.2	0.7709	0.9661	1.0235	1.0719	1.0984	1.1224	1.1434	1.1588	1.1737	1.1863
353.2	0.7354	0.9353	0.9934	1.0425	1.0706	1.0947	1.1265	1.1321	1.1468	1.1597
393.2	0.6921	0.9038	0.9642	1.0138	1.0423	1.0672	1.0981	1.1059	1.1213	1.1343
433.2	0.6367	0.8703	0.9336	0.9854	1.0151	1.0407	1.0643	1.0799	1.0958	1.1112

(continued)

	x_2									
T/K	0.0000	0.2000	0.3000	0.3999	0.4998	0.6000	0.6996	0.8001	0.8998	1.0000
p = 1.80 MPa										
313.2	0.7713	0.9669	1.0238	1.0725	1.0991	1.1227	1.1438	1.1594	1.1742	1.1866
353.2	0.7362	0.9360	0.9938	1.0428	1.0709	1.0949	1.1267	1.1325	1.1472	1.1600
393.2	0.6931	0.9043	0.9645	1.0141	1.0429	1.0677	1.0986	1.1065	1.1217	1.1346
433.2	0.6381	0.8707	0.9339	0.9858	1.0159	1.0409	1.0644	1.0803	1.0961	1.1113
p = 2.00 MPa										
313.2	0.7717	0.9673	1.0241	1.0729	1.1016	1.1230	1.1443	1.1596	1.1746	1.1867
353.2	0.7380	0.9362	0.9940	1.0429	1.0710	1.0951	1.1269	1.1327	1.1475	1.1603
393.2	0.6934	0.9045	0.9649	1.0145	1.0431	1.0679	1.0988	1.1067	1.1218	1.1350
433.2	0.6384	0.8710	0.9342	0.9860	1.0161	1.0412	1.0646	1.0806	1.0963	1.1117

	x_2							
T/K	0.2000	0.3000	0.3999	0.4998	0.6000	0.6996	0.8001	0.8998
V^E/cm^3 · mol^{-1}								
p = 0.20 MPa								
313.2	−0.982	−1.248	−2.013	−1.345	−0.949	−0.800	−0.386	−0.306
353.2	−1.409	−1.647	−2.327	−1.718	−1.245	−2.456	−0.569	−0.313
p = 0.60 MPa								
313.2	−1.010	−1.235	−1.994	−1.335	−0.960	−0.799	−0.351	−0.291
353.2	−1.398	−1.627	−2.337	−1.742	−1.227	−2.442	−0.602	−0.309
393.2	−2.347	−2.642	−3.238	−2.438	−1.814	−2.752	−0.895	−0.478
p = 1.00 MPa								
313.2	−0.999	−1.223	−1.958	−1.300	−0.990	−0.835	−0.387	−0.251
353.2	−1.378	−1.596	−2.321	−1.724	−1.234	−2.453	−0.583	−0.261
393.2	−2.300	−2.623	−3.216	−2.404	−1.827	−2.776	−0.861	−0.464
p = 1.40 MPa								
313.2	−0.975	−1.230	−1.943	−1.297	−0.980	−0.852	−0.375	−0.211
353.2	−1.356	−1.602	−2.313	−1.701	−1.286	−2.515	−0.588	−0.294
393.2	−2.307	−2.608	−3.202	−2.397	−1.844	−2.800	−0.880	−0.466
433.2	−4.272	−4.495	−4.983	−3.869	−2.968	−2.462	−1.224	−0.475
p = 1.80 MPa								
313.2	−1.011	−1.221	−1.958	−1.333	−0.975	−0.848	−0.414	−0.239
353.2	−1.359	−1.582	−2.285	−1.679	−1.250	−2.483	−0.593	−0.289
393.2	−2.272	−2.560	−3.168	−2.393	−1.841	−2.808	−0.919	−0.480
433.2	−4.183	−4.417	−4.929	−3.887	−2.924	−2.422	−1.242	−0.489
p = 2.00 MPa								
313.2	−1.022	−1.226	−1.979	−1.584	−0.987	−0.896	−0.429	−0.273
353.2	−1.251	−1.480	−2.187	−1.593	−1.189	−2.434	−0.551	−0.282
393.2	−2.255	−2.558	−3.158	−2.373	−1.815	−2.780	−0.883	−0.435
433.2	−4.180	−4.407	−4.901	−3.857	−2.916	−2.394	−1.227	−0.458

Ref. [135]

x_2	η/(mPa · s)	x_2	η/(mPa · s)	x_2	η/(mPa · s)
298.15 K					
0.00	1.087	0.3520	6.848	0.6137	19.756
0.0545	1.453	0.3963	8.200	0.6870	27.000
0.0898	1.811	0.4321	10.800	0.7705	38.560
0.1993	3.432	0.4622	16.700	0.8633	58.400
0.2882	5.276	0.5803	17.500	1.00	110.308

(2) Excess Enthalpy

Ref. [132]

x_2	H^E/ J · mol^{-1}	x_2	H^E/ J · mol^{-1}	x_2	H^E/ J · mol^{-1}	x_2	H^E/ J · mol^{-1}
T = 303.15 K							
0.0201	360	0.1962	1956	0.5001	2509	0.7422	1668
0.0573	852	0.3371	2472	0.5874	2279	0.8291	1159
0.1005	1285	0.4225	2609	0.6333	2049	0.9257	625

(3) Electrical Conductivity

Ref. [135]

x_2	$10^3\kappa$ (S · cm^{-1})	x_2	$10^3\kappa$ (S · cm^{-1})	x_2	$10^3\kappa$ (S · cm^{-1})
298.15 K					
0.0545	6.47	0.3963	13.77	0.6870	9.46
0.0898	9.09	0.4321	13.13	0.7705	6.06
0.1993	14.84	0.4622	12.60	0.8633	4.62
0.2882	15.13	0.5803	10.61	1.00	2.48
0.3520	15.95	0.6137	9.58		

Mixture 3: CH_3NO_2 (1) + [Bmim][BF_4] (2)
(1) Volumetric Properties and Refractive Index

Ref. [131]

x_2	ρ (g · cm^{-3})	n_D	x_2	ρ (g · cm^{-3})	n_D	x_2	ρ (g · cm^{-3})	n_D
T = 298.15 K								
0.0512	1.14818	1.3889	0.3565	1.18514	1.4112	0.6863	1.19669	1.4188
0.0991	1.15864	1.3952	0.3972	1.18727	1.4125	0.8073	1.19886	1.4202
0.1392	1.16519	1.3990	0.4553	1.18989	1.4142	0.8634	1.19969	1.4207
0.2026	1.17312	1.4038	0.5231	1.19230	1.4158			
0.2986	1.18143	1.4089	0.5949	1.19448	1.4172			

Ref. [132]

	T/K					T/K			
x_2	293.15	303.15	313.15	323.15	x_2	293.15	303.15	313.15	323.15
ρ (g · cm^{-3})									
0	1.13798	1.12439	1.11068	1.09688	0.4553	1.19413	1.18565	1.17721	1.16883
0.0512	1.15427	1.14208	1.12984	1.11757	0.5231	1.19642	1.18818	1.17999	1.17187
0.0991	1.16429	1.15298	1.14166	1.13032	0.5949	1.19848	1.19048	1.18253	1.17465
0.1392	1.17056	1.15982	1.14908	1.13835	0.6863	1.20057	1.19282	1.18511	1.17748
0.2026	1.17815	1.16809	1.15806	1.14805	0.8073	1.20260	1.19513	1.18769	1.18033
0.2986	1.18608	1.17678	1.16752	1.15831	0.8634	1.20337	1.19601	1.18869	1.18143
0.3565	1.18962	1.18067	1.17176	1.16290	1	1.20499	1.19789	1.19081	1.18378
0.3972	1.19164	1.18290	1.17420	1.16555					

(2) Excess Enthalpy

Ref. [132]

x_2	H^E/J · mol^{-1}	x_2	H^E/J · mol^{-1}	x_2	H^E/J · mol^{-1}	x_2	H^E/J · mol^{-1}
$T = 303.15$ K							
0.0016	−7	0.1131	−287	0.5067	−452	0.8827	−159
0.0273	−92	0.202	−406	0.6037	−409		
0.0558	−182	0.3621	−487	0.7226	−316		

Mixture 4: 1,3-$ClCH_2CH_2CH_2Cl$ (1) + [Bmim][BF_4] (2)
(1) Volumetric Properties and Refractive Index

Ref. [131]

x_2	ρ (g · cm^{-3})	n_D	x_2	ρ (g · cm^{-3})	n_D	x_2	ρ (g · cm^{-3})	n_D
$T = 298.15$ K								
0.0536	1.18413	1.4448	0.3528	1.19653	1.4359	0.5955	1.19976	1.4296
0.1135	1.18780	1.4430	0.4042	1.19751	1.4345	0.6786	1.20034	1.4276
0.1516	1.18978	1.4418	0.4626	1.19841	1.4330	0.7863	1.20089	1.4255
0.2101	1.19232	1.4401	0.5254	1.19914	1.4314	0.9028	1.20120	1.4234
0.2817	1.19471	1.4381	0.5466	1.19938	1.4309			

Mixture 5: $HOCH_2CH_2OCH_2CH_2OC_2H_5$ (1) + [Bmim][BF_4] (2)
(1) Volumetric Properties and Refractive Index

Ref. [131]

x_2	ρ (g · cm^{-3})	n_D	x_2	ρ (g · cm^{-3})	n_D	x_2	ρ (g · cm^{-3})	n_D
$T = 298.15$ K								
0.0544	1.00336	1.4268	0.3215	1.07837	1.4277	0.6998	1.15507	1.4247
0.1016	1.01820	1.4274	0.3932	1.09493	1.4273	0.8304	1.17650	1.4235
0.1492	1.03287	1.4277	0.5108	1.11982	1.4265	0.9551	1.19504	1.4223
0.1941	1.04510	1.4278	0.6016	1.13749	1.4257			
0.2686	1.06522	1.4278	0.6657	1.14916	1.4250			

Mixture 6: 1,2-$HOCH_2CH_2OH$ (1) + [Bmim][BF_4] (2)
(1) Volumetric Properties and Refractive Index

Ref. [131]

x_2	ρ (g · cm^{-3})	n_D	x_2	ρ (g · cm^{-3})	n_D	x_2	ρ (g · cm^{-3})	n_D
T = 298.15 K								
0.0517	1.12311	1.4287	0.3324	1.16186	1.4231	0.6320	1.18329	1.4215
0.1380	1.13878	1.4261	0.4020	1.16789	1.4226	0.6960	1.18682	1.4215
0.1928	1.14651	1.4250	0.4162	1.16898	1.4225	0.8045	1.19248	1.4215
0.2567	1.15422	1.4240	0.4997	1.17510	1.4219	0.9083	1.19733	1.4217
0.3000	1.15872	1.4235	0.5803	1.18022	1.4216			

Ref. [136]

	ρ/(g · cm^{-3})	ρ/(g · cm^{-3})	ρ/(g · cm^{-3})	n_D
x_2	T = 298.15 K	T = 308.15 K	T = 318.15 K	T = 298.15 K
0.0000	1.1100	1.1029	1.0958	1.4304
0.0266	1.1167	1.1094	1.1022	1.4290
0.0766	1.1273	1.1199	1.1125	1.4274
0.1017	1.1319	1.1244	1.1169	1.4266
0.1678	1.1424	1.1347	1.1269	1.4248
0.2425	1.1521	1.1441	1.1362	1.4234
0.3825	1.1660	1.1578	1.1497	1.4217
0.4389	1.1705	1.1624	1.1543	1.4213
0.5207	1.1764	1.1683	1.1603	1.4209
0.6049	1.1817	1.1737	1.1658	1.4207
0.7196	1.1881	1.1804	1.1727	1.4206
0.8106	1.1927	1.1851	1.1776	1.4207
0.8978	1.1968	1.1894	1.1821	1.4208
1.0000	1.2012	1.1940	1.1869	1.4208

Mixture 7: $HCOOCH_3$ (1) + [Bmim][BF_4] (2)
(1) Volumetric Properties and Transport Properties

Ref. [137]

x_2	w_2	ρ (g · cm^{-3})	η (mPa · s)	V_m^E (cm^3 · mol^{-1})	$\Delta\eta$ (mPa · s)
T = 298.15 K					
0	0	0.96478	0.3	0	0
0.0500	0.1653	1.01172	0.5	−0.905	−5.3
0.1011	0.2974	1.04555	0.9	−1.335	−10.5
0.2030	0.4895	1.09140	1.9	−1.602	−20.7
0.2993	0.6165	1.12147	3.5	−1.632	−29.7
0.3395	0.6592	1.13120	4.4	−1.576	−33.2
0.3994	0.7145	1.14364	6.3	−1.453	−38.0
0.4498	0.7547	1.15273	8.2	−1.337	−41.6
0.4997	0.7899	1.16043	10.6	−1.183	−44.7

(continued)

x_2	w_2	ρ (g · cm^{-3})	η (mPa · s)	V_m^E (cm^3 · mol^{-1})	$\Delta\eta$ (mPa · s)
0.5482	0.8204	1.16719	13.7	−1.032	−46.9
0.5994	0.8492	1.17337	17.12	−0.841	−49.0
0.6999	0.8977	1.18234	28.3	−0.265	−49.0
0.7996	0.9376	1.19278	44.2	−0.099	−44.1
0.8993	0.9711	1.20231	68.4	−0.019	−30.8
1	1	1.21105	110.3	0	0

Mixture 8: CH_3COOCH_3 (1) + [BMIM][BF_4] (2)
(1) Volumetric Properties and Transport Properties

Ref. [137]

x_2	w_2	ρ (g · cm^{-3})	η (mPa · s)	V_m^E (cm^3 · mol^{-1})	$\Delta\eta$ (mPa · s)
T = 298.15 K					
0	0	0.91597	0.4	0	0
0.0501	0.1386	0.96450	0.6	−1.477	−5.3
0.1004	0.2540	0.99604	0.9	−1.804	−10.5
0.1999	0.4326	1.04598	2.0	−2.155	−20.4
0.3031	0.5703	1.08458	3.6	−2.166	−30.1
0.3504	0.6220	1.09890	4.7	−2.072	−34.2
0.3995	0.6699	1.11224	6.2	−1.947	−38.1
0.4511	0.7149	1.12480	8.2	−1.786	−41.8
0.4993	0.7526	1.13601	10.2	−1.689	−45.1
0.5484	0.7875	1.14494	13.1	−1.394	−47.6
0.5995	0.8204	1.15395	16.3	−1.144	−50.0
0.6996	0.8766	1.16955	25.8	−0.635	−51.5
0.8000	0.9243	1.18288	40.8	−0.094	−47.5
0.8999	0.9648	1.19770	63.7	−0.033	−35.6
1	1	1.21105	110.3	0	0

Mixture 9: $HCOOC_2H_5$ (1) + [Bmim][BF_4] (2)
(1) Volumetric Properties and Transport Properties

Ref. [137]

x_2	w_2	ρ (g · cm^{-3})	η (mPa · s)	V_m^E (cm^3 · mol^{-1})	$\Delta\eta$ (mPa · s)
T = 298.15 K					
0	0	0.92848	0.4	0	0
0.0500	0.1384	0.96120	0.6	−0.156	−5.3
0.1004	0.2540	0.98920	1.0	−0.203	−10.4
0.1986	0.4306	1.04081	2.0	−0.839	−20.2
0.3047	0.5721	1.08216	3.8	−1.106	−30.1
0.3560	0.6278	1.09826	5.1	−1.120	−34.4
0.4027	0.6729	1.11127	6.5	−1.091	−38.1
0.4523	0.7159	1.12369	8.5	−1.029	−41.6
0.4971	0.7510	1.13373	10.6	−0.937	−44.4

(*continued*)

x_2	w_2	ρ (g · cm^{-3})	η (mPa · s)	V_m^E (cm^3 · mol^{-1})	$\Delta\eta$ (mPa · s)
0.5443	0.7847	1.14334	13.3	−0.817	−46.9
0.5987	0.8199	1.15332	16.9	−0.649	−49.3
0.6976	0.8756	1.16916	26.6	−0.303	−50.5
0.7973	0.9231	1.18378	43.7	−0.057	−44.3
0.8996	0.9647	1.19830	69.0	−0.018	−30.3
1	1	1.21105	110.3	0	0

Mixture 10: CH_3COCH_3 (1) + [Bmim][BF_4] (2)
(1) Volumetric Properties and Transport Properties

Ref. [137]

x_2	w_2	ρ (g · cm^{-3})	η (mPa · s)	V_m^E (cm^3 · mol^{-1})	$\Delta\eta$ (mPa · s)
T = 298.15 K					
0	0	0.78536	0.3	0	0
0.0508	0.1724	0.84615	0.5	−0.958	−5.3
0.1006	0.3033	0.89439	0.7	−1.461	−10.6
0.2012	0.4950	0.97029	1.5	−1.942	−20.6
0.3022	0.6276	1.02698	3.0	−2.033	−30.8
0.3501	0.6770	1.04891	4.0	−1.976	−35.5
0.3975	0.7197	1.06829	5.2	−1.887	−39.4
0.4491	0.7603	1.08616	6.9	−1.616	−43.1
0.4965	0.7933	1.10246	9.0	−1.584	−46.0
0.5507	0.8267	1.11843	12.0	−1.384	−48.1
0.5998	0.8536	1.13138	15.1	−1.169	−51.1
0.7037	0.9024	1.15529	25.5	−0.679	−51.5
0.7982	0.9390	1.17432	40.1	−0.284	−47.9
0.8949	0.9707	1.19301	64.7	−0.131	−34.6
1	1	1.21105	110.3	0	0

Mixture 11: H_2O (1) + [bmim][BF_4] (2)
(1) Volumetric Properties, Transport Properties, Surface Tension, Refractive Index, and Conductivity

Ref. [138]

x_2	ρ/ (g · cm^{-3})	η (average value)/ (mPa · s)	Standard deviation	$\Delta\eta$/ (mPa · s)	x_2	n_D	Δn_D	R_m/ (cm^3 · mol^{-1})
T = 298.15 K								
0.0000	0.997043	0.89	0.002	0.00	0.0000	1.333	0.0000	3.716
0.0360	1.050069	1.51	0.001	−2.97	0.0440	1.36	0.0231	5.663
0.0440	1.058546	1.67	0.002	−3.61	0.1267	1.384	0.0397	9.324
0.1267	1.112514	3.22	0.003	−10.31	0.3134	1.405	0.0441	17.602
0.3134	1.158587	7.20	0.005	−24.96	0.4984	1.412	0.0346	25.708
0.4984	1.177854	13.63	0.008	−36.99	0.5802	1.415	0.0304	29.350
0.5802	1.183492	19.25	0.017	−39.53	0.7488	1.418	0.0184	36.736

(*continued*)

x_2	ρ/ ($g \cdot cm^{-3}$)	η (average value)/ (mPa · s)	Standard deviation	$\Delta\eta$/ (mPa · s)	x_2	n_D	Δn_D	R_m/ ($cm^3 \cdot mol^{-1}$)
0.8089	1.194472	46.55	0.044	−35.04	0.8089	1.419	0.0140	39.384
0.9719	1.199541	88.87	0.084	−8.99	0.9719	1.421	0.0015	46.550
1.0000	1.200297	100.66	0.205	0.00	1	1.422	0.0000	47.856

Ref. [139]

x_2	m/($mol \cdot g^{-1}$)	C/($mol \cdot L^{-1}$)	ρ/($kg \cdot dm^{-3}$)	$\varphi_{V,2}$/($cm^3 \cdot mol^{-1}$)
298 K				
0			0.997043	
0.000587	0.0326	0.0323	0.998273	188.51
0.000650	0.0361	0.0358	0.998405	188.51
0.001001	0.0556	0.0549	0.999130	188.58
0.001583	0.0881	0.0864	1.000317	188.69
0.001889	0.1051	0.1028	1.000931	188.76

Ref. [139]

m_1/($mol \cdot g^{-1}$)	ρ/($kg \cdot dm^{-3}$)	$\varphi_{V,1}$/($cm^3 \cdot mol^{-1}$)
298 K		
0.3678	1.192922	18.99
0.2159	1.193752	19.01
0.4894	1.192229	19.02
0.6261	1.191489	19.01
0.8059	1.190502	19.02
0.9897	1.189527	19.00

Ref. [139]

x_2	C/($mol \cdot L^{-1}$)	ρ/($kg \cdot dm^{-3}$)	V^E_{mol}/($cm^3 \cdot mol^{-1}$)
298 K			
0	0	0.997043	0.000
0.0499	1.8700	1.064225	0.118
0.1237	3.1418	1.111069	0.246
0.2342	4.0187	1.145010	0.342
0.3016	4.3204	1.156777	0.395
0.4073	4.6373	1.169740	0.423
0.5268	4.8723	1.180153	0.367
0.6698	5.0587	1.188391	0.303
0.7818	5.1649	1.193431	0.199
0.8960	5.2486	1.197370	0.098
0.9228	5.2657	1.198200	0.070
0.9682	5.2924	1.199432	0.035
1	5.3101	1.200297	0.000

Ref. [139]

m_u/ (mol · g^{-1})	ρ/ (kg · dm^{-3})	$\varphi_{1,x2}$/ (cm^3 · mol^{-1})	m_u/ (mol · g^{-1})	ρ/ (kg · dm^{-3})	$\varphi_{1,x2}$/ (cm^3 · mol^{-1})	m_u/ (mol · g^{-1})	ρ/ (kg · dm^{-3})	$\varphi_{1,x2}$/ (cm^3 · mol^{-1})
298 K								
	$x_2 = 0.0591$			$x_2 = 0.1099$			$x_2 = 0.1631$	
0.1891	1.069881	18.19	0.0965	1.105046	18.07	0.1941	1.126047	18.10
0.2933	1.069706	18.23	0.1340	1.104965	18.08	0.3511	1.125628	18.10
0.4628	1.069419	18.27	0.3243	1.104554	18.08	0.7760	1.124502	18.11
0.3962	1.069532	18.26	0.5252	1.104120	18.07	0.9777	1.123966	18.12
0.8024	1.068771	18.38	0.7968	1.103538	18.09	1.1248	1.123573	18.13
	$x_2 = 0.4183$			$x_2 = 0.7140$			$x_2 = 0.9510$	
0.2634	1.169039	18.42	0.1806	1.188860	18.90	0.2772	1.197389	19.23
0.3117	1.168829	18.45	0.2437	1.188445	19.14	0.4557	1.196345	19.20
0.4099	1.168429	18.44	0.3280	1.187990	19.1	0.5312	1.195898	19.21
0.4600	1.168206	18.47	0.4257	1.187648	18.77	0.6377	1.195299	19.18
0.8413	1.166547	18.55	0.8242	1.185634	18.79	0.7591	1.194598	19.18

Ref. [139]

x_2	$C/(\text{mol} \cdot \text{L}^{-1})$	$V'_1/(\text{cm}^3 \cdot \text{mol}^{-1})$	$V'_2/(\text{cm}^3 \cdot \text{mol}^{-1})$
298 K			
0.0591	2.0873	18.14	190.35
0.1099	2.9719	18.07	190.14
0.1631	3.5376	18.09	189.85
0.4183	4.6603	18.36	189.05
0.7140	5.1010	19.10	188.39
0.9510	5.2829	19.27	188.30
1.000	5.3101	19.01	188.32

Ref. [140]

	T/K							
x_1	303.15	308.15	313.15	318.15	323.15	333.15	343.15	353.15
$\rho/\text{g} \cdot \text{cm}^{-3}$								
0.0000	1.1984	1.1954	1.1922	1.1890	1.1860	1.1798	1.1735	1.1674
0.1002	1.1948	1.1917	1.1884	1.1853	1.1823	1.1758	1.1694	1.1631
0.2090	1.1915	1.1883	1.1852	1.1819	1.1787	1.1724	1.1658	1.1595
0.2984	1.1881	1.1849	1.1815	1.1780	1.1748	1.1682	1.1614	1.1553
0.4001	1.1833	1.1801	1.1767	1.1730	1.1698	1.1630	1.1561	1.1491
0.5014	1.1762	1.1731	1.1698	1.1662	1.1628	1.1555	1.1487	1.1417
0.6004	1.1670	1.1638	1.1605	1.1572	1.1537	1.1462	1.1389	1.1315
0.7000	1.1548	1.1519	1.1487	1.1448	1.1414	1.1331	1.1247	1.1163
0.8000	1.1347	1.1319	1.1289	1.1250	1.1217	1.1142	1.1068	1.0981
0.8999	1.0995	1.0955	1.0919	1.0899	1.0865	1.0789	1.0717	1.0647
1.0000	0.9957	0.9940	0.9922	0.9902	0.9881	0.9832	0.9778	0.9718
$\eta/\text{mPa} \cdot \text{s}$								
0.0000	74.21	58.18	46.51	37.79	31.08	21.52	15.61	11.49
0.1002	47.61	38.12	30.72	25.72	21.49	15.42	11.46	8.63
0.2090	34.89	28.35	23.50	19.64	16.74	12.02	8.98	6.79
0.2984	23.32	19.42	16.15	14.07	11.67	8.59	6.46	4.96
0.4001	16.57	13.57	11.44	9.72	8.41	6.16	4.68	3.56
0.5014	12.05	10.50	8.98	7.77	6.80	5.28	4.24	3.44
0.6004	10.05	8.69	7.40	6.40	5.69	4.75	3.75	3.19
0.7000	6.70	5.85	5.35	4.53	4.05	3.29	2.73	2.31
0.8000	4.36	3.80	3.36	2.99	2.68	2.19	1.82	1.54
0.8999	2.53	2.21	1.95	1.73	1.54	1.28	1.07	0.91
1.0000	0.80	0.72	0.65	0.60	0.55	0.47	0.40	0.35
$V^E/\text{cm}^3 \cdot \text{mol}^{-1}$								
0.1002	0.21	0.23	0.25	0.23	0.24	0.29	0.31	0.34
0.2090	0.24	0.27	0.27	0.29	0.32	0.34	0.39	0.42
0.2984	0.27	0.30	0.34	0.38	0.41	0.47	0.54	0.54
0.4001	0.30	0.33	0.36	0.42	0.45	0.53	0.60	0.71
0.5014	0.38	0.41	0.43	0.48	0.52	0.64	0.70	0.79

(*continued*)

	T/K							
x_1	303.15	308.15	313.15	318.15	323.15	333.15	343.15	353.15
0.6004	0.43	0.46	0.49	0.51	0.56	0.68	0.77	0.88
0.7000	0.39	0.40	0.42	0.48	0.52	0.67	0.82	0.98
0.8000	0.35	0.36	0.37	0.42	0.45	0.53	0.60	0.74
0.8999	0.16	0.22	0.25	0.23	0.26	0.33	0.38	0.41
$\Delta\eta$/mPa · s								
0.1002	−19.3	−14.3	−11.2	−8.4	−6.5	−4.0	−2.6	−1.8
0.2090	−24.0	−17.8	−13.4	−10.4	−8.0	−5.1	−3.5	−2.4
0.2984	−29.0	−21.1	−16.1	−12.6	−10.3	−6.7	−4.6	−3.2
0.4001	−28.3	−21.6	−16.7	−13.2	−10.5	−6.9	−4.8	−3.5
0.5014	−25.4	−18.6	−14.5	−11.4	−9.0	−5.7	−3.8	−2.5
0.6004	−20.1	−15.0	−11.6	−9.1	−7.1	−4.1	−2.7	−1.6
0.7000	−16.1	−12.1	−9.1	−7.2	−5.7	−3.5	−2.2	−1.4
0.8000	−11.1	−8.4	−6.5	−5.1	−4.0	−2.5	−1.6	−1.0
0.8999	−5.6	−4.3	−3.3	−2.6	−2.1	−1.3	−0.9	−0.6

Ref. [34]

x_2	*c* (mol/L)	ρ (g/cm^3)	x_2	*c* (mol/L)	ρ (g/cm^3)
298.15 K					
1.0000	5.312	1.2012	0.2963	4.301	1.1562
0.9005	5.252	1.1978	0.2520	4.107	1.1482
0.8167	5.187	1.1937	0.1688	3.584	1.1280
0.7065	5.099	1.1911	0.1398	3.318	1.1180
0.6922	5.082	1.1894	0.1072	2.935	1.1039
0.6326	5.022	1.1879	0.0736	2.394	1.0841
0.6000	4.977	1.1850	0.0502	1.878	1.0652
0.5723	4.933	1.1818	0.0335	1.405	1.0482
0.5304	4.881	1.1815	0.0197	1.381	1.0306
0.5252	4.863	1.1787	0.0092	0.904	1.0144
0.4946	4.818	1.1779	0.0069	0.467	1.0119
0.4408	4.707	1.1719	0.0052	0.359	1.0085
0.4067	4.634	1.1694	0.0035	0.275	1.0048
0.3734	4.552	1.1669	0.0017	0.186	1.0009

Ref. [34]

x_2	*c* (mol/L)	σ (mN/m)	$\delta\sigma$ (mN/m)	x_2	*c* (mol/L)	σ (mN/m)	$\delta\sigma$ (mN/m)
298.15 K							
1.0000	5.312	45.33	0.00	0.0998	2.835	46.19	−22.56
0.9005	5.252	45.41	−2.51	0.0799	2.509	45.99	−23.27
0.8003	5.185	45.52	−5.01	0.0698	2.317	46.14	−23.39
0.7065	5.099	45.68	−7.29	0.0586	2.077	46.25	−23.57
0.6326	5.022	45.81	−9.08	0.0498	1.862	46.27	−23.78

(*continued*)

x_2	c (mol/L)	σ (mN/m)	$\delta\sigma$ (mN/m)	x_2	c (mol/L)	σ (mN/m)	$\delta\sigma$ (mN/m)
0.5304	4.881	46.02	−11.53	0.0404	1.610	46.02	−24.27
0.4067	4.634	46.32	−14.45	0.0200	0.925	46.93	−23.90
0.2963	4.300	46.25	−17.39	0.0094	0.476	50.63	−20.47
0.2128	3.904	45.62	−20.19	0.0074	0.384	51.77	−19.38
0.1818	3.696	45.91	−20.71	0.0058	0.305	54.34	−16.85
0.1788	3.674	46.06	−20.64	0.0039	0.207	57.30	−13.94
0.1621	3.539	46.16	−20.97	0.0010	0.055	51.62	−19.70
0.1390	3.320	46.22	−21.51	0.0003	0.017	61.50	−9.84
0.1081	2.952	46.23	−22.31	0.0001	0.006	61.54	−9.80
0.1013	2.857	45.63	−23.08	0.0000	0.000	71.35	0.00

Ref. [44]

	$\rho/\mathrm{g \cdot cm^{-3}}$		
x_1	298.15 K	303.15 K	333.15 K
0	1.2011	1.1974	1.1754
0.0895	1.1983	1.1946	1.1728
0.1950	1.1944	1.1909	1.1690
0.3080	1.1893	1.1858	1.1639
0.4996	1.1778	1.1739	1.1515
0.7500	1.1479	1.1438	1.1210
0.8990	1.1007	1.0972	1.0764
x_1	$V^E_m/\mathrm{cm^3 \cdot mol^{-1}}$		
0.0895	0.129	0.133	0.123
0.1950	0.267	0.251	0.280
0.3080	0.392	0.384	0.439
0.4996	0.477	0.519	0.659
0.7500	0.399	0.454	0.644
0.8990	0.203	0.236	0.360

Ref. [135]

x_2	η/(mPa · s)	x_2	η/(mPa · s)	x_2	η/(mPa · s)
298.15 K					
0.00	0.894	0.3463	8.868	0.5987	22.383
0.0510	1.845	0.4050	11.041	0.6877	31.171
0.0951	2.728	0.4509	12.908	0.7947	47.513
0.2103	5.083	0.5044	15.654	0.8914	70.099
0.2978	7.383	0.5399	17.871	1.00	110.308

Ref. [141]

T/(K)	$x_2 = 0.20$	$x_2 = 0.40$	$x_2 = 0.60$	$x_2 = 0.80$
σ (S/m)				
293.2	3.530	2.250	1.401	0.897
298.2	4.100	2.610	1.613	1.026
303.2	4.670	2.950	1.832	1.158
308.2	5.280	3.280	2.040	1.297
313.2	5.900	3.690	2.270	1.450
318.2	6.550	4.130	2.530	1.594
323.2	7.220	4.570	2.780	1.754
333.2	8.540	5.310	3.360	2.100
343.2	9.890	6.270	3.870	2.490
353.2	11.510	7.440	4.560	2.940

Ref. [135]

x_2	$10^3\kappa$ (S · cm^{-1})	x_2	$10^3\kappa$ (S · cm^{-1})	x_2	$10^3\kappa$ (S · cm^{-1})
298.15 K					
0.0473	27.91	0.2496	32.20	0.5917	15.08
0.0589	32.10	0.3108	28.10	0.7361	10.30
0.0753	35.32	0.3730	24.08	0.8525	7.07
0.0993	37.81	0.4433	21.41	1.00	2.48
0.1705	37.55	0.4735	19.90		

Ref. [35]

x_2	$10^{-3}c_2$/mol · m^{-3}	ρ/kg · m^{-3}	κ/S · cm^{-1}	$10^4\Lambda$/S m^2 · mol^{-1}
298.15 K				
0.005053	0.2669	1007.00	1.720	64.3
0.008751	0.4471	1013.47	2.47	55.2
0.01927	0.9018	1030.78	3.59	39.8
0.03304	1.391	1047.81	4.46	32.1
0.04969	1.867	1065.19	4.90	26.3
0.07385	2.400	1084.47	5.17	21.5
0.1053	2.910	1103.47	5.21	17.92
0.1566	3.482	1124.77	4.93	14.15
0.1991	3.810	1137.30	4.55	11.94
0.2824	4.247	1154.46	3.71	8.73
0.4140	4.656	1171.08	2.57	5.51
0.4919	4.815	1177.86	2.04	4.23
0.5749	4.946	1183.94	1.574	3.18
0.6838	5.078	1190.14	1.098	2.16
0.7594	5.150	1193.53	0.846	1.643
0.8931	5.253	1198.74	0.527	1.003
1	5.319	1202.19	0.353	0.664

(2) Boiling Point

Ref. [32]

w_2	T/K	γ_1	w_2	T/K	γ_1
100 kPa					
0.8474	379.19	1.152	0.5892	374.71	1.040
0.7979	377.19	1.125	0.4905	374.27	1.021
0.7324	375.98	1.087	0.3711	373.86	1.007
0.6725	375.45	1.058			

(3) Osmotic Coefficient

Ref. [119]

m_2	Φ	m_2	Φ	m_2	Φ
0.0101	1.0050	0.152	0.8370	0.5632	0.5753
0.0355	0.9797	0.1977	0.7851	0.799	0.5058
0.044	0.9388	0.2891	0.7362	1.0232	0.4142
0.0701	0.9042	0.3968	0.6947	1.4685	0.2878
0.0971	0.8712	0.4678	0.6586	2.4171	1.0243

Ref. [83]

	Φ			
m_2 (mol · kg^{-1})	T = 298.15 K	T = 308.15 K	T = 318.15 K	T = 328.15 K
0.0120	1.023	1.009	1.006	0.994
0.0223	0.994	0.974	0.986	0.978
0.0369	0.987	0.959	0.955	0.965
0.0450	0.956	0.956	0.960	0.959
0.0652	0.952	0.926	0.937	0.920
0.0855	0.862	0.879	0.892	0.900
0.1058	0.854	0.872	0.887	0.883
0.1629	0.825	0.823	0.826	0.839
0.2108	0.804	0.787	0.802	0.797
0.2695	0.749	0.766	0.766	0.772
0.3284	0.699	0.708	0.748	0.761
0.4307	0.677	0.685	0.693	0.726
0.5330	0.648	0.666	0.673	0.685
0.6375	0.619	0.622	0.634	0.647
0.7233	0.575	0.587	0.610	0.632
0.9322	0.523	0.545	0.564	0.589
1.2923	0.462	0.495	0.508	0.517
2.0659	0.392	0.396	0.412	0.415

(4) Phase Equilibrium

Ref. [142]

w_1 SLE	T/K	w_1 LLE	T/K
0.0379	245.6	0.2995	274.8
0.1127	265.1	0.3491	277.0
0.1527	269.0	0.4090	278.2
0.1784	270.0	0.4518	278.4
0.1946	270.6	0.4963	278.6[a]
0.3056	270.9[b]	0.5551	278.5
0.3147	270.9[b]	0.5844	278.3
0.3344	270.9[b]	0.6525	277.6
0.3956	270.9[b]	0.6986	276.3
0.4105	270.9[b]	0.7490	273.3
0.5111	270.9[b]		
0.5691	270.9[b]		
0.6510	270.9[b]		
0.6889	270.9[b]		
0.6983	270.9[b]		
0.7501	271.3		
0.7789	271.3		
0.8509	271.4		
0.9502	272.5		

[a]Upper critical solution point
[b]At SLLE

(5) Vapor Pressure

Ref. [122]

X_2 = 20 mol %		X_2 = 40 mol %		X_2 = 60 mol %		X_2 = 80 mol %	
T/K	P/kPa	T/K	P/kPa	T/K	P/kPa	T/K	P/kPa
311.4	3.9	320.4	3.5	342.0	3.9	375.0	4.3
318.0	6.1	332.0	6.5	353.2	7.2	385.0	6.8
327.8	10.5	341.6	10.2	359.8	10.0	401.0	12.2
335.8	16.0	351.4	15.9	367.4	13.7	413.0	17.8
344.0	22.3	360.4	22.6	374.2	17.6	424.2	24.1
350.8	29.7	369.4	32.1	380.6	22.2	431.4	29.6
358.0	40.6	377.0	41.8	389.0	29.4	435.4	35.5
364.4	53.3	385.2	56.0	396.0	37.3	442.6	43.1
368.4	66.9	389.8	68.2	403.6	45.8	448.4	51.8
376.4	82.4	395.0	82.1	409.2	55.0	451.8	59.0
380.6	102.6	398.4	92.0	414.6	63.0	457.6	68.9
				419.4	74.6	464.8	80.7
				425.0	86.6	475.2	101.6
				430.8	102.2		

(6) Heat Capacity

Ref. [141]

	$x_2 = 0.20$		$x_2 = 0.40$		$x_2 = 0.60$		$x_2 = 0.80$	
T/(K)	C_P/(J/(mol · K))	C_P^E/(J/(mol · K))	C_P/(J/(mol · K))	C_P^E/(J/(mol · K))	C_P/(J/(mol · K))	C_P^E/(J/(mol · K))	C_P/(J/(mol · K))	C_P^E/(J/(mol · K))
303.2	137.5	3.6	200.1	7.6	259.4	8.2	314.2	4.4
308.2	138.4	4.0	201.7	8.1	261.5	8.8	316.6	4.7
313.2	139.3	4.5	203.2	8.8	263.2	9.2	318.6	5.1
318.2	140.5	5.2	204.5	9.1	265.1	9.7	320.9	5.4
323.2	141.2	5.4	206.2	9.8	267.0	10.1	323.2	5.7
328.2	141.8	5.5	207.4	10.2	268.8	10.7	325.2	6.1
333.2	142.9	6.2	208.8	10.8	270.5	11.3	327.0	6.4
338.2	143.8	6.8	210.1	11.4	272.1	11.8	328.7	6.7
343.2	144.7	7.0	211.7	11.9	274.3	12.2	331.3	7.0
348.2	145.3	7.1	213.2	12.4	276.2	12.7	333.4	7.3
353.2	146.2	7.6	214.6	12.9	278.0	13.2	335.7	7.9

Mixture 12: CH_3CN (1) + [BMIM][BF_4] (2)
(1) Volumetric Properties, Speed of Sound, Transport Properties, Compressibility Properties and Conductivity

Ref. [143]

x_2	ρ/g · cm^{-3}	V^E/cm^3 · mol^{-1}	x_2	ρ/g · cm^{-3}	V^E/cm^3 · mol^{-1}
T = 293.15 K					
0.0000	0.78230	0.0000	0.6015	1.13887	−0.8730
0.1000	0.91354	−0.9460	0.6998	1.15673	−0.6406
0.1997	0.99350	−1.2439	0.7990	1.17164	−0.4010
0.3000	1.04776	−1.2984	0.9012	1.18523	−0.2554
0.4005	1.08695	−1.2417	1.0000	1.19571	0.0000
0.4995	1.11642	−1.1530			
T = 298.15 K					
0.0000	0.77693	0.0000	0.6015	1.13503	−0.9123
0.1000	0.90878	−0.9887	0.6998	1.15298	−0.6740
0.1997	0.98903	−1.2947	0.7990	1.16796	−0.4272
0.3000	1.04355	−1.3537	0.9012	1.18158	−0.2700
0.4005	1.08289	−1.2932	1.0000	1.19207	0.0000
0.4995	1.11251	−1.2027			
T = 303.15 K					
0.0000	0.77144	0.0000	0.6015	1.13108	−0.9486
0.1000	0.90394	−1.0353	0.6998	1.14911	−0.7020
0.1997	0.98451	−1.3514	0.7990	1.16416	−0.4467
0.3000	1.03925	−1.4112	0.9012	1.17780	−0.2748
0.4005	1.07874	−1.3460	1.0000	1.18837	0.0000
0.4995	1.10844	−1.2454			
T = 308.15 K					
0.0000	0.76597	0.0000	0.6015	1.12717	−0.9854
0.1000	0.89910	−1.0813	0.6998	1.14529	−0.7313
0.1997	0.98003	−1.4104	0.7990	1.16039	−0.4642
0.3000	1.03499	−1.4707	0.9012	1.17409	−0.2831
0.4005	1.07462	−1.3994	1.0000	1.18472	0.0000
0.4995	1.10446	−1.2944			
T = 313.15 K					
0.0000	0.76068	0.0000	0.6015	1.12364	−1.0290
0.1000	0.89451	−1.1300	0.6998	1.14182	−0.7625
0.1997	0.97582	−1.4715	0.7990	1.15700	−0.4868
0.3000	1.03110	−1.5390	0.9012	1.17073	−0.2909
0.4005	1.07084	−1.4574	1.0000	1.18143	0.0000
0.4995	1.10065	−1.3281			
T = 323.15 K					
0.0000	0.74973	0.0000	0.6015	1.11623	−1.1232
0.1000	0.88505	−1.2386	0.6998	1.13455	−0.8334
0.1997	0.96707	−1.6049	0.7990	1.14988	−0.5378

(*continued*)

x_2	ρ/g · cm^{-3}	V^E/cm^3 · mol^{-1}	x_2	ρ/g · cm^{-3}	V^E/cm^3 · mol^{-1}
0.3000	1.02286	−1.6756	0.9012	1.16380	−0.3308
0.4005	1.06298	−1.5876	1.0000	1.17449	0.0000
0.4995	1.09319	−1.4598			
T = 333.15 K					
0.0000	0.73819	0.0000	0.6015	1.10814	−1.2221
0.1000	0.87491	−1.2386	0.6998	1.12663	−0.9111
0.1997	0.95771	−1.7520	0.7990	1.14209	−0.5911
0.3000	1.01398	−1.8242	0.9012	1.15607	−0.3513
0.4005	1.05445	−1.7266	1.0000	1.16689	0.0000
0.4995	1.08493	−1.5749			
T = 343.15 K					
0.0000	0.72694	0.0000	0.6015	1.10067	−1.3273
0.1000	0.86642	−1.5800	0.6998	1.11931	−0.9884
0.1997	0.94885	−1.9111	0.7990	1.13492	−0.6427
0.3000	1.00539	−1.9598	0.9012	1.14906	−0.3814
0.4005	1.04632	−1.8569	1.0000	1.15997	0.0000
0.4995	1.07727	−1.7182			

Ref. [128]

m_2 (mol · kg^{-1})	$10^3\rho$ (kg · m^{-3})	u (m · s^{-1})	$10^6 V_\varphi$ (m^3 · mol^{-1})	$10^{12}\kappa_s$ (Pa^{-1})	$10^{14}\kappa_\varphi$ (m^3 · mol^{-1} · Pa^{-1})
T = 298.15 K					
0.0269	0.778582	1280.40	166.54	783.44	−5.31
0.0578	0.780850	1281.84	167.19	779.41	−4.51
0.0851	0.782806	1282.97	168.05	776.09	−3.89
0.1274	0.785804	1284.71	168.85	771.04	−3.39
0.1488	0.787297	1285.61	169.13	768.50	−3.26
0.2335	0.793124	1289.28	169.87	758.52	−2.97
0.2616	0.794982	1289.81	170.30	756.12	−2.46
0.2819	0.796334	1290.52	170.45	754.01	−2.34
0.3755	0.802412	1293.74	171.22	744.58	−1.90
0.5204	0.811448	1298.45	172.18	730.95	−1.35
0.6652	0.819953	1302.92	173.23	718.42	−0.88

Ref. [130]

x_2	$T = 298.15$ K		$T = 303.15$ K		$T = 308.15$ K		$T = 313.15$ K		$T = 318.15$ K	
	ρ (g · cm^{-3})	u (m · s^{-1})	ρ (g · cm^{-3})	u (m · s^{-1})	ρ (g · cm^{-3})	u (m · s^{-1})	ρ (g · cm^{-3})	u (m · s^{-1})	ρ (g · cm^{-3})	u (m · s^{-1})
0	0.776533	1278.62	0.771116	1258.73	0.76566	1238.66	0.760162	1218.52	0.754462	1198.32
0.0103	0.794416	1289.15	0.789093	1269.35	0.783745	1249.62	0.778368	1229.96	0.772971	1210.35
0.0442	0.845059	1315.67	0.839982	1296.95	0.834883	1278.18	0.829765	1259.48	0.82463	1240.84
0.1032	0.913689	1354.8	0.908908	1337.19	0.904123	1319.67	0.899327	1302.27	0.894531	1284.94
0.2015	0.994016	1407.74	0.989579	1391.76	0.985147	1375.82	0.980715	1359.59	0.976287	1344.09
0.2931	1.045577	1445.25	1.041358	1430.21	1.037146	1415.25	1.032939	1400.46	1.028745	1385.81
0.3935	1.086978	1476.75	1.08293	1462.66	1.078891	1448.62	1.074862	1434.64	1.070843	1420.75
0.5352	1.128517	1509.22	1.12464	1495.94	1.120775	1482.62	1.11692	1469.74	1.113084	1456.87
0.7043	1.162306	1535.91	1.158572	1532.48	1.154845	1511.09	1.151133	1498.78	1.147435	1486.56
1	1.20142	1576.05	1.197884	1555.47	1.194346	1543.92	1.190815	1532.52	1.187294	1521.19

Ref. [134]

	x_2										
T/K	0.0000	0.1000	0.2000	0.3000	0.4000	0.5000	0.6000	0.6999	0.7998	0.8993	1.0000
ρ/g · cm^{-3}											
p = 0.20 MPa											
313.2	0.7548	0.8931	0.9800	1.0403	1.0770	1.1046	1.1268	1.1462	1.1612	1.1747	1.1852
353.2	0.7097	0.8567	0.9464	1.0078	1.0466	1.0756	1.0978	1.1177	1.1330	1.1475	1.1587
p = 0.60 MPa											
313.2	0.7553	0.8939	0.9809	1.0408	1.0774	1.1054	1.1270	1.1464	1.1616	1.1748	1.1856
353.2	0.7103	0.8572	0.9471	1.0083	1.0469	1.0759	1.0981	1.1181	1.1332	1.1475	1.1590
393.2	0.6629	0.8197	0.9135	0.9773	1.0170	1.0467	1.0710	1.0907	1.1067	1.1216	1.1337
p = 1.00 MPa											
313.2	0.7560	0.8945	0.9814	1.0410	1.0780	1.1056	1.1275	1.1467	1.1618	1.1751	1.1860
353.2	0.7105	0.8575	0.9476	1.0089	1.0473	1.0763	1.0987	1.1184	1.1334	1.1479	1.1595
393.2	0.6636	0.8203	0.9138	0.9775	1.0174	1.0470	1.0714	1.0912	1.1070	1.1217	1.1341
433.2	0.6102	0.7804	0.8802	0.9461	0.9879	1.0194	1.0441	1.0646	1.0805	1.0965	1.1111
p = 1.40 MPa											
313.2	0.7562	0.8949	0.9819	1.0412	1.0784	1.1060	1.1279	1.1469	1.1620	1.1753	1.1863
353.2	0.7111	0.8579	0.9479	1.0093	1.0477	1.0765	1.0993	1.1187	1.1336	1.1483	1.1597
393.2	0.6646	0.8207	0.9142	0.9781	1.0178	1.0476	1.0717	1.0915	1.1071	1.1219	1.1343
433.2	0.6112	0.7814	0.8806	0.9466	0.9884	1.0196	1.0442	1.0649	1.0807	1.0967	1.1112
p = 1.80 MPa											
313.2	0.7568	0.8953	0.9824	1.0413	1.0790	1.1063	1.1283	1.1471	1.1625	1.1754	1.1866
353.2	0.7119	0.8583	0.9485	1.0097	1.0481	1.0768	1.0997	1.1195	1.1339	1.1486	1.1600
393.2	0.6653	0.8209	0.9145	0.9783	1.0182	1.0478	1.0721	1.0918	1.1073	1.1221	1.1346
433.2	0.6128	0.7822	0.8810	0.9470	0.9888	1.0198	1.0448	1.0653	1.0813	1.0970	1.1113
473.2	0.5492	0.7402	0.8456	0.9157	0.9587	0.9911	1.0167	1.0388	1.0556	1.0708	1.0847

(*continued*)

	x_2										
T/K	0.0000	0.1000	0.2000	0.3000	0.4000	0.5000	0.6000	0.6999	0.7998	0.8993	1.0000
p = 2.00 MPa											
313.2	0.7570	0.8955	0.9828	1.0415	1.0792	1.1066	1.1286	1.1474	1.1627	1.1762	1.1867
353.2	0.7125	0.8585	0.9487	1.0099	1.0483	1.0770	1.0999	1.1198	1.1342	1.1488	1.1603
393.2	0.6658	0.8210	0.9149	0.9785	1.0184	1.0482	1.0723	1.0922	1.1076	1.1228	1.1350
433.2	0.6135	0.7825	0.8812	0.9474	0.9890	1.0200	1.0452	1.0655	1.0815	1.0974	1.1117
473.2	0.5507	0.7405	0.8462	0.9160	0.9590	0.9916	1.0170	1.0391	1.0558	1.0710	1.0849

	x_2								
T/K	0.1000	0.2000	0.3000	0.4000	0.5000	0.6000	0.6999	0.7998	0.8993
$V^E/cm^3 \cdot mol^{-1}$									
p = 0.20 MPa									
313.2	−1.344	−2.008	−2.482	−2.097	−1.658	−1.251	−1.033	−0.661	−0.431
353.2	−2.052	−2.825	−3.210	−2.815	−2.307	−1.691	−1.327	−0.788	−0.510
p = 0.60 MPa									
313.2	−1.364	−2.049	−2.481	−2.094	−1.687	−1.229	−1.005	−0.652	−0.385
353.2	−2.050	−2.828	−3.213	−2.799	−2.280	−1.677	−1.335	−0.761	−0.462
393.2	−3.024	−3.971	−4.367	−3.776	−3.057	−2.429	−1.780	−1.069	−0.616
p = 1.00 MPa									
313.2	−1.357	−2.031	−2.448	−2.099	−1.664	−1.226	−0.981	−0.631	−0.374
353.2	−2.048	−2.843	−3.229	−2.793	−2.278	−1.696	−1.312	−0.728	−0.440
393.2	−3.008	−3.944	−4.326	−3.759	−3.041	−2.423	−1.786	−1.054	−0.567
433.2	−4.592	−5.843	−6.078	−5.283	−4.357	−3.349	−2.400	−1.253	−0.570
p = 1.40 MPa									
313.2	−1.369	−2.048	−2.435	−2.110	−1.672	−1.236	−0.968	−0.610	−0.344
353.2	−2.030	−2.831	−3.225	−2.793	−2.266	−1.728	−1.323	−0.726	−0.473
393.2	−2.962	−3.897	−4.308	−3.734	−3.041	−2.396	−1.780	−1.024	−0.544
433.2	−4.587	−5.783	−6.050	−5.278	−4.320	−3.320	−2.402	−1.262	−0.586

$p = 1.80$ MPa									
313.2	−1.352	−2.045	−2.395	−2.115	−1.648	−1.235	−0.940	−0.624	−0.311
353.2	−2.005	−2.820	−3.204	−2.774	−2.247	−1.726	−1.371	−0.709	−0.468
393.2	−2.907	−3.864	−4.264	−3.716	−3.012	−2.390	−1.769	−1.005	−0.538
433.2	−4.498	−5.684	−5.958	−5.211	−4.250	−3.308	−2.394	−1.296	−0.598
473.2	−7.664	−9.180	−9.404	−8.201	−6.830	−5.386	−4.135	−2.586	−1.228
$p = 2.00$ MPa									
313.2	−1.350	−2.060	−2.402	−2.121	−1.662	−1.253	−0.962	−0.635	−0.413
353.2	−1.978	−2.794	−3.176	−2.753	−2.220	−1.707	−1.367	−0.702	−0.457
393.2	−2.870	−3.840	−4.235	−3.681	−2.992	−2.355	−1.753	−0.978	−0.579
433.2	−4.453	−5.631	−5.936	−5.161	−4.203	−3.292	−2.346	−1.257	−0.585
473.2	−7.506	−9.076	−9.285	−8.096	−6.777	−5.318	−4.081	−2.546	−1.223

Ref. [135]

x_2	η/(mPa · s)	x_2	η/(mPa · s)	x_2	η/(mPa · s)
298.15 K					
0.00	0.346	0.3219	3.886	0.5567	14.013
0.0335	0.481	0.3679	4.871	0.6523	22.416
0.0905	0.778	0.4282	7.057	0.7487	35.711
0.1787	1.529	0.4782	9.278	0.8473	56.112
0.2717	2.854	0.5195	11.546	1.00	110.308

Ref. [135]

x_2	$10^3\kappa$ (S · cm^{-1})	x_2	$10^3\kappa$ (S · cm^{-1})	x_2	$10^3\kappa$ (S · cm^{-1})
298.15 K					
0.0335	18.90	0.3679	16.11	0.6523	7.07
0.0905	21.80	0.4282	13.06	0.7487	4.88
0.1787	23.29	0.4782	10.20	0.8473	3.27
0.2717	20.25	0.5195	10.22	1.0000	2.48
0.3219	16.60	0.5567	9.37		

Ref. [35]

x_2	$10^{-3}c_2$/ mol · m^{-3}	$10^{-3}c_2{}^a$/ mol · m^{-3}	ρ/ kg · m^{-3}	ρ^a/ kg · m^{-3}	κ/ S · cm^{-1}	$10^4\Lambda$/ S m^2 · mol^{-1}	$10^4\Lambda^a$/ S m^2 · mol^{-1}
298.15 K							
0.009414	0.1744	0.1744	792.87	792.87	1.629	93.4	93.4
0.01981	0.3588	0.3588	809.82	809.81	2.71	75.4	75.4
0.03142	0.5545	0.5545	827.13	827.12	3.51	63.3	63.3
0.04389	0.7538	0.7538	844.54	844.54	4.12	54.7	54.7
0.07179	1.162	1.162	879.61	879.61	4.88	42.0	42.0
0.1077	1.622	1.622	918.10	918.10	5.19	32.0	32.0
0.1538	2.123	2.123	959.09	959.08	5.04	23.8	23.8
0.2149	2.666	2.666	1002.40	1002.38	4.50	16.89	16.89
0.2966	3.239	3.239	1047.32	1047.29	3.62	11.16	11.16
0.3900	3.741	3.741	1085.90	1085.85	2.72	7.27	7.27
0.6126	4.549	4.548	1146.22	1146.07	1.285	2.82	2.82
0.7544	4.893	4.892	1171.38	1171.11	0.802	1.639	1.640
0.8943	5.157	5.155	1190.63	1190.20	0.500	0.970	0.970
1	5.319	5.316	1202.19	1201.64	0.353	0.664	0.664

[a]Corrected for the damping effect in vibrating tube Volumetric Properties measurements

(2) Osmotic Coefficient

Ref. [119]

m_2	Φ	m_2	Φ	m_2	Φ
0.0093	0.9062	0.0795	0.2729	0.1472	0.4621
0.0203	0.6481	0.1043	0.3139	0.1631	0.5535
0.0341	0.4619	0.1163	0.3445	0.1791	0.6460
0.0631	0.2932	0.1359	0.4120	0.2091	0.8720

Mixture 13: CH_3SOCH_3 (1) + [BMIm][BF_4] (2)

(1) Volumetric Properties, Speed of Sound, Compressibility Properties and Conductivity

Ref. [128]

m_2 (mol · kg^{-1})	$10^3\rho$ (kg · m^{-3})	u (m · s^{-1})	10^6V_φ (m^3 · mol^{-1})	$10^{12}\kappa_S$ (Pa^{-1})	$10^{14}\kappa_\varphi$ (m^3 · mol^{-1} · Pa^{-1})
T = 298.15 K					
0.0162	1.095695	1485.68	184.07	413.49	4.94
0.0289	1.096029	1486.09	184.14	413.13	4.99
0.0523	1.096640	1486.83	184.16	412.49	5.03
0.0685	1.097059	1487.31	184.20	412.07	5.07
0.0982	1.097818	1488.18	184.24	411.30	5.10
0.1164	1.098270	1488.73	184.31	410.83	5.11
0.1628	1.099414	1490.04	184.41	409.68	5.15
0.2046	1.100409	1491.15	184.53	408.70	5.19
0.2621	1.101750	1492.90	184.64	407.24	5.18

Ref. [35]

x_2	$10^{-3}c_2$/mol · m^{-3}	ρ/kg · m^{-3}	κ/S · cm^{-1}	$10^4\Lambda$/S m^2 · mol^{-1}
298.15 K				
0.01925	0.2618	1101.56	0.687	26.2
0.03427	0.4556	1106.21	1.029	22.6
0.07837	0.9763	1117.80	1.558	15.96
0.1247	1.457	1128.31	1.715	11.77
0.1866	2.011	1139.50	1.672	8.31
0.2554	2.536	1150.71	1.502	5.92
0.3251	2.986	1159.40	1.310	4.39
0.4423	3.611	1171.71	1.015	2.81
0.5246	3.969	1178.20	0.854	2.15
0.6132	4.301	1184.27	0.711	1.653
0.7494	4.728	1192.13	0.543	1.149

Mixture 14: $CH_3CON(CH_3)_2$ (1) + $[BMIm][BF_4]$ (2)
(1) Volumetric Properties, Speed of Sound and Compressibility Properties

Ref. [128]

m_2 ($mol \cdot kg^{-1}$)	$10^3\rho$ ($kg \cdot m^{-3}$)	u ($m \cdot s^{-1}$)	10^6V_φ ($m^3 \cdot mol^{-1}$)	$10^{12}\kappa_s$ (Pa^{-1})	$10^{14}\kappa_\varphi$ ($m^3 \cdot mol^{-1} \cdot Pa^{-1}$)
$T = 298.15$ K					
0.0172	0.937286	1457.32	173.53	502.36	1.20
0.0222	0.937582	1457.65	173.50	501.98	1.02
0.0370	0.938456	1458.45	173.53	500.96	1.15
0.0672	0.940182	1459.98	174.21	498.99	1.41
0.0870	0.941300	1460.88	174.50	497.79	1.58
0.1353	0.943917	1462.95	175.46	495.00	1.92
0.1650	0.945493	1464.10	175.90	493.40	2.09
0.2144	0.948055	1465.93	176.48	490.84	2.32
0.2521	0.949998	1467.28	176.69	488.94	2.44
0.3439	0.954527	1470.26	177.37	484.64	2.72
0.4517	0.959626	1473.47	177.99	479.97	2.96
0.5318	0.963257	1475.75	178.37	476.69	3.10

Mixture 15: $OCH_2CH_2CH_2CH_2$ (1) + $[BMIm][BF_4]$ (2)
(1) Volumetric Properties, Speed of Sound and Compressibility Properties

Ref. [128]

m_2 ($mol \cdot kg^{-1}$)	$10^3\rho$ ($kg \cdot m^{-3}$)	u ($m \cdot s^{-1}$)	10^6V_φ ($m^3 \cdot mol^{-1}$)	$10^{12}\kappa_s$ (Pa^{-1})	$10^{14}\kappa_\varphi$ ($m^3 \cdot mol^{-1} \cdot Pa^{-1}$)
$T = 298.15$ K					
0.0095	0.882621	1278.6	173.25	693.04	4.70
0.0200	0.883210	1278.9	178.80	692.25	4.45
0.0317	0.883326	1278.98	202.65	692.07	8.40
0.0827	0.883900	1279.07	226.40	691.53	12.75
0.1562	0.884926	1279.13	231.51	690.66	13.82
0.1967	0.885110	1279.10	235.18	690.55	14.45
0.2644	0.885264	1279.13	239.56	690.40	15.14

Mixture 16: $[C_4mim][BF_4]$ (1) + $[C_6mim][BF_4]$ (2)
(1) Volumetric Properties and Transport Properties

Ref. [27]

η/mPa · s

x_2	T = 298.15 K	T = 299.15 K	T = 301.15 K	T = 302.15 K	T = 303.15 K	T = 304.15 K	T = 305.15 K	T = 307.15 K	T = 308.15 K
0	106.39	101.03	91.07	86.71	82.58	78.91	75.15	68.48	65.39
0.1083	115.27	109.48	98.64	93.85	89.30	85.00	80.98	73.62	70.30
0.1910	122.27	115.71	104.43	99.21	94.45	89.86	85.58	77.74	74.51
0.4020	142.45	134.69	121.10	114.88	108.62	103.58	98.54	89.33	85.21
0.4450	144.00	135.86	122.23	115.94	109.98	104.44	99.25	89.99	85.88
0.4953	149.83	144.00	129.15	122.41	116.08	109.97	104.67	94.56	90.03
0.5617	158.65	148.97	133.37	126.30	119.65	113.72	107.75	97.41	92.82
0.6398	165.72	156.73	140.45	132.93	126.19	119.77	113.55	102.55	97.30
0.7005	167.28	161.30	141.59	134.16	127.15	120.63	114.56	103.53	98.47
0.7924	172.40	161.30	144.97	137.35	130.18	122.77	117.32	106.12	101.04
0.8970	195.09	184.18	164.51	156.76	147.13	139.21	131.96	118.99	112.99
0.9460	200.19	188.67	168.26	160.41	147.51	142.49	135.66	123.85	117.41
1	209.27	197.25	175.84	165.95	157.01	148.58	140.70	126.74	119.98

Ref. [28]

ρ/g · cm^{-3}

x_1	T/K = 298.15	T/K = 299.15	T/K = 301.15	T/K = 302.15	T/K = 303.15	T/K = 304.15	T/K = 305.15	T/K = 307.15	T/K = 308.15
0	1.2017	1.2008	1.1994	1.1987	1.1980	1.1973	1.1966	1.1952	1.1945
0.1083	1.1946	1.1937	1.1923	1.1916	1.1909	1.1903	1.1896	1.1882	1.1875
0.1910	1.1895	1.1887	1.1873	1.1866	1.1859	1.1852	1.1845	1.1831	1.1824
0.4020	1.1770	1.1761	1.1747	1.1740	1.1733	1.1726	1.1719	1.1705	1.1698
0.4450	1.1746	1.1738	1.1724	1.1717	1.1710	1.1703	1.1696	1.1682	1.1675
0.4953	1.1717	1.1709	1.1695	1.1688	1.1681	1.1674	1.1667	1.1653	1.1646
0.5617	1.1680	1.1671	1.1657	1.1650	1.1643	1.1636	1.1629	1.1615	1.1608
0.6398	1.1636	1.1628	1.1614	1.1607	1.1600	1.1594	1.1587	1.1573	1.1566
0.7924	1.1559	1.1550	1.1536	1.1529	1.1522	1.1516	1.1509	1.1495	1.1488
0.8970	1.1506	1.1498	1.1484	1.1477	1.1470	1.1463	1.1457	1.1443	1.1436
0.9460	1.1481	1.1473	1.1459	1.1452	1.1445	1.1438	1.1432	1.1418	1.1411
1	1.1453	1.1445	1.1431	1.1424	1.1417	1.1411	1.1404	1.1390	1.1383

Ref. [28]

x_1	α_p/k K^{-1}	x_1	α_p/k K^{-1}
303.15 K			
0	0.5937	0.5617	0.6076
0.1083	0.5888	0.6398	0.5955
0.1910	0.5957	0.7924	0.6106
0.4020	0.6027	0.8970	0.6070
0.4450	0.6015	0.9460	0.6059
0.4953	0.6027	1	0.6035

(2) Excess Molar Enthalpy

Ref. [28]

x_1	H^E/J · mol^{-1}	x_1	H^E/J · mol^{-1}
303.15 K			
0.1071	17	0.5499	67
0.2572	62	0.6548	58
0.3454	66	0.7573	41
0.4453	62	0.8925	16
0.5099	61	-	-

Mixture 17: $[C_4mim][MeSO_4]$ (1) + $[C_4mim][BF_4]$ (2)
(1) Volumetric Properties and Transport Properties

Ref. [27]

η/mPa · s

x_2	$T = 298.15$ K	$T = 299.15$ K	$T = 301.15$ K	$T = 302.15$ K	$T = 303.15$ K	$T = 304.15$ K	$T = 305.15$ K	$T = 307.15$ K	$T = 308.15$ K
0	215.16	202.90	181.25	171.38	162.32	156.80	145.67	131.22	124.53
0.1209	192.53	181.76	162.84	153.94	145.84	138.21	131.11	118.21	112.42
0.2142	188.24	177.62	158.65	147.05	142.18	134.78	127.80	115.20	109.48
0.3057	180.18	170.33	152.31	144.09	136.68	129.64	123.17	111.22	105.94
0.4060	168.13	160.67	142.45	134.97	127.87	121.40	114.96	103.97	98.70
0.4358	161.70	151.31	136.38	129.32	122.40	116.01	109.99	99.42	94.49
0.5118	158.90	147.44	132.08	125.12	118.60	112.70	107.02	96.76	91.97
0.5412	149.54	143.58	128.59	122.17	115.75	109.66	103.98	93.91	89.59
0.6026	146.16	138.39	124.09	117.54	111.40	105.67	100.35	90.91	86.80
0.6480	141.56	133.79	119.95	113.77	108.05	102.41	97.07	87.85	83.57
0.6934	137.53	130.05	117.04	110.81	105.11	99.94	95.06	86.24	81.99
0.7813	128.44	121.50	109.46	103.90	98.68	93.79	89.35	81.17	77.38
0.9005	117.61	111.47	100.65	95.99	90.74	86.89	82.78	75.43	72.60
0.9447	110.84	105.31	95.10	90.67	86.22	82.03	78.24	71.12	68.00
1	108.92	103.49	93.24	88.70	84.54	80.52	76.80	69.89	66.81

Ref. [28]

ρ/g · cm^{-3}									
x_2	T/K = 298.15	T/K = 299.15	T/K = 301.15	T/K = 302.15	T/K = 303.15	T/K = 304.15	T/K = 305.15	T/K = 307.15	T/K = 308.15
0	1.2079	1.2073	1.2060	1.2053	1.2046	1.2040	1.2033	1.2020	1.2014
0.1209	1.2074	1.2067	1.2054	1.2047	1.2040	1.2034	1.2027	1.2013	1.2006
0.2166	1.2069	1.2063	1.2049	1.2042	1.2036	1.2029	1.2022	1.2008	1.2001
0.3057	1.2064	1.2057	1.2043	1.2037	1.2030	1.2023	1.2016	1.2003	1.1996
0.4060	1.2058	1.2051	1.2038	1.2031	1.2024	1.2017	1.2010	1.1997	1.1990
0.4358	1.2056	1.2050	1.2036	1.2029	1.2022	1.2015	1.2009	1.1995	1.1988
0.5118	1.2052	1.2045	1.2031	1.2024	1.2017	1.2011	1.2004	1.1990	1.1983
0.5412	1.2050	1.2043	1.2029	1.2022	1.2015	1.2009	1.2002	1.1988	1.1981
0.6026	1.2045	1.2039	1.2025	1.2018	1.2011	1.2004	1.1997	1.1983	1.1976
0.6480	1.2044	1.2036	1.2022	1.2015	1.2008	1.2001	1.1994	1.1981	1.1973
0.6934	1.2038	1.2031	1.2017	1.2011	1.2004	1.1997	1.1990	1.1976	1.1969
0.7813	1.2032	1.2025	1.2011	1.2004	1.1998	1.1991	1.1984	1.1970	1.1963
0.9005	1.2022	1.2016	1.2002	1.1995	1.1988	1.1981	1.1974	1.1960	1.1953
0.9363	1.2020	1.2012	1.1998	1.1991	1.1984	1.1977	1.1971	1.1957	1.1950
0.9683	1.2019	1.2011	1.1997	1.1990	1.1983	1.1976	1.1969	1.1957	1.1948
1	1.2014	1.2007	1.1993	1.1986	1.1979	1.1972	1.1966	1.1952	1.1945

Ref. [28]

x_2	α_p/kK^{-1}	x_2	α_p/kK^{-1}
303.15 K			
0	0.5454	0.6026	0.5755
0.1209	0.5610	0.6480	0.5811
0.2142	0.5657	0.6934	0.5723
0.3057	0.5623	0.7813	0.5745
0.4060	0.5691	0.9005	0.5823
0.4358	0.5714	0.9363	0.5828
0.5118	0.5710	0.9683	0.5817
0.5412	0.5769	1	0.5762

(2) Excess Molar Enthalpy

Ref. [28]

x_2	H^E/J · mol^{-1}	x_2	H^E/J · mol^{-1}
303.15 K			
0.0780	−73	0.5475	−367
0.2142	−203	0.7088	−260
0.3533	−297	0.7019	−286
0.4511	−343	0.8096	−201
0.5674	−362	0.9006	−100

Mixture 18: [C_4mim][PF_6] (1) + [C_4mim][BF_4] (2)
(1) Volumetric Properties and Transport Properties

Ref. [27]

η/mPa · s

x_1	T = 298.15 K	T = 299.15 K	T = 301.15 K	T = 302.15 K	T = 303.15 K	T = 304.15 K	T = 305.15 K	T = 307.15 K	T = 308.15 K
0	107.21	101.94	92.00	87.52	83.36	79.57	75.85	69.07	66.06
0.0568	106.81	101.35	91.60	87.14	82.86	78.95	75.28	68.50	65.39
0.1034	116.13	110.18	99.53	94.41	89.68	85.51	81.45	74.26	70.87
0.2095	128.14	121.29	109.47	103.94	98.85	94.17	89.46	81.25	77.35
0.3012	137.43	130.09	116.96	111.11	105.39	100.22	95.34	86.45	82.42
0.3825	146.67	139.86	125.59	118.81	112.97	107.12	102.08	92.49	88.09
0.4647	157.66	145.55	131.81	124.40	118.14	111.83	106.86	96.35	91.90
0.4999	169.30	159.74	142.82	135.79	128.19	121.55	115.26	103.88	98.44
0.5385	168.47	159.64	142.92	134.84	128.05	121.63	114.89	104.11	98.80
0.5974	182.46	172.40	156.70	145.41	137.60	130.55	123.92	111.75	105.99
0.6928	201.44	190.14	169.85	160.86	155.35	144.05	136.46	123.33	116.90
0.8330	226.69	214.02	190.21	179.51	169.68	160.47	154.46	136.09	129.23
0.8877	237.48	223.51	198.98	187.72	177.28	167.51	159.03	142.25	134.82
0.9450	249.19	234.50	208.03	196.24	185.36	175.33	166.08	149.71	142.47
1	278.58	258.97	231.96	218.40	205.96	194.78	183.78	164.46	155.99

Ref. [28]

ρ/g · cm^{-3}									
x_1	T/K = 298.15	T/K = 299.15	T/K = 301.15	T/K = 302.15	T/K = 303.15	T/K = 304.15	T/K = 305.15	T/K = 307.15	T/K = 308.15
0	1.2014	1.2006	1.1993	1.1987	1.1981	1.1973	1.1966	1.1952	1.1946
0.0568	1.2116	1.2109	1.2095	1.2089	1.2082	1.2075	1.2068	1.2053	1.2046
0.1034	1.2202	1.2195	1.2181	1.2174	1.2167	1.2159	1.2152	1.2138	1.2131
0.2095	1.2391	1.2383	1.2369	1.2362	1.2355	1.2347	1.2340	1.2326	1.2318
0.3012	1.2548	1.2540	1.2526	1.2519	1.2512	1.2504	1.2497	1.2482	1.2475
0.3825	1.2687	1.2679	1.2665	1.2657	1.2650	1.2642	1.2635	1.2620	1.2612
0.4647	1.2826	1.2817	1.2803	1.2796	1.2789	1.2781	1.2773	1.2758	1.2750
0.4999	1.2883	1.2875	1.2861	1.2853	1.2846	1.2838	1.2830	1.2815	1.2807
0.5974	1.3045	1.3035	1.3021	1.3014	1.3006	1.2998	1.2990	1.2975	1.2967
0.6928	1.3198	1.3189	1.3174	1.3166	1.3159	1.3150	1.3142	1.3126	1.3119
0.8330	1.3423	1.3414	1.3399	1.3391	1.3383	1.3375	1.3366	1.3350	1.3342
0.8877	1.3507	1.3497	1.3482	1.3474	1.3466	1.3458	1.3450	1.3433	1.3425
0.9450	1.3594	1.3585	1.3570	1.3562	1.3554	1.3545	1.3537	1.3521	1.3512
1	1.3679	1.3669	1.3654	1.3647	1.3639	1.3630	1.3622	1.3605	1.3597

Ref. [28]

x_1	α_p/kK^{-1}	x_2	α_p/kK^{-1}
303.15 K			
0	0.5686	0.4999	0.5921
0.0568	0.5775	0.5974	0.5922
0.1034	0.5859	0.6928	0.5971
0.2095	0.5848	0.8330	0.6059
0.3012	0.5827	0.8877	0.6031
0.3825	0.5881	0.9450	0.6026
0.4647	0.5910	1	0.6012

Ref. [57]

	V (pure component) or V^E/cm^3 · mol^{-1}	
x_1	T = 298.15 K	T = 333.15 K
1	207.986 (207.8)	212.470 (212.3)
0.600	0.13_9	0.13_7
0.500	0.11_8	0.11_0
0.400	0.13_3	0.12_8
0	187.634 (187.6)	191.558 (191.4)
1	207.826 (207.8)	212.308 (212.3)
0.500	0.12_5	0.11_6
0.250	0.09_2	0.09_0
0	187.634 (187.6)	191.558 (191.6)

(2) Excess Molar Enthalpy

Ref. [28]

x_2	H^E/J · mol^{-1}	x_2	H^E/J · mol^{-1}
303.15 K			
0.1066	−37	0.6024	−136
0.2935	−68	0.6470	−103
0.4424	−105	0.7973	−76
0.5030	−125	0.8968	−30
0.5316	−131	–	–

Mixture 19: $CF_3CHFCHFCF_2CF_3$ *threo*-H (1) + [bmim][BF_4] (2)
(1) Phase Equilibrium

Ref. [30]

T/K	x'_1/mol%	x_1/mol%	V'^a/ cm^3 · mol^{-1}	V^a/ cm^3 · mol^{-1}	$V^{ex/b}$/ cm^3 · mol^{-1}	V^{exb}/ cm^3 · mol^{-1}
313.6 ± 0.2	95.4 ± 0.8	40.6 ± 1.5	162.8 ± 0.6	172.2 ± 1.0	−1.9 ± 0.6	−6.8 ± 1.0
323.7 ± 0.2	98.1 ± 0.6	27.1 ± 1.1	165.5 ± 0.6	177.7 ± 1.1	−1.7 ± 0.6	−6.6 ± 1.1
334.1 ± 0.2	98.5 ± 0.6	21.8 ± 0.9	168.8 ± 0.4	181.3 ± 1.3	−2.0 ± 0.4	−6.0 ± 1.3

[a]Observed molar volume
[b]Volumetric Properties

(2) Cloud Point

Ref. [30]

T/K	x_1/mol %	T/K	x_1/mol %
311 ± 1	50.0 ± 0.5	307 ± 1	79.5 ± 0.5
309 ± 1	59.1 ± 0.5	308 ± 1	91.0 ± 0.5
308 ± 1	69.8 ± 0.5		

Mixture 20: $CF_3CHFCHFCF_2CF_3$ *threo*-D (1) + [bmim][BF_4] (2)
(1) Phase Equilibrium

Ref. [30]

T/K	x'_1/mol%	x_1/mol%	V'^a/ $cm^3 \cdot mol^{-1}$	V^a/ $cm^3 \cdot mol^{-1}$	$V^{ex/b}$/ $cm^3 \cdot mol^{-1}$	V^{exb}/ $cm^3 \cdot mol^{-1}$
313.6 ± 0.2	93.9 ± 0.7	46.8 ± 1.4	162.5 ± 0.7	170.7 ± 0.8	−3.1 ± 0.7	−7.0 ± 0.8
323.7 ± 0.2	97.5 ± 0.5	29.0 ± 1.0	164.9 ± 0.7	177.2 ± 0.9	−3.1 ± 0.7	−6.9 ± 0.9
334.1 ± 0.2	98.1 ± 0.6	23.1 ± 0.7	169.2 ± 0.8	180.8 ± 0.7	−2.5 ± 0.8	−6.4 ± 0.7

[a]Observed molar volume
[b]Volumetric Properties

(2) Cloud Point

Ref. [30]

T/K	x_1/mol %	T/K	x_1/mol %
311 ± 1	59.9 ± 0.5	309 ± 1	80.1 ± 0.5
310 ± 1	69.4 ± 0.5	310 ± 1	89.3 ± 0.5

Mixture 21: $CF_3CHFCHFCF_2CF_3$ *erythro*-H (1) + [bmim][BF_4] (2)
(1) Phase Equilibrium

Ref. [30]

T/K	x'_1/mol%	x_1/mol%	V'^a/ $cm^3 \cdot mol^{-1}$	V^a/ $cm^3 \cdot mol^{-1}$	$V^{ex/b}$/ $cm^3 \cdot mol^{-1}$	V^{exb}/ $cm^3 \cdot mol^{-1}$
283.6 ± 0.2	99.2 ± 0.7	7.0 ± 1.1	157.4 ± 0.7	180.3 ± 0.7	−0.3 ± 0.7	−4.1 ± 0.7
293.6 ± 0.2	99.5 ± 0.8	5.5 ± 0.9	159.5 ± 0.7	182.1 ± 0.9	−0.9 ± 0.7	−3.8 ± 0.9
303.7 ± 0.2	99.4 ± 0.8	6.2 ± 1.3	163.0 ± 0.6	182.6 ± 1.1	−0.3 ± 0.6	−4.4 ± 1.1
313.7 ± 0.2	99.4 ± 0.8	5.4 ± 1.3	166.2 ± 0.5	184.4 ± 1.0	−0.3 ± 0.5	−4.0 ± 1.0
325.4 ± 0.2	99.0 ± 0.9	5.6 ± 1.1	170.1 ± 0.6	185.3 ± 0.9	−0.4 ± 0.6	−4.6 ± 0.9
333.5 ± 0.2	98.8 ± 1.0	5.4 ± 1.1	172.6 ± 0.5	186.7 ± 0.8	−0.9 ± 0.5	−4.2 ± 0.8

[a]Observed molar volume
[b]Volumetric Properties

Mixture 22: $CF_3CHFCHFCF_2CF_3$ *erythro*-D (1) + [bmim][BF_4] (2)
(1) Phase Equilibrium

Ref. [30]

T/K	x'_1/mol%	x_1/mol%	V'^a/ $cm^3 \cdot mol^{-1}$	V^a/ $cm^3 \cdot mol^{-1}$	$V^{ex/b}$/ $cm^3 \cdot mol^{-1}$	V^{exb}/ $cm^3 \cdot mol^{-1}$
283.5 ± 0.2	99.2 ± 0.7	7.1 ± 1.2	157.1 ± 0.6	179.6 ± 1.2	−0.6 ± 0.6	−4.7 ± 1.2
294.1 ± 0.2	99.4 ± 0.7	6.5 ± 1.6	159.9 ± 0.5	181.6 ± 1.0	−0.5 ± 0.5	−4.1 ± 1.0
303.9 ± 0.2	99.0 ± 0.9	6.3 ± 1.3	163.0 ± 0.5	182.6 ± 0.9	−0.4 ± 0.5	−4.4 ± 0.9
313.5 ± 0.2	98.9 ± 0.8	5.5 ± 1.5	165.7 ± 0.6	184.1 ± 1.0	−0.9 ± 0.6	−4.3 ± 1.0
323.4 ± 0.2	98.8 ± 0.8	5.3 ± 1.1	169.2 ± 0.4	185.7 ± 0.9	−1.4 ± 0.4	−4.0 ± 0.9
334.0 ± 0.2	98.7 ± 0.9	5.6 ± 1.3	172.3 ± 0.5	185.9 ± 0.8	−1.2 ± 0.5	−5.1 ± 0.8

[a]Observed molar volume
[b]Volumetric Properties

Mixture 23: $CF_3CHFCHFCF_2CF_3$ [*threo*-H(87.12 ± 0.6 %) + *erythro*-H (11.89 ± 0.6 %)] (1) + [bmim][BF_4] (2)
(1) Phase Equilibrium

Ref. [30]

T/K	x'_1/mol%	x_1/mol%	V'^a/ $cm^3 \cdot mol^{-1}$	V^a/ $cm^3 \cdot mol^{-1}$	$V^{ex/b}$/ $cm^3 \cdot mol^{-1}$	V^{exb}/ $cm^3 \cdot mol^{-1}$
313.7 ± 0.2	99.5 ± 0.5	24.5 ± 2.2	160.4 ± 0.6	177.8 ± 1.1	−3.4 ± 0.6	−5.5 ± 1.1
325.6 ± 0.2	98.6 ± 0.6	20.4 ± 0.8	165.6 ± 0.6	179.8 ± 1.2	−2.4 ± 0.6	−6.5 ± 1.2
333.5 ± 0.2	98.6 ± 0.6	17.6 ± 0.8	168.1 ± 0.4	183.7 ± 0.9	−2.7 ± 0.4	−4.4 ± 0.9

[a]Observed molar volume
[b]Volumetric Properties

Mixture 24: C_6H_6 (1) + [BMIM][BF_4] (2)
(1) Volumetric Properties

Ref. [143]

x_2	ρ/g · cm^{-3}	$V^E/cm^3 \cdot mol^{-1}$	x_2	ρ/g · cm^{-3}	$V^E/cm^3 \cdot mol^{-1}$
T = 293.15 K					
0.0000	0.87823	0.0000	0.8000	1.16873	−0.9308
0.4871	1.10520	−1.8411	0.8990	1.18280	−0.4588
0.5499	1.12099	−1.7394	0.9498	1.18944	−0.2228
0.5999	1.13231	−1.6357	1.0000	1.19571	0.0000
0.7001	1.15184	−1.2957			
T = 298.15 K					
0.0000	0.87311	0.0000	0.8000	1.16502	−0.9616
0.4871	1.10128	−1.9063	0.8990	1.17915	−0.4780
0.5499	1.11714	−1.7888	0.9498	1.18585	−0.2408
0.5999	1.12850	−1.6931	1.0000	1.19207	0.0000
0.7001	1.14813	−1.3569			
T = 303.15 K					
0.0000	0.86759	0.0000	0.8000	1.16124	−0.9988

(*continued*)

x_2	ρ/g · cm^{-3}	V^E/cm^3 · mol^{-1}	x_2	ρ/g · cm^{-3}	V^E/cm^3 · mol^{-1}
0.4871	1.09721	−1.9802	0.8990	1.17538	−0.4920
0.5499	1.11320	−1.8642	0.9498	1.18212	−0.2485
0.5999	1.12457	−1.7584	1.0000	1.18837	0.0000
0.7001	1.14426	−1.4059			
T = 308.15 K					
0.0000	0.86221	0.0000	0.8000	1.15748	−1.0304
0.4871	1.09323	−2.0558	0.8990	1.17167	−0.5069
0.5499	1.10930	−1.9356	0.9498	1.17845	−0.2574
0.5999	1.12071	−1.8236	1.0000	1.18472	0.0000
0.7001	1.14047	−1.4571			
T = 313.15 K					
0.0000	0.85682	0.0000	0.8000	1.15413	−1.0768
0.4871	1.08958	−2.1469	0.8990	1.16835	−0.5301
0.5499	1.10572	−2.0186	0.9498	1.17516	−0.2713
0.5999	1.11722	−1.9053	1.0000	1.18143	0.0000
0.7001	1.13704	−1.5196			
T = 323.15 K					
0.0000	0.84605	0.0000	0.8000	1.14703	−1.1597
0.4871	1.08190	−2.3149	0.8990	1.16136	−0.5762
0.5499	1.09823	−2.1787	0.9498	1.16819	−0.2934
0.5999	1.10981	−2.0507	1.0000	1.17449	0.0000
0.7001	1.12983	−1.6399			
T = 333.15 K					
0.0000	0.83576	0.0000	0.8000	1.13928	−1.2248
0.4871	1.07373	−2.4527	0.8990	1.15364	−0.6016
0.5499	1.09017	−2.3058	0.9498	1.16054	−0.3075
0.5999	1.10175	−2.1576	1.0000	1.16689	0.0000
0.7001	1.12195	−1.7292			
T = 343.15 K					
0.0000	0.82489	0.0000	0.8000	1.13217	−1.3127
0.4871	1.06572	−2.5966	0.8990	1.14664	−0.6478
0.5499	1.08251	−2.4602	0.9498	1.15358	−0.3304
0.5999	1.09435	−2.3196	1.0000	1.15997	0.0000
0.7001	1.11470	−1.8548			

Ref. [134]

	x_2					
T/	0.5000	0.5999	0.7001	0.8000	0.8999	1.0000
ρ/g · cm^{-3}						
p = 0.20 MPa						
313.2	1.1104	1.1289	1.1447	1.1576	1.1717	1.1852
353.2	1.0815	1.1017	1.1170	1.1302	1.1446	1.1587

(*continued*)

	x_2					
$T/$	0.5000	0.5999	0.7001	0.8000	0.8999	1.0000
p = 0.60 MPa						
313.2	1.1106	1.1298	1.1451	1.1578	1.1720	1.1856
353.2	1.0818	1.1021	1.1173	1.1308	1.1447	1.1590
393.2	1.0540	1.0754	1.0909	1.1046	1.1189	1.1337
p = 1.00 MPa						
313.2	1.1107	1.1303	1.1455	1.1581	1.1725	1.1860
353.2	1.0823	1.1022	1.1176	1.1309	1.1449	1.1595
393.2	1.0540	1.0756	1.0910	1.1047	1.1190	1.1341
p = 1.40 MPa						
313.2	1.1109	1.1305	1.1460	1.1583	1.1727	1.1863
353.2	1.0827	1.1029	1.1179	1.1311	1.1453	1.1597
393.2	1.0542	1.0762	1.0913	1.1049	1.1194	1.1343
433.2	1.0271	1.0488	1.0650	1.0790	1.0945	1.1112
p = 1.80 MPa						
313.2	1.1112	1.1312	1.1466	1.1586	1.1728	1.1866
353.2	1.0831	1.1032	1.1181	1.1315	1.1456	1.1600
393.2	1.0551	1.0766	1.0915	1.1050	1.1199	1.1346
433.2	1.0275	1.0492	1.0654	1.0793	1.0948	1.1113
p = 2.00 MPa						
313.2	1.1113	1.1315	1.1467	1.1590	1.1730	1.1867
353.2	1.0833	1.1035	1.1183	1.1316	1.1457	1.1603
393.2	1.0553	1.0768	1.0918	1.1052	1.1201	1.1350
433.2	1.0278	1.0496	1.0657	1.0797	1.0951	1.1117

	x_2				
T/K	0.5000	0.5999	0.7001	0.8000	0.8999
$V^E/cm^3 \cdot mol^{-1}$					
p = 0.20 MPa					
313.2	−4.220	−3.269	−2.294	−1.199	−0.531
353.2	−5.130	−4.150	−2.843	−1.530	−0.662
p = 0.60 MPa					
313.2	−4.159	−3.310	−2.278	−1.146	−0.503
353.2	−5.127	−4.155	−2.838	−1.568	−0.615
393.2	−6.420	−5.279	−3.648	−2.059	−0.847
p = 1.00 MPa					
313.2	−4.120	−3.326	−2.278	−1.148	−0.515
353.2	−5.074	−4.061	−2.779	−1.486	−0.564
393.2	−6.328	−5.208	−3.574	−1.993	−0.790
p = 1.40 MPa					
313.2	−4.075	−3.283	−2.281	−1.111	−0.485
353.2	−5.098	−4.133	−2.799	−1.489	−0.606
393.2	−6.274	−5.235	−3.568	−1.969	−0.811
433.2	−8.011	−6.448	−4.434	−2.415	−0.951

(continued)

	x_2					
T/	0.5000	0.5999	0.7001	0.8000	0.8999	1.0000
p = 1.80 MPa						
313.2	−4.068	−3.325	−2.313	−1.103	−0.462	
353.2	−5.078	−4.107	−2.754	−1.490	−0.585	
393.2	−6.305	−5.201	−3.522	−1.928	−0.842	
433.2	−8.000	−6.439	−4.446	−2.420	−0.980	
p = 2.00 MPa						
313.2	−4.049	−3.339	−2.304	−1.133	−0.466	
353.2	−5.050	−4.099	−2.734	−1.457	−0.560	
393.2	−6.267	−5.168	−3.499	−1.891	−0.804	
433.2	−7.971	−6.430	−4.432	−2.422	−0.959	

(2) Vapor Pressure and Activity Coefficient
Ref. [144]

T (K)	p^{exp} (kPa)	p^{nrtl} (kPa)	RD (p)	$\gamma^{exp}{}_1$	$\gamma^{NRTL}{}_1$
x_1 = 0.5129					
303.37	15.7	15.4	1.74	1.8985	1.8714
308.46	19.0	19.2	1.01	1.8378	1.8669
313.27	23.9	23.4	2.08	1.8977	1.8623
318.11	28.7	28.4	1.19	1.8746	1.8575
322.90	35.0	34.1	2.69	1.9007	1.8525
328.16	41.1	41.3	0.56	1.8300	1.8468
333.19	49.1	49.4	0.57	1.8244	1.8412
338.17	57.8	58.5	1.25	1.8060	1.8355
343.15	68.2	68.9	1.09	1.8034	1.8296
x_1 = 0.4499					
303.98	15.1	15.0	0.75	2.0095	2.0174
308.46	18.2	18.2	0.19	1.9981	2.0142
313.44	22.3	22.3	0.07	1.9917	2.0102
318.16	26.7	26.9	0.83	1.9736	2.0061
322.98	32.6	32.4	0.67	2.0024	2.0017
328.19	38.9	39.2	0.87	1.9661	1.9966
333.10	47.4	46.7	1.46	2.0107	1.9915
338.14	55.9	55.5	0.73	1.9900	1.9860
343.17	66.0	65.5	0.75	1.9851	1.9804
x_1 = 0.4003					
303.43	13.7	13.8	0.70	2.0801	2.1384
308.27	17.1	17.0	0.59	2.1157	2.1357
313.25	20.9	20.9	0.00	2.1046	2.1324
318.09	26.0	25.3	2.50	2.1623	2.1288
323.18	30.5	30.8	1.03	2.0803	2.1246
328.36	36.4	37.3	2.48	2.0469	2.1200
333.17	43.3	44.3	2.21	2.0497	2.1154

(*continued*)

T (K)	p^{exp} (kPa)	p^{nrtl} (kPa)	RD (p)	$\gamma^{exp}{}_1$	$\gamma^{NRTL}{}_1$
338.18	52.8	52.5	0.49	2.1039	2.1104
343.16	63.7	61.9	2.77	2.1506	2.1051
x_1 = 0.2999					
303.27	11.9	11.5	3.16	2.3926	2.4013
308.37	14.5	14.4	0.87	2.3502	2.3999
313.46	17.6	17.8	0.91	2.3162	2.3979
318.10	21.4	21.4	0.09	2.3462	2.3956
323.12	25.7	25.9	0.94	2.3241	2.3926
328.23	31.0	31.3	1.12	2.3207	2.3892
333.28	37.2	37.5	0.89	2.3256	2.3853
338.25	44.0	44.5	1.18	2.3169	2.3811
343.15	52.2	52.4	0.33	2.3347	2.3766
x_1 = 0.2002					
303.34	8.8	8.7	1.58	2.5722	2.6952
308.35	11.3	10.8	4.74	2.6989	2.6950
313.38	13.6	13.3	2.38	2.6516	2.6941
318.16	16.4	16.1	1.95	2.6524	2.6928
323.10	19.6	19.5	0.72	2.6269	2.6908
328.23	23.2	23.5	1.48	2.5734	2.6882
333.20	27.8	28.1	1.16	2.5847	2.6852
338.29	32.9	33.5	1.87	2.5673	2.6816
343.16	39.0	39.4	1.03	2.5888	2.6778
x_1 = 0.0997					
303.46	5.1	4.9	4.21	2.8857	3.0372
308.48	6.2	6.1	2.03	2.8776	3.0375
313.34	7.5	7.4	0.81	2.8772	3.0372
318.39	9.1	9.1	0.10	2.8748	3.0363
323.17	11.2	11.0	2.18	2.9596	3.0348
328.22	13.8	13.2	4.21	3.0352	3.0327
333.26	15.9	15.8	0.42	2.9248	3.0300
338.25	19.0	18.8	1.01	2.9470	3.0268
343.17	22.8	22.2	2.83	3.0054	3.0232

Mixture 25: 1-C_3H_7OH (1) + [BMIM][BF_4] (2)
(1) Volumetric Properties

Ref. [143]

x_2	ρ/g · cm^{-3}	V^E/cm^3 · mol^{-1}	x_2	ρ/g · cm^{-3}	V^E/cm^3 · mol^{-1}
T = 293.15 K					
0.0000	0.80497	0.0000	0.6978	1.14141	−0.3752
0.3999	1.05466	−0.4988	0.8002	1.16300	−0.3388
0.4498	1.07268	−0.4986	0.9003	1.18065	−0.1978
0.5001	1.08908	−0.4819	1.0000	1.19571	0.0000
0.6001	1.11770	−0.4372			

(continued)

x_2	ρ/g · cm^{-3}	V^E/cm^3 · mol^{-1}	x_2	ρ/g · cm^{-3}	V^E/cm^3 · mol^{-1}
$T = 298.15$ K					
0.0000	0.80101	0.0000	0.6978	1.13771	−0.3884
0.3999	1.05103	−0.5370	0.8002	1.15935	−0.3523
0.4498	1.06890	−0.5171	0.9003	1.17699	−0.2026
0.5001	1.08528	−0.4991	1.0000	1.19207	0.0000
0.6001	1.11399	−0.4518			
$T = 303.15$ K					
0.0000	0.79689	0.0000	0.6978	1.13391	−0.3995
0.3999	1.04717	−0.5624	0.8002	1.15561	−0.3638
0.4498	1.06503	−0.5380	0.9003	1.17325	−0.2054
0.5001	1.08139	−0.5142	1.0000	1.18837	0.0000
0.6001	1.11015	−0.4654			
$T = 308.15$ K					
0.0000	0.79283	0.0000	0.6978	1.13015	−0.4090
0.3999	1.04337	−0.5887	0.8002	1.15192	−0.3752
0.4498	1.06122	−0.5599	0.9003	1.16956	−0.2081
0.5001	1.07757	−0.5314	1.0000	1.18472	0.0000
0.6001	1.10639	−0.4827			
$T = 313.15$ K					
0.0000	0.78907	0.0000	0.6978	1.12678	−0.4235
0.3999	1.03988	−0.6112	0.8002	1.14861	−0.3900
0.4498	1.05772	−0.5775	0.9003	1.16626	−0.2156
0.5001	1.07409	−0.5475	1.0000	1.18143	0.0000
0.6001	1.10301	−0.5037			
$T = 323.15$ K					
0.0000	0.78074	0.0000	0.6978	1.11964	−0.4625
0.3999	1.03241	−0.6711	0.8002	1.14154	−0.4170
0.4498	1.05032	−0.6356	0.9003	1.15925	−0.2282
0.5001	1.06673	−0.6001	1.0000	1.17449	0.0000
0.6001	1.09573	−0.5452			
$T = 333.15$ K					
0.0000	0.77175	0.0000	0.6978	1.11183	−0.5046
0.3999	1.02428	−0.7382	0.8002	1.13381	−0.4473
0.4498	1.04229	−0.7044	0.9003	1.15158	−0.2425
0.5001	1.05877	−0.6669	1.0000	1.16689	0.0000
0.6001	1.08782	−0.5963			
$T = 343.15$ K					
0.0000	0.76286	0.0000	0.6978	1.10468	−0.5613
0.3999	1.01675	−0.8307	0.8002	1.12678	−0.4920
0.4498	1.03483	−0.7916	0.9003	1.14459	−0.2624
0.5001	1.05138	−0.7487	1.0000	1.15997	0.0000
0.6001	1.08057	−0.6678			

(2) Boiling Point
Ref. [32]

w_2	T/K	γ_1	w_2	T/K	γ_1
100 kPa					
0.8190	380.08	1.514	0.4853	371.82	1.160
0.7526	376.19	1.429	0.3683	370.99	1.106
0.6704	373.93	1.322	0.2769	370.74	1.065
0.5857	372.57	1.241	0.2077	370.60	1.039

Mixture 26: 1,3-HOC_3H_6OH (1) + [bmim][BF_4] (2)
(1) Phase Equilibrium

Ref. [31]

T/K	$100x'_1$	$100x_1$	V'^{a}/ $cm^3 \cdot mol^{-1}$	V^{b}/ $cm^3 \cdot mol^{-1}$	$V^{E\prime c}$/ $cm^3 \cdot mol^{-1}$	V^{Ed}/ $cm^3 \cdot mol^{-1}$
286.7 ± 0.2	36.8 ± 0.2	95.8 ± 0.8	144.0 ± 0.2	75.2 ± 0.2	−0.5 ± 0.2	−1.6 ± 0.2
287.9 ± 0.2	38.4 ± 0.4	95.2 ± 0.2	141.8 ± 0.4	75.7 ± 0.2	−1.0 ± 0.4	−1.8 ± 0.2
289.1 ± 0.2	39.7 ± 0.4	94.9 ± 0.2	140.7 ± 0.5	76.0 ± 0.3	−0.6 ± 0.5	−1.8 ± 0.3
291.9 ± 0.2	47.0 ± 0.5	93.1 ± 0.1	131.6 ± 0.5	78.9 ± 0.1	−1.6 ± 0.5	−1.2 ± 0.1

[a] V', observed molar volume lower phase
[b] V, observed molar volume upper phase
[c] $V^{E\prime}$, Volumetric Properties lower phase
[d] V^{E}, Volumetric Properties upper phase

Mixture 27: 2-CH_3CHCH_3OH (1) + [BMIM][BF_4] (2)
(1) Boiling Point

Ref. [32]

w_2	T/K	γ_1	w_2	T/K	γ_1
100 kPa					
0.8302	363.56	1.646	0.5376	356.54	1.228
0.7909	361.40	1.557	0.4971	356.31	1.195
0.7300	359.19	1.453	0.4593	356.10	1.170
0.7033	358.43	1.420	0.4234	355.98	1.146
0.6790	358.07	1.380	0.3860	355.88	1.124
0.6554	357.64	1.352	0.3475	355.80	1.102
0.6249	357.40	1.308	0.3006	355.69	1.081
0.5797	356.90	1.264			

Mixture 28: $HOOC(CH_2)_4COOH$ (1) + [Bmim][BF_4] (2)
(1) Phase Equilibrium

Ref. [142]

x_1	T/K	x_1	T/K	x_1	T/K	x_1	T/K
0.0331	327.3	0.0986	355.2	0.4984	405.3	0.8779	421.3
0.0675	345.0	0.3034	389.3	0.6340	413.4		

Mixture 29: $HOOC(CH_2)_3COOH$ (1) + [Bmim][BF_4] (2)
(1) Phase Equilibrium

Ref. [142]

x_1	T/K	x_1	T/K	x_1	T/K	x_1	T/K
0.1052	299.0	0.1991	320.5	0.4115	341.8	0.6074	355.9
0.7965	364.6						

Mixture 30: $HOOC(CH_2)_2COOH$ (1) + [BMIM][BF_4] (2)
(1) Phase Equilibrium

Ref. [142]

x_1	T/K	x_1	T/K	x_1	T/K	x_1	T/K
0.0596	338.6	0.2137	383.1	0.5922	429.9	0.7946	444.8
0.1137	360.3	0.4094	409.9				

Mixture 31: 1-C_4H_9OH (1) + [bmim][BF_4] (2)
(1) Miscibility Temperatures

Ref. [145]

x_2	T/K	x_2	T/K	x_2	T/K
0.036	322.15	0.112	334.35	0.227	335.11
0.046	327.12	0.120	334.54	0.289	333.92
0.053	331.56	0.146	334.55	0.329	332.17
0.070	332.65	0.170	334.95	0.405	327.75
0.103	334.15	0.208	335.00	0.453	324.40

Mixture 32: $(CH_3)_2CHCH_2OH$ (1) + [bmim][BF_4] (2)
(1) Miscibility Temperatures

Ref. [145]

x_2	T/K	x_2	T/K	x_2	T/K
0.031	317.35	0.246	332.15	0.474	318.55
0.045	326.55	0.289	331.75	0.510	315.95
0.070	330.95	0.334	329.55	0.573	311.15
0.129	332.05	0.400	323.15	0.608	307.75
0.180	332.55	0.450	319.95	0.623	306.75
0.218	332.55				

Mixture 33: 2-$CH_3CH(OH)CH_2CH_3$ (1) + [bmim][BF_4] (2)
(1) Miscibility Temperatures

Ref. [145]

x_2	T/K	x_2	T/K	x_2	T/K
0.031	329.25	0.120	332.95	0.301	326.85
0.062	332.15	0.153	332.75	0.326	325.15

(continued)

x_2	T/K	x_2	T/K	x_2	T/K
0.075	332.75	0.183	332.45	0.361	320.95
0.100	332.95	0.242	330.45	0.464	311.35

Mixture 34: $(CH_3)_3COH$ (1) + [Bmim][BF_4] (2)
(1) Miscibility Temperatures

Ref. [145]

x_2	T/K	x_2	T/K	x_2	T/K
0.076	291.75	0.155	320.15	0.320	315.05
0.082	303.15	0.179	320.95	0.498	301.35
0.098	314.75	0.198	321.15	0.518	300.15
0.118	318.95	0.244	320.55	0.600	294.35

Mixture 35: 1-C_4H_9OD (1) + [bmim][BF_4] (2)
(1) Miscibility Temperatures

Ref. [145]

x_2	T/K	x_2	T/K	x_2	T/K
0.035	327.25	0.174	330.45	0.289	329.05
0.050	328.75	0.191	330.45	0.302	328.65
0.101	329.55	0.210	330.35	0.339	326.65
0.110	329.75	0.226	330.35	0.416	320.15
0.161	330.25	0.246	329.95	0.493	310.15

Mixture 36: $(CH_3)_2CHCH_2OD$ (1) + [bmim][BF_4] (2)
(1) Miscibility Temperatures

Ref. [145]

x_2	T/K	x_2	T/K	x_2	T/K
0.040	325.35	0.143	328.00	0.358	325.15
0.081	326.75	0.182	328.35	0.392	323.45
0.083	326.75	0.221	328.25	0.466	320.00
0.112	327.45	0.286	327.45		

Mixture 37: 2-$CH_3CH(OD)CH_2CH_3$ (1) + [bmim][BF_4] (2)
(1) Miscibility Temperatures

Ref. [145]

x_2	T/K	x_2	T/K	x_2	T/K
0.065	295.15	0.120	327.45	0.283	316.95
0.077	311.15	0.159	327.15	0.323	307.25
0.088	324.45	0.212	326.15	0.353	301.15
0.114	327.05	0.244	321.95		

Mixture 38: $(CH_3)_3COD$ (1) + [bmim][BF_4] (2)
(1) Miscibility Temperatures

Ref. [145]

x_2	T/K	x_2	T/K	x_2	T/K
0.110	310.15	0.200	314.15	0.372	307.15
0.128	312.35	0.220	313.95	0.432	303.75
0.136	313.15	0.262	312.95	0.473	301.55
0.169	313.55	0.322	309.55	0.549	295.15
0.154	313.75				

Mixture 39: $HCON(CH_3)_2$ (1) + $bmimBF_4$ (2)
(1) Transport Properties

Ref. [146]

x_1	η/mPa · s	x_1	η/mPa · s	x_1	η/mPa · s
295.2 K					
0.00	122.9	0.33	26.3	0.61	9.8
0.14	65.6	0.44	16.1	0.73	4.7
0.23	45.4	0.53	11.2		

(2) Electrical Conductivity

Ref. [146]

x_1	σ/mS · cm^{-1}	x_1	σ/mS · cm^{-1}	x_1	σ/mS · cm^{-1}	x_1	σ/mS · cm^{-1}
295.2 K							
0.000	3.75	0.366	8.71	0.658	17.24	0.906	24.36
0.088	4.64	0.434	10.33	0.728	20.11	0.957	18.27
0.161	5.51	0.490	11.69	0.793	23.15	0.973	13.92
0.224	6.29	0.535	12.96	0.828	24.03	0.981	11.28
0.324	7.90	0.606	15.49	0.852	24.37		

Mixture 40: $3\text{-}CH_3C_5H_4N$ (3-Picoline) (1) + $bmimBF_4$ (2)
(1) Electrical Conductivity

Ref. [146]

x_1	σ/mS · cm^{-1}	x_1	σ/mS · cm^{-1}	x_1	σ/mS · cm^{-1}	x_1	σ/mS · cm^{-1}
295.2 K							
0.000	3.78	0.244	5.88	0.568	8.45	0.838	8.38
0.061	4.40	0.279	6.16	0.659	9.15	0.866	7.47
0.114	4.76	0.340	6.74	0.721	9.28	0.906	5.90
0.162	5.20	0.392	7.20	0.763	9.36	0.928	4.74
0.205	5.55	0.474	8.02	0.798	9.40		

Mixture 41: CO_2 (1) + [bmim][BF_4] (2)
(1) Henry's Law Constant and Thermodynamic Properties

Ref. [147]

T (K)	K_H (MPa)	T (K)	K_H (MPa)
322.15	2.35	312.45	1.91
317.45	2.11	307.55	1.71

Ref. [147]

$\Delta_{sil}H^o$ (kJ · mol^{-1})	$\Delta_{sil}G^o$ (kJ · mol^{-1})	$\Delta_{sil}S^o$ (J · K^{-1} · mol^{-1})
−15.8	0.94	−55.8

(2) Solubility and Diffusivity

Ref. [148]

T/K	P/MPa	C_0/(mass%)	C_s/(mass%)	$D \times 10^{-11}$/m^2 · s^{-1}	$x_{1,calc}$/mol %	$x_{1,meas}$/mol %
282.75	0.01017	0.00	0.04	4.0	0.2	0.2
282.95	0.05015	0.07	0.24	3.4	1.3	1.3
283.05	0.10010	0.26	0.46	4.8	2.3	2.6
283.05	0.40010	0.97	1.88	3.8	9.0	9.6
283.25	0.69962	2.40	3.25	4.5	14.7	15.9
283.05	0.99967	3.69	4.64	5.5	20.2	21.6
283.05	1.30019	5.12	5.93	6.9	24.9	26.5
283.15	1.50006	6.28	6.79	6.7	27.8	29.5
283.05	2.00019	7.89	8.97	8.0	34.6	36.4
297.95	0.00973	–	–	–	–	0.2
298.15	0.05002	0.08	0.19	6.6	1.1	1.0
298.05	0.10013	0.23	0.36	7.3	2.0	1.9
298.15	0.39956	0.81	1.42	6.4	6.9	6.9
298.15	0.70022	1.87	2.45	7.1	11.6	11.6
297.95	0.99967	2.88	3.46	7.7	15.8	15.8
298.15	1.30022	3.88	4.44	9.3	19.7	19.7
297.95	1.50008	4.71	5.09	8.2	22.2	22.1
298.05	2.00020	6.03	6.66	10.0	27.7	27.7
323.25	0.01015	0.02	0.06	21.0	0.3	0.3
323.15	0.05011	0.08	0.13	25.0	0.6	0.6
323.15	0.10014	0.18	0.23	12.0	1.2	1.2
323.15	0.39970	0.62	0.88	16.0	4.3	4.4
323.15	0.69962	1.23	1.49	17.0	7.2	7.3
323.15	1.00020	1.88	2.09	14.0	9.9	10.0
323.15	1.29965	2.48	2.69	13.0	12.4	12.7
323.15	1.50011	2.94	3.07	17.0	14.0	14.3
323.15	2.00000	3.82	4.02	15.0	17.7	18.2
348.15	0.01020	0.02	0.04	33.0	0.2	0.2
348.15	0.05013	0.04	0.07	34.0	0.3	0.3

(continued)

T/K	P/MPa	C_0/(mass%)	C_s/(mass%)	$D \times 10^{-11}/m^2 \cdot s^{-1}$	$x_{1,calc}$/mol %	$x_{1,meas}$/mol %
348.05	0.10018	0.10	0.15	33.0	0.8	0.7
348.05	0.40018	0.48	0.60	30.0	3.0	3.0
348.15	0.69956	0.91	1.01	24.0	5.0	5.1
347.95	1.00027	1.30	1.43	33.0	6.9	7.1
348.15	1.30000	1.72	1.83	24.0	8.7	8.9
348.15	1.50024	2.01	2.11	29.0	10.0	10.3
348.05	1.99990	2.68	2.77	46.0	12.7	13.3

Ref. [148]

283.15 K	298.15 K	323.15 K	348.15 K
Diffusivity ($m^2 \cdot s^{-1}$) (over Pressures (0.1–1.0 MPa))			
$(5.3 \pm 2.3) \times 10^{-11}$	$(7.8 \pm 1.8) \times 10^{-11}$	$(1.7 \pm 0.6) \times 10^{-10}$	$(3.2 \pm 1.1) \times 10^{-10}$

(3) Bubble Point

Ref. [149]

x_1	T/K	p/MPa	T/K	p/MPa	T/K	p/MPa
0.1022	293.13	0.587	323.11	1.044	348.12	1.539
	298.14	0.659	328.14	1.143	353.16	1.664
	303.15	0.740	333.16	1.239	358.17	1.769
	308.15	0.803	338.16	1.344	363.16	1.875
	313.19	0.885	343.14	1.438	368.14	1.989
	318.11	0.968				
0.1998	293.17	1.243	323.15	2.269	348.15	3.378
	298.17	1.388	328.12	2.466	353.16	3.629
	303.19	1.542	333.18	2.686	358.16	3.884
	308.19	1.710	338.12	2.909	363.16	4.150
	313.15	1.886	343.13	3.137	368.22	4.418
	318.12	2.074				
0.3035	278.53	1.547	313.39	3.332	342.73	5.577
	283.61	1.742	318.28	3.667	347.95	6.027
	288.47	1.957	323.16	4.002	353.11	6.512
	293.50	2.192	328.12	4.367	357.96	7.012
	298.47	2.447	332.82	4.757	362.91	7.522
	303.40	2.722	337.75	5.157	367.96	8.047
	308.36	3.012				
0.4005	278.68	2.233	313.00	4.953	343.12	8.748
	283.52	2.508	318.13	5.513	348.08	9.473
	288.51	2.838	323.22	6.098	352.82	10.233
	293.33	3.198	328.12	6.703	357.69	11.063
	298.15	3.573	333.19	7.343	363.02	11.968
	303.11	3.993	338.28	8.053	367.92	12.888
	308.10	4.458				
0.4825	278.47	2.926	303.16	5.496	323.10	8.921

(*continued*)

x_1	T/K	p/MPa	T/K	p/MPa	T/K	p/MPa
	283.39	3.346	308.16	6.206	327.99	10.071
	288.55	3.801	313.27	7.046	333.10	11.406
	293.32	4.301	318.16	7.931	338.17	12.856
	298.28	4.861				
0.5507	313.12	14.44	332.94	26.12	347.66	33.82
	318.07	17.46	337.86	28.46	352.69	36.58
	322.95	20.22	342.78	31.12	357.90	38.86
	327.96	23.06				
0.6017	313.12	36.58	332.85	50.24	352.77	62.16
	318.08	39.66	337.85	53.70	357.71	64.82
	322.94	43.24	342.94	56.74	362.58	67.62
	327.91	46.72	347.98	59.46		

Mixture 42: $[C_4mim][BF_4]$ (1) + $[C_4mim][NTf_2]$ (2)
(1) Volumetric Properties

Ref. [57]

	V (pure component) or V^E/cm^3 · mol^{-1}	
x_1	T = 298.15 K	T = 333.15 K
1	188.042 (188.0)	191.367 (191.4)
0.977	0.10_2	0.10_1
0.899	0.14_7	0.21_8
0.699	0.27_5	0.28_0
0.450	0.28_7	0.25_7
0.301	0.28_9	0.25_0
0.102	0.14_0	0.11_8
0	292.794 (292.8)	299.005 (299.0)

Mixture 43: CF_3CH_2OH (1) + $[BMIm][BF_4]$ (2)
(1) Vapor Pressure

Ref. [123]

T (K)	P (kPa)	T (K)	P (kPa)	T (K)	P (kPa)	T (K)	P (kPa)
X_2 = 40 %		X_2 = 60 %		X_2 = 80 %		X_2 = 90 %	
315.2	10.5	321.6	6.8	353.8	7.1	430.4	27.4
325.2	17.7	341.0	15.3	376.0	15.3	442.4	36.0
334.0	26.1	350.6	22.5	387.8	23.1	451.6	44.2
340.8	34.1	361.4	35.5	396.8	30.8	461.2	54.4
346.8	43.7	367.2	41.8	407.2	40.0	465.2	63.7
351.8	53.2	373.0	50.6	414.4	49.7	469.4	73.8
357.2	64.2	379.0	60.6	418.4	59.0		
362.6	77.8	384.4	70.5	425.8	69.4		
370.2	100.5	388.4	82.1	432.8	85.6		
		395.4	100.3	439.0	99.8		

Mixture 44: 1H,1H,2H,2H-$C_8F_{13}H_4OH$ (1) + $[C_4mim][BF_4]$ (2)
(1) Cloud Point

Ref. [150]

x_2	T/K	x_2	T/K	x_2	T/K	x_2	T/K
0.313	284.3	0.601	302.8	0.986	301.3	0.959	303.0
0.371	287.7	0.621	303.7	0.942	307.5	0.924	304.9
0.404	290.1	0.609	303.5	0.794	307.1	0.691	304.2
0.439	291.9	0.645	304.8	0.818	307.1	0.479	294.2
0.480	294.8	0.696	306.5	0.851	307.3	0.591	300.5
0.508	297.0	0.744	307.8	0.890	306.4	0.405	293.2
0.538	299.1	0.805	308.7	0.982	296.1	0.645	306.6
0.572	301.2	0.846	309.4				

Mixture 45: $C_4H_6O_3$ (propylene carbonate) [PC] (1) + $BMIBF_4$ (2)
(1) Volumetric Properties and Conductivity

Ref. [151]

Composition (mol%)		
Solvent	Ionic liquid	σ (mS · cm^{-1}) at 298.15 K
80 % PC	20 % $BMIBF_4$	11.07

Ref. [35]

x_2	$10^{-3}c_2$/mol · m^{-3}	ρ/kg · m^{-3}	κ/S · cm^{-1}	$10^4\Lambda$/S · m^2 · mol^{-1}
298.15 K				
0.02318	0.2653	1201.32	0.537	20.2
0.04836	0.5380	1202.51	0.873	16.22
0.1028	1.078	1204.30	1.230	11.41
0.1601	1.583	1205.43	1.346	8.51
0.2277	2.107	1206.17	1.332	6.32
0.3056	2.634	1206.51	1.229	4.67
0.4070	3.219	1206.44	1.055	3.28
0.5126	3.733	1206.03	0.875	2.34
0.6192	4.174	1205.48	0.715	1.713
0.7173	4.523	1204.39	0.593	1.310
0.8497	4.930	1203.39	0.463	0.939

Mixture 46: SCH = CHCH = CH (thiophene) (1) + $[BMIM][BF_4]$ (2)

(1) Vapor Pressure and Activity Coefficient

Ref. [144]

T (K)	p^{exp} (kPa)	p^{nrtl} (kPa)	RD (p)	$\gamma^{exp}{}_1$	$\gamma^{NRTL}{}_1$
$x_1 = 0.5998$					
303.43	13.5	13.4	0.63	1.6611	1.6512
308.38	16.7	16.7	0.20	1.6474	1.6458
313.34	20.6	20.5	0.29	1.6432	1.6401
318.07	24.6	24.9	1.15	1.6127	1.6346
322.85	29.8	30.0	0.65	1.6156	1.6290
328.12	36.3	36.6	0.74	1.6078	1.6228
333.07	43.5	43.7	0.57	1.6048	1.6168
337.96	52.0	51.9	0.21	1.6118	1.6109
342.99	61.7	61.5	0.37	1.6082	1.6048
$x_1 = 0.3997$					
303.38	11.2	10.8	3.43	2.0314	2.0023
308.28	13.6	13.4	1.11	1.9878	2.0016
313.26	16.4	16.6	1.46	1.9391	2.0005
318.24	19.9	20.4	2.59	1.9192	1.9988
323.12	24.8	24.8	0.16	1.9759	1.9968
328.25	29.8	30.1	0.99	1.9520	1.9942
333.18	36.2	36.0	0.42	1.9794	1.9913
338.12	43.2	42.9	0.65	1.9822	1.9881
343.24	51.3	51.1	0.41	1.9745	1.9845
$x_1 = 0.1997$					
303.40	6.0	6.2	2.97	2.0782	2.2873
308.28	7.7	7.7	0.03	2.1788	2.2945
313.27	9.5	9.6	0.70	2.1894	2.3010
318.14	11.7	11.7	0.20	2.2188	2.3067
323.04	14.3	14.3	0.17	2.2413	2.3116
328.52	17.4	17.6	1.39	2.2177	2.3163
333.18	21.2	21.0	1.04	2.2815	2.3197
338.16	25.2	25.1	0.45	2.2746	2.3226
343.14	29.7	29.8	0.35	2.2613	2.3250

Mixture 47: C_6H_5CHO (1) + [bmim][BF_4] (2)
(1) Volumetric Properties

Ref. [152]

x_2	ρ/ ($g \cdot cm^{-3}$)	V^E_m/ ($cm^3 \cdot mol^{-1}$)	$\overline{V}_1$/ ($cm^3 \cdot mol^{-1}$)	$\overline{V}_2$/ ($cm^3 \cdot mol^{-1}$)	$V_{\Phi 1}$/ ($cm^3 \cdot mol^{-1}$)	$V_{\Phi 2}$/ ($cm^3 \cdot mol^{-1}$)	$\overline{V}^E_1$/ ($cm^3 \cdot mol^{-1}$)	$\overline{V}^E_2$/ ($cm^3 \cdot mol^{-1}$)
$T = 298.15$ K								
0.10009	1.07864	−0.6272	182.01	101.45	180.37	100.93	−4.6294	−0.1817
0.19983	1.10545	−0.9440	183.96	101.11	181.91	100.45	−2.6772	−0.5136
0.29972	1.12730	−1.0913	184.97	100.78	182.99	100.07	−1.6648	−0.8435
0.40003	1.14556	−1.1278	185.58	100.46	183.82	99.75	−1.0567	−1.1680
0.49993	1.16076	−1.0626	186.04	100.08	184.51	99.50	−0.5960	−1.5450
0.59982	1.17377	−0.9361	186.42	99.61	185.07	99.29	−0.2104	−2.0162
0.69989	1.18470	−0.7136	186.70	99.11	185.61	99.25	0.0632	−2.5191
0.79711	1.19380	−0.4362	186.80	98.83	186.09	99.48	0.1632	−2.7983
0.89906	1.20244	−0.1520	186.73	99.29	186.47	100.12	0.0937	−2.3385
$T = 303.15$ K								
0.10009	1.07278	−0.5919	184.05	101.88	182.84	101.36	−4.6991	−0.1366
0.19983	1.09860	−0.9453	185.79	101.58	184.02	100.84	−2.9598	−0.4365
0.29972	1.11946	−1.1164	186.88	101.22	185.03	100.42	−1.8682	−0.7946
0.40003	1.13676	−1.1678	187.58	100.85	185.83	100.07	−1.1737	−1.1644
0.49993	1.15114	−1.1205	188.03	100.48	186.51	99.78	−0.7196	−1.5333
0.59982	1.16340	−1.0095	188.34	100.11	187.07	99.50	−0.4093	−1.9106
0.69989	1.17386	−0.8307	188.55	99.72	187.56	99.25	−0.1973	−2.3021
0.79711	1.18259	−0.6023	188.68	99.34	188.00	99.05	−0.0699	−2.6781
0.89906	1.19051	−0.3107	188.74	99.01	188.41	98.94	−0.0095	−3.0050

T = 308.15 K								
0.10009	1.06860	−0.5656	184.41	102.32	183.55	101.79	−4.7941	−0.0961
0.19983	1.09481	−0.9452	185.96	102.05	184.47	101.24	−3.2458	−0.3674
0.29972	1.11611	−1.1499	187.12	101.67	185.36	100.78	−2.0793	−0.7520
0.40003	1.13367	−1.2176	187.90	101.25	186.16	100.39	−1.3004	−1.1663
0.49993	1.14832	−1.1901	188.36	100.88	186.82	100.04	−0.8437	−1.5349
0.59982	1.16074	−1.0888	188.61	100.58	187.39	99.70	−0.5928	−1.8377
0.69989	1.17152	−0.9433	188.77	100.28	187.85	99.27	−0.4306	−2.1396
0.79711	1.18059	−0.7525	188.92	99.81	188.26	98.71	−0.2766	−2.6081
0.89906	1.18853	−0.4565	189.10	98.78	188.69	97.90	−0.1007	−3.6338
T = 313.15 K								
0.10009	1.06470	−0.5840	184.91	102.72	184.03	102.17	−4.9486	−0.0981
0.19983	1.09093	−0.9715	186.49	102.44	185.00	101.60	−3.3688	−0.3749
0.29972	1.11230	−1.1905	187.69	102.05	185.89	101.11	−2.1752	−0.7686
0.40003	1.12992	−1.2693	188.49	101.62	186.69	100.70	−1.3721	−1.1959
0.49993	1.14450	−1.2364	188.97	101.23	187.39	100.34	−0.8939	−1.5822
0.59982	1.15692	−1.1369	189.24	100.91	187.97	99.97	−0.6245	−1.9075
0.69989	1.16764	−0.9855	189.41	100.58	188.45	99.53	−0.4476	−2.2366
0.79711	1.17660	−0.7807	189.58	100.08	188.88	98.97	−0.2838	−2.7346
0.89906	1.18448	−0.4746	189.76	99.02	189.33	98.11	−0.1022	−3.7914

Mixture 48: CH_2Cl_2 (1) + $[C_4mim][BF_4]$ (2)

(1) Volumetric Properties, Transport Properties and Electrical Conductivity

Ref. [135]

x_2	η/(mPa · s)	x_2	η/(mPa · s)	x_2	η/(mPa · s)
298.15 K					
0.00	0.434	0.3131	5.214	0.5728	18.650
0.0713	0.883	0.3724	7.224	0.6658	27.054
0.0923	1.078	0.4226	9.371	0.7595	36.725
0.1827	2.254	0.4649	11.429	0.8511	55.483
0.2861	4.456	0.5107	14.157	1.0000	110.308

Ref. [135]

x_2	$10^3\kappa$ (S · cm^{-1})	x_2	$10^3\kappa$ (S · cm^{-1})	x_2	$10^3\kappa$ (S · cm^{-1})
298.15 K					
0.0713	5.42	0.3724	10.67	0.6658	6.64
0.0923	7.01	0.4226	10.11	0.7595	5.78
0.1827	10.51	0.4649	9.43	0.8511	4.35
0.2861	11.22	0.5107	8.76	1.00	2.48
0.3131	11.24	0.5728	7.71		

Ref. [35]

x_2	$10^{-3}c_2$/mol · m^{-3}	ρ/kg · m^{-3}	κ/S · cm^{-1}	$10^4\Lambda$/S · m^2 · mol^{-1}
298.15 K				
0.007550	0.1156	1316.60	0.0602	5.21
0.01921	0.2883	1314.92	0.200	6.92
0.02714	0.4016	1313.40	0.310	7.72
0.03878	0.5627	1311.80	0.436	7.76
0.07987	1.083	1304.97	0.980	9.05
0.1395	1.726	1294.06	1.427	8.27
0.1940	2.219	1284.29	1.550	6.99
0.2674	2.773	1271.96	1.607	5.80
0.3355	3.199	1261.43	1.530	4.78
0.4787	3.902	1242.95	1.209	3.10
0.6019	4.360	1230.33	0.930	2.13
0.7730	4.846	1216.18	0.620	1.278
0.8496	5.024	1211.19	0.511	1.018
0.9107	5.152	1207.46	0.441	0.856
1	5.319	1202.19	0.353	0.664

Mixture 49: $CH_3N(CH_2)_3CO$ (*N*-methyl-2-pyrrolidinone) (1) + $[bmim][BF_4]$ (2)
(1) Volumetric Properties

Ref. [153]

x_2	ρ/ ($g \cdot cm^{-3}$)	V^E_m/ ($cm^3 \cdot mol^{-1}$)	$\overline{V}_2$/ ($cm^3 \cdot mol^{-1}$)	$\overline{V}_1$/ ($cm^3 \cdot mol^{-1}$)	$V_{\varphi 2}$/ ($cm^3 \cdot mol^{-1}$)	$V_{\varphi 1}$/ ($cm^3 \cdot mol^{-1}$)	$\overline{V}^E_2$/ ($cm^3 \cdot mol^{-1}$)	$\overline{V}^E_1$/ ($cm^3 \cdot mol^{-1}$)
T = 298.15 K								
0.10050	1.06641	−0.5561	183.48	96.14	181.10	95.78	−3.155	−0.266
0.20027	1.09471	−0.7064	185.52	95.80	183.11	95.52	−1.113	−0.602
0.30970	1.11966	−0.7133	186.03	95.64	184.33	95.37	−0.606	−0.763
0.39986	1.13716	−0.6916	186.17	95.56	184.90	95.25	−0.465	−0.841
0.49920	1.15377	−0.6271	186.48	95.30	185.38	95.15	−0.150	−1.106
0.60001	1.16795	−0.4808	186.96	94.71	185.83	95.20	0.329	−1.697
0.69931	1.17948	−0.2242	187.34	94.02	186.31	95.66	0.705	−2.384
0.79967	1.18939	0.0993	187.36	94.03	186.76	96.90	0.721	−2.372
0.89838	1.19889	0.3045	186.98	96.34	186.97	99.40	0.345	−0.060
T = 303.15 K								
0.10050	1.06141	−0.5735	185.26	96.49	183.04	96.10	−3.491	−0.247
0.20027	1.08859	−0.7558	187.33	96.14	184.98	95.79	−1.421	−0.591
0.30970	1.11240	−0.7916	188.00	95.93	186.20	95.59	−0.753	−0.807
0.39986	1.12887	−0.7687	188.21	95.81	186.83	95.45	−0.544	−0.921
0.49920	1.14452	−0.7122	188.49	95.58	187.32	95.31	−0.260	−1.158
0.60001	1.15788	−0.5788	188.88	95.10	187.79	95.29	0.126	−1.635
0.69931	1.16884	−0.3551	189.19	94.53	188.24	95.55	0.438	−2.206
0.79967	1.17831	−0.0738	189.24	94.42	188.66	96.37	0.487	−2.312
0.89838	1.18708	0.1371	188.99	95.95	188.90	98.08	0.242	−0.785

(continued)

x_2	ρ/ (g · cm^{-3})	V^E_m/ (cm^3 · mol^{-1})	$\overline{V}_2$/ (cm^3 · mol^{-1})	$\overline{V}_1$/ (cm^3 · mol^{-1})	$V_{\varphi 2}$/ (cm^3 · mol^{-1})	$V_{\varphi 1}$/ (cm^3 · mol^{-1})	$\overline{V}^E_2$/ (cm^3 · mol^{-1})	$\overline{V}^E_1$/ (cm^3 · mol^{-1})
T = 308.15 K								
0.10050	1.05684	−0.5864	185.46	96.97	183.37	96.55	−3.737	−0.235
0.20027	1.08456	−0.7983	187.56	96.62	185.21	96.21	−1.642	−0.585
0.30970	1.10876	−0.8478	188.33	96.37	186.46	95.98	−0.875	−0.835
0.39986	1.12548	−0.8316	188.56	96.24	187.12	95.82	−0.638	−0.963
0.49920	1.14137	−0.7801	188.81	96.04	187.64	95.65	−0.396	−1.164
0.60001	1.15509	−0.6711	189.11	95.66	188.08	95.53	−0.088	−1.545
0.69931	1.16647	−0.4853	189.38	95.16	188.51	95.59	0.183	−2.044
0.79967	1.17633	−0.2411	189.48	94.91	188.90	96.00	0.277	−2.300
0.89838	1.18510	−0.0153	189.35	95.71	189.18	97.06	0.152	−1.495
T = 313.15 K								
0.10050	1.05264	−0.6019	185.87	97.40	183.87	96.95	−3.995	−0.222
0.20027	1.08056	−0.8356	188.00	97.04	185.69	96.57	−1.864	−0.581
0.30970	1.10492	−0.9040	188.88	96.75	186.94	96.31	−0.978	−0.871
0.39986	1.12167	−0.8909	189.16	96.60	187.63	96.13	−0.704	−1.018
0.49920	1.13763	−0.8489	189.35	96.44	188.16	95.92	−0.507	−1.180
0.60001	1.15146	−0.7549	189.58	96.17	188.60	95.73	−0.286	−1.453
0.69931	1.16310	−0.6045	189.79	95.76	189.00	95.61	−0.069	−1.858
0.79967	1.17319	−0.3959	189.93	95.38	189.37	95.64	0.064	−2.243
0.89838	1.18193	−0.1684	189.92	95.45	189.67	95.96	0.058	−2.164

Ref. [153]

T/K	V_2^0/ $(cm^3 \cdot mol^{-1})$	$\overline{V}_1^0$/ $(cm^3 \cdot mol^{-1})$	$\overline{V}_2^{\infty a}$/ $(cm^3 \cdot mol^{-1})$	$\overline{V}_1^{\infty b}$/ $(cm^3 \cdot mol^{-1})$	$\overline{V}_2^{\infty c}$/ $(cm^3 \cdot mol^{-1})$	$\overline{V}_2^{\infty d}$/ $(cm^3 \cdot mol^{-1})$
298.15	186.63	96.40	177.63	103.89	178.88	102.64
303.15	188.75	96.73	179.90	101.26	180.86	100.30
308.15	189.20	97.21	180.43	99.11	181.18	98.36
313.15	189.86	97.62	181.17	96.86	181.68	96.33

[a]From equation $\overline{V}_1^{\infty} = V_1^0 + \sum_{i=0}^{i=n} A_i$

[b]From equation $\overline{V}_2^{\infty} = V_2^0 + \sum_{i=0}^{i=n} A_i(-1)^i$

[c]From equation $V_{\Phi 1} = V_1^0 - V_m^E/x$

[d]From equation $V_{\Phi 2} = V_2^0 - V_m^E/(1-x)$

2.2.31.2 Ternary Mixtures

Mixture 1: H_2O (1) + C_2H_5OH (2) + [bmim][BF_4] (3)

(1) Phase Equilibrium

Ref. [24]

x'_2 [a]	w_3 [a]	y_2	T/K	γ_1	γ_2	$\alpha_{2,1}$ [b]
$p = 100$ kPa						
0.9502	0.9084	0.9818	372.15	1.10	1.40	2.83
0.9501	0.8000	0.9780	360.01	1.27	1.31	2.34
0.9501	0.7000	0.9745	356.15	1.39	1.23	2.01
0.9501	0.6000	0.9701	354.65	1.54	1.15	1.70
0.9501	0.4999	0.9663	353.49	1.68	1.11	1.51
0.9500	0.3999	0.9629	352.87	1.79	1.07	1.37
0.9500	0.3001	0.9596	352.35	1.91	1.04	1.25
0.9500	0.2000	0.9552	351.91	2.08	1.02	1.12
0.9499	0.1000	0.9518	351.45	2.22	1.01	1.04

[a]Compositions reported are of equilibrium liquid

[b]$\alpha_{2,1} = (y_2/x'_2)/(y_1/x'_1)$

Ref. [36]

x_2	y_2	x_2	y_2	x_2	y_2
$T = 337.15$ K					
$x_3 = 0.1$		$x_3 = 0.3$		$x_3 = 0.5$	
0.388	0.635	0.388	0.660	0.388	0.677
0.504	0.696	0.504	0.735	0.504	0.759
0.636	0.764	0.636	0.813	0.636	0.834
0.725	0.810	0.725	0.857	0.725	0.880
0.766	0.837	0.766	0.876	0.766	0.897
0.858	0.893	0.858	0.922	0.858	0.938

(continued)

x_2	y_2	x_2	y_2	x_2	y_2
0.898	0.918	0.898	0.943	0.898	0.955
0.953	0.957	0.953	0.969	0.953	0.983
0.980	0.985	0.980	0.991	0.980	0.994

(2) Transport Properties

Ref. [135]

x_3	η/(mPa · s)	x_3	η/(mPa · s)	x_3	η/(mPa · s)
298.15 K, the molar ratio of EtOH to H_2O is 3:1					
0.00	1.437	0.2824	6.318	0.5740	18.956
0.0308	1.799	0.3311	7.654	0.6817	29.532
0.0965	2.716	0.4309	11.502	0.7743	42.012
0.1873	4.287	0.4834	13.716	0.8725	63.266
0.2487	5.538	0.5379	16.332	1.00	110.308
298.15 K, the molar ratio of EtOH to H_2O is 1:1					
0.00	1.932	0.2625	7.112	0.5961	24.314
0.0339	2.392	0.3033	8.265	0.6508	28.986
0.0977	3.274	0.4379	13.352	0.7708	46.686
0.1944	5.189	0.4922	16.513	0.8700	67.556
0.2792	7.812	0.5538	20.614	1.00	110.038
298.15 K, the molar ratio of EtOH to H_2O is 1:3					
0.00	2.383	0.2867	10.117	0.5276	21.226
0.0480	3.552	0.3337	11.486	0.6419	33.110
0.0848	4.456	0.3989	14.501	0.7534	48.352
0.1421	6.012	0.4263	15.598	0.8615	70.870
0.2488	9.176	0.4716	18.256	1.00	110.308

(3) Electrical Conductivity

Ref. [135]

x_3	$10^3\kappa$ (S · cm^{-1})	x_3	$10^3\kappa$ (S · cm^{-1})	x_3	$10^3\kappa$ (S · cm^{-1})
298.15 K, the molar ratio of EtOH to H_2O is 3:1					
0.0308	7.74	0.3311	15.37	0.6817	7.71
0.0965	15.17	0.4309	13.86	0.7743	5.91
0.1873	18.22	0.4834	12.97	0.8725	4.29
0.2487	18.09	0.5379	11.22	1.0000	2.48
0.2824	16.67	0.5740	10.18		
298.15 K, the molar ratio of EtOH to H_2O is 1:1					
0.0339	12.28	0.3033	20.33	0.7708	6.10
0.0977	21.66	0.4379	14.59	0.6508	7.62
0.1944	23.62	0.4922	12.35	0.8700	4.90
0.2792	21.24	0.5538	10.56	1.0000	2.48
0.2625	21.99	0.5961	9.55		
298.15 K, the molar ratio of EtOH to H_2O is 1:3					

(continued)

x_3	$10^3\kappa$ (S · cm^{-1})	x_3	$10^3\kappa$ (S · cm^{-1})	x_3	$10^3\kappa$ (S · cm^{-1})
0.0482	19.60	0.3337	20.79	0.6419	8.62
0.0848	25.25	0.3989	18.51	0.7534	6.53
0.1421	27.21	0.4263	16.52	0.8615	4.90
0.2488	24.82	0.4716	14.49	1.0000	2.48
0.2867	23.08	0.5276	12.34		

Mixture 2: $C_6H_5COCH_3$ (acetophenone) (1) + CO_2 (2) + bmim[BF_4] (3)
(1) Bubble Point

Ref. [154]

x_2	T (K)	P (MPa)	T (K)	P (MPa)	T (K)	P (MPa)	T (K)	P (MPa)
0.1010	278.94	0.487	308.93	0.882	338.83	1.437	358.17	1.867
	288.75	0.597	318.56	1.047	348.32	1.642	367.20	2.082
	299.39	0.737	328.21	1.227				
0.2213	278.98	1.101	308.87	2.103	338.41	3.507	356.79	4.562
	289.06	1.383	318.31	2.518	348.12	4.047	365.89	5.122
	298.57	1.698	328.12	2.978				
0.2971	279.63	1.631	308.60	3.106	338.18	5.296	356.97	6.966
	288.87	2.031	318.62	3.766	347.88	6.136	366.78	7.966
	298.63	2.526	328.50	4.496				
0.3989	278.84	2.367	308.87	4.733	337.92	8.358	357.63	11.658
	288.72	3.022	318.16	5.744	348.16	9.988	365.12	12.993
	299.19	3.857	328.25	7.004				
0.5014	279.44	3.433	298.35	5.613	318.41	9.723	324.22	11.743
	288.80	4.378	308.46	7.308				

Mixture 3: $C_6H_5CHOHCH_3$ (1-phenylethanol) (1) + CO_2 (2) + bmim[BF_4] (3)
(1) Bubble Point

Ref. [154]

x_2	T (K)	P (MPa)	T (K)	P (MPa)	T (K)	P (MPa)	T (K)	P (MPa)
0.1192	285.35	0.718	313.20	1.207	334.15	1.683	355.14	2.228
	292.37	0.828	320.19	1.383	341.24	1.853	361.96	2.412
	299.27	0.958	327.24	1.518	348.16	2.028	366.23	2.522
	306.34	1.088						
0.2047	278.86	1.038	303.34	1.713	330.33	2.802	353.31	3.957
	285.37	1.162	308.32	1.898	338.34	3.188	360.30	4.353
	291.40	1.333	316.31	2.203	345.28	3.547	367.28	4.738
	296.90	1.503	323.29	2.497				
0.3009	279.35	1.648	304.00	2.888	333.33	4.992	355.30	6.983
	286.36	1.958	311.34	3.378	340.30	5.603	361.30	7.583
	293.38	2.298	318.29	3.839	348.24	6.328	367.32	8.243
	298.34	2.578	325.32	4.352				

(*continued*)

x_2	T (K)	P (MPa)	T (K)	P (MPa)	T (K)	P (MPa)	T (K)	P (MPa)
0.4019	279.36	2.693	305.39	4.773	326.32	7.269	347.25	10.548
	285.39	3.103	312.37	5.553	333.43	8.328	354.24	11.822
	291.90	3.553	319.32	6.328	340.31	9.363	359.31	12.762
	297.79	4.063						
0.4995	279.36	3.579	293.37	5.189	307.35	7.403	319.15	10.694
	286.33	4.294	300.30	6.259	314.29	9.224	323.09	12.209

Mixture 4: CO_2 (1) + $(CH_3)_2CHCH_2C_6H_4COCH_3$ (4-isobutylacetophenone) (2) + bmim[BF_4] (3)

(1) Liquid–vapor boundaries data

Ref. [155]

x_1	T (K)	P (MPa)	T (K)	P (MPa)	T (K)	P (MPa)
0.0971	285.39	0.583	319.23	1.072	354.19	1.743
	292.31	0.663	326.26	1.187	360.24	1.864
	299.28	0.763	333.24	1.318	367.17	2.014
	306.29	0.853	340.12	1.434		
	313.27	0.954	347.23	1.588		
0.2067	279.15	1.000	310.94	1.925	339.49	3.081
	285.50	1.149	315.91	2.116	344.63	3.331
	291.27	1.300	318.99	2.230	348.92	3.541
	296.36	1.459	324.73	2.456	353.41	3.755
	300.76	1.584	329.58	2.656	359.71	4.083
	305.97	1.750	334.50	2.871	366.11	4.420
0.2960	279.27	1.598	314.30	3.403	349.04	6.039
	286.35	1.918	321.14	3.848	356.16	6.674
	293.21	2.228	328.17	4.343	362.21	7.204
	300.20	2.588	335.23	4.899	367.12	7.654
	307.26	2.973	342.17	5.464		
0.4004	280.84	2.498	314.50	5.199	349.82	9.739
	286.90	2.854	319.65	5.708	354.55	10.509
	292.59	3.254	325.75	6.444	360.06	11.359
	297.35	3.629	332.55	7.274	366.21	12.399
	303.14	4.104	338.28	8.035		
	308.66	4.604	344.18	8.885		
0.5037	279.34	3.378	300.31	5.718	321.13	9.759
	286.37	4.043	307.24	6.803	326.11	11.193
	293.17	4.803	314.16	8.069	330.15	12.563

Molar fraction of 4-isobutylacetophenone in [bmim][BF_4] was kept constant at 0.09425
Mixture 5: H_2O (1) + 1-C_3H_7OH (2) + [bmim][BF_4] (3)
(1) Phase Equilibrium

Ref. [32]

$x'_2{}^{a,b}$	y_2	T/K	$x_1{}^{a,c}$	$x_2{}^{a,c}$	γ_1	γ_2	$x'_2{}^{a,b}$	y_2	T/K	$x_1{}^{a,c}$	$x_2{}^{a,c}$	γ_1	γ_2
100 kPa													
$w_3 = 0.300$													
0.9805	0.9564	370.39	0.0176	0.8813	2.71	1.06	0.5655	0.4787	362.76	0.4025	0.5239	1.87	1.20
0.9602	0.9121	369.76	0.0359	0.8643	2.73	1.06	0.4573	0.4491	362.21	0.5068	0.4271	1.61	1.42
0.9298	0.8543	368.86	0.0634	0.8388	2.65	1.06	0.3589	0.4345	362.04	0.6032	0.3377	1.39	1.74
0.8599	0.7410	367.11	0.1270	0.7797	2.51	1.05	0.2647	0.4247	362.18	0.6967	0.2508	1.22	2.28
0.7726	0.6366	365.32	0.2075	0.7050	2.30	1.07	0.1748	0.4264	362.67	0.7873	0.1668	1.06	3.38
0.6694	0.5428	363.75	0.3039	0.6155	2.10	1.12	0.1139	0.4178	363.70	0.8494	0.1091	0.96	4.85
$w_3 = 0.500$													
0.9776	0.9613	371.51	0.0178	0.7747	2.28	1.16	0.4115	0.4584	363.73	0.5088	0.3558	1.48	1.63
0.9038	0.8424	369.78	0.0771	0.7241	2.28	1.17	0.3195	0.4408	363.58	0.5972	0.2804	1.31	2.00
0.8107	0.7186	367.92	0.1538	0.6587	2.18	1.17	0.2214	0.4195	363.93	0.6945	0.1975	1.16	2.67
0.7067	0.6134	366.37	0.2421	0.5833	2.02	1.20	0.1394	0.4028	365.14	0.7783	0.1260	1.01	3.83
0.6138	0.5569	365.23	0.3234	0.5140	1.81	1.29	0.0913	0.3977	366.32	0.8285	0.0832	0.92	5.47
0.5078	0.5141	364.35	0.4191	0.4324	1.58	1.47							
$w_3 = 0.700$													
0.9956	0.9950	374.81	0.0027	0.6149	1.71	1.34	0.4226	0.5014	367.81	0.4214	0.3084	1.42	1.76
0.8640	0.8473	372.97	0.0870	0.5532	1.74	1.36	0.3314	0.4511	367.41	0.5025	0.2491	1.33	1.99
0.7790	0.7531	371.55	0.1449	0.5108	1.78	1.38	0.2456	0.4307	367.32	0.5834	0.1899	1.19	2.50
0.6902	0.6658	370.36	0.2084	0.4643	1.75	1.41	0.1596	0.3978	367.95	0.6692	0.1271	1.07	3.36
0.6031	0.6021	369.38	0.2740	0.4163	1.64	1.47	0.1046	0.3352	369.17	0.7269	0.0849	1.04	4.05
0.5143	0.5371	368.43	0.3445	0.3647	1.57	1.55	0.0702	0.2889	370.56	0.7641	0.0577	1.01	4.87

[a]Compositions reported are of equilibria liquid
[b]Mole fraction on IL-free basis
[c]Mole fraction on IL-containing basis

Mixture 6: $HOOC(CH_2)_4COOH$ (1) + H_2O (2) + [Bmim][BF_4] (3)
(1) Phase Equilibrium

Ref. [142]

x_1	T/K
$HOOC(CH_2)_4COOH$ (1) + (H_2O:[Bmim][BF_4] = 0.689:0.311)[a] (2)	
0.0	270.0
0.0592	331.3
0.1952	374.1
0.3614	396.8
0.5690	407.3
0.8359	419.0
$HOOC(CH_2)_4COOH$ (1) + (H_2O:[Bmim][BF_4] = 0.984:0.016)[a] (2)	
0.0	271.9
0.0136	321.1
0.0452	343.1
0.0793	354.3
0.2600	382.6
0.3313	390.5
0.3736	393.6
[Bmim][BF_4] (1) + (H_2O:$HOOC(CH_2)_4COOH$ = 0.5:0.5)[a] (2)	
0.0	398.4
0.2689	394.2
0.3909	389.9
0.5984	378.2
0.7902	358.7

[a]Molar ratio

Mixture 7: $HOOC(CH_2)_3COOH$ (1) + H_2O (2) + [Bmim][BF_4] (3)
(1) Phase Equilibrium

Ref. [142]

x_1	T/K
[Bmim][BF_4] (1) + (H_2O:$HOOC(CH_2)_3COOH$ = 0.5:0.5)[a] (2)	
0.0	345.0
0.1989	338.2
0.3996	329.6
0.6008	317.5
0.7973	307.1

[a]Molar ratio

Mixture 8: $HOOC(CH_2)_2COOH$ (1) + H_2O (2) + [Bmim][BF_4] (3)
(1) Phase Equilibrium

Ref. [142]

x_1	T/K
[Bmim][BF_4] (1) + (H_2O: $HOOC(CH_2)_2COOH$ = 0.5:0.5)[a] (2)	
0.0	428.5
0.2005	426.4
0.4381	402.1
0.6020	385.3
0.7989	358.8

[a]Molar ratio

Mixture 9: $2\text{-}CH_3CHCH_3OH$ (1) + H_2O (2) + [bmim][BF_4] (3)

(1) Phase Equilibrium

Ref. [156]

$x_1^{a,b}$	y_1	T/K	$x_2^{a,c}$	$x_1^{a,c}$	γ_2	γ_1	$x_1^{a,b}$	y_1	T/K	$x_2^{a,c}$	$x_1^{a,c}$	γ_2	γ_1
100 kPa													
$w_3 = 0.300$													
0.99803	0.99753	355.70	0.00177	0.8960	2.68	1.08	0.5511	0.6812	354.22	0.4164	0.5111	1.54	1.37
0.9535	0.9495	355.38	0.0418	0.8588	2.35	1.09	0.4529	0.6379	354.48	0.5111	0.4231	1.41	1.54
0.9172	0.9146	355.14	0.0748	0.8282	2.24	1.10	0.3579	0.6224	354.89	0.6042	0.3367	1.22	1.86
0.8864	0.8899	354.95	0.1028	0.8022	2.11	1.11	0.2652	0.6094	355.55	0.6963	0.2513	1.07	2.38
0.8212	0.8334	354.62	0.1626	0.7467	2.04	1.13	0.1827	0.5929	356.63	0.7793	0.1742	0.95	3.20
0.7373	0.7796	354.34	0.2403	0.6745	1.84	1.18	0.1108	0.5561	358.49	0.8525	0.1063	0.88	4.59
0.6492	0.7262	354.23	0.3230	0.5978	1.71	1.25							
$w_3 = 0.500$													
0.99708	0.99748	356.29	0.00231	0.7878	2.05	1.20	0.5323	0.6916	355.38	0.3966	0.4515	1.50	1.51
0.9770	0.9799	356.23	0.0183	0.7742	2.07	1.20	0.4287	0.6511	355.82	0.4926	0.3697	1.34	1.71
0.9173	0.9305	355.92	0.0661	0.7334	2.00	1.22	0.3329	0.6278	356.39	0.5842	0.2915	1.18	2.04
0.8496	0.8720	355.62	0.1215	0.6862	2.02	1.24	0.2345	0.6077	357.46	0.6813	0.2087	1.02	2.65
0.7372	0.7864	355.29	0.2159	0.6056	1.92	1.28	0.1395	0.5606	360.27	0.7782	0.1261	0.90	3.65
0.6405	0.7339	355.18	0.2998	0.5341	1.72	1.36							
$w_3 = 0.700$													
0.99488	0.99703	358.72	0.00316	0.6144	1.61	1.40	0.4457	0.6822	358.74	0.4015	0.3229	1.34	1.83
0.9596	0.9761	358.65	0.0252	0.5983	1.63	1.42	0.3538	0.6282	359.85	0.4821	0.2639	1.25	1.98
0.9320	0.9602	358.62	0.0427	0.5854	1.60	1.42	0.2821	0.5957	360.43	0.5483	0.2155	1.16	2.25
0.8663	0.9205	358.53	0.0855	0.5541	1.60	1.45	0.2010	0.5447	361.97	0.6271	0.1578	1.08	2.66
0.7569	0.8574	358.35	0.1604	0.4993	1.53	1.51	0.1256	0.4897	364.57	0.7046	0.1012	0.98	3.40
0.6527	0.8016	358.31	0.2362	0.4439	1.45	1.59	0.0850	0.4231	366.60	0.7479	0.0695	0.96	3.98
0.5488	0.7385	358.42	0.3165	0.3850	1.41	1.68							

[a]Compositions reported are of equilibrium liquid. For simplicity subscript l is omitted
[b]Mole fraction on IL-free basis
[c]Mole fraction on IL-containing basis

Ref. [157]

$100x_3$	T/K	x_1	y_1	γ_1	γ_2	α_{12}
P = 101.32 kPa						
9.951	367.98	0.017	0.170	6.831	1.128	11.8433
10.000	364.43	0.062	0.316	3.976	1.112	6.989
10.001	359.75	0.142	0.423	2.779	1.226	4.430
9.975	357.17	0.245	0.524	2.207	1.272	3.392
9.999	356.46	0.285	0.531	1.978	1.361	2.840
10.035	355.44	0.451	0.620	1.520	1.496	1.986
10.035	355.06	0.599	0.673	1.262	1.790	1.378
9.940	355.07	0.699	0.734	1.177	1.937	1.188
10.014	355.25	0.766	0.787	1.145	1.982	1.129
9.987	355.57	0.877	0.852	1.068	2.586	0.807
9.971	355.90	0.946	0.924	1.060	2.985	0.694
19.734	369.80	0.035	0.187	3.830	1.181	6.342
19.801	365.23	0.127	0.337	2.254	1.261	3.494
19.666	361.84	0.294	0.440	1.443	1.495	1.887
19.702	358.56	0.409	0.528	1.414	1.710	1.616
19.982	358.10	0.474	0.586	1.383	1.722	1.571
19.961	356.46	0.561	0.632	1.345	1.956	1.344
20.014	356.28	0.633	0.698	1.326	1.935	1.340
19.953	356.41	0.705	0.760	1.289	1.902	1.325
19.959	356.65	0.863	0.889	1.220	1.876	1.271
19.755	356.97	0.918	0.932	1.174	2.114	1.224
29.934	370.94	0.055	0.214	3.066	1.281	4.678
29.961	367.36	0.126	0.382	2.724	1.242	4.288
29.728	363.51	0.202	0.476	2.437	1.328	3.589
29.915	361.06	0.272	0.571	2.391	1.312	3.562
29.697	359.97	0.372	0.655	2.085	1.272	3.205
29.990	359.74	0.473	0.710	1.801	1.291	2.728
29.666	358.86	0.562	0.768	1.689	1.280	2.580
29.953	358.50	0.750	0.857	1.438	1.407	1.998
29.834	359.15	0.784	0.868	1.356	1.463	1.812
29.922	359.34	0.933	0.951	1.241	1.740	1.394

Mixture 10: $(C_5H_5)_2Fe$ (1) + $HCON(CH_3)_2$ (2) + $bmimBF_4$ (3)
(1) Diffusion Coefficient

Ref. [146]

(2) + (3)			
x_2	$10^6\ D_{Fc}$/cm$^2 \cdot$ s^{-1}	x_2	$10^6\ D_{Fc}$/cm$^2 \cdot$ s^{-1}
295.2 K			
0.00	0.04	0.60	0.56
0.10	0.05	0.70	0.88
0.20	0.08	0.80	1.40

(*continued*)

(2) + (3)			
x_2	$10^6\ D_{Fc}/cm^2 \cdot s^{-1}$	x_2	$10^6\ D_{Fc}/cm^2 \cdot s^{-1}$
0.30	0.14	0.90	2.40
0.40	0.22	0.95	3.20
0.50	0.34	0.97	3.40

Mixture 11: H_2O (1) + $C_{12}H_{22}O_{11}$ (Sucrose) (2) + [Bmim]BF_4 (3)

(1) Phase Equilibrium

Ref. [158]

278 K		298 K		308 K	
100 w_3	100 w_2	100 w_3	100 w_2	100 w_3	100 w_2
66.25	0.68	72.44	1.96	61.58	3.22
61.15	1.14	67.91	2.22	57.12	3.84
56.36	1.63	62.24	2.86	52.05	4.55
52.22	2.15	58.55	3.33	48.36	5.3
48.89	2.53	54.62	3.74	44.65	6.02
45.35	3.11	50.96	4.1	40.33	7.11
42.26	3.64	46.02	4.94	38.51	8.96
38.11	4.28	41.69	5.59	36.47	10.32
35.32	4.86	36.85	6.51	34.03	12.75
31.24	5.67	32.32	7.62	30.63	15.24
26.45	6.21	28.26	8.88	28.67	18.67
21.42	6.85	24.97	9.79	24.32	21.33
15.33	7.88	21.33	10.55	19.4	24.38
12.87	10.24	19.04	12.44	16.58	27.88
10.55	11.57	16.92	14.18	13.15	30.34
9.38	13.68	14.56	15.71	10.2	34.48
8.67	16.44	12.29	16.68	9.14	38.3
6.21	19.96	10.86	18.4	7.37	42.12
5.75	23.55	9.57	21.25	6.3	46.92
5.05	27.63	8.14	24.34	5.84	50.48
4.26	32.14	7.05	27.37	3.83	52.45
3.62	37.24	4.96	30.46		
3.18	41.37	3.81	35.37		
2.41	46.22	3.21	39.65		
2.03	50.05	2.75	43.27		
1.66	54.25	2.45	48.24		
1.24	60.35	2.08	53.12		
0.98	64.12	1.76	58.46		

Ref. [159]

At room temperature		At room temperature		At room temperature	
100 w_3	100 w_2	100 w_3	100 w_2	100 w_3	100 w_2
72.44	1.96	28.26	8.88	7.05	27.37
67.91	2.22	24.97	9.79	4.96	30.46
62.24	2.86	21.33	10.55	3.81	35.37
58.55	3.33	19.04	12.44	3.21	39.65
54.62	3.74	16.92	14.18	2.75	43.27
50.96	4.10	14.56	15.71	2.45	48.24
46.02	4.94	12.29	16.68	2.08	53.12
41.69	5.59	10.86	18.40	1.76	58.46
36.85	6.51	9.57	21.25	1.23	53.05
32.32	7.62	8.14	24.34		

Mixture 12: CO_2 (1) + 1-(4-$(CH_3)_3C$-C_6H_4)-CH_3CHOH (1-(4-isobutyl phenyl)-ethanol) (2) + bmim[BF_4] (3)
(1) Bubble Point

Ref. [160]

x_{CO2}	T (K)	P (MPa)	T (K)	P (MPa)	T (K)	P (MPa)
The concentration of 1-(4-isobutylphenyl)-ethanol in bmim[BF_4] was kept constant at 9.636 mol%						
0.1116	292.98	0.693	353.26	1.864	328.23	1.295
	303.17	0.855	363.14	2.099	298.25	0.754
	313.24	1.025	278.28	0.501	318.26	1.095
	323.25	1.210	283.34	0.556	348.23	1.730
	333.26	1.415	368.22	2.231	358.36	1.980
	343.22	1.629	308.28	0.924	363.29	4.643
0.2144	293.14	1.433	328.09	2.813	279.24	1.032
	300.10	1.668	335.15	3.138	286.24	1.217
	307.26	1.928	342.22	3.503	368.24	4.947
	314.18	2.199	348.97	3.868	358.16	7.025
	321.19	2.499	356.05	4.243	279.42	1.549
0.3027	293.18	2.158	328.18	4.361	286.15	1.834
	300.15	2.528	335.14	4.893	367.37	7.909
	307.17	2.933	342.09	5.478	356.09	11.315
	314.17	3.372	351.25	6.268	363.09	12.605
	321.22	3.856	328.05	6.951	286.12	2.761
0.4066	293.13	3.299	335.16	7.955	279.05	2.300
	300.33	3.899	342.14	8.984	286.22	3.954
	307.26	4.551	349.22	10.159	313.17	7.994
	314.27	5.291	321.17	10.084	323.18	10.714
	321.29	6.095	327.27	12.189		
0.5058	293.19	4.777	279.21	3.269		
	300.09	5.703				
	307.14	6.842				
	314.16	8.242				

Mixture 13: H_2O (1) + $C_5H_{10}O_5$ (xylose) (2) + [Bmim]BF_4 (3)
(1) Phase Equilibrium

Ref. [159]

At room temperature		At room temperature		At room temperature	
100 w_3	100 w_2	100 w_3	100 w_2	100 w_3	100 w_2
66.24	4.22	36.58	13.62	8.26	34.33
62.91	4.78	31.37	15.83	7.14	37.54
58.24	5.63	26.9	18.32	6.02	40.74
56.37	6.36	23.53	20.04	4.98	45.38
53.45	7.33	19.18	23.69	3.72	50.67
49.86	8.22	16.45	26.05	3.07	55.42
45.57	9.84	13.63	29.66	2.53	62.56
40.42	11.25	10.18	32.25	2.03	67.22

Mixture 14: H_2O (1) + $C_6H_{12}O_6$ (glucose) (2) + [Bmim]BF_4 (3)
(1) Phase Equilibrium

Ref. [159]

At room temperature		At room temperature		At room temperature	
100 w_3	100 w_2	100 w_3	100 w_2	100 w_3	100 w_2
61.58	3.27	36.47	8.95	13.15	24.23
57.12	4.01	34.03	9.66	10.20	28.85
52.05	4.83	30.63	10.17	9.14	32.19
48.36	5.56	28.67	11.42	7.37	35.38
44.65	6.22	24.32	13.39	6.30	37.32
40.33	7.02	19.40	17.57	5.84	41.73
38.51	7.89	16.58	20.25	3.83	45.80

Mixture 15: H_2O (1) + $C_6H_{12}O_6$ (fructose) (2) + [Bmim]BF_4 (3)
(1) Phase Equilibrium

Ref. [159]

At room temperature		At room temperature		At room temperature	
100 w_3	100 w_2	100 w_3	100 w_2	100 w_3	100 w_2
74.69	3.26	39.70	14.67	12.05	36.70
70.92	4.24	35.93	16.38	9.52	40.93
67.77	5.12	30.30	18.91	7.51	45.75
63.73	6.38	25.48	21.45	6.21	51.09
58.97	7.91	21.30	23.93	4.75	56.70
55.53	8.63	18.04	27.84	4.55	62.13
49.21	10.78	15.26	31.93	4.37	67.26
43.96	12.56				

Mixture 16: C_6H_{14} (1) + CH_3CH_2COOH (2) + [bmim][BF_4] (3)
(1) Selectivity

Ref. [75]

Hexane-rich phase		RTIL-rich phase		
x_3'	x_2'	x_3''	x_2''	Selectivity (S)
298.2 K				
0.0027	0.8755	0.1436	0.8561	397
0.0018	0.7152	0.2804	0.7194	1423
0.0011	0.5188	0.4464	0.5534	2561
0.0008	0.2663	0.6356	0.3642	5012
0.0001	0.1504	0.8239	0.1760	9941

Mixture 17: C_6H_{14} (1) + $CH_3CH_2COOCH = CH_2$ (Vinyl propionate) (2) + [bmim][BF_4] (3)
(1) Selectivity

Ref. [75]

Hexane-rich phase		RTIL-rich phase		
x_3'	x_2'	x_3''	x_2''	Selectivity (S)
298.2 K				
0.0110	0.8694	0.6226	0.3774	–
0.0100	0.7128	0.7917	0.2083	–
0.0009	0.5185	0.8432	0.1568	–
0.0007	0.2721	0.8708	0.1292	–
0.0005	0.1326	0.9086	0.0914	–

Mixture 18: C_6H_{14} (1) + $CH_3CH(Ph)COOCH_2CH_3$ (*rac*-1-Phenylethyl propionate) (2) + [bmim][BF_4] (3)
(1) Selectivity

Ref. [75]

Hexane-rich phase		RTIL-rich phase		
x_3'	x_2'	x_3''	x_2''	Selectivity (S)
298.2 K				
0.0045	0.7403	0.2619	0.7379	1272
0.0030	0.5100	0.4687	0.5312	5072
0.0020	0.3132	0.6581	0.3419	–
0.0012	0.1253	0.7879	0.2121	–
0.0000	0.0511	0.8731	0.1269	–

Mixture 19: C_6H_{14} (1) + $CH_3CH(Ph)OH$ (*rac*-1-Phenylethanol) (2) + [bmim][BF_4] (3)
(1) Selectivity

Ref. [75]

Hexane-rich phase		RTIL-rich phase		
x_3'	x_2'	x_3''	x_2''	Selectivity (*S*)
298.2 K				
0.0000	0.4335	0.1650	0.7000	7
0.0000	0.1785	0.3150	0.6000	33
0.0000	0.0641	0.5500	0.3900	95
0.0000	0.0183	0.7000	0.2550	304

Mixture 20: $C_4H_6O_3$ (propylene carbonates) (1) + $C_3H_4O_3$ (ethylene carbonates) (2) + $BMIBF_4$ (3)
(1) Electrical Conductivity

Ref. [151]

Composition (mol%)		
Solvent	Ionic liquid	σ (mS · cm^{-1}) at 298.15 K
40 % PC, 40 % EC	20 % $BMIBF_4$	14.05

Mixture 21: H_2O (1) + $OCH_2CH_2CH_2CH_2$ (2) + [BMIM][BF_4] (3)
(1) Phase Equilibrium

Ref. [36]

x_2	y_2	x_2	y_2	x_2	y_2
T = 337.15 K					
x_3 = 0.1		x_3 = 0.3		x_3 = 0.5	
0.356	0.788	0.356	0.795	0.356	0.811
0.428	0.801	0.428	0.826	0.428	0.833
0.622	0.833	0.622	0.879	0.700	0.922
0.700	0.850	0.700	0.900	0.792	0.947
0.792	0.878	0.792	0.925	0.851	0.965
0.851	0.901	0.851	0.947	0.900	0.978
0.900	0.924	0.900	0.963	0.948	0.987
0.948	0.955	0.948	0.980	0.980	0.995
0.980	0.982	0.980	0.992		

Mixture 22: CH_3CN (1) + CH_2Cl_2 (2) + [C_4mim][BF_4] (3)
(1) Transport Properties

Ref. [135]

x_3	η/(mPa · s)	x_3	η/(mPa · s)	x_3	η/(mPa · s)
298.15 K, the molar ratio of CH_3CN to CH_2Cl_2 is 1:3					
0.00	0.424	0.2950	4.281	0.6032	20.744
0.0451	0.644	0.3632	6.351	0.6715	27.994

(*continued*)

x_3	η/(mPa · s)	x_3	η/(mPa · s)	x_3	η/(mPa · s)
0.1015	1.073	0.3880	7.266	0.8165	51.288
0.1201	1.235	0.4313	8.989	0.8953	68.285
0.2169	2.594	0.5446	14.896	1.0000	107.320

Ref. [135]

x_3	η/(mPa · s)	x_3	η/(mPa · s)	x_3	η/(mPa · s)
298.15 K, the molar ratio of CH_3CN to CH_2Cl_2 is 1:1					
0.00	0.405	0.3452	5.301	0.6105	19.712
0.0691	0.759	0.3760	6.313	0.6981	28.225
0.1000	0.985	0.4427	9.082	0.7760	39.447
0.2152	2.347	0.4945	11.690	0.8816	65.160
0.2969	3.979	0.5378	14.438	1.00	110.308

Ref. [135]

x_3	η/(mPa · s)	x_3	η/(mPa · s)	x_3	η/(mPa · s)
298.15 K, the molar ratio of CH_3CN to CH_2Cl_2 is 3:1					
0.00	0.373	0.3397	4.731	0.5733	16.078
0.0639	0.680	0.3806	5.979	0.6531	23.128
0.0855	0.816	0.4316	7.926	0.7788	38.843
0.1916	1.836	0.4891	10.694	0.8600	55.221
0.2803	3.301	0.5249	12.727	1.00	110.308

(2) Electrical Conductivity

Ref. [135]

x_3	$10^3\kappa$ (S · cm^{-1})	x_3	$10^3\kappa$ (S · cm^{-1})	x_3	$10^3\kappa$ (S · cm^{-1})
298.15 K, the molar ratio of CH_3CN to CH_2Cl_2 is 1:3					
0.0451	10.27	0.3632	14.70	0.6715	7.85
0.1015	15.90	0.3880	13.19	0.8165	5.31
0.1201	16.84	0.4313	12.90	0.8953	4.46
0.2169	17.68	0.5446	10.98	1.0000	2.48
0.2950	15.78	0.6032	9.24		
298.15 K, the molar ratio of CH_3CN to CH_2Cl_2 is 1:1					
0.0691	12.75	0.3760	16.97	0.6981	7.39
0.1000	18.31	0.4427	14.51	0.7760	6.32
0.2152	19.79	0.4945	12.38	0.8816	4.32
0.2969	20.38	0.5378	11.50	1.0000	2.48
0.3452	18.00	0.6105	9.69		
298.15 K, the molar ratio of CH_3CN to CH_2Cl_2 is 3:1					
0.0639	16.76	0.3806	14.27	0.6531	7.87
0.0855	17.96	0.4316	13.80	0.7788	6.54
0.1916	20.46	0.4891	11.02	0.8600	5.75
0.2803	19.17	0.5249	10.96	1.0000	2.48
0.3397	17.65	0.5733	9.96		

2.2.32 02011-03001: 1-Butyl-3-methylimidazolium bis((trifluoromethyl)sulfonyl)imide

Abbreviation:	[C_4MIm][NTf_2], [BMIM][NTf_2],[bmim][NTf_2], [C_4MIM][NTf_2], [C_4mim][NTf_2], [BMIM][$(CF_3SO_2)_2N$], [BMIM][BTI], [bmim][Tf_2N], [C_1C_4Im][NTf_2]
Molecular Formula:	$C_{10}H_{15}F_6N_3O_4S_2$
Molar Mass:	419.37
Structure:	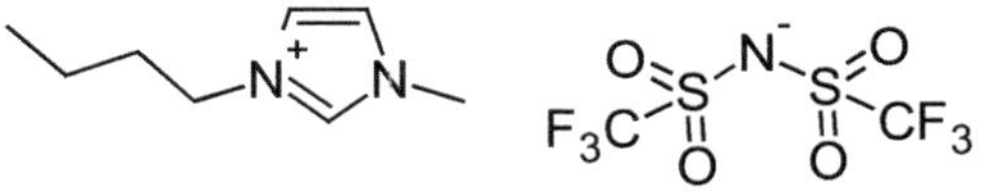

2.2.32.1 Binary Mixtures

Mixture 1: 1-C_3H_7OH (1) + [C_4MIM][NTf_2] (2)

(1) Volumetric Properties and Surface Tension

Ref. [161]

x_2	ρ/g · cm^{-3}	σ/mN · m^{-1}	V^E/cm^3 · mol^{-1}
0.000	0.7990	23.34	0
0.035	0.8782	23.60	−0.096
0.131	1.0362	24.16	−0.154
0.202	1.1165	24.73	−0.143
0.296	1.1941	25.77	−0.027
0.373	1.2435	26.47	0.007
0.517	1.3118	28.60	0.128
0.664	1.3619	29.89	0.177
0.685	1.3680	30.54	0.167
0.884	1.4220	32.49	–
1.000	1.4364	32.80	0

(2) Phase Equilibrium

Ref. [162]

	T = 298.15 K		T = 303.15 K		T = 308.15 K		T = 313.15 K	
x_1	p (Pa)	γ_1	p (Pa)	γ_1	p (Pa)	γ_1	p (Pa)	γ_1
0.0000	0	2.707	0	2.437	0	2.314	0	2.241
0.1155	789	2.420	976	2.192	1241	2.042	1594	1.960
0.2119	1309	2.215	1633	2.021	1994	1.868	2543	1.785
0.3339	1849	1.986	2373	1.833	2912	1.693	3714	1.615
0.4673	2356	1.761	3001	1.652	3790	1.540	4875	1.471
0.5733	2582	1.592	3361	1.517	4331	1.435	5560	1.374
0.6716	2703	1.438	3589	1.392	4702	1.340	6066	1.290
0.7370	2746	1.335	3696	1.306	4879	1.276	6332	1.234

(continued)

	T = 298.15 K		T = 303.15 K		T = 308.15 K		T = 313.15 K	
x_1	p (Pa)	γ_1	p (Pa)	γ_1	p (Pa)	γ_1	p (Pa)	γ_1
0.8178	2772	1.209	3772	1.198	5031	1.190	6584	1.159
0.8675	2779	1.134	3803	1.130	5102	1.132	6714	1.110
0.8910	2782	1.101	3812	1.099	5132	1.104	6771	1.086
0.9358	2784	1.043	3820	1.044	5168	1.050	6866	1.041
1.0000	2786	1.000	3826	1.000	5199	1.000	6986	1.000

(3) Activity Coefficient at Infinite Dilution and Partial Molar Excess Enthalpy at Infinite Dilution

Ref. [163]

Solute i	Temp range/K	γ_i^{∞} (298 K)	$H_i^{E,\infty}$/kJ · mol^{-1}
1-propanol	312.2–354.2	2.68	12.6

Mixture 2: 1-C_4H_9OH (1) + [C_4MIM][NTf_2] (2)
(1) Volumetric Properties and Surface Tension

Ref. [161]

x_2	ρ/g · cm^{-3}	σ/mN · m^{-1}	V^E/cm^3 · mol^{-1}
0.0000	0.8053	23.96	
0.0941	0.9622	23.98	−0.074
0.2025	1.0871	23.88	0.026
0.2045	1.0890	23.80	−0.089
0.3060	1.1726	24.16	0.088
0.3064	1.1725	24.25	0.101
0.3964	1.2305	25.17	0.134
0.5002	1.2836	26.79	0.227
0.5977	1.3235	28.32	0.405
0.6848	1.3548	29.56	-
0.8016	1.3886	30.96	0.404
0.8986	1.4134	31.94	0.271
1.0000	1.4364	32.80	0

(2) Activity Coefficient at Infinite Dilution and Partial Molar Excess Enthalpy at Infinite Dilution

Ref. [163]

Solute i	Temp range/K	γ_i^{∞} (298 K)	$H_i^{E,\infty}$/kJ · mol^{-1}
1-butanol	312.1–343.6	2.96	9.8

Mixture 3: H_2O (1) + $[C_4mim][NTf_2]$ (2)
(1) Volumetric Properties

Ref. [44]

x_1	$\rho/g \cdot cm^{-3}$	x_1	$\rho/g \cdot cm^{-3}$
298.21 K		323.21 K	
0	1.4371	0	1.4138
0.0308	1.4357	0.0482	1.4118
0.0753	1.4340	0.0816	1.4105
0.1194	1.4322	0.1190	1.4088
0.1731	1.4300	0.1597	1.4071
0.2391	1.4267	0.2141	1.4045
		0.2583	1.4020
x_1	$V^E_m/cm^3 \cdot mol^{-1}$	x_1	$V^E_m/cm^3 \cdot mol^{-1}$
298.21 K		323.21 K	
0	0	0	0
0.0308	0.103	0.0482	0.141
0.0753	0.170	0.0816	0.196
0.1194	0.223	0.1190	0.283
0.1731	0.245	0.1597	0.325
0.2391	0.314	0.2141	0.389
		0.2583	0.459

(2) Solubility

Ref. [45]

T/K	$x_1 \pm \sigma$	$10^4 (x_2 \pm \sigma)$
288.15	0.2307 ± 0.0003	2.86 ± 0.01
293.15	0.2443 ± 0.0006	2.92 ± 0.02
298.15	0.2568 ± 0.0006	3.07 ± 0.01
303.15	0.2715 ± 0.0007	3.18 ± 0.01
308.15	0.2853 ± 0.0007	3.44 ± 0.02
313.15	0.2989 ± 0.0004	3.94 ± 0.01
318.15	0.3136 ± 0.0009	4.70 ± 0.01

Ref. [44]

T/K	x_1	T/K	x_1	T/K	x_1	T/K	x_1
290.13	0.2431	332.89	0.3571	360.05	0.4465	320.64	0.9994
299.32	0.2618	345.19	0.3971	357.32	0.9983	310.38	0.9996
307.12	0.2891	350.14	0.4123	345.83	0.9989	290.27	0.9998
319.76	0.3201						

(3) Thermodynamic Properties

Ref. [45]

$(\Delta_{sol}H_m^0 \pm \sigma^a)$/ kJ · mol^{-1}	$(\Delta_{sol}G_m^0 \pm \sigma^a)$/ kJ · mol^{-1}	$(\Delta_{sol}S_m^0 \pm \sigma^a)$/ J · K^{-1} · mol^{-1}	$(\Delta_{svt}H_m^0 \pm \sigma^a)$/ kJ · mol^{-1}	$(\Delta_{svt}Gm^0 \pm \sigma^a)$/ kJ · mol^{-1}	$(\Delta_{svt}S_m^0 \pm \sigma^a)$/ J · K^{-1} · mol^{-1}	$(\Delta_{svt}H_m^* \pm \sigma^a)$/ kJ · mol^{-1}	$(\Delta_{svt}G_m^* \pm \sigma^a)$/ kJ · mol^{-1}	$(\Delta_{svt}S_m^* \pm \sigma^a)$/ J · K^{-1} · mol^{-1}
298.15 K								
7.1 ± 1.5	20.053 ± 0.008	−43.4 ± 5.1	−147.9 ± 6.2	−60.301 ± 0.008	−294 ± 21	−145.7 ± 6.2	−78.209 ± 0.008	−226 ± 21

[a]Standard deviation

(4) Phase Equilibrium

Ref. [18]

x_1	P (kPa)	x_1	P (kPa)	x_1	P (kPa)	x_1	P (kPa)
353.15 K							
0.049	5.32	0.458	43.68	0.839	47.00	0.950	47.12
0.097	10.18	0.563	46.31	0.874	47.05	0.960	47.13
0.192	20.07	0.657	46.66	0.902	47.07	0.967	47.15
0.264	27.00	0.733	46.82	0.923	47.09	0.972	47.15
0.341	34.49	0.793	46.94	0.939	47.11	0.977	47.16

Ref. [46]

x_1	P (mbar)	x_1	P (mbar)	x_1	P (mbar)	x_1	P (mbar)
353.15 K							
0.0490	53.17	0.4577	436.78	0.8391	470.01	0.9504	471.21
0.0969	101.75	0.5626	463.10	0.8739	470.48	0.9595	471.35
0.1920	200.70	0.6572	466.60	0.9022	470.75	0.9666	471.45
0.2636	269.95	0.7333	468.21	0.9229	470.91	0.9723	471.52
0.3415	344.95	0.7932	469.40	0.9386	471.07	0.9767	471.56

(5) Activity Coefficient at Infinite Dilution and Partial Molar Excess Enthalpy at Infinite Dilution

Ref. [18]

Solute	$\gamma^{\infty}{}_1$
353.15 K	
Water	2.27

Ref. [4]

Solute (1)	$H_{12}{}^{E,\infty}$/kJ · mol^{-1}
Water	10.542

Mixture 4: C_8H_{18} (1) + [BMIM][NTf$_2$] (2)

(1) Phase Equilibrium

Ref. [50]

x_1	P (kPa)	x_1	P (kPa)	x_1	P (kPa)	x_1	P (kPa)
353.15 K							
0.0065	3.15	0.0194	8.613	0.0498	19.79	1.0000	23.16
0.0134	6.07	0.0323	13.53				
363.15 K							
0.0107	4.741	0.0274	11.54	0.0635	24.72	1.0000	38.51
0.0186	7.939	0.0419	17.07	0.0941	33.87		

(2) Activity Coefficient at Infinite Dilution and Partial Molar Excess Enthalpy at Infinite Dilution

Ref. [163]

Solute *i*	Temp range/K	γ_i^{∞} (298 K)	$H_i^{E,\infty}$/kJ · mol^{-1}
1-octane	302.0–333.6	33.8	7.7

Mixture 5: C_9H_{20} (1) + [BMIM][NTf$_2$] (2)
(1) Phase Equilibrium

Ref. [50]

x_1	P (kPa)	x_1	P (kPa)	x_1	P (kPa)	x_1	P (kPa)
353.15 K							
0.0047	1.462	0.0145	3.962	0.0262	6.736	1.0000	9.650
0.0096	2.732						

(2) Activity Coefficient at Infinite Dilution and Partial Molar Excess Enthalpy at Infinite Dilution

Ref. [163]

Solute *i*	Temp range/K	γ_i^{∞} (298 K)	$H_i^{E,\infty}$/kJ · mol^{-1}
1-nonane	302.0–333.6	49.6	8.5

Mixture 6: $C_6H_5CH_3$ (1) + [BMIM][NTf$_2$] (2)
(1) Phase Equilibrium

Ref. [50]

x_1	P (kPa)	x_1	P (kPa)	x_1	P (kPa)	x_1	P (kPa)
363.15 K							
0.0245	1.823	0.0656	4.953	0.1166	8.897	0.4485	36.03
0.0340	2.541	0.0773	5.856	0.1617	12.47	0.6306	49.56
0.0408	3.077	0.0899	6.804	0.2210	17.24		
0.0528	4.010	0.1010	7.667	0.3151	24.99		

(2) Excess Molar Enthalpy

Ref. [50]

x_1	H^E (J · mol^{-1})	x_1	H^E (J · mol^{-1})	x_1	H^E (J · mol^{-1})	x_1	H^E (J · mol^{-1})
363.15 K							
0.0653	−96.09	0.2803	−380.0	0.5389	−579.9	0.8911	−187.2
0.1255	−182.1	0.3249	−422.4	0.6451	−564.8	0.9608	−71.58
0.1811	−261.2	0.4054	−508.5	0.7317	−451.9		
0.2325	−322.9	0.4761	−555.8	0.8035	−344.6		

(3) Activity Coefficient at Infinite Dilution and Partial Molar Excess Enthalpies at Infinite Dilution

Ref. [163]

Solute i	Temp range/K	γ_i^∞ (298 K)	$H_i^{E,\infty}$/kJ · mol^{-1}
Toluene	312.2–343.7	1.40	5.6

Ref. [4]

Solute (1)	$H_{12}^{E,\infty}$/kJ · mol^{-1}
Toluene	−1.260

Mixture 7: C_6H_{14} (1) + [BMIM][NTf_2] (2)
(1) Excess Molar Enthalpy

Ref. [50]

x_1	H^E (J · mol^{-1})	x_1	H^E (J · mol^{-1})	x_1	H^E (J · mol^{-1})	x_1	H^E (J · mol^{-1})
353.15 K							
0.0200	55.9	0.1503	397.2	0.3004	393.8	0.7004	159.2
0.0501	147.9	0.2003	454.9	0.5004	277.0	0.9002	48.7
0.1002	288.1						

(2) Activity Coefficient at Infinite Dilution and Partial Molar Excess Enthalpies at Infinite Dilution

Ref. [163]

Solute i	Temp range/K	γ_i^∞ (298 K)	$H_i^{E,\infty}$/kJ · mol^{-1}
1-hexane	302.0–333.6	15.4	6.3

Ref. [4]

Solute (1)	$H_{12}^{E,\infty}$/kJ · mol^{-1}
n-hexane	4.486

Mixture 8: C_6H_6 (1) + [C_4MIm][NTf_2] (2)
(1) Excess Molar Enthalpy

Ref. [50]

x_1	H^E (J · mol^{-1})	x_1	H^E (J · mol^{-1})	x_1	H^E (J · mol^{-1})	x_1	H^E (J · mol^{-1})
363.15 K							
0.0498	−98.2	0.1994	−366.2	0.5991	−863.7	0.8495	−470.1
0.0997	−188.6	0.2992	−529.5	0.6992	−829.9	0.8997	−299.5
0.1495	−283.8	0.4491	−735.8	0.7994	−613.1	0.9498	−147.6
413.15 K							
0.1462	−288.0	0.3647	−695.6	0.5823	−920.2	0.8299	−595.6

(*continued*)

x_1	H^E (J · mol^{-1})	x_1	H^E (J · mol^{-1})	x_1	H^E (J · mol^{-1})	x_1	H^E (J · mol^{-1})
0.2087	−399.3	0.4485	−820.3	0.6844	−909.7	0.9070	−331.5
0.2654	−500.4	0.5202	−885.0	0.7648	−778.6	0.9670	−131.0
0.3172	−594.1						

(2) Phase Equilibrium

Ref. [162]

	T = 298.15 K		T = 303.15 K		T = 308.15 K		T = 313.15 K	
x_1	p (Pa)	γ_1	p (Pa)	γ_1	p (Pa)	γ_1	p (Pa)	γ_1
0.0000	0	0.635	0	0.634	0	0.633	0	0.630
0.2189	2348	0.853	2924	0.848	3576	0.842	4394	0.836
0.2351	2571	0.870	3187	0.865	3890	0.858	4761	0.853
0.3353	4218	0.974	5208	0.967	6320	0.957	7681	0.950
0.4401	6024	1.074	7462	1.064	9087	1.052	11133	1.043
0.5478	7935	1.151	9843	1.139	12014	1.125	14741	1.113
0.5970	8794	1.173	10901	1.160	13264	1.146	16221	1.134
0.6761	10088	1.185	12514	1.173	15275	1.158	18669	1.146
0.7116	10693	1.180	13219	1.168	16113	1.155	19693	1.143
0.7478	11288	1.168	14000	1.157	17061	1.145	20874	1.134
1.0000	12704	1.000	15919	1.000	19659	1.000	24331	1.000

Ref. [3]

x_1	P (kPa)	x_1	P (kPa)	x_1	P (kPa)	x_1	P (kPa)
353.15 K							
0.0117	1.40	0.3475	38.06	0.7775	97.46	0.9845	100.83
0.0231	2.53	0.4059	45.65	0.7862	98.51	0.9900	100.83
0.0357	3.79	0.4662	53.91	0.8043	99.47	0.9939	100.82
0.0479	5.00	0.5251	62.46	0.8420	100.86	0.9961	100.82
0.0602	6.22	0.5797	70.67	0.8871	100.86	0.9975	100.81
0.0944	9.63	0.6293	78.24	0.9216	100.86	0.9986	100.81
0.1355	13.84	0.6733	84.79	0.9479	100.86	0.9993	100.81
0.1835	18.93	0.7124	90.21	0.9657	100.85	0.9998	100.80
0.2354	24.62	0.7342	93.19	0.9766	100.84	1.0000	100.81
0.2893	30.90	0.7471	94.43				

(3) Activity Coefficient at Infinite Dilution and Partial Molar Excess Enthalpies at Infinite Dilution

Ref. [163]

Solute i	Temp range/K	γ_i^∞ (298 K)	$H_i^{E,\infty}$/kJ · mol^{-1}
Benzene	322.8–354.5	0.88	1.0

Ref. [4]

Solute (1)	$H_{12}^{E,\infty}$/kJ · mol^{-1}
Benzene	−0.920

Mixture 9: $CH_3CH(CH_2)_4CH_2$ (1) + [BMIM][BTI] (2)
(1) Excess Molar Enthalpy
Ref. [50]

x_1	H^E (J · mol^{-1})	x_1	H^E (J · mol^{-1})	x_1	H^E (J · mol^{-1})	x_1	H^E (J · mol^{-1})
363.15 K							
0.1067	463.8	0.2860	574.1	0.4932	414.3	0.7730	183.8
0.2014	619.7	0.3621	516.0	0.6022	323.4	0.8720	108.1
0.2449	599.9	0.4308	463.4	0.6942	245.4		

Mixture 10: CH_3COCH_3 (1) + [BMIM][BTI] (2)
(1) Excess Molar Enthalpy

Ref. [50]

x_1	H^E (J · mol^{-1})	x_1	H^E (J · mol^{-1})	x_1	H^E (J · mol^{-1})	x_1	H^E (J · mol^{-1})
353.15 K							
0.0500	−129.9	0.1999	−512.0	0.5998	−1217	0.8499	−984.6
0.0999	−254.5	0.2998	−730.2	0.6998	−1239	0.8999	−778.2
0.1499	−384.9	0.4498	−1037	0.7999	−1110	0.9500	−492.5

(2) Phase Equilibrium

Ref. [18]

x_1	P (kPa)	x_1	P (kPa)	x_1	P (kPa)	x_1	P (kPa)
353.15 K							
0.056	4.92	0.575	80.91	0.872	172.53	0.950	199.85
0.089	7.27	0.638	96.67	0.892	179.77	0.955	201.60
0.146	12.82	0.693	112.26	0.907	185.16	0.973	207.00
0.196	18.59	0.741	126.60	0.909	185.73	0.990	211.93
0.241	24.18	0.783	140.78	0.922	190.54	0.996	213.96
0.340	36.46	0.819	153.05	0.934	194.48	0.999	214.72
0.422	49.81	0.848	163.69	0.941	196.86	1.000	215.04
0.504	65.45	0.865	170.13	0.943	197.64		

Ref. [46]

x_1	P (mbar)	x_1	P (mbar)	x_1	P (mbar)	x_1	P (mbar)
353.15 K							
0.0562	49.21	0.5745	809.05	0.8721	1725.3	0.9495	1998.5
0.0889	72.71	0.6384	966.68	0.8921	1797.7	0.9548	2016.0
0.1459	128.24	0.6935	1122.6	0.9071	1851.6	0.9728	2070.0

(continued)

x_1	P (mbar)	x_1	P (mbar)	x_1	P (mbar)	x_1	P (mbar)
0.1963	185.89	0.7407	1266.0	0.9086	1857.3	0.9895	2119.3
0.2411	241.81	0.7831	1407.8	0.9223	1905.4	0.9964	2139.6
0.3400	364.56	0.8186	1530.5	0.9335	1944.8	0.9990	2147.2
0.4221	498.12	0.8480	1636.9	0.9410	1968.6	1.0000	2150.4
0.5043	654.45	0.8653	1701.3	0.9429	1976.4		

(3) Activity Coefficient at Infinite Dilution and Partial Molar Excess Enthalpy at Infinite Dilution

Ref. [163]

Solute i	Temp range/K	γ_i^∞ (298 K)	$H_i^{E,\infty}$/kJ · mol^{-1}
Acetone	312.3–343.7	0.37	−0.42

Ref. [18]

Solute	$\gamma^\infty{}_1$
353.15 K	
Acetone	0.406

Mixture 11: 2-$CH_3COC_2H_5$ (1) + [BMIM][BTI] (2)
(1) Excess Molar Enthalpy

Ref. [50]

x_1	H^E (J · mol^{-1})	x_1	H^E (J · mol^{-1})	x_1	H^E (J · mol^{-1})	x_1	H^E (J · mol^{-1})
353.15 K							
0.0508	−122.0	0.2026	−491.9	0.6039	−1179	0.8521	−958.5
0.1015	−247.5	0.3035	−721.4	0.7034	−1208	0.9015	−758.9
0.1521	−379.0	0.4541	−1002	0.8026	−1082	0.9508	−470.7

(2) Partial Molar Excess Enthalpies at Infinite Dilution

Ref. [4]

Solute (1)	$H_{12}{}^{E,\infty}$/kJ · mol^{-1}
2-Butanone	−2.188

Mixture 12: 1-$C_6H_{13}OH$ (1) + [C_4MIM][NTf_2] (2)
(1) Cloud Point

Ref. [164]

w_2	T/K	w_2	T/K
0.1587	327.13	0.1752	326.55
0.2872	331.37	0.2226	328.39
0.6199	330.16	0.2851	329.51
0.1185	323.77	0.2919	329.52
0.5077	331.81	0.3171	329.72

(continued)

w_2	T/K	w_2	T/K
0.7449	323.32	0.3779	329.67
0.4088	332.00	0.4349	329.43
0.4670	332.05	0.5114	328.80
0.2475	329.21	0.5711	327.99
0.2926	329.61	0.5998	327.36
0.3780	329.88	0.6307	326.35
0.4702	329.23	0.7103	322.00
0.5140	329.05	0.7832	314.50
0.1010	319.90	0.3441	329.75

(2) Activity Coefficient at Infinite Dilution and Partial Molar Excess Enthalpy at Infinite Dilution

Ref. [163]

Solute i	Temp range/K	γ_i^∞ (298 K)	$H_i^{E,\infty}$/kJ · mol^{-1}
1-hexanol	354.0–385.6	4.00	9.9

Mixture 13: 1,2-$C_6H_{12}(OH)_2$ (1) + [C_4MIM][NTf_2] (2)
(1) Densities of the Liquid-Liquid Equilibria and Interfacial Tensions

Ref. [164]

T/K	ρ_l/g · cm^{-3}	ρ_u/g · cm^{-3}	$\Delta\rho$/g · cm^{-3}	Γ/mN · m^{-1}
293.75	0.9712	1.3763	0.4051	1.27
298.14	0.9727	1.3609	0.3882	1.00
303.18	0.9751	1.3426	0.3675	0.813
305.80	0.9785	1.3315	0.3530	0.759
308.17	0.9760	1.3245	0.3485	0.612
310.60	0.9832	1.3082	0.3250	0.529
312.56	0.9828	1.3016	0.3188	0.422
318.20	0.9961	1.2641	0.2680	0.253
323.20	1.0190	1.2144	0.1954	0.073

(2) Cloud Point

Ref. [51]

w_2	T/K	w_2	T/K
First run		Second run	
0.1131	312.4	0.0802	304.5
0.1726	320.3	0.2845	326.5
0.2376	324.6	0.5157	328.2
0.3584	327.9	0.5675	328.0
0.4500	328.2	0.6606	326.5
0.5916	327.9	0.8688	307.1
0.6830	325.9		
0.7221	324.1		

(continued)

w_2	T/K	w_2	T/K
First run		Second run	
0.7781	320.0		
0.8380	312.7		

Mixture 14: CH_3OH (1) + [BMIM][NTf_2] (2)
(1) Phase Equilibrium

Ref. [162]

	T = 298.15 K		T = 303.15 K		T = 308.15 K		T = 313.15 K	
x_1	p (Pa)	γ_1	p (Pa)	γ_1	p (Pa)	γ_1	p (Pa)	γ_1
0.0000	0	0.945	0	0.911	0	0.850	0	0.806
0.0454	748	0.964	928	0.931	1113	0.873	1299	0.830
0.0816	1356	0.980	1670	0.946	1989	0.891	2399	0.849
0.0992	1588	0.987	1972	0.954	2404	0.899	2844	0.859
0.1400	2332	1.004	2899	0.970	3482	0.919	4174	0.880
0.2537	4287	1.047	5303	1.014	6421	0.970	7780	0.937
0.3349	6126	1.072	7574	1.041	9382	1.002	11439	0.973
0.4658	8826	1.101	11085	1.073	13737	1.042	17045	1.022
0.5852	10875	1.109	13765	1.085	17180	1.061	21347	1.049
0.6878	12730	1.098	16060	1.079	20065	1.060	24766	1.054
0.7840	14086	1.071	17878	1.059	22370	1.045	27947	1.043
0.8526	15001	1.045	19045	1.038	24104	1.029	30239	1.029
0.9007	15609	1.026	19953	1.021	25260	1.016	31701	1.016
0.9321	16001	1.014	20491	1.012	26003	1.009	32702	1.009
0.9786	16597	1.002	21342	1.001	27198	1.001	34391	1.001
0.9900	16772	1.000	21607	1.000	27540	1.000	34850	1.000
1.0000	16958	1.000	21880	1.000	27960	1.000	35450	1.000

(2) Activity Coefficient at Infinite Dilution and Partial Molar Excess Enthalpy at Infinite Dilution

Ref. [163]

Solute i	Temp range/K	γ_i^{∞} (298 K)	$H_i^{E,\infty}$/kJ · mol^{-1}
Methanol	312.2–354.2	1.30	8.8

Mixture 15: C_2H_5OH (1) + [BMIM][NTf_2] (2)
(1) Phase Equilibrium

Ref. [162]

	T = 298.15 K		T = 303.15 K		T = 308.15 K		T = 313.15 K	
x_1	p (Pa)	γ_1	p (Pa)	γ_1	p (Pa)	γ_1	p (Pa)	γ_1
0.0000	0	1.819	0	1.556	0	1.359	0	1.209
0.0631	852	1.756	1006	1.541	1141	1.374	1334	1.244
0.0824	1136	1.738	1325	1.536	1552	1.378	1827	1.254
0.1995	2626	1.638	3223	1.501	3993	1.390	4850	1.301
0.2690	3392	1.584	4247	1.476	5202	1.389	6342	1.317

(*continued*)

	$T = 298.15$ K		$T = 303.15$ K		$T = 308.15$ K		$T = 313.15$ K	
x_1	p (Pa)	γ_1	p (Pa)	γ_1	p (Pa)	γ_1	p (Pa)	γ_1
0.3622	4294	1.517	5471	1.438	6797	1.375	8331	1.322
0.5088	5621	1.418	7242	1.363	9196	1.322	11603	1.288
0.5584	6059	1.384	7789	1.332	9952	1.296	12510	1.265
0.6777	6979	1.298	8937	1.248	11431	1.216	14336	1.191
0.7116	7154	1.272	9169	1.222	11773	1.190	14767	1.166
0.8008	7546	1.193	9742	1.145	12527	1.117	15875	1.099
0.8768	7733	1.115	10108	1.077	13051	1.057	16679	1.046
0.9365	7785	1.048	10262	1.028	13337	1.019	17193	1.014
1.0000	7876	1.000	10468	1.000	13768	1.000	17928	1.000

(2) Activity Coefficient at Infinite Dilution and Partial Molar Excess Enthalpy at Infinite Dilution

Ref. [163]

Solute i	Temp range/K	γ_i^{∞} (298 K)	$H_i^{E,\infty}$/kJ · mol^{-1}
Ethanol	312.2–343.7	1.88	9.3

Mixture 16: [EMIM][BTI] (1) + [BMIM][BTI] (2)
(1) Volumetric Properties

Ref. [4]
These physicochemical properties data are listed in a mixture 31 of IL 02003-03001.
Mixture 17: 1-$C_5H_{11}OH$ (1) + [C_4MIM][NTf_2] (2)
(1) Activity Coefficient at Infinite Dilution and Partial Molar Excess Enthalpy at Infinite Dilution

Ref. [163]

Solute i	Temp range/K	γ_i^{∞} (298 K)	$H_i^{E,\infty}$/kJ · mol^{-1}
1-Pentanol	354.0–385.6	3.15	9.7

Mixture 18: 2-CH_3CHCH_3OH (1) + [BMIM][$(CF_3SO_2)_2N$] (2)
(1) Phase Equilibrium

Ref. [18]

x_1	P (kPa)	x_1	P (kPa)	x_1	P (kPa)	x_1	P (kPa)
353.15 K							
0.037	3.85	0.505	56.32	0.862	86.25	0.945	89.72
0.074	7.25	0.580	64.08	0.879	86.98	0.951	89.93
0.115	11.58	0.644	70.28	0.898	87.83	0.957	90.16
0.149	15.65	0.699	75.21	0.905	88.12	0.972	90.60
0.180	19.24	0.750	79.20	0.914	88.52	0.989	91.33
0.262	28.28	0.792	82.15	0.927	89.00	0.996	91.71
0.340	37.42	0.827	84.29	0.937	89.41	0.999	91.92
0.424	47.22	0.856	85.84	0.940	89.44	1.000	91.99

Ref. [46]

x_1	P (mbar)	x_1	P (mbar)	x_1	P (mbar)	x_1	P (mbar)
353.15 K							
0.0374	38.48	0.5047	563.19	0.8624	862.46	0.9449	897.15
0.0744	72.46	0.5804	640.81	0.8790	869.81	0.9508	899.27
0.1147	115.76	0.6437	702.80	0.8982	878.30	0.9569	901.62
0.1492	156.48	0.6994	752.14	0.9051	881.22	0.9721	906.03
0.1800	192.40	0.7501	792.00	0.9139	885.18	0.9892	913.29
0.2616	282.75	0.7921	821.48	0.9267	890.01	0.9962	917.06
0.3396	374.24	0.8272	842.92	0.9374	894.09	0.9989	919.19
0.4245	472.21	0.8557	858.41	0.9398	894.42	1.0000	919.92

(2) Activity Coefficient at Infinite Dilution and Partial Molar Excess Enthalpy at Infinite Dilution

Ref. [163]

Solute i	Temp range/K	γ_i^∞ (298 K)	$H_i^{E,\infty}$/kJ · mol^{-1}
2-Propanol	312.2–343.7	2.07	9.2

Ref. [18]

Solute	γ^∞_1
353.15 K	
2-Propanol	1.10

Mixture 19: $[C_4mim][NTf_2]$ (1) + $[C_{10}mim][NTf_2]$ (2)
(1) Volumetric Properties

Ref. [57]

	V (pure component) or V^E/cm^3 · mol^{-1}	
x_1	T = 298.15 K	T = 333.15 K
1	292.490 (292.4)	299.414 (299.3)
0.500	0.22_1	0.23_1
0.250	0.16_0	0.17_7
0	404.351 (404.0)	413.872 (413.5)

Mixture 20: $[C_4mim][NTf_2]$ (1) + $[C_8mim][NTf_2]$ (2)
(1) Volumetric Properties

Ref. [57]

	V (pure component) or V^E/cm^3 · mol^{-1}	
x_1	T = 298.15 K	T = 333.15 K
1	292.349 (292.4)	299.309 (299.3)
0.472	0.13_9	0.11_9
0.471	0.13_9	0.11_9
0	362.141 (361.8)	370.874 (370.5)

Mixture 21: $[C_4mim][PF_6]$ (1) + $[C_4mim][NTf_2]$ (2)
(1) Volumetric Properties
Ref. [57]

	V (pure component) or $V^E/cm^3 \cdot mol^{-1}$	
x_1	$T = 298.15$ K	$T = 333.15$ K
1	207.811 (207.8)	212.291 (212.3)
0.750	0.09_5	0.07_6
0.500	0.12_0	0.10_7
0.250	0.13_1	0.10_9
0	292.453 (292.4)	299.375 (299.0)

Mixture 22: $[C_4mim][BF_4]$ (1) + $[C_4mim][NTf_2]$ (2)
(1) Volumetric Properties

Ref. [57]
These physicochemical properties data are listed in a mixture 42 of IL 02011-02001.
Mixture 23: $CH_2(CH_2)_4CH_2$ (1) + [BMIM]$[NTf_2]$ (2)
(1) Phase Equilibrium

Ref. [3]

x_1	P (kPa)	x_1	P (kPa)	x_1	P (kPa)	x_1	P (kPa)
353.15 K							
0.0089	5.70	0.2061	98.98	0.6941	98.99	0.9653	98.95
0.0171	10.69	0.2586	99.09	0.6949	99.03	0.9786	98.95
0.0269	16.40	0.3149	99.09	0.7314	99.01	0.9861	98.94
0.0364	21.85	0.3783	99.07	0.7637	99.01	0.9916	98.94
0.0452	26.90	0.4402	99.07	0.7683	98.98	0.9948	98.94
0.0709	41.37	0.5001	99.06	0.8240	98.99	0.9971	98.95
0.0956	54.65	0.5558	99.06	0.8714	98.98	0.9987	98.94
0.1265	69.82	0.6070	99.06	0.9136	98.97	0.9995	98.94
0.1621	85.76	0.6534	99.05	0.9441	98.96	1.0000	98.94

(2) Activity Coefficient at Infinite Dilution and Partial Molar Excess Enthalpies at Infinite Dilution

Ref. [163]

Solute i	Temp range/K	γ_i^{∞} (298 K)	$H_i^{E,\infty}/kJ \cdot mol^{-1}$
Cyclohexane	301.9–333.6	9.32	5.6

Ref. [4]

Solute (1)	${H_{12}}^{E,\infty}/kJ \cdot mol^{-1}$
Cyclohexane	4.502

Mixture 24: C_5H_{12} (1) + [BMIM][NTf_2] (2)
(1) Activity Coefficient at Infinite Dilution and Partial Molar Excess Enthalpies at Infinite Dilution

Ref. [163]

Solute i	Temp range/K	γ_i^∞ (298 K)	$H_i^{E,\infty}$/kJ · mol^{-1}
1-Pentane	302.0–333.6	10.1	6.0

Ref. [4]

Solute (1)	$H_{12}^{E,\infty}$/kJ · mol^{-1}
n-Pentane	4.332

Mixture 25: C_7H_{16} (1) + [BMIM][NTf_2] (2)
(1) Activity Coefficient at Infinite Dilution and Partial Molar Excess Enthalpies at Infinite Dilution

Ref. [163]

Solute i	Temp range/K	γ_i^∞ (298 K)	$H_i^{E,\infty}$/kJ · mol^{-1}
1-Heptane	302.0–333.6	23.0	7.1

Ref. [4]

Solute (1)	$H_{12}^{E,\infty}$/kJ · mol^{-1}
n-Heptane	5.207

Mixture 26: $C_{10}H_{22}$ (1) + [BMIM][NTf_2] (2)
(1) Activity Coefficient at Infinite Dilution and Partial Molar Excess Enthalpy at Infinite Dilution

Ref. [163]

Solute i	Temp range/K	γ_i^∞ (298 K)	$H_i^{E,\infty}$/kJ · mol^{-1}
1-Decane	312.3–354.3	64.5	9.0

Mixture 27: $C_{11}H_{24}$ (1) + [BMIM][NTf_2] (2)
(1) Activity Coefficient at Infinite Dilution and Partial Molar Excess Enthalpy at Infinite Dilution

Ref. [163]

Solute i	Temp range/K	γ_i^∞ (298 K)	$H_i^{E,\infty}$/kJ · mol^{-1}
1-Undecane	312.3–354.3	92.6	9.9

Mixture 28: $C_{12}H_{26}$ (1) + [BMIM][NTf$_2$] (2)
(1) Activity Coefficient at Infinite Dilution and Partial Molar Excess Enthalpy at Infinite Dilution

Ref. [163]

Solute *i*	Temp range/K	γ_i^{∞} (298 K)	$H_i^{E,\infty}$/kJ · mol^{-1}
1-Dodecane	322.8–354.3	127.3	10.9

Mixture 29: 1-C_5H_{10} (1) + [BMIM][NTf$_2$] (2)
(1) Activity Coefficient at Infinite Dilution and Partial Molar Excess Enthalpies at Infinite Dilution

Ref. [163]

Solute *i*	Temp range/K	γ_i^{∞} (298 K)	$H_i^{E,\infty}$/kJ · mol^{-1}
1-Pentene	301.9–333.7	5.00	3.3

Ref. [4]

Solute (1)	$H_{12}^{E,\infty}$/kJ · mol^{-1}
1-Pentene	0.989

Mixture 30: 1-C_6H_{12} (1) + [BMIM][NTf$_2$] (2)
(1) Activity Coefficient at Infinite Dilution and Partial Molar Excess Enthalpies at Infinite Dilution

Ref. [163]

Solute *i*	Temp range/K	γ_i^{∞} (298 K)	$H_i^{E,\infty}$/kJ · mol^{-1}
1-Hexene	301.9–333.7	7.67	3.5

Ref. [4]

Solute (1)	$H_{12}^{E,\infty}$/kJ · mol^{-1}
1-Hexene	2.096

Mixture 31: 1-C_7H_{14} (1) + [BMIM][NTf$_2$] (2)
(1) Activity Coefficient at Infinite Dilution and Partial Molar Excess Enthalpies at Infinite Dilution

Ref. [163]

Solute *i*	Temp range/K	γ_i^{∞} (298 K)	$H_i^{E,\infty}$/kJ · mol^{-1}
1-Heptene	301.9–333.6	11.4	3.8

Ref. [4]

Solute (1)	$H_{12}^{E,\infty}$/kJ · mol^{-1}
1-Heptene	3.509

Mixture 32: 1-C_8H_{16} (1) + [BMIM][NTf_2] (2)
(1) Activity Coefficient at Infinite Dilution and Partial Molar Excess Enthalpies at Infinite Dilution

Ref. [79]

Solute i	Temp range/K	γ_i^∞ (298 K)	$H_i^{E,\infty}$/kJ · mol^{-1}
1-Octene	301.9–333.6	16.8	4.5

Mixture 33: 1-C_9H_{18} (1) + [BMIM][NTf_2] (2)
(1) Activity Coefficient at Infinite Dilution and Partial Molar Excess Enthalpy at Infinite Dilution

Ref. [163]

Solute i	Temp range/K	γ_i^∞ (298 K)	$H_i^{E,\infty}$/kJ · mol^{-1}
1-Nonene	301.9–333.6	24.7	5.1

Mixture 34: 1-$C_{10}H_{20}$ (1) + [BMIM][NTf_2] (2)
(1) Activity Coefficient at Infinite Dilution and Partial Molar Excess Enthalpy at Infinite Dilution

Ref. [163]

Solute i	Temp range/K	γ_i^∞ (298 K)	$H_i^{E,\infty}$/kJ · mol^{-1}
1-Decene	312.2–354.3	34.4	7.4

Mixture 35: 1-$C_{11}H_{22}$ (1) + [BMIM][NTf_2] (2)
(1) Activity Coefficient at Infinite Dilution and Partial Molar Excess Enthalpy at Infinite Dilution

Ref. [163]

Solute i	Temp range/K	γ_i^∞ (298 K)	$H_i^{E,\infty}$/kJ · mol^{-1}
1-Undecene	312.2–354.3	50.6	8.7

Mixture 36: 1-$C_{12}H_{24}$ (1) + [BMIM][NTf_2] (2)
(1) Activity Coefficient at Infinite Dilution and Partial Molar Excess Enthalpy at Infinite Dilution

Ref. [163]

Solute i	Temp range/K	γ_i^∞ (298 K)	$H_i^{E,\infty}$/kJ · mol^{-1}
1-Dodecene	312.2–354.3	73.6	9.9

Mixture 37: $CH = CHCH_2CH_2CH_2CH_2$ (1) + [BMIM][NTf_2] (2)
(1) Activity Coefficient at Infinite Dilution and Partial Molar Excess Enthalpies at Infinite Dilution

Ref. [163]

Solute i	Temp range/K	γ_i^∞ (298 K)	$H_i^{E,\infty}$/kJ · mol^{-1}
Cyclohexene	301.9–333.6	5.29	3.8

Ref. [4]

Solute (1)	$H_{12}^{E,\infty}$/kJ · mol^{-1}
Cyclohexene	3.716

Mixture 38: CH = CHCH = $CHCH_2CH_2$ (1) + [BMIM][NTf_2] (2)
(1) Activity Coefficient at Infinite Dilution and Partial Molar Excess Enthalpy at Infinite Dilution
Ref. [163]

Solute *i*	Temp range/K	γ_i^{∞} (298 K)	$H_i^{E,\infty}$/kJ · mol^{-1}
1,3-Cyclohexadiene	301.9–333.6	2.62	2.0

Mixture 39: $C_6H_5C_2H_5$ (1) + [BMIM][NTf_2] (2)
(1) Activity Coefficient at Infinite Dilution and Partial Molar Excess Enthalpy at Infinite Dilution
Ref. [163]

Solute *i*	Temp range/K	γ_i^{∞} (298 K)	$H_i^{E,\infty}$/kJ · mol^{-1}
Ethyl benzene	354.2–385.6	2.23	2.1

Mixture 40: $C_6H_5C_3H_7$ (1) + [BMIM][NTf_2] (2)
(1) Activity Coefficient at Infinite Dilution and Partial Molar Excess Enthalpy at Infinite Dilution
Ref. [163]

Solute *i*	Temp range/K	γ_i^{∞} (298 K)	$H_i^{E,\infty}$/kJ · mol^{-1}
Propyl benzene	354.2–385.6	3.23	2.7

Mixture 41: $C_6H_5C_4H_9$ (1) + [BMIM][NTf_2] (2)
(1) Activity Coefficient at Infinite Dilution and Partial Molar Excess Enthalpy at Infinite Dilution
Ref. [163]

Solute *i*	Temp range/K	γ_i^{∞} (298 K)	$H_i^{E,\infty}$/kJ · mol^{-1}
Butyl benzene	354.2–385.6	4.66	3.5

Mixture 42: $C_6H_5C_5H_9$ (1) + [BMIM][NTf_2] (2)
(1) Activity Coefficient at Infinite Dilution and Partial Molar Excess Enthalpy at Infinite Dilution
Ref. [163]

Solute *i*	Temp range/K	γ_i^{∞} (298 K)	$H_i^{E,\infty}$/kJ · mol^{-1}
Pentyl benzene	354.2–385.6	6.64	4.1

Mixture 43: 1-$C_7H_{15}OH$ (1) + [BMIM][NTf_2] (2)
(1) Activity Coefficient at Infinite Dilution and Partial Molar Excess Enthalpy at Infinite Dilution

Ref. [163]

Solute *i*	Temp range/K	γ_i^∞ (298 K)	$H_i^{E,\infty}$/kJ · mol^{-1}
1-Heptanol	354.0–385.6	5.33	10.7

Mixture 44: $(CH_3)_2CHCH_2OH$ (1) + [BMIM][NTf_2] (2)
(1) Activity Coefficient at Infinite Dilution, Partial Molar Excess Enthalpies at Infinite Dilution and Partial Molar Enthalpies at Infinite Dilution

Ref. [163]

Solute *i*	Temp range/K	γ_i^∞ (298 K)	$H_i^{E,\infty}$/kJ · mol^{-1}
iso-butanol	312.1–343.6	2.88	10.8

Ref. [53]

$H_1^{E,\infty}$(J/mol)
+13119

(2) Cloud Point

Ref. [53]

p (bar)	*T* (K)	$10^2 w_2$	$10^2 dT/dp$ (K/bar)
1.00	285.02	5.29	
2.47	297.96	14.43	−0.33
9.20	297.93		
21.81	297.89		
30.06	297.86		
41.33	297.83		
2.41	301.36	21.14	−0.24
9.73	301.35		
21.10	301.32		
31.85	301.29		
41.60	301.27		
2.23	303.51	30.67	0.08
6.86	303.57		
10.02	303.56		
20.89	303.54		
30.62	303.55		
43.84	303.58		
2.06	303.68	33.77	−0.40
9.51	303.63		
20.10	303.60		
29.84	303.56		

(continued)

p (bar)	T (K)	$10^2 w_2$	$10^2 \mathrm{d}T/\mathrm{d}p$ (K/bar)
43.34	303.51		
2.89	303.35	50.86	−0.26
12.65	303.34		
21.73	303.30		
30.97	303.27		
43.57	303.25		
2.10	302.33	58.71	−0.12
21.00	302.31		
31.87	302.30		
44.24	302.28		
2.14	293.05	75.26	−0.18
21.13	293.01		
42.33	292.98		
29.26	292.99		
1.00	292.45	8.58	–
1.00	303.55	41.96	–
1.00	303.55	48.97	
1.00	303.55	50.38	
1.00	300.65	64.86	
1.00	284.95	81.15	

Mixture 45: $CH_3CH_2CH(OH)CH_3$ (1) + [BMIM][NTf_2] (2)
(1) Activity Coefficient at Infinite Dilution and Partial Molar Excess Enthalpy at Infinite Dilution

Ref. [163]

Solute i	Temp range/K	γ_i^∞ (298 K)	$H_i^{E,\infty}$/kJ · mol^{-1}
sec-butanol	312.1–343.6	2.49	9.8

Mixture 46: $(CH_3)_3COH$ (1) + [BMIM][NTf_2] (2)
(1) Activity Coefficient at Infinite Dilution and Partial Molar Excess Enthalpy at Infinite Dilution

Ref. [163]

Solute i	Temp range/K	γ_i^∞ (298 K)	$H_i^{E,\infty}$/kJ · mol^{-1}
tert-butyl alcohol	312.1–343.6	1.90	8.3

Mixture 47: $CH_3CH_2C(CH_3)_2OH$ (1) + [BMIM][NTf_2] (2)
(1) Activity Coefficient at Infinite Dilution and Partial Molar Excess Enthalpy at Infinite Dilution

Ref. [163]

Solute i	Temp range/K	γ_i^∞ (298 K)	$H_i^{E,\infty}$/kJ · mol^{-1}
tert-pentanol	312.5–344.0	1.98	6.6

Mixture 48: $(CH_2)_5CHOH$ (1) + [BMIM][NTf_2] (2)
(1) Activity Coefficient at Infinite Dilution and Partial Molar Excess Enthalpy at Infinite Dilution

Ref. [163]

Solute *i*	Temp range/K	γ_i^∞ (298 K)	$H_i^{E,\infty}$/kJ · mol^{-1}
Cyclohexanol	354.3–385.8	3.35	12.5

Mixture 49: CH_3CH_2CHO (1) + [BMIM][NTf_2] (2)
(1) Activity Coefficient at Infinite Dilution and Partial Molar Excess Enthalpy at Infinite Dilution

Ref. [163]

Solute *i*	Temp range/K	γ_i^∞ (298 K)	$H_i^{E,\infty}$/kJ · mol^{-1}
Propanal	312.3–343.8	0.50	0.49

Mixture 50: $CH_3(CH_2)_2CHO$ (1) + [BMIM][NTf_2] (2)
(1) Activity Coefficient at Infinite Dilution and Partial Molar Excess Enthalpy at Infinite Dilution

Ref. [163]

Solute *i*	Temp range/K	γ_i^∞ (298 K)	$H_i^{E,\infty}$/kJ · mol^{-1}
Butanal	312.3–343.8	0.68	0.11

Mixture 51: $CH_3(CH_2)_3CHO$ (1) + [BMIM][NTf_2] (2)
(1) Activity Coefficient at Infinite Dilution and Partial Molar Excess Enthalpy at Infinite Dilution

Ref. [163]

Solute *i*	Temp range/K	γ_i^∞ (298 K)	$H_i^{E,\infty}$/kJ · mol^{-1}
Pentanal	312.3–343.8	1.39	1.4

Mixture 52: $CH_3(CH_2)_4CHO$ (1) + [BMIM][NTf_2] (2)
(1) Activity Coefficient at Infinite Dilution and Partial Molar Excess Enthalpy at Infinite Dilution

Ref. [163]

Solute *i*	Temp range/K	γ_i^∞ (298 K)	$H_i^{E,\infty}$/kJ · mol^{-1}
Hexanal	322.8–375.1	1.48	−0.78

Mixture 53: $CH_3(CH_2)_5CHO$ (1) + [BMIM][NTf_2] (2)
(1) Activity Coefficient at Infinite Dilution and Partial Molar Excess Enthalpy at Infinite Dilution

Ref. [163]

Solute *i*	Temp range/K	γ_i^∞ (298 K)	$H_i^{E,\infty}$/kJ · mol^{-1}
Heptanal	354.3–385.7	1.81	6.0

Mixture 54: $CH_3(CH_2)_6CHO$ (1) + [BMIM][NTf_2] (2)
(1) Activity Coefficient at Infinite Dilution and Partial Molar Excess Enthalpy at Infinite Dilution

Ref. [163]

Solute *i*	Temp range/K	γ_i^∞ (298 K)	$H_i^{E,\infty}$/kJ · mol^{-1}
Octanal	354.3–385.7	2.52	−0.73

Mixture 55: $CH_3OOCCH_2CH_3$ (1) + [BMIM][NTf_2] (2)
(1) Activity Coefficient at Infinite Dilution and Partial Molar Excess Enthalpy at Infinite Dilution

Ref. [79]

Solute *i*	Temp range/K	γ_i^∞ (298 K)	$H_i^{E,\infty}$/kJ · mol^{-1}
Methyl propanoate	312.3–343.7	0.80	0.24

Mixture 56: $CH_3OOCCH_2CH_2CH_3$ (1) + [BMIM][NTf_2] (2)
(1) Activity Coefficient at Infinite Dilution and Partial Molar Excess Enthalpy at Infinite Dilution

Ref. [163]

Solute *i*	Temp range/K	γ_i^∞(298 K)	$H_i^{E,\infty}$/kJ · mol^{-1}
Methyl butanoate	312.3–343.7	1.06	0.18

Mixture 57: $CH_3OOCCH_2CH_2CH_2CH_3$ (1) + [BMIM][NTf_2] (2)
(1) Activity Coefficient at Infinite Dilution and Partial Molar Excess Enthalpy at Infinite Dilution

Ref. [163]

Solute *i*	Temp range/K	γ_i^∞ (298 K)	$H_i^{E,\infty}$/kJ · mol^{-1}
Methyl pentanoate	354.3–385.7	1.52	4.6

Mixture 58: $CH_3OOCCH_2CH_2CH_2CH_2CH_3$ (1) + [BMIM][NTf_2] (2)
(1) Activity Coefficient at Infinite Dilution and Partial Molar Excess Enthalpy at Infinite Dilution

Ref. [163]

Solute *i*	Temp range/K	γ_i^∞ (298 K)	$H_i^{E,\infty}$/kJ · mol^{-1}
Methyl hexanoate	354.3–385.7	2.23	4.6

Mixture 59: $CH_3OOCCH_2CH_2CH_2CH_2CH_2CH_3$ (1) + [BMIM][NTf_2] (2)
(1) Activity Coefficient at Infinite Dilution and Partial Molar Excess Enthalpy at Infinite Dilution

Ref. [163]

Solute *i*	Temp range/K	γ_i^∞ (298 K)	$H_i^{E,\infty}$/kJ · mol^{-1}
Methyl heptanoate	354.3–385.7	2.83	5.6

Mixture 60: CH_3CN (1) + [BMIM][NTf_2] (2)
(1) Activity Coefficient at Infinite Dilution and Partial Molar Excess Enthalpy at Infinite Dilution

Ref. [163]

Solute i	Temp range/K	γ_i^∞ (298 K)	$H_i^{E,\infty}$/kJ · mol^{-1}
Acetonitrile	312.3–343.7	0.48	2.4

Mixture 61: $CH_3N(CH_2)_3CO$ (1) + [BMIM][NTf_2] (2)
(1) Activity Coefficient at Infinite Dilution and Partial Molar Excess Enthalpy at Infinite Dilution

Ref. [163]

Solute i	Temp range/K	γ_i^∞ (298 K)	$H_i^{E,\infty}$/kJ · mol^{-1}
N-methylpyrrolidone	312.4–343.9	123.7	19.3

Mixture 62: 1,4-$CH_2CH_2OCH_2CH_2O$ (1) + [BMIM][NTf_2] (2)
(1) Activity Coefficient at Infinite Dilution and Partial Molar Excess Enthalpy at Infinite Dilution

Ref. [163]

Solute i	Temp range/K	γ_i^∞ (298 K)	$H_i^{E,\infty}$/kJ · mol^{-1}
1,4-Dioxane	312.4–343.9	0.52	−1.1

Mixture 63: $CH = CHCH_2CH_2CH_2$ (1) + [BMIM][BTI] (2)
(1) Partial Molar Excess Enthalpy at Infinite Dilution

Ref. [4]

Solute (1)	$H_{12}^{E,\infty}$/kJ · mol^{-1}
Cyclopentene	1.830

Mixture 64: $CH_2(CH_2)_3CH_2$ (1) + [BMIM][BTI] (2)
(1) Partial Molar Excess Enthalpy at Infinite Dilution

Ref. [4]

Solute (1)	$H_{12}^{E,\infty}$/kJ · mol^{-1}
Cyclopentane	3.575

Mixture 65: $CH_2(CH_2)_4CH_2$ (1) + [BMIM][NTf_2] (2)
(1) Activity Coefficient at Infinite Dilution and Partial Molar Excess Enthalpies at Infinite Dilution

Ref. [163]

Solute i	Temp range/K	γ_i^∞ (298 K)	$H_i^{E,\infty}$/kJ · mol^{-1}
Cyclohexane	301.9–333.6	9.32	5.6

Ref. [4]

Solute (1)	$H_{12}^{E,\infty}$/kJ · mol^{-1}
Cyclohexane	4.502

Mixture 66: CO_2 (1) + [bmim][Tf_2N] (2)
(1) Bubble Point

Ref. [165]

x_1	T/K	P/MPa	T/K	P/MPa	T/K	P/MPa
0.0847	313.55	0.422	323.08	0.488	332.73	0.558
	342.43	0.648	352.13	0.723	361.82	0.813
	371.50	0.903	381.17	1.003	390.67	1.098
	400.58	1.198	410.32	1.303	420.12	1.408
	429.98	1.503	439.54	1.603	449.29	1.703
0.1886	313.60	1.002	323.08	1.172	332.74	1.357
	342.42	1.552	352.11	1.757	361.76	1.972
	371.43	2.202	380.99	2.433	390.86	2.668
	400.64	2.912	410.37	3.158	420.16	3.398
	429.89	3.648	439.64	3.888	449.38	4.133
0.2926	313.58	1.724	323.08	2.024	332.75	2.350
	342.55	2.700	352.10	3.105	361.77	3.500
	371.56	3.900	381.22	4.304	390.88	4.729
	400.62	5.150	410.37	5.600	420.21	6.035
	429.87	6.475	439.60	6.900	449.36	7.325
0.3818	313.50	2.473	323.08	2.938	332.77	3.423
	342.47	3.943	352.12	4.497	361.86	5.082
	371.50	5.682	381.25	6.303	390.91	6.943
	400.63	7.593	410.52	8.248	420.12	8.933
	429.86	9.548	439.65	10.198	449.41	10.833
0.4866	313.57	3.799	332.76	5.289	342.53	6.137
	352.10	7.197	371.46	9.172	390.87	11.273
	410.37	13.473	415.25	14.023		
0.5257	314.65	4.448	323.15	5.174	332.79	6.099
	342.45	7.074	352.11	8.124	361.76	9.250
	371.43	10.425	381.11	11.625	390.95	12.874
	400.59	14.099				
0.5852	313.65	5.483	323.13	6.588	332.73	7.828
	342.43	9.213	352.11	10.713	361.93	12.304
	371.59	13.959	313.65	5.527		

(2) Solubility

Ref. [165]

T/K	313.15	333.15	353.15	373.15	393.15	413.15	433.15	453.15
x_1	P/MPa	P/MPa	P/MPa	P/MPa	P/MPa	P/MPa	P/MPa	P/MPa
0.0847	0.420	0.564	0.734	0.922	1.123	1.331	1.539	1.741
0.1886	0.994	1.363	1.783	2.241	2.725	3.225	3.729	4.224
0.2926	1.723	2.379	3.132	3.957	4.829	5.724	6.617	7.483
0.3818	2.467	3.437	4.557	5.789	7.094	8.435	9.774	11.073
0.4866	3.758	5.339	7.234	9.341	11.558	13.782		
0.5257	4.331	6.119	8.254	10.632	13.151			
0.5852	5.437	7.907	10.849	14.261				

2.2.32.2 Ternary Mixtures

Mixture 1: CH_3COCH_3 (1) + 2-CH_3CHCH_3OH (2) + [BMIM][$(CF_3(SO_2)_2N$] (3)
(1) Phase Equilibrium

Ref. [46]

x_1	x_2	P (mbar)	x_1	x_2	P (mbar)
353.15 K					
19.10 mol.% acetone			59.85 mol.% acetone		
0.0040	0.0250	29.07	0.0282	0.0191	45.76
0.0076	0.0472	54.80	0.0573	0.0387	89.88
0.0118	0.0730	85.72	0.0811	0.0547	131.38
0.0160	0.0980	115.38	0.1035	0.0698	171.80
0.0235	0.1408	168.02	0.1237	0.0834	210.32
0.0354	0.2078	262.14	0.1787	0.1201	322.59
0.0486	0.2762	361.98	0.2289	0.1535	440.21
0.0645	0.3512	475.34	0.2791	0.1866	568.65
0.0796	0.4173	581.90	0.3248	0.2165	698.67
0.0949	0.4785	682.18	0.3662	0.2433	826.15
0.1094	0.5324	771.94	0.4023	0.2665	947.81
0.1223	0.5775	847.75	0.4338	0.2866	1062.5
0.1342	0.6171	914.44	0.4626	0.3047	1173.0
0.1443	0.6489	968.82	0.4864	0.3196	1269.6
0.1528	0.6750	1014.0	0.5062	0.3319	1352.9
0.1598	0.6961	1051.7	0.5224	0.3418	1422.6
0.1657	0.7134	1083.6	0.5357	0.3500	1481.1
0.1706	0.7274	1110.8	0.5467	0.3567	1530.1
0.1747	0.7390	1134.1	0.5558	0.3622	1570.8
0.1786	0.7498	1157.5	0.5633	0.3667	1604.4
0.1819	0.7587	1177.8	0.5695	0.3704	1632.5
0.1845	0.7659	1195.1	0.5751	0.3738	1658.2
38.69 mol.% acetone			0.5771	0.3765	1680.7

(*continued*)

x_1	x_2	P (mbar)	x_1	x_2	P (mbar)
0.0132	0.0234	36.80	80.25 mol.% acetone		
0.0345	0.0607	93.29	0.0525	0.0116	57.41
0.0545	0.0957	151.54	0.0781	0.0173	87.14
0.0723	0.1263	205.68	0.1251	0.0277	144.63
0.0883	0.1538	258.35	0.1651	0.0367	198.36
0.1179	0.2036	359.16	0.2018	0.0449	251.79
0.1475	0.2524	463.47	0.2732	0.0609	369.18
0.1768	0.2989	572.29	0.3408	0.0763	494.53
0.2042	0.3415	677.83	0.4054	0.0912	633.91
0.2304	0.3810	782.40	0.4635	0.1048	776.22
0.2540	0.4156	878.62	0.5147	0.1170	919.29
0.2749	0.4455	966.29	0.5591	0.1278	1055.8
0.2946	0.4729	1051.0	0.5971	0.1371	1183.8
0.3109	0.4952	1122.9	0.6320	0.1459	1309.6
0.3247	0.5135	1185.0	0.6608	0.1532	1419.2
0.3359	0.5284	1238.2	0.6847	0.1594	1513.7
0.3451	0.5402	1282.5	0.7043	0.1646	1592.8
0.3527	0.5500	1321.3	0.7204	0.1689	1658.2
0.3591	0.5581	1354.3	0.7337	0.1724	1712.2
0.3644	0.5648	1382.5	0.7448	0.1754	1756.5
0.3689	0.5703	1406.9	0.7538	0.1779	1792.5
0.3729	0.5753	1429.5	0.7614	0.1800	1822.4
0.3747	0.5803	1446.8	0.7668	0.1814	1843.5
			0.7696	0.1840	1859.4

Mixture 2: 2-CH_3CHCH_3OH (1) + H_2O (2) + [BMIM][($CF_3(SO_2)_2N$] (3)
(1) Phase Equilibrium

Ref. [46]

x_1	x_2	P (mbar)	x_1	x_2	P (mbar)
353.15 K					
21.28 mol.% 2-propanol			59.25 mol.% 2-propanol		
0.0077	0.0412	50.95	0.0279	0.0199	49.38
0.0137	0.0737	94.62	0.0517	0.0367	91.63
0.0220	0.1153	146.99	0.0799	0.0567	143.95
0.0371	0.1863	236.09	0.1138	0.0807	211.42
0.0502	0.2442	309.72	0.1497	0.1060	283.83
0.0754	0.3468	438.16	0.1969	0.1390	376.00
0.0991	0.4343	538.23	0.2462	0.1732	469.94
0.1215	0.5108	611.57	0.2980	0.2090	568.55
0.1399	0.5697	655.93	0.3471	0.2425	656.07
0.1546	0.6158	691.42	0.3881	0.2704	723.93
0.1661	0.6515	721.97	0.4221	0.2934	776.05

(continued)

x_1	x_2	P (mbar)	x_1	x_2	P (mbar)
0.1747	0.6777	746.36	0.4513	0.3131	818.00
0.1822	0.7002	768.99	0.4775	0.3307	853.43
0.1879	0.7173	786.10	0.4985	0.3447	880.40
0.1926	0.7310	799.28	0.5154	0.3561	901.40
0.1961	0.7414	808.80	0.5289	0.3651	917.74
0.1989	0.7495	821.48	0.5397	0.3723	930.72
0.2011	0.7560	833.41	0.5484	0.3781	941.21
0.2029	0.7612	844.25	0.5555	0.3828	949.77
0.2044	0.7654	854.07	0.5612	0.3866	956.72
0.2055	0.7688	863.14	0.5659	0.3897	962.69
0.2066	0.7718	872.09	0.5701	0.3925	967.76
40.47 mol.% 2-propanol			80.54 mol.% 2-propanol		
0.0173	0.0289	47.52	0.0475	0.0103	55.77
0.0329	0.0550	92.30	0.0854	0.0186	104.49
0.0496	0.0822	135.38	0.1254	0.0274	156.36
0.0654	0.1079	180.40	0.1731	0.0380	219.78
0.0915	0.1498	256.41	0.2087	0.0460	270.02
0.1354	0.2185	382.88	0.2657	0.0589	354.02
0.1777	0.2824	498.39	0.3260	0.0729	443.76
0.2185	0.3419	599.29	0.3910	0.0883	537.48
0.2529	0.3906	675.62	0.4512	0.1030	622.52
0.2815	0.4303	731.58	0.5064	0.11678	695.88
0.3049	0.4622	773.42	0.5552	0.1290	756.80
0.3228	0.4864	803.93	0.594	0.1390	801.73
0.3386	0.5076	830.28	0.6307	0.1485	840.70
0.3509	0.5239	851.22	0.6608	0.1564	870.35
0.3607	0.5370	868.51	0.6855	0.1629	892.27
0.3685	0.5473	882.97	0.7055	0.1683	909.52
0.3748	0.5555	895.05	0.7218	0.1726	922.99
0.3798	0.5621	905.30	0.7351	0.1762	933.52
0.3838	0.5674	914.05	0.7460	0.1791	942.12
0.3871	0.5717	921.45	0.7549	0.1814	949.03
0.3898	0.5752	928.11	0.7623	0.1834	954.65
0.3925	0.5787	938.03	0.7689	0.1852	960.28
0.3942	0.5814	943.40			

Mixture 3: H_2O (1) + 1-C_4H_9OH (2) + [bmim][Tf_2N] (3)
(1) Distribution Ratio and Selectivity

Ref. [166]

H_2O-rich phase			[bmim][Tf_2N]-rich phase				
w_1	w_2	w_3	w_1	w_2	w_3	S	β_{BuOH}
298.15 K							
0.9396	0.0604	0.0000	0.0418	0.2674	0.6908	99.49	4.43
0.9314	0.0686	0.0000	0.0715	0.3317	0.5968	62.99	4.84
0.9177	0.0682	0.0141	0.1131	0.4520	0.4350	53.77	6.62
0.9071	0.0801	0.0128	0.0802	0.3237	0.5961	45.70	4.04
0.9267	0.0733	0.0000	0.1278	0.4610	0.4111	45.62	6.29
0.9303	0.0697	0.0000	0.1379	0.4698	0.3922	45.49	6.74
0.9267	0.0733	0.0000	0.1623	0.5134	0.3243	39.99	7.00
0.9294	0.0706	0.0000	0.1725	0.5237	0.3038	39.94	7.41
0.9337	0.0663	0.0000	0.1112	0.3118	0.5770	39.49	4.70
0.9268	0.0732	0.0000	0.1734	0.5124	0.3142	37.43	7.00
0.9047	0.0953	0.0000	0.1007	0.3903	0.5090	36.80	4.09
0.9306	0.0694	0.0000	0.1409	0.3613	0.4978	34.37	5.20
0.8662	0.0797	0.0541	0.0733	0.2210	0.7057	32.78	2.77
0.9876	0.0124	0.0000	0.5717	0.1744	0.2540	24.36	14.10
0.9951	0.0049	0.0000	0.5613	0.0591	0.3796	21.25	11.98
0.9358	0.0642	0.0000	0.1399	0.1739	0.6862	18.13	2.71
0.9405	0.0595	0.0000	0.1581	0.1635	0.6783	16.34	2.75
0.9571	0.0430	0.0000	0.1306	0.0936	0.7758	15.97	2.18
0.9794	0.0206	0.0000	0.1010	0.0278	0.8712	13.09	1.35
0.9559	0.0441	0.0000	0.1181	0.0617	0.8202	11.31	1.40
0.9527	0.0473	0.0000	0.1264	0.0669	0.8067	10.65	1.41

(2) Cloud Point

Ref. [167]

w_3	x_1	p/bar	T/K	$dT/dp \times 10^3$/K bar^{-1}
x_1 is the mole fraction of water in the water–1-butanol mixture				
0.1312	0.5319	1	302	–
0.1333	0.5101	1	276	–
0.1527	0.2220	1	254	–
0.1554	0.1646	1	261	–
0.1571	0.1241	1	272	–
0.1596	0.0585	1	283	–
0.1616	0	1	294	–
0.2028	0	1	296	–
		2.2	297.12	
		10.7	297.09	
0.2220	0	19.9	297.06	−3.37

(*continued*)

w_3	x_1	p/bar	T/K	$dT/dp \times 10^3$/K bar^{-1}
		30.1	297.02	
		41.4	296.99	
0.2361	0	1	297	–
0.2813	0	1	298	–
0.2937	0.8016	1	355	–
		2.9	282.32	
		12.9	282.28	
0.2960	0.0990	21.6	282.27	−3.45
		31.1	282.22	
		40.2	282.19	
0.2986	0.7938	1	346	–
0.3017	0.7888	1	340	–
0.3045	0.7840	1	334	–
0.3066	0.7805	1	330	–
0.3084	0	1	299	–
0.3128	0.7693	1	310	–
0.3144	0.7664	1	304	–
0.3336	0	1	299	–
0.3353	0.0452	1	292	–
0.3371	0.0768	1	287	–
0.3390	0.1078	1	282	–
0.3414	0.1431	1	276	–
0.3420	0.1458	1	275	–
0.3623	0.6496	1	259	–
0.3888	0	1	299	–
0.4177	0.3793	1	259	–
0.4197	0.3644	1	260	–
0.4215	0.0000	1	299	–
0.4236	0.3327	1	262	–
0.4287	0.2875	1	264	–
0.4337	0.2377	1	269	–
0.4386	0.1831	1	276	–
0.4407	0	1	299	–
0.4432	0	1	299	–
0.4456	0.0933	1	287	–
0.4473	0.0684	1	290	–
0.4494	0.0378	1	294	–
0.4509	0.0135	1	298	–
0.4517	0	1	299	–
0.4928	0	1	299	–
		2.0	299.09	
		14.8	299.03	
0.5107	0	25.7	299.00	−3.73
		41.3	298.95	

(*continued*)

w_3	x_1	p/bar	T/K	$dT/dp \times 10^3$/K bar^{-1}
		42.6	298.95	
0.5259	0	1	299	–
0.6004	0	1	299	–
0.7607	0	1	291	–

Mixture 4: H_2O (1) + 1-$C_6H_{13}OH$ (2) + $[C_4MIM][NTf_2]$ (3)
(1) Cloud Point

Ref. [164]

10^{-3} w_{H_2O}/ppm	T/K	10^{-3} w_{H_2O}/ppm	T/K	10^{-3} w_{H_2O}/ppm	T/K
(2) + (3) = 1 (w_3 = 34.41 mass%)					
0.163	329.8	2.761	327.3	6.417	323.9
1.965	328.1	4.554	325.6	7.942	322.4

Mixture 5: H_2O (1) + 1,2-$C_6H_{12}(OH)_2$ (2) + $[C_4MIM][NTf_2]$ (3)
(1) Cloud Point

Ref. [164]

10^{-3} w_{H_2O}/ppm	T/K	10^{-3} w_{H_2O}/ppm	T/K	10^{-3} w_{H_2O}/ppm	T/K
(2) + (3) = 1 (w_3 = 35.18 mass%)					
0.456	325.5	9.565	322.5	14.759	321.0
3.883	324.5	11.138	322.1	15.960	320.7
6.856	323.5				

Mixture 6: H_2O (1) + $CH_2(CH_2)_4CHOH$ (2) + $[C_4MIM][NTf_2]$ (3)
(1) Cloud Point

Ref. [164]

10^{-3} w_{H_2O}/ppm	T/K	10^{-3} w_{H_2O}/ppm	T/K	10^{-3} w_{H_2O}/ppm	T/K
(2) + (3) = 1 (w_3 = 32.00 mass%)					
0.285	316.9	5.568	311.6	11.038	306.3
1.387	315.8	7.324	309.8	12.950	304.6
3.515	313.6	9.052	308.1	14.898	302.9

Mixture 7: C_6H_{14} (1) + [EMIM][BTI] (2) + [BMIM][BTI] (3) (w_2:w_3 = 50:50 weight percent)
(1) Activity Coefficient at Infinite Dilution

Ref. [4]
These physicochemical properties data are listed in a mixture 6 of IL 02003-03001.
Mixture 8: C_7H_{16} (1) + [EMIM][BTI] (2) + [BMIM][BTI] (3) (w_2:w_3 = 50:50 weight percent)
(1) Activity Coefficient at Infinite Dilution

Ref. [4]
These physicochemical properties data are listed in a mixture 7 of IL 02003-03001.
Mixture 9: C_8H_{18} (1) + [EMIM][BTI] (2) + [BMIM][BTI] (3) (w_2:w_3 = 50:50 weight percent)
(1) Activity Coefficient at Infinite Dilution

Ref. [4]
These physicochemical properties data are listed in a mixture 8 of IL 02003-03001.
Mixture 10: 1-C_5H_{10} (1) + [EMIM][BTI] (2) + [BMIM][BTI] (3) (w_2:w_3 = 50:50 weight percent)
(1) Activity Coefficient at Infinite Dilution

Ref. [4]
These physicochemical properties data are listed in a mixture 9 of IL 02003-03001.
Mixture 11: 2-$CH_3COC_2H_5$ (1) + [EMIM][BTI] (2) + [BMIM][BTI] (3) (w_2:w_3 = 50:50 weight percent)
(1) Activity Coefficient at Infinite Dilution

Ref. [4]
These physicochemical properties data are listed in a mixture 10 of IL 02003-03001.
Mixture 12: 2-$CH_3COC_3H_7$ (1) + [EMIM][BTI] (2) + [BMIM][BTI] (3) (w_2:w_3 = 50:50 weight percent)
(1) Activity Coefficient at Infinite Dilution

Ref. [4]
These physicochemical properties data are listed in a mixture 11 of IL 02003-03001.
Mixture 13: H_2O (1) + [EMIM][BTI] (2) + [BMIM][BTI] (3) (w_2:w_3 = 50:50 weight percent)
(1) Activity Coefficient at Infinite Dilution

Ref. [4]
These physicochemical properties data are listed in a mixture 12 of IL 02003-03001.
Mixture 14: [C_2mim][NTf_2] (1) + [C_4mim][NTf_2] (2) + [$P_{666,14}$][NTf_2] (3)
(1) Phase Equilibrium

Ref. [60]
These physicochemical properties data are listed in a mixture 13 of IL 02003-03001.
Mixture 15: C_6H_{14} (1) + C_6H_6 (2) + [C_4mim][NTf_2] (3)
(1) Distribution Ratio and Selectivity

Ref. [168]

Organic-rich phase			Ionic-liquid-rich phase				
x_1	x_2	x_3	x_1	x_2	x_3	β	S
298.15 K							
1.000	0.000	0.000	0.093	0.000	0.907		
0.962	0.038	0.000	0.099	0.068	0.833	1.79	17.39
0.900	0.100	0.000	0.094	0.166	0.740	1.66	15.89
0.846	0.154	0.000	0.087	0.232	0.681	1.51	14.65
0.777	0.223	0.000	0.086	0.319	0.595	1.43	12.92

(continued)

Organic-rich phase			Ionic-liquid-rich phase				
x_1	x_2	x_3	x_1	x_2	x_3	β	S
0.681	0.319	0.000	0.083	0.401	0.516	1.26	10.31
0.556	0.444	0.000	0.068	0.486	0.446	1.09	8.95
0.458	0.542	0.000	0.065	0.554	0.381	1.02	7.20
0.353	0.647	0.000	0.062	0.616	0.322	0.95	5.42
0.195	0.805	0.000	0.047	0.699	0.254	0.87	3.60
0.073	0.927	0.000	0.022	0.762	0.216	0.82	2.73
0.000	1.000	0.000	0.000	0.816	0.184	0.82	

(2) Selectivity at Infinite Dilution and Capacities at Infinite Dilution

Ref. [169]

Solvent	S^{∞}_{ij}	k_j^{∞}
$T = 298.15$ K	*n*-Hexane (1)/benzene (2)	Benzene
[BMIM][NTf_2]	16.7[a]	1.14[a]

[a]Extrapolated value

Mixture 16: H_2O (1) + $(CH_3)_2CHCH_2OH$ (2) + [bmim][NTf_2] (3)
(1) Cloud Point

Ref. [53]

p (bar)	T (K)	10^2w_3	10^2w_1
1.00	292.95	49.79	2.31
1.00	280.45	49.18	4.66
1.00	270.65	48.73	6.38
1.00	264.15	48.19	8.36
1.00	260.15	47.66	10.31
1.00	258.15	47.11	12.27
1.00	258.15	46.23	15.30
1.00	271.15	45.66	17.24
1.00	290.95	45.13	18.98
1.00	303.85	44.54	20.89
1.00	307.15	43.96	22.71
1.00	263.15	45.62	16.92
1.00	287.15	44.70	19.94
1.00	283.25	50.01	3.90
1.00	262.15	48.47	9.64
1.00	257.15	47.14	14.32
1.00	266.15	46.26	17.30

Mixture 17: $CH_2(CH_2)_4CH_2$ (1) + C_6H_6 (2) + [BMIM][NTf_2] (3)
(1) Selectivity at Infinite Dilution and Capacities at Infinite Dilution

Ref. [169]

Solvent	S^{∞}_{ij}	k_j^{∞}
T = 298.15 K	Cyclohexane (1)/benzene (2)	Benzene
[BMIM][NTf_2]	10.2[a]	1.14[a]

[a]Extrapolated value

2.2.32.3 Quaternary Mixtures

Mixture 1: H_2O (1) + 1-C_4H_9OH (2) + [C_4mim][NTf_2] (3) + CO_2 (4)
(1) Cloud Point

Ref. [167]

	Estimated composition			Cloud-point data	
No	x_3	x_1	$x_{4\ calc.}$	T/K	p/bar
1	0.046	0.257	0.628	302.25	68.5
				313.15	96.5
				323.15	137.9
				333.15	158.6
2	0.025	0.295	0.587	295.6	213.7
3	0.016	0.087	0.868	296.70	218.4
				298.65	417.1
				298.75	448.2
				299.13	489.5
				299.35	517.1
				300.67	558.5
				301.10	596.4
				301.44	701.5
				301.60	723.9
				313.15	965.3
4	0.016	0.078	0.859	296.15	806.7
5	0	0.363	0.526	297.15	59.6
6	0	0.343	0.509	297.15	641.2
7	0	0.352	0.517	297.15	339.6

2.2.33 02011-03004: 1-Butyl-3-methylimidazolium dicyanamide

Abbreviation:	[C_4MIm][dca], [BMIM][dca], [bmim][dca], [bmim][$N(CN)_2$]
Molecular Formula:	$C_{10}H_{15}N_5$
Molar Mass:	205.26
Structure:	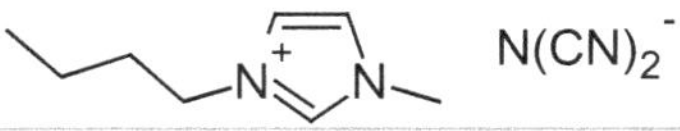

2.2.33.1 Binary Mixtures

Mixture 1: H_2O (1) + [bmim][DCA] (2)

(1) Volumetric Properties and Conductivity

Ref. [35]

x_2	$10^{-3}c_2$/mol · m^{-3}	ρ/kg · m^{-3}	κ/S · cm^{-1}	$10^4\Lambda$/S · m^2 · mol^{-1}
298.15 K				
0.009639	0.4879	1003.22	2.02	41.5
0.02791	1.214	1011.25	3.49	28.7
0.05432	1.965	1019.63	4.43	22.5
0.07995	2.485	1025.43	4.80	19.3
0.1112	2.952	1031.05	4.90	16.60
0.1933	3.708	1040.01	4.47	12.04
0.3214	4.304	1047.31	3.55	8.25
0.4686	4.661	1051.86	2.71	5.82
0.6386	4.898	1055.38	2.00	4.09
0.8742	5.092	1058.33	1.366	2.68
1	5.160	1059.12	1.139	2.21

2.2.33.2 Ternary Mixtures

Mixture 1: H_2O (1) + C_2H_5OH (2) + [bmim][N(CN)$_2$] (3)

(1) Phase Equilibrium

Ref. [24]

x'_2[a]	w_3[a]	y_2	T/K	γ_1	γ_2	α_{21}[b]
p = 100 kPa						
0.9501	0.7996	0.9802	373.95	0.67	0.82	2.71
0.9500	0.7000	0.9776	365.25	0.89	0.91	2.29
0.9500	0.5999	0.9748	360.00	1.10	0.96	1.98
0.9500	0.5000	0.9709	356.85	1.31	0.99	1.73
0.9500	0.3999	0.9671	354.82	1.52	1.00	1.51
0.9499	0.3001	0.9633	353.41	1.74	1.01	1.33
0.9499	0.2000	0.9598	352.40	1.94	1.01	1.19
0.9499	0.1000	0.9550	351.70	2.16	1.00	1.07

[a]Compositions reported are of equilibrium liquid

[b]$\alpha_{21} = (y_2/x'_2)/(y_1/x'_1)$

2.2.34 02011-03009: 1-Butyl-3-methylimidazolium nitrate

Abbreviation:	$[C_4MIm][NO_3]$, $[BMIM][NO_3]$, $[bmim][NO_3]$, $BMImNO_3$
Molecular Formula:	$C_8H_{15}N_3O_3$
Molar Mass:	201.23
Structure:	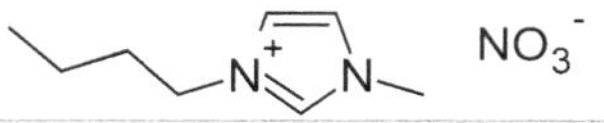

2.2.34.1 Binary Mixtures

Mixture 1: H_2O (1) + $BMImNO_3$ (2)

(1) Volumetric Properties

Ref. [170]

	ρ_2 (g/cm^3)			
	x_1			
T (K)	0.00 %	13.05 %	19.46 %	27.75 %
292	1.1573	1.1547	1.1529	1.1500
302	1.1508	1.1481	1.1462	1.1432
312	1.1445	1.1417	1.1399	1.1369
322	1.1382	1.1354	1.1335	1.1305
332	1.1320	1.1292	1.1274	1.1244
342	1.1260	1.1231	1.1212	1.1181

Mixture 2: CO_2 (1) + $BMImNO_3$ (2)

(1) Phase Equilibrium

Ref. [170]

x_1 = 3.82 %		x_1 = 9.99		x_1 = 14.91 %		x_1 = 20.00 %		x_1 = 0.2491		x_1 = 29.97 %	
P (MPa)	T (K)	P (MPa)	T (K)	P (MPa)	T (K)	P (MPa)	T (K)	P (MPa)	T (K)	P (MPa)	T (K)
0.368	293.20	0.938	293.16	1.285	293.13	1.911	293.2	2.535	293.17	3.162	293.18
0.413	300.63	1.048	300.71	1.515	300.64	2.251	300.71	2.995	300.64	3.757	300.68
0.463	308.22	1.203	308.21	1.770	308.44	2.616	308.13	3.515	308.17	4.417	308.15
0.523	315.68	1.373	315.72	2.035	315.72	3.036	315.66	4.074	315.68	5.202	315.68
0.587	323.17	1.558	323.19	2.324	323.11	3.486	323.18	4.700	323.14	6.047	323.17
0.662	330.68	1.758	330.66	2.625	330.69	3.970	330.7	5.390	330.72	6.982	330.68
0.742	338.14	1.973	338.21	2.955	338.12	4.471	338.21	6.105	338.22	8.017	338.14
0.822	345.72	2.198	345.64	3.305	345.68	5.020	345.71	6.875	345.69	9.112	345.64
0.902	353.17	2.433	353.22	3.675	353.17	5.580	353.18	7.690	353.18	10.297	353.18
0.992	360.69	2.678	360.69	4.055	360.72	6.175	360.69	8.534	360.73	11.542	360.74
1.077	368.21	2.928	368.18	4.445	368.24	6.790	368.22	9.419	368.18	12.832	368.19

Mixture 3: C_2H_5OH (1) + [Bmim][NO_3] (2)
(1) Volumetric Properties and Transport Properties

Ref. [171]

x_1	$\rho/(g \cdot cm^{-3})$	$\eta/(mPa \cdot s)$	$V^E/(cm^3 \cdot mol^{-1})$	$\Delta\eta/(mPa \cdot s)$
$T = 283.15$ K				
0.110	1.1498	259.88	−0.207	−146.26
0.254	1.1297	120.38	−0.538	−220.28
0.352	1.1123	65.96	−0.666	−230.13
0.447	1.0916	35.48	−0.717	−217.41
0.556	1.0622	21.24	−0.729	−182.08
0.663	1.0257	10.88	−0.723	−143.79
0.748	0.9891	7.45	−0.709	−108.57
0.852	0.9301	3.90	−0.604	−64.83
$T = 288.15$ K				
0.110	1.1466	189.76	−0.244	−91.00
0.254	1.1264	90.72	−0.565	−144.82
0.352	1.1089	51.47	−0.689	−153.29
0.447	1.0881	28.43	−0.738	−146.50
0.556	1.0587	17.96	−0.762	−122.75
0.663	1.0221	9.58	−0.757	−97.52
0.748	0.9856	6.39	−0.758	−74.02
0.852	0.9263	3.47	−0.640	−44.28
$T = 293.15$ K				
0.110	1.1434	139.32	−0.267	−60.47
0.254	1.1231	67.91	−0.592	−99.75
0.352	1.1055	41.09	−0.713	−104.69
0.447	1.0846	23.18	−0.761	−101.40
0.556	1.0551	15.21	−0.787	−85.05
0.663	1.0184	8.84	−0.788	−67.53
0.748	0.9817	5.61	−0.782	−51.80
0.852	0.9225	3.08	−0.672	−31.12
$T = 298.15$ K				
0.110	1.1402	105.51	−0.270	−40.55
0.254	1.1197	53.62	−0.581	−68.97
0.352	1.1020	32.83	−0.705	−73.79
0.447	1.0812	19.16	−0.765	−71.99
0.556	1.0516	13.06	−0.786	−60.33
0.663	1.0151	7.62	−0.806	−48.34
0.748	0.9782	4.83	−0.784	−37.28
0.852	0.9186	2.75	−0.653	−22.42
$T = 303.15$ K				
0.110	1.1370	86.79	−0.282	−23.24
0.254	1.1164	43.08	−0.610	−49.31
0.352	1.0986	26.28	−0.731	−54.09

(continued)

x_1	$\rho/(g \cdot cm^{-3})$	$\eta/(mPa \cdot s)$	$V^E/(cm^3 \cdot mol^{-1})$	$\Delta\eta/(mPa \cdot s)$
0.447	1.0776	16.15	−0.794	−52.58
0.556	1.0480	11.34	−0.832	−44.03
0.663	1.0114	6.71	−0.867	−35.56
0.748	0.9745	4.28	−0.866	−27.56
0.852	0.9148	2.46	−0.737	−16.64
$T = 308.15$ K				
0.110	1.1338	59.59	−0.290	−25.64
0.254	1.1131	35.32	−0.614	−36.26
0.352	1.0952	21.40	−0.742	−40.89
0.447	1.0742	13.73	−0.817	−39.56
0.556	1.0447	9.83	−0.883	−33.13
0.663	1.0081	5.77	−0.929	−27.05
0.748	0.9710	3.76	−0.921	−21.00
0.852	0.9109	2.31	−0.768	−12.60
$T = 313.15$ K				
0.110	1.1306	47.55	−0.299	−15.80
0.254	1.1099	29.20	−0.638	−24.02
0.352	1.0919	17.78	−0.771	−28.57
0.447	1.0710	11.88	−0.861	−27.79
0.556	1.0413	9.05	−0.915	−22.96
0.663	1.0046	5.00	−0.959	−19.49
0.748	0.9673	3.38	−0.935	−15.14
0.852	0.9070	2.13	−0.780	−9.08
$T = 318.15$ K				
0.110	1.1274	38.81	−0.311	−11.55
0.254	1.1065	24.53	−0.643	−17.81
0.352	1.0885	15.48	−0.785	−21.39
0.447	1.0677	10.26	−0.906	−21.32
0.556	1.0383	8.02	−1.008	−17.48
0.663	1.0014	4.49	−1.054	−15.04
0.748	0.9638	3.08	−1.010	−11.71
0.852	0.9032	1.90	−0.833	−7.10
$T = 323.15$ K				
0.110	1.1243	32.11	−0.323	−8.57
0.254	1.1032	20.84	−0.656	−13.37
0.352	1.0852	14.06	−0.809	−15.75
0.447	1.0642	8.91	−0.927	−16.62
0.556	1.0346	6.96	−1.022	−13.68
0.663	0.9975	4.04	−1.059	−11.79
0.748	0.9603	2.81	−1.063	−9.20
0.852	0.8994	1.75	−0.877	−5.58
$T = 328.15$ K				
0.110	1.1211	26.90	−0.337	−6.56
0.254	1.1000	17.90	−0.684	−10.25

(*continued*)

x_1	$\rho/(\mathrm{g\cdot cm^{-3}})$	$\eta/(\mathrm{mPa\cdot s})$	$V^E/(\mathrm{cm^3\cdot mol^{-1}})$	$\Delta\eta/(\mathrm{mPa\cdot s})$
0.352	1.0817	12.51	−0.812	−12.02
0.447	1.0606	7.87	−0.937	−13.16
0.556	1.0309	6.08	−1.032	−10.93
0.663	0.9938	3.70	−1.087	−9.36
0.748	0.9563	2.57	−1.080	−7.36
0.852	0.8958	1.65	−0.937	−4.44
$T = 333.15$ K				
0.110	1.1181	22.91	−0.368	−4.99
0.254	1.0967	15.52	−0.691	−7.96
0.352	1.0784	10.88	−0.833	−9.59
0.447	1.0571	7.70	−0.946	−9.86
0.556	1.0276	5.49	−1.078	−8.72
0.663	0.9906	3.39	−1.155	−7.54
0.748	0.9536	2.38	−1.203	−5.94
0.852	0.8928	1.59	−1.038	−3.54

Mixture 4: 1-C_3H_7OH (1) + [Bmim][NO_3] (2)
(1) Volumetric Properties and Transport Properties

Ref. [171]

x_1	$\rho/(\mathrm{g\cdot cm^{-3}})$	$\eta/(\mathrm{mPa\cdot s})$	$V^E/(\mathrm{cm^3\cdot mol^{-1}})$	$\Delta\eta/(\mathrm{mPa\cdot s})$
$T = 283.15$ K				
0.121	1.1442	243.50	−0.139	−157.90
0.257	1.1200	125.76	−0.338	−214.07
0.364	1.0974	72.52	−0.470	−218.87
0.471	1.0707	42.71	−0.575	−200.24
0.554	1.0461	27.88	−0.596	−177.49
0.673	1.0041	15.61	−0.613	−135.90
0.755	0.9686	10.26	−0.561	−104.13
0.857	0.9146	6.16	−0.438	−62.05
$T = 288.15$ K				
0.121	1.1408	176.09	−0.140	−101.40
0.257	1.1167	95.71	−0.360	−139.27
0.364	1.0941	56.76	−0.496	−144.78
0.471	1.0673	34.53	−0.602	−133.57
0.554	1.0427	23.00	−0.624	−119.16
0.673	1.0005	13.20	−0.637	−91.77
0.755	0.9649	8.83	−0.581	−70.51
0.857	0.9109	5.29	−0.457	−42.17
$T = 293.15$ K				
0.121	1.1376	130.88	−0.161	−66.61
0.257	1.1134	68.18	−0.387	−99.13
0.364	1.0907	45.10	−0.524	−98.46
0.471	1.0639	28.35	−0.633	−91.47

(continued)

x_1	$\rho/(g \cdot cm^{-3})$	$\eta/(mPa \cdot s)$	$V^E/(cm^3 \cdot mol^{-1})$	$\Delta\eta/(mPa \cdot s)$
0.554	1.0392	19.19	−0.655	−82.20
0.673	0.9970	11.32	−0.670	−63.66
0.755	0.9613	7.60	−0.619	−49.19
0.857	0.9071	4.64	−0.479	−29.51
T = 298.15 K				
0.121	1.1344	99.82	−0.174	−44.57
0.257	1.1102	53.67	−0.405	−68.71
0.364	1.0874	36.43	−0.551	−68.63
0.471	1.0605	23.48	−0.660	−64.26
0.554	1.0358	16.29	−0.684	−58.02
0.673	0.9934	9.79	−0.699	−45.25
0.755	0.9577	6.56	−0.647	−35.21
0.857	0.9034	4.05	−0.501	−21.21
T = 303.15 K				
0.121	1.1312	74.02	−0.180	−34.77
0.257	1.1069	43.04	−0.421	−49.21
0.364	1.0841	29.92	−0.571	−49.32
0.471	1.0570	19.71	−0.673	−46.51
0.554	1.0323	13.94	−0.712	−42.19
0.673	0.9898	8.54	−0.728	−33.11
0.755	0.9540	5.77	−0.665	−25.90
0.857	0.8996	3.59	−0.525	−15.67
T = 308.15 K				
0.121	1.1280	58.48	−0.191	−25.80
0.257	1.1037	35.02	−0.441	−36.48
0.364	1.0808	24.93	−0.593	−36.51
0.471	1.0537	16.82	−0.703	−34.56
0.554	1.0289	12.03	−0.745	−31.56
0.673	0.9863	7.43	−0.760	−24.97
0.755	0.9503	5.04	−0.698	−19.65
0.857	0.8959	3.21	−0.553	−11.90
T = 313.15 K				
0.121	1.1248	47.07	−0.205	−15.59
0.257	1.1003	28.93	−0.450	−24.27
0.364	1.0773	21.01	−0.605	−24.75
0.471	1.0503	14.44	−0.737	−23.87
0.554	1.0255	10.50	−0.783	−22.05
0.673	0.9827	6.49	−0.795	−17.77
0.755	0.9468	4.49	−0.745	−14.07
0.857	0.8921	2.87	−0.582	−8.60
T = 318.15 K				
0.121	1.1217	38.59	−0.220	−11.24
0.257	1.0972	24.21	−0.485	−18.12
0.364	1.0741	17.89	−0.644	−18.54

(continued)

x_1	$\rho/(g \cdot cm^{-3})$	$\eta/(mPa \cdot s)$	$V^E/(cm^3 \cdot mol^{-1})$	$\Delta\eta/(mPa \cdot s)$
0.471	1.0468	12.54	−0.761	−18.00
0.554	1.0221	9.21	−0.823	−16.75
0.673	0.9791	5.78	−0.829	−13.63
0.755	0.9431	3.97	−0.775	−10.91
0.857	0.8883	2.59	−0.615	−6.67
$T = 323.15$ K				
0.121	1.1186	31.91	−0.243	−8.35
0.257	1.0940	20.58	−0.521	−13.65
0.364	1.0709	15.41	−0.685	−14.06
0.471	1.0436	10.97	−0.814	−13.75
0.554	1.0186	8.16	−0.866	−12.88
0.673	0.9755	5.09	−0.866	−10.67
0.755	0.9395	3.58	−0.818	−8.55
0.857	0.8845	2.35	−0.651	−5.25
$T = 328.15$ K				
0.121	1.1155	26.75	−0.269	−6.36
0.257	1.0909	17.65	−0.556	−10.52
0.364	1.0677	13.41	−0.725	−10.86
0.471	1.0403	9.68	−0.860	−10.70
0.554	1.0152	7.20	−0.916	−10.17
0.673	0.9720	4.58	−0.922	−8.46
0.755	0.9359	3.24	−0.879	−6.82
0.857	0.8807	2.14	−0.693	−4.21
$T = 333.15$ K				
0.121	1.1125	22.73	−0.305	−4.89
0.257	1.0877	15.31	−0.586	−8.20
0.364	1.0644	11.77	−0.764	−8.51
0.471	1.0368	8.61	−0.896	−8.44
0.554	1.0119	6.37	−0.971	−8.17
0.673	0.9687	4.14	−1.004	−6.81
0.755	0.9323	2.94	−0.935	−5.54
0.857	0.8768	1.96	−0.739	−3.44

Mixture 5: 1-C_4H_9OH (1) + [Bmim][NO_3] (2)
(1) Volumetric Properties and Transport Properties

Ref. [171]

x_1	$\rho/(g \cdot cm^{-3})$	$\eta/(mPa \cdot s)$	$V^E/(cm^3 \cdot mol^{-1})$	$\Delta\eta/(mPa \cdot s)$
$T = 283.15$ K				
0.129	1.1390	232.09	−0.156	−165.73
0.260	1.1115	124.98	−0.331	−213.59
0.361	1.0875	75.49	−0.482	−217.40
0.456	1.0620	47.92	−0.602	−202.01
0.565	1.0284	29.00	−0.682	−171.63

(*continued*)

x_1	$\rho/(g \cdot cm^{-3})$	$\eta/(mPa \cdot s)$	$V^E/(cm^3 \cdot mol^{-1})$	$\Delta\eta/(mPa \cdot s)$
0.650	0.9982	19.86	−0.699	−142.32
0.753	0.9554	11.42	−0.590	−104.17
0.848	0.9088	7.82	−0.393	−64.81
$T = 288.15$ K				
0.129	1.1356	170.62	−0.162	−104.45
0.260	1.1081	94.83	−0.343	−139.38
0.361	1.0841	58.80	−0.497	−143.90
0.456	1.0586	38.34	−0.628	−134.73
0.565	1.0251	23.86	−0.723	−115.22
0.650	0.9947	16.11	−0.724	−96.45
0.753	0.9518	9.74	−0.612	−70.69
0.848	0.9051	6.78	−0.399	−44.02
$T = 293.15$ K				
0.129	1.1324	126.89	−0.184	−68.90
0.260	1.1048	68.45	−0.362	−98.35
0.361	1.0807	46.54	−0.519	−97.89
0.456	1.0553	31.10	−0.658	−92.32
0.565	1.0215	19.88	−0.744	−79.41
0.650	0.9912	13.67	−0.755	−66.81
0.753	0.9483	8.41	−0.648	−49.27
0.848	0.9015	5.88	−0.420	−30.77
$T = 298.15$ K				
0.129	1.1292	96.98	−0.191	−46.18
0.260	1.1014	53.94	−0.370	−68.08
0.361	1.0773	37.64	−0.533	−68.08
0.456	1.0519	25.72	−0.683	−64.67
0.565	1.0181	16.79	−0.772	−56.01
0.650	0.9876	11.74	−0.770	−47.35
0.753	0.9447	7.28	−0.670	−35.19
0.848	0.8978	5.12	−0.440	−22.01
$T = 303.15$ K				
0.129	1.1260	70.83	−0.199	−37.04
0.260	1.0981	43.31	−0.377	−48.69
0.361	1.0740	30.76	−0.545	−49.00
0.456	1.0484	21.48	−0.692	−46.76
0.565	1.0146	14.34	−0.791	−40.69
0.650	0.9842	10.17	−0.801	−34.57
0.753	0.9412	6.35	−0.696	−25.90
0.848	0.8942	4.52	−0.463	−16.22
$T = 308.15$ K				
0.129	1.1228	56.32	−0.214	−27.26
0.260	1.0948	35.31	−0.390	−36.01
0.361	1.0706	25.60	−0.560	−36.27
0.456	1.0450	18.23	−0.714	−34.75

(continued)

x_1	$\rho/(g \cdot cm^{-3})$	$\eta/(mPa \cdot s)$	$V^E/(cm^3 \cdot mol^{-1})$	$\Delta\eta/(mPa \cdot s)$
0.565	1.0111	12.37	−0.815	−30.41
0.650	0.9807	8.91	−0.830	−25.91
0.753	0.9376	5.59	−0.724	−19.59
0.848	0.8905	3.98	−0.489	−12.31
$T = 313.15$ K				
0.129	1.1196	45.28	−0.231	−16.87
0.260	1.0916	29.24	−0.410	−23.84
0.361	1.0673	21.55	−0.586	−24.53
0.456	1.0417	15.59	−0.745	−23.92
0.565	1.0078	10.76	−0.855	−21.20
0.650	0.9771	7.80	−0.861	−18.28
0.753	0.9340	4.91	−0.756	−14.03
0.848	0.8869	3.56	−0.518	−8.82
$T = 318.15$ K				
0.129	1.1165	37.11	−0.250	−12.31
0.260	1.0883	24.53	−0.433	−17.71
0.361	1.0639	18.41	−0.603	−18.30
0.456	1.0382	13.52	−0.760	−17.97
0.565	1.0042	9.44	−0.880	−16.07
0.650	0.9736	6.94	−0.892	−13.91
0.753	0.9303	4.38	−0.779	−10.83
0.848	0.8832	3.19	−0.552	−6.81
$T = 323.15$ K				
0.129	1.1134	30.72	−0.272	−9.21
0.260	1.0851	20.83	−0.460	−13.32
0.361	1.0606	15.85	−0.637	−13.84
0.456	1.0349	11.81	−0.801	−13.70
0.565	1.0009	8.33	−0.935	−12.37
0.650	0.9701	6.15	−0.940	−10.79
0.753	0.9268	3.92	−0.832	−8.48
0.848	0.8796	2.87	−0.589	−5.33
$T = 328.15$ K				
0.129	1.1103	25.86	−0.297	−6.99
0.260	1.0819	17.89	−0.490	−10.22
0.361	1.0573	13.75	−0.658	−10.71
0.456	1.0313	10.37	−0.807	−10.65
0.565	0.9971	7.37	−0.928	−9.71
0.650	0.9664	5.52	−0.954	−8.50
0.753	0.9230	3.50	−0.852	−6.78
0.848	0.8759	2.60	−0.631	−4.25
$T = 333.15$ K				
0.129	1.1072	21.95	−0.318	−5.45
0.260	1.0788	15.46	−0.530	−8.01
0.361	1.0541	12.03	−0.703	−8.40
0.456	1.0279	9.18	−0.842	−8.40

(continued)

x_1	$\rho/(\text{g}\cdot\text{cm}^{-3})$	$\eta/(\text{mPa}\cdot\text{s})$	$V^E/(\text{cm}^3\cdot\text{mol}^{-1})$	$\Delta\eta/(\text{mPa}\cdot\text{s})$
0.565	0.9935	6.56	−0.958	−7.74
0.650	0.9627	4.93	−0.973	−6.82
0.753	0.9193	3.17	−0.888	−5.49
0.848	0.8722	2.37	−0.676	−3.44

2.2.34.2 Ternary Mixtures

Mixture 1: CO_2 (1) + H_2O (2) + $BMImNO_3$ (3)
(1) Phase Equilibrium

Ref. [170]

X_2 = 1.99 %, X_1 = 10.00 %, x_2 = 1.76 %, x_1 = 8.93 %		X_2 = 9.18 %, X_1 = 10.00 %, x_2 = 7.60 %, x_1 = 8.33 %		X_2 = 21.26 %, X_1 = 10.00 %, x_2 = 16.06 %, x_1 = 7.30 %		X_2 = 1.99 %, X_1 = 25.00 %, x_2 = 1.47 %, x_1 = 19.68 %		X_2 = 9.18 %, X_1 = 25.00 %, x_2 = 6.44 %, x_1 = 18.50 %		X_2 = 21.26 %, X_1 = 25.00 %, x_2 = 13.75 %, x_1 = 16.45 %	
P (MPa)	T (K)	P (MPa)	T (K)	P (MPa)	T (K)	P (MPa)	T (K)	P (MPa)	T (K)	P (MPa)	T (K)
0.894	293.13	0.853	293.15	0.930	293.21	2.418	293.16	2.534	293.18	2.671	293.18
1.029	300.66	0.998	300.67	1.070	300.66	2.858	300.65	2.994	300.66	3.136	300.69
1.179	308.13	1.151	308.13	1.175	308.14	3.328	308.25	3.514	308.34	3.646	308.15
1.349	315.66	1.321	315.62	1.345	315.69	3.857	315.66	4.054	315.66	4.226	315.66
1.529	323.21	1.508	323.17	1.525	323.21	4.428	323.17	4.658	323.18	4.846	323.20
1.714	330.64	1.708	330.66	1.720	330.73	5.037	330.71	5.308	330.70	5.541	330.69
1.914	338.16	1.913	338.20	1.925	338.21	5.707	338.17	5.998	338.20	6.251	338.17
2.124	345.65	2.118	345.69	2.135	345.72	6.437	345.71	6.723	345.66	6.991	345.65
2.349	353.16	2.338	353.21	2.355	353.20	7.212	353.18	7.503	353.20	7.791	353.17
2.579	360.68	2.563	360.65	2.585	360.68	8.022	360.68	8.313	360.65	8.616	360.67
2.824	368.20	2.803	368.18	2.830	368.20	8.887	368.22	9.183	368.21	9.471	368.20

$x_i = n_i/(n_{CO2} + n_{H2O} + n_{IL})$; $X_i = n_i/(n_i + n_{IL})$; i = CO_2, H_2O; n = number of moles

2.2.35 02011-04003: 1-Butyl-3-methylimidazolium trifluoromethanesulfonate

Abbreviation:	$[C_4MIm][TfO]$, [BMIM][TfO], [bmim][TfO], [bmim][triflate], [BMIM] $[CF_3SO_3]$, $[C_4mim][OTf]$
Molecular Formula:	$C_9H_{15}F_3N_2O_3S$
Molar Mass:	288.29
Structure:	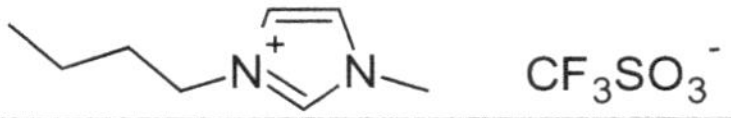

2.2.35.1 Binary Mixtures

Mixture 1: C_2H_5OH (1) + [bmim][triflate] (2)

(1) Excess Molar Enthalpy

Ref. [65]

x_2	H^E (J · mol^{-1})	x_2	H^E (J · mol^{-1})	x_2	H^E (J · mol^{-1})
T = 303.15 K					
0.0987	963	0.4016	1773	0.6832	1217
0.2029	1518	0.4928	1700	0.8026	833
0.2948	1742	0.5952	1479	0.8759	539

(2) Activity Coefficient at Infinite Dilution and Partial Molar Excess Enthalpy at Infinite Dilution

Ref. [169]

Solute	γ_{12}^{∞} (gas–liquid chromatography)	$\Delta H_1^{E,\infty}$ (kJ · mol^{-1})
298.15 K		
Ethanol	1.07	4.3

Mixture 2: H_2O (1) + [BMIM][CF_3SO_3] (2)

(1) Volumetric Properties and Transport Properties

Ref. [172]

x_1	303.15 K	308.15 K	313.15 K	318.15 K	323.15 K	333.15 K	343.15 K
	ρ/g · cm^{-3}						
0.0000	1.2984	1.2949	1.2914	1.2878	1.2844	1.2774	1.2705
0.1032	1.2945	1.2910	1.2875	1.2838	1.2803	1.2731	1.2660
0.1939	1.2912	1.2877	1.2841	1.2804	1.2768	1.2697	1.2626
0.2987	1.2865	1.2830	1.2794	1.2758	1.2722	1.2650	1.2579
0.3999	1.2808	1.2773	1.2738	1.2702	1.2665	1.2593	1.2522
0.5010	1.2728	1.2693	1.2657	1.2620	1.2583	1.2511	1.2440
0.5923	1.2630	1.2595	1.2559	1.2522	1.2485	1.2413	1.2339
0.7023	1.2462	1.2425	1.2388	1.2349	1.2311	1.2234	1.2156
0.7990	1.2201	1.2166	1.2129	1.2089	1.2052	1.1972	1.1893
0.9010	1.1646	1.1606	1.1567	1.1530	1.1500	1.1426	1.1343
1.0000	0.9957	0.9940	0.9922	0.9902	0.9881	0.9832	0.9778
	η/mPa · s						
0	63.190	51.729	42.458	35.174	29.553	21.609	16.376
0.1032	45.696	37.227	30.927	25.041	21.507	16.069	12.443
0.1939	35.415	29.270	24.463	20.791	18.108	13.507	10.545
0.2987	28.969	24.276	20.532	17.549	15.173	11.613	9.1368
0.3999	23.248	19.951	16.880	14.377	12.618	9.8800	7.8156
0.5010	19.187	16.805	13.948	12.001	10.423	8.1008	6.6122
0.5923	15.467	13.202	11.210	9.9072	8.5157	6.7021	5.3842
0.7023	11.556	10.130	8.6049	7.4650	6.5763	5.1369	4.2397

(continued)

x_1	303.15 K	308.15 K	313.15 K	318.15 K	323.15 K	333.15 K	343.15 K
0.7990	8.2063	6.9320	5.9890	5.3393	4.5704	3.6181	2.9364
0.9010	4.5521	3.8692	3.3759	3.1563	3.0054	2.0992	1.7141
1.0000	0.7977	0.7190	0.6532	0.6001	0.5470	0.4665	0.4040
	$V^E/cm^3 \cdot mol^{-1}$						
0.1032	0.1693	0.1732	0.1770	0.1965	0.2155	0.2534	0.2908
0.1939	0.1958	0.2026	0.2234	0.2441	0.2783	0.3023	0.3393
0.2987	0.2059	0.2158	0.2380	0.2471	0.2930	0.3189	0.3551
0.3999	0.2253	0.2382	0.2506	0.2624	0.3058	0.3448	0.3801
0.5010	0.2567	0.2729	0.2979	0.3223	0.3634	0.4033	0.4385
0.5923	0.2711	0.2902	0.3165	0.3420	0.3809	0.4213	0.4810
0.7023	0.2167	0.2516	0.2856	0.3251	0.3677	0.4412	0.5217
0.7990	0.2075	0.2327	0.2666	0.3092	0.3429	0.4241	0.4983
0.9010	0.1632	0.2084	0.2491	0.2785	0.2863	0.3348	0.4090
	$\Delta\eta/mPa \cdot s$						
0.1032	−11.055	−9.2381	−7.2166	−6.5649	−5.0524	−3.3579	−2.2836
0.1939	−15.678	−12.568	−9.8881	−7.6789	−5.8211	−4.0031	−2.7329
0.2987	−15.584	−12.217	−9.4376	−7.2974	−5.7166	−3.6816	−2.4669
0.3999	−14.993	−11.380	−8.8587	−6.9704	−5.3366	−3.2754	−2.1714
0.5010	−12.746	−9.3686	−7.5642	−5.8513	−4.5989	−2.9175	−1.7596
0.5923	−10.770	−8.3141	−6.4846	−4.7886	−3.8587	−2.3863	−1.5291
0.7023	−7.8175	−5.7750	−4.4912	−3.4277	−2.6078	−1.6262	−0.9162
0.7990	−5.1340	−4.0409	−3.0644	−2.2101	−1.8091	−1.1008	−0.6747
0.9010	−2.4245	−1.9007	−1.4131	−0.8665	−0.4158	−0.4636	−0.2675

(2) Thermal Conductivity

Ref. [82]

w_1	$\lambda/W \cdot m^{-1} \cdot K^{-1}$	w_1	$\lambda/W \cdot m^{-1} \cdot K^{-1}$
293 K			
0	0.147	0.20	0.221
0.01	0.150	0.50	0.356
0.05	0.157	0.70	0.440
0.10	0.173	1.00	0.607

(3) Heat Capacity

Ref. [173]

	$x_2 = 0.20$		$x_2 = 0.40$		$x_2 = 0.60$		$x_2 = 0.80$	
T/(K)	C_p ($J \cdot mol^{-1} \cdot K^{-1}$)	C_p^E ($J \cdot mol^{-1} \cdot K^{-1}$)	C_p ($J \cdot mol^{-1} \cdot K^{-1}$)	C_p^E ($J \cdot mol^{-1} \cdot K^{-1}$)	C_p ($J \cdot mol^{-1} \cdot K^{-1}$)	C_p^E ($J \cdot mol^{-1} \cdot K^{-1}$)	C_p ($J \cdot mol^{-1} \cdot K^{-1}$)	C_p^E ($J \cdot mol^{-1} \cdot K^{-1}$)
303.2	149	−0.2	223	0.6	295	−0.7	366	−2.4
308.2	150	0.6	225	1.8	298	1.0	370	−1.1
313.2	151	1.4	227	3.2	301	2.5	372	−0.2
318.2	152	1.7	229	4.2	304	4.2	376	1.0
323.2	153	2.3	231	5.4	307	5.6	378	2.0
328.2	154	3.1	234	7.1	310	7.2	381	3.1
333.2	155	3.7	236	8.6	313	8.9	385	4.6
338.2	157	4.5	238	9.5	316	10.7	388	5.4
343.2	158	5.1	241	10.9	319	12.2	392	7.1
348.2	159	6.0	243	12.1	323	13.8	395	8.1
353.2	161	6.7	246	13.8	325	15.2	398	9.3

(4) Excess Molar Enthalpy

Ref. [71]

x_2	$H^E/(J \cdot mol^{-1})$	x_2	$H^E/(J \cdot mol^{-1})$	x_2	$H^E/(J \cdot mol^{-1})$
303.15 K					
0.1007	656	0.3944	1389	0.6624	1023
0.1989	1074	0.4985	1327	0.8201	569
0.2954	1304	0.5757	1224		

Mixture 3: C_6H_{14} (1) + [BMIM][CF_3SO_3] (2)
(1) Activity Coefficient at Infinite Dilution and Partial Molar Excess Enthalpy at Infinite Dilution

Ref. [169]

Solute	γ_{12}^{∞} (gas–liquid chromatography)	$\Delta H_1^{E,\infty}$ (kJ · mol^{-1})
298.15 K		
Hexane	41.6	7.0

Mixture 4: $CH_2(CH_2)_4CH_2$ (1) + [BMIM][CF_3SO_3] (2)
(1) Activity Coefficient at Infinite Dilution and Partial Molar Excess Enthalpy at Infinite Dilution

Ref. [169]

Solute	γ_{12}^{∞} (gas–liquid chromatography)	$\Delta H_1^{E,\infty}$ (kJ · mol^{-1})
298.15 K		
Cyclohexane	20.6	6.5

Mixture 5: 1-C_6H_{12} (1) + [BMIM][CF_3SO_3] (2)
(1) Activity Coefficient at Infinite Dilution and Partial Molar Excess Enthalpy at Infinite Dilution

Ref. [169]

Solute	γ_{12}^{∞} (gas–liquid chromatography)	$\Delta H_1^{E,\infty}$ (kJ · mol^{-1})
298.15 K		
Hex-1-ene	17.6	4.1

Mixture 6: $CH \equiv CCH_2CH_2CH_2CH_3$ (1) + $[BMIM][CF_3SO_3]$ (2)
(1) Activity Coefficient at Infinite Dilution and Partial Molar Excess Enthalpy at Infinite Dilution

Ref. [169]

Solute	γ_{12}^{∞} (gas–liquid chromatography)	$\Delta H_1^{E,\infty}$ (kJ · mol^{-1})
298.15 K		
Hex-1-yne	3.68	−1.5

Mixture 7: C_6H_6 (1) + $[BMIM][CF_3SO_3]$ (2)
(1) Activity Coefficient at Infinite Dilution and Partial Molar Excess Enthalpy at Infinite Dilution

Ref. [169]

Solute	γ_{12}^{∞} (gas–liquid chromatography)	$\Delta H_1^{E,\infty}$ (kJ · mol^{-1})
298.15 K		
Benzene	1.55	−0.7

Mixture 8: CH_3NO_2 (1) + [bmim][triflate] (2)
(1) Excess Molar Enthalpy

Ref. [74]

x_2	H^E/(J · mol^{-1})	x_2	H^E/(J · mol^{-1})	x_2	H^E/(J · mol^{-1})
303.15 K					
0.1022	39	0.4044	59	0.7030	31
0.1994	57	0.5067	56	0.7814	5
0.3029	62	0.5931	45	0.8933	−9

2.2.35.2 Ternary Mixtures

Mixture 1: C_6H_{14} (1) + C_6H_6 (2) + $[BMIM][CF_3SO_3]$ (3)
(1) Selectivity at Infinite Dilution and Capacities at Infinite Dilution

Ref. [169]

Solvent	S^{∞}_{ij}	k_j^{∞}
$T = 298.15$ K	*n*-Hexane (1)/benzene (2)	Benzene
$[BMIM][CF_3SO_3]$	26.8	0.65

Mixture 2: $CH_2(CH_2)_4CH_2$ (1) + C_6H_6 (2) + $[BMIM][CF_3SO_3]$ (3)
(1) Selectivity at Infinite Dilution and Capacities at Infinite Dilution

Ref. [169]

Solvent	S^{∞}_{ij}	k_j^{∞}
$T = 298.15$ K	Cyclohexane (1)/benzene (2)	Benzene
$[BMIM][CF_3SO_3]$	13.3	0.65

Mixture 3: C_6H_{14} (1) + SCH = CHCH = CH (thiophene) (2) + [BMIM][CF_3SO_3] (3)
(1) Selectivity at Infinite Dilution and Capacities at Infinite Dilution

Ref. [169]

Solvent	$S^{\infty}{}_{12}$	
T = 298.15 K	n-Hexane (1)/thiophene (2)	k_2^{∞}
[BMIM][CF_3SO_3]	36.5	0.88

Mixture 4: $CH_2(CH_2)_4CH_2$ (1) + SCH = CHCH = CH (thiophene) (2) + [BMIM][CF_3SO_3] (3)
(1) Selectivity at Infinite Dilution and Capacities at Infinite Dilution

Ref. [169]

Solvent	$S^{\infty}{}_{12}$	
T = 298.15 K	Cyclohexane (1)/thiophene (2)	k_2^{∞}
[BMIM][CF_3SO_3]	18.1	0.88

Mixture 5: C_6H_6 (1) + SCH = CHCH = CH (thiophene) (2) + [BMIM][CF_3SO_3] (3)
(1) Selectivity at Infinite Dilution and Capacities at Infinite Dilution

Ref. [169]

Solvent	$S^{\infty}{}_{12}$	
T = 298.15 K	Benzene (1)/thiophene (2)	k_2^{∞}
[BMIM][CF_3SO_3]	1.4	0.88

2.2.36 02011-04007: 1-Butyl-3-methylimidazolium tosylate

Abbreviation:	[C_4mim][Tos]
Molecular Formula:	$C_{15}H_{22}N_2O_3S$
Molar Mass:	310.42
Structure:	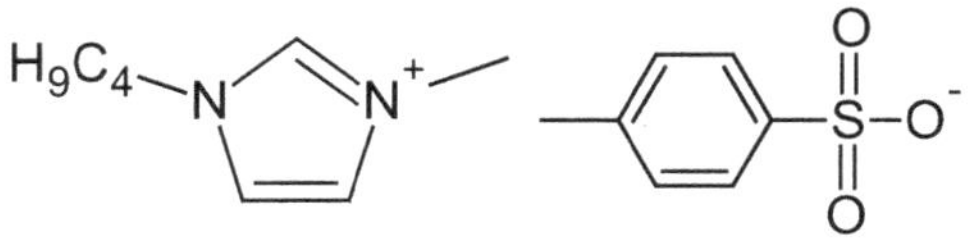

2.2.36.1 Binary Mixtures

Mixture 1: H_2O (1) + [C_4mim][Tos] (2)
(1) Activity Coefficient

Ref. [174]

x_2	T_{sat}/K (real solution)	γ_i
0.960	343.3	0.991
0.860	333.2	0.927

(*continued*)

x_2	T_{sat}/K (real solution)	γ_i
0.754	324.0	0.870
0.707	318.9	0.834
0.629	310.7	0.793
0.592	305.7	0.762
0.523	296.7	0.726
0.501	292.2	0.696
0.477	285.8	0.650
0.442	280.0	0.631
0.408	272.2	0.599

Mixture 2: $NH(CH_2)_5CO$ (1) + [C_4mim][Tos] (2)
(1) Activity Coefficient

Ref. [174]

x_2	T_{sat}/K (real solution)	γ_i
0.960	343.3	1.033
0.824	334.7	0.999
0.684	326.7	1.014
0.568	317.7	1.013
0.537	315.7	1.027
0.482	311.7	1.056
0.392	313.0	1.019
0.351	316.2	1.005
0.286	322.2	1.007
0.227	327.0	1.006
0.171	331.7	1.012
0.114	335.9	1.014
0.063	339.2	1.013

2.2.37 02011-04008: 1-Butyl-3-methylimidazolium octylsulfate

Abbreviation:	[bmim][$OcOSO_3$], [bmim][C_8OSO_3], [BMIM][$OcSO_4$]
Molecular Formula:	$C_{16}H_{32}N_2O_4S$
Molar Mass:	348.51
Structure:	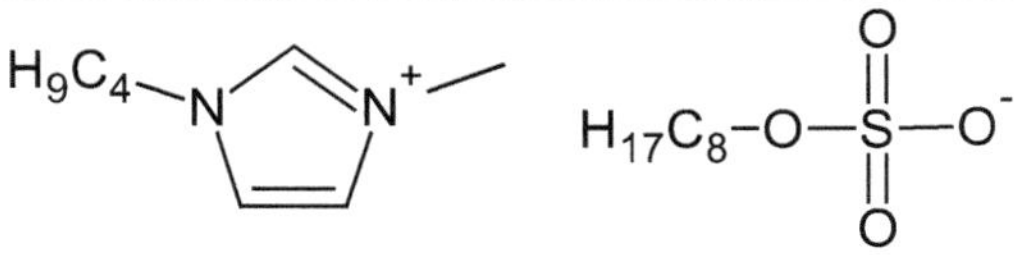

2.2.37.1 Binary Mixtures

Mixture 1: H_2O (1) + [bmim][$OcOSO_3$] (2)

(1) Volumetric Properties

Ref. [175]

T/K	m_2/ $(\mathrm{mol \cdot kg^{-1}})$	ρ/ $(\mathrm{kg \cdot m^{-3}})$	V/ $(\mathrm{cm^3 \cdot mol^{-1}})$	V_Φ/ $(\mathrm{cm^3 \cdot mol^{-1}})$	$u(V_\Phi)$/ $(\mathrm{cm^3 \cdot mol^{-1}})$
278.15	0.000	999.97	18.017		
	0.629	1015.71	21.382	318.48	0.08
	1.519	1030.10	26.037	319.07	0.06
	2.395	1039.66	30.476	319.28	0.06
	3.321	1046.92	35.030	319.39	0.07
	4.251	1052.18	39.465	319.55	0.08
288.15	0.000	999.09	18.032		
	0.640	1013.50	21.495	321.65	0.08
	1.515	1026.30	26.108	322.06	0.06
	2.397	1035.05	30.622	322.20	0.06
	3.311	1041.51	35.161	322.31	0.07
	4.246	1046.54	39.657	322.33	0.08
298.15	0.000	997.05	18.069		
	0.639	1010.20	21.559	324.57	0.08
	1.502	1021.80	26.155	324.91	0.06
	2.394	1029.92	30.758	325.01	0.06
	3.321	1035.98	35.402	325.06	0.07
	4.238	1040.38	39.853	325.13	0.08
308.15	0.000	994.05	18.124		
	0.647	1006.28	21.686	327.42	0.08
	1.499	1016.86	26.265	327.68	0.06
	2.393	1024.44	30.922	327.72	0.07
	3.311	1030.03	35.553	327.73	0.07
	4.274	1034.42	40.260	327.72	0.08
318.15	0.000	990.22	18.194		
	0.636	1001.43	21.729	330.23	0.08
	1.487	1011.32	26.343	330.44	0.06
	2.400	1018.63	31.134	330.40	0.07
	3.297	1023.73	35.697	330.40	0.07
	4.253	1027.86	40.412	330.37	0.08
328.15	0.000	985.73	18.277		
	0.639	996.28	21.857	332.99	0.08
	1.504	1005.67	26.583	333.17	0.06
	2.404	1012.35	31.350	333.17	0.07
	3.314	1017.28	36.014	333.10	0.07
	4.258	1021.29	40.697	332.95	0.08

(2) Theoretical Debye-HüCkel Limiting Slope and Volumetric Properties

Ref. [175]

T/K	$S^{DH}{}_V$/(cm^3 · kg$^{1/2}$ · mol$^{-3/2}$)	$\overline{V}^{\infty}$/(cm^3 · mol^{-1})	b_V/(cm^3 · kg · mol^{-2})
278.15	1.572	317.5	−0.272
288.15	1.716	320.6	−0.414
298.15	1.875	323.4	−0.507
308.15	2.054	326.1	−0.634
318.15	2.259	328.8	−0.753
328.15	2.493	331.5	−0.877

Mixture 2: 1-C_3H_7OH (1) + [bmim][$OcOSO_3$] (2)
(1) Volumetric Properties

Ref. [175]

T/K	m_2/ (mol · kg^{-1})	ρ/ (kg · m^{-3})	V/ (cm^3 · mol^{-1})	V_Φ/ (cm^3 · mol^{-1})	$u(V_\Phi)$/ (cm^3 · mol^{-1})
278.15	0.000	815.98	73.649		
	0.196	829.35	76.514	319.5	0.3
	0.468	846.13	80.348	318.57	0.13
	0.726	859.61	83.934	319.74	0.09
	1.000	872.81	87.585	319.49	0.08
	1.302	885.46	91.513	319.76	0.07
288.15	0.000	807.54	74.419		
	0.188	820.73	77.155	319.0	0.3
	0.447	836.66	80.852	320.17	0.14
	0.742	852.43	84.935	320.90	0.09
	0.997	864.61	88.362	321.09	0.08
	1.287	876.57	92.171	321.78	0.07
298.15	0.000	799.54	75.164		
	0.175	812.03	77.699	319.1	0.3
	0.443	828.68	81.547	321.28	0.14
	0.720	843.94	85.385	321.58	0.09
	0.991	856.55	89.085	322.86	0.08
	1.306	869.95	93.212	323.11	0.07
308.15	0.000	791.44	75.933		
	0.196	805.47	78.781	320.4	0.3
	0.459	821.77	82.558	322.55	0.14
	0.756	838.02	86.673	323.00	0.09
	0.960	847.43	89.473	324.25	0.08
	1.272	861.12	93.573	324.34	0.07
318.15	0.000	783.22	76.729		
	0.199	797.70	79.611	320.5	0.3
	0.454	813.46	83.298	323.95	0.14

(*continued*)

T/K	m_2/ (mol · kg^{-1})	ρ/ (kg · m^{-3})	V/ (cm^3 · mol^{-1})	V_Φ/ (cm^3 · mol^{-1})	$u(V_\Phi)$/ (cm^3 · mol^{-1})
	0.738	828.90	87.277	325.11	0.09
	1.003	841.94	90.851	325.15	0.08
	1.279	853.78	94.506	325.72	0.07

(2) Theoretical Debye-HüCkel Limiting Slope and Volumetric Properties

Ref. [175]

T/K	$S^{DH}{}_V$/(cm^3 · kg$^{1/2}$ · mol$^{-3/2}$)	$\overline{V}^\infty$/(cm^3 · mol^{-1})	b_V/(cm^3 · kg · mol^{-2})
278.15	19.88	311.8	−11.86
288.15	23.38	310.5	−12.34
298.15	27.31	309.4	−13.85
308.15	37.14	308.9	−16.45
318.15	36.76	307.6	−18.71

(3) Relative Permittivity and Isothermal Compressibility

Ref. [175]

T/K	ϵ_r [a]	$(\partial \ln \epsilon_r/\partial P)_T$ [b]/GPa^{-1}	k_T [c]/GPa^{-1}
278.15	23.33	0.884	0.87
288.15	21.85	0.958	0.94
298.15	20.44	1.033	1.01
308.15	19.09	1.108	1.08
318.15	17.80	1.183	1.16

[a] From Barthel et al.
[b] From a polynomial fitting of Chen et al. data
[c] From a polynomial fitting of Diaz-Peña and Tardajos data

(4) Vapor Pressure and Activity Coefficient

Ref. [176]

	298.15 K		303.15 K		308.15 K		313.15 K			298.15 K		303.15 K		308.15 K		313.15 K	
x_1	p/Pa	γ_1	p/Pa	γ_1	p/Pa	γ_1	p/Pa	γ_1	x_1	p/Pa	γ_1	p/Pa	γ_1	p/Pa	γ_1	p/Pa	γ_1
0.0000	0	0.325	0	0.281	0	0.251	0	0.250	0.59067	1400	0.851	1861	0.823	2430	0.799	3130	0.776
0.1218	146	0.395	177	0.353	218	0.326	277	0.317	0.72583	1982	1.024	2608	0.982	3402	0.944	4386	0.943
0.1464	171	0.411	210	0.369	260	0.343	322	0.332	0.79989	2491	1.096	3225	1.039	4070	0.995	5220	1.008
0.1724	197	0.429	248	0.388	304	0.362	372	0.350	0.85309	2730	1.121	3560	1.054	4600	1.013	5944	1.032
0.31478	454	0.539	596	0.503	778	0.482	987	0.459	0.90911	2754	1.106	3674	1.042	4875	1.014	6422	1.030
0.38852	680	0.611	902	0.577	1175	0.558	1508	0.531	1.00000	2786	1.000	3826	1.000	5199	1.000	6986	1.000
0.56081	1290	0.812	1705	0.784	2240	0.762	2865	0.736									

Mixture 3: H_2 (1) + $[C_4C_1Im][C_8SO_4]$ (2)
(1) Henry's Law Constant

Ref. [91]

T/K	$p/10^2$ Pa	$K_H/10^5$ Pa	$x_1/10^{-4}$	Dev%
$P_0 = 0.1$ MPa				
313.33	921.66	1892	5.288	+0.1
313.33	832.13	1891	5.292	−0.0
323.03	857.54	2062	4.853	+0.3
323.20	950.28	2047	4.887	−0.5
332.87	883.34	2279	4.390	+0.2
342.69	909.05	2531	3.953	−0.9
342.98	1007.71	2584	3.873	+0.8

(2) Thermodynamic Properties

Ref. [91]

T/K	$\Delta_{solv}G$/kJ · mol^{-1}	$\Delta_{solv}H$/kJ · mol^{-1}	$\Delta_{solv}S$/J · mol^{-1} · K^{-1}
$P_0 = 0.1$ MPa			
313	19.63	−6.08	−82.1
323	20.48	−8.24	−88.9
333	21.41	−10.3	−95.1
343	22.38	−12.2	−101

Mixture 4: CO_2 (1) + $[C_4C_1Im][C_8SO_4]$ (2)
(1) Henry's Law Constant

Ref. [91]

T/K	$p/10^2$ Pa	$K_H/10^5$ Pa	$x_1/10^{-4}$	Dev%
$P_0 = 0.1$ MPa				
313.15	682.01	64.84	153.6	+0.9
313.33	702.76	63.71	156.3	−1.0
322.31	725.63	70.81	140.7	−1.1
322.81	705.87	73.09	136.3	+1.5
332.27	728.35	80.51	123.8	−1.1
332.28	752.10	81.89	121.7	+0.6
342.23	778.22	94.57	105.4	+1.1
343.22	755.86	94.03	106.0	−0.9

(2) Thermodynamic Properties

Ref. [91]

T/K	$\Delta_{solv}G$/kJ · mol^{-1}	$\Delta_{solv}H$/kJ · mol^{-1}	$\Delta_{solv}S$/J · mol^{-1} · K^{-1}
$P_0 = 0.1$ MPa			
313	10.83	−9.05	−63.5
323	11.49	−10.8	−69.1
333	12.21	−12.5	−74.1
343	12.97	−14.0	−78.8

Mixture 5: 1, 2-$HOCH_2CH_2OH$ (1) + [bmim][C_8OSO_3] (2)
(1) Volumetric Properties and Refractive Index

Ref. [136]

	ρ/(g · cm^{-3})	ρ/(g · cm^{-3})	ρ/(g · cm^{-3})	n_D
x_2	$T = 298.15$ K	$T = 308.15$ K	$T = 318.15$ K	$T = 298.15$ K
0.0000	1.1100	1.1029	1.0956	1.4304
0.0003	1.1099	1.1028	1.0955	1.4304
0.0008	1.1098	1.1026	1.0953	1.4305
0.0012	1.1096	1.1025	1.0952	1.4306
0.0017	1.1095	1.1024	1.0951	1.4307
0.0020	1.1094	1.1023	1.0950	1.4307
0.0038	1.1088	1.1017	1.0944	1.4308
0.0070	1.1078	1.1007	1.0934	1.4313
0.0119	1.1063	1.0992	1.0920	1.4320
0.0174	1.1047	1.0977	1.0905	1.4329
0.0258	1.1025	1.0955	1.0882	1.4342
0.0508	1.0968	1.0898	1.0826	1.4378
0.1900	1.0799	1.0729	1.0659	1.4509
0.3010	1.0742	1.0673	1.0604	1.4566
0.3712	1.0722	1.0653	1.0584	1.4593
0.4591	1.0705	1.0636	1.0568	1.4619
0.5435	1.0694	1.0627	1.0559	1.4638
0.6225	1.0687	1.0621	1.0554	1.4652
0.7366	1.0682	1.0615	1.0549	1.4670
0.8073	1.0679	1.0613	1.0548	1.4680
0.9007	1.0677	1.0612	1.0546	1.4689
1.0000	1.0676	1.0610	1.0545	1.4699

Mixture 6: CH_3OH (1) + [BMIM][$OcSO_4$] (2)
(1) Volumetric Properties

Ref. [177]

p/(MPa)	ρ/(kg · m^{-3})	V_φ/(cm^3 · mol^{-1})	p/(MPa)	ρ/(kg · m^{-3})	V_φ/(cm^3 · mol^{-1})	p/(MPa)	ρ/(kg · m^{-3})	V_φ/(cm^3 · mol^{-1})
Panel A: x_2 = 0.0 (pure methanol)			x_2 = 0.0388			x_2 = 0.1351		
T = 298.15 K								
0.24	785.98	–	1.19	865.73	311.25	0.58	955.08	318.95
5.00	790.69	–	5.38	869.16	311.15	5.39	958.39	318.47
10.00	795.40	–	11.56	873.89	311.08	10.69	961.82	317.98
15.01	799.88	–	15.42	876.83	310.83	15.97	965.26	317.39
19.99	804.10	–	20.33	880.16	310.89	20.44	967.95	316.97
25.01	808.17	–	25.75	883.80	310.78	25.56	970.87	316.53
30.00	812.04	–	30.57	887.24	310.18	31.22	974.06	316.00
34.99	815.76	–	35.23	890.20	309.99	34.74	975.93	315.70
39.98	819.34	–	39.56	892.79	309.81	39.75	978.54	315.28
T = 323.15 K								
0.27	762.14	–	1.03	843.45	314.73	1.03	936.98	322.24
4.99	767.53	–	5.41	847.36	315.00	5.08	939.90	321.97
9.98	772.87	–	10.03	851.32	315.15	10.61	943.73	321.56
14.98	777.91	–	15.47	855.79	315.15	15.09	946.69	321.22
19.99	782.69	–	20.31	859.57	315.12	20.41	950.05	320.81
24.99	787.21	–	25.19	863.21	315.07	25.09	952.85	320.47
30.00	791.53	–	30.51	866.98	315.02	30.74	956.05	320.09
34.99	795.64	–	35.62	870.40	315.03	35.12	958.39	319.83
39.99	799.60	–	39.97	873.16	315.10	39.97	960.84	319.56

(*continued*)

p/(MPa)	ρ/(kg · m^{-3})	V_φ/(cm^3 · mol^{-1})	p/(MPa)	ρ/(kg · m^{-3})	V_φ/(cm^3 · mol^{-1})	p/(MPa)	ρ/(kg · m^{-3})	V_φ/(cm^3 · mol^{-1})
T = 348.15 K								
0.366	737.29	–	0.77	820.93	316.61	1.52	919.20	324.93
4.987	743.46	–	6.32	826.78	316.98	5.91	922.89	324.60
9.986	749.65	–	10.23	830.82	316.85	10.73	926.46	324.39
14.991	755.42	–	15.23	835.08	317.66	15.44	929.89	324.08
20.004	760.84	–	20.31	839.59	317.65	19.77	932.75	323.87
25.02	765.94	–	25.07	843.72	317.42	25.53	936.41	323.57
30.013	770.76	–	30.28	847.59	317.84	30.27	939.26	323.32
34.989	775.32	–	35.24	851.10	318.20	35.19	942.16	323.02
40.000	779.70	–	39.53	854.39	317.76	39.72	944.76	322.73
T = 373.15 K								
0.534	710.55	–	2.04	800.53	314.52	1.08	901.34	326.17
4.983	717.67	–	5.74	804.93	315.43	5.07	904.88	326.25
9.99	724.99	–	10.03	809.80	316.09	10.06	909.13	326.17
14.998	731.69	–	15.21	815.36	316.55	15.21	913.30	325.98
19.985	737.88	–	20.64	820.81	316.83	20.08	917.06	325.72
24.994	743.68	–	25.91	825.73	317.06	25.09	920.72	325.44
29.997	749.13	–	30.46	829.69	317.31	30.14	924.22	325.14
35.001	754.27	–	35.72	833.93	317.71	35.09	927.44	324.87
39.996	759.14	–	39.38	836.66	317.99	39.94	930.42	324.63
T = 398.15 K								
0.934	680.93	–	1.70	778.33	305.18	0.57	883.56	325.03
4.984	689.00	–	5.92	784.21	307.70	5.92	888.97	325.91
9.998	697.91	–	10.61	790.40	309.45	11.36	894.23	326.15
14.986	705.88	–	15.26	796.18	310.53	15.85	898.38	326.13
19.987	713.16	–	21.34	803.19	311.44	21.06	902.99	325.95

24.992	719.87	–	26.08	808.24	311.99	25.86	907.03	325.69
29.991	726.11	–	30.05	812.19	312.42	30.31	910.61	325.41
35.005	731.99	–	35.22	816.93	313.11	36.23	915.12	325.00
40.004	737.47	–	40.08	820.99	313.90	39.76	917.66	324.77
$x_2 = 0.2785$			$x_2 = 0.4961$			$x_2 = 0.7497$		
$T = 298.15$ K								
0.32	1007.47	322.78	0.34	1040.32	324.91	0.45	1057.86	325.96
5.04	1010.37	322.16	5.03	1042.94	324.23	5.12	1060.31	325.25
10.23	1013.45	321.48	10.06	1045.68	323.51	10.34	1062.99	324.48
15.75	1016.60	320.78	15.12	1048.35	322.80	15.82	1065.72	323.70
20.64	1019.29	320.18	20.23	1050.96	322.11	20.03	1067.77	323.11
25.87	1022.04	319.57	25.06	1053.35	321.48	25.04	1070.16	322.42
30.32	1024.30	319.07	30.03	1055.73	320.86	30.12	1072.51	321.75
35.84	1026.98	318.47	35.07	1058.05	320.25	35.21	1074.80	321.09
39.96	1028.89	318.06	39.98	1060.24	319.67	39.93	1076.87	320.51
$T = 323.15$ K								
0.75	991.06	326.66	0.56	1024.37	329.34	0.56	1041.97	330.72
5.82	994.27	326.04	5.09	1027.02	328.66	5.02	1044.45	329.99
10.83	997.32	325.44	10.03	1029.80	327.94	10.08	1047.18	329.19
15.83	1000.24	324.84	15.05	1032.53	327.23	15.06	1049.78	328.43
20.98	1003.11	324.25	20.14	1035.20	326.53	20.21	1052.38	327.67
25.71	1005.62	323.74	25.09	1037.69	325.88	25.31	1054.87	326.94
30.73	1008.16	323.22	30.06	1040.08	325.25	30.06	1057.11	326.29
35.98	1010.68	322.71	35.92	1042.78	324.55	35.02	1059.36	325.63
39.95	1012.49	322.34	39.94	1044.55	324.09	39.98	1061.53	325.00
$T = 348.15$ K								
0.72	975.08	330.14	0.76	1009.49	333.41	0.98	1027.58	335.09
4.89	977.83	329.71	5.23	1012.10	332.78	4.92	1029.73	334.46

(*continued*)

p/(MPa)	ρ/(kg · m^{-3})	V_φ/(cm^3 · mol^{-1})	p/(MPa)	ρ/(kg · m^{-3})	V_φ/(cm^3 · mol^{-1})	p/(MPa)	ρ/(kg · m^{-3})	V_φ/(cm^3 · mol^{-1})
10.43	981.36	329.11	10.05	1014.84	332.10	10.03	1032.46	333.65
15.86	984.68	328.51	15.02	1017.58	331.40	15.21	1035.16	332.85
20.32	987.29	328.02	19.98	1020.24	330.72	19.98	1037.59	332.13
25.81	990.37	327.44	24.97	1022.84	330.05	25.03	1040.11	331.39
29.94	992.59	327.01	30.05	1025.39	329.39	30.09	1042.58	330.66
35.96	995.68	326.41	35.91	1028.24	328.64	35.02	1044.92	329.96
39.96	997.63	326.03	39.95	1030.13	328.15	39.98	1047.22	329.28
T = 373.15 K								
0.84	959.27	333.21	0.43	994.16	337.52	0.92	1012.72	339.67
4.92	962.23	332.85	5.06	997.08	336.86	5.08	1015.16	338.95
9.74	965.61	332.38	10.12	1000.20	336.12	10.12	1018.05	338.09
15.72	969.65	331.73	15.32	1003.31	335.35	14.89	1020.72	337.29
20.09	972.49	331.24	19.98	1006.02	334.67	20.12	1023.59	336.43
25.42	975.83	330.63	25.62	1009.22	333.84	25.02	1026.22	335.64
29.97	978.57	330.11	29.97	1011.61	333.22	30.12	1028.90	334.83
35.62	981.83	329.48	35.02	1014.31	332.51	35.03	1031.41	334.08
39.96	984.23	329.01	39.96	1016.87	331.83	39.97	1033.88	333.33
T = 398.15 K								
0.74	943.53	335.44	0.87	979.60	341.22	0.76	998.18	344.15
5.07	947.05	335.24	5.32	982.70	340.59	4.92	1000.85	343.38
10.93	951.67	334.76	10.03	985.91	339.88	10.04	1004.06	342.43
15.72	955.33	334.26	15.05	989.25	339.09	15.09	1007.16	341.50
20.84	959.11	333.66	19.98	992.46	338.30	20.13	1010.18	340.59
25.74	962.63	333.04	25.03	995.66	337.48	25.07	1013.06	339.71
29.97	965.56	332.50	29.94	998.69	336.70	30.07	1015.91	338.84
35.21	969.09	331.81	35.62	1002.09	335.80	35.02	1018.66	338.00

39.19	971.68	331.29	39.95	1004.62	335.12	39.98	1021.35	337.18
$x_2 = 0.9312$			$x_2 = 1.0$ (pure [BMIM][OcSO$_4$])					
$T = 298.15$ K								
0.32	1065.12	326.42	0.29	1067.24	–			
5.02	1067.51	325.70	5.71	1069.98	–			
10.45	1070.22	324.88	10.52	1072.36	–			
15.32	1072.60	324.17	15.07	1074.57	–			
20.21	1074.94	323.47	20.13	1076.98	–			
25.32	1077.32	322.77	25.11	1079.30	–			
29.93	1079.43	322.14	30.40	1081.70	–			
35.02	1081.70	321.47	35.24	1083.85	–			
39.95	1083.85	320.84	40.01	1085.91	–			
$T = 323.15$ K								
0.65	1049.36	331.27	0.28	1051.28	–			
5.03	1051.74	330.53	5.06	1053.88	–			
10.34	1054.54	329.67	10.21	1056.60	–			
15.92	1057.39	328.80	15.64	1059.37	–			
20.43	1059.63	328.11	20.03	1061.55	–			
25.83	1062.23	327.32	25.08	1063.98	–			
30.32	1064.32	326.68	30.18	1066.35	–			
35.73	1066.76	325.95	35.34	1068.67	–			
39.94	1068.60	325.39	39.94	1070.67	–			
$T = 348.15$ K								
0.32	1034.68	335.91	0.48	1036.91	–			
5.02	1037.16	335.12	6.34	1039.98	–			
10.32	1039.92	334.25	10.35	1042.05	–			
15.84	1042.73	333.37	15.72	1044.77	–			
20.65	1045.12	332.62	21.53	1047.66	–			

(continued)

p/(MPa)	ρ/(kg · m^{-3})	V_φ/(cm^3 · mol^{-1})	p/(MPa)	ρ/(kg · m^{-3})	V_φ/(cm^3 · mol^{-1})	p/(MPa)	ρ/(kg · m^{-3})	V_φ/(cm^3 · mol^{-1})
25.62	1047.56	331.86	25.99	1049.84	–			
29.95	1049.64	331.21	30.61	1052.06	–			
35.92	1052.45	330.34	36.07	1054.63	–			
39.96	1054.31	329.77	39.95	1056.43	–			
T = 373.15 K								
0.76	1020.19	340.61	0.11	1022.04	–			
5.62	1022.93	339.72	5.95	1025.31	–			
10.43	1025.60	338.86	10.82	1027.98	–			
15.23	1028.22	338.01	15.34	1030.41	–			
20.54	1031.05	337.11	20.73	1033.26	–			
25.31	1033.55	336.31	25.58	1035.77	–			
29.94	1035.92	335.55	30.85	1038.45	–			
35.06	1038.50	334.73	36.22	1041.11	–			
39.97	1040.91	333.97	40.01	1042.96	–			
T = 398.15 K								
0.87	1005.97	345.32	0.33	1007.09	–			
5.21	1008.65	344.44	6.18	1011.44	–			
10.32	1011.74	343.42	10.81	1014.19	–			
15.92	1015.04	342.33	15.66	1017.00	–			
20.43	1017.63	341.48	21.15	1020.12	–			
25.84	1020.67	340.49	25.48	1022.53	–			
30.43	1023.19	339.67	30.95	1025.51	–			
35.92	1026.12	338.71	35.97	1028.18	–			
39.95	1028.22	338.03	40.03	1030.29	–			

(2) Volumetric Properties

Ref. [177]

x_2	p = 0.1 MPa V^E_m (cm^3 · mol^{-1})		p = 10 MPa V^E_m (cm^3 · mol^{-1})		p = 20 MPa V^E_m (cm^3 · mol^{-1})		p = 30 MPa V^E_m (cm^3 · mol^{-1})		p = 40 MPa V^E_m (cm^3 · mol^{-1})	
T = 323.15 K										
0.0	0.0		0.0		0.0		0.0		0.0	
0.0388	−0.6860		−0.5931		−0.5152		−0.4526		−0.3993	
0.1351	−1.2752		−1.1370		−1.0123		−0.9011		−0.7979	
0.2785	−1.3591		−1.2258		−1.1017		−0.9870		−0.8771	
0.4961	−1.0731		−0.9710		−0.8768		−0.7861		−0.7031	
0.7497	−0.5657		−0.5112		−0.4618		−0.4154		−0.3735	
0.9312	−0.1660		−0.1412		−0.1266		−0.1130		−0.1098	
1.0	0.0		0.0		0.0		0.0		0.0	
	$\overline{V}_1$	$\overline{V}_2$	$\overline{V}_1$	$\overline{V}_2$	$\overline{V}_1$	$\overline{V}_2$	$\overline{V}_1$	$\overline{V}_2$	$\overline{V}_1$	$\overline{V}_2$
T = 323.15 K										
0.0	42.05	324.42	41.46	323.44	40.94	322.52	40.48	321.68	40.08	320.90
0.0388	41.62	324.60	41.10	323.64	40.63	322.72	40.21	321.86	39.84	321.06
0.1351	41.45	325.97	40.93	324.86	40.47	323.80	40.07	322.83	39.72	321.92
0.2785	41.43	328.26	40.91	326.90	40.44	325.63	40.04	324.45	39.69	323.37
0.4961	40.58	330.88	40.14	329.27	39.75	327.75	39.42	326.35	39.13	325.04
0.7497	37.84	332.19	37.67	330.46	37.54	328.83	37.45	327.31	37.39	325.90
0.9312	34.65	331.91	34.81	330.21	34.98	328.61	35.17	327.12	35.36	325.73
1.0	33.18	331.54	33.48	329.88	33.79	328.31	34.10	326.85	34.42	325.50

Ref. [177]

T (K)	V^0_φ (cm^3 · mol^{-1})	T (K)	V^0_φ (cm^3 · mol^{-1})	T (K)	V^0_φ (cm^3 · mol^{-1})
p = 0.1 MPa		p = 20 MPa		p = 40 MPa	
298.15	298.68	298.15	300.13	298.15	300.83
323.15	301.10	323.15	304.81	323.15	307.08
348.15	301.20	348.15	305.94	348.15	308.37
373.15	293.32	373.15	300.73	373.15	305.57
398.15	271.69	398.15	286.01	398.15	294.58

(3) Vapor Pressure and Activity Coefficient

Ref. [176]

	298.15 K		303.15 K		308.15 K		313.15 K			298.15 K		303.15 K		308.15 K		313.15 K	
x_1	p/Pa	γ_1	p/Pa	γ_1	p/Pa	γ_1	p/Pa	γ_1	x_1	p/Pa	γ_1	p/Pa	γ_1	p/Pa	γ_1	p/Pa	γ_1
0.0000	0	0.343	0	0.382	0	0.365	0	0.316	0.7638	10155	0.910	12880	0.910	16058	0.901	19880	0.888
0.1083	1142	0.405	1417	0.440	1739	0.420	2101	0.37	0.8792	13903	0.974	17732	0.973	22429	0.970	28172	0.967
0.2222	1708	0.48	2102	0.509	2544	0.488	3101	0.438	0.9235	15713	0.989	19952	0.989	24961	0.987	31222	0.986
0.2629	2066	0.51	2570	0.536	3130	0.514	3813	0.465	0.9392	16150	0.993	20583	0.993	25914	0.992	32289	0.991
0.4629	3612	0.668	4482	0.682	5507	0.661	6673	0.621	0.9754	16621	0.999	21348	0.999	27136	0.999	34240	0.998
0.6156	6101	0.797	7603	0.802	9404	0.786	11402	0.758	0.9874	16682	1	21485	1	27405	1	34710	1
0.6549	7046	0.829	8769	0.833	10796	0.818	13260	0.793	1.0000	16958	1	21880	1	27960	1	35450	1

Mixture 7: C_2H_5OH (1) + [BMIM][OctS] (2)
(1) Vapor Pressure and Activity Coefficient

Ref. [176]

	298.15 K		303.15 K		308.15 K		313.15 K			298.15 K		303.15 K		308.15 K		313.15 K	
x_1	p/Pa	γ_1	p/Pa	γ_1	p/Pa	γ_1	p/Pa	γ_1	x_1	p/Pa	γ_1	p/Pa	γ_1	p/Pa	γ_1	p/Pa	γ_1
0.0000	0	0.455	0	0.413	0	0.389	0	0.392	0.6364	5550	1.012	6659	0.912	7730	0.855	9007	0.822
0.0617	304	0.493	369	0.444	428	0.416	495	0.417	0.6992	6405	1.071	7862	0.975	9454	0.922	11168	0.887
0.1429	571	0.548	707	0.489	863	0.456	1058	0.453	0.7299	6789	1.096	8477	1.004	10306	0.954	12313	0.918
0.1794	752	0.575	921	0.512	1124	0.476	1360	0.471	0.7831	7359	1.13	9398	1.046	11688	1.005	14342	0.97
0.2873	1330	0.663	1683	0.586	2082	0.543	2567	0.531	0.8343	7668	1.146	10004	1.072	12722	1.043	16002	1.01
0.3544	1864	0.724	2346	0.639	2940	0.591	3610	0.575	0.87474	7824	1.141	10275	1.078	13240	1.061	16842	1.031
0.3861	2142	0.754	2723	0.666	3357	0.616	4062	0.598	0.95435	7851	1.065	10406	1.036	13625	1.037	17669	1.023
0.4892	3326	0.858	3976	0.761	4818	0.706	5817	0.68	1	7876	1	10468	1	13768	1	17928	1
0.5515	4164	0.924	4949	0.824	5892	0.766	6900	0.737									

Mixture 8: C_6H_6 (1) + [BMIM][OctS] (2)
(1) Vapor Pressure and Activity Coefficient

Ref. [176]

	298.15 K		303.15 K		308.15 K		313.15 K			298.15 K		303.15 K		308.15 K		313.15 K	
x_1	p/Pa	γ_1	p/Pa	γ_1	p/Pa	γ_1	p/Pa	γ_1	x_1	p/Pa	γ_1	p/Pa	γ_1	p/Pa	γ_1	p/Pa	γ_1
0.0000	0	1.660	0	1.492	0	1.444	0	1.436	0.6988	12101	1.334	14847	1.312	17951	1.288	21735	1.266
0.0624	1335	1.635	1480	1.487	1736	1.439	2109	1.425	0.7843	12529	1.262	15541	1.247	19012	1.232	23260	1.217
0.1498	2958	1.600	3463	1.479	4114	1.430	4904	1.410	0.8183	12620	1.227	15706	1.214	19264	1.204	23655	1.193
0.2467	4760	1.562	5700	1.465	6805	1.418	8205	1.393	0.87018	12635	1.166	15845	1.157	19500	1.152	24050	1.147
0.3758	6963	1.509	8320	1.440	9948	1.396	12032	1.366	0.91993	12651	1.096	15861	1.091	19581	1.091	24200	1.091
0.4165	7725	1.491	9274	1.430	11082	1.387	13340	1.357	0.959	12679	1.038	15872	1.036	19586	1.037	24224	1.039
0.5480	10143	1.428	12191	1.388	14479	1.352	17484	1.322	0.97415	12692	1.019	15877	1.017	19594	1.018	24239	1.02
0.6275	11365	1.382	13780	1.353	16496	1.322	19824	1.296	1	12704	1	15919	1	19659	1	24331	1

2.2.38 02011-04017: 1-Butyl-3-methyl imidazolium methylsulfate

Abbreviation:	[BMIM][MS], [BMIM][$MeSO_4$], [Bmim][$MeSO_4$], [bmim][CH_3SO_4], $BMimMSO_4$
Molecular Formula:	$C_9H_{18}N_2O_4S$
Molar Mass:	250.31
Structure:	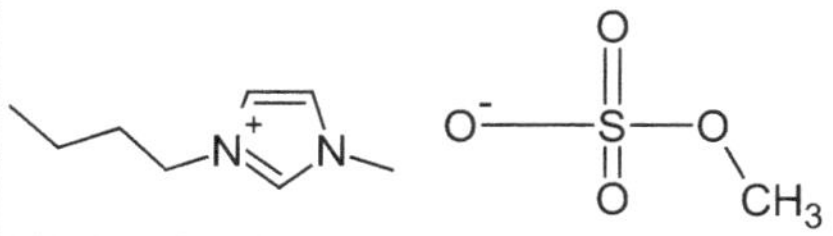

2.2.38.1 Binary Mixtures

Mixture 1: H_2O (1) + $BmimMeSO_4$ (2)

(1) Osmotic Coefficient, Activity and Vapor Pressure

Ref. [5]

$m/(mol \cdot kg^{-1})$	Φ_1	a_1	p/kPa	$m/(mol \cdot kg^{-1})$	Φ_1	a_1	p/kPa
$T = 313.15$ K				$T = 333.15$ K			
0.0519	0.995	0.9981	7.364	0.0519	0.989	0.9982	19.887
0.2039	0.952	0.9930	7.326	0.2039	0.930	0.9932	19.788
0.2934	0.915	0.9904	7.306	0.2934	0.900	0.9905	19.735
0.4033	0.901	0.9870	7.281	0.4033	0.888	0.9872	19.667
0.4947	0.888	0.9843	7.261	0.4947	0.873	0.9846	19.615
0.7132	0.852	0.9783	7.217	0.7132	0.840	0.9787	19.497
0.9875	0.829	0.9709	7.162	0.9875	0.810	0.9716	19.356
1.2867	0.791	0.9640	7.111	1.2867	0.786	0.9642	19.207
1.6992	0.767	0.9542	7.038	1.6992	0.748	0.9552	19.027
2.1321	0.742	0.9446	6.967	2.1321	0.733	0.9453	18.828
2.4949	0.730	0.9365	6.908	2.4949	0.716	0.9376	18.675
2.8891	0.730	0.9269	6.836	2.8891	0.714	0.9284	18.491
3.2941	0.733	0.9166	6.761	3.2941	0.710	0.9192	18.307
3.6976	0.722	0.9082	6.699	3.6976	0.694	0.9117	18.157
4.0910	0.720	0.8993	6.632	4.0910	0.695	0.9026	17.975

Ref. [83]

	Φ			
m_2 $(mol \cdot kg^{-1})$	$T = 298.15$ K	$T = 308.15$ K	$T = 318.15$ K	$T = 328.15$ K
0.0132	0.923	0.937	0.968	0.987
0.0225	0.895	0.921	0.958	0.977
0.0328	0.884	0.897	0.94	0.966
0.0434	0.863	0.885	0.896	0.959
0.0634	0.837	0.872	0.882	0.896

(*continued*)

	Φ			
m_2 (mol · kg^{-1})	T = 298.15 K	T = 308.15 K	T = 318.15 K	T = 328.15 K
0.078	0.821	0.836	0.877	0.887
0.1205	0.809	0.813	0.839	0.874
0.1570	0.791	0.809	0.839	0.858
0.1944	0.780	0.802	0.814	0.857
0.2801	0.767	0.779	0.802	0.840
0.3552	0.741	0.758	0.775	0.81
0.4687	0.721	0.749	0.762	0.800
0.7057	0.710	0.740	0.756	0.786
0.9552	0.698	0.716	0.732	0.76
1.4403	0.692	0.708	0.730	0.752
2.1770	0.693	0.700	0.726	0.743

(2) Volumetric Properties, Transport Properties, Refractive Index and Thermodynamic Properties

Ref. [178]

x_1	ρ/ (g · cm^{-3})	n_D	η/ (mPa · s)	V^E/ (cm^3 · mol^{-1})	Δn_D	$\Delta\eta$/ (mPa · s)	ΔG^{*E}/ (J · mol^{-1})
T = 298.15 K							
0.0000	1.21343	1.47792	182.3	0.000	0.0000	0.000	0.0
0.0325	1.21291	1.47746	177.0	−0.020	0.0043	0.634	477.3
0.1703	1.21105	1.47584	137.0	−0.208	0.0227	−14.423	2144.3
0.2989	1.20833	1.47341	95.4	−0.334	0.0389	−32.720	3344.4
0.4104	1.20462	1.47032	65.5	−0.387	0.0521	−42.349	4181.6
0.4774	1.20172	1.46814	50.20	−0.417	0.0596	−45.497	4554.2
0.5886	1.19468	1.46300	31.93	−0.426	0.0707	−43.596	5077.1
0.6859	1.18523	1.45603	20.28	−0.425	0.0779	−37.587	5293.0
0.7856	1.16821	1.44395	11.42	−0.366	0.0803	−28.358	5094.3
0.8899	1.13052	1.41842	5.071	−0.230	0.0699	−15.795	4073.5
0.9466	1.08581	1.38937	2.597	−0.101	0.0491	−7.977	2714.1
1.0000	0.9972	1.33251	0.890	0.000	0.0000	0.000	0.0
T = 313.15 K							
0.0000	1.20348		84.8	0.000		0.000	0.0
0.0325	1.20296		82.8	−0.015		0.756	477.8
0.1703	1.20102		68.9	−0.184		−1.562	2255.4
0.2989	1.19830		51.50	−0.302		−8.115	3551.0
0.4104	1.19457		35.77	−0.347		−14.469	4327.8
0.4774	1.19155		27.55	−0.361		−17.053	4653.6
0.5886	1.18443		18.40	−0.358		−16.847	5199.8
0.6859	1.17480		12.03	−0.341		−15.033	5387.7
0.7856	1.15761		7.027	−0.271		−11.651	5157.7
0.8899	1.12017		3.296	−0.142		−6.617	4105.0
0.9466	1.07683		1.763	−0.045		−3.379	2721.4

(*continued*)

x_1	ρ/ (g · cm^{-3})	n_D	η/ (mPa · s)	V^E/ (cm^3 · mol^{-1})	Δn_D	$\Delta\eta$/ (mPa · s)	ΔG^{*E}/ (J · mol^{-1})
1.0000	0.99221		0.653	0.000		0.000	0.0
T = 328.15 K							
0.0000	1.19359		45.35	0.000		0.000	0.0
0.0325	1.19303		44.80	−0.009		0.917	500.0
0.1703	1.19099		41.09	−0.157		3.382	2491.4
0.2989	1.18825		33.85	−0.271		1.909	3984.9
0.4104	1.18454		23.52	−0.315		−3.422	4688.9
0.4774	1.18133		18.25	−0.308		−5.683	4984.9
0.5886	1.17425		11.98	−0.307		−6.973	5397.9
0.6859	1.16426		7.862	−0.264		−6.726	5509.3
0.7856	1.14683		4.738	−0.185		−5.378	5255.0
0.8899	1.10950		2.306	−0.065		−3.135	4150.0
0.9466	1.06717		1.281	0.005		−1.617	2744.1
1.0000	0.98569		0.504	0.000		0.000	0.0

(3) Phase Equilibrium

Ref. [179]

T/K	x_1	γ_1	T/K	x_1	γ_1
101.3 kPa					
373.15	1.000	1.000	373.90	0.704	1.382
373.48	0.931	1.061	373.93	0.664	1.465
373.59	0.901	1.092	373.94	0.650	1.495
373.74	0.848	1.154	374.09	0.605	1.598
373.83	0.817	1.194	374.09	0.550	1.759
373.84	0.758	1.288	374.10	0.507	1.908
373.89	0.740	1.317	374.12	0.455	2.125

(4) Heat Capacity

Ref. [173]

	$x_2 = 0.20$		$x_2 = 0.40$		$x_2 = 0.60$		$x_2 = 0.80$	
T/(K)	C_p (J · mol^{-1} · K^{-1})	C_p^E (J · mol^{-1} · K^{-1})	C_p (J · mol^{-1} · K^{-1})	C_p^E (J · mol^{-1} · K^{-1})	C_p (J · mol^{-1} · K^{-1})	C_p^E (J · mol^{-1} · K^{-1})	C_p (J · mol^{-1} · K^{-1})	C_p^E (J · mol^{-1} · K^{-1})
303.2	140	1.9	203	1.9	265	0.03	326	−1.9
308.2	142	2.8	205	2.7	267	1.3	328	−0.7
313.2	142	3.3	207	3.7	269	2.4	331	0.6
318.2	143	3.8	209	4.9	272	3.8	334	1.8
323.2	145	4.5	210	5.7	275	5.5	337	2.8
328.2	146	5.2	212	6.7	277	6.5	340	4.2
333.2	147	5.8	214	7.8	280	7.8	342	5.1
338.2	148	6.7	216	8.7	282	8.9	346	6.7
343.2	149	7.4	218	9.5	285	10.7	349	8.1
348.2	151	8.2	220	10.8	287	11.4	352	9.2
353.2	152	8.8	222	11.8	291	13.2	355	10.8

(5) Excess Molar Enthalpy

Ref. [71]

x_2	$H^E/(J \cdot mol^{-1})$	x_2	$H^E/(J \cdot mol^{-1})$	x_2	$H^E/(J \cdot mol^{-1})$
303.15 K					
0.0506	−283	0.2018	−522	0.5447	−413
0.1003	−404	0.2990	−537	0.6795	−316
0.1461	−482	0.4021	−491	0.8434	−173

Mixture 2: C_2H_5OH (1) + [Bmim][$MeSO_4$] (2)
(1) Volumetric Properties, Transport Properties, Refractive Index, Compressibility Properties and Thermodynamic Properties

Ref. [178]

x_1	$\rho/(g \cdot cm^{-3})$	n_D	$\eta/(mPa \cdot s)$	$V^E/(cm^3 \cdot mol^{-1})$	Δn_D	$\Delta\eta/(mPa \cdot s)$	$\Delta G^{*E}/(J \cdot mol^{-1})$
T = 298.15 K							
0.0000	1.21343	1.47792	182.3	0.000	0.0000	0.000	0.0
0.0651	1.20577	1.47590	156.7	−0.107	0.0056	−13.797	535.4
0.0922	1.20249	1.47500	131.9	−0.172	0.0079	−33.739	484.2
0.2011	1.18741	1.47110	88.6	−0.381	0.0168	−57.244	1003.0
0.3091	1.16923	1.46633	56.32	−0.563	0.0248	−69.974	1350.4
0.4020	1.15009	1.46114	37.85	−0.682	0.0305	−71.600	1612.0
0.4910	1.12758	1.45512	25.04	−0.744	0.0350	−68.276	1762.2
0.5788	1.10078	1.44780	16.22	−0.833	0.0380	−61.187	1819.2
0.7142	1.04370	1.43243	7.427	−0.776	0.0386	−45.449	1575.4
0.8070	0.98909	1.41714	4.349	−0.715	0.0342	−31.707	1354.7
0.8950	0.91729	1.39734	2.465	−0.548	0.0247	−17.653	941.4
0.9445	0.86352	1.38212	1.733	−0.365	0.0154	−9.414	598.8
1.0000	0.78546	1.36023	1.082	0.000	0.0000	0.000	0.0
T = 313.15 K							
0.0000	1.20348		84.7	0.000		0.000	0.0
0.0651	1.19571		74.9	−0.113		−4.427	547.8
0.0922	1.19240		64.1	−0.185		−12.918	503.9
0.2011	1.17724		46.28	−0.426		−21.590	1092.9
0.3091	1.15888		31.64	−0.625		−27.168	1506.4
0.4020	1.13956		22.35	−0.761		−28.657	1789.4
0.4910	1.11704		15.62	−0.857		−27.918	1973.1
0.5788	1.08992		10.58	−0.948		−25.589	2035.5
0.7142	1.03243		5.271	−0.901		−19.541	1822.7
0.8070	0.97750		3.197	−0.843		−13.826	1562.3
0.8950	0.90522		1.863	−0.660		−7.779	1089.2
0.9445	0.85111		1.319	−0.456		−4.168	686.5
1.0000	0.77200		0.827	0.000		0.000	0.0

(continued)

x_1	ρ/ (g · cm^{-3})	n_D	η/ (mPa · s)	V^E/ (cm^3 · mol^{-1})	Δn_D	$\Delta\eta$/ (mPa · s)	ΔG^{*E}/ (J · mol^{-1})
T = 328.15 K							
0.0000	1.19359		45.35	0.000		0.000	0.0
0.0651	1.18573		40.76	−0.125		−1.672	554.9
0.0922	1.18240		35.90	−0.205		−5.323	557.8
0.2011	1.16712		27.16	−0.475		−9.196	1189.7
0.3091	1.14861		19.89	−0.701		−11.636	1699.8
0.4020	1.12914		14.41	−0.855		−12.968	1968.4
0.4910	1.10655		10.53	−0.982		−12.861	2192.5
0.5788	1.07914		7.263	−1.076		−12.206	2216.6
0.7142	1.02120		4.019	−1.037		−9.399	2143.0
0.8070	0.96592		2.503	−0.982		−6.766	1848.6
0.8950	0.89302		1.432	−0.769		−3.905	1216.2
0.9445	0.83848		1.033	−0.535		−2.090	797.9
1.0000	0.75855		0.641	0.000		0.000	0.0

Ref. [131]

x_2	ρ (g · cm^{-3})	n_D	x_2	ρ (g · cm^{-3})	n_D	x_2	ρ (g · cm^{-3})	n_D
T = 298.15 K								
0.0459	0.85062	1.3791	0.2733	1.03354	1.4314	0.6742	1.16039	1.4667
0.0987	0.90924	1.3956	0.3516	1.06963	1.4414	0.7665	1.17677	1.4715
0.1420	0.94807	1.4067	0.4381	1.10130	1.4503	0.8482	1.18891	1.4745
0.1905	0.98434	1.4170	0.5335	1.12908	1.4581	0.9248	1.19903	1.4772
0.2357	1.01298	1.4253	0.5712	1.13845	1.4609			

Ref. [132]

x_2	ρ (g · cm^{-3})				x_2	ρ (g · cm^{-3})			
	293.15 K	303.15 K	313.15 K	323.15 K		293.15 K	303.15 K	313.15 K	323.15 K
0	0.78953	0.78094	0.77220	0.76328	0.4381	1.10490	1.09771	1.09057	1.08348
0.0459	0.85475	0.84648	0.83814	0.82969	0.5335	1.13261	1.12556	1.11855	1.11160
0.0987	0.91325	0.90523	0.89717	0.88906	0.5712	1.14196	1.13495	1.12799	1.12109
0.1420	0.95200	0.94414	0.93626	0.92836	0.6742	1.16383	1.15695	1.15010	1.14331
0.1905	0.98820	0.98049	0.97279	0.96508	0.7665	1.18016	1.17337	1.16661	1.15990
0.2357	1.01676	1.00919	1.00163	0.99407	0.8482	1.19227	1.18555	1.17885	1.17221
0.2733	1.03727	1.02980	1.02232	1.01488	0.9248	1.20235	1.19572	1.18908	1.18249
0.3516	1.07329	1.06597	1.05867	1.05141	1	1.21117	1.20448	1.19792	1.19139

Ref. [7]

x_2	ρ/ (g · cm^{-3})	$V^E{}_m$/ (cm^3 · mol^{-1})	n_D	Δn_D	u/ (m · s^{-1})	K_s/ (T · Pa^{-1})	ΔK_s/ (T · Pa^{-1})
T = 293.15 K							
0	0.7893	0	1.36138	0	1160	941	0
0.0501	0.8602	−0.304	1.38149	0.0141	1218	783	−126
0.1015	0.9169	−0.489	1.39827	0.0248	1277	669	−207
0.1961	0.9937	−0.629	1.42043	0.0356	1358	546	−268
0.2990	1.0522	−0.659	1.43625	0.0392	1425	468	−280
0.3914	1.0913	−0.648	1.44713	0.0390	1479	419	−269
0.5025	1.1273	−0.598	1.45714	0.0358	1538	375	−241
0.6073	1.1535	−0.501	1.46429	0.0304	1576	349	−200
0.6982	1.1720	−0.401	1.46926	0.0245	1598	334	−156
0.8002	1.1892	−0.260	1.47391	0.0170	1626	318	−106
0.8992	1.2031	−0.096	1.47757	0.0088	1650	305	−55
0.9498	1.2095	−0.028	1.47901	0.0042	1659	300	−27
1	1.2156	0	1.48077	0	1671	295	0
T = 298.15 K							
0	0.7851	0	1.35929	0	1143	974	0
0.0501	0.8564	−0.345	1.37970	0.0144	1203	806	−134
0.1015	0.9130	−0.523	1.39667	0.0252	1262	688	−218
0.1961	0.9900	−0.675	1.41884	0.0360	1343	560	−282
0.2990	1.0486	−0.706	1.43478	0.0396	1412	479	−294
0.3914	1.0878	−0.697	1.44560	0.0393	1466	428	−283
0.5025	1.1237	−0.624	1.45586	0.0362	1525	383	−253
0.6073	1.1500	−0.529	1.46284	0.0306	1563	356	−209
0.6982	1.1685	−0.418	1.46780	0.0246	1584	341	−163
0.8002	1.1856	−0.252	1.47249	0.0171	1612	325	−110
0.8992	1.1997	−0.107	1.47623	0.0089	1636	312	−57
0.9498	1.2061	−0.034	1.47766	0.0043	1645	306	−28
1	1.2122	0	1.47942	0	1658	300	0
T = 303.15 K							
0	0.7807	0	1.35725	0	1127	1009	0
0.0501	0.8522	−0.358	1.37885	0.0155	1189	830	−144
0.1015	0.9090	−0.551	1.39541	0.0259	1248	706	−231
0.1961	0.9862	−0.714	1.41656	0.0356	1324	579	−292
0.2990	1.0449	−0.746	1.43335	0.0400	1392	494	−304
0.3914	1.0841	−0.727	1.44408	0.0396	1446	441	−292
0.5025	1.1201	−0.653	1.45509	0.0371	1508	392	−263
0.6073	1.1465	−0.558	1.46136	0.0308	1553	362	−220
0.6982	1.1650	−0.438	1.46639	0.0248	1574	347	−171
0.8002	1.1822	−0.274	1.47111	0.0172	1602	330	−117
0.8992	1.1963	−0.119	1.47485	0.0090	1626	316	−60
0.9498	1.2027	−0.040	1.47622	0.0042	1633	312	−29
1	1.2088	0	1.47806	0	1645	306	0

(2) Excess Enthalpy

Ref. [132]

x_2	H^E/ J · mol^{-1}	x_2	H^E/ J · mol^{-1}	x_2	H^E/ J · mol^{-1}	x_2	H^E/ J · mol^{-1}
T = 303.15 K							
0.016	211	0.1897	1157	0.5068	1070	0.7889	510
0.0338	368	0.2821	1323	0.5946	958	0.9083	220
0.1032	862	0.3379	1277	0.7119	696		

(3) Osmotic Coefficient, Activity and Vapor Pressure

Ref. [9]

m/(mol · kg^{-1})	Φ_1	a_1	p/kPa
T = 323.15 K			
0.0495	0.780	0.996	29.371
0.0783	0.716	0.995	29.323
0.1760	0.646	0.990	29.166
0.3283	0.573	0.983	28.965
0.3907	0.550	0.980	28.893
0.5167	0.529	0.975	28.737
0.6846	0.508	0.968	28.537
0.9851	0.496	0.956	28.166
1.3181	0.499	0.941	27.724
1.7272	0.480	0.927	27.289
2.0926	0.474	0.913	26.876
2.5070	0.465	0.898	26.446
2.9138	0.448	0.887	26.104
3.3109	0.424	0.879	25.866
3.6970	0.408	0.870	25.617

(4) Phase Equilibrium

Ref. [179]

T/K	x_1	γ_1	T/K	x_1	γ_1
101.3 kPa					
351.44	1.000	1.000	354.43	0.862	1.032
351.67	0.980	1.012	355.37	0.835	1.028
351.88	0.966	1.018	356.85	0.804	1.009
352.14	0.943	1.032	360.43	0.735	0.964
352.56	0.923	1.037	362.98	0.693	0.930
352.97	0.904	1.042	368.33	0.629	0.845
353.66	0.884	1.037	373.70	0.565	0.780

Mixture 3: CH_3NO_2 (1) + [Bmim][$MeSO_4$] (2)

(1) Volumetric Properties and Refractive Index

Ref. [131]

x_2	ρ (g · cm^{-3})	n_D	x_2	ρ (g · cm^{-3})	n_D	x_2	ρ (g · cm^{-3})	n_D
T = 298.15 K								
0.0398	1.14667	1.3953	0.2444	1.18095	1.4382	0.5652	1.19852	1.4648
0.1021	1.16166	1.4128	0.2998	1.18558	1.4449	0.7127	1.20266	1.4714
0.1433	1.16887	1.4218	0.3979	1.19169	1.4541	0.8212	1.20474	1.4748
0.1988	1.17621	1.4316	0.4786	1.19547	1.4596	0.8546	1.20540	1.4758

Ref. [132]

	ρ (g · cm^{-3})					ρ (g · cm^{-3})			
x_2	293.15 K	303.15 K	313.15 K	323.15 K	x_2	293.15 K	303.15 K	313.15 K	323.15 K
0	1.13792	1.12432	1.11062	1.09682	0.3979	1.19581	1.18757	1.17937	1.17123
0.0398	1.15279	1.14052	1.12819	1.11581	0.4786	1.19941	1.19153	1.18369	1.17591
0.1021	1.16713	1.15618	1.14520	1.13422	0.5652	1.20231	1.19473	1.18719	1.17971
0.1433	1.17403	1.16371	1.15339	1.14309	0.7127	1.20624	1.19908	1.19195	1.18488
0.1988	1.18104	1.17138	1.16173	1.15211	0.8212	1.20819	1.20128	1.19438	1.18753
0.2444	1.18556	1.17634	1.16713	1.15797	0.8546	1.20881	1.20197	1.19514	1.18836
0.2998	1.18998	1.18118	1.17241	1.16368	1	1.21116	1.20445	1.19788	1.19135

(2) Excess Enthalpy

Ref. [132]

x_2	H^E/J · mol^{-1}	x_2	H^E/J · mol^{-1}	x_2	H^E/J · mol^{-1}
T = 303.15 K					
0.0804	−159	0.3940	−466	0.6675	−370
0.1617	−285	0.4351	−451	0.7671	−273
0.2832	−396	0.5566	−418	0.8761	−130

Mixture 4: 1,3-$ClCH_2CH_2CH_2Cl$ (1) + [Bmim][$MeSO_4$] (2)

(1) Volumetric Properties and Refractive Index

Ref. [131]

x_2	ρ (g · cm^{-3})	n_D	x_2	ρ (g · cm^{-3})	n_D	x_2	ρ (g · cm^{-3})	n_D
T = 298.15 K								
0.0429	1.18363	1.4499	0.2496	1.19576	1.4624	0.6797	1.20461	1.4747
0.0674	1.18562	1.4518	0.3166	1.19802	1.4654	0.7887	1.20573	1.4765
0.1040	1.18824	1.4543	0.3917	1.20002	1.4677	0.8987	1.20682	1.4784
0.1376	1.19037	1.4564	0.4881	1.20198	1.4706			
0.1894	1.19317	1.4596	0.5686	1.20319	1.4725			

Mixture 5: $HOCH_2CH_2OCH_2CH_2OC_2H_5$ (1) + $[Bmim][MeSO_4]$ (2)
(1) Volumetric Properties and Refractive Index

Ref. [131]

x_2	ρ $(g \cdot cm^{-3})$	n_D	x_2	ρ $(g \cdot cm^{-3})$	n_D	x_2	ρ $(g \cdot cm^{-3})$	n_D
$T = 298.15$ K								
0.0517	1.00336	1.4307	0.3194	1.08250	1.4507	0.6960	1.16173	1.4697
0.0891	1.01608	1.4339	0.3902	1.09982	1.4548	0.8065	1.17990	1.4735
0.1483	1.03502	1.4388	0.4732	1.11860	1.4593	0.8947	1.19318	1.4763
0.2233	1.05710	1.4443	0.5364	1.13183	1.4626			
0.2901	1.07498	1.4488	0.6235	1.14855	1.4663			

Mixture 6: 1,2-$HOCH_2CH_2OH$ (1) + $[Bmim][MeSO_4]$ (2)
(1) Volumetric Properties and Refractive Index

Ref. [131]

x_2	ρ $(g \cdot cm^{-3})$	n_D	x_2	ρ $(g \cdot cm^{-3})$	n_D	x_2	ρ $(g \cdot cm^{-3})$	n_D
$T = 298.15$ K								
0.0854	1.11582	1.4373	0.5061	1.15015	1.4616	0.8055	1.18208	1.4735
0.1946	1.12361	1.4449	0.7170	1.15576	1.4703	0.5299	1.18848	1.4625
0.2999	1.12601	1.4511	0.3530	1.16456	1.4540	0.0341	1.19509	1.4334
0.4399	1.13805	1.4582	0.2446	1.17078	1.4479	0.9070	1.20187	1.4768
0.6360	1.14398	1.4672	0.1079	1.17289	1.4390			

Mixture 7: $[C_4mim][MeSO_4]$ (1) + $[C_4mim][BF_4]$ (2)
(1) Volumetric Properties and Transport Properties

Ref. [27]
These physicochemical properties data are listed in a mixture 17 of IL 02011-02001.

Ref. [28]
These physicochemical properties data are listed in a mixture 17 of IL 02011-02001.
(2) Volumetric Properties

Ref. [28]
These physicochemical properties data are listed in a mixture 17 of IL 02011-02001.
(3) Excess Molar Enthalpy

Ref. [28]
These physicochemical properties data are listed in a mixture 17 of IL 02011-02001.
Mixture 8: $CH_3CH_2CH_2OCH_2CH_2CH_3$ (1) + $[bmim][CH_3SO_4]$ (2)
(1) Phase Equilibrium

Ref. [13]

x_2	T/K	x_2	T/K	x_2	T/K	x_2	T/K
0.7366	356.05	0.8195	330.27	0.8684	315.75	0.8964	306.05
0.7772	340.95						

Mixture 9: $CH_3(CH_2)_3O(CH)_3CH_3$ (1) + [bmim][CH_3SO_4] (2)
(1) Phase Equilibrium

Ref. [13]

x_2	T/K	x_2	T/K	x_2	T/K	x_2	T/K
0.6816	407.85	0.7307	380.42	0.7658	363.62	0.8688	321.88
0.6904	398.18	0.7439	374.37	0.7999	349.37	0.8821	317.59
0.7053	392.06	0.7545	369.73	0.8280	337.66	0.8974	311.66
0.7175	385.78						

Mixture 10: $(CH_3)_3COCH_3$ (1) + [bmim][CH_3SO_4] (2)
(1) Phase Equilibrium

Ref. [13]

x_2	T/K	x_2	T/K	x_2	T/K	x_2	T/K
0.6844	352.10	0.7164	329.55	0.7342	308.72	0.7414	297.78
0.6947	346.95	0.7219	323.20	0.7386	303.17	0.7449	292.44
0.7065	339.59	0.7295	313.76				

Mixture 11: $CH_3OC(CH_3)_2C_2H_5$ (1) + [bmim][CH_3SO_4] (2)
(1) Phase Equilibrium

Ref. [13]

x_2	T/K	x_2	T/K	x_2	T/K	x_2	T/K
0.4200	467.14	0.5073	458.10	0.6782	415.29	0.8546	342.51
0.4481	465.33	0.5515	448.57	0.7168	396.50	0.8865	324.08
0.4731	461.71	0.5707	443.80	0.7506	381.71	0.9251	297.06
0.4892	460.39	0.6525	421.87	0.8057	363.59		

Mixture 12: 2-$CH_3COC_3H_7$ (1) + [C_4MIm][$MeSO_4$] (2)
(1) Phase Equilibrium

Ref. [13]

x_2	T/K	x_2	T/K	x_2	T/K	x_2	T/K
0.4323	377.09	0.4566	339.82	0.4747	313.81	0.4884	293.35
0.4388	365.50	0.4646	327.30	0.4818	302.42	0.4982	281.20
0.4499	350.92	0.4704	318.83				

Mixture 13: 3-$C_2H_5COC_2H_5$ (1) + [bmim][CH_3SO_4] (2)
(1) Phase Equilibrium

Ref. [13]

x_2	T/K	x_2	T/K	x_2	T/K	x_2	T/K
0.4320	372.70	0.4510	349.70	0.4761	313.19	0.4861	292.20
0.4435	360.52	0.4652	326.60	0.4782	306.70	0.5022	276.50

Mixture 14: 2-$CH_3COC_4H_9$ (1) + [bmim][CH_3SO_4] (2)
(1) Phase Equilibrium

Ref. [13]

x_2	T/K	x_2	T/K	x_2	T/K	x_2	T/K
0.5182	436.17	0.5544	373.48	0.5881	330.79	0.6184	294.23
0.5395	399.32	0.5759	345.10	0.6071	305.90		

Mixture 15: 4-$C_3H_7COC_3H_7$ (1) + [bmim][CH_3SO_4] (2)
(1) Phase Equilibrium

Ref. [13]

x_2	T/K	x_2	T/K	x_2	T/K	x_2	T/K
0.7196	451.30	0.7649	385.06	0.8259	322.61	0.8892	295.76
0.7306	420.37	0.7939	363.00				

Mixture 16: C_5H_{12} (1) + [bmim][CH_3SO_4] (2)
(1) Phase Equilibrium

Ref. [14]

x_2	T/K	x_2	T/K
0.8640	325.00	0.9411	298.47
0.8910	316.90	0.9456	294.88
0.9110	310.89	0.9499	289.96
0.9300	304.37	0.9522	288.48

Mixture 17: C_6H_{14} (1) + [bmim][CH_3SO_4] (2)
(1) Phase Equilibrium

Ref. [14]

x_2	T/K	x_2	T/K
0.8048	346.77	0.9450	312.54
0.8821	325.96	0.9770	302.52

Mixture 18: C_7H_{16} (1) + [bmim][CH_3SO_4] (2)
(1) Phase Equilibrium

Ref. [14]

x_2	T/K	x_2	T/K
0.4433	442.73	0.8855	343.24
0.4697	439.12	0.8936	341.88
0.4962	430.22	0.8963	340.42
0.5396	417.45	0.9010	338.59
0.6283	394.32	0.9101	336.96
0.6748	384.35	0.9230	334.69
0.7734	363.55	0.9291	332.08
0.7941	359.25	0.9343	330.24
0.8233	355.75	0.9443	328.17
0.8451	352.65	0.9487	326.69
0.8642	348.05	0.9883	299.85
0.8855	343.24		

Mixture 19: C_8H_{18} (1) + [bmim][CH_3SO_4] (2)
(1) Phase Equilibrium

Ref. [14]

x_2	T/K	x_2	T/K
0.6109	401.69	0.7982	366.05
0.6286	397.66	0.8181	362.57
0.6494	394.65	0.8386	358.56
0.6682	389.67	0.8593	353.37
0.6941	385.66	0.8727	349.89
0.7160	382.65	0.8995	342.37
0.7338	378.18	0.9291	335.42
0.7535	374.88	0.9451	329.51
0.7773	370.04	0.9655	319.49

Mixture 20: $C_{10}H_{22}$ (1) + [bmim][CH_3SO_4] (2)
(1) Phase Equilibrium

Ref. [14]

x_2	T/K	x_2	T/K
0.5612	445.70	0.8442	380.92
0.6187	433.52	0.8622	375.53
0.6611	424.02	0.8970	365.75
0.7026	412.30	0.9114	358.06
0.7242	408.42	0.9333	349.67
0.7511	403.03	0.9548	342.22
0.7821	394.41	0.9625	332.69
0.8135	388.06	0.9726	325.77
0.8306	383.22	0.9912	320.76

Mixture 21: $CH_2(CH_2)_4CH_2$ (1) + [bmim][CH_3SO_4] (2)
(1) Phase Equilibrium

Ref. [14]

x_2	T/K	x_2	T/K
0.6627	353.35	0.8340	318.55
0.7132	344.55	0.8627	313.86
0.7421	340.75	0.8915	302.75
0.7706	333.11	0.9136	296.25
0.8076	325.62		

Mixture 22: $CH_2(CH_2)_5CH_2$ (1) + [bmim][CH_3SO_4] (2)
(1) Phase Equilibrium

Ref. [14]

x_2	T/K	x_2	T/K
0.7342	406.33	0.8958	350.42
0.7702	399.99	0.9090	342.01
0.8078	384.42	0.9257	328.59
0.8573	364.80	0.9416	318.11
0.8774	357.10		

Mixture 23: C_6H_6 (1) + [bmim][CH_3SO_4] (2)
(1) Phase Equilibrium

Ref. [14]

x_2	T/K	x_2	T/K
0.2767	348.37	0.3558	315.75
0.2925	337.71	0.3708	313.58
0.3015	329.18	0.3857	311.96
0.3220	322.44	0.4258	308.57
0.3416	317.96	0.4533	303.12

Mixture 24: $C_6H_5CH_3$ (1) + [bmim][CH_3SO_4] (2)
(1) Phase Equilibrium

Ref. [14]

x_2	T/K	x_2	T/K
0.4153	381.40	0.5577	320.66
0.4620	344.87	0.5919	313.37
0.5223	331.41	0.6044	311.33
0.5231	331.27	0.6440	302.77
0.5521	322.39	0.6740	295.76

Mixture 25: $C_6H_5C_2H_5$ (1) + [bmim][CH_3SO_4] (2)
(1) Phase Equilibrium
Ref. [14]

x_2	T/K	x_2	T/K
0.5038	408.81	0.6222	350.88
0.5051	407.27	0.6326	348.07
0.5251	388.49	0.6422	343.18
0.5342	387.57	0.6542	337.53
0.5432	381.59	0.6664	333.76
0.5463	381.65	0.6664	335.48
0.5598	376.98	0.6856	330.56
0.5784	367.57	0.6952	328.86
0.5800	366.24	0.7228	320.96
0.5843	365.19	0.7415	317.19
0.6089	355.77		

Mixture 26: $C_6H_5C_3H_7$ (1) + [bmim][CH_3SO_4] (2)
(1) Phase Equilibrium

Ref. [14]

x_2	T/K	x_2	T/K
0.6457	430.20	0.8392	347.96
0.6720	414.83	0.8614	336.96
0.6931	405.76	0.8867	320.37
0.7074	397.96	0.9145	299.80
0.7495	384.57	0.9174	290.40
0.7824	371.44	0.9278	275.50
0.8159	356.29		

Mixture 27: o-$C_6H_4(CH_3)_2$ (1) + [bmim][CH_3SO_4] (2)
(1) Phase Equilibrium

Ref. [14]

x_2	T/K	x_2	T/K
0.5870	412.47	0.7190	344.86
0.6044	406.56	0.7521	328.70
0.6454	384.88	0.7852	317.67
0.6856	365.96	0.8094	307.93

Mixture 28: m-$C_6H_4(CH_3)_2$ (1) + [bmim][CH_3SO_4] (2)
(1) Phase Equilibrium

Ref. [14]

x_2	T/K	x_2	T/K
0.6075	410.08	0.7170	347.83
0.6246	399.82	0.7530	332.36

(*continued*)

x_2	T/K	x_2	T/K
0.6355	391.75	0.7771	325.21
0.6497	385.67	0.8274	306.96
0.6554	382.09	0.8322	305.77
0.6635	377.18		

Mixture 29: p-$C_6H_4(CH_3)_2$ (1) + [bmim][CH_3SO_4] (2)
(1) Phase Equilibrium

Ref. [14]

x_2	T/K	x_2	T/K
0.5750	411.18	0.7056	343.78
0.5882	405.75	0.7263	332.54
0.6020	400.82	0.7422	328.26
0.6116	396.56	0.7611	323.17
0.6212	385.79	0.7811	313.18
0.6383	376.27	0.7973	309.19
0.6465	373.49	0.8012	308.26
0.6782	354.97	0.8129	305.73
0.6980	345.45	0.8218	302.37

Mixture 30: CH_2F_2 (1) + [bmim][$MeSO_4$] (2)
(1) Solubility and Diffusivity

Ref. [55]

T/K	P/MPa	100 x_1	$10^{11}D/m^2 \cdot s^{-1}$	T/K	P/MPa	100 x_1	$10^{11}D/m^2 \cdot s^{-1}$
298.15	0.0099	1.2	2.1	298.25	0.5501	30.5	5.2
298.15	0.1001	6.8	2.7	298.15	0.6995	37.4	6.2
298.15	0.2498	15.6	3.0	298.15	0.8500	43.2	9.9
298.15	0.3995	23.4	4.2	298.15	1.0006	48.9	12

Mixture 31: 1-C_3H_7OH (1) + $BMimMSO_4$ (2)
(1) Osmotic Coefficient, Activity and Vapor Pressure

Ref. [9]

$m/(mol \cdot kg^{-1})$	Φ_1	a_1	p/kPa
T = 323.15 K			
0.0437	0.873	0.995	12.046
0.0547	0.829	0.995	12.036
0.2041	0.691	0.983	11.896
0.3571	0.639	0.973	11.771
0.5002	0.628	0.963	11.648
0.6627	0.628	0.951	11.504
0.8487	0.629	0.938	11.341
1.0071	0.629	0.927	11.204

(continued)

$m/(\text{mol} \cdot \text{kg}^{-1})$	Φ_1	a_1	p/kPa
1.2796	0.628	0.908	10.974
1.6654	0.620	0.883	10.673
2.1435	0.599	0.857	10.351
2.9365	0.581	0.815	9.836

Mixture 32: 2-CH_3CHCH_3OH (1) + $BMimMSO_4$ (2)
(1) Osmotic Coefficient, Activity and Vapor Pressure

Ref. [9]

$m/(\text{mol} \cdot \text{kg}^{-1})$	Φ_1	a_1	p/kPa
T = 323.15 K			
0.0524	0.950	0.994	23.448
0.0842	0.905	0.991	23.372
0.1084	0.871	0.989	23.321
0.2087	0.799	0.980	23.115
0.3036	0.733	0.974	22.959
0.4106	0.685	0.967	22.794
0.4869	0.648	0.963	22.700
0.6820	0.588	0.953	22.463
0.9823	0.523	0.940	22.157
1.3202	0.493	0.925	21.791
1.6922	0.488	0.905	21.329
2.0498	0.502	0.884	20.808
2.4462	0.480	0.868	20.445
2.8295	0.455	0.856	20.160
3.3106	0.451	0.836	19.668

2.2.38.2 Ternary Mixtures

Mixture 1: C_2H_5OH (1) + H_2O (2) + [Bmim][$MeSO_4$] (3)
(1) Volumetric Properties, Transport Properties, Refractive Index and Thermodynamic Properties

Ref. [178]

x_1	x_2	$\rho/(\text{g} \cdot \text{cm}^{-3})$	n_D	$\eta/(\text{mPa} \cdot \text{s})$	$V^E/(\text{cm}^3 \cdot \text{mol}^{-1})$	Δn_D	$\Delta\eta/(\text{mPa} \cdot \text{s})$	$\Delta G^{*E}/(\text{J} \cdot \text{mol}^{-1})$
T = 298.15 K								
0.0555	0.9174	1.03064	1.37206	2.190	−0.319	0.03407	−3.631	2336.5
0.2592	0.7195	0.95633	1.37513	2.678	−0.981	0.03234	−2.121	2825.1
0.3461	0.6352	0.92740	1.37415	2.518	−1.048	0.02932	−1.845	2649.0
0.4416	0.5424	0.89964	1.37260	2.324	−1.071	0.02552	−1.560	2405.4
0.6769	0.3138	0.84417	1.36776	1.739	−0.897	0.01514	−0.964	1514.5
0.8250	0.1700	0.81468	1.36440	1.420	−0.540	0.00829	−0.540	872.5
0.0661	0.8839	1.05876	1.38998	2.956	−0.422	0.04836	−7.020	3022.7

(continued)

x_1	x_2	ρ/ $(g \cdot cm^{-3})$	n_D	η/ $(mPa \cdot s)$	V^E/ $(cm^3 \cdot mol^{-1})$	Δn_D	$\Delta\eta$/ $(mPa \cdot s)$	ΔG^{*E}/ $(J \cdot mol^{-1})$
0.1379	0.8159	1.02961	1.38927	3.131	−0.682	0.04623	−6.161	3151.0
0.2136	0.7443	0.99968	1.38752	3.110	−0.860	0.04297	−5.461	3116.0
0.2971	0.6652	0.96891	1.38502	2.962	−0.989	0.03880	−4.814	2968.1
0.3912	0.5762	0.93671	1.38185	2.678	−1.046	0.03376	−4.201	2680.5
0.5975	0.3810	0.87681	1.37453	2.076	−0.999	0.02232	−2.839	1933.0
0.0475	0.9409	1.00575	1.35525	1.689	−0.260	0.01973	−1.316	1682.1
0.0844	0.9044	0.99314	1.35878	2.011	−0.458	0.02230	−0.920	2130.8
0.2286	0.7619	0.94433	1.36577	2.502	−0.933	0.02555	−0.138	2704.2
0.4771	0.5165	0.87559	1.36681	2.101	−1.078	0.02015	−0.037	2195.0
0.0894	0.8190	1.08893	1.41128	4.506	−0.545	0.06297	−13.027	3821.1
0.1833	0.7345	1.05117	1.40658	4.234	−0.781	0.05703	−11.602	3642.6
0.2777	0.6496	1.01443	1.40138	3.816	−0.934	0.05060	−10.315	3360.1
0.3752	0.5619	0.97824	1.39582	3.359	−1.021	0.04377	−9.010	3017.5
0.4749	0.4723	0.94318	1.39008	2.883	−1.045	0.03672	−7.686	2610.0
0.5749	0.3823	0.90995	1.38459	2.466	−1.008	0.02993	−6.295	2191.1
0.1145	0.7501	1.10742	1.42498	6.379	−0.630	0.06961	−19.082	4287.3
0.2255	0.6561	1.06553	1.41779	5.345	−0.848	0.06182	−17.060	3853.9
0.3319	0.5660	1.02544	1.41044	4.504	−0.977	0.05388	−14.971	3437.0
0.4361	0.4777	0.98644	1.40296	3.613	−1.026	0.04583	−12.994	2901.2
0.5417	0.3883	0.94801	1.39537	3.081	−1.038	0.03766	−10.620	2516.2
0.7310	0.2279	0.87985	1.38111	2.109	−0.835	0.02236	−6.379	1602.6
0.8227	0.1502	0.84786	1.37409	1.705	−0.664	0.01483	−4.258	1089.4
0.0971	0.5483	1.16962	1.45656	21.91	−0.591	0.06979	−43.325	4718.6
0.1994	0.4862	1.14228	1.45015	16.17	−0.730	0.06639	−41.795	4288.4
0.2976	0.4266	1.11342	1.44324	12.07	−0.834	0.06237	−38.914	3871.1
0.3991	0.3649	1.08060	1.43537	8.886	−0.917	0.05748	−34.893	3424.1
0.5979	0.2442	1.00519	1.41668	4.848	−0.952	0.04463	−24.804	2525.5
0.6989	0.1829	0.95962	1.40526	3.465	−0.858	0.03618	−19.010	1995.5
0.0971	0.3675	1.18696	1.46632	46.76	−0.525	0.05326	−51.276	3931.8
0.1968	0.3269	1.16628	1.46116	32.91	−0.653	0.05394	−54.420	3708.2
0.2969	0.2862	1.14244	1.45511	22.80	−0.754	0.05374	−53.780	3439.0
0.3990	0.2446	1.11435	1.44801	15.54	−0.834	0.05262	−50.083	3130.4
0.4992	0.2038	1.08219	1.43961	10.584	−0.887	0.05008	−44.279	2796.8
0.5993	0.1631	1.04401	1.42960	7.095	−0.884	0.04592	−37.010	2410.5
0.6966	0.1235	0.99987	1.41799	4.782	−0.840	0.04001	−28.881	2005.0
0.7977	0.0823	0.94417	1.40306	3.089	−0.723	0.03099	−19.718	1500.1
0.1282	0.1706	1.19197	1.47059	78.10	−0.465	0.03256	−50.054	2493.8
0.3090	0.1352	1.15811	1.46183	41.44	−0.694	0.03993	−60.344	2756.3
0.4085	0.1157	1.13426	1.45547	27.31	−0.770	0.04245	−59.967	2710.0

(2) Phase Equilibrium

Ref. [179]

T/K	x_1	x_2	y_1	γ_1	γ_2
101.3 kPa					
357.00	0.174	0.818	0.567	2.622	0.973
357.56	0.167	0.818	0.536	2.532	1.018
359.28	0.150	0.817	0.508	2.504	1.012
360.19	0.141	0.817	0.509	2.583	0.976
361.25	0.128	0.817	0.501	2.684	0.951
362.24	0.119	0.811	0.496	2.767	0.931
363.16	0.106	0.811	0.461	2.778	0.962
365.20	0.070	0.811	0.459	3.888	0.894
361.67	0.122	0.759	0.545	3.014	0.919
359.30	0.183	0.707	0.600	2.422	0.951
356.79	0.270	0.629	0.655	1.971	1.017
358.39	0.256	0.619	0.656	1.957	0.965
360.30	0.237	0.611	0.655	1.971	0.912
362.19	0.209	0.601	0.661	2.099	0.847
363.24	0.197	0.585	0.682	2.206	0.785
365.97	0.175	0.553	0.694	2.297	0.719
362.42	0.261	0.487	0.741	1.868	0.791
359.82	0.375	0.403	0.787	1.519	0.868
358.07	0.492	0.334	0.808	1.271	1.012
355.84	0.589	0.263	0.842	1.205	1.154
354.62	0.678	0.201	0.864	1.125	1.368
355.63	0.369	0.563	0.666	1.533	1.149
357.45	0.227	0.729	0.586	2.043	1.026
359.12	0.153	0.814	0.525	2.547	0.986
360.21	0.118	0.855	0.482	2.923	0.981
361.61	0.090	0.888	0.430	3.245	0.985
362.88	0.070	0.911	0.393	3.628	0.974
364.84	0.051	0.934	0.325	3.853	0.982
360.35	0.097	0.888	0.455	3.340	0.989
355.85	0.239	0.748	0.574	2.020	1.095
355.16	0.293	0.695	0.591	1.745	1.164
356.45	0.275	0.684	0.596	1.783	1.109
355.41	0.330	0.630	0.625	1.621	1.166
354.60	0.391	0.573	0.653	1.477	1.225
353.78	0.470	0.498	0.684	1.328	1.326
353.00	0.563	0.411	0.724	1.210	1.448
352.47	0.643	0.335	0.762	1.137	1.570
353.31	0.630	0.320	0.782	1.153	1.451
354.40	0.605	0.308	0.801	1.179	1.312

(*continued*)

T/K	x_1	x_2	y_1	γ_1	γ_2
354.98	0.423	0.511	0.691	1.423	1.204
354.27	0.470	0.466	0.711	1.354	1.268
352.38	0.515	0.428	0.740	1.385	1.343

Mixture 2: $C_6H_5CH_3$ (1) + C_7H_{16} (2) + [bmim][CH_3SO_4] (3)
(1) Selectivity

Ref. [16]

n-Heptane-rich phase			[bmim][CH_3SO_4] -rich phase			Experimental Selectivity
x_1	x_2	x_3	x_1	x_2	x_3	
0.1 MPa						
313.2 K						
0.0000	1.0000	0.0000	0.0000	0.0128	0.9872	–
0.0152	0.9848	0.0000	0.0041	0.0106	0.9854	24.86
0.0553	0.9447	0.0000	0.0191	0.0093	0.9716	35.02
0.0863	0.9137	0.0000	0.0289	0.0098	0.9613	31.17
0.1288	0.8649	0.0063	0.0440	0.0092	0.9468	32.13
0.1451	0.8549	0.0000	0.0482	0.0090	0.9428	31.75
0.2969	0.7030	0.0001	0.0960	0.0087	0.8953	26.26
0.4327	0.5672	0.0001	0.1350	0.0076	0.8574	23.19
0.5458	0.4531	0.0011	0.1673	0.0072	0.8255	19.26
0.6927	0.3059	0.0014	0.2185	0.0057	0.7758	16.86
0.8207	0.1792	0.0001	0.2518	0.0044	0.7439	12.61
0.9976	0.0024	0.0000	0.3474	0.0001	0.6526	12.08
1.0000	0.0000	0.0000	0.3222	0.0000	0.6778	–
348.2 K						
0.0000	1.0000	0.0000	0.0000	0.0117	0.9883	–
0.0178	0.9822	0.0000	0.0041	0.0116	0.9843	19.40
0.0562	0.9437	0.0000	0.0189	0.0107	0.9704	29.53
0.0918	0.9082	0.0000	0.0290	0.0103	0.9608	27.96
0.1359	0.8615	0.0026	0.0426	0.0110	0.9463	24.53
0.1528	0.8472	0.0000	0.0476	0.0101	0.9423	26.08
0.3008	0.6992	0.0000	0.0926	0.0099	0.8975	21.79
0.4366	0.5634	0.0000	0.1313	0.0092	0.8595	18.44
0.5494	0.4441	0.0065	0.1684	0.0096	0.8220	14.17
0.7075	0.2886	0.0038	0.2245	0.0071	0.7684	12.94
0.8243	0.1754	0.0003	0.2387	0.0044	0.7569	11.54
0.9967	0.0033	0.0000	0.3431	0.0001	0.6567	9.21
1.0000	0.0000	0.0000	0.3166	0.0000	0.6834	–

(2) Solubility

Ref. [16]

Temperature					
Toluene in IL, maximum		*n*-Heptane in IL, average		IL in *n*-Heptane, average	
313.2 K	348.2 K	313.2 K	348.2 K	313.2 K	348.2 K
0.32	0.32	0.0074	0.0088	0.0009	0.0015

Mixture 3: C_7H_{16} (1) + C_2H_5OH (2) + [BMIM][$MeSO_4$] (3)
(1) Distribution Ratio and Selectivity

Ref. [180]

Hydrocarbon-rich phase		Ionic liquid-rich phase			
w^I_1	w^I_2	w^{II}_1	w^{II}_2	β_2	S
298.15 K					
0.998	0.002	0.001	0.041	19.10	26795
0.996	0.004	0.001	0.066	16.26	22326
0.995	0.005	0.001	0.104	21.52	17389
0.993	0.007	0.003	0.133	19.06	6891.1
0.992	0.008	0.009	0.157	20.21	2347.7
0.990	0.010	0.019	0.181	18.32	973.4
0.990	0.010	0.022	0.225	21.48	950.9
0.985	0.015	0.034	0.277	18.41	526.2
0.972	0.028	0.046	0.326	11.80	251.1
0.959	0.041	0.058	0.372	9.13	150.4
0.951	0.048	0.082	0.402	8.32	96.6
0.927	0.073	0.105	0.440	6.06	53.5
0.892	0.108	0.131	0.471	4.36	29.6
0.884	0.116	0.173	0.496	4.28	21.9
0.876	0.124	0.217	0.504	4.07	16.4
0.856	0.144	0.261	0.512	3.56	11.7

Mixture 4: C_6H_{14} (1) + C_2H_5OH (2) + [BMIM][$MeSO_4$] (3)
(1) Distribution Ratio and Selectivity

Ref. [181]

Hydrocarbon-rich phase		Ionic liquid-rich phase			
$w_1^{HC\text{-}phase}$	$w_2^{HC\text{-}phase}$	$w_1^{IL\text{-}phase}$	$w_2^{IL\text{-}phase}$	β_2	S
298.15 K					
0.994	0.005	0.013	0.030	5.51	432.0
0.994	0.006	0.016	0.045	7.64	482.1
0.993	0.007	0.020	0.070	10.27	523.0
0.992	0.008	0.024	0.123	15.98	651.7
0.991	0.009	0.033	0.156	18.28	553.9
0.988	0.012	0.040	0.213	17.62	437.3
0.987	0.012	0.048	0.282	23.31	483.2

(*continued*)

Hydrocarbon-rich phase		Ionic liquid-rich phase			
$w_1^{\text{HC-phase}}$	$w_2^{\text{HC-phase}}$	$w_1^{\text{IL-phase}}$	$w_2^{\text{IL-phase}}$	β_2	S
0.979	0.019	0.076	0.334	17.89	231.0
0.962	0.035	0.101	0.381	11.01	105.0
0.951	0.044	0.128	0.409	9.30	68.9
0.934	0.060	0.160	0.441	7.31	42.6
0.905	0.088	0.199	0.466	5.27	24.0

2.2.39 02011-04019: 1-Butyl-3-methylimidazolium diethyleneglycolmonomethylethersulphate

Abbreviation:	$[BMIM][CH_3(OCH_2CH_2)_2OSO_3]$
Molecular Formula:	$C_{13}H_{26}N_2O_6S$
Molar Mass:	338.42
Structure:	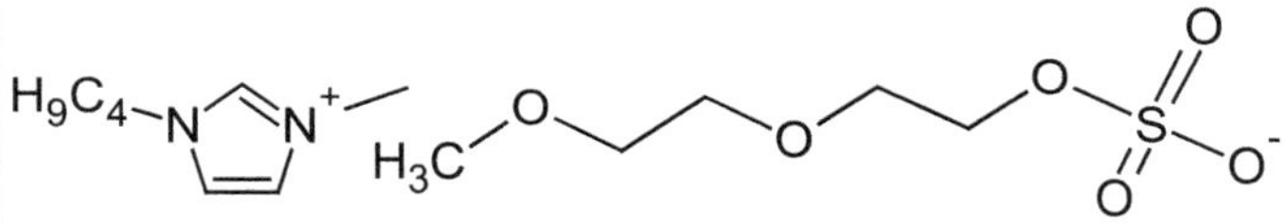

2.2.39.1 Binary Mixtures

Mixture 1: CH_3OH (1) + $[BMIM][CH_3(OCH_2CH_2)_2OSO_3]$ (2)

(1) Volumetric Properties

Ref. [102]

x_1	ρ (g · cm^{-3})	V^E_m (cm^3 · mol^{-1})	x_1	ρ (g · cm^{-3})	V^E_m (cm^3 · mol^{-1})
$T = 298.15$ K					
0.055	1.1873	−0.015	0.533	1.1381	−0.586
0.093	1.1850	−0.048	0.649	1.1125	−0.725
0.136	1.1821	−0.090	0.711	1.0929	−0.777
0.199	1.1774	−0.143	0.775	1.0661	−0.786
0.265	1.1718	−0.192	0.877	0.9972	−0.757
0.335	1.1651	−0.303	0.912	0.9593	−0.672
0.388	1.1592	−0.364	0.951	0.9027	−0.511
0.481	1.1467	−0.513	0.979	0.8438	−0.284
$T = 303.15$ K					
0.055	1.1837	−0.019	0.533	1.1345	−0.631
0.093	1.1814	−0.056	0.649	1.1088	−0.768
0.136	1.1785	−0.101	0.711	1.0891	−0.815
0.199	1.1738	−0.160	0.775	1.0622	−0.822
0.265	1.1682	−0.214	0.877	0.9932	−0.796

(continued)

x_1	ρ (g · cm^{-3})	V^E_m (cm^3 · mol^{-1})	x_1	ρ (g · cm^{-3})	V^E_m (cm^3 · mol^{-1})
0.335	1.1615	−0.331	0.912	0.9551	−0.701
0.388	1.1556	−0.396	0.951	0.8984	−0.536
0.481	1.1431	−0.554	0.979	0.8394	−0.304
$T = 313.15$ K					
0.055	1.1767	−0.052	0.533	1.1275	−0.737
0.093	1.1744	−0.094	0.649	1.1018	−0.890
0.136	1.1715	−0.146	0.711	1.0818	−0.914
0.199	1.1669	−0.235	0.775	1.0549	−0.932
0.265	1.1614	−0.317	0.877	0.9856	−0.899
0.335	1.1546	−0.425	0.912	0.9469	−0.767
0.388	1.1488	−0.513	0.951	0.8901	−0.600
0.481	1.1361	−0.652	0.979	0.8303	−0.321

02011-04021: 1-Butyl-3-methylimidazolium hydrogen sulfate

Abbreviation:	[BMIM][HSO_4]
Molecular Formula:	$C_8H_{16}N_2O_4S$
Molar Mass:	236.29
Structure:	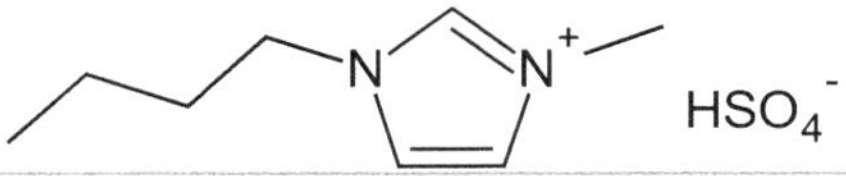

2.2.39.2 Ternary Mixtures

Mixture 1: $CH_3COOCH_2CH_3$ (1) + C_2H_5OH (2) + [BMIM][HSO_4] (3)

(1) Volumetric Properties

Ref. [104]

ester-rich phase			[BMIM][HSO_4]-rich phase		
w_1	w_2	ρ/(g · cm^{-3})	w_1	w_2	ρ/(g · cm^{-3})
313 K					
0.998	0.000	0.87700	0.057	0.000	1.24238
0.927	0.069	0.86779	0.097	0.076	1.17467
0.848	0.143	0.86092	0.151	0.143	1.10987
0.752	0.211	0.86291	0.215	0.187	1.05863
0.688	0.235	0.87131	0.258	0.208	1.03065
0.604	0.242	0.89436	0.372	0.249	0.96708
0.556	0.251	0.90516	0.128	0.124	1.12802
0.555	0.250	0.90641	0.179	0.167	1.08246
0.816	0.169	0.85967			
0.773	0.199	0.86079			
0.675	0.235	0.87490			
0.599	0.242	0.89543			

Ref. [104]

ester-rich phase			[BMIM][HSO$_4$]-rich phase		
w_1	w_2	ρ/(g · cm^{-3})	w_1	w_2	ρ/(g · cm^{-3})
313 K					
0.915	0.081	0.86660	0.120	0.109	1.14401
0.877	0.116	0.86324	0.147	0.138	1.11498
0.670	0.240	0.87434	0.339	0.238	0.98470
0.826	0.158	–	0.200	0.180	–
0.753	0.21	–	0.281	0.214	–

Mixture 2: $CH_3COOCH_2CH_3$ (1) + CH_3COOH (2) + [BMIM][HSO$_4$] (3)
(1) Volumetric Properties

Ref. [104]

ester-rich phase			[BMIM][HSO$_4$]-rich phase		
w_1	w_2	ρ/(g · cm^{-3})	w_1	w_2	ρ/(g · cm^{-3})
313 K					
0.998	0.000	0.87700	0.057	0.000	1.24238
0.894	0.099	0.89277	0.098	0.083	1.19892
0.811	0.169	0.90748	0.109	0.109	1.18664
0.736	0.212	0.92434	0.142	0.145	1.16514
0.645	0.229	0.95164	0.264	0.218	1.10135
0.550	0.238	0.98270	0.403	0.242	1.03697
0.757	0.203	0.91908			
0.895	0.100	0.89289			
0.647	0.237	0.95175			

Ref. [104]

ester-rich phase			[BMIM][HSO$_4$]-rich phase		
w_1	w_2	ρ/(g · cm^{-3})	w_1	w_2	ρ/(g · cm^{-3})
313 K					
0.955	0.044	0.88307	0.094	0.079	1.20116
0.857	0.133	0.89884	0.209	0.197	1.12719
0.782	0.186	0.91377	0.320	0.230	1.07622
0.917	0.081	0.88881	0.138	0.142	1.16741
0.764	0.196	0.91783	0.346	0.234	1.06426

2.2.40 02011-04028: 1-Butyl-3-methylimidazolium 1,1,2-trifluoro-2-(perfluoroethoxy) ethanesulfonate

Abbreviation:	[bmim][TPES]
Molecular Formula:	$C_{12}H_{16}F_8N_2O_4S$
Molar Mass:	436.32
Structure:	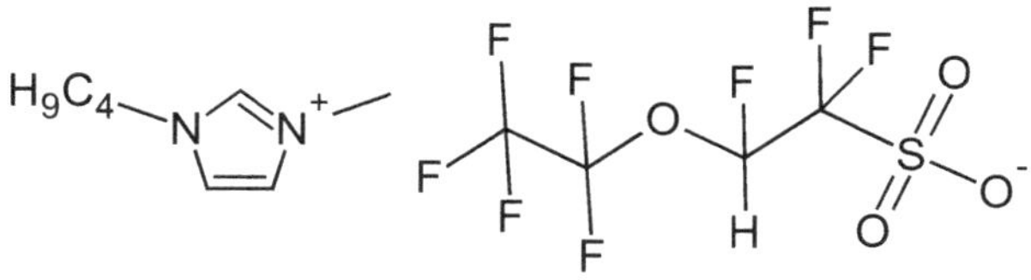

2.2.40.1 Binary Mixtures

Mixture 1: 1,1,1,2-$C_2H_2F_4$ (1) + [bmim][TPES] (2)
(1) Solubility and Diffusivity

Ref. [64]

T/K	P/MPa	$100x_1$	10^{11} D/m^2 · s^{-1}	T/K	P/MPa	$100x_1$	10^{11} D/m^2 · s^{-1}
283.15	0.0102	2.8	0.9	323.15	0.0104	1.1	–
283.10	0.0504	13.2	1.2	323.15	0.0503	4.5	6.0
283.15	0.1004	25.3	1.7	323.15	0.1002	8.7	5.7
283.10	0.1503	36.3	2.5	323.15	0.1502	12.9	6.6
283.05	0.2002	46.6	3.7	323.15	0.2002	16.8	6.9
283.10	0.2502	56.8	5.4	323.15	0.2504	20.6	7.4
283.10	0.3004	67.3	7.1	323.10	0.3004	24.1	8.6
283.10	0.3504	79.8	–	323.15	0.3503	27.6	8.0
298.05	0.0103	1.8	1.8	348.10	0.0105	0.6	11
298.10	0.0503	8.4	2.3	348.10	0.0505	2.7	12
298.10	0.1002	16.2	2.6	348.10	0.1005	5.3	12
298.10	0.1502	23.6	3.2	348.10	0.1505	7.8	–
298.05	0.2003	30.4	4.1	347.90	0.2006	10.3	11
298.10	0.2504	36.8	5.0	348.10	0.2505	12.6	15
298.10	0.3003	43.1	5.7	348.10	0.3004	14.9	13
298.10	0.3501	49.4	4.9	348.10	0.3505	17.1	14

Mixture 2: CH_2F_2 (1) + [bmim][TPES] (2)
(1) Solubility and Diffusivity

Ref. [55]

T/K	P/MPa	100 x_1	$10^{11}D/$ $m^2 \cdot s^{-1}$	T/K	P/MPa	100 x_1	$10^{11}D/$ $m^2 \cdot s^{-1}$
298.15	0.0095	1.0	4.5	298.15	0.5494	44.5	12
298.15	0.1001	10.2	5.3	298.15	0.6994	52.9	15
298.15	0.2496	23.7	6.7	298.15	0.8495	60.5	18
298.15	0.3995	34.8	9.6	298.15	0.9994	67.4	21

2.2.41 02011-04029: 1-Butyl-3-methylimidazolium 1,1,2,3,3,3-hexafluoropropanesulfonate

Abbreviation:	[bmim][HFPS]
Molecular Formula:	$C_{11}H_{16}F_6N_2O_3S$
Molar Mass:	370.31
Structure:	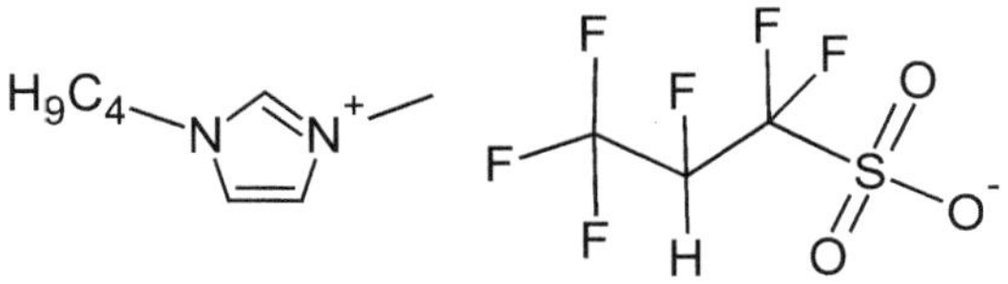

2.2.41.1 Binary Mixtures

Mixture 1: 1,1,1,2-$C_2H_2F_4$ (1) + [bmim][HFPS] (2)
(1) Solubility and Diffusivity

Ref. [64]

T/K	P/MPa	100 x_1	10^{11} $D/m^2 \cdot s^{-1}$	T/K	P/MPa	100 x_1	10^{11} $D/m^2 \cdot s^{-1}$
283.10	0.0099	1.5	–	323.15	0.0101	0.6	3.9
283.10	0.0502	8.2	0.9	323.15	0.0500	2.9	3.5
283.10	0.1000	17.1	1.1	323.15	0.1000	5.8	2.8
283.15	0.1500	26.4	1.4	323.10	0.1500	8.8	3.3
283.10	0.2001	35.0	–	323.15	0.1993	11.7	-
283.15	0.2500	44.7	2.4	323.15	0.2504	14.5	4.2
283.15	0.3002	55.0	4.8	323.15	0.3004	17.2	4.8
283.20	0.3506	67.7	7.1	323.15	0.3502	19.8	4.8
298.00	0.0101	1.0	1.6	348.05	0.0099	0.3	5.1
298.10	0.0501	5.5	–	348.05	0.0501	1.7	–
298.10	0.1000	10.9	–	348.05	0.1000	3.5	6.7
298.05	0.1499	16.3	1.5	348.10	0.1499	5.3	7.6

(*continued*)

T/K	*P*/MPa	100 x_1	10^{11} D/m^2 · s^{-1}	*T*/K	*P*/MPa	100 x_1	10^{11} D/m^2 · s^{-1}
298.10	0.1994	21.9	1.8	348.10	0.2000	7.0	6.7
298.15	0.2504	26.9	2.7	348.05	0.2495	8.7	8.1
298.10	0.3000	32.1	2.5	348.05	0.3001	10.3	8.1
298.00	0.3504	37.4	3.1	348.05	0.3494	12.0	–

Mixture 2: CH_2F_2 (1) + [bmim][HFPS] (2)
(1) Solubility and Diffusivity

Ref. [55]

T/K	*P*/MPa	100 x_1	$10^{11}$$D$/m^2 · s^{-1}	*T*/K	*P*/MPa	100 x_1	$10^{11}$$D$/m^2 · s^{-1}
298.15	0.0095	1.0	3.3	298.15	0.5495	44.6	7.0
298.15	0.1000	10.4	3.2	298.15	0.6998	52.7	11
298.15	0.2498	24.2	3.9	298.15	0.8494	60.4	10
298.15	0.3995	35.0	7.0	298.15	1.0004	67.0	17

2.2.42 02011-04030: 1-Butyl-3-methylimidazolium 1,1,2,2-tetrafluoroethanesulfonate

Abbreviation:	[bmim][TFES]
Molecular Formula:	$C_{10}H_{16}F_4N_2O_3S$
Molar Mass:	320.30
Structure:	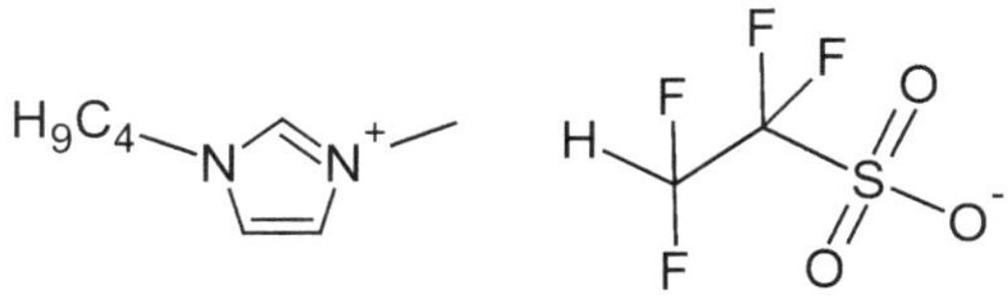

2.2.42.1 Binary Mixtures

Mixture 1: CH_2F_2 (1) + [bmim][TFES] (2)
(1) Solubility and Diffusivity

Ref. [55]

T/K	*P*/MPa	100 x_1	$10^{11}$$D$/m^2 · s^{-1}	*T*/K	*P*/MPa	100 x_1	$10^{11}$$D$/m^2 · s^{-1}
298.15	0.0097	0.7		298.15	0.5484	34.5	4.8
298.15	0.0999	7.2	1.5	298.15	0.6995	42.0	7.8
298.15	0.2500	17.2	2.7	298.15	0.8481	49.0	8.7
298.15	0.3972	26.3	4.0	298.15	0.9989	55.6	14

2.2.43 02011-04031: 1-Butyl-3-methylimidazolium 2-(1,2,2,2-tetrafluoroethoxy)-1,1,2,2-tetrafluoroethanesulfonate

Abbreviation:	[bmim][FS]
Molecular Formula:	$C_{12}H_{16}F_8N_2O_4S$
Molar Mass:	436.32
Structure:	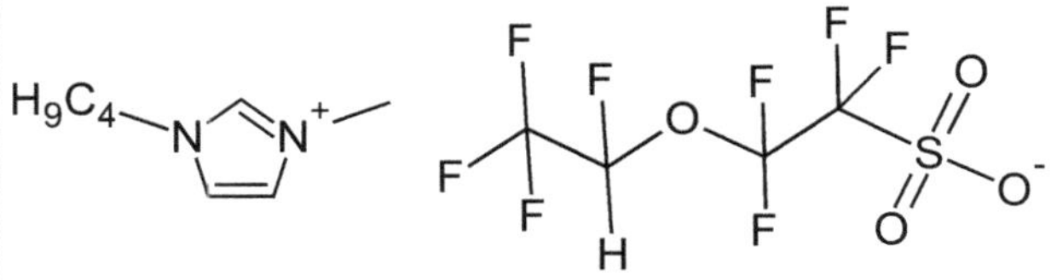

2.2.43.1 Binary Mixtures

Mixture 1: CH_2F_2 (1) + [bmim][FS] (2)

(1) Solubility and Diffusivity

Ref. [55]

T/K	P/MPa	100 x_1	$10^{11}D$/ $m^2 \cdot s^{-1}$	T/K	P/MPa	100 x_1	$10^{11}D$/ $m^2 \cdot s^{-1}$
298.15	0.0100	0.9	4.3	298.15	0.5500	41.5	10
298.15	0.0997	9.2	4.3	298.15	0.6996	49.7	15
298.15	0.2501	21.4	5.4	298.15	0.8494	57.1	
298.15	0.4004	31.9	9.1	298.15	1.0005	63.8	28

2.2.44 02011-04033: 1-Butyl-3-methylimidazolium 1,1,2-trifluoro-2-(trifluoromethoxy) ethanesulfonate

Abbreviation:	[bmim][TTES]
Molecular Formula:	$C_{11}H_{16}F_6N_2O_4S$
Molar Mass:	386.31
Structure:	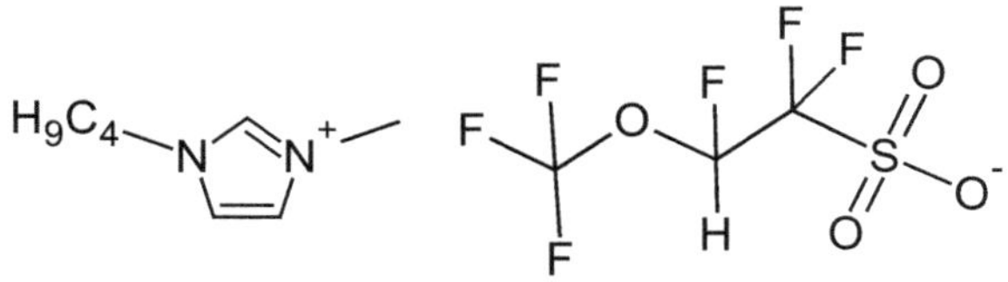

2.2.44.1 Binary Mixtures

Mixture 1: 1,1,1,2-$C_2H_2F_4$ (1) + [bmim][TTES] (2)

(1) Solubility and Diffusivity

Ref. [64]

T/K	*P*/MPa	100 x_1	10^{11} *D*/ $m^2 \cdot s^{-1}$	*T*/K	*P*/MPa	100 x_1	10^{11} *D*/ $m^2 \cdot s^{-1}$
283.20	0.0103	2.6	1.0	323.15	0.0105	0.9	5.8
283.20	0.0500	11.7	1.3	323.15	0.0503	3.9	5.4
283.15	0.1003	22.5	1.6	323.15	0.1002	7.6	5.8
283.15	0.1502	32.7	2.2	323.15	0.1502	11.2	6.4
282.10	0.2002	42.6	3.3	323.15	0.2003	14.6	6.6
283.10	0.2505	51.6	5.3	323.15	0.2503	18.0	7.0
283.10	0.2995	63.0	5.3	323.15	0.3004	21.3	7.2
283.10	0.3505	74.8	–	323.15	0.3502	24.5	6.9
298.10	0.0102	1.7	1.2	348.15	0.0103	0.5	10
298.10	0.0503	7.5	2.2	348.10	0.0505	2.3	12
298.20	0.1002	14.4	2.6	348.10	0.1005	4.5	12
298.00	0.1502	20.9	3.0	348.05	0.1503	6.7	12
298.10	0.2003	27.3	3.3	348.05	0.2004	8.8	13
298.15	0.2504	33.3	4.4	348.10	0.2504	10.9	13
298.10	0.3001	39.2	5.1	348.10	0.3002	12.9	14
297.90	0.3501	45.3	4.5	348.10	0.3504	14.9	–

Mixture 2: CH_2F_2 (1) + [bmim][TTES] (2)

(1) Solubility and Diffusivity

Ref. [55]

T/K	*P*/MPa	100 x_1	10^{11}*D*/ $m^2 \cdot s^{-1}$	*T*/K	*P*/MPa	100 x_1	10^{11}*D*/ $m^2 \cdot s^{-1}$
298.15	0.0095	1.0	4.3	298.15	0.5500	42.8	10
298.15	0.1003	9.6	4.5	298.15	0.6997	51.0	12
298.15	0.2503	22.3	5.8	298.15	0.8495	58.3	16
298.15	0.3996	33.4	7.1	298.15	0.9992	65.0	25

2.2.45 02011-04036: 1-Butyl-3-methylimidazolium ethylenglykolmonomethylethersulfate

Abbreviation:	[bmim][MDEGSO$_4$]
Molecular Formula:	$C_{12}H_{24}N_2O_6S$
Molar Mass:	324.40
Structure:	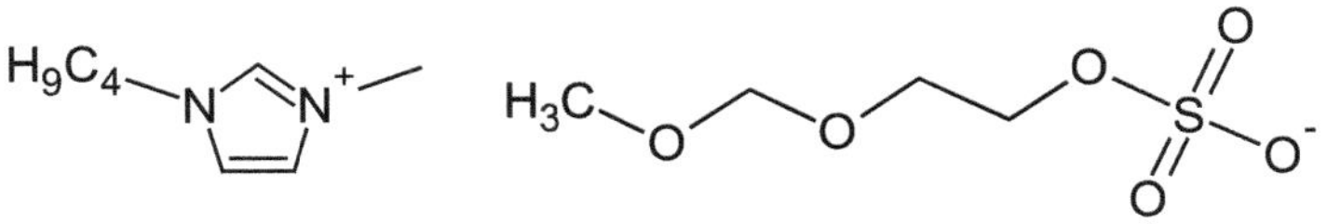

2.2.45.1 Ternary Mixtures

Mixture 1: C_6H_{14} (1) + CH_3CH_2COOH (2) + [bmim][MDEGSO$_4$] (3)
(1) Selectivity

Ref. [75]

Hexane-rich phase		RTIL-rich phase		
x_3'	x_2'	x_3''	x_2''	Selectivity (S)
0.0000	0.5198	0.0650	0.7850	5
0.0000	0.2136	0.1400	0.7850	39
0.0000	0.0465	0.3000	0.6200	159
0.0000	0.0074	0.4750	0.4600	949

Mixture 2: C_6H_{14} (1) + $CH_3CH_2COOCH = CH_2$ (Vinyl propionate) (2) + [bmim][MDEGSO$_4$] (3)
(1) Selectivity

Ref. [75]

Hexane-rich phase		RTIL-rich phase		
x_3'	x_2'	x_3''	x_2''	Selectivity (S)
0.0010	0.8638	0.3190	0.6810	–
0.0009	0.6952	0.4581	0.5419	–
0.0008	0.5013	0.6413	0.3587	–
0.0007	0.2633	0.7772	0.2228	–
0.0005	0.1350	0.8786	0.1214	–

Mixture 3: C_6H_{14} (1) + $CH_3CH(Ph)COOCH_2CH_3$ (*rac*-1-Phenylethyl propionate) (2) + [bmim][$MDEGSO_4$] (3)
(1) Selectivity

Ref. [75]

Hexane-rich phase		RTIL-rich phase		
x_3'	x_2'	x_3''	x_2''	Selectivity (S)
0.0050	0.7428	0.2059	0.7939	1348
0.0031	0.4822	0.3285	0.6710	1433
0.0020	0.2775	0.4941	0.5058	13133
0.0010	0.1360	0.7512	0.2487	15782
0.0000	0.0669	0.8770	0.1230	–

Mixture 4: C_6H_{14} (1) + $CH_3CH(Ph)OH$ (*rac*-1-Phenylethanol) (2) + [bmim][$MDEGSO_4$] (3)
(1) Selectivity

Ref. [75]

Hexane-rich phase		RTIL-rich phase		
x_3'	x_2'	x_3''	x_2''	Selectivity (S)
0.0000	0.0618	0.0900	0.7200	58
0.0000	0.0249	0.1900	0.6600	172
0.0000	0.0092	0.3950	0.4950	485
0.0000	0.0055	0.5900	0.3300	746

2.2.46 02011-05001: 1-Butyl-3-methylimidazolium hexafluorophosphate

Abbreviation:	[C_4MIm][PF_6], [BMIM][PF_6], [bmim][PF_6], [C_4mim][PF_6], $BMIMPF_6$, $BMIPF_6$, C_4mimPF_6, [C_1C_4Im][PF_6]
Molecular Formula:	$C_8H_{15}F_6N_2P$
Molar Mass:	284.18
Structure:	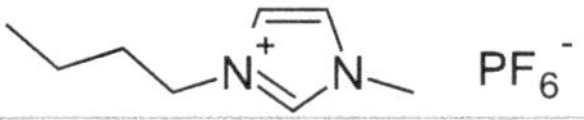

2.2.46.1 Binary Mixtures

Mixture 1: $HOCH_2CH_2OCH_3$ (1) + [bmim][PF_6] (2)
(1) Volumetric Properties and Refractive Index

Ref. [182]

	$\rho/(g \cdot cm^{-3})$			n_D
x_2	$T = 298.15$ K	$T = 308.15$ K	$T = 318.15$ K	$T = 298.15$ K
0.0000	0.96005	0.95081	0.94145	1.4004
0.0037	0.96455	0.96253	0.94596	1.4006
0.0078	0.96943	0.96734	0.95088	1.4010
0.0120	0.97436	0.97219	0.95582	1.4013
0.0354	1.00013	0.99766	0.98170	1.4030
0.0801	1.04278	1.03997	1.02448	1.4050
0.1468	1.09424	1.09112	1.07609	1.4065
0.2542	1.15735	1.15341	1.13953	1.4078
0.3960	1.21908	1.21360	1.20172	1.4083
0.6104	1.28464	1.27788	1.26748	1.4086
0.8127	1.32813	1.32065	1.31110	1.4086
0.8961	1.34477	1.33668	1.32790	1.4089
0.9345	1.35285	1.34449	1.33600	1.4090
0.9749	1.36190	1.35335	1.34499	1.4090
1.0000	1.36788	1.35933	1.35087	1.4089

Ref. [182]

T/K	$V^{\infty}_{\varphi,i}/(cm^3 \cdot mol^{-1})$	$\overline{V}^{\infty}_{m,i}/(cm^3 \cdot mol^{-1})$	$\overline{V}^{E,\infty}_{m,i}/(cm^3 \cdot mol^{-1})$
298.15	194.83	194.94	−12.81

Mixture 2: $CH_3OCH_2CH_2OCH_3$ (1) + [bmim][PF_6] (2)
(1) Volumetric Properties and Refractive Index

Ref. [182]

	$\rho/(g \cdot cm^{-3})$			n_D
x_2	$T = 298.15$ K	$T = 308.15$ K	$T = 318.15$ K	$T = 298.15$ K
0.0000	0.86447	0.85347	0.84227	1.3790
0.0041	–	–	–	1.3781
0.0125	0.87928	0.86836	0.85725	1.3798
0.0416	0.91209	0.90137	0.89047	1.3859
0.1055	0.97653	0.96623	0.95579	1.3926
0.1219	0.99151	0.98131	0.97100	1.3941
0.2474	1.08914	1.07957	1.06997	1.4009
0.3744	1.16422	1.15509	1.14598	1.4043
0.4937	1.21989	1.21103	1.20222	1.4062
0.7092	1.29636	1.28772	1.27925	1.4076
0.9329	1.35378	1.34527	1.33685	1.4087
1.0000	1.36788	1.35933	1.35087	1.4089

Ref. [182]

T/K	$V^{\infty}_{\varphi,i}$/(cm^3 · mol^{-1})	$\overline{V}^{\infty}_{m,i}$/(cm^3 · mol^{-1})	$\overline{V}^{E,\infty}_{m,i}$/(cm^3 · mol^{-1})
298.15	183.66	183.63	−24.13

Mixture 3: $CH_3(OCH_2CH_2)_2OCH_3$ (1) + [bmim][PF_6] (2)
(1) Volumetric Properties and Refractive Index

Ref. [182]

x_2	ρ/(g · cm^{-3})			n_D
	T = 298.15 K	T = 308.15 K	T = 318.15 K	T = 298.15 K
0.0000	0.93910	0.92960	0.91930	1.4058
0.0039	0.94216	0.93257	0.92234	1.4063
0.0076	0.94503	0.93537	0.92520	1.4065
0.0127	0.94897	0.93922	0.92913	1.407
0.0304	0.96227	0.95230	0.94243	1.4081
0.0505	0.97674	0.96673	0.95703	1.4092
0.1003	1.01023	1.00046	0.99093	1.4113
0.1617	1.04781	1.03853	1.02898	1.4141
0.2070	1.07360	1.06459	1.05502	1.4149
0.3024	1.12455	1.11544	1.10604	1.4148
0.4018	1.17396	1.16458	1.15565	1.4142
0.5023	1.21947	1.21050	1.20205	1.4138
0.5623	1.24360	1.23527	1.22697	1.4123
0.7197	1.29224	1.28605	1.27769	1.4098
0.9477	1.34516	1.33693	1.32847	1.4094
0.9842	1.35924	1.35131	1.34285	1.4089
0.9952	1.36536	1.35687	1.34847	-
1.0000	1.36788	1.35933	1.35087	1.4058

Ref. [182]

T/K	$V^{\infty}_{\varphi,i}$/(cm^3 · mol^{-1})	$\overline{V}^{\infty}_{m,i}$/(cm^3 · mol^{-1})	$\overline{V}^{E,\infty}_{m,i}$/(cm^3 · mol^{-1})
298.15	182.64	182.61	−25.14

Mixture 4: $HCON(CH_3)_2$ (1) + $[C_4mim][PF_6]$ (2)
(1) Volumetric Properties and Transport Properties

Ref. [183]

x_1	ρ/g · cm^{-3}						η/mPa · s					
	293.15 K	298.15 K	303.15 K	308.15 K	313.15 K	318.15 K	293.15 K	298.15 K	303.15 K	308.15 K	313.15 K	318.15 K
0.0000	1.3689	1.3635	1.3592	1.3555	1.3520	1.3474	265.19	205.57	165.71	133.89	110.79	95.75
0.1002	1.3493	1.3454	1.3409	1.3367	1.3323	1.3283	151.11	114.49	88.20	68.75	54.88	44.63
0.2000	1.3298	1.3256	1.3213	1.3170	1.3127	1.3086	94.60	73.44	58.02	45.91	37.38	31.22
0.3002	1.3095	1.3053	1.3008	1.2965	1.2923	1.2881	59.64	47.42	38.13	30.91	25.68	21.55
0.3999	1.2899	1.2856	1.2812	1.2769	1.2723	1.2683	39.70	32.09	26.34	21.74	18.30	15.57
0.5000	1.2628	1.2586	1.2541	1.2498	1.2453	1.2411	23.41	19.44	16.30	13.74	11.82	10.24
0.6000	1.2244	1.2200	1.2155	1.2111	1.2067	1.2023	12.37	10.57	9.11	7.89	6.94	6.16
0.7000	1.1819	1.1778	1.1732	1.1689	1.1643	1.1599	6.86	6.02	5.32	4.71	4.22	3.82
0.7999	1.1292	1.1247	1.1201	1.1154	1.1110	1.1065	3.71	3.35	3.03	2.73	2.51	2.30
0.8999	1.0550	1.0505	1.0457	1.0409	1.0363	1.0318	1.90	1.77	1.62	1.50	1.40	1.31
1.0000	0.9491	0.9449	0.9402	0.9355	0.9307	0.9259	0.91	0.87	0.82	0.78	0.75	0.72

Mixture 5: CH_3OH (1) + [bmim][PF_6] (2)
(1) Volumetric Properties, Transport Properties, Refractive Index, Speed of Sound and Compressibility Properties

Ref. [184]

p (MPa)	ρ ($kg \cdot m^{-3}$)	φ_V ($cm^3 \cdot mol^{-1}$)	p (MPa)	ρ ($kg \cdot m^{-3}$)	φ_V ($cm^3 \cdot mol^{-1}$)	p (MPa)	ρ ($kg \cdot m^{-3}$)	φ_V ($cm^3 \cdot mol^{-1}$)
$x_2 = 0.0$ (Pure methanol)			$x_2 = 0.02297$			$x_2 = 0.08317$		
$T = 298.15$ K								
0.1	785.85	–	0.22	857.21	188.46	0.36	985.28	197.88
5	790.77	–	5.02	861.80	188.37	4.82	989.26	197.73
10	795.51	–	11.48	867.54	188.39	9.41	993.08	197.61
15	799.99	–	15.62	870.87	188.74	14.93	997.17	197.65
20	804.24	–	20.41	874.60	189.04	19.44	1000.73	197.43
25	808.29	–	25.83	878.69	189.24	25.34	1004.89	197.35
30	812.15	–	29.96	881.70	189.35	29.55	1007.68	197.32
35	815.85	–	35.77	885.67	189.72	35.11	1011.29	197.24
40	819.40	–	39.52	887.91	190.31	39.22	1014.28	196.90
$T = 323.15$ K								
0.1	761.97	–	1.34	833.67	191.79	1.02	961.09	200.67
5	767.60	–	5.14	837.57	192.30	5.04	964.84	200.77
10	772.97	–	10.47	842.79	192.80	10.32	969.57	200.80
15	778.02	–	15.31	847.27	193.17	15.47	973.97	200.78
20	782.79	–	20.01	851.39	193.52	20.06	977.71	200.74
25	787.31	–	25.09	855.59	193.95	25.41	981.86	200.69
30	791.61	–	30.41	859.69	194.59	30.06	985.29	200.66
35	795.72	–	35.08	863.05	195.29	35.12	988.82	200.65
40	799.64	–	39.97	866.33	196.12	39.98	992.03	200.63
$T = 348.15$ K								
0.18	737.16	–	1.15	808.17	194.01	2.03	938.17	202.26
5	743.58	–	5.69	813.81	194.29	6.56	943.22	202.28
10	749.76	–	10.67	819.31	195.13	11.11	947.99	202.26
15	755.52	–	15.31	824.02	196.03	15.47	951.72	202.63
20	760.91	–	20.11	829.05	195.84	20.53	956.28	202.68
25	766.00	–	25.96	834.30	196.84	25.24	960.30	202.71
30	770.81	–	30.03	837.78	197.45	30.76	964.68	202.79
35	775.38	–	34.31	841.14	198.41	35.31	968.16	202.82
40	779.74	–	39.35	845.18	198.81	39.95	971.75	202.69
$T = 373.15$ K								
0.39	710.54	–	1.84	783.65	191.26	1.03	914.74	200.77
5	717.84	–	5.74	789.05	192.66	5.12	919.61	201.45
10	725.10	–	10.58	795.40	193.69	10.04	925.20	201.98
15	731.78	–	15.06	800.93	194.29	15.31	930.89	202.32
20	737.98	–	20.41	807.11	194.81	20.06	935.74	202.52
25	743.78	–	25.31	812.35	195.35	25.14	940.64	202.67
30	749.23	–	30.08	817.08	196.01	30.12	945.15	202.79
35	754.37	–	35.73	822.20	197.13	35.02	949.30	202.93
40	759.24	–	39.93	825.66	198.29	39.97	953.22	203.08

(continued)

p (MPa)	ρ (kg · m^{-3})	φ_V (cm^3 · mol^{-1})	p (MPa)	ρ (kg · m^{-3})	φ_V (cm^3 · mol^{-1})	p (MPa)	ρ (kg · m^{-3})	φ_V (cm^3 · mol^{-1})
T = 398.15 K								
0.76	680.87	–	1.70	757.50	178.78	2.11	894.76	195.82
5	689.18	–	5.62	763.96	182.07	5.55	899.39	197.05
10	697.99	–	11.06	772.39	184.69	10.58	905.87	198.29
15	705.96	–	15.97	779.47	186.00	16.31	912.84	199.18
20	713.25	–	21.63	787.02	186.97	20.96	918.17	199.67
25	719.99	–	26.33	792.78	187.72	26.67	924.32	200.09
30	726.26	–	30.12	797.10	188.39	30.22	927.92	200.30
35	732.13	–	35.17	802.39	189.61	35.34	932.81	200.59
40	737.65	–	39.74	806.73	190.93	39.89	936.87	200.84
x_2 = 0.26147			x_2 = 0.49343			x_2 = 0.75255		
T = 298.15 K								
0.17	1171.68	204.71	0.14	1275.52	206.77	0.14	1332.67	207.76
5.06	1175.25	204.43	5.06	1278.75	206.39	5.07	1335.75	207.33
10.04	1178.79	204.12	10.32	1282.12	205.99	10.02	1338.76	206.91
15.62	1182.63	203.78	15.62	1285.44	205.59	15.61	1342.07	206.45
20.08	1185.62	203.50	20.03	1288.13	205.26	20.04	1344.61	206.09
25.45	1189.11	203.16	24.59	1290.85	204.92	25.84	1347.83	205.65
30.09	1192.03	202.88	29.86	1293.91	204.54	30.01	1350.08	205.33
35.61	1195.40	202.54	34.62	1296.61	204.21	35.41	1352.91	204.94
39.98	1197.98	202.27	39.98	1299.56	203.84	39.94	1355.21	204.62
T = 323.15 K								
0.14	1150.15	207.11	0.17	1253.49	209.84	0.11	1309.84	211.20
4.98	1153.93	206.88	4.96	1256.97	209.44	5.04	1313.28	210.71
10.24	1157.95	206.58	10.03	1260.57	209.01	9.86	1316.54	210.24
14.89	1161.42	206.30	15.23	1264.17	208.57	14.89	1319.85	209.77
20.45	1165.47	205.95	20.03	1267.40	208.17	20.02	1323.13	209.30
25.08	1168.76	205.64	25.61	1271.07	207.71	25.62	1326.59	208.80
30.41	1172.44	205.29	30.06	1273.92	207.35	30.01	1329.21	208.43
35.01	1175.54	204.98	35.41	1277.25	206.93	35.41	1332.33	207.98
39.94	1178.78	204.65	39.96	1280.00	206.58	39.95	1334.87	207.62
T = 348.15 K								
0.24	1127.98	209.53	0.24	1231.81	212.84	0.22	1288.16	214.52
5.62	1132.63	209.31	4.96	1235.51	212.44	4.98	1291.71	214.01
9.94	1136.28	209.09	10.01	1239.39	211.99	10.02	1295.39	213.48
14.26	1139.86	208.83	14.95	1243.12	211.54	15.42	1299.25	212.92
20.08	1144.57	208.45	20.05	1246.90	211.08	19.96	1302.42	212.45
25.14	1148.55	208.09	24.87	1250.40	210.64	25.41	1306.15	211.90
29.98	1152.27	207.74	30.98	1254.75	210.08	30.02	1309.23	211.45
35.06	1156.08	207.36	35.12	1257.64	209.70	35.41	1312.75	210.93
39.94	1159.65	206.98	39.85	1260.88	209.27	39.78	1315.55	210.52
T = 373.15 K								
0.42	1105.35	211.71	0.47	1210.19	215.76	0.42	1266.72	217.86
5.06	1109.81	211.63	5.04	1214.05	215.39	5.02	1270.40	217.33
10.03	1114.49	211.44	10.06	1218.21	214.95	10.02	1274.32	216.76
14.85	1118.93	211.17	14.89	1222.16	214.49	15.62	1278.63	216.13

(continued)

p (MPa)	ρ ($kg \cdot m^{-3}$)	φ_V ($cm^3 \cdot mol^{-1}$)	p (MPa)	ρ ($kg \cdot m^{-3}$)	φ_V ($cm^3 \cdot mol^{-1}$)	p (MPa)	ρ ($kg \cdot m^{-3}$)	φ_V ($cm^3 \cdot mol^{-1}$)
19.74	1123.34	210.85	20.51	1226.68	213.95	20.03	1281.95	215.63
25.65	1128.53	210.42	25.15	1230.35	213.49	24.89	1285.54	215.10
29.47	1131.80	210.12	30.04	1234.16	213.00	29.96	1289.21	214.55
35.16	1136.57	209.65	35.61	1238.42	212.45	35.41	1293.06	213.97
39.92	1140.46	209.24	39.97	1241.70	212.01	39.98	1296.23	213.49
$T = 398.15$ K								
0.89	1082.27	213.38	0.95	1186.54	218.97	0.86	1243.51	221.55
4.78	1086.55	213.48	4.87	1190.40	218.64	4.98	1247.34	221.01
9.98	1092.14	213.42	10.05	1195.39	218.14	10.03	1251.91	220.34
15.02	1097.39	213.23	15.42	1200.42	217.59	15.75	1256.95	219.58
20.04	1102.46	212.94	20.06	1204.66	217.09	19.92	1260.51	219.04
25.74	1108.03	212.54	24.87	1208.95	216.56	25.41	1265.08	218.34
29.98	1112.05	212.22	29.96	1213.36	216.00	29.96	1268.75	217.78
35.15	1116.79	211.79	35.45	1217.98	215.39	35.63	1273.19	217.10
39.98	1121.08	211.38	39.92	1221.63	214.91	39.93	1276.45	216.60

p (MPa)	ρ ($kg \cdot m^{-3}$)	φ_V ($cm^3 \cdot mol^{-1}$)	p (MPa)	ρ ($kg \cdot m^{-3}$)
$x_2 = 0.90612$			$x_2 = 1.0$ (pure $BMIMPF_6$)	
$T = 298.15$ K				
0.12	1354.02	208.12	1.22	1365.43
5.03	1357.04	207.68	5.34	1367.76
10.32	1360.19	207.21	10.36	1370.59
15.62	1363.25	206.76	15.91	1373.69
20.25	1365.84	206.38	20.37	1376.28
25.64	1368.75	205.95	25.18	1378.89
30.06	1371.06	205.62	30.93	1381.99
35.41	1373.76	205.22	35.94	1384.47
39.98	1375.99	204.90	39.51	1386.19
$T = 323.15$ K				
0.17	1330.88	211.68	1.01	1341.59
4.98	1334.18	211.18	5.06	1344.34
10.03	1337.53	210.67	10.31	1347.79
15.62	1341.13	210.12	15.41	1351.04
20.95	1344.43	209.62	20.78	1354.34
25.43	1347.11	209.22	25.06	1356.88
30.14	1349.85	208.81	30.74	1360.13
35.61	1352.90	208.35	35.62	1362.82
39.94	1355.24	208.00	39.93	1365.11
$T = 348.15$ K				
0.21	1309.09	215.13	1.36	1320.13
5.06	1312.64	214.58	5.97	1323.35
10.06	1316.22	214.02	11.02	1327.03
15.61	1320.10	213.41	15.64	1330.17
20.04	1323.12	212.94	20.53	1333.42
25.54	1326.78	212.37	25.44	1336.71
30.08	1329.73	211.92	30.54	1339.89

(*continued*)

p (MPa)	ρ (kg · m^{-3})	φ_V (cm^3 · mol^{-1})	p (MPa)	ρ (kg · m^{-3})	φ_V (cm^3 · mol^{-1})	p (MPa)	ρ (kg · m^{-3})	φ_V (cm^3 · mol^{-1})
35.74	1333.32		211.36		35.87		1343.24	
39.97	1335.93		210.96		39.67		1345.56	
T = 373.15 K								
0.42	1287.73		218.61		0.87		1298.41	
5.78	1291.93		217.94		5.06		1301.65	
10.08	1295.23		217.41		10.06		1305.45	
15.94	1299.63		216.70		15.41		1309.41	
20.08	1302.66		216.22		20.09		1312.79	
25.64	1306.65		215.58		25.74		1316.77	
30.09	1309.76		215.09		30.31		1319.91	
35.64	1313.54		214.49		35.01		1323.07	
39.87	1316.36		214.04		39.98		1326.32	
T = 398.15 K								
0.96	1264.88		222.44		0.73		1278.21	
5.98	1269.43		221.69		6.58		1282.89	
10.36	1273.30		221.05		10.42		1285.92	
14.64	1276.98		220.45		15.54		1289.90	
19.86	1281.35		219.73		20.39		1293.61	
25.08	1285.57		219.04		26.31		1298.06	
30.48	1289.80		218.35		31.34		1301.77	
35.06	1293.27		217.78		36.62		1305.59	
39.86	1296.80		217.21		40.07		1308.05	

Ref. [184]

	V^E_m (cm^3 · mol^{-1})	V^E_m (cm^3 · mol^{-1})	V^E_m (cm^3 · mol^{-1})
p = 0.1 MPa			
x_2	T = 298.15 K	T = 323.15 K	T = 398.15 K
0	0	0	–
0.02297	−0.4778	−0.4979	–
0.08317	−0.8759	−0.9636	–
0.26147	−0.8908	−1.2641	–
0.49343	−0.6450	−1.0092	–
0.75255	−0.2215	−0.4987	–
0.90612	−0.0718	−0.1498	–
1	0	0	–
p = 10 MPa			
0	0	0	0
0.02297	−0.4362	−0.5652	−0.8447
0.08317	−0.8098	−1.3190	−1.9042
0.26147	−0.8551	−1.6490	−1.9987
0.49343	−0.6776	−1.3672	−1.4399

(*continued*)

	V^E_m (cm^3 · mol^{-1})	V^E_m (cm^3 · mol^{-1})	V^E_m (cm^3 · mol^{-1})
0.75255	−0.3525	−0.7155	−0.5476
0.90612	−0.1298	−0.2678	−0.0376
1	0	0	0
P = 30 MPa			
0	0	0	0
0.02297	−0.3635	−0.4223	−0.6520
0.08317	−0.6943	−1.0234	−1.4755
0.26147	−0.7329	−1.3623	−1.6101
0.49343	−0.5781	−1.1443	−1.2081
0.75255	−0.2881	−0.6037	−0.5227
0.90612	−0.0877	−0.2282	−0.0664
1	0	0	0

Ref. [184]

	$\overline{V}_1$ (cm^3 · mol^{-1})	$\overline{V}_2$ (cm^3 · mol^{-1})	$\overline{V}_1$ (cm^3 · mol^{-1})	$\overline{V}_2$ (cm^3 · mol^{-1})	$\overline{V}_1$ (cm^3 · mol^{-1})	$\overline{V}_2$ (cm^3 · mol^{-1})
p = 0.1 MPa						
x_2	T = 298.15 K		T = 373.15 K		T = 398.15 K	
0	40.77	202.53	–	–	–	–
0.02297	40.77	202.81	44.36	212.55	45.87	215.63
0.08317	40.73	203.52	44.31	213.37	45.82	216.54
0.26147	40.35	205.36	43.87	215.50	45.33	218.90
0.49343	39.26	207.16	42.61	217.58	43.93	221.20
0.75255	37.24	208.38	40.29	218.98	41.35	222.76
0.90612	35.66	208.71	38.45	219.36	39.32	223.18
1	34.54	208.76	37.16	219.42	37.89	223.25
p = 10 MPa						
0	40.28	202.04	44.19	208.29	45.91	209.08
0.02297	40.28	202.31	44.19	208.76	45.90	209.69
0.08317	40.24	202.98	44.12	209.93	45.81	211.22
0.26147	39.88	204.72	43.49	212.96	44.99	215.19
0.49343	38.84	206.42	41.69	215.93	42.63	219.07
0.75255	36.94	207.57	38.37	217.94	38.29	221.69
0.90612	35.44	207.88	35.76	218.48	34.87	222.40
1	34.39	207.93	33.92	218.57	32.47	222.52
p = 30 MPa						
0	39.45	201.14	42.77	207.73	44.12	208.88
0.02297	39.45	201.37	42.76	208.10	44.11	209.37
0.08317	39.42	201.95	42.71	209.05	44.05	210.59
0.26147	39.11	203.44	42.20	211.49	43.39	213.74
0.49343	38.22	204.91	40.75	213.89	41.52	216.83
0.75255	36.58	205.90	38.07	215.51	38.07	218.91
0.90612	35.29	206.16	35.96	215.95	35.35	219.47
1	34.39	206.21	34.47	216.02	33.43	219.57

Ref. [185]

x_2	η (mPa · s)	$\Delta\eta$ (mPa · s)	n_D	R_m ($m^3 \cdot mol^{-1}$)	$10^6 \cdot \Delta R$ ($m^3 \cdot mol^{-1}$)
298.15 K					
0.0503	0.913	−10.258	1.3480	10.389	−7.051
0.0993	1.483	−20.058	1.3636	12.580	−11.234
0.1517	2.331	−30.312	1.3710	14.766	−14.120
0.1969	3.059	−39.138	1.3768	16.690	−15.592
0.3024	6.057	−58.462	1.3872	21.253	−16.775
0.3531	8.072	−67.170	1.3892	23.373	−16.696
0.4106	11.027	−76.381	1.3941	25.955	−16.044
0.4510	13.523	−82.424	1.3961	27.707	−15.454
0.5020	17.293	−89.450	1.3986	29.943	−14.508
0.6015	27.411	−100.388	1.4015	34.225	−12.296
0.7082	45.953	−104.410	1.4045	38.861	−9.396
0.7989	66.789	−102.766	1.4066	42.809	−6.646
0.8889	107.939	−80.655	1.4078	46.668	−3.785

Ref. [186]

m_2/ (mol · kg^{-1})	10^3 ρ/ (kg · m^{-3})	u/ (m · s^{-1})	10^6 V_φ/ ($m^3 \cdot mol^{-1}$)	10^{12} k_s/ Pa^{-1}	10^{14} k_φ/ ($m^3 \cdot mol^{-1} \cdot Pa^{-1}$)
T = 298.15 K					
0.0133	0.788011	1103.75	179.14	1041.66	−21.7
0.0288	0.789686	1104.88	182.99	1037.33	−18.71
0.0424	0.791120	1105.70	184.92	1033.91	−16.73
0.0739	0.794425	1107.66	186.45	1025.97	−15.12
0.0966	0.796760	1108.94	187.47	1020.60	−14.13
0.1419	0.801334	1111.23	189.02	1010.60	−12.49
0.1992	0.807045	1113.98	189.78	998.50	−11.29
0.2412	0.811062	1115.85	190.84	990.23	−10.44
0.2796	0.814795	1117.46	190.98	982.85	−9.88
0.404	0.826259	1122.26	192.56	960.94	−8.23
0.461	0.831366	1124.25	193.05	951.66	−7.61
0.6132	0.844434	1129.26	194.32	928.64	−6.26

Ref. [187]

x_2	ρ/g · cm^{-3}	u/m · s^{-1}	$10^6\ V^E$/m^3 · mol^{-1}	ρ/g · cm^{-3}	u/m · s^{-1}	$10^6\ V^E$/m^3 · mol^{-1}	ρ/g · cm^{-3}	u/m · s^{-1}	$10^6\ V^E$/m^3 · mol^{-1}
	T = 298.15 K			T = 303.15 K			T = 308.15 K		
1	1366.657	1442.41	0	1362.401	1430.58	0	1358.304	1418.95	0
0.9208	1357.705	1435.22	−0.095	1353.548	1423.47	−0.115	1349.387	1411.94	−0.112
0.7736	1337.498	1421.10	−0.292	1333.239	1409.46	−0.308	1328.992	1397.95	−0.307
0.6222	1308.958	1400.51	−0.439	1304.585	1388.86	−0.453	1300.243	1377.38	−0.455
0.4572	1263.599	1366.68	−0.594	1259.044	1355.14	−0.603	1254.805	1343.61	−0.631
0.3573	1223.908	1340.69	−0.691	1219.386	1329.07	−0.711	1214.871	1317.49	−0.724
0.2156	1136.500	1280.66	−0.761	1131.894	1268.92	−0.784	1127.297	1257.18	−0.805
0.1869	1111.366	1261.62	−0.754	1106.761	1249.80	−0.779	1102.148	1237.97	−0.800
0.0786	972.943	1180.74	−0.578	968.272	1167.69	−0.598	963.583	1154.62	−0.618
0.0484	914.856	1155.52	−0.467	910.156	1141.54	−0.483	905.442	1127.55	−0.498
0.0223	852.499	1132.08	−0.285	847.778	1116.93	−0.294	843.039	1101.80	−0.303
0	786.531	1102.68	0	781.808	1086.37	0	777.066	1070.11	0
	T = 313.15 K			T = 318.15 K					
1	1354.195	1407.53	0	1350.081	1396.27	0			
0.9208	1345.226	1400.58	−0.111	1341.078	1389.38	−0.113			
0.7736	1324.76	1386.56	−0.310	1320.543	1375.31	−0.316			
0.6222	1295.964	1365.94	−0.467	1291.779	1354.65	−0.491			
0.4572	1250.504	1332.11	−0.656	1246.033	1320.70	−0.667			
0.3573	1210.364	1305.95	−0.741	1205.873	1294.50	−0.760			
0.2156	1122.699	1245.49	−0.828	1118.121	1233.85	−0.855			
0.1869	1097.528	1226.13	−0.824	1092.914	1214.27	−0.851			
0.0786	958.877	1141.58	−0.639	954.071	1128.58	−0.656			
0.0484	900.703	1113.58	−0.515	895.952	1099.67	−0.533			
0.0223	838.277	1086.69	−0.313	833.490	1071.68	−0.324			
0	772.283	1053.96	0	767.474	1037.91	0			

Mixture 6: $[C_4mim][PF_6]$ (1) + $[C_4mim][BF_4]$ (2)
(1) Volumetric Properties and Transport Properties

Ref. [27]
These physicochemical properties data are listed in a mixture 18 of IL 02011-02001.

Ref. [28]
These physicochemical properties data are listed in a mixture 18 of IL 02011-02001.

Ref. [28]
These physicochemical properties data are listed in a mixture 18 of IL 02011-02001.

Ref. [57]
These physicochemical properties data are listed in a mixture 18 of IL 02011-02001.
(2) Excess Molar Enthalpy

Ref. [28]
These physicochemical properties data are listed in a mixture 18 of IL 02011-02001.
Mixture 7: $[C_4mim][PF_6]$ (1) + $[C_4mim][NTf_2]$ (2)
(1) Volumetric Properties

Ref. [57]
These physicochemical properties data are listed in a mixture 21 of IL 02011-03001.
Mixture 8: $CF_3CHFCHFCF_2CF_3$ *threo*-H (1) + $[bmim][PF_6]$ (2)
(1) Phase Equilibrium

Ref. [30]

T/K	x'_1/mol%	x_1/mol%	V'^a/ $cm^3 \cdot mol^{-1}$	V^a/ $cm^3 \cdot mol^{-1}$	$V^{ex/b}$/ $cm^3 \cdot mol^{-1}$	V^{exb}/ $cm^3 \cdot mol^{-1}$
293.4 ± 0.2	97.1 ± 1.0	27.8 ± 1.0	156.0 ± 1.1	189.0 ± 1.3	−2.8 ± 1.1	−4.3 ± 1.3
303.3 ± 0.2	97.8 ± 0.8	24.2 ± 1.2	158.5 ± 1.0	192.7 ± 1.6	−2.8 ± 1.0	−4.1 ± 1.6
313.1 ± 0.2	98.0 ± 1.0	21.7 ± 1.2	161.5 ± 1.0	195.1 ± 1.4	−2.7 ± 1.0	−4.4 ± 1.4
323.3 ± 0.2	98.2 ± 1.0	19.5 ± 1.6	164.6 ± 1.0	198.1 ± 1.6	−2.9 ± 1.0	−4.3 ± 1.6
334.1 ± 0.2	98.2 ± 1.1	17.9 ± 1.8	169.4 ± 1.0	199.9 ± 1.4	−2.8 ± 1.0	−4.9 ± 1.4

[a]Observed molar volume
[b]Volumetric Properties

(2) Cloud Point

Ref. [30]

T/K	x_1/mol %	T/K	x_1/mol %
270 ± 1	40.6 ± 0.5	265 ± 1	69.7 ± 0.5
269 ± 1	50.8 ± 0.5	264 ± 1	80.0 ± 0.5
268 ± 1	60.4 ± 0.5	276 ± 1	90.0 ± 0.5

Mixture 9: $CF_3CHFCHFCF_2CF_3$ *threo*-D (1) + [bmim][PF_6] (2)
(1) Phase Equilibrium

Ref. [30]

T/K	x'_1/mol%	x_1/mol%	V'^{a}/ $cm^3 \cdot mol^{-1}$	V^{a}/ $cm^3 \cdot mol^{-1}$	$V^{ex/b}$/ $cm^3 \cdot mol^{-1}$	V^{exb}/ $cm^3 \cdot mol^{-1}$
293.4 ± 0.2	97.6 ± 0.6	28.9 ± 1.0	157.0 ± 1.0	188.0 ± 1.5	−2.3 ± 1.0	−4.9 ± 1.5
303.3 ± 0.2	98.2 ± 0.8	25.0 ± 1.2	159.7 ± 1.0	191.6 ± 1.4	−2.1 ± 1.0	−4.9 ± 1.4
313.1 ± 0.2	98.4 ± 1.0	22.3 ± 1.4	162.5 ± 1.0	194.2 ± 1.6	−1.6 ± 1.0	−5.1 ± 1.6
323.3 ± 0.2	98.5 ± 1.1	20.2 ± 1.2	165.6 ± 1.0	196.9 ± 1.9	−2.4 ± 1.0	−5.3 ± 1.9
334.1 ± 0.2	98.7 ± 1.0	18.1 ± 1.8	169.3 ± 1.0	198.9 ± 1.4	−2.5 ± 1.0	−6.0 ± 1.4

[a]Observed molar volume
[b]Volumetric Properties

(2) Cloud Point

Ref. [30]

T/K	x_1/mol %	T/K	x_1/mol %
272 ± 1	40.0 ± 0.5	267 ± 1	69.8 ± 0.5
271 ± 1	50.0 ± 0.5	267 ± 1	79.7 ± 0.5
271 ± 1	59.7 ± 0.5	272 ± 1	88.5 ± 0.5

Mixture 10: $CF_3CHFCHFCF_2CF_3$ *erythro*-H (1) + [bmim][PF_6] (2)
(1) Phase Equilibrium

Ref. [30]

T/K	x'_1/mol%	x_1/mol%	V'^{a}/ $cm^3 \cdot mol^{-1}$	V^{a}/ $cm^3 \cdot mol^{-1}$	$V^{ex/b}$/ $cm^3 \cdot mol^{-1}$	V^{exb}/ $cm^3 \cdot mol^{-1}$
282.6 ± 0.2	99.5 ± 0.5	8.2 ± 1.2	157.6 ± 0.7	199.7 ± 1.0	0.2 ± 0.7	−2.0 ± 1.0
294.8 ± 0.2	99.4 ± 0.6	8.3 ± 1.1	161.1 ± 0.6	201.1 ± 0.9	0.2 ± 0.6	−2.3 ± 0.9
303.5 ± 0.2	99.2 ± 0.5	8.6 ± 1.1	163.8 ± 0.4	202.2 ± 1.0	0.3 ± 0.4	−2.4 ± 1.0
313.7 ± 0.2	98.7 ± 0.6	8.6 ± 1.4	166.6 ± 0.4	203.5 ± 1.1	−0.3 ± 0.4	−2.5 ± 1.1
323.8 ± 0.2	98.3 ± 0.7	7.6 ± 1.5	170.2 ± 0.5	205.4 ± 1.2	−0.2 ± 0.5	−2.5 ± 1.2
333.6 ± 0.2	98.2 ± 0.7	7.4 ± 1.2	173.4 ± 0.5	206.5 ± 1.0	−0.6 ± 0.5	−3.0 ± 1.0

[a]Observed molar volume
[b]Volumetric Properties

Mixture 11: $CF_3CHFCHFCF_2CF_3$ *erythro*-D (1) + [bmim][PF_6] (2)
(1) Phase Equilibrium

Ref. [30]

T/K	x'_1/mol%	x_1/mol%	V'^{a}/ $cm^3 \cdot mol^{-1}$	V^{a}/ $cm^3 \cdot mol^{-1}$	$V^{ex/b}$/ $cm^3 \cdot mol^{-1}$	V^{exb}/ $cm^3 \cdot mol^{-1}$
283.7 ± 0.2	99.7 ± 0.3	8.6 ± 1.7	157.4 ± 0.9	198.2 ± 1.3	−0.2 ± 0.9	−3.5 ± 1.3
294.0 ± 0.2	99.6 ± 0.4	8.2 ± 1.7	159.8 ± 0.8	199.6 ± 1.2	−0.7 ± 0.8	−3.8 ± 1.2
303.9 ± 0.2	99.0 ± 0.5	9.3 ± 1.6	162.9 ± 0.9	200.1 ± 1.5	−0.8 ± 0.9	−4.2 ± 1.5

(*continued*)

T/K	x'_1/mol%	x_1/mol%	V'^{a}/ $cm^3 \cdot mol^{-1}$	V^{a}/ $cm^3 \cdot mol^{-1}$	$V^{ex\prime b}$/ $cm^3 \cdot mol^{-1}$	V^{exb}/ $cm^3 \cdot mol^{-1}$
313.5 ± 0.2	99.0 ± 0.6	9.5 ± 1.8	166.2 ± 0.8	201.2 ± 1.5	−0.6 ± 0.8	−4.4 ± 1.5
323.5 ± 0.2	98.7 ± 0.6	8.7 ± 1.8	169.4 ± 0.8	203.3 ± 1.3	−0.8 ± 0.8	−4.1 ± 1.3
333.9 ± 0.2	98.8 ± 0.6	9.3 ± 1.9	172.9 ± 0.7	203.4 ± 1.1	−1.0 ± 0.7	−5.3 ± 1.1

[a]Observed molar volume
[b]Volumetric Properties

Mixture 12: $CF_3CHFCHFCF_2CF_3$ [*threo*-H (87.12 ± 0.6 %) + *erythro*-H (11.89 ± 0.6 %)] (1) + [bmim][PF_6] (2)
(1) Phase Equilibrium

Ref. [30]

T/K	x'_1/mol%	x_1/mol%	V'^{a}/ $cm^3 \cdot mol^{-1}$	V^{a}/ $cm^3 \cdot mol^{-1}$	$V^{ex\prime b}$/ $cm^3 \cdot mol^{-1}$	V^{exb}/ $cm^3 \cdot mol^{-1}$
286.6 ± 0.2	98.3 ± 0.7	21.3 ± 0.4	153.4 ± 1.3	189.0 ± 0.2	−3.2 ± 1.3	−6.5 ± 0.2
291.9 ± 0.2	98.8 ± 0.5	19.6 ± 0.5	156.0 ± 1.0	192.6 ± 0.3	−2.0 ± 1.0	−4.6 ± 0.3
298.3 ± 0.2	98.7 ± 0.5	18.9 ± 0.6	156.8 ± 0.3	191.9 ± 0.3	−2.9 ± 0.3	−6.7 ± 0.3
303.2 ± 0.2	98.5 ± 0.5	17.9 ± 0.7	157.6 ± 0.5	193.6 ± 0.3	−3.6 ± 0.7	−6.1 ± 0.3
313.5 ± 0.2	99.0 ± 0.6	16.6 ± 0.7	161.8 ± 0.8	195.4 ± 0.3	−2.3 ± 0.8	−6.6 ± 0.3
323.4 ± 0.2	98.9 ± 0.6	15.4 ± 0.7	165.0 ± 0.7	197.8 ± 0.3	−2.3 ± 0.7	−6.4 ± 0.3

[a]Observed molar volume
[b]Volumetric Properties

Mixture 13: H_2O (1) + [C_4MIm][PF_6] (2)
(1) Volumetric Properties

Ref. [44]

x_1	$\rho/g \cdot cm^{-3}$	x_1	$\rho/g \cdot cm^{-3}$
298.21 K		323.21 K	
0	1.3683	0	1.3477
0.0431	1.3665	0.0431	1.3456
0.0982	1.3635	0.0982	1.3422
0.1222	1.3621	0.1222	1.3406
0.1566	1.3601	0.1566	1.3385
0.1873	1.3585	0.1873	1.3364
0.2548	1.3555	0.2548	1.3323
0.2971	1.3541	0.2971	1.3299
x_1	$V^E_m/cm^3 \cdot mol^{-1}$	x_1	$V^E_m/cm^3 \cdot mol^{-1}$
0	0	0	0
0.0431	0.051	0.0431	0.106
0.0982	0.182	0.0982	0.307
0.1222	0.238	0.1222	0.394
0.1566	0.301	0.1566	0.475
0.1873	0.317	0.1873	0.559
0.2548	0.244	0.2548	0.616

(*continued*)

x_1	$V^E{}_m$/cm^3 · mol^{-1}	x_1	$V^E{}_m$/cm^3 · mol^{-1}
0.2971	0.115	0.2971	0.591

(2) Solubility

Ref. [188]

T/K	$x_1 \pm \sigma^a$	$10^4 (x_2 \pm \sigma^a)$
288.15	0.229 ± 0.003	1.01 ± 0.06
293.15	0.251 ± 0.001	1.10 ± 0.01
298.15	0.272 ± 0.003	1.21 ± 0.01
303.15	0.293 ± 0.004	1.29 ± 0.01
308.15	0.303 ± 0.004	1.43 ± 0.04
313.15	0.333 ± 0.004	1.58 ± 0.01
318.15	0.348 ± 0.004	1.76 ± 0.02

[a]Standard deviation

Ref. [44]

T/K	x_1	T/K	x_1	T/K	x_1	T/K	x_1
295.76	0.2832	318.62	0.3897	347.00	0.5597	324.72	0.9974
298.15	0.2970	324.82	0.4231	356.41	0.6563	314.24	0.9979
303.27	0.3135	333.76	0.4765	354.97	0.9954	303.76	0.9982
309.97	0.3401	340.76	0.5084	343.75	0.9966	298.15	0.9984

(3) Thermodynamic Properties

Ref. [188]

$(\Delta_{sol}H_m{}^0 \pm \sigma^a)$/kJ · mol^{-1}	$(\Delta_{sol}G_m{}^0 \pm \sigma^a)$/kJ · mol^{-1}	$(\Delta_{sol}S_m{}^0 \pm \sigma^a)$/kJ · mol^{-1}
298.15 K		
12.6 ± 1.5	16.66 ± 0.01	−13.5 ± 5.1

[a]Standard deviation

(4) Cloud Point

Ref. [53]

p (bar)	T (K)	$10^2\ w_2$	$10^2\ w_1$	$10^2 dT/dp$ (K/bar)
1.00	284.75	1.53	100	
1.00	291.15	1.73	100	
1.00	297.25	1.94	100	
1.00	307.55	2.35	100	
1.00	317.95	2.99	100	
1.00	331.45	3.98	100	
1.00	354.65	7.07	100	
1.00	288.35	98.24	100	
1.00	294.85	97.87	100	
1.00	298.75	97.51	100	
1.00	305.05	97.06	100	
1.00	316.15	96.24	100	
1.00	360.35	90.78	100	
2.61	310.59	96.27	100	0.50
5.02	310.60			
11.49	310.62			
28.29	310.70			
35.92	310.76			
41.90	310.78			
1.00	323.65	94.49	100	

(5) Heat Capacity

Ref. [141]

	$x_2 = 0.80$		$x_2 = 0.85$		$x_2 = 0.90$		$x_2 = 0.95$	
T(K)	C_p/(J/(mol · K))	C_p^E/(J/(mol · K))	C_p/(J/(mol · K))	C_p^E/(J/(mol · K))	C_p/(J/(mol · K))	C_p^E/(J/(mol · K))	C_p/(J/(mol · K))	C_p^E/(J/(mol · K))
303.2	342.2	−3.4	360.0	−2.5	377.7	−1.7	395.5	−0.7
308.2	344.5	−3.7	362.6	−2.7	380.5	−1.9	398.6	−0.8
313.2	346.3	−4.0	364.5	−3.0	382.7	−2.0	401.0	−0.9
318.2	348.2	−4.5	366.7	−3.3	385.1	−2.3	403.7	−1.0
323.2	349.7	−4.6	368.1	−3.7	386.8	−2.4	405.6	−1.1
328.2	351.6	−4.9	370.1	−4.0	389.0	−2.7	408.1	−1.1
333.2	354.2	−5.0	372.7	−4.3	391.9	−2.8	411.2	−1.3
338.2	355.9	−5.4	374.6	−4.6	394.0	−3.0	413.6	−1.3
343.2	358.1	−5.8	377.0	−4.9	396.7	−3.2	416.5	−1.4
348.2	360.5	−6.1	379.6	−5.2	399.5	−3.5	419.7	−1.5
353.2	362.9	−6.2	382.0	−5.5	402.3	−3.5	422.6	−1.6

(6) Electrical Conductivity

Ref. [141]

T/(K)	$x_2 = 0.80$	$x_2 = 0.85$	$x_2 = 0.90$	$x_2 = 0.95$
σ (S/m)				
293.2	0.290	0.235	0.199	0.181
298.2	0.364	0.361	0.268	0.231
303.2	0.448	0.382	0.331	0.291
308.2	0.546	0.469	0.401	0.351
313.2	0.652	0.561	0.488	0.423
318.2	0.775	0.675	0.563	0.491
323.2	0.911	0.792	0.678	0.592
333.2	1.060	0.923	0.797	0.701
343.2	1.230	1.060	0.928	0.833
353.2	1.621	1.405	1.217	1.061

Mixture 14: $OCH_2CH_2CH_2CH_2$ (1) + [BMIM][PF_6] (2)
(1) Volumetric Properties, Transport Properties, Refractive Index, Speed of Sound and Compressibility Properties

Ref. [185]

x_2	η (mPa · s)	$\Delta\eta$ (mPa · s)	n_D	R_m ($m^3 \cdot mol^{-1}$)	$10^6 \cdot \Delta R$ ($m^3 \cdot mol^{-1}$)
298.15 K					
0.0502	0.750	−10.332	1.4084	21.468	−2.302
0.0967	1.140	−19.784	1.4117	22.930	−3.844
0.1524	1.732	−30.971	1.4125	24.593	−5.317
0.1999	2.476	−40.294	1.4126	26.039	−6.224
0.2969	4.937	−58.344	1.4130	29.079	−7.246
0.3484	6.979	−67.219	1.4128	30.703	−7.462
0.3994	10.065	−74.917	1.4127	32.315	−7.489
0.4484	14.539	−80.813	1.4126	33.869	−7.371
0.5028	18.503	−88.364	1.4117	35.532	−7.153
0.6047	33.632	−94.796	1.4115	38.797	−6.260
0.7016	52.844	−96.107	1.4110	41.928	−5.054
0.8050	90.817	−80.019	1.4105	45.281	−3.473
0.9595	186.856	−16.674	1.4095	50.183	−0.790

Ref. [185]

x_2	$10^{-3} \cdot \rho$ ($kg \cdot m^{-3}$)	$10^6 V^E_m$ ($m^3 \cdot mol^{-1}$)	u ($m \cdot s^{-1}$)	$10^{12} \cdot \kappa_s$ (Pa^{-1})	$10^{12} \cdot \Delta\kappa_s$ (Pa^{-1})
298.15 K					
0.0199	0.911565	−0.537	1279.16	670.444	−16.656
0.0376	0.936570	−0.994	1279.92	651.770	−29.249
0.0737	0.979097	−1.447	1287.89	615.767	−52.879
0.1500	1.055633	−2.257	1312.64	549.789	−92.744

(continued)

x_2	$10^{-3} \cdot \rho$ (kg · m^{-3})	$10^6 V^E{}_m$ (m^3 · mol^{-1})	u (m · s^{-1})	$10^{12} \cdot \kappa_s$ (Pa^{-1})	$10^{12} \cdot \Delta\kappa_s$ (Pa^{-1})
0.2003	1.095843	−2.485	1329.15	516.540	−108.754
0.2486	1.129411	−2.618	1343.72	490.377	−118.343
0.3004	1.159966	−2.605	1357.31	467.947	−123.057
0.3491	1.185700	−2.585	1368.62	450.256	−124.058
0.4634	1.235907	−2.410	1390.52	418.466	−116.675
0.5808	1.276688	−2.137	1406.94	395.698	−99.223
0.6941	1.306088	−1.491	1418.73	380.389	−75.725
0.7996	1.329247	−0.906	1427.83	369.013	−50.948
0.8991	1.348144	−0.367	1435.73	359.847	−26.031

Ref. [186]

m_2/ (mol · kg^{-1})	10^3 ρ/ (kg · m^{-3})	u/ (m · s^{-1})	$10^6 \cdot V_\varphi$/ (m^3 · mol^{-1})	10^{12} k_s/ Pa^{-1}	10^{14} k_φ/ (m^3 · mol^{-1} · Pa^{-1})
T = 298.15 K					
0.0100	0.883070	1278.69	183.52	692.59	0.66
0.0218	0.884323	1278.92	184.40	691.36	0.83
0.0306	0.885225	1279.01	185.63	690.56	1.34
0.0578	0.887952	1279.15	188.38	688.28	2.44
0.0731	0.889468	1279.18	189.14	687.08	2.8
0.1018	0.892277	1279.19	190.12	684.90	3.28
0.1228	0.894293	1279.25	190.75	683.30	3.47
0.1561	0.897462	1279.26	191.48	680.87	3.76
0.2045	0.901980	1279.29	192.22	677.43	4.03
0.2466	0.905831	1279.30	192.73	674.54	4.21
0.2811	0.908968	1279.31	192.92	672.20	4.32
0.4163	0.921020	1279.56	193.15	663.15	4.50

Mixture 15: CH_3SOCH_3 (1) + [BMIM][PF_6] (2)
(1) Volumetric Properties, Transport Properties, Refractive Index, Speed of Sound and Compressibility Properties

Ref. [185]

x_2	η (mPa · s)	$\Delta\eta$ (mPa · s)	n_D	R_m (m^3 · mol^{-1})	$10^6 \cdot \Delta R$ (m^3 · mol^{-1})
298.15 K					
0.0500	2.749	−9.721	1.4676	21.707	−2.616
0.0997	3.698	−19.211	1.4606	23.284	−4.515
0.1477	4.962	−28.046	1.4546	24.798	−5.874
0.1977	6.489	−37.010	1.4491	26.366	−6.887
0.2995	11.954	−52.939	1.4396	29.536	−8.004
0.3524	15.998	−60.007	1.4361	31.221	−8.151
0.3997	21.172	−64.793	1.4331	32.719	−8.117
0.4328	24.610	−68.293	1.4309	33.745	−8.027

(continued)

x_2	η (mPa · s)	$\Delta\eta$ (mPa · s)	n_D	R_m (m^3 · mol^{-1})	$10^6 \cdot \Delta R$ (m^3 · mol^{-1})
0.5018	36.976	−70.444	1.4278	35.957	−7.572
0.5980	56.306	−71.317	1.4226	38.915	−6.701
0.7043	83.729	−66.229	1.4182	42.222	−5.319
0.7812	117.542	−48.582	1.4158	44.655	−4.086
0.8930	173.317	−16.298	1.4125	48.150	−2.099

Ref. [185]

x_2	$10^{-3} \cdot \rho$ (kg · m^{-3})	$10^6 V^E{}_m$ (m^3 · mol^{-1})	u (m · s^{-1})	$10^{12} \cdot \kappa_s$ (Pa^{-1})	$10^{12} \cdot \Delta\kappa_s$ (Pa^{-1})
298.15 K					
0.0011	1.096211	−0.003	1484.84	413.758	−0.199
0.0024	1.097258	−0.006	1484.88	413.341	−0.536
0.0187	1.109730	−0.011	1483.45	409.484	−3.368
0.0321	1.119442	−0.017	1482.37	406.523	−5.488
0.0552	1.135088	−0.027	1480.53	401.917	−8.648
0.0635	1.140411	−0.028	1479.89	400.387	−9.654
0.0996	1.161891	−0.041	1477.26	394.385	−13.393
0.1530	1.189469	−0.058	1473.65	387.131	−17.297
0.2381	1.225522	−0.094	1469.37	377.935	−21.154
0.3372	1.258279	−0.115	1464.46	370.568	−22.303
0.5586	1.309441	−0.125	1454.94	360.765	−18.222
0.7791	1.342894	−0.108	1448.17	355.074	−10.075
0.9250	1.359297	−0.081	1444.74	352.457	−3.541

Ref. [186]

m_2/ (mol · kg^{-1})	10^3 ρ/ (kg · m^{-3})	u/ (m · s^{-1})	10^6 V_φ/ (m^3 · mol^{-1})	10^{12} k_s/ Pa^{-1}	10^{14} k_φ/ (m^3 · mol^{-1} · Pa^{-1})
T = 298.15 K					
0.0112	1.095976	1485.1	206.73	413.7	6.47
0.0243	1.096798	1485.05	206.76	413.42	6.53
0.044	1.098015	1484.94	206.9	413.02	6.60
0.0716	1.099708	1484.78	207	412.48	6.65
0.0889	1.100744	1484.6	207.08	412.19	6.72
0.1233	1.102805	1484.35	207.11	411.56	6.75
0.147	1.104201	1484.19	207.12	411.12	6.76
0.238	1.109444	1483.57	207.14	409.52	6.78
0.3189	1.113934	1483.03	207.14	408.17	6.80
0.3955	1.118048	1482.55	207.14	406.93	6.81
0.5188	1.124403	1481.76	207.14	405.06	6.83

Mixture 16: CH_3CN (1) + [BMIM][PF_6] (2)
(1) Volumetric Properties, Transport Properties, Refractive Index, Speed of Sound, Compressibility Properties and Conductivity

Ref. [185]

x_2	η (mPa · s)	$\Delta\eta$ (mPa · s)	n_D	R_m ($m^3 \cdot mol^{-1}$)	$10^6 \cdot \Delta R$ ($m^3 \cdot mol^{-1}$)
298.15 K					
0.0496	0.556	−10.281	1.3575	13.085	−4.923
0.1016	0.864	−20.989	1.3687	15.187	−8.375
0.1470	1.187	−30.292	1.3754	16.993	−10.451
0.2022	1.866	−41.282	1.3819	19.223	−12.06
0.2995	3.792	−59.979	1.3902	23.190	−13.255
0.3524	5.308	−69.649	1.3934	25.333	−13.308
0.4036	7.674	−78.136	1.3959	27.402	−13.070
0.4337	9.153	−83.023	1.3974	28.631	−12.804
0.4955	14.443	−90.834	1.3998	31.128	−12.065
0.5892	23.841	−101.269	1.4024	34.892	−10.522
0.6857	40.692	−104.855	1.4048	38.799	−8.480
0.8046	81.532	−89.185	1.4068	43.576	−5.565
0.8055	86.271	−84.650	1.4069	43.622	−5.532
0.9176	154.061	−40.601	1.4084	48.137	−2.448

Ref. [143]

x_2	$\rho/g \cdot cm^{-3}$	$V^E/cm^3 \cdot mol^{-1}$	x_2	$\rho/g \cdot cm^{-3}$	$V^E/cm^3 \cdot mol^{-1}$
$T = 293.15$ K					
0.0000	0.78230	0.0000	0.5996	1.29348	−0.8020
0.1000	0.97406	−0.8408	0.7001	1.31851	−0.5637
0.1997	1.08947	−1.1234	0.8000	1.33925	−0.3641
0.3001	1.16716	−1.2178	0.8992	1.35681	−0.2139
0.4002	1.22132	−1.1104	1.0000	1.37157	0.0000
0.4995	1.26157	−0.9524			
$T = 298.15$ K					
0.0000	0.77693	0.0000	0.5996	1.28908	−0.8495
0.1000	0.96905	−0.8862	0.7001	1.31415	−0.6024
0.1997	1.08437	−1.1585	0.8000	1.33492	−0.3935
0.3001	1.16248	−1.2762	0.8992	1.35254	−0.2390
0.4002	1.21677	−1.1681	1.0000	1.36722	0.0000
0.4995	1.25710	−1.0052			
$T = 303.15$ K					
0.0000	0.77144	0.0000	0.5996	1.28455	−0.8876
0.1000	0.9639	−0.9307	0.7001	1.30969	−0.6332
0.1997	1.07944	−1.2148	0.8000	1.33047	−0.4104
0.3001	1.15773	−1.3365	0.8992	1.34819	−0.2554
0.4002	1.21210	−1.2217	1.0000	1.36286	0.0000

(continued)

x_2	ρ/g · cm^{-3}	V^E/cm^3 · mol^{-1}	x_2	ρ/g · cm^{-3}	V^E/cm^3 · mol^{-1}
0.4995	1.25255	−1.0559			
$T = 308.15$ K					
0.0000	0.76597	0.0000	0.5996	1.28012	−0.9322
0.1000	0.95879	−0.9773	0.7001	1.30551	−0.6928
0.1997	1.07457	−1.2745	0.8000	1.32623	−0.4484
0.3001	1.15301	−1.3973	0.8992	1.34383	−0.2622
0.4002	1.20752	−1.2807	1.0000	1.35856	0.0000
0.4995	1.24807	−1.1101			
$T = 313.15$ K					
0.0000	0.76068	0.0000	0.5996	1.27612	−1.0184
0.1000	0.95396	−1.0334	0.7001	1.30133	−0.7454
0.1997	1.07004	−1.3511	0.8000	1.32221	−0.5093
0.3001	1.14867	−1.4820	0.8992	1.33974	−0.3016
0.4002	1.20329	−1.3648	1.0000	1.35430	0.0000
0.4995	1.24397	−1.1962			
$T = 323.15$ K					
0.0000	0.74973	0.0000	0.5996	1.26765	−1.1335
0.1000	0.94395	−1.1458	0.7001	1.29268	−0.8036
0.1997	1.06053	−1.4899	0.8000	1.31373	−0.5546
0.3001	1.13957	−1.6302	0.8992	1.33133	−0.3230
0.4002	1.19450	−1.5105	1.0000	1.34598	0.0000
0.4995	1.23537	−1.3292			
$T = 333.15$ K					
0.0000	0.73819	0.0000	0.5996	1.25864	−1.2047
0.1000	0.93335	−1.2604	0.7001	1.28391	−0.8589
0.1997	1.05042	−1.6227	0.8000	1.30514	−0.5912
0.3001	1.12980	−1.7562	0.8992	1.32302	−0.3593
0.4002	1.18499	−1.6173	1.0000	1.33770	0.0000
0.4995	1.22609	−1.4148			
$T = 343.15$ K					
0.0000	0.72694	0.0000	0.5996	1.25062	−1.2617
0.1000	0.92324	−1.3789	0.7001	1.27641	−0.9258
0.1997	1.04099	−1.7598	0.8000	1.29784	−0.6345
0.3001	1.12072	−1.8734	0.8992	1.31594	−0.3863
0.4002	1.17643	−1.7273	1.0000	1.33076	0.0000
0.4995	1.21782	−1.4989			

Ref. [186]

m_2/ (mol · kg^{-1})	$10^3\rho$/ (kg · m^{-3})	u/ (m · s^{-1})	10^6 V_φ/ (m^3 · mol^{-1})	10^{12} k_s/ Pa^{-1}	10^{14} · k_φ/ (m^3 · mol^{-1} · Pa^{-1})
T = 298.15 K					
0.0452	0.781327	1279.76	189.75	781.47	−2.28
0.0689	0.783783	1280.19	190.17	778.49	−1.98
0.0835	0.785289	1280.49	190.27	776.64	−1.94
0.1021	0.78719	1280.72	190.57	774.48	−1.63
0.1088	0.787875	1280.83	190.56	773.68	−1.59
0.1409	0.791131	1281.3	190.84	769.93	−1.34
0.1939	0.796403	1282.07	191.27	763.91	−1.04
0.2643	0.803205	1282.96	192.13	756.39	−0.61
0.3566	0.811886	1284.06	192.85	747.02	−0.2
0.4576	0.821093	1285.28	193.44	737.25	0.13

Ref. [187]

x_2	ρ/g · cm^{-3}	u/m · s^{-1}	10^6 V^E/m^3 · mol^{-1}	ρ/g · cm^{-3}	u/m · s^{-1}	10^6 V^E/m^3 · mol^{-1}	ρ/g · cm^{-3}	u/m · s^{-1}	10^6 V^E/m^3 · mol^{-1}
	T = 298.15 K			T = 303.15 K			T = 308.15 K		
1	1366.657	1442.41	0	1362.401	1430.58	0	1358.304	1418.95	0
0.8925	1350.778	1436.54	−0.234	1346.331	1424.49	−0.223	1342.199	1412.62	−0.235
0.7886	1332.552	1427.94	−0.466	1328.297	1415.92	−0.497	1324.054	1404.07	−0.512
0.7285	1320.819	1423.93	−0.657	1316.586	1411.85	−0.700	1312.315	1399.90	−0.722
0.6186	1294.947	1413.51	−0.948	1290.953	1401.05	−1.034	1286.592	1388.69	−1.066
0.5140	1263.518	1402.00	−1.178	1258.712	1389.19	−1.190	1254.638	1376.46	−1.270
0.4241	1228.439	1388.99	−1.277	1223.914	1375.61	−1.333	1219.397	1362.34	−1.382
0.2966	1159.895	1366.45	−1.296	1155.226	1352.23	−1.355	1150.557	1338.10	−1.411
0.2007	1085.082	1341.88	−1.185	1080.247	1326.18	−1.242	1075.415	1310.91	−1.298
0.0935	959.103	1308.88	−0.860	954.020	1291.76	−0.903	948.924	1274.60	−0.947
0.0453	878.471	1293.53	−0.622	873.235	1275.06	−0.650	866.821	1256.66	−0.600
0	776.533	1278.62	0	771.116	1258.73	0	765.660	1238.66	0
	T = 313.15 K			T = 318.15 K					
1	1354.195	1407.53	0	1350.081	1396.27	0			
0.8925	1338.047	1401.01	−0.247	1333.888	1389.50	−0.260			
0.7886	1319.812	1392.36	−0.530	1315.594	1380.84	−0.555			
0.7285	1308.063	1388.01	−0.750	1303.820	1376.25	−0.783			
0.6186	1282.249	1376.50	−1.102	1277.921	1364.43	−1.148			
0.5140	1250.219	1363.79	−1.317	1245.806	1351.22	−1.373			
0.4241	1214.893	1349.24	−1.436	1210.405	1336.27	−1.502			
0.2966	1145.899	1324.03	−1.473	1141.242	1310.05	−1.546			
0.2007	1070.583	1295.79	−1.358	1065.762	1280.89	−1.433			
0.0935	943.817	1257.52	−0.994	938.706	1240.52	−1.056			
0.0453	862.713	1238.37	−0.713	857.428	1220.16	−0.759			
0	760.162	1218.52	0	754.462	1198.32	0			

Ref. [35]

x_2	$10^{-3}c_2$/mol · m^{-3}	ρ/kg · m^{-3}	κ/S · cm^{-1}	$10^4\Lambda$/S m^2 · mol^{-1}
298.15 K				
0.007571	0.1404	795.63	1.485	105.8
0.01576	0.2863	815.17	2.54	88.7
0.02489	0.4415	835.56	3.39	76.8
0.03448	0.5970	855.89	4.05	67.9
0.05838	0.9529	901.77	5.00	52.5
0.08779	1.338	950.71	5.39	40.3
0.1261	1.766	1004.47	5.27	29.8
0.1772	2.238	1062.70	4.68	20.9
0.2521	2.780	1128.61	3.63	13.04
0.3649	3.374	1199.71	2.30	6.82
0.5591	4.038	1278.14	0.995	2.47
0.7291	4.415	1321.97	0.469	1.062
0.8729	4.652	1349.76	0.253	0.543
1	4.815	1368.32	0.1469	0.305

(2) Osmotic Coefficient

Ref. [119]

m_2	Φ	m_2	Φ	m_2	Φ
T = 318.15 K					
0.0116	0.7601	0.0544	0.3229	0.1285	0.4155
0.0190	0.6364	0.0691	0.2780	0.1496	0.5146
0.0321	0.4771	0.0918	0.2867	0.1789	0.6924
0.0366	0.4358	0.1015	0.3139	0.2032	0.8570

Mixture 17: $CFCl_3$ (1) + [bmim][PF_6] (2)
(1) Phase Equilibrium

Ref. [189]

T(K)	x'_1 (mol%)	x_1 (mol%)	V'^{a} (cm^3 · mol^{-1})	V^{a} (cm^3 · mol^{-1})	$V^{ex\prime\ b}$ (cm^3 · mol^{-1})	$V^{ex\ b}$ (cm^3 · mol^{-1})
282.7 ± 0.2	99.0 ± 0.5	21.0 ± 3.0	91.6 ± 1.2	177.8 ± 3.2	−0.4 ± 1.2	−3.8 ± 3.2
293.5 ± 0.2	99.5 ± 0.5	20.6 ± 1.2	92.5 ± 0.6	179.0 ± 2.5	−0.4 ± 0.6	−4.5 ± 2.5
303.7 ± 0.2	98.0 ± 0.4	20.5 ± 1.6	96.0 ± 0.5	180.5 ± 2.4	−0.2 ± 0.5	−4.4 ± 2.4
313.8 ± 0.2	98.4 ± 0.4	20.4 ± 1.6	97.3 ± 0.4	181.9 ± 2.4	0.0 ± 0.4	−4.4 ± 2.4
323.9 ± 0.2	100.0–0.4	19.3 ± 2.4	96.9 ± 0.4	184.9 ± 2.6	−0.4 ± 0.4	−4.2 ± 2.6
333.6 ± 0.2	100.0–0.4	17.8 ± 2.4	97.3 ± 0.4	188.6 ± 2.6	−0.3 ± 0.4	−3.5 ± 2.6

[a]Observed molar volume [cm^3 · mol^{-1}]
[b]Volumetric Properties [cm^3 · mol^{-1}]

Mixture 18: $CHCl_3$ (1) + [bmim][PF_6] (2)

(1) Phase Equilibrium

Ref. [189]

T(K)	x'_1 (mol%)	x_1 (mol%)	V'^a ($cm^3 \cdot mol^{-1}$)	V^a ($cm^3 \cdot mol^{-1}$)	$V^{ex\prime\ b}$ ($cm^3 \cdot mol^{-1}$)	$V^{ex\ b}$ ($cm^3 \cdot mol^{-1}$)
283.1 ± 0.2	100.0-0.5	73.5 ± 2.6	79.0 ± 1.2	107.8 ± 4.4	0.0 ± 1.2	−4.9 ± 4.4
293.2 ± 0.2	100.0-0.5	72.6 ± 2.6	79.8 ± 1.2	109.7 ± 3.5	−0.3 ± 1.2	−5.2 ± 3.5
303.2 ± 0.2	100.0-0.5	71.9 ± 2.5	80.9 ± 1.0	111.8 ± 3.1	−0.2 ± 1.0	−5.1 ± 3.1
313.0 ± 0.2	99.6 ± 0.4	72.2 ± 2.4	82.6 ± 1.2	112.8 ± 3.1	−0.1 ± 1.2	−4.9 ± 3.1
323.3 ± 0.2	99.4 ± 0.6	71.9 ± 2.4	84.0 ± 1.2	114.2 ± 3.0	0.0 ± 1.2	−5.0 ± 3.0
333.7 ± 0.2	99.6 ± 0.4	70.6 ± 1.6	85.1 ± 1.2	116.0 ± 3.0	0.1 ± 1.2	−6.0 ± 3.0

[a]Observed molar volume [$cm^3\ mol^{-1}$]
[b]Volumetric Properties [$cm^3\ mol^{-1}$]

Mixture 19: $CDCl_3$ (1) + [bmim][PF_6] (2)

(1) Phase Equilibrium

Ref. [189]

T(K)	x'_1 (mol%)	x_1 (mol%)	V'^a ($cm^3 \cdot mol^{-1}$)	V^a ($cm^3 \cdot mol^{-1}$)	$V^{ex\prime\ b}$ ($cm^3 \cdot mol^{-1}$)	$V^{ex\ b}$ ($cm^3 \cdot mol^{-1}$)
283.1 ± 0.2	99.6 ± 0.4	74.2 ± 1.4	79.6 ± 1.2	107.9 ± 2.8	−0.1 ± 1.2	−3.9 ± 2.8
293.2 ± 0.2	99.4 ± 0.5	73.5 ± 1.4	80.3 ± 1.4	110.3 ± 2.5	−0.3 ± 1.4	−3.3 ± 2.5
303.2 ± 0.2	99.5 ± 0.4	72.2 ± 2.0	81.3 ± 1.2	111.8 ± 3.0	−0.1 ± 1.2	−3.4 ± 3.0
313.1 ± 0.2	98.9 ± 0.4	73.3 ± 1.6	83.1 ± 0.7	112.7 ± 3.4	−0.2 ± 0.7	−3.3 ± 3.4
323.3 ± 0.2	99.2 ± 0.4	71.3 ± 2.0	84.1 ± 1.2	115.6 ± 3.0	0.2 ± 1.2	−3.9 ± 3.0
333.9 ± 0.2	99.5 ± 0.5	70.8 ± 0.9	85.3 ± 1.4	116.3 ± 2.8	0.4 ± 1.4	−5.3 ± 2.8

[a]Observed molar volume [$cm^3\ mol^{-1}$]
[b]Volumetric Properties [$cm^3\ mol^{-1}$]

Mixture 20: $CFCl_2CF_2Cl$ (1) + [bmim][PF_6] (2)

(1) Phase Equilibrium

Ref. [189]

T(K)	x'_1 (mol%)	x_1 (mol%)	V'^a ($cm^3 \cdot mol^{-1}$)	V^a ($cm^3 \cdot mol^{-1}$)	$V^{ex\prime\ b}$ ($cm^3 \cdot mol^{-1}$)	$V^{ex\ b}$ ($cm^3 \cdot mol^{-1}$)
282.9 ± 0.2	98.3 ± 1.2	9.7 ± 1.2	117.8 ± 1.2	194.2 ± 2.0	−0.9 ± 1.2	−3.0 ± 2.0
288.5 ± 0.2	98.6 ± 1.0	8.4 ± 1.2	118.1 ± 1.0	196.9 ± 2.0	−1.2 ± 1.0	−2.2 ± 2.0
293.4 ± 0.2	98.7 ± 1.0	7.6 ± 1.7	119.0 ± 1.0	197.9 ± 2.0	−1.2 ± 1.0	−2.5 ± 2.0
303.3 ± 0.2	98.1 ± 1.2	8.7 ± 2.0	121.3 ± 1.0	197.5 ± 2.4	−1.1 ± 1.0	−3.3 ± 2.4
313.6 ± 0.2	98.8 ± 1.0	7.2 ± 2.2	123.0 ± 1.0	200.6 ± 3.0	−0.8 ± 1.0	−2.9 ± 3.0
323.6 ± 0.2	98.5 ± 0.8	8.8 ± 2.7	124.9 ± 0.9	200.7 ± 3.0	−1.2 ± 0.9	−2.7 ± 3.0
333.7 ± 0.2	99.1 ± 0.9	10.7 ± 2.8	126.6 ± 0.8	200.4 ± 3.0	−1.2 ± 0.8	−2.8 ± 3.0

[a]Observed molar volume [$cm^3\ mol^{-1}$]
[b]Volumetric Properties [$cm^3\ mol^{-1}$]

Mixture 21: $CHCl_2CF_3$ (1) + [bmim][PF_6] (2)
(1) Phase Equilibrium

Ref. [189]

T(K)	x'_1 (mol%)	x_1 (mol%)	V'^a ($cm^3 \cdot mol^{-1}$)	V^a ($cm^3 \cdot mol^{-1}$)	$V^{ex\prime\ b}$ ($cm^3 \cdot mol^{-1}$)	$V^{ex\ b}$ ($cm^3 \cdot mol^{-1}$)
283.0 ± 0.2	99.4 ± 0.5	77.2 ± 1.0	102.3 ± 0.6	121.1 ± 1.0	−0.1 ± 0.6	−4.4 ± 1.0
288.5 ± 0.2	99.2 ± 0.8	76.1 ± 0.8	103.6 ± 0.6	122.8 ± 0.8	0.1 ± 0.6	−4.8 ± 0.8
293.4 ± 0.2	99.3 ± 0.7	74.9 ± 1.1	104.8 ± 0.7	124.3 ± 0.8	0.5 ± 0.7	−5.3 ± 0.8
303.4 ± 0.2	72.6 ± 0.8	99.7 ± 0.3	127.7 ± 0.6	105.9 ± 0.8	−6.0 ± 0.8	0.2 ± 0.6
313.7 ± 0.2	70.4 ± 0.8	99.8 ± 0.2	130.9 ± 0.6	108.1 ± 0.7	−6.9 ± 0.7	0.4 ± 0.6
323.7 ± 0.2	67.7 ± 1.1	99.8 ± 0.2	134.6 ± 0.8	110.1 ± 0.7	−7.7 ± 0.7	0.5 ± 0.8
333.7 ± 0.2	65.1 ± 1.2	99.9 ± 0.1	138.4 ± 0.9	112.2 ± 0.6	−8.4 ± 0.6	0.4 ± 0.9

[a]Observed molar volume [$cm^3 \cdot mol^{-1}$]
[b]Volumetric Properties [$cm^3 \cdot mol^{-1}$]

Mixture 22: C_6H_6 (1) + [BMIM][PF_6] (2)
(1) Volumetric Properties

Ref. [143]

x_2	ρ/g · cm^{-3}	V^E/$cm^3 \cdot mol^{-1}$	x_2	ρ/g · cm^{-3}	V^E/$cm^3 \cdot mol^{-1}$
T = 293.15 K					
0.0000	0.87823	0.0000	0.7000	1.30429	−1.2330
0.3400	1.16634	−2.1019	0.8003	1.32979	−0.8227
0.3995	1.19759	−2.2168	0.9001	1.35151	−0.3427
0.4999	1.23920	−1.8918	1.0000	1.37157	0.0000
0.5999	1.27425	−1.5676			
T = 298.15 K					
0.0000	0.87311	0.0000	0.7000	1.29995	−1.2817
0.3400	1.16180	−2.1780	0.8003	1.32597	−0.8600
0.3995	1.19315	−2.2668	0.9001	1.34729	−0.3696
0.4999	1.23486	−1.9633	1.0000	1.36722	0.0000
0.5999	1.26999	−1.6338			
T = 303.15 K					
0.0000	0.86759	0.0000	0.7000	1.29560	−1.3436
0.3400	1.15716	−2.2683	0.8003	1.32158	−0.8963
0.3995	1.18862	−2.3615	0.9001	1.34285	−0.3780
0.4999	1.23038	−2.0462	1.0000	1.36286	0.0000
0.5999	1.26561	−1.7107			
T = 308.15 K					
0.0000	0.86221	0.0000	0.7000	1.29134	−1.4079
0.3400	1.15256	−2.3524	0.8003	1.31728	−0.9356
0.3995	1.18416	−2.4538	0.9001	1.33855	−0.3975
0.4999	1.22601	−2.1323	1.0000	1.35856	0.0000
0.5999	1.26130	−1.7866			

(continued)

x_2	ρ/g · cm^{-3}	V^E/cm^3 · mol^{-1}	x_2	ρ/g · cm^{-3}	V^E/cm^3 · mol^{-1}
T = 313.15 K					
0.0000	0.85682	0.0000	0.7000	1.28735	−1.5058
0.3400	1.14835	−2.4810	0.8003	1.31309	−0.9866
0.3995	1.18004	−2.5856	0.9001	1.33445	−0.4415
0.4999	1.22202	−2.2638	1.0000	1.35430	0.0000
0.5999	1.25739	−1.9120			
T = 323.15 K					
0.0000	0.84605	0.0000	0.7000	1.27903	−1.6383
0.3400	1.13950	−2.6858	0.8003	1.30476	−1.0732
0.3995	1.17145	−2.8031	0.9001	1.32620	−0.4960
0.4999	1.21365	−2.4703	1.0000	1.34598	0.0000
0.5999	1.24914	−2.0946			
T = 333.15 K					
0.0000	0.83576	0.0000	0.7000	1.27076	−1.7626
0.3400	1.12994	−2.7802	0.8003	1.29668	−1.1839
0.3995	1.16213	−2.9088	0.9001	1.31785	−0.5263
0.4999	1.20495	−2.6144	1.0000	1.33770	0.0000
0.5999	1.24067	−2.2308			
T = 343.15 K					
0.0000	0.82489	0.0000	0.7000	1.26366	−1.9244
0.3400	1.12087	−2.9086	0.8003	1.28979	−1.3135
0.3995	1.15345	−3.0513	0.9001	1.31077	−0.5659
0.4999	1.19696	−2.7805	1.0000	1.33076	0.0000
0.5999	1.23313	−2.3938			

(2) Vapor Pressure and Activity Coefficient

Ref. [144]

T (K)	p^{exp} (kPa)	p^{nrtl} (kPa)	RD (p)	γ_1^{exp}	γ_1^{NRTL}
x_1 = 0.6600					
303.56	15.1	15.2	0.43	1.3984	1.4176
308.28	18.5	18.6	0.37	1.3984	1.4143
313.17	22.6	22.7	0.54	1.3939	1.4108
317.93	27.6	27.5	0.53	1.4075	1.4073
323.10	33.8	33.5	1.01	1.4116	1.4035
328.36	40.5	40.6	0.25	1.3894	1.3995
333.34	49.2	48.4	1.57	1.4128	1.3957
338.31	57.1	57.4	0.50	1.3787	1.3919
343.42	67.8	67.9	0.15	1.3799	1.3879
x_1 = 0.6000					
303.53	14.7	14.6	0.95	1.4946	1.4992
308.57	17.9	18.1	1.01	1.4637	1.4963
313.26	21.9	21.9	0.22	1.4761	1.4935
318.17	26.5	26.7	0.69	1.4674	1.4904

(continued)

T (K)	p^{exp} (kPa)	p^{nrtl} (kPa)	RD (p)	γ_1^{exp}	γ_1^{NRTL}
323.11	32.7	32.2	1.40	1.4984	1.4871
328.28	39.5	39.0	1.23	1.4930	1.4836
333.19	46.8	46.4	0.76	1.4826	1.4801
338.25	55.5	55.2	0.49	1.4752	1.4765
343.05	64.3	64.7	0.65	1.4544	1.4729
x_1 = 0.5001					
303.44	13.2	13.1	0.50	1.5967	1.6289
308.47	16.0	16.3	2.02	1.5609	1.6276
313.36	20.5	20.0	2.45	1.6421	1.6260
318.17	24.5	24.2	1.07	1.6192	1.6242
323.05	29.3	29.2	0.18	1.6043	1.6221
328.25	35.8	35.5	0.94	1.6169	1.6197
333.18	42.1	42.3	0.43	1.5926	1.6173
338.19	49.5	50.2	1.48	1.5738	1.6146
343.20	58.8	59.3	0.88	1.5816	1.6117
x_1 = 0.3999					
303.37	11.2	11.3	0.57	1.6706	1.7520
308.29	13.6	14.0	2.58	1.6507	1.7526
313.25	17.4	17.2	1.35	1.7312	1.7528
318.10	20.9	20.9	0.19	1.7159	1.7526
323.20	25.4	25.4	0.04	1.7159	1.7521
328.25	30.5	30.7	0.55	1.7096	1.7513
333.27	36.7	36.7	0.04	1.7200	1.7502
338.19	43.8	43.5	0.64	1.7327	1.7489
343.17	51.6	51.4	0.42	1.7286	1.7473
x_1 = 0.3004					
303.08	9.1	8.9	1.75	1.7919	1.8693
308.05	11.3	11.2	0.64	1.7975	1.8721
313.17	14.1	13.7	2.58	1.8520	1.8742
317.90	16.8	16.8	0.21	1.8174	1.8759
322.80	20.2	20.3	0.36	1.8151	1.8772
328.13	24.5	24.6	0.50	1.8189	1.8782
333.03	29.9	29.5	1.27	1.8573	1.8788
338.05	35.1	35.1	0.00	1.8362	1.8791
342.95	40.8	41.3	1.13	1.8173	1.8790
x_1 = 0.2004					
303.20	6.5	6.3	2.48	1.8727	1.9839
308.11	8.1	7.9	2.64	1.9146	1.9886
313.08	10.1	9.7	3.52	1.9592	1.9928
317.97	12.2	11.9	2.46	1.9554	1.9965
322.83	14.5	14.4	0.82	1.9355	1.9997
328.13	17.6	17.7	0.64	1.9181	2.0029
333.05	21.3	21.1	0.96	1.9576	2.0052
338.04	24.8	25.1	1.07	1.9236	2.0072

(*continued*)

T (K)	p^{exp} (kPa)	p^{nrtl} (kPa)	RD (p)	$\gamma_1{}^{exp}$	$\gamma_1{}^{NRTL}$
342.97	29.5	29.7	0.52	1.9393	2.0088
$x_1 = 0.1001$					
303.10	3.5	3.3	4.73	1.9730	2.0973
308.26	4.3	4.2	2.63	1.9843	2.1044
313.36	5.3	5.2	1.95	2.0067	2.1107
318.10	6.4	6.3	1.50	2.0217	2.1162
323.11	7.7	7.7	0.34	2.0171	2.1215
328.20	9.4	9.3	1.05	2.0473	2.1263
333.17	11.2	11.1	0.48	2.0474	2.1306
338.19	13.4	13.3	0.80	2.0638	2.1345
343.11	15.5	15.7	1.31	2.0282	2.1378

Mixture 23: 1-C_3H_7OH (1) + [BMIM][PF_6] (2)
(1) Phase Equilibrium

Ref. [190]

T/K	x_2 (IL-poor phase)	x_2 (IL-rich phase)	T/K	x_2 (IL-poor phase)	x_2 (IL-rich phase)
278.15	0.0004	0.7864	313.15	0.0015	0.6033
283.15	0.0005	0.7667	318.15	0.0018	0.5760
288.15	0.0006	0.7437	323.15	0.0024	0.5497
293.15	0.0007	0.7198	328.15	0.0032	0.5303
298.15	0.0008	0.6946	333.15	0.0049	0.5085
303.15	0.0010	0.6664	338.15	0.0090	0.4842
308.15	0.0012	0.6380	343.15	0.0101	0.4706

Ref. [191]

	[bmim][PF_6]-rich phase			1-propanol-rich phase		
T/K	$x_1 \pm \Delta x_1$	$w_1 \pm \Delta w_1$		$x_2 \pm \Delta x_2$	$w_2 \pm \Delta w_2$	
$p = 0.1$ MPa						
293.2	0.2755 ± 0.0022	0.0744 ± 0.0020	CP			
293.2	0.2721 ± 0.0022	0.0732 ± 0.0020	CP			
303.2	0.3285 ± 0.0022	0.0938 ± 0.0020	CP			
303.2	0.3293 ± 0.0022	0.0941 ± 0.0020	CP			
313.1	0.3838 ± 0.0022	0.1164 ± 0.0020	CP			
313.1	0.3831 ± 0.0022	0.1161 ± 0.0020	CP			
323.1	0.4517 ± 0.0022	0.1484 ± 0.0020	CP			
323.1	0.4508 ± 0.0022	0.1479 ± 0.0020	CP			
333.0	0.5279 ± 0.0022	0.1912 ± 0.0020	CP	0.0076 ± 0.0011	0.0349 ± 0.0045	CP
333.0	0.5243 ± 0.0022	0.1890 ± 0.0020	CP	0.0084 ± 0.0011	0.0384 ± 0.0045	CP
343.0	0.6006 ± 0.0022	0.2413 ± 0.0020	CP	0.0147 ± 0.0011	0.0660 ± 0.0045	CP
343.0	0.5992 ± 0.0022	0.2402 ± 0.0020	CP	0.0169 ± 0.0011	0.0751 ± 0.0045	CP
352.4	0.6439 ± 0.0022	0.2766 ± 0.0020	CP	0.0288 ± 0.0011	0.1230 ± 0.0045	CP

(*continued*)

	[bmim][PF_6]-rich phase			1-propanol-rich phase		
T/K	$x_1 \pm \Delta x_1$	$w_1 \pm \Delta w_1$		$x_2 \pm \Delta x_2$	$w_2 \pm \Delta w_2$	
352.4	0.6483 ± 0.0022	0.2805 ± 0.0020	CP	0.0297 ± 0.0011	0.1262 ± 0.0045	CP
361.9	0.7498 ± 0.0022	0.3879 ± 0.0020	CP			
361.9	0.7520 ± 0.0022	0.3907 ± 0.0020	CP			
262.6				0.0004 ± 0.0001	0.0020 ± 0.0003	UV
262.6				0.0004 ± 0.0001	0.0020 ± 0.0003	UV
262.6				0.0004 ± 0.0001	0.0020 ± 0.0003	UV
267.6				0.0006 ± 0.0001	0.0029 ± 0.0004	UV
267.6				0.0005 ± 0.0001	0.0026 ± 0.0004	UV
267.6				0.0005 ± 0.0001	0.0026 ± 0.0004	UV
271.4				0.0007 ± 0.0001	0.0031 ± 0.0005	UV
271.4				0.0006 ± 0.0001	0.0030 ± 0.0005	UV
271.4				0.0006 ± 0.0001	0.0030 ± 0.0005	UV
274.3				0.0007 ± 0.0001	0.0035 ± 0.0005	UV
274.3				0.0007 ± 0.0001	0.0033 ± 0.0005	UV
274.3				0.0007 ± 0.0001	0.0034 ± 0.0005	UV
278.2				0.0008 ± 0.0001	0.0036 ± 0.0006	UV
278.2				0.0008 ± 0.0001	0.0037 ± 0.0006	UV
278.2				0.0007 ± 0.0001	0.0034 ± 0.0005	UV
284.1				0.0009 ± 0.0002	0.0045 ± 0.0007	UV
284.1				0.0009 ± 0.0002	0.0044 ± 0.0007	UV
284.1				0.0009 ± 0.0001	0.0044 ± 0.0007	UV
289.0				0.0012 ± 0.0002	0.0055 ± 0.0009	UV
289.0				0.0012 ± 0.0002	0.0055 ± 0.0009	UV
289.0				0.0012 ± 0.0002	0.0056 ± 0.0009	UV
294.0				0.0016 ± 0.0002	0.0074 ± 0.0011	UV
294.0				0.0015 ± 0.0002	0.0068 ± 0.0011	UV
294.0				0.0014 ± 0.0002	0.0068 ± 0.0010	UV
298.9				0.0019 ± 0.0003	0.0088 ± 0.0014	UV
298.9				0.0019 ± 0.0003	0.0088 ± 0.0014	UV
298.9				0.0019 ± 0.0003	0.0088 ± 0.0014	UV
303.9				0.0023 ± 0.0004	0.0107 ± 0.0016	UV
303.9				0.0023 ± 0.0004	0.0110 ± 0.0017	UV
303.9				0.0023 ± 0.0004	0.0108 ± 0.0017	UV
308.8				0.0025 ± 0.0004	0.0117 ± 0.0018	UV
308.8				0.0026 ± 0.0004	0.0122 ± 0.0019	UV
308.8				0.0026 ± 0.0004	0.0122 ± 0.0019	UV
313.8				0.0032 ± 0.0005	0.0151 ± 0.0023	UV
313.8				0.0031 ± 0.0005	0.0145 ± 0.0022	UV
313.8				0.0033 ± 0.0005	0.0152 ± 0.0024	UV

Mixture 24: 1-C4H9OH (1) + [bmim][PF6] (2)

(1) Phase Equilibrium

Ref. [190]

T/K	x_2 (IL-poor phase)	x_2 (IL-rich phase)	T/K	x_2 (IL-poor phase)	x_2 (IL-rich phase)
278.15	0.0003	0.8730	313.15	0.0010	0.7597
283.15	0.0004	0.8610	318.15	0.0013	0.7387
288.15	0.0004	0.8465	323.15	0.0016	0.7164
293.15	0.0005	0.8299	328.15	0.0020	0.7069
298.15	0.0007	0.8124	333.15	0.0025	0.6980
303.15	0.0008	0.7932	338.15	0.0032	0.6893
308.15	0.0009	0.7718	343.15	0.0040	0.6785

Ref. [191]

	[bmim][PF_6]-rich phase			1-butanol-rich phase		
T/K	$x_1 \pm \Delta x_1$	$w_1 \pm \Delta w_1$		$x_2 \pm \Delta x_2$	$w_2 \pm \Delta w_2$	
$p = 0.1$ MPa						
283.2	0.1371 ± 0.0017	0.0398 ± 0.0006	CP			
283.2	0.1404 ± 0.0017	0.0409 ± 0.0006	CP			
288.2	0.1621 ± 0.0017	0.0480 ± 0.0006	CP			
288.2	0.1600 ± 0.0017	0.0473 ± 0.0006	CP			
291.2	0.1686 ± 0.0017	0.0502 ± 0.0006	CP			
291.2	0.1670 ± 0.0017	0.0497 ± 0.0006	CP			
293.2	0.1779 ± 0.0017	0.0534 ± 0.0006	CP			
293.2	0.1755 ± 0.0017	0.0526 ± 0.0006	CP			
298.2	0.2107 ± 0.0017	0.0651 ± 0.0006	CP			
298.2	0.2108 ± 0.0017	0.0651 ± 0.0006	CP			
308.1	0.2357 ± 0.0017	0.0745 ± 0.0006	CP			
308.1	0.2352 ± 0.0017	0.0743 ± 0.0006	CP			
318.1	0.3038 ± 0.0017	0.1022 ± 0.0006	CP			
318.1	0.3036 ± 0.0017	0.1021 ± 0.0006	CP			
333.0	0.3509 ± 0.0017	0.1236 ± 0.0006	CP			
333.0	0.3514 ± 0.0017	0.1238 ± 0.0006	CP			
347.2	0.4088 ± 0.0017	0.1528 ± 0.0006	CP			
347.2	0.4094 ± 0.0017	0.1531 ± 0.0006	CP			
357.1	0.4447 ± 0.0017	0.1728 ± 0.0006	CP			
357.1	0.4438 ± 0.0017	0.1722 ± 0.0006	CP			
361.9	0.4876 ± 0.0017	0.1989 ± 0.0006	CP	0.0127 ± 0.0003	0.0469 ± 0.0010	CP
361.9	0.4868 ± 0.0017	0.1983 ± 0.0006	CP	0.0131 ± 0.0003	0.0485 ± 0.0010	CP
278.2				0.0004 ± 0.0001	0.0016 ± 0.0002	UV

(*continued*)

	[bmim][PF$_6$]-rich phase		1-butanol-rich phase		
T/K	$x_1 \pm \Delta x_1$	$w_1 \pm \Delta w_1$	$x_2 \pm \Delta x_2$	$w_2 \pm \Delta w_2$	
278.2			0.0004 ± 0.0001	0.0015 ± 0.0002	UV
278.2			0.0004 ± 0.0001	0.0016 ± 0.0002	UV
281.1			0.0005 ± 0.0001	0.0017 ± 0.0003	UV
281.1			0.0005 ± 0.0001	0.0018 ± 0.0003	UV
281.1			0.0005 ± 0.0001	0.0018 ± 0.0003	UV
284.1			0.0004 ± 0.0001	0.0017 ± 0.0003	UV
284.1			0.0004 ± 0.0001	0.0017 ± 0.0003	UV
284.1			0.0004 ± 0.0001	0.0017 ± 0.0003	UV
289.0			0.0005 ± 0.0001	0.0020 ± 0.0003	UV
289.0			0.0005 ± 0.0001	0.0020 ± 0.0003	UV
289.0			0.0005 ± 0.0001	0.0020 ± 0.0003	UV
293.9			0.0007 ± 0.0001	0.0025 ± 0.0004	UV
293.9			0.0007 ± 0.0001	0.0025 ± 0.0004	UV
293.9			0.0007 ± 0.0001	0.0025 ± 0.0004	UV
298.9			0.0008 ± 0.0001	0.0030 ± 0.0005	UV
298.9			0.0008 ± 0.0001	0.0030 ± 0.0005	UV
298.9			0.0008 ± 0.0001	0.0030 ± 0.0005	UV
303.8			0.0011 ± 0.0002	0.0040 ± 0.0006	UV
303.8			0.0011 ± 0.0002	0.0041 ± 0.0006	UV
303.8			0.0010 ± 0.0002	0.0040 ± 0.0006	UV
308.8			0.0012 ± 0.0002	0.0045 ± 0.0007	UV
308.8			0.0012 ± 0.0002	0.0046 ± 0.0007	UV
308.8			0.0012 ± 0.0002	0.0045 ± 0.0007	UV
313.8			0.0016 ± 0.0002	0.0060 ± 0.0009	UV
313.8			0.0016 ± 0.0002	0.0059 ± 0.0009	UV
313.8			0.0016 ± 0.0002	0.0060 ± 0.0009	UV
323.7			0.0024 ± 0.0004	0.0091 ± 0.0014	UV
323.7			0.0024 ± 0.0004	0.0093 ± 0.0014	UV
323.7			0.0024 ± 0.0004	0.0092 ± 0.0014	UV
333.7			0.0040 ± 0.0006	0.0150 ± 0.0023	UV
333.7			0.0039 ± 0.0006	0.0148 ± 0.0023	UV
333.7			0.0038 ± 0.0006	0.0146 ± 0.0023	UV
333.7			0.0038 ± 0.0006	0.0144 ± 0.0022	UV

Ref. [192]

T/K	x_2' (IL phase)	x_2'' (alcohol phase)	T/K	x_2' (IL phase)	x_2'' (alcohol phase)
288.15	0.854	0.0021	318.15	0.738	0.0026
298.15	0.820	0.0024	328.15	0.684	0.0037
308.15	0.780	0.0024			

Ref. [193]

T/K	x_{11}/mol % liquid $i = 1$	x_{11}/ mol % liquid $i = 1^a$	x_{12}/mol % liquid $i = 2$	$\overline{V}_1{}^b$ liquid $i = 1$	$\overline{V}_2{}^b$ liquid $i = 2$	$\overline{V}_1^{exc}$ liquid $i = 1$	$\overline{V}_2^{exc}$ liquid $i = 2$
285.7	20.4 ± 1.5	21.4	99.7 ± 0.3	177.8 ± 1.6	89.6 ± 0.5	−4.8 ± 1.6	−1.4 ± 0.5
294.9	22.9 ± 1.5	24.5	99.4 ± 0.3	174.4 ± 1.6	91.5 ± 0.5	−6.4 ± 1.6	−0.9 ± 0.5
313.1	28.0 ± 1.5	31.9	99.6 ± 0.3	171.9 ± 1.6	92.7 ± 0.5	−5.2 ± 1.6	−1.4 ± 0.5
332.0	35.9 ± 1.5	40.3	99.6 ± 0.3	167.5 ± 1.6	94.2 ± 0.5	−3.0 ± 1.6	−2.1 ± 0.5

[a]Interpolated mol % from the cloud-point measurements in Ref. [6]
[b]Observed molar volume [cm^3 mol^{-1}]
[c]Volumetric Properties [cm^3 mol^{-1}]

(2) Cloud Point

Ref. [192]

x_2	T/K	x_2	T/K	x_2	T/K	x_2	T/K
0.0056	343.87	0.0850	382.65	0.1872	383.00	0.3770	369.86
0.0078	350.83	0.0956	382.94	0.1951	382.81	0.4643	361.61
0.0285	375.40	0.1133	383.17	0.2082	382.48	0.5171	354.48
0.0376	378.39	0.1271	383.19	0.2167	382.21	0.5982	341.43
0.0493	379.97	0.1457	383.18	0.2615	380.52	0.6822	326.14
0.0641	381.69	0.1535	383.21	0.3058	376.03	0.7261	317.12
0.0762	382.39	0.1661	383.08	0.3523	373.59	0.8190	296.47

Mixture 25: 1-$C_5H_{11}OH$ (1) + $BMIMPF_6$ (2)
(1) Phase Equilibrium

Ref. [190]

T/K	x_2 (IL-poor phase)	x_2 (IL-rich phase)	T/K	x_2 (IL-poor phase)	x_2 (IL-rich phase)
278.15	0.0007	0.9193	313.15	0.0013	0.8379
283.15	0.0007	0.9157	318.15	0.0015	0.8212
288.15	0.0007	0.9066	323.15	0.0018	0.8044
293.15	0.0007	0.8960	328.15	0.0021	0.7872
298.15	0.0008	0.8832	333.15	0.0025	0.7695
303.15	0.0009	0.8714	338.15	0.0030	0.7584
308.15	0.0010	0.8568	343.15	0.0035	0.7470

Mixture 26: 1,3-HOC_3H_6OH (1) + [bmim][PF_6] (2)
(1) Phase Equilibrium

Ref. [31]

T/K	$100x'_1$	$100x_1$	V'/ $cm^3 \cdot mol^{-1}$	V/ $cm^3 \cdot mol^{-1}$	$V^{E\prime}$/ $cm^3 \cdot mol^{-1}$	V^E/ $cm^3 \cdot mol^{-1}$
286.6 ± 0.2	10.9 ± 1.7	99.4 ± 0.1	184.5 ± 2.2	71.5 ± 0.2	−7.1 ± 2.2	−1.3 ± 0.2
287.8 ± 0.2	11.4 ± 2.3	99.4 ± 0.1	184.2 ± 3.0	71.6 ± 0.1	−6.9 ± 3.0	−1.2 ± 0.1

(continued)

T/K	$100x'_1$	$100x_1$	V'/ $cm^3 \cdot mol^{-1}$	V/ $cm^3 \cdot mol^{-1}$	$V^{E\prime}$/ $cm^3 \cdot mol^{-1}$	V^E/ $cm^3 \cdot mol^{-1}$
288.2 ± 0.2	11.8 ± 1.6	99.4 ± 0.1	183.4 ± 2.1	71.8 ± 0.2	−7.2 ± 2.1	−1.1 ± 0.2
291.9 ± 0.2	13.3 ± 1.9	99.1 ± 0.1	182.6 ± 2.4	72.3 ± 0.1	−6.5 ± 2.4	−1.0 ± 0.1
298.3 ± 0.2	15.0 ± 2.4	99.0 ± 0.1	181.1 ± 3.1	72.6 ± 0.1	−6.4 ± 3.1	−1.2 ± 0.1
303.3 ± 0.2	15.5 ± 1.5	98.8 ± 0.1	180.9 ± 2.0	73.0 ± 0.2	−6.4 ± 2.0	−1.3 ± 0.2
313.5 ± 0.2	19.0 ± 1.9	98.5 ± 0.1	176.8 ± 2.4	74.1 ± 0.1	−7.0 ± 2.4	−1.0 ± 0.1
323.4 ± 0.2	24.3 ± 1.5	97.8 ± 0.1	171.2 ± 1.9	75.5 ± 0.1	−6.3 ± 1.9	−1.1 ± 0.1

V', observed molar volume lower phase; V, observed molar volume upper phase; $V^{E\prime}$, Volumetric Properties lower phase; V^E, Volumetric Properties upper phase

Mixture 27: C_2H_5OH (1) + [bmim][PF_6] (2)
(1) Volumetric Properties, Refractive Index, Speed of Sound and Compressibility Properties

Ref. [7]

x_2	ρ/ $(g \cdot cm^{-3})$	V^E_m/ $(cm^3 \cdot mol^{-1})$	n_D	Δn_D	u/ $(m \cdot s^{-1})$	κ_s/ $(T \cdot Pa^{-1})$	$\Delta\kappa_s$/ $(T \cdot Pa^{-1})$
T = 293.15 K							
0	0.7893	0	1.36138	0	1160	941	0
0.0009	0.7915	−0.024	1.36162	0.0002	1163	935	−6
0.0021	0.7943	−0.051	1.36196	0.0005	1162	932	−8
0.0033	0.7970	−0.061	1.36222	0.0007	1164	927	−13
0.0056	0.8023	−0.114	1.36281	0.0012	1166	917	−21
0.4960	1.2467	−0.499	1.40171	0.0159	1371	427	−219
0.5032	1.2494	−0.482	1.40193	0.0157	1373	424	−216
0.5981	1.2820	−0.357	1.40432	0.0135	1397	400	−185
0.7045	1.3116	−0.193	1.40639	0.0103	1417	380	−141
0.8003	1.3339	−0.078	1.40790	0.0071	1431	366	−97
0.9002	1.3541	−0.026	1.40930	0.0035	1443	355	−49
0.9834	1.3688	0.001	1.41052	0.0006	1454	346	−8
1	1.3716	0	1.41069	0	1455	344	0
T = 298.15 K							
0	0.7851	0	1.35929	0	1143	974	0
0.0009	0.7874	−0.038	1.35956	0.0002	1144	971	−3
0.0021	0.7901	−0.058	1.35990	0.0005	1145	965	−8
0.0033	0.7928	−0.069	1.36018	0.0007	1146	961	−11
0.0056	0.7981	−0.124	1.36080	0.0012	1147	952	−19
0.4960	1.2422	−0.502	1.40034	0.0162	1359	436	−229
0.5032	1.2450	−0.496	1.40055	0.0161	1361	434	−227
0.5981	1.2776	−0.365	1.40294	0.0137	1384	408	−193
0.7045	1.3073	−0.206	1.40507	0.0105	1404	388	−147
0.8003	1.3296	−0.085	1.40656	0.0072	1418	374	−102
0.9002	1.3499	−0.041	1.40799	0.0036	1430	362	−51
0.9834	1.3647	−0.025	1.40923	0.0007	1440	353	−8

(continued)

x_2	ρ/ (g · cm^{-3})	V^E_m/ (cm^3 · mol^{-1})	n_D	Δn_D	u/ (m · s^{-1})	κ_s/ (T · Pa^{-1})	$\Delta\kappa_s$/ (T · Pa^{-1})
1	1.3673	0	1.40937	0	1443	351	0
T = 303.15 K							
0	0.7807	0	1.35725	0	1127	1009	0
0.0009	0.7831	−0.040	1.35751	0.0002	1126	1008	0
0.0021	0.7858	−0.061	1.35781	0.0005	1128	1000	−7
0.0033	0.7885	−0.072	1.35812	0.0007	1128	996	−10
0.0056	0.7938	−0.128	1.35871	0.0012	1130	987	−18
0.4960	1.2379	−0.529	1.39940	0.0169	1347	445	−241
0.5032	1.2407	−0.522	1.39959	0.0168	1349	443	−238
0.5981	1.2734	−0.398	1.40176	0.0141	1372	417	−203
0.7045	1.3030	−0.221	1.40376	0.0107	1391	396	−154
0.8003	1.3254	−0.106	1.40522	0.0073	1405	382	−106
0.9002	1.3457	−0.056	1.40666	0.0037	1418	370	−54
0.9834	1.3605	−0.036	1.40795	0.0007	1428	360	−9
1	1.3630	0	1.40808	0	1431	358	0

(2) Phase Equilibrium

Ref. [7]

T/K	x_2 Ethanol-rich phase	x_2 Ionic liquid-rich phase
278.15	0.003	0.592
283.15	0.004	0.563
288.15	0.005	0.530
293.15	0.006	0.493
298.15	0.008	0.455
303.15	0.010	0.414
308.15	0.012	0.368
313.15	0.019	0.331
318.15	0.028	0.264
323.15	0.036	0.206
325.15	0.047	0.179

Ref. [191]

	[bmim][PF_6]-rich phase			Ethanol-rich phase		
T/K	$x_1 \pm \Delta x_1$	$w_1 \pm \Delta w_1$		$x_2 \pm \Delta x_2$	$w_2 \pm \Delta w_2$	
p = 0.1 MPa						
288.2	0.4835 ± 0.0038	0.1318 ± 0.0019	CP	0.0047 ± 0.0008	0.0285 ± 0.0021	CP
288.2	0.4789 ± 0.0038	0.1297 ± 0.0019	CP	0.0047 ± 0.0008	0.0285 ± 0.0021	CP
288.2	0.4837 ± 0.0038	0.1318 ± 0.0019	CP			
293.2	0.5048 ± 0.0038	0.1418 ± 0.0019	CP	0.0063 ± 0.0008	0.0375 ± 0.0021	CP
293.2	0.5090 ± 0.0038	0.1439 ± 0.0019	CP	0.0065 ± 0.0008	0.0389 ± 0.0021	CP
295.2	0.5180 ± 0.0038	0.1484 ± 0.0019	CP			
295.2	0.5255 ± 0.0038	0.1522 ± 0.0019	CP			

(*continued*)

	$[bmim][PF_6]$-rich phase			Ethanol-rich phase		
T/K	$x_1 \pm \Delta x_1$	$w_1 \pm \Delta w_1$		$x_2 \pm \Delta x_2$	$w_2 \pm \Delta w_2$	
298.2	0.5489 ± 0.0038	0.1648 ± 0.0019	CP	0.0071 ± 0.0008	0.0423 ± 0.0021	CP
298.2	0.5459 ± 0.0038	0.1631 ± 0.0019	CP	0.0070 ± 0.0008	0.0416 ± 0.0021	CP
308.1	0.6320 ± 0.0038	0.2178 ± 0.0019	CP	0.0112 ± 0.0008	0.0696 ± 0.0021	CP
308.1	0.6329 ± 0.0038	0.2185 ± 0.0019	CP	0.0127 ± 0.0008	0.0737 ± 0.0021	CP
318.1	0.7149 ± 0.0038	0.2890 ± 0.0019	CP	0.0231 ± 0.0008	0.1273 ± 0.0021	CP
318.1	0.7141 ± 0.0038	0.2882 ± 0.0019	CP	0.0234 ± 0.0008	0.1288 ± 0.0021	CP
328.0	0.8279 ± 0.0038	0.4382 ± 0.0019	CP	0.0636 ± 0.0008	0.2951 ± 0.0021	CP
328.0	0.8267 ± 0.0038	0.4361 ± 0.0019	CP	0.0644 ± 0.0008	0.2982 ± 0.0021	CP
279.1				0.0041 ± 0.0007	0.0249 ± 0.0038	UV
279.1				0.0039 ± 0.0007	0.0233 ± 0.0036	UV
279.1				0.0040 ± 0.0006	0.0241 ± 0.0037	UV
279.1				0.0039 ± 0.0006	0.0234 ± 0.0036	UV
279.1				0.0037 ± 0.0006	0.0222 ± 0.0034	UV
279.1				0.0039 ± 0.0006	0.0236 ± 0.0036	UV
284.0				0.0045 ± 0.0007	0.0271 ± 0.0042	UV
284.0				0.0049 ± 0.0008	0.0297 ± 0.0046	UV
284.0				0.0046 ± 0.0007	0.0276 ± 0.0043	UV
284.0				0.0048 ± 0.0008	0.0288 ± 0.0044	UV
284.0				0.0048 ± 0.0008	0.0288 ± 0.0045	UV
284.0				0.0047 ± 0.0007	0.0283 ± 0.0044	UV
289.0				0.0056 ± 0.0009	0.0335 ± 0.0052	UV
289.0				0.0057 ± 0.0009	0.0339 ± 0.0052	UV
289.0				0.0061 ± 0.0010	0.0366 ± 0.0057	UV
289.0				0.0060 ± 0.0010	0.0358 ± 0.0055	UV
289.0				0.0060 ± 0.0010	0.0360 ± 0.0055	UV
289.0				0.0058 ± 0.0009	0.0350 ± 0.0054	UV
293.9				0.0066 ± 0.0011	0.0395 ± 0.0061	UV
293.9				0.0066 ± 0.0011	0.0392 ± 0.0060	UV
293.9				0.0068 ± 0.0011	0.0402 ± 0.0062	UV
298.8				0.0083 ± 0.0014	0.0493 ± 0.0076	UV
298.8				0.0080 ± 0.0013	0.0472 ± 0.0073	UV
298.8				0.0083 ± 0.0014	0.0492 ± 0.0076	UV
303.9				0.0099 ± 0.0016	0.0580 ± 0.0090	UV
303.9				0.0099 ± 0.0016	0.0581 ± 0.0090	UV
303.9				0.0100 ± 0.0016	0.0584 ± 0.0090	UV
308.8				0.0162 ± 0.0027	0.0919 ± 0.014	UV
308.8				0.0167 ± 0.0028	0.0949 ± 0.015	UV
308.8				0.0154 ± 0.0026	0.0881 ± 0.014	UV
308.8				0.0142 ± 0.0024	0.0817 ± 0.013	UV
308.8				0.0148 ± 0.0025	0.0846 ± 0.013	UV
308.8				0.0167 ± 0.0028	0.0947 ± 0.015	UV
308.8				0.0169 ± 0.0029	0.0960 ± 0.015	UV
308.8				0.0145 ± 0.0024	0.0831 ± 0.013	UV
308.8				0.0142 ± 0.0024	0.0816 ± 0.013	UV

(*continued*)

	[bmim][PF_6]-rich phase			Ethanol-rich phase		
T/K	$x_1 \pm \Delta x_1$	$w_1 \pm \Delta w_1$		$x_2 \pm \Delta x_2$	$w_2 \pm \Delta w_2$	
313.8				0.0190 ± 0.0033	0.1069 ± 0.016	UV
313.8				0.0197 ± 0.0034	0.1101 ± 0.017	UV
313.8				0.0189 ± 0.0033	0.1061 ± 0.016	UV

(3) Cloud Point

Ref. [53]

p (bar)	T (K)	$10^2\ w_2$	$10^2 dT/dp$ (K/bar)
[bmim][PF_6] + ethanol			
108.50	288.66	3.30	−1.51
201.30	287.35		
307.90	285.65		
401.70	284.25		
1.00	298.55	4.78	–
2.46	313.43	12.23	−1.88
16.37	313.23		
25.94	313.02		
35.46	312.85		
42.99	312.67		
2.08	322.58	26.59	−1.84
12.69	322.41		
24.95	322.18		
31.60	322.09		
38.31	321.92		
48.90	321.72		
2.18	323.80	30.38	−1.88
12.72	323.63		
25.02	323.40		
37.52	323.15		
48.44	322.94		
2.06	325.20	36.13	−1.86
9.36	325.03		
15.49	324.98		
25.84	324.77		
34.79	324.59		
49.29	324.31		
2.18	325.10	43.24	−2.00
12.57	324.88		
25.18	324.62		
37.43	324.40		
47.75	324.18		
1.00	325.18	43.83	
1.96	325.33	48.96	−1.84

(*continued*)

p (bar)	T (K)	$10^2\ w_2$	$10^2 dT/dp$ (K/bar)
13.88	325.18		
24.95	324.96		
37.82	324.70		
48.02	324.50		
2.06	321.48	59.77	−1.59
12.90	321.32		
19.79	321.25		
40.66	320.90		
48.66	320.74		
2.14	312.83	72.21	−1.22
13.91	312.79		
24.88	312.61		
37.95	312.43		
46.77	312.32		
1.00	296.55	83.45	
1.00	294.55	84.0	
[bmim][PF_6] + ethanol ($-d_1$)			
2.03	323.76	31.08	
1.81	324.40	38.28	
1.95	323.60	42.64	
1.20	324.13	45.90	
2.06	324.22	48.59	−2.19
10.25	324.05		
19.03	323.86		
31.24	323.55		
43.94	323.32		
2.06	324.02	49.85	−2.04
9.94	323.91		
19.86	323.80		
29.64	323.54		
42.57	323.19		

(4) Partial Molar Enthalpies at Infinite Dilution

Ref. [53]

$H_1^{E,\infty}$(J/mol)
+10397

Mixture 28: 2-$CH_3COC_2H_5$ (1) + $BMIMPF_6$ (2)
(1) Volumetric Properties, Transport Properties, Refractive Index, Speed of Sound and Compressibility Properties

Ref. [12]

x_2	ρ/ $g \cdot cm^{-3}$	V_m^E/ $cm^3 \cdot mol^{-1}$	n_D	Δn_D	u/ $m \cdot s^{-1}$	κ_S/ $T \cdot Pa^{-1}$	$\Delta\kappa_S$/ $T \cdot Pa^{-1}$
$T = 293.15$ K							
0	0.8056	0	1.37879	0	1212	845	0
0.0533	0.8777	−0.742	1.38660	0.0061	1242	739	−80
0.1058	0.9388	−1.249	1.39154	0.0094	1265	666	−126
0.2081	1.0362	−1.816	1.39807	0.0126	1301	570	−171
0.2992	1.1045	−1.984	1.40183	0.0135	1331	511	−184
0.4044	1.1679	−1.929	1.40475	0.0131	1360	463	−180
0.4926	1.2116	−1.742	1.40661	0.0121	1380	433	−165
0.5943	1.2541	−1.456	1.40795	0.0102	1399	407	−140
0.7085	1.2939	−1.036	1.40900	0.0076	1418	385	−106
0.8093	1.3239	−0.647	1.40971	0.0051	1433	368	−72
0.9013	1.3483	−0.333	1.41019	0.0026	1444	356	−38
0.9491	1.3597	−0.141	1.41042	0.0014	1449	350	−20
1	1.3716	0	1.41069	0	1455	344	0
$T = 298.15$ K							
0	0.7997	0	1.37618	0	1192	880	0
0.0533	0.8726	−0.846	1.38487	0.0069	1224	765	−87
0.1058	0.9340	−1.390	1.39004	0.0103	1248	687	−136
0.2081	1.0313	−1.940	1.39642	0.0133	1286	586	−184
0.2992	1.0995	−2.082	1.39987	0.0138	1317	524	−197
0.4044	1.1632	−2.038	1.40296	0.0134	1348	473	−193
0.4926	1.2073	−1.877	1.40505	0.0125	1369	442	−177
0.5943	1.2501	−1.603	1.40645	0.0105	1390	414	−152
0.7085	1.2898	−1.140	1.40757	0.0079	1408	391	−114
0.8093	1.3198	−0.723	1.40833	0.0053	1421	375	−77
0.9013	1.3440	−0.356	1.40888	0.0028	1433	362	−41
0.9491	1.3555	−0.165	1.40911	0.0014	1438	357	−22
1	1.3673	0	1.40937	0	1443	351	0
$T = 303.15$ K							
0	0.7949	0	1.37355	0	1163	931	0
0.0533	0.8676	−0.845	1.38273	0.0073	1195	807	−93
0.1058	0.9291	−1.412	1.38810	0.0109	1220	724	−147
0.2081	1.0268	−2.013	1.39466	0.0139	1260	614	−198
0.2992	1.0953	−2.185	1.39814	0.0143	1292	547	−212
0.4044	1.1590	−2.133	1.40149	0.0140	1322	494	−206
0.4926	1.2030	−1.950	1.40357	0.0130	1349	457	−192
0.5943	1.2457	−1.650	1.40517	0.0111	1374	426	−165
0.7085	1.2855	−1.183	1.40617	0.0082	1393	401	−124

(*continued*)

x_2	ρ/ g · cm^{-3}	V_m^E/ cm^3 · mol^{-1}	n_D	Δn_D	u/ m · s^{-1}	κ_S/ T · Pa^{-1}	$\Delta\kappa_S$/ T · Pa^{-1}
0.8093	1.3156	−0.764	1.40668	0.0052	1408	384	−84
0.9013	1.3397	−0.367	1.40753	0.0029	1419	370	−44
0.9491	1.3512	−0.168	1.40779	0.0015	1425	364	−23
1	1.3630	0	1.40808	0	1431	358	0

Ref. [194]

x_2	ρ (kg · m^{-3})	η (mPa · s)	$V^E{}_m$ (10^{-6}m^3 · mol^{-1})	(lnη)^E (mPa · s)
298.15 K				
0.00	799.65	0.388	0.00	0.00
0.0513	872.33	0.606	−1.075	0.104
0.0973	925.05	0.861	−1.364	0.162
0.1996	1024.05	1.861	−1.893	0.263
0.2977	1098.25	3.701	−2.035	0.308
0.3483	1131.68	5.150	−2.140	0.307
0.4012	1161.35	7.271	−1.995	0.306
0.4494	1186.14	9.819	−1.873	0.290
0.4998	1210.23	13.549	−1.777	0.282
0.5515	1232.63	18.183	−1.635	0.238
0.5976	1250.77	23.549	−1.463	0.195
0.6984	1286.51	43.329	−1.115	0.145
0.7993	1317.51	81.624	−0.765	0.118
0.9005	1343.93	144.875	−0.299	0.029
1.00	1367.88	269.959	0.00	0.00

Ref. [135]

x_2	η/(mPa · s)	x_2	η/(mPa · s)	x_2	η/(mPa · s)
298.15 K					
0.00	0.388	0.3483	5.150	0.5976	23.549
0.0513	0.606	0.4012	7.271	0.6984	43.329
0.0973	0.861	0.4494	9.819	0.7993	81.624
0.1996	1.861	0.4998	13.549	0.9005	144.875
0.2977	3.701	0.5515	18.183	1.00	269.959

(2) Electrical Conductivity

Ref. [135]

x_2	10$^3\kappa$ (S · cm^{-1})	x_2	10$^3\kappa$ (S · cm^{-1})	x_2	10$^3\kappa$ (S · cm^{-1})
298.15 K					
0.0516	14.13	0.4012	13.52	0.6984	5.39
0.1043	18.52	0.4494	11.90	0.7993	3.59
0.2084	19.28	0.4998	10.26	0.9005	2.42

(continued)

x_2	$10^3\kappa$ (S · cm^{-1})	x_2	$10^3\kappa$ (S · cm^{-1})	x_2	$10^3\kappa$ (S · cm^{-1})
0.2977	17.28	0.5515	8.81	1.00	1.55
0.3483	15.43	0.5976	7.69		

Mixture 29: $CH_3COOCH_2CH_3$ (1) + BMIM PF_6 (2)
(1) Volumetric Properties, Transport Properties, Refractive Index, Speed of Sound and Compressibility Properties

Ref. [12]

x_2	ρ/ g · cm^{-3}	V_m^E/ cm^3 · mol^{-1}	n_D	Δn_D	u/ m · s^{-1}	K_s/ T · Pa^{-1}	ΔK_s/ T · Pa^{-1}
T = 293.15 K							
0	0.9005	0	1.37241	0	1162	822	0
0.0520	0.9563	−0.734	1.37948	0.0051	1182	749	−49
0.0979	0.9990	−1.134	1.38419	0.0080	1200	695	−80
0.1980	1.0776	−1.702	1.39206	0.0121	1240	603	−124
0.2939	1.1382	−1.944	1.39736	0.0137	1280	536	−145
0.3968	1.1914	−1.980	1.40129	0.0137	1319	482	−150
0.4883	1.2307	−1.853	1.40391	0.0128	1350	446	−143
0.5936	1.2687	−1.567	1.40599	0.0109	1379	414	−124
0.7000	1.3010	−1.171	1.40788	0.0087	1399	393	−95
0.8146	1.3309	−0.710	1.40911	0.0055	1421	372	−61
0.8997	1.3506	−0.379	1.40991	0.0031	1438	358	−34
0.9499	1.3615	−0.206	1.41046	0.0017	1447	351	−17
1	1.3716	0	1.41069	0	1455	344	0
T = 298.15 K							
0	0.8944	0	1.36977	0	1141	859	0
0.0520	0.9506	−0.783	1.37726	0.0054	1162	780	−53
0.0979	0.9934	−1.193	1.38208	0.0084	1180	723	−86
0.1980	1.0722	−1.772	1.39005	0.0124	1228	618	−140
0.2939	1.1331	−2.029	1.39538	0.0140	1272	545	−164
0.3968	1.1868	−2.099	1.39946	0.0140	1310	491	−166
0.4883	1.2266	−2.011	1.40230	0.0132	1338	455	−156
0.5936	1.2647	−1.712	1.40451	0.0112	1366	424	−134
0.7000	1.2970	−1.290	1.40638	0.0089	1391	399	−105
0.8146	1.3266	−0.757	1.40768	0.0057	1412	378	−67
0.8997	1.3463	−0.402	1.40855	0.0032	1427	365	−37
0.9499	1.3573	−0.229	1.40915	0.0018	1436	358	−19
1	1.3673	0	1.40937	0	1443	351	0
T = 303.15 K							
0	0.8882	0	1.36712	0	1111	912	0
0.0520	0.9446	−0.812	1.37520	0.0059	1137	819	−64
0.0979	0.9877	−1.254	1.38016	0.0090	1160	752	−106
0.1980	1.0670	−1.875	1.38819	0.0130	1207	643	−159
0.2939	1.1280	−2.124	1.39346	0.0143	1247	570	−180

(continued)

x_2	ρ/ g · cm^{-3}	V_m^E/ cm^3 · mol^{-1}	n_D	Δn_D	u/ m · s^{-1}	K_s/ T · Pa^{-1}	ΔK_s/ T · Pa^{-1}
0.3968	1.1820	−2.204	1.39774	0.0144	1291	508	−185
0.4883	1.2220	−2.116	1.40065	0.0135	1323	468	−174
0.5936	1.2603	−1.812	1.40301	0.0116	1354	433	−151
0.7000	1.2926	−1.360	1.40490	0.0091	1378	407	−117
0.8146	1.3222	−0.792	1.40632	0.0058	1401	386	−76
0.8997	1.3420	−0.427	1.40723	0.0033	1415	372	−42
0.9499	1.3529	−0.224	1.40793	0.0019	1420	367	−20
1	1.3630	0	1.40808	0	1431	358	0

Ref. [194]

x_2	ρ (kg · m^{-3})	η (mPa · s)	$V^E{}_m$ (10^{-6}m^3 · mol^{-1})	$(\ln\eta)^E$ (mPa · s)
298.15 K				
0.00	894.60	0.439	0.00	0.00
0.0496	947.52	0.699	−0.661	0.145
0.1020	996.53	1.083	−1.157	0.247
0.1963	1070.84	2.208	−1.728	0.354
0.2908	1130.97	4.210	−1.941	0.393
0.3476	1162.91	6.216	−2.122	0.418
0.3961	1186.49	8.449	−2.081	0.413
0.4471	1209.21	11.522	−2.012	0.396
0.5033	1231.70	16.160	−1.865	0.373
0.5515	1249.51	20.543	−1.691	0.304
0.6056	1267.08	27.516	−1.409	0.249
0.7082	1297.87	48.203	−0.992	0.151
0.8086	1325.64	87.906	−0.838	0.107
0.9063	1347.90	149.310	−0.382	0.009
1.00	1367.88	269.959	0.00	0.00

Mixture 30: 2-CH_3CHCH_3OH (1) + [BMIM][PF_6] (2)
(1) Volumetric Properties, Refractive Index, Speed of Sound and Compressibility Properties

Ref. [12]

x_2	ρ/ g · cm^{-3}	V_m^E/ cm^3 · mol^{-1}	n_D	Δn_D	u/ m · s^{-1}	κ_s/ T · Pa^{-1}	$\Delta\kappa_s$/ T · Pa^{-1}
T = 293.15 K							
0	0.7850	0	1.37707	0	1156	953	0
0.0008	0.7863	−0.004	1.37717	0.0001	1156	951	−1
0.0011	0.7869	−0.012	1.37726	0.0002	1157	950	−2
0.7136	1.2985	−0.350	1.40727	0.0062	1409	388	−131
0.8175	1.3282	−0.183	1.40833	0.0038	1429	369	−86
0.8945	1.3476	−0.075	1.40869	0.0015	1441	358	−51
0.9542	1.3615	−0.020	1.40976	0.0006	1449	350	−22

(continued)

x_2	ρ/ $g \cdot cm^{-3}$	V_m^E/ $cm^3 \cdot mol^{-1}$	n_D	Δn_D	u/ $m \cdot s^{-1}$	κ_s/ $T \cdot Pa^{-1}$	$\Delta\kappa_s$/ $T \cdot Pa^{-1}$
1	1.3716	0	1.41069	0	1455	344	0
$T = 298.15$ K							
0	0.7810	0	1.37496	0	1139	986	0
0.0008	0.7823	−0.008	1.37509	0.0001	1139	985	−1
0.0011	0.7829	−0.014	1.37521	0.0002	1140	984	−2
0.7136	1.2943	−0.370	1.40584	0.0063	1397	396	−138
0.8175	1.3240	−0.206	1.40689	0.0038	1417	376	−91
0.8945	1.3434	−0.092	1.40738	0.0016	1429	365	−54
0.9542	1.3572	−0.027	1.40844	0.0007	1437	357	−24
1	1.3673	0	1.40937	0	1443	351	0
$T = 303.15$ K							
0	0.7766	0	1.37278	0	1121	1025	0
0.0008	0.7780	−0.009	1.37301	0.0002	1121	1022	−2
0.0011	0.7785	−0.017	1.37312	0.0003	1122	1021	−3
0.7136	1.2900	−0.401	1.40450	0.0065	1385	404	−145
0.8175	1.3198	−0.236	1.40562	0.0040	1405	384	−96
0.8945	1.3393	−0.120	1.40595	0.0016	1417	372	−57
0.9542	1.3530	−0.039	1.40721	0.0008	1425	364	−25
1	1.3630	0	1.40808	0	1431	358	0

(2) Phase Equilibrium

Ref. [12]

	x_2			x_2	
T/K	IL-poor phase	IL-rich phase	T/K	IL-poor phase	IL-rich phase
278.15	0.0008	0.7516	313.15	0.0027	0.5592
283.15	0.0008	0.7309	318.15	0.0036	0.5335
288.15	0.0009	0.7070	323.15	0.0044	0.4905
293.15	0.0012	0.6820	328.15	0.0059	0.4661
298.15	0.0014	0.6517	333.15	0.0081	0.4339
303.15	0.0017	0.6238	338.15	0.0109	0.4057
308.15	0.0022	0.5910	343.15	0.0133	0.3678

Mixture 31: $C_6H_5CH_2OH$ (1) + [bmim][PF_6] (2)
(1) Volumetric Properties

Ref. [195]

x_2	$\rho/(kg \cdot m^{-3})$	$V^E_m/(cm^3 \cdot mol^{-1})$	x_2	$\rho/(kg \cdot m^{-3})$	$V^E_m/(cm^3 \cdot mol^{-1})$
$T = 298.15$ K			$T = 308.15$ K		
0.0000	1044.85	0.0000	0.0000	1037.09	0.0000
0.0502	1079.03	−0.3081	0.0502	1071.16	−0.3107
0.0989	1108.64	−0.5333	0.0989	1100.70	−0.5403
0.1998	1160.82	−0.8133	0.1998	1152.74	−0.8225

(*continued*)

x_2	ρ/(kg · m^{-3})	V^E_m/(cm^3 · mol^{-1})	x_2	ρ/(kg · m^{-3})	V^E_m/(cm^3 · mol^{-1})
0.3001	1203.01	−0.8941	0.3001	1194.89	−0.9097
0.4012	1238.30	−0.8468	0.4012	1230.15	−0.8659
0.5012	1267.82	−0.7201	0.5012	1259.64	−0.7402
0.6014	1293.35	−0.5567	0.6014	1285.12	−0.5731
0.7003	1315.45	−0.3847	0.7003	1307.18	−0.3979
0.8005	1335.42	−0.2236	0.8005	1327.14	−0.2371
0.9016	1353.60	−0.0919	0.9016	1345.29	−0.1019
1.0000	1369.74	0.0000	1.0000	1361.36	0.0000
T = 303.15 K			T = 313.15 K		
0.0000	1040.97	0.0000	0.0000	1033.17	0.0000
0.0502	1075.09	−0.3089	0.0502	1067.20	−0.3136
0.0989	1104.66	−0.5367	0.0989	1096.72	−0.5473
0.1998	1156.76	−0.8169	0.1998	1148.71	−0.8321
0.3001	1198.94	−0.9021	0.3001	1190.83	−0.9211
0.4012	1234.23	−0.8586	0.4012	1226.07	−0.8774
0.5012	1263.74	−0.7341	0.5012	1255.53	−0.7503
0.6014	1289.24	−0.5697	0.6014	1280.98	−0.5801
0.7003	1311.33	−0.3977	0.7003	1303.04	−0.4047
0.8005	1331.26	−0.2337	0.8005	1322.96	−0.2389
0.9016	1349.39	−0.0942	0.9016	1341.08	−0.0989
1.0000	1365.51	0.0000	1.0000	1357.17	0.0000

Mixture 32: C_6H_5CHO (1) + [bmim][PF_6] (2)
(1) Volumetric Properties

Ref. [195]

x_2	ρ/(kg · m^{-3})	V^E_m/(cm^3 · mol^{-1})	x_2	ρ/(kg · m^{-3})	V^E_m/(cm^3 · mol^{-1})
T = 298.15 K			T = 308.15 K		
0.0000	1043.68	0.0000	0.0000	1035.35	0.0000
0.0523	1081.36	−0.4643	0.0523	1072.95	−0.4689
0.1034	1113.62	−0.7913	0.1034	1105.14	−0.7993
0.2003	1164.90	−1.1541	0.2003	1156.46	−1.1751
0.2996	1207.37	−1.2954	0.2996	1198.97	−1.3258
0.4026	1243.63	−1.2954	0.4026	1235.23	−1.3259
0.5118	1275.65	−1.1943	0.5118	1267.23	−1.2198
0.5983	1297.30	−1.0537	0.5983	1288.90	−1.0779
0.7011	1319.60	−0.8271	0.7011	1311.29	−0.8587
0.7977	1337.76	−0.5653	0.7977	1329.56	−0.6059
0.8996	1354.63	−0.2611	0.8996	1346.50	−0.3049
1.0000	1369.74	0.0000	1.0000	1361.36	0.0000
T = 303.15 K			T = 313.15 K		
0.0000	1039.64	0.0000	0.0000	1031.00	0.0000
0.0523	1077.27	−0.4667	0.0523	1068.58	−0.4723
0.1034	1109.46	−0.7931	0.1034	1100.81	−0.8106

(continued)

x_2	ρ/(kg · m^{-3})	V^E_m/(cm^3 · mol^{-1})	x_2	ρ/(kg · m^{-3})	V^E_m/(cm^3 · mol^{-1})
0.2003	1160.74	−1.1628	0.2003	1152.16	−1.1931
0.2996	1203.19	−1.3057	0.2996	1194.73	−1.3498
0.4026	1239.44	−1.3069	0.4026	1231.01	−1.3511
0.5118	1271.44	−1.2048	0.5118	1263.03	−1.2432
0.5983	1293.14	−1.0692	0.5983	1284.74	−1.1029
0.7011	1315.55	−0.8581	0.7011	1307.11	−0.8767
0.7977	1333.78	−0.6043	0.7977	1325.40	−0.6209
0.8996	1350.65	−0.3006	0.8996	1342.31	−0.3109
1.0000	1365.51	0.0000	1.0000	1357.17	0.0000

Mixture 33: CHF_3 (1) + [bmim][PF_6] (2)
(1) Phase Equilibrium

Ref. [196]

T/K	x_1/mol %	x'_1/mol %	$\overline{V}$/ cm^3 · mol^{-1}	$\overline{V}'$/ cm^3 · mol^{-1}	$\overline{V}^{ex}$/ cm^3 · mol^{-1}	$\overline{V}^{ex\prime}$/ cm^3 · mol^{-1}
293.6	77.3 ± 2.0	100.1 ± 0.3	87.8 ± 3.0	86.2 ± 3.0	−26.2 ± 3.0	−0.2 ± 3.0
285.6	81.2 ± 2.0	99.7 ± 3.0	86.5 ± 3.0	73.6 ± 3.0	−14.0 ± 3.0	−2.8 ± 3.0
280.2	85.3 ± 2.0	99.6 ± 0.3	79.4 ± 3.0	70.6 ± 3.0	−12.0 ± 3.0	−1.7 ± 3.0

x_1: CHF_3 composition in one liquid, x_1': CHF_3 composition in another equilibrium liquid; $\overline{V}$: molar volume of one liquid, $\overline{V}'$: molar volume of another equilibrium liquid; $\overline{V}^{ex}$: Volumetric Properties of one liquid, $\overline{V}^{ex\prime}$: Volumetric Properties of another equilibrium liquid

Ref. [197]

T/K	100 x_1'	100 x_1	V'/ cm^3 · mol^{-1}	V/ cm^3 · mol^{-1}	$V^{E\prime}$/ cm^3 · mol^{-1}	V^E/ cm^3 · mol^{-1}
280.2	85.3 ± 2.0	99.6 ± 0.3	79.4 ± 3.0	70.6 ± 3.0	−12.0 ± 3.0	−1.7 ± 3.0
285.6	81.2 ± 2.0	99.7 ± 0.3	86.5 ± 3.0	73.6 ± 3.0	−14.0 ± 3.0	−2.8 ± 3.0
293.6	77.3 ± 2.0	100.1 ± 0.3	87.8 ± 3.0	86.2 ± 3.0	−26.2 ± 3.0	−0.2 ± 3.0

x_1', mole fraction in lower liquid phase; x_1, mole fraction in upper liquid phase; V', molar volume in lower liquid phase; V, molar volume in upper liquid phase; $V^{E\prime}$, Volumetric Properties in lower liquid phase; V^E, Volumetric Properties in upper liquid phase

Ref. [198]

x_1*	T (K)	P (MPa)	T (K)	P (MPa)	T (K)	P (MPa)
0.102(b.p.)	303.20	0.58	328.24	0.88	353.23	1.24
	308.19	0.63	333.22	0.94	358.22	1.33
	313.20	0.69	338.17	1.02	363.25	1.41
	318.18	0.74	343.25	1.09		
	323.26	0.82	348.22	1.17		
0.203(b.p.)	303.29	1.16	328.22	1.81	353.27	2.62
	308.21	1.28	333.30	1.96	358.23	2.80
	313.25	1.40	338.23	2.11	363.39	2.98
	318.24	1.54	343.29	2.28		

(*continued*)

x_1*	T (K)	P (MPa)	T (K)	P (MPa)	T (K)	P (MPa)
	323.31	1.67	348.26	2.45		
0.302(b.p.)	303.28	1.78	328.26	2.85	353.32	4.17
	308.30	1.97	333.28	3.08	358.33	4.49
	313.28	2.18	338.29	3.35	363.31	4.79
	318.25	2.38	343.32	3.62		
	323.32	2.61	348.29	3.89		
0.400(b.p.)	303.38	2.43	328.33	3.97	353.36	5.99
	308.33	2.69	333.39	4.33	358.22	6.43
	313.32	2.97	338.30	4.71	363.15	6.91
	318.24	3.27	343.36	5.13		
	323.33	3.61	348.27	5.55		
0.483(b.p.)	303.24	3.01	328.33	5.11	353.25	7.99
	308.22	3.37	333.29	5.65	358.30	8.67
	313.24	3.75	338.31	6.19	363.42	9.41
	318.34	4.17	343.32	6.75		
	323.28	4.61	348.25	7.37		
0.552(b.p.)	313.15	4.56	327.79	6.36	342.66	8.76
	318.09	5.12	332.77	7.14	347.62	9.68
	322.77	5.68	337.73	7.94	352.57	10.60
0.621(b.p.)	312.75	5.28	332.63	8.92	347.51	12.54
	317.76	6.02	337.65	10.10	352.43	13.80
	322.77	6.88	342.56	11.30	357.26	15.06
	327.70	7.84				
0.700(b.p.)	312.95	7.84	332.80	14.64	347.43	19.70
	317.84	9.50	337.71	16.36	352.30	21.32
	322.81	11.22	342.62	18.04	357.12	22.88
	327.80	12.92				
0.780(b.p.)	312.90	14.20	327.88	20.52	342.80	26.44
	317.98	16.30	332.85	22.50	347.69	28.32
	322.93	18.44	337.82	24.48	352.73	30.28
0.850(b.p.)	313.11	19.44	332.79	27.72	347.53	33.62
	317.86	21.50	337.72	29.64	352.39	35.56
	322.90	23.64	342.58	31.62	357.42	37.54
	327.79	25.68				
0.900(b.p.)	312.85	20.60	332.80	29.54	347.92	35.98
	318.36	23.12	337.79	31.70	352.56	37.90
	323.73	25.54	342.73	33.82	358.16	40.16
	327.87	27.38				
0.925(b.p.)	312.95	21.60	332.88	30.52	347.99	36.94
	317.86	23.84	337.83	32.66	352.66	38.86
	322.95	26.14	342.69	34.72	357.50	40.82
	327.79	28.28				
0.956(d.p.)	312.94	21.76	333.82	31.06	352.67	38.98
	320.01	24.96	340.74	34.02	357.64	41.00

(*continued*)

x_1*	T (K)	P (MPa)	T (K)	P (MPa)	T (K)	P (MPa)
	326.87	28.02	347.69	36.94		
0.980(d.p.)	313.48	21.62	332.83	30.12	347.68	36.34
	317.90	23.60	337.83	32.26	352.68	38.36
	322.92	25.82	342.69	34.28	357.56	40.30
	327.91	28.00				
0.990(d.p.)	312.86	19.48	328.30	26.00	343.48	32.08
	317.93	21.66	332.84	27.82	348.15	33.94
	322.89	23.74	338.66	30.18	352.78	35.74

**b.p.* Bubble Point measurements; *d.p.* dew point measurements

Ref. [198]

T (K)	P (MPa)	x_1
315	22.77	0.9486
320	25.04	0.9479
325	27.27	0.9473
330	29.48	0.9467
335	31.65	0.9462
340	33.79	0.9456
345	35.90	0.9452
350	37.97	0.9447
355	40.02	0.9443

Ref. [198]

T (K)	P (MPa)	T (K)	P (MPa)
288.22	3.68	296.13	4.47
290.20	3.86	296.16	4.47
292.14	4.05	297.20	4.58
293.21	4.16	298.18	4.69
294.15	4.26	298.70	4.75
294.20	4.26	299.08	4.80
295.18	4.37	299.20	4.81

Mixture 34: 1-C_6H_{12} (1) + [BMIM][PF_6] (2)
(1) Phase Equilibrium

Ref. [199]

283.15 K		293.15 K		303.15 K	
x_1	P (kPa)	x_1	P (kPa)	x_1	P (kPa)
0.001882	0.4317	0.001870	0.5456	0.001857	0.6867
0.004559	1.006	0.004537	1.275	0.004504	1.596
0.008411	1.775	0.008379	2.303	0.008321	2.803
0.01309	2.605	0.01304	3.448	0.01295	4.242
0.01910	3.589	0.01902	4.886	0.01889	5.920

(*continued*)

283.15 K		293.15 K		303.15 K	
x_1	P (kPa)	x_1	P (kPa)	x_1	P (kPa)
0.02162	3.972	0.02154	5.444	0.02141	6.649
0.02586	4.502	0.02578	6.345	0.02563	7.756
0.03547	5.847	0.03536	8.317	0.03516	10.14
0.04270	6.702	0.04259	9.759	0.04236	11.78
0.05004	7.565	0.04991	11.05	0.04967	13.34
0.05740	8.557	0.05725	12.26	0.05697	14.54
0.05971	8.793	0.05956	12.57	0.05929	15.02
0.06744	9.746	0.06727	13.80	0.06697	16.65
0.07879	10.55	0.07856	15.65	0.07821	19.09
0.08498	10.71	0.08472	16.59	0.08436	20.23
0.09046	11.12	0.09018	17.40	0.08982	21.17
0.09384	11.34	0.09356	17.80	0.09321	21.77
0.1042	11.58	0.1039	18.62	0.1035	23.66
0.1269	11.87	0.1265	19.81	0.1259	26.77
0.1557	11.86	0.1553	19.80	0.1548	28.02
0.2208	11.91	0.2204	19.72	0.2199	28.14
0.3191	12.03	0.3188	19.79	0.3184	28.91

(2) Solubility

Ref. [199]

T (K)	283.15	293.15	303.15
x_1	0.07888	0.1008	0.1189

Mixture 35: *trans*-3-C_6H_{12} (1) + [BMIM][PF_6] (2)
(1) Phase Equilibrium

Ref. [199]

283.15 K		293.15 K		303.15 K	
x_1	P (kPa)	x_1	P (kPa)	x_1	P (kPa)
0.004494	1.284	0.004489	1.650	0.004464	2.143
0.008940	2.513	0.008931	3.203	0.008897	4.079
0.01741	4.848	0.01740	6.096	0.01735	7.562
0.02557	7.052	0.02557	8.732	0.02550	10.73
0.03447	9.385	0.03441	11.44	0.03432	13.79
0.04171	9.823	0.04153	13.32	0.04141	16.22
0.05066	10.19	0.05031	15.58	0.05016	19.30
0.06112	10.22	0.06069	16.45	0.06042	22.61
0.06742	10.25	0.06701	16.47	0.06669	24.44
0.07549	10.19	0.07505	16.81	0.07475	25.47
0.08199	10.21	0.08156	17.02	0.08125	25.93
0.1134	10.29	0.1129	16.97	0.1127	26.11
0.1350	10.29	0.1346	16.82	0.1343	26.20

(*continued*)

283.15 K		293.15 K		303.15 K	
x_1	P (kPa)	x_1	P (kPa)	x_1	P (kPa)
0.1547	10.25	0.1543	16.74	0.1540	26.29

(2) Solubility

Ref. [199]

T (K)	283.15	293.15	303.15
x_1	0.03729	0.05183	0.06801

Mixture 36: CH_3F (1) + [bmim][PF_6] (2)

(1) Phase Equilibrium

Ref. [197]

T/K	100 x_1'	100 x_1	V'/ $cm^3 \cdot mol^{-1}$	V/ $cm^3 \cdot mol^{-1}$	$V^{E\prime}$/ $cm^3 \cdot mol^{-1}$	V^E/ $cm^3 \cdot mol^{-1}$
243.2	81.8 ± 1.0	100.0 ± 1.0	67.6 ± 1.0	43.8 ± 1.0	−5.3 ± 1.0	−0.6 ± 1.0
263.2	80.7 ± 1.0	99.9 ± 1.0	71.2 ± 1.0	47.8 ± 1.0	−6.8 ± 1.0	−0.3 ± 1.0
294.1	74.6 ± 1.0	100.0 ± 0.3	82.7 ± 1.0	55.9 ± 1.0	−12.6 ± 1.0	−1.4 ± 1.0

x_1', mole fraction in lower liquid phase; x_1, mole fraction in upper liquid phase; V', molar volume in lower liquid phase; V, molar volume in upper liquid phase; $V^{E\prime}$, Volumetric Properties in lower liquid phase; V^E, Volumetric Properties in upper liquid phase

Mixture 37: 1,1,1,2-$C_2H_2F_4$ (1) + [bmim][PF_6] (2)

(1) Phase Equilibrium

Ref. [197]

T/K	100 x_1'	100 x_1	V'/ $cm^3 \cdot mol^{-1}$	V/ $cm^3 \cdot mol^{-1}$	$V^{E\prime}$/ $cm^3 \cdot mol^{-1}$	V^E/ $cm^3 \cdot mol^{-1}$
318.2	69.5 ± 2.0	101.2 ± 1.5	114.7 ± 5.0	88.5 ± 5.0	−12.4 ± 5.0	−2.1 ± 5.0
333.8	63.8 ± 2.0	99.4 ± 0.3	127.4 ± 3.0	96.6 ± 1.5	−11.4 ± 3.0	−1.3 ± 1.5
355.0	57.5 ± 2.0	99.1 ± 1.0	139.0 ± 5.0	110.7 ± 5.0	−16.7 ± 5.0	−1.9 ± 5.0

x_1', mole fraction in lower liquid phase; x_1, mole fraction in upper liquid phase; V', molar volume in lower liquid phase; V, molar volume in upper liquid phase; $V^{E\prime}$, Volumetric Properties in lower liquid phase; V^E, Volumetric Properties in upper liquid phase

Mixture 38: 1,1,1-$C_2H_3F_3$ (1) + [bmim][PF_6] (2)

(1) Phase Equilibrium

Ref. [197]

T/K	100 x_1'	100 x_1	V'/ $cm^3 \cdot mol^{-1}$	V/ $cm^3 \cdot mol^{-1}$	$V^{E\prime}$/ $cm^3 \cdot mol^{-1}$	V^E/ $cm^3 \cdot mol^{-1}$
291.8	29.0 ± 1.0	100.7 ± 1.0	163.1 ± 1.0	89.2 ± 1.0	−9.2 ± 1.0	1.3 ± 1.0
324.6	25.8 ± 1.0	100.1 ± 1.0	178.8 ± 1.0	105.6 ± 1.0	−5.2 ± 1.0	−0.3 ± 1.0
338.6	24.0 ± 1.0	100.7 ± 1.0	181.3 ± 1.0	124.1 ± 1.0	−10.5 ± 1.0	−0.7 ± 1.0

x_1', mole fraction in lower liquid phase; x_1, mole fraction in upper liquid phase; V', molar volume in lower liquid phase; V, molar volume in upper liquid phase; $V^{E\prime}$, Volumetric Properties in lower liquid phase; V^E, Volumetric Properties in upper liquid phase

Mixture 39: 1,1-$C_2H_4F_2$ (1) + [bmim][PF_6] (2)
(1) Phase Equilibrium

Ref. [197]

T/K	100 x_1'	100 x_1	V'/ cm^3 · mol^{-1}	V/ cm^3 · mol^{-1}	$V^{E\prime}$/ cm^3 · mol^{-1}	V^E/ cm^3 · mol^{-1}
318.3	77.4 ± 1.0	96.1 ± 1.0	80.5 ± 5.0	78.0 ± 5.0	−27.5 ± 5.0	−5.2 ± 5.0
333.8	72.2 ± 1.0	95.0 ± 1.0	93.2 ± 3.0	76.6 ± 5.0	−25.6 ± 3.0	−12.7 ± 5.0
354.9	67.6 ± 1.0	98.7 ± 1.0	118.2 ± 5.0	78.6 ± 5.0	−13.7 ± 5.0	−15.1 ± 5.0

x_1', mole fraction in lower liquid phase; x_1, mole fraction in upper liquid phase; V', molar volume in lower liquid phase; V, molar volume in upper liquid phase; $V^{E\prime}$, Volumetric Properties in lower liquid phase; V^E, Volumetric Properties in upper liquid phase

Mixture 40: C_2H_5F (1) + [bmim][PF_6] (2)
(1) Phase Equilibrium

Ref. [197]

T/K	100 x_1'	100 x_1	V'/ cm^3 · mol^{-1}	V/ cm^3 · mol^{-1}	$V^{E\prime}$/ cm^3 · mol^{-1}	V^E/ cm^3 · mol^{-1}
243.2	76.1 ± 2.0	102.1 ± 2.5	72.3 ± 3.0	49.7 ± 2.0	−21.4 ± 3.0	−10.4 ± 2.5
263.2	71.8 ± 1.0	102.0 ± 2.0	78.0 ± 1.0	52.6 ± 1.0	−24.5 ± 1.0	−10.3 ± 2.0
293.7	68.9 ± 1.0	100.2 ± 0.3	83.1 ± 1.0	59.7 ± 1.0	−28.4 ± 1.0	−8.5 ± 1.0
324.6	62.5 ± 1.0	100.0 ± 2.0	104.6 ± 5.0	73.6 ± 5.0	−22.3 ± 5.0	−2.3 ± 5.0
338.6	60.8 ± 1.0	100.8 ± 2.0	120.4 ± 15.0	81.7 ± 5.0	−12.5 ± 15.0	0.5 ± 10.0

x_1', mole fraction in lower liquid phase; x_1, mole fraction in upper liquid phase; V', molar volume in lower liquid phase; V, molar volume in upper liquid phase; $V^{E\prime}$, Volumetric Properties in lower liquid phase; V^E, Volumetric Properties in upper liquid phase

Mixture 41: C_2HF_5 (1) + [bmim][PF_6] (2)
(1) Phase Equilibrium

Ref. [197]

T/K	100 x_1'	100 x_1	V'/ cm^3 · mol^{-1}	V/ cm^3 · mol^{-1}	$V^{E\prime}$/ cm^3 · mol^{-1}	V^E/ cm^3 · mol^{-1}
285.8	69.5 ± 2.0	99.1 ± 0.2	118.9 ± 2.5	96.0 ± 0.5	−10.3 ± 2.5	−0.4 ± 0.5
294.8	61.8 ± 2.0	99.5 ± 0.2	129.2 ± 2.5	100.1 ± 0.5	−11.0 ± 2.5	−0.2 ± 0.5
313.1	48.3 ± 2.0	99.7 ± 0.2	147.5 ± 2.5	110.5 ± 0.5	−14.1 ± 2.5	0 ± 0.5

x_1', mole fraction in lower liquid phase; x_1, mole fraction in upper liquid phase; V', molar volume in lower liquid phase; V, molar volume in upper liquid phase; $V^{E\prime}$, Volumetric Properties in lower liquid phase; V^E, Volumetric Properties in upper liquid phase

Ref. [193]

T/K	x_{11}/mol % liquid i = 1	x_{12}/mol % liquid i = 2	$\overline{V}_1$[a] liquid i = 1	$\overline{V}_2$[a] liquid i = 2	$\overline{V}_1^{ex}$[b] liquid i = 1	$\overline{V}_2^{ex}$[b] liquid i = 2
285.8	69.5 ± 2.0	99.1 ± 0.2	118.9 ± 2.5	96.0 ± 0.5	−10.3 ± 2.5	−0.4 ± 0.5
294.8	61.8 ± 2.0	99.5 ± 0.2	129.2 ± 2.5	100.1 ± 0.5	−11.0 ± 2.5	−0.2 ± 0.5
313.1	48.3 ± 2.0	99.7 ± 0.2	147.5 ± 2.5	110.5 ± 0.5	−14.1 ± 2.5	0 ± 0.5

[a]Observed molar volume [cm^3 mol^{-1}]
[b]Volumetric Properties [cm^3 mol^{-1}]

Mixture 42: $CH_3CON(CH_3)_2$ (1) + [BMIM][PF_6] (2)
(1) Volumetric Properties and Compressibility Properties

Ref. [186]

m_2/ (mol · kg^{-1})	10^3 ρ/ (kg · m^{-3})	u/ (m · s^{-1})	10^6 V_φ/ (m^3 · mol^{-1})	10^{12} · κ_s/ Pa^{-1}	10^{14} κ_φ/ (m^3 · mol^{-1} · Pa^{-1})
T = 298.15 K					
0.0125	0.937454	1456.71	195.18	502.69	2.42
0.0255	0.938668	1456.95	196.13	501.88	2.83
0.0621	0.942010	1457.6	196.94	499.65	3.13
0.0894	0.944466	1457.93	197.24	498.13	3.34
0.1235	0.947495	1458.62	197.54	496.07	3.32
0.1351	0.948511	1458.44	197.58	495.66	3.55
0.1613	0.950795	1458.88	197.74	494.17	3.55
0.1905	0.953306	1459.19	197.95	492.66	3.64
0.3037	0.962705	1460.19	198.63	487.18	3.92
0.3683	0.967921	1460.61	198.79	484.27	4.04
0.4697	0.975843	1461.11	199.01	480.01	4.20

Mixture 43: O_2 (1) + [BMIM][PF_6] (2)
(1) Solubility

Ref. [200]

T/K	m_1/ mol · kg^{-1}	Δm_1/ mol · kg^{-1}	p/ MPa	Δp/ MPa	(f_1/m_1) MPa/mol · kg^{-1}	Δ (f_1/m_1) MPa/mol · kg^{-1}
293.2	0.02991	0.00010	1.568	0.027	51.89	0.93
	0.04486	0.00012	2.351	0.029	51.62	0.65
	0.05249	0.00013	2.779	0.037	52.00	0.71
	0.05994	0.00014	3.165	0.037	51.74	0.64
	0.07725	0.00050	4.176	0.056	52.62	0.80
	0.09408	0.00052	5.167	0.056	53.12	0.67
	0.11236	0.00053	6.205	0.057	53.06	0.57
	0.13000	0.00055	7.315	0.058	53.67	0.51
	0.15837	0.00058	9.055	0.060	53.93	0.43
313.15	0.02243	0.00008	1.164	0.027	51.63	1.21
	0.05189	0.00013	2.746	0.037	52.23	0.72
	0.06610	0.00015	3.539	0.038	52.64	0.59
	0.08280	0.00018	4.498	0.039	53.16	0.48
	0.09894	0.00021	5.434	0.041	53.51	0.43
	0.11880	0.00025	6.569	0.043	53.58	0.38
	0.13900	0.00029	7.799	0.045	54.04	0.34
	0.15927	0.00033	9.086	0.047	54.61	0.32
333.15	0.02371	0.00008	1.229	0.027	51.61	1.14
	0.04171	0.00011	2.178	0.028	51.82	0.69
	0.05540	0.00013	2.915	0.037	52.07	0.67
	0.08036	0.00046	4.338	0.054	53.15	0.74

(continued)

T/K	m_1/ mol · kg^{-1}	Δm_1/ mol · kg^{-1}	p/ MPa	Δp/ MPa	(f_1/m_1) MPa/mol · kg^{-1}	$\Delta\ (f_1/m_1)$ MPa/mol · kg^{-1}
	0.10781	0.00049	5.887	0.055	53.48	0.56
	0.12979	0.00052	7.194	0.056	54.03	0.49
	0.14824	0.00053	8.250	0.057	54.05	0.43
	0.16164	0.00054	9.116	0.058	54.60	0.41
353.2	0.02612	0.00008	1.343	0.027	51.24	1.04
	0.04100	0.00011	2.118	0.028	51.37	0.69
	0.05692	0.00013	2.982	0.037	51.99	0.66
	0.07996	0.00044	4.279	0.053	52.93	0.72
	0.10377	0.00046	5.581	0.054	53.02	0.57
	0.12323	0.00048	6.681	0.055	53.29	0.49
	0.14241	0.00051	7.756	0.056	53.39	0.44
	0.16243	0.00053	8.962	0.057	53.92	0.39
373.1	0.02708	0.00008	1.371	0.027	50.49	1.00
	0.04718	0.00011	2.433	0.028	51.34	0.61
	0.06687	0.00042	3.449	0.051	51.25	0.83
	0.08717	0.00043	4.567	0.052	51.96	0.65
	0.10810	0.00045	5.702	0.053	52.21	0.54
	0.12642	0.00047	6.740	0.054	52.67	0.47
	0.14491	0.00049	7.742	0.055	52.69	0.42
	0.16338	0.00052	8.822	0.056	53.14	0.38

(2) Henry's Law Constant

Ref. [200]

T/K	$k^{(0)}_{H,O_2}$/MPa	$\Delta k^{(0)}_{H,O_2}$/MPa
293.2	51.09	0.68
313.15	51.32	0.52
333.15	51.13	0.63
353.2	50.92	0.63
373.1	50.27	0.62

Mixture 44: CO_2 (1) + [bmim][PF_6] (2)
(1) Solubility, Diffusivity and Transport Properties

Ref. [148]

T/K	P/MPa	C_0/(mass %)	C_s/(mass %)	$D \times 10^{-11}$/ m^2 · s^{-1}	$x_{1,calc}$/ mol %	$x_{1,meas}$/ mol %
283.05	0.00969	–	–	–	–	0.4
283.05	0.05009	0.10	0.24	2.0	1.7	1.6
283.05	0.10018	0.29	0.46	2.2	3.0	2.9
283.55	0.39956	0.85	1.72	2.6	10.1	10.2
283.75	0.69959	2.15	2.97	3.0	16.5	16.7
283.65	0.99996	3.43	4.22	3.6	22.1	22.4

(continued)

T/K	P/MPa	C_0/(mass %)	C_s/(mass %)	$D \times 10^{-11}$/ $m^2 \cdot s^{-1}$	$x_{1,calc}$/ mol %	$x_{1,meas}$/ mol %
282.05	1.30027	–	–	–	–	28.4
283.05	1.49980	–	–	–	–	30.9
283.05	1.99975	7.51	8.35	5.2	37.5	37.9
298.05	0.01015	–	–	–	–	0.1
298.05	0.05015	0.03	0.14	5.1	1.1	0.9
298.05	0.10020	0.16	0.29	4.8	2.0	1.8
298.15	0.39963	0.65	1.19	4.5	7.2	7.2
298.15	0.70000	1.55	2.07	5.4	12.2	12.2
298.05	0.99941	2.45	2.95	5.7	16.7	16.7
298.05	1.29990	3.31	3.80	6.8	20.8	20.8
298.05	1.49942	4.05	4.41	7.7	23.8	23.3
298.05	1.99919	5.19	5.80	8.1	28.5	29.1
323.25	0.01020	0.01	0.03	11.0	0.2	0.2
323.15	0.05026	0.04	0.09	11.0	0.6	0.6
323.15	0.10020	0.12	0.19	9.7	1.2	1.2
323.25	0.39961	0.50	0.76	10.0	4.7	4.7
323.15	0.70004	1.07	1.29	10.0	7.8	7.9
323.15	0.99979	1.59	1.81	13.0	10.7	10.9
323.15	1.30023	2.13	2.34	13.0	13.4	13.6
323.25	1.50027	2.53	2.70	14.0	15.2	15.5
323.25	1.99978	3.27	3.56	15.0	19.2	19.7
348.15	0.01021	–	–	–	–	0.1
348.05	0.05008	0.02	0.04	24.0	0.2	0.2
348.05	0.10003	0.07	0.11	21.0	0.7	0.7
348.05	0.39969	0.40	0.51	28.0	3.2	3.2
348.05	0.70001	0.78	0.89	24.0	5.5	5.6
347.95	1.00023	–	–	–	–	7.8
348.15	1.30032	–	–	–	–	9.9
348.05	1.49985	–	–	–	–	11.3
348.25	1.99951	2.36	2.45	28.0	14.0	14.9

Ref. [148]

283.15 K	298.15 K	323.15 K	348.15 K
diffusivity ($m^2 \cdot s^{-1}$) (over Pressures (0.1–1.0 MPa))			
$(2.9 \pm 1.9) \times 10^{-11}$	$(6.0 \pm 1.8) \times 10^{-11}$	$(1.2 \pm 0.3) \times 10^{-10}$	$(2.4 \pm 0.8) \times 10^{-10}$

Ref. [201]

$T = 313.15$ K			$T = 323.15$ K			$T = 333.15$ K		
P/MPa	x_2	η/cp	P/MPa	x_2	η/cp	P/MPa	x_2	η/cp
0.10	0.0	92.3	0.10	0.0	60.4	0.10	0.0	44.1
0.93	0.122	50.8	0.57	0.061	48.0	1.50	0.154	39.4
1.98	0.242	43.8	1.54	0.176	41.5	2.48	0.266	36.4

(*continued*)

$T = 313.15$ K			$T = 323.15$ K			$T = 333.15$ K		
P/MPa	x_2	η/cp	P/MPa	x_2	η/cp	P/MPa	x_2	η/cp
2.43	0.275	40.6	2.38	0.266	38.3	4.43	0.398	33.0
3.06	0.332	38.5	3.83	0.369	34.8	6.69	0.475	31.2
3.95	0.389	36.0	5.08	0.433	33.0	8.74	0.523	29.7
4.70	0.430	34.3	6.71	0.494	31.3	9.90	0.541	29.3
5.78	0.489	32.7	7.29	0.518	31.2	10.69	0.562	29.0
6.09	0.503	32.0	8.25	0.542	30.0	11.58	0.571	28.5
6.94	0.543	31.5	10.20	0.586	29.1	12.93	0.583	28.0
8.02	0.576	30.5	11.58	0.604	28.5			
8.53	0.591	30.6						
9.87	0.619	29.5						
10.93	0.639	29.1						

(2) Phase Equilibrium

Ref. [198]

x_1	T (K)	P (MPa)	T (K)	P (MPa)	T (K)	P (MPa)
0.100	298.29	0.59	323.45	0.93	348.38	1.34
	303.35	0.65	328.43	1.01	353.25	1.43
	308.31	0.71	333.39	1.09	358.16	1.53
	313.35	0.78	338.66	1.17	363.45	1.63
	318.35	0.85	343.51	1.26		
0.203	303.21	1.45	328.23	2.24	353.24	3.25
	308.23	1.58	333.25	2.43	358.17	3.45
	313.27	1.74	338.13	2.61	363.27	3.69
	318.24	1.91	343.29	2.83		
	323.19	2.08	348.20	3.07		
0.250	293.32	1.53	318.62	2.50	343.48	3.79
	298.32	1.69	323.39	2.77	348.59	4.05
	303.33	1.88	328.41	2.97	353.31	4.31
	308.34	2.07	333.46	3.25	358.44	4.61
	313.35	2.29	338.59	3.49	363.28	4.93
0.351	293.29	2.43	318.35	4.09	343.39	6.22
	298.28	2.73	323.31	4.47	348.41	6.72
	303.29	3.09	328.28	4.85	353.50	7.22
	308.33	3.37	333.39	5.29	358.40	7.70
	313.32	3.71	338.32	5.76	363.46	8.24
0.399	293.56	2.97	318.31	5.01	343.46	7.73
	298.52	3.29	323.35	5.47	348.44	8.37
	303.28	3.69	328.31	6.01	353.42	9.05
	308.23	4.13	333.38	6.55	358.35	9.67
	313.21	4.51	338.47	7.15	363.54	10.43
0.501	312.97	6.88	327.99	9.72	342.90	13.46
	318.02	7.72	332.73	10.84	347.58	14.80

(*continued*)

x_1	T (K)	P (MPa)	T (K)	P (MPa)	T (K)	P (MPa)
	323.00	8.68	337.82	12.10	352.81	16.30
0.598	313.00	25.36	327.94	31.70	342.82	39.60
	317.99	26.74	332.67	34.50	347.57	42.18
	322.96	29.20	337.81	37.10	352.53	43.90
0.650	313.03	52.66	327.91	61.36	342.52	68.78
	318.03	55.60	332.87	63.90	347.49	71.14
	322.99	58.62	337.79	66.30	352.60	73.50

Ref. [198]

T (K)	P (MPa)	T (K)	P (MPa)
293.14	5.73	298.23	6.43
293.16	5.73	299.20	6.59
294.38	5.89	299.25	6.59
295.62	6.07	300.23	6.74
295.73	6.09	301.14	6.88
296.18	6.14	302.14	7.04
296.68	6.22	302.20	7.04
297.16	6.29	303.15	7.20
297.46	6.33	304.16	7.36
297.70	6.36	304.20	7.37
298.14	6.43	304.23	7.38

Mixture 45: CH_3COCH_3 (1) + $[C_4mim][PF_6]$ (2)

(1) Volumetric Properties and Transport Properties

Ref. [194]

x_2	ρ (kg · m^{-3})	η (mPa · s)	V^E_m (10^{-6} m^3 · mol^{-1})	$(\ln\eta)^E$ (mPa · s)
298.15 K				
0.00	784.85	0.316	0.00	0.00
0.0519	874.30	0.516	−1.086	0.139
0.1044	946.81	0.810	−1.691	0.235
0.1942	1044.05	1.648	−2.290	0.339
0.3011	1127.27	3.690	−2.358	0.424
0.3481	1156.19	4.986	−2.253	0.408
0.3993	1184.71	6.964	−2.177	0.397
0.4505	1209.43	9.680	−2.014	0.380
0.5007	1231.10	13.406	−1.835	0.367
0.5515	1251.55	18.463	−1.726	0.344
0.5965	1267.23	23.998	−1.523	0.303
0.7077	1302.50	46.683	−1.271	0.218
0.8020	1326.74	84.774	−0.817	0.178
0.8853	1345.33	136.974	−0.453	0.096
1.00	1367.88	269.959	0.00	0.00

Ref. [135]

x_2	η/(mPa · s)	x_2	η/(mPa · s)	x_2	η/(mPa · s)
298.15 K					
0.00	0.316	0.3481	4.986	0.5965	23.998
0.0519	0.516	0.3993	6.964	0.7077	46.683
0.1044	0.810	0.4505	9.680	0.8020	84.774
0.1942	1.648	0.5007	13.406	0.8853	136.974
0.3011	3.690	0.5515	18.463	1.00	269.959

(2) Electrical Conductivity

Ref. [135]

x_2	$10^3\kappa$ (S · cm^{-1})	x_2	$10^3\kappa$ (S · cm^{-1})	x_2	$10^3\kappa$ (S · cm^{-1})
298.15 K					
0.0519	25.69	0.3993	16.97	0.7077	5.46
0.1067	31.21	0.4505	14.47	0.8020	3.75
0.2130	28.77	0.5007	12.30	0.8853	2.61
0.3011	21.41	0.5515	9.96	1.00	1.55
0.3481	19.85	0.5965	8.38		

Mixture 46: 3-$C_2H_5COC_2H_5$ (1) + [C_4mim][PF_6] (2)
(1) Volumetric Properties and Transport Properties

Ref. [194]

x_2	ρ (kg · m^{-3})	η (mPa · s)	V^E_m (10^{-6} m^3 · mol^{-1})	$(\ln\eta)^E$ (mPa · s)
298.15 K				
0.00	810.86	0.458	0.00	0.00
0.0520	870.89	0.690	−0.774	0.078
0.0981	917.86	0.972	−1.170	0.126
0.1993	1006.35	2.022	−1.641	0.214
0.3017	1080.18	4.033	−1.793	0.251
0.3480	1109.30	5.360	−1.773	0.240
0.3990	1138.66	7.366	−1.684	0.224
0.4497	1165.69	10.023	−1.582	0.217
0.4991	1190.78	13.360	−1.548	0.189
0.5509	1214.94	17.939	−1.453	0.154
0.6014	1236.56	24.258	−1.302	0.133
0.6986	1273.62	42.696	−0.887	0.079
0.8073	1311.36	81.387	−0.581	0.030
0.8912	1336.51	136.738	−0.198	0.011
1.00	1367.88	269.959	0.00	0.00

Mixture 47: $(CH_2)_4CO$ (1) + $[C_4mim][PF_6]$ (2)
(1) Volumetric Properties and Transport Properties

Ref. [194]

x_2	ρ (kg · m^{-3})	η (mPa · s)	V^E_m (10^{-6}m^3 · mol^{-1})	$(\ln\eta)^E$ (mPa · s)
298.15 K				
0.00	943.86	1.074	0.00	0.00
0.0482	991.01	1.566	−0.224	0.111
0.1018	1039.43	2.274	−0.676	0.188
0.1959	1106.01	4.188	−0.870	0.278
0.2989	1163.92	7.890	−0.932	0.343
0.3498	1188.58	11.006	−0.968	0.394
0.3966	1210.86	14.519	−0.932	0.412
0.4486	1230.69	19.231	−0.867	0.406
0.4975	1246.27	24.709	−0.781	0.386
0.5525	1264.18	33.036	−0.690	0.373
0.5988	1278.14	42.163	−0.579	0.361
0.6907	1302.86	66.012	−0.435	0.301
0.8117	1330.69	114.768	−0.166	0.185
0.9064	1350.22	174.670	−0.048	0.082
1.00	1367.88	269.959	0.00	0.00

Mixture 48: DODCI (3,3′-Di iethyloxadicarbocyanine iodide) (1) + $[bmim][PF_6]$ (2)
(1) Emission Properties

Ref. [202]

Medium	ϕ_f	τ_f (ns)	τ_{rot} (ns)	$k_r \times 10^{-9}$ s^{-1}	$k_{nr} \times 10^{-9}$ s^{-1}
$[bmim][PF_6]$	0.93	3.64	2.56	0.256	0.019

Mixture 49: $C_4H_6O_3$ (propylene carbonate) (1) + $BMIPF_6$ (2)
(1) Electrical Conductivity

Ref. [151]

Composition (mol%)		
Solvent	Ionic liquid	σ (mS · cm^{-1}) at 298.15 K
80 % PC	20 % $BMIPF_6$	9.37

(2) Potential Limits

Ref. [151]

Sample	Cathodic limit (V)	Anodic limit (V)
$BMIPF_6$-PC	−1.8	2.8

Mixture 50: $[CH_2\text{-}CHCN]_n$ (poly(acrylonitrile)) (1) + $BMIPF_6$ (2)
(1) Electrical Conductivity

Ref. [38]

Composition (wt.%)		
Polymer	Ionic liquid	σ (mS/cm) at 298.15 K
62.7 % PAN	62.7 % $BMIPF_6$	0.0003
29.3 % PAN	70.7 % $BMIPF_6$	0.02

Mixture 51: SCH = CHCH = CH (thiophene) (1) + $[BMIM][PF_6]$ (2)
(1) Vapor Pressure and Activity Coefficient

Ref. [144]

T (K)	p^{exp} (kPa)	p^{nrtl} (kPa)	$RD\ (p)$	γ_1^{exp}	γ_1^{NRTL}
$x_1 = 0.7501$					
303.77	12.5	12.5	0.13	1.1986	1.2131
308.47	15.2	15.4	1.22	1.1839	1.2100
313.24	19.5	18.8	3.47	1.2442	1.2070
317.95	23.1	22.8	1.26	1.2123	1.2040
322.96	27.8	27.8	0.10	1.1950	1.2009
328.21	33.8	33.9	0.20	1.1885	1.1977
332.93	39.6	40.2	1.58	1.1692	1.1949
338.12	47.8	48.3	1.01	1.1735	1.1919
342.84	56.8	56.7	0.22	1.1860	1.1891
$x_1 = 0.5997$					
303.62	11.2	11.1	0.95	1.3382	1.3540
308.44	14.1	13.7	2.66	1.3672	1.3519
313.10	16.9	16.7	1.00	1.3438	1.3498
317.96	20.3	20.4	0.57	1.3219	1.3475
323.09	24.9	25.0	0.37	1.3241	1.3449
328.19	30.6	30.3	0.90	1.3406	1.3423
333.12	36.7	36.3	1.07	1.3413	1.3397
338.00	43.5	43.1	0.87	1.3365	1.3370
342.88	51.0	50.9	0.17	1.3246	1.3344

(continued)

T (K)	p^{exp} (kPa)	p^{nrtl} (kPa)	$RD\ (p)$	γ_1^{exp}	γ_1^{NRTL}
$x_1 = 0.4001$					
303.13	7.7	7.9	2.11	1.3681	1.4708
308.32	9.8	9.9	1.28	1.3981	1.4731
313.33	11.9	12.3	3.51	1.3778	1.4750
318.17	14.8	15.1	1.71	1.4108	1.4765
322.94	17.8	18.2	2.32	1.4076	1.4777
328.22	22.4	22.3	0.39	1.4518	1.4787
333.15	26.6	26.8	0.67	1.4391	1.4793
338.14	31.8	32.0	0.62	1.4421	1.4797
342.99	37.6	37.8	0.57	1.4443	1.4799
$x_1 = 0.1997$					
303.24	4.2	4.0	3.92	1.4415	1.5049
308.18	5.3	5.1	4.59	1.4858	1.5129
313.34	6.4	6.3	0.92	1.4536	1.5208
318.07	7.8	7.7	0.75	1.4680	1.5275
323.14	9.6	9.5	0.92	1.4850	1.5342
328.24	11.7	11.6	0.77	1.4947	1.5404
333.12	14.1	14.0	1.05	1.5087	1.5459
338.32	16.9	16.8	0.30	1.5059	1.5514
343.05	20.2	19.9	1.56	1.5324	1.5560

Mixture 52: $CH_2 = C(CH_3)COOCH_3$ (1) + $[BMIM][PF_6]$ (2)
(1) Volumetric Properties and Transport Properties

Ref. [203]

x_1	283.15 K	288.15 K	293.15 K	298.15 K	303.15 K	308.15 K	313.15 K	308.15 K	323.15 K	328.15 K	333.15 K	343.15 K	353.15 K
$\rho/(g \cdot cm^{-3})$													
0.0986	1.360727	1.356381	1.352041	1.347842	1.343673	1.339494	1.335321	1.331153	1.326996	1.322846	1.318711	1.310465	1.302231
0.2094	1.335266	1.330982	1.326717	1.322446	1.318175	1.313914	1.309659	1.305410	1.301169	1.296917	1.292662	1.284513	1.275985
0.3073	1.310417	1.306048	1.301685	1.297323	1.292970	1.288620	1.284277	1.279937	1.275582	1.271138	1.267356	1.259315	1.250714
0.4053	1.282194	1.277721	1.273259	1.268808	1.264369	1.259943	1.255531	1.251130	1.246737	1.242359	1.237992	1.229290	1.220625
0.4977	1.250448	1.245869	1.241301	1.236747	1.232201	1.227671	1.223150	1.218637	1.214134	1.209647	1.205167	1.196230	1.187324
0.5993	1.211208	1.206490	1.201584	1.197089	1.192403	1.187726	1.183054	1.178390	1.173733	1.169083	1.164441	1.155167	1.145906
0.7009	1.164351	1.159456	1.154070	1.149686	1.144808	1.139934	1.135064	1.130194	1.125325	1.120457	1.115589	1.105843	1.096084
0.8003	1.109097	1.103970	1.098448	1.093722	1.088593	1.083459	1.078316	1.073169	1.068012	1.062848	1.057670	1.047287	1.036871
0.9007	1.042117	1.036709	1.031096	1.025867	1.020427	1.014968	1.009493	1.003999	0.998489	0.992980	0.987485	0.976413	0.965130
$\eta/(mPa \cdot s)$													
0.0986	434.9	308.3	224.9	168.1	128.5	100.5	80.38	62.98	51.32	42.36	35.43	25.61	19.25
0.2094	218.0	161.2	122.3	94.88	75.44	58.33	47.27	38.91	32.45	27.39	23.38	17.56	13.65
0.3073	118.6	91.29	71.80	55.22	44.49	36.45	30.27	25.48	21.70	18.69	16.24	12.57	10.01
0.4053	64.04	50.28	40.30	32.87	27.22	22.86	19.43	16.70	14.51	12.71	11.22	8.922	7.259
0.4977	34.64	28.16	23.27	19.54	16.62	14.29	12.43	10.90	9.640	8.584	7.691	6.278	5.224
0.5993	17.78	14.99	12.81	11.07	9.662	8.505	7.550	6.745	6.066	5.498	4.991	4.187	3.579
0.7009	8.859	7.717	6.781	6.014	5.374	4.838	4.379	3.990	3.653	3.362	3.111	2.708	2.415
0.8003	3.909	3.485	3.136	2.836	2.583	2.369	2.175	2.007	1.868	1.743	1.629	1.432	1.272
0.9007	1.719	1.575	1.450	1.342	1.247	1.165	1.092	1.026	0.9687	0.9201	0.9150	0.8545	0.7810

x_1	283.15 K	288.15 K	293.15 K	298.15 K	303.15 K	308.15 K	313.15 K	308.15 K	323.15 K	328.15 K	333.15 K	343.15 K	353.15 K
$V^E/(cm^3 \cdot mol^{-1})$													
0.0986	−0.4099	−0.4220	−0.4345	−0.4682	−0.4443	−0.5192	−0.5371	−0.5568	−0.5790	−0.6013	−0.6259	−0.6767	−0.7277
0.2094	−0.7624	−0.8078	−0.8565	−0.9057	−0.9014	−0.9874	−1.0245	−1.0642	−1.1071	−1.1485	−1.1908	−1.3304	−1.4189
0.3073	−1.1353	−1.1909	−1.2486	−1.3083	−1.3231	−1.4170	−1.4694	−1.5247	−1.5807	−1.6252	−1.7659	−1.9863	−2.1382
0.4053	−1.4802	−1.5437	−1.6106	−1.6813	−1.7149	−1.8187	−1.8892	−1.9642	−2.0441	−2.1281	−2.2174	−2.4083	−2.6212
0.4977	−1.5972	−1.6661	−1.7389	−1.8162	−1.8628	−1.9702	−2.0504	−2.1352	−2.2259	−2.3214	−2.4229	−2.6391	−2.8815
0.5993	−1.7840	−1.8566	−1.9097	−2.0153	−2.0739	−2.1817	−2.2691	−2.3613	−2.4599	−2.5629	−2.6730	−2.9071	−3.1699
0.7009	−1.8045	−1.8749	−1.8916	−2.0284	−2.0910	−2.1913	−2.2782	−2.3692	−2.4665	−2.5679	−2.6762	−2.9061	−3.1654
0.8003	−1.6199	−1.6789	−1.6968	−1.8083	−1.8644	−1.9464	−2.0201	−2.0975	−2.1800	−2.2660	−2.3576	−2.5527	−2.7772
0.9007	−1.2784	−1.3169	−1.3363	−1.4014	−1.4401	−1.4917	−1.5407	−1.5912	−1.6460	−1.7051	−1.7733	−1.9144	−2.0632

Mixture 53: $CH_3N(CH_2)_3CO$ (N-methyl-2-pyrrolidinone) (1) + [bmim][PF_6] (2)
(1) Volumetric Properties

Ref. [153]

x_2	ρ/ ($g \cdot cm^{-3}$)	V^E_m/ ($cm^3 \cdot mol^{-1}$)	$\overline{V}_2$/ ($cm^3 \cdot mol^{-1}$)	$\overline{V}_1$/ ($cm^3 \cdot mol^{-1}$)	$V_{\varphi 2}$/ ($cm^3 \cdot mol^{-1}$)	$V_{\varphi 1}$/ ($cm^3 \cdot mol^{-1}$)	$\overline{V}^E_2$/ ($cm^3 \cdot mol^{-1}$)	$\overline{V}^E_1$/ ($cm^3 \cdot mol^{-1}$)
T = 298.15 K								
0.09532	1.09591	−0.4898	204.12	96.27	202.85	95.86	−3.870	−0.134
0.19100	1.14965	−0.7463	205.84	95.99	204.08	95.48	−2.155	−0.412
0.28837	1.19383	−0.8470	206.78	95.70	205.05	95.21	−1.213	−0.702
0.38693	1.23081	−0.8655	207.23	95.48	205.75	94.99	−0.764	−0.925
0.48512	1.26178	−0.8260	207.41	95.34	206.29	94.80	−0.579	−1.064
0.58725	1.28938	−0.7698	207.51	95.23	206.68	94.54	−0.486	−1.172
0.68840	1.31291	−0.6881	207.60	95.05	206.99	94.19	−0.387	−1.349
0.78484	1.33234	−0.5717	207.74	94.67	207.26	93.75	−0.254	−1.733
0.89380	1.35113	−0.3577	207.91	93.76	207.59	93.03	−0.086	−2.646
T = 303.15 K								
0.09532	1.09257	−0.5034	204.38	96.60	203.12	96.18	−4.022	−0.133
0.19100	1.14652	−0.7749	206.11	96.32	204.35	95.78	−2.293	−0.414
0.28837	1.19089	−0.8877	207.08	96.02	205.33	95.49	−1.321	−0.713
0.38693	1.22797	−0.9086	207.56	95.78	206.06	95.25	−0.841	−0.953
0.48512	1.25909	−0.8787	207.78	95.62	206.59	95.03	−0.630	−1.112
0.58725	1.28669	−0.8138	207.89	95.49	207.02	94.76	−0.518	−1.241
0.68840	1.31028	−0.7318	208.00	95.29	207.34	94.39	−0.407	−1.441
0.78484	1.32970	−0.6047	208.14	94.89	207.64	93.92	−0.265	−1.848
0.89380	1.34847	−0.3771	208.32	93.94	207.98	93.18	−0.089	−2.800
T = 308.15 K								
0.09532	1.08771	−0.5263	204.99	97.07	203.66	96.62	−4.193	−0.140

x_2	ρ/ (g · cm^{-3})	V^E_m/ (cm^3 · mol^{-1})	$\overline{V}_2$/ (cm^3 · mol^{-1})	$\overline{V}_1$/ (cm^3 · mol^{-1})	$V_{\varphi 2}$/ (cm^3 · mol^{-1})	$V_{\varphi 1}$/ (cm^3 · mol^{-1})	$\overline{V}^E_2$/ (cm^3 · mol^{-1})	$\overline{V}^E_1$/ (cm^3 · mol^{-1})
0.19100	1.14168	−0.8066	206.78	96.77	204.96	96.21	−2.396	−0.432
0.28837	1.18609	−0.9282	207.78	96.47	205.96	95.90	−1.399	−0.738
0.38693	1.22323	−0.9588	208.27	96.22	206.70	95.64	−0.909	−0.982
0.48512	1.25428	−0.9223	208.49	96.06	207.28	95.42	−0.690	−1.148
0.58725	1.28193	−0.8631	208.62	95.91	207.71	95.12	−0.563	−1.295
0.68840	1.30547	−0.7748	208.75	95.68	208.05	94.72	−0.434	−1.526
0.78484	1.32488	−0.6449	208.90	95.23	208.36	94.21	−0.277	−1.977
0.89380	1.34352	−0.3976	209.09	94.22	208.74	93.46	−0.091	−2.982
T = 313.15 K								
0.09532	1.08351	−0.5424	205.45	97.48	204.11	97.02	−4.348	−0.142
0.19100	1.13760	−0.8372	207.28	97.18	205.42	96.58	−2.526	−0.437
0.28837	1.18212	−0.9692	208.30	96.87	206.44	96.26	−1.506	−0.752
0.38693	1.21931	−1.0028	208.81	96.61	207.21	95.98	−0.992	−1.008
0.48512	1.25046	−0.9753	209.05	96.42	207.79	95.72	−0.746	−1.195
0.58725	1.27811	−0.9143	209.21	96.25	208.24	95.40	−0.594	−1.370
0.68840	1.30161	−0.8156	209.36	95.98	208.62	95.00	−0.446	−1.636
0.78484	1.32097	−0.6754	209.52	95.50	208.94	94.48	−0.278	−2.116
0.89380	1.33955	−0.4120	209.71	94.49	209.34	93.74	−0.090	−3.131

Ref. [153]

T/K	V_2^0/ ($cm^3 \cdot mol^{-1}$)	$\overline{V}_1^0$/ ($cm^3 \cdot mol^{-1}$)	$\overline{V}_2^{\infty}$ [a]/ ($cm^3 \cdot mol^{-1}$)	$\overline{V}_1^{\infty}$ [b]/ ($cm^3 \cdot mol^{-1}$)	$\overline{V}_2^{\infty}$ [c]/ ($cm^3 \cdot mol^{-1}$)	$\overline{V}_2^{\infty}$ [d]/ ($cm^3 \cdot mol^{-1}$)
298.15	207.99	96.40	201.26	92.10	201.29	92.06
303.15	208.41	96.73	201.54	92.21	201.57	92.18
308.15	209.18	97.21	201.99	92.47	202.04	92.43
313.15	209.80	97.62	202.42	92.78	202.48	92.72

[a]From equation $\overline{V}_1^{\infty} = V_1^0 + \sum_{i=0}^{i=n} A_i$

[b]From equation $\overline{V}_2^{\infty} = V_2^0 + \sum_{i=0}^{i=n} A_i(-1)^i$

[c]From equation $V_{\Phi 1} = V_1^0 - V_m^E/x$

[d]From equation $V_{\Phi 2} = V_2^0 - V_m^E/(1-x)$

2.2.46.2 Ternary Mixtures

Mixture 1: $HOC_6H_4C(CH_3)_2C_6H_4OH$ (1) + H_2O (2) + $[C_4mim][PF_6]$ (3)
(1) Distribution Ratio

Ref. [204]

pH	$T = 297.15 \pm 1$ K	pH	$T = 297.15 \pm 1$ K	pH	$T = 297.15 \pm 1$ K	pH	$T = 297.15 \pm 1$ K
1.86	389.0	4.95	382.3	9.17	225.0	12.01	0.18
3.26	389.0	7.03	416.1	11.61	0.60		

Mixture 2: C_6H_5OH (1) + H_2O (2) + $[C_4MIm][PF_6]$ (3)
(1) Distribution Ratio

Ref. [204]

pH	$T = 297.15 \pm 1$ K	pH	$T = 297.15 \pm 1$ K	pH	$T = 297.15 \pm 1$ K
3.23	11.4	7.16	11.2	11.68	1.9
5.25	11.3	9.52	7.8		

Mixture 3: $4\text{-}C_9H_{19}C_6H_4OH$ (1) + H_2O (2) + $[C_4mim][PF_6]$ (3)
(1) Distribution Ratio

Ref. [204]

pH	$T = 297.15 \pm 1$ K	pH	$T = 297.15 \pm 1$ K	pH	$T = 297.15 \pm 1$ K	pH	$T = 297.15 \pm 1$ K
3.23[a]	200.0	7.09	200.0	10.20	93.2	12.61	0.49
5.19	210.0	8.43	156.3	11.03	38.6		

[a]$V_W:V_{IL}$ = 25:1 was used at these pHs because of the low solubility of 4-nonylphenol l in water and their very low concentrations in aqueous phase after extraction

Mixture 4: 4-$C_8H_{17}C_6H_4OH$ (1) + H_2O (2) + [C_4mim][PF_6] (3)
(1) Distribution Ratio

Ref. [204]

pH	$T = 297.15 \pm 1$ K	pH	$T = 297.15 \pm 1$ K	pH	$T = 297.15 \pm 1$ K
3.23[a]	124.2	7.04	160.8	12.07	0.81
5.27	123.3	10.22	35.5		

[a]V_W:V_{IL} = 25:1 was used at these pHs because of the low solubility of 4-octylphenol in water and their very low concentrations in aqueous phase after extraction

Mixture 5: C_6Cl_5OH (1) + H_2O (2) + [C_4mim][PF_6] (3)
(1) Distribution Ratio

Ref. [204]

pH	$T = 297.15 \pm 1$ K	pH	$T = 297.15 \pm 1$ K	pH	$T = 297.15 \pm 1$ K
3.16[a]	58.9	7.03	27.4	12.73	14.3
5.12	54.6	10.25	21.5		

[a]V_W:V_{IL} = 25:1 was used at these pHs because of the low solubility of pentachlorophenol in water and their very low concentrations in aqueous phase after extraction

Mixture 6: CO_2 (1) + [bmim][PF_6] (2) + $C_{10}H_8$ (naphthalene) (3)
(1) Infinite dilution partition coefficients

Ref. [205]

313.2 K		333.2 K		353.2 K	
P (MPa)	$K^{\infty}{}_{w3}$	P (MPa)	$K^{\infty}{}_{w3}$	P (MPa)	$K^{\infty}{}_{w3}$
7.74	0.014	5.12	0.093	11.24	0.044
8.63	0.049	7.07	0.116	13.00	0.070
9.04	0.073	9.06	0.157	14.90	0.106
9.84	0.107	10.83	0.206		
11.68	0.164	12.81	0.334		
13.53	0.190	14.65	0.477		

(2) Solubility

Ref. [206]

m_3 (mol/kg) in the liquid phase	x_3 in the liquid phase	x_1 in the liquid phase	c_3 (g/L) in the CO_2-rich phase
$T = 313$ K, $P = 8$ MPa			
0.375	0.0482	0.500	0.322
0.186	0.0245	0.512	0.227
0.0932	0.0124	0.518	0.110
$T = 313$ K, $P = 9$ MPa			
0.363	0.0439	0.53	1.07
0.175	0.0217	0.542	0.912
0.0851	0.0107	0.548	0.595

(*continued*)

m_3 (mol/kg) in the liquid phase	x_3 in the liquid phase	x_1 in the liquid phase	c_3 (g/L) in the CO_2-rich phase
$T = 313$ K, $P = 10$ MPa			
0.315	0.0370	0.550	3.83
0.163	0.0195	0.560	1.58
0.0828	0.0100	0.565	0.724
$T = 313$ K, $P = 12$ MPa			
0.302	0.0311	0.607	4.60
0.150	0.0157	0.616	2.36
0.0777	0.00819	0.621	1.03
$T = 313$ K, $P = 15$ MPa			
0.297	0.0294	0.622	4.91
$T = 313$ K, $P = 20$ MPa			
0.289	0.0281	0.630	5.36
$T = 323$ K, $P = 8$ MPa			
0.378	0.0529	0.454	0.182
0.189	0.0272	0.466	0.0865
0.0943	0.0138	0.472	0.0471
0.0471	0.00692	0.476	0.0333
0.0234	0.00345	0.477	0.0245
$T = 323$ K, $P = 9$ MPa			
0.374	0.0505	0.474	0.411
0.187	0.0259	0.486	0.206
0.0932	0.0131	0.492	0.108
0.0463	0.00656	0.496	0.0658
0.0231	0.00328	0.497	0.0401
$T = 323$ K, $P = 10$ MPa			
0.357	0.0465	0.495	1.41
0.180	0.0240	0.507	0.593
0.0895	0.0121	0.513	0.337
0.0445	0.00605	0.516	0.176
0.0218	0.00298	0.517	0.114
$T = 323$ K, $P = 12$ MPa			
0.337	0.0411	0.530	2.53
0.167	0.0208	0.541	1.35
0.0814	0.0102	0.547	0.816
0.0406	0.00513	0.550	0.412
0.0195	0.00247	0.552	0.252
$T = 323$ K, $P = 15$ MPa			
0.305	0.0359	0.550	4.43
0.153	0.0184	0.560	2.17
0.0747	0.00905	0.565	1.20
0.0370	0.00450	0.567	0.629
0.0177	0.00216	0.569	0.356

(continued)

m_3 (mol/kg) in the liquid phase	x_3 in the liquid phase	x_1 in the liquid phase	c_3 (g/L) in the CO_2-rich phase
$T = 323$ K, $P = 20$ MPa			
0.291	0.0328	0.571	5.27
0.147	0.0168	0.580	2.54
0.07	0.00809	0.585	1.47
0.0348	0.00404	0.588	0.747
0.0169	0.00196	0.589	0.404
$T = 333$ K, $P = 8$ MPa			
0.378	0.0571	0.411	0.162
0.189	0.0294	0.423	0.0954
0.0947	0.0149	0.429	0.0383
0.0471	0.00749	0.433	0.0306
0.0235	0.00376	0.434	0.0165
$T = 333$ K, $P = 9$ MPa			
0.376	0.0546	0.435	0.261
0.188	0.02800	0.447	0.150
0.0939	0.0142	0.453	0.0673
0.0471	0.00717	0.457	0.0358
0.0234	0.00358	0.458	0.0250
$T = 333$ K, $P = 10$ MPa			
0.372	0.0520	0.456	0.519
0.186	0.0266	0.468	0.278
0.0928	0.0135	0.475	0.141
0.0463	0.00679	0.478	0.0662
0.0231	0.00339	0.479	0.0435
$T = 333$ K, $P = 12$ MPa			
0.352	0.0467	0.486	1.70
0.176	0.0239	0.498	0.859
0.0869	0.0120	0.504	0.483
0.0445	0.00616	0.507	0.186
0.0217	0.00301	0.508	0.124
$T = 323$ K, $P = 15$ MPa			
0.318	0.0398	0.519	3.66
0.158	0.0202	0.530	1.88
0.0777	0.0100	0.536	1.03
0.0395	0.00513	0.538	0.474
0.0196	0.00255	0.540	0.245
$T = 323$ K, $P = 20$ MPa			
0.294	0.0352	0.544	5.04
0.145	0.0176	0.554	2.67
0.0700	0.00860	0.559	1.47
0.0359	0.00443	0.562	0.683
0.0177	0.00219	0.563	0.355

Mixture 7: CO_2 (1) + [bmim][PF_6] (2) + C_6H_6 (3)
(1) Infinite Dilution Partition Coefficients

Ref. [205]

313.2 K		333.2 K	
P (MPa)	K^{∞}_{w3}	*P* (MPa)	K^{∞}_{w3}
5.47	0.126	5.53	0.254
7.32	0.223	7.32	0.277
9.31	0.316	9.28	0.297
11.17	0.782	11.16	0.339
13.02	1.046	12.98	0.487
14.83	1.061	14.80	0.559

Mixture 8: CO_2 (1) + [bmim][PF_6] (2) + $C_6H_5CH_3$ (3)
(1) Infinite Dilution Partition Coefficients

Ref. [205]

313.2 K		333.2 K	
P (MPa)	K^{∞}_{w3}	*P* (MPa)	K^{∞}_{w3}
5.37	0.119	5.31	0.185
7.19	0.189	7.18	0.218
9.20	0.308	9.14	0.258
11.11	0.875	10.94	0.298
12.88	0.878	12.84	0.410
14.80	0.899	14.61	0.518

Mixture 9: CO_2 (1) + [bmim][PF_6] (2) + C_6H_5Cl (3)
(1) Infinite Dilution Partition Coefficients

Ref. [205]

313.2 K		333.2 K	
P (MPa)	K^{∞}_{w3}	*P* (MPa)	K^{∞}_{w3}
5.25	0.055	5.12	0.093
7.14	0.098	7.07	0.116
9.09	0.208	9.06	0.157
10.90	0.583	10.83	0.206
12.80	0.930	12.81	0.334
14.50	0.944	14.65	0.477

Mixture 10: C_2H_5OH (1) + 2-$CH_3COC_2H_5$ (2) + [bmim][PF_6] (3)
(1) Distribution Ratio and Selectivity

Ref. [15]

Organic-rich phase		Ionic-liquid-rich phase			
x_1^I	x_2^I	x_1^{II}	x_2^{II}	β	S
298.15 K					
0.973	0.019	0.533	0.025	1.33	2.42
0.965	0.024	0.543	0.041	1.68	2.99
0.957	0.031	0.568	0.054	1.78	3.00
0.946	0.037	0.585	0.067	1.79	2.90
0.910	0.062	0.647	0.087	1.40	1.96
0.869	0.076	0.718	0.104	1.36	1.64

Mixture 11: 2-C_3H_7OH (1) + 2-$CH_3COC_2H_5$ (2) + [bmim][PF_6] (3)
(1) Distribution Ratio and Selectivity

Ref. [15]

Organic-rich phase		Ionic-liquid-rich phase			
x_1^I	x_2^I	x_1^{II}	x_2^{II}	β	S
298.15 K					
0.967	0.033	0.344	0.041	1.27	3.58
0.949	0.049	0.347	0.070	1.43	3.93
0.938	0.059	0.351	0.104	1.76	4.71
0.928	0.069	0.355	0.131	1.90	4.97
0.910	0.086	0.362	0.165	1.92	4.83
0.889	0.105	0.365	0.179	1.70	4.14
0.871	0.122	0.376	0.218	1.80	4.16
0.851	0.140	0.407	0.243	1.73	3.61
0.821	0.159	0.435	0.273	1.71	3.23
0.776	0.185	0.489	0.280	1.51	2.40
0.718	0.221	0.561	0.281	1.27	1.63

Mixture 12: 1-C_6H_{12} (1) + *trans*-3-C_6H_{12} (2) + [BMIM][PF_6] (3)
(1) Phase Equilibrium

Ref. [199]

		y'_1		
x_1	x_2	283.15 K	293.15 K	303.15 K
0.003465	0.03144	0.07435	0.07787	0.08154
0.006920	0.02791	0.1833	0.1894	0.1914
0.01039	0.02444	0.2870	0.2889	0.2920
0.01395	0.02110	0.3781	0.3816	0.3873
0.01717	0.01731	0.4641	0.4713	0.4811
0.02092	0.01406	0.5545	0.5701	0.5786
0.02444	0.01056	0.6457	0.6602	0.6740

(continued)

x_1	x_2	y'_1		
		283.15 K	293.15 K	303.15 K
0.02772	0.006986	0.7239	0.7342	0.7441
0.03110	0.003484	0.8046	0.8108	0.8111

Mixture 13: C_2H_5OH (1) + H_2O (2) + [BMIM][PF_6] (3)
(1) Phase Equilibrium

Ref. [21]

wt.% of IL	T (K)	x_1	y_1	γ_1	γ_2	α_{12}
101.32 kPa						
About 10 %						
9.810	351.41	0.903	0.910	1.025	2.181	1.077
10.442	351.54	0.847	0.866	1.036	2.030	1.169
10.685	351.71	0.788	0.827	1.055	1.886	1.281
9.960	351.93	0.726	0.794	1.090	1.713	1.456
10.504	352.21	0.661	0.748	1.115	1.675	1.524
10.687	352.62	0.627	0.733	1.135	1.584	1.638
10.628	353.03	0.564	0.713	1.205	1.437	1.915
11.140	353.42	0.509	0.696	1.283	1.329	2.205
11.504	353.83	0.448	0.676	1.396	1.237	2.576
11.146	354.34	0.384	0.663	1.562	1.131	3.147
11.083	354.95	0.327	0.657	1.778	1.026	3.943
11.111	355.66	0.264	0.643	2.094	0.949	5.014
10.778	356.47	0.220	0.625	2.372	0.909	5.924
10.683	357.70	0.164	0.597	2.890	0.868	7.540
About 20 %						
20.818	352.01	0.925	0.936	1.030	2.000	1.179
21.134	352.18	0.874	0.897	1.037	1.902	1.247
21.302	352.38	0.815	0.853	1.048	1.826	1.313
21.589	352.70	0.739	0.811	1.085	1.635	1.517
21.801	353.20	0.659	0.774	1.136	1.467	1.769
21.739	353.62	0.603	0.742	1.170	1.411	1.892
22.203	354.01	0.539	0.719	1.250	1.298	2.196
22.749	354.53	0.475	0.704	1.357	1.179	2.622

(2) Cloud Point

Ref. [53]

p (bar)	T (K)	$10^2\ w_3$	$10^2\ w_2$	$10^2 dT/dp$ (K/bar)
1.00	324.15	43.57	1.06	
1.00	322.95	43.38	1.82	
1.00	321.65	43.27	2.24	
1.00	320.15	43.14	2.76	
1.00	317.85	42.88	3.79	
1.00	314.65	42.61	4.84	
1.00	311.45	42.32	5.98	

(*continued*)

p (bar)	T (K)	$10^2\ w_3$	$10^2\ w_2$	$10^2 dT/dp$ (K/bar)
1.00	305.95	41.57	8.84	
1.00	296.55	40.77	11.79	
1.00	293.05	39.90	14.64	
1.00	289.15	38.45	19.67	
1.00	288.35	37.08	24.22	
1.00	289.65	35.22	30.11	
1.00	294.05	32.55	37.96	
1.00	301.45	29.02	47.61	
1.00	302.65	28.54	48.81	
1.00	306.45	27.05	52.50	
1.00	310.05	25.39	56.40	
1.00	317.35	22.08	63.69	
1.00	292.75	82.78	4.63	
1.00	288.35	81.96	9.86	
1.00	286.05	81.53	12.47	
1.00	284.25	81.04	15.24	
1.00	283.25	80.59	17.63	
1.00	283.35	80.22	19.59	
1.00	284.05	79.76	21.83	
1.00	285.05	79.28	24.09	
1.00	287.65	78.56	27.32	
1.00	293.15	77.35	32.28	
1.00	314.95	93.66	86.13	
1.00	307.05	91.50	62.73	
1.00	303.45	90.19	53.59	
1.00	289.05	84.50	31.79	
1.00	285.15	82.31	27.12	
1.00	285.85	76.86	19.37	
1.00	286.75	76.01	18.47	
1.00	289.65	73.10	15.85	
1.00	292.75	70.54	13.96	
1.00	295.45	67.82	12.29	
1.00	300.15	63.20	10.01	
1.00	304.45	58.33	8.16	
1.00	307.95	53.36	6.67	
2.04	288.38	37.15	24.19	−0.05
20.71	288.37			
31.05	288.37			
42.76	288.36			
1.90	288.47	37.90	24.51	
45.70	288.46			
158.70	288.50			
244.40	288.53			

w_3 represents the mass fraction of RTIL, with respect to the entire solution; w_2 represents the mass fraction of water in the mixed solvent

(3) Electrical Conductivity and Transport Properties

Ref. [207]

T/K	x_2	x_1	10^4 k/ $S \cdot cm^{-1}$	η/ $mPa \cdot s$	T/K	x_2	x_1	10^4 k/ $S \cdot cm^{-1}$	η/ $mPa \cdot s$
288.15	0.20	0.00			298.15	0.31	0.31	122.0	
	0.00	0.20	30.2	200.6		0.35	0.35	142.2	
	0.14	0.06	29.8			0.00	0.48	79.8	30.1
	0.06	0.14	30.3			0.24	0.24	87.1	34.8
	0.10	0.10	29.5			0.34	0.14	85.5	
	0.20	0.20	60.6			0.14	0.34	86.4	
	0.31	0.31	101.8		303.15	0.20	0.00	42.9	99.2
	0.35	0.35				0.00	0.20	43.0	88.4
	0.00	0.48	68.0	48.0		0.14	0.06	42.3	
	0.24	0.24	73.3	54.7		0.06	0.14	43.4	
	0.34	0.14				0.10	0.10	42.1	
	0.14	0.34	72.6			0.20	0.20	80.3	
293.15	0.20	0.00	33.0	166.4		0.31	0.31	132.8	
	0.00	0.20	33.6	148.1		0.35	0.35	153.8	
	0.14	0.06	33.4			0.00	0.48	86.3	24.8
	0.06	0.14	34.6			0.24	0.24	95.5	27.2
	0.10	0.10	33.0			0.34	0.14	96.3	
	0.20	0.20	67.7			0.14	0.34	93.8	
	0.31	0.31	111.7		308.15	0.20	0.00	47.8	78.2
	0.35	0.35	130.9			0.00	0.20	47.5	71.5
	0.00	0.48	73.3	37.7		0.14	0.06	46.8	
	0.24	0.24	80.9	44.2		0.06	0.14	48.5	
	0.34	0.14				0.10	0.10	47.1	
	0.14	0.34	79.5			0.20	0.20	88.0	
298.15	0.20	0.00	37.2	128.7		0.31	0.31	140.5	
	0.00	0.20	38.0	113.7		0.35	0.35	162.5	
	0.14	0.06	36.8			0.00	0.48	93.3.	21.0
	0.06	0.14	38.7			0.24	0.24	103.1	23.3
	0.10	0.10	37.6			0.34	0.14	106.1	
	0.20	0.20	73.5			0.14	0.34	99.8	

Mixture 14: DODCI (3,3′-diethyloxadicarbocyanine iodide) (1) + H_2O (2) + [bmim][PF_6] (3)
(1) Emission Properties and Transport Properties

Ref. [202]

Medium	ϕ_f	τ_f (ns)	τ_{rot} (ns)	$k_r \times 10^{-9}$ s^{-1}	$k_{nr} \times 10^{-9}$ s^{-1}	Calculated Transport Properties[a] (cP)	Measured Transport Properties (cP)
[bmim][PF_6] + 0.18 % w/w water	0.80	3.46	2.25	0.231	0.058	56	136
[bmim][PF_6] + 0.71 % w/w water	0.76	3.12	2.14	0.244	0.077	42	92
[bmim][PF_6] + 1.41 % w/w water	0.68	2.94	1.80	0.231	0.108	30	76

[a]Calculated using isomerisation rate

Mixture 15: C_6H_{14} (1) + CH_3CH_2COOH (2) + [bmim][PF_6] (3)
(1) Selectivity

Ref. [75]

Hexane-rich phase		RTIL-rich phase		
x_3'	x_2'	x_3''	x_2''	Selectivity (S)
298.2 K				
0.0013	0.7700	0.7300	0.2600	8
0.0012	0.5900	0.7600	0.2350	33
0.0011	0.3400	0.8100	0.1900	–
0.0010	0.1900	0.9400	0.0600	–

Mixture 16: C_6H_{14} (1) + $CH_3CH_2COOCH = CH_2$ (Vinyl propionate) (2) + [bmim][PF_6] (3)
(1) Selectivity

Ref. [75]

Hexane-rich phase		RTIL-rich phase		
x_3'	x_2'	x_3''	x_2''	Selectivity (S)
298.2 K				
0.0010	0.8289	0.1293	0.8707	–
0.0006	0.5800	0.2000	0.8000	–
0.0005	0.3200	0.3100	0.6900	–
0.0004	0.1800	0.5800	0.4200	–
0.0001	0.0900	0.7700	0.2300	–

Mixture 17: C_6H_{14} (1) + $CH_3CH(Ph)COOCH_2CH_3$ (*rac*-1-Phenylethyl propionate) (2) + [bmim][PF_6] (3)
(1) Selectivity

Ref. [75]

Hexane-rich phase		RTIL-rich phase		Selectivity (*S*)
x_3'	x_2'	x_3''	x_2''	
298.2 K				
0.0033	0.7510	0.2625	0.7373	1206
0.0031	0.5104	0.4264	0.5735	5467
0.0023	0.2711	0.5265	0.4734	12688
0.0012	0.1293	0.7648	0.2351	15810
0.0000	0.0758	0.9227	0.0773	–

Mixture 18: C_6H_{14} (1) + $CH_3CH(Ph)OH$ (*rac*-1-Phenylethanol) (2) + [bmim][PF_6] (3)
(1) Selectivity

Ref. [75]

Hexane-rich phase		RTIL-rich phase		Selectivity (*S*)
x_3'	x_2'	x_3''	x_2''	
298.2 K				
0.0025	0.7918	0.0647	0.9351	1215
0.0018	0.5314	0.2359	0.7640	–
0.0016	0.2875	0.3609	0.6390	15801
0.0010	0.2534	0.4099	0.5901	–
0.0000	0.0751	0.5922	0.4077	50

Mixture 19: 2-CH_3CHCH_3OH (1) + $CH_3COOCH_2CH_3$ (2) + BMIM PF_6(3)
(1) Distribution Ratio and Selectivity

Ref. [17]

Organic-rich phase		Ionic liquid-rich phase		β	*S*
x^I_1	x^I_2	x^{II}_1	x^{II}_2		
298.15 K					
0.954	0.045	0.348	0.081	1.80	4.93
0.926	0.073	0.363	0.112	1.53	3.91
0.897	0.101	0.375	0.139	1.38	3.29
0.864	0.132	0.385	0.172	1.30	2.92
0.853	0.142	0.389	0.194	1.36	2.99
0.836	0.159	0.394	0.210	1.32	2.80
0.809	0.183	0.404	0.242	1.32	2.65
0.781	0.209	0.416	0.276	1.32	2.48
0.764	0.222	0.430	0.297	1.34	2.38
0.730	0.247	0.452	0.315	1.28	2.06
0.701	0.267	0.479	0.330	1.24	1.81
0.669	0.290	0.493	0.346	1.19	1.62

Mixture 20: CO_2 (1) + CH_3COCH_3 (2) + [bmim][PF_6] (3)
(1) Phase Equilibrium

Ref. [208]

P (MPa)	x_1	x_2	x_3	y_1	y_2	y_3	K_1	K_2
313.15 K								
3.0	0.319	0	0.681	1.000	0	–	3.13	–
3.0	0.317	0.172	0.511	0.994	0.0065	–	3.14	0.038
3.0	0.317	0.330	0.353	0.985	0.0152	–	3.11	0.046
3.0	0.352	0.523	0.125	0.970	0.0299	–	2.76	0.057
5.0	0.422	0	0.578	1.000	0	–	2.37	–
5.0	0.423	0.123	0.454	0.994	0.0059	–	2.35	0.048
5.0	0.426	0.327	0.247	0.985	0.0144	–	2.31	0.044
7.0	0.518	0	0.482	1.000	0	–	1.93	–
7.0	0.493	0.127	0.380	0.991	0.0091	–	2.01	0.072
7.0	0.518	0.294	0.188	0.815	0.185	3.5×10^{-4}	1.57	0.629
7.0	0.555	0.375	0.070	0.669	0.312	0.019	1.21	0.832
10.0	0.546	0	0.454	1.000	0	–	1.83	–
10.0	0.537	0.088	0.375	0.971	0.029	–	1.81	0.33
10.0	0.526	0.199	0.275	0.853	0.147	–	1.62	0.74
10.0	0.526	0.318	0.155	0.746	0.254	–	1.42	0.80
10.0	0.597	0.345	0.058	0.652	0.321	0.027	1.09	0.93
15.0	0.599	0	0.401	1.000	0	3.2×10^{-7a}	1.67	–
15.0	0.596	0.078	0.326	0.922	0.078	3.7×10^{-6a}	1.55	1.00
15.0	0.549	0.203	0.249	0.840	0.160	1.8×10^{-4a}	1.53	0.79
15.0	0.545	0.319	0.136	0.759	0.241	6.3×10^{-4a}	1.39	0.76
15.0	0.600	0.335	0.065	0.656	0.319	0.025	1.09	0.95

[a]Concentration of the IL in the top phase were obtained by correlating the data in our previous work

Ref. [208]

P (MPa)	x_2	x_3	z_2	z_3	y_2	K_2	K_{2m}	K_{3m}
313.15 K								
5.0	0.397	0.061	0.395	0.024	0.0185	0.046	0.995	0.393
7.0	0.228	0.270	0.120	4.3×10^{-5}	0.0203	0.089	0.526	1.6×10^{-4}

Mixture 21: $C_4H_6O_3$ (propylene carbonates) (1) + $C_3H_4O_3$ (ethylene carbonates) (2) + $BMIPF_6$ (3)
(1) Electrical Conductivity

Ref. [151]

Composition (mol%)		
Solvent	Ionic liquid	k (mS · cm^{-1}) at 298.15 K
40 % PC, 40 % EC	20 % $BMIPF_6$	8.79

(2) Potential Limits

Ref. [151]

Sample	Cathodic limit (V)	Anodic limit (V)
$BMIPF_6$-PC-EC	−1.8	2.9

Mixture 22: PEOEMA (Poly(2-ethoxyethyl methacrylate)) (1) + $C_4H_6O_3$ (propylene carbonates) (2) + $BMIPF_6$ (3)
(1) Electrical Conductivity

Ref. [151]

Composition (mol%)			
Polymer	Solvent	Ionic liquid	k (mS · cm^{-1}) at 298.15 K
50 % PEOEMA	40 % PC	10 % $BMIPF_6$	0.50
55 % PEOEMA	37 % PC	8 % BMIPF6	0.38
60 % PEOEMA	32 % PC	5 % BMIPF6	0.18

(2) Potential Limits

Ref. [151]

Sample	Cathodic limit (V)	Anodic limit (V)
PEOEMA-PC-$BMIPF_6$	−1.6	2.8

Mixture 23: $[CH_2\text{-}CHCN]_n$ (poly(acrylonitrile)) (1) + $CH_2CH_2CH_2CH_2SO_2$ (sulpholane) (2) + $BMIPF_6$ (3)
(1) Electrical Conductivity

Ref. [38]

Composition (wt.%)			
Polymer	Ionic liquid	TMS (%)	k (mS/cm) at 298.15 K
12.5 % PAN	68.0 % $BMIPF_6$	19.5	1.0
12.6 % PAN	47.3 % $BMIPF_6$	40.1	2.7
12.3 % PAN	33.2 % $BMIPF_6$	54.5	6.3

Mixture 24: CH_3COCH_3 (1) + H_2O (2) + [bmim][PF_6] (3)
(1) Electrical Conductivity and Transport Properties

Ref. [207]

T/K	x_2	x_1	10^4 k/ S · cm^{-1}	η/ mPa · s	T/K	x_2	x_1	10^4 k/ S · cm^{-1}	η/ mPa · s
288.15	0.20	0.00			298.15	0.00	0.48	101.7	18.2
	0.00	0.20	29.8	165.0		0.14	0.34	108.0	
	0.10	0.10	31.5			0.24	0.24	169.1	25.8
	0.14	0.06	28.7			0.34	0.14	94.0	
	0.06	0.14	32.2			0.17	0.41	166.6	
	0.00	0.48	92.5	25.1	303.15	0.20	0.00	42.9	99.2
	0.24	0.24	157.7	37.3		0.00	0.20	42.9	79.8

(*continued*)

T/K	x_2	x_1	10^4 k/ S · cm^{-1}	η/ mPa · s	T/K	x_2	x_1	10^4 k/ S · cm^{-1}	η/ mPa · s
	0.34	0.14	74.8			0.10	0.10	44.2	
	0.14	0.34	96.0			0.14	0.06	41.5	
	0.17	0.41	160.4			0.06	0.14	44.3	
293.15	0.20	0.00	33.0	166.4		0.00	0.48	106.1	16.0
	0.00	0.20	34.2	127.0		0.14	0.34	114.0	
	0.10	0.10	35.5			0.24	0.24	174.1	21.4
	0.14	0.06	33.6			0.34	0.14	106.9	
	0.06	0.14	35.9			0.17	0.41	170.3	
	0.00	0.48	96.5	21.2	308.15	0.20	0.00	47.8	78.2
	0.14	0.37	100.8			0.00	0.20	46.7	64.1
	0.24	0.24	165.0	31.0		0.10	0.10	49.5	
	0.34	0.14	86.0			0.14	0.06	47.0	
	0.17	0.41	162.9			0.06	0.14	48.7	
298.15	0.20	0.00	37.2	128.7		0.00	0.48	110.3	13.8
	0.00	0.20	37.5	100.2		0.24	0.24	178.9	18.6
	0.10	0.10	37.5			0.34	0.14	114.8	
	0.14	0.06	37.3			0.14	0.34	118.2	
	0.06	0.14	39.4			0.17	0.41	173.7	

Mixture 25: CO_2 (1) + CH_3OH (2) + [C_4mim][PF_6] (3)
(1) Phase Equilibrium, Fugacity Coefficients and Transport Properties

Ref. [201]

x_1	x_2	x_3	y_1	y_2	η/cp	K_1	K_2	φ^l_1	φ^l_2	φ^v_1	φ^v_2
313.15 K, 7.15 MPa											
0.570	0.087	0.343	0.983	0.017	27.5	1.72	0.20	1.19	0.0411	0.689	0.210
0.489	0.194	0.317	0.979	0.021	27.6	2.00	0.11	1.38	0.0221	0.690	0.205
0.433	0.265	0.302	0.977	0.023	26.8	2.26	0.09	1.56	0.0175	0.690	0.202
0.403	0.360	0.237	0.975	0.025	26.1	2.41	0.07	1.67	0.0138	0.690	0.199
0.307	0.556	0.137	0.972	0.028	25.2	3.17	0.05	2.19	0.0097	0.691	0.194
313.15 K, 10.00 MPa											
0.554	0.124	0.322	0.927	0.073	28.0	1.67	0.59	0.973	0.0144	0.581	0.0245
0.529	0.162	0.309	0.923	0.077	27.6	1.74	0.48	1.017	0.0113	0.583	0.0237
0.456	0.276	0.268	0.906	0.094	26.8	1.98	0.34	1.171	0.0072	0.589	0.0210
0.317	0.487	0.196	0.886	0.114	25.4	2.79	0.23	1.671	0.0044	0.598	0.0185

2.2.46.3 Quaternary Mixtures

Mixture 1: PEOEMA (Poly(2-ethoxyethyl methacrylate)) (1) + $C_4H_6O_3$ (propylene carbonates) [PC] (2) + $C_3H_4O_3$ (ethylene carbonates) [EC] (3) + $BMIPF_6$ (4)
(1) Electrical Conductivity

Ref. [151]

Composition (mol%)			
Polymer	Solvent	Ionic liquid	k (mS · cm^{-1}) at 298.15 K
50 % PEOEMA	20 % PC, 20 % EC	10 % $BMIPF_6$	0.35

(2) Potential Limits

Ref. [151]

Sample	Cathodic limit (V)	Anodic limit (V)
PEOEMA-PC/EC-$BMIPF_6$	−1.6	2.8

Mixture 2: PEOEMA (Poly(2-ethoxyethyl methacrylate)) (1) + $C_4H_6O_3$ (propylene carbonates) [PC] (2) + $LiPF_6$ (3) + $BMIPF_6$ (4)
(1) Electrical Conductivity

Ref. [151]

Composition (mol%)				
Polymer	Solvent	Ionic liquid	Lithium salt	k (mS · cm^{-1}) at 298.15 K
50 % PEOEMA	40 % PC	9.8 % $BMIPF_6$	1.8 % $LiPF_6$	0.61

(2) Potential Limits

Ref. [151]

Sample	Cathodic limit (V)	Anodic limit (V)
PEOEMA-PC-$BMIPF_6$-$LiPF_6$	−1.5	2.8

Mixture 3: PEOEMA (Poly(2-ethoxyethyl methacrylate)) (1) + $C_4H_6O_3$ (propylene carbonates) [PC] (2) + $LiClO_4$ (3) + $BMIPF_6$ (4)
(1) Electrical Conductivity

Ref. [151]

Composition (mol%)				k (mS · cm^{-1}) at 298.15 K
Polymer	Solvent	Ionic liquid	Lithium salt	
48.4 % PEOEMA	38.8 % PC	9.7 % $BMIPF_6$	3.1 % $LiClO_4$	0.61

(2) Potential Limits

Ref. [151]

Sample	Cathodic limit (V)	Anodic limit (V)
PEOEMA-PC-$BMIPF_6$-$LiClO_4$	−0.9 (−1.4)[a]	2.8

[a]Lower anodic limit value involves the effect of present water traces.

2.2.46.4 Quinary Mixtures

Mixture 1: PEOEMA (Poly(2-ethoxyethyl methacrylate)) (1) + $C_4H_6O_3$ (propylene carbonates) [PC] (2) + $C_3H_4O_3$ (ethylene carbonates) [EC] (3) + $LiPF_6$ (4) + $BMIPF_6$ (5)
(1) Electrical Conductivity

Ref. [151]

Composition (mol%)				
Polymer	Solvent	Ionic liquid	Lithium salt	k (mS · cm^{-1}) at 298.15 K
49.2 % PEOEMA	19.6 % PC, 19.6 % EC	9.8 % $BMIPF_6$	1.8 % $LiPF_6$	0.94

Mixture 2: PEOEMA (Poly(2-ethoxyethyl methacrylate)) (1) + $C_4H_6O_3$ (propylene carbonates) [PC] (2) + $C_3H_4O_3$ (ethylene carbonates) [EC] (3) + $LiClO_4$ (4) + $BMIPF_6$ (5)
(1) Electrical Conductivity

Ref. [151]

Composition (mol%)				
Polymer	Solvent	Ionic liquid	Lithium salt	k (mS · cm^{-1}) at 298.15 K
47.9 % PEOEMA	19.6 % PC, 19.6 % EC	10.3 % $BMIPF_6$	2.6 % $LiClO_4$	0.64

2.2.47 02011-05013: 1-Butyl-3-methylimidazolium dibutylphosphate

Abbreviation:	[BMIM][DBP]
Molecular Formula:	$C_{16}H_{33}N_2O_4P$
Molar Mass:	348.42
Structure:	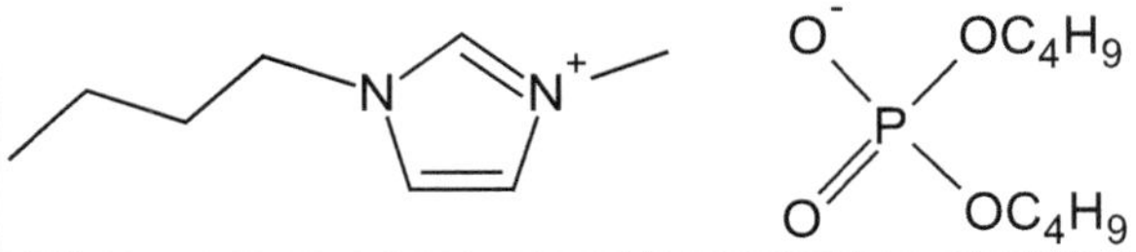

2.2.47.1 Binary Mixtures

Mixture 1: H_2O (1) + [BMIM][DBP] (2)

(1) Vapor Pressure

Ref. [19]

T (K)	P^{exp} (kPa)	P^{NRTL} (kPa)	γ_1^{exp}	γ_1^{NRTL}
IL-content: 20 wt.%				
317.98	9.347	9.339	0.999	0.998
327.23	14.87	14.810	1.002	0.998
335.39	21.746	21.703	1.000	0.998
341.27	28.264	28.231	1.000	0.998
346.32	35.133	35.105	0.999	0.999
351.12	42.896	42.889	0.999	0.999
355.52	51.267	51.259	0.999	0.999
359.23	59.443	59.343	1.000	0.999
362.85	68.319	68.252	1.000	0.999
366.44	78.095	78.138	0.998	0.999
IL-content: 30 wt.%				
318.28	9.426	9.371	1.001	0.995
327.75	15.056	15.002	0.999	0.995
335.18	21.351	21.247	1.000	0.996
341.57	28.398	28.268	1.000	0.996
346.94	35.761	35.597	1.000	0.996
352.11	44.258	44.125	0.999	0.996
356.79	53.410	53.264	0.998	0.996
361.32	63.772	63.601	0.998	0.996
364.91	73.029	72.907	0.997	0.996
367.94	81.63	81.643	0.996	0.996
IL-content: 40 wt.%				
317.98	9.184	9.060	1.003	0.989
328.67	15.542	15.392	0.999	0.989
336.62	22.470	22.270	0.998	0.989
342.86	29.643	29.363	0.999	0.989
348.18	37.166	36.831	0.998	0.989
352.58	44.579	44.171	0.999	0.989
357.21	53.665	53.197	0.998	0.990
361.04	62.318	61.778	0.998	0.990
364.52	71.260	70.554	0.999	0.990
367.81	80.600	79.801	1.000	0.990
IL-content: 70 wt.%				
330.60	13.806	13.912	0.877	0.884
338.90	20.205	20.374	0.878	0.885
344.99	26.505	26.597	0.883	0.886
350.48	33.486	33.519	0.886	0.887
355.36	40.979	40.904	0.889	0.887

(continued)

T (K)	P^{exp} (kPa)	P^{NRTL} (kPa)	γ_1^{exp}	γ_1^{NRTL}
359.92	49.182	48.992	0.891	0.888
364.26	58.075	57.905	0.891	0.888
368.07	67.007	66.821	0.891	0.889

Mixture 2: CH_3OH (1) + [BMIM][DBP] (2)
(1) Vapor Pressure

Ref. [19]

T (K)	P^{exp} (kPa)	P^{NRTL} (kPa)	γ_1^{exp}	γ_1^{NRTL}
IL-content: 20 wt.%				
286.92	8.845	8.818	0.983	0.980
296.97	15.260	15.286	0.980	0.982
304.16	21.981	22.091	0.978	0.983
309.76	28.961	29.048	0.981	0.984
314.51	36.135	36.328	0.980	0.985
318.60	43.721	43.770	0.984	0.985
322.59	52.228	52.240	0.986	0.986
325.89	59.928	60.246	0.981	0.986
329.29	69.181	69.537	0.982	0.987
332.34	78.595	78.898	0.983	0.987
IL-content: 30 wt.%				
286.82	8.431	8.345	0.957	0.947
297.07	14.713	14.674	0.955	0.953
304.71	21.627	21.728	0.952	0.956
309.80	27.865	27.894	0.957	0.958
314.96	35.489	35.577	0.958	0.960
319.24	43.224	43.269	0.961	0.962
322.88	51.125	50.846	0.968	0.963
326.23	58.474	58.790	0.959	0.964
329.29	66.561	66.910	0.960	0.965
332.14	75.006	75.328	0.962	0.966
IL-content: 40 wt.%				
288.72	8.748	8.573	0.911	0.893
298.87	15.142	14.957	0.914	0.903
306.32	21.969	21.901	0.912	0.909
312.08	29.130	29.003	0.918	0.914
317.47	37.666	37.337	0.926	0.918
321.56	45.422	44.930	0.930	0.920
325.37	53.571	53.117	0.931	0.923
328.82	62.340	61.604	0.936	0.925
332.13	71.022	70.775	0.930	0.927
335.10	80.194	79.946	0.931	0.928
IL-content: 60 wt.%				
300.79	12.031	12.014	0.706	0.705

(continued)

T (K)	P^{exp} (kPa)	P^{NRTL} (kPa)	γ_1^{exp}	γ_1^{NRTL}
307.90	17.535	17.441	0.723	0.719
313.87	23.595	23.497	0.733	0.730
320.15	31.621	31.695	0.739	0.741
325.54	40.282	40.545	0.745	0.750
330.02	48.791	49.409	0.747	0.757
334.02	57.845	58.643	0.752	0.763
337.39	66.635	67.509	0.757	0.767

Mixture 3: C_2H_5OH (1) + [BMIM][DBP] (2)
(1) Vapor Pressure

Ref. [19]

T (K)	P^{exp} (kPa)	P^{NRTL} (kPa)	γ_1^{exp}	γ_1^{NRTL}
IL-content: 10 wt.%				
301.83	9.504	9.539	0.993	0.997
310.71	15.437	15.488	0.994	0.997
318.13	22.588	22.662	0.994	0.997
323.39	29.387	29.310	1.000	0.997
327.76	35.946	36.039	0.995	0.997
331.83	43.306	43.428	0.994	0.997
335.71	51.436	51.632	0.993	0.997
339.18	59.886	60.051	0.994	0.997
342.37	68.560	68.762	0.994	0.997
345.13	77.029	77.134	0.996	0.997
IL-content: 20 wt.%				
298.42	7.606	7.608	0.984	0.984
308.89	13.645	13.639	0.985	0.985
316.81	20.568	20.587	0.984	0.985
322.67	27.531	27.484	0.987	0.985
327.64	34.697	34.794	0.983	0.986
332.17	42.736	42.823	0.984	0.986
335.90	50.433	50.579	0.983	0.986
339.38	58.652	58.828	0.983	0.986
342.65	67.448	67.600	0.984	0.986
345.72	76.972	76.792	0.989	0.987
IL-content: 30 wt.%				
301.19	8.480	8.473	0.957	0.956
311.58	15.037	14.956	0.963	0.958
318.85	21.779	21.691	0.963	0.959
324.97	29.202	29.226	0.959	0.960
329.89	36.805	36.787	0.961	0.961
334.17	44.617	44.652	0.960	0.961
338.07	53.059	53.008	0.963	0.962
341.35	61.147	61.025	0.964	0.962

(continued)

T (K)	P^{exp} (kPa)	P^{NRTL} (kPa)	γ_1^{exp}	γ_1^{NRTL}
344.41	69.684	69.402	0.966	0.962
347.07	77.658	77.445	0.965	0.963
IL-content: 60 wt.%				
315.47	11.263	11.321	0.669	0.673
322.24	16.027	15.996	0.681	0.679
328.82	22.057	22.035	0.686	0.685
334.93	29.319	29.266	0.692	0.691
339.44	35.811	35.814	0.695	0.695
344.62	44.862	44.845	0.699	0.699
350.02	55.861	56.205	0.699	0.704
355.12	69.426	69.066	0.711	0.708

2.2.47.2 Ternary Mixtures

Mixture 1: C_2H_5OH (1) + CH_3OH (2) + [BMIM][DBP] (3)
(1) Vapor Pressure

Ref. [20]

T/K	p/kPa	y_1	T/K	p/kPa	y_1
$x_1 = 0.1349, x_2 = 0.7759$					
302.98	14.95	0.1445	326.03	44.61	0.1557
310.24	21.52	0.1483	330.36	53.74	0.1576
316.32	28.82	0.1513	334.08	62.99	0.1591
321.55	36.67	0.1538			
$x_1 = 0.3361, x_2 = 0.5673$					
298.90	10.25	0.4075	325.51	37.54	0.4279
307.25	15.72	0.4148	330.20	46.25	0.4307
314.17	22.12	0.4202	334.46	55.62	0.4331
320.33	29.60	0.4245	338.24	65.18	0.4350
$x_1 = 0.4962, x_2 = 0.4013$					
300.69	9.73	0.6203	327.30	36.51	0.6361
309.65	15.68	0.6265	331.43	43.85	0.6379
316.21	21.73	0.6304	335.63	52.86	0.6397
322.29	28.99	0.6337	339.47	62.14	0.6411
$x_1 = 0.6771, x_2 = 0.2137$					
302.47	9.00	0.8285	328.77	33.81	0.8353
311.79	14.82	0.8314	333.76	42.19	0.8362
318.49	20.72	0.8331	338.02	50.95	0.8369
324.11	27.22	0.8344	341.82	60.16	0.8374
$x_1 = 0.8349, x_2 = 0.05$					
305.84	9.24	0.9667	333.10	36.02	0.9677
315.23	15.31	0.9671	337.44	43.91	0.9677
322.12	21.62	0.9674	341.73	53.32	0.9678
327.78	28.27	0.9676	345.42	62.59	0.9679

2.2.48 02011-06003: 1-Butyl-3-methylimidazolium acetate

Abbreviation:	[C_4MIm][OAc], [BMIM][OAc], [bmim][OAc], [bmim][Ac]
Molecular Formula:	$C_{10}H_{18}N_2O_2$
Molar Mass:	198.27
Structure:	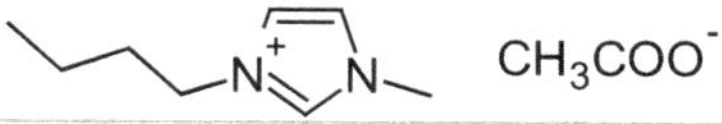

2.2.48.1 Binary Mixtures

Mixture 1: CH_2F_2 (1) + [bmim][Ac] (2)
(1) Solubility and Diffusivity

Ref. [55]

T/K	P/MPa	100 x_1	$10^{11}D$/m$^2 \cdot$ s^{-1}	T/K	P/MPa	100 x_1	$10^{11}D$/m$^2 \cdot$ s^{-1}
298.25	0.0099	1.0	2.2	298.15	0.5498	33.6	10
298.15	0.0997	7.7	2.6	298.15	0.6997	40.1	15
298.15	0.2498	17.4	4.1	298.05	0.8502	46.2	18
298.15	0.4004	26.0	5.6	298.15	1.0004	51.8	21

2.2.48.2 Ternary Mixtures

Mixture 1: H_2O (1) + C_2H_5OH (2) + [bmim][OAc] (3)
(1) Phase Equilibrium

Ref. [24]

x'_2 [a]	w_3 [a]	y_2	T/K	γ_1	γ_2	α_{21} [b]
p = 100 kPa						
0.9501	0.5998	0.9740	377.15	0.60	0.53	1.96
0.9500	0.5000	0.9718	367.48	0.84	0.68	1.81
0.9500	0.4000	0.9691	360.87	1.11	0.81	1.65
0.9500	0.3000	0.9652	356.73	1.39	0.89	1.46
0.9500	0.2000	0.9609	353.72	1.70	0.96	1.29
0.9499	0.0999	0.9550	351.87	2.05	1.00	1.12

[a] Compositions reported are of equilibrium liquid
[b] $\alpha_{21} = (y_2/x'_2)/(y_1/x'_1)$

Mixture 2: H_2O (1) + $(CH_3)_3COH$ (2) + [bmim][OAc] (3)
(1) Phase Equilibrium

Ref. [25]

x'_2 [a]	w_3 [a]	y_2	T/K	γ_1	γ_2	α_{21} [b]
p = 100 kPa						
0.9499	0.6008	0.9822	381.22	0.41	0.63	2.91
0.9499	0.5498	0.9809	376.11	0.49	0.69	2.72
0.9499	0.5001	0.9800	371.77	0.56	0.76	2.58
0.9499	0.4496	0.9784	367.26	0.68	0.85	2.38
0.9500	0.4000	0.9767	363.97	0.80	0.91	2.20
0.9501	0.3002	0.9700	359.80	1.12	0.98	1.70
0.9503	0.1999	0.9620	357.35	1.49	1.01	1.32
0.9502	0.1000	0.9444	355.38	2.24	1.02	0.89
0.9501	0.0450	0.9326	354.83	2.71	1.01	0.73

[a]Compositions reported are of equilibrium liquid
[b]$\alpha_{21} = (y_2/x'_2)/(y_1/x'_1)$

2.2.49 02011-07002: 1-Butyl-3-methylimidazolium tricyanomethane

Abbreviation:	$[C_4mim][C(CN)_3]$
Molecular Formula:	$C_{12}H_{15}N_5$
Molar Mass:	229.28
Structure:	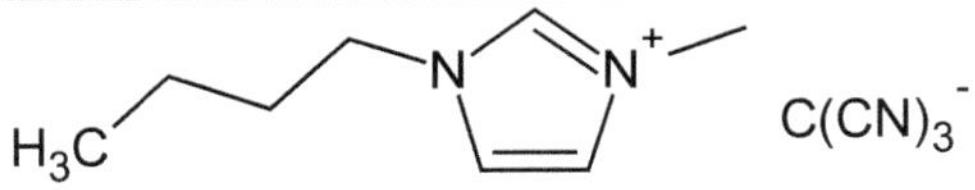

2.2.49.1 Binary Mixtures

Mixture 1: H_2O (1) + $[C_4mim][C(CN)_3]$ (2)
(1) Solubility

Ref. [188]

T/K	$x_1 \pm \sigma^a$	$10^4\ (x_2 \pm \sigma^a)$
288.15	0.844 ± 0.005	5.62 ± 0.01
293.15	0.848 ± 0.003	5.75 ± 0.07
298.15	0.857 ± 0.009	6.22 ± 0.03
303.15	0.863 ± 0.005	6.83 ± 0.01
308.15	0.870 ± 0.002	7.38 ± 0.01
313.15	0.879 ± 0.005	7.99 ± 0.02
318.15	0.889 ± 0.004	9.56 ± 0.05

[a]Standard deviation

(2) Thermodynamic Properties

Ref. [188]

$(\Delta_{sol}H_m{}^0 \pm \sigma^a)$/kJ · mol^{-1}	$(\Delta_{sol}G_m{}^0 \pm \sigma^a)$/kJ · mol^{-1}	$(\Delta_{sol}S_m{}^0 \pm \sigma^a)$/kJ · mol^{-1}
298.15 K		
10.1 ± 1.5	12.59 ± 0.01	−8.5 ± 5.1

[a]Standard deviation

2.2.50 02011-15003: 1-Butyl-3-methylimidazolium thiocyanate

Abbreviation:	[bmim][SCN], [BMIM][SCN]
Molecular Formula:	$C_9H_{15}N_3S$
Molar Mass:	197.30
Structure:	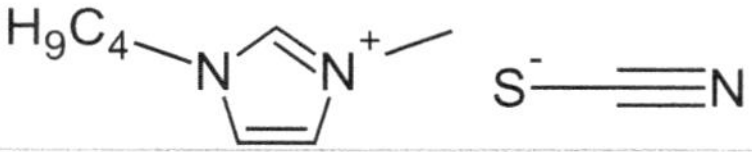

2.2.50.1 Binary Mixtures

Mixture 1: CH_2F_2 (1) + [bmim][SCN] (2)
(1) Solubility and Diffusivity

Ref. [55]

T/K	P/MPa	100 x_1	$10^{11}D$/m^2 · s^{-1}	T/K	P/MPa	100 x_1	$10^{11}D$/m^2 · s^{-1}
298.15	0.0095	0.4	8.1	298.15	0.5500	21.4	
298.15	0.1003	4.1	8.6	298.15	0.6997	27.0	14
298.15	0.2503	10.0	10	298.15	0.8495	32.4	18
298.15	0.3996	15.8	11	298.15	0.9992	37.9	21

Mixture 2: SCH = CHCH = CH (thiophene) (1) + [BMIM][SCN] (2)
(1) Phase Equilibrium

Ref. [110]

x_2	T^{LLE}/K	x_2	T^{LLE}/K
0.2542	354.0	0.2152	271.9
0.2520	349.4	0.2145	274.6
0.2493	341.2	0.2048	262.6
0.2470	332.9	0.1981	249.8
0.2437	326.2	0.1947	244.7
0.2404	320.5	0.1921	233.0
0.2389	315.3	$5.28 \cdot 10^{-4}$	341.1[a]
0.2356	310.3	$3.62 \cdot 10^{-4}$	319.9[a]
0.2317	303.3	$3.44 \cdot 10^{-4}$	313.3[a]

(*continued*)

x_2	T^{LLE}/K	x_2	T^{LLE}/K
0.2300	301.5	$3.26 \cdot 10^{-4}$	308.1[a]
0.2281	299.4	$3.06 \cdot 10^{-4}$	303.3[a]
0.2250	295.8	$3.06 \cdot 10^{-4}$	285.3[a]
0.2233	291.6	$3.26 \cdot 10^{-4}$	283.9[a]
0.2211	287.2	$3.44 \cdot 10^{-4}$	282.5[a]
0.2191	281.5	$3.62 \cdot 10^{-4}$	281.4[a]

[a]Thiophene rich phase

Mixture 3: CH_3OH (1) + [BMIM][SCN] (2)
(1) Volumetric Properties and Transport Properties

Ref. [209]

x_2	298.15 K	308.15 K	318.15 K	328.15 K
$\rho/g \cdot cm^{-3}$				
1.0000	1.06979	1.06389	1.05802	1.05219
0.9046	1.06423	1.05822	1.05227	1.04637
0.8006	1.05746	1.05139	1.04536	1.03938
0.7114	1.05029	1.04413	1.03801	1.03194
0.6227	1.04188	1.03564	1.02943	1.02326
0.5083	1.02810	1.02168	1.01530	1.00899
0.3602	1.00145	0.99474	0.98806	0.98143
0.2076	0.95494	0.94769	0.94047	0.93323
0.1443	0.92463	0.91704	0.90943	0.90181
0.0964	0.89265	0.88464	0.87663	0.86859
0.0531	0.85466	0.84616	0.83763	0.82905
0.0154	0.81000	0.80087	0.79170	0.78241
0.0000	0.78664	0.77715	0.76759	0.75783
$V^E_m/cm^3 \cdot mol^{-1}$				
1.0000	0.0000	0.0000	0.0000	0.0000
0.9046	−0.1416	−0.1455	−0.1552	−0.1684
0.8006	−0.3578	−0.3785	−0.4029	−0.4328
0.7114	−0.5143	−0.5450	−0.5804	−0.6228
0.6227	−0.6878	−0.7319	−0.7803	−0.8361
0.5083	−0.9013	−0.9565	−1.0182	−1.0929
0.3602	−1.0557	−1.1255	−1.2023	−1.2924
0.2076	−1.0830	−1.1579	−1.2409	−1.3341
0.1443	−1.0237	−1.0966	−1.1745	−1.2644
0.0964	−0.8408	−0.9021	−0.9698	−1.0485
0.0531	−0.6038	−0.6494	−0.6990	−0.7586
0.0154	−0.2448	−0.2635	−0.2847	−0.3112
0.0000	0.0000	0.0000	0.0000	0.0000
$\eta/mPa \cdot s$				
1.0000	51.7	34.9	24.2	16.7
0.9046	34.1	23.3	16.9	12.78

(*continued*)

x_2	298.15 K	308.15 K	318.15 K	328.15 K
0.8006	25.4	18.0	13.4	10.32
0.7114	19.4	14.1	10.7	8.40
0.6227	14.5	10.8	8.40	6.68
0.5083	9.75	7.55	6.03	4.85
0.3602	5.15	4.13	3.33	2.82
0.2076	2.34	1.96	1.67	1.45
0.1443	1.62	1.39	1.20	1.06
0.0964	1.18	1.02	0.892	0.792
0.0531	0.871	0.765	0.681	0.613
0.0154	0.665	0.593	0.536	0.490
0.0000	0.577	0.525	0.481	0.443
$\Delta\eta$/mPa · s				
1.0000	0.00	0.00	0.00	0.00
0.9046	−12.72	−8.32	−5.04	−2.37
0.8006	−16.11	−10.05	−6.07	−3.14
0.7114	−17.55	−10.88	−6.65	−3.61
0.6227	−17.91	−11.13	−6.85	−3.89
0.5083	−16.81	−10.45	−6.51	−3.86
0.3602	−13.84	−8.78	−5.69	−3.48
0.2076	−8.85	−5.70	−3.74	−2.37
0.1443	−6.33	−4.10	−2.70	−1.73
0.0964	−4.33	−2.82	−1.88	−1.22
0.0531	−2.42	−1.59	−1.06	−0.69
0.0154	−0.70	−0.46	−0.31	−0.20
0.0000	0.00	0.00	0.00	0.00

Mixture 4: C_2H_5OH (1) + [BMIM][SCN] (2)
(1) Volumetric Properties and Transport Properties

Ref. [209]

x_2	298.15 K	308.15 K	318.15 K	328.15 K	338.15 K	348.15 K
ρ/g · cm^{-3}						
1.0000	1.06979	1.06389	1.05088	1.05219	1.04643	1.04070
0.9174	1.06278	1.05682	1.05088	1.04499	1.03915	1.03334
0.8114	1.05290	1.04684	1.04084	1.03486	1.02895	1.02305
0.6960	1.03972	1.03357	1.02747	1.02140	1.01536	1.00940
0.5306	1.01459	1.00822	1.00190	0.99561	0.98936	0.98314
0.3637	0.97805	0.97137	0.96472	0.95811	0.95147	0.94488
0.2943	0.95741	0.95056	0.94372	0.93687	0.93003	0.92318
0.1877	0.91669	0.90946	0.90220	0.89491	0.88757	0.88022
0.1129	0.87641	0.86874	0.86106	0.85327	0.84544	0.83747
0.0473	0.82974	0.82161	0.81339	0.80505	0.79653	0.78785
0.0000	0.78517	0.77649	0.76767	0.75865	0.74935	0.73975
$V^E{}_m$/cm^3 · mol^{-1}						

(continued)

x_2	298.15 K	308.15 K	318.15 K	328.15 K	338.15 K	348.15 K
1.0000	0.0000	0.0000	0.0000	0.0000	0.0000	0.0000
0.9174	−0.1508	−0.1626	−0.1744	−0.1900	−0.2053	−0.2237
0.8114	−0.4140	−0.4397	−0.4737	−0.5097	−0.5526	−0.5994
0.6960	−0.6545	−0.6997	−0.7539	−0.8146	−0.8808	−0.9659
0.5306	−0.9028	−0.9668	−1.0424	−1.1290	−1.2290	−1.3465
0.3637	−1.0706	−1.1492	−1.2399	−1.3469	−1.4657	−1.6100
0.2943	−1.0789	−1.1610	−1.2542	−1.3601	−1.4847	−1.6306
0.1877	−1.0563	−1.1353	−1.2234	−1.3246	−1.4425	−1.5850
0.1129	−0.8230	−0.8842	−0.9567	−1.0368	−1.1365	−1.2507
0.0473	−0.4779	−0.5164	−0.5599	−0.6106	−0.6703	−0.7433
0.0000	0.0000	0.0000	0.0000	0.0000	0.0000	0.0000
η/mPa · s						
1.0000	51.7	34.9	24.2	16.7	12.73	9.98
0.9174	36.2	24.7	17.8	13.4	10.5	8.39
0.8114	27.4	19.2	14.2	10.9	8.64	7.01
0.6960	19.7	14.3	10.9	8.54	6.84	5.58
0.5306	11.7	8.89	7.00	5.97	4.56	3.80
0.3637	6.36	4.95	3.94	3.24	2.75	1.18
0.2943	4.70	3.77	3.10	2.59	2.20	1.90
0.1877	2.95	2.41	2.01	1.71	1.47	1.28
0.1129	2.07	1.71	1.44	1.23	1.06	0.926
0.0473	1.48	1.23	1.05	0.896	0.780	0.344
0.0000	1.09	0.912	0.774	0.664	0.589	0.512
$\Delta\eta$/mPa · s						
1.0000	0.00	0.00	0.00	0.00	0.00	0.00
0.9174	−11.32	−7.36	−4.46	−1.98	−1.27	−0.80
0.8114	−14.77	−9.24	−5.57	−2.76	−1.80	−1.19
0.6960	−16.64	−10.20	−6.17	−3.29	−2.21	−1.51
0.5306	−16.26	−10.04	−6.23	−3.21	−2.47	−1.73
0.3637	−13.15	−8.31	−5.35	−3.26	−2.25	−1.46
0.2943	−11.29	−7.14	−4.57	−2.79	−1.96	−1.40
0.1877	−7.65	−4.87	−3.16	−1.97	−1.40	−1.01
0.1129	−4.74	−3.03	−1.98	−1.25	−0.90	−0.65
0.0473	−2.01	−1.28	−0.84	−0.53	−0.38	−0.32
0.0000	0.00	0.00	0.00	0.00	0.00	0.00

Mixture 5: 1-C_3H_7OH (1) + [BMIM][SCN] (2)
(1) Volumetric Properties and Transport Properties

Ref. [209]

x_2	298.15 K	308.15 K	318.15 K	328.15 K	338.15 K	348.15 K
ρ/g · cm^{-3}						
1.0000	1.06979	1.06389	1.05802	1.05219	1.04643	1.04070
0.9342	1.06297	1.05702	1.05109	1.04522	1.03940	1.03361

(*continued*)

x_2	298.15 K	308.15 K	318.15 K	328.15 K	338.15 K	348.15 K
0.8050	1.04814	1.04209	1.03608	1.03011	1.02420	1.01832
0.6791	1.03055	1.02439	1.01827	1.01218	1.00614	1.00013
0.5868	1.01547	1.00921	1.00301	0.99684	0.99070	0.98460
0.4577	0.98894	0.98250	0.97607	0.96967	0.96330	0.95697
0.3429	0.95902	0.95234	0.94566	0.93900	0.93234	0.92571
0.2538	0.93041	0.92349	0.91658	0.90963	0.90270	0.89573
0.1849	0.90316	0.89599	0.88881	0.88157	0.87427	0.86694
0.0833	0.85416	0.84663	0.83900	0.83126	0.82339	0.81539
0.0267	0.81906	0.81118	0.80318	0.79500	0.78660	0.77798
0.0000	0.79952	0.79141	0.78315	0.77466	0.76591	0.75684
$V^E{}_m/cm^3 \cdot mol^{-1}$						
1.0000	0.0000	0.0000	0.0000	0.0000	0.0000	0.0000
0.9342	−0.1198	−0.1290	−0.1380	−0.1526	−0.1666	−0.1836
0.8050	−0.4094	−0.4383	−0.4734	−0.5152	−0.5635	−0.6210
0.6791	−0.6391	−0.6865	−0.7430	−0.8089	−0.8853	−0.9766
0.5868	−0.8171	−0.8767	−0.9502	−1.0364	−1.1341	−1.2526
0.4577	−0.9069	−0.9783	−1.0595	−1.1575	−1.2725	−1.4137
0.3429	−0.9114	−0.9846	−1.0690	−1.1724	−1.2937	−1.4445
0.2538	−0.8799	−0.9499	−1.0338	−1.1324	−1.2547	−1.4017
0.1849	−0.7391	−0.7988	−0.8713	−0.9582	−1.0620	−1.1938
0.0833	−0.5275	−0.5740	−0.6255	−0.6887	−0.7647	−0.8608
0.0267	−0.2365	−0.2573	−0.2814	−0.3107	−0.3443	−0.3891
0.0000	0.0000	0.0000	0.0000	0.0000	0.0000	0.0000
$\eta/mPa \cdot s$						
1.0000	51.7	34.9	24.2	16.7	12.7	9.98
0.9342	38.6	26.2	18.7	14.0	10.9	8.69
0.8050	25.8	18.1	13.4	10.4	8.19	6.62
0.6791	20.5	14.8	11.2	8.7	6.97	5.64
0.5868	16.7	12.3	9.37	7.3	5.89	4.81
0.4577	10.9	8.22	6.43	5.1	4.18	3.46
0.3429	7.42	5.71	4.49	3.6	3.01	2.56
0.2538	5.34	4.21	3.40	2.8	2.35	2.00
0.1849	4.27	3.39	2.75	2.27	1.90	1.62
0.0833	2.87	2.30	1.87	1.55	1.30	1.10
0.0267	2.26	1.80	1.46	1.20	1.00	0.848
0.0000	1.94	1.54	1.24	1.02	0.842	0.710
$\Delta\eta/mPa \cdot s$						
1.0000	0.00	0.00	0.00	0.00	0.00	0.00
0.9342	−9.91	−6.50	−3.97	−1.64	−1.06	−0.67
0.8050	−16.24	−10.26	−6.27	−3.29	−2.22	−1.55
0.6791	−15.22	−9.36	−5.65	−2.95	−1.95	−1.36
0.5868	−14.48	−8.83	−5.34	−2.88	−1.93	−1.34
0.4577	−13.88	−8.57	−5.31	−3.07	−2.11	−1.49
0.3429	−11.60	−7.27	−4.63	−2.78	−1.91	−1.33

(*continued*)

x_2	298.15 K	308.15 K	318.15 K	328.15 K	338.15 K	348.15 K
0.2538	−9.24	−5.79	−3.67	−2.20	−1.51	−1.06
0.1849	−6.88	−4.32	−2.74	−1.65	−1.14	−0.81
0.0833	−3.22	−2.02	−1.28	−0.78	−0.54	−0.38
0.0267	−1.01	−0.63	−0.39	−0.24	−0.16	−0.11
0.0000	0.00	0.00	0.00	0.00	0.00	0.00

Mixture 6: 1-C_4H_9OH (1) + [BMIM][SCN] (2)
(1) Volumetric Properties and Transport Properties

Ref. [210]

x_2	298.15 K	308.15 K	318.15 K	328.15 K	338.15 K	348.15 K
ρ/g · cm^{-3}						
1.0000	1.06979	1.06389	1.05802	1.05219	1.04643	1.04070
0.9089	1.05853	1.05257	1.04666	1.04080	1.03499	1.02920
0.7758	1.04000	1.03394	1.02794	1.02200	1.01609	1.01019
0.6683	1.02196	1.01582	1.00972	1.00368	0.99766	0.99168
0.5448	0.99765	0.99139	0.98518	0.97899	0.97282	0.96671
0.4404	0.97354	0.96711	0.96072	0.95435	0.94801	0.94169
0.3778	0.95704	0.95050	0.94400	0.93751	0.93103	0.92456
0.3166	0.93922	0.93255	0.92590	0.91921	0.91260	0.90590
0.2100	0.90340	0.89646	0.88954	0.88259	0.87561	0.86858
0.1513	0.88020	0.87310	0.86598	0.85880	0.85157	0.84426
0.1025	0.85883	0.85155	0.84426	0.83688	0.82939	0.82180
0.0262	0.82088	0.81332	0.80565	0.79784	0.78983	0.78160
0.0000	0.80578	0.79806	0.79021	0.78216	0.77388	0.76532
$V^E{}_m$/cm^3 · mol^{-1}						
1.0000	0.0000	0.0000	0.0000	0.0000	0.0000	0.0000
0.9089	−0.2174	−0.2617	−0.2327	−0.2785	−0.2537	−0.3012
0.7758	−0.5458	−0.5248	−0.5824	−0.5626	−0.6296	−0.6103
0.6683	−0.6859	−0.6707	−0.7394	−0.7230	−0.8026	−0.7865
0.5448	−0.7861	−0.7867	−0.8545	−0.8522	−0.9370	−0.9304
0.4404	−0.8402	−0.8406	−0.9119	−0.9129	−0.9984	−0.9997
0.3778	−0.8364	−0.8472	−0.9090	−0.9209	−0.9977	−1.0101
0.3166	−0.8154	−0.8285	−0.8860	−0.9010	−0.9711	−0.9897
0.2100	−0.7194	−0.7146	−0.7803	−0.7771	−0.8576	−0.8555
0.1513	−0.5925	−0.5929	−0.6455	−0.6446	−0.7108	−0.7103
0.1025	−0.4553	−0.4508	−0.4953	−0.4899	−0.5492	−0.5402
0.0262	−0.1811	−0.1375	−0.1985	−0.1493	−0.2185	−0.1647
0.0000	0.0000	0.0000	0.0000	0.0000	0.0000	0.0000
η/mPa · s						
1.0000	51.7	34.9	24.2	16.7	12.7	9.98
0.9089	39.2	26.5	19.0	14.2	11.0	8.82
0.6683	21.2	15.2	11.5	8.92	7.12	5.78
0.5448	15.3	11.3	8.70	6.86	5.51	4.51

(continued)

x_2	298.15 K	308.15 K	318.15 K	328.15 K	338.15 K	348.15 K
0.4404	11.2	8.44	6.55	5.16	4.19	3.50
0.3778	9.23	7.04	5.50	4.39	3.60	3.02
0.3166	7.53	5.80	4.59	3.72	3.08	2.59
0.2100	5.28	4.12	3.30	2.69	2.24	1.89
0.1513	4.37	3.43	2.75	2.25	1.87	1.58
0.1025	3.74	2.94	2.36	1.92	1.60	1.34
0.0262	2.90	2.27	1.81	1.46	1.20	1.00
0.0000	2.52	1.97	1.56	1.25	1.02	0.846

Mixture 7: 1-$C_5H_{11}OH$ (1) + [BMIM][SCN] (2)
(1) Volumetric Properties and Transport Properties

Ref. [210]

x_2	298.15 K	308.15 K	318.15 K	328.15 K	338.15 K	348.15 K
ρ/g · cm^{-3}						
1.0000	1.06979	1.06389	1.05802	1.05219	1.04643	1.04070
0.9807	1.06714	1.06123	1.05535	1.04952	1.04374	1.03801
0.8557	1.04845	1.04247	1.03653	1.03064	1.02478	1.01898
0.7249	1.02599	1.01994	1.01391	1.00794	1.00201	0.99612
0.6157	1.00435	0.99820	0.99210	0.98601	0.97997	0.97397
0.5130	0.98156	0.97529	0.96905	0.96284	0.95668	0.95053
0.4443	0.96472	0.95835	0.95201	0.94571	0.93941	0.93313
0.3840	0.94869	0.94223	0.93579	0.92937	0.92298	0.91658
0.2698	0.91510	0.90842	0.90176	0.89508	0.88839	0.88169
0.2086	0.89501	0.88821	0.88139	0.87455	0.86768	0.86076
0.1440	0.87176	0.86481	0.85783	0.85079	0.84368	0.83650
0.0865	0.84933	0.84222	0.83506	0.82781	0.82047	0.81298
0.0000	0.81094	0.80352	0.79598	0.78827	0.78036	0.77221
V^E_m/cm^3 · mol^{-1}						
1.0000	0.0000	0.0000	0.0000	0.0000	0.0000	0.0000
0.9807	−0.0534	−0.0572	−0.0615	−0.0682	−0.0724	−0.0811
0.8557	−0.3423	−0.3690	−0.4014	−0.4409	−0.4805	−0.5325
0.7249	−0.5613	−0.6118	−0.6682	−0.7380	−0.8160	−0.9094
0.6157	−0.6447	−0.7072	−0.7831	−0.8683	−0.9683	−1.0890
0.5130	−0.6978	−0.7673	−0.8495	−0.9477	−1.0653	−1.2036
0.4443	−0.7051	−0.7758	−0.8609	−0.9657	−1.0852	−1.2301
0.3840	−0.6813	−0.7524	−0.8376	−0.9415	−1.0672	−1.2162
0.2698	−0.6230	−0.6869	−0.7672	−0.8637	−0.9804	−1.1252
0.2086	−0.5697	−0.6286	−0.6994	−0.7880	−0.8962	−1.0286
0.1440	−0.4589	−0.5079	−0.5684	−0.6424	−0.7325	−0.8458
0.0865	−0.3458	−0.3812	−0.4259	−0.4812	−0.5513	−0.6342
0.0000	0.0000	0.0000	0.0000	0.0000	0.0000	0.0000
η/mPa · s						
1.0000	51.7	34.9	24.2	16.7	12.7	9.98

(continued)

x_2	298.15 K	308.15 K	318.15 K	328.15 K	338.15 K	348.15 K
0.9807	46.9	31.1	21.9	16.2	12.5	9.88
0.8557	34.2	23.4	17.0	12.8	9.98	7.97
0.7249	26.0	18.3	13.5	10.4	8.20	6.60
0.6157	20.0	14.4	10.9	8.49	6.75	5.49
0.5130	15.5	11.4	8.74	6.84	5.48	4.47
0.4443	12.7	9.50	7.31	5.74	4.61	3.78
0.3840	10.8	8.15	6.63	5.05	4.13	3.42
0.2698	7.74	5.89	4.62	3.71	3.04	2.54
0.2086	6.49	4.97	3.92	3.16	2.60	2.17
0.1440	5.38	4.14	3.27	2.63	2.16	1.80
0.0865	4.56	3.15	2.77	2.22	1.81	1.50
0.0000	3.37	2.57	2.00	1.58	1.27	1.03

Mixture 8: 1-$C_6H_{13}OH$ (1) + [BMIM][SCN] (2)
(1) Volumetric Properties and Transport Properties

Ref. [210]

x_2	298.15 K	308.15 K	318.15 K	328.15 K	338.15 K	348.15 K
ρ/g · cm^{-3}						
1.0000	1.06979	1.06389	1.05802	1.05219	1.04643	1.04070
0.9576	1.06302	1.05709	1.05122	1.04538	1.03959	1.03384
0.8495	1.04439	1.03841	1.03248	1.02660	1.02076	1.01496
0.6907	1.01312	1.00702	1.00093	0.99490	0.98897	0.98313
0.5470	0.98120	0.97499	0.96881	0.96266	0.95654	0.95041
0.4227	0.95044	0.94406	0.93770	0.93137	0.92505	0.91873
0.3693	0.93618	0.92971	0.92328	0.91685	0.91044	0.90401
0.3038	0.91771	0.91116	0.90460	0.89805	0.89148	0.88490
0.2417	0.89921	0.89253	0.88585	0.87915	0.87243	0.86569
0.1488	0.86952	0.86268	0.85582	0.84890	0.84190	0.83485
0.0898	0.84902	0.84207	0.83507	0.82795	0.82077	0.81343
0.0000	0.81494	0.80772	0.80039	0.79291	0.78524	0.77736
V^E_m/cm^3 · mol^{-1}						
1.0000	0.0000	0.0000	0.0000	0.0000	0.0000	0.0000
0.9576	−0.1163	−0.1233	−0.1368	−0.1503	−0.1624	−0.1784
0.8495	−0.3346	−0.3632	−0.3995	−0.4434	−0.4898	−0.5458
0.6907	−0.4605	−0.5105	−0.5644	−0.6340	−0.7244	−0.8411
0.5470	−0.5170	−0.5832	−0.6619	−0.7558	−0.8657	−0.9912
0.4227	−0.5281	−0.5950	−0.6757	−0.7756	−0.8941	−1.0355
0.3693	−0.5189	−0.5827	−0.6648	−0.7629	−0.8836	−1.0261
0.3038	−0.4894	−0.5537	−0.6296	−0.7250	−0.8392	−0.9796
0.2417	−0.4552	−0.5104	−0.5801	−0.6663	−0.7736	−0.9077
0.1488	−0.3797	−0.4251	−0.4834	−0.5542	−0.6398	−0.7502
0.0898	−0.2765	−0.3122	−0.3569	−0.4057	−0.4743	−0.5509
0.0000	0.0000	0.0000	0.0000	0.0000	0.0000	0.0000

(continued)

x_2	298.15 K	308.15 K	318.15 K	328.15 K	338.15 K	348.15 K
η/mPa · s						
1.0000	51.7	34.9	24.2	16.7	12.7	9.98
0.9576	44.4	29.7	21.0	15.6	12.0	9.56
0.8495	36.2	24.6	17.7	13.3	10.3	8.26
0.6907	26.1	18.3	13.5	10.3	8.15	6.54
0.5470	18.2	13.1	9.87	7.65	6.07	4.92
0.4227	13.8	10.2	7.71	6.00	4.76	3.88
0.3693	12.1	9.01	6.91	5.39	4.30	3.51
0.3038	10.3	7.67	5.87	4.59	3.67	3.04
0.2417	8.77	6.54	4.98	3.92	3.18	2.65
0.1488	6.83	5.15	3.99	3.17	2.57	2.10
0.0898	5.87	4.42	3.41	2.69	2.17	1.77
0.0000	4.48	3.33	2.55	1.98	1.57	1.28

2.2.50.2 Ternary Mixtures

Mixture 1: C_6H_{14} (1) + SCH = CHCH = CH (thiophene) (2) + [BMIM][SCN] (3)
(1) Selectivity at Infinite Dilution and Capacities at Infinite Dilution

Ref. [169]

Solvent	$S^{\infty}{}_{12}$	k_2^{∞}
T = 298.15 K	n-Hexane (1)/thiophene (2)	
[BMIM][SCN]	186	0.82

Mixture 2: $CH_2(CH_2)_4CH_2$ (1) + SCH = CHCH = CH (thiophene) (2) + [BMIM][SCN] (3)
(1) Selectivity at Infinite Dilution and Capacities at Infinite Dilution

Ref. [169]

Solvent	$S^{\infty}{}_{12}$	k_2^{∞}
T = 298.15 K	Cyclohexane (1)/thiophene (2)	
[BMIM][SCN]	51.3	0.82

Mixture 3: C_6H_6 (1) + SCH = CHCH = CH (thiophene) (2) + [BMIM][SCN] (3)
(1) Selectivity at Infinite Dilution and Capacities at Infinite Dilution

Ref. [169]

Solvent	$S^{\infty}{}_{12}$	k_2^{∞}
T = 298.15 K	Benzene (1)/thiophene (2)	
[BMIM][SCN]	1.7	0.82

2.2.51 02016-01001: 1-Pentyl-3-methylimidazolium chloride

Abbreviation:	[PnMIm]Cl
Molecular Formula:	$C_9H_{17}N_2Cl$
Molar Mass:	188.70
Structure:	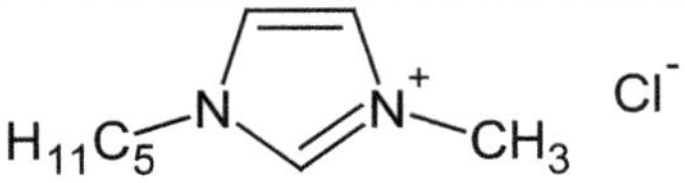

2.2.51.1 Binary Mixtures

Mixture 1: H_2O (1) + [PnMIm]Cl (2)

(1) Osmotic Coefficient

Ref. [211]

m_2(mol · kg^{-1})	T = 298.15 K	T = 308.15 K	T = 318.15 K	T = 328.15 K
Φ				
0.0140	0.931	0.938	0.941	0.949
0.0263	0.889	0.905	0.911	0.923
0.0391	0.882	0.887	0.903	0.908
0.0518	0.871	0.878	0.885	0.897
0.0634	0.863	0.868	0.878	0.886
0.0840	0.852	0.858	0.865	0.870
0.1007	0.841	0.845	0.854	0.866
0.1369	0.821	0.829	0.845	0.849
0.1784	0.805	0.812	0.831	0.837
0.2198	0.797	0.803	0.822	0.828
0.2765	0.790	0.795	0.803	0.809
0.3517	0.781	0.788	0.794	0.799
0.4043	0.768	0.780	0.784	0.790
0.5014	0.750	0.753	0.760	0.771
0.5758	0.735	0.742	0.748	0.756
0.8464	0.722	0.726	0.735	0.739
1.0016	0.710	0.715	0.723	0.730
1.3758	0.697	0.704	0.711	0.718
1.7972	0.687	0.691	0.697	0.704

(2) Electrical Conductivity

Ref. [211]

298.15 K		308.15 K		318.15 K		328.15 K	
c	Λ	c	Λ	c	Λ	c	Λ
0.6930	88.50	0.7520	99.32	0.2730	115.47	0.5880	118.70
1.4530	88.01	1.6430	97.49	0.5920	113.69	1.0760	117.62

(*continued*)

298.15 K		308.15 K		318.15 K		328.15 K	
c	Λ	c	Λ	c	Λ	c	Λ
2.0970	87.28	2.6860	96.35	1.2070	112.21	1.6120	116.75
2.7320	86.94	3.4700	95.61	2.0020	111.25	2.1090	116.41
4.6370	86.38	4.1910	95.95	2.7570	109.82	2.9490	115.95
6.2720	85.78	5.0810	95.32	3.4830	108.02	3.7210	115.57
7.7540	85.18	6.2150	94.45	4.4580	107.34	4.6080	114.63
10.2260	84.05	7.6460	93.70	5.6010	106.49	5.8970	113.41
12.8330	83.24	9.1720	92.68	7.3480	104.92	7.0870	111.27
17.7490	81.65	11.0210	91.68	8.9040	103.80	8.6320	109.73
21.8570	80.44	12.6850	91.16	10.5170	102.96	10.3430	107.84
27.0420	79.33	14.4700	90.62	12.4040	101.97	12.2920	105.69
32.0540	78.28	16.2500	90.17	14.7780	100.62	15.3870	104.78
		17.9940	89.50	16.6060	99.96	17.8760	103.59
		19.8590	88.95	19.5950	98.89	20.1130	102.93
		22.5340	88.06	22.4110	97.76	22.7300	101.07
		25.4910	87.30	25.4460	96.63		
		28.8450	86.30	28.9690	95.27		
		32.4530	85.58				

(3) Refractive Index

Ref. [211]

	n_D			
m_2 (mol · kg^{-1})	298.15 K	308.15 K	318.15 K	328.15 K
0.0140	1.3331	1.3318	1.3302	1.3282
0.0263	1.3334	1.3321	1.3305	1.3284
0.0391	1.3338	1.3324	1.3308	1.3288
0.0518	1.3342	1.3328	1.3313	1.3291
0.0634	1.3344	1.3331	1.3316	1.3294
0.0840	1.3351	1.3339	1.3322	1.3305
0.1007	1.3358	1.3344	1.3327	1.3309
0.1369	1.3367	1.3357	1.3336	1.3316
0.1784	1.3379	1.3367	1.3349	1.3328
0.2198	1.3392	1.3378	1.3359	1.3339
0.2765	1.3408	1.3397	1.3377	1.3356
0.3517	1.3430	1.3416	1.3397	1.3376
0.4043	1.3443	1.3430	1.3411	1.3389
0.5014	1.3470	1.3457	1.3437	1.3419
0.5758	1.3489	1.3475	1.3457	1.3440
0.8464	1.3558	1.3547	1.3523	1.3505
1.0016	1.3596	1.3579	1.3557	1.3537
1.3758	1.3679	1.3658	1.3639	1.3619
1.7972	1.3764	1.3745	1.3721	1.3697

2.2.52 02016-01002: 3-Methyl-1-pentylimidazolium bromide

Abbreviation:	[C_5MIm]Br, [PnMIm]Br
Molecular Formula:	$C_9H_{17}N_2Br$
Molar Mass:	233.15
Structure:	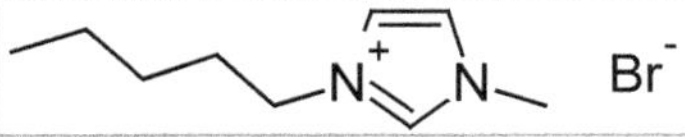

2.2.52.1 Binary Mixtures

Mixture 1: [C_5mim]Br (1) + [C_5mim][NTf_2] (2)
(1) Volumetric Properties

Ref. [212]

[C_5mim]Br/[C_5mim][NTf_2]	$T = 298$ K		
x_{Br}	$\overline{V}_{mix}^{exp\,a}$ (experimental)	$\overline{V}_{mix}^{0\ b}$ (calculated)	$\overline{V}_E^{\ c}$
0.25	278.2	277.3	0.9
0.50	246.8	246.3	0.5
0.75	215.7	215.2	0.5

[a]Experimental molar volume
[b]Calculated molar volume based on eq.
[c]Volumetric Properties

Mixture 2: H_2O (1) + [PnMIm]Br (2)
(1) Osmotic Coefficient

Ref. [112]

	Φ			
$m/(\text{mol} \cdot \text{kg}^{-1})$	298.15 K	308.15 K	318.15 K	328.15 K
0.0179	0.911	0.987	0.940	1.008
0.0250	0.925	0.964	0.924	0.996
0.0301	0.906	0.933	0.912	0.982
0.0518	0.878	0.892	0.895	0.955
0.0722	0.861	0.872	0.882	0.920
0.0904	0.852	0.856	0.871	0.910
0.1155	0.826	0.843	0.857	0.852
0.1697	0.808	0.811	0.831	0.843
0.2113	0.793	0.798	0.816	0.827
0.2308	0.785	0.786	0.813	0.811
0.3311	0.745	0.760	0.776	0.782
0.3787	0.726	0.745	0.762	0.766
0.4716	0.704	0.727	0.749	0.754
0.5627	0.694	0.709	0.720	0.734
0.6532	0.679	0.689	0.697	0.706

(*continued*)

	Φ			
$m/(\text{mol} \cdot \text{kg}^{-1})$	298.15 K	308.15 K	318.15 K	328.15 K
0.7246	0.669	0.682	0.692	0.697
0.8353	0.657	0.666	0.680	0.681
1.0258	0.618	0.637	0.632	0.649
1.5089	0.583	0.588	0.594	0.608
2.0049	0.561	0.565	0.568	0.594

2.2.53 02016-03001: 1-Amyl-3-methylimidazolium bis((trifluoromethyl)sulfonyl)imide

Abbreviation:	$[C_5MIm][NTf_2]$, $[C_5mim][NTf_2]$, $[C_5mim][Tf_2N]$
Molecular Formula:	$C_{11}H_{17}F_6N_3O_4S_2$
Molar Mass:	433.39
Structure:	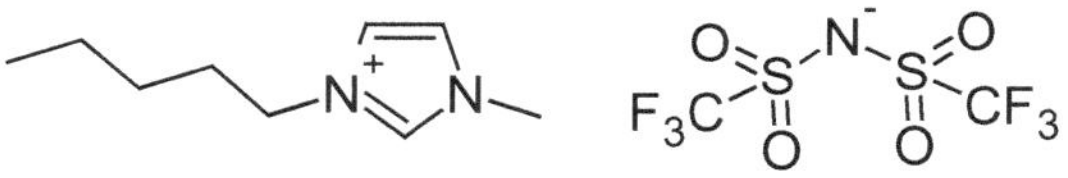

2.2.53.1 Binary Mixtures

Mixture 1: $[C_5mim]Br$ (1) + $[C_5mim][NTf_2]$ (2)

(1) Volumetric Properties

Ref. [212]

These physicochemical properties data are listed in a mixture 1 of IL 02016-01002.

Mixture 2: H_2O (1) + $[C_5mim][NTf_2]$ (2)

(1) Solubility

Ref. [45]

T/K	$x_1 \pm \sigma^a$	$10^4\,(x_2 \pm \sigma)$
288.15	0.2052 ± 0.0005	1.84 ± 0.02
293.15	0.2127 ± 0.0004	1.88 ± 0.03
298.15	0.2211 ± 0.0005	1.96 ± 0.01
303.15	0.2324 ± 0.0004	2.01 ± 0.01
308.15	0.2449 ± 0.0001	2.18 ± 0.01
313.15	0.2570 ± 0.0009	2.32 ± 0.04
318.15	0.2669 ± 0.0007	2.53 ± 0.01

[a]Standard deviation

(2) Thermodynamic Properties

Ref. [45]

$(\Delta_{sol}H_m^{0} \pm \sigma^{a})$/ kJ · mol^{-1}	$(\Delta_{sol}G_m^{0} \pm \sigma^{a})$/ kJ · mol^{-1}	$(\Delta_{sol}S_m^{0} \pm \sigma^{a})$/ J · K^{-1} · mol^{-1}	$(\Delta_{svt}H_m^{0} \pm \sigma^{a})$/ kJ · mol^{-1}	$(\Delta_{svt}Gm^{0} \pm \sigma^{a})$/ kJ · mol^{-1}	$(\Delta_{svt}S_m^{0} \pm \sigma^{a})$/ J · K^{-1} · mol^{-1}	$(\Delta_{svt}H_m^{*} \pm \sigma^{a})$/ kJ · mol^{-1}	$(\Delta_{svt}G_m^{*} \pm \sigma^{a})$/ kJ · mol^{-1}	$(\Delta_{svt}S_m^{*} \pm \sigma^{a})$/ J · K^{-1} · mol^{-1}
298.15 K								
5.9 ± 1.5	21.169 ± 0.009	−51.0 ± 5.0	−156.1 ± 6.2	−59.975 ± 0.009	−322 ± 21	−153.9 ± 6.2	−77.883 ± 0.009	−255 ± 21

[a]Standard deviation

Mixture 3: CO_2 (1) + [C_5mim][Tf_2N] (2)
(1) Bubble Point

Ref. [49]

x_1	T/K	p/MPa	x_1	T/K	p/MPa	x_1	T/K	p/MPa	x_1	T/K	p/MPa	x_1	T/K	p/MPa
0.212	298.65	0.720	0.351	298.77	1.517	0.390	298.65	1.736	0.453	298.70	2.152	0.619	298.55	3.915
0.212	303.61	0.799	0.351	303.34	1.675	0.390	303.39	1.915	0.453	303.38	2.395	0.619	303.33	4.360
0.212	313.15	1.044	0.351	313.21	2.031	0.390	312.99	2.334	0.453	312.85	2.927	0.619	312.90	5.426
0.212	323.35	1.252	0.351	323.05	2.468	0.390	322.52	2.785	0.453	322.62	3.512	0.619	322.72	6.645
0.212	333.25	1.525	0.351	332.85	2.894	0.390	332.54	3.312	0.453	333.07	4.218	0.619	332.50	7.984
0.212	343.00	1.780	0.351	343.19	3.398	0.390	342.47	3.858	0.453	342.53	4.839	0.619	343.04	9.650
0.212	353.49	2.124	0.351	353.26	3.907	0.390	352.85	4.430	0.453	352.65	5.598	0.619	352.86	11.212
0.212	363.29	2.370	0.351	362.68	4.400	0.390	363.05	5.050	0.453	363.05	6.430	0.619	362.95	12.990
0.212	293.75	0.618	0.351	293.65	1.322	0.390	293.73	1.564	0.453	293.75	1.943	0.619	293.75	3.486
0.654	298.47	4.460	0.701	298.45	5.245	0.751	298.42	8.365	0.802	298.37	30.098			
0.654	303.32	5.030	0.701	303.49	5.990	0.751	303.38	10.625	0.802	303.38	32.881			
0.654	313.00	6.344	0.701	313.19	7.732	0.751	313.57	15.435	0.802	293.30	27.460			
0.654	322.96	7.870	0.701	322.85	10.198	0.751	323.34	19.995	0.802	313.44	38.299			
0.654	333.56	9.778	0.701	333.28	13.225	0.751	333.11	24.280	0.802	323.19	43.225			
0.654	343.37	11.776	0.701	343.08	16.438	0.751	343.05	28.266	0.802	333.08	47.960			
0.654	353.30	13.980	0.701	353.06	19.692	0.751	353.20	32.086	0.802	343.14	52.431			
0.654	362.95	16.246	0.701	363.12	22.730	0.751	363.11	35.340	0.802	353.13	56.146			
0.654	293.76	3.974	0.701	293.43	4.617	0.751	293.42	5.609	0.802	363.29	59.805			

(2) Thermodynamic Properties

Ref. [49]

T/K	$\Delta_{sol}H$/J · mol^{-1}	$T\Delta_{sol}S$/(J · mol^{-1})
298.15	−14621.1	−14549.7
303.15	−14446.9	−14429.9
313.15	−14115.2	−14123.1
323.15	−13804.0	−13830.8
333.15	−13511.5	−13492.6
343.15	−13236.1	−13142.6

2.2.54 02016-05001: 1-Amyl-3-methylimidazolium hexafluorophosphate

Abbreviation:	$[C_5MIm][PF_6]$
Molecular Formula:	$C_9H_{17}F_6N_2P$
Molar Mass:	298.21
Structure:	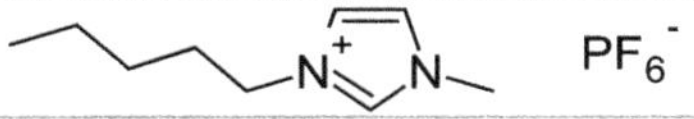

2.2.54.1 Binary Mixtures

Mixture 1: $[C_5mim][TFA]$ (1) + $[C_5mim][PF_6]$ (2)

(1) Volumetric Properties

Ref. [212]

$[C_5mim][PF_6]/[C_5mim][TFA]$	T = 298 K		
x_{PF6}	$\overline{V}_{mix}^{exp\,a}$ (experimental)	$\overline{V}_{mix}^{0\ b}$ (calculated)	$\overline{V}_E^{\ c}$
0.25	223.2	223.1	0.1
0.50	223.3	223.4	−0.1
0.75	223.6	223.7	−0.1

[a]Experimental molar volume
[b]Calculated molar volume based on eq.
[c]Volumetric Properties

2.2.55 02016-06001: 3-methyl-1-pentylimidazolium trifluoroacetate

Abbreviation:	[C_5MIm][TA], [C_5mim][TFA]
Molecular Formula:	$C_{11}H_{17}F_3N_2O_2$
Molar Mass:	266.27
Structure:	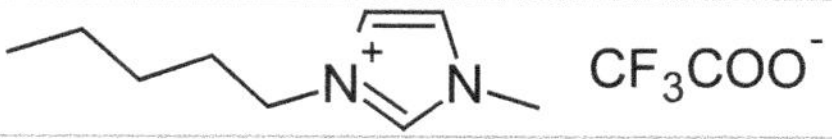

2.2.55.1 Binary Mixtures

Mixture 1: [C_5mim][TFA] (1) + [C_5mim][PF_6] (2)

(1) Volumetric Properties

Ref. [212]

These physicochemical properties data are listed in a mixture 1 of IL 02016-05001.

2.2.56 02021-01001: 1-Hexyl-3-methylimidazolium chloride

Abbreviation:	[C_6MIm]Cl, [Hmim]Cl, [C_6mim]Cl, HmimCl, [hmim]Cl
Molecular Formula:	$C_{10}H_{19}N_2Cl$
Molar Mass:	202.73
Structure:	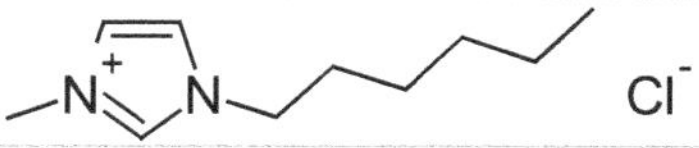

2.2.56.1 Binary Mixtures

Mixture 1: H_2O (1) + [C_6mim]Cl (2)

(1) Volumetric Properties, Refractive Index and Transport Properties

Ref. [213]

x_1	ρ (g · cm^{-3})	n_D	V^E (cm^3 · mol^{-1})	Δn_D
T = 298.15 K				
0.0000	1.0404	1.52182	0.000	0.0000
0.1032	1.0411	1.51718	−0.196	0.0149
0.2001	1.0420	1.51563	−0.395	0.0317
0.3013	1.0422	1.51066	−0.470	0.0459
0.3890	1.0423	1.50890	−0.521	0.0608
0.5125	1.0422	1.49981	−0.564	0.0751
0.5926	1.0412	1.49319	−0.514	0.0836
0.6972	1.0387	1.48167	−0.407	0.0919
0.8007	1.0342	1.46350	−0.285	0.0933
0.9088	1.0236	1.42700	−0.132	0.0773
1.0000	0.9972	1.33240	0.000	0.0000

Ref. [115]

x_1	ρ/ g · cm^{-3}	V^E/ cm^3 · mol^{-1}	$10^3\eta$/ Pa · s	$10^3\Delta\eta$/ Pa · s	x_1	ρ/ g · cm^{-3}	V^E/ cm^3 · mol^{-1}	$10^3\eta$/ Pa · s	$10^3\Delta\eta$/ Pa · s
T = 298.15 K									
0.0000	1.03967	0.000	18089	0.000	0.7127	1.03877	−0.469	60.75	−5108
0.1274	1.04045	−0.224	7667	−8129	0.7830	1.03572	−0.367	33.79	−3929
0.3173	1.04070	−0.373	1020	−11268	0.8935	1.02528	−0.154	8.688	−1931
0.3906	1.04095	−0.444	499.4	−10597	0.9417	1.01817	−0.113	3.505	−1019
0.5594	1.04140	−0.574	201.2	−7748	1.0000	0.99703	0.000	0.895	0.000
T = 313.15 K									
0.0000	1.03113	0.000	3429	0.000	0.7127	1.02980	−0.400	30.04	−948.0
0.1274	1.03189	−0.216	1746	−1245	0.7830	1.02667	−0.292	17.49	−730.4
0.3173	1.03222	−0.366	393.5	−1955	0.8935	1.01620	−0.076	5.191	−363.9
0.3906	1.03263	−0.452	188.9	−1891	0.9417	1.00943	−0.045	2.323	−194.8
0.5594	1.03276	−0.536	88.11	−1431	1.0000	0.99220	0.000	0.658	0.000
T = 328.15 K									
0.0000	1.02269	0.000	942.1	0.000	0.7127	1.02075	−0.340	16.83	−251.99
0.1274	1.02344	−0.214	516.4	−304.55	0.7830	1.01717	−0.210	10.17	−193.95
0.3173	1.02380	−0.363	165.7	−485.28	0.8935	1.00679	−0.009	3.367	−98.265
0.3906	1.02430	−0.460	86.76	−477.48	0.9417	1.00013	0.012	1.649	−54.158
0.5594	1.02414	−0.508	44.87	−375.91	1.0000	0.98568	0.000	0.519	0.000
T = 343.15 K									
0.0000	1.01434	0.000	339.5	0.000	0.7127	1.01145	−0.274	10.39	−86.478
0.1274	1.01506	−0.211	207.4	−88.424	0.7830	1.00771	−0.144	6.542	−67.131
0.3173	1.01550	−0.373	79.07	−156.38	0.8935	0.99700	0.052	2.358	−34.542
0.3906	1.01577	−0.440	45.52	−156.98	0.9417	0.99026	0.064	1.224	−19.167
0.5594	1.01540	−0.474	25.37	−126.923	1.0000	0.97775	0.000	0.419	0.000

(2) Osmotic Coefficient, Activity and Vapor Pressure

Ref. [5]

m_1/(mol · kg^{-1})	Φ_1	a_1	p/kPa	m_1/(mol · kg^{-1})	Φ_1	a_1	p/kPa
T = 313.15 K				T = 333.15 K			
0.0932	0.985	0.9967	7.353	0.0932	0.960	0.9968	19.860
0.2730	0.939	0.9908	7.309	0.2730	0.903	0.9912	19.747
0.4844	0.860	0.9851	7.267	0.4844	0.831	0.9856	19.636
0.6500	0.824	0.9809	7.236	0.6500	0.783	0.9818	19.560
1.0049	0.765	0.9727	7.175	1.0049	0.732	0.9739	19.401
1.1991	0.737	0.9687	7.146	1.1991	0.703	0.9701	19.325
1.3652	0.723	0.9651	7.119	1.3652	0.691	0.9666	19.255
1.5875	0.701	0.9607	7.087	1.5875	0.659	0.9630	19.183
1.7918	0.689	0.9565	7.055	1.7918	0.643	0.9593	19.110
1.9290	0.681	0.9538	7.035	1.9290	0.642	0.9564	19.051
2.2578	0.667	0.9472	6.987	2.2578	0.627	0.9502	18.928
2.6240	0.648	0.9406	6.938	2.6240	0.612	0.9438	18.799

(continued)

m_1/(mol · kg^{-1})	Φ_1	a_1	p/kPa	m_1/(mol · kg^{-1})	Φ_1	a_1	p/kPa
2.9150	0.637	0.9353	6.899	2.9150	0.609	0.9380	18.684
3.2198	0.631	0.9294	6.855	3.2198	0.603	0.9324	18.571
3.5333	0.628	0.9232	6.809	3.5333	0.600	0.9265	18.452
3.9846	0.615	0.9155	6.753	3.9846	0.588	0.9191	18.304
4.1289	0.614	0.9127	6.732	4.1289	0.593	0.9155	18.233

(3) Phase Equilibrium

Ref. [214]

T(K)	x_1	γ_1
101.300 kPa		
373.15	1.000	1.000
373.30	0.998	0.997
373.38	0.996	0.996
373.48	0.994	0.994
373.59	0.990	0.994
373.69	0.986	0.995
374.39	0.965	0.992
374.51	0.961	0.992
374.83	0.952	0.990
374.98	0.947	0.990
375.37	0.941	0.982
375.70	0.934	0.978
375.97	0.931	0.972
376.39	0.925	0.964
376.90	0.917	0.955
377.80	0.909	0.934
379.15	0.894	0.907
384.09	0.847	0.809
389.30	0.803	0.719
396.05	0.740	0.629
407.59	0.640	0.514
415.51	0.580	0.452

(4) Boiling Point

Ref. [215]

w_2	T/K	γ_1
100 kPa		
0.8007	394.82	0.648
0.6999	384.41	0.807
0.5999	379.47	0.896
0.5003	376.62	0.950
0.3997	375.05	0.977

(*continued*)

w_2	T/K	γ_1
0.2998	374.2	0.986
0.2000	373.62	0.992
0.1000	373.19	0.995

Mixture 2: C_2H_5OH (1) + [Hmim]Cl (2)
(1) Volumetric Properties and Refractive Index

Ref. [213]

x_1	ρ (g · cm^{-3})	n_D	V^E (cm^3 · mol^{-1})	Δn_D
T = 298.15 K				
0.0000	1.0404	1.52182	0.000	0.0000
0.0980	1.0328	1.52022	−0.084	0.0143
0.1960	1.0240	1.51336	−0.170	0.0234
0.2898	1.0140	1.50603	−0.229	0.0313
0.4075	0.9997	1.49228	−0.424	0.0367
0.5680	0.9727	1.47741	−0.557	0.0480
0.7044	0.9405	1.45530	−0.696	0.0480
0.7955	0.9097	1.43465	−0.651	0.0422
0.9047	0.8570	1.40175	−0.460	0.0271
0.9445	0.8315	1.38605	−0.355	0.0178
1.0000	0.7854	1.35920	0.000	0.0000

(2) Phase Equilibrium

Ref. [214]

T(K)	x_1	γ_1
101.300 kPa		
351.44	1.000	1.000
351.74	0.992	0.996
351.91	0.985	0.997
352.18	0.975	0.996
352.50	0.965	0.994
352.95	0.952	0.990
353.76	0.935	0.977
354.51	0.919	0.965
355.40	0.903	0.949
357.63	0.875	0.900
358.77	0.862	0.874
360.16	0.847	0.845
364.98	0.808	0.742
367.56	0.785	0.696
371.45	0.751	0.634
377.59	0.706	0.547
379.97	0.687	0.520

(*continued*)

T(K)	x_1	γ_1
386.47	0.643	0.450
391.16	0.618	0.404
397.00	0.584	0.359
405.53	0.540	0.302
416.78	0.503	0.238

(3) Osmotic Coefficient, Activity and Vapor Pressure

Ref. [84]

$m/(\text{mol} \cdot \text{kg}^{-1})$	Φ_1	α_1	p/kPa	$m/(\text{mol} \cdot \text{kg}^{-1})$	Φ_1	α_1	p/kPa
T = 313.15 K				T = 333.15 K			
0.0583	0.955	0.9949	17.784	0.0583	0.7423	0.9960	46.721
0.0975	0.920	0.9918	17.728	0.0975	0.7140	0.9936	46.605
0.4274	0.735	0.9715	17.360	0.4274	0.6079	0.9763	45.776
0.6999	0.695	0.9562	17.083	0.6999	0.5835	0.9631	45.138
1.0138	0.677	0.9387	16.767	1.0138	0.5679	0.9483	44.432
1.3536	0.662	0.9207	16.441	1.3536	0.5613	0.9324	43.667
1.5970	0.659	0.9076	16.204	1.5970	0.5578	0.9212	43.131
1.9237	0.652	0.8909	15.901	1.9237	0.5512	0.9069	42.448
2.2111	0.648	0.8763	15.638	2.2111	0.5532	0.8934	41.802
2.5789	0.636	0.8597	15.339	2.5789	0.5444	0.8787	41.096
2.8590	0.635	0.8460	15.091	2.8590	0.5443	0.8664	40.512
3.2716	0.628	0.8276	14.758	3.2716	0.5347	0.8512	39.783

Mixture 3: [C_6mim]Cl (1) + [C_6mim][NTf_2] (2)
(1) Thermal Conductivity

Ref. [82]

w_1	$\lambda/\text{W} \cdot \text{m}^{-1} \cdot \text{K}^{-1}$	w_1	$\lambda/\text{W} \cdot \text{m}^{-1} \cdot \text{K}^{-1}$
293 K			
0	0.128	0.20	0.132
0.01	0.128	0.50	0.139
0.05	0.129	0.70	0.151
0.10	0.130	1.00	0.172

2.2.56.2 Ternary Mixtures

Mixture 1: C_2H_5OH (1) + H_2O (2) + [C_6mim]Cl (3)
(1) Volumetric Properties and Refractive Index

Ref. [213]

x_1	x_2	ρ (g · cm^{-3})	n_D	V^E (cm^3 · mol^{-1})	Δn_D
T = 298.15 K					
0.0549	0.7567	1.0216	1.45831	−0.418	0.0906
0.1054	0.7163	1.0098	1.45364	−0.525	0.0861

(continued)

x_1	x_2	ρ (g · cm^{-3})	n_D	V^E (cm^3 · mol^{-1})	Δn_D
0.1932	0.6460	0.9892	1.44553	−0.668	0.0779
0.0520	0.8615	1.0084	1.42339	−0.336	0.0718
0.1008	0.8173	0.9943	1.41988	−0.505	0.0676
0.1461	0.7761	0.9814	1.41663	−0.636	0.0637
0.1118	0.4552	1.0255	1.49314	−0.575	0.0745
0.0575	0.5585	1.0319	1.48868	−0.542	0.0821
0.1637	0.4956	1.0150	1.47932	−0.598	0.0791
0.0542	0.6594	1.0287	1.47701	−0.487	0.0889
0.1077	0.6221	1.0183	1.47228	−0.550	0.0863
0.1502	0.5925	1.0101	1.46777	−0.593	0.0839
0.7887	0.0847	0.8849	1.41520	−0.685	0.0372
0.8212	0.0594	0.8779	1.41213	−0.639	0.0346
0.8572	0.0519	0.8611	1.40220	−0.569	0.0285
0.6070	0.2006	0.9292	1.43738	−0.778	0.0529
0.6051	0.0999	0.9504	1.45576	−0.716	0.0529
0.6528	0.0296	0.9488	1.45733	−0.680	0.0497
0.6557	0.0517	0.9441	1.45352	−0.701	0.0503
0.0467	0.5540	1.0337	1.49097	−0.541	0.0814
0.4178	0.2450	0.9791	1.46858	−0.675	0.0620
0.5638	0.1836	0.9480	1.45208	−0.746	0.0565
0.6365	0.1529	0.9302	1.44037	−0.762	0.0518
0.7320	0.1128	0.9027	1.42465	−0.738	0.0433
0.3855	0.3435	0.9749	1.46070	−0.699	0.0676
0.5031	0.2778	0.9499	1.44715	−0.758	0.0602
0.6273	0.2083	0.9191	1.43047	−0.791	0.0500
0.7205	0.1562	0.8929	1.41624	−0.762	0.0406
0.2691	0.5313	0.9818	1.45207	−0.703	0.0750
0.4188	0.4225	0.9485	1.43647	−0.791	0.0620
0.5023	0.3618	0.9284	1.42682	−0.825	0.0543
0.6146	0.2801	0.9008	1.41394	−0.839	0.0436
0.6105	0.2931	0.8972	1.41089	−0.849	0.0422
0.6546	0.1484	0.9242	1.43663	−0.764	0.0503
0.0408	0.9476	0.9890	1.35378	−0.237	0.0194
0.9030	0.0000	0.8575	1.40198	−0.485	0.0273
0.0504	0.9440	0.9833	1.34716	−0.284	0.0137
0.0502	0.9468	0.9817	1.34374	−0.282	0.0107
0.2244	0.3265	1.0119	1.48997	−0.586	0.0644
0.3065	0.2919	0.9992	1.48176	−0.619	0.0638
0.3032	0.3581	0.9943	1.47485	−0.644	0.0697
0.3996	0.3085	0.9761	1.46417	−0.692	0.0659
0.1104	0.5270	1.0239	1.48387	−0.571	0.0809
0.3436	0.1977	0.9992	1.48705	0.592	0.0545

(2) Phase Equilibrium

Ref. [214]

T(K)	x_1	x_2	y_1	γ_1	γ_2
101.300 kPa					
372.45	0.026	0.884	0.264	17.43	1.160
367.09	0.091	0.815	0.399	6.135	1.530
362.41	0.255	0.637	0.513	2.580	2.336
359.76	0.423	0.460	0.575	1.717	3.582
357.89	0.531	0.354	0.627	1.467	5.016
358.72	0.403	0.506	0.572	1.875	3.396
359.77	0.286	0.646	0.530	2.537	2.552
360.64	0.212	0.737	0.498	3.319	2.162
361.45	0.155	0.805	0.454	4.411	1.918
362.69	0.120	0.847	0.394	5.444	1.739
363.48	0.100	0.871	0.398	6.351	1.640
364.74	0.081	0.894	0.357	7.442	1.525
366.38	0.066	0.911	0.311	8.649	1.406
362.34	0.116	0.862	0.418	5.713	1.731
360.31	0.161	0.817	0.467	4.423	1.976
358.68	0.227	0.749	0.507	3.329	2.297
357.65	0.269	0.704	0.540	2.919	2.542
356.81	0.328	0.644	0.563	2.477	2.876
356.01	0.402	0.568	0.585	2.085	3.367
354.83	0.512	0.456	0.609	1.711	4.391
351.47	0.873	0.125	0.865	1.144	18.34
351.86	0.903	0.080	0.889	1.090	28.36
352.40	0.918	0.047	0.898	1.050	47.32
353.11	0.855	0.089	0.817	1.096	24.17
353.89	0.855	0.069	0.820	1.062	30.32
355.49	0.837	0.054	0.824	1.021	36.25
356.13	0.788	0.089	0.789	1.057	21.38
358.66	0.710	0.130	0.766	1.066	13.19
362.51	0.583	0.217	0.732	1.124	6.830
366.79	0.460	0.314	0.692	1.220	4.013
372.99	0.389	0.341	0.690	1.161	2.953
382.32	0.317	0.358	0.677	1.044	2.026
387.31	0.157	0.521	0.482	1.797	1.180
378.48	0.093	0.706	0.348	4.041	1.174
376.09	0.036	0.830	0.255	11.22	1.086
368.92	0.115	0.748	0.434	4.509	1.558
364.52	0.274	0.581	0.540	2.228	2.363
364.77	0.170	0.735	0.434	3.559	1.851
365.27	0.111	0.819	0.400	5.354	1.631
366.29	0.067	0.876	0.352	8.579	1.467

(continued)

T(K)	x_1	x_2	y_1	γ_1	γ_2
362.48	0.147	0.794	0.445	4.452	1.870
364.47	0.088	0.866	0.383	6.907	1.589
366.63	0.066	0.901	0.308	8.505	1.409
368.63	0.053	0.927	0.229	9.978	1.271
370.51	0.040	0.949	0.136	12.24	1.159

Ref. [215]

x'_1	w_3	y_1	T/K	γ_2	γ_1	$\alpha_{1,2}$
100 kPa						
x'_1 = 0.1						
0.0980	0.7987	0.2946	390.15	0.61	1.08	3.84
0.0995	0.6993	0.3213	377.82	0.78	1.51	4.29
0.0990	0.6007	0.3389	371.10	0.90	1.88	4.66
0.0990	0.5005	0.3570	367.10	0.97	2.18	5.05
0.0989	0.4000	0.3740	364.16	1.02	2.46	5.44
0.0991	0.3005	0.3916	362.58	1.03	2.66	5.85
0.0992	0.2000	0.4136	361.27	1.02	2.89	6.41
0.0994	0.1002	0.4261	359.88	1.04	3.09	6.73
x'_1 = 0.2						
0.1963	0.8002	0.4489	388.60	0.59	0.90	3.34
0.1960	0.7010	0.4645	376.17	0.75	1.21	3.56
0.1992	0.5987	0.4786	367.89	0.91	1.50	3.69
0.1986	0.5008	0.4861	363.51	1.01	1.70	3.82
0.1984	0.4005	0.5006	360.69	1.05	1.88	4.05
0.1984	0.3005	0.5107	358.82	1.08	2.00	4.22
0.1990	0.2001	0.5251	357.52	1.08	2.11	4.45
0.1992	0.1002	0.5333	356.28	1.10	2.21	4.59
x'_1 = 0.4						
0.3983	0.7959	0.6749	390.37	0.46	0.67	3.14
0.4013	0.6967	0.6575	374.75	0.71	0.92	2.86
0.4007	0.6002	0.6448	366.15	0.92	1.11	2.71
0.3989	0.5002	0.6380	361.05	1.06	1.25	2.66
0.3992	0.4004	0.6318	357.83	1.18	1.33	2.58
0.3991	0.3002	0.6277	355.81	1.25	1.39	2.54
0.3995	0.2001	0.6215	354.58	1.30	1.40	2.47
0.3997	0.1002	0.6185	353.80	1.33	1.41	2.43
x'_1 = 0.6						
0.5985	0.7987	0.8054	393.73	0.40	0.52	2.78
0.5992	0.7012	0.7772	376.79	0.68	0.71	2.33
0.5999	0.6005	0.7655	366.76	0.92	0.89	2.18
0.6003	0.5002	0.7508	360.72	1.14	1.01	2.01
0.6003	0.4005	0.7395	356.96	1.32	1.09	1.89
0.6001	0.3003	0.7288	354.66	1.45	1.13	1.79
0.6000	0.2000	0.7210	353.23	1.53	1.15	1.72

(*continued*)

x'_1	w_3	y_1	T/K	γ_2	γ_1	$\alpha_{1,2}$
0.6003	0.1003	0.7129	352.39	1.59	1.15	1.65
$x'_1 = 0.8$						
0.7999	0.8005	0.9211	396.94	0.32	0.43	2.92
0.8007	0.7008	0.8999	379.64	0.58	0.59	2.24
0.8005	0.6001	0.8849	368.35	0.88	0.75	1.92
0.7997	0.4999	0.8734	361.55	1.15	0.88	1.73
0.7995	0.4000	0.8601	358.05	1.37	0.93	1.54
0.8000	0.3003	0.8511	354.23	1.63	1.02	1.43
0.8003	0.2002	0.8416	352.90	1.78	1.03	1.33
0.8004	0.1000	0.8256	351.65	2.00	1.03	1.18
$x'_1 = 0.98$						
0.9798	0.8008	0.9935	400.40	0.25	0.36	3.15
0.9799	0.7006	0.9917	381.77	0.46	0.51	2.46
0.9801	0.6006	0.9895	370.39	0.77	0.66	1.92
0.9802	0.5006	0.9876	363.04	1.10	0.78	1.61
0.9803	0.4004	0.9860	357.93	1.43	0.89	1.41
0.9805	0.3003	0.9846	354.80	1.71	0.95	1.27
0.9804	0.2002	0.9827	352.90	1.98	0.98	1.14
0.9802	0.1002	0.9804	351.77	2.27	1.00	1.01

Mixture 2: H_2O (1) + $(CH_3)_3COH$ (2) + [hmim]Cl (3)
(1) Phase Equilibrium

Ref. [25]

x'_2 [a]	w_3 [a]	y_2	T/K	γ_1	γ_2	$\alpha_{2,1}$ [b]
p = 100 kPa						
0.9501	0.5915	0.9778	374.26	0.64	0.77	2.31
0.9500	0.5501	0.9764	370.65	0.73	0.83	2.18
0.9500	0.4999	0.9750	366.44	0.85	0.91	2.05
0.9500	0.4499	0.9713	363.55	1.04	0.96	1.78
0.9500	0.4002	0.9695	361.50	1.15	0.99	1.67
0.9500	0.2999	0.9631	358.10	1.47	1.04	1.37
0.9501	0.2000	0.9567	356.55	1.74	1.04	1.16
0.9500	0.1000	0.9415	354.94	2.39	1.04	0.85
0.9499	0.0494	0.9334	354.61	2.70	1.02	0.74

[a]Compositions reported are of equilibrium liquid
[b]$\alpha_{21} = (y_2/x'_2)/(y_1/x'_1)$

2.2.57 02021-01002: 1-Hexyl-3-methylimidazolium bromide

Abbreviation:	[C_6MIm]Br, [HMIm]Br
Molecular Formula:	$C_{10}H_{19}N_2Br$
Molar Mass:	247.18
Structure:	

2.2.57.1 Binary Mixtures

Mixture 1: H_2O (1) + [HMIm]Br (2)

(1) Volumetric Properties, Speed of Sound and Compressibility Properties

Ref. [216]

m_2/(mol · kg^{-1})	$10^{-3} \cdot \rho$/ (kg · m^{-3})	u/(m · s^{-1})	$10^6 \cdot V_\varphi$/ (m^3 · mol^{-1})	$10^{12} \cdot k_s$/ Pa^{-1}	$10^{14} \cdot k_\varphi$/ (m^3 · mol^{-1} · Pa^{-1})
T = 283.15 K					
0.0042	0.999919	1448.70	195.20	476.52	−10.46
0.0048	0.999941	1448.79	196.66	476.45	−9.54
0.0067	1.000033	1449.21	197.34	476.13	−8.89
0.0081	1.000096	1449.29	197.85	476.04	−6.83
0.0139	1.000379	1450.30	198.00	475.25	−5.79
0.0295	1.001151	1452.94	197.72	473.16	−4.87
0.0369	1.001520	1453.35	197.55	472.72	−3.22
0.0487	1.002104	1455.35	197.35	471.14	−3.46
0.0640	1.002842	1457.93	197.51	469.13	−3.58
0.0730	1.003271	1459.38	197.61	468.00	−3.56
0.0940	1.004288	1462.85	197.52	465.31	−3.62
0.1183	1.005446	1466.89	197.49	462.22	−3.67
0.1423	1.006600	1470.86	197.35	459.20	−3.70
0.1639	1.007599	1474.04	197.46	456.77	−3.55
0.2206	1.010253	1482.68	197.30	450.27	−3.40
0.2581	1.011963	1488.08	197.28	446.25	−3.25
0.3034	1.014040	1494.64	197.13	441.44	−3.14
0.3543	1.016301	1501.62	197.09	436.37	−2.97
0.3874	1.017818	1506.11	196.89	433.13	−2.89
0.4492	1.020542	1513.65	196.73	427.68	−2.65
0.6269	1.028050	1535.23	196.43	412.70	−2.21
0.8614	1.037256	1557.70	196.25	397.33	−1.50
1.0007	1.042275	1568.46	196.32	390.01	−1.07
1.3210	1.053013	1586.34	196.40	377.38	−0.16
1.7388	1.065229	1600.03	196.65	366.69	0.85
2.0867	1.074156	1607.38	196.89	360.33	1.48

(continued)

m_2/(mol · kg^{-1})	$10^{-3} \cdot \rho$/ (kg · m^{-3})	u/(m · s^{-1})	$10^6 \cdot V_\varphi$/ (m^3 · mol^{-1})	$10^{12} \cdot k_s$/ Pa^{-1}	$10^{14} \cdot k_\varphi$/ (m^3 · mol^{-1} · Pa^{-1})
T = 288.15 K					
0.0042	0.999312	1467.02	196.95	464.97	−4.74
0.0048	0.999334	1467.07	198.22	464.93	−3.94
0.0067	0.999423	1467.40	198.92	464.68	−3.90
0.0081	0.999485	1467.43	199.31	464.63	−2.27
0.0139	0.999762	1468.41	199.32	463.88	−2.87
0.0295	1.000518	1470.81	198.94	462.02	−2.81
0.0369	1.000880	1471.09	198.73	461.68	−1.34
0.0487	1.001447	1472.96	198.62	460.25	−1.78
0.0640	1.002170	1475.26	198.74	458.48	−1.95
0.0730	1.002592	1476.58	198.80	457.47	−1.99
0.0940	1.003585	1479.77	198.72	455.05	−2.15
0.1183	1.004717	1483.49	198.70	452.26	−2.27
0.1423	1.005846	1487.10	198.55	449.56	−2.33
0.1639	1.006823	1490.04	198.65	447.35	−2.23
0.2206	1.009416	1497.86	198.50	441.56	−2.13
0.2581	1.011087	1502.72	198.48	437.98	−2.00
0.3034	1.013114	1508.68	198.34	433.66	−1.92
0.3543	1.015322	1515.02	198.31	429.1	−1.79
0.3874	1.016804	1519.09	198.11	426.18	−1.73
0.4492	1.019461	1525.87	197.96	421.3	−1.52
0.6269	1.026781	1545.27	197.69	407.86	−1.15
0.8614	1.035745	1565.11	197.54	394.15	−0.51
1.0007	1.040636	1574.56	197.6	387.60	−0.14
1.3210	1.051103	1590.30	197.68	376.18	0.66
1.7388	1.063056	1602.46	197.89	366.33	1.54
2.0867	1.071818	1608.87	198.07	360.44	2.10
T = 293.15 K					
0.0042	0.998411	1483.21	198.27	455.29	−1.22
0.0048	0.998432	1483.24	199.61	455.26	−0.55
0.0067	0.998518	1483.54	200.41	455.04	−1.10
0.0081	0.998579	1483.57	200.69	454.99	0.05
0.0139	0.99885	1484.47	200.62	454.31	−1.05
0.0295	0.999592	1486.67	200.1	452.63	−1.42
0.0369	0.999946	1486.81	199.91	452.39	0.01
0.0487	1.000504	1488.54	199.73	451.09	−0.52
0.0640	1.001211	1490.65	199.87	449.49	−0.76
0.0730	1.001625	1491.86	199.92	448.58	−0.83
0.0940	1.002596	1494.78	199.86	446.40	−1.01
0.1183	1.003707	1498.15	199.81	443.90	−1.15
0.1423	1.004812	1501.44	199.68	441.47	−1.22
0.1639	1.005768	1504.11	199.79	439.48	−1.14
0.2206	1.008306	1511.19	199.64	434.28	−1.07

(continued)

m_2/(mol · kg^{-1})	$10^{-3} \cdot \rho$/ (kg · m^{-3})	u/(m · s^{-1})	$10^6 \cdot V_\varphi$/ (m^3 · mol^{-1})	$10^{12} \cdot k_s$/ Pa^{-1}	$10^{14} \cdot k_\varphi$/ (m^3 · mol^{-1} · Pa^{-1})
0.2581	1.009943	1515.58	199.62	431.07	−0.96
0.3034	1.011926	1521.01	199.49	427.16	−0.91
0.3543	1.014084	1526.74	199.47	423.05	−0.80
0.3874	1.015533	1530.40	199.27	420.43	−0.75
0.4492	1.018130	1536.42	199.13	416.08	−0.55
0.6269	1.025278	1553.85	198.89	403.96	−0.24
0.8614	1.034026	1571.30	198.75	391.70	0.34
1.0007	1.038798	1579.58	198.83	385.82	0.67
1.3210	1.049027	1593.40	198.89	375.46	1.38
1.7388	1.060744	1604.06	199.06	366.39	2.15
2.0867	1.069356	1609.56	199.20	360.96	2.64
T = 298.15 K					
0.0042	0.997246	1497.46	199.63	447.19	0.26
0.0048	0.997266	1497.48	201.05	447.16	0.89
0.0067	0.997351	1497.73	201.64	446.98	0.42
0.0081	0.997413	1497.74	201.62	446.94	1.44
0.0139	0.997679	1498.58	201.60	446.32	0.11
0.0295	0.998405	1500.55	201.20	444.83	−0.31
0.0369	0.998754	1500.60	200.96	444.64	1.04
0.0487	0.999300	1502.18	200.83	443.47	0.49
0.064	0.999994	1504.08	200.95	442.04	0.25
0.0730	1.000398	1505.16	201.03	441.23	0.19
0.0940	1.001354	1507.83	200.93	439.25	−0.03
0.1183	1.002444	1510.92	200.88	436.98	−0.19
0.1423	1.003527	1513.91	200.76	434.78	−0.27
0.1639	1.004465	1516.32	200.86	433.00	−0.21
0.2206	1.006953	1522.75	200.73	428.29	−0.16
0.2581	1.008555	1526.73	200.72	425.38	−0.08
0.3034	1.010498	1531.62	200.59	421.85	−0.03
0.3543	1.012613	1536.76	200.57	418.16	0.07
0.3874	1.014031	1540.06	200.39	415.79	0.11
0.4492	1.016573	1545.40	200.26	411.89	0.29
0.6269	1.023564	1561.01	200.04	400.93	0.56
0.8614	1.032118	1576.29	199.92	389.94	1.09
1.0007	1.036783	1583.54	199.99	384.64	1.39
1.3210	1.046799	1595.58	200.04	375.23	2.02
1.7388	1.058306	1604.80	200.17	366.90	2.69
2.0867	1.066779	1609.39	200.29	361.91	3.13
T = 303.15 K					
0.0042	0.995843	1509.84	201.03	440.50	1.24
0.0048	0.995864	1509.82	202.11	440.50	2.22
0.0067	0.995948	1510.06	202.61	440.33	1.49
0.0081	0.996006	1510.09	202.97	440.28	2.21

(*continued*)

m_2/(mol · kg^{-1})	10^{-3} · ρ/ (kg · m^{-3})	u/(m · s^{-1})	10^6 · V_φ/ (m^3 · mol^{-1})	10^{12} · k_s/ Pa^{-1}	10^{14} · k_φ/ (m^3 · mol^{-1} · Pa^{-1})
0.0139	0.996267	1510.84	202.84	439.73	1.01
0.0295	0.996983	1512.62	202.24	438.38	0.56
0.0369	0.997324	1512.54	202.06	438.28	1.94
0.0487	0.997863	1514.00	201.86	437.2	1.35
0.0640	0.998545	1515.72	201.98	435.91	1.09
0.0730	0.998942	1516.73	202.06	435.15	1.00
0.0940	0.999878	1519.17	201.99	433.35	0.77
0.1183	1.000950	1522.00	201.93	431.28	0.60
0.1423	1.002014	1524.70	201.81	429.30	0.53
0.1639	1.002936	1526.91	201.91	427.66	0.57
0.2206	1.005377	1532.67	201.78	423.42	0.62
0.2581	1.006949	1536.25	201.78	420.79	0.70
0.3034	1.008855	1540.65	201.66	417.60	0.73
0.3543	1.010929	1545.26	201.65	414.26	0.82
0.3874	1.012318	1548.20	201.47	412.12	0.86
0.4492	1.014808	1552.93	201.35	408.61	1.03
0.6269	1.021657	1566.84	201.15	398.70	1.27
0.8614	1.030031	1580.17	201.05	388.81	1.75
1.0007	1.034602	1586.48	201.12	384.02	2.02
1.3210	1.044425	1596.90	201.16	375.46	2.58
1.7388	1.055742	1604.73	201.25	367.82	3.19
2.0867	1.064092	1608.43	201.33	363.26	3.58

Ref. [216]

T/K	283.15	288.15	293.15	298.15	303.15
$E^0{}_\Phi$/cm^3 · mol^{-1} · K^{-1}	0.2892	0.2692	0.2492	0.2292	0.2092

(2) Molar Conductivity

Ref. [216]

$T = 283.15$ K		$T = 288.15$ K		$T = 293.15$ K		$T = 298.15$ K		$T = 303.15$ K	
c_2/ $(mol \cdot m^{-3})$	Λ/ $(S \cdot cm^2 \cdot mol^{-1})$	c_2/ $(mol \cdot m^{-3})$	Λ/ $(S \cdot cm^2 \cdot mol^{-1})$	c_2/ $(mol \cdot m^{-3})$	Λ/ $(S \cdot cm^2 \cdot mol^{-1})$	c_2/ $(mol \cdot m^{-3})$	Λ/ $(S \cdot cm^2 \cdot mol^{-1})$	c_2/ $(mol \cdot m^{-3})$	Λ/ $(S \cdot cm^2 \cdot mol^{-1})$
0.71	97.26	0.89	103.05	0.60	106.27	0.76	106.63	0.64	109.17
1.38	96.91	2.42	101.88	1.37	105.63	1.57	106.28	1.45	107.75
2.41	96.09	2.95	101.47	2.18	105.58	2.39	106.20	2.28	107.08
3.23	96.09	3.75	101.25	2.82	104.86	3.32	105.73	3.14	106.47
4.08	95.46	5.44	100.47	4.05	103.71	3.94	105.36	4.11	105.62
4.96	95.41	6.89	100.18	5.93	103.24	4.62	105.19	5.65	105.25
5.96	94.62	9.37	99.00	7.59	102.38	5.24	104.65	7.37	104.43
6.83	94.60	11.85	97.97	9.37	101.13	6.68	103.56	9.98	103.30
7.54	94.50	14.44	97.00	10.25	100.53	8.02	102.70	12.35	102.47
9.33	94.14	17.59	95.69	12.60	99.10	9.95	101.66	15.04	101.30
11.66	93.16	20.53	94.65	14.68	97.84	12.15	100.36	18.88	99.99
14.10	92.63	24.47	93.22	17.10	96.92	15.12	98.89	22.32	98.72
16.70	91.89	28.82	91.63	19.49	95.86	17.97	97.65	26.57	97.00
21.00	90.58	32.68	90.36	21.85	94.96	21.82	95.90	31.69	95.38
24.46	89.56	36.91	89.06	24.61	93.98	25.91	94.26	36.29	93.63
28.21	88.51	41.83	87.66	27.44	92.88	30.63	92.33	40.92	92.45
33.35	86.93	46.40	86.45	30.87	91.62	35.36	90.65	44.95	91.28
37.71	85.74	52.14	84.91	34.66	90.41	39.61	89.16	50.10	89.88
43.69	84.15	58.97	83.22	38.02	89.21	44.40	87.67	63.84	85.38
45.83	83.64			41.20	88.28	49.44	86.14		
				45.50	86.96	59.14	83.39		
				51.39	85.28				

(3) Electrical Properties

Ref. [216]

T/K	K_a/(dm^3 · mol^{-1})	Λ_0/(S · cm^2 · mol^{-1})	10^{10} R/m	σ (Λ)
c_2/<10^2 mol · m^{-3}				
283.15	2.94	97.98	6.00	0.04
288.15	3.73	103.90	6.00	0.03
293.15	4.84	107.27	10.02	0.08
298.15	4.90	108.36	5.14	0.20
303.15	4.93	109.46	9.89	0.17

(4) Osmotic Coefficient

Ref. [112]

	Φ			
m/(mol · kg^{-1})	298.15 K	308.15 K	318.15 K	328.15 K
0.0155	0.876	0.889	0.910	0.935
0.0210	0.869	0.870	0.876	0.911
0.0305	0.854	0.858	0.866	0.869
0.0467	0.835	0.838	0.848	0.861
0.0633	0.815	0.811	0.816	0.821
0.0819	0.761	0.763	0.800	0.815
0.1210	0.741	0.742	0.775	0.790
0.1580	0.703	0.732	0.754	0.760
0.1845	0.708	0.727	0.742	0.750
0.2503	0.670	0.690	0.702	0.719
0.3312	0.650	0.659	0.672	0.670
0.4111	0.602	0.605	0.625	0.630
0.5114	0.554	0.585	0.589	0.592
0.5898	0.540	0.546	0.555	0.569
0.6829	0.517	0.513	0.537	0.544
0.7428	0.488	0.489	0.509	0.514
1.1189	0.384	0.397	0.413	0.429
1.6856	0.312	0.322	0.352	0.362
2.3976	0.271	0.292	0.309	0.328

2.2.58 02021-02001: 1-Hexyl-3-methylimidazolium tetrafluoroborate

Abbreviation:	$[C_6MIm][BF_4]$, $[hmim][BF_4]$, $[HMIM][BF_4]$, $[C_6mim][BF_4]$, $HMIMBF_4$
Molecular Formula:	$C_{10}H_{19}BF_4N_2$
Molar Mass:	254.08
Structure:	

2.2.58.1 Binary Mixtures

Mixture 1: C_2H_5OH (1) + $[hmim][BF_4]$ (2)

(1) Excess Molar Enthalpy

Ref. [65]

x_2	H^E (J · mol^{-1})	x_2	H^E (J · mol^{-1})	x_2	H^E (J · mol^{-1})
T = 303.15 K					
0.1010	1346	0.3977	2567	0.6901	1855
0.2037	2046	0.4988	2479	0.7932	1314
0.3018	2408	0.6045	2184	0.8733	848

(2) Volumetric Properties and Surface Tension

Ref. [34]

x_2	c (mol/L)	ρ (g/cm^3)	x_2	c (mol/L)	ρ (g/cm^3)
298.15 K					
1.0000	4.451	1.1455	0.5079	3.594	1.0737
0.9000	4.377	1.1347	0.3991	3.234	1.0461
0.8030	4.234	1.1239	0.3000	2.798	1.0119
0.7020	4.051	1.1088	0.1998	2.199	0.9646
0.6123	3.866	1.0955	0.0980	1.318	0.8939

Ref. [34]

x_2	c (mol/L)	σ (mN/m)	δσ (mN/m)	x_2	c (mol/L)	σ (mN/m)	δσ (mN/m)
298.15 K							
1.0000	4.451	37.33	0.00	0.5079	3.594	28.82	−0.97
0.9000	4.377	35.75	−0.05	0.3991	3.234	27.09	−1.03
0.8030	4.234	34.22	−0.09	0.3000	2.798	25.62	−0.98
0.7020	4.051	32.47	−0.30	0.1998	2.199	24.21	−0.85
0.6123	3.866	30.74	−0.65	0.0980	1.318	23.11	−0.40

Mixture 2: CH_3OH (1) + $HMIMBF_4$ (2)

(1) Heat Capacity

Ref. [217]

x_2	$C_p/(J \cdot mol^{-1} \cdot K^{-1})$	$C_p^E/(J \cdot mol^{-1} \cdot K^{-1})$	$C_p/(J \cdot mol^{-1} \cdot K^{-1})$	$C_p^E/(J \cdot mol^{-1} \cdot K^{-1})$	$C_p/(J \cdot mol^{-1} \cdot K^{-1})$	$C_p^E/(J \cdot mol^{-1} \cdot K^{-1})$
	$T = 283.15$ K		$T = 288.15$ K		$T = 293.15$ K	
0	78.6	0.00	79.5	0.00	80.4	0.00
0.0751	108.0	3.51	108.7	3.26	109.5	3.02
0.1511	135.5	4.80	136.4	4.63	137.4	4.43
0.2973	187.2	6.17	188.1	5.62	188.9	5.08
0.3970	220.7	5.25	221.7	4.75	222.8	4.25
0.6423	301.9	2.01	303.6	1.68	305.3	1.36
0.8759	380.4	−0.05	382.7	−0.14	385.0	−0.22
1	423.2	0.00	425.8	0.00	428.4	0.00
	$T = 298.15$ K		$T = 303.15$ K		$T = 308.15$ K	
0	81.3	0.00	82.3	0.00	83.4	0.00
0.0751	110.3	2.78	111.2	2.53	112.2	2.28
0.1511	138.3	4.20	139.3	3.93	140.3	3.63
0.2973	189.9	4.55	190.8	4.02	191.8	3.49
0.3970	223.9	3.76	225.1	3.27	226.3	2.79
0.6423	307.0	1.05	308.7	0.75	310.5	0.45
0.8759	387.4	−0.28	389.7	−0.33	392.1	−0.37
1	431.0	0.00	433.6	0.00	436.2	0.00
	$T = 313.15$ K		$T = 318.15$ K		$T = 323.15$ K	
0	84.5	0.00	85.7	0.00	87.0	0.00
0.0751	113.2	2.03	114.2	1.78	115.3	1.52
0.1511	141.3	3.29	142.4	2.91	143.4	2.50
0.2973	192.9	2.97	193.9	2.46	195.1	1.95
0.3970	227.5	2.32	228.8	1.86	230.1	1.40
0.6423	312.3	0.17	314.1	−0.11	315.9	−0.39
0.8759	394.5	−0.39	396.9	−0.40	399.3	−0.39
1	438.8	0.00	441.4	0.00	444.0	0.00

Mixture 3: $[C_2mim][BF_4]$ (1) + $[C_6mim][BF_4]$ (2)
(1) Volumetric Properties and Transport Properties

Ref. [27]
These physicochemical properties data are listed in a mixture 1 of IL 02003-02001.

Ref. [28]
These physicochemical properties data are listed in a mixture 1 of IL 02003-02001.

Ref.[28]
These physicochemical properties data are listed in a mixture 1 of IL 02003-02001.
(2) Excess Molar Enthalpy

Ref.[28]
These physicochemical properties data are listed in a mixture 1 of IL 02003-02001.
Mixture 4: $[C_4mim][BF_4]$ (1) + $[C_6mim][BF_4]$ (2)
(1) Volumetric Properties and Transport Properties

Ref. [27]
These physicochemical properties data are listed in a mixture 16 of IL 02011-02001.

Ref. [28]
These physicochemical properties data are listed in a mixture 16 of IL 02011-02001.

Ref. [28]
These physicochemical properties data are listed in a mixture 16 of IL 02011-02001.
(2) Excess Molar Enthalpy

Ref. [28]
These physicochemical properties data are listed in a mixture 16 of IL 02011-02001.
Mixture 5: CO_2 (1) + $[C_6MIm][BF_4]$ (2)
(1) Henry's Law Constant and Thermodynamic Properties

Ref. [147]

T (K)	K_H (MPa)	T (K)	K_H (MPa)
322.15	2.19	312.45	1.75
317.45	2.01	307.55	1.59

Ref. [147]

$\Delta_{sil}H^o$ ($kJ \cdot mol^{-1}$)	$\Delta_{sil}G^o$ ($kJ \cdot mol^{-1}$)	$\Delta_{sil}S^o$ ($J \cdot K^{-1} \cdot mol^{-1}$)
−17.3	0.72	−60.0

(2) Bubble Point

Ref. [218]

x_1	T/K	p/MPa	T/K	p/MPa	T/K	p/MPa
0.103	293.18	0.54	323.14	0.90	348.10	1.28
	298.18	0.57	328.13	0.97	353.13	1.37
	303.14	0.63	333.28	1.05	358.15	1.47
	308.14	0.69	338.10	1.11	363.19	1.57
	313.12	0.77	343.16	1.20	368.14	1.65
	318.17	0.83				
0.201	293.18	1.09	323.06	1.91	348.15	2.80
	298.14	1.19	328.14	2.08	353.13	2.99
	303.15	1.32	333.11	2.25	358.12	3.20
	308.17	1.45	338.15	2.43	363.13	3.40
	313.19	1.62	343.10	2.61	368.16	3.62
	318.13	1.76				
0.299	302.37	2.17	327.96	3.49	347.21	4.74
	308.66	2.46	328.04	3.52	351.75	5.05
	313.48	2.69	332.64	3.82	357.27	5.42
	318.40	2.96	337.39	4.09	363.33	5.84
	323.29	3.23	342.41	4.40	367.81	6.18
0.404	303.39	3.44	328.25	5.56	352.98	8.28
	308.37	3.81	333.25	6.06	358.23	8.92
	313.39	4.20	338.16	6.58	362.98	9.52
	318.33	4.63	343.13	7.12	367.82	10.16
	323.27	5.08	348.10	7.70		
0.498	303.63	4.81	323.07	7.31	337.87	9.77
	308.28	5.32	327.84	8.04	342.82	10.72
	313.39	5.96	332.91	8.89	348.15	11.77
	318.51	6.64				
0.602	313.07	10.46	332.80	19.82	352.87	28.78
	317.88	12.76	338.05	22.14	357.77	30.88
	322.98	15.22	342.88	24.34	362.68	32.92
	327.97	17.52	347.81	26.54		
0.653	312.95	24.52	332.97	36.28	352.94	46.56
	318.08	27.70	337.93	38.96	358.04	48.98
	323.07	30.84	342.80	41.58	362.94	51.32
	327.91	33.54	347.88	44.26		
0.703	313.03	56.74	333.04	69.40	352.92	81.68
	318.04	59.94	337.84	72.56	357.64	84.46
	323.02	63.12	342.93	75.52	362.58	86.60
	327.95	66.26	347.82	78.72		

(3) Phase Equilibrium

Ref. [218]

T/K	p/MPa	T/K	p/MPa	T/K	p/MPa
278.70	4.04	298.49	6.49	303.86	7.34
283.54	4.56	302.46	7.11	303.99	7.36
288.49	5.14	303.51	7.28	304.09	7.37
293.61	5.80	303.75	7.32		

Mixture 6: 1-$C_5H_{11}OH$ (1) + C_6mim BF_4 (2)
(1) Transport Properties

Ref. [219]

T/K	η/cp (experimental dynamic Transport Properties)	η_b/cP (interpolated background Transport Properties)	τ (reduced temperature) $\times 10^{-4}$	η/η_b
w_c (critical mass fraction) = 0.31034, T (critical temperature) = 314.91 K				
328.15	3.0474	3.0600	420.44	0.9959
325.69	3.2616	3.2634	342.21	0.9995
323.16	3.5122	3.5062	261.87	1.0017
321.14	3.7440	3.7296	197.94	1.0039
319.34	3.9844	3.9577	140.52	1.0068
318.02	4.1937	4.1428	98.76	1.0122
317.23	4.3456	4.2629	73.67	1.0194
316.73	4.4596	4.3426	57.79	1.0269
316.42	4.5367	4.3931	48.06	1.0327
316.22	4.5953	4.4272	41.60	1.0380
316.07	4.6429	4.4521	36.94	1.0428
315.98	4.6694	4.4681	33.98	1.0451
315.94	4.6845	4.4750	32.71	1.0468
315.91	4.6931	4.4802	31.76	1.0475
315.88	4.7010	4.4848	30.91	1.0482
315.86	4.7090	4.4889	30.17	1.0490
315.84	4.7137	4.4923	29.53	1.0493
315.79	4.7264	4.5011	27.94	1.0501
315.75	4.7445	4.5087	26.57	1.0523
315.71	4.7624	4.5157	25.30	1.0546
315.63	4.7835	4.5292	22.86	1.0562
315.58	4.8038	4.5375	21.38	1.0587
315.54	4.8256	4.5452	20.01	1.0617
315.50	4.8487	4.5523	18.74	1.0651
315.44	4.8722	4.5631	16.83	1.0677
315.39	4.9010	4.5721	15.24	1.0719
315.35	4.9306	4.5799	13.86	1.0766
315.30	4.9631	4.5890	12.28	1.0815
315.24	4.9989	4.5993	10.48	1.0869

(continued)

T/K	η/cp (experimental dynamic Transport Properties)	η_b/cP (interpolated background Transport Properties)	τ (reduced temperature) $\times$ 10^{-4}	η/η_b
315.20	5.0379	4.6072	9.10	1.0935
315.15	5.0915	4.6167	7.46	1.1028
315.10	5.1411	4.6256	5.93	1.1114
315.04	5.2162	4.6355	4.23	1.1253
315.00	5.3054	4.6435	2.86	1.1425
314.94	5.4676	4.6541	1.06	1.1748

(2) Critical Temperatures of Phase Separation

Ref. [220]

$100w_2$ (mass%)	T_S/K	$100w_2$ (mass%)	T_S/K
10.00	312.46	32.04	315.20
20.02	314.98	44.92	314.56
31.20	315.20	60.02	309.77
31.44	315.20		
Critical data			
31.03	315.20		

Ref. [220]

Solvent	T/K
1-pentanol	315.55

Mixture 7: H_2O (1) + $[C_6mim][BF_4]$ (2)
(1) Critical Temperatures of Phase Separation

Ref. [220]

$100w_2$ (mass%)	T_S/K	$100w_2$ (mass%)	T_S/K
9.609	317.22	39.70	331.29
19.83	329.14	60.08	328.77
28.28	331.00		
Critical data			
40.97[a]	331.31[a]		

[a]Supposed critical composition and temperature

Ref. [220]

Solvent	T/K
H_2O	327.05

(2) Volumetric Properties, Surface Tension and Electrical Conductivity

Ref. [34]

x_2	c (mol/L)	ρ (g/cm^3)	x_2	c (mol/L)	ρ (g/cm^3)
298.15 K					
1.0000	4.507	1.1455	0.6061	4.265	1.1338
0.9002	4.458	1.1426	0.5058	4.153	1.1292
0.8030	4.407	1.1403	0.4059	4.008	1.1239
0.7057	4.344	1.1374	0.3032	3.776	1.1158

Ref. [34]

x_2	c (mol/L)	σ (mN/m)	$\delta\sigma$ (mN/m)	x_2	c (mol/L)	σ (mN/m)	$\delta\sigma$ (mN/m)
298.15 K							
1.0000	4.507	37.33	0.00	0.6061	4.265	37.52	−13.21
0.9002	4.458	37.33	−3.40	0.5058	4.153	37.65	−16.50
0.8030	4.407	37.35	−6.68	0.4059	4.008	37.79	−19.75
0.7057	4.344	37.40	−9.94	0.3032	3.776	37.93	−23.10

Ref. [35]

x_2	$10^{-3}c_2$/mol · m^{-3}	ρ/kg · m^{-3}	κ/S · cm^{-1}	$10^4\Lambda$/S · m^2 · mol^{-1}
298.15 K				
0.0007116	0.03906	998.06[a]	0.327	83.8
0.001428	0.07778	999.52	0.567	72.9
0.001482	0.08064	999.64[a]	0.608	75.4
0.002238	0.1208	1001.24[a]	0.854	70.7
0.002960	0.1568	1002.63	1.049	66.9
0.002960	0.1586	1002.70[a]	1.064	67.1
0.003642	0.1937	1003.94[a]	1.225	63.3
0.004470	0.2356	1005.24	1.411	59.9
0.004747	0.2495	1005.61[a]	1.466	58.8
0.2816	3.712	1113.63[a]	2.03	5.48
0.3140	3.806	1116.76	1.829	4.81
0.3441	3.881	1119.43[a]	1.655	4.27
0.3813	3.962	1122.43	1.456	3.68
0.4148	4.024	1124.75[a]	1.293	3.21
0.5139	4.170	1130.58	0.898	2.16
0.5161	4.173	1130.73[a]	0.885	2.12
0.7069	4.354	1138.70[a]	0.417	0.958
1	4.512	1146.34	0.1228	0.272

[a]Interpolated values

Mixture 8: 1-C_3H_7OH (1) + $[C_6mim][BF_4]$ (2)
(1) Critical Temperatures of Phase Separation

Ref. [220]

$100w_2$ (mass%)	T_S/K	$100w_2$ (mass%)	T_S/K
9.987	284.45	32.17	287.70
19.98	287.33	40.02	287.61
29.99	287.67	59.97	283.83
Critical data			
33.51	287.71		

Ref. [220]

Solvent	T/K
1-propanol	287.25

Mixture 9: 1-C_4H_9OH (1) + $[C_6mim][BF_4]$ (2)
(1) Critical Temperatures of Phase Separation

Ref. [220]

$100w_2$ (mass%)	T_S/K	$100w_2$ (mass%)	T_S/K
11.08	300.82	31.64	303.70
20.00	303.39	32.10	303.70
31.57	303.70	45.10	303.13
31.62	303.70	60.04	298.66
Critical data			
37.71	303.70		

Ref. [220]

Solvent	T/K
1-butanol	304.15

Mixture 10: 2-$CH_3CH(OH)CH_2CH_3$ (1) + $[C_6mim][BF_4]$ (2)
(1) Critical Temperatures of Phase Separation

Ref. [220]

$100w_2$ (mass%)	T_S/K	$100w_2$ (mass%)	T_S/K
10.00	301.39	30.67	303.37
19.98	303.25	45.01	302.30
28.56	303.37	59.95	297.04
Critical data			
29.26	303.37		

Ref. [220]

Solvent	T/K
2-butanol	302.85

Mixture 11: 2-$CH_3CH(OH)(CH_2)_2CH_3$ (1) + [C_6mim][BF_4] (2)
(1) Critical Temperatures of Phase Separation

Ref. [220]

$100w_2$ (mass%)	T_S/K	$100w_2$ (mass%)	T_S/K
9.992	315.39	29.10	317.66
20.05	317.47	30.10	317.66
27.01	317.66	45.04	316.35
28.06	317.66	60.07	310.95
Critical data			
27.90	317.66		

Ref. [220]

Solvent	T/K
2-pentanol	317.45

Mixture 12: 1-$C_6H_{13}OH$ (1) + [C_6mim][BF_4] (2)
(1) Critical Temperatures of Phase Separation

Ref. [220]

$100w_2$ (mass%)	T_S/K	$100w_2$ (mass%)	T_S/K
10.01	323.66	39.93	326.15
20.27	326.17	60.08	320.56
29.99	326.38		
Critical data			
30.31	326.38		

Ref. [220]

Solvent	T/K
1-Hexanol	325.45

Mixture 13: 1,1,1,2-$C_2H_2F_4$ (1) + [HMIm][BF_4] (2)
(1) Phase Equilibrium

Ref. [221]

T(K)	P(bar)	x_1	$\Delta V/V_0$[a]	V^L(cm^3/mol)
	0.81	0.0438 ± 0.0024	0.0169 ± 0.0009	226.05 ± 0.14
	1.56	0.1305 ± 0.0020	0.0444 ± 0.0009	211.09 ± 0.13
298.15	2.42	0.2236 ± 0.0016	0.0846 ± 0.0010	195.75 ± 0.12
	4.22	0.4156 ± 0.0010	0.2026 ± 0.0010	163.38 ± 0.09

(*continued*)

T(K)	P(bar)	x_1	$\Delta V/V_0$[a]	V^L(cm^3/mol)
	5.12	0.5222 ± 0.0007	0.3362 ± 0.0011	148.44 ± 0.07
	0.86	0.0308 ± 0.0023	0.0021 ± 0.0009	227.35 ± 0.15
	1.70	0.0761 ± 0.0022	0.0055 ± 0.0009	217.45 ± 0.14
	3.50	0.1768 ± 0.0018	0.0467 ± 0.0010	201.70 ± 0.13
	5.30	0.2734 ± 0.0015	0.1024 ± 0.0010	187.48 ± 0.11
323.15	7.68	0.3953 ± 0.0012	0.2009 ± 0.0010	169.98 ± 0.09
	10.60	0.5737 ± 0.0007	0.4459 ± 0.0012	144.27 ± 0.07
	12.27	0.7071 ± 0.0003	0.8315 ± 0.0014	125.57 ± 0.05
	12.75^{vll}	0.7766 ± 0.0002		
	46.4^{CP}	0.8904 ± 0.0002		
	6.12	0.0000 ± 0.0049	0.012 ± 0.002	209.3 ± 0.3
	9.86	0.2006 ± 0.0043	0.062 ± 0.002	188.1 ± 0.3
	13.44	0.3155 ± 0.0036	0.124 ± 0.002	166.0 ± 0.2
348.15	16.62	0.4292 ± 0.0030	0.213 ± 0.002	149.5 ± 0.2
	19.70	0.5235 ± 0.0025	0.333 ± 0.002	136.7 ± 0.1
	23.24^{Vll}	0.8291 ± 0.0015		
	143.30^{CP}	0.9631 ± 0.0003		

[a]$\Delta V/V_0 = (V_m(T, P, x_i) - V_{IL}(T, P = 1\ \text{bar}))/(V_{IL}(T, P = 1\ \text{bar}))$; cp: mixture critical point; vll: vapor–liquid–liquid equilibrium points for the IL-rich phase

Mixture 14: 2-$CH_3COC_2H_5$ (1) + [C_6mim][BF_4] (2)

(1) Transport Properties

Ref. [135]

x_2	η/(mPa · s)	x_2	η/(mPa · s)	x_2	η/(mPa · s)
298.15 K					
0.00	0.388	0.3482	5.578	0.6057	24.459
0.0492	0.600	0.3972	7.507	0.7054	43.382
0.0988	0.914	0.4605	11.080	0.7955	69.399
0.2011	2.059	0.4981	13.673	0.8931	110.337
0.2989	4.063	0.5500	18.570	1.00	195.367

(2) Electrical Conductivity

Ref. [135]

x_2	$10^3\kappa$ (S · cm^{-1})	x_2	$10^3\kappa$ (S · cm^{-1})	x_2	$10^3\kappa$ (S · cm^{-1})
298.15 K					
0.0492	7.32	0.3972	7.82	0.7054	3.14
0.0988	9.07	0.4605	6.43	0.7955	2.38
0.2011	10.09	0.4981	5.87	0.8931	1.73
0.2989	9.13	0.5500	5.23	1.00	1.11
0.3482	8.46	0.6057	4.43		

Mixture 15: $CH_3COOCH_2CH_3$ (1) + $[C_6mim][BF_4]$ (2)

(1) Transport Properties

Ref. [135]

x_2	η/(mPa · s)	x_2	η/(mPa · s)	x_2	η/(mPa · s)
298.15 K					
0.00	0.437	0.3475	6.846	0.6022	28.479
0.0498	0.738	0.3983	9.328	0.7075	48.111
0.1000	1.188	0.4488	12.307	0.8066	80.426
0.2027	2.674	0.5010	16.471	0.8908	119.875
0.3020	5.221	0.5487	21.492	1.00	195.367

(2) Electrical Conductivity

Ref. [135]

x_2	$10^3\kappa$ (S · cm^{-1})	x_2	$10^3\kappa$ (S · cm^{-1})	x_2	$10^3\kappa$ (S · cm^{-1})
298.15 K					
0.0498	0.78	0.3983	4.59	0.7075	2.54
0.1000	2.10	0.4488	4.43	0.8066	1.87
0.2027	4.11	0.5010	4.00	0.8908	1.48
0.3020	4.76	0.5487	3.59	1.00	1.11
0.3475	4.82	0.6022	3.24		

Mixture 16: CH_3CN (1) + [hmim]$[BF_4]$ (2)

(1) Volumetric Properties and Electrical Conductivity

Ref. [35]

x_2	$10^{-3}c_2$/mol · m^{-3}	ρ/kg · m^{-3}	κ/S · cm^{-1}	$10^4\Lambda$/S · m^2 · mol^{-1}
298.15 K				
0.004117	0.07700	784.24	0.838	108.8
0.008445	0.1560	791.63	1.442	92.5
0.01744	0.3142	806.26	2.35	74.7
0.03816	0.6489	836.42	3.51	54.2
0.06486	1.028	869.76	4.13	40.2
0.09763	0.1427	903.86	4.25	29.8
0.1399	1.855	939.67	3.96	21.4
0.1943	2.300	975.94	3.37	14.67
0.2712	2.785	1014.86	2.53	9.08
0.3907	3.322	1056.74	1.543	4.65
0.5902	3.894	1100.45	0.656	1.685
0.7435	4.183	1122.12	0.344	0.823
0.8889	4.287	1137.20	0.192	0.437
1	4.512	1146.34	0.1228	0.272

The standard uncertainty of ρ is 0.05 kg · m^{-3}; the relative standard uncertainty of κ and Λ is 0.5 %.

Mixture 17: CH_3NO_2 (1) + [hmim][BF_4] (2)
(1) Excess Molar Enthalpy

Ref. [74]

x_2	$H^E/(J \cdot mol^{-1})$	x_2	$H^E/(J \cdot mol^{-1})$	x_2	$H^E/(J \cdot mol^{-1})$
303.15 K					
0.1000	−73	0.4096	−160	0.7148	−113
0.2074	−124	0.5095	−159	0.7891	−94
0.3028	−146	0.6072	−143	0.8821	−50

2.2.58.2 Ternary Mixtures

Mixture 1: $HOC_6H_4C(CH_3)_2C_6H_4OH$ (1) + H_2O (2) + [C_6mim][BF_4] (3)
(1) Distribution Ratio

Ref. [204]

pH	$T = 297.15 \pm 1$ K	pH	$T = 297.15 \pm 1$ K	pH	$T = 297.15 \pm 1$ K	pH	$T = 297.15 \pm 1$ K
1.86	759.1	4.95	808.1	9.17	562.5	12.01	2.07
3.26	747.7	7.03	962.1	11.61	21.2		

Mixture 2: C_6H_5OH (1) + H_2O (2) + [C_6mim][BF_4] (3)
(1) Distribution Ratio

Ref. [204]

pH	$T = 297.15 \pm 1$ K	pH	$T = 297.15 \pm 1$ K	pH	$T = 297.15 \pm 1$ K
3.23	20.2	7.16	20.1	11.68	4.1
5.25	20.8	9.52	14.1		

Mixture 3: 4-$C_9H_{19}C_6H_4OH$ (1) + H_2O (2) + [C_6mim][BF_4] (3)
(1) Distribution Ratio

Ref. [204]

pH	$T = 297.15 \pm 1$ K	pH	$T = 297.15 \pm 1$ K	pH	$T = 297.15 \pm 1$ K	pH	$T = 297.15 \pm 1$ K
3.23[a]	ND[b]	7.09	ND	10.20	368.8	12.61	162.7
5.19	ND	8.43	375.0	11.03	327.8		

[a] V_W:V_{IL} = 25:1 was used at these pHs because of the low solubility of 4-nonylphenol in water and their very low concentrations in aqueous phase after extraction
[b] Not detectable because of low concentration of the compound in aqueous phase after extraction

Mixture 4: 4-$C_8H_{17}C_6H_4OH$ (1) + H_2O (2) + [C_6mim][BF_4] (3)
(1) Distribution Ratio

Ref. [204]

pH	$T = 297.15 \pm 1$ K	pH	$T = 297.15 \pm 1$ K	pH	$T = 297.15 \pm 1$ K
3.23[a]	362.5	7.04	397.2	12.07	62.4
5.27	383.4	10.22	118.8		

[a]V_W:V_{IL} = 25:1 was used at these pHs because of the low solubility of 4-octylphenol in water and their very low concentrations in aqueous phase after extraction.

Mixture 5: C_6Cl_5OH (1) + H_2O (2) + [C_6mim][BF_4] (3)
(1) Distribution Ratio

Ref. [204]

pH	$T = 297.15 \pm 1$ K	pH	$T = 297.15 \pm 1$ K	pH	$T = 297.15 \pm 1$ K
3.16[a]	430.0	7.03	408.0	12.73	121.0
5.12	424.9	10.25	387.8		

[a]V_W:V_{IL} = 25:1 was used at these pHs because of the low solubility of pentachlorophenol in water and their very low concentrations in aqueous phase after extraction.

Mixture 6: C_7H_{16} (1) + C_6H_6 (2) + [hmim][BF_4] (3)
(1) Phase Equilibrium

Ref. [222]

x_3	x_2	x_3	x_2	x_3	x_2
298.2 K					
0.0010	0.6220	0.0060	0.7530	0.6240	0.3200
0.0020	0.4640	0.2540	0.7460	0.7300	0.2000
0.0020	0.0000	0.5330	0.4210	0.8200	0.1030
0.0050	0.9950	0.5730	0.3870	0.9020	0.0000

Ref. [222]

[hmim]BF_4-rich layer		Alkane-rich layer		
x'_3	x'_2	x''_3	x''_2	S
298.2 K				
0.2798	0.7179	0.0030	0.9438	17.59
0.2894	0.7044	0.0032	0.9333	7.73
0.3155	0.6769	0.0022	0.8881	11.00
0.3844	0.5961	0.0016	0.8067	7.29
0.4562	0.5140	0.0020	0.7095	7.01
0.4818	0.4852	0.0022	0.6495	7.88
0.5370	0.4180	0.0032	0.4600	10.84
0.5738	0.3800	0.0022	0.4799	8.84
0.6000	0.3480	0.0018	0.4199	9.22
0.6792	0.2600	0.0016	0.3203	9.05

(*continued*)

[hmim]BF_4-rich layer		Alkane-rich layer		
x'_3	x'_2	x''_3	x''_2	S
0.7499	0.1783	0.0011	0.2286	8.37
0.8108	0.1043	0.0016	0.1280	8.35

Mixture 7: $C_{12}H_{26}$ (1) + C_6H_6 (2) + [hmim][BF_4] (3)
(1) Phase Equilibrium

Ref. [222]

x_3	x_2	x_3	x_2	x_3	x_2
298.2 K					
0.0001	0.0000	0.2540	0.7460	0.6658	0.3051
0.0008	0.8748	0.3513	0.6440	0.8366	0.1321
0.0017	0.7580	0.4759	0.5013	0.8752	0.0919
0.0050	0.9950	0.5946	0.3836	0.9638	0.0000

Ref. [222]

[hmim]BF_4-rich layer		Alkane-rich layer		
x'_3	x'_2	x''_3	x''_2	S
298.2 K				
0.3399	0.6507	0.0074	0.9413	3.77
0.3775	0.6127	0.0060	0.8554	10.13
0.4799	0.5001	0.0052	0.7121	9.93
0.5310	0.4476	0.0048	0.6200	12.66
0.6211	0.3517	0.0042	0.5157	12.04
0.7395	0.2306	0.0035	0.3516	14.15
0.8680	0.1002	0.0021	0.2060	12.11

Mixture 8: $C_{16}H_{34}$ (1) + C_6H_6 (2) + [hmim][BF_4] (3)
(1) Phase Equilibrium

Ref. [222]

x_3	x_2	x_3	x_2	x_3	x_2
298.2 K					
0.0001	0.0000	0.0050	0.9950	0.5348	0.4535
0.0007	0.8603	0.2540	0.7460	0.7334	0.2517
0.0016	0.5736	0.2989	0.7002	0.9090	0.0761
0.0027	0.2988	0.4332	0.5652	0.9808	0.0000

Ref. [222]

[hmim]BF_4-rich layer		Alkane-rich layer		
x'_3	x'_2	x''_3	x''_2	S
298.2 K				
0.3348	0.6616	0.0007	0.9950	8.08
0.3701	0.6240	0.0007	0.9441	5.95
0.4992	0.4917	0.0006	0.8103	12.48
0.6145	0.3758	0.0005	0.6558	20.20
0.6919	0.2957	0.0003	0.5201	21.93
0.8142	0.1701	0.0003	0.4051	15.82
0.8812	0.1001	0.0001	0.3350	10.60
0.9162	0.0659	0.0001	0.2485	11.15
0.9542	0.0263	0.0000	0.0972	12.49

Mixture 9: 1-C_6H_{12} (1) + C_2H_5OH (2) + [C_6mim]BF_4 (3)
(1) Phase Equilibrium

Ref. [223]

x_3	x_2	x_3	x_2	x_3	x_2
298.2 K					
0.0008	0.0000	0.0492	0.6222	0.3413	0.5109
0.0012	0.1744	0.0500	0.6329	0.4365	0.4270
0.0023	0.2973	0.1612	0.6487	0.6716	0.2099
0.0039	0.4230	0.1898	0.6407	0.8235	0.0488
0.0142	0.5878	0.2172	0.6310	0.8688	0.0000
0.0205	0.6021	0.2624	0.5909		

Ref. [223]

[C_6mim]BF_4-rich layer		Alkene-rich layer		
x'_3	x'_2	x''_3	x''_2	S
298.2 K				
0.8242	0.0428	0.0001	0.0054	59.17
0.6810	0.1990	0.0002	0.0158	103.28
0.5452	0.3257	0.0008	0.0557	42.73
0.5422	0.3082	0.0001	0.0311	64.18
0.5086	0.3602	0.0003	0.0654	39.23
0.4856	0.3813	0.0004	0.0968	26.72
0.4492	0.4144	0.0012	0.1191	22.44
0.4082	0.4538	0.0023	0.1643	16.27
0.3388	0.5207	0.0014	0.2526	10.95
0.2804	0.5771	0.0018	0.3300	8.20
0.1502	0.6504	0.0039	0.4239	4.40
0.0512	0.6323	0.0129	0.5328	1.70

Mixture 10: C_7H_{14} (1) + C_2H_5OH (2) + [C_6mim][BF_4] (3)
(1) Phase Equilibrium

Ref. [223]

x_3	x_2	x_3	x_2	x_3	x_2
298.2 K					
0.0013	0.0000	0.0522	0.6874	0.3506	0.5425
0.0041	0.4993	0.0663	0.6926	0.4555	0.4543
0.0126	0.6011	0.1045	0.6950	0.5525	0.3539
0.0228	0.6582	0.1523	0.6866	0.7104	0.2029
0.0289	0.6643	0.2230	0.6472	0.9118	0.0000
0.0329	0.6780	0.3204	0.5717		

Ref. [223]

[C_6mim]BF_4-rich layer		Alkene-rich layer		
x'_3	x'_2	x''_3	x''_2	S
298.2 K				
0.7761	0.1405	0.0006	0.0065	257.34
0.5942	0.3193	0.0007	0.0309	115.69
0.4328	0.4747	0.0011	0.1359	32.59
0.4108	0.4966	0.0009	0.1748	25.29
0.3878	0.5181	0.0020	0.2067	21.12
0.3198	0.5759	0.0054	0.3122	12.07
0.1994	0.6602	0.0051	0.3998	7.00
0.0799	0.6930	0.0090	0.5153	2.82

2.2.59 02021-03001: 1-Hexyl-3-methylimidazolium bis((trifluoromethyl)sulfonyl)imide

Abbreviation:	[C_6MIm][NTf_2], [C_6C_1Im][Ntf_2], [HMIM][BTI], [C_6mim][Tf_2N], [C_6mim][NTf_2], [C_6MIM][NTf_2], [HMIM][NTf_2], [hmim][Tf_2N]
Molecular Formula:	$C_{12}H_{19}F_6N_3O_4S_2$
Molar Mass:	447.42
Structure:	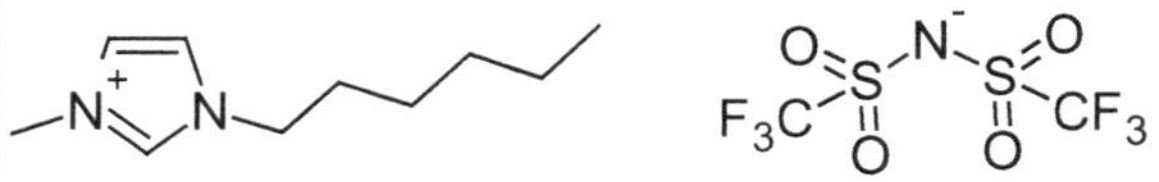

2.2.59.1 Binary Mixtures

Mixture 1: CF_4 (1) + $[C_6C_1Im][Ntf_2]$ (2)

(1) Phase Equilibrium

Ref. [224]

T/K	$p/10^2$ Pa	$K_H/10^5$ Pa	$x_1/10^{-3}$	dev %
CF_4 (Partial Pressure of Solute is 0.1 Mpa)				
303.42	1146.54	665.69	1.50	+0.2
313.40	1184.57	721.00	1.39	−0.4
323.44	1222.83	783.18	1.28	−0.3
333.45	1260.93	849.06	1.18	+0.8
343.44	1299.08	946.33	1.06	−0.4

(2) Thermodynamic Properties

Ref. [224]

T/K	$\Delta_{solv}G^{\infty}$/kJ · mol^{-1}	$\Delta_{solv}H^{\infty}$/kJ · mol^{-1}	$\Delta_{solv}S^{\infty}$/J · mol^{-1} · K^{-1}
CF_4 in $[C_6C_1Im][Ntf_2]$			
p^0 = 0.1 MPa			
303	16.38	−5.38	−71.8
313	17.11	−6.38	−75.1
323	17.88	−7.48	−78.5
333	18.68	−8.68	−82.2
343	19.52	−9.99	−86.0

Mixture 2: $CH_3CH(CH_2)_4CH_2$ (1) + [HMIM][BTI] (2)

(1) Excess Molar Enthalpy

Ref. [78]

x_1	H^E (J · mol^{-1})	x_1	H^E (J · mol^{-1})	x_1	H^E (J · mol^{-1})	x_1	H^E (J · mol^{-1})
363.15 K, 1617 kPa							
0.0612	192.0	0.2663	735.3	0.5213	502.5	0.8840	112.5
0.1180	385.0	0.3096	723.4	0.6288	394.6		
0.1708	553.3	0.3885	643.7	0.7176	302.1		
0.2202	695.4	0.4586	569.3	0.7922	219.8		

(2) Activity Coefficient at Infinite Dilution

Ref. [78]

	$\gamma^{\infty}{}_1$			
Solvent	303.15 K	313.15 K	323.15 K	333.15 K
[HMIM][BTI]	6.83	6.59	6.30	6.06

Mixture 3: C_8H_{18} (1) + [HMIM][BTI] (2)
(1) Excess Molar Enthalpy

Ref. [78]

x_1	H^E (J · mol^{-1})	x_1	H^E (J · mol^{-1})	x_1	H^E (J · mol^{-1})	x_1	H^E (J · mol^{-1})
363.15 K, 1548 kPa							
0.0486	203.6	0.2217	462.9	0.4608	331.0	0.8568	95.1
0.0950	369.2	0.2603	443.1	0.5707	266.3		
0.1392	449.8	0.3327	401.1	0.6660	202.2		
0.1814	483.5	0.3993	367.0	0.7495	160.8		

(2) Activity Coefficient at Infinite Dilution and Partial Molar Excess Enthalpy at Infinite Dilution

Ref. [4]

	$\gamma^{\infty}{}_1$ (dilutor technique)			
Solute (1)	303.15 K	313.15 K	323.15 K	333.15 K
n-Octane	14.4	14.1	13.1	12.8

Ref. [4]

	$\gamma^{\infty}{}_1$ (gas–liquid chromatography)			
Solute (1)	303.15 K	313.15 K	323.15 K	333.15 K
n-Octane	15.2	14.3	13.5	13. 0

Ref. [4]

Solute (1)	$H_{12}{}^{E,\infty}$/kJ · mol^{-1}
n-Octane	4.031

Mixture 4: $C_6H_5CH_3$ (1) + [HMIM][BTI] (2)
(1) Activity Coefficient at Infinite Dilution and Partial Molar Excess Enthalpy at Infinite Dilution

Ref. [78]

	$\gamma^{\infty}{}_1$			
Solvent	303.15 K	313.15 K	323.15 K	333.15 K
[HMIM][BTI]	0.85	0.95	1.05	1.14

Ref. [4]

	$\gamma^{\infty}{}_1$ (gas–liquid chromatography)			
Solute (1)	303.15 K	313.15 K	323.15 K	333.15 K
Toluene	1.01	1.03	1.05	1.08

Ref. [4]

Solute (1)	$H_{12}^{E,\infty}$/kJ · mol^{-1}
Toluene	−1.843

(2) Excess Molar Enthalpy

Ref. [50]

x_1	H^E (J · mol^{-1})	x_1	H^E (J · mol^{-1})	x_1	H^E (J · mol^{-1})	x_1	H^E (J · mol^{-1})
363.15 K							
0.0501	−91.4	0.2002	−362.2	0.6003	−765.4	0.8501	−363.0
0.1001	−184.4	0.3002	−523.1	0.7002	−685.5	0.9001	−239.0
0.1501	−269.7	0.4503	−707.4	0.8002	−487.4	0.9501	−120.3

Mixture 5: H_2O (1) + [C_6mim][Tf_2N] (2)
(1) Solubility

Ref. [45]

T/K	$x_1 \pm \sigma^a$	10^4 ($x_2 \pm \sigma$)
288.15	0.1906 ± 0.0006	8.90 ± 0.03
293.15	0.1978 ± 0.0006	8.96 ± 0.07
298.15	0.2076 ± 0.0006	9.56 ± 0.02
303.15	0.2187 ± 0.0005	9.73 ± 0.01
308.15	0.2303 ± 0.0006	10.2 ± 0.1
313.15	0.2424 ± 0.0006	11.7 ± 0.2
318.15	0.2556 ± 0.0007	13.2 ± 0.2

[a]Standard deviation

(2) Thermodynamic Properties

Ref. [45]

$(\Delta_{sol}H_m^0 \pm \sigma^a)$/ kJ · mol^{-1}	$(\Delta_{sol}G_m^0 \pm \sigma^a)$/ kJ · mol^{-1}	$(\Delta_{sol}S_m^0 \pm \sigma^a)$/ J · K^{-1} · mol^{-1}	$(\Delta_{svt}H_m^0 \pm \sigma^a)$/ kJ · mol^{-1}	$(\Delta_{svt}Gm^0 \pm \sigma^a)$/ kJ · mol^{-1}	$(\Delta_{svt}S_m^0 \pm \sigma^a)$/ J · K^{-1} · mol^{-1}	$(\Delta_{svt}H_m^* \pm \sigma^a)$/ kJ · mol^{-1}	$(\Delta_{svt}G_m^* \pm \sigma^a)$/ kJ · mol^{-1}	$(\Delta_{svt}S_m^* \pm \sigma^a)$/ J · K^{-1} · mol^{-1}
298.15 K								
6.0 ± 1.5	22.969 ± 0.004	−56.8 ± 5.1	−167.0 ± 6.2	−58.830 ± 0.004	−363 ± 21	−164.8 ± 6.2	−76.738 ± 0.004	−295 ± 21

[a]Standard deviation

(3) Phase Equilibrium

Ref. [50]

x_1	P (kPa)	x_1	P (kPa)	x_1	P (kPa)	x_1	P (kPa)
353.15 K							
0.1016	12.18	0.1708	20.02	0.2122	24.66	0.2971	33.32
0.3997	43.08						

Mixture 6: [C_6mim]Cl (1) + [C_6mim][NTf_2] (2)
(1) Thermal Conductivity

Ref. [82]
These physicochemical properties data are listed in a mixture 3 of IL 02021-01001.
Mixture 7: C_6H_{14} (1) + [HMIM][BTI] (2)
(1) Phase Equilibrium

Ref. [50]

x_1	P (kPa)	x_1	P (kPa)	x_1	P (kPa)	x_1	P (kPa)
353.15 K							
0.0112	9.561	0.0436	35.92	0.0839	66.92	0.1944	133.0
0.0243	20.22	0.0578	46.98	0.1380	102.6		

(2) Excess Molar Enthalpy

Ref. [50]

x_1	H^E ($J \cdot mol^{-1}$)	x_1	H^E ($J \cdot mol^{-1}$)	x_1	H^E ($J \cdot mol^{-1}$)	x_1	H^E ($J \cdot mol^{-1}$)
363.15 K							
0.0200	51.9	0.2003	516.7	0.6004	297.4	0.9501	33.5
0.0501	146.3	0.3003	527.0	0.7003	213.8		
0.1002	305.4	0.4004	446.9	0.8002	137.3		
0.1503	434.2	0.5004	367.2	0.9001	65.7		

(3) Activity Coefficient at Infinite Dilution and Partial Molar Excess Enthalpy at Infinite Dilution

Ref. [4]

	$\gamma^{\infty}{}_1$ (dilutor technique)			
Solute (1)	303.15 K	313.15 K	323.15 K	333.15 K
n-Hexane	7.82	7.55	6.96	6.55

Ref. [4]

	$\gamma^{\infty}{}_1$ (gas–liquid chromatography)			
Solute (1)	303.15 K	313.15 K	323.15 K	333.15 K
n-Hexane	8.23	7.82	7.48	7.27

Ref. [4]

Solute (1)	$H_{12}^{E,\infty}$/kJ · mol^{-1}
n-Hexane	3.525

Mixture 8: 1-C_8H_{16} (1) + [HMIM][BTI] (2)
(1) Excess Molar Enthalpy

Ref. [50]

x_1	H^E (J · mol^{-1})	x_1	H^E (J · mol^{-1})	x_1	H^E (J · mol^{-1})	x_1	H^E (J · mol^{-1})
413.15 K							
0.0504	168.9	0.2281	596.1	0.4700	429.5	0.8613	114.0
0.0982	305.1	0.2675	588.4	0.5797	341.0		
0.1437	426.9	0.3409	543.0	0.6742	258.2		
0.1869	546.7	0.4082	484.4	0.7563	197.3		

Mixture 9: C_6H_6 (1) + [HMIM][BTI] (2)
(1) Excess Molar Enthalpy

Ref. [50]

x_1	H^E (J · mol^{-1})	x_1	H^E (J · mol^{-1})	x_1	H^E (J · mol^{-1})	x_1	H^E (J · mol^{-1})
363.15 K							
0.0854	−158.1	0.3421	−593.2	0.6094	−897.3	0.9161	−285.6
0.1608	−291.9	0.3912	−681.3	0.7082	−862.1	0.9704	−88.75
0.2279	−415.3	0.4765	−798.2	0.7845	−765.5		
0.2880	−502.8	0.5482	−879.4	0.8452	−576.2		

(2) Activity Coefficient at Infinite Dilution and Partial Molar Excess Enthalpy at Infinite Dilution

Ref. [4]

	$\gamma^{\infty}{}_1$ (gas–liquid chromatography)			
Solute (1)	303.15 K	313.15 K	323.15 K	333.15 K
Benzene	0.75	0.76	0.78	0.79

Ref. [4]

Solute (1)	$H_{12}^{E,\infty}$/kJ · mol^{-1}
Benzene	−1.511

Mixture 10: C_2H_5OH (1) + [HMIM][BTI] (2)
(1) Excess Molar Enthalpy

Ref. [50]

x_1	H^E ($J \cdot mol^{-1}$)	x_1	H^E ($J \cdot mol^{-1}$)	x_1	H^E ($J \cdot mol^{-1}$)	x_1	H^E ($J \cdot mol^{-1}$)
353.15 K							
0.0506	270.9	0.1517	846.3	0.4534	2133	0.9012	1074
0.1012	551.5	0.2022	1127	0.6033	2305	0.9506	577.8

(2) Volumetric Properties

Ref. [225]

x_2	$10^6 \cdot V^E/m^3 \cdot mol^{-1}$ 293.15 K	$10^6 \cdot V^E/m^3 \cdot mol^{-1}$ 298.15 K	$10^6 \cdot V^E/m^3 \cdot mol^{-1}$ 303.15 K
0.1 MPa			
0.0354	−0.23	−0.24	−0.24
0.1970	−0.34	−0.35	−0.35
0.3950	−0.24	−0.24	−0.24
0.4977	−0.18	−0.17	−0.17
0.5942	−0.12	−0.11	−0.11
0.7946	−0.04	−0.04	−0.04
0.0998	−0.33	−0.34	−0.35
0.1480	−0.34	−0.34	−0.35
0.2232	−0.31	−0.32	−0.32
0.3009	−0.28	−0.28	−0.28
0.6939	−0.06	−0.05	−0.05
0.4330	−0.22	−0.22	−0.22

(3) Phase Equilibrium

Ref. [4]

x_1	P/kPa	x_1	P/kPa	x_1	P/kPa	x_1	P/kPa
353.16 K							
0.062	9.01	0.568	70.33	0.860	97.19	0.980	106.21
0.118	16.55	0.650	78.62	0.878	98.68	0.988	106.86
0.168	23.12	0.725	85.67	0.885	99.21	0.994	107.35
0.214	29.23	0.784	90.87	0.908	100.89	0.996	107.58
0.260	34.96	0.810	93.34	0.933	102.76	0.998	107.72
0.375	48.83	0.828	94.54	0.954	104.26	0.999	107.81
0.476	60.49	0.845	96.15	0.969	105.38	1.000	107.91

(4) Activity Coefficient at Infinite Dilution and Partial Molar Excess Enthalpy at Infinite Dilution

Ref. [4]

	$\gamma^{\infty}{}_1$ (gas–liquid chromatography)			
Solute (1)	303.15 K	313.15 K	323.15 K	333.15 K
Ethanol	1.60	1.47	1.36	1.27

Ref. [4]

Solute (1)	$H_{12}{}^{E,\infty}$/kJ · mol^{-1}
Ethanol	6.478

Mixture 11: CH_3COCH_3 (1) + [HMIM][BTI] (2)
(1) Excess Molar Enthalpy

Ref. [50]

x_1	H^E (J · mol^{-1})	x_1	H^E (J · mol^{-1})	x_1	H^E (J · mol^{-1})	x_1	H^E (J · mol^{-1})
353.15 K							
0.0200	−44.80	0.1998	−444.3	0.5996	−1088	0.9499	−488.6
0.0499	−105.9	0.2997	−661.3	0.6997	−1123		
0.0999	−214.4	0.3996	−846.2	0.7998	−1045		
0.1498	−332.8	0.4996	−998.5	0.8999	−760.3		

(2) Activity Coefficient at Infinite Dilution and Partial Molar Excess Enthalpy at Infinite Dilution

Ref. [4]

	$\gamma^{\infty}{}_1$ (dilutor technique)			
Solute (1)	303.15 K	313.15 K	323.15 K	333.15 K
Acetone	0.31	0.32	0.34	0.34

Ref. [4]

Solute (1)	$H_{12}{}^{E,\infty}$/kJ · mol^{-1}
Acetone	−2.665

Mixture 12: 2-$CH_3COC_2H_5$ (1) + [HMIM][BTI] (2)
(1) Excess Molar Enthalpy

Ref. [50]

x_1	H^E (J · mol^{-1})	x_1	H^E (J · mol^{-1})	x_1	H^E (J · mol^{-1})	x_1	H^E (J · mol^{-1})
353.15 K							
0.0200	−52.4	0.2001	−502.0	0.6001	−1145	0.9500	−478.5
0.0500	−123.8	0.3001	−713.3	0.7001	−1165		
0.1000	−242.0	0.4001	−900.3	0.8001	−1067		
0.1500	−377.5	0.5001	−1047	0.9000	−757.6		

(2) Activity Coefficient at Infinite Dilution

Ref. [4]

	$\gamma^{\infty}{}_1$ (gas–liquid chromatography)			
Solute (1)	303.15 K	313.15 K	323.15 K	333.15 K
2-Butanone	0.41	0.42	0.44	0.44

Mixture 13: 3-$C_2H_5COC_2H_5$ (1) + [HMIM][BTI] (2)
(1) Excess Molar Enthalpy

Ref. [50]

x_1	H^E (J · mol^{-1})	x_1	H^E (J · mol^{-1})	x_1	H^E (J · mol^{-1})	x_1	H^E (J · mol^{-1})
353.15 K							
0.0503	−103.5	0.2009	−448.2	0.5013	−1006	0.8009	−1042
0.1005	−220.5	0.3011	−653.9	0.6013	−1113	0.9005	−741.5
0.1507	−332.4	0.4013	−829.0	0.7011	−1155	0.9503	−454.8

Mixture 14: $NH(CH_2)_5C = O$ (1) + [HMIM][NTf_2] (2)
(1) Saturation Temperatures, Activity Coefficient, Molar Enthalpies of Mixing for the Solute and Vapor Pressure

Ref. [226]

T^{SLE}/K	x_1^{SLE}	γ_i^{SLE}(exp tl)	γ_i^{SLE}(calcd)	Δ[a]	$\Delta_{mix}H_{m,i}(T_{tp,i})$/kJ · mol^{-1}
Caprolactam					
342.2[b]	1.00	1.00	1	0.0	−6.9
331.9	0.8867	0.95	0.93	2.1	
320.4	0.7756	0.89	0.87	2.2	
315.4	0.7475	0.84	0.85	−1.2	
311.9	0.7117	0.83	0.83	0.0	
305.9	0.6520	0.82	0.81	1.2	
303.8	0.6457	0.79	0.81	−2.5	
298.4	0.5911	0.78	0.79	−1.3	
281.4	0.4389	0.76	0.76	0.0	
264.1[c]	0.312[c]	0.76	0.76	0.0	
[C_6MIm][NTf_2]					
264.1[c]	0.312[c]	1.01	1.01	0.0	
265.6	0.2692	1.02	1.00	2.0	
267.2	0.1934	0.99	1.00	−1.0	
269.3	0.1034	0.98	1.00	−2.0	
272.0[b]	0.00	1.00	1.00	0.0	−0.1

[a]$\Delta = 100\cdot(\gamma_i^{SLE}$ (exptl) $-\gamma_i^{SLE}$(calcd))/(γ_i^{SLE} (exptl))
[b]Triple-point temperature of the pure solute
[c]Extrapolated values, eutectic point

Ref. [226]

x_1	p^{exptl}/ Pa	p^{NRTL}/ Pa	γ_1^{NRTL}	x_1	p^{exptl}/ Pa	p^{NRTL}/ Pa	γ_1^{NRTL}	x_1	p^{exptl}/ Pa	p^{NRTL}/ Pa	γ_1^{NRTL}
315.3 K				320.3 K				331.9 K			
1.00	2.86	2.86	1.00	1.00	4.42	4.42	1.00	1	11.44	11.44	1.00
0.748	(1.82[a])	1.80	0.84 (0.85[a])	0.776	(3.05[a])	2.98	0.88 (0.87[a])	0.887	(9.44[a])	9.96	0.98 (0.93[a])
0.682	1.52	1.58	0.81	0.718	2.72	2.76	0.87	0.853	9.25	9.56	0.98
0.623	1.36	1.41	0.79	0.652	2.48	2.45	0.85	0.796	9.19	8.92	0.98
0.569	1.27	1.25	0.77	0.488	1.71	1.73	0.80	0.723	7.74	8.02	0.97
0.548	1.23	1.19	0.76	0.262	0.70	0.75	0.65	0.595	6.46	6.40	0.94
0.458	0.94	0.94	0.72	0.228	0.60	0.62	0.62	0.453	4.94	4.61	0.89
0.397	0.81	0.77	0.68	0.193	0.49	0.49	0.58	0.421	4.41	4.24	0.88
0.349	0.61	0.65	0.65	0.154	0.38	0.36	0.53	0.171	1.17	1.29	0.66
0.223	0.36	0.34	0.54	0.139	0.29	0.31	0.51	0.124	0.82	0.85	0.6
0.193	0.30	0.28	0.51	0.081	0.17	0.15	0.42	0.066	0.40	0.39	0.51
0.121	0.15	0.15	0.42	0.061	0.10	0.11	0.39	0.038	0.22	0.20	0.46
0.00			0.24	0.00			0.28	0			0.40

[a]Vapor pressure was calculated as $p^{exptl} = \gamma_{CL}^{SLE} \cdot P_0 \cdot x_{CL}$

(2) Activity Coefficient at Infinite Dilution, Partial Molar Excess Enthalpies at Infinite Dilution and The Excess Partial Molar Heat Capacity at Infinite Dilution

Ref. [226]

substance	T/K	$\gamma\infty$	<T>/K	$\overline{H}_i^{E,\infty}(\langle T\rangle)$/ kJ · mol^{-1}	$\overline{C}_{p,m,i}^{E,\infty}(\langle T\rangle)$/ J · K · mol^{-1}
NH $(CH_2)_5CO$	427.5	0.49	–	–	–

Mixture 15: $CH_2(CH_2)_3CH_2C = NOH$ (1) + [HMIM][NTf_2] (2)

(1) Saturation Temperatures, Activity Coefficient and Molar Enthalpies of Mixing for the Solute

Ref. [226]

T^{SLE}/K	x_1^{SLE}	γ_1^{SLE}(exp tl)	$\gamma_1^{SLE}(calcd)$	Δ[a]	$\Delta_{mix}H_{m,1}(T_{tp,i})$/kJ · mol^{-1}
310.0	0.2127	2.50	2.73	−9.1	
338.2	0.4087	1.84	1.50	18.1	
362.5[b]	1.0	1.00	1.00	0.0	16.4

[a]$\Delta = 100\cdot(\gamma_i^{SLE}\ (\text{exptl}) - \gamma_i^{SLE}(\text{calcd}))/(\gamma_i^{SLE}\ (\text{exptl}))$

[b]Triple-point temperature of the pure cyclohexanone oxime

(2) Activity Coefficient at Infinite Dilution, Partial Molar Excess Enthalpies at Infinite Dilution and Excess Partial Molar Heat Capacity at Infinite Dilution

Ref. [226]

substance	T/K	$\gamma\infty$	<T>/K	$\overline{H}_i^{E,\infty}(\langle T\rangle)$/ kJ · mol^{-1}	$\overline{C}_{p,m,i}^{E,\infty}(\langle T\rangle)$/ J · K · mol^{-1}
cyclohexanone oxime	364.9	2.61	403.3	13.3 ± 0.3	175 ± 29
	375.4	2.41			
	386.0	2.22			
	396.5	2.03			
	407.0	1.84			
	417.1	1.67			
	417.6	1.61			
	427.5	1.51			
	438.0	1.28			

Mixture 16: 1-$C_6H_{13}OH$ (1) + [C_6MIM][NTf_2] (2)
(1) Cloud Point

Ref. [164]

w_2	T/K	w_2	T/K	w_2	T/K	w_2	T/K
0.0568	288.19	0.2764	305.46	0.5254	305.75	0.7114	299.71
0.0633	289.30	0.2764	305.46	0.5816	305.00	0.7424	297.28
0.1331	299.83	0.3832	305.94	0.6175	304.22	0.7770	293.61
0.1656	302.40	0.3956	306.06	0.6325	303.78	0.8185	286.77
0.2224	304.54	0.3956	306.06	0.6663	302.20	0.8279	285.39
0.2675	305.33	0.4629	306.08				

(2) Phase Equilibrium

Ref. [225]

x_2	w_2	T/K	x_2	w_2	T/K	x_2	w_2	T/K
0.1 MPa								
0.01188	0.05001	288.41	0.10888	0.34853	306.21	0.29698	0.64909	303.35
0.02489	0.10054	297.40	0.12973	0.39494	306.30	0.34941	0.70164	300.33
0.03031	0.12040	299.50	0.15293	0.44150	306.32[a]	0.37049	0.72044	299.17
0.03868	0.14980	301.76	0.15745	0.45003	306.33	0.40548	0.74915	295.76
0.05384	0.19947	303.79	0.18893	0.50495	306.15	0.47827	0.80056	290.24
0.07020	0.24844	304.99	0.21692	0.54811	305.84	0.55978	0.84774	281.78
0.08838	0.29801	305.75	0.25160	0.59548	305.11			

[a]Values obtained from extrapolation of $T = f(p)$ to $p = 0.1$ MPa

Ref. [225]

w_2	x_2	p/MPa	T/K	T/K ($p = 0.1$ MPa)	dT/dp (K/MPa)
0.44150	0.15293	10.055	306.53	306.32	0.023
		14.881	306.65		
		20.176	306.84		
		24.967	306.85		
		30.110	307.01		
		35.077	307.18		
		40.046	307.25		
		45.079	307.36		
		49.926	307.46		
		55.067	307.56		

(3) Excess Molar Partial Enthalpies

Ref. [227]

x_1	$H^E{}_1$/ (J · mol^{-1})	x_1	$H^E{}_1$/ (J · mol^{-1})	x_1	$H^E{}_1$/ (J · mol^{-1})	x_1	$H^E{}_1$/ (J · mol^{-1})
298.15 K							
0.0152	8719	0.0371	8099	0.0549	7847	0.0851	7175
0.0226	8540	0.0444	8106	0.0652	7410	0.0947	7051
0.0299	8372	0.0512	8078	0.0752	7253	0.1041	6663

(4) Activity Coefficient at Infinite Dilution and Partial Molar Excess Enthalpies at Infinite Dilution

Ref. [227]

	Number of moles of the stationary phase component on the column/10^{-3} mol						
T/K	2.01 (32 %)[a]	2.01 (32 %)[a]	3.36 (48 %)[a]	3.36 (48 %)[a]	3.36 (48 %)[a]	7.61 (48 %)[a]	5.83 (60 %)[a]
333.4		2.57					
343.9	2.32	2.25			2.37		
354.4	2.03	1.98	2.13	2.18	2.15	2.14	2.19
364.8	1.83	1.80	1.97	1.97	1.95	1.95	1.99
375.2	1.67		1.80	1.82	1.77	1.82	1.85
386.0			1.67	1.67		1.69	1.70
396.0				1.56			
$\gamma^{\infty}{}_1$ (298 K)[b]	4.21	4.22	3.79	3.91	4.05	3.67	3.90
$H^{E/\infty}{}_1$/ (kJ · mol^{-1})	11.2	11.6	8.9	9.3	10.0	8.5	9.1

[a]Weight per cent IL of the mass of support material
[b]Extrapolated to $T = 298.15$ K

Ref. [227]

	$H^{E/\infty}{}_1$/(kJ · mol^{-1}) at T = 298.15 K		
Solute	Calorimetric		From $\gamma^{\infty}{}_1$
1-Hexanol	9.04 ± 0.07	(9.8 ± 0.8)[a]	12.1[a]

[a]Derived from g.l.c. data

Mixture 17: 1-C_3H_7OH (1) + [C_6mim][NTf_2] (2)
(1) Volumetric Properties

Ref. [225]

	$10^6 \cdot V^E$/m^3 · mol^{-1}	$10^6 \cdot V^E$/m^3 · mol^{-1}	$10^6 \cdot V^E$/m^3 · mol^{-1}
x_2	293.15 K	298.15 K	303.15 K
0.1 MPa			
0.1769	−0.14	−0.135	−0.135
0.3724	−0.08	−0.068	−0.064
0.4923	−0.00	0.003	0.005
0.5971	0.04	0.053	0.056
0.7899	0.05	0.052	0.053
0.2008	−0.14	−0.134	−0.134
0.6764	0.04	0.046	0.050

(2) Activity Coefficient at Infinite Dilution and Partial Molar Excess Enthalpy at Infinite Dilution

Ref. [4]

	$\gamma^{\infty}{}_1$ (gas–liquid chromatography)			
Solute (1)	303.15 K	313.15 K	323.15 K	333.15 K
1-Propanol	1.90	1.71	1.57	1.45

Ref. [4]

Solute (1)	$H_{12}{}^{E,\infty}$/kJ · mol^{-1}
1-Propanol	7.536

Mixture 18: 1-C_4H_9OH (1) + [C_6MIm][NTf_2] (2)
(1) Volumetric Properties

Ref. [225]

	$10^6 \cdot V^E$/m^3 · mol^{-1}	$10^6 \cdot V^E$/m^3 · mol^{-1}	$10^6 \cdot V^E$/m^3 · mol^{-1}
x_2	293.15 K	298.15 K	303.15 K
0.1 MPa			
0.2003	−0.01	−0.01	−0.01
0.3968	0.06	0.07	0.08
0.5031	0.10	0.11	0.11
0.5963	0.11	0.12	0.12

(continued)

x_2	$10^6 \cdot V^E/m^3 \cdot mol^{-1}$ 293.15 K	$10^6 \cdot V^E/m^3 \cdot mol^{-1}$ 298.15 K	$10^6 \cdot V^E/m^3 \cdot mol^{-1}$ 303.15 K
0.7693	0.11	0.11	0.12
0.0511	−0.06	−0.06	−0.06
0.1023	−0.05	−0.05	−0.05
0.1523	−0.03	−0.03	−0.03
0.3013	0.03	0.04	0.05
0.6851	0.11	0.12	0.12

(2) Phase Equilibrium

Ref. [225]

x_2	w_2	T/K	x_2	w_2	T/K	x_2	w_2	T/K
0.1 MPa								
0.01829	0.10107	260.11	0.08011	0.34453	269.20	0.16393	0.54202	269.01
0.02813	0.14872	264.52	0.09883	0.39830	269.44	0.19937	0.60048	268.17
0.03987	0.20039	266.93	0.10685	0.41931	269.45[a]	0.23802	0.65343	266.90
0.05236	0.25011	267.97	0.11947	0.45024	269.46	0.27730	0.69843	265.04
0.06599	0.29898	268.69	0.14224	0.50024	269.30	0.32046	0.74002	262.00

[a]Values obtained from extrapolation of $T = f(p)$ to $p = 0.1$ MPa

Ref. [225]

w_2	x_2	p/MPa	T/K	T/K (p = 0.1 MPa)	dT/dp (K/MPa)
0.41931	0.10685	10.129	269.16	269.45	−0.027
		20.101	268.90		
		30.054	268.66		
		40.026	268.36		
		50.258	268.06		

(3) Excess Molar Partial Enthalpies

Ref. [227]

x_1	H^E_1/ (J · mol^{-1})	x_1	H^E_1/ (J · mol^{-1})	x_1	H^E_1/ (J · mol^{-1})	x_1	H^E_1/ (J · mol^{-1})
298.15 K							
0.0132	9141	0.0386	8893	0.0567	8659	0.0742	8436
0.0197	9063	0.0447	8782	0.0622	8560	0.0786	8359
0.0260	9034	0.0452	8730	0.0626	8558	0.0799	8325
0.0276	8962	0.0508	8706	0.0685	8459	0.0856	8254
0.0323	8901	0.0538	8654	0.0705	8453	0.0866	8236
0.0365	8851						

(4) Activity Coefficient at Infinite Dilution and Partial Molar Excess Enthalpies at Infinite Dilution

Ref. [227]

	Number of moles of the stationary phase component on the column/10^{-3} mol						
T/K	2.01 (32 %)[a]	2.01 (32 %)[a]	3.36 (48 %)[a]	3.36 (48 %)[a]	3.36 (48 %)[a]	7.61 (48 %)[a]	5.83 (60 %)[a]
323.1	1.99	1.98	1.99	2.06	1.99		
333.5	1.78	1.77	1.78	1.83	1.78		1.87
344.0	1.60	1.57	1.60	1.67	1.61	1.60	1.68
354.5	1.45	1.42	1.45	1.53	1.46	1.46	1.52
364.9	1.34	1.30	1.34		1.32	1.32	1.37
375.4	1.23		1.23			1.23	1.28
386.1						1.14	
γ^{∞}_{1} (298 K)[b]	2.73	2.69	2.65	2.70	2.68	2.59	2.82
$H^{E/\infty}_{1}$/(kJ mol^{-1})	10.8	9.9	9.3	9.0	9.6	9.0	9.6

[a]Weight per cent IL of the mass of support material
[b]Extrapolated to T = 298.15 K

Ref. [227]

	$H^{E/\infty}_{1}$/(kJ · mol^{-1})		
Solute	Calorimetric		From γ^{∞}_{i}
1-Butanol	9.32 ± 0.01	(9.6 ± 0.8)[a]	9.8[b]

[a]This work
[b]Derived from g.l.c. data

Mixture 19: 1-$C_5H_{11}OH$ (1) + [C_6mim][NTf_2] (2)
(1) Volumetric Properties

Ref. [225]

	$10^6 \cdot V^E/m^3 \cdot mol^{-1}$	$10^6 \cdot V^E/m^3 \cdot mol^{-1}$	$10^6 \cdot V^E/m^3 \cdot mol^{-1}$
x_2	293.15 K	298.15 K	303.15 K
0.1 MPa			
0.1011	0.03	0.03	0.03
0.2010	0.08	0.09	0.09
0.4967	0.20	0.21	0.21
0.3987	0.18	0.19	0.20
0.5988	0.20	0.21	0.21
0.7995	0.16	0.16	0.16

(2) Phase Equilibrium

Ref. [225]

x_2	w_2	T/K	x_2	w_2	T/K	x_2	w_2	T/K
0.1 MPa								
0.01142	0.05536	272.94	0.09594	0.35006	289.44	0.22635	0.59758	288.46
0.02147	0.10018	281.13	0.11610	0.40000	289.62	0.26737	0.64940	287.04
0.02724	0.12444	283.31	0.11622	0.40028	289.62[a]	0.31615	0.70117	284.50
0.03372	0.15047	285.08	0.13966	0.45172	289.64	0.34027	0.72359	282.94
0.04706	0.20040	287.30	0.16444	0.49972	289.52	0.37491	0.75273	280.58
0.06095	0.24781	288.49	0.19134	0.54564	289.20	0.44041	0.79978	275.02
0.07807	0.30061	289.10	0.19400	0.54989	288.94[a]			

[a]Values obtained from extrapolation of $T = f(p)$ to $p = 0.1$ MPa

Ref. [225]

w_2	x_2	p/MPa	T/K	T/K (p = 0.1 MPa)	dT/dp (K/MPa)
0.40028	0.11622	10.148	289.67	289.62	0.004
		20.089	289.70		
		24.965	289.72		
		30.055	289.74		
		39.508	289.76		
		49.981	289.83		
0.54989	0.19400	10.240	288.96	288.94	0.007
		20.316	289.01		
		30.245	289.09		
		40.251	289.18		
		50.141	289.24		

Mixture 20: 1-$C_8H_{17}OH$ (1) + [C_6MIm][NTf_2] (2)
(1) Phase Equilibrium

Ref. [225]

x_2	w_2	T/K	x_2	w_2	T/K	x_2	w_2	T/K
0.1 MPa								
0.01488	0.04933	316.25	0.13627	0.35149	333.50	0.34523	0.64431	329.32
0.03134	0.10003	325.97	0.16148	0.39816	333.62	0.40333	0.69900	325.95
0.04779	0.14707	329.66	0.19309	0.45118	333.46	0.47166	0.75411	321.33
0.06762	0.19945	331.91	0.22455	0.49871	333.14	0.62114	0.84923	305.79
0.08787	0.24865	332.86	0.26092	0.54809	332.48	0.71942	0.89805	287.75
0.10969	0.29739	333.32	0.30860	0.60528	331.23			

Mixture 21: $CH_2(CH_2)_4CH_2$ (1) + [HMIM][BTI] (2)
(1) Phase Equilibrium

Ref. [4]

x_1	P/kPa	x_1	P/kPa	x_1	P/kPa	x_1	P/kPa
353.23 K							
0.0310	9.94	0.1252	47.12	0.2939	97.39	0.7903	99.97
0.0641	21.92	0.1502	57.12	0.6908	99.72	0.9993	99.81
0.0951	34.51	0.2165	79.44	0.7408	99.83	1.0000	99.58

(2) Activity Coefficient at Infinite Dilution and Partial Molar Excess Enthalpy at Infinite Dilution
Ref. [4]

	$\gamma^{\infty}{}_1$ (dilutor technique)			
Solute (1)	303.15 K	313.15 K	323.15 K	333.15 K
Cyclohexane	5.29	5.01	4.78	4.68

	$\gamma^{\infty}{}_1$ (gas–liquid chromatography)			
Solute (1)	303.15 K	313.15 K	323.15 K	333.15 K
Cyclohexane	5.48	5.15	4.90	4.75

Ref. [4]

Solute (1)	$H_{12}{}^{E,\infty}$/kJ · mol^{-1}
Cyclohexane	3.751

Mixture 22: $CH = CHCH_2CH_2CH_2CH_2$ (1) + [HMIM][BTI] (2)
(1) Phase Equilibrium

Ref. [4]

x_1	P/kPa	x_1	P/kPa	x_1	P/kPa	x_1	P/kPa
353.41 K							
0.0432	10.81	0.394	89.84	0.856	94.01	0.993	94.07
0.0624	16.67	0.485	94.21	0.890	94.02	0.996	94.07
0.0821	22.36	0.576	94.21	0.920	94.03	0.998	94.07
0.102	28.19	0.653	94.13	0.945	94.04	0.999	94.07
0.139	38.15	0.717	94.03	0.963	94.05	0.999	94.06
0.228	59.54	0.768	93.93	0.976	94.05	1.000	94.04
0.309	76.27	0.818	93.99	0.986	94.06		

(2) Activity Coefficient at Infinite Dilution and Partial Molar Excess Enthalpy at Infinite Dilution

Ref. [4]

	$\gamma^{\infty}{}_1$ (dilutor technique)			
Solute (1)	303.15 K	313.15 K	323.15 K	333.15 K
Cyclohexene	3.20	3.15	3.07	3.01

Ref. [4]

	$\gamma^{\infty}{}_1$ (gas–liquid chromatography)			
Solute (1)	303.15 K	313.15 K	323.15 K	333.15 K
Cyclohexene	3.42	3.28	3.18	3.14

Ref. [4]

Solute (1)	$H_{12}{}^{E,\infty}$/kJ · mol^{-1}
Cyclohexene	0.440

Mixture 23: CH_3OH (1) + [HMIM][BTI] (2)
(1) Phase Equilibrium

Ref. [4]

x_1	P/kPa	x_1	P/kPa	x_1	P/kPa	x_1	P/kPa
353.11 K							
0.136	20.72	0.749	137.10	0.902	163.43	0.978	175.78
0.232	38.03	0.803	146.63	0.912	164.46	0.986	177.07
0.306	51.92	0.847	154.18	0.920	166.42	0.992	178.15
0.373	65.10	0.860	155.85	0.920	166.41	0.996	179.10
0.422	75.05	0.860	155.84	0.934	168.13	0.998	179.50
0.529	96.18	0.860	155.84	0.952	171.27	0.999	179.79
0.613	112.05	0.879	159.48	0.967	173.85	1.000	179.93
0.685	125.68	0.887	160.36				

(2) Excess Molar Partial Enthalpies

Ref. [227]

x_1	$H^E{}_1$/ (J · mol^{-1})	x_1	$H^E{}_1$/ (J · mol^{-1})	x_1	$H^E{}_1$/ (J · mol^{-1})	x_1	$H^E{}_1$/ (J · mol^{-1})
298.15 K							
0.0242	7200	0.0693	6550	0.1104	6231	0.1480	5802
0.0359	6941	0.0799	6609	0.1201	6103	0.1569	5722
0.0473	6939	0.0903	6454	0.1296	6091	0.1656	5651
0.0584	6732	0.1004	6298	0.1389	5875	0.1741	5573

(3) Activity Coefficient at Infinite Dilution and Partial Molar Excess Enthalpies at Infinite Dilution

Ref. [227]

	Number of moles of the stationary phase component on the column/10^{-3} mol				
T/K	2.01 (32 %)[a]	2.01 (32 %)[a]	3.36 (48 %)[a]	5.83 (60 %)[a]	(30 %)[b]
301.8	1.43	1.50	1.46	1.46	1.34 (298.2)[b]
312.4	1.27	1.35	1.34	1.29	1.17 (313.2)[b]
322.9	1.15	1.25	1.22	1.15	
333.2	1.05	1.17	1.10	1.07	1.00 (333.2)[b]
344.0		1.02	0.99	0.98	
354.3	0.87		0.93	0.90	
$\gamma^{\infty}{}_1$(298 K)[c]	1.49	1.53	1.53	1.50	1.34
$H^{E/\infty}{}_1$/ (kJ · mol^{-1})	8.4	6.4	7.8	8.4	6.9

[a]Weight per cent IL of the mass of support material
[b]Derived from data
[c]Extrapolated to $T = 298.15$ K

Ref. [227]

	$H_i^{E,\infty}$/(kJ · mol^{-1})		
Solute	Calorimetric		From $\gamma^{\infty}{}_i$
Methanol	7.38 ± 0.03	(7.8 ± 1.0)[a]	6.9[b], 6.8[b], 4.3[c]

[a]This work
[b]Derived from g.l.c. data.
[c]Derived from dilutor technique data

Ref. [4]

	$\gamma^{\infty}{}_1$ (dilutor technique)			
Solute (1)	303.15 K	313.15 K	323.15 K	333.15 K
Methanol		1.16	1.14	1.05

Ref. [4]

	$\gamma^{\infty}{}_1$ (gas–liquid chromatography)			
Solute (1)	303.15 K	313.15 K	323.15 K	333.15 K
Methanol	1.30	1.19	1.10	1.02

Ref. [4]

Solute (1)	$H_{12}{}^{E,\infty}$/kJ · mol^{-1}
Methanol	6.208

Mixture 24: $[C_6mim][NTf_2]$ (1) + $[C_{10}mim][NTf_2]$ (2)
(1) Volumetric Properties

Ref. [57]

	V (pure component) or V^E/cm^3 · mol^{-1}	
x_1	T = 298.15 K	T = 333.15 K
1	326.222 (326.2)	334.058 (334.1)
0.836	0.05_3	0.06_1
0.660	0.09_1	0.09_3
0.556	0.11_0	0.11_5
0.463	0.15_9	0.11_1
0.246	0.07_1	0.10_6
0	404.021 (404.0)	413.736 (413.5)

Mixture 25: C_5H_{12} (1) + [HMIM][BTI] (2)
(1) Activity Coefficient at Infinite Dilution and Partial Molar Excess Enthalpy at Infinite Dilution

Ref. [4]

	$\gamma^{\infty}{}_1$ (dilutor technique)			
Solute (1)	303.15 K	313.15 K	323.15 K	333.15 K
n-Pentane	5.93	5.67	5.48	5.27

Ref. [4]

	$\gamma^{\infty}{}_1$ (gas–liquid chromatography)			
Solute (1)	303.15 K	313.15 K	323.15 K	333.15 K
n-Pentane	5.95	5.76	5.65	5.52

Ref. [4]

Solute (1)	$H_{12}{}^{E,\infty}$/kJ · mol^{-1}
n-Pentane	2.663

Mixture 26: C_7H_{16} (1) + [HMIM][BTI] (2)
(1) Activity Coefficient at Infinite Dilution and Partial Molar Excess Enthalpy at Infinite Dilution

Ref. [4]

	$\gamma^{\infty}{}_1$ (dilutor technique)			
Solute (1)	303.15 K	313.15 K	323.15 K	333.15 K
n-Heptane	11.0	10.6	10.1	9.77

Ref. [4]

	$\gamma^{\infty}{}_1$ (gas–liquid chromatography)			
Solute (1)	303.15 K	313.15 K	323.15 K	333.15 K
n-Heptane	11.2	10.6	10. 0	9.67

Ref. [4]

Solute (1)	$H_{12}{}^{E,\infty}$/kJ · mol^{-1}
n-Heptane	3.844

Mixture 27: 1-C_5H_{10} (1) + [HMIM][BTI] (2)
(1) Activity Coefficient at Infinite Dilution and Partial Molar Excess Enthalpy at Infinite Dilution

Ref. [4]

	$\gamma^{\infty}{}_1$ (dilutor technique)			
Solute (1)	303.15 K	313.15 K	323.15 K	333.15 K
1-Pentene	3.36	3.33	3.30	3.23

Ref. [4]

	$\gamma^{\infty}{}_1$ (gas–liquid chromatography)			
Solute (1)	303.15 K	313.15 K	323.15 K	333.15 K
1-Pentene	3.47	3.40	3.30	3.31

Ref. [4]

Solute (1)	$H_{12}{}^{E,\infty}$/kJ · mol^{-1}
1-Pentene	1.220

Mixture 28: 1-C_6H_{12} (1) + [HMIM][BTI] (2)
(1) Activity Coefficient at Infinite Dilution and Partial Molar Excess Enthalpy at Infinite Dilution

Ref. [4]

	$\gamma^{\infty}{}_1$ (dilutor technique)			
Solute (1)	303.15 K	313.15 K	323.15 K	333.15 K
1-Hexene	4.64	4.56	4.48	4.39

Ref. [4]

	$\gamma^{\infty}{}_1$ (gas–liquid chromatography)			
Solute (1)	303.15 K	313.15 K	323.15 K	333.15 K
1-Hexene	4.71	4.61	4.48	4.44

Ref. [4]

Solute (1)	$H_{12}^{E,\infty}$/kJ · mol^{-1}
1-Hexene	1.672

Mixture 29: 1-C_7H_{14} (1) + [HMIM][BTI] (2)
(1) Activity Coefficient at Infinite Dilution and Partial Molar Excess Enthalpy at Infinite Dilution

Ref. [4]

	$\gamma^{\infty}{}_1$ (dilutor technique)			
Solute (1)	303.15 K	313.15 K	323.15 K	333.15 K
1-Heptene	6.28	6.12	5.96	5.86

Ref. [4]

	$\gamma^{\infty}{}_1$ (gas–liquid chromatography)			
Solute (1)	303.15 K	313.15 K	323.15 K	333.15 K
1-Heptene	6.46	6.28	6.09	5.96

Ref. [4]

Solute (1)	$H_{12}^{E,\infty}$/kJ · mol^{-1}
1-Heptene	2.133

Mixture 30: 1-C_8H_{16} (1) + [HMIM][BTI] (2)
(1) Activity Coefficient at Infinite Dilution and Partial Molar Excess Enthalpy at Infinite Dilution

Ref. [4]

	$\gamma^{\infty}{}_1$ (dilutor technique)			
Solute (1)	303.15 K	313.15 K	323.15 K	333.15 K
1-Octene	8.28	8.08	7.92	7.77

Ref. [4]

	$\gamma^{\infty}{}_1$ (gas–liquid chromatography)			
Solute (1)	303.15 K	313.15 K	323.15 K	333.15 K
1-Octene	8.81	8.47	8.18	7.90

Ref. [4]

Solute (1)	$H_{12}^{E,\infty}$/kJ · mol^{-1}
1-Octene	2.369

Mixture 31: $CH_2(CH_2)_3CH_2$ (1) + [HMIM][BTI] (2)
(1) Activity Coefficient at Infinite Dilution and Partial Molar Excess Enthalpy at Infinite Dilution

Ref. [4]

	$\gamma^{\infty}{}_1$ (dilutor technique)			
Solute (1)	303.15 K	313.15 K	323.15 K	333.15 K
Cyclopentane	3.64	3.48	3.27	3.26

Ref. [4]

Solute (1)	$H_{12}{}^{E,\infty}/kJ \cdot mol^{-1}$
Cyclopentane	3.393

Mixture 32: *m*-$C_6H_4(CH_3)_2$ (1) + [HMIM][BTI] (2)
(1) Activity Coefficient at Infinite Dilution and Partial Molar Excess Enthalpy at Infinite Dilution
Ref. [4]

	$\gamma^{\infty}{}_1$ (gas–liquid chromatography)			
Solute (1)	303.15 K	313.15 K	323.15 K	333.15 K
m-Xylene	1.40	1.41	1.46	1.50

Ref. [4]

Solute (1)	$H_{12}{}^{E,\infty}/kJ \cdot mol^{-1}$
m-Xylene	−1.649

Mixture 33: *p*-$C_6H_4(CH_3)_2$ (1) + [HMIM][BTI] (2)
(1) Activity Coefficient at Infinite Dilution and Partial Molar Excess Enthalpy at Infinite Dilution

Ref. [4]

	$\gamma^{\infty}{}_1$ (gas–liquid chromatography)			
Solute (1)	303.15 K	313.15 K	323.15 K	333.15 K
p-Xylene	1.42	1.42	1.48	1.51

Ref. [4]

Solute (1)	$H_{12}{}^{E,\infty}/kJ \cdot mol^{-1}$
p-Xylene	−2.020

Mixture 34: 2-$CH_3COC_3H_7$ (1) + [HMIM][BTI] (2)

(1) Activity Coefficient at Infinite Dilution and Partial Molar Excess Enthalpy at Infinite Dilution

Ref. [4]

	$\gamma^{\infty}{}_1$ (gas–liquid chromatography)			
Solute (1)	303.15 K	313.15 K	323.15 K	333.15 K
2-Pentanone	0.51	0.52	0.54	0.55

Ref. [4]

Solute (1)	$H_{12}{}^{E,\infty}$/kJ · mol^{-1}
2-Pentanone	−2.338

Mixture 35: 2-CH_3CHCH_3OH (1) + [HMIM][BTI] (2)

(1) Activity Coefficient at Infinite Dilution and Partial Molar Excess Enthalpy at Infinite Dilution

Ref. [4]

	$\gamma^{\infty}{}_1$ (dilutor technique)			
Solute (1)	303.15 K	313.15 K	323.15 K	333.15 K
2-Propanol	1.72	1.48	1.37	1.23

Ref. [4]

	$\gamma^{\infty}{}_1$ (gas–liquid chromatography)			
Solute (1)	303.15 K	313.15 K	323.15 K	333.15 K
2-Propanol	1.71	1.55	1.44	1.33

Ref. [4]

Solute (1)	$H_{12}{}^{E,\infty}$/kJ · mol^{-1}
2-Propanol	8.071

Mixture 36: $OCH_2CH_2CH_2CH_2$ (1) + [HMIM][BTI] (2)

(1) Activity Coefficient at Infinite Dilution and Partial Molar Excess Enthalpy at Infinite Dilution

Ref. [4]

	$\gamma^{\infty}{}_1$ (gas–liquid chromatography)			
Solute (1)	303.15 K	313.15 K	323.15 K	333.15 K
Tetrahydrofuran	0.49	0.51	0.53	0.55

Ref. [4]

Solute (1)	$H_{12}^{E,\infty}$/kJ · mol^{-1}
Tetrahydrofuran	−3.149

Mixture 37: $(CH_3)_2CHOCH(CH_3)_2$ (1) + [HMIM][BTI] (2)
(1) Activity Coefficient at Infinite Dilution

Ref. [4]

	γ^{∞}_1 (gas–liquid chromatography)			
Solute (1)	303.15 K	313.15 K	323.15 K	333.15 K
Diisopropyl ether	2.76	2.82	2.85	2.92

Mixture 38: $(CH_3)_3COC_2H_5$ (1) + [HMIM][BTI] (2)
(1) Activity Coefficient at Infinite Dilution and Partial Molar Excess Enthalpy at Infinite Dilution

Ref. [4]

	γ^{∞}_1 (gas–liquid chromatography)			
Solute (1)	303.15 K	313.15 K	323.15 K	333.15 K
Ethyl tert-butyl ether	2.69	2.74	2.76	2.82

Ref. [4]

Solute (1)	$H_{12}^{E,\infty}$/kJ · mol^{-1}
Ethyl tert-butyl ether	−1.333

Mixture 39: $CH_3OC(CH_3)_2C_2H_5$ (1) + [HMIM][BTI] (2)
(1) Activity Coefficient at Infinite Dilution and Partial Molar Excess Enthalpy at Infinite Dilution

Ref. [4]

	γ^{∞}_1 (gas–liquid chromatography)			
Solute (1)	303.15 K	313.15 K	323.15 K	333.15 K
Methyl *tert*-amyl ether	1.83	1.90	1.94	2.02

Ref. [4]

Solute (1)	$H_{12}^{E,\infty}$/kJ · mol^{-1}
Methyl *tert*-amyl ether	−2.657

Mixture 40: SO_2 (1) + [hmim][Tf_2N] (2)
(1) Solubility

Ref. [228]

T/K	*P*/MPa	$100x_1$
283.2	0.0107	17.1
282.4	0.0503	51.2
283.5	0.0753	61.6
282.8	0.1002	70.6
283.3	0.1252	76.7
283.4	0.1501	82.5
283.5	0.1750	87.5
298.1	0.0109	12.4
298.1	0.0505	37.7
298.1	0.0753	47.7
298.2	0.1003	55.3
298.1	0.1504	66.7
298.1	0.2004	75.2
298.1	0.2502	82.0
298.1	0.3002	88.1
298.1	0.2505	82.1[a]
298.1	0.2005	75.5[a]
298.2	0.1503	67.0[a]
298.1	0.1004	55.5[a]
298.1	0.0761	48.1[a]
298.1	0.0502	37.9[a]
298.1	0.0111	12.9[a]
323.1	0.0109	4.9
323.1	0.0505	19.2
323.1	0.0754	26.6
323.1	0.1003	32.9
323.1	0.1502	43.1
323.1	0.2001	51.3
323.2	0.2503	58.1
323.1	0.3002	63.9
348.2	0.0110	3.7
348.1	0.0504	12.4
348.1	0.0756	16.9
348.1	0.1003	21.0
348.2	0.1508	28.0
348.2	0.2003	34.3
348.1	0.2500	40.1
348.1	0.3003	45.7

[a]Desorption measurements

Mixture 41: 1,1,1,2-$C_2H_2F_4$ (1) + [HMIm][Tf_2N] (2)
(1) Phase Equilibrium

Ref. [221]

T(K)	P(bar)	x_1	$\Delta V/V_0$	V^L(cm^3/mol)
	0.42	0.0247 ± 0.0019	0.0007 ± 0.0006	335.48 ± 0.14
	0.75	0.0761 ± 0.0017	0.0085 ± 0.0006	320.28 ± 0.13
	1.17	0.1442 ± 0.0015	0.0273 ± 0.0006	302.21 ± 0.12
	1.81	0.2360 ± 0.0012	0.0569 ± 0.0006	277.58 ± 0.11
	2.51	0.3416 ± 0.0009	0.1060 ± 0.0006	250.29 ± 0.09
298.15	3.46	0.4609 ± 0.0006	0.1812 ± 0.0006	218.88 ± 0.08
	4.02	0.5264 ± 0.0005	0.2414 ± 0.0007	202.09 ± 0.07
	4.55	0.5893 ± 0.0004	0.3219 ± 0.0007	186.60 ± 0.06
	5.23	0.6724 ± 0.0002	0.4668 ± 0.0007	165.19 ± 0.05
	5.86	0.7578 ± 0.0001	0.7304 ± 0.0008	144.08 ± 0.03
	0.53	0.0282 ± 0.0014	0.0015 ± 0.0005	336.77 ± 0.12
	1.04	0.0652 ± 0.0013	0.0093 ± 0.0005	326.46 ± 0.12
	1.55	0.1100 ± 0.0012	0.0194 ± 0.0005	313.96 ± 0.11
	2.13	0.1604 ± 0.0011	0.0346 ± 0.0005	300.56 ± 0.11
	2.86	0.2123 ± 0.0010	0.0537 ± 0.0005	287.21 ± 0.10
	3.92	0.2860 ± 0.0008	0.0839 ± 0.0005	267.80 ± 0.09
	5.06	0.3577 ± 0.0007	0.1225 ± 0.0006	249.48 ± 0.08
323.15	6.12	0.4223 ± 0.0006	0.1649 ± 0.0006	232.85 ± 0.07
	7.48	0.5019 ± 0.0005	0.2329 ± 0.0006	212.50 ± 0.06
	8.66	0.5702 ± 0.0004	0.3109 ± 0.0006	194.98 ± 0.06
	9.72	0.6267 ± 0.0003	0.3981 ± 0.0006	180.57 ± 0.05
	10.62	0.6807 ± 0.0003	0.5119 ± 0.0007	167.04 ± 0.04
	11.43	0.7334 ± 0.0002	0.6685 ± 0.0007	153.94 ± 0.03
	11.92	0.7642 ± 0.0002	0.7894 ± 0.0008	146.01 ± 0.03
	5.02	0.2429 ± 0.0044	0.050 ± 0.002	293.7 ± 0.3
	8.74	0.3916 ± 0.0033	0.113 ± 0.002	250.1 ± 0.2
	12.11	0.5145 ± 0.0025	0.189 ± 0.002	213.3 ± 0.2
348.15	15.59	0.6205 ± 0.0019	0.310 ± 0.002	183.7 ± 0.2
	18.78	0.7157 ± 0.0013	0.498 ± 0.002	157.3 ± 0.1
	22.92[a]	0.9107 ± 0.0010		
	43.2[b]	0.9600 ± 0.0010		

[a]Vapor–liquid–liquid equilibrium points for the IL-rich phase
[b]Mixture critical point

2.2.59.2 Ternary Mixtures

Mixture 1: $CH_3CH(CH_2)_4CH_2$ (1) + $C_6H_5CH_3$ (2) + [HMIM][BTI] (3)
(1) Selectivities and capacities at infinite dilution $\gamma^{\infty}{}_1$ for the separation of methylcyclohexane from toluene for ionic liquids

Ref. [78]

Separation problem		$[C_6MIm][NTf_2]$	
Methylcyclohexane (1)/toluene (2)	T (K)	S^{∞}_{12}	k^{∞}_{2}
	303.15	8.07	1.18
	313.15	6.97	1.06
	323.15	6.03	0.96
	333.15	5.34	0.88

(2) Phase Equilibrium Separation Factors

Ref. [78]

80 mol% IL; 324.95 K			90 mol% IL; 324.95 K			80 mol% IL; 333.65 K		
$x_{1,\mathrm{corr.}}$	y_1	α_{12}	$x_{1,\mathrm{corr.}}$	y_1	α_{12}	$x_{1,\mathrm{corr.}}$	y_1	α_{12}
0.0556	0.3558	9.38	0.0559	0.3718	10.00	0.0454	0.2947	8.79
0.0873	0.4578	8.83	0.1072	0.5241	9.18	0.0932	0.4614	8.33
0.1840	0.6397	7.88	0.1149	0.5449	9.22	0.1364	0.5565	7.95
0.2773	0.7541	7.99	0.1743	0.6507	8.82	0.1804	0.6279	7.67
0.3722	0.8169	7.53	0.2313	0.7253	8.77	0.2989	0.7584	7.37
0.4745	0.8686	7.32	0.2630	0.7550	8.63	0.2731	0.7352	7.39
0.5844	0.9109	7.27	0.3285	0.8066	8.52	0.3351	0.7763	6.89
0.6816	0.9369	6.94	0.3734	0.8334	8.40	0.3687	0.8059	7.11
0.8329	0.9782	9.00	0.4292	0.8631	8.38	0.4218	0.8358	6.98
			0.4669	0.8777	8.19	0.4747	0.8624	6.93
			0.5746	0.9164	8.12	0.5631	0.8956	6.65
			0.6694	0.9421	8.04	0.6670	0.9282	6.46
			0.7740	0.9647	7.98	0.7792	0.9571	6.32
			0.9205	0.9894	8.04	0.8711	0.9781	6.60

Mixture 2: 1-C_8H_{16} (1) + C_8H_{18} (2) + [HMIM][BTI](3)
(1) Phase Equilibrium Separation Factors

Ref. [78]

$x_{1,\mathrm{corr.}}$	y_1	$1/\alpha_{12}$
95 mol%$[C_6MIm][NTf_2]$		
342.45 K		
0.0725	0.0609	1.21
0.0846	0.0754	1.33
0.1846	0.1513	1.27
0.2100	0.1706	1.29
0.3297	0.2715	1.32
0.4229	0.3501	1.36
0.4995	0.4231	1.36
0.6081	0.5266	1.39
0.7079	0.6302	1.42
0.7891	0.7208	1.45
0.8978	0.8511	1.54

Mixture 3: H_2O (1) + 1-C_4H_9OH (2) + [hmim][Tf_2N] (3)
(1) Distribution Ratio and Selectivity

Ref. [166]

H_2O-rich phase			[hmim][Tf_2N]-rich phase				
w_1	w_2	w_3	w_1	w_2	w_3	S	β_{BuOH}
298.15 K							
0.9342	0.0658	0.0000	0.0997	0.5286	0.3717	75.27	8.04
0.9370	0.0630	0.0000	0.0518	0.2341	0.7141	67.27	3.72
0.9301	0.0699	0.0000	0.1239	0.4545	0.4216	48.81	6.50
0.9282	0.0718	0.0000	0.1419	0.5227	0.3353	47.59	7.28
0.9389	0.0611	0.0000	0.1102	0.3399	0.5499	47.42	5.56
0.9292	0.0708	0.0000	0.1292	0.4649	0.4059	47.23	6.57
0.9290	0.0710	0.0000	0.1299	0.4555	0.4146	45.93	6.42
0.9294	0.0706	0.0000	0.1643	0.5060	0.3297	40.55	7.17
0.9403	0.0598	0.0000	0.1084	0.2516	0.6400	36.52	4.21
0.9258	0.0742	0.0000	0.1585	0.4243	0.4173	33.43	5.72
0.9276	0.0724	0.0000	0.1238	0.3172	0.5589	32.84	4.38
0.9502	0.0498	0.0000	0.1362	0.1853	0.6784	25.97	3.72
0.9349	0.0651	0.0000	0.1311	0.2310	0.6379	25.30	3.55
0.9629	0.0371	0.0000	0.1208	0.0651	0.8141	14.00	1.75
0.9638	0.0362	0.0000	0.2054	0.0573	0.7373	7.42	1.58
0.7993	0.2007	0.0000	0.0522	0.0790	0.8689	6.03	0.39
0.9390	0.0610	0.0000	0.3560	0.0360	0.6080	1.56	0.59
0.9239	0.0761	0.0000	0.8728	0.0452	0.0821	0.63	0.59

Mixture 4: CO_2 (1) + SO_2 (2) + [hmim][Tf_2N] (3)
(1) Phase Equilibrium

Ref. [228]

Feed CO_2/ mol %	Feed SO_2/ mol %	Feed [hmim] [Tf_2N]/ mol %	T/K	P/ MPa	Liquid SO_2 calculated/ mol %	Liquid [hmim] [Tf_2N] calculated/ mol %	Vapor SO_2 calculated/ mol %	Vapor SO_2 measured/ mol %
8.15	66.73	25.13	295.2	0.143	60.2	38.8	78.7	79.0[a]
6.62	54.25	39.13	296.7	0.115	50.0	49.1	71.2	71.7[a]
6.03	51.83	42.13	294.8	0.132	50.7	47.8	59.9	62.3[a]
4.51	43.67	51.81	294.5	0.105	42.8	55.9	54.7	52.9[a]
8.15	66.73	25.13	322.1	0.172	44.1	55.4	85.5	86.1[a]
6.62	54.25	39.13	322.1	0.168	41.9	57.4	80.6	80.2[a]
6.03	51.83	42.13	322.1	0.201	45.0	54.1	75.9	75.1[a]
42.20	42.63	15.18	295.7	0.391	59.3	35.8	30.4	31.0[b]
32.81	33.14	34.05	294.7	0.263	40.5	54.9	21.1	20.0[b]
27.78	28.06	44.16	295.2	0.253	32.9	62.0	16.2	16.0[b]
7.99	8.08	83.93	295.3	0.119	8.2	88.0	6.3	6.0[b]
27.78	28.06	44.16	320.6	0.298	29.1	67.3	26.1	27.1[b]
7.99	8.08	83.93	320.6	0.132	7.8	89.8	12.4	12.4[b]
76.06	6.80	17.14	296.0	0.384	14.9	74.5	4.4	4.3[c]

(continued)

Feed CO_2/ mol %	Feed SO_2/ mol %	Feed [hmim] [Tf_2N]/ mol %	T/K	P/ MPa	Liquid SO_2 calculated/ mol %	Liquid [hmim] [Tf_2N] calculated/ mol %	Vapor SO_2 calculated/ mol %	Vapor SO_2 measured/ mol %
62.26	5.57	32.17	296.0	0.363	9.6	79.5	2.8	2.8[c]
47.40	4.24	48.36	296.0	0.358	6.0	82.6	1.8	2.3[c]
34.42	3.08	62.50	296.0	0.356	3.8	84.5	1.1	1.6[c]
22.49	2.01	75.50	296.1	0.350	2.2	86.0	0.7	1.0[c]
76.06	6.80	17.14	321.5	0.413	10.1	82.3	5.9	7.0[c]
62.26	5.57	32.17	321.5	0.398	7.4	84.9	4.4	5.0[c]
47.40	4.24	48.36	321.1	0.377	5.1	87.3	3.1	3.9[c]
34.42	3.08	62.50	320.2	0.412	3.5	87.8	2.0	2.4[c]
22.49	2.01	75.50	320.1	0.419	2.2	88.8	1.2	1.2[c]

[a]CO_2/SO_2 gas mixture: (9.4/90.6) mole %
[b](49.7/50.3) mole %
[c](91.8/8.2) mole %. Liquid CO_2 mole% = (100−Liquid SO_2 mole%−Liquid [hmim][Tf_2N] mole %). Vapor CO_2 mole % = (100−Vapor SO_2 mole %, and Vapor [hmim][Tf_2N] = 0 mol %)

2.2.60 02021-04003: 1-Hexyl-3-methylimidazolium trifluoromethanesulfonate

Abbreviation:	[HMIM][TfO], [HMIM][OTF]
Molecular Formula:	$C_{11}H_{19}F_3N_2O_3S$
Molar Mass:	316.34
Structure:	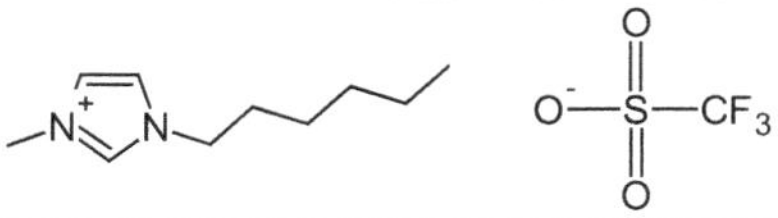

2.2.60.1 Binary Mixtures

Mixture 1: $C_6H_5CH_3$ (1) + [HMIM][OTF] (2)
(1) Activity Coefficient at Infinite Dilution

Ref. [78]

	$\gamma^{\infty}{}_1$			
Solvent	303.15 K	313.15 K	323.15 K	333.15 K
[HMIM][OTF]	1.38	1.52	1.65	1.75

Mixture 2: $CH_3CH(CH_2)_4CH_2$ (1) + [HMIM][OTF] (2)
(1) Activity Coefficient at Infinite Dilution

Ref. [78]

	$\gamma^{\infty}{}_1$			
Solvent	303.15 K	313.15 K	323.15 K	333.15 K
[HMIM][OTF]	10.53	10.2	9.77	9.49

Mixture 3: C_2H_5OH (1) + [HMIM][OTF] (2)
(1) Activity Coefficient at Infinite Dilution

Ref. [78]

	$\gamma^{\infty}{}_1$					
Solvent	303.15 K	313.15 K	323.15 K	333.15 K	343.15 K	353.15 K
[HMIM][OTF]	1.05	1.00	0.94	0.86	–	–

Mixture 4: H_2O (1) + [HMIM][OTF] (2)
(1) Activity Coefficient at Infinite Dilution

Ref. [78]

	$\gamma^{\infty}{}_1$				
Solvent	303.15 K	313.15 K	323.15 K	333.15 K	343.15 K
[HMIM][OTF]	1.31	1.04	0.82	0.64	–

2.2.60.2 Ternary Mixtures

Mixture 1: $CH_3CH(CH_2)_4CH_2$ (1) + $C_6H_5CH_3$ (2) + [HMIM][OTF] (3)
(1) Selectivities and capacities at infinite dilution $\gamma^{\infty}{}_1$ for the separation of methylcyclohexane from toluene for ionic liquids

Ref. [78]

Separation problem		[HMIM][OTF]	
Methylcyclohexane (1)/toluene (2)	T (K)	$S^{\infty}{}_{12}$	$k^{\infty}{}_2$
	303.15	7.63	0.73
	313.15	6.69	0.66
	323.15	5.91	0.61
	333.15	5.43	0.57

Mixture 2: C_2H_5OH (1) + H_2O (2) + [HMIM][OTF] (3)
(1) Selectivities and capacities at infinite dilution $\gamma^{\infty}{}_1$ for the separation of ethanol from water in ionic liquids

Ref. [78]

Separation problem		[HMIM][OTF]	
Ethanol (1)/water (2)	T (K)	$S^{\infty}{}_{12}$	$k^{\infty}{}_2$
	303.15	0.80	0.76
	313.15	0.96	0.96
	323.15	1.14	1.22
	333.15	1.37	1.56
	343.15	–	–

Mixture 3: C_6H_{14} (1) + C_6H_6 (2) + [HMIM][CF_3SO_3] (3)
(1) Selectivity at Infinite Dilution and Capacities at Infinite Dilution

Ref. [169]

Solvent	S^{∞}_{ij}	k_j^{∞}
$T = 298.15$ K	*n*-Hexane (1)/benzene (2)	Benzene
[HMIM][CF_3SO_3]	15.2	0.71

Mixture 4: $CH_2(CH_2)_4CH_2$ (1) + C_6H_6 (2) + [HMIM][CF_3SO_3] (3)
(1) Selectivity at Infinite Dilution and Capacities at Infinite Dilution

Ref. [169]

Solvent	S^{∞}_{ij}	k_j^{∞}
$T = 298.15$ K	Cyclohexane (1)/benzene (2)	Benzene
[HMIM][CF_3SO_3]	7.8	0.71

2.2.61 02021-04021: 1-Hexyl-3-methylimidazolium hydrogen sulfate

Abbreviation:	[HMIM][HSO_4]
Molecular Formula:	$C_{10}H_{20}N_2O_4S$
Molar Mass:	264.34
Structure:	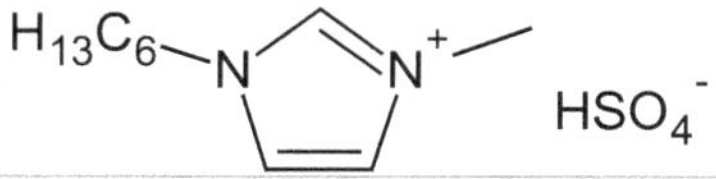

2.2.61.1 Ternary Mixtures

Mixture 1: $CH_3COOCH_2CH_3$ (1) + C_2H_5OH (2) + [HMIM][HSO_4] (3)
(1) Volumetric Properties

Ref. [104]

ester-rich phase			[HMIM][HSO_4]-rich phase		
313 K					
w_1	w_2	$\rho/(g \cdot cm^{-3})$	w_1	w_2	$\rho/(g \cdot cm^{-3})$
0.998	0.000	0.87639	0.010	0.000	1.46294
0.913	0.085	0.86580	0.019	0.052	1.40296
0.850	0.148	0.85793	0.029	0.128	1.32482
0.833	0.164	0.85670	0.034	0.133	1.31709
0.762	0.235	0.84892	0.031	0.149	1.29952
0.749	0.248	0.84726	0.041	0.211	1.24011
0.651	0.342	0.83914	0.043	0.219	1.23135
0.646	0.348	0.83753	0.047	0.225	1.22522

(*continued*)

ester-rich phase			[HMIM][HSO$_4$]-rich phase		
0.572	0.416	0.83373	0.059	0.286	1.16334
0.540	0.444	0.83109	0.078	0.358	1.09549
0.502	0.478	0.82940	0.141	0.512	0.96152
0.451	0.519	0.82924	0.170	0.565	0.91877
0.417	0.546	0.83046	0.182	0.577	0.90707
0.363	0.579	0.83410	0.194	0.595	0.89457
0.340	0.589	0.83738	0.221	0.586	0.88552
0.321	0.602	0.83991	0.211	0.600	0.88295
0.317	0.602	0.84016	0.235	0.615	0.86652
0.282	0.614	0.84803			
0.265	0.618	0.85306			

Ref. [104]

Ester-rich phase			[HMIM][HSO$_4$]-rich phase		
313 K					
w_1	w_2	ρ/(g · cm^{-3})	w_1	w_2	ρ/(g · cm^{-3})
0.741	0.256	0.84627	0.025	0.108	1.34578
0.204	0.590	0.89092	0.060	0.296	1.15479
0.896	0.102	0.86389	0.016	0.056	1.40256
0.331	0.595	–	0.045	0.213	–
0.938	0.060	0.86914	0.014	0.036	1.42326
0.818	0.180	0.85454	0.022	0.094	1.36170

Mixture 2: $CH_3COOCH_2CH_3$ (1) + CH_3COOH (2) + [HMIM][HSO$_4$] (3)
(1) Volumetric Properties

Ref. [104]

Ester-rich phase			[HMIM][HSO$_4$]-rich phase		
313 K					
w_1	w_2	ρ/(g · cm^{-3})	w_1	w_2	ρ/(g · cm^{-3})
0.999	0.000	0.87646	0.010	0.000	1.46294
0.897	0.101	0.89151	0.025	0.071	1.41143
0.819	0.178	0.90294	0.025	0.088	1.40977
0.735	0.261	0.91510	0.037	0.173	1.36451
0.665	0.330	0.92572	0.048	0.232	1.32863
0.586	0.406	0.93855	0.061	0.284	1.29483
0.543	0.446	0.94579	0.090	0.365	1.24029
0.481	0.499	0.95755	0.126	0.435	1.18679
0.420	0.549	0.97214	0.171	0.507	1.12841
0.339	0.589	1.00190	0.224	0.546	1.07974
			0.282	0.579	1.03632

Ref. [104]

Ester-rich phase			[HMIM][HSO$_4$]-rich phase		
313 K					
w_1	w_2	ρ/(g · cm^{-3})	w_1	w_2	ρ/(g · cm^{-3})
0.841	0.156	0.89972	0.027	0.125	1.39255
0.586	0.405	0.93836	0.063	0.287	1.29295
0.667	0.328	0.92542	0.048	0.231	1.32848
0.494	0.489	0.95520	0.087	0.359	1.24435
0.403	0.560	0.97667	0.125	0.438	1.18569
0.745	0.251	0.91373	0.038	0.182	1.35827
0.591	0.401	0.93762	0.063	0.286	1.29384
0.943	0.055	0.88474	0.019	0.053	1.42911

2.2.62 02021-05001: 1-Hexyl-3-methylimidazolium hexafluorophosphate

Abbreviation:	[C$_6$MIm][PF$_6$], HMIM PF$_6$, [C$_6$mim][PF$_6$], C$_6$mimPF$_6$, [hmim][PF$_6$], [hmim]PF$_6$, [C$_6$mim]PF$_6$
Molecular Formula:	$C_{10}H_{19}F_6N_2P$
Molar Mass:	312.24
Structure:	

2.2.62.1 Binary Mixtures

Mixture 1: $CH_3COOCH_2CH_3$ (1) + HMIM PF$_6$ (2)

(1) Volumetric Properties, Refractive Index, Speed of Sound and Compressibility Properties

Ref. [229]

x_2	ρ/ g · cm^{-3}	V_m^E/ cm^3 · mol^{-1}	n_D	Δn_D	u/ m · s^{-1}	κ_s/ T · Pa^{-1}	$\Delta\kappa_s$/ T · Pa^{-1}
T = 293.15 K							
0	0.9005	0	1.37241	0	1162	822	0
0.0506	0.9526	−0.66	1.38064	0.0059	1185	748	−52
0.1040	0.9987	−1.13	1.38740	0.0101	1210	684	−91
0.1978	1.0638	−1.57	1.39645	0.0148	1252	600	−133
0.2924	1.1146	−1.74	1.40200	0.0159	1288	541	−150
0.4549	1.1797	−1.66	1.40845	0.0147	1335	476	−142
0.4974	1.1933	−1.58	1.40969	0.0140	1346	463	−136
0.5904	1.2195	−1.35	1.41237	0.0123	1366	440	−117

(continued)

x_2	ρ/ g · cm^{-3}	V_m^E/ cm^3 · mol^{-1}	n_D	Δn_D	u/ m · s^{-1}	κ_S/ T · Pa^{-1}	$\Delta\kappa_S$/ T · Pa^{-1}
0.7015	1.2458	−1.03	1.41438	0.0091	1387	417	−89
0.7926	1.2638	−0.69	1.41592	0.0064	1403	402	−64
0.8941	1.2816	−0.35	1.41731	0.0030	1420	387	−33
0.9459	1.2898	−0.17	1.41797	0.0013	1429	380	−17
1	1.2979	0	1.41923	0	1438	372	0
T = 298.15 K							
0	0.8944	0	1.36977	0	1141	859	0
0.0506	0.9468	−0.70	1.37838	0.0062	1163	780	−54
0.1040	0.9932	−1.18	1.38537	0.0106	1189	712	−97
0.1978	1.0586	−1.66	1.39415	0.0149	1233	621	−143
0.2924	1.1098	−1.83	1.40007	0.0162	1271	558	−161
0.4549	1.1752	−1.77	1.40690	0.0153	1320	488	−153
0.4974	1.1889	−1.68	1.40821	0.0145	1331	475	−146
0.5904	1.2152	−1.44	1.41083	0.0127	1352	450	−126
0.7015	1.2415	−1.10	1.41296	0.0094	1374	427	−97
0.7926	1.2598	−0.77	1.41446	0.0066	1390	411	−70
0.8941	1.2776	−0.42	1.41594	0.0032	1407	395	−36
0.9459	1.2858	−0.22	1.41662	0.0014	1415	388	−19
1	1.2937	0	1.41787	0	1424	381	0
T = 303.15 K							
0	0.8882	0	1.36712	0	1111	912	0
0.0506	0.9410	−0.74	1.37606	0.0064	1142	815	−71
0.1040	0.9876	−1.25	1.38312	0.0109	1169	741	−117
0.1978	1.0534	−1.75	1.39194	0.0151	1215	643	−165
0.2924	1.1049	−1.93	1.39821	0.0167	1254	576	−184
0.4549	1.1705	−1.84	1.40523	0.0157	1305	502	−173
0.4974	1.1842	−1.74	1.40664	0.0150	1317	487	−165
0.5904	1.2107	−1.50	1.40893	0.0127	1338	461	−142
0.7015	1.2372	−1.15	1.41150	0.0098	1361	437	−109
0.7926	1.2556	−0.83	1.41307	0.0069	1378	420	−78
0.8941	1.2738	−0.49	1.41454	0.0033	1395	404	−41
0.9459	1.2820	−0.30	1.41526	0.0015	1403	396	−21
1	1.2896	0	1.41643	0	1411	389	0

Ref. [230]

x_2	ρ/(g · cm^{-3})	V^E/(cm^3 · mol^{-1})	n_D	Δn_D
298.15 K				
0	0.8944	0	1.36977	0
0.0585	0.9540	−0.768	1.37943	0.0068
0.1018	0.9913	−1.151	1.38485	0.0102
0.2029	1.0620	−1.709	1.39463	0.0151
0.3005	1.1138	−1.860	1.40092	0.0167

(continued)

x_2	$\rho/(g \cdot cm^{-3})$	$V^E/(cm^3 \cdot mol^{-1})$	n_D	Δn_D
0.3997	1.1555	−1.824	1.40541	0.0164
0.4917	1.1871	−1.691	1.40863	0.0152
0.5924	1.2157	−1.439	1.41140	0.0131
0.7028	1.2418	−1.103	1.41357	0.0100
0.8020	1.2617	−0.767	1.41517	0.0068
0.8994	1.2786	−0.414	1.41657	0.0035
0.9477	1.2861	−0.218	1.41713	0.0018
1	1.2937	0	1.41787	0

Mixture 2: H_2O (1) + [C_6mim][PF_6] (2)
(1) Solubility

Ref. [188]

T/K	$x_1 \pm \sigma^a$	$10^4\ (x_2 \pm \sigma^a)$
288.15	0.196 ± 0.003	3.54 ± 0.01
293.15	0.209 ± 0.001	3.92 ± 0.01
298.15	0.229 ± 0.003	4.34 ± 0.02
303.15	0.250 ± 0.004	4.68 ± 0.03
308.15	0.267 ± 0.004	5.04 ± 0.02
313.15	0.288 ± 0.004	5.59 ± 0.01
318.15	0.303 ± 0.004	6.78 ± 0.01

[a]Standard deviation

(2) Thermodynamic Properties

Ref. [188]

$(\Delta_{sol}H_m{}^0 \pm \sigma^a)$/kJ · mol^{-1}	$(\Delta_{sol}G_m{}^0 \pm \sigma^a)$/kJ · mol^{-1}	$(\Delta_{sol}S_m{}^0 \pm \sigma^a)$/kJ · mol^{-1}
298.15 K		
12.7 ± 1.5	19.19 ± 0.01	−21.5 ± 5.1

[a]Standard deviation

Mixture 3: 1-C_3H_7OH (1) + HMIM PF_6 (2)
(1) Phase Equilibrium

Ref. [190]

T/K	x_2 (IL-poor phase)	x_2 (IL-rich phase)	T/K	x_2 (IL-poor phase)	x_2 (IL-rich phase)
278.15	0.0018	0.6973	308.15	0.0078	0.5072
283.15	0.0022	0.6674	313.15	0.0106	0.4646
288.15	0.0027	0.6365	318.15	0.0175	0.4346
293.15	0.0035	0.6023	323.15	0.0482	0.4120
298.15	0.0044	0.5691	328.15	0.0785	0.3819
303.15	0.0057	0.5331			

Mixture 4: 1-C_4H_9OH (1) + HMIM PF_6 (2)
(1) Phase Equilibrium

Ref. [190]

T/K	x_2 (IL-poor phase)	x_2 (IL-rich phase)	T/K	x_2 (IL-poor phase)	x_2 (IL-rich phase)
278.15	0.0009	0.7691	313.15	0.0043	0.5864
283.15	0.0012	0.7501	318.15	0.0056	0.5576
288.15	0.0015	0.7255	323.15	0.0073	0.5314
293.15	0.0018	0.6991	328.15	0.0101	0.5017
298.15	0.0021	0.6706	333.15	0.0169	0.4824
303.15	0.0026	0.6401	338.15	0.0275	0.4581
308.15	0.0034	0.6128	343.15	0.0407	0.4298

Mixture 5: 1-$C_5H_{11}OH$ (1) + HMIM PF_6 (2)
(1) Phase Equilibrium

Ref. [190]

T/K	x_2 (IL-poor phase)	x_2 (IL-rich phase)	T/K	x_2 (IL-poor phase)	x_2 (IL-rich phase)
278.15	0.0010	0.8402	313.15	0.0019	0.6963
283.15	0.0010	0.8261	318.15	0.0026	0.6779
288.15	0.0010	0.8130	323.15	0.0029	0.6555
293.15	0.0010	0.7931	328.15	0.0036	0.6326
298.15	0.0013	0.7708	333.15	0.0046	0.6079
303.15	0.0014	0.7474	338.15	0.0057	0.5858
308.15	0.0016	0.7233	343.15	0.0091	0.5654

Mixture 6: C_2H_5OH (1) + HMIM PF_6 (2)
(1) Volumetric Properties, Refractive Index, Speed of Sound and Compressibility Properties

Ref. [7]

x_2	ρ/ (g · cm^{-3})	V^E_m/ (cm^3 · mol^{-1})	n_D	Δn_D	u/ (m · s^{-1})	κ_s/ T · Pa^{-1}	$\Delta\kappa_s$/ T · Pa^{-1}
T = 293.15 K							
0	0.7893	0	1.36138	0	1160	941	0
0.0198	0.8300	−0.117	1.36597	0.0034	1164	889	−41
0.2033	1.0556	−0.505	1.39358	0.0204	1248	608	−218
0.3120	1.1256	−0.504	1.40092	0.0215	1300	526	−238
0.3897	1.1618	−0.438	1.40471	0.0208	1329	488	−232
0.5002	1.2014	−0.340	1.40893	0.0186	1360	450	−207
0.5982	1.2284	−0.239	1.41163	0.0156	1380	427	−174
0.6961	1.2502	−0.150	1.41394	0.0123	1396	410	−135
0.7902	1.2676	−0.076	1.41562	0.0085	1410	397	−95
0.9033	1.2852	−0.030	1.41784	0.0042	1423	384	−43

(continued)

x_2	ρ/ (g · cm^{-3})	V^E_m/ (cm^3 · mol^{-1})	n_D	Δn_D	u/ (m · s^{-1})	κ_S/ T · Pa^{-1}	$\Delta\kappa_S$/ T · Pa^{-1}
0.9920	1.2970	−0.016	1.41884	0.0001	1435	374	−3
1	1.2979	0	1.41923	0	1438	372	0
T = 298.15 K							
0	0.7851	0	1.35929	0	1143	974	0
0.0198	0.8258	−0.130	1.36482	0.0044	1150	916	−47
0.0465	0.8735	−0.261	1.37109	0.0091	1160	851	−95
0.0991	0.9487	−0.423	1.38024	0.0151	1185	751	−165
0.2033	1.0516	−0.556	1.39210	0.0209	1237	622	−232
0.3120	1.1215	−0.547	1.39964	0.0221	1288	537	−252
0.3897	1.1577	−0.480	1.40388	0.0218	1317	498	−245
0.5002	1.1972	−0.366	1.40810	0.0195	1348	460	−218
0.5982	1.2244	−0.288	1.41098	0.0166	1368	436	−183
0.6961	1.2462	−0.196	1.41313	0.0131	1384	419	−142
0.7902	1.2635	−0.104	1.41472	0.0091	1397	405	−100
0.9033	1.2811	−0.052	1.41645	0.0042	1410	392	−46
0.9920	1.2928	−0.017	1.41775	0.0003	1421	383	−3
1	1.2937	0	1.41787	0	1424	381	0
T = 303.15 K							
0	0.7807	0	1.35725	0	1127	1009	0
0.0198	0.8218	−0.161	1.36297	0.0046	1134	947	−50
0.0465	0.8698	−0.321	1.36941	0.0094	1144	878	−102
0.0991	0.9448	−0.476	1.37870	0.0156	1169	775	−172
0.2033	1.0475	−0.598	1.39057	0.0213	1226	635	−248
0.3120	1.1173	−0.577	1.39816	0.0224	1277	549	−267
0.3897	1.1536	−0.517	1.40239	0.0221	1306	509	−259
0.5002	1.1933	−0.423	1.40693	0.0201	1336	469	−230
0.5982	1.2204	−0.328	1.40986	0.0172	1357	445	−193
0.6961	1.2422	−0.231	1.41197	0.0135	1372	427	−150
0.7902	1.2595	−0.134	1.41366	0.0096	1386	413	−106
0.9033	1.2770	−0.059	1.41525	0.0045	1399	400	−49
0.9920	1.2887	−0.018	1.41635	0.0004	1409	391	−4
1	1.2896	0	1.41643	0	1411	389	0

(2) Phase Equilibrium

Ref. [7]

T/K	x_2 Ethanol-rich phase	x_2 Ionic liquid-rich phase
278.15	0.010	0.366
283.15	0.013	0.318
288.15	0.019	0.263
293.15	0.036	0.183

Mixture 7: 2-CH_3CHCH_3OH (1) + HMIM PF_6 (2)
(1) Volumetric Properties, Refractive Index, Speed of Sound and Compressibility Properties

Ref. [12]

x_2	ρ/ $g \cdot cm^{-3}$	V_m^E/ $cm^3 \cdot mol^{-1}$	n_D	Δn_D	u/ $m \cdot s^{-1}$	κ_S/ $T \cdot Pa^{-1}$	$\Delta\kappa_S$/ $T \cdot Pa^{-1}$
T = 293.15 K							
0	0.7850	0	1.37707	0	1156	953	0
0.0011	0.7870	−0.031	1.37727	0.0002	1157	949	−3
0.0019	0.7884	−0.041	1.37764	0.0005	1158	946	−6
0.6123	1.2135	−0.230	1.41255	0.0097	1370	439	−158
0.6958	1.2360	−0.118	1.41425	0.0078	1388	420	−129
0.8110	1.2618	0.119	1.41602	0.0048	1408	400	−83
0.9260	1.2839	0.231	1.41766	0.0015	1427	383	−33
0.9933	1.2962	0.109	1.41900	0.0001	1437	373	−3
1	1.2979	0	1.41923	0	1438	372	0
T = 298.15 K							
0	0.7810	0	1.37496	0	1139	986	0
0.0011	0.7828	−0.010	1.37513	0.0001	1140	983	−2
0.0019	0.7842	−0.022	1.37534	0.0003	1141	979	−6
0.0024	0.7854	−0.043	1.37542	0.0004	1141	978	−7
0.4893	1.1687	−0.402	1.40818	0.0122	1325	487	−203
0.6123	1.2094	−0.262	1.41101	0.0098	1358	448	−167
0.6958	1.2320	−0.156	1.41282	0.0080	1375	429	−136
0.8110	1.2577	0.087	1.41451	0.0047	1396	408	−87
0.9260	1.2799	0.193	1.41628	0.0016	1413	391	−35
0.9933	1.2922	0.073	1.41766	0.0001	1423	382	−3
1	1.2937	0	1.41787	0	1424	381	0
T = 303.15 K							
0	0.7766	0	1.37278	0	1121	1025	0
0.0011	0.7785	−0.023	1.37296	0.0001	1122	1020	−4
0.0019	0.7800	−0.036	1.37319	0.0003	1124	1015	−8
0.0024	0.7811	−0.061	1.37337	0.0005	1124	1014	−9
0.4893	1.1646	−0.454	1.40669	0.0126	1313	498	−216
0.6123	1.2053	−0.300	1.40962	0.0101	1346	458	−178
0.6958	1.2279	−0.192	1.41139	0.0082	1363	438	−145
0.8110	1.2538	0.047	1.41312	0.0049	1383	417	−93
0.9260	1.2759	0.176	1.41490	0.0017	1401	399	−37
0.9933	1.2883	0.036	1.41624	0.0001	1410	390	−3
1	1.2896	0	1.41643	0	1411	389	0

(2) Phase Equilibrium

Ref. [12]

	x_2			x_2	
T/K	IL-poor phase	IL-rich phase	T/K	IL-poor phase	IL-rich phase
278.15	0.0013	0.6218	308.15	0.0047	0.4054
283.15	0.0014	0.5904	313.15	0.0064	0.3709
288.15	0.0016	0.5574	318.15	0.0103	0.3141
293.15	0.0024	0.5208	323.15	0.0215	0.2821
298.15	0.0027	0.4856	328.15	0.0377	0.2106
303.15	0.0035	0.4458			

Mixture 8: $(CH_3O)_2C = O$ (1) + HMIM PF_6 (2)
(1) Volumetric Properties and Refractive Index

Ref. [230]

x_2	ρ/(g · cm^{-3})	V^E/(cm^3 · mol^{-1})	n_D	Δn_D
298.15 K				
0	1.0633	0	1.36640	0
0.0524	1.1017	−0.594	1.37624	0.0071
0.1026	1.1312	−1.007	1.38348	0.0118
0.1915	1.1704	−1.397	1.39262	0.0164
0.2935	1.2021	−1.513	1.39977	0.0183
0.4000	1.2261	−1.427	1.40466	0.0177
0.5043	1.2442	−1.262	1.40837	0.0160
0.5985	1.2573	−1.066	1.41086	0.0137
0.6999	1.2690	−0.834	1.41318	0.0108
0.8033	1.2788	−0.548	1.41505	0.0073
0.9001	1.2866	−0.271	1.41653	0.0038
0.9500	1.2902	−0.123	1.41717	0.0019
1	1.2937	0	1.41787	0

Mixture 9: $(CH_3CH_2O)_2C = O$ [DEC] (1) + HMIM PF_6 (2)
(1) Volumetric Properties and Refractive Index
Ref. [230]

x_2	ρ/(g · cm^{-3})	V^E/(cm^3 · mol^{-1})	n_D	Δn_D
298.15 K				
0	0.9692	0	1.38280	0
0.0098	0.9761	−0.084	1.38348	0.0003
0.1298	1.0521	−1.163	1.39410	0.0067
0.1454	1.0607	−1.277	1.39532	0.0074
0.1989	1.0882	−1.615	1.39905	0.0093
0.2988	1.1317	−1.948	1.40412	0.0108
0.3995	1.1675	−2.011	1.40802	0.0112
0.4975	1.1965	−1.881	1.41068	0.0104

(continued)

x_2	ρ/(g · cm^{-3})	V^E/(cm^3 · mol^{-1})	n_D	Δn_D
0.6052	1.2236	−1.635	1.41299	0.0090
0.7019	1.2443	−1.309	1.41462	0.0072
0.8014	1.2629	−0.911	1.41594	0.0050
0.8997	1.2788	−0.431	1.41692	0.0026
0.9498	1.2862	−0.170	1.41740	0.0013
1	1.2937	0	1.41787	0

(2) Phase Equilibrium

Ref. [230]

T/K	DEC -rich phase	HMIM PF_6-rich phase
278.15	0.0081	0.1347
283.15	0.0086	0.1330
288.15	0.0090	0.1313
293.15	0.0092	0.1295
298.15	0.0094	0.1273
303.15	0.0096	0.1258
308.15	0.0101	0.1230
313.15	0.0112	0.1208
318.15	0.0136	0.1191
323.15	0.0152	0.1173
328.15	0.0187	0.1155
333.15	0.0225	0.1126

Mixture 10: CH_3COCH_3 (1) + HMIM PF_6 (2)
(1) Volumetric Properties and Refractive Index

Ref. [230]

x_2	ρ/(g · cm^{-3})	V^E/(cm^3 · mol^{-1})	n_D	Δn_D
298.15 K				
0	0.78440	0	1.35566	0
0.0462	0.8641	−0.974	1.36998	0.0114
0.0911	0.9254	−1.500	1.37940	0.0181
0.1933	1.0276	−2.055	1.39202	0.0243
0.2775	1.0873	−2.188	1.39899	0.0261
0.3621	1.1328	−2.111	1.40383	0.0256
0.4680	1.1768	−1.913	1.40814	0.0234
0.5483	1.2032	−1.687	1.41040	0.0206
0.6029	1.2186	−1.511	1.41169	0.0185
0.7127	1.2449	−1.126	1.41389	0.0139
0.8174	1.2654	−0.727	1.41565	0.0091
0.9110	1.2809	−0.356	1.41684	0.0045
0.9518	1.2868	−0.163	1.41731	0.0024
1	1.2937	0	1.41787	0

Mixture 11: 2-$CH_3COC_2H_5$ (1) + HMIM PF_6 (2)
(1) Volumetric Properties and Refractive Index

Ref. [230]

x_2	ρ/(g · cm^{-3})	V^E/(cm^3 · mol^{-1})	n_D	Δn_D
298.15 K				
0	0.7997	0	1.37604	0
0.0579	0.8783	−0.990	1.38671	0.0082
0.0977	0.9226	−1.349	1.39160	0.0115
0.1906	1.0063	−1.849	1.39962	0.0156
0.2929	1.0750	−1.949	1.40509	0.0168
0.4037	1.1320	−1.867	1.40900	0.0161
0.4995	1.1709	−1.664	1.41154	0.0146
0.6006	1.2048	−1.411	1.41341	0.0122
0.7012	1.2325	−1.058	1.41483	0.0095
0.8036	1.2565	−0.697	1.41607	0.0064
0.9012	1.2764	−0.380	1.41701	0.0033
0.9483	1.2852	−0.248	1.41741	0.0017
1	1.2937	0	1.41787	0

Mixture 12: 2-$CH_3COC_3H_7$ (1) + HMIM PF_6 (2)
(1) Volumetric Properties and Refractive Index

Ref. [230]

x_2	ρ/(g · cm^{-3})	V^E/(cm^3 · mol^{-1})	n_D	Δn_D
298.15 K				
0	0.8014	0	1.38823	0
0.0538	0.8635	−0.845	1.39480	0.0050
0.0932	0.9028	−1.210	1.39817	0.0072
0.1960	0.9884	−1.734	1.40449	0.0104
0.2934	1.0525	−1.890	1.40850	0.0116
0.3985	1.1082	−1.803	1.41141	0.0114
0.4946	1.1505	−1.628	1.41337	0.0105
0.5983	1.1891	−1.374	1.41465	0.0087
0.7039	1.2223	−1.033	1.41579	0.0067
0.7993	1.2483	−0.707	1.41661	0.0047
0.8981	1.2719	−0.336	1.41719	0.0023
0.9519	1.2838	−0.171	1.41763	0.0012
1	1.2937	0	1.41787	0

Mixture 13: CH_3COOCH_3 (1) + HMIM PF_6 (2)

(1) Volumetric Properties and Refractive Index

Ref. [230]

x_2	$\rho/(g \cdot cm^{-3})$	$V^E/(cm^3 \cdot mol^{-1})$	n_D	Δn_D
298.15 K				
0	0.9269	0	1.35881	0
0.0411	0.9764	−0.660	1.36891	0.0077
0.0967	1.0297	−1.225	1.37917	0.0147
0.2066	1.1051	−1.715	1.39227	0.0213
0.2880	1.1446	−1.769	1.39832	0.0225
0.4075	1.1881	−1.667	1.40488	0.0220
0.5423	1.2236	−1.370	1.40950	0.0187
0.6247	1.2409	−1.176	1.41170	0.0160
0.7020	1.2548	−0.971	1.41331	0.0130
0.8043	1.2703	−0.653	1.41503	0.0087
0.9002	1.2826	−0.332	1.41652	0.0045
0.9499	1.2883	−0.162	1.41723	0.0023
1	1.2937	0	1.41787	0

Mixture 14: $CH_3COOC_4H_9$ (1) + HMIM PF_6 (2)

(1) Volumetric Properties and Refractive Index

Ref. [230]

x_2	$\rho/(g \cdot cm^{-3})$	$V^E/(cm^3 \cdot mol^{-1})$	n_D	Δn_D
298.15 K				
0	0.8763	0	1.39190	0
0.0365	0.9052	−0.296	1.39472	0.0019
0.0913	0.9452	−0.661	1.39831	0.0040
0.2054	1.0178	−1.205	1.40418	0.0069
0.2945	1.0660	−1.460	1.40759	0.0080
0.4109	1.1195	−1.534	1.41084	0.0083
0.5066	1.1571	−1.439	1.41282	0.0078
0.5784	1.1823	−1.309	1.41392	0.0070
0.6918	1.2178	−1.052	1.41542	0.0056
0.7965	1.2467	−0.767	1.41643	0.0038
0.8940	1.2705	−0.419	1.41715	0.0020
0.9496	1.2830	−0.207	1.41753	0.0010
1	1.2937	0	1.41787	0

Mixture 15: CO_2 (1) + [hmim][PF_6] (2)

(1) Bubble Point

Ref. [231]

x_1	T (K)	P (MPa)	T (K)	P (MPa)	T (K)	P (MPa)
0.098	308.50	0.64	328.58	0.89	348.52	1.15
	313.47	0.70	333.55	0.95	353.53	1.23
	318.54	0.75	338.52	1.02	358.53	1.31

(continued)

x_1	T (K)	P (MPa)	T (K)	P (MPa)	T (K)	P (MPa)
	323.46	0.81	343.58	1.08	363.58	1.41
0.198	298.45	1.15	323.43	1.80	348.42	2.57
	303.37	1.27	328.40	1.94	353.49	2.75
	308.43	1.40	333.45	2.10	358.45	2.92
	313.49	1.55	338.47	2.26	363.30	3.09
	318.41	1.65	343.53	2.43		
0.299	303.34	2.13	328.44	3.31	348.46	4.56
	308.35	2.34	333.36	3.60	353.48	4.79
	313.36	2.57	338.43	3.86	358.65	5.09
	318.39	2.80	343.63	4.17	363.37	5.43
	323.41	3.06				
0.410	298.31	3.05	323.57	4.94	348.63	7.37
	303.48	3.37	328.58	5.38	353.59	7.85
	308.48	3.80	333.54	5.85	358.62	8.41
	313.52	4.12	338.55	6.31	363.37	9.03
	318.55	4.54	343.58	6.77		
0.501	313.05	4.48	328.07	6.02	342.93	7.88
	318.13	4.98	333.05	6.60	347.90	8.52
	323.13	5.52	337.88	7.22	352.56	9.20
0.599	313.03	8.32	327.97	12.58	342.70	17.36
	318.11	9.34	332.90	14.12	347.65	18.46
	322.97	11.10	337.82	15.94		
0.650	313.14	21.80	323.02	25.78	332.87	31.10
	318.04	23.16	327.88	28.84	337.83	33.44
0.702	313.19	53.74	327.89	60.74	337.84	65.70
	318.07	55.16	332.87	63.34	342.79	68.18
	323.05	57.90				
0.711	313.05	59.10	327.74	66.88	342.72	74.50
	318.03	62.18	332.79	69.26	347.70	76.90
	322.99	64.50	337.82	71.20		
0.727	313.08	81.20	327.93	87.58	337.89	92.50
	318.03	82.80	332.94	90.26	342.79	94.60
	323.00	85.00				

Mixture 16: 1,1,1,2-$C_2H_2F_4$ (1) + [HMIm][PF_6] (2)
(1) Phase Equilibrium

Ref. [221]

T(K)	P(bar)	x_1	$\Delta V/V_0$	V^L(cm^3/mol)
	0.81	0.0594 ± 0.0042	0.010 ± 0.001	244.03 ± 0.23
	1.33	0.1242 ± 0.0037	0.023 ± 0.001	230.31 ± 0.22
	2.2	0.2116 ± 0.0031	0.060 ± 0.001	214.82 ± 0.19
298.15	3.04	0.3093 ± 0.0024	0.081 ± 0.001	191.84 ± 0.17
	4.29	0.4529 ± 0.0016	0.179 ± 0.001	165.74 ± 0.13

(continued)

T(K)	P(bar)	x_1	$\Delta V/V_0$	V^L(cm^3/mol)
	5.76	0.6507 ± 0.0007	0.508 ± 0.001	135.42 ± 0.09
	0.89	0.0295 ± 0.0027	0.003 ± 0.001	257.07 ± 0.17
	1.77	0.0739 ± 0.0025	0.009 ± 0.001	246.77 ± 0.16
	3.40	0.1659 ± 0.0021	0.032 ± 0.001	227.48 ± 0.15
	5.12	0.2723 ± 0.0017	0.079 ± 0.001	207.45 ± 0.13
	6.65	0.3544 ± 0.0015	0.115 ± 0.001	190.15 ± 0.11
323.15	8.10	0.4238 ± 0.0013	0.170 ± 0.001	178.12 ± 0.10
	9.21	0.4809 ± 0.0011	0.226 ± 0.001	168.13 ± 0.09
	10.77	0.5856 ± 0.0008	0.370 ± 0.001	150.00 ± 0.07
	11.93	0.6683 ± 0.0005	0.559 ± 0.001	136.66 ± 0.06
	12.83	0.7677 ± 0.0002	0.978 ± 0.001	121.40 ± 0.04
	6.23	0.1997 ± 0.0045	0.005 ± 0.002	238.9 ± 0.3
	10.11	0.3231 ± 0.0038	0.059 ± 0.002	212.8 ± 0.2
	13.48	0.4330 ± 0.0032	0.119 ± 0.002	188.4 ± 0.2
348.15	17.13	0.5437 ± 0.0025	0.210 ± 0.002	164.0 ± 0.2
	20.57	0.6368 ± 0.0020	0.360 ± 0.002	146.6 ± 0.1
	23.22[a]	0.8262 ± 0.0012		
	128.8[a]	0.9572 ± 0.0015		

[a]Vapor–liquid–liquid equilibrium points for the IL-rich phase
[b]Mixture critical point

2.2.62.2 Ternary Mixtures

Mixture 1: $CH_2(CH_2)_4CH_2$ (1) + 2-$CH_3COC_2H_5$ (2) + HMIM PF_6 (3)
(1) Distribution Ratio and Selectivity

Ref. [232]

Organic-rich phase		Ionic liquid-rich phase			
x_1^I	x_2^I	x_1^{II}	x_2^{II}	β	S
298.15 K					
0.985	0.015	0.071	0.128	8.31	114.85
0.961	0.039	0.079	0.217	5.56	67.99
0.947	0.053	0.083	0.293	5.53	63.40
0.919	0.081	0.093	0.366	4.51	44.52
0.867	0.132	0.110	0.416	3.14	24.68
0.815	0.185	0.135	0.476	2.58	15.52
0.788	0.211	0.154	0.514	2.43	12.44
0.750	0.249	0.180	0.558	2.24	9.33
0.695	0.304	0.199	0.587	1.93	6.74
0.660	0.338	0.215	0.600	1.77	5.45
0.629	0.369	0.237	0.606	1.64	4.36

Mixture 2: $HOC_6H_4C(CH_3)_2C_6H_4OH$ (Bisphenol A) (1) + H_2O (2) + $[C_6mim][PF_6]$ (3)
(1) Distribution Ratio

Ref. [204]

Bisphenol A		V_2/V_3					
pH	T	1	5	10	12	25	50
7.03	297.15 ± 1 K	490.5	473.1	455.7	445.1	432.3	414.1

Ref. [204]

pH	T = 297.15 ± 1 K	pH	T = 297.15 ± 1 K	pH	T = 297.15 ± 1 K	pH	T = 297.15 ± 1 K
1.86	412.0	4.95	414.8	9.17	233.6	12.01	0.22
3.26	412.0	7.03	432.3	11.61	0.76		

Mixture 3: C_6H_5OH (1) + H_2O (2) + $[C_6mim][PF_6]$ (3)
(1) Distribution Ratio

Ref. [204]

Phenol		V_2/V_3					
pH	T	1	5	10	12	25	50
7.16	297.15 ± 1 K	21.6	18.3	18.0	17.1	16.5	15.9

Ref. [204]

pH	T = 297.15 ± 1 K	pH	T = 297.15 ± 1 K	pH	T = 297.15 ± 1 K
3.23	17.4	7.16	15.9	11.68	2.1
5.25	16.7	9.52	10.6		

Mixture 4: $4\text{-}C_9H_{19}C_6H_4OH$ (1) + H_2O (2) + $[C_6mim][PF_6]$ (3)
(1) Distribution Ratio

Ref. [204]

pH	T = 97.15 ± 1 K	pH	T = 297.15 ± 1 K	pH	T = 297.15 ± 1 K	pH	T = 297.15 ± 1 K
3.23[a]	350.0	7.09	338.9	10.20	212.5	12.61	156.7
5.19	347.2	8.43	283.3	11.03	192.0		

[a]$V_W:V_{IL}$ = 25:1 was used at these pHs because of the low solubility of 4-nonylphenol in water and their very low concentrations in aqueous phase after extraction

Mixture 5: $4\text{-}C_8H_{17}C_6H_4OH$ (1) + H_2O (2) + $[C_6mim][PF_6]$ (3)
(1) Distribution Ratio

Ref. [204]

pH	$T = 297.15 \pm 1$ K	pH	$T = 297.15 \pm 1$ K	pH	$T = 297.15 \pm 1$ K
3.23[a]	211.9	7.04	237.5	12.07	53.3
5.27	231.7	10.22	111.5		

[a] $V_W{:}V_{IL} = 25{:}1$ was used at these pHs because of the low solubility of 4-octylphenol in water and their very low concentrations in aqueous phase after extraction

Mixture 6: C_6Cl_5OH (1) + H_2O (2) + $[C_6mim][PF_6]$ (3)
(1) Distribution Ratio

Ref. [204]

pH	$T = 297.15 \pm 1$ K	pH	$T = 297.15 \pm 1$ K	pH	$T = 297.15 \pm 1$ K
3.16	176.5	7.03	162.5	12.73	120.3
5.12	171.4	10.25	153.2		

[a] $V_W{:}V_{IL} = 25{:}1$ was used at these pHs because of the low solubility of pentachlorophenol in water and their very low concentrations in aqueous phase after extraction

Mixture 7: C_6H_{14} (1) + $CH_3COOCH_2CH_3$ (2) + HMIM PF_6 (3)
(1) Distribution Ratio and Selectivity

Ref. [229]

Organic-rich phase		Ionic liquid-rich phase			
x_1^I	x_2^I	x_1^{II}	x_2^{II}	β	S
$T = 298.15$ K					
0.940	0.060	0.019	0.218	3.66	186.19
0.877	0.123	0.033	0.302	2.45	66.01
0.816	0.184	0.039	0.384	2.08	43.06
0.743	0.258	0.051	0.415	1.61	23.62
0.648	0.352	0.064	0.482	1.37	13.96
0.582	0.417	0.076	0.520	1.25	9.57
0.501	0.498	0.091	0.578	1.16	6.37
0.427	0.571	0.108	0.631	1.11	4.40
0.371	0.626	0.111	0.658	1.05	3.52
0.326	0.670	0.125	0.675	1.01	2.63
0.286	0.710	0.134	0.696	0.98	2.10
0.244	0.747	0.141	0.720	0.96	1.68
0.193	0.782	0.159	0.745	0.95	1.16

Mixture 8: 2-CH_3CHCH_3OH (1) + $CH_3COOCH_2CH_3$ (2) + HMIM PF_6 (3)
(1) Distribution Ratio and Selectivity

Ref. [17]

Organic-rich phase		Ionic liquid-rich phase			
x^{I}_{1}	x^{I}_{2}	x^{II}_{1}	x^{II}_{2}	β	S
298.15 K					
0.988	0.010	0.516	0.008	0.82	1.58
0.987	0.010	0.520	0.012	1.21	2.30
0.984	0.012	0.527	0.035	2.97	5.54
0.979	0.015	0.528	0.065	4.21	7.80
0.975	0.019	0.532	0.084	4.46	8.17
0.955	0.038	0.540	0.106	2.84	5.01
0.927	0.065	0.555	0.132	2.03	3.40
0.895	0.093	0.574	0.158	1.70	2.65
0.882	0.104	0.585	0.171	1.64	2.47
0.845	0.129	0.609	0.188	1.46	2.02
0.804	0.158	0.650	0.200	1.27	1.57

Mixture 9: C_7H_{16} (1) + C_6H_6 (2) + [hmim]PF_6 (3)
(1) Phase Equilibrium

Ref. [222]

x_3	x_2	x_3	x_2	x_3	x_2
298.2 K					
0.0002	0.8962	0.0040	0.9992	0.6598	0.2965
0.0004	0.8762	0.2165	0.7835	0.8034	0.1234
0.0007	0.9993	0.4100	0.5688	0.8859	0.1141
0.0015	0.6902	0.5288	0.4402	0.9161	0.0000
0.0016	0.0000				

Ref. [222]

[hmim]PF_6-rich layer		Alkane-rich layer		
x'_3	x'_2	x''_3	x''_2	S
298.2 K				
0.2692	0.7239	0.0030	0.9495	5.25
0.3055	0.6857	0.0022	0.8501	13.54
0.3270	0.6630	0.0020	0.8390	12.56
0.4070	0.5737	0.0014	0.7307	10.90
0.4250	0.5548	0.0012	0.6882	12.22
0.4308	0.5461	0.0012	0.6722	11.49
0.4782	0.4954	0.0007	0.5689	14.20
0.4900	0.4801	0.0006	0.5506	13.06
0.5872	0.3755	0.0008	0.4197	13.90
0.6756	0.2783	0.0007	0.3020	13.94

(continued)

[hmim]PF$_6$-rich layer		Alkane-rich layer		
x'_3	x'_2	x''_3	x''_2	S
0.7918	0.1387	0.0006	0.1512	11.20
0.8904	0.0288	0.0005	0.0414	13.09

Mixture 10: $C_{12}H_{26}$ (1) + C_6H_6 (2) + [hmim]PF$_6$ (3)
(1) Phase Equilibrium

Ref. [222]

x_3	x_2	x_3	x_2	x_3	x_2
298.2 K					
0.0004	0.0000	0.0040	0.9960	0.5999	0.3800
0.0008	0.8854	0.2165	0.7835	0.7308	0.2300
0.0013	0.7713	0.3809	0.6108	0.8463	0.1061
0.0018	0.4960	0.5343	0.4448	0.9471	0.0000
0.0035	0.1503				

Ref. [222]

[hmim]PF$_6$-rich layer		Alkane-rich layer		
x'_3	x'_2	x''_3	x''_2	S
298.2 K				
0.3308	0.6603	0.0038	0.9428	4.20
0.3597	0.6306	0.0032	0.8279	13.26
0.4309	0.5590	0.0028	0.7446	18.78
0.4920	0.4912	0.0024	0.6562	15.21
0.6448	0.3274	0.0021	0.4602	13.76
0.8299	0.1275	0.0018	0.2536	8.79
0.8912	0.0590	0.001	0.1451	6.98

Mixture 11: $C_{16}H_{34}$ (1) + C_6H_6 (2) + [hmim]PF$_6$ (3)
(1) Phase Equilibrium

Ref. [222]

x_3	x_2	x_3	x_2	x_3	x_2
298.2 K					
0.0001	0.0000	0.2165	0.7835	0.7365	0.2489
0.0003	0.5452	0.3724	0.6262	0.9080	0.0715
0.0004	0.8631	0.4791	0.5181	0.9783	0.0000
0.0040	0.9960	0.6166	0.3786		

Ref. [222]

[hmim]PF$_6$-rich layer		Alkane-rich layer		
x'_3	x'_2	x''_3	x''_2	S
298.2 K				
0.3132	0.6840	0.0004	0.9450	14.11
0.3829	0.6136	0.0004	0.8509	30.64

(*continued*)

[hmim]PF_6-rich layer		Alkane-rich layer		
x'_3	x'_2	x''_3	x''_2	S
0.6240	0.3683	0.0001	0.5448	39.94
0.7501	0.2348	0.0001	0.3950	23.81
0.9062	0.0745	0.0000	0.1722	18.51

Mixture 12: 1-C_6H_{12} (1) + C_2H_5OH (2) + [C_6mim]PF_6 (3)
(1) Phase Equilibrium

Ref. [223]

x_3	x_2	x_3	x_2	x_3	x_2
298.2 K					
0.0021	0.0000	0.0122	0.8703	0.2619	0.6681
0.0028	0.5531	0.0378	0.9033	0.4489	0.4641
0.0038	0.4378	0.0677	0.8927	0.7866	0.0998
0.0040	0.7309	0.1384	0.8111	0.8770	0.0000
0.0052	0.7749	0.2145	0.7241		

Ref. [223]

[C_6mim]PF_6-rich layer		Alkene-rich layer		
x'_3	x'_2	x''_3	x''_2	S
298.2 K				
0.6802	0.2120	0.0065	0.2305	6.51
0.6718	0.2204	0.0030	0.2746	5.38
0.6200	0.2801	0.0045	0.3608	4.93
0.5238	0.3831	0.0054	0.4808	4.40
0.5108	0.3978	0.0043	0.5103	2.39
0.4859	0.4251	0.0068	0.5565	3.75
0.4108	0.5062	0.0065	0.6452	3.29
0.3830	0.5365	0.0069	0.6765	3.12
0.3692	0.5515	0.0064	0.7196	2.65
0.3672	0.5523	0.0088	0.7821	1.83
0.3648	0.5549	0.0099	0.8359	1.28
0.2972	0.6303	0.0120	0.8705	1.17
0.2698	0.6603	0.0122	0.8889	0.78

Mixture 13: C_7H_{14} (1) + C_2H_5OH (2) + [C_6mim]PF_6 (3)
(1) Phase Equilibrium

Ref. [223]

x_3	x_2	x_3	x_2	x_3	x_2
298.2 K					
0.0001	0.0000	0.0157	0.8946	0.1429	0.8189
0.0012	0.5060	0.0174	0.9015	0.2023	0.7507
0.0019	0.6111	0.0195	0.9115	0.5868	0.3262

(continued)

x_3	x_2	x_3	x_2	x_3	x_2
0.0033	0.7235	0.0248	0.9171	0.7391	0.1590
0.0072	0.8438	0.0718	0.8944	0.8772	0.0000
0.0079	0.7990				

Ref. [223]

$[C_6mim]PF_6$-rich layer		Alkene-rich layer		
x'_3	x'_2	x''_3	s''_2	S
298.2 K				
0.7919	0.0988	0.0001	0.1118	7.18
0.5629	0.3545	0.0020	0.4148	6.03
0.4941	0.4311	0.0029	0.5129	5.42
0.4499	0.4822	0.0019	0.5711	5.32
0.4459	0.4840	0.0027	0.5946	4.68
0.4361	0.4951	0.0015	0.6001	4.78
0.4221	0.5099	0.0030	0.6808	3.48
0.4096	0.5255	0.0030	0.7229	3.07
0.4061	0.5267	0.0057	0.7721	2.26
0.3932	0.5431	0.0073	0.8059	1.98
0.3731	0.5663	0.0099	0.8442	1.62
0.3580	0.5818	0.0098	0.8886	1.12
0.3091	0.6348	0.0382	0.9171	0.55

2.2.63 02021-06003: 1-Hexyl-3-methylimidazolium acetate

Abbreviation:	$[C_6MIm][OAc]$, [hmim][OAc]
Molecular Formula:	$C_{12}H_{22}N_2O_2$
Molar Mass:	226.31
Structure:	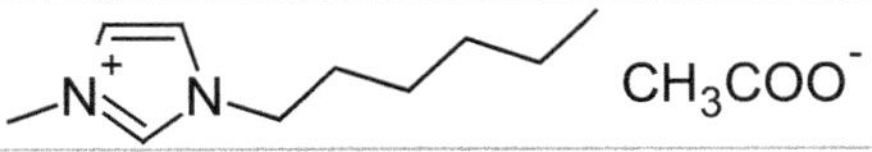

2.2.63.1 Ternary Mixtures

Mixture 1: H_2O (1) + $(CH_3)_3COH$ (2) + [hmim][OAc] (3)

(1) Phase Equilibrium

Ref. [25]

x'_2 [a]	w_3 [a]	y_2	T/K	γ_1	γ_2	$\alpha_{2,1}$ [b]
p = 100 kPa						
0.9501	0.6166	0.9823	378.35	0.44	0.68	2.92
0.9501	0.5502	0.9798	372.03	0.58	0.77	2.55
0.9501	0.5000	0.9776	368.36	0.69	0.83	2.30

(continued)

x'_2[a]	w_3[a]	y_2	T/K	γ_1	γ_2	$\alpha_{2,1}$[b]
0.9501	0.4502	0.9752	365.60	0.81	0.87	2.06
0.9501	0.4001	0.9729	362.90	0.95	0.92	1.88
0.9501	0.3000	0.9670	359.80	1.22	0.97	1.54
0.9503	0.2002	0.9580	357.07	1.65	1.01	1.19
0.9501	0.1000	0.9410	355.30	2.37	1.02	0.84
0.9501	0.0500	0.9315	354.95	2.73	1.01	0.72

[a]Compositions reported are of equilibrium liquid
[b]$\alpha_{2,1} = (y_2/x'_2)/(y_1/x'_1)$

2.2.64 02021-15003: 1-Hexyl-3-methylimidazolium thiocyanate

Abbreviation:	[HMIM][SCN]
Molecular Formula:	$C_{11}H_{19}N_3S$
Molar Mass:	225.36
Structure:	

2.2.64.1 Binary Mixtures

Mixture 1: SCH = CHCH = CH (thiophene) (1) + [HMIM][SCN] (2)
(1) Phase Equilibrium

Ref. [110]

x_2	T^{LLE}/K	x_2	T^{LLE}/K
0.1640	364.7	0.1426	303.8
0.1625	362.5	0.1413	290.8
0.1607	355.4	0.1395	279.1
0.1585	352.4	0.1376	265.8
0.1553	346.5	0.1367	253.3
0.1456	320.6	0.1341	247.8
0.1453	333.4		

2.2.65 02024-03001: 1-Heptyl-3-methylimidazolium bis((trifluoromethyl)sulfonyl)imide

Abbreviation:	[C_7MIm][NTf_2], [C_7mim][Tf_2N]
Molecular Formula:	$C_{13}H_{21}F_6N_3O_4S_2$
Molar Mass:	461.45
Structure:	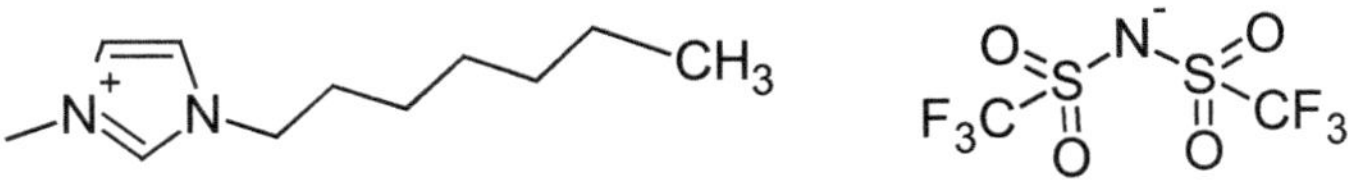

2.2.65.1 Binary Mixtures

Mixture 1: H_2O (1) + [C_7mim][Tf_2N] (2)
(1) Solubility

Ref. [45]

T/K	$x_1 \pm \sigma^a$	$10^5\ (x_2 \pm \sigma^a)$
288.15	0.1806 ± 0.0009	4.85 ± 0.02
293.15	0.1879 ± 0.0006	4.98 ± 0.04
298.15	0.1971 ± 0.0004	5.09 ± 0.02
303.15	0.2081 ± 0.0005	5.28 ± 0.09
308.15	0.2190 ± 0.0006	5.81 ± 0.09
313.15	0.2334 ± 0.0005	6.87 ± 0.01
318.15	0.2447 ± 0.0009	7.93 ± 0.06

[a]Standard deviation

(2) Thermodynamic Properties

Ref. [45]

$(\Delta_{sol}H_m^0 \pm \sigma^a)$/ $kJ \cdot mol^{-1}$	$(\Delta_{sol}G_m^0 \pm \sigma^a)$/ $kJ \cdot mol^{-1}$	$(\Delta_{sol}S_m^0 \pm \sigma^a)$/ $J \cdot K^{-1} \cdot mol^{-1}$	$(\Delta_{svt}H_m^0 \pm \sigma^a)$/ $kJ \cdot mol^{-1}$	$(\Delta_{svt}Gm^0 \pm \sigma^a)$/ $kJ \cdot mol^{-1}$	$(\Delta_{svt}S_m^0 \pm \sigma^a)$/ $J \cdot K^{-1} \cdot mol^{-1}$	$(\Delta_{svt}H_m^* \pm \sigma^a)$/ $kJ \cdot mol^{-1}$	$(\Delta_{svt}G_m^* \pm \sigma^a)$/ $kJ \cdot mol^{-1}$	$(\Delta_{svt}S_m^* \pm \sigma^a)$/ $J \cdot K^{-1} \cdot mol^{-1}$
298.15 K								
7.1 ± 1.5	24.504 ± 0.008	−58.4 ± 5.0	−172.9 ± 6.2	−58.561 ± 0.008	−383 ± 21	−170.7 ± 6.2	−76.469 ± 0.008	−316 ± 21

[a]Standard deviation

2.2.66 02024-04030: 1-Heptyl-3-methylimidazolium 1,1,2,2-tetrafluoroethanesulfonate

Abbreviation:	[hmim][TFES]
Molecular Formula:	$C_{13}H_{22}F_4N_2O_3S$
Molar Mass:	362.38
Structure:	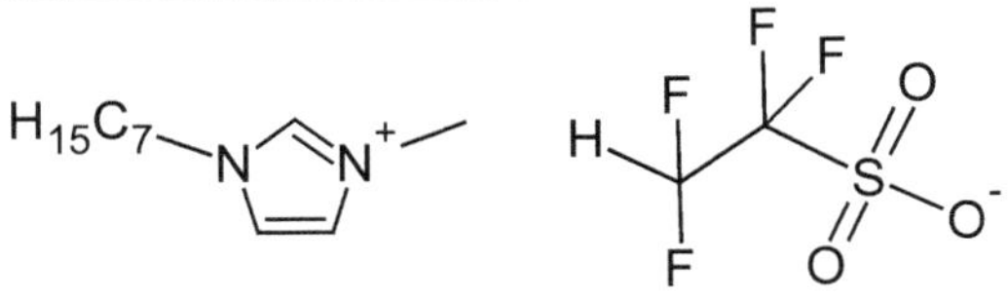

2.2.66.1 Binary Mixtures

Mixture 1: CH_2F_2 (1) + [hmim][TFES] (2)
(1) Solubility and Diffusivity

Ref. [55]

T/K	P/MPa	100 x_1	$10^{11}D/$ $m^2 \cdot s^{-1}$	T/K	P/MPa	100 x_1	$10^{11}D/$ $m^2 \cdot s^{-1}$
298.15	0.0099	0.8	3.9	298.15	0.5497	37.4	6.4
298.15	0.1002	8.1	3.9	298.15	0.6987	45.2	9.0
298.15	0.2510	19.1	4.4	298.15	0.8479	52.3	13
298.15	0.3988	28.8	5.8	298.15	0.9980	59.2	16

2.2.67 02029-01001: 1-Octyl-3-methylimidazolium chloride

Abbreviation:	$[C_8MIm]Cl$, $[C_8mim]Cl$, MeOctImCl
Molecular Formula:	$C_{12}H_{23}N_2Cl$
Molar Mass:	230.78
Structure:	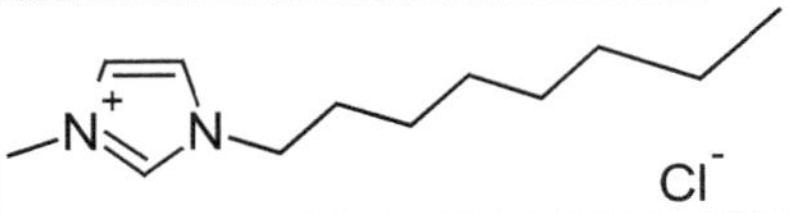

2.2.67.1 Binary Mixtures

Mixture 1: H_2O (1) + $[C_8mim]Cl$ (2)
(1) Volumetric Properties, Refractive Index, Speed of Sound, Compressibility Properties and Transport Properties

Ref. [95]

x_1	n_D	u/m · s^{-1}	κ_S/T · Pa^{-1}	Δn_D	$\Delta\kappa_S$/T · Pa^{-1}
298.15 K					
0.0000	1.51051	1721.0	334.7	0.0000	0.0
0.1484	1.50760	1666.7	356.6	0.0235	20.4
0.2030	1.50625	1655.1	361.4	0.0319	24.5
0.4076	1.49957	1633.8	370.3	0.0616	29.8
0.5334	1.49329	1632.2	370.6	0.0777	26.6
0.6807	1.48143	1639.3	367.6	0.0921	16.6
0.8185	1.45934	1650.7	363.7	0.0945	−0.6
0.8803	1.44036	1651.4	364.3	0.0865	−11.9
0.9307	1.41432	1639.6	370.6	0.0695	−22.1
0.9635	1.38658	1611.3	384.8	0.0476	−26.2
1.0000	1.33251	1497.0	447.6	0.0000	0.0

Ref. [115]

x_1	ρ/ g · cm^{-3}	V^E/ cm^3 · mol^{-1}	$10^3\eta$/ Pa · s	$10^3\Delta\eta$/ Pa · s	x_1	ρ/ g · cm^{-3}	V^E/ cm^3 · mol^{-1}	$10^3\eta$/ Pa · s	$10^3\Delta\eta$/ Pa · s
T = 298.15 K									
0.0000	1.00882	0.000	20883	0.000	0.8185	1.00916	−0.192	193.4	−3599
0.1484	1.00961	−0.185	5570	−12214	0.8803	1.00663	−0.092	80.77	−2420
0.2030	1.00999	−0.258	3815	−12829	0.9307	1.00369	−0.031	22.48	−1426
0.4076	1.01183	−0.510	1325	−11046	0.9635	1.00110	−0.007	5.714	−756.8
0.5334	1.01270	−0.558	921.1	−8824	1.0000	0.99703	0.000	0.895	0.000
0.6807	1.01237	−0.442	433.0	−6236					
T = 313.15 K									
0.0000	1.00020	0.000	4066	0.000	0.8185	1.00009	−0.113	83.87	−654.9
0.1484	1.00098	−0.177	1470	−1992	0.8803	0.99760	−0.015	38.02	−449.4
0.2030	1.00135	−0.246	1097	−2144	0.9307	0.99512	0.032	13.27	−269.2
0.4076	1.00320	−0.491	387.9	−2021	0.9635	0.99358	0.032	4.248	−144.7
0.5334	1.00376	−0.494	337.9	−1560	1.0000	0.99220	0.000	0.658	0.000
0.6807	1.00316	−0.353	175.5	−1123					
T = 328.15 K									
0.0000	0.99167	0.000	1091	0.000	0.8185	0.99076	−0.038	42.22	−156.3
0.1484	0.99242	−0.167	484.6	−444.6	0.8803	0.98816	0.059	20.45	−110.6
0.2030	0.99277	−0.231	393.5	−476.1	0.9307	0.98589	0.091	8.178	−67.92
0.4076	0.99440	−0.444	170.8	−475.7	0.9635	0.98506	0.068	3.295	−36.99
0.5334	0.99481	−0.432	147.7	−361.7	1.0000	0.98568	0.000	0.519	0.000
0.6807	0.99398	−0.276	83.26	−265.5					
T = 343.15 K									
0.0000	0.98324	0.000	387.2	0.000	0.8185	0.98111	0.041	24.22	−46.42
0.1484	0.98387	−0.146	200.1	−129.7	0.8803	0.97839	0.129	12.17	−34.56
0.2030	0.98421	−0.210	165.4	−143.3	0.9307	0.97615	0.146	5.329	−21.90
0.4076	0.98560	−0.394	84.75	−144.8	0.9635	0.97571	0.103	2.647	−11.88
0.5334	0.98596	−0.385	74.31	−106.6	1.0000	0.97775	0.000	0.419	0.000
0.6807	0.98480	−0.209	44.23	−79.69					

Mixture 2: CH_3OH (1) + [C_8mim]Cl (2)
(1) Volumetric Properties, Speed of Sound, Refractive Index, Transport Properties and Compressibility Properties

Ref. [233]

x_1	ρ/ g · cm^{-3}	u/ m · s^{-1}	$10^3\eta$/ Pa · s	n_D	V^E/ cm^3 · mol^{-1}	k_s/ T · Pa^{-1}	Δk_s/ T · Pa^{-1}	$\Delta\eta$/ Pa · s
T = 298.15 K								
0.0000	1.00882	1721	20868	1.51051	0.000	335	0.00	0.000
0.0565	1.00693	1683	10353	1.50817	−0.098	351	−24.0	−9336
0.1686	1.00242	1638	3456	1.50362	−0.264	372	−83.0	−13894
0.2638	0.99772	1608	1489	1.49975	−0.400	388	−134	−13874
0.3525	0.99206	1581	534	1.49421	−0.471	403	−182	−12978
0.4395	0.98537	1559	237.23	1.48828	−0.559	418	−229	−11460
0.5589	0.97332	1526	64.10	1.47696	−0.686	441	−290	−9141
0.7536	0.93925	1447	8.427	1.44786	−0.812	508	−361	−5134
0.8231	0.91805	1397	4.193	1.43053	−0.798	558	−361	−3688
0.8713	0.89737	1348	2.390	1.41277	−0.723	613	−340	−2684
0.8813	0.89218	1335	2.205	1.40840	−0.696	629	−332	−2475
0.9236	0.86628	1274	1.416	1.38754	−0.583	711	−279	−1593
0.9436	0.85042	1237	1.135	1.37577	−0.479	768	−236	−1176
0.9803	0.81347	1157	0.729	1.34680	−0.235	918	−113	−410.9
1.0000	0.78664	1103	0.543	1.32687	0.000	1045	0.00	0.000
T = 313.15 K								
0.0000	1.00020	1624	4062		0.000	379	0	0.000
0.0565	0.99827	1602	2450		−0.109	390	−33	−1382
0.1686	0.99368	1578	1002		−0.299	404	−108	−2375
0.2638	0.98889	1558	499.91		−0.453	417	−170	−2491
0.3525	0.98319	1536	211.59		−0.551	431	−225	−2419
0.4395	0.97646	1518	106.68		−0.666	444	−280	−2170
0.5589	0.96407	1486	34.968		−0.796	470	−349	−1757
0.7536	0.92918	1406	5.817		−0.930	544	−427	−995.3
0.8231	0.9075	1354	3.580		−0.915	601	−425	−715.3
0.8713	0.88632	1304	1.854		−0.831	664	−401	−521.3
0.8813	0.88100	1292	1.722		−0.802	680	−392	−480.8
0.9236	0.85444	1229	1.133		−0.673	775	−331	−309.6
0.9436	0.83816	1192	0.914		−0.556	840	−281	−228.6
0.9803	0.80013	1110	0.594		−0.274	1014	−136	−79.86
1.0000	0.77238	1054	0.444		0.000	1165	0	0.000
T = 328.15 K								
0.0000	0.99167	1560	1091		0.000	414	0	0.000
0.0565	0.98971	1548	757		−0.125	422	−43	−272.6
0.1686	0.98504	1531	364.92		−0.342	433	−130	−542.2
0.2638	0.98025	1514	205.49		−0.535	445	−203	−597.8
0.3525	0.97448	1496	98.97		−0.655	459	−268	−607.6

(*continued*)

x_1	ρ/ g · cm^{-3}	u/ m · s^{-1}	$10^3\eta$/ Pa · s	n_D	V^E/ cm^3 · mol^{-1}	k_s/ T · Pa^{-1}	Δk_s/ T · Pa^{-1}	$\Delta\eta$/ Pa · s
0.4395	0.96751	1478	55.20		−0.770	473	−330	−556.5
0.5589	0.95487	1447	21.442		−0.920	500	−408	−460.0
0.7536	0.91917	1365	4.388		−1.069	584	−497	−264.7
0.8231	0.89697	1313	2.496		−1.051	647	−496	−190.8
0.8713	0.87526	1263	1.491		−0.958	716	−469	−139.2
0.8813	0.86980	1250	1.388		−0.925	736	−458	−128.4
0.9236	0.84252	1186	0.93		−0.777	844	−387	−82.76
0.9436	0.82578	1147	0.754		−0.645	920	−328	−61.13
0.9803	0.78658	1070	0.497		−0.319	1110	−171	−21.36
1.0000	0.75785	1008	0.369		0.000	1299	0	0.000

Mixture 3: C_2H_5OH (1) + [C_8mim]Cl (2)
(1) Volumetric Properties, Speed of Sound, Refractive Index, Transport Properties and Compressibility Properties

Ref. [233]

x_1	ρ/ g · cm^{-3}	u/ m · s^{-1}	$10^3\eta$/ Pa · s	n_D	V^E/ cm^3 · mol^{-1}	k_s/ T · Pa^{-1}	Δk_s/ T · Pa^{-1}	$\Delta\eta$/ Pa · s
T = 298.15 K								
0.0000	1.00882	1721	20868	1.51051	0.000	335	0	0.000
0.0449	1.00627	1692	12571	1.50835	−0.025	347	−16	−7360
0.1305	1.00097	1646	5426	1.50474	−0.091	369	−49	−12719
0.2969	0.98854	1579	1127	1.49563	−0.285	406	−119	−13546
0.4753	0.96979	1518	185.58	1.48233	−0.480	447	−192	−10764
0.6858	0.93354	1437	21.263	1.45682	−0.602	519	−255	−6536
0.8809	0.86864	1304	3.501	1.41305	−0.574	677	−222	−2483
0.9170	0.84907	1264	2.551	1.40102	−0.481	737	−185	−1730
0.9550	0.82408	1215	1.754	1.38404	−0.342	822	−124	−938.3
0.9794	0.80461	1179	1.396	1.37207	−0.197	894	−68	−429.5
1.0000	0.78511	1143	1.085	1.35960	0.000	975	0	0.000
T = 313.15 K								
0.0000	1.00020	1624	4062		0.000	379	0	0.000
0.0449	0.99761	1606	2717		−0.034	389	−22	−1163
0.1305	0.99224	1578	1411		−0.118	405	−66	−2121
0.2969	0.97963	1530	393.63		−0.347	436	−152	−2463
0.4753	0.96076	1477	87.97		−0.600	477	−237	−2044
0.6858	0.92378	1395	14.598		−0.722	556	−306	−1262
0.8809	0.85766	1258	2.601		−0.678	737	−263	−481.9
0.9170	0.83768	1218	1.919		−0.570	−221	−221	−336.0
0.9550	0.81212	1169	1.341		−0.406	−151	−151	−182.2
0.9794	0.79215	1131	1.061		−0.233	−83	−83	−83.42
1.0000	0.77207	1093	0.826		0.000	0	0	0.000

(*continued*)

x_1	ρ/ g · cm^{-3}	u/ m · s^{-1}	$10^3\eta$/ Pa · s	n_D	V^E/ cm^3 · mol^{-1}	k_s/ T · Pa^{-1}	Δk_s/ T · Pa^{-1}	$\Delta\eta$/ Pa · s
T = 328.15 K								
0.0000	0.99167	1560	1091		0.000	414	0	0.000
0.0449	0.98905	1547	827		−0.047	422	−28	−214.9
0.1305	0.98362	1527	481.47		−0.157	436	−82	−467.2
0.2969	0.97096	1487	164.99		−0.453	466	−185	−602.3
0.4753	0.95169	1438	45.212		−0.726	508	−285	−527.5
0.6858	0.91405	1353	9.120		−0.873	598	−364	−334.1
0.8809	0.84668	1214	1.997		−0.818	801	−315	−128.5
0.917	0.82615	1175	1.488		−0.683	877	−269	−89.65
0.955	0.79994	1122	1.038		−0.486	993	−183	−48.67
0.9794	0.77938	1090	0.819		−0.279	1080	−115	−22.28
1.0000	0.75861	1043	0.641		0.000	1212	0	0.000

Mixture 4: 1-C_3H_7OH (1) + [C_8MIm]Cl (2)
(1) Volumetric Properties, Speed of Sound, Refractive Index, Transport Properties and Compressibility Properties

Ref. [233]

x_1	ρ/ g · cm^{-3}	u/ m · s^{-1}	$10^3\eta$/ Pa · s	n_D	V^E/ cm^3 · mol^{-1}	k_s/ T · Pa^{-1}	Δk_s/ T · Pa^{-1}	$\Delta\eta$/ Pa · s
T = 298.15 K								
0.0000	1.00882	1721	20868	1.51051	0.000	335	0	0.000
0.1071	1.0011	1659	7238	1.50524	−0.046	363	−28	−11395
0.2407	0.98968	1593	2056	1.49750	−0.118	398	−63	−13789
0.3036	0.98357	1566	912	1.49364	−0.182	415	−80	−13621
0.4070	0.97162	1531	385.53	1.48636	−0.227	439	−110	−11990
0.4887	0.9606	1504	178.51	1.47959	−0.290	460	−131	−10492
0.6209	0.93812	1456	50.81	1.46578	−0.362	502	−159	−7861
0.7248	0.91481	1413	19.246	1.45126	−0.399	548	−168	−5725
0.8516	0.87563	1342	6.342	1.42753	−0.404	634	−148	−3092
0.9024	0.85486	1304	4.597	1.41523	−0.374	688	−121	−2034
0.9400	0.83647	1271	3.238	1.40435	−0.297	740	−89	−1251
0.9729	0.8178	1238	2.648	1.39350	−0.190	798	−48	−564.7
1.0000	0.79948	1205	1.951	1.38302	0.000	861	0	0.000
T = 313.15 K								
0.0000	1.00020	1624	4062		0.000	379	0.00	0.000
0.1071	0.99242	1584	1720		−0.075	402	−39	−1907
0.2407	0.98087	1539	649		−0.174	430	−87	−2435
0.3036	0.9747	1517	334.46		−0.253	445	−108	−2494
0.4070	0.96254	1487	159.63		−0.304	470	−142	−2250
0.4887	0.9515	1461	84.10		−0.401	492	−167	−1993
0.6209	0.92861	1414	28.813		−0.470	538	−197	−1512
0.7248	0.90492	1369	12.215		−0.507	590	−205	−1107

(*continued*)

x_1	ρ/ g · cm^{-3}	u/ m · s^{-1}	$10^3\eta$/ Pa · s	n_D	V^E/ cm^3 · mol^{-1}	k_s/ T · Pa^{-1}	Δk_s/ T · Pa^{-1}	$\Delta\eta$/ Pa · s
0.8516	0.86508	1296	4.401		−0.498	688	−179	−599.6
0.9024	0.84394	1257	3.230		−0.455	750	−146	−394.5
0.9400	0.82519	1223	2.293		−0.361	811	−107	−242.7
0.9729	0.80609	1188	1.878		−0.227	879	−58	−109.5
1.0000	0.78728	1155	1.383		0.000	952	0.00	0.000
T = 328.15 K								
0.0000	0.99167	1560	1091		0.000	414	0	0.000
0.1071	0.98383	1531	561		−0.111	433	−50	−413.4
0.2407	0.9721	1494	230.72		−0.236	461	−108	−597.9
0.3036	0.96591	1475	144.13		−0.342	476	−134	−616.1
0.4070	0.95364	1445	77.39		−0.424	502	−174	−570.0
0.4887	0.9424	1421	42.72		−0.529	525	−203	−515.6
0.6209	0.91912	1373	17.726		−0.609	577	−236	−396.5
0.7248	0.89505	1326	8.254		−0.651	635	−245	−292.7
0.8516	0.85446	1250	3.189		−0.626	748	−214	−159.6
0.9024	0.83287	1211	2.352		−0.564	819	−176	−105.0
0.9400	0.81368	1175	1.679		−0.445	890	−129	−64.73
0.9729	0.79406	1140	1.372		−0.277	970	−70	−29.17
1.0000	0.77464	1105	1.007		0.000	1057	0	0.000

Mixture 5: C_7H_{16} (1) + MeOctImCl (2)
(1) Solubility

Ref. [234]

Binary mixtures	Mole fraction	Binary mixtures	Mole fraction
298.2 K			
$C_{12}H_{23}N_2 \cdot Cl$ in C_7H_{16}	0.007	C_7H_{16} in $C_{12}H_{23}N_2 \cdot Cl$	0.065

Mixture 6: $C_{12}H_{26}$ (1) + MeOctImCl (2)
(1) Solubility

Ref. [234]

Binary mixtures	Mole fraction	Binary mixtures	Mole fraction
298.2 K			
$C_{12}H_{23}N_2 \cdot Cl$ in $C_{12}H_{26}$	0.009	$C_{12}H_{26}$ in $C_{12}H_{23}N_2 \cdot Cl$	0.024

Mixture 7: $C_{16}H_{34}$ (1) + MeOctImCl (2)
(1) Solubility

Ref. [234]

Binary mixtures	Mole fraction	Binary mixtures	Mole fraction
298.2 K			
$C_{12}H_{23}N_2 \cdot Cl$ in $C_{16}H_{34}$	0.010	$C_{16}H_{34}$ in $C_{12}H_{23}N_2 \cdot Cl$	0.019

2.2.67.2 Ternary Mixtures

Mixture 1: C_2H_5OH (1) + H_2O (2) + [C_8mim]Cl (3)

(1) Volumetric Properties, Speed of Sound, Refractive Index, Compressibility Properties and Transport Properties

Ref. [95]

x_1	x_2	ρ/g · cm^{-3}	u/m · s^{-1}	n_D	κ_S/T · Pa^{-1}	$10^3\eta$/Pa · s
298.15 K						
0.0467	0.9348	0.9837	1590.1	1.36831	402.1	2.514
0.0962	0.8862	0.9688	1587.6	1.37031	409.6	2.748
0.2254	0.7596	0.9315	1514.6	1.37306	468.0	2.603
0.3057	0.6808	0.9103	1458.9	1.37294	516.1	2.662
0.5377	0.4533	0.8588	1329.3	1.36990	659.0	1.472
0.0826	0.9100	0.9706	1600.1	1.35765	402.4	2.007
0.1329	0.8601	0.9555	1587.3	1.36124	415.4	2.881
0.2008	0.7927	0.9353	1540.2	1.36427	450.7	2.822
0.2733	0.7208	0.9147	1482.5	1.36592	497.5	2.591
0.3651	0.6297	0.8913	1418.2	1.36663	557.8	2.282
0.6077	0.3891	0.8421	1293.0	1.36557	710.3	1.864
0.0558	0.9143	0.9824	1593.2	1.38190	401.0	3.908
0.1867	0.7875	0.9455	1535.3	1.38199	448.7	3.207
0.2604	0.7162	0.9259	1488.1	1.38087	487.8	3.164
0.3386	0.6405	0.9063	1439.5	1.37926	532.5	2.729
0.4437	0.5387	0.8822	1380.4	1.37657	594.9	2.553
0.0679	0.8854	0.9816	1594.6	1.39833	400.7	7.008
0.2958	0.6690	0.9221	1466.3	1.39021	504.4	3.560
0.3954	0.5743	0.8986	1410.6	1.38593	559.3	2.873
0.5841	0.3951	0.8588	1319.1	1.37780	669.2	2.156
0.7005	0.2845	0.8370	1269.7	1.37290	741.1	1.923
0.8470	0.1454	0.8108	1208.9	1.36639	843.9	1.508
0.0766	0.8576	0.9823	1594.2	1.41129	400.6	10.50
0.3335	0.6189	0.9191	1449.6	1.39734	517.8	3.791
0.4166	0.5418	0.9001	1406.3	1.39250	561.7	3.353
0.5440	0.4234	0.8728	1344.9	1.38532	633.4	2.522
0.1218	0.7443	0.9818	1572.8	1.44104	411.8	18.90
0.2235	0.6582	0.9604	1520.2	1.43225	450.6	10.37
0.3218	0.5748	0.9396	1473.1	1.42378	490.4	7.072
0.4363	0.4778	0.9150	1419.7	1.41372	542.2	4.956
0.5320	0.3966	0.8942	1375.7	1.40523	590.9	3.823
0.6374	0.3073	0.8711	1327.4	1.39554	651.5	2.912
0.0929	0.8165	0.9822	1586.9	1.42562	404.3	13.64
0.1807	0.7375	0.9615	1537.1	1.41834	440.2	7.695
0.3069	0.6239	0.9325	1469.9	1.40940	496.3	5.009
0.3714	0.5658	0.9179	1437.3	1.40940	527.4	4.285
0.4443	0.5002	0.9019	1401.7	1.39964	564.4	3.630

(continued)

x_1	x_2	ρ/g · cm^{-3}	u/m · s^{-1}	n_D	κ_S/T · Pa^{-1}	$10^3\eta$/Pa · s
0.6743	0.2932	0.8527	1295.5	1.38333	698.8	2.278
0.2624	0.3089	0.9837	1544.6	1.48118	426.1	119.1
0.3599	0.2680	0.9709	1514.0	1.47289	449.3	58.73
0.4432	0.2332	0.9586	1487.6	1.46531	471.4	33.75
0.5558	0.1860	0.9386	1446.8	1.45230	509.0	16.16
0.1594	0.6721	0.9791	1554.6	1.44928	422.6	21.90
0.3263	0.5386	0.9463	1478.2	1.43391	483.6	9.470
0.3660	0.5069	0.9385	1460.8	1.43040	499.3	8.153
0.4617	0.4303	0.9185	1418.9	1.42100	540.8	5.846
0.5878	0.3295	0.8908	1361.8	1.40841	605.3	3.918
0.7599	0.1919	0.8502	1278.6	1.38964	719.4	2.375
0.2008	0.4721	0.9852	1553.2	1.47378	420.8	68.12
0.2777	0.4267	0.9741	1525.8	1.46709	441.0	41.41
0.3806	0.3659	0.9577	1489.5	1.45750	470.7	24.18
0.5079	0.2907	0.9345	1442.0	1.44384	514.6	11.59
0.6004	0.2361	0.9149	1402.5	1.43252	555.7	7.378
0.6991	0.1777	0.8913	1355.3	1.41902	610.8	4.710
0.7997	0.1184	0.8623	1296.9	1.40247	689.5	2.964
0.0984	0.1869	1.0029	1617.4	1.50191	381.2	2330
0.1968	0.1665	0.9943	1583.3	1.49589	401.2	855.1
0.2833	0.1486	0.9860	1556.1	1.49005	418.8	402.0
0.6985	0.0625	0.9199	1408.2	1.44546	548.2	11.08
0.7937	0.0428	0.8922	1351.9	1.42733	613.3	5.553

(2) Volumetric Properties, Transport Properties, Refractive Index and Thermodynamic Properties

Ref. [95]

x_1	x_2	V^E/ cm^3 · mol^{-1}	Δn_D	$\Delta\kappa_S$/ T · Pa^{-1}	$10^3\Delta\eta$/ Pa · s	ΔG^{*E}/ J · mol^{-1}
298.15 K						
0.0467	0.9348	−0.215	0.03122	−67.8	−384.481	2502.8
0.0962	0.8862	−0.435	0.03201	−86.3	−364.204	2745.7
0.2254	0.7596	−0.821	0.03163	−95.4	−312.075	2643.1
0.3057	0.6808	−0.953	0.02956	−89.4	−279.476	2698.3
0.5377	0.4533	−1.016	0.02088	−67.9	−186.760	1153.9
0.0826	0.9100	−0.408	0.02152	−87.4	−154.916	2053.8
0.1329	0.8601	−0.621	0.02379	−100.7	−145.496	2975.2
0.2008	0.7927	−0.824	0.02503	−100.9	−134.013	2946.1
0.2733	0.7208	−0.953	0.02478	−92.0	−121.936	2744.1
0.3651	0.6297	−1.037	0.02308	−79.5	−106.653	2417.5
0.6077	0.3891	−0.959	0.01564	−53.8	−65.859	1781.5
0.0558	0.9143	−0.225	0.04252	−72.3	−620.966	3482.8
0.1867	0.7875	−0.683	0.03972	−93.3	−535.170	3025.0

(continued)

x_1	x_2	V^E/ $cm^3 \cdot mol^{-1}$	Δn_D	$\Delta\kappa_S$/ $T \cdot Pa^{-1}$	$10^3\Delta\eta$/ $Pa \cdot s$	ΔG^{*E}/ $J \cdot mol^{-1}$
0.2604	0.7162	−0.843	0.03697	−93.0	−486.519	2997.5
0.3386	0.6405	−0.947	0.03363	−89.3	−435.305	2628.1
0.4437	0.5387	−1.008	0.02863	−82.1	−366.041	2443.6
0.0679	0.8854	−0.248	0.05562	−77.1	−967.999	4714.5
0.2958	0.6690	−0.829	0.04322	−93.4	−733.371	3081.4
0.3954	0.5743	−0.934	0.03707	−91.0	−629.936	2554.9
0.5841	0.3951	−0.944	0.02539	−80.6	−433.511	1828.5
0.7005	0.2845	−0.842	0.01830	−70.0	−312.133	1520.4
0.8470	0.1454	−0.521	0.00904	−44.4	−159.494	877.9
0.0766	0.8576	−0.260	0.06493	−79.5	−1364.76	5422.1
0.3335	0.6189	−0.834	0.04712	−98.4	−989.078	2989.5
0.4166	0.5418	−0.908	0.04103	−98.4	−865.916	2708.8
0.5440	0.4234	−0.944	0.03194	−94.2	−677.089	2033.2
0.1218	0.7443	−0.377	0.08132	−84.2	−2775.19	5540.8
0.2235	0.6582	−0.568	0.07248	−100.2	−2460.42	4224.0
0.3218	0.5748	−0.719	0.06395	−113.3	−2150.885	3441.4
0.4363	0.4778	−0.834	0.05382	−123.2	−1788.809	2756.2
0.5320	0.3966	−0.879	0.04528	−126.0	−1485.570	2278.7
0.6374	0.3073	−0.873	0.03552	−122.2	−1151.402	1788.6
0.0929	0.8165	−0.307	0.07441	−81.5	−1877.36	5626.5
0.1807	0.7375	−0.526	0.06626	−92.4	−1700.466	4272.1
0.3069	0.6239	−0.762	0.05606	−103.5	−1440.107	3300.5
0.3714	0.5658	−0.839	0.05542	−106.8	−1306.479	2960.8
0.4443	0.5002	−0.906	0.04494	−108.6	−1155.242	2601.7
0.6743	0.2932	−0.860	0.02634	−96.7	−677.409	1611.4
0.2624	0.3089	−0.519	0.06508	−109.9	−8828.4	2515.0
0.3599	0.2680	−0.574	0.06417	−143.9	−7706.57	2025.8
0.4432	0.2332	−0.648	0.06290	−170.7	−6721.29	1722.0
0.5558	0.1860	−0.690	0.05842	−199.2	−5372.67	1326.1
0.1594	0.6721	−0.455	0.08234	−89.1	−3496.78	5088.1
0.3263	0.5386	−0.690	0.06830	−118.9	−2810.485	3466.6
0.3660	0.5069	−0.749	0.06511	−124.7	−2646.029	3203.6
0.4617	0.4303	−0.829	0.05648	−135.4	−2247.645	2643.4
0.5878	0.3295	−0.861	0.04489	−139.4	−1721.958	2002.3
0.7599	0.1919	−0.740	0.02750	−119.0	−1003.192	1246.8
0.2008	0.4721	−0.495	0.07748	−94.6	−6758.10	3955.4
0.2777	0.4267	−0.560	0.07426	−118.0	−6127.85	3348.0
0.3806	0.3659	−0.646	0.06932	−146.8	−5266.05	2846.7
0.5079	0.2907	−0.735	0.06141	−175.1	−4191.82	2042.4
0.6004	0.2361	−0.758	0.05425	−186.5	−3406.476	1655.7
0.6991	0.1777	−0.774	0.04521	−187.4	−2565.692	1318.2
0.7997	0.1184	−0.658	0.03319	−165.7	−1709.028	953.2
0.0984	0.1869	−0.306	0.03946	−37.1	−12584	2438.9

(*continued*)

x_1	x_2	V^E/ $cm^3 \cdot mol^{-1}$	Δn_D	$\Delta\kappa_S$/ $T \cdot Pa^{-1}$	$10^3\Delta\eta$/ $Pa \cdot s$	ΔG^{*E}/ $J \cdot mol^{-1}$
0.1968	0.1665	−0.291	0.04459	−77.1	−12432.0	1870.5
0.2833	0.1486	−0.346	0.04856	−112.3	−11454.4	1670.3
0.6985	0.0625	−0.609	0.05105	−236.6	−4976.53	552.5
0.7937	0.0428	−0.589	0.04371	−229.7	−3407.771	542.8

Mixture 2: C_7H_{16} (1) + CH_3OH (2) + MeOctImCl (3)
(1) The Composition of Points on the Coexistence Curve, Compositions of the Conjugate Solutions and the Selectivities

Ref. [234]

x_3	x_2	x_3	x_2	x_3	x_2
T = 298.2 K					
0.000	0.902	0.282	0.641	0.551	0.346
0.007	0.000	0.384	0.533	0.622	0.274
0.090	0.850	0.463	0.447	0.737	0.179
0.183	0.728	0.466	0.439	0.935	0.000
0.235	0.697	0.546	0.365		

Ref. [234]

x'_3	x'_2	x''_3	x''_2	S
T = 298.2 K				
0.183	0.728	0.000	0.020	401
0.463	0.447	0.000	0.001	5000
0.546	0.365	0.001	0.0005	8200

The superscript ′ indicates the ionic liquid rich phase, and the ″ indicates the alkane rich phase

Mixture 3: $C_{16}H_{34}$ (1) + CH_3OH (2) + MeOctImCl (3)
(1) The Composition of Points on the Coexistence Curve, Compositions of the Conjugate Solutions and the Selectivities

Ref. [234]

x_3	x_2	x_3	x_2	x_3	x_2
T = 298.2 K					
0.000	0.008	0.295	0.700	0.712	0.278
0.010	0.000	0.411	0.582	0.873	0.111
0.100	0.898	0.492	0.501	0.903	0.078
0.139	0.859	0.690	0.297	0.981	0.000
0.245	0.750				

Ref. [234]

x'_3	x'_2	x''_3	x''_2	S
$T = 298.2$ K				
0.873	0.111	0.001	0.0003	23100
0.690	0.297	0.001	0.0003	76100
0.411	0.582	0.001	0.0003	27700
0.245	0.750	0.001	0.0003	500000
0.139	0.859	0.0004	0.0006	715000

The superscript ′ indicates the ionic liquid rich phase, and the ″ indicates the alkane rich phase

Mixture 4: C_7H_{16} (1) + C_2H_5OH (2) + MeOctImCl (3)
(1) The Composition of Points on the Coexistence Curve, Compositions of the Conjugate Solutions and the Selectivities

Ref. [234]

x_3	x_2	x_3	x_2	x_3	x_2
$T = 298.2$ K					
0.007	0.000	0.062	0.678	0.454	0.400
0.007	0.007	0.264	0.574	0.493	0.371
0.007	0.003	0.306	0.536	0.634	0.241
0.029	0.489	0.360	0.489	0.935	0.000

Ref. [234]

x'_3	x'_2	x''_3	x''_2	S
$T = 298.2$ K				
0.493	0.371	0.007	0.007	384
0.360	0.489	0.007	0.007	457
0.264	0.574	0.007	0.007	503

The superscript ′ indicates the ionic liquid rich phase, and the ″ indicates the alkane rich phase

Mixture 5: $C_{12}H_{26}$ (1) + C_2H_5OH (2) + MeOctImCl (3)
(1) The Composition of Points on the Coexistence Curve, Compositions of the Conjugate Solutions and the Selectivities

Ref. [234]

x_3	x_2	x_3	x_2	x_3	x_2
$T = 298.2$ K					
0.009	0.000	0.270	0.684	0.500	0.478
0.008	0.535	0.286	0.672	0.694	0.294
0.027	0.875	0.315	0.651	0.890	0.009
0.077	0.860	0.430	0.532	0.923	0.064
0.098	0.859	0.481	0.482	0.976	0.000
0.140	0.810				

Ref. [234]

x'_3	x'_2	x''_3	x''_2	S
$T = 298.2$ K				
0.923	0.064	0.001	0.006	814
0.430	0.532	0.001	0.007	2000
0.270	0.684	0.001	0.007	2110
0.140	0.810	0.002	0.008	2000

The superscript ′ indicates the ionic liquid rich phase, and the ″ indicates the alkane rich phase

Mixture 6: $C_{16}H_{34}$ (1) + C_2H_5OH (2) + MeOctImCl (3)

(1) The Composition of Points on the Coexistence Curve, Compositions of the Conjugate Solutions and the Selectivities

Ref. [234]

x_3	x_2	x_3	x_2	x_3	x_2
$T = 298.2$ K					
0.000	0.956	0.279	0.709	0.480	0.513
0.010	0.000	0.291	0.693	0.543	0.449
0.097	0.876	0.306	0.679	0.586	0.407
0.125	0.846	0.451	0.538	0.599	0.396
0.221	0.758	0.459	0.532	0.981	0.000

Ref. [234]

x'_3	x'_2	x''_3	x''_2	S
$T = 298.2$ K				
0.125	0.846	0.001	0.002	14500
0.221	0.758	0.000	0.002	18000
0.306	0.679	0.001	0.001	45200
0.460	0.532	0.001	0.000	33200
0.587	0.407	0.001	0.000	33900

The superscript ′ indicates the ionic liquid rich phase, and the ″ indicates the alkane rich phase

Mixture 7: C_7H_{16} (1) + C_6H_6 (2) + MeOctImCl (3)

(1) The Composition of Points on the Coexistence Curve

Ref. [235]

x_3	x_2	x_3	x_2	x_3	x_2
$T = 298.2$ K					
0.0034	0.9015	0.3270	0.6729	0.5952	0.3621
0.0039	0.1490	0.4708	0.4994	0.6728	0.3017
0.0069	0.0000	0.5456	0.4132	0.8006	0.1879
0.0072	0.7202	0.5702	0.4031	0.9347	0.0000
0.0073	0.9927				

(2) Tie-line Compositions and the Selectivities

Ref. [235]

Tie-line composition				Selectivity (*S*)	
MeOctImCl-rich phase		Alkane-rich phase			
x'_2	x'_1	x''_2	x''_1	Exp.	Calc.
T = 298.2 K					
0.5528	0.0226	0.9600	0.0335	0.85	0.68
0.5360	0.0249	0.6613	0.0130	0.42	0.64
0.4301	0.0342	0.9438	0.0507	0.67	0.79
0.4002	0.0356	0.7166	0.2834	4.48	4.33
0.1538	0.0702	0.2656	0.7344	7.70	5.68

Mixture 8: $C_{12}H_{26}$ (1) + C_6H_6 (2) + MeOctImCl (3)
(1) The Composition of Points on the Coexistence Curve

Ref. [235]

x_3	x_2	x_3	x_2	x_3	x_2
298.2 K					
0.0012	0.3720	0.0072	0.9927	0.5902	0.3939
0.0020	0.7387	0.0095	0.0000	0.6134	0.3704
0.0027	0.5078	0.0111	0.9186	0.7060	0.2781
0.0047	0.4025	0.3270	0.6729	0.9146	0.0694
0.0054	0.0054	0.3835	0.5978	0.9758	0.0000
0.0065	0.8008	0.5031	0.4832		

(2) Tie-line Compositions and the Selectivities

Ref. [235]

Tie-line composition				Selectivity (*S*)	
MeOctImCl-rich phase		Alkane-rich phase			
x'_2	x'_1	x''_2	x''_1	Exp.	Calc.
T = 298.2 K					
0.4078	0.0170	0.6712	0.3254	11.66	11.39
0.3141	0.0156	0.4199	0.5774	27.78	22.89
0.2925	0.0155	0.4028	0.5944	27.92	34.08
0.2162	0.0170	0.2493	0.7463	38.27	37.32

Mixture 9: $C_{16}H_{34}$ (1) + C_6H_6 (2) + MeOctImCl (3)
(1) The Composition of Points on the Coexistence Curve

Ref. [235]

x_3	x_2	x_3	x_2	x_3	x_2
298.2 K					
0.0011	0.7074	0.0491	0.9508	0.5743	0.4214
0.0013	0.5300	0.0072	0.9927	0.6831	0.3152

(*continued*)

x_3	x_2	x_3	x_2	x_3	x_2
0.0015	0.2021	0.0073	0.5658	0.7170	0.2753
0.0019	0.6949	0.3270	0.6729	0.7552	0.2379
0.0034	0.9458	0.4173	0.5755	0.0096	0.0000
0.0035	0.0227	0.5507	0.4455	0.9812	0.0000

(2) Tie-line Compositions and the Selectivities

Ref. [235]

Tie-line composition				Selectivity (S)	
MeOctImCl-rich phase		Alkane-rich phase			
x'_2	x'_1	x''_2	x''_1	Exp.	Calc.
$T = 298.2$ K					
0.4589	0.0099	0.8974	0.1026	5.32	5.30
0.3674	0.0088	0.7849	0.2151	11.39	11.32
0.3054	0.0077	0.5881	0.4119	27.89	28.15
0.1564	0.0081	0.3087	0.6913	43.26	43.24

2.2.68 02029-01002: 1-Octyl-3-methylimidazolium bromide

Abbreviation:	$[C_8MIm]Br$, $[C_8mim]Br$
Molecular Formula:	$C_{12}H_{23}N_2Br$
Molar Mass:	275.23
Structure:	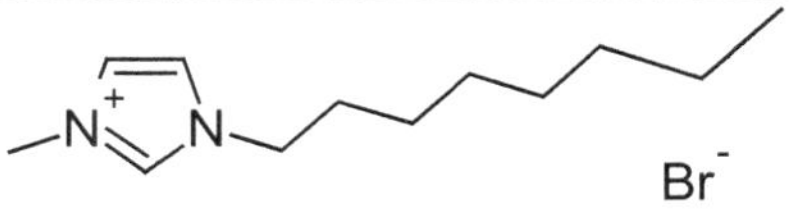

2.2.68.1 Binary Mixtures

Mixture 1: CH_3SOCH_3 (1) + $[C_8mim]Br$ (2)

(1) Volumetric Properties

Ref. [236]

$10^3 \cdot m_2$/ $(mol \cdot g^{-1})$	ρ/g · cm^{-3}	$V_{\Phi,2}$/ $cm^3 \cdot mol^{-1}$	$10^3 \cdot m_2$/ $(mol \cdot g^{-1})$	ρ/ g · cm^{-3}	$V_{\Phi,2}$/ $cm^3 \cdot mol^{-1}$	$10^3 \cdot m_2$/ $(mol \cdot g^{-1})$	ρ/ g · cm^{-3}	$V_{\Phi,2}$/ $cm^3 \cdot mol^{-1}$
298.15 K								
0	1.095332	–	0.01568	1.095742	229.31	0.03185	1.096150	229.61
0.003697	1.095431	228.84	0.01953	1.095840	229.39	0.03581	1.096248	229.67
0.007705	1.095536	229.07	0.02373	1.095946	229.49	0.03956	1.096340	229.73
0.01147	1.095634	229.17	0.02785	1.096051	229.52			

(2) Electrical Conductivity

Ref. [236]

10^4c/ $(mol \cdot L^{-1})$	$k/(mS \cdot cm^{-1})$	$\Lambda_m/(S \cdot cm^2 \cdot mol^{-1})$	10^4c/ $(mol \cdot L^{-1})$	$k/(mS \cdot cm^{-1})$	$\Lambda_m/(S \cdot cm^2 \cdot mol^{-1})$	10^4c/ $(mol \cdot L^{-1})$	$k/(mS \cdot cm^{-1})$	$\Lambda_m/(S \cdot cm^2 \cdot mol^{-1})$
298.15 K								
3.084	10.999	35.66	5.853	20.652	35.29	8.527	29.848	35.01
3.997	14.193	35.51	6.583	23.178	35.21	9.411	32.876	34.93
4.924	17.433	35.40	7.657	26.873	35.10	10.33	35.981	34.85

Mixture 2: 1,2-$HOCH_2CH_2OH$ (1) + [C_8mim]Br (2)
(1) Volumetric Properties

Ref. [236]

$10^3 \cdot m_2/$ $(mol \cdot g^{-1})$	$\rho/g \cdot cm^{-3}$	$V_{\Phi,2}/$ $cm^3 \cdot mol^{-1}$	$10^3 \cdot m_2/$ $(mol \cdot g^{-1})$	$\rho/g \cdot cm^{-3}$	$V_{\Phi,2}/$ $cm^3 \cdot mol^{-1}$	$10^3 \cdot m_2/$ $(mol \cdot g^{-1})$	$\rho/g \cdot cm^{-3}$	$V_{\Phi,2}/$ $cm^3 \cdot mol^{-1}$
298.15 K								
0	1.109957	–	0.02074	1.110217	237.64	0.03492	1.110387	237.79
0.006765	1.110043	237.54	0.02364	1.110252	237.68	0.03798	1.110425	237.77
0.01325	1.110124	237.60	0.03075	1.110338	237.73	0.04271	1.110480	237.82

(2) Electrical Conductivity

Ref. [236]

10^4c/ $(mol \cdot L^{-1})$	k/ $(mS \cdot cm^{-1})$	Λ_m/ $(S \cdot cm^2 \cdot mol^{-1})$	10^4c/ $(mol \cdot L^{-1})$	k/ $(mS \cdot cm^{-1})$	Λ_m/ $(S \cdot cm^2 \cdot mol^{-1})$	10^4c/ $(mol \cdot L^{-1})$	k/ $(mS \cdot cm^{-1})$	Λ_m/ $(S \cdot cm^2 \cdot mol^{-1})$
298.15 K								
3.384	2.2948	6.78	6.367	4.2872	6.73	8.592	5.7569	6.70
4.129	2.7954	6.77	7.095	4.7682	6.72	9.615	6.4291	6.69
4.841	3.2701	6.75	7.842	5.2632	6.71	10.54	7.0333	6.68
5.608	3.7807	6.74						

Mixture 3: $HCON(CH_3)_2$ (1) + $[C_8mim]Br$ (2)
(1) Volumetric Properties

Ref. [236]

$10^3 \cdot m_2/$ $(mol \cdot g^{-1})$	$\rho/$ $g \cdot cm^{-3}$	$V_{\Phi,2}/$ $cm^3 \cdot mol^{-1}$	$10^3 \cdot m_2/$ $(mol \cdot g^{-1})$	$\rho/$ $g \cdot cm^{-3}$	$V_{\Phi,2}/$ $cm^3 \cdot mol^{-1}$	$10^3 \cdot m_2/$ $(mol \cdot g^{-1})$	$\rho/$ $g \cdot cm^{-3}$	$V_{\Phi,2}/$ $cm^3 \cdot mol^{-1}$
298.15 K								
0	0.944305	–	0.01485	0.945316	214.71	0.03009	0.946260	218.03
0.003600	0.944562	210.99	0.01862	0.945557	215.65	0.03685	0.946653	219.36
0.007420	0.944824	212.80	0.02243	0.945792	216.65	0.04079	0.946877	220.04
0.01114	0.945075	213.68	0.02634	0.946030	217.50			

(2) Electrical Conductivity

Ref. [236]

10^4c/ (mol · L^{-1})	k/ (mS · cm^{-1})	Λ_m/ (S · cm^2 · mol^{-1})	10^4c/ (mol · L^{-1})	k/ (mS · cm^{-1})	Λ_m/ (S · cm^2 · mol^{-1})	10^4c/ (mol · L^{-1})	k/ (mS · cm^{-1})	Λ_m/ (S · cm^2 · mol^{-1})
298.15 K								
2.813	22.058	78.41	5.571	42.898	77.00	7.743	58.939	76.12
3.412	26.643	78.08	6.412	49.137	76.63	8.340	63.303	75.90
3.997	31.079	77.75	7.099	54.221	76.38	8.925	67.545	75.68
4.885	37.776	77.32						

2.2.69 02029-02001: 1-Octyl-3-methylimidazolium tetrafluoroborate

Abbreviation:	$[C_8MIm][BF_4]$, $[Omim][BF_4]$, $OMImBF_4$, $[omim][BF_4]$, $[OMIM][BF_4]$, $[C_8mim][BF_4]$, C_8mimBF_4
Molecular Formula:	$C_{12}H_{23}BF_4N_2$
Molar Mass:	282.13
Structure:	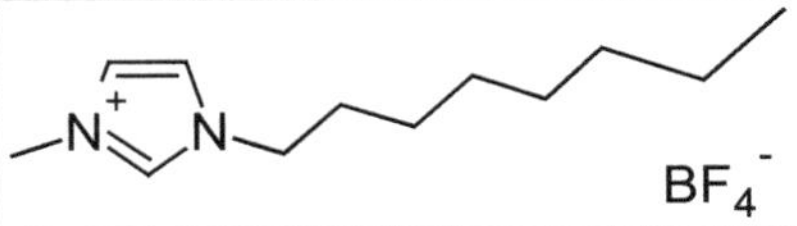

2.2.69.1 Binary Mixtures

Mixture 1: $HOCH_2CH_2OCH_3$ (1) + $[Omim][BF_4]$ (2)

(1) Volumetric Properties and Refractive Index

Ref. [237]

	$\rho/(g \cdot cm^{-3})$			n_D
x_2	$T = 298.15$ K	$T = 308.15$ K	$T = 318.15$ K	$T = 298.15$ K
0	0.96005	0.95081	0.94145	1.4004
0.0106	0.96572	0.95659	0.94731	1.4015
0.0477	0.98308	0.97421	0.96556	1.4057
0.1083	1.00486	0.99643	0.98784	1.4108
0.1495	1.01631	1.00797	0.99962	1.4134
0.2062	1.02920	1.02109	1.01299	1.4163
0.2827	1.04298	1.03509	1.02725	1.4193
0.3204	1.04869	1.04105	1.03314	1.4208
0.5420	1.07351	1.06616	1.05883	1.4258
0.7354	1.08707	1.07983	1.07261	1.4285
0.9632	1.09890	1.09203	1.08514	1.4312
0.9929	1.09825	1.09142	1.08447	1.4318
1	1.10108	1.09413	1.08724	1.4319

Ref. [237]

T/K	$V^{\infty}_{\phi,i}/(cm^3 \cdot mol^{-1})$	$\overline{V}^{\infty}_{m,i}/(cm^3 \cdot mol^{-1})$	$\overline{V}^{E,\infty}_{m,i}/(cm^3 \cdot mol^{-1})$
298.15	248.4011	248.4093	−7.7950

Mixture 2: $HO(CH_2)_2O(CH_2)_2OCH_3$ (1) + $[C_8MIm][BF_4]$ (2)
(1) Volumetric Properties and Refractive Index

Ref. [237]

	$\rho/(g \cdot cm^{-3})$			n_D
x_2	$T = 298.15$ K	$T = 308.15$ K	$T = 318.15$ K	$T = 298.15$ K
0	1.01487	1.00599	0.97705	1.4242
0.0088	1.01727	1.00844	0.99955	1.4227
0.0165	1.01928	1.01049	1.00165	1.4251
0.0745	1.03237	1.02386	1.01533	1.4271
0.1611	1.04673	1.03856	1.03036	1.4288
0.2555	1.05805	1.05015	1.04223	1.4298
0.3459	1.06648	1.05876	1.05105	1.4304
0.4603	1.07527	1.06773	1.06025	1.4308
0.6468	1.08677	1.07950	1.07231	1.4312
0.8198	1.09483	1.08777	1.08074	1.4314
0.9227	1.09858	1.09161	1.08465	1.4317
0.9912	1.10080	1.09385	1.08695	1.4318

Ref. [237]

T/K	$V_{\phi,i}^{\infty}/(cm^3 \cdot mol^{-1})$	$\overline{V}_{m,i}^{\infty}/(cm^3 \cdot mol^{-1})$	$\overline{V}_{m,i}^{E,\infty}/(cm^3 \cdot mol^{-1})$
298.15	245.6700	245.6803	−10.5422

Mixture 3: $HO(CH_2)_2O(CH_2)_2O(CH_2)_2OCH_3$ (1) + $[C_8MIm][BF_4]$ (2)
(1) Volumetric Properties and Refractive Index

Ref. [237]

	$\rho/(g \cdot cm^{-3})$			n_D
x_2	$T = 298.15$ K	$T = 308.15$ K	$T = 318.15$ K	$T = 298.15$ K
0	1.04491	1.03611	1.02728	1.4362
0.0122	1.04670	1.03794	1.02917	1.4364
0.0223	1.04813	1.03941	1.03069	1.4366
0.0994	1.05770	1.04921	1.04084	1.4372
0.2073	1.06789	1.05971	1.05167	1.4373
0.3192	1.07571	1.06781	1.06000	1.4369
0.4200	1.08120	1.07351	1.06588	1.4361
0.5301	1.08614	1.07864	1.07118	1.4352
0.7153	1.09301	1.08574	1.07858	1.4334
0.8449	1.09708	1.08995	1.08296	1.4325
0.9395	1.09966	1.09263	1.08572	1.4321
0.9920	1.10090	1.09394	1.08705	1.4320
1	1.10108	1.09413	1.08724	1.4319

Ref. [237]

T/K	$V_{\phi,i}^{\infty}/(cm^3 \cdot mol^{-1})$	$\overline{V}_{m,i}^{\infty}/(cm^3 \cdot mol^{-1})$	$\overline{V}_{m,i}^{E,\infty}/(cm^3 \cdot mol^{-1})$
298.15	247.5055	247.5231	−8.7072

Mixture 4: CH_3OH (1) + $OMImBF_4$ (2)
(1) Heat Capacity and Derived Properties

Ref. [217]

x_2	C_p/(J · mol^{-1} · K^{-1})	C_p^E/(J · mol^{-1} · K^{-1})	C_p/(J · mol^{-1} · K^{-1})	C_p^E/(J · mol^{-1} · K^{-1})	C_p/(J · mol^{-1} · K^{-1})	C_p^E/(J · mol^{-1} · K^{-1})
	T = 283.15 K		T = 288.15 K		T = 293.15 K	
0	78.6	0.00	79.5	0.00	80.4	0.00
0.0736	111.8	2.92	112.7	2.77	113.6	2.62
0.1652	151.5	4.93	152.6	4.77	153.7	4.58
0.3441	226.4	6.18	227.4	5.54	228.5	4.95
0.4516	269.5	5.00	270.8	4.41	272.1	3.88
0.6526	349.3	2.17	351.3	1.77	353.3	1.42
0.8472	427.6	0.41	430.3	0.34	433.0	0.28
1	490.1	0.00	493.2	0.00	496.3	0.00
	T = 298.15 K		T = 303.15 K		T = 308.15 K	
0	81.3	0.00	82.3	0.00	83.4	0.00
0.0736	114.6	2.46	115.6	2.31	116.6	2.14
0.1652	154.7	4.36	155.9	4.12	157.0	3.85
0.3441	229.6	4.42	230.8	3.94	232.1	3.51
0.4516	273.5	3.39	275.0	2.95	276.6	2.55
0.6526	355.3	1.14	357.4	0.91	359.6	0.73
0.8472	435.7	0.22	438.4	0.17	441.1	0.12
1	499.4	0.00	502.5	0.00	505.4	0.00
	T = 313.15 K		T = 318.15 K		T = 323.15 K	
0	84.5	0.00	85.7	0.00	87.0	0.00
0.0736	117.7	1.98	118.9	1.81	120.1	1.63
0.1652	158.1	3.56	159.3	3.23	160.4	2.88
0.3441	233.5	3.14	235.0	2.82	236.6	2.56
0.4516	278.2	2.21	279.9	1.91	281.6	1.65
0.6526	361.8	0.62	364.0	0.56	366.4	0.56
0.8472	443.7	0.08	446.3	0.05	448.9	0.02
1	508.4	0.00	511.3	0.00	514.2	0.00

(2) Vapor Pressure and Activity Coefficient

Ref. [176]

	298.15 K		303.15 K		308.15 K		313.15 K			298.15 K		303.15 K		308.15 K		313.15 K	
x_1	p/Pa	γ_1	p/Pa	γ_1	p/Pa	γ_1	p/Pa	γ_1	x_1	p/Pa	γ_1	p/Pa	γ_1	p/Pa	γ_1	p/Pa	γ_1
0.0000	0	0.691	0	0.66	0	0.626	0	0.597	0.4444	8577	1.125	10779	1.103	13434	1.077	16527	1.064
0.0388	515	0.736	626	0.705	745	0.670	875	0.641	0.5473	10465	1.157	13243	1.14	16523	1.121	20554	1.116
0.0657	851	0.767	1054	0.736	1282	0.700	1515	0.671	0.6278	12120	1.155	15300	1.141	19076	1.129	23759	1.128
0.0931	1139	0.799	1408	0.768	1724	0.732	2086	0.703	0.7474	14063	1.113	17912	1.105	22469	1.100	27997	1.103
0.1740	2548	0.891	3128	0.861	3787	0.825	4544	0.797	0.8860	15858	1.036	20185	1.034	25398	1.033	31626	1.035
0.2228	3586	0.944	4406	0.915	5302	0.879	6403	0.853	0.9400	16254	1.012	20770	1.011	26263	1.011	33026	1.012
0.3539	6602	1.067	8239	1.041	10137	1.009	12553	0.99	1	16958	1	21880	1	27960	1	35450	1
0.4243	8160	1.115	10258	1.091	12744	1.064	15682	1.049									

Mixture 5: C_2H_5OH (1) + [Omim][BF_4] (2)
(1) Volumetric Properties, Transport Properties, Refractive Index and Surface Tension

Ref. [126]

x_1	T/K	ρ/g · cm^{-3}	η/mPa · s	n_D	V^E/cm^3 · mol^{-1}	$\Delta\eta$/mPa · s	Δn_D
0.081	283.15	1.1229	718.47	1.4335	−0.941	−234.32	0.0031
0.081	288.15	1.1194	507.18	1.4323	−0.952	−119.81	0.0033
0.081	293.15	1.1159	366.59	1.4310	−0.966	−56.27	0.0034
0.081	298.15	1.1124	268.93	1.4295	−0.971	−35.42	0.0034
0.081	303.15	1.1089	203.35	1.4284	−0.989	−21.62	0.0036
0.081	308.15	1.1055	156.29	1.4268	−1.017	−14.73	0.0040
0.081	313.15	1.1022	123.27	1.4255	−1.038	−8.01	0.0042
0.081	318.15	1.0988	97.69	1.4240	−1.057	−4.80	0.0042
0.081	323.15	1.0954	79.43	1.4225	−1.073	−1.85	0.0038
0.081	328.15	1.0921	65.03	1.4213	−1.090	−0.83	0.0042
0.081	333.15	1.0887	54.15	1.4205	−1.108	0.53	0.0047
0.081	338.15	1.0854	45.46	1.4190	−1.126	1.08	0.0045
0.081	343.15	1.0821	39.03	1.4175	−1.145	1.85	0.0045
0.161	283.15	1.1180	492.73	1.4315	−1.361	−377.52	0.0069
0.161	288.15	1.1145	354.89	1.4305	−1.377	−217.81	0.0074
0.161	293.15	1.1110	261.54	1.4292	−1.393	−124.73	0.0076
0.161	298.15	1.1075	194.74	1.4280	−1.396	−83.30	0.0079
0.161	303.15	1.1040	150.36	1.4270	−1.442	−55.17	0.0082
0.161	308.15	1.1007	117.36	1.4255	−1.483	−38.90	0.0087
0.161	313.15	1.0973	93.20	1.4240	−1.505	−26.76	0.0088
0.161	318.15	1.0939	75.37	1.4225	−1.531	−18.29	0.0088
0.161	323.15	1.0905	61.66	1.4210	−1.552	−12.62	0.0085
0.161	328.15	1.0871	51.12	1.4200	−1.575	−9.09	0.0091
0.161	333.15	1.0837	42.99	1.4188	−1.597	−6.02	0.0093
0.161	338.15	1.0803	36.47	1.4173	−1.620	−4.11	0.0091
0.161	343.15	1.0770	31.51	1.4160	−1.653	−2.49	0.0094
0.315	283.15	1.0987	167.49	1.4300	−0.879	−543.29	0.0167
0.315	288.15	1.0952	127.69	1.4288	−0.914	−340.13	0.0170
0.315	293.15	1.0918	98.92	1.4273	−0.949	−216.66	0.0171
0.315	298.15	1.0883	78.63	1.4260	−0.959	−148.56	0.0174
0.315	303.15	1.0848	63.33	1.4245	−1.009	−104.66	0.0173
0.315	308.15	1.0813	51.59	1.4230	−1.042	−76.15	0.0178
0.315	313.15	1.0778	42.48	1.4219	−1.057	−55.61	0.0184
0.315	318.15	1.0743	35.84	1.4205	−1.084	−40.77	0.0186
0.315	323.15	1.0709	30.38	1.4190	−1.105	−30.40	0.0185
0.315	328.15	1.0674	26.04	1.4178	−1.129	−23.23	0.0189
0.315	333.15	1.0640	22.54	1.4160	−1.149	−17.59	0.0186
0.315	338.15	1.0605	19.73	1.4145	−1.171	−13.50	0.0186
0.315	343.15	1.0571	17.42	1.4132	−1.199	−10.42	0.0190

(*continued*)

x_1	T/K	ρ/g · cm^{-3}	η/mPa · s	n_D	V^E/cm^3 · mol^{-1}	$\Delta\eta$/mPa · s	Δn_D
0.511	283.15	1.0647	51.57	1.4265	−0.461	−456.23	0.0275
0.511	288.15	1.0613	41.43	1.4250	−0.513	−292.89	0.0276
0.511	293.15	1.0578	35.28	1.4237	−0.551	−190.33	0.0280
0.511	298.15	1.0542	28.87	1.4230	−0.555	−133.62	0.0291
0.511	303.15	1.0506	23.94	1.4208	−0.611	−96.25	0.0285
0.511	308.15	1.0469	20.13	1.4195	−0.627	−71.31	0.0292
0.511	313.15	1.0432	17.09	1.4180	−0.618	−53.16	0.0294
0.511	318.15	1.0393	14.67	1.4168	−0.608	−40.23	0.0300
0.511	323.15	1.0349	12.72	1.4150	−0.508	−30.86	0.0297
0.511	328.15	1.0303	11.10	1.4137	−0.382	−24.25	0.0301
0.511	333.15	1.0287	9.77	1.4125	−0.686	−19.04	0.0306
0.511	338.15	1.0252	8.65	1.4118	−0.728	−15.22	0.0315
0.511	343.15	1.0213	7.74	1.4110	−0.705	−12.28	0.0325
0.612	283.15	1.0397	27.12	1.4225	−0.244	−341.92	0.0333
0.612	288.15	1.0360	22.05	1.4212	−0.257	−221.01	0.0337
0.612	293.15	1.0322	18.24	1.4200	−0.271	−145.87	0.0343
0.612	298.15	1.0284	15.35	1.4185	−0.249	−102.90	0.0346
0.612	303.15	1.0247	12.98	1.4172	−0.302	−74.54	0.0350
0.612	308.15	1.0209	11.16	1.4160	−0.319	−55.46	0.0358
0.612	313.15	1.0172	9.61	1.4143	−0.316	−41.61	0.0359
0.612	318.15	1.0134	8.39	1.4130	−0.340	−31.67	0.0364
0.612	323.15	1.0097	7.39	1.4115	−0.349	−24.43	0.0366
0.612	328.15	1.0059	6.55	1.4100	−0.363	−19.29	0.0369
0.612	333.15	1.0022	5.80	1.4085	−0.372	−15.28	0.0372
0.612	338.15	0.9985	5.18	1.4078	−0.382	−12.30	0.0382
0.612	343.15	0.9947	4.70	1.4065	−0.394	−9.97	0.0387
0.753	283.15	0.9913	12.45	1.4132	−0.137	−212.65	0.0341
0.753	288.15	0.9875	10.36	1.4125	−0.154	−138.03	0.0352
0.753	293.15	0.9836	8.79	1.4110	−0.169	−91.51	0.0356
0.753	298.15	0.9797	7.53	1.4098	−0.140	−64.83	0.0363
0.753	303.15	0.9757	6.51	1.4085	−0.195	−47.12	0.0368
0.753	308.15	0.9718	5.66	1.4072	−0.204	−35.22	0.0375
0.753	313.15	0.9680	4.96	1.4070	−0.208	−26.52	0.0392
0.753	318.15	0.9641	4.38	1.4065	−0.237	−20.28	0.0406
0.753	323.15	0.9602	3.89	1.4055	−0.256	−15.74	0.0414
0.753	328.15	0.9564	3.50	1.4045	−0.275	−12.47	0.0423
0.753	333.15	0.9525	3.17	1.4040	−0.283	−9.88	0.0436
0.753	338.15	0.9486	2.89	1.4033	−0.296	−7.96	0.0448
0.753	343.15	0.9447	2.65	1.4020	−0.300	−6.48	0.0453

Ref. [34]

x_2	c (mol/L)	ρ (g/cm^3)	x_2	c (mol/L)	ρ (g/cm^3)
298.15 K					
1.0000	3.913	1.1042	0.5125	3.214	1.0478
0.9014	3.814	1.0956	0.4096	2.944	1.0264
0.8150	3.720	1.0888	0.3000	2.564	0.9966
0.7207	3.597	1.0793	0.2034	2.071	0.9579
0.6156	3.425	1.0652	0.1012	1.292	0.8930

Ref. [34]

x_2	c (mol/L)	σm (mN/m)	δσN (mN/m)	x_2	c (mol/L)	σm (mN/m)	δσN (mN/m)
298.15 K							
1.0000	3.913	32.82	0.00	0.5125	3.214	27.90	0.35
0.9014	3.814	31.84	0.09	0.4096	2.944	27.28	0.85
0.8150	3.720	30.86	0.04	0.3000	2.564	25.99	0.74
0.7207	3.597	29.90	0.10	0.2034	2.071	24.70	0.49
0.6156	3.425	28.88	0.22	0.1012	1.292	23.33	0.22

(2) Vapor Pressure and Activity Coefficient

Ref. [176]

	298.15 K		303.15 K		308.15 K		313.15 K			298.15 K		303.15 K		308.15 K		313.15 K	
x_1	p/Pa	γ_1	p/Pa	γ_1	p/Pa	γ_1	p/Pa	γ_1	x_1	p/Pa	γ_1	p/Pa	γ_1	p/Pa	γ_1	p/Pa	γ_1
0.0000	0	1.763	0	1.552	0	1.380	0	1.225	0.5721	6001	1.331	7705	1.285	9834	1.245	12283	1.196
0.0838	1103	1.676	1336	1.513	1575	1.372	1863	1.252	0.6202	6379	1.303	8189	1.258	10433	1.222	13086	1.172
0.1226	1565	1.64	1876	1.495	2263	1.367	2742	1.261	0.7258	7043	1.235	9043	1.193	11618	1.164	14643	1.113
0.2113	2610	1.565	3237	1.455	4010	1.353	4908	1.274	0.8596	7652	1.128	9921	1.093	12890	1.074	16352	1.04
0.2510	3068	1.535	3813	1.437	4667	1.346	5756	1.276	0.9196	7750	1.067	10150	1.044	13200	1.033	16930	1.015
0.4046	4524	1.431	5801	1.368	7235	1.307	9089	1.258	0.9346	7760	1.051	10180	1.032	13278	1.024	17089	1.01
0.4883	5293	1.38	6801	1.328	8631	1.279	10795	1.232	1.0000	7876	1	10468	1	13768	1	17928	1

Mixture 6: CO_2 (1) + [omim][BF_4] (2)
(1) Bubble Point

Ref. [238]

x_1	T (K)	P (MPa)	T (K)	P (MPa)	T (K)	P (MPa)
0.1005	308.22	0.571	328.14	0.796	348.20	1.054
	313.11	0.621	333.12	0.857	353.17	1.123
	318.13	0.677	338.16	0.921	358.13	1.193
	323.23	0.736	343.11	0.986	363.12	1.264
0.200	308.20	1.254	328.00	1.756	348.13	2.347
	313.25	1.373	333.05	1.896	353.19	2.502
	318.13	1.495	338.18	2.045	358.19	2.665
	323.14	1.625	343.08	2.194	363.21	2.827
0.2992	308.26	2.123	333.40	3.209	363.10	4.789
	313.24	2.299	338.18	3.449	343.13	3.709
	318.39	2.514	343.13	3.729	353.09	4.229
	323.62	2.748	348.11	3.969		
	328.19	2.954	358.27	4.529		
0.4149	309.16	3.452	333.12	5.271	353.30	7.09
	313.18	3.753	338.12	5.691	358.25	7.57
	318.11	4.092	343.13	6.15	363.29	8.09
	323.25	4.471	343.12	6.19	348.24	6.59
	328.13	4.871	348.12	6.59	336.47	5.531
0.5050	308.33	4.707	333.31	7.53	358.37	11.13
	313.42	5.227	338.15	8.19	315.35	5.39
	318.39	5.747	343.13	8.89	363.14	11.93
	323.27	6.29	348.31	9.63		
	328.11	6.89	353.34	10.39		
0.6019	313.16	7.92	332.79	12.48	352.46	18.28
	318.03	8.84	337.69	13.88	357.50	19.80
	323.05	9.94	342.59	15.32	307.79	6.98
	327.93	11.16	347.81	16.88	362.51	21.30
0.7024	313.08	29.50	342.61	43.56	308.25	26.80
	318.03	32.04	347.68	45.68	362.35	51.50
	322.92	34.48	352.57	47.68	362.44	51.54
	327.87	36.88	357.52	49.64	322.90	34.50
	332.79	39.20	322.87	34.40		
	337.69	41.40	322.86	34.44		
0.7523	312.98	67.60	335.29	77.60	357.41	85.80
	320.42	71.04	342.66	80.48		
	327.89	74.54	350.08	83.20		

(2) Phase Equilibrium

Ref. [238]

T (K)	P (MPa)	T (K)	P (MPa)
293.20	5.75	301.11	6.89
295.15	6.01	303.17	7.21
297.15	6.29	304.19	7.37
299.18	6.59		

(3) Henry's Law Constant and Thermodynamic Properties

Ref. [147]

T (K)	K_H (MPa)	T (K)	K_H (MPa)
322.15	2.06	312.45	1.69
317.45	1.85	307.55	1.50

Ref. [147]

$\Delta_{sil}H^o$ (kJ · mol^{-1})	$\Delta_{sil}G^o$ (kJ · mol^{-1})	$\Delta_{sil}S^o$ (J · K^{-1} · mol^{-1})
−18.3	0.64	−63.0

Mixture 7: 1-C_3H_7OH (1) + [OMIM][BF_4] (2)
(1) Vapor Pressure and Activity Coefficient

Ref. [176]

	298.15 K		303.15 K		308.15 K		313.15 K			298.15 K		303.15 K		308.15 K		313.15 K	
x_1	p/Pa	γ_1	p/Pa	γ_1	p/Pa	γ_1	p/Pa	γ_1	x_1	p/Pa	γ_1	p/Pa	γ_1	p/Pa	γ_1	p/Pa	γ_1
0.0000	0	2.153	0	1.990	0	1.867	0	1.852	0.7281	2679	1.319	3577	1.282	4717	1.242	6057	1.189
0.1742	954	1.959	1211	1.822	1528	1.706	1955	1.626	0.8026	2739	1.217	3695	1.194	4917	1.169	6384	1.134
0.3484	1730	1.771	2235	1.662	2872	1.562	3591	1.463	0.8407	2753	1.164	3727	1.148	4975	1.130	6526	1.104
0.5068	2286	1.594	2971	1.513	3781	1.435	4759	1.344	0.8656	2758	1.130	3742	1.118	4998	1.104	6602	1.084
0.6075	2510	1.474	3278	1.413	4226	1.351	5361	1.274	0.9228	2765	1.057	3764	1.052	5056	1.047	6753	1.039
0.7041	2654	1.351	3522	1.309	4622	1.265	5929	1.207	1.0000	2786	1.000	3826	1.000	5199	1.000	6986	1.000

Mixture 8: C_6H_6 (1) + [OMIM][BF_4] (2)
(1) Vapor Pressure and Activity Coefficient

Ref. [176]

	298.15 K		303.15 K		308.15 K		313.15 K			298.15 K		303.15 K		308.15 K		313.15 K	
x_1	p/Pa	γ_1	p/Pa	γ_1	p/Pa	γ_1	p/Pa	γ_1	x_1	p/Pa	γ_1	p/Pa	γ_1	p/Pa	γ_1	p/Pa	γ_1
0.0000	0	1.342	0	1.356	0	1.383	0	1.411	0.6905	11226	1.288	13881	1.268	16646	1.234	19816	1.190
0.1839	3209	1.350	3981	1.346	4868	1.345	5949	1.344	0.7464	12006	1.263	14830	1.247	17701	1.214	21120	1.168
0.2640	4398	1.350	5458	1.340	6663	1.330	8146	1.319	0.7586	12189	1.257	15036	1.241	17954	1.209	21432	1.163
0.3583	5995	1.347	7427	1.331	9048	1.311	11053	1.290	0.7620	12223	1.255	15083	1.240	18070	1.207	21532	1.162
0.5299	9006	1.329	11072	1.308	13321	1.276	15966	1.241	0.78731	12452	1.24	15476	1.227	18670	1.196	22072	1.15
0.5660	9517	1.322	11702	1.301	14105	1.268	16893	1.231	0.83959	12578	1.201	15700	1.193	19099	1.166	22760	1.122
0.6417	10581	1.304	13050	1.283	15720	1.249	18719	1.207	1	12704	1.000	15919	1	19659	1	24331	1

Mixture 9: 1,2-$HOCH_2CH_2OH$ (1) + C_8mimBF_4 (2)
(1) Volumetric Properties and Refractive Index

Ref. [136]

x_2	$\rho/(g \cdot cm^{-3})$ $T = 298.15$ K	$\rho/(g \cdot cm^{-3})$ $T = 308.15$ K	$\rho/(g \cdot cm^{-3})$ $T = 318.15$ K	n_D $T = 298.15$ K
0.0000	1.1100	1.1029	1.0958	1.4304
0.0027	1.1099	1.1028	1.0957	1.4303
0.0046	1.1097	1.1027	1.0954	1.4302
0.0081	1.1095	1.1024	1.0952	1.4301
0.0151	1.1090	1.1019	1.0946	1.4301
0.0316	1.1080	1.1008	1.0934	1.4299
0.0614	1.1064	1.0992	1.0917	1.4296
0.0830	1.1054	1.0980	1.0906	1.4296
0.1341	1.1035	1.0960	1.0886	1.4296
0.2026	1.1017	1.0940	1.0868	1.4296
0.2730	1.1004	1.0927	1.0853	1.4297
0.4125	1.0992	1.0915	1.0842	1.4300
0.5198	1.0990	1.0913	1.0841	1.4302
0.6306	1.0993	1.0916	1.0845	1.4306
0.7919	1.1003	1.0926	1.0855	1.4312
0.9583	1.1016	1.0940	1.0871	1.4318
1.0000	1.1019	1.09432	1.0875	1.4319

Mixture 10: 2-$CH_3COC_2H_5$ (1) + $[C_8mim][BF_4]$ (2)
(1) Transport Properties

Ref. [135]

x_2	$\eta/(mPa \cdot s)$	x_2	$\eta/(mPa \cdot s)$	x_2	$\eta/(mPa \cdot s)$
298.15 K					
0.00	0.388	0.3558	7.539	0.6009	34.241
0.0482	0.625	0.3994	10.257	0.7287	69.932
0.1010	1.026	0.4490	13.820	0.7910	100.303
0.1967	2.287	0.4968	18.400	0.8990	187.796
0.2977	5.037	0.5495	24.781	1.00	336.322

(2) Electrical Conductivity

Ref. [135]

x_2	$10^3\kappa$ $(S \cdot cm^{-1})$	x_2	$10^3\kappa$ $(S \cdot cm^{-1})$	x_2	$10^3\kappa$ $(S \cdot cm^{-1})$
298.15 K					
0.0482	4.95	0.3994	3.90	0.7287	1.56
0.1010	6.24	0.4490	3.47	0.7910	1.11
0.1967	6.33	0.4968	3.00	0.8990	0.74
0.2977	4.98	0.5495	2.52	1.00	0.49
0.3558	4.34	0.6009	2.13		

Mixture 11: CH_3COOCH_3 (1) + $[C_8mim][BF_4]$ (2)

(1) Transport Properties

Ref. [135]

x_2	η/(mPa · s)	x_2	η/(mPa · s)	x_2	η/(mPa · s)
298.15 K					
0.00	0.376	0.3481	8.679	0.5937	37.958
0.0485	0.670	0.3944	11.69	0.7036	71.839
0.1011	1.189	0.4488	16.507	0.8010	117.296
0.1990	2.854	0.5036	23.382	0.8827	183.897
0.2973	6.093	0.5489	30.370	1.00	336.322

(2) Electrical Conductivity

Ref. [135]

x_2	$10^3\kappa$ (S · cm^{-1})	x_2	$10^3\kappa$ (S · cm^{-1})	x_2	$10^3\kappa$ (S · cm^{-1})
298.15 K					
0.0485	2.27	0.3944	4.09	0.7036	1.42
0.1011	4.41	0.4488	3.65	0.8010	1.10
0.199	5.61	0.5036	2.95	0.8827	0.86
0.2973	5.08	0.5489	2.55	1.00	0.49
0.3481	4.56	0.5937	2.28		

Mixture 12: $CH_3COOCH_2CH_3$ (1) + $[C_8mim][BF_4]$ (2)

(1) Transport Properties

Ref. [135]

x_2	η/(mPa · s)	x_2	η/(mPa · s)	x_2	η/(mPa · s)
298.15 K					
0.00	0.437	0.3760	10.930	0.6014	41.354
0.0470	0.739	0.4016	12.720	0.7047	69.749
0.0995	1.281	0.4493	17.011	0.7960	115.491
0.1983	3.095	0.5001	23.129	0.8908	181.150
0.2988	6.562	0.5539	31.639	1.00	336.322

(2) Electrical Conductivity

Ref. [135]

x_2	$10^3\kappa$ (S · cm^{-1})	x_2	$10^3\kappa$ (S · cm^{-1})	x_2	$10^3\kappa$ (S · cm^{-1})
298.15 K					
0.0470	0.68	0.4016	2.73	0.7047	1.36
0.0995	1.72	0.4493	2.52	0.7960	0.99
0.1983	2.87	0.5001	2.24	0.8908	0.80
0.2988	3.06	0.5539	2.00	1.00	0.49
0.3760	2.87	0.6014	1.75		

2.2.69.2 Ternary Mixtures

Mixture 1: $HOC_6H_4C(CH_3)_2C_6H_4OH$ (1) + H_2O (2) + $[C_8mim][BF_4]$ (3)
(1) Distribution Ratio

Ref. [204]

pH	$T = 297.15 \pm 1$ K	pH	$T = 297.15 \pm 1$ K	pH	$T = 297.15 \pm 1$ K	pH	$T = 297.15 \pm 1$ K
1.86	1043.3	4.95	1882.6	9.17	802.9	12.01	31.8
3.26	1882.6	7.03	2465.1	11.61	40.3		

Mixture 2: C_6H_5OH (1) + H_2O (2) + $[C_8mim][BF_4]$ (3)
(1) Distribution Ratio

Ref. [204]

pH	$T = 297.15 \pm 1$ K	pH	$T = 297.15 \pm 1$ K	pH	$T = 297.15 \pm 1$ K
3.23	37.5	7.16	36.0	11.68	7.5
5.25	37.4	9.52	32.7		

Mixture 3: $4\text{-}C_9H_{19}C_6H_4OH$ (1) + H_2O (2) + $[C_8mim][BF_4]$ (3)
(1) Distribution Ratio

Ref. [204]

pH	$T = 297.15 \pm 1$ K	pH	$T = 297.15 \pm 1$ K	pH	$T = 297.15 \pm 1$ K	pH	$T = 297.15 \pm 1$ K
3.23[a]	ND[b]	7.09	ND[b]	10.20	700.0	12.61	233.3
5.19	ND[b]	8.43	725.0	11.03	595.0		

[a] $V_W:V_{IL} = 25:1$ was used at these pHs because of the low solubility of 4-octylphenol in water and their very low concentrations in aqueous phase after extraction
[b] Not detectable because of low concentration of the compound in aqueous phase after extraction

Mixture 4: $4\text{-}C_8H_{17}C_6H_4OH$ (1) + H_2O (2) + $[C_8mim][BF_4]$ (3)
(1) Distribution Ratio

Ref. [204]

pH	$T = 297.15 \pm 1$ K	pH	$T = 297.15 \pm 1$ K	pH	$T = 297.15 \pm 1$ K
3.23[a]	656.3	7.04	687.5	12.07	75.8
5.27	645.8	10.22	132.8		

[a] $V_W:V_{IL} = 25:1$ was used at these pHs because of the low solubility of 4-octylphenol in water and their very low concentrations in aqueous phase after extraction

Mixture 5: C_6Cl_5OH (1) + H_2O (2) + [C_8mim][BF_4] (3)
(1) Distribution Ratio

Ref. [204]

pH	$T = 297.15 \pm 1$ K	pH	$T = 297.15 \pm 1$ K	pH	$T = 297.15 \pm 1$ K
3.16[a]	1175.0	7.03	1096.7	12.73	544.9
5.12	1141.7	10.25	922.4		

[a]V_W:V_{IL} = 25:1 was used at these pHs because of the low solubility of 4-octylphenol in water and their very low concentrations in aqueous phase after extraction

Mixture 6: C_7H_{16} (1) + SCH = CHCH = CH (thiophene) (2) + [C_8mim][BF_4] (3)
(1) Distribution Ratio and Selectivity

Ref. [239]

IL-rich phase			Hydrocarbon-rich phase				
x^{II}_3	x^{II}_2	x^{II}_1	x^{I}_3	x^{I}_2	x^{I}_1	β	S
298.15 K							
0.862	0.000	0.138	0.000	0.000	1.000		
0.710	0.152	0.139	0.004	0.094	0.902	1.61	10.48
0.558	0.306	0.137	0.004	0.226	0.770	1.35	7.63
0.445	0.422	0.134	0.004	0.370	0.626	1.14	5.34
0.362	0.508	0.130	0.003	0.510	0.487	1.00	3.72
0.312	0.563	0.126	0.000	0.651	0.349	0.86	2.40
0.260	0.627	0.113	0.000	0.771	0.229	0.81	1.65
0.224	0.679	0.097	0.000	0.846	0.154	0.80	1.27
0.122	0.836	0.042	0.000	0.954	0.046	0.88	0.95
0.100	0.900	0.000	0.000	1.000	0.000	0.90	

Mixture 7: $C_{12}H_{26}$ (1) + SCH = CHCH = CH (thiophene) (2) + [C_8mim][BF_4] (3)
(1) Distribution Ratio and Selectivity

Ref. [239]

IL-rich phase			Hydrocarbon-rich phase				
x^{II}_3	x^{II}_2	x^{II}_1	x^{I}_3	x^{I}_2	x^{I}_1	β	S
298.15 K							
0.956	0.000	0.044	0.000	0.000	1.000		
0.834	0.116	0.050	0.000	0.117	0.883	1.00	17.62
0.691	0.254	0.055	0.000	0.273	0.727	0.93	12.29
0.550	0.389	0.061	0.000	0.452	0.548	0.86	7.76
0.437	0.494	0.069	0.005	0.631	0.364	0.78	4.10
0.417	0.512	0.071	0.002	0.659	0.339	0.78	3.72
0.344	0.580	0.076	0.005	0.769	0.226	0.75	2.23
0.299	0.621	0.080	0.006	0.856	0.138	0.73	1.25
0.255	0.675	0.070	0.000	0.902	0.098	0.75	1.05
0.207	0.730	0.063	0.000	0.933	0.067	0.78	0.83
0.152	0.811	0.037	0.000	0.963	0.037	0.84	0.84
0.100	0.900	0.000	0.000	1.000	0.000	0.90	

Mixture 8: $C_{16}H_{34}$ (1) + SCH = CHCH = CH (thiophene) (2) + $[C_8mim][BF_4]$ (3)
(1) Distribution Ratio and Selectivity

Ref. [239]

IL-rich phase			Hydrocarbon-rich phase			β	S
x^{II}_3	x^{II}_2	x^{II}_1	x^{I}_3	x^{I}_2	x^{I}_1		
298.15 K							
0.979	0.000	0.021	0.000	0.000	1.000		
0.886	0.093	0.021	0.000	0.145	0.855	0.64	26.11
0.773	0.202	0.025	0.000	0.317	0.683	0.64	17.41
0.649	0.319	0.032	0.000	0.510	0.490	0.63	9.58
0.521	0.431	0.048	0.000	0.686	0.314	0.63	4.11
0.416	0.528	0.056	0.000	0.834	0.166	0.63	1.88
0.358	0.585	0.057	0.000	0.902	0.098	0.65	1.12
0.295	0.648	0.057	0.000	0.948	0.052	0.68	0.62
0.225	0.748	0.027	0.000	0.987	0.013	0.76	0.36
0.201	0.792	0.007	0.000	0.997	0.003	0.79	0.34
0.100	0.900	0.000	0.000	1.000	0.000	0.90	

Mixture 9: C_8H_{18} (*i*-octane) (1) + SCH = CHCH = CH (thiophene) (2) + $[C_8mim][BF_4]$ (3)
(1) Distribution Ratio and Selectivity

Ref. [240]

Hydrocarbon-rich Phase			IL-rich Phase			β	S
x_3	x_2	x_1	x_3	x_2	x_1		
298.15 K							
0.000	0.000	1.000	0.919	0.000	0.081	–	–
0.000	0.060	0.940	0.790	0.134	0.076	2.23	27.62
0.000	0.124	0.876	0.661	0.258	0.081	2.08	22.50
0.000	0.239	0.762	0.514	0.410	0.076	1.72	17.20
0.000	0.356	0.644	0.418	0.518	0.064	1.46	14.64
0.000	0.516	0.484	0.316	0.615	0.068	1.19	8.48
0.000	0.663	0.337	0.213	0.729	0.058	1.10	6.39
0.000	0.764	0.236	0.155	0.791	0.054	1.04	4.52
0.000	0.829	0.172	0.135	0.818	0.046	0.99	3.69
0.000	0.940	0.060	0.112	0.864	0.024	0.92	2.30
0.000	1.000	0.000	0.100	0.900	0.000	0.90	–

Mixture 10: $C_6H_5CH_3$ (1) + SCH = CHCH = CH (thiophene) (2) + [C_8mim] [BF_4] (3)
(1) Distribution Ratio and Selectivity

Ref. [240]

Hydrocarbon-rich Phase			IL-rich Phase				
x_3	x_2	x_1	x_3	x_2	x_1	β	S
298.15 K							
0.000	0.000	1.000	0.287	0.000	0.713	–	–
0.003	0.124	0.873	0.255	0.139	0.606	1.12	1.61
0.003	0.235	0.762	0.234	0.218	0.549	0.93	1.29
0.003	0.362	0.635	0.205	0.331	0.464	0.92	1.25
0.003	0.547	0.450	0.174	0.498	0.328	0.91	1.25
0.003	0.671	0.326	0.149	0.605	0.246	0.90	1.19
0.003	0.758	0.239	0.137	0.680	0.182	0.90	1.18
0.003	0.820	0.177	0.127	0.736	0.137	0.90	1.16
0.003	0.889	0.108	0.118	0.786	0.095	0.88	1.01
0.003	0.929	0.068	0.112	0.821	0.066	0.88	0.91
0.003	0.997	0.000	0.100	0.900	0.000	0.90	–

Mixture 11: C_6H_{14} (1) + $CH_2(CH_2)_4CH_2$ (2) + [C_8mim][BF_4] (3)
(1) Distribution Ratio and Selectivity

Ref. [240]

Hydrocarbon-rich Phase			IL-rich Phase				
x_3	x_2	x_1	x_3	x_2	x_1	β	S
298.15 K							
0.000	0.000	1.000	0.855	0.000	0.146	–	–
0.000	0.101	0.899	0.840	0.023	0.137	0.23	1.49
0.000	0.246	0.754	0.825	0.054	0.121	0.22	1.37
0.000	0.323	0.677	0.816	0.074	0.110	0.23	1.41
0.000	0.457	0.543	0.803	0.102	0.095	0.22	1.28
0.000	0.586	0.414	0.795	0.133	0.073	0.23	1.29
0.000	0.680	0.321	0.793	0.155	0.052	0.23	1.41
0.001	0.837	0.162	0.792	0.179	0.030	0.21	1.15
0.021	0.886	0.094	0.790	0.195	0.016	0.22	1.29
0.052	0.948	0.000	0.792	0.208	0.000	0.22	–

Mixture 12: H_2O (1) + $OCH_2CH_2CH_2CH_2$ (2) + [OMIM][BF_4] (3)
(1) Phase Equilibrium

Ref. [36]

x_2	y_2	x_2	y_2	x_2	y_2
T = 337.15 K					
x_3 = 0.1		x_3 = 0.3		x_3 = 0.5	
0.428	0.769	0.428	0.757	0.622	0.840

(*continued*)

x_2	y_2	x_2	y_2	x_2	y_2
0.622	0.819	0.622	0.837	0.700	0.885
0.700	0.839	0.700	0.869	0.794	0.922
0.794	0.872	0.794	0.906	0.852	0.948
0.852	0.896	0.852	0.932	0.899	0.965
0.899	0.922	0.899	0.953	0.949	0.984
0.949	0.956	0.949	0.976	0.979	0.994
0.979	0.982	0.979	0.991		

2.2.70 02029-03001: 1-Octyl-3-methylimidazolium bis((trifluoromethyl)sulfonyl)imide

Abbreviation:	$[C_8MIm][NTf_2]$, $[C_8mim][Tf_2N]$, $[C_8mim][NTf_2]$, $C_8mimNTf_2$, [OMIM][BTI]
Molecular Formula:	$C_{14}H_{23}F_6N_3O_4S_2$
Molar Mass:	475.47
Structure:	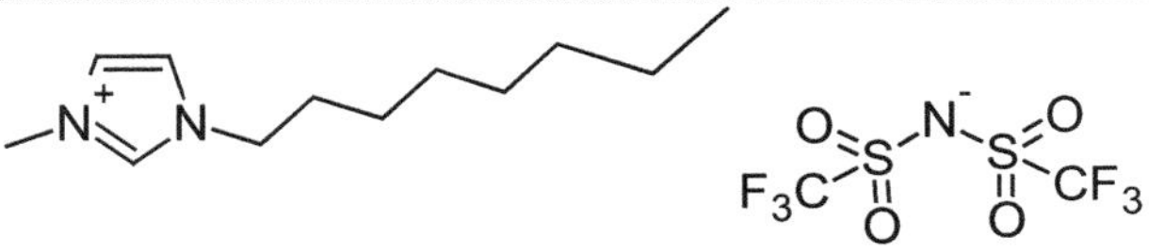

2.2.70.1 Binary Mixtures

Mixture 1: H_2O (1) + $[C_8mim][Tf_2N]$ (2)
(1) Solubility

Ref. [45]

T/K	$x_1 \pm \sigma^a$	$10^4 (x_2 \pm \sigma^a)$
288.15	0.1715 ± 0.0006	3.17 ± 0.01
293.15	0.1781 ± 0.0003	3.23 ± 0.01
298.15	0.1868 ± 0.0003	3.36 ± 0.01
303.15	0.1949 ± 0.0005	3.55 ± 0.01
308.15	0.2054 ± 0.0006	3.68 ± 0.02
313.15	0.2149 ± 0.0004	3.94 ± 0.06
318.15	0.2241 ± 0.0005	4.55 ± 0.07

[a]Standard deviation

(2) Thermodynamic Properties

Ref. [45]

$(\Delta_{sol}H_m{}^0 \pm \sigma^a)$/ $kJ \cdot mol^{-1}$	$(\Delta_{sol}G_m{}^0 \pm \sigma^a)$/ $kJ \cdot mol^{-1}$	$(\Delta_{sol}S_m{}^0 \pm \sigma^a)$/ $J \cdot K^{-1} \cdot mol^{-1}$	$(\Delta_{svt}H_m{}^0 \pm \sigma^a)$/ $kJ \cdot mol^{-1}$	$(\Delta_{svt}Gm^0 \pm \sigma^a)$/ $kJ \cdot mol^{-1}$	$(\Delta_{svt}S_m{}^0 \pm \sigma^a)$/ $J \cdot K^{-1} \cdot mol^{-1}$	$(\Delta_{svt}H_m{}^* \pm \sigma^a)$/ $kJ \cdot mol^{-1}$	$(\Delta_{svt}G_m{}^* \pm \sigma^a)$/ $kJ \cdot mol^{-1}$	$(\Delta_{svt}S_m{}^* \pm \sigma^a)$/ $J \cdot K^{-1} \cdot mol^{-1}$
298.15 K								
5.6 ± 1.5	25.537 ± 0.002	−66.7 ± 5.1	−186.4 ± 6.2	−60.643 ± 0.002	−422 ± 21	−184.2 ± 6.2	−78.552 ± 0.002	−354 ± 21

[a]Standard deviation

(3) Activity Coefficient at Infinite Dilution and Partial Molar Excess Enthalpy at Infinite Dilution

Ref. [4]

	$\gamma^{\infty}{}_1$ (dilutor technique)			
Solute (1)	303.15 K	313.15 K	323.15 K	333.15 K
Water	3.76	3.51	3.29	–

Ref. [4]

	$\gamma^{\infty}{}_1$ (gas-liquid chromatography)			
Solute (1)	303.15 K	313.15 K	323.15 K	333.15 K
Water	3.89	3.52	3.03	2.58

Ref. [4]

Solute (1)	$H_{12}{}^{E,\infty}$/kJ · mol^{-1}
Water	−10.093

Mixture 2: C_6H_{14} (1) + [OMIM][BTI] (2)
(1) Phase Equilibrium

Ref. [50]

x_1	P (kPa)	x_1	P (kPa)	x_1	P (kPa)	x_1	P (kPa)
353.15 K							
0.0154	9.398	0.0513	30.07	0.0893	50.66	0.2257	114.4
0.0337	19.96	0.0704	40.38	0.1595	86.09	0.3050	139.8

(2) Activity Coefficient at Infinite Dilution and Partial Molar Excess Enthalpy at Infinite Dilution

Ref. [4]

	$\gamma^{\infty}{}_1$ (dilutor technique)			
Solute (1)	303.15 K	313.15 K	323.15 K	333.15 K
n-Hexane	5.44	5.26	5.13	4.97

Ref. [4]

	$\gamma^{\infty}{}_1$ (gas-liquid chromatography)			
Solute (1)	303.15 K	313.15 K	323.15 K	333.15 K
n-Hexane	5.32	5.13	4.92	4.80

Ref. [4]

Solute (1)	$H_{12}{}^{E,\infty}$/kJ · mol^{-1}
n-Hexane	2.744

Mixture 3: $[C_8mim][NTf_2]$ (1) + $[C_{10}mim][NTf_2]$ (2)
(1) Volumetric Properties

Ref. [57]

	V (pure component) or V^E/cm^3 · mol^{-1}	
x_1	$T = 298.15$ K	$T = 333.15$ K
1	361.784 (361.8)	370.508 (370.5)
0.501	0.08_4	0.07_7
0	406.689 (404.0)	416.532 (413.5)

Mixture 4: $[C_4mim][NTf_2]$ (1) + $[C_8mim][NTf_2]$ (2)
(1) Volumetric Properties

Ref. [57]
These physicochemical properties data are listed in a mixture 20 of IL 02011-03001.
Mixture 5: $[C_2mim][NTf_2]$ (1) + $[C_8mim][NTf_2]$ (2)
(1) Volumetric Properties

Ref. [57]
These physicochemical properties data are listed in a mixture 38 of IL 02003-03001.
Mixture 6: 1H,1H,2H,2H-$C_8F_{13}H_4OH$ (1) + $C_8mimNTf_2$ (2)
(1) Cloud Point

Ref. [150]

x_2	T/K	x_2	T/K	x_2	T/K	x_2	T/K
0.155	263.9	0.302	268.3	0.670	264.5	0.466	269.2
0.220	266.5	0.384	268.3	0.569	268.5	0.346	268.9

Mixture 7: C_5H_{12} (1) + [OMIM][BTI] (2)
(1) Activity Coefficient at Infinite Dilution and Partial Molar Excess Enthalpy at Infinite Dilution

Ref. [4]

	$\gamma^{\infty}{}_1$ (dilutor technique)			
Solute (1)	303.15 K	313.15 K	323.15 K	333.15 K
n-Pentane	3.98	3.92	3.87	3.84

Ref. [4]

	$\gamma^{\infty}{}_1$ (gas-liquid chromatography)			
Solute (1)	303.15 K	313.15 K	323.15 K	333.15 K
n-Pentane	4.07	3.94	3.78	3.60

Ref. [4]

Solute (1)	$H_{12}^{E,\infty}$/kJ · mol^{-1}
n-Pentane	2.212

Mixture 8: C_7H_{16} (1) + [OMIM][BTI] (2)
(1) Activity Coefficient at Infinite Dilution and Partial Molar Excess Enthalpy at Infinite Dilution

Ref. [4]

	$\gamma^{\infty}{}_1$ (dilutor technique)			
Solute (1)	303.15 K	313.15 K	323.15 K	333.15 K
n-Heptane	7.01	6.81	6.57	6.40

Ref. [4]

	$\gamma^{\infty}{}_1$ (gas-liquid chromatography)			
Solute (1)	303.15 K	313.15 K	323.15 K	333.15 K
n-Heptane	6.82	6.55	6.38	6.21

Ref. [4]

Solute (1)	$H_{12}^{E,\infty}$/kJ · mol^{-1}
n-Heptane	2.576

Mixture 9: C_8H_{18} (1) + [OMIM][BTI] (2)
(1) Activity Coefficient at Infinite Dilution and Partial Molar Excess Enthalpy at Infinite Dilution

Ref. [4]

	$\gamma^{\infty}{}_1$ (gas-liquid chromatography)			
Solute (1)	303.15 K	313.15 K	323.15 K	333.15 K
n-Octane	8.92	8.43	8.14	7.82

Ref. [4]

Solute (1)	$H_{12}^{E,\infty}$/kJ · mol^{-1}
n-Octane	3.547

Mixture 10: 1-C_5H_{10} (1) + [OMIM][BTI] (2)
(1) Activity Coefficient at Infinite Dilution and Partial Molar Excess Enthalpy at Infinite Dilution

Ref. [4]

	$\gamma^{\infty}{}_1$ (dilutor technique)			
Solute (1)	303.15 K	313.15 K	323.15 K	333.15 K
1-Pentene	2.54	2.53	2.53	2.52

Ref. [4]

	γ^{∞}_1 (gas-liquid chromatography)			
Solute (1)	303.15 K	313.15 K	323.15 K	333.15 K
1-Pentene	2.52	2.53	2.46	2.45

Ref. [4]

Solute (1)	$H_{12}^{E,\infty}$/kJ · mol^{-1}
1-Pentene	0.570

Mixture 11: 1-C_6H_{12} (1) + [OMIM][BTI] (2)
(1) Activity Coefficient at Infinite Dilution and Partial Molar Excess Enthalpy at Infinite Dilution

Ref. [4]

	γ^{∞}_1 (dilutor technique)			
Solute (1)	303.15 K	313.15 K	323.15 K	333.15 K
1-Hexene	3.33	3.30	3.27	3.23

Ref. [4]

	γ^{∞}_1 (gas-liquid chromatography)			
Solute (1)	303.15 K	313.15 K	323.15 K	333.15 K
1-Hexene	3.29	3.27	3.18	3.18

Ref. [4]

Solute (1)	$H_{12}^{E,\infty}$/kJ · mol^{-1}
1-Hexene	0.974

Mixture 12: 1-C_7H_{14} (1) + [OMIM][BTI] (2)
(1) Activity Coefficient at Infinite Dilution and Partial Molar Excess Enthalpy at Infinite Dilution

Ref. [4]

	γ^{∞}_1 (dilutor technique)			
Solute (1)	303.15 K	313.15 K	323.15 K	333.15 K
1-Heptene	4.33	4.26	4.20	4.15

Ref. [4]

	γ^{∞}_1 (gas-liquid chromatography)			
Solute (1)	303.15 K	313.15 K	323.15 K	333.15 K
1-Heptene	4.29	4.23	4.13	4.13

Ref. [4]

Solute (1)	$H_{12}^{E,\infty}$/kJ · mol^{-1}
1-Heptene	1.180

Mixture 13: 1-C_8H_{16} (1) + [OMIM][BTI] (2)
(1) Activity Coefficient at Infinite Dilution and Partial Molar Excess Enthalpy at Infinite Dilution

Ref. [4]

	γ^{∞}_1 (gas-liquid chromatography)			
Solute (1)	303.15 K	313.15 K	323.15 K	333.15 K
1-Octene	5.57	5.48	5.32	5.28

Ref. [4]

Solute (1)	$H_{12}^{E,\infty}$/kJ · mol^{-1}
1-Octene	1.582

Mixture 14: $CH_2(CH_2)_3CH_2$ (1) + [OMIM][BTI] (2)
(1) Activity Coefficient at Infinite Dilution and Partial Molar Excess Enthalpy at Infinite Dilution

Ref. [4]

	γ^{∞}_1 (dilutor technique)			
Solute (1)	303.15 K	313.15 K	323.15 K	333.15 K
Cyclopentane	2.68	2.64	2.59	2.56

Ref. [4]

Solute (1)	$H_{12}^{E,\infty}$/kJ · mol^{-1}
Cyclopentane	1.388

Mixture 15: $CH_2(CH_2)_4CH_2$ (1) + [OMIM][BTI] (2)
(1) Activity Coefficient at Infinite Dilution and Partial Molar Excess Enthalpy at Infinite Dilution

Ref. [4]

	γ^{∞}_1 (dilutor technique)			
Solute (1)	303.15 K	313.15 K	323.15 K	333.15 K
Cyclohexane	3.67	3.59	3.48	3.39

Ref. [4]

	γ^{∞}_1 (gas-liquid chromatography)			
Solute (1)	303.15 K	313.15 K	323.15 K	333.15 K
Cyclohexane	3.64	3.52	3.36	3.29

Ref. [4]

Solute (1)	$H_{12}^{E,\infty}$/kJ · mol^{-1}
Cyclohexane	2.625

Mixture 16: CH = $CHCH_2CH_2CH_2$ (1) + [OMIM][BTI] (2)
(1) Activity Coefficient at Infinite Dilution and Partial Molar Excess Enthalpy at Infinite Dilution

Ref. [4]

	$\gamma^{\infty}{}_1$ (dilutor technique)			
Solute (1)	303.15 K	313.15 K	323.15 K	333.15 K
Cyclopentene	1.86	1.84	1.81	1.78

Ref. [4]

Solute (1)	$H_{12}^{E,\infty}$/kJ · mol^{-1}
Cyclopentene	1.348

Mixture 17: CH = $CHCH_2CH_2CH_2CH_2$ (1) + [OMIM][BTI] (2)
(1) Activity Coefficient at Infinite Dilution and Partial Molar Excess Enthalpy at Infinite Dilution

Ref. [4]

	$\gamma^{\infty}{}_1$ (dilutor technique)			
Solute (1)	303.15 K	313.15 K	323.15 K	333.15 K
Cyclohexene	2.44	2.38	2.33	2.28

Ref. [4]

	$\gamma^{\infty}{}_1$ (gas-liquid chromatography)			
Solute (1)	303.15 K	313.15 K	323.15 K	333.15 K
Cyclohexene	2.41	2.38	2.31	2.29

Ref. [4]

Solute (1)	$H_{12}^{E,\infty}$/kJ · mol^{-1}
Cyclohexene	1.664

Mixture 18: C_6H_6 (1) + [OMIM][BTI] (2)
(1) Activity Coefficient at Infinite Dilution

Ref. [4]

	$\gamma^{\infty}{}_1$ (gas-liquid chromatography)			
Solute (1)	303.15 K	313.15 K	323.15 K	333.15 K
Benzene	0.63	0.65	0.65	0.66

Mixture 19: $C_6H_5CH_3$ (1) + [OMIM][BTI] (2)
(1) Activity Coefficient at Infinite Dilution and Partial Molar Excess Enthalpy at Infinite Dilution

Ref. [4]

	$\gamma^{\infty}{}_1$ (gas-liquid chromatography)			
Solute (1)	303.15 K	313.15 K	323.15 K	333.15 K
Toluene	0.82	0.83	0.85	0.85

Ref. [4]

Solute (1)	$H_{12}{}^{E,\infty}$/kJ · mol^{-1}
Toluene	−0.932

Mixture 20: m-$C_6H_4(CH_3)_2$ (1) + [OMIM][BTI] (2)
(1) Activity Coefficient at Infinite Dilution and Partial Molar Excess Enthalpy at Infinite Dilution

Ref. [4]

	$\gamma^{\infty}{}_1$ (gas-liquid chromatography)			
Solute (1)	303.15 K	313.15 K	323.15 K	333.15 K
m-Xylene	1.13	1.15	1.17	1.19

Ref. [4]

Solute (1)	$H_{12}{}^{E,\infty}$/kJ · mol^{-1}
m-Xylene	−1.447

Mixture 21: o-$C_6H_4(CH_3)_2$ (1) + [OMIM][BTI] (2)
(1) Activity Coefficient at Infinite Dilution

Ref. [4]

	$\gamma^{\infty}{}_1$ (gas-liquid chromatography)			
Solute (1)	303.15 K	313.15 K	323.15 K	333.15 K
o-Xylene	0.98	1.03	1.03	1.05

Mixture 22: p-$C_6H_4(CH_3)_2$ (1) + [OMIM][BTI] (2)
(1) Activity Coefficient at Infinite Dilution and Partial Molar Excess Enthalpy at Infinite Dilution

Ref. [4]

	$\gamma^{\infty}{}_1$ (gas-liquid chromatography)			
Solute (1)	303.15 K	313.15 K	323.15 K	333.15 K
p-Xylene	1.11	1.13	1.16	1.16

Ref. [4]

Solute (1)	$H_{12}^{E,\infty}$/kJ · mol^{-1}
p-Xylene	−1.336

Mixture 23: CH_3OH (1) + [OMIM][BTI] (2)
(1) Activity Coefficient at Infinite Dilution and Partial Molar Excess Enthalpy at Infinite Dilution

Ref. [4]

	γ^{∞}_1 (gas-liquid chromatography)			
Solute (1)	303.15 K	313.15 K	323.15 K	333.15 K
Methanol	1.23	1.14	1.03	0.96

Ref. [4]

Solute (1)	$H_{12}^{E,\infty}$/kJ · mol^{-1}
Methanol	6.965

(2) Volumetric Properties, Refractive Index and Transport Properties

Ref. [241]

x_1	ρ/(g · cm^{-3})	n_D	η/(mPa · s)	V^E/(cm^3 · mol^{-1})	ΔR	$\Delta\eta$/(mPa · s)
T = 298.15 K, atmospheric pressure						
0.0654	1.31650	1.43207	78.34	−0.026	−0.014	−8.151
0.0863	1.31510	1.43188	74.57	−0.039	0.002	−10.00
0.2134	1.30518	1.42995	53.65	−0.108	0.012	−19.23
0.3138	1.29508	1.42809	40.77	−0.151	0.035	−22.88
0.3873	1.28591	1.42619	32.92	−0.177	0.026	−23.96
0.5038	1.26704	1.42266	22.27	−0.231	0.05	−23.90
0.6155	1.24079	1.41741	14.57	−0.272	0.043	−21.33
0.7018	1.21098	1.41157	9.493	−0.300	0.046	−18.46
0.8050	1.15404	1.40017	4.919	−0.307	0.035	−13.55
0.9020	1.05254	1.38017	2.144	−0.294	0.031	−7.403

Mixture 24: C_2H_5OH (1) + [OMIM][BTI] (2)
(1) Activity Coefficient at Infinite Dilution and Partial Molar Excess Enthalpy at Infinite Dilution

Ref. [4]

	γ^{∞}_1 (gas-liquid chromatography)			
Solute (1)	303.15 K	313.15 K	323.15 K	333.15 K
Ethanol	1.46	1.34	1.22	1.12

Ref. [4]

Solute (1)	$H_{12}^{E,\infty}$/kJ · mol^{-1}
Ethanol	7.509

(2) Volumetric Properties, Refractive Index and Transport Properties

Ref. [242]

x_1	ρ/ (g · cm^{-3})	n_D	η/ (mPa · s)	V^E/ (cm^3 · mol^{-1})	ΔR	$\Delta\eta$/ (mPa · s)
T = 298.15 K, atmospheric pressure						
0.0959	1.31129	1.43164	71.28	0.051	0.004	−12.46
0.1162	1.30917	1.43133	67.63	0.047	0.001	−14.25
0.1935	1.30029	1.43016	55.53	0.036	0.013	−19.28
0.2269	1.29602	1.42965	51.54	0.029	0.026	−20.21
0.3304	1.28078	1.42766	39.23	−0.009	0.039	−23.06
0.4164	1.26519	1.42549	29.89	−0.070	0.028	−24.54
0.5412	1.23517	1.42143	19.92	−0.148	0.024	−23.09
0.6012	1.21621	1.41886	15.86	−0.181	0.022	−21.67
0.7033	1.17346	1.41319	10.25	−0.264	0.023	−17.93
0.8067	1.10685	1.40447	5.830	−0.316	0.034	−12.90
0.8967	1.01050	1.39142	3.079	−0.315	0.025	−7.426

Mixture 25: 1-C_3H_7OH (1) + [OMIM][BTI] (2)
(1) Activity Coefficient at Infinite Dilution and Partial Molar Excess Enthalpy at Infinite Dilution

Ref. [4]

	$\gamma^{\infty}{}_1$ (gas-liquid chromatography)			
Solute (1)	303.15 K	313.15 K	323.15 K	333.15 K
1-Propanol	1.67	1.51	1.35	1.24

Ref. [4]

Solute (1)	$H_{12}^{E,\infty}$/kJ · mol^{-1}
1-Propanol	8.445

Mixture 26: 2-CH_3CHCH_3OH (1) + [OMIM][BTI] (2)
(1) Partial Molar Excess Enthalpy at Infinite Dilution

Ref. [4]

Solute (1)	$H_{12}^{E,\infty}$/kJ · mol^{-1}
2-Propanol	8.025

Mixture 27: CH_3COCH_3 (1) + [OMIM][BTI] (2)
(1) Partial Molar Excess Enthalpy at Infinite Dilution

Ref. [4]

Solute (1)	$H_{12}^{E,\infty}$/kJ · mol^{-1}
Acetone	−1.793

Mixture 28: 2-$CH_3COC_2H_5$ (1) + [OMIM][BTI] (2)
(1) Partial Molar Excess Enthalpy at Infinite Dilution

Ref. [4]

Solute (1)	$H_{12}^{E,\infty}$/kJ · mol^{-1}
2-Butanone	−2.343

Mixture 29: 2-$CH_3COC_3H_7$ (1) + [OMIM][BTI] (2)
(1) Partial Molar Excess Enthalpy at Infinite Dilution

Ref. [4]

Solute (1)	$H_{12}^{E,\infty}$/kJ · mol^{-1}
2-Pentanone	−1.982

Mixture 30: $OCH_2CH_2CH_2CH_2$ (1) + [OMIM][BTI] (2)
(1) Partial Molar Excess Enthalpy at Infinite Dilution

Ref. [4]

Solute (1)	$H_{12}^{E,\infty}$/kJ · mol^{-1}
Tetrahydrofuran	−2.118

Mixture 31: $(CH_3)_2CHOCH(CH_3)_2$ (1) + [OMIM][BTI] (2)
(1) Partial Molar Excess Enthalpy at Infinite Dilution

Ref. [4]

Solute (1)	$H_{12}^{E,\infty}$/kJ · mol^{-1}
Diisopropyl ether	−1.510

Mixture 32: $(CH_3)_3COC_2H_5$ (1) + [OMIM][BTI] (2)
(1) Partial Molar Excess Enthalpy at Infinite Dilution

Ref. [4]

Solute (1)	$H_{12}^{E,\infty}$/kJ · mol^{-1}
Ethyl *tert*-butyl ether	−0.856

Mixture 33: $CH_3OC(CH_3)_2C_2H_5$ (1) + [OMIM][BTI] (2)
(1) Partial Molar Excess Enthalpy at Infinite Dilution

Ref. [4]

Solute (1)	$H_{12}^{E,\infty}$/kJ · mol^{-1}
Methyl *tert*-amyl ether	−2.039

Mixture 34: CH_3COOCH_3 (1) + [C_8mim][NTf_2] (2)
(1) Volumetric Properties, Refractive Index and Transport Properties

Ref. [241]

x_1	ρ/(g · cm^{-3})	n_D	η/(mPa · s)	V^E/ (cm^3 · mol^{-1})	ΔR	$\Delta\eta$/ (mPa · s)
T = 298.15 K, atmospheric pressure						
0.0699	1.31477	1.43190	75.62	−0.164	−0.013	−10.45
0.1317	1.30887	1.43089	63.25	−0.269	−0.004	−17.12
0.1748	1.30445	1.43009	56.12	−0.367	−0.009	−20.28
0.2177	1.29957	1.42927	49.13	−0.431	0.000	−23.32
0.3226	1.28583	1.42694	33.76	−0.599	0.018	−29.03
0.4105	1.27188	1.42453	23.79	−0.763	0.024	−30.90
0.5011	1.25368	1.42124	15.98	−0.845	0.025	−30.35
0.6001	1.22833	1.41684	9.815	−0.947	0.041	−27.40
0.6992	1.19395	1.41055	5.527	−1.002	0.034	−22.55
0.7872	1.15129	1.40296	3.066	−0.984	0.049	−16.91
0.8984	1.06739	1.38700	1.206	−0.764	0.028	−8.514

Mixture 35: $CH_3COOCH_2CH_3$ (1) + [C_8mim][NTf_2] (2)
(1) Volumetric Properties, Refractive Index and Transport Properties

Ref. [242]

x_1	ρ/(g · cm^{-3})	n_D	η/(mPa · s)	V^E/ (cm^3 · mol^{-1})	ΔR	$\Delta\eta$/ (mPa · s)
T = 298.15 K, atmospheric pressure						
0.0493	1.31493	1.43216	78.89	−0.077	−0.005	−9.081
0.1112	1.30734	1.43115	65.93	−0.215	−0.003	−16.35
0.1546	1.30155	1.43042	58.25	−0.312	0.005	−20.02
0.2125	1.29310	1.42930	48.82	−0.425	0.011	−24.13
0.3102	1.27660	1.42726	34.78	−0.589	0.045	−29.17
0.4040	1.25753	1.42453	24.15	−0.730	0.030	−31.16
0.4779	1.23988	1.42219	17.54	−0.879	0.035	−30.96
0.5967	1.20386	1.41721	9.795	−1.027	0.035	−27.77
0.7035	1.16022	1.41123	5.384	−1.120	0.044	−22.35
0.8233	1.08962	1.40097	2.284	−1.049	0.034	−14.41
0.9016	1.02264	1.39100	1.233	−0.815	0.037	−8.254

2.2.70.2 Ternary Mixtures

Mixture 1: $C_{12}H_{26}$ (1) + SCH = CHCH = CH (thiophene) (2) + [C_8mim][NTf_2] (3)
(1) Distribution Ratio and Selectivity

Ref. [243]

IL-rich phase			Hydrocarbon-rich phase				
x_3	x_2	x_1	x_3	x_2	x_1	β	S
298.15 K							
0.911	0.000	0.089	0.000	0.000	1.000	–	–
0.843	0.064	0.093	0.003	0.046	0.951	1.39	14.23
0.692	0.213	0.095	0.002	0.174	0.824	1.22	10.62
0.538	0.363	0.099	0.003	0.307	0.691	1.18	8.25
0.498	0.405	0.097	0.001	0.375	0.624	1.08	6.95
0.455	0.446	0.099	0.003	0.426	0.570	1.05	6.03
0.413	0.485	0.102	0.000	0.457	0.543	1.06	5.65
0.368	0.529	0.104	0.006	0.513	0.482	1.03	4.78
0.336	0.559	0.105	0.000	0.597	0.403	0.94	3.59
0.292	0.604	0.104	0.006	0.667	0.328	0.91	2.86
0.218	0.676	0.106	0.004	0.797	0.198	0.85	1.58
0.163	0.741	0.096	0.002	0.871	0.127	0.85	1.13
0.059	0.900	0.041	0.002	0.949	0.048	0.95	1.11
0.054	0.946	0.000	0.000	1.000	0.000	0.95	–

Mixture 2: $CH_2(CH_2)_4CH_2$ (1) + SCH = CHCH = CH (thiophene) (2) + [C_8mim][NTf_2] (3)
(1) Distribution Ratio and Selectivity

Ref. [243]

IL-rich phase			Hydrocarbon-rich phase				
x_3	x_2	x_1	x_3	x_2	x_1	β	S
298.15 K							
0.646	0.000	0.354	0.003	0.000	0.997	–	–
0.488	0.195	0.317	0.002	0.087	0.911	2.24	6.44
0.419	0.272	0.309	0.004	0.127	0.869	2.14	6.02
0.301	0.427	0.272	0.003	0.271	0.726	1.58	4.19
0.210	0.552	0.238	0.003	0.455	0.542	1.21	2.76
0.166	0.629	0.205	0.003	0.587	0.410	1.07	2.14
0.133	0.698	0.169	0.003	0.712	0.285	0.98	1.65
0.093	0.806	0.101	0.003	0.838	0.159	0.96	1.51
0.080	0.851	0.068	0.001	0.906	0.092	0.94	1.28
0.054	0.946	0.000	0.001	1	0.000	0.95	–

Mixture 3: C_6H_{14} (1) + SCH = CHCH = CH (thiophene) (2) + [C_8mim][NTf_2] (3)
(1) Distribution Ratio and Selectivity

Ref. [244]

IL-rich phase			Hydrocarbon-rich phase				
x^{II}_3	x^{II}_2	x^{II}_1	x^I_3	x^I_2	x^I_1	β	S
298.15 K							
0.711	0.000	0.289	0.000	0.000	1.000		
0.547	0.190	0.264	0.000	0.087	0.913	2.18	7.55
0.363	0.412	0.225	0.001	0.267	0.732	1.54	5.02
0.274	0.518	0.208	0.000	0.399	0.601	1.30	3.75
0.215	0.593	0.192	0.000	0.545	0.455	1.09	2.58
0.168	0.660	0.172	0.000	0.620	0.380	1.06	2.35
0.143	0.701	0.156	0.000	0.695	0.305	1.01	1.97
0.118	0.768	0.114	0.000	0.800	0.200	0.96	1.68
0.100	0.809	0.091	0.000	0.858	0.142	0.94	1.47
0.065	0.936	0.000	0.000	1.000	0.000	0.94	–

Mixture 4: C_7H_{16} (1) + SCH = CHCH = CH (thiophene) (2) + [C_8mim][NTf_2] (3)
(1) Distribution Ratio and Selectivity

Ref. [244]

IL-rich phase			Hydrocarbon-rich phase				
x^{II}_3	x^{II}_2	x^{II}_1	x^I_3	x^{II}_2	x^{II}_1	β	S
298.15 K							
0.755	0.000	0.245	0.000	0.000	1.000		
0.560	0.225	0.215	0.001	0.098	0.901	2.30	9.62
0.399	0.409	0.192	0.001	0.232	0.767	1.76	7.04
0.273	0.552	0.175	0.000	0.405	0.595	1.36	4.63
0.181	0.665	0.154	0.000	0.634	0.366	1.05	2.49
0.140	0.732	0.129	0.000	0.782	0.218	0.94	1.58
0.117	0.788	0.095	0.000	0.872	0.128	0.90	1.22
0.088	0.862	0.050	0.000	0.950	0.050	0.91	0.91
0.065	0.936	0.000	0.000	1.000	0.000	0.94	

Mixture 5: $C_{16}H_{34}$ (1) + SCH = CHCH = CH (thiophene) (2) + [C_8mim][NTf_2] (3)
(1) Distribution Ratio and Selectivity

Ref. [244]

IL-rich phase			Hydrocarbon-rich phase				
x^{II}_3	x^{II}_2	x^{II}_1	x^I_3	x^{II}_2	x^{II}_1	β	S
298.15 K							
0.953	0.000	0.047	0.009	0.000	0.991		
0.845	0.106	0.049	0.004	0.143	0.852	0.74	12.89
0.697	0.242	0.062	0.007	0.297	0.696	0.81	9.15
0.592	0.343	0.065	0.008	0.469	0.523	0.73	5.88

(*continued*)

IL-rich phase			Hydrocarbon-rich phase				
x^{II}_3	x^{II}_2	x^{II}_1	x^{I}_3	x^{II}_2	x^{II}_1	β	S
0.471	0.450	0.079	0.001	0.628	0.362	0.72	3.28
0.407	0.515	0.078	0.009	0.742	0.250	0.69	2.22
0.313	0.599	0.088	0.005	0.878	0.115	0.68	0.89
0.226	0.702	0.072	0.007	0.952	0.041	0.74	0.42
0.065	0.936	0.000	0.000	1.000	0.000	0.94	

Mixture 6: C_6H_{14} (1) + C_6H_6 (2) + [C_8mim][NTf_2] (3)
(1) Distribution Ratio and Selectivity

Ref. [168]

organic-rich phase			ionic-liquid-rich phase				
x_1	x_2	x_3	x_1	x_2	x_3	β	S
298.15 K							
1.000	0.000	0.000	0.295	0.000	0.705		
0.883	0.117	0.000	0.228	0.222	0.550	1.90	7.35
0.799	0.201	0.000	0.215	0.337	0.448	1.68	6.23
0.672	0.328	0.000	0.189	0.465	0.346	1.42	5.04
0.446	0.554	0.000	0.163	0.613	0.224	1.11	3.03
0.317	0.683	0.000	0.138	0.696	0.166	1.02	2.34
0.186	0.814	0.000	0.097	0.768	0.135	0.94	1.81
0.077	0.922	0.001	0.047	0.850	0.103	0.92	1.51
0.000	0.999	0.001	0.000	0.915	0.085	0.92	

Mixture 7: CH_3COOCH_3 (1) + CH_3OH (2) + [C_8mim][NTf_2] (3)
(1) Volumetric Properties, Refractive Index and Transport Properties

Ref. [241]

x_1	x_2	ρ/ (g · cm^{-3})	n_D	η/ (mPa · s)	V^E/ (cm^3 · mol^{-1})	ΔR	$\Delta\eta$/ (mPa · s)
T = 298.15 K, atmospheric pressure							
0.8612	0.1250	0.94561	1.36274	0.456	−0.224	0.012	−1.205
0.7744	0.2031	0.95684	1.36488	0.608	−0.311	0.016	−1.864
0.6883	0.2805	0.96805	1.36728	0.670	−0.386	0.033	−2.609
0.5755	0.3821	0.98244	1.36966	0.777	−0.436	0.030	−3.559
0.4944	0.4551	0.99278	1.37136	0.794	−0.457	0.030	−4.301
0.3837	0.5547	1.00685	1.37393	1.005	−0.462	0.043	−5.126
0.2875	0.6413	1.01915	1.37556	1.155	−0.451	0.032	−5.877
0.1967	0.7230	1.03063	1.37721	1.414	−0.419	0.030	−6.468
0.1482	0.7667	1.03683	1.37814	1.533	−0.401	0.030	−6.803
0.0972	0.8126	1.04333	1.37934	1.756	−0.376	0.040	−7.057
0.0440	0.8604	1.05012	1.37995	1.953	−0.347	0.026	−7.358
0.8760	0.0989	0.96672	1.36706	0.522	−0.315	0.017	−2.168
0.7914	0.1664	0.99135	1.37194	0.747	−0.452	0.026	−3.532

(continued)

x_1	x_2	ρ/ $(g \cdot cm^{-3})$	n_D	η/ (mPa · s)	V^E/ $(cm^3 \cdot mol^{-1})$	ΔR	$\Delta\eta$/ (mPa · s)
0.6849	0.2513	1.02006	1.37741	0.915	−0.567	0.034	−5.366
0.5963	0.3220	1.04206	1.38168	1.063	−0.609	0.049	−6.883
0.4928	0.4045	1.06594	1.38589	1.377	−0.618	0.051	−8.513
0.4050	0.4746	1.08486	1.38895	1.784	−0.600	0.041	−9.755
0.2826	0.5722	1.10973	1.39315	2.411	−0.569	0.037	−11.43
0.2047	0.6344	1.12444	1.39560	3.024	−0.523	0.036	−12.28
0.1684	0.6633	1.13084	1.39676	3.307	−0.480	0.044	−12.68
0.1206	0.7014	1.13943	1.39828	3.832	−0.453	0.048	−13.05
0.0544	0.7542	1.15075	1.39995	4.467	−0.396	0.037	−13.66
0.8870	0.0795	0.98130	1.37008	0.616	−0.389	0.021	−2.846
0.7857	0.1508	1.02208	1.37796	0.811	−0.572	0.029	−5.427
0.6962	0.2138	1.05350	1.38416	1.132	−0.662	0.049	−7.564
0.5976	0.2831	1.08394	1.38950	1.512	−0.698	0.041	−9.888
0.5049	0.3484	1.10931	1.39398	2.047	−0.692	0.041	−11.90
0.4017	0.4209	1.13454	1.39837	2.740	−0.659	0.038	−14.03
0.3124	0.4838	1.15405	1.40172	3.680	−0.595	0.037	−15.54
0.2174	0.5506	1.17301	1.40527	4.871	−0.523	0.056	−16.96
0.1752	0.5803	1.18106	1.40642	5.504	−0.507	0.038	−17.48
0.1216	0.6180	1.19059	1.40815	6.515	−0.457	0.046	−17.94
0.0531	0.6662	1.20196	1.40985	7.961	−0.375	0.031	−18.38
0.8973	0.0616	0.99342	1.37240	0.667	−0.440	0.017	−3.490
0.7848	0.1291	1.05009	1.38355	1.032	−0.680	0.037	−7.286
0.6870	0.1877	1.08960	1.39073	1.566	−0.757	0.033	−10.37
0.5947	0.2432	1.12088	1.39647	2.220	−0.766	0.040	−13.13
0.5042	0.2974	1.14726	1.40125	3.060	−0.749	0.044	−15.63
0.4026	0.3584	1.17270	1.40580	4.207	−0.690	0.049	−18.24
0.3188	0.4086	1.19100	1.40893	5.676	−0.618	0.045	−19.87
0.2104	0.4736	1.21191	1.41261	7.984	−0.521	0.048	−21.57
0.1609	0.5034	1.22057	1.41420	9.363	−0.475	0.054	−22.02
0.1059	0.5364	1.22958	1.41543	11.21	−0.417	0.027	−22.21
0.0506	0.5695	1.23817	1.41693	13.07	−0.372	0.025	−22.39
0.8913	0.0543	1.01217	1.37607	0.773	−0.517	0.019	−4.607
0.7958	0.1020	1.06746	1.38676	1.188	−0.735	0.030	−8.603
0.6991	0.1503	1.11105	1.39485	1.936	−0.818	0.033	−12.32
0.6063	0.1967	1.14459	1.40103	2.773	−0.811	0.042	−15.77
0.5062	0.2468	1.17443	1.40635	4.233	−0.768	0.042	−18.94
0.4122	0.2937	1.19783	1.41065	5.949	−0.701	0.054	−21.56
0.3026	0.3485	1.22079	1.41448	8.719	−0.595	0.038	−23.85
0.2420	0.3788	1.23192	1.41671	10.66	−0.533	0.062	−24.72
0.1583	0.4206 ·	1.24597	1.41901	14.06	−0.473	0.043	−25.18
0.1108	0.4443	1.25304	1.42013	16.13	−0.406	0.034	−25.30
0.0415	0.4790	1.26264	1.42171	20.00	−0.311	0.025	−24.63
0.8939	0.0428	1.02420	1.37845	0.848	−0.586	0.018	−5.351

(*continued*)

x_1	x_2	ρ/ $(g \cdot cm^{-3})$	n_D	η/ $(mPa \cdot s)$	V^E/ $(cm^3 \cdot mol^{-1})$	ΔR	$\Delta\eta$/ $(mPa \cdot s)$
0.7940	0.0831	1.08725	1.39048	1.440	−0.793	0.027	−10.25
0.6965	0.1225	1.13302	1.39919	2.476	−0.852	0.047	−14.59
0.5995	0.1617	1.16868	1.40540	3.752	−0.879	0.026	−18.65
0.5071	0.1989	1.19579	1.41049	5.644	−0.834	0.043	−21.84
0.4081	0.2389	1.21954	1.41464	8.298	−0.745	0.039	−24.63
0.3216	0.2738	1.23678	1.41758	11.36	−0.626	0.033	−26.32
0.2304	0.3106	1.25260	1.42041	15.45	−0.523	0.036	−27.25
0.1474	0.3441	1.26519	1.42254	20.25	−0.433	0.025	−27.02
0.0912	0.3668	1.27289	1.42387	24.00	−0.373	0.020	−26.37
0.0423	0.3866	1.27884	1.42493	27.81	−0.275	0.026	−25.25
0.9034	0.0289	1.02989	1.37958	0.882	−0.607	0.019	−5.720
0.8020	0.0592	1.10088	1.39322	1.648	−0.833	0.035	−11.51
0.7197	0.0838	1.14263	1.40102	2.701	−0.895	0.043	−15.78
0.6067	0.1176	1.18556	1.40911	4.790	−0.877	0.069	−21.00
0.5114	0.1461	1.21328	1.41375	7.387	−0.825	0.042	−24.56
0.4453	0.1659	1.22923	1.41645	9.760	−0.773	0.027	−26.46
0.3245	0.2020	1.25333	1.42073	15.48	−0.663	0.023	−28.55
0.2485	0.2247	1.26598	1.42297	20.32	−0.599	0.018	−28.63
0.1936	0.2411	1.27393	1.42434	24.23	−0.515	0.016	−28.26
0.1512	0.2538	1.27944	1.42541	27.75	−0.419	0.032	−27.48
0.0864	0.2731	1.28763	1.42670	33.95	−0.360	0.008	−25.48
0.8965	0.0208	1.04749	1.38303	1.026	−0.681	0.022	−6.961
0.7990	0.0403	1.11776	1.39645	1.978	−0.880	0.037	−13.20
0.6965	0.0609	1.16852	1.40574	3.818	−0.942	0.034	−18.91
0.6164	0.0769	1.19792	1.41109	5.764	−0.907	0.038	−22.87
0.4919	0.1019	1.23259	1.41749	10.40	−0.801	0.057	−27.42
0.4285	0.1146	1.24651	1.41972	13.57	−0.721	0.033	−28.91
0.3079	0.1388	1.26861	1.42386	21.67	−0.611	0.053	−29.70
0.2350	0.1534	1.27965	1.42553	27.87	−0.536	0.014	−28.87
0.1783	0.1648	1.28702	1.42684	33.60	−0.424	0.018	−27.32
0.1455	0.1714	1.29119	1.42762	37.22	−0.404	0.020	−26.12
0.0441	0.1917	1.30220	1.42946	49.70	−0.221	0.007	−21.12
0.8921	0.0111	1.06236	1.38595	1.174	−0.731	0.026	−8.102
0.7982	0.0208	1.13178	1.39912	2.318	−0.928	0.037	−14.72
0.6966	0.0313	1.18203	1.40823	4.578	−0.951	0.032	−20.86
0.5872	0.0426	1.22004	1.41512	8.328	−0.898	0.032	−26.16
0.5013	0.0514	1.24247	1.41909	12.57	−0.810	0.027	−29.02
0.4038	0.0615	1.26278	1.42279	19.30	−0.708	0.033	−30.35
0.3093	0.0712	1.27863	1.42544	27.85	−0.592	0.009	−29.60
0.2026	0.0822	1.29313	1.42795	39.99	−0.425	−0.001	−26.29
0.1446	0.0882	1.29978	1.42927	47.27	−0.308	0.022	−23.81
0.0981	0.0930	1.30477	1.43009	53.99	−0.240	0.011	−20.93
0.0073	0.1024	1.31339	1.43154	70.21	−0.087	−0.003	−12.21

Mixture 8: $CH_3COOCH_2CH_3$ (1) + C_2H_5OH (2) + [C_8mim][NTf_2] (3)
(1) Volumetric Properties, Refractive Index and Transport Properties

Ref. [242]

x_1	x_2	ρ/ (g · cm^{-3})	n_D	η/ (mPa · s)	V^E/ (cm^3 · mol^{-1})	ΔR	$\Delta\eta$/ (mPa · s)
T = 298.15 K, atmospheric pressure							
0.8849	0.1035	0.90663	1.37241	0.498	−0.081	0.013	−1.062
0.8735	0.1138	0.90787	1.37266	0.503	−0.090	0.013	−1.170
0.7954	0.1840	0.91647	1.37435	0.555	−0.156	0.014	−1.887
0.6802	0.2876	0.92911	1.37686	0.679	−0.223	0.020	−2.898
0.5940	0.3650	0.93865	1.37889	0.773	−0.261	0.033	−3.654
0.5043	0.4457	0.94876	1.38068	0.928	−0.300	0.028	−4.383
0.4189	0.5225	0.95846	1.38257	1.076	−0.327	0.034	−5.076
0.3104	0.6200	0.97093	1.38481	1.371	−0.351	0.032	−5.850
0.2274	0.6947	0.98048	1.38642	1.610	−0.353	0.028	−6.429
0.1905	0.7278	0.98468	1.38705	1.824	−0.345	0.023	−6.578
0.1248	0.7869	0.99241	1.38831	2.113	−0.345	0.018	−6.937
0.0800	0.8272	0.99776	1.38928	2.461	−0.347	0.019	−7.031
0.8748	0.0997	0.92911	1.37615	0.588	−0.250	0.015	−2.252
0.7846	0.1715	0.95243	1.38022	0.733	−0.362	0.024	−3.846
0.6915	0.2456	0.97543	1.38412	0.936	−0.453	0.030	−5.438
0.5910	0.3256	0.99895	1.38800	1.199	−0.511	0.035	−7.113
0.4913	0.4050	1.02098	1.39149	1.570	−0.528	0.038	−8.663
0.3835	0.4908	1.04363	1.39496	2.037	−0.520	0.036	−10.28
0.2797	0.5734	1.06426	1.39804	2.726	−0.483	0.034	−11.59
0.1952	0.6406	1.08046	1.40051	3.493	−0.453	0.035	−12.45
0.1462	0.6797	1.08956	1.40165	4.020	−0.428	0.022	−12.87
0.1078	0.7103	1.09663	1.40272	4.540	−0.415	0.022	−13.09
0.0502	0.7561	1.10682	1.40418	5.392	−0.373	0.019	−13.35
0.8768	0.0857	0.94728	1.37908	0.685	−0.355	0.019	−3.252
0.7742	0.1570	0.98577	1.38552	0.930	−0.528	0.033	−5.931
0.6981	0.2099	1.01159	1.38960	1.261	−0.608	0.035	−7.767
0.5954	0.2813	1.04336	1.39451	1.663	−0.662	0.037	−10.29
0.4921	0.3531	1.07192	1.39884	2.392	−0.644	0.044	−12.51
0.3889	0.4249	1.09793	1.40273	3.145	−0.606	0.049	−14.70
0.2817	0.4994	1.12262	1.40640	4.576	−0.554	0.054	−16.32
0.1993	0.5567	1.14002	1.40852	5.829	−0.485	0.027	−17.41
0.1489	0.5917	1.15003	1.40998	6.763	−0.429	0.031	−17.92
0.0995	0.6261	1.15956	1.41124	7.881	−0.382	0.024	−18.21
0.0424	0.6658	1.17020	1.41273	9.352	−0.337	0.021	−18.36
0.8932	0.0638	0.95585	1.38071	0.743	−0.428	0.036	−3.675
0.7892	0.1261	1.00541	1.38845	1.085	−0.635	0.032	−7.227
0.6998	0.1795	1.04198	1.39430	1.610	−0.737	0.047	−10.04
0.6078	0.2345	1.07466	1.39903	2.286	−0.758	0.040	−12.81
0.5162	0.2893	1.10346	1.40329	3.107	−0.748	0.043	−15.41

(continued)

x_1	x_2	ρ/ (g · cm^{-3})	n_D	η/ (mPa · s)	V^E/ (cm^3 · mol^{-1})	ΔR	$\Delta\eta$/ (mPa · s)
0.4132	0.3509	1.13182	1.40722	4.382	−0.676	0.035	−17.99
0.3042	0.4160	1.15833	1.41110	6.305	−0.581	0.045	−20.15
0.2069	0.4742	1.17936	1.41388	8.621	−0.475	0.031	−21.47
0.1578	0.5036	1.18933	1.41526	10.05	−0.439	0.026	−21.88
0.0987	0.5389	1.20038	1.41676	11.96	−0.350	0.025	−22.17
0.0576	0.5635	1.20775	1.41770	13.66	−0.294	0.018	−22.01
0.8999	0.0505	0.96521	1.38193	0.800	−0.482	0.023	−4.223
0.8028	0.0995	1.02001	1.39061	1.212	−0.713	0.031	−8.276
0.7065	0.1480	1.06452	1.39739	1.953	−0.808	0.037	−11.96
0.5911	0.2063	1.10845	1.40394	2.980	−0.813	0.048	−16.23
0.4990	0.2528	1.13790	1.40803	4.430	−0.763	0.039	−19.02
0.3979	0.3037	1.16575	1.41194	6.374	−0.675	0.038	−21.72
0.3119	0.3471	1.18642	1.41486	8.577	−0.566	0.043	−23.47
0.2071	0.4000	1.20888	1.41787	12.06	−0.461	0.029	−24.80
0.1685	0.4195	1.21641	1.41886	13.74	−0.413	0.023	−24.90
0.1118	0.4481	1.22676	1.42028	16.28	−0.328	0.023	−24.96
0.0645	0.4719	1.23502	1.42156	18.90	−0.273	0.036	−24.51
0.8901	0.0443	0.98547	1.38518	0.931	−0.598	0.028	−5.564
0.7900	0.0846	1.04730	1.39468	1.547	−0.829	0.029	−10.47
0.7002	0.1208	1.09139	1.40130	2.524	−0.906	0.036	−14.46
0.5947	0.1634	1.13322	1.40733	3.856	−0.887	0.037	−18.95
0.4877	0.2065	1.16773	1.41238	6.229	−0.801	0.055	−22.49
0.3874	0.2469	1.19476	1.41595	9.092	−0.692	0.037	−25.17
0.2689	0.2947	1.22172	1.41984	13.86	−0.559	0.050	−26.94
0.1848	0.3286	1.23808	1.42183	18.21	−0.429	0.023	−27.23
0.1753	0.3324	1.23985	1.42207	18.85	−0.420	0.022	−27.12
0.1317	0.3499	1.24764	1.42315	21.47	−0.373	0.020	−26.91
0.0370	0.3881	1.26257	1.42512	28.31	−0.162	0.021	−25.30
0.8981	0.0294	0.99407	1.38673	0.992	−0.648	0.043	−6.137
0.8137	0.0536	1.05407	1.39582	1.727	−0.871	0.042	−10.95
0.7173	0.0814	1.10678	1.40358	2.906	−0.944	0.046	−16.11
0.5899	0.1181	1.15950	1.41099	5.526	−0.915	0.037	−21.86
0.4936	0.1458	1.19054	1.41530	8.229	−0.827	0.034	−25.49
0.3880	0.1762	1.21845	1.41909	12.79	−0.693	0.025	−27.88
0.2964	0.2026	1.23884	1.42194	17.52	−0.569	0.028	−29.16
0.2087	0.2279	1.25550	1.42415	23.50	−0.403	0.023	−28.95
0.1575	0.2426	1.26441	1.42541	27.51	−0.340	0.025	−28.31
0.1042	0.2580	1.27292	1.42652	32.57	−0.253	0.017	−26.75
0.0538	0.2725	1.28042	1.42758	37.90	−0.170	0.021	−24.74
0.8952	0.0203	1.00754	1.38867	1.126	−0.715	0.037	−7.092
0.7968	0.0394	1.07980	1.39946	2.230	−0.956	0.031	−13.31
0.6913	0.0598	1.13610	1.40764	4.029	−1.024	0.031	−19.35
0.5970	0.0781	1.17423	1.41310	6.502	−0.952	0.042	−23.89

(*continued*)

x_1	x_2	ρ/ (g · cm^{-3})	n_D	η/ (mPa · s)	V^E/ (cm^3 · mol^{-1})	ΔR	$\Delta\eta$/ (mPa · s)
0.4956	0.0978	1.20682	1.41760	10.28	−0.842	0.039	−27.65
0.4014	0.1160	1.23148	1.42096	15.35	−0.721	0.033	−29.59
0.3052	0.1347	1.25239	1.42365	21.83	−0.560	0.012	−30.26
0.2045	0.1542	1.27074	1.42629	30.54	−0.365	0.033	−29.03
0.1600	0.1628	1.27808	1.42715	35.37	−0.308	0.009	−27.51
0.0967	0.1751	1.28749	1.42850	42.55	−0.178	0.021	−25.04
0.0507	0.1840	1.29387	1.42930	47.86	−0.095	0.010	−23.15
0.9031	0.0093	1.01113	1.38906	1.138	−0.745	0.026	−7.356
0.7979	0.0194	1.09370	1.40184	2.769	−1.004	0.060	−14.49
0.7077	0.0281	1.14388	1.40879	4.389	−1.028	0.041	−20.38
0.5961	0.0388	1.19009	1.41523	8.007	−0.951	0.037	−26.06
0.5102	0.0471	1.21765	1.41901	12.10	−0.850	0.033	−29.12
0.4059	0.0571	1.24463	1.42271	18.76	−0.687	0.032	−31.15
0.2967	0.0676	1.26733	1.42574	28.46	−0.493	0.023	−30.55
0.2012	0.0768	1.28374	1.42795	39.19	−0.298	0.022	−27.77
0.1527	0.0815	1.29110	1.42893	45.00	−0.198	0.019	−25.99
0.0984	0.0867	1.29872	1.43013	52.99	−0.087	0.046	−22.53
0.0545	0.0909	1.30443	1.43071	60.11	0.001	0.014	−19.07

2.2.71 02029-04019: 1-Methyl-3-octylimidazolium diethyleneglycolmonomethylethersulphate

Abbreviation:	[MOIM][$CH_3(OCH_2CH_2)_2OSO_3$]
Molecular Formula:	$C_{17}H_{34}N_2O_6S$
Molar Mass:	394.53
Structure:	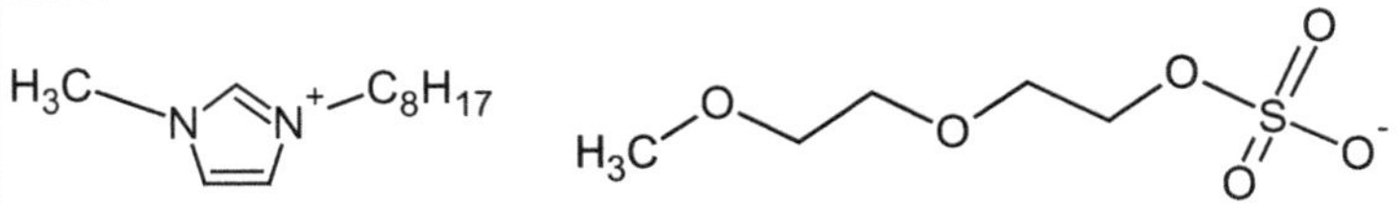

2.2.71.1 Binary Mixtures

Mixture 1: CH_3OH (1) + [MOIM][$CH_3(OCH_2CH_2)_2OSO_3$] (2)

(1) Volumetric Properties

Ref. [102]

x_1	ρ (g · cm^{-3})	V^E_m (cm^3 · mol^{-1})	x_1	ρ (g · cm^{-3})	V^E_m (cm^3 · mol^{-1})
T = 298.15 K					
0.110	1.1123	−0.098	0.608	1.0709	−0.649
0.307	1.1022	−0.374	0.803	1.0182	−0.721
0.447	1.0910	−0.502	0.909	0.9471	−0.502

(continued)

x_1	ρ (g · cm^{-3})	V^E_m (cm^3 · mol^{-1})	x_1	ρ (g · cm^{-3})	V^E_m (cm^3 · mol^{-1})
T = 303.15 K					
0.110	1.1087	−0.106	0.608	1.0672	−0.681
0.307	1.0986	−0.399	0.803	1.0143	−0.752
0.447	1.0874	−0.538	0.909	0.9431	−0.538
T = 313.15 K					
0.110	1.1019	−0.155	0.608	1.0601	−0.755
0.307	1.0918	−0.477	0.803	1.0066	−0.804
0.447	1.0805	−0.616	0.909	0.9350	−0.589

2.2.72 02029-05001: 1-Octyl-3-methylimidazoliumhexafluorophosphate

Abbreviation:	[C_8MIm][PF_6], OMIM PF_6, C_8mimPF$_6$, [C_8mim][PF_6]
Molecular Formula:	$C_{12}H_{23}F_6N_2P$
Molar Mass:	340.29
Structure:	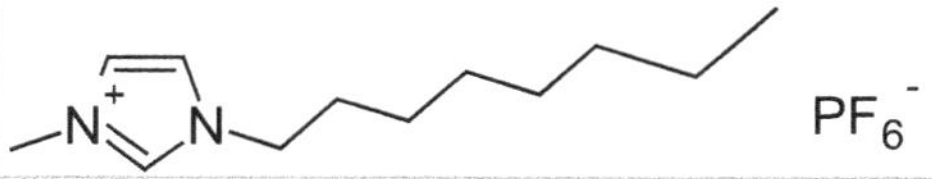

2.2.72.1 Binary Mixtures

Mixture 1: 2-$CH_3COC_2H_5$ (1) + OMIM PF_6 (2)

(1) Volumetric Properties, Refractive Index, Speed of Sound and Compressibility Properties

Ref. [232]

x_2	ρ/ (g · cm^{-3})	V^E_m/ (cm^3 · mol^{-1})	n_D	Δn_D	u/ (m · s^{-1})	k_s/ (T · Pa^{-1})	Δk_s/ (T · Pa^{-1})
T = 293.15 K							
0	0.8056	0	1.37879	0	1212	845	0
0.0475	0.8698	−0.748	1.38881	0.0079	1243	744	−80
0.0972	0.9237	−1.204	1.39567	0.0124	1268	674	−128
0.1918	1.0020	−1.689	1.40468	0.0171	1303	588	−171
0.2947	1.0631	−1.870	1.41058	0.0184	1331	531	−182
0.3949	1.1079	−1.897	1.41468	0.0179	1352	494	−175
0.4983	1.1436	−1.786	1.41774	0.0162	1368	467	−155
0.5752	1.1649	−1.606	1.41934	0.0143	1378	452	−136
0.6815	1.1895	−1.318	1.42126	0.0114	1389	436	−105
0.7683	1.2058	−0.983	1.42239	0.0086	1399	424	−78
0.8315	1.2164	−0.749	1.42355	0.0068	1406	416	−57
0.9315	1.2309	−0.326	1.42474	0.0035	1425	400	−29
1	1.2396	0	1.42440	0	1424	398	0
T = 298.15 K							

(continued)

x_2	ρ/ (g · cm^{-3})	$V^E{}_m$/ (cm^3 · mol^{-1})	n_D	Δn_D	u/ (m · s^{-1})	k_s/ (T · Pa^{-1})	Δk_s/ (T · Pa^{-1})
0	0.7997	0	1.37618	0	1192	880	0
0.0475	0.8643	−0.800	1.38663	0.0082	1224	773	−85
0.0972	0.9189	−1.329	1.39445	0.0137	1249	698	−136
0.1918	0.9976	−1.848	1.40289	0.0177	1286	606	−183
0.2947	1.0588	−1.999	1.40877	0.0188	1315	546	−195
0.3949	1.1037	−2.015	1.41302	0.0183	1337	507	−187
0.4983	1.1394	−1.890	1.41614	0.0166	1354	479	−166
0.5752	1.1609	−1.710	1.41779	0.0147	1364	463	−146
0.6815	1.1855	−1.394	1.41979	0.0117	1376	446	−113
0.7683	1.2020	−1.060	1.42095	0.0088	1386	433	−84
0.8315	1.2128	−0.850	1.42198	0.0069	1392	425	−62
0.9315	1.2275	−0.442	1.42337	0.0036	1410	410	−31
1	1.2357	0	1.42302	0	1408	408	0
$T = 303.15$ K							
0	0.7949	0	1.37355	0	1163	931	0
0.0475	0.8596	−0.824	1.38505	0.0092	1195	814	−92
0.0972	0.9141	−1.340	1.39284	0.0146	1229	724	−157
0.1918	0.9932	−1.909	1.40124	0.0185	1269	625	−207
0.2947	1.0546	−2.071	1.40709	0.0194	1300	561	−218
0.3949	1.0995	−2.072	1.41133	0.0188	1323	520	−208
0.4983	1.1353	−1.935	1.41466	0.0171	1340	490	−185
0.5752	1.1569	−1.753	1.41645	0.0152	1351	474	−162
0.6815	1.1815	−1.430	1.41825	0.0119	1363	456	−126
0.7683	1.1983	−1.126	1.41949	0.0090	1372	443	−94
0.8315	1.2091	−0.907	1.42051	0.0070	1379	435	−70
0.9315	1.2239	−0.494	1.42199	0.0036	1395	420	−34
1	1.2319	0	1.42165	0	1393	418	0

Mixture 2: C_6H_{14} (1) + OMIM PF_6 (2)
(1) Volumetric Properties, Refractive Index, Speed of Sound and Compressibility Properties

Ref. [245]

x_2	ρ (g · cm^{-3})	$V^E{}_m$ (cm^3 · mol^{-1})	n_D	Δn_D	u (m · s^{-1})	k_s (T · Pa^{-1})	Δk_s (T · Pa^{-1})
$T = 293.15$ K							
0	0.6595	0	1.37495	0	1100	1252	0
0.8732	1.2053	−0.694	1.42206	0.0039	1380	436	−70
0.9107	1.2161	−0.523	1.42276	0.0028	1394	423	−51
0.9288	1.2210	−0.400	1.42314	0.0023	1401	418	−41
0.9532	1.2274	−0.244	1.42356	0.0015	1408	411	−27
1	1.2396	0	1.42440	0	1424	398	0
$T = 298.15$ K							
0	0.6549	0	1.37217	0	1078	1314	0

(*continued*)

x_2	ρ (g · cm^{-3})	V^E_m (cm^3 · mol^{-1})	n_D	Δn_D	u (m · s^{-1})	k_s (T · Pa^{-1})	Δk_s (T · Pa^{-1})
0.8732	1.2015	−0.728	1.42057	0.0040	1364	447	−76
0.9107	1.2122	−0.548	1.42133	0.0029	1378	434	−55
0.9288	1.2171	−0.423	1.42174	0.0023	1385	429	−44
0.9532	1.2236	−0.259	1.42216	0.0015	1393	421	−29
1	1.2357	0	1.42302	0	1408	408	0
T = 303.15 K							
0	0.6504	0	1.36936	0	1055	1380	0
0.8732	1.1976	−0.761	1.41939	0.0044	1349	459	−82
0.9107	1.2084	−0.571	1.41996	0.0030	1364	445	−59
0.9288	1.2133	−0.442	1.42042	0.0025	1370	439	−48
0.9532	1.2197	−0.267	1.42078	0.0016	1378	432	−32
1	1.2319	0	1.42165	0	1393	418	0

(2) Phase Equilibrium

Ref. [245]

T (K)	x_2 Poor phase	x_2 Rich phase
278.15	0.0001	0.8852
283.15	0.0001	0.8827
288.15	0.0001	0.8754
293.15	0.0002	0.8721
298.15	0.0002	0.8690
303.15	0.0002	0.8669
308.15	0.0002	0.8659
313.15	0.0002	0.8650
318.15	0.0002	0.8646
323.15	0.0002	0.8636
328.15	0.0002	0.8615

Mixture 3: C_7H_{16} (1) + OMIM PF_6 (2)
(1) Volumetric Properties, Refractive Index, Speed of Sound and Compressibility Properties

Ref. [245]

x_2	ρ (g · cm^{-3})	V^E_m (cm^3 · mol^{-1})	n_D	Δn_D	u (m · s^{-1})	k_s (T · Pa^{-1})	Δk_s (T · Pa^{-1})
T = 293.15 K							
0	0.6837	0	1.38776	0	1153	1101	0
0.9197	1.2161	−0.278	1.42315	0.0017	1396	422	−33
0.9484	1.2242	−0.072	1.42362	0.0011	1406	413	−21
0.9926	1.2374	−0.004	1.42426	0.0001	1421	400	−3
1	1.2396	0	1.42440	0	1424	398	0
T = 298.15 K							
0	0.6795	0	1.38517	0	1131	1150	0

(continued)

x_2	ρ (g · cm^{-3})	V^E_m (cm^3 · mol^{-1})	n_D	Δn_D	u (m · s^{-1})	k_s (T · Pa^{-1})	Δk_s (T · Pa^{-1})
0.9197	1.2122	−0.299	1.42173	0.0018	1381	433	−35
0.9484	1.2204	−0.088	1.42221	0.0011	1391	424	−23
0.9926	1.2335	−0.005	1.42290	0.0002	1406	410	−4
1	1.2357	0	1.42302	0	1408	408	0
$T = 303.15$ K							
0	0.6752	0	1.38258	0	1110	1202	0
0.9197	1.2084	−0.316	1.42033	0.0018	1366	443	−38
0.9484	1.2165	−0.092	1.42079	0.0012	1376	434	−25
0.9926	1.2297	−0.006	1.42152	0.0002	1391	420	−4
1	1.2319	0	1.42165	0	1393	418	0

(2) Phase Equilibrium

Ref. [245]

T (K)	x_2 Poor phase	x_2 Rich phase
278.15	0.0001	0.9124
283.15	0.0001	0.9034
288.15	0.0001	0.9035
293.15	0.0001	0.9027
298.15	0.0001	0.9014
303.15	0.0001	0.9003
308.15	0.0002	0.8997
313.15	0.0002	0.8981
318.15	0.0002	0.8975
323.15	0.0002	0.8951
328.15	0.0002	0.8937
333.15	0.0002	0.8928
338.15	0.0002	0.8919
343.15	0.0002	0.8915

Mixture 4: 2-CH_3CHCH_3OH (1) + OMIM PF_6 (2)
(1) Volumetric Properties, Refractive Index and Compressibility Properties

Ref. [245]

x_2	ρ (g · cm^{-3})	V^E_m (cm^3 · mol^{-1})	n_D	Δn_D	u (m · s^{-1})	k_s (T · Pa^{-1})	Δk_s (T · Pa^{-1})
$T = 293.15$ K							
0	0.7850	0	1.37707	0	1156	953	0
0.0004	0.7857	−0.004	1.37729	0.0002	1156	952	−1
0.0009	0.7866	−0.010	1.37736	0.0002	1157	950	−2
0.0014	0.7875	−0.018	1.37743	0.0003	1157	948	−4
0.0024	0.7891	−0.020	1.37752	0.0003	1158	945	−7
0.0030	0.7901	−0.024	1.37772	0.0005	1158	944	−7
0.3513	1.0885	−0.465	1.41129	0.0176	1293	550	−208

(continued)

x_2	ρ (g · cm^{-3})	V^E_m (cm^3 · mol^{-1})	n_D	Δn_D	u (m · s^{-1})	k_s (T · Pa^{-1})	Δk_s (T · Pa^{-1})
0.4011	1.1093	−0.467	1.41393	0.0179	1310	526	−205
0.5057	1.1451	−0.440	1.41743	0.0164	1340	486	−186
0.5991	1.1705	−0.396	1.41896	0.0135	1364	459	−161
0.6960	1.1920	−0.322	1.42004	0.0100	1379	441	−125
0.7988	1.2109	−0.230	1.42137	0.0065	1395	424	−85
0.9035	1.2271	−0.140	1.42302	0.0032	1412	409	−43
0.9931	1.2388	−0.024	1.42426	0.0002	1422	399	−3
1	1.2396	0	1.42440	0	1424	398	0
T = 298.15 K							
0	0.7810	0	1.37496	0	1139	986	0
0.0004	0.7817	−0.010	1.37519	0.0002	1140	985	−1
0.0009	0.7826	−0.017	1.37520	0.0002	1140	983	−3
0.0014	0.7835	−0.024	1.37546	0.0004	1140	981	−4
0.0024	0.7851	−0.026	1.37562	0.0005	1141	978	−7
0.0030	0.7861	−0.029	1.37559	0.0005	1141	976	−8
0.0053	0.7900	−0.047	1.37596	0.0007	1142	970	−13
0.3036	1.0617	−0.467	1.40657	0.0170	1261	592	−219
0.3513	1.0845	−0.476	1.40992	0.0181	1279	563	−220
0.4011	1.1053	−0.487	1.41265	0.0184	1296	539	−216
0.5057	1.1412	−0.467	1.41622	0.0170	1327	498	−196
0.5991	1.1665	−0.415	1.41775	0.0140	1351	469	−171
0.6960	1.1881	−0.337	1.41888	0.0105	1366	451	−133
0.7988	1.2071	−0.251	1.42042	0.0071	1382	434	−91
0.9035	1.2233	−0.159	1.42216	0.0038	1397	419	−46
0.9931	1.2350	−0.033	1.42309	0.0004	1407	409	−3
1	1.2357	0	1.42302	0	1408	408	0
T = 303.15 K							
0	0.7766	0	1.37278	0	1121	1025	0
0.0004	0.7773	−0.011	1.37302	0.0002	1121	1023	−1
0.0009	0.7782	−0.019	1.37301	0.0002	1122	1021	−3
0.0014	0.7791	−0.026	1.37322	0.0004	1122	1019	−4
0.0024	0.7807	−0.028	1.37336	0.0005	1123	1016	−7
0.0030	0.7818	−0.042	1.37353	0.0006	1123	1014	−9
0.0053	0.7857	−0.062	1.37386	0.0008	1125	1006	−16
0.0098	0.7932	−0.107	1.37475	0.0015	1128	991	−28
0.2531	1.0297	−0.474	1.40116	0.0160	1235	637	−234
0.3036	1.0575	−0.485	1.40528	0.0177	1252	604	−237
0.3513	1.0803	−0.495	1.40861	0.0187	1267	576	−235
0.4011	1.1011	−0.500	1.41146	0.0191	1283	552	−230
0.5057	1.1371	−0.482	1.41501	0.0175	1315	509	−209
0.5991	1.1626	−0.438	1.41680	0.0147	1339	480	−182
0.6960	1.1842	−0.362	1.41792	0.0111	1353	461	−141
0.7988	1.2032	−0.266	1.41950	0.0077	1368	444	−96
0.9035	1.2195	−0.175	1.42128	0.0044	1384	428	−49
0.9931	1.2312	−0.036	1.42192	0.0006	1393	419	−4
1	1.2319	0	1.42165	0	1393	418	0

(2) Phase Equilibrium

Ref. [245]

T (K)	x_2 Poor phase	x_2 Rich phase
278.15	0.0023	0.4606
283.15	0.0028	0.4229
288.15	0.0035	0.3827
293.15	0.0049	0.3393
298.15	0.0073	0.2929
303.15	0.0113	0.2357
308.15	0.0162	0.1674
313.15	0.0289	0.0985

Mixture 5: H_2O (1) + [C_8mim][PF_6] (2)
(1) Solubility

Ref. [188]

T/K	$x_1 \pm \sigma^a$	10^4 ($x_2 \pm \sigma^a$)
288.15	0.172 ± 0.003	1.13 ± 0.01
293.15	0.185 ± 0.002	1.20 ± 0.02
298.15	0.205 ± 0.004	1.27 ± 0.03
303.15	0.217 ± 0.004	1.46 ± 0.04
308.15	0.229 ± 0.006	1.59 ± 0.02
313.15	0.250 ± 0.004	1.76 ± 0.01
318.15	0.257 ± 0.004	2.03 ± 0.02

[a]Standard deviation

(2) Thermodynamic Properties

Ref. [188]

($\Delta_{sol}H_m{}^0 \pm \sigma^a$)/kJ · mol^{-1}	($\Delta_{sol}G_m{}^0 \pm \sigma^a$)/kJ · mol^{-1}	($\Delta_{sol}S_m{}^0 \pm \sigma^a$)/kJ · mol^{-1}
298.15 K		
12.6 ± 1.5	22.24 ± 0.05	−32.2 ± 5.2

[a]Standard deviation

Mixture 6: 1-C_3H_7OH (1) + OMIM PF_6 (2)
(1) Phase Equilibrium

Ref. [190]

T/K	x_2 (IL-poor phase)	x_2 (IL-rich phase)	T/K	x_2 (IL-poor phase)	x_2 (IL-rich phase)
278.15	0.0024	0.4646	298.15	0.0088	0.2910
283.15	0.0030	0.4278	303.15	0.0151	0.2335
288.15	0.0040	0.3870	305.15	0.0171	0.2026
293.15	0.0058	0.3422	307.15	0.0269	0.1566

Mixture 7: $1\text{-}C_4H_9OH$ (1) + OMIM PF_6 (2)
(1) Phase Equilibrium

Ref. [190]

T/K	x_2 (IL-poor phase)	x_2 (IL-rich phase)	T/K	x_2 (IL-poor phase)	x_2 (IL-rich phase)
278.15	0.0013	0.6184	303.15	0.0040	0.4395
283.15	0.0014	0.5903	308.15	0.0055	0.3963
288.15	0.0018	0.5549	313.15	0.0082	0.3390
293.15	0.0022	0.5202	318.15	0.0138	0.2924
298.15	0.0030	0.4820	323.15	0.0284	0.2341

Mixture 8: $1\text{-}C_5H_{11}OH$ (1) + OMIM PF_6 (2)
(1) Phase Equilibrium

Ref. [190]

T/K	x_2 (IL-poor phase)	x_2 (IL-rich phase)	T/K	x_2 (IL-poor phase)	x_2 (IL-rich phase)
278.15	0.0004	0.7551	308.15	0.0027	0.5370
283.15	0.0006	0.7158	313.15	0.0034	0.5001
288.15	0.0008	0.6787	318.15	0.0048	0.4523
293.15	0.0012	0.6384	323.15	0.0068	0.4175
298.15	0.0016	0.6068	328.15	0.0106	0.3807
303.15	0.0022	0.5730	333.15	0.0290	0.3482

Mixture 9: C_2H_5OH (1) + OMIM PF_6 (2)
(1) Volumetric Properties, Refractive Index, Speed of Sound and Compressibility Properties

Ref. [7]

x_2	ρ/ ($g \cdot cm^{-3}$)	V^E_m/ ($cm^3 \cdot mol^{-1}$)	n_D	Δn_D	u/ ($m \cdot s^{-1}$)	k_s ($T \cdot Pa^{-1}$)	Δk_s ($T \cdot Pa^{-1}$)
T = 293.15 K							
0	0.7893	0	1.36138	0	1160	941	0
0.0447	0.8739	−0.261	1.37484	0.0106	1180	822	−95
0.0982	0.9465	−0.389	1.38484	0.0173	1206	727	−161
0.1976	1.0358	−0.475	1.39750	0.0237	1257	611	−223
0.2897	1.0895	−0.461	1.40449	0.0249	1295	547	−237
0.3897	1.1304	−0.417	1.41009	0.0242	1327	502	−227
0.4990	1.1628	−0.342	1.41448	0.0216	1352	470	−200
0.5896	1.1832	−0.268	1.41703	0.0185	1369	451	−170
0.7080	1.2043	−0.175	1.41965	0.0136	1386	432	−124
0.8037	1.2179	−0.099	1.42150	0.0095	1398	420	−85
0.9040	1.2299	−0.046	1.42304	0.0047	1411	408	−42
0.9502	1.2347	−0.018	1.42375	0.0025	1416	404	−21
1	1.2396	0	1.42440	0	1424	398	0

(continued)

x_2	ρ/ (g · cm^{-3})	V^E_m/ (cm^3 · mol^{-1})	n_D	Δn_D	u/ (m · s^{-1})	k_s (T · Pa^{-1})	Δk_s (T · Pa^{-1})
T = 298.15 K							
0	0.7851	0	1.35929	0	1143	974	0
0.0447	0.8697	−0.275	1.37299	0.0108	1165	847	−102
0.0982	0.9423	−0.409	1.38315	0.0176	1193	746	−173
0.1976	1.0317	−0.494	1.39590	0.0240	1245	625	−237
0.2897	1.0853	−0.474	1.40291	0.0252	1283	560	−251
0.3897	1.1263	−0.428	1.40855	0.0244	1315	514	−240
0.4990	1.1588	−0.359	1.41289	0.0218	1340	481	−211
0.5896	1.1793	−0.282	1.41553	0.0187	1357	461	−180
0.7080	1.2004	−0.194	1.41817	0.0138	1373	442	−132
0.8037	1.2140	−0.118	1.42004	0.0095	1385	429	−90
0.9040	1.2260	−0.057	1.42165	0.0048	1397	418	−45
0.9502	1.2309	−0.029	1.42231	0.0025	1401	414	−23
1	1.2357	0	1.42302	0	1408	408	0
T = 303.15 K							
0	0.7807	0	1.35725	0	1127	1009	0
0.0447	0.8654	−0.288	1.37124	0.0111	1151	872	−110
0.0982	0.9381	−0.428	1.38286	0.0193	1180	766	−185
0.1976	1.0275	−0.512	1.39478	0.0248	1233	641	−251
0.2897	1.0811	−0.482	1.40181	0.0259	1271	573	−265
0.3897	1.1222	−0.438	1.40752	0.0252	1303	525	−253
0.4990	1.1548	−0.368	1.41174	0.0224	1328	491	−223
0.5896	1.1753	−0.294	1.41412	0.0189	1344	471	−190
0.7080	1.1965	−0.207	1.41682	0.0140	1360	452	−139
0.8037	1.2102	−0.131	1.41858	0.0096	1372	439	−95
0.9040	1.2222	−0.070	1.42026	0.0048	1383	428	−47
0.9502	1.2271	−0.040	1.42112	0.0027	1387	424	−24
1	1.2319	0	1.42165	0	1393	418	0

Mixture 10: $CH_3COOCH_2CH_3$ (1) + OMIM PF_6 (2)
(1) Volumetric Properties, Refractive Index, Speed of Sound and Compressibility Properties

Ref. [12]

x_2	ρ/ g · cm^{-3}	V_m^E/ cm^3 · mol^{-1}	n_D	Δn_D	u/ m · s^{-1}	k_s (T · Pa^{-1})	Δk_s (T · Pa^{-1})
T = 293.15 K							
0	0.9005	0	1.37241	0	1162	822	0
0.0547	0.9543	−0.729	1.38271	0.0075	1189	741	−58
0.0964	0.9878	−1.063	1.38868	0.0112	1209	692	−89
0.1996	1.0523	−1.540	1.39940	0.0166	1255	604	−134
0.2941	1.0955	−1.673	1.40596	0.0183	1288	550	−147
0.3939	1.1306	−1.641	1.41079	0.0179	1315	511	−144

(continued)

x_2	ρ/ g · cm^{-3}	V_m^E/ cm^3 · mol^{-1}	n_D	Δn_D	u/ m · s^{-1}	k_s (T · Pa^{-1})	Δk_s (T · Pa^{-1})
0.4944	1.1583	−1.482	1.41443	0.0163	1337	483	−130
0.6046	1.1827	−1.249	1.41759	0.0137	1358	458	−108
0.7033	1.2001	−0.907	1.41968	0.0107	1375	441	−83
0.8019	1.2153	−0.625	1.42153	0.0074	1391	425	−57
0.9033	1.2287	−0.336	1.42308	0.0037	1408	411	−28
0.9952	1.2390	0.001	1.42431	0.0002	1423	398	−2
1	1.2396	0	1.42440	0	1424	398	0
T = 298.15 K							
0	0.8944	0	1.36977	0	1141	859	0
0.0547	0.9486	−0.762	1.38058	0.0079	1168	773	−61
0.0964	0.9823	−1.114	1.38636	0.0115	1189	720	−95
0.1996	1.0472	−1.614	1.39757	0.0172	1236	625	−144
0.2941	1.0908	−1.754	1.40405	0.0186	1271	567	−159
0.3939	1.1261	−1.719	1.40900	0.0183	1299	526	−155
0.4944	1.1539	−1.554	1.41271	0.0166	1323	495	−141
0.6046	1.1785	−1.313	1.41599	0.0140	1344	470	−117
0.7033	1.1962	−0.983	1.41820	0.0110	1361	451	−91
0.8019	1.2113	−0.649	1.42008	0.0076	1377	435	−62
0.9033	1.2248	−0.351	1.42168	0.0038	1393	421	−31
0.9952	1.2352	−0.008	1.42294	0.0002	1408	409	−2
1	1.2357	0	1.42302	0	1408	408	0
T = 303.15 K							
0	0.8882	0	1.36712	0	1111	912	0
0.0547	0.9429	−0.805	1.37906	0.0090	1147	807	−78
0.0964	0.9768	−1.174	1.38508	0.0127	1169	750	−115
0.1996	1.0422	−1.698	1.39557	0.0176	1218	647	−167
0.2941	1.0860	−1.843	1.40226	0.0191	1253	586	−181
0.3939	1.1216	−1.807	1.40728	0.0187	1284	541	−177
0.4944	1.1496	−1.634	1.41112	0.0170	1309	507	−161
0.6046	1.1744	−1.381	1.41437	0.0143	1330	481	−133
0.7033	1.1922	−1.044	1.41663	0.0112	1348	462	−103
0.8019	1.2074	−0.694	1.41862	0.0078	1364	445	−71
0.9033	1.2209	−0.363	1.42027	0.0039	1379	431	−36
0.9952	1.2313	−0.009	1.42155	0.0002	1393	419	−2
1	1.2319	0	1.42165	0	1393	418	0

Mixture 11: 1H,1H,2H,2H-$C_8F_{13}H_4OH$ (1) + C_8mimPF_6 (2)
(1) Cloud Point

Ref. [150]

x_2	T/K	x_2	T/K	x_2	T/K	x_2	T/K
0.041	314.3	0.092	323.5	0.145	326.0	0.232	327.6
0.065	320.0	0.115	325.1	0.183	327.0	0.284	328.1

2.2.72.2 Ternary Mixtures

Mixture 1: $CH_2(CH_2)_4CH_2$ (1) + 2-$CH_3COC_2H_5$ (2) + OMIM PF_6 (3)
(1) Distribution Ratio and Selectivity

Ref. [232]

Organic-rich phase		Ionic liquid-rich phase			
x_1^I	x_1^I	x_1^{II}	x_2^{II}	β	S
$T = 298.15$ K					
0.991	0.009	0.209	0.108	11.66	55.30
0.974	0.026	0.209	0.177	6.77	31.56
0.959	0.041	0.210	0.240	5.90	26.93
0.942	0.058	0.215	0.303	5.23	22.88
0.922	0.078	0.221	0.362	4.62	19.27
0.897	0.103	0.236	0.409	3.98	15.13
0.852	0.148	0.261	0.461	3.12	10.19
0.778	0.221	0.296	0.503	2.27	5.97
0.708	0.290	0.328	0.530	1.83	3.94
0.640	0.356	0.357	0.538	1.51	2.71

Mixture 2: $HOC_6H_4C(CH_3)_2C_6H_4OH$ (1) + H_2O (2) + $[C_8mim][PF_6]$ (3)
(1) Distribution Ratio

Ref. [204]

pH	$T = 297.15 \pm 1$ K	pH	$T = 297.15 \pm 1$ K	pH	$T = 297.15 \pm 1$ K	pH	$T = 297.15 \pm 1$ K
1.86	447.1	4.95	503.6	9.17	432.0	12.01	0.45
3.26	447.0	7.03	492.4	11.61	20.8		

Mixture 3: C_6H_5OH (1) + H_2O (2) + $[C_8mim][PF_6]$ (3)
(1) Distribution Ratio

Ref. [204]

pH	$T = 297.15 \pm 1$ K	pH	$T = 297.15 \pm 1$ K	pH	$T = 297.15 \pm 1$ K
3.23	23.3	7.16	22.8	11.68	3.8
5.25	23.4	9.52	11.8		

Mixture 4: 4-$C_9H_{19}C_6H_4OH$ (1) + H_2O (2) + $[C_8mim][PF_6]$ (3)
(1) Distribution Ratio

Ref. [204]

pH	$T = 297.15 \pm 1$ K	pH	$T = 297.15 \pm 1$ K	pH	$T = 297.15 \pm 1$ K	pH	$T = 297.15 \pm 1$ K
3.23[a]	550.0	7.09	450.0	10.20	235.7	12.61	163.8
5.19	527.8	8.43	316.7	11.03	227.1		

[a] $V_W:V_{IL} = 25:1$ was used at these pHs because of the low solubility of 4-nonylphenol, 4-octylphenol and pentachlorophenol in water and their very low concentrations in aqueous phase after extraction

Mixture 5: 4-$C_8H_{17}C_6H_4OH$ (1) + H_2O (2) + [C_8mim][PF_6] (3)
(1) Distribution Ratio

Ref. [204]

pH	$T = 297.15 \pm 1$ K	pH	$T = 297.15 \pm 1$ K	pH	$T = 297.15 \pm 1$ K
3.23[a]	243.3	7.04	278.6	12.07	62.5
5.27	275.0	10.22	125.0		

[a] V_W:V_{IL} = 25:1 was used at these pHs because of the low solubility of 4-nonylphenol, 4-octylphenol and pentachlorophenol in water and their very low concentrations in aqueous phase after extraction

Mixture 6: C_6Cl_5OH (1) + H_2O (2) + [C_8mim][PF_6] (3)
(1) Distribution Ratio

Ref. [204]

pH	$T = 297.15 \pm 1$ K	pH	$T = 297.15 \pm 1$ K	pH	$T = 297.15 \pm 1$ K
3.16[a]	541.1	7.03	505.6	12.73	462.8
5.12	537.5	10.25	496.5		

[a] V_W:V_{IL} = 25:1 was used at these pHs because of the low solubility of 4-nonylphenol, 4-octylphenol and pentachlorophenol in water and their very low concentrations in aqueous phase after extraction

Mixture 7: C_6H_{14} (1) + $CH_3COOCH_2CH_3$ (2) + OMIM PF_6 (3)
(1) Distribution Ratio and Selectivity

Ref. [229]

Organic-rich phase		Ionic liquid-rich phase			
x_1^I	x_2^I	x_1^{II}	x_2^{II}	β	S
$T = 298.15$ K					
0.972	0.028	0.087	0.209	7.45	83.29
0.920	0.080	0.099	0.267	3.34	31.03
0.889	0.111	0.116	0.332	3.00	23.09
0.825	0.175	0.132	0.393	2.24	14.07
0.764	0.236	0.147	0.444	1.88	9.77
0.698	0.302	0.158	0.490	1.62	7.18
0.653	0.345	0.163	0.544	1.58	6.31
0.592	0.406	0.173	0.600	1.48	5.07
0.532	0.466	0.178	0.620	1.33	3.96
0.492	0.505	0.195	0.650	1.29	3.26
0.462	0.534	0.204	0.670	1.26	2.85
0.396	0.598	0.228	0.681	1.14	1.98
0.350	0.639	0.262	0.681	1.07	1.42

2.2.73 02039-01001: 1-Decyl-3-methylimidazolium chloride

Abbreviation:	$[C_{10}MIm]Cl$, $[C_{10}mim]Cl$
Molecular Formula:	$C_{14}H_{27}N_2Cl$
Molar Mass:	258.83
Structure:	

2.2.73.1 Binary Mixtures

Mixture 1: C_2H_5OH (1) + $[C_{10}mim]Cl$ (2)
(1) Phase Equilibrium

Ref. [246]

x_2	$T_{\alpha 1}$ or $T_{\alpha 2}$/ K	γ_2	x_2	$T_{\alpha 1}$ or $T_{\alpha 2}$/ K	γ_2	x_2	$T_{\alpha 1}$ or $T_{\alpha 2}$/ K	γ_2
0.3711	275.96(α_2)	0.59	0.5743	298.02(α_2)	1.03	0.8834	310.67(α_2)	1.11
0.4177	279.89(α_2)	0.63	0.6181	302.59(α_2)	1.15	0.9498	311.15(α_2)	1.05
0.4385	283.30(α_2)	0.70	0.6953	304.98(α_2)	1.13	1.0000	311.17(α_2)	1.00
0.4615	287.99(α_2)	0.83	0.7358	306.51(α_2)	1.13			
0.5299	293.88(α_2)	0.93	0.7878	308.37(α_2)	1.14			

T, Liquidus Curves α_1, α_2, Respectively.

Mixture 2: $1\text{-}C_4H_9OH$ (1) + $[C_{10}mim]Cl$ (2)
(1) Phase Equilibrium

Ref. [246]

x_2	$T_{\alpha 1}$ or $T_{\alpha 2}$/ K	γ_2	x_2	$T_{\alpha 1}$ or $T_{\alpha 2}$/ K	γ_2	x_2	$T_{\alpha 1}$ or $T_{\alpha 2}$/ K	γ_2
0.1704	274.53(α_2)	1.19	0.4614	293.01(α_2)	1.03	0.7024	309.14(α_2)	1.32
0.2180	278.55(α_2)	1.13	0.5121	295.39(α_2)	1.03	0.7353	309.84(α_2)	1.29
0.2569	280.56(α_2)	1.06	0.5566	298.56(α_2)	1.08	0.7945	310.36(α_2)	1.22
0.3188	283.73(α_2)	0.99	0.5823	299.76(α_2)	1.09	0.8924	310.88(α_2)	1.11
0.3252	284.17(α_2)	0.99	0.6001	301.66(α_2)	1.14	0.9459	310.94(α_2)	1.05
0.3691	287.09(α_2)	0.99	0.6235	304.01(α_2)	1.21	1.0000	311.17(α_2)	1.00
0.3918	289.60(α_2)	1.05	0.6656	307.14(α_2)	1.28			
0.4165	290.85(α_2)	1.04	0.6825	308.28(α_2)	1.31			

T, Liquidus Curves α_1, α_2, Respectively.

Mixture 3: 2-$CH_3CH(OH)CH_2CH_3$ (1) + [C_{10}mim]Cl (2)
(1) Phase Equilibrium

Ref. [246]

x_2	$T_{\alpha 1}$ or $T_{\alpha 2}$/K	γ_2	x_2	$T_{\alpha 1}$ or $T_{\alpha 2}$/K	γ_2	x_2	$T_{\alpha 1}$ or $T_{\alpha 2}$/K	γ_2
0.1597	273.85(α_2)	1.23	0.4178	299.65(α_2)	1.51	0.7662	310.57(α_2)	1.28
0.2056	278.50(α_2)	1.20	0.4831	305.81(α_2)	1.68	0.8806	311.04(α_2)	1.13
0.2740	283.74(α_2)	1.15	0.5484	309.47(α_2)	1.71	1.0000	311.17(α_2)	1.00
0.3346	290.10(α_2)	1.25	0.5718	310.20(α_2)	1.68			
0.3694	293.77(α_2)	1.33	0.6509	310.38(α_2)	1.49			

T, Liquidus Curves α_1, α_2, Respectively.

Mixture 4: $(CH_3)_3COH$ (1) + [C_{10}mim]Cl (2)
(1) Phase Equilibrium

Ref. [246]

x_2	$T_{\alpha 1}$ or $T_{\alpha 2}$/K	γ_2	x_2	$T_{\alpha 1}$ or $T_{\alpha 2}$/K	γ_2	x_2	$T_{\alpha 1}$ or $T_{\alpha 2}$/K	γ_2
0.0000	297.99(α_1)	-	0.2274	281.26(α_2)	1.23	0.5941	298.64(α_2)	1.02
0.0205	293.16(α_1)	-	0.2654	282.87(α_2)	1.14	0.6531	300.99(α_2)	1.02
0.0411	288.12(α_1)	-	0.2815	285.04(α_2)	1.19	0.7120	303.02(α_2)	1.02
0.0680	282.52(α_1)	-	0.3356	286.38(α_2)	1.06	0.7710	305.85(α_2)	1.05
0.0930	274.81(α_1)	-	0.3702	287.69(α_2)	1.02	0.8132	307.03(α_2)	1.05
0.1157	273.26(α_2)	1.65	0.4013	289.28(α_2)	1.01	0.8932	310.27(α_2)	1.08
0.1299	274.29(α_2)	1.54	0.4334	290.66(α_2)	0.99	1.0000	311.17(α_2)	1.00
0.1381	275.13(α_2)	1.38	0.4843	293.69(α_2)	1.01			
0.1894	278.94(α_2)	1.33	0.5352	296.20(α_2)	1.02			

T, Liquidus Curves α_1, α_2, Respectively.

(2) Eutectic Temperature

Ref. [246]

$T_{2,e}$/K	$x_{2,e}$
272.91	0.1008

Mixture 5: 1-$C_6H_{13}OH$ (1) + [C_{10}mim]Cl (2)
(1) Phase Equilibrium

Ref. [246]

x_2	$T_{\alpha 1}$ or $T_{\alpha 2}$/K	γ_2	x_2	$T_{\alpha 1}$ or $T_{\alpha 2}$/K	γ_2	x_2	$T_{\alpha 1}$ or $T_{\alpha 2}$/K	γ_2
0.1013	277.27(α_2)	2.29	0.4835	301.46(α_2)	1.41	0.7805	310.08(α_2)	1.23
0.1898	284.49(α_2)	1.72	0.5498	303.61(α_2)	1.35	0.8480	310.59(α_2)	1.15
0.2493	288.08(α_2)	1.54	0.6072	305.74(α_2)	1.33	0.9006	311.12(α_2)	1.11

(*continued*)

x_2	$T_{\alpha1}$ or $T_{\alpha2}$/K	γ_2	x_2	$T_{\alpha1}$ or $T_{\alpha2}$/K	γ_2	x_2	$T_{\alpha1}$ or $T_{\alpha2}$/K	γ_2
0.3022	291.60(α_2)	1.48	0.6455	307.63(α_2)	1.35	0.9451	311.13(α_2)	1.06
0.3723	294.27(α_2)	1.35	0.6914	309.16(α_2)	1.34	1.0000	311.17(α_2)	1.00
0.4321	298.59(α_2)	1.40	0.7427	309.70(α_2)	1.27			

T, Liquidus Curves α_1, α_2, Respectively.

Mixture 6: 1-$C_8H_{17}OH$ (1) + [C_{10}mim]Cl (2)

(1) Phase Equilibrium

Ref. [246]

x_2	$T_{\alpha1}$ or $T_{\alpha2}$/K	γ_2	x_2	$T_{\alpha1}$ or $T_{\alpha2}$/K	γ_2	x_2	$T_{\alpha1}$ or $T_{\alpha2}$/K	γ_2
0.1464	279.91(α_2)	1.80	0.4977	306.81(α_2)	1.70	0.8109	310.56(α_2)	1.20
0.2021	286.73(α_2)	1.79	0.5397	307.99(α_2)	1.64	0.8765	310.65(α_2)	1.12
0.2546	292.14(α_2)	1.80	0.5978	309.16(α_2)	1.55	0.9250	310.80(α_2)	1.07
0.3095	296.85(α_2)	1.81	0.6548	309.88(α_2)	1.45	0.9632	311.01(α_2)	1.03
0.3716	301.10(α_2)	1.80	0.7082	310.27(α_2)	1.36	1.0000	311.17(α_2)	1.00
0.4386	304.57(α_2)	1.76	0.7603	310.46(α_2)	1.28			

T, Liquidus Curves α_1, α_2, Respectively.

Mixture 7: 1-$C_{10}H_{21}OH$ (1) + [C_{10}mim]Cl (2)

(1) Phase Equilibrium

Ref. [246]

x_2	$T_{\alpha1}$ or $T_{\alpha2}$/K	γ_2	x_2	$T_{\alpha1}$ or $T_{\alpha2}$/K	γ_2	x_2	$T_{\alpha1}$ or $T_{\alpha2}$/K	γ_2
0.0000	280.50(α_1)	-	0.2084	290.00(α_2)	2.00	0.7113	311.14(α_2)	1.40
0.0295	277.44(α_1)	-	0.2695	295.88(α_2)	2.00	0.7785	311.15(α_2)	1.28
0.0536	276.65(α_2)	4.20	0.3266	301.42(α_2)	2.08	0.8548	311.14(α_2)	1.17
0.0759	279.13(α_2)	3.34	0.4110	306.88(α_2)	2.06	0.9109	311.17(α_2)	1.10
0.0950	281.09(α_2)	2.93	0.4990	308.76(α_2)	1.83	1.0000	311.17(α_2)	1.00
0.1231	282.49(α_2)	2.41	0.5806	311.12(α_2)	1.72			
0.1558	285.30(α_2)	2.17	0.6416	311.13(α_2)	1.56			

(2) Eutectic Temperature

Ref. [246]

$T_{2,e}$/K	$x_{2,e}$
275.44	0.0488

Mixture 8: 1-$C_{12}H_{25}OH$ (1) + [C_{10}mim]Cl (2)
(1) Phase Equilibrium

Ref. [246]

x_2	$T_{\alpha1}$ or $T_{\alpha2}$/K	γ_2	x_2	$T_{\alpha1}$ or $T_{\alpha2}$/K	γ_2	x_2	$T_{\alpha1}$ or $T_{\alpha2}$/K	γ_2
0.0000	296.84(α_1)	–	0.3030	307.43(α_2)	2.85	0.6730	311.16(α_2)	1.49
0.0188	295.71(α_1)	–	0.3647	308.90(α_2)	2.51	0.7325	311.16(α_2)	1.36
0.0574	297.31(α_2)	9.98	0.3742	309.23(α_2)	2.48	0.7917	311.16(α_2)	1.26
0.0936	299.23(α_2)	6.63	0.4284	311.01(α_2)	2.32	0.8493	311.15(α_2)	1.18
0.1348	301.33(α_2)	5.02	0.5008	311.12(α_2)	1.99	0.9015	311.16(α_2)	1.11
0.1900	303.16(α_2)	3.84	0.5515	311.15(α_2)	1.81	0.9520	311.17(α_2)	1.05
0.2529	306.08(α_2)	3.24	0.5841	311.16(α_2)	1.71	1.0000	311.17(α_2)	1.00

T, Liquidus Curves α_1, α_2, Respectively.

(2) Eutectic Temperature

Ref. [246]

$T_{2,e}$/K	$x_{2,e}$
295.57	0.0211

2.2.74 02039-01002: 1-Decyl-3-methylimidazolium bromide

Abbreviation:	[DMIm]Br, [C_{10}mim]Br
Molecular Formula:	$C_{14}H_{27}N_2Br$
Molar Mass:	303.28
Structure:	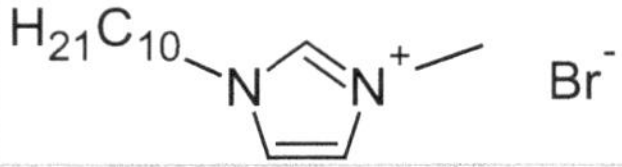

2.2.74.1 Binary Mixtures

Mixture 1: H_2O (1) + [DMIm]Br (2)
(1) Volumetric Properties

Ref. [121]

m_2/(mol · kg^{-1})	ρ/(kg · dm^{-3})	φ_2: Apparent molar volume (cm^3 · mol^{-1})	$V_2^\infty = \varphi - 1.86\sqrt{m}$ (cm^3 · mol^{-1})
0.0040	0.99720	264.8	264.7
0.0086	0.99738	264.9	264.7
0.0193	0.99779	265.0	264.7
0.0291	0.99816	265.0	264.7
	cmc		

(continued)

m_2/(mol · kg^{-1})	ρ/(kg · dm^{-3})	φ_2: Apparent molar volume (cm^3 · mol^{-1})	$V_2^\infty = \varphi - 1.86\sqrt{m}$ (cm^3 · mol^{-1})
0.0539	0.99904	266.1	
0.0671	0.99946	267.0	
0.0788	0.99983	267.6	
0.0994	1.00048	268.2	
0.2565	1.00513	270.0	

Ref. [121]

Molality of the salt (mol · kg^{-1})	X_2: Molar fraction of the salt	ρ (kg · dm^{-3})	φ_2/ (cm^3 · mol^{-1})	V'_2/ (cm^3 · mol^{-1})
1.530	0.0267	1.033747	269.8	269.3
3.640	0.0616	1.059110	269.9	269.5
pur	1	1.126172	–	269.0

Mixture 2: CH_3SOCH_3 (1) + [C_{10}mim]Br (2)
(1) Volumetric Properties

Ref. [236]

$10^3 \cdot m_2/$ (mol · g^{-1})	$\rho/$ g · cm^{-3}	$V_{\Phi,2}/$ cm^3 · mol^{-1}	$10^3 \cdot m_2/$ (mol · g^{-1})	$\rho/$ g · cm^{-3}	$V_{\Phi,2}/$ cm^3 · mol^{-1}	$10^3 \cdot m_2/$ (mol · g^{-1})	$\rho/$ g · cm^{-3}	$V_{\Phi,2}/$ cm^3 · mol^{-1}
298.15 K								
0	1.095332	–	0.01682	1.095585	264.31	0.04096	1.096000	263.15
0.005147	1.095402	265.55	0.02087	1.095652	264.05	0.04462	1.096070	262.94
0.008680	1.095455	265.06	0.02779	1.095770	263.66	0.05299	1.096220	262.72
0.01336	1.095529	264.57	0.03347	1.095870	263.38			

(2) Electrical Conductivity

Ref. [236]

$10^4 c_2$/ (mol · L^{-1})	k_2/ (μS · cm^{-1})	Λ_{m2}/ (S · cm^2 · mol^{-1})	$10^4 c_2$/ (mol · L^{-1})	k_2/ (μS · cm^{-1})	Λ_{m2}/ (S · cm^2 · mol^{-1})	$10^4 c_2$/ (mol · L^{-1})	k_2/ (μS · cm^{-1})	Λ_{m2}/ (S · cm^2 · mol^{-1})
298.15 K								
2.071	7.2672	35.08	5.770	19.959	34.59	8.974	30.743	34.26
3.447	12.013	34.85	6.934	23.891	34.46	10.43	35.611	34.13
4.551	15.802	34.72	8.066	27.695	34.33	12.34	41.931	33.97

Mixture 3: 1,2-$HOCH_2CH_2OH$ (1) + [C_{10}mim]Br (2)
(1) Volumetric Properties

Ref. [236]

$10^3 \cdot m_2$/ $(mol \cdot g^{-1})$	ρ/$g \cdot cm^{-3}$	$V_{\Phi,2}$/ $cm^3 \cdot mol^{-1}$	$10^3 \cdot m_2$/ $(mol \cdot g^{-1})$	ρ/$g \cdot cm^{-3}$	$V_{\Phi,2}$/ $cm^3 \cdot mol^{-1}$	$10^3 \cdot m_2$/ $(mol \cdot g^{-1})$	ρ/$g \cdot cm^{-3}$	$V_{\Phi,2}$/ $cm^3 \cdot mol^{-1}$
298.15 K								
0	1.109977	–	0.01114	1.110009	270.91	0.01950	1.110022	271.36
0.002798	1.109988	270.06	0.01398	1.110014	271.09	0.02259	1.110024	271.55
0.005415	1.109996	270.40	0.01720	1.110019	271.26	0.02552	1.110026	271.68
0.008253	1.110003	270.69						

(2) Electrical Conductivity

Ref. [236]

10^4c_2/ (mol · L^{-1})	k_2/ (μS · cm^{-1})	Λ_{m2}/ (S · cm^2 · mol^{-1})	10^4c_2/ (mol · L^{-1})	k_2/ (μS · cm^{-1})	Λ_{m2}/ (S · cm^2 · mol^{-1})	10^4c_2/ (mol · L^{-1})	k_2/ (μS · cm^{-1})	Λ_{m2}/ (S · cm^2 · mol^{-1})
298.15 K								
3.546	2.3639	6.67	6.739	4.4585	6.61	9.056	5.9611	6.58
4.359	2.9024	6.66	7.633	5.0409	6.60	9.847	6.4719	6.57
5.147	3.4170	6.64	8.415	5.5471	6.59	10.65	6.9876	6.56
5.865	3.8881	6.63						

Mixture 4: $HCON(CH_3)_2$ (1) + $[C_{10}mim]Br$ (2)
(1) Volumetric Properties

Ref. [236]

$10^3 \cdot m_2/$ $(mol \cdot g^{-1})$	$\rho/g \cdot cm^{-3}$	$V_{\Phi,2}/$ $cm^3 \cdot mol^{-1}$	$10^3 \cdot m_2/$ $(mol \cdot g^{-1})$	$\rho/g \cdot cm^{-3}$	$V_{\Phi,2}/$ $cm^3 \cdot mol^{-1}$	$10^3 \cdot m_2/$ $(mol \cdot g^{-1})$	$\rho/g \cdot cm^{-3}$	$V_{\Phi,2}/$ $cm^3 \cdot mol^{-1}$
298.15 K								
0	0.944305	–	0.01889	0.945493	250.33	0.03813	0.946640	251.87
0.004548	0.944601	248.02	0.02351	0.945774	250.70	0.04346	0.946942	252.43
0.009331	0.944904	248.99	0.02843	0.946065	251.30	0.04852	0.947237	252.63
0.01416	0.945206	249.58	0.03329	0.946353	251.63			

(2) Electrical Conductivity

Ref. [236]

10^4c_2/ (mol · L^{-1})	k_2/ (μS · cm^{-1})	Λ_{m2}/ (S · cm^2 · mol^{-1})	10^4c_2/ (mol · L^{-1})	k_2/ (μS · cm^{-1})	Λ_{m2}/ (S · cm^2 · mol^{-1})	10^4c_2/ (mol · L^{-1})	k_2/ (μS · cm^{-1})	Λ_{m2}/ (S · cm^2 · mol^{-1})
298.15 K								
1.732	13.502	77.94	4.585	35.017	76.36	6.636	50.097	75.49
2.484	19.233	77.42	5.275	40.130	76.07	7.275	54.729	75.23
3.119	24.037	77.08	5.939	44.991	75.75	7.761	58.221	75.02
3.859	29.603	76.72						

2.2.75 02039-03001: 1-Decyl-3-methylimidazolium bis((trifluoromethyl)sulfonyl)imide

Abbreviation:	$[C_{10}MIm][NTf_2]$, $[C_{10}mim][NTf_2]$, $C_{10}mimNTf_2$
Molecular Formula:	$C_{16}H_{27}F_6N_3O_4S_2$
Molar Mass:	503.53
Structure:	503.53 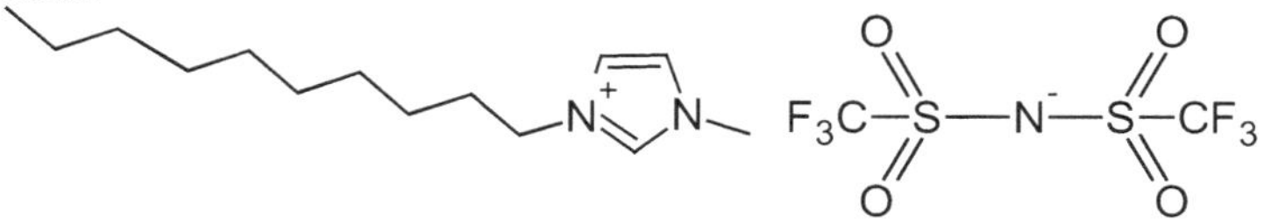

2.2.75.1 Binary Mixtures

Mixture 1: CH_3OH (1) + $[C_{10}mim][NTf_2]$ (2)
(1) Transport Properties

Ref. [247]

v_1:v_2 = 2:3	η/mPa · s			
	293.15 K	303.15 K	313.15 K	323.15 K
	8.09	7.84	7.08	6.52

Mixture 2: $[C_2mim][NTf_2]$ (1) + $[C_{10}mim][NTf_2]$ (2)
(1) Volumetric Properties

Ref. [57]
These physicochemical properties data are listed in a mixture 37 of IL 02003-03001
Mixture 3: $[C_4mim][NTf_2]$ (1) + $[C_{10}mim][NTf_2]$ (2)
(1) Volumetric Properties

Ref. [57]
These physicochemical properties data are listed in a mixture 19 of IL 02011-03001.
Mixture 4: $[C_6mim][NTf_2]$ (1) + $[C_{10}mim][NTf_2]$ (2)
(1) Volumetric Properties

Ref. [57]
These physicochemical properties data are listed in a mixture 24 of IL 02021-03001.
Mixture 5: $[C_8mim][NTf_2]$ (1) + $[C_{10}mim][NTf_2]$ (2)
(1) Volumetric Properties

Ref. [57]
These physicochemical properties data are listed in a mixture 3 of IL 02029-03001.
Mixture 6: 1H,1H,2H,2H,3H,3H -$C_9F_{13}H_6OH$ (1) + $C_{10}mimNTf_2$ (2)
(1) Cloud Point

Ref. [150]

x_2	T/K	x_2	T/K	x_2	T/K	x_2	T/K
0.093	280.8	0.273	288.3	0.062	274.3	0.451	284.9
0.118	283.3	0.350	287.8	0.078	278.1	0.501	283.0
0.141	285.5	0.406	286.1	0.291	288.2	0.573	279.3
0.174	287.1	0.048	270.3	0.329	288.0	0.648	273.0
0.222	287.9	0.072	276.6	0.388	286.6	0.703	269.3

Mixture 7: 1-C_3H_7OH (1) + C_{10}mimNTf$_2$ (2)
(1) Cloud Point

Ref. [150]

x_2	T/K	x_2	T/K	x_2	T/K	x_2	T/K
0.038	252.3	0.045	256.9	0.185	274.4	0.436	275.6
0.052	261.2	0.067	262.9	0.219	275.0	0.480	274.1
0.083	267.5	0.080	266.0	0.261	275.6	0.527	273.1
0.102	270.4	0.094	268.6	0.313	276.0	0.565	272.2
0.760	259.9	0.121	270.8	0.260	275.9	0.605	270.6
0.690	266.2	0.139	272.2	0.329	276.1	0.653	268.3
0.625	270.7	0.168	274.1	0.382	275.8	0.730	263.0

2.2.75.2 Ternary Mixtures

Mixture 1: C_6H_{14} (1) + C_6H_6 (2) + [C_{10}mim][NTf$_2$] (3)
(1) Distribution Ratio and Selectivity

Ref. [168]

Organic-rich phase			Ionic-liquid-rich phase			β	S
x_1	x_2	x_3	x_1	x_2	x_3		
298.15 K							
1.000	0.000	0.000	0.419	0.000	0.581		
0.962	0.038	0.000	0.390	0.084	0.526	2.21	5.45
0.895	0.105	0.000	0.346	0.199	0.455	1.90	4.90
0.801	0.199	0.000	0.306	0.337	0.357	1.69	4.43
0.731	0.269	0.000	0.280	0.411	0.309	1.53	3.99
0.542	0.458	0.000	0.240	0.553	0.207	1.21	2.73
0.363	0.637	0.000	0.194	0.667	0.139	1.05	1.96
0.269	0.731	0.000	0.163	0.728	0.109	1.00	1.64
0.184	0.815	0.001	0.121	0.792	0.087	0.97	1.48
0.057	0.940	0.003	0.046	0.900	0.054	0.96	1.19
0.000	0.997	0.003	0.000	0.957	0.043	0.96	

2.2.76 02042-01002: *n*-Acryloyloxypropyl-*N*-methylimidazolium bromide

Abbreviation:	$[C_6H_9O_2MIM]Br$
Molecular Formula:	$C_{10}H_{15}N_2O_2Br$
Molar Mass:	275.14
Structure:	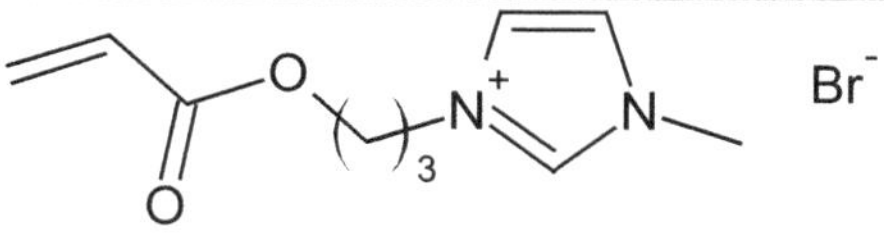

2.2.76.1 Ternary Mixtures

Mixture 1: C_6H_{14} (1) + SCH = CHCH = CH (thiophene) (2) + $[C_6H_9O_2MIM]Br$ (3)
(1) Selectivity at Infinite Dilution and Capacities at Infinite Dilution

Ref. [169]

Solvent	$S^{\infty}{}_{12}$	
$T = 298.15$ K	*n*-Hexane (1)/thiophene (2)	k_2^{∞}
$[C_6H_9O_2MIM]Br$	85.3	0.46

Mixture 2: $CH_2(CH_2)_4CH_2$ (1) + SCH = CHCH = CH (thiophene) (2) + $[C_6H_9O_2MIM]Br$ (3)
(1) Selectivity at Infinite Dilution and Capacities at Infinite Dilution

Ref. [169]

Solvent	$S^{\infty}{}_{12}$	
$T = 298.15$ K	Cyclohexane (1)/thiophene (2)	k_2^{∞}
$[C_6H_9O_2MIM]Br$	80.9	0.46

Mixture 3: C_6H_6 (1) + SCH = CHCH = CH (thiophene) (2) + $[C_6H_9O_2MIM]Br$ (3)
(1) Selectivity at Infinite Dilution and Capacities at Infinite Dilution

Ref. [169]

Solvent	$S^{\infty}{}_{12}$	
$T = 298.15$ K	Benzene (1)/thiophene (2)	k_2^{∞}
$[C_6H_9O_2MIM]Br$	3.4	0.46

2.2.77 02046-01002: 1-Dodecyl-3-methylimidazolium bromide

Abbreviation:	$[C_{12}MIm]Br$, $[C_{12}mim]Br$
Molecular Formula:	$C_{16}H_{31}N_2Br$
Molar Mass:	331.34
Structure:	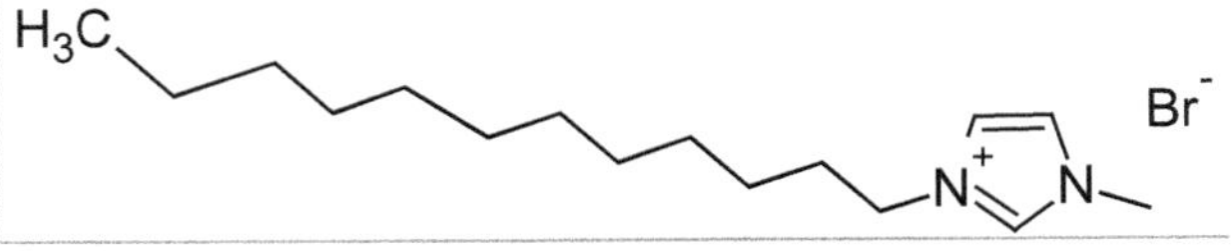

2.2.77.1 Binary Mixtures

Mixture 1: CH_3SOCH_3 (1) + $[C_{12}mim]Br$ (2)

(1) Volumetric Properties

Ref. [236]

$10^3 \cdot m_2$/ $(mol \cdot g^{-1})$	ρ/ $g \cdot cm^{-3}$	$V_{\Phi,2}$/ $cm^3 \cdot mol^{-1}$	$10^3 \cdot m_2$/ $(mol \cdot g^{-1})$	ρ/ $g \cdot cm^{-3}$	$V_{\Phi,2}$/ $cm^3 \cdot mol^{-1}$	$10^3 \cdot m_2$/ $(mol \cdot g^{-1})$	ρ/ $g \cdot cm^{-3}$	$V_{\Phi,2}$/ $cm^3 \cdot mol^{-1}$
298.15 K								
0	1.095347	–	0.01038	1.095384	299.48	0.02082	1.095428	299.20
0.002680	1.095355	299.97	0.01304	1.095395	299.38	0.02457	1.095445	299.11
0.004967	1.095363	299.77	0.01600	1.095407	299.32	0.02736	1.095458	299.05
0.007868	1.095374	299.59	0.01873	1.095419	299.24			

(2) Electrical Conductivity

Ref. [236]

$10^4 c_2$/ (mol · L^{-1})	k/ (mS · cm^{-1})	Λ_m/ (S · cm^2 · mol^{-1})	$10^4 c_2$/ (mol · L^{-1})	k/ (mS · cm^{-1})	Λ_m/ (S · cm^2 · mol^{-1})	$10^4 c_2$/ (mol · L^{-1})	k/ (mS · cm^{-1})	Λ_m/ (S · cm^2 · mol^{-1})
298.15 K								
3.026	10.415	34.42	5.343	18.199	34.06	7.464	25.234	33.81
3.758	12.885	34.29	6.078	20.646	33.97	8.059	27.189	33.74
4.555	15.563	34.16	6.771	22.943	33.89	9.021	30.353	33.65

Mixture 2: 1,2-$HOCH_2CH_2OH$ (1) + [C_{12}mim]Br (2)
(1) Volumetric Properties

Ref. [236]

$10^3 \cdot m_2$/ $(mol \cdot g^{-1})$	ρ/ $g \cdot cm^{-3}$	$V_{\Phi,2}$/ $cm^3 \cdot mol^{-1}$	$10^3 \cdot m_2$/ $(mol \cdot g^{-1})$	ρ/ $g \cdot cm^{-3}$	$V_{\Phi,2}$/ $cm^3 \cdot mol^{-1}$	$10^3 \cdot m_2$/ $(mol \cdot g^{-1})$	ρ/ $g \cdot cm^{-3}$	$V_{\Phi,2}$/ $cm^3 \cdot mol^{-1}$
298.15 K								
0	1.109947	–	0.01013	1.109847	306.52	0.02023	1.109746	306.60
0.003873	1.109909	306.46	0.01330	1.109815	306.57	0.03123	1.109635	306.68
0.007663	1.109872	306.50	0.01680	1.109780	306.58	0.03429	1.109604	306.70

(2) Electrical Conductivity

Ref. [236]

10^4c_2/ $(mol \cdot L^{-1})$	k/ $(mS \cdot cm^{-1})$	Λ_m/ $(S \cdot cm^2 \cdot mol^{-1})$	10^4c_2/ $(mol \cdot L^{-1})$	k/ $(mS \cdot cm^{-1})$	Λ_m/ $(S \cdot cm^2 \cdot mol^{-1})$	10^4c_2/ $(mol \cdot L^{-1})$	k/ $(mS \cdot cm^{-1})$	Λ_m/ $(S \cdot cm^2 \cdot mol^{-1})$
298.15 K								
3.480	2.2762	6.54	6.521	4.2308	6.49	8.687	5.6080	6.45
4.233	2.7632	6.53	7.213	4.6736	6.48	9.337	6.0180	6.44
4.995	3.2552	6.52	8.076	5.2208	6.46	10.12	6.5106	6.43
5.785	3.7608	6.50						

Mixture 3: $HCON(CH_3)_2$ (1) + [C_{12}mim]Br (2)
(1) Volumetric Properties

Ref. [236]

$10^3 \cdot m_2$/ $(mol \cdot g^{-1})$	ρ/ $g \cdot cm^{-3}$	$V_{\Phi,2}$/ $cm^3 \cdot mol^{-1}$	$10^3 \cdot m_2$/ $(mol \cdot g^{-1})$	ρ/ $g \cdot cm^{-3}$	$V_{\Phi,2}$/ $cm^3 \cdot mol^{-1}$	$10^3 \cdot m_2$/ $(mol \cdot g^{-1})$	ρ/ $g \cdot cm^{-3}$	$V_{\Phi,2}$/ $cm^3 \cdot mol^{-1}$
298.15 K								
0	0.944305	–	0.01627	0.945277	283.49	0.03337	0.946237	285.31
0.003861	0.944545	280.91	0.02032	0.945508	284.07	0.03725	0.946450	285.62
0.008055	0.944798	282.03	0.02446	0.945745	284.37	0.04099	0.946651	285.94
0.01230	0.945047	282.90	0.02703	0.945890	284.59			

(2) Electrical Conductivity

Ref. [236]

10^4c_2/ $(mol \cdot L^{-1})$	k/ $(mS \cdot cm^{-1})$	Λ_m/ $(S \cdot cm^2 \cdot mol^{-1})$	10^4c_2/ $(mol \cdot L^{-1})$	k/ $(mS \cdot cm^{-1})$	Λ_m/ $(S \cdot cm^2 \cdot mol^{-1})$	10^4c_2/ $(mol \cdot L^{-1})$	k/ $(mS \cdot cm^{-1})$	Λ_m/ $(S \cdot cm^2 \cdot mol^{-1})$
298.15 K								
1.357	10.437	76.91	3.169	24.024	75.81	4.399	33.111	75.27
1.826	13.983	76.56	3.578	27.066	75.64	4.725	35.485	75.10
2.263	17.270	76.31	4.043	30.484	75.41	5.219	39.114	74.94
2.718	20.667	76.05						

2.2.78 02046-03001: 1-Dodecyl-3-methylimidazolium bis((trifluoromethyl)sulfonyl)imide

Abbreviation:	$[C_{12}MIm][NTf_2]$, $C_{12}mimNTf_2$, $[C_{12}mim][NTf_2]$
Molecular Formula:	$C_{18}H_{31}F_6N_3O_4S_2$
Molar Mass:	531.58
Structure:	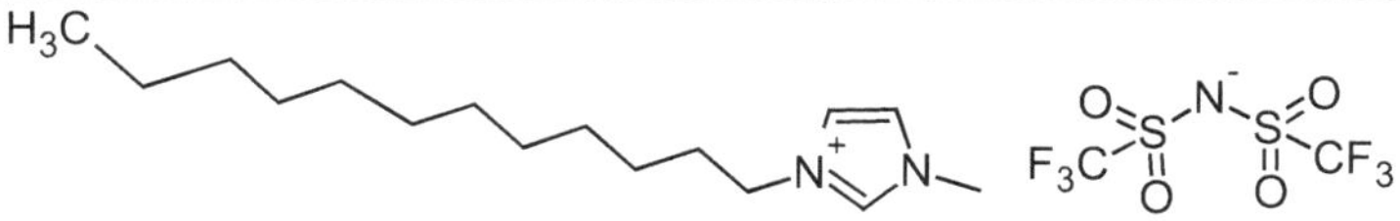

2.2.78.1 Binary Mixtures

Mixture 1: C_2H_5OH (1) + $C_{12}mimNTf_2$ (2)
(1) Cloud Point

Ref. [150]

x_2	T/K	x_2	T/K	x_2	T/K	x_2	T/K
0.273	285.8	0.393	285.2	0.533	282.0	0.060	273.7
0.136	283.2	0.199	284.8	0.739	267.3	0.623	277.2

Mixture 2: 1H,1H,2H,2H-$C_8F_{13}H_4OH$ (1) + $C_{12}mimNTf_2$ (2)
(1) Cloud Point

Ref. [150]

x_2	T/K	x_2	T/K	x_2	T/K	x_2	T/K
0.533	282.0	0.136	283.2	0.739	267.3	0.623	277.2
0.060	273.7	0.199	284.8				

2.2.78.2 Ternary Mixtures

Mixture 1: C_6H_{14} (1) + C_6H_6 (2) + $[C_{12}mim][NTf_2]$ (3)
(1) Distribution Ratio and Selectivity

Ref. [168]

Organic-rich phase			Ionic-liquid-rich phase				
x_1	x_2	x_3	x_1	x_2	x_3	β	S
298.15 K							
1.000	0.000	0.000	0.551	0.000	0.449		
0.908	0.092	0.000	0.468	0.168	0.364	1.83	3.54
0.773	0.227	0.000	0.383	0.345	0.272	1.52	3.07
0.697	0.303	0.000	0.353	0.416	0.231	1.37	2.71
0.496	0.504	0.000	0.284	0.573	0.143	1.14	1.99
0.385	0.615	0.000	0.245	0.641	0.114	1.04	1.64

(*continued*)

Organic-rich phase			Ionic-liquid-rich phase				
x_1	x_2	x_3	x_1	x_2	x_3	β	S
0.251	0.748	0.001	0.189	0.748	0.063	1.00	1.33
0.177	0.820	0.003	0.151	0.809	0.040	0.99	1.16

2.2.79 02046-04030: 1-Dodecyl-3-methylimidazolium 1,1,2,2-tetrafluoroethanesulfonate

Abbreviation:	[dmim][TFES]
Molecular Formula:	$C_{18}H_{32}N_2F_4O_3S$
Molar Mass:	432.52
Structure:	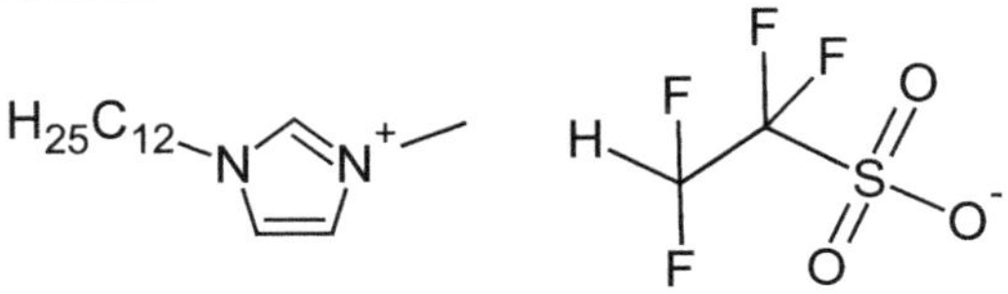

2.2.79.1 Binary Mixtures

Mixture 1: CH_2F_2 (1) + [dmim][TFES] (2)
(1) Solubility and Diffusivity

Ref. [55]

T/K	P/MPa	100 x_1	$10^{11}D$/m$^2\cdot$s^{-1}	T/K	P/MPa	100 x_1	$10^{11}D$/m$^2\cdot$s^{-1}
298.15	0.0096	0.6		298.15	0.5481	35.7	4.7
298.15	0.0995	7.4	4.7	298.15	0.7001	43.3	6.8
298.15	0.2510	17.9	5.1	298.15	0.8500	50.3	8.2
298.15	0.3997	27.3	3.0	298.15	1.0010	56.9	13

2.2.80 02059-03001: 1-Octadecyl-3-methylimidazolium bis((trifluoromethyl)sulfonyl)imide

Abbreviation:	[C_{18}MIm][NTf_2], C_{18}mimNTF$_2$
Molecular Formula:	$C_{24}H_{43}F_6N_3O_4S_2$
Molar Mass:	615.75
Structure:	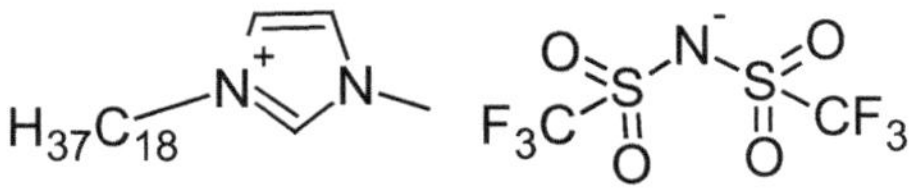

2.2.80.1 Binary Mixtures

Mixture 1: $CH_2(CH_2)_4CH_2$ (1) + $C_{18}mimNTF_2$ (2)
(1) Solubility

Ref. [248]

x_2	T/K	x_2	T/K	x_2	T/K
0.0246	304.75	0.0172	305.14	0.0145	305.44
0.0234	304.87	0.0165	305.19	0.0128	305.41
0.0201	305.09	0.0162	305.22	0.0105	305.35
0.0190	305.10	0.0158	305.15	0.0091	305.09
0.0186	305.15	0.0149	305.41	0.0078	304.90
0.0177	305.10				

Mixture 2: $CH_3CH(CH_2)_4CH_2$ (1) + $C_{18}mimNTF_2$ (2)
(1) Solubility

Ref. [248]

x_2	T/K	x_2	T/K	x_2	T/K
0.0115	301.48	0.0180	301.91	0.0208	301.90
0.0126	301.79	0.0185	301.92	0.0227	301.84
0.0143	301.87	0.0191	301.92	0.0254	301.65
0.0161	301.90				

Mixture 3: $C_{10}H_{18}$ (decalin) (1) + $C_{18}mimNTF_2$ (2)
(1) Solubility

Ref. [248]

x_2	T/K	x_2	T/K	x_2	T/K
0.0102	329.66	0.0212	330.57	0.0350	330.42
0.0118	329.94	0.0236	330.55	0.0376	330.33
0.0144	330.28	0.0268	330.53	0.0400	330.17
0.0174	330.52	0.0307	330.50	0.0429	330.01

2.2.81 02085-04017: 3-Methyl-1-vinylimidazolium methyl sulfate

Abbreviation:	[VMIM][$MeSO_4$]
Molecular Formula:	$C_7H_{12}N_2O_4S$
Molar Mass:	220.26
Structure:	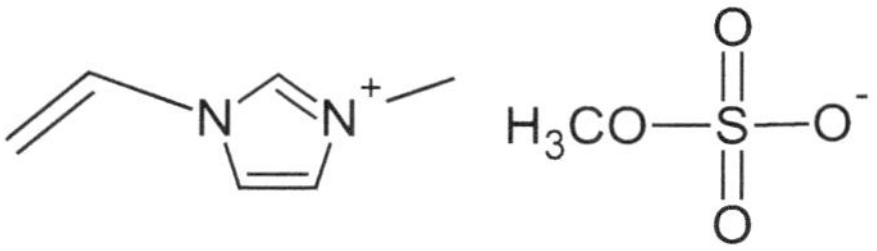

2.2.81.1 Binary Mixtures

Mixture 1: H_2O (1) + [VMIM][$MeSO_4$] (2)
(1) Activity

Ref. [249]

$T = 298$ K		$T = 338$ K	
w_2	a_1	w_2	a_1
0.1055	0.9856	0.1257	0.9806
0.1169	0.9822	0.1303	0.9807
0.1561	0.9772	0.1476	0.9767
0.1772	0.9732	0.2277	0.9638
0.1893	0.9694	0.2749	0.9533
0.2179	0.9677	0.3155	0.9446
0.2272	0.9644	0.3482	0.9364
0.2469	0.9578		
0.2478	0.9599		
0.2690	0.9555		
0.2904	0.9509		
0.2954	0.9511		
0.3388	0.9405		
0.3584	0.9286		
0.3853	0.9269		
0.4015	0.9226		

2.2.81.2 Ternary Mixtures

Mixture 1: H_2O (1) + Na_2SO_4 (2) + [VMIM][$MeSO_4$] (3)
(1) Cloud Point

Ref. [249]

$T = 298$ K		$T = 338$ K	
w_3	w_2	w_3	w_2
0.4320	0.0443	0.4253	0.0588
0.3733	0.0658	0.3755	0.0745
0.3206	0.0748	0.3103	0.1033
0.2728	0.0943	0.1960	0.1685
0.1780	0.1306	0.1627	0.1900
0.1325	0.1540	0.1156	0.2092
0.1165	0.1803	0.0951	0.2383
0.0844	0.2074	0.0794	0.2381
0.0322	0.2595	0.0327	0.2951

(2) Phase Equilibrium

Ref. [249]

Feed solution		Liquid phase	
w_3	w_2	w_3	w_2
T = 298 K (solid phase = sodium sulfate decahydrate)			
0.1311	0.2362	0.2304	0.0934
0.1656	0.1509	0.2040	0.1059
0.1880	0.1828	0.2717	0.0773
0.2682	0.1318	0.3333	0.0633
T = 338 K (solid phase = sodium sulfate)			
0.0984	0.2816	0.1125	0.2133
0.1466	0.2399	0.1583	0.1606
0.2010	0.1913	0.2048	0.1337
0.2041	0.2219	0.2246	0.1331
0.2622	0.2171	0.3215	0.0887

2.2.82 02118-02001: 1, 3-Dihexyloxymethylimidazolium tetrafluoroborate

Abbreviation:	$[(C_6Om)_2Im][BF_4]$, $[(C_6H_{13}OCH_2)_2IM][BF_4]$
Molecular Formula:	$C_{17}H_{33}N_2O_2BF_4$
Molar Mass:	384.26
Structure:	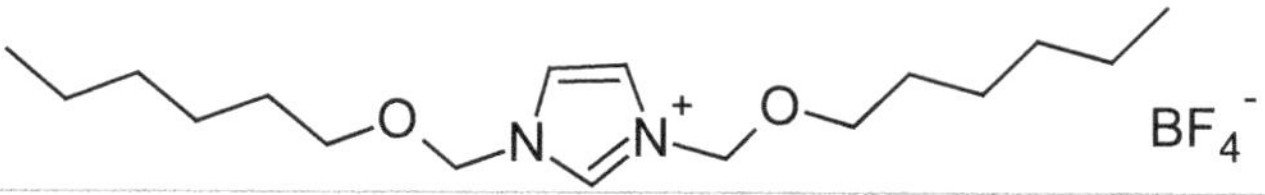

2.2.82.1 Binary Mixtures

Mixture 1: CH_3OH (1) + $[(C_6H_{13}OCH_2)_2IM][BF_4]$ (2)

(1) Volumetric Properties and Surface Tension

Ref. [8]

x_2	ρ (g · cm^{-3})	σ (mN · m^{-1})	x_2	ρ (g · cm^{-3})	σ (mN · m^{-1})
308.15 K			318.15 K		
0.0019	0.7741	21.40	0.0019	0.7678	20.86
0.0014	0.7730	21.30	0.0012	0.7670	20.77
0.0009	0.7720	21.21	0.0007	0.7669	20.78
0.0000	0.7713	21.09	0.0000	0.7664	20.55

Mixture 2: 1-C_4H_9OH (1) + [$(C_6H_{13}OCH_2)_2IM$][BF_4] (2)
(1) Volumetric Properties and Surface Tension

Ref. [8]

x_2	ρ (g · cm^{-3})	σ (mN · m^{-1})	x_2	ρ (g · cm^{-3})	σ (mN · m^{-1})
308.15 K			318.15 K		
0.0038	0.8045	23.17	0.0037	0.7930	22.37
0.0027	0.8030	23.18	0.0028	0.7921	22.42
0.0012	0.7998	23.10	0.0016	0.7905	22.39
0.0000	0.7978	22.89	0.0000	0.7897	22.30

Mixture 3: 1-$C_6H_{13}OH$ (1) + [$(C_6H_{13}OCH_2)_2IM$][BF_4] (2)
(1) Volumetric Properties and Surface Tension

Ref. [8]

x_2	ρ (g · cm^{-3})	σ (mN · m^{-1})	x_2	ρ (g · cm^{-3})	σ (mN · m^{-1})
308.15 K			318.15 K		
0.0055	0.8123	23.52	0.0061	0.8048	22.92
0.0035	0.8110	23.56	0.0048	0.8036	23.02
0.0020	0.8095	23.67	0.0034	0.8020	22.78
0.0000	0.8081	23.77	0.0000	0.7992	23.02

(2) Phase Equilibrium

Ref. [250]

x_2	T (K)	x_2	T (K)	x_2	T (K)
0.0082	280.98	0.1182	285.56	0.4494	290.27
0.0149	282.67	0.1883	286.04	0.5377	292.96
0.0254	283.84	0.2313	286.40	0.6045	295.55
0.0365	285.01	0.2885	287.18	0.7147	299.26
0.0541	285.13	0.3451	287.99	1.0000	309.46
0.0801	285.33	0.3904	288.83		

Mixture 4: 1-$C_8H_{17}OH$ (1) + [$(C_6H_{13}OCH_2)_2IM$][BF_4] (2)
(1) Phase Equilibrium

Ref. [250]

x_2	T (K)	x_2	T (K)	x_2	T (K)
0.0067	285.79	0.1638	289.11	0.4230	291.57
0.0145	287.29	0.1754	289.20	0.5442	294.63
0.0259	288.15	0.2067	289.46	0.6820	298.69
0.0425	288.37	0.2301	289.65	0.7943	303.35
0.0659	288.50	0.2719	289.87	1.0000	309.46
0.1023	288.71	0.2996	290.02		
0.1334	288.88	0.3567	290.53		

Mixture 5: 1-$C_{10}H_{21}OH$ (1) + [$(C_6H_{13}OCH_2)_2$IM][BF_4] (2)
(1) Phase Equilibrium

Ref. [250]

x_2	T (K)	x_2	T (K)	x_2	T (K)
0.0060	289.63	0.1075	294.39	0.5123	296.25
0.0097	291.90	0.1524	294.39	0.6141	298.53
0.0260	294.20	0.2229	294.39	0.7250	301.67
0.0347	294.35	0.3026	294.41	0.8445	305.71
0.0574	294.35	0.4023	294.97	1.0000	309.46

2.2.82.2 Ternary Mixtures

Mixture 1: C_6H_{14} (1) + p-$C_6H_4(CH_3)_2$ (2) + [$(C_6H_{13}OCH_2)_2$Im][BF_4] (3)
(1) Distribution Ratio and Selectivity

Ref. [251]

Ionic liquid-rich layer		Hexane-rich layer			
x_3^{II}	x_2^{II}	x_3^{I}	x_2^{I}	β	S
298.15 K					
0.44	0.00	0.00	0.00		
0.40	0.14	0.00	0.20	0.68	1.2
0.37	0.25	0.00	0.34	0.74	1.3
0.34	0.33	0.00	0.41	0.80	1.4
0.34	0.34	0.00	0.44	0.77	1.4
0.34	0.35	0.00	0.45	0.78	1.0
0.31	0.43	0.00	0.55	0.78	1.4
0.29	0.47	0.00	0.60	0.78	1.3
0.27	0.50	0.00	0.63	0.79	1.3
0.20	0.63	0.00	0.74	0.85	1.4

2.2.83 02118-03001: 1, 3-Dihexyloxymethylimidazolium bis((trifluoromethyl)sulfonyl)imide

Abbreviation:	[$(C_6Om)_2$Im][NTf_2], [$(C_6H_{13}OCH_2)_2$Im][Tf_2N]
Molecular Formula:	$C_{19}H_{33}F_6N_3O_6S_2$
Molar Mass:	577.6
Structure:	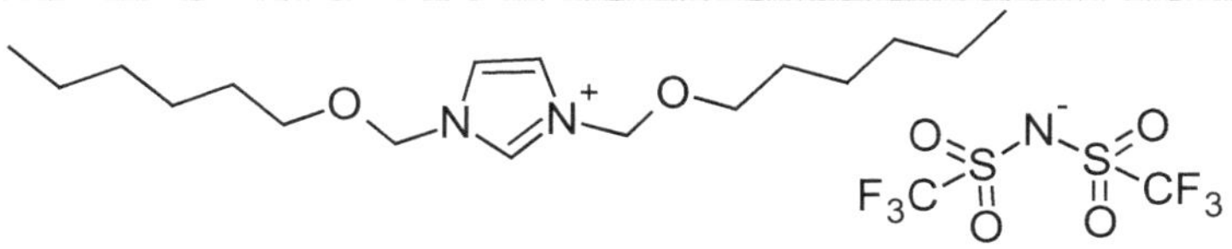

2.2.83.1 Binary Mixtures

Mixture 1: H_2O (1) + $[(C_6H_{13}OCH_2)_2Im][Tf_2N]$ (2)
(1) Phase Equilibrium

Ref. [250]

x_2	T_d (K)	T_a (K)	x_2	T_d (K)	T_a (K)
0.5333	306.16	305.08	0.3960	341.25	340.25
0.4933	316.74	315.63	0.3570	350.32	349.76
0.4374	331.05	330.05	0.3190	363.90	363.42

T_d (disappearance of the second liquid phase on heating) and temperature, T_a (appearance of the second liquid phase on cooling)

Mixture 2: 1-C_6H_{14} (1) + $[(C_6H_{13}OCH_2)_2Im][Tf_2N]$ (2)
(1) Phase Equilibrium

Ref. [250]

x_2	T_d (K)	T_a (K)	x_2	T_d (K)	T_a (K)
0.5012	300.05	297.35	0.4659	324.45	322.05
0.4885	307.75	305.35	0.4544	330.95	328.75
0.4757	316.35	314.20	0.4430	341.25	339.45

T_d (disappearance of the second liquid phase on heating) and temperature, T_a (appearance of the second liquid phase on cooling)

Mixture 3: $CH_2(CH_2)_4CH_2$ (1) + $[(C_6H_{13}OCH_2)_2Im][Tf_2N]$ (2)
(1) Phase Equilibrium

Ref. [250]

x_2	T_d (K)	T_a (K)	x_2	T_d (K)	T_a (K)
0.3704	300.65	298.75	0.3053	331.45	329.95
0.3582	306.35	304.65	0.2852	339.45	338.15
0.3440	312.85	311.25	0.2717	345.65	344.45
0.3344	317.35	315.65	0.2605	352.75	351.55
0.3212	324.25	322.75			

T_d (disappearance of the second liquid phase on heating) and temperature, T_a (appearance of the second liquid phase on cooling)

2.2.84 02125-02001: 1-Hydroxyethyl-3-methylimidazolium tetrafluoroborate

Abbreviation:	$[HEMIm][BF_4]$, $[hydemim][BF_4]$, $[C_2OHmim][BF_4]$
Molecular Formula:	$C_6H_{11}BF_4N_2O$
Molar Mass:	213.97
Structure:	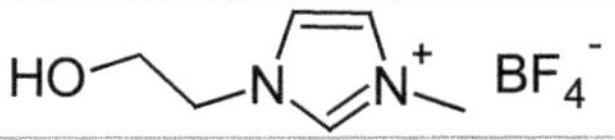

2.2.84.1 Binary Mixtures

Mixture 1: H_2O (1) + [hydemim][BF_4] (2)
(1) Vapor Pressure

Ref. [122]

X_2 = 20 mol %		X_2 = 40 mol %		X_2 = 60 mol %		X_2 = 80 mol %	
T/K	P/kPa	T/K	P/kPa	T/K	P/kPa	T/K	P/kPa
325.8	9.2	326.2	5.5	337.2	4.6	352.8	3.3
335.6	15.1	346.2	14.5	367.4	16.3	392.6	12.2
343.2	20.9	353.8	20.1	378.0	21.9	402.0	16.5
350.4	28.9	361.6	26.7	388.2	30.3	420.0	27.5
357.8	38.8	368.0	34.2	395.0	38.5	425.8	32.8
364.2	51.2	375.4	43.8	403.0	48.3	432.2	39.2
370.2	63.5	380.0	55.3	409.8	58.5	435.0	44.2
375.2	76.5	391.0	71.9	416.2	69.4	439.6	50.2
380.4	88.9	396.8	84.5	420.2	80.9	443.8	56.3
383.8	98.8	402.2	100.0	427.4	100.8	448.0	62.6
						454.0	69.7
						458.4	75.8
						462.2	81.9
						464.8	87.9

Mixture 2: CH_2Cl_2 (1) + [C_2OHmim][BF_4] (2)
(1) Cloud Point

Ref. [252]

x_2	T/K	x_2	T/K
0.1 MPa			
0.4876	298.2	0.0099	298.2

2.2.84.2 Ternary Mixtures

Mixture 1: 2-CH_3CHCH_3OH (1) + CH_2Cl_2 (2) + [C_2OHmim][BF_4] (3)
(1) Cloud Point

Ref. [252]

x_3	x_2	x_3	x_2
298.15 K at atmospheric pressure			
0.4876	0.5124	0.0410	0.8157
0.0099	0.9901	0.0316	0.8318
0.5906	0.0000	0.2455	0.3005
0.0010	0.0000	0.1596	0.3354
0.1233	0.7148	0.1048	0.3402
0.2885	0.6143	0.0119	0.1962
0.2242	0.6497	0.0211	0.2938
0.0932	0.7608	0.0304	0.3159
0.0669	0.7745		

Ref. [252]

x_3	T/K	x_3	T/K
0.1 MPa The composition of the (CH_2Cl_2 + 2-propanol) solvent mixture in mole fractions is 0.5425 of CH_2Cl_2 and 0.4575 of 2-propanol. x_3 represents mole fraction of ionic liquid in pseudo-Binary Mixtures			
0.1808	274.4	0.2941	282.0
0.2172	277.4	0.3430	282.7
0.2301	278.4	0.4601	278.9
0.0978	264.2	0.1634	273.4
0.1530	272.8	0.4214	281.5
0.2421	280.1	0.5203	273.2

2.2.85 02125-03001: 1-Hydroxyethyl-3-methylimidazolium bis((trifluoromethyl)sulfonyl)imide

Abbreviation:	[HEMIm][NTf_2], [C_2OHmim][$(CF_3SO_2)_2N$]
Molecular Formula:	$C_8H_{11}F_6N_3O_5S_2$
Molar Mass:	407.31
Structure:	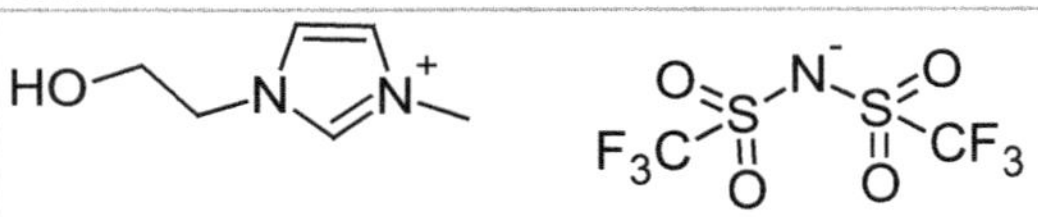

2.2.85.1 Binary Mixtures

Mixture 1: CH_2Cl_2 (1) + [C_2OHmim][$(CF_3SO_2)_2N$] (2)

(1) Cloud Point

Ref. [252]

x_2	T/K	x_2	T/K
0.1 MPa			
0.0219	298.2	0.0683	333.8
0.0272	327.1	0.0754	331.3
0.0344	331.1	0.1025	323.2
0.0404	333.2	0.1214	315.2
0.0466	334.2	0.1417	298.2
0.0595	333.9		

2.2.85.2 Ternary Mixtures

Mixture 1: 2-CH_3CHCH_3OH (1) + CH_2Cl_2 (2) + [C_2OHmim][($CF_3SO_2)_2N$] (3)
(1) Cloud Point

Ref. [252]

x_3	x_2	x_3	x_2
298.15 K at atmospheric pressure			
0.1417	0.8583	0.0700	0.9120
0.1154	0.8604	0.0984	0.8744
0.0219	0.9781	0.0499	0.9311

Mixture 2: 2-CH_3CHCH_3OH (1) + 1,2-$C_2H_4Cl_2$ (2) + [C_2OHmim][($CF_3SO_2)_2N$] (3)
(1) Cloud Point

Ref. [252]

x_3	x_2	x_3	x_2
298.15 K at atmospheric pressure			
0.1503	0.8497	0.0498	0.9032
0.0987	0.8521	0.0312	0.9389
0.0750	0.8748	0.0107	0.9832

2.2.86 02125-04003: 1-Hydroxyethyl-3-methylimidazolium triflate

Abbreviation:	[C_2OHmim][CF_3SO_3]
Molecular Formula:	$C_7H_{11}F_3N_2O_4S$
Molar Mass:	276.23
Structure:	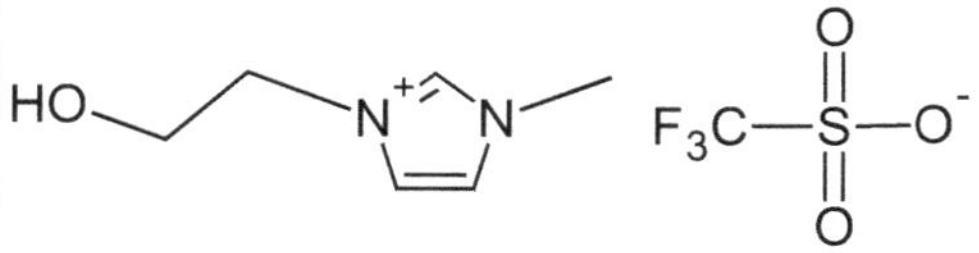

2.2.86.1 Binary Mixtures

Mixture 1: CH_2Cl_2 (1) + [C_2OHmim][CF_3SO_3] (2)
(1) Cloud Point

Ref. [252]

x_2	T/K	x_2	T/K
0.1 MPa			
0.0011	298.2	0.0012	263.2
0.4224	298.2	0.3925	331.2

2.2.86.2 Ternary Mixtures

Mixture 1: 2-CH_3CHCH_3OH (1) + CH_2Cl_2 (2) + [C_2OHmim][CF_3SO_3] (3)
(1) Cloud Point

Ref. [252]

x_3	x_2	x_3	x_2
298.15 K at atmospheric pressure			
0.0011	0.9989	0.1912	0.6898
0.0035	0.9748	0.2610	0.6550
0.0377	0.8024	0.3038	0.6349
0.0598	0.7744	0.4224	0.5776
0.1160	0.7460		

Mixture 2: 2-CH_3CHCH_3OH (1) + 1,2-$C_2H_4Cl_2$ (2) + [C_2OHmim][CF_3SO_3] (3)
(1) Cloud Point

Ref. [252]

x_3	x_2	x_3	x_2
298.15 K at atmospheric pressure			
0.4108	0.5892	0.0524	0.8002
0.3370	0.5707	0.0034	0.9966
0.2058	0.6190	0.0997	0.7267
0.0764	0.7702	0.0113	0.9690

2.2.87 02125-05001: 1-Hydroxyethyl-3-methylimidazolium hexafluorophosphate

Abbreviation:	[HEMIm][PF_6], [C_2OHmim][PF_6]
Molecular Formula:	$C_6H_{11}F_6N_2OP$
Molar Mass:	272.13
Structure:	

2.2.87.1 Binary Mixtures

Mixture 1: CH_2Cl_2 (1) + [C_2OHmim][PF_6] (2)
(1) Cloud Point

Ref. [252]

x_2	T/K	x_2	T/K
0.1 MPa			
0.0015	298.2	0.3076	298.2

2.2.87.2 Ternary Mixtures

Mixture 1: 2-CH_3CHCH_3OH (1) + CH_2Cl_2 (2) + [C_2OHmim][PF_6] (3)
(1) Cloud Point

Ref. [252]

x_3	x_2	x_3	x_2
298.15 K at atmospheric pressure			
0.0015	0.9985	0.3385	0.2994
0.3076	0.6924	0.2996	0.3194
0.6340	0.0000	0.2127	0.3658
0.0008	0.0000	0.1691	0.3793
0.2108	0.6895	0.1352	0.3857
0.1250	0.7153	0.0923	0.3810
0.0803	0.7585	0.0756	0.3797
0.0763	0.7648	0.0035	0.1104
0.0666	0.7772	0.0173	0.2955
0.0098	0.9214	0.0280	0.3402
0.0178	0.8961	0.3385	0.2994
0.0526	0.8166	0.2996	0.3194
0.4217	0.2104		

Ref. [252]

x_3	T/K	x_3	T/K
0.1 MPa The composition of the (CH_2Cl_2 + 2-propanol) solvent mixture in mole fractions is 0.5425 of CH_2Cl_2 and 0.4575 of 2-propanol. x_3 represents mole fraction of ionic liquid in pseudo-Binary Mixtures			
0.2109	277.9	0.2914	278.6
0.1449	272.6	0.3207	274.6
0.1143	269.3	0.3942	266.3
0.0644	261.8	0.1810	276.0
0.2328	279.1		

2.2.88 02199-01001: 1-Allyl-3-methyl-imidazolium chloride

Abbreviation:	[CC = CMIm]Cl, [amim]Cl
Molecular Formula:	$C_7H_{11}N_2Cl$
Molar Mass:	158.63
Structure:	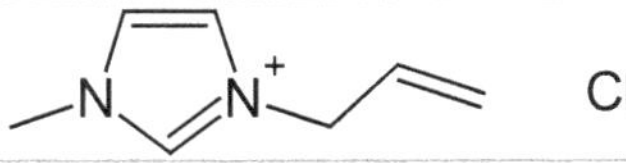

2.2.88.1 Binary Mixtures

Mixture 1: H_2O (1) + [amim]Cl (2)

(1) Electrical Conductivity and Transport Properties

Ref. [253]

T/K	x_1	10^4 *k*/ S · cm^{-1}	*η*/ mPa · s	*T*/K	x_1	10^4 *k*/ S · cm^{-1}	*η*/ mPa · s
293.15	0.0	38.2	66.0	293.15	0.65	155.5	29.1
303.15		44.0	44.7	303.15		160.9	24.2
313.15		50.1	25.3	313.15		166.2	20.9
323.15		56.2	20.6	323.15		171.6	17.6
333.15		62.4	16.2	333.15		177.0	14.3
293.15	0.30	81.7	51.3	293.15	0.70	173.0	26.1
303.15		87.4	40.7	303.15		178.4	22.0
313.15		93.0	24.1	313.15		183.7	19.5
323.15		98.5	19.6	323.15		189.0	16.4
333.15		104.1	15.0	333.15		194.2	13.0
293.15	0.45	111.4	42.7	293.15	0.75	184.2	22.6
303.15		117.1	31.4	303.15		189.4	18.7
313.15		122.5	22.9	313.15		194.8	15.4
323.15		128.0	18.2	323.15		199.9	12.2
333.15		133.4	15.1	333.15		205.1	10.1
293.15	0.59	136.2	34.4	293.15	0.80	193.5	19.3
303.15		141.9	27.5	303.15		198.6	15.6
313.15		147.3	22.7	313.15		203.5	12.7
323.15		152.7	18.3	323.15		208.6	11.1
333.15		158.0	14.9	333.15		213.5	9.2

Mixture 2: C_2H_5OH (1) + [amim]Cl (2)

(1) Electrical Conductivity and Transport Properties

Ref. [253]

T/K	x_1	10^4 *k*/ S · cm^{-1}	*η*/ mPa · s	*T*/K	x_1	10^4 *k*/ S · cm^{-1}	*η*/ mPa · s
293.15	0.00	38.2	66.0	293.15	0.35	80.0	30.5
303.15		44.0	44.7	303.15		85.5	25.1
313.15		50.1	25.3	313.15		91.0	19.4
323.15		56.2	20.6	323.15		96.9	16.0
333.15		62.4	16.2	333.15		102.4	14.4
293.15	0.10	48.5	54.6	293.15	0.45	88.4	26.4
303.15		54.3	44.1	303.15		94.3	22.1
313.15		60.4	24.6	313.15		99.9	20.1
323.15		66.0	20.0	323.15		105.6	16.7
333.15		72.2	16.0	333.15		111.0	13.1
293.15	0.20	62.7	45.7	293.15	0.50	95.2	23.2
303.15		68.2	42.2	303.15		100.6	18.1

(*continued*)

T/K	x_1	10^4 k/ S · cm^{-1}	η/ mPa · s	T/K	x_1	10^4 k/ S · cm^{-1}	η/ mPa · s
313.15		74.0	24.1	313.15		106.0	15.3
323.15		80.1	19.5	323.15		111.3	13.6
333.15		85.9	15.4	333.15		116.6	11.2
293.15	0.30	70.3	39.1	293.15	0.55	103.3	20.4
303.15		75.9	30.9	303.15		108.9	16.3
313.15		81.8	22.6	313.15		114.3	14.0
323.15		87.6	18.7	323.15		119.7	12.5
333.15		93.5	15.1	333.15		125.0	10.1

2.2.89 02236-01001: 1-Methyl-3-ethylimidazolium chloride

Abbreviation:	$[C_1EIm]Cl$
Molecular Formula:	$C_6H_{11}N_2Cl$
Molar Mass:	146.62
Structure:	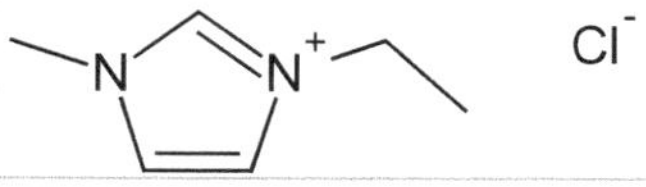

2.2.89.1 Binary Mixtures

Mixture 1: $AlCl_3$ (1) + $[C_1EIm]Cl$ (2)

(1) Volumetric Properties

Ref. [254]

x_1	$\rho/(g \cdot cm^{-3})$									
	288.15 K	298.15 K	308.15 K	318.15 K	328.15 K	338.15 K	348.15 K	358.15 K	368.15 K	378.15 K
0.5000	1.30469	1.29671	1.28873	1.28074	1.27276	1.26477	1.25679	1.24881	1.24082	1.23284
0.5500	1.32910	1.32029	1.31290	1.30390	1.29625	1.28730	1.27870	1.27073	1.26240	1.25402
0.6000	1.35796	1.34880	1.34068	1.33162	1.32376	1.31450	1.30514	1.29694	1.28793	1.27938
0.6667	1.40102	1.39064	1.38189	1.37216	1.36271	1.35356	1.34351	1.33365	1.32428	1.31491

2.2.89.2 Ternary Mixtures

Mixture 1: $AlCl_3$ (1) + $[C_1EIm]Cl$ (2) + NaCl (3)
(1) Volumetric Properties

Ref. [254]

x_1	x_2	$\rho/(g \cdot cm^{-3})$									
		288.15 K	298.15 K	308.15 K	318.15 K	328.15 K	338.15 K	348.15 K	358.15 K	368.15 K	378.15 K
0.5230	0.4270	1.34146	1.33266	1.32544	1.31651	1.30904	1.30019	1.29093	1.28354	1.27553	1.26699
0.5940	0.3960	1.36093	1.35194	1.34369	1.33417	1.32626	1.31694	1.30755	1.29904	1.29036	1.28159
0.5700	0.3800	1.37042	1.36118	1.35357	1.34404	1.33612	1.32672	1.31754	1.30931	1.30079	1.29190
0.5460	0.3640	1.39055	1.37192	1.36379	1.35463	1.34677	1.33748	1.32839	1.32005	1.31147	1.30288
0.6616	0.3284	1.40475	1.39530	1.38585	1.37640	1.36695	1.35749	1.34804	1.33859	1.32914	1.31969
0.5896	0.2997	1.42058	1.41079	1.40288	1.39305	1.38472	1.37497	1.36517	1.35669	1.34765	1.33842
0.5000	0.2500	1.46958	1.46016	1.45247	1.44320	1.43506	1.42569	1.41616	1.40745	1.39945	1.39027

Mixture 2: $AlCl_3$ (1) + $[C_1EIm]Cl$ (2) + KCl (3)
(1) Volumetric Properties

Ref. [254]

		$\rho/(g \cdot cm^{-3})$									
x_1	x_2	288.15 K	298.15 K	308.15 K	318.15 K	328.15 K	338.15 K	348.15 K	358.15 K	368.15 K	378.15 K
0.5953	0.3969	1.36071	1.35153	1.34340	1.33440	1.32571	1.31700	1.30801	1.29899	1.29036	1.28163
0.5850	0.3900	1.36647	1.35716	1.34786	1.33996	1.33025	1.32245	1.31365	1.30484	1.29604	1.28724
0.5700	0.3806	1.37415	1.34500	1.35503	1.34766	1.33857	1.33051	1.32138	1.31234	1.30396	1.29540

Mixture 3: $AlCl_3$ (1) + [C_1EIm]Cl (2) + LiCl (3)
(1) Volumetric Properties

Ref. [254]

x_1	x_2	$\rho/(g \cdot cm^{-3})$									
		288.15 K	298.15 K	308.15 K	318.15 K	328.15 K	338.15 K	348.15 K	358.15 K	368.15 K	378.15 K
0.5942	0.3961	1.36088	1.35203	1.34319	1.33434	1.32550	1.31666	1.30781	1.29897	1.29013	1.28128
0.5602	0.3734	1.36843	1.35917	1.35161	1.34236	1.33433	1.32517	1.31586	1.30748	1.29924	1.29036
0.5100	0.3400	1.38060	1.37136	1.36391	1.35470	1.34676	1.33831	1.32871	1.32021	1.31158	1.30295
0.5091	0.3394	1.38220	1.37559	1.36502	1.35640	1.34851	1.33920	1.33060	1.32201	1.31320	1.30480

Mixture 4: $AlCl_3$ (1) + [C_1EIm]Cl (2) + RbCl (3)
(1) Volumetric Properties

Ref. [254]

		ρ/(g · cm^{-3})									
x_1	x_2	288.15 K	298.15 K	308.15 K	318.15 K	328.15 K	338.15 K	348.15 K	358.15 K	368.15 K	378.15 K
0.5991	0.3994	1.36084	1.35147	1.34391	1.33440	1.32643	1.31727	1.30786	1.29915	1.29103	1.28201
0.5991	0.3960	1.36464	1.35521	1.34765	1.33807	1.33005	1.32055	1.31186	1.30281	1.29425	1.28546

Mixture 5: $AlCl_3$ (1) + $[C_1EIm]Cl$ (2) + CsCl (3)

(1) Volumetric Properties

Ref. [254]

x_1	x_2	$\rho/(g \cdot cm^{-3})$									
		288.15 K	298.15 K	308.15 K	318.15 K	328.15 K	338.15 K	348.15 K	358.15 K	368.15 K	378.15 K
0.5989	0.3993	1.36119	1.35181	1.34414	1.33484	1.32687	1.31760	1.30814	1.29971	1.29125	1.28240
0.5969	0.3979	1.36455	1.35521	1.34754	1.30807	1.32021	1.32088	1.31148	1.30315	1.29447	1.28574

Mixture 6: $AlCl_3$ (1) + [C_1EIm]Cl (2) + NaBr (3)
(1) Volumetric Properties

Ref. [254]

x_1	x_2	$\rho/(g \cdot cm^{-3})$									
		288.15 K	298.15 K	308.15 K	318.15 K	328.15 K	338.15 K	348.15 K	358.15 K	368.15 K	378.15 K
0.5579	0.3719	1.40518	1.39649	1.38668	1.37878	1.37012	1.36179	1.35116	1.34421	1.33451	1.32568
0.5190	0.3460	1.45031	1.44109	1.43277	1.42313	1.41523	1.40573	1.39621	1.38796	1.37890	1.36996

Mixture 7: $AlCl_3$ (1) + [C_1EIm]Cl (2) + NaI (3)
(1) Volumetric Properties

Ref. [254]

x_1	x_2	ρ/(g · cm^{-3})									
		288.15 K	298.15 K	308.15 K	318.15 K	328.15 K	338.15 K	348.15 K	358.15 K	368.15 K	378.15 K
0.5766	0.3862	1.38935	1.37993	1.37219	1.36258	1.35493	1.34518	1.33575	1.32769	1.31865	1.30988
0.5056	0.3374	1.51311	1.50351	1.49558	1.48527	1.47752	1.46781	1.45799	1.44932	1.44050	1.43127

2.2.90 02237-02001: 1-Hexyloxymethyl-3-methylimidazolium tetrafluoroborate

Abbreviation:	$[C_6H_{13}OCH_2MIM][BF_4]$
Molecular Formula:	$C_{11}H_{21}N_2OBF_4$
Molar Mass:	284.11
Structure:	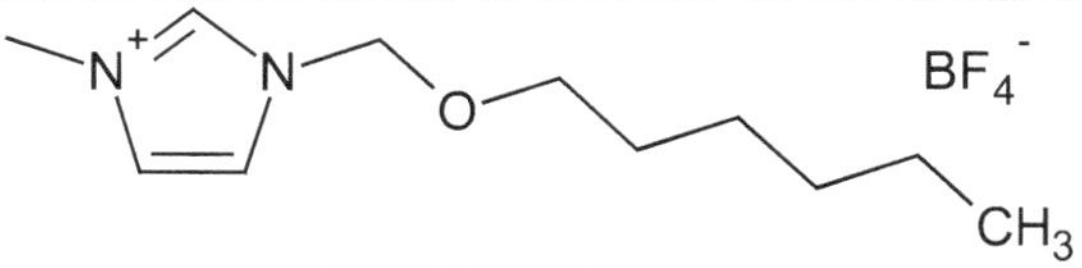

2.2.90.1 Binary Mixtures

Mixture 1: CH_3OH (1) + $[C_6H_{13}OCH_2MIM][BF_4]$ (2)
(1) Volumetric Properties and Surface Tension

Ref. [8]

x_2	ρ (g · cm^{-3})	σ (mN · m^{-1})	x_2	ρ (g · cm^{-3})	σ (mN · m^{-1})
308.15 K			318.15 K		
0.0037	0.7734	21.13	0.0044	0.7668	20.73
0.0032	0.7729	21.08	0.0036	0.7662	20.69
0.0021	0.7723	21.04	0.0026	0.7653	20.71
0.0000	0.7713	21.09	0.0000	0.7664	20.55

Mixture 2: $1\text{-}C_4H_9OH$ (1) + $[C_6H_{13}OCH_2MIM][BF_4]$ (2)
(1) Volumetric Properties and Surface Tension

Ref. [8]

x_2	ρ (g · cm^{-3})	σ (mN · m^{-1})	x_2	ρ (g · cm^{-3})	σ (mN · m^{-1})
308.15 K			318.15 K		
0.0040	0.7999	23.19	0.0064	0.7938	22.43
0.0029	0.7990	23.13	0.0046	0.7922	22.50
0.0014	0.7982	23.01	0.0040	0.7916	22.31
0.0000	0.7978[a]	22.89[a]	0.0000	0.7897	22.30

[a]Volumetric Properties is 0.79807 g · cm^{-3} and surface tension is 22.27 mN · m^{-1} at 308.15 K

(2) Phase Equilibrium

Ref. [250]

x_2 (ionic liquid rich phase)	x_2 (solvent rich phase)	T_d (K)	T_a (K)	x_2 (ionic liquid rich phase)	x_2 (solvent rich phase)	T_d (K)	T_a (K)
0.0162	0.2505[a]	288.61	288.43	0.0428[a]	0.1385	292.06	291.82
0.0178[a]	0.2419	289.23	289.11	0.0437	0.1326[a]	292.11	292.04

(continued)

x_2 (ionic liquid rich phase)	x_2 (solvent rich phase)	T_d (K)	T_a (K)	x_2 (ionic liquid rich phase)	x_2 (solvent rich phase)	T_d (K)	T_a (K)
0.0217	0.2213[a]	290.14	289.95	0.0493[a]	0.1215	292.21	292.04
0.0279	0.1933	291.07	290.94	0.0503	0.1190[a]	292.24	292.16
0.0285[a]	0.1921	291.13	290.85	0.0537[a]	0.1066	292.32	292.14
0.0325	0.1772[a]	291.47	291.36	0.0565	0.1024[a]	292.37	292.28
0.0372[a]	0.1672	291.77	291.47	0.0619[a]	0.0912	292.50	292.33
0.0374	0.1591[a]	291.81	291.73	0.0665	0.0772	292.55	292.46

T_d (disappearance of the second liquid phase on heating) and temperature, T_a (appearance of the second liquid phase on cooling)
[a]Values calculated at constant temperature for the experimental point in the ionic liquid rich phase, or solvent rich phase

Mixture 3: 1-$C_6H_{13}OH$ (1) + [$C_6H_{13}OCH_2MIM$][BF_4] (2)

(1) Volumetric Properties and Surface Tension

Ref. [8]

x_2	ρ ($g \cdot cm^{-3}$)	σ ($mN \cdot m^{-1}$)	x_2	ρ ($g \cdot cm^{-3}$)	σ ($mN \cdot m^{-1}$)
308.15 K			318.15 K		
0.0061	0.8122	23.95	0.0110	0.8064	23.44
0.0041	0.8109	23.81	0.0083	0.8045	23.44
0.0023	0.8095	23.99	0.0061	0.8031	23.19
0.0000	0.8081[a]	23.77[b]	0.0000	0.7992	23.02

[a]Volumetric Properties is 0.8080 $g \cdot cm^{-3}$ and surface tension is 24.19 $mN \cdot m^{-1}$ at 308.15 K

(2) Phase Equilibrium

Ref. [250]

x_2 (ionic liquid rich phase)	x_2 (solvent rich phase)	T_d (K)	T_a (K)	x_2 (ionic liquid rich phase)	x_2 (solvent rich phase)	T_d (K)	T_a (K)
0.0083[a]	0.5071	286.65	284.85	0.0176[a]	0.1535	308.49	308.18
0.0100[a]	0.4041	297.49	297.01	0.0178	0.1526[a]	308.51	307.84
0.0114	0.3547	300.66	300.18	0.0208[a]	0.0985	309.99	309.57
0.0133[a]	0.3011	303.71	303.13	0.0277[a]	0.0579	311.04	310.49
0.0151	0.2484	306.04	305.56	0.0282	0.0560[a]	311.10	310.54
0.0168[a]	0.1911	307.62	307.23	0.0368	0.0370	311.46	311.22

T_d (disappearance of the second liquid phase on heating) and temperature, T_a (appearance of the second liquid phase on cooling)
[a]Values calculated at constant temperature for the experimental point in the ionic liquid rich phase, or solvent rich phase

Mixture 4: 1-$C_8H_{17}OH$ (1) + [$C_6H_{13}OCH_2MIM$][BF_4] (2)
(1) Phase Equilibrium

Ref. [250]

x_2 (ionic liquid rich phase)	x_2 (solvent rich phase)	T_d (K)	T_a (K)	x_2 (ionic liquid rich phase)	x_2 (solvent rich phase)	T_d (K)	T_a (K)
0.0011[a]	0.5683	295.87	294.00	0.0345[a]	0.2905	325.25	324.83
0.0019[a]	0.5468	303.21	302.63	0.0391	0.2770[a]	325.75	325.39
0.0030[a]	0.4850	309.22	308.55	0.0410[a]	0.2700	325.95	325.47
0.0046	0.4634[a]	312.93	311.98	0.0453[a]	0.2441	326.40	326.06
0.0056[a]	0.4489	314.52	314.11	0.0477	0.2310[a]	326.61	326.33
0.0058	0.4477[a]	314.71	313.84	0.0520[a]	0.2232	326.75	326.50
0.0108	0.4305[a]	316.95	315.93	0.0561	0.2139[a]	326.92	326.63
0.0133	0.4148[a]	318.41	317.26	0.0594[a]	0.2016	327.09	326.85
0.0158[a]	0.3960	319.96	319.30	0.0637[a]	0.1846	327.35	327.20
0.0162	0.3955[a]	320.18	319.57	0.0642	0.1820[a]	327.39	327.23
0.0204[a]	0.3773	321.85	321.15	0.0731[a]	0.1676	327.59	327.41
0.0208	0.3760[a]	321.98	321.65	0.0740	0.1670[a]	327.61	327.44
0.0236[a]	0.3538	323.06	322.41	0.0823	0.1598[a]	327.73	327.51
0.0264[a]	0.3294	323.93	323.60	0.0940	0.1553[a]	327.78	327.61
0.0280[a]	0.3146	324.35	324.10	0.0968[a]	0.1542	327.80	327.53
0.0282	0.3131[a]	324.41	324.08	0.1055	0.1430	327.87	327.68
0.0339	0.2961[a]	324.98	324.76	0.1216	0.1298	327.97	327.74

T_d (disappearance of the second liquid phase on heating) and temperature, T_a (appearance of the second liquid phase on cooling)
[a]Values calculated at constant temperature for the experimental point in the ionic liquid rich phase, or solvent rich phase

Mixture 5: C_5H_{12} (1) + [$C_6H_{13}OCH_2MIM$][BF_4] (2)
(1) Phase Equilibrium

Ref. [255]

x_2 (ionic liquid rich phase)	T/K	x_2 (ionic liquid rich phase)	T/K
0.7427	309.05	0.8009	301.24
0.7627	306.34	0.8153	299.05
0.7798	303.95		

Mixture 6: C_6H_{14} (1) + [$C_6H_{13}OCH_2MIM$][BF_4] (2)
(1) Phase Equilibrium

Ref. [255]

x_2 (ionic liquid rich phase)	T/K	x_2 (ionic liquid rich phase)	T/K
0.7862	341.32	0.8080	333.46
0.7955	339.05	0.8233	322.98
0.8050	334.68	0.8443	307.65

Mixture 7: C_7H_{16} (1) + [$C_6H_{13}OCH_2MIM$][BF_4] (2)
(1) Phase Equilibrium

Ref. [255]

x_2 (ionic liquid rich phase)	T/K	x_2 (ionic liquid rich phase)	T/K
0.8387	366.55	0.8735	330.35
0.8425	362.45	0.8834	319.85
0.8500	354.05	0.8945	306.75
0.8609	343.85		

Mixture 8: C_8H_{18} (1) + [$C_6H_{13}OCH_2MIM$][BF_4] (2)
(1) Phase Equilibrium

Ref. [255]

x_2 (ionic liquid rich phase)	T/K	x_2 (ionic liquid rich phase)	T/K
0.8508	396.95	0.8968	360.85
0.8634	388.95	0.9171	338.45
0.8759	377.35	0.9319	308.95
0.8933	363.95		

Mixture 9: $CH_2(CH_2)_4CH_2$ (1) + [$C_6H_{13}OCH_2MIM$][BF_4] (2)
(1) Phase Equilibrium

Ref. [255]

x_2 (ionic liquid rich phase)	T/K	x_2 (ionic liquid rich phase)	T/K
0.6609	353.20	0.6914	333.95
0.6788	341.84	0.7073	317.99
0.6908	334.56	0.7187	300.38

Mixture 10: C_6H_6 (1) + $[C_6H_{13}OCH_2MIM][BF_4]$ (2)
(1) Phase Equilibrium

Ref. [255]

x_2 (ionic liquid rich phase)	x^*_2 (hydrocarbon rich phase)	T/K	x_2 (ionic liquid rich phase)	x^*_2 (hydrocarbon rich phase)	T/K
0.1870[a]	0.0023	351.25	0.1934	0.0013	312.65
0.1874	0.0022[a]	348.00	0.1951[a]	0.0012	305.35
0.1895[a]	0.0017	332.85	0.1975	0.0010[a]	297.00
0.1898	0.0017[a]	332.21	0.1991[a]	0.0010	292.65
0.1916	0.0014	320.42			

[a]Values calculated at constant temperature for the experimental point in the ionic liquid rich phase, or hydrocarbon rich phase

Mixture 11: $C_6H_5CH_3$ (1) + $[C_6H_{13}OCH_2MIM][BF_4]$ (2)
(1) Phase Equilibrium

Ref. [255]

x_2 (ionic liquid rich phase)	T/K	x_2 (ionic liquid rich phase)	T/K
0.3137	382.95	0.3403	321.65
0.3191	372.25	0.3436	311.65
0.3260	357.95	0.3530	288.15
0.3338	339.55		

Mixture 12: $C_6H_5C_2H_5$ (1) + $[C_6H_{13}OCH_2MIM][BF_4]$ (2)
(1) Phase Equilibrium

Ref. [255]

x_2 (ionic liquid rich phase)	T/K	x_2 (ionic liquid rich phase)	T/K
0.4348	408.35	0.4756	348.35
0.4432	402.55	0.4847	327.85
0.4538	389.45	0.4937	306.85
0.4651	371.95		

Mixture 13: *o*-$C_6H_4(CH_3)_2$ (1) + $[C_6H_{13}OCH_2MIM][BF_4]$ (2)
(1) Phase Equilibrium

Ref. [255]

x_2 (ionic liquid rich phase)	T/K	x_2 (ionic liquid rich phase)	T/K
0.3785	416.05	0.4499	342.85
0.3919	403.55	0.4577	330.75
0.4049	389.95	0.4631	322.85

(continued)

x_2 (ionic liquid rich phase)	T/K	x_2 (ionic liquid rich phase)	T/K
0.4189	378.45	0.4804	304.95
0.4340	364.35	0.4895	294.75
0.4431	351.95		

Mixture 14: m-$C_6H_4(CH_3)_2$ (1) + [$C_6H_{13}OCH_2MIM$][BF_4] (2)
(1) Phase Equilibrium

Ref. [255]

x_2 (ionic liquid rich phase)	T/K	x_2 (ionic liquid rich phase)	T/K
0.4496	408.35	0.4890	365.05
0.4692	390.45	0.4902	363.05
0.4802	377.95	0.5032	341.45
0.4874	367.25	0.5148	310.85

Mixture 15: p-$C_6H_4(CH_3)_2$ (1) + [$C_6H_{13}OCH_2MIM$][BF_4] (2)
(1) Phase Equilibrium

Ref. [255]

x_2 (ionic liquid rich phase)	T/K	x_2 (ionic liquid rich phase)	T/K
0.4335	410.95	0.4817	366.55
0.4471	402.05	0.4949	345.65
0.4618	390.45	0.4970	340.95
0.4786	369.75	0.5084	298.55

2.2.91 02237-03001: 1-Hexyloxymethyl-3-methylimidazolium bis((trifluoromethyl)sulfonyl)imide

Abbreviation:	[$C_6H_{13}OCH_2MIM$][Tf_2N]
Molecular Formula:	$C_{13}H_{21}F_6N_3O_5S_2$
Molar Mass:	477.45
Structure:	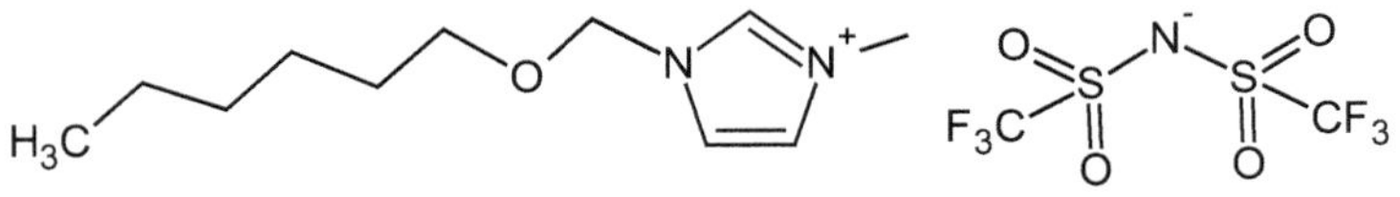

2.2.91.1 Binary Mixtures

Mixture 1: H_2O (1) + [$C_6H_{13}OCH_2MIM$][Tf_2N] (2)
(1) Phase Equilibrium

Ref. [250]

x_2 (ionic liquid rich phase)	x_2 (solvent rich phase)	T_d (K)	T_a (K)	x_2 (ionic liquid rich phase)	x_2 (solvent rich phase)	T_d (K)	T_a (K)
0.00010	0.6100[a]	307.20	305.05	0.00016[a]	0.4186	342.25	341.35
0.00011[a]	0.5817	312.35	310.65	0.00018	0.3958[a]	347.65	345.33
0.00013[a]	0.5224	324.65	323.15	0.00018[a]	0.3942	348.05	347.05
0.00013[a]	0.4961	328.35	326.95	0.00021[a]	0.3589	357.25	356.35
0.00014[a]	0.4650	334.75	333.75	0.00024	0.3465[a]	362.45	360.55
0.00015	0.4492[a]	336.55	334.95	0.00028[a]	0.3289	369.95	369.05

T_d (disappearance of the second liquid phase on heating) and temperature, T_a (appearance of the second liquid phase on cooling)
[a]Values calculated at constant temperature for the experimental point in the ionic liquid rich phase, or solvent rich phase

Mixture 2: 1-$C_6H_{13}OH$ (1) + [$C_6H_{13}OCH_2MIM$][Tf_2N] (2)
(1) Phase Equilibrium

Ref. [250]

x_2 (ionic liquid rich phase)	x_2 (solvent rich phase)	T_d (K)	T_a (K)	x_2 (ionic liquid rich phase)	x_2 (solvent rich phase)	T_d (K)	T_a (K)
0.0194[a]	0.3930	277.87	277.63	0.0568[a]	0.1979	284.98	284.81
0.0198	0.3900[a]	278.00	277.30	0.0572	0.1920[a]	285.08	284.81
0.0242[a]	0.3507	279.72	279.52	0.0654[a]	0.1643	285.47	285.31
0.0265	0.3359[a]	280.50	279.88	0.0709	0.1469[a]	285.72	285.49
0.0342[a]	0.2984	282.05	281.85	0.0717[a]	0.1445	285.73	285.57
0.0370	0.2866[a]	282.49	281.93	0.0767[a]	0.1221	285.83	285.67
0.0445[a]	0.2599	283.52	283.31	0.0813	0.1109	285.94	285.82
0.0485	0.2403[a]	284.04	283.66	0.0974	0.0974	285.97	285.88
0.0519[a]	0.2241	284.49	284.30				

T_d (disappearance of the second liquid phase on heating) and temperature, T_a (appearance of the second liquid phase on cooling)
[a]Values calculated at constant temperature for the experimental point in the ionic liquid rich phase, or solvent rich phase

Mixture 3: 1-$C_8H_{17}OH$ (1) + [$C_6H_{13}OCH_2MIM$][Tf_2N] (2)
(1) Phase Equilibrium

Ref. [250]

x_2 (ionic liquid rich phase)	x_2 (solvent rich phase)	T_d (K)	T_a (K)	x_2 (ionic liquid rich phase)	x_2 (solvent rich phase)	T_d (K)	T_a (K)
0.0145	0.7150[a]	297.61	297.10	0.0467[a]	0.2806	311.48	311.15
0.0173[a]	0.6181	301.12	300.45	0.0522	0.2550[a]	312.05	311.85
0.0194	0.5610[a]	303.24	302.95	0.0537[a]	0.2493	312.15	311.95
0.0230[a]	0.5225	304.94	304.46	0.0633	0.2153	312.87	312.55
0.0280[a]	0.4550	306.90	306.48	0.0660	0.1980[a]	313.05	312.83
0.0311	0.4080[a]	308.15	307.92	0.0730[a]	0.1735	313.30	313.05
0.0322[a]	0.3950	308.48	308.10	0.0810[a]	0.1386	313.50	313.26
0.0385[a]	0.3305	310.05	309.66	0.0858	0.1106	313.58	313.39
0.0415	0.3070[a]	310.80	310.61	0.0919	0.0919	313.65	313.52

Td (disappearance of the second liquid phase on heating) and temperature, Ta (appearance of the second liquid phase on cooling)

[a]Values calculated at constant temperature for the experimental point in the ionic liquid rich phase, or solvent rich phase

Mixture 4: C_6H_{14} (1) + [$C_6H_{13}OCH_2MIM$][Tf_2N] (2)
(1) Phase Equilibrium

Ref. [255]

x_2 (ionic liquid rich phase)	T/K	x_2 (ionic liquid rich phase)	T/K
0.6823	341.45	0.7505	312.95
0.7095	328.49	0.7605	308.28
0.7299	320.65	0.7722	303.10
0.7357	318.45	0.7819	299.95

Mixture 5: C_7H_{16} (1) + [$C_6H_{13}OCH_2MIM$][Tf_2N] (2)
(1) Phase Equilibrium

Ref. [255]

x_2 (ionic liquid rich phase)	T/K	x_2 (ionic liquid rich phase)	T/K
0.7368	368.95	0.8002	322.05
0.7520	357.95	0.8096	315.10
0.7647	347.95	0.8185	309.38
0.7749	340.85	0.8274	303.25
0.7912	328.27		

Mixture 6: $CH_2(CH_2)_4CH_2$ (1) + $[C_6H_{13}OCH_2MIM][Tf_2N]$ (2)
(1) Phase Equilibrium

Ref. [255]

x_2 (ionic liquid rich phase)	T/K	x_2 (ionic liquid rich phase)	T/K
0.5105	352.65	0.5928	319.35
0.5308	344.65	0.6152	310.35
0.5498	336.25	0.6279	305.75
0.5669	329.05		

Mixture 7: C_6H_6 (1) + $[C_6H_{13}OCH_2MIM][Tf_2N]$ (2)
(1) Phase Equilibrium

Ref. [255]

x_2 (ionic liquid rich phase)	x^*_2 (hydrocarbon rich phase)	T/K	x_2 (ionic liquid rich phase)	x^*_2 (hydrocarbon rich phase)	T/K
0.0201[a]	0.0189	351.70	0.0604[a]	0.0056	328.65
0.0253	0.0165[a]	351.55	0.0637	0.0049[a]	324.05
0.0269[a]	0.0149	351.37	0.0663[a]	0.0044	319.75
0.0314	0.0130[a]	350.59	0.0693	0.0038[a]	315.10
0.0386[a]	0.0105	348.70	0.0730[a]	0.0029	308.93
0.0390	0.0104[a]	348.61	0.0734	0.0030[a]	308.35
0.0437[a]	0.0086	345.77	0.0766	0.0024[a]	303.29
0.0460	0.0082[a]	344.25	0.0775[a]	0.0024	302.25
0.0525[a]	0.0070	338.35	0.0800	0.0019[a]	298.45
0.0564	0.0063[a]	333.58			

[a]Values calculated at constant temperature for the experimental point in the ionic liquid rich phase, or hydrocarbon rich phase

Mixture 8: $C_6H_5CH_3$ (1) + $[C_6H_{13}OCH_2MIM][Tf_2N]$ (2)
(1) Phase Equilibrium

Ref. [255]

x_2 (ionic liquid rich phase)	x^*_2 (hydrocarbon rich phase)	T/K	x_2 (ionic liquid rich phase)	x^*_2 (hydrocarbon rich phase)	T/K
0.0116[a]	0.0835	383.42	0.0048[a]	0.1141	340.23
0.0105	0.0885[a]	379.50	0.0044	0.1177	334.10
0.0097	0.0912[a]	376.40	0.0041[a]	0.1193	331.30
0.0086	0.0959[a]	369.80	0.0035[a]	0.1243	323.15
0.0997	0.0078[a]	364.33	0.1266	0.0031[a]	319.56
0.1017	0.0074[a]	362.14	0.1301[a]	0.0026	313.10

(continued)

x_2 (ionic liquid rich phase)	x^*_2 (hydrocarbon rich phase)	T/K	x_2 (ionic liquid rich phase)	x^*_2 (hydrocarbon rich phase)	T/K
0.1056	0.0067[a]	355.10	0.1312	0.0025[a]	311.90
0.1089[a]	0.0059	349.10	0.1349	0.0019[a]	304.56
0.1116	0.0052[a]	344.25	0.1374[a]	0.0016	298.13

[a]Values calculated at constant temperature for the experimental point in the ionic liquid rich phase, or hydrocarbon rich phase

Mixture 9: $C_6H_5C_2H_5$ (1) + [$C_6H_{13}OCH_2MIM$][Tf_2N] (2)
(1) Phase Equilibrium

Ref. [255]

x_2 (ionic liquid rich phase)	x^*_2 (hydrocarbon rich phase)	T/K	x_2 (ionic liquid rich phase)	x^*_2 (hydrocarbon rich phase)	T/K
0.1136	0.0095[a]	407.35	0.1699[a]	0.0039	353.35
0.1165[a]	0.0093	405.45	0.1749	0.0034[a]	345.35
0.1232	0.0087[a]	400.45	0.1790[a]	0.0029	337.45
0.1336[a]	0.0075	391.65	0.1802	0.0028[a]	334.25
0.1409	0.0067[a]	384.55	0.1853	0.0023[a]	324.25
0.1509	0.0058[a]	375.45	0.1877[a]	0.0021	319.65
0.1538[a]	0.0056	373.25	0.1897	0.0018[a]	312.95
0.1594	0.0050[a]	367.95	0.1943	0.0014[a]	302.85
0.1682	0.0042	356.75	0.1950[a]	0.0013	299.95

[a]Values calculated at constant temperature for the experimental point in the ionic liquid rich phase, or hydrocarbon rich phase

Mixture 10: o-$C_6H_4(CH_3)_2$ (1) + [$C_6H_{13}OCH_2MIM$][Tf_2N] (2)
(1) Phase Equilibrium

Ref. [255]

x_2 (ionic liquid rich phase)	x^*_2 (hydrocarbon rich phase)	T/K	x_2 (ionic liquid rich phase)	x^*_2 (hydrocarbon rich phase)	T/K
0.1435	0.0099[a]	416.89	0.1826	0.0043[a]	357.12
0.1485[a]	0.0092	412.90	0.1886	0.0036[a]	340.80
0.1543	0.0089[a]	409.60	0.1906[a]	0.0034	336.90
0.1576	0.0083[a]	405.40	0.1930	0.0032[a]	329.76
0.1628	0.0078[a]	400.15	0.1941	0.0031	325.14
0.1658[a]	0.0072	395.70	0.1970	0.0026[a]	316.98
0.1690	0.0067[a]	390.00	0.1989	0.0025[a]	310.40
0.1735[a]	0.0055	378.17	0.2006	0.0019[a]	304.38
0.1799	0.0047[a]	363.90	0.2018[a]	0.0018	298.58

[a]Values calculated at constant temperature for the experimental point in the ionic liquid rich phase, or hydrocarbon rich phase

Mixture 11: m-$C_6H_4(CH_3)_2$ (1) + [$C_6H_{13}OCH_2MIM$][Tf_2N] (2)
(1) Phase Equilibrium

Ref. [255]

x_2 (ionic liquid rich phase)	x^*_2 (hydrocarbon rich phase)	T/K	x_2 (ionic liquid rich phase)	x^*_2 (hydrocarbon rich phase)	T/K
0.1500[a]	0.0094	411.15	0.1886	0.0043[a]	357.55
0.1542	0.0088[a]	409.95	0.1886[a]	0.0043	357.95
0.1637[a]	0.0075	402.45	0.1895[a]	0.0042	356.65
0.1660	0.0072[a]	398.25	0.1923	0.0038[a]	346.95
0.1681[a]	0.0068	395.75	0.1939[a]	0.0035	340.65
0.1731	0.0062[a]	388.35	0.1958	0.0032[a]	333.85
0.1755[a]	0.0060	385.65	0.1972	0.0029[a]	323.45
0.1788	0.0056[a]	379.95	0.1985[a]	0.0026	316.15
0.1840[a]	0.0051	370.25	0.1990	0.0023[a]	308.95
0.1841	0.0050	368.95	0.2007[a]	0.0021	301.95

[a]Values calculated at constant temperature for the experimental point in the ionic liquid rich phase, or hydrocarbon rich phase

Mixture 12: p-$C_6H_4(CH_3)_2$ (1) + [$C_6H_{13}OCH_2MIM$][Tf_2N] (2)
(1) Phase Equilibrium

Ref. [255]

x_2 (ionic liquid rich phase)	x^*_2 (hydrocarbon rich phase)	T/K	x_2 (ionic liquid rich phase)	x^*_2 (hydrocarbon rich phase)	T/K
0.1613[a]	0.0071	411.55	0.1891[a]	0.0033	349.55
0.1620	0.0070[a]	410.45	0.1913	0.0030[a]	342.95
0.1686[a]	0.0060	397.15	0.1921[a]	0.0030	340.65
0.1693	0.0059[a]	396.05	0.1950	0.0025[a]	329.95
0.1744	0.0051[a]	384.65	0.1971[a]	0.0022	322.35
0.1793	0.0045[a]	373.55	0.1987	0.0020[a]	315.55
0.1809[a]	0.0043	369.95	0.2022[a]	0.0014	295.90
0.1837	0.0040[a]	364.05	0.2023	0.0014[a]	295.25
0.1874	0.0035[a]	354.05			

[a]Values calculated at constant temperature for the experimental point in the ionic liquid rich phase, or hydrocarbon rich phase

2.2.92 02238-02001: 1-Isobutenyl-3-methylimidazolium tetrafluoroborate

Abbreviation:	[*i*BeMIM][BF_4]
Molecular Formula:	$C_8H_{13}N_2BF_4$
Molar Mass:	224.01
Structure:	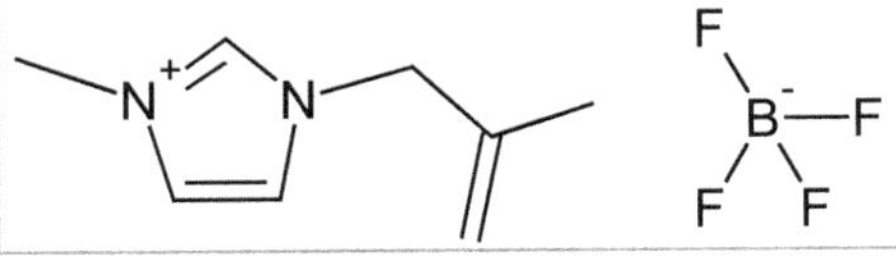

2.2.92.1 Binary Mixtures

Mixture 1: 1-C_6H_{12} (1) + [*i*BeMIM][BF_4]-$AgBF_4$ (2)

(1) Phase Equilibrium

Ref. [199]

283.15 K		293.15 K		303.15 K	
x_1	*P* (kPa)	x_1	*P* (kPa)	x_1	*P* (kPa)
0.003833	1.826	0.003801	2.322	0.003761	3.150
0.008064	3.741	0.007998	4.735	0.007910	6.477
0.01113	5.046	0.01104	6.426	0.01094	8.759
0.01526	6.792	0.01514	8.695	0.01500	11.70
0.01871	8.229	0.01857	10.49	0.01842	14.04
0.02368	10.10	0.02349	13.02	0.02329	17.42
0.02494	10.49	0.02476	13.62	0.02458	18.24
0.02981	11.15	0.02951	15.83	0.02929	21.15
0.03488	11.80	0.03451	17.23	0.03418	23.87
0.03793	11.84	0.03751	17.94	0.03726	24.95
0.04451	11.91	0.04404	18.70	0.04380	25.92
0.04700	12.05	0.04657	18.70	0.04633	26.49
0.05566	12.10	0.05522	18.72	0.05495	27.25
0.05818	12.34	0.05779	18.84	0.05754	27.67
0.06469	12.37	0.06429	19.04	0.06406	28.60
0.07100	12.49	0.07059	19.53	0.07037	29.34
0.1035	12.53	0.1021	19.73	0.1011	29.90

(2) Solubility

Ref. [199]

T (K)	283.15	293.15	303.15
x_1	0.02616	0.03168	0.03473

Mixture 2: *trans*-3-C_6H_{12} (1) + [*i*BeMIM][BF_4]-$AgBF_4$ (2)
(1) Phase Equilibrium

Ref. [199]

283.15 K		293.15 K		303.15 K	
x_1	P (kPa)	x_1	P (kPa)	x_1	P (kPa)
0.004017	3.182	0.004000	3.838	0.003953	4.025
0.007839	5.664	0.007791	7.134	0.007720	7.286
0.01157	7.716	0.01148	9.973	0.01139	10.23
0.01612	9.655	0.01599	12.84	0.01586	13.32
0.01992	9.638	0.01974	14.31	0.01948	15.63
0.02367	9.678	0.02347	15.15	0.02310	17.67
0.02756	9.696	0.02727	15.97	0.02694	18.56
0.03212	9.992	0.03175	16.10	0.03149	18.87
0.03880	9.956	0.03852	15.86	0.03821	18.67
0.06188	10.42	0.06163	16.12	0.06126	19.91

(2) Solubility

Ref. [199]

T (K)	283.15	293.15	303.15
x_1	0.01569	0.01986	0.02246

2.2.92.2 Ternary Mixtures

Mixture 1: 1-C_6H_{12} (1) + *trans*-3-C_6H_{12} (2) + [*i*BeMIM][BF_4]-$AgBF_4$ (3)
(1) Phase Equilibrium

Ref. [199]

		y'_2		
x_1	x_2	283.15 K	293.15 K	303.15 K
0.001860	0.01500	0.08858	0.09427	0.1012
0.002780	0.01402	0.1323	0.1386	0.1476
0.005580	0.01125	0.2693	0.2749	0.2921
0.008370	0.008439	0.4145	0.4188	0.4438
0.01119	0.005639	0.5739	0.5762	0.6015
0.01393	0.002808	0.7642	0.7651	0.7800
0.01488	0.001875	0.8285	0.8309	0.8369
0.01581	0.0009377	0.9063	0.9082	0.9085

2.2.93 02239-03009: 1-Hydroxy-1-propyl-3-methylimidazolium nitrate

Abbreviation:	[HOPMIm][NO_3]
Molecular Formula:	$C_7H_{13}N_3O_4$
Molar Mass:	203.19
Structure:	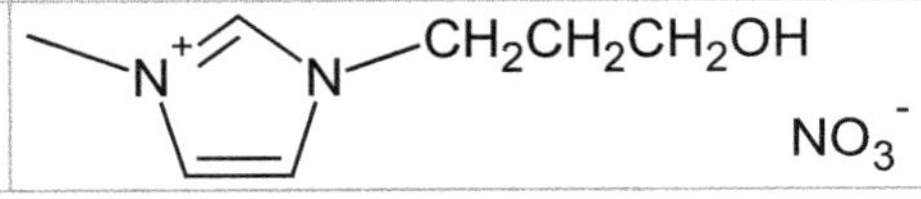

2.2.93.1 Binary Mixtures

Mixture 1: H_2O (1) + [HOPMIm][NO_3] (2)

(1) Volumetric Properties

Ref. [170]

	ρ_2 (g/cm^3)				
	x_1				
T (K)	0.00 %	10.08 %	14.62 %	21.21 %	29.31 %
292	1.2734	1.2703	1.2683	1.2648	1.2600
302	1.2668	1.2637	1.2616	1.2581	1.2533
312	1.2604	1.2572	1.2553	1.2516	1.2469
322	1.2544	1.2511	1.2490	1.2455	1.2408
332	1.2486	1.2450	1.2428	1.2394	1.2348
342	1.2427	1.2391	1.2369	1.2334	1.2287

Mixture 2: CO_2 (1) + [HOPMIm][NO_3] (2)

(1) Phase Equilibrium

Ref. [170]

P (MPa)	T (K)	P (MPa)	T (K)	P (MPa)	T (K)	P (MPa)	T (K)	P (MPa)	T (K)	P (MPa)	T (K)
x_1 = 5.28 %		x_1 = 10.04 %		x_1 = 14.33 %		x_1 = 20.74 %		x_1 = 24.84 %		x_1 = 30.81 %	
0.851	303.36	1.784	303.93	2.595	303.51	3.921	303.63	5.545	303.39	5.976	293.47
1.021	313.36	2.140	314.59	3.165	313.43	4.884	313.37	6.975	313.11	8.101	298.57
1.206	323.24	2.564	325.73	3.815	323.56	6.110	323.43	7.915	318.42	9.925	300.86
1.416	333.60	3.184	338.60	4.485	333.27	6.702	328.20	8.790	323.10		
1.650	343.33	4.114	355.95	5.280	342.82	7.415	333.43	10.015	328.42		
1.920	353.34	4.609	363.55	6.205	353.38	9.076	344.17	11.305	333.15		
				7.205	363.33	10.647	353.42				
						11.621	358.63				

P: absolute pressure; $x_{CO2} = n_{CO2}/(n_{CO2} + n_{IL})$; n: number of moles

2.2.93.2 Ternary Mixtures

Mixture 1: CO_2 (1) + H_2O (2) + [HOPMIm][NO_3] (3)

(1) Phase Equilibrium

Ref. [170]

P (MPa)	T (K)	P (MPa)	T (K)	P (MPa)	T (K)	P (MPa)	T (K)
X_2 = 0.77 %, X_1 = 5.43 %, x_2 = 0.73 %, x_1 = 5.39 %		X_2 = 4.11 %, X_1 = 9.90 %, x_2 = 3.72 %, x_1 = 9.53 %		X_2 = 5.54 %, X_1 = 14.71 %, x_2 = 4.76 %, x_1 = 14.01 %		X_2 = 3.99 %, X_1 = 19.32 %, x_2 = 3.25 %, x_1 = 18.69 %	
0.842	303.50	1.786	303.37	2.626	303.50	3.685	303.21
1.023	313.46	2.131	313.48	3.241	313.52	4.616	313.13
1.233	323.50	2.331	318.45	3.633	318.34	5.151	318.17
1.423	333.55	2.531	323.46	3.946	323.45	5.681	323.17
1.668	343.55	2.951	333.49	4.731	333.51	6.981	333.20
1.950	353.54	3.421	343.54	5.632	343.35	8.361	343.13
		3.926	353.52	6.633	353.43		
X_2 = 13.37 %, X_1 = 6.02 %, x_2 = 12.66 %, x_1 = 5.26 %		X_2 = 12.91 %, X_1 = 11.47 %, x_2 = 11.47 %, x_1 = 10.14 %		X_2 = 15.10 %, X_1 = 17.74 %, x_2 = 12.76 %, x_1 = 15.47 %		X_2 = 11.15 %, X_1 = 21.30 %, x_2 = 8.99 %, x_1 = 19.38 %	
0.953	303.18	2.152	303.42	3.516	303.49	4.557	303.40
1.159	313.31	2.542	313.43	4.206	313.27	5.697	313.33
1.388	323.08	3.027	323.44	5.206	323.40	6.406	318.40
1.526	328.24	3.552	333.37	6.151	333.38	7.052	323.49
1.644	333.34	4.187	343.43	7.371	343.37	8.676	333.50
1.926	343.11	4.811	352.97	8.496	352.92	9.536	338.43
2.250	352.85	5.536	363.34	10.072	362.89		
X_2 = 34.11 %, X_1 = 7.33 %, x_2 = 32.42 %, x_1 = 4.95 %		X_2 = 39.80 %, X_1 = 13.79 %, x_2 = 36.30 %, x_1 = 8.78 %		X_2 = 34.84 %, X_1 = 18.28 %, x_2 = 30.41 %, x_1 = 12.72 %		X_2 = 33.96 %, X_1 = 22.17 %, x_2 = 28.59 %, x_1 = 15.83 %	
1.237	303.28	2.476	303.15	3.596	303.34	5.153	303.23
1.481	313.28	2.986	313.35	4.382	313.30	6.293	313.18
1.655	318.38	3.280	318.20	4.914	318.20	7.878	323.29
1.768	323.39	3.571	323.16	5.346	323.24	9.606	333.22
2.071	333.37	3.900	328.29	6.427	333.26	11.338	341.35
2.403	343.34	4.307	333.32	7.552	342.92		
2.776	353.35	4.924	343.24				
		5.774	353.22				

P: absolute pressure; $x_i = n_i/(n_{CO2} + n_{H2O} + n_{IL})$; $X_i = n_i/(n_i + n_{IL})$; i = CO_2, H_2O; n = number of moles

2.3 Mixtures Containing Tri-alkyl Imidazolium Ionic Liquids

2.3.1 03007-03001: 1, 2-Dimethyl-3-propylimidazolium bis((trifluoromethyl)sulfonyl)imide

Abbreviation:	[DMPIM][TFSI], [dmpim][BMeI]
Molecular Formula:	$C_{10}H_{15}F_6N_3O_4S_2$
Molar Mass:	419.37
Structure:	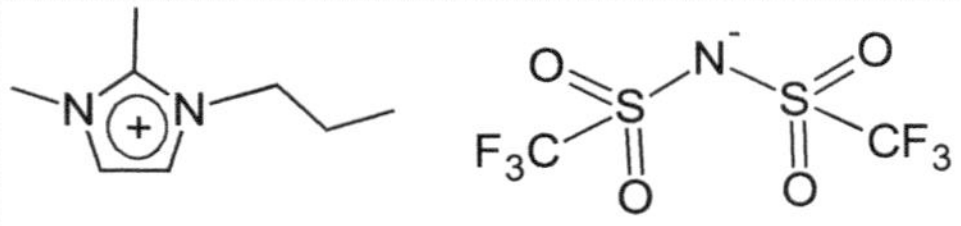

2.3.1.1 Binary Mixtures

Mixture 1: CH_2F_2 (1) + [dmpim][BMeI] (2)
(1) Solubility and Diffusivity

Ref. [55]

T/K	P/MPa	100 x_1	$10^{11}D$/ $m^2 \cdot s^{-1}$	T/K	P/MPa	100 x_1	$10^{11}D$/ $m^2 \cdot s^{-1}$
298.15	0.0099	0.8	6.3	298.15	0.5501	41.5	15
298.15	0.0995	9.0	6.6	298.15	0.7017	50.1	18
298.15	0.2495	21.3	8.9	298.15	0.8513	57.8	22
298.15	0.3976	31.9	12	298.15	1.0011	65.1	23

2.3.2 03007-07001: 1, 2-Dimethyl-3-propylimidazolium tri (trifluoromethylsulfonyl)methide

Abbreviation:	[DMPIM][Me], [dmpim][TMeM]
Molecular Formula:	$C_{12}H_{15}F_9N_2O_6S_3$
Molar Mass:	550.44
Structure:	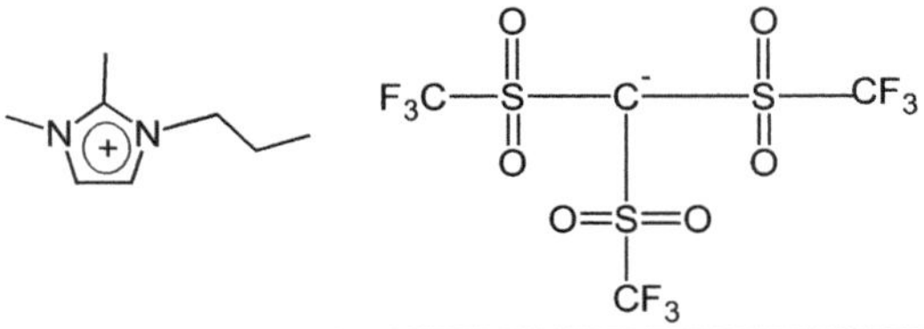

2.3.2.1 Binary Mixtures

Mixture 1: CH_2F_2 (1) + [dmpim][TMeM] (2)

(1) Solubility and Diffusivity

Ref. [55]

T/K	P/MPa	100 x_1	$10^{11}D$/m$^2 \cdot$ s^{-1}	T/K	P/MPa	100 x_1	$10^{11}D$/m$^2 \cdot$ s^{-1}
283.15	0.0101	1.2		323.15	0.1000	5.7	6.7
283.15	0.0996	13.6	1.4	323.15	0.2494	14.1	7.9
283.15	0.2497	31.2	2.5	323.15	0.3995	21.5	9.6
283.15	0.3997	46.0	5.0	323.15	0.5495	28.1	12
283.15	0.5495	58.6	8.9	323.15	0.6997	34.1	13
283.15	0.6997	70.0	14	323.15	0.8494	39.7	15
283.15	0.8496	80.5	16	323.15	0.9995	44.8	17
298.05	0.0100	1.0	2.7	348.05	0.0094	0.0	
298.05	0.1000	9.6	2.5	348.15	0.1002	3.1	
298.05	0.2496	22.6		348.05	0.2504	8.8	12
298.05	0.3996	33.7	5.5	348.05	0.3996	13.9	13
298.05	0.5493	43.5	7.9	348.05	0.5494	18.6	16
298.05	0.6995	52.0	11	348.05	0.7005	23.1	
298.05	0.8495	59.8	13	348.05	0.8495	27.2	
298.05	1.0000	66.7	18	348.05	1.0005	31.2	21
323.15	0.0099	0.4	6.4				

2.3.3 03009-02001: 1-Propyl-2,3-dimethylimidazolium tetrafluoroborate

Abbreviation:	[PDMIM][BF_4]
Molecular Formula:	$C_8H_{15}N_2BF_4$
Molar Mass:	226.02
Structure:	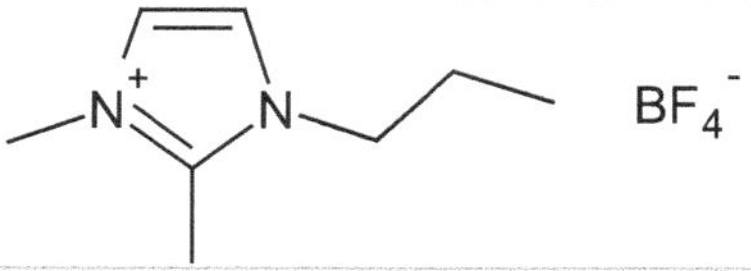

2.3.3.1 Binary Mixtures

Mixture 1: H_2O (1) + [PDMIM][BF_4] (2)

(1) Volumetric Properties and Transport Properties

Ref. [256]

x_1	298.15 K	303.15 K	308.15 K	313.15 K	318.15 K	323.15 K	333.15 K	343.15 K
ρ/g · cm^{-3}								
0.0000	1.2251	1.2218	1.2188	1.2174	1.2162	1.2154	1.2134	1.2119
0.0939	1.2219	1.2179	1.2148	1.2132	1.2119	1.2110	1.2089	1.2071
0.1932	1.2180	1.2140	1.2108	1.2093	1.2079	1.2068	1.2044	1.2026
0.2833	1.2136	1.2097	1.2065	1.2048	1.2032	1.2021	1.1997	1.1978
0.3944	1.2074	1.2037	1.2002	1.1982	1.1968	1.1956	1.1932	1.1912
0.4987	1.1998	1.1960	1.1926	1.1906	1.1890	1.1878	1.1853	1.1831
0.5995	1.1896	1.1861	1.1825	1.1805	1.1789	1.1776	1.1747	1.1723
0.6985	1.1757	1.1717	1.1683	1.1665	1.1650	1.1637	1.1607	1.1579
0.8009	1.1535	1.1499	1.1469	1.1447	1.1431	1.1417	1.1384	1.1343
0.8988	1.1130	1.1098	1.1074	1.1053	1.1033	1.1016	1.0978	1.0942
1.0000	0.9970	0.9957	0.9940	0.9922	0.9902	0.9881	0.9832	0.9778
η/mPa · s								
0.0000	377.28	271.90	201.99	152.20	117.44	92.29	60.45	40.74
0.0939	224.97	167.61	127.89	99.34	78.56	63.30	42.52	29.96
0.1932	134.95	103.29	81.12	64.54	52.07	42.80	34.62	21.84
0.2833	85.25	66.81	53.23	43.20	35.71	29.65	21.37	16.02
0.3944	50.58	40.44	32.95	27.18	22.79	19.32	14.38	11.04
0.4987	31.29	25.50	21.07	17.63	14.97	12.87	9.78	7.66
0.5995	19.71	16.21	13.58	11.52	9.88	8.58	6.65	5.30
0.6985	12.03	10.04	8.53	7.31	6.34	5.55	4.37	3.53
0.8009	6.69	5.67	4.90	4.23	3.71	3.29	2.62	2.15
0.8988	3.23	2.78	2.42	2.14	1.89	1.69	1.39	1.16
1.0000	0.89	0.80	0.72	0.65	0.60	0.55	0.47	0.40
V^E/cm^3 · mol^{-1}								
0.0939	0.13	0.22	0.24	0.27	0.28	0.29	0.31	0.34
0.1932	0.24	0.33	0.36	0.37	0.40	0.43	0.47	0.50
0.2833	0.33	0.41	0.44	0.48	0.53	0.55	0.59	0.61
0.3944	0.39	0.45	0.50	0.57	0.58	0.61	0.63	0.66
0.4987	0.42	0.49	0.53	0.59	0.61	0.62	0.65	0.68
0.5995	0.45	0.48	0.54	0.58	0.61	0.62	0.66	0.69
0.6985	0.42	0.48	0.52	0.54	0.55	0.57	0.59	0.62
0.8009	0.31	0.36	0.38	0.41	0.41	0.42	0.44	0.50
0.8988	0.19	0.21	0.22	0.23	0.24	0.25	0.26	0.27
$\Delta\eta$/mPa · s								
0.0939	−117.0	−78.8	−55.2	−38.6	−27.9	−20.4	−12.3	−7.0
0.1932	−169.6	−116.2	−82.0	−58.4	−42.8	−31.8	−17.2	−11.1
0.2833	−185.4	−128.3	−91.7	−66.1	−48.6	−36.7	−22.1	−13.3
0.3944	−178.3	−124.5	−89.7	−65.2	−48.6	−36.8	−22.4	−13.8

(*continued*)

x_1	298.15 K	303.15 K	308.15 K	313.15 K	318.15 K	323.15 K	333.15 K	343.15 K
0.4987	−158.3	−111.2	−80.5	−59.0	−44.2	−33.7	−20.8	−13.0
0.5995	−131.9	−93.2	−67.8	−49.8	−37.5	−28.7	−17.8	−11.3
0.6985	−102.3	−72.5	−52.9	−39.0	−29.5	−22.7	−14.2	−9.0
0.8009	−69.1	−49.1	−35.9	−26.6	−20.2	−15.5	−9.8	−6.3
0.8988	−35.8	−25.5	−18.7	−13.8	−10.5	−8.1	−5.1	−3.3

2.3.4 03010-02001: 1-Butyl-2,3-dimethylimidazolium tetrafluoroborate

Abbreviation:	[BDMIM][BF_4], bdmimBF_4
Molecular Formula:	$C_9H_{17}N_2BF_4$
Molar Mass:	240.05
Structure:	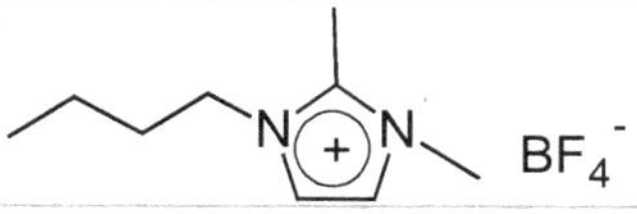

2.3.4.1 Binary Mixtures

Mixture 1: H_2O (1) + bdmimBF_4 (2)

(1) Volumetric Properties, Transport Properties and Refractive Index

Ref. [138]

x_2	ρ/ $(g \cdot cm^{-3})$	η (average value)/ (mPa · s)	Standard deviation	$\Delta\eta$/ (mPa · s)	x_2	n_D	Δn_D	R_m/ $(cm^3 \cdot mol^{-1})$
T = 298.15 K								
0.0000	0.997043	0.89	0.002	0.00	0	1.333	0.0000	3.716
0.0085	1.012564	1.06	0.009	−3.70	0.0397	1.362	0.0250	5.649
0.0397	1.053332	1.83	0.005	−17.12	0.1068	1.387	0.0433	8.927
0.1068	1.100329	3.96	0.008	−45.52	0.1997	1.404	0.0510	13.469
0.1997	1.132177	7.34	0.005	−84.40	0.2577	1.411	0.0522	16.325
0.2577	1.144223	10.07	0.003	−108.05	0.3222	1.417	0.0518	19.514
0.3222	1.154043	12.62	0.022	−134.85	0.4983	1.423	0.0402	27.979
0.7058	1.182144	94.63	0.083	−227.34	0.7058	1.429	0.0254	38.104
0.7325	1.183223	104.97	0.077	−229.15	0.7325	1.43	0.0238	39.441
0.8659	1.187778	214.82	0.324	−179.99	0.8659	1.432	0.0124	45.917
1.0000	1.1912	455.81	0.482	0.00	1.000	1.433	0.0000	52.374

Ref. [139]

x_2	m/(mol · kg^{-1})	C/(mol · L^{-1})	ρ/(kg · dm^{-3})	$\varphi_{V,2}$/(cm^3 · mol^{-1})
298 K				
0.000339	0.0188	0.0187	0.997743	203.22
0.000762	0.0423	0.0418	0.998600	203.43
0.001181	0.0657	0.0646	0.999440	203.55
0.001432	0.0796	0.0781	0.999935	203.62
0.001477	0.0821	0.0805	1.000024	203.63
0.001903	0.1059	0.1033	1.000856	203.75

Ref. [139]

m_1/(mol · g^{-1})	ρ/(kg · dm^{-3})	$\varphi_{V,1}$/(cm^3 · mol^{-1})
298 K		
0.2116	1.190599	18.98
0.4529	1.189231	19.08
0.7871	1.187243	19.22
0.9494	1.186326	19.23
1.1738	1.185125	19.20

Ref. [139]

x_2	C/(mol · L^{-1})	ρ/(kg · dm^{-3})	V^E_{mol}/(cm^3 · mol^{-1})
298 K			
0	0	0.997043	0.000
0.0071	0.3656	1.010238	0.022
0.0200	0.9167	1.029861	0.067
0.0289	1.2328	1.041283	0.094
0.0686	2.2242	1.077836	0.193
0.2076	3.6734	1.134407	0.361
0.2204	3.7431	1.136955	0.386
0.2979	4.0735	1.150711	0.419
0.4286	4.4128	1.165249	0.434
0.4983	4.5352	1.170897	0.396
0.6317	4.7049	1.178812	0.312
0.7598	4.8192	1.184289	0.207
0.8480	4.8801	1.187227	0.132
1	4.9623	1.1912	0.000

Ref. [139]

m_u/ (mol · g^{-1})	ρ/ (kg · dm^{-3})	$\varphi_{1,x2}$/ (cm^3 · mol^{-1})	m_u/ (mol · g^{-1})	ρ/ (kg · dm^{-3})	$\varphi_{1,x2}$/ (cm^3 · mol^{-1})	m_u/ (mol · g^{-1})	ρ/ (kg · dm^{-3})	$\varphi_{1,x2}$/ (cm^3 · mol^{-1})
298 K								
	x_2 = 0.0189			x_2 = 0.0368			x_2 = 0.0757	
0.3515	1.028268	18.07	0.1536	1.050346	18.07	0.1868	1.081493	18.11
0.7068	1.028057	18.08	0.2433	1.050255	18.07	0.3747	1.081183	18.09
0.9237	1.027929	18.08	0.6065	1.049887	18.08	0.5824	1.080855	18.07
1.2612	1.027730	18.08	0.8312	1.049661	18.08	0.9357	1.080328	18.03
1.3288	1.027690	18.08	1.0007	1.049490	18.08	1.0900	1.080047	18.06
	x_2 = 0.1197			x_2 = 0.1525			x_2 = 0.5807	
0.2495	1.106891	18.10	0.1876	1.118998	18.14	0.2605	1.173711	18.90
0.3582	1.106651	18.09	0.3475	1.118581	18.16	0.3544	1.173280	18.85
0.4264	1.106537	18.02	0.5386	1.118072	18.20	0.4408	1.172874	18.84
0.4431	1.106471	18.08	0.7732	1.117454	18.21	0.6036	1.172152	18.78
0.9270	1.105420	18.08	0.9830	1.116880	18.24	0.9345	1.170648	18.76
1.0486	1.105159	18.08						

Ref. [139]

x_2	C/(mol · L^{-1})	V'_1/(cm^3 · mol^{-1})	V'_2/(cm^3 · mol^{-1})
298 K			
0.0189	0.8746	18.07	204.37
0.0368	1.4766	18.07	204.28
0.0757	2.3520	18.11	204.05
0.1197	2.9730	18.08	203.39
0.1525	3.2912	18.12	203.14
0.5807	4.6432	18.93	201.70
1.000	4.9623	19.03	201.52

2.3.5 03010-05001: 1-Butyl-2,3-dimethylimidazolium hexafluorophosphate

Abbreviation:	[BDMIM][PF_6], [C_4C_1mim][PF_6]
Molecular Formula:	$C_9H_{17}N_2PF_6$
Molar Mass:	298.21
Structure:	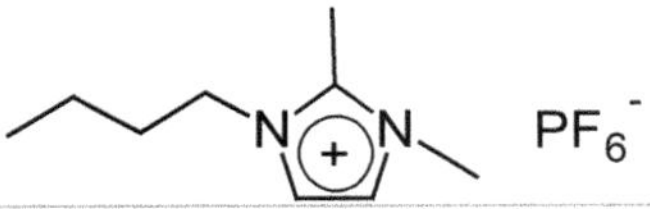

2.3.5.1 Binary Mixtures

Mixture 1: H_2O (1) + [C_4C_1mim][PF_6] (2)

(1) Solubility

Ref. [188]

T/K	$x_1 \pm \sigma^a$	10^4 ($x_2 \pm \sigma$)
288.15		7.18 ± 0.01
293.15		7.59 ± 0.02
298.15		8.17 ± 0.04
303.15	0.219 ± 0.003	8.74 ± 0.06
308.15	0.237 ± 0.007	9.97 ± 0.04
313.15	0.252 ± 0.006	10.84 ± 0.03
318.15	0.263 ± 0.004	11.51 ± 0.08

[a]Standard deviation

(2) Thermodynamic Properties

Ref. [188]

$(\Delta_{sol}H_m{}^0 \pm \sigma^a)$/kJ · mol^{-1}	$(\Delta_{sol}G_m{}^0 \pm \sigma^a)$/kJ · mol^{-1}	$(\Delta_{sol}S_m{}^0 \pm \sigma^a)$/kJ · mol^{-1}
298.15 K		
11.9 ± 1.5	17.63 ± 0.01	−19.2 ± 5.1

[a]Standard deviation

2.3.6 03017-01002: 1-Decyl-2,3-dimethylimidazolium bromide

Abbreviation:	[ddmim]Br
Molecular Formula:	$C_{15}H_{29}N_2Br$
Molar Mass:	317.31
Structure:	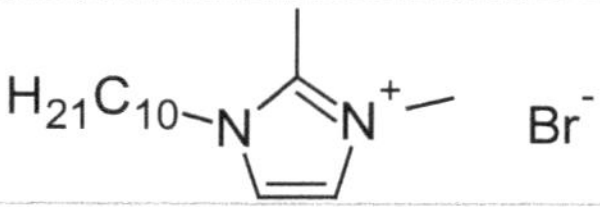

2.3.6.1 Ternary Mixtures

Mixture 1: H_2O (1) + EA (ethanolamine) (2) + [ddmim]Br (3)

(1) Surface Tension, Molar Volume and Gordon Parameter

Ref. [257]

wt$_2$% ($w_1 + w_2 = 1$)	σ (mN · m^{-1})	V (cm^3 · mol^{-1})	G (Jm^{-3})
298.15 K			
0	72.4	18.06	2.76
9.93	67.9	19.35	2.53
19.95	65.5	20.85	2.38
28.96	64.1	22.41	2.27
39.79	62.2	24.64	2.14

(2) Volumetric Properties

Ref. [257]

		Below the CMC		Above the CMC	
wt$_2$% ($w_1 + w_2 = 1$)	ρ_0 (g · cm^{-3})	d_ρ/d_c	υ (cm^3 · g^{-1})	d_ρ/d_c	υ (cm^3 · g^{-1})
298.15 K					
0	0.997043	0.1218	0.8808	0.1125	0.8901
9.93	1.000775	0.1254	0.8739	0.1110	0.8883
19.95	1.005342	0.131	0.8644	0.1190	0.8763
28.96	1.009880	0.1303	0.8612	0.1085	0.8828
39.79	1.015869	0.1159	0.8703	0.1000	0.8859

References

1. Gardas RL, Dagade DH, Terdale SS, Coutinho JAP, Patil KJ (2008) Acoustic and volumetric properties of aqueous solutions of imidazolium based ionic liquids at 298.15K. J Chem Thermodyn 40(4):695–701
2. Dong L, Zheng DX, Wei Z, Wu XH (2009) Synthesis of 1,3-dimethylimidazolium chloride and volumetric property investigations of its aqueous solution. Int J Thermophys 30 (5):1480–1490
3. Kato R, Krummen M, Gmehling J (2004) Measurement and correlation of vapor–liquid equilibria and excess enthalpies of binary systems containing ionic liquids and hydrocarbons. Fluid Phase Equilib 224(1):47–54
4. Kato R, Gmehling J (2005) Systems with ionic liquids: measurement of VLE and γ^{∞} data and prediction of their thermodynamic behavior using original UNIFAC, mod. UNIFAC(Do) and COSMO-RS(Ol). J Chem Thermodyn 37(6):603–619
5. González B, Calvar N, Domínguez Á, Macedo EA (2008) Osmotic coefficients of aqueous solutions of four ionic liquids at T = (313.15 and 333.15) K. J Chem Thermodyn 40 (9):1346–1351
6. Gómez E, González B, Calvar N, Domínguez Á (2008) Excess molar properties of ternary system (ethanol + water + 1,3-dimethylimidazolium methylsulphate) and its binary mixtures at several temperatures. J Chem Thermodyn 40(8):1208–1216
7. Pereiro AB, Rodríguez A (2007) Study on the phase behaviour and thermodynamic properties of ionic liquids containing imidazolium cation with ethanol at several temperatures. J Chem Thermodyn 39(6):978–989
8. Domańska U, Pobudkowska A, Rogalski M (2008) Surface tension of binary mixtures of imidazolium and ammonium based ionic liquids with alcohols, or water: cation, anion effect. J Colloid Interface Sci 322(1):342–350
9. Calvar N, González B, Domínguez Á, Macedo EA (2009) Osmotic coefficients of binary mixtures of 1-butyl-3-methylimidazolium methylsulfate and 1,3-dimethylimidazolium methylsulfate with alcohols at T = 323.15K. J Chem Thermodyn 41(5):617–622
10. Gołdon A, Dabrowska K, Hofman T (2007) Densities and excess volumes of the 1,3-dimethylimidazolium methylsulfate + methanol system at temperatures from (313.15 to 333.15) K and pressures from (0.1 to 25) MPa. J Chem Eng Data 52(5):1830–1837
11. Pereiro AB, Canosa J, Rodríguez A (2007) Liquid–liquid equilibria of 1,3-dimethylimidazolium methyl sulfate with ketones, dialkyl carbonates and acetates. Fluid Phase Equilib 254(1–2):150–157
12. Pereiro AB, Rodríguez A (2007) Thermodynamic properties of ionic liquids in organic solvents from (293.15 to 303.15) K. J Chem Eng Data 52(2):600–608
13. Domańska U, Pobudkowska A, Eckert F (2006) (Liquid + liquid) phase equilibria of 1-alkyl-3-methylimidazolium methylsulfate with alcohols, or ethers, or ketones. J Chem Thermodyn 38(6):685–695
14. Domańska U, Pobudkowska A, Eckert F (2006) Liquid–liquid equilibria in the binary systems (1,3-dimethylimidazolium, or 1-butyl-3-methylimidazolium methylsulfate + hydrocarbons). Green Chem 8(3):268–276
15. Pereiro AB, Rodríguez A (2007) Ternary liquid – liquid equilibria ethanol + 2-butanone + 1-butyl-3-methylimidazolium hexafluorophosphate, 2-propanol + 2-butanone + 1-butyl-3-methylimidazolium hexafluorophosphate, and 2-butanone + 2-propanol + 1,3-dimethylimidazolium methyl sulfate at 298.15 K. J Chem Eng Data 52(6):2138–2142
16. Meindersma GW, Podt AJG, de Haan AB (2006) Ternary liquid–liquid equilibria for mixtures of toluene + *n*-heptane + an ionic liquid. Fluid Phase Equilib 247(1–2):158–168
17. Pereiro AB, Rodríguez A (2007) Ternary (liquid + liquid) equilibria of the azeotrope (ethyl acetate + 2-propanol) with different ionic liquids at T = 298.15K. J Chem Thermodyn 39 (12):1608–1613

18. Kato R, Gmehling J (2005) Measurement and correlation of vapor–liquid equilibria of binary systems containing the ionic liquids [EMIM][$(CF_3SO_2)_2N$], [BMIM][$(CF_3SO_2)_2N$], [MMIM][$(CH_3)_2PO_4$] and oxygenated organic compounds respectively water. Fluid Phase Equilib 231 (1):38–43
19. Zhao J, Jiang XC, Li CX, Wang ZH (2006) Vapor pressure measurement for binary and ternary systems containing a phosphoric ionic liquid. Fluid Phase Equilib 247(1–2):190–198
20. Zhao J, Li CX, Wang ZH (2006) Vapor pressure measurement and prediction for ethanol + methanol and ethanol + water systems containing ionic liquids. J Chem Eng Data 51 (5):1755–1760
21. Zhao J, Dong CC, Li CX, Meng H, Wang ZH (2006) Isobaric vapor–liquid equilibria for ethanol–water system containing different ionic liquids at atmospheric pressure. Fluid Phase Equilib 242(2):147–153
22. Tsuda T, Tomioka T, Hussey CL (2008) Physicochemical properties of highly conductive urea-EtMeImCl melts. Chem Commun 25:2908–2910
23. Rodriguez H, Francisco M, Rahman M, Sun N, Rogers RD (2009) Biphasic liquid mixtures of ionic liquids and polyethylene glycols. Phys Chem Chem Phys 11(46):10916–10922
24. Ge Y, Zhang L, Yuan X, Geng W, Ji J (2008) Selection of ionic liquids as entrainers for separation of (water + ethanol). J Chem Thermodyn 40(8):1248–1252
25. Zhang L, Qiao B, Ge Y, Deng D, Ji J (2009) Effect of ionic liquids on (vapor + liquid) equilibrium behavior of (water + 2-methyl-2-propanol). J Chem Thermodyn 41(1):138–143
26. Gardas RL, Dagade DH, Coutinho JAP, Patil KJ (2008) Thermodynamic studies of ionic interactions in aqueous solutions of imidazolium-based ionic liquids [Emim][Br] and [Bmim][Cl]. J Phys Chem B 112(11):3380–3389
27. Navia P, Troncoso J, Romaní L (2008) Viscosities for ionic liquid binary mixtures with a common ion. J Solut Chem 37(5):677–688
28. Navia P, Troncoso J, Romaní L (2007) Excess magnitudes for ionic liquid binary mixtures with a common ion. J Chem Eng Data 52(4):1369–1374
29. Finotello A, Bara JE, Narayan S, Camper D, Noble RD (2008) Ideal gas solubilities and solubility selectivities in a binary mixture of room-temperature ionic liquids. J Phys Chem B 112(8):2335–2339
30. Shiflett MB, Yokozeki A (2008) Phase equilibria of hydrofluorocarbon-4310mee mixtures with ionic liquids: miscibility of *threo*- and *erythro*-diastereomers in ionic liquids. Ind Eng Chem Res 47(3):926–934
31. Shiflett MB, Yokozeki A (2007) Liquid – liquid equilibria in binary mixtures of 1,3-propanediol + ionic liquids [bmim][PF_6], [bmim][BF_4], and [emim][BF_4]. J Chem Eng Data 52(4):1302–1306
32. Zhang L, Han J, Wang R, Qiu X, Ji J (2007) Isobaric vaporric 7. equilibria for three ternary systems: water + 2-propanol + 1-ethyl-3-methylimidazolium tetrafluoroborate, water + 1-propanol + 1-ethyl-3-methylimidazolium tetrafluoroborate, and water + 1-propanol + 1-butyl-3-methylimidazolium tetrafluoroborate. J Chem Eng Data 52(4):1401–1407
33. Zhang S, Li X, Chen H, Wang J, Zhang J, Zhang M (2004) Determination of physical properties for the binary system of 1-ethyl-3-methylimidazolium tetrafluoroborate + H_2O. J Chem Eng Data 49(4):760–764
34. Rilo E, Pico J, García-Garabal S, Varela LM, Cabeza O (2009) Density and surface tension in binary mixtures of CnMIM-BF_4 ionic liquids with water and ethanol. Fluid Phase Equilib 285 (1–2):83–89
35. Stoppa A, Hunger J, Buchner R (2009) Conductivities of binary mixtures of ionic liquids with polar solvents. J Chem Eng Data 54(2):472–479
36. Jork C, Seiler M, Beste YA, Arlt W (2004) Influence of ionic liquids on the phase behavior of aqueous azeotropic systems. J Chem Eng Data 49(4):852–857
37. Hayamizu K, Aihara Y, Nakagawa H, Nukuda T, Price WS (2004) Ionic conduction and ion diffusion in binary room-temperature ionic liquids composed of [emim][BF_4] and $LiBF_4$. J Phys Chem B 108(50):19527–19532

38. Lewandowski A, Świderska A (2004) New composite solid electrolytes based on a polymer and ionic liquids. Solid State Ion 169(1–4):21–24
39. Li Q, Zhang J, Lei Z, Zhu J, Xing F (2009) Isobaric vaporric – equilibrium for ethyl acetate + ethanol + 1-ethyl-3-methylimidazolium tetrafluoroborate. J Chem Eng Data 54 (2):193–197
40. Wachter P, Zistler M, Schreiner C, Fleischmann M, Gerhard D, Wasserscheid P, Barthel J, Gores HJ (2009) Temperature dependence of the non-Stokesian charge transport in binary blends of ionic liquids. J Chem Eng Data 54(2):491–497
41. Shiflett MB, Yokozeki A (2008) Liquid – liquid equilibria in binary mixtures containing fluorinated benzenes and ionic liquid 1-ethyl-3-methylimidazolium bis(trifluoromethylsulfonyl)imide. J Chem Eng Data 53(11):2683–2691
42. Yokozeki A, Shiflett MB (2008) Binary and ternary phase diagrams of benzene, hexafluorobenzene, and ionic liquid [emim][Tf_2N] using equations of state. Ind Eng Chem Res 47 (21):8389–8395
43. Shiflett MB, Yokozeki A (2008) Binary vapor–liquid and vapor–liquid–liquid equilibria of hydrofluorocarbons (HFC-125 and HFC-143a) and hydrofluoroethers (HFE-125 and HFE-143a) with ionic liquid [emim][Tf_2N]. J Chem Eng Data 53(2):492–497
44. Wang S, Jacquemin J, Husson P, Hardacre C, Costa Gomes MF (2009) Liquid–liquid miscibility and volumetric properties of aqueous solutions of ionic liquids as a function of temperature. J Chem Thermodyn 41(11):1206–1214
45. Freire MG, Carvalho PJ, Gardas RL, Marrucho IM, Santos LMNBF, Coutinho JAP (2008) Mutual solubilities of water and the [$C_n$mim][Tf_2N] hydrophobic ionic liquids. J Phys Chem B 112(6):1604–1610
46. Döker M, Gmehling J (2005) Measurement and prediction of vapor–liquid equilibria of ternary systems containing ionic liquids. Fluid Phase Equilib 227(2):255–266
47. Shiflett MB, Yokozeki A (2007) Liquid – liquid equilibria of hydrofluoroethers and ionic liquid 1-ethyl-3-methylimidazolium bis(trifluoromethylsulfonyl)imide. J Chem Eng Data 52 (6):2413–2418
48. Schilderman AM, Raeissi S, Peters CJ (2007) Solubility of carbon dioxide in the ionic liquid 1-ethyl-3-methylimidazolium bis(trifluoromethylsulfonyl)imide. Fluid Phase Equilib 260 (1):19–22
49. Carvalho PJ, Álvarez VH, Machado JJB, Pauly J, Daridon J-L, Marrucho IM, Aznar M, Coutinho JAP (2009) High pressure phase behavior of carbon dioxide in 1-alkyl-3-methylimidazolium bis(trifluoromethylsulfonyl)imide ionic liquids. J Supercrit Fluids 48(2):99–107
50. Nebig S, Bölts R, Gmehling J (2007) Measurement of vapor–liquid equilibria (VLE) and excess enthalpies (H^E) of binary systems with 1-alkyl-3-methylimidazolium bis(trifluoromethylsulfonyl)imide and prediction of these properties and γ^∞ using modified UNIFAC (Dortmund). Fluid Phase Equilib 258(2):168–178
51. Heintz A, Lehmann JK, Wertz C, Jacquemin J (2005) Thermodynamic properties of mixtures containing ionic liquids. 4. LLE of binary mixtures of [C_2MIM][NTf_2] with propan-1-ol, butan-1-ol, and pentan-1-ol and [C_4MIM][NTF_2] with cyclohexanol and 1,2-hexanediol including studies of the influence of small amounts of water. J Chem Eng Data 50(3):956–960
52. Heintz A, Lehmann JK, Wertz C (2003) Thermodynamic properties of mixtures containing ionic liquids. 3. Liquid – liquid equilibria of binary mixtures of 1-ethyl-3-methylimidazolium bis(trifluoromethylsulfonyl)imide with propan-1-ol, butan-1-ol, and pentan-1-ol. J Chem Eng Data 48(3):472–474
53. Najdanovic-Visak V, Esperança JMSS, Rebelo LPN, Nunes da Ponte M, Guedes HJR, Seddon KR, de Sousa HC, Szydlowski J (2003) Pressure, isotope, and water co-solvent effects in liquid – liquid equilibria of (ionic liquid + alcohol) systems. J Phys Chem B 107 (46):12797–12807
54. Hong G, Jacquemin J, Husson P, Costa Gomes MF, Deetlefs M, Nieuwenhuyzen M, Sheppard O, Hardacre C (2006) Effect of acetonitrile on the solubility of carbon dioxide in

1-ethyl-3-methylimidazolium bis(trifluoromethylsulfonyl)amide. Ind Eng Chem Res 45 (24):8180–8188

55. Shiflett MB, Harmer MA, Junk CP, Yokozeki A (2006) Solubility and diffusivity of difluoromethane in room-temperature ionic liquids. J Chem Eng Data 51(2):483–495
56. Vasiltsova TV, Verevkin SP, Bich E, Heintz A, Bogel-Lukasik R, Domańska U (2006) Thermodynamic properties of mixtures containing ionic liquids. 7. Activity coefficients of aliphatic and aromatic esters and benzylamine in 1-methyl-3-ethylimidazolium bis(trifluoromethylsulfonyl) imide using the transpiration method. J Chem Eng Data 51(1):213–218
57. Canongia Lopes JN, Cordeiro TC, Esperanca JM, Guedes HJ, Huq S, Rebelo LP, Seddon KR (2005) Deviations from ideality in mixtures of two ionic liquids containing a common ion. J Phys Chem B 109(8):3519–3525
58. Vasiltsova TV, Verevkin SP, Bich E, Heintz A, Bogel-Lukasik R, Domanska U (2005) Thermodynamic properties of mixtures containing ionic liquids. Activity coefficients of ethers and alcohols in 1-methyl-3-ethyl-imidazolium bis(trifluoromethyl-sulfonyl) imide using the transpiration method. J Chem Eng Data 50(1):142–148
59. Verevkin SP, Vasiltsova TV, Bich E, Heintz A (2004) Thermodynamic properties of mixtures containing ionic liquids activity coefficients of aldehydes and ketones in 1-methyl-3-ethyl-imidazolium bis(trifluoromethyl-sulfonyl) imide using the transpiration method. Fluid Phase Equilib 218(2):165–175
60. Arce A, Earle MJ, Katdare SP, Rodríguez H, Seddon KR (2007) Phase equilibria of mixtures of mutually immiscible ionic liquids. Fluid Phase Equilib 261(1–2):427–433
61. Shiflett MB, Niehaus AMS, Yokozeki A (2009) Liquid – liquid equilibria in binary mixtures containing chlorobenzene, bromobenzene, and iodobenzene with ionic liquid 1-ethyl-3-methylimidazolim bis(trifluoromethylsulfonyl)imide. J Chem Eng Data 54(7):2090–2094
62. Arce A, Earle MJ, Rodríguez H, Seddon KR, Soto A (2008) 1-ethyl-3-methylimidazolium bis {(trifluoromethyl)sulfonyl}amide as solvent for the separation of aromatic and aliphatic hydrocarbons by liquid extraction *xt*extension to C_7- and C_8-fractions. Green Chem 10 (12):1294–1300
63. Deenadayalu N, Bhujrajh P (2008) Density, speed of sound, and derived thermodynamic properties of ionic liquids $[EMIM]^+[BETI]^-$ or ($[EMIM]^+[CH_3(OCH_2CH_2)_2OSO_3]^-$ + methanol or + acetone) at T = (298.15 or 303.15 or 313.15) K. J Chem Eng Data 53(5):1098–1102
64. Shiflett MB, Harmer MA, Junk CP, Yokozeki A (2006) Solubility and diffusivity of 1,1,1,2-tetrafluoroethane in room-temperature ionic liquids. Fluid Phase Equilib 242(2):220–232
65. García-Miaja G, Troncoso J, Romaní L (2008) Excess properties for binary systems ionic liquid + ethanol: experimental results and theoretical description using the ERAS model. Fluid Phase Equilib 274(1–2):59–67
66. Orchillés AV, Miguel PJ, Vercher E, Martínez-Andreu A (2007) Isobaric vaporric And equilibria for ethyl acetate + ethanol + 1-ethyl-3-methylimidazolium trifluoromethanesulfonate at 100 kPa. J Chem Eng Data 52(6):2325–2330
67. Vercher E, Orchillés AV, Miguel PJ, Martínez-Andreu A (2007) Volumetric and ultrasonic studies of 1-ethyl-3-methylimidazolium trifluoromethanesulfonate ionic liquid with methanol, ethanol, 1-propanol, and water at several temperatures. J Chem Eng Data 52(4):1468–1482
68. Orchillés AV, Miguel PJ, Vercher E, Martínez-Andreu A (2008) Isobaric vaporric And and liquid – liquid equilibria for chloroform + ethanol + 1-ethyl-3-methylimidazolium trifluoromethanesulfonate at 100 kPa. J Chem Eng Data 53(11):2642–2648
69. Orchillés AV, Miguel PJ, Vercher E, Martínez-Andreu A (2008) Isobaric vaporric And equilibria for 1-propanol + water + 1-ethyl-3-methylimidazolium trifluoromethanesulfonate at 100 kPa. J Chem Eng Data 53(10):2426–2431
70. Rodríguez H, Brennecke JF (2006) Temperature and composition dependence of the density and viscosity of binary mixtures of water + ionic liquid. J Chem Eng Data 51(6):2145–2155
71. García-Miaja G, Troncoso J, Romaní L (2009) Excess enthalpy, density, and heat capacity for binary systems of alkylimidazolium-based ionic liquids + water. J Chem Thermodyn 41 (2):161–166

72. Orchillés AV, Miguel PJ, Vercher E, Martínez-Andreu A (2007) Ionic liquids as entrainers in extractive distillation: isobaric vaporic quid equilibria for acetone + methanol + 1-ethyl-3-methylimidazolium trifluoromethanesulfonate. J Chem Eng Data 52(1):141–147
73. Orchillés AV, Miguel PJ, Vercher E, Martínez-Andreu A (2007) Isobaric vaporric And equilibria for methyl acetate + methanol + 1-ethyl-3-methylimidazolium trifluoromethanesulfonate at 100 kPa. J Chem Eng Data 52(3):915–920
74. García-Miaja G, Troncoso J, Romaní L (2009) Excess molar properties for binary systems of alkylimidazolium-based ionic liquids + nitromethane. Experimental results and ERAS-model calculations. J Chem Thermodyn 41(3):334–341
75. Hernández-Fernández FJ, de los Ríos AP, Gómez D, Rubio M, Tomás-Alonso F, Víllora G (2008) Ternary liquid–liquid equilibria for mixtures of an ionic liquid + *n*-hexane + an organic compound involved in the kinetic resolution of *rac*-1-phenyl ethanol (*rac*-1-phenyl ethanol, vinyl propionate, *rac*-1-phenylethyl propionate or propionic acid) at 298.2 K and atmospheric pressure. Fluid Phase Equilib 263(2):190–198
76. Gómez E, González B, Calvar N, Tojo E, Domínguez Á (2006) Physical properties of pure 1-ethyl-3-methylimidazolium ethylsulfate and its binary mixtures with ethanol and water at several temperatures. J Chem Eng Data 51(6):2096–2102
77. Calvar N, González B, Gómez E, Domínguez Á (2008) Vapor–liquid equilibria for the ternary system ethanol + water + 1-ethyl-3-methylimidazolium ethylsulfate and the corresponding binary systems containing the ionic liquid at 101.3 kPa. J Chem Eng Data 53(3):820–825
78. Liebert V, Nebig S, Gmehling J (2008) Experimental and predicted phase equilibria and excess properties for systems with ionic liquids. Fluid Phase Equilib 268(1–2):14–20
79. Sumartschenkowa IA, Verevkin SP, Vasiltsova TV, Bich E, Heintz A, Shevelyova MP, Kabo GJ (2006) Experimental study of thermodynamic properties of mixtures containing ionic liquid 1-ethyl-3-methylimidazolium ethyl sulfate using gaseriment chromatography and transpiration method. J Chem Eng Data 51(6):2138–2144
80. Calvar N, González B, Domínguez Á, Macedo EA (2009) Vapour pressures and osmotic coefficients of binary mixtures of 1-ethyl-3-methylimidazolium ethylsulfate and 1-ethyl-3-methylpyridinium ethylsulfate with alcohols at T = 323.15 K. J Chem Thermodyn 41 (12):1439–1445
81. Wang JF, Li CX, Wang ZH (2007) Measurement and prediction of vapor pressure of binary and ternary systems containing 1-ethyl-3-methylimidazolium ethyl sulfate. J Chem Eng Data 52(4):1307–1312
82. Ge R, Hardacre C, Nancarrow P, Rooney DW (2007) Thermal conductivities of ionic liquids over the temperature range from 293 K to 353 K. J Chem Eng Data 52(5):1819–1823
83. Shekaari H, Mousavi SS (2009) Measurement and modeling of osmotic coefficients of aqueous solution of ionic liquids using vapor pressure osmometry method. Fluid Phase Equilib 279(1):73–79
84. Calvar N, González B, Domínguez Á, Macedo EA (2009) Osmotic coefficients of binary mixtures of four ionic liquids with ethanol or water at T = (313.15 and 333.15) K. J Chem Thermodyn 41(1):11–16
85. Domańska U, Laskowska M, Marciniak A (2008) Phase equilibria of (1-ethyl-3-methylimidazolium ethylsulfate + hydrocarbon, + ketone, and + ether) binary systems. J Chem Eng Data 53(2):498–502
86. Bendová M, Wagner Z (2009) Thermodynamic description of liquid–liquid equilibria in systems 1-ethyl-3-methylimidazolium ethylsulfate + C_7-hydrocarbons by polymer-solution models. Fluid Phase Equilib 284(2):80–85
87. Yokozeki A, Shiflett MB (2007) Vapor–liquid equilibria of ammonia + ionic liquid mixtures. Appl Energy 84(12):1258–1273
88. Hofman T, Gołdon A, Nevines A, Letcher TM (2008) Densities, excess volumes, isobaric expansivity, and isothermal compressibility of the (1-ethyl-3-methylimidazolium ethylsulfate + methanol) system at temperatures (283.15 to 333.15) K and pressures from (0.1 to 35) MPa. J Chem Thermodyn 40(4):580–591

89. González EJ, González B, Calvar N, Domínguez Á (2007) Physical properties of binary mixtures of the ionic liquid 1-ethyl-3-methylimidazolium ethyl sulfate with several alcohols at T = (298.15, 313.15, and 328.15) K and atmospheric pressure. J Chem Eng Data 52 (5):1641–1648
90. Deenadayalu N, Sen S, Sibiya PN (2009) Application of the PFV EoS correlation to excess molar volumes of (1-ethyl-3-methylimidazolium ethylsulfate + alkanols) at different temperatures. J Chem Thermodyn 41(4):538–548
91. Jacquemin J, Husson P, Majer V, Padua AAH, Costa Gomes MF (2008) Thermophysical properties, low pressure solubilities and thermodynamics of solvation of carbon dioxide and hydrogen in two ionic liquids based on the alkylsulfate anion. Green Chem 10(9):944–950
92. Hunger J, Stoppa A, Buchner R, Hefter G (2009) Dipole correlations in the ionic liquid 1-N-ethyl-3-N-methylimidazolium ethylsulfate and its binary mixtures with dichloromethane. J Phys Chem B 113(28):9527–9537
93. Alonso L, Arce A, Francisco M, Soto A (2008) Thiophene separation from aliphatic hydrocarbons using the 1-ethyl-3-methylimidazolium ethylsulfate ionic liquid. Fluid Phase Equilib 270(1–2):97–102
94. Alonso L, Arce A, Francisco M, Soto A (2008) Liquid-liquid equilibria of ([C_2mim][$EtSO_4$] + thiophene + 2,2,4-trimethylpentane) and ([C_2mim][$EtSO_4$] + thiophene + toluene): experimental data and correlation. J Solut Chem 37(10):1355–1363
95. Calvar N, Gómez E, González B, Domínguez Á (2007) Experimental determination, correlation, and prediction of physical properties of the ternary mixtures ethanol + water with 1-octyl-3-methylimidazolium chloride and 1-ethyl-3-methylimidazolium ethylsulfate. J Chem Eng Data 52(6):2529–2535
96. González B, Calvar N, González E, Domínguez Á (2008) Density and viscosity experimental data of the ternary mixtures 1-propanol or 2-propanol + water + 1-ethyl-3-methylimidazolium ethylsulfate. Correlation and prediction of physical properties of the ternary systems. J Chem Eng Data 53(3):881–887
97. Arce A, Rodil E, Soto A (2006) Volumetric and viscosity study for the mixtures of 2-ethoxy-2-methylpropane, ethanol, and 1-ethyl-3-methylimidazolium ethyl sulfate ionic liquid. J Chem Eng Data 51(4):1453–1457
98. Arce A, Rodríguez H, Soto A (2007) Use of a green and cheap ionic liquid to purify gasoline octane boosters. Green Chem 9(3):247–253
99. Arce A, Pobudkowska A, Rodríguez O, Soto A (2007) Citrus essential oil terpenless by extraction using 1-ethyl-3-methylimidazolium ethylsulfate ionic liquid: effect of the temperature. Chem Eng J 133(1–3):213–218
100. Arce A, Rodríguez O, Soto A (2006) A comparative study on solvents for separation of tert-amyl ethyl ether and ethanol mixtures. New experimental data for 1-ethyl-3-methyl imidazolium ethyl sulfate ionic liquid. Chem Eng Sci 61(21):6929–6935
101. García J, Fernández A, Torrecilla JS, Oliet M, Rodríguez F (2009) Liquid–liquid equilibria for {hexane + benzene + 1-ethyl-3-methylimidazolium ethylsulfate} at (298.2, 313.2 and 328.2) K. Fluid Phase Equilib 282(2):117–120
102. Bhujrajh P, Deenadayalu N (2007) Liquid densities and excess molar volumes for binary systems (ionic liquids + methanol or water) at 298.15, 303.15 and 313.15 K, and at atmospheric pressure. J Solut Chem 36(5):631–642
103. Deenadayalu N, Kumar S, Bhujrajh P (2007) Liquid densities and excess molar volumes for (ionic liquids + methanol + water) ternary system at atmospheric pressure and at various temperatures. J Chem Thermodyn 39(9):1318–1324
104. Naydenov D, Bart H-J (2007) Ternary liquid – liquid equilibria for six systems containing ethylacetate + ethanol or acetic acid + an imidazolium-based ionic liquid with a hydrogen sulfate anion at 313.2 K. J Chem Eng Data 52(6):2375–2381
105. Naydenov D, Bart H-J (2009) Ternary liquid – liquid equilibria for systems containing alcohol or acetic acid + ester + 1-ethyl-3-methylimidazolium hydrogen sulfate at 313.2 K using headspace gas chromatography. J Chem Eng Data 54(1):43–47

106. Sifaoui H, Ait-Kaci A, Modarressi A, Rogalski M (2007) Solid–liquid equilibria of three binary systems: {1-ethyl-3-methylimidazolium hexafluorophosphate + 2-phenylimidazole, or 4,5-diphenylimidazole or 2,4,5-triphenylimidazole}. Thermochim Acta 456(2):114–119
107. Shariati A, Peters CJ (2004) High-pressure phase behavior of systems with ionic liquids: II. The binary system carbon dioxide + 1-ethyl-3-methylimidazolium hexafluorophosphate. J Supercrit Fluids 29(1–2):43–48
108. Wang JF, Li CX, Wang ZH, Li ZJ, Jiang YB (2007) Vapor pressure measurement for water, methanol, ethanol, and their binary mixtures in the presence of an ionic liquid 1-ethyl-3-methylimidazolium dimethylphosphate. Fluid Phase Equilib 255(2):186–192
109. Shiflett MB, Yokozeki A (2009) Phase behavior of carbon dioxide in ionic liquids: [emim][acetate], [emim][trifluoroacetate], and [emim][acetate] + [emim][trifluoroacetate] mixtures. J Chem Eng Data 54(1):108–114
110. Domańska U, Królikowski M, Ślesińska K (2009) Phase equilibria study of the binary systems (ionic liquid + thiophene): desulphurization process. J Chem Thermodyn 41(11):1303–1311
111. Jiang XC, Wang JF, Li CX, Wang LM, Wang ZH (2007) Vapour pressure measurement for binary and ternary systems containing water methanol ethanol and an ionic liquid 1-ethyl-3-ethylimidazolium diethylphosphate. J Chem Thermodyn 39(6):841–846
112. Shekaari H, Mousavi SS (2009) Influence of alkyl chain on the thermodynamic properties of aqueous solutions of ionic liquids 1-alkyl-3-methylimidazolium bromide at different temperatures. J Chem Thermodyn 41(1):90–96
113. Sadeghi R, Shekaari H, Hosseini R (2009) Volumetric and isentropic compressibility behavior of ionic liquid, 1-propyl-3-methylimidazolium bromide in acetonitrile, dimethylformamide, and dimethylsulfoxide at T = (288.15 to 308.15) K. Int J Thermophys 30(5):1491–1509
114. Calvar N, González B, Domínguez A, Tojo J (2006) Physical properties of the ternary mixture ethanol + water + 1-butyl-3-methylimidazolium chloride at 298.15 K. J Solut Chem 35 (9):1217–1225
115. Gómez E, González B, Domínguez Á, Tojo E, Tojo J (2006) Dynamic viscosities of a series of 1-alkyl-3-methylimidazolium chloride ionic liquids and their binary mixtures with water at several temperatures. J Chem Eng Data 51(2):696–701
116. Shekaari H, Mousavi SS (2009) Osmotic coefficients and refractive indices of aqueous solutions of ionic liquids containing 1-butyl-3-methylimidazolium halide at T = (298.15 to 328.15) K. J Chem Eng Data 54(3):823–829
117. Calvar N, González B, Gómez E, Domínguez Á (2006) Vaporguez1. equilibria for the ternary system ethanol + water + 1-butyl-3-methylimidazolium chloride and the corresponding binary systems at 101.3 kPa. J Chem Eng Data 51(6):2178–2181
118. Sedláková Z, Sauton H, Hynek V, Malijevská I (2008) Solidevská– equilibrium in the systems with an ionic liquid. Collect Czechoslov Chem Commun 73(5):657–664
119. Shekaari H, Zafarani-Moattar MT (2007) Osmotic coefficients of some imidazolium based ionic liquids in water and acetonitrile at temperature 318.15 K. Fluid Phase Equilib 254 (1–2):198–203
120. Zafarani-Moattar MT, Shekaari H (2005) Apparent molar volume and isentropic compressibility of ionic liquid 1-butyl-3-methylimidazolium bromide in water, methanol, and ethanol at T = (298.15 to 318.15)K. J Chem Thermodyn 37(10):1029–1035
121. Gaillon L, Sirieix-Plenet J, Letellier P (2004) Volumetric study of binary solvent mixtures constituted by amphiphilic ionic liquids at room temperature (1-alkyl-3-methylimidazolium bromide) and water. J Solut Chem 33(11):1333–1347
122. Kim K-S, Park S-Y, Choi S, Lee H (2004) Vapor pressures of the 1-butyl-3-methylimidazolium bromide + water, 1-butyl-3-methylimidazolium tetrafluoroborate + water, and 1-(2-hydroxyethyl)-3-methylimidazolium tetrafluoroborate + water systems. J Chem Eng Data 49(6):1550–1553
123. Kim K-S, Shin B-K, Lee H, Ziegler F (2004) Refractive index and heat capacity of 1-butyl-3-methylimidazolium bromide and 1-butyl-3-methylimidazolium tetrafluoroborate, and vapor pressure of binary systems for 1-butyl-3-methylimidazolium bromide + trifluoroethanol and

1-butyl-3-methylimidazolium tetrafluoroborate + trifluoroethanol. Fluid Phase Equilib 218 (2):215–220
124. Zafarani-Moattar MT, Hamzehzadeh S (2007) Liquid – liquid equilibria of aqueous two-phase systems containing 1-butyl-3-methylimidazolium bromide and potassium phosphate or dipotassium hydrogen phosphate at 298.15 K. J Chem Eng Data 52(5):1686–1692
125. Zafarani-Moattar MT, Hamzehzadeh S (2009) Phase diagrams for the aqueous two-phase ternary system containing the ionic liquid 1-butyl-3-methylimidazolium bromide and tri-potassium citrate at T = (278.15, 298.15, and 318.15) K. J Chem Eng Data 54(3):833–841
126. Mokhtarani B, Mojtahedi MM, Mortaheb HR, Mafi M, Yazdani F, Sadeghian F (2008) Densities, refractive indices, and viscosities of the ionic liquids 1-methyl-3-octylimidazolium tetrafluoroborate and 1-methyl-3-butylimidazolium perchlorate and their binary mixtures with ethanol at several temperatures. J Chem Eng Data 53(3):677–682
127. Abdulagatov IM, Tekin A, Safarov J, Shahverdiyev A, Hassel E (2008) High-pressure densities and derived volumetric properties (excess, apparent, and partial molar volumes) of binary mixtures of {methanol (1) + [BMIM][BF_4] (2)}. J Chem Thermodyn 40(9):1386–1401
128. Shekaari H, Zafarani-Moattar MT (2008) Volumetric properties of the ionic liquid, 1-butyl-3-methylimidazolium tetrafluoroborate, in organic solvents at T = 298.15 K. Int J Thermophys 29(2):534–545
129. Iglesias-Otero MÁ, Troncoso J, Carballo E, Romaní L (2007) Density and refractive index for binary systems of the ionic liquid [Bmim][BF_4] with methanol, 1,3-dichloropropane, and dimethyl carbonate. J Sol Chem 36(10):1219–1230
130. Zafarani-Moattar MT, Shekaari H (2006) Application of Prigogine–Flory–Patterson theory to excess molar volume and speed of sound of 1-*n*-butyl-3-methylimidazolium hexafluorophosphate or 1-*n*-butyl-3-methylimidazolium tetrafluoroborate in methanol and acetonitrile. J Chem Thermodyn 38(11):1377–1384
131. Iglesias-Otero MA, Troncoso J, Carballo E, Romaní L (2008) Density and refractive index in mixtures of ionic liquids and organic solvents: correlations and predictions. J Chem Thermodyn 40(6):949–956
132. Iglesias-Otero MA, Troncoso J, Carballo E, Romaní L (2008) Densities and excess enthalpies for ionic liquids + ethanol or + nitromethane. J Chem Eng Data 53(6):1298–1301
133. Abdulagatov IM, Tekin A, Safarov J, Shahverdiyev A, Hassel E (2008) Densities and excess, apparent, and partial molar volumes of binary mixtures of $BMIMBF_4$ + ethanol as a function of temperature, pressure, and concentration. Int J Thermophys 29(2):505–533
134. Han C, Xia S, Ma P, Zeng F (2009) Densities of ionic liquid [BMIM][BF_4] + ethanol, + benzene, and + acetonitrile at different temperature and pressure. J Chem Eng Data 54 (10):2971–2977
135. Zhu A, Wang J, Han L, Fan M (2009) Measurements and correlation of viscosities and conductivities for the mixtures of imidazolium ionic liquids with molecular solutes. Chem Eng J 147(1):27–35
136. Singh T, Kumar A, Kaur M, Kaur G, Kumar H (2009) Non-ideal behaviour of imidazolium based room temperature ionic liquids in ethylene glycol at T = (298.15 to 318.15) K. J Chem Thermodyn 41(6):717–723
137. Tian Y, Wang X, Wang J (2008) Densities and viscosities of 1-butyl-3-methylimidazolium tetrafluoroborate + molecular solvent binary mixtures. J Chem Eng Data 53(9):2056–2059
138. Bou Malham I, Turmine M (2008) Viscosities and refractive indices of binary mixtures of 1-butyl-3-methylimidazolium tetrafluoroborate and 1-butyl-2,3-dimethylimidazolium tetrafluoroborate with water at 298 K. J Chem Thermodyn 40(4):718–723
139. Bou Malham I, Letellier P, Mayaffre A, Turmine M (2007) Part I: thermodynamic analysis of volumetric properties of concentrated aqueous solutions of 1-butyl-3-methylimidazolium tetrafluoroborate, 1-butyl-2,3-dimethylimidazolium tetrafluoroborate, and ethylammonium nitrate based on pseudo-lattice theory. J Chem Thermodyn 39(8):1132–1143

140. Zhou Q, Wang L-S, Chen H-P (2006) Densities and viscosities of 1-butyl-3-methylimidazolium tetrafluoroborate + H_2O binary mixtures from (303.15 to 353.15) K. J Chem Eng Data 51(3):905–908
141. Yu Y-H, Soriano AN, Li M-H (2009) Heat capacity and electrical conductivity of aqueous mixtures of [Bmim][BF_4] and [Bmim][PF_6]. J Taiwan Inst Chem Eng 40(2):205–212
142. Lin HM, Tien HY, Hone YT, Lee MJ (2007) Solubility of selected dibasic carboxylic acids in water, in ionic liquid of [Bmim][BF_4], and in aqueous [Bmim][BF_4] solutions. Fluid Phase Equilib 253(2):130–136
143. Huo Y, Xia S, Ma P (2007) Densities of ionic liquids, 1-butyl-3-methylimidazolium hexafluorophosphate and 1-butyl-3-methylimidazolium tetrafluoroborate, with benzene, acetonitrile, and 1-propanol at T = (293.15 to 343.15) K. J Chem Eng Data 52(5):2077–2082
144. Huo Y, Xia S, Yi S, Ma P (2009) Measurement and correlation of vapor pressure of benzene and thiophene with [BMIM][PF_6] and [BMIM][BF_4] ionic liquids. Fluid Phase Equilib 276 (1):46–52
145. Makowska A, Siporska A, Szydlowski J (2006) Isotope effects on miscibility of 1-butyl-3-methylimidazolium tetrafluoroborate with butanols. J Phys Chem B 110(34):17195–17199
146. Comminges C, Barhdadi R, Laurent M, Troupel M (2006) Determination of viscosity, ionic conductivity, and diffusion coefficients in some binary systems: ionic liquids + molecular solvents. J Chem Eng Data 51(2):680–685
147. Chen Y, Zhang S, Yuan X, Zhang Y, Zhang X, Dai W, Mori R (2006) Solubility of CO_2 in imidazolium-based tetrafluoroborate ionic liquids. Thermochim Acta 441(1):42–44
148. Shiflett MB, Yokozeki A (2005) Solubilities and diffusivities of carbon dioxide in ionic liquids: [bmim][PF_6] and [bmim][BF_4]. Ind Eng Chem Res 44(12):4453–4464
149. Kroon MC, Shariati A, Costantini M, van Spronsen J, Witkamp G-J, Sheldon RA, Peters CJ (2005) High-pressure phase behavior of systems with ionic liquids: part V. The binary system carbon dioxide + 1-butyl-3-methylimidazolium tetrafluoroborate. J Chem Eng Data 50 (1):173–176
150. Ferreira R, Blesic M, Trindade J, Marrucho I, Lopes JNC, Rebelo LPN (2008) Solubility of fluorinated compounds in a range of ionic liquids. Cloud-point temperature dependence on composition and pressure. Green Chem 10(9):918–928
151. Reiter J, Vondrák J, Michálek J, Mička Z (2006) Ternary polymer electrolytes with 1-methylimidazole based ionic liquids and aprotic solvents. Electrochim Acta 52 (3):1398–1408
152. Gao H, Qi F, Wang H (2009) Densities and volumetric properties of binary mixtures of the ionic liquid 1-butyl-3-methylimidazolium tetrafluoroborate with benzaldehyde at T = (298.15 to 313.15) K. J Chem Thermodyn 41(7):888–892
153. Qi F, Wang H (2009) Application of Prigogine–Flory–Patterson theory to excess molar volume of mixtures of 1-butyl-3-methylimidazolium ionic liquids with N-methyl-2-pyrrolidinone. J Chem Thermodyn 41(2):265–272
154. Kühne E, Santarossa S, Perez E, Witkamp GJ, Peters CJ (2008) New approach in the design of reactions and separations using an ionic liquid and carbon dioxide as solvents: phase equilibria in two selected ternary systems. J Supercrit Fluids 46(2):93–98
155. Kühne E, Perez E, Witkamp GJ, Peters CJ (2008) Solute influence on the high-pressure phase equilibrium of ternary systems with carbon dioxide and an ionic liquid. J Supercrit Fluids 45 (1):27–31
156. Zhang LZ, Deng DS, Han JZ, Ji DX, Ji JB (2007) Isobaric vaporric – equilibria for water + 2-propanol + 1-butyl-3-methylimidazolium tetrafluoroborate. J Chem Eng Data 52(1):199–205
157. Li Q, Zhang J, Lei Z, Zhu J, Wang B, Huang X (2009) Isobaric vaporric 8. equilibrium for (propan-2-ol + water + 1-butyl-3-methylimidazolium tetrafluoroborate). J Chem Eng Data 54 (9):2785–2788
158. Wu B, Zhang Y, Wang H, Yang L (2008) Temperature dependence of phase behavior for ternary systems composed of ionic liquid + sucrose + water. J Phys Chem B 112 (41):13163–13165

159. Wu B, Zhang Y, Wang H (2008) Phase behavior for ternary systems composed of ionic liquid + saccharides + water. J Phys Chem B 112(20):6426–6429
160. Kühne E, Calvo ES, Witkamp GJ, Peters CJ (2008) Fluid phase behaviour of the ternary system bmim[BF_4] + 1-(4-isobutylphenyl)-ethanol + carbon dioxide. J Supercrit Fluids 45 (3):293–297
161. Wandschneider A, Lehmann JK, Heintz A (2008) Surface tension and density of pure ionic liquids and some binary mixtures with 1-propanol and 1-butanol. J Chem Eng Data 53 (2):596–599
162. Verevkin SP, Safarov J, Bich E, Hassel E, Heintz A (2005) Thermodynamic properties of mixtures containing ionic liquids vapor pressures and activity coefficients of *n*-alcohols and benzene in binary mixtures with 1-methyl-3-butyl-imidazolium bis(trifluoromethyl-sulfonyl) imide. Fluid Phase Equilib 236(1–2):222–228
163. Heintz A, Casás LM, Nesterov IA, Emel'yanenko VN, Verevkin SP (2005) Thermodynamic properties of mixtures containing ionic liquids. 5. Activity coefficients at infinite dilution of hydrocarbons, alcohols, esters, and aldehydes in 1-methyl-3-butyl-imidazolium bis (trifluoromethyl-sulfonyl) imide using gasrmodyna chromatography. J Chem Eng Data 50 (5):1510–1514
164. Wertz C, Tschersich A, Lehmann JK, Heintz A (2007) Liquid–liquid equilibria and liquid–liquid interfacial tension measurements of mixtures containing ionic liquids. J Mol Liq 131–132:2–6
165. Raeissi S, Peters CJ (2009) Carbon dioxide solubility in the homologous 1-alkyl-3-methylimidazolium bis (trifluoromethylsulfonyl)imide family. J Chem Eng Data 54(2):382–386
166. Davis SE, Morton SA (2008) Investigation of ionic liquids for the separation of butanol and water. Sep Sci Technol 43(9–10):2460–2472
167. Najdanovic-Visak V, Rebelo LPN, Nunes da Ponte M (2005) Liquid–liquid behaviour of ionic liquid–1-butanol–water and high pressure CO_2-induced phase changes. Green Chem 7 (6):443–450
168. Arce A, Earle MJ, Rodríguez H, Seddon KR (2007) Separation of benzene and hexane by solvent extraction with 1-alkyl-3-methylimidazolium bis{(trifluoromethyl)sulfonyl}amide ionic liquids: effect of the alkyl-substituent length. J Phys Chem B 111(18):4732–4736
169. Domańska U, Redhi GG, Marciniak A (2009) Activity coefficients at infinite dilution measurements for organic solutes and water in the ionic liquid 1-butyl-1-methylpyrrolidinium trifluoromethanesulfonate using GLC. Fluid Phase Equilib 278(1–2):97–102
170. Bermejo MD, Montero M, Saez E, Florusse LJ, Kotlewska AJ, Cocero MJ, van Rantwijk F, Peters CJ (2008) Liquidjka541 equilibrium of the systems butylmethylimidazolium nitrate-CO_2 and hydroxypropylmethylimidazolium nitrate-CO_2 at high pressure: influence of water on the phase behavior. J Phys Chem B 112(43):13532–13541
171. Mokhtarani B, Sharifi A, Mortaheb HR, Mirzaei M, Mafi M, Sadeghian F (2009) Density and viscosity of 1-butyl-3-methylimidazolium nitrate with ethanol, 1-propanol, or 1-butanol at several temperatures. J Chem Thermodyn 41(12):1432–1438
172. Ge ML, Zhao RS, Yi YF, Zhang Q, Wang LS (2008) Densities and viscosities of 1-butyl-3-methylimidazolium trifluoromethanesulfonate + H_2O binary mixtures at T = (303.15 to 343.15) K. J Chem Eng Data 53(10):2408–2411
173. Lin PY, Soriano AN, Caparanga AR, Li MH (2009) Molar heat capacity and electrolytic conductivity of aqueous solutions of [Bmim][$MeSO_4$] and [Bmim][triflate]. Thermochim Acta 496(1–2):105–109
174. Strechan AA, Paulechka YU, Kabo AG, Blokhin AV, Kabo GJ (2007) 1-butyl-3-methylimidazolium tosylate ionic liquid: heat capacity, thermal stability, and phase equilibrium of its binary mixtures with water and caprolactam. J Chem Eng Data 52(5):1791–1799
175. Orchillés AV, González-Alfaro V, Miguel PJ, Vercher E, Martínez-Andreu A (2006) Volumetric properties of binary mixtures of ionic liquid 1-butyl-3-methylimidazolium octylsulfate with water or propanol in the temperature range of 278.15 K to 328.15 K. J Chem Thermodyn 38 (9):1124–1129

176. Safarov J, Verevkin SP, Bich E, Heintz A (2006) Vapor pressures and activity coefficients ofn-alcohols and benzene in binary mixtures with 1-methyl-3-butylimidazolium octyl sulfate and 1-methyl-3-octylimidazolium tetrafluoroborate. J Chem Eng Data 51(2):518–525
177. Abdulagatov IM, Safarov J, Guliyev T, Shahverdiyev A, Hassel E (2009) High temperature and high pressure volumetric properties of (methanol + [BMIM+][$OcSO_4^-$]) mixtures. Phys Chem Liq 47(1):9–34
178. González B, Calvar N, Gómez E, Domínguez Á (2008) Physical properties of the ternary system (ethanol + water + 1-butyl-3-methylimidazolium methylsulphate) and its binary mixtures at several temperatures. J Chem Thermodyn 40(8):1274–1281
179. Calvar N, Ba G, Gómez E, An D (2009) Vaporguez8. equilibria for the ternary system ethanol + water + 1-butyl-3-methylimidazolium methylsulfate and the corresponding binary systems at 101.3 kPa. J Chem Eng Data 54(3):1004–1008
180. Pereiro AB, Rodríguez A (2008) Azeotrope-breaking using [BMIM] [$MeSO_4$] ionic liquid in an extraction column. Sep Purif Technol 62(3):733–738
181. Pereiro AB, Rodriguez A (2009) Purification of hexane with effective extraction using ionic liquid as solvent. Green Chem 11(3):346–350
182. Kumar A, Singh T, Gardas RL, Coutinho JAP (2008) Non-ideal behaviour of a room temperature ionic liquid in an alkoxyethanol or poly ethers at T = (298.15 to 318.15) K. J Chem Thermodyn 40(1):32–39
183. Geng Y, Wang T, Yu D, Peng C, Liu H, Hu Y (2008) Densities and viscosities of the ionic liquid [C_4mim][PF_6] + N, N-dimethylformamide binary mixtures at 293.15 K to 318.15 K. Chin J Chem Eng 16(2):256–262
184. Abdulagatov IM, Tekin A, Safarov J, Shahverdiyev A, Hassel E (2008) High-pressure densities and derived volumetric properties (excess, apparent and partial molar volumes) of binary mixtures of methanol + [BMIM][PF_6]. J Solut Chem 37(6):801–833
185. Zafarani-Moattar MT, Majdan-Cegincara R (2007) Viscosity, density, speed of sound, and refractive index of binary mixtures of organic solvent + ionic liquid, 1-butyl-3-methylimidazolium hexafluorophosphate at 298.15 K. J Chem Eng Data 52(6):2359–2364
186. Zafarani-Moattar MT, Shekaari H (2006) Volumetric and compressibility behaviour of ionic liquid, 1-*n*-butyl-3-methylimidazolium hexafluorophosphate and tetrabutylammonium hexafluorophosphate in organic solvents at T = 298.15 K. J Chem Thermodyn 38(5):624–633
187. Zafarani-Moattar MT, Shekaari H (2005) Volumetric and speed of sound of ionic liquid, 1-butyl-3-methylimidazolium hexafluorophosphate with acetonitrile and methanol at T = (298.15 to 318.15) K. J Chem Eng Data 50(5):1694–1699
188. Freire MG, Neves CMSS, Carvalho PJ, Gardas RL, Fernandes AM, Marrucho IM, Santos LMNBF, Coutinho JAP (2007) Mutual solubilities of water and hydrophobic ionic liquids. J Phys Chem B 111(45):13082–13089
189. Shiflett MB, Yokozeki A (2007) Hydrogen substitution effect on the solubility of perhalogenated compounds in ionic liquid [bmim][PF_6]. Fluid Phase Equilib 259(2):210–217
190. Pereiro AB, Rodríguez A (2007) Experimental liquid – liquid equilibria of 1-alkyl-3-methylimidazolium hexafluorophosphate with 1-alcohols. J Chem Eng Data 52(4):1408–1412
191. Sahandzhieva K, Tuma D, Breyer S, Pérez-Salado Kamps Á, Maurer G (2006) Liquid – liquid equilibrium in mixtures of the ionic liquid 1-*n*-butyl-3-methylimidazolium hexafluorophosphate and an alkanol. J Chem Eng Data 51(5):1516–1525
192. Bendová M, Wagner Z (2006) Liquid – liquid equilibrium in binary system [bmim] [PF_6] + 1-butanol. J Chem Eng Data 51(6):2126–2131
193. Shiflett MB, Yokozeki A (2006) Vaporeki4443-liquid equilibria of pentafluoroethane and ionic liquid [bmim][PF_6] mixtures studied with the volumetric method. J Phys Chem B 110 (29):14436–14443
194. Wang J, Zhu A, Zhao Y, Zhuo K (2005) Excess molar volumes and excess logarithm viscosities for binary mixtures of the ionic liquid 1-butyl-3-methylimidazolium hexafluorophosphate with some organic compounds. J Solut Chem 34(5):585–596

195. Zhong Y, Wang H, Diao K (2007) Densities and excess volumes of binary mixtures of the ionic liquid 1-butyl-3-methylimidazolium hexafluorophosphate with aromatic compound at T = (298.15 to 313.15) K. J Chem Thermodyn 39(2):291–296
196. Yokozeki A, Shiflett MB (2006) Global phase behaviors of trifluoromethane in ionic liquid [bmim][PF_6]. AIChE J 52(11):3952–3957
197. Shiflett MB, Yokozeki A (2006) Vaporeki39. – liquid equilibria of hydrofluorocarbons + 1-butyl-3-methylimidazolium hexafluorophosphate. J Chem Eng Data 51(5):1931–1939
198. Shariati A, Gutkowski K, Peters CJ (2005) Comparison of the phase behavior of some selected binary systems with ionic liquids. AIChE J 51(5):1532–1540
199. Zhu J, Chen J, Li C, Fei W (2006) Study on the separation of 1-hexene and *trans*-3-hexene using ionic liquids. Fluid Phase Equilib 247(1–2):102–106
200. Kumełan J, Kamps ÁP-S, Urukova I, Tuma D, Maurer G (2005) Solubility of oxygen in the ionic liquid [bmim][PF_6]: experimental and molecular simulation results. J Chem Thermodyn 37(6):595–602
201. Liu Z, Wu W, Han B, Dong Z, Zhao G, Wang J, Jiang T, Yang G (2003) Study on the phase behaviors, viscosities, and thermodynamic properties of CO_2/[C_4mim][PF_6]/methanol system at elevated pressures. Chem–Eur J 9(16):3897–3903
202. Chakrabarty D, Chakraborty A, Hazra P, Seth D, Sarkar N (2004) Dynamics of photoisomerisation and rotational relaxation of 3,3′-diethyloxadicarbocyanine iodide in room temperature ionic liquid and binary mixture of ionic liquid and water. Chem Phys Lett 397(1–3):216–221
203. Fan W, Zhou Q, Sun J, Zhang S (2009) Density, excess molar volume, and viscosity for the methyl methacrylate + 1-butyl-3-methylimidazolium hexafluorophosphate ionic liquid binary system at atmospheric pressure. J Chem Eng Data 54(8):2307–2311
204. Fan J, Fan Y, Pei Y, Wu K, Wang J, Fan M (2008) Solvent extraction of selected endocrine-disrupting phenols using ionic liquids. Sep Purif Technol 61(3):324–331
205. Machida H, Sato Y, Smith RL (2008) Measurement and correlation of infinite dilution partition coefficients of aromatic compounds in the ionic liquid 1-butyl-3-methyl-imidazolium hexafluorophosphate ([bmim][PF_6])–CO_2 system at temperatures from 313 to 353 K and at pressures up to 16 MPa. J Supercrit Fluids 43(3):430–437
206. Dongbao F, Xuewen S, Yanhua Q, Xiaohui J, Suoqi Z (2007) High-pressure phase behavior of the ternary system CO_2 + ionic liquid [bmim][PF_6] + naphthalene. Fluid Phase Equilib 251 (2):114–120
207. Zhang J, Wu W, Jiang T, Gao H, Liu Z, He J, Han B (2003) Conductivities and viscosities of the ionic liquid [bmim][PF_6] + water + ethanol and [bmim][PF_6] + water + acetone ternary mixtures. J Chem Eng Data 48(5):1315–1317
208. Zhang Z, Wu W, Wang B, Chen J, Shen D, Han B (2007) High-pressure phase behavior of CO_2/acetone/ionic liquid system. J Supercrit Fluids 40(1):1–6
209. Domańska U, Laskowska M (2009) Temperature and composition dependence of the density and viscosity of binary mixtures of {1-butyl-3-methylimidazolium thiocyanate + 1-alcohols}. J Chem Eng Data 54(7):2113–2119
210. Domańska U, Laskowska M (2009) Effect of temperature and composition on the density and viscosity of binary mixtures of ionic liquid with alcohols. J Solut Chem 38(6):779–799
211. Shekaari H, Mousavi SS, Mansoori Y (2009) Thermophysical properties of ionic liquid, 1-pentyl-3-methylimidazolium chloride in water at different temperatures. Inter J Thermophys 30(2):499–514
212. Xiao D, Rajian JR, Hines LG Jr, Li S, Bartsch RA, Quitevis EL (2008) Nanostructural organization and anion effects in the optical Kerr effect spectra of binary ionic liquid mixtures. J Phys Chem B 112(42):13316–13325
213. Gómez E, Calvar N, Domínguez I, Domínguez A (2006) Physical properties of the ternary mixture ethanol + water + 1-hexyl-3-methylimidazolium chloride at 298.15 K. Phys Chem Liq 44(4):409–417

214. Calvar N, González B, Gómez E, Domínguez A (2007) Study of the behaviour of the azeotropic mixture ethanol–water with imidazolium-based ionic liquids. Fluid Phase Equilib 259(1):51–56
215. Zhang L, Ge Y, Ji D, Ji J (2009) Experimental measurement and modeling of vaporimental equilibrium for ternary systems containing ionic liquids: a case study for the system water + ethanol + 1-hexyl-3-methylimidazolium chloride. J Chem Eng Data 54(8):2322–2329
216. Shekaari H, Mansoori Y, Sadeghi R (2008) Density, speed of sound, and electrical conductance of ionic liquid 1-hexyl-3-methyl-imidazolium bromide in water at different temperatures. J Chem Thermodyn 40(5):852–859
217. Waliszewski D (2008) Heat capacities of the mixtures of ionic liquids with methanol at temperatures from 283.15 K to 323.15 K. J Chem Thermodyn 40(2):203–207
218. Costantini M, Toussaint VA, Shariati A, Peters CJ, Kikic I (2005) High-pressure phase behavior of systems with ionic liquids: part iv. Binary system carbon dioxide + 1-hexyl-3-methylimidazolium tetrafluoroborate. J Chem Eng Data 50(1):52–55
219. Wagner M, Stanga O, Schröer W (2004) Critical viscosity near the liquid? Liquid phase transition in the solution of the ionic liquid 1-methyl-3-hexylimidazolium tetrafluoroborate in 1-pentanol. Phys Chem Chem Phys 6(8):1750–1757
220. Wagner M, Stanga O, Schröer W (2003) Corresponding states analysis of the critical points in binary solutions of room temperature ionic liquids. Phys Chem Chem Phys 5(18):3943–3950
221. Ren W, Scurto AM (2009) Phase equilibria of imidazolium ionic liquids and the refrigerant gas, 1,1,1,2-tetrafluoroethane (R-134a). Fluid Phase Equilib 286(1):1–7
222. Letcher TM, Reddy P (2005) Ternary (liquid + liquid) equilibria for mixtures of 1-hexyl-3-methylimidazolium (tetrafluoroborate or hexafluorophosphate) + benzene + an alkane at $T = 298.2$ K and $p = 0.1$ MPa. J Chem Thermodyn 37(5):415–421
223. Letcher TM, Reddy P (2004) Ternary liquid–liquid equilibria for mixtures of 1-hexyl-3-methylimidozolium (tetrafluoroborate or hexafluorophosphate) + ethanol + an alkene at $T = 298.2$ K. Fluid Phase Equilib 219(2):107–112
224. Pison L, Canongia Lopes JN, Rebelo LPN, Padua AAH, Costa Gomes MF (2008) Interactions of fluorinated gases with ionic liquids: solubility of CF_4, C_2F_6, and C_3F_8 in trihexyltetradecylphosphonium bis(trifluoromethylsulfonyl)amide. J Phys Chem B 112(39):12394–12400
225. Łachwa J, Morgado P, Esperança JMSS, Guedes HJR, Canongia Lopes JN, Rebelo LPN (2006) Fluid-phase behavior of {1-hexyl-3-methylimidazolium bis(trifluoromethylsulfonyl) imide, $[C_6mim][NTf_2]$, + $C_2 - C_8$ *n*-alcohol} mixtures: liquid – liquid equilibrium and excess volumes. J Chem Eng Data 51(6):2215–2221
226. Shevelyova MP, Zaitsau DH, Paulechka YU, Blokhin AV, Kabo GJ, Verevkin SP, Heintz A (2007) Solidzinaa. equilibrium and activity coefficients for caprolactam + 1-hexyl-3-methylimidazolium bis(trifluoromethylsulfonyl)imide and cyclohexanone oxime + 1-hexyl-3-methylimidazolium bis(trifluoromethylsulfonyl)imide. J Chem Eng Data 52(4):1360–1365
227. Heintz A, Verevkin SP, Lehmann JK, Vasiltsova TV, Ondo D (2007) Activity coefficients at infinite dilution and enthalpies of solution of methanol, 1-butanol, and 1-hexanol in 1-hexyl-3-methyl-imidazolium bis(trifluoromethyl-sulfonyl) imide. J Chem Thermodyn 39(2):268–274
228. Yokozeki A, Shiflett MB (2009) Separation of carbon dioxide and sulfur dioxide gases using room-temperature ionic liquid $[hmim][Tf_2N]$. Energy Fuel 23(9):4701–4708
229. Pereiro AB, Rodríguez A (2008) Phase equilibria of the azeotropic mixture hexane + ethyl acetate with ionic liquids at 298.15 K. J Chem Eng Data 53(6):1360–1366
230. Pereiro AB, Tojo E, Rodrıguez A, Canosa J, Tojo J (2006) Properties of ionic liquid $HMIMPF_6$ with carbonates, ketones and alkyl acetates. J Chem Thermodyn 38(6):651–661
231. Shariati A, Peters CJ (2004) High-pressure phase behavior of systems with ionic liquids. Part III. The binary system carbon dioxide + 1-hexyl-3-methylimidazolium hexafluorophosphat. J Supercrit Fluids 30(2):139–144
232. Pereiro AB, Rodríguez A (2008) Measurement and correlation of (liquid + liquid) equilibrium of the azeotrope (cyclohexane + 2-butanone) with different ionic liquids at $T = 298.15$ K. J Chem Thermodyn 40(8):1282–1289

233. González EJ, Alonso L, Domínguez Á (2006) Physical properties of binary mixtures of the ionic liquid 1-methyl-3-octylimidazolium chloride with methanol, ethanol, and 1-propanol at T = (298.15, 313.15, and 328.15) K and at P = 0.1 MPa. J Chem Eng Data 51(4):1446–1452
234. Letcher TM, Deenadayalu N, Soko B, Ramjugernath D, Naicker PK (2003) Ternary liquid – liquid equilibria for mixtures of 1-methyl-3-octylimidazolium chloride + an alkanol + an alkane at 298.2 K and 1 bar. J Chem Eng Data 48(4):904–907
235. Letcher TM, Deenadayalu N (2003) Ternary liquid–liquid equilibria for mixtures of 1-methyl-3-octyl-imidazolium chloride + benzene + an alkane at T =298.2 K and 1 atm. J Chem Thermodyn 35(1):67–76
236. Wang J, Zhang S, Wang H, Pei Y (2009) Apparent molar volumes and electrical conductance of ionic liquids [Cnmim]Br (n = 8, 10, 12) in ethylene glycol, N, N-dimethylformamide, and dimethylsulfoxide at 298.15 K. J Chem Eng Data 54(12):3252–3258
237. Singh T, Kumar A (2008) Physical and excess properties of a room temperature ionic liquid (1-methyl-3-octylimidazolium tetrafluoroborate) with n-alkoxyethanols (C_1E_m, m = 1 to 3) at T = (298.15 to 318.15) K. J Chem Thermodyn 40(3):417–423
238. Gutkowski KI, Shariati A, Peters CJ (2006) High-pressure phase behavior of the binary ionic liquid system 1-octyl-3-methylimidazolium tetrafluoroborate + carbon dioxide. J Supercrit Fluids 39(2):187–191
239. Alonso L, Arce A, Francisco M, Soto A (2008) Solvent extraction of thiophene from *n*-alkanes (C_7, C_{12}, and C_{16}) using the ionic liquid [C_8mim][BF4]. J Chem Thermodyn 40(6):966–972
240. Alonso L, Arce A, Francisco M, Rodríguez O, Soto A (2007) Gasoline desulfurization using extraction with [C_8mim][BF_4] ionic liquid. AIChE J 53(12):3108–3115
241. Andreatta AE, Arce A, Rodil E, Soto A (2009) Physical and excess properties of (methyl acetate + methanol + 1-octyl-3-methyl-imidazolium bis(trifluoromethylsulfonyl)imide) and its binary mixtures at T = 298.15 K and atmospheric pressure. J Chem Thermodyn 41 (12):1317–1323
242. Andreatta AE, Arce A, Rodil E, Soto A (2009) Physical properties of binary and ternary mixtures of ethyl acetate, ethanol, and 1-octyl-3-methyl-imidazolium bis(trifluoromethyl-sulfonyl)imide at 298.15 K. J Chem Eng Data 54(3):1022–1028
243. Alonso L, Arce A, Francisco M, Soto A (2008) (Liquid + liquid) equilibria of [C_8mim][NTf_2] ionic liquid with a sulfur-component and hydrocarbons. J Chem Thermodyn 40(2):265–270
244. Alonso L, Arce A, Francisco M, Soto A (2008) Phase behaviour of 1-methyl-3-octylimidazolium bis[trifluoromethylsulfonyl]imide with thiophene and aliphatic hydrocarbons: the influence of *n*-alkane chain length. Fluid Phase Equilib 263(2):176–181
245. Pereiro AB, Rodríguez A (2008) Binary mixtures containing OMIM PF_6: density, speed of sound, refractive index and LLE with hexane, heptane and 2-propanol at several temperatures. Phys Chem Liq 46(2):162–174
246. Domańska U, Bogel-Łukasik E (2003) Measurements and correlation of the (solid + liquid) equilibria of [1-decyl-3-methylimidazolium chloride + alcohols (C_2 – C_{12})]. Ind Eng Chem Res 42(26):6986–6992
247. Gan Q, Xue M, Rooney D (2006) A study of fluid properties and microfiltration characteristics of room temperature ionic liquids [C_{10}-min][NTf_2] and N_{8881}[NTf_2] and their polar solvent mixtures. Sep Purif Technol 51(2):185–192
248. Schröer W, Vale VR (2009) Liquid-liquid phase separation in solutions of ionic liquids: phase diagrams, corresponding state analysis and comparison with simulations of the primitive model. J Phys Condens Matter 21(42):1–21
249. Fedicheva N, Ninni L, Maurer G (2007) Phase equilibrium of aqueous systems containing 3-methyl-1-vinylimidazolium methyl sulfate. J Chem Eng Data 52(6):2228–2232
250. Domańska U, Marciniak A (2007) Phase behaviour of 1-hexyloxymethyl-3-methyl-imidazolium and 1,3-dihexyloxymethyl-imidazolium based ionic liquids with alcohols, water, ketones and hydrocarbons: the effect of cation and anion on solubility. Fluid Phase Equilib 260(1):9–18

251. Domańska U, Pobudkowska A, Żołek-Tryznowska Z (2007) Effect of an ionic liquid (IL) cation on the ternary system (IL + p-xylene + hexane) at T = 298.15 K. J Chem Eng Data 52(6):2345–2349
252. Najdanovic-Visak V, Rodriguez A, Visak ZP, Rosa JN, Afonso CAM, Nunes da Ponte M, Rebelo LPN (2007) Co-solvent effects in LLE of 1-hydroxyethyl-3-methylimidazolium based ionic liquids + 2-propanol + dichloromethane or 1,2-dichloroethane. Fluid Phase Equilib 254 (1–2):35–41
253. Xu H, Zhao D, Xu P, Liu F, Gao G (2005) Conductivity and viscosity of 1-allyl-3-methyl-imidazolium chloride + water and + ethanol from 293.15 K to 333.15 K. J Chem Eng Data 50 (1):133–135
254. Elias AM, Wilkes JS (1994) Densities, molar volumes, and thermal expansivities of 1-methyl-3-ethylimidazolium chloride + aluminum chloride + alkali-metal halide molten salts. J Chem Eng Data 39(1):79–82
255. Domańska U, Marciniak A (2005) Liquid phase behaviour of 1-hexyloxymethyl-3-methyl-imidazolium-based ionic liquids with hydrocarbons: the influence of anion. J Chem Thermodyn 37(6):577–585
256. Ge ML, Ren XG, Song YJ, Wang LS (2009) Densities and viscosities of 1-propyl-2,3-dimethylimidazolium tetrafluoroborate + H_2O at T = (298.15 to 343.15) K. J Chem Eng Data 54(4):1400–1402
257. Bou Malham I, Letellier P, Turmine M (2008) Synthesis and micellar properties of 1-decyl-2,3-dimethylimidazolium bromide surfactant in water and water-ethanolamine mixtures at 298.15 K. J Colloid Interface Sci 328(1):166–171

Mixtures Containing Triazolium Ionic Liquids

3

3.1 07068-05001: 1-Ethyl-3-butylbenzotriazolium Hexafluorophosphate

Abbreviation:	C_2BtPF_6
Molecular formula:	$C_{12}H_{18}N_3PF_6$
Molar mass:	349.25
Structure:	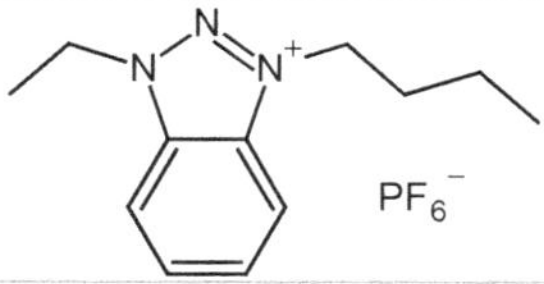

3.1.1 Binary Mixtures

Mixture 1: IM (1) + C_2BtPF_6 (2)
(1) Melting Point and Transport Properties

Ref. [1]

Molar ratio (C_2BtPF_6:IM)	T_m(K)	η/(cp) 343.15 K	353.15 K	363.15 K	373.15 K	383.15 K	393.15 K	403.15 K
1: 4	330.15	11.0	7.7	5.6	5.2	4.2	3.6	3.1

Ref. [1]

Shear rate (s^{-1})	η/(cp)	Shear rate (s^{-1})	η/(cp)
343.15 K Molar ratio (C_2BtPF_6:IM=1:4)			
836	11	1855	10.8
1006	11	2028	9.7
1092	10.9		

S. Zhang et al., *Physicochemical Properties of Ionic Liquid Mixtures*,
DOI 10.1007/978-94-017-7573-1_3

(2) Electrical Conductivity

Ref. [1]

Molar ratio (C_2BtPF_6:IM)	Conductivity ($S{\cdot}cm^{-1}$)						
	343.15 K	353.15 K	363.15 K	373.15 K	383.15 K	393.15 K	403.15 K
1: 4	1.1×10^{-2}	1.4×10^{-2}	1.8×10^{-2}	2.2×10^{-2}	2.9×10^{-2}	3.2×10^{-2}	3.8×10^{-2}

3.1.2 Ternary Mixtures

Mixture 1: IM (1) + C_2BtPF_6 (2) + Solvents (3)
(1) Solubility in common solvents

Ref. [1]

Molar ratio (C_2BtPF_6: IM)	Solubility (wt.%)										
	Water	Acetonitrile	Ethanol	Dichloro-methane	Chloroform	Acetone	Ethyl acetate	Toluene	Benzene	Ethyl ether	Hexane
1:4	0.41	54.33	1.31	24.76	13.46	66.28	39.93	im	im	im	im

im immiscible

Mixture 2: IM (1) + C_2BtPF_6 (2) + Compounds (3)
(1) Solubility of some compounds in Binary Mixtures

Ref. [1]

Molar ratio (C_2BtPF_6:IM)	Solubility (wt.%) (343.15 K)							
	CuO	Fe_2O_3	$MgCl_2$	LiCl	$ZnCl_2$	Glucose	Citric acid	Benzoic acid
1:4	0.05	0.01	1.62	5.07	0.49	4.60	7.68	23.19

References

1. Hou Y, Gu Y, Zhang S, Yang F, Ding H, Shan Y (2008) Novel binary eutectic mixtures based on imidazole. J Mol Liq 143(2–3):154–159

Mixtures Containing Pyrrolidinium Ionic Liquids

4

4.1 08003-03001: *N*-methyl-*N*-propyl-pyrrolidinium bis ((trifluoromethyl)sulfonyl)imide

Abbreviation:	[P13][TFSI], $[C_3mpyr][Tf_2N]$
Molecular formula:	$C_{10}H_{18}F_6N_2O_4S_2$
Molar mass:	408.39
Structure:	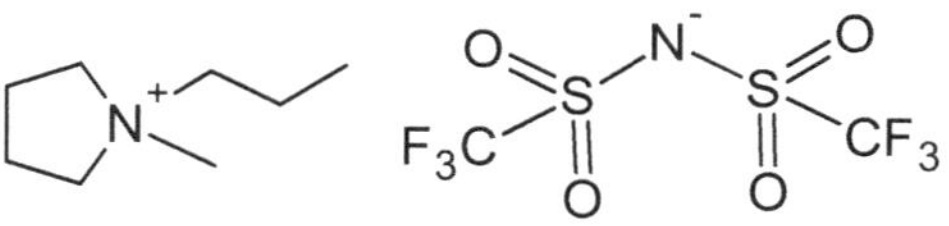

4.1.1 Binary Mixtures

Mixture 1: H_2O (1) + $[C_3mpyr][Tf_2N]$ (2)
(1) Solubility

Ref. [1]

T/K	$x_1 \pm \sigma^a$	*T*/K	$x_1 \pm \sigma^a$
288.15	0.200 ± 0.004	308.15	0.268 ± 0.004
293.15	0.220 ± 0.001	313.15	0.290 ± 0.004
298.15	0.233 ± 0.004	318.15	0.299 ± 0.007
303.15	0.254 ± 0.001		

[a]Standard deviation

S. Zhang et al., *Physicochemical Properties of Ionic Liquid Mixtures*,
DOI 10.1007/978-94-017-7573-1_4

4.2 08004-03001: *N*-methyl-*N*-butyl-pyrrolidinium bis ((trifluoromethyl)sulfonyl)imide

Abbreviation:	[P14][TFSI], [C_4mpyr][Tf_2N], [bmpy][Tf_2N], [BMPYR][BTI], [C_1C_4Pyrro][NTf_2], BMPIm
Molecular formula:	$C_{11}H_{20}F_6N_2O_4S_2$
Molar mass:	422.41
Structure:	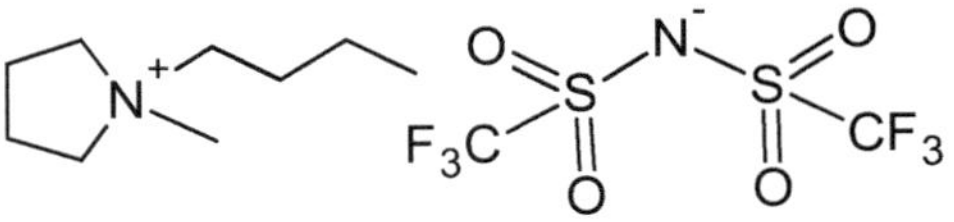

4.2.1 Binary Mixtures

Mixture 1: H_2O (1) + [C_4mpyr][Tf_2N] (2)
(1) Volumetric Properties

Ref. [2]

x_1	ρ/g · cm^{-3}	x_1	ρ/g · cm^{-3}
298.21 K		323.21 K	
0	1.3953	0	1.3736
0.0435	1.3940	0.0298	1.3725
0.0654	1.3932	0.0673	1.3711
0.0860	1.3923	0.1012	1.3699
0.0913	1.3921	0.1406	1.3683
0.1262	1.3907	0.2033	1.3653
0.1396	1.3901		
0.1909	1.3881		
0.2228	1.3871		
x_1	V^E_m/cm^3 · mol^{-1}	x_1	V^E_m/cm^3 · mol^{-1}
298.21 K		323.21 K	
0	0	0	0
0.0435	0.058	0.0298	0.090
0.0654	0.100	0.0673	0.172
0.0860	0.148	0.1012	0.241
0.0913	0.169	0.1406	0.320
0.1262	0.224	0.2033	0.467
0.1396	0.255		
0.1909	0.304		
0.2228	0.265		

(2) Solubility

Ref. [1]

T/K	$x_1 \pm \sigma^a$	T/K	$x_1 \pm \sigma^a$
288.15	0.186 ± 0.004	308.15	0.244 ± 0.005
293.15	0.199 ± 0.004	313.15	0.263 ± 0.003
298.15	0.211 ± 0.001	318.15	0.282 ± 0.004
303.15	0.233 ± 0.004		

[a]Standard deviation

Ref. [2]

T/K	x_1	T/K	x_1	T/K	x_1	T/K	x_1
278.46	0.1800	312.37	0.2921	345.11	0.3902	358.06	0.9994
290.35	0.2260	319.21	0.3050	350.14	0.3996	338.19	0.9995
297.17	0.2430	327.37	0.3250	357.17	0.4270	318.24	0.9997
303.45	0.2650	333.21	0.3450	365.12	0.4390	290.11	0.9998
308.01	0.2765	338.25	0.3685				

Mixture 2: $HCON(CH_3)_2$ (1) + [bmpy][Tf_2N] (2)

(1) Transport Properties

Ref. [3]

x_1	η/mPa · s	x_1	η/mPa · s	x_1	η/mPa · s
295.2 K					
0.00	80.0	0.33	21.7	0.64	6.6
0.13	40.3	0.45	14.4	0.74	3.7
0.24	29.6	0.54	9.9		

(2) Electrical Conductivity

Ref. [3]

x_1	k/mS · cm^{-1}	x_1	k/mS · cm^{-1}	x_1	k/mS · cm^{-1}	x_1	k/mS · cm^{-1}
295.2 K							
0.000	2.80	0.457	7.80	0.773	18.49	0.931	19.83
0.144	3.87	0.519	9.81	0.835	20.95	0.953	17.04
0.252	4.95	0.574	11.05	0.871	21.65	0.971	13.12
0.336	6.12	0.637	13.17	0.894	21.70	0.981	10.10
0.403	7.40	0.716	16.01	0.910	21.20		

Mixture 3: $3\text{-}CH_3C_5H_4N$ (3-Picoline) (1) + [bmpy][Tf_2N] (2)
(1) Electrical Conductivity

Ref. [3]

x_1	k/mS · cm^{-1}	x_1	k/mS · cm^{-1}	x_1	k/mS · cm^{-1}	x_1	k/mS · cm^{-1}
295.2 K							
0.000	2.80	0.194	4.45	0.490	6.35	0.743	8.14
0.029	3.53	0.249	4.67	0.540	6.86	0.794	8.08
0.057	3.35	0.338	5.22	0.591	7.26	0.845	7.71
0.095	3.67	0.387	5.78	0.640	7.61	0.896	6.65
0.153	4.16	0.439	6.06	0.691	7.93		

Mixture 4: C_5H_{12} (1) + [BMPYR][BTI] (2)
(1) Activity Coefficient at Infinite Dilution and Partial Molar Excess Enthalpy at Infinite Dilution

Ref. [4]

Solute (1)	$\gamma^{\infty}{}_1$ (dilutor technique)				$H_{12}{}^{E,\infty}$/kJ · mol^{-1}
	303.15 K	313.15 K	323.15 K	333.15 K	
n-Pentane	9.22	9.18	8.62	8.29	3.180

Mixture 5: C_6H_{14} (1) + [BMPYR][BTI] (2)
(1) Activity Coefficient at Infinite Dilution, Partial Molar Excess Enthalpy at Infinite Dilution and Excess Molar Enthalpy at Infinite Dilution

Ref. [4]

Solute (1)	$\gamma^{\infty}{}_1$ (dilutor technique)				$H_{12}{}^{E,\infty}$/kJ · mol^{-1}
	303.15 K	313.15 K	323.15 K	333.15 K	
n-Hexane	13.8	13.3	12.1	11.0	6.822

Ref. [5]

Solute	$\gamma_{12}{}^{\infty}$ (gas–liquid chromatography)	$\Delta H_1{}^{E,\infty}$ (kJ · mol^{-1})
298.15 K		
n-Hexane	14.7	6.5

Mixture 6: C_7H_{16} (1) + [BMPYR][BTI] (2)
(1) Activity Coefficient at Infinite Dilution and Partial Molar Excess Enthalpy at Infinite Dilution

Ref. [4]

Solute (1)	$\gamma^{\infty}{}_1$ (dilutor technique)				$H_{12}{}^{E,\infty}$/kJ · mol^{-1}
	303.15 K	313.15 K	323.15 K	333.15 K	
n-Heptane	20.7	19.3	18.2	17.2	5.172

Mixture 7: C_8H_{18} (1) + [BMPYR][BTI] (2)

(1) Activity Coefficient at Infinite Dilution and Partial Molar Excess Enthalpy at Infinite Dilution

Ref. [4]

Solute (1)	$\gamma^{\infty}{}_1$ (dilutor technique)				$H_{12}{}^{E,\infty}$/kJ · mol^{-1}
	303.15 K	313.15 K	323.15 K	333.15 K	
n-Octane	28.3	26.5	25.2	23.7	4.890

Mixture 8: $CH_2(CH_2)_3CH_2$ (1) + [BMPYR][BTI] (2)

(1) Activity Coefficient at Infinite Dilution and Partial Molar Excess Enthalpy at Infinite Dilution

Ref. [4]

Solute (1)	$\gamma^{\infty}{}_1$ (dilutor technique)				$H_{12}{}^{E,\infty}$/kJ · mol^{-1}
	303.15 K	313.15 K	323.15 K	333.15 K	
Cyclopentane	6.14	5.70	5.41	5.12	5.019

Mixture 9: $CH_2(CH_2)_4CH_2$ (1) + [BMPYR][BTI] (2)

(1) Activity Coefficient at Infinite Dilution, Partial Molar Excess Enthalpy at Infinite Dilution and Excess Molar Enthalpy at Infinite Dilution

Ref. [4]

Solute (1)	$\gamma^{\infty}{}_1$ (dilutor technique)				$H_{12}{}^{E,\infty}$/kJ · mol^{-1}
	303.15 K	313.15 K	323.15 K	333.15 K	
Cyclohexane	8.75	8.43	7.91	7.42	4.667

Ref. [5]

Solute	$\gamma_{12}{}^{\infty}$ (gas–liquid chromatography)	$\Delta H_1{}^{E,\infty}$ (kJ · mol^{-1})
298.15 K		
Cyclohexane	9.11	4.7

Mixture 10: C_6H_6 (1) + [BMPYR][BTI] (2)

(1) Activity Coefficient at Infinite Dilution, Partial Molar Excess Enthalpy at Infinite Dilution and Excess Molar Enthalpy at Infinite Dilution

Ref. [4]

Solute (1)	$\gamma^{\infty}{}_1$ (dilutor technique)				$H_{12}{}^{E,\infty}$/kJ · mol^{-1}
	303.15 K	313.15 K	323.15 K	333.15 K	
Benzene	0.84	0.86	0.88	0.89	−1.654

Ref. [5]

Solute	$\gamma_{12}{}^{\infty}$ (gas–liquid chromatography)	$\Delta H_1{}^{E,\infty}$ (kJ · mol^{-1})
298.15 K		
Benzene	0.83	−1.7

Mixture 11: $1\text{-}C_5H_{10}$ (1) + [BMPYR][BTI] (2)
(1) Activity Coefficient at Infinite Dilution and Partial Molar Excess Enthalpy at Infinite Dilution

Ref. [4]

Solute (1)	$\gamma^{\infty}{}_1$ (dilutor technique)				$H_{12}{}^{E,\infty}$/kJ · mol^{-1}
	303.15 K	313.15 K	323.15 K	333.15 K	
1-Pentene	4.95	4.91	4.81	4.80	0.950

Mixture 12: $1\text{-}C_6H_{12}$ (1) + [BMPYR][BTI] (2)
(1) Activity Coefficient at Infinite Dilution, Partial Molar Excess Enthalpy at Infinite Dilution and Excess Molar Enthalpy at Infinite Dilution

Ref. [4]

Solute (1)	$\gamma^{\infty}{}_1$ (dilutor technique)				$H_{12}{}^{E,\infty}$/kJ · mol^{-1}
	303.15 K	313.15 K	323.15 K	333.15 K	
1-Hexene	7.08	6.92	6.82	6.64	1.735

Ref. [5]

Solute	$\gamma_{12}{}^{\infty}$ (gas–liquid chromatography)	$\Delta H_1{}^{E,\infty}$ (kJ · mol^{-1})
298.15 K		
Hex-1-ene	7.17	1.7

(2) Excess Enthalpy

Ref. [5]

x_1	H^E/J · mol^{-1}	x_1	H^E/J · mol^{-1}	x_1	H^E/J · mol^{-1}
363.15 K, 2239 kPa					
0.0587	139.1	0.3781	459.2	0.7086	222.9
0.1135	267.7	0.4477	412.8	0.7849	166.9
0.1647	349.4	0.5104	361.4	0.8795	84.91
0.2579	435.0	0.6185	278.6		

Mixture 13: $1\text{-}C_7H_{14}$ (1) + [BMPYR][BTI] (2)
(1) Activity Coefficient at Infinite Dilution and Partial Molar Excess Enthalpy at Infinite Dilution

Ref. [4]

Solute (1)	$\gamma^{\infty}{}_1$ (dilutor technique)				$H_{12}{}^{E,\infty}$/kJ · mol^{-1}
	303.15 K	313.15 K	323.15 K	333.15 K	
1-Heptene	10.4	9.97	9.69	9.37	2.867

(2) Excess Enthalpy

Ref. [6]

x_1	H^E/J · mol^{-1}	x_1	H^E/J · mol^{-1}	x_1	H^E/J · mol^{-1}
413.15 K, 2239 kPa					
0.0525	134.8	0.3509	418.6	0.6838	209.8
0.1022	259.7	0.4189	381.9	0.7643	146.5
0.1492	375.8	0.4810	337.8	0.8664	85.18
0.2762	470.9	0.5904	260.4	0.9511	38.13

Mixture 14: 1-C_8H_{16} (1) + [BMPYR][BTI] (2)
(1) Activity Coefficient at Infinite Dilution and Partial Molar Excess Enthalpy at Infinite Dilution

Ref. [4]

Solute (1)	γ^{∞}_1 (dilutor technique)				$H_{12}^{E,\infty}$/kJ · mol^{-1}
	303.15 K	313.15 K	323.15 K	333.15 K	
1-Octene	15.0	13.9	13.3	13.2	3.627

Mixture 15: 2-$CH_3COC_2H_5$ (1) + [BMPYR][BTI] (2)
(1) Activity Coefficient at Infinite Dilution and Partial Molar Excess Enthalpy at Infinite Dilution

Ref. [4]

Solute (1)	γ^{∞}_1 (dilutor technique)				$H_{12}^{E,\infty}$/kJ · mol^{-1}
	303.15 K	313.15 K	323.15 K	333.15 K	
2-Butanone	0.52	0.53	0.55	0.56	−2.177

Mixture 16: 2-$CH_3COC_3H_7$ (1) + [BMPYR][BTI] (2)
(1) Activity Coefficient at Infinite Dilution and Partial Molar Excess Enthalpy at Infinite Dilution

Ref. [4]

Solute (1)	γ^{∞}_1 (dilutor technique)				$H_{12}^{E,\infty}$/kJ · mol^{-1}
	303.15 K	313.15 K	323.15 K	333.15 K	
2-Pentanone	0.63	0.66	0.68	0.71	−3.261

Mixture 17: C_2H_5OH (1) + [BMPYR][BTI] (2)
(1) Activity Coefficient at Infinite Dilution and Partial Molar Excess Enthalpy at Infinite Dilution

Ref. [4]

Solute (1)	γ^{∞}_1 (dilutor technique)				$H_{12}^{E,\infty}$/kJ · mol^{-1}
	303.15 K	313.15 K	323.15 K	333.15 K	
Ethanol	1.84	1.73	1.63	1.50	5.635

Ref. [5]

Solute	γ_{12}^{∞} (gas–liquid chromatography)	$\Delta H_1^{E,\infty}$ (kJ · mol^{-1})
298.15 K		
Ethanol	1.92	5.6

Mixture 18: 2-CH_3CHCH_3OH (1) + [BMPYR][BTI] (2)
(1) Activity Coefficient at Infinite Dilution and Partial Molar Excess Enthalpy at Infinite Dilution

Ref. [4]

Solute (1)	γ^{∞}_1 (dilutor technique)				$H_{12}^{E,\infty}$/kJ · mol^{-1}
	303.15 K	313.15 K	323.15 K	333.15 K	
2-Propanol	2.11	1.92	1.77	1.63	7.194

Mixture 19: $(CH_3)_2CHOCH(CH_3)_2$ (1) + [BMPYR][BTI] (2)
(1) Partial Molar Excess Enthalpy at Infinite Dilution

Ref. [4]

Solute (1)	$H_{12}^{E,\infty}$/kJ · mol^{-1}
Diisopropyl ether	−2.197

Mixture 20: $[CH_2\text{-}CH_2\text{-}O]_n$ (poly(ethyleneoxide)) (1) + BMPIm (2)
(1) Electrical Conductivity

Ref. [7]

Composition (wt.%)		k (mS/cm), 298.15 K
Polymer	Ionic liquid	
64.8 % PEO	35.2 % BMPIm	0.008
38.8 % PEO	61.2 % BMPIm	0.05
25.4 % PEO	74.6 % BMPIm	0.3

4.2.2 Ternary Mixtures

Mixture 1: $(C_5H_5)_2Fe$ (1) + $HCON(CH_3)_2$ (2) + [bmpy][Tf_2N] (3)
(1) Diffusion Coefficient

Ref. [3]

(2) + (3)		(2) + (3)	
x_2	$10^6\ D_{Fc}$/cm^2 · s^{-1}	x_2	$10^6\ D_{Fc}$/cm^2 · s^{-1}
295.2 K			
0.00	0.08	0.60	0.81
0.10	0.12	0.70	1.20

(*continued*)

(2) + (3)		(2) + (3)	
x_2	$10^6\ D_{Fc}/cm^2 \cdot s^{-1}$	x_2	$10^6\ D_{Fc}/cm^2 \cdot s^{-1}$
0.20	0.18	0.80	1.80
0.30	0.27	0.90	2.60
0.40	0.40	0.94	3.00
0.50	0.59	0.97	3.10

Mixture 2: C_6H_{14} (1) + C_6H_6 (2) + [BMPYR][NTf$_2$] (3)
(1) Selectivity at Infinite Dilution and Capacities at Infinite Dilution

Ref. [5]

Solvent	S^{∞}_{ij}	k_j^{∞}
T = 298.15 K	n-Hexane (1)/benzene (2)	Benzene
[BMPYR][NTf$_2$]	17.7[a]	1.20[a]

[a]Extrapolated value

Mixture 3: $CH_2(CH_2)_4CH_2$ (1) + C_6H_6 (2) + [BMPYR][NTf$_2$] (3)
(1) Selectivity at Infinite Dilution and Capacities at Infinite Dilution

Ref. [5]

Solvent	S^{∞}_{ij}	k_j^{∞}
T = 298.15 K	Cyclohexane (1)/benzene (2)	Benzene
[BMPYR][NTf$_2$]	11.0	1.20[a]

[a]Extrapolated value

4.3 08004-04003: 1-Butyl-1-methyl-pyrrolidinium trifluoromethanesulfonate

Abbreviation:	[BMPYR][CF_3SO_3], [MBPYR][OTF]
Molecular formula:	$C_{10}H_{20}F_3NO_3S$
Molar mass:	291.33
Structure:	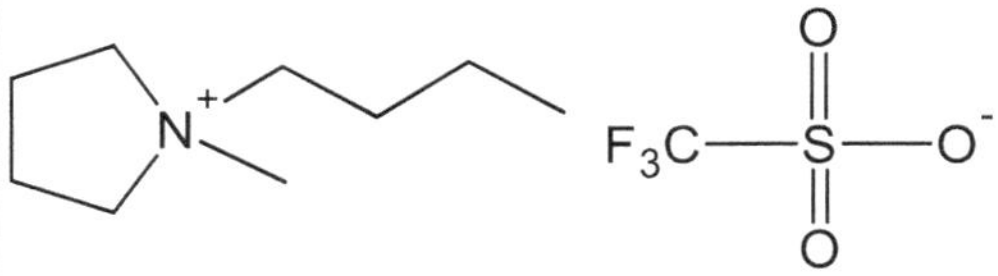

4.3.1 Binary Mixtures

Mixture 1: C_5H_{12} (1) + [BMPYR][CF_3SO_3] (2)
(1) Activity Coefficient at Infinite Dilution and Excess Molar Enthalpy at Infinite Dilution

Ref. [5]

Solute	γ_{12}^{∞} (gas–liquid chromatography)								$\Delta H_1^{E,\infty}$
	298.15 K	308.15 K	318.15 K	328.15 K	338.15 K	348.15 K	358.15 K	368.15 K	(kJ · mol^{-1})
n-Pentane	52.1	46.1	40.3	35.9	32.2	28.9	26.1		10.3

Mixture 2: C_6H_{14} (1) + [BMPYR][CF_3SO_3] (2)
(1) Activity Coefficient at Infinite Dilution and Excess Molar Enthalpy at Infinite Dilution

Ref. [5]

Solute	γ_{12}^{∞} (gas–liquid chromatography)								$\Delta H_1^{E,\infty}$
	298.15 K	308.15 K	318.15 K	328.15 K	338.15 K	348.15 K	358.15 K	368.15 K	(kJ · mol^{-1})
n-Hexane	55.8	50.6	46.4	42.2	38.8	35.8	33.3	31.3	7.6

Mixture 3: C_7H_{16} (1) + [BMPYR][CF_3SO_3] (2)
(1) Activity Coefficient at Infinite Dilution and Excess Molar Enthalpy at Infinite Dilution

Ref. [5]

Solute	γ_{12}^{∞} (gas–liquid chromatography)								$\Delta H_1^{E,\infty}$
	298.15 K	308.15 K	318.15 K	328.15 K	338.15 K	348.15 K	358.15 K	368.15 K	(kJ · mol^{-1})
n-Heptane	71.5	64.7	59.3	54.2	50.4	46.7	43.3	40.8	7.3

Mixture 4: C_8H_{18} (1) + [BMPYR][CF_3SO_3] (2)
(1) Activity Coefficient at Infinite Dilution and Excess Molar Enthalpy at Infinite Dilution

Ref. [5]

Solute	γ_{12}^{∞} (gas–liquid chromatography)								$\Delta H_1^{E,\infty}$
	298.15 K	308.15 K	318.15 K	328.15 K	338.15 K	348.15 K	358.15 K	368.15 K	(kJ · mol^{-1})
n-Octane	95.2	86.9	79.4	73.1	67.2	62.9	58.4	54.4	7.3

Mixture 5: C_9H_{20} (1) + [BMPYR][CF_3SO_3] (2)
(1) Activity Coefficient at Infinite Dilution and Excess Molar Enthalpies at Infinite Dilution

Ref. [5]

Solute	γ_{12}^{∞} (gas–liquid chromatography)								$\Delta H_1^{E,\infty}$
	298.15 K	308.15 K	318.15 K	328.15 K	338.15 K	348.15 K	358.15 K	368.15 K	(kJ · mol^{-1})
n-Nonane	135	122	111	100	92.4	85.2	78.9	73.3	8.0

Mixture 6: $C_{10}H_{22}$ (1) + [BMPYR][CF_3SO_3] (2)

(1) Activity Coefficient at Infinite Dilution and Excess Molar Enthalpy at Infinite Dilution

Ref. [5]

Solute	γ_{12}^{∞} (gas–liquid chromatography)								$\Delta H_1^{E,\infty}$
	298.15 K	308.15 K	318.15 K	328.15 K	338.15 K	348.15 K	358.15 K	368.15 K	(kJ · mol^{-1})
n-Decane	195	174	157	143	130	120	110	101	8.5

Mixture 7: $CH_2(CH_2)_3CH_2$ (1) + [BMPYR][CF_3SO_3] (2)

(1) Activity Coefficient at Infinite Dilution and Excess Molar Enthalpy at Infinite Dilution

Ref. [5]

Solute	γ_{12}^{∞} (gas–liquid chromatography)								$\Delta H_1^{E,\infty}$
	298.15 K	308.15 K	318.15 K	328.15 K	338.15 K	348.15 K	358.15 K	368.15 K	(kJ · mol^{-1})
Cyclopentane	19.2	17.4	15.9	14.6	13.5	12.6	11.7	11.0	7.3

Mixture 8: $CH_2(CH_2)_4CH_2$ (1) + [BMPYR][CF_3SO_3] (2)

(1) Activity Coefficient at Infinite Dilution and Excess Molar Enthalpy at Infinite Dilution

Ref. [5]

Solute	γ_{12}^{∞} (gas–liquid chromatography)								$\Delta H_1^{E,\infty}$
	298.15 K	308.15 K	318.15 K	328.15 K	338.15 K	348.15 K	358.15 K	368.15 K	(kJ · mol^{-1})
Cyclohexane	25.7	23.4	21.4	19.7	18.1	16.8	15.6	14.6	7.4

Mixture 9: $CH_2(CH_2)_5CH_2$ (1) + [BMPYR][CF_3SO_3] (2)

(1) Activity Coefficient at Infinite Dilution and Excess Molar Enthalpy at Infinite Dilution

Ref. [5]

Solute	γ_{12}^{∞} (gas–liquid chromatography)								$\Delta H_1^{E,\infty}$
	298.15 K	308.15 K	318.15 K	328.15 K	338.15 K	348.15 K	358.15 K	368.15 K	(kJ · mol^{-1})
Cycloheptane	29.3	27.0	24.9	23.1	21.6	20.1	18.9	17.9	6.5

Mixture 10: $CH_2(CH_2)_6CH_2$ (1) + [BMPYR][CF_3SO_3] (2)

(1) Activity Coefficient at Infinite Dilution and Excess Molar Enthalpy at Infinite Dilution

Ref. [5]

Solute	γ_{12}^{∞} (gas–liquid chromatography)								$\Delta H_1^{E,\infty}$
	298.15 K	308.15 K	318.15 K	328.15 K	338.15 K	348.15 K	358.15 K	368.15 K	(kJ · mol^{-1})
Cyclooctane	37.4	34.0	31.4	29.1	26.9	25.1	23.5	22.2	6.8

Mixture 11: 1-C_5H_{10} (1) + [BMPYR][CF_3SO_3] (2)

(1) Activity Coefficient at Infinite Dilution and Excess Molar Enthalpy at Infinite Dilution

Ref. [5]

Solute	γ_{12}^{∞} (gas–liquid chromatography)								$\Delta H_1^{E,\infty}$
	298.15 K	308.15 K	318.15 K	328.15 K	338.15 K	348.15 K	358.15 K	368.15 K	(kJ · mol^{-1})
Pent-1-ene	17.6	16.3	15.3	14.4	13.5	12.8	12.2		5.4

Mixture 12: 1-C_6H_{12} (1) + [BMPYR][CF_3SO_3] (2)

(1) Activity Coefficient at Infinite Dilution and Excess Molar Enthalpy at Infinite Dilution

Ref. [5]

Solute	γ_{12}^{∞} (gas–liquid chromatography)								$\Delta H_1^{E,\infty}$
	298.15 K	308.15 K	318.15 K	328.15 K	338.15 K	348.15 K	358.15 K	368.15 K	(kJ · mol^{-1})
Hex-1-ene	21.6	20.2	19.2	18.1	17.3	16.4	15.7	15.0	4.8

Mixture 13: 1-C_7H_{14} (1) + [BMPYR][CF_3SO_3] (2)

(1) Activity Coefficient at Infinite Dilution and Excess Molar Enthalpy at Infinite Dilution

Ref. [5]

Solute	γ_{12}^{∞} (gas–liquid chromatography)								$\Delta H_1^{E,\infty}$
	298.15 K	308.15 K	318.15 K	328.15 K	338.15 K	348.15 K	358.15 K	368.15 K	(kJ · mol^{-1})
Hept-1-ene	29.0	27.5	26.1	24.8	23.6	22.6	21.6	20.8	4.4

Mixture 14: 1-C_8H_{16} (1) + [BMPYR][CF_3SO_3] (2)

(1) Activity Coefficient at Infinite Dilution and Excess Molar Enthalpy at Infinite Dilution

Ref. [5]

Solute	γ_{12}^{∞} (gas–liquid chromatography)								$\Delta H_1^{E,\infty}$
	298.15 K	308.15 K	318.15 K	328.15 K	338.15 K	348.15 K	358.15 K	368.15 K	(kJ · mol^{-1})
Oct-1-ene	40.9	38.4	36.2	34.3	32.7	31.2	29.8	28.6	4.7

Mixture 15: $CH \equiv CCH_2CH_2CH_3$ (1) + [BMPYR][CF_3SO_3] (2)

(1) Activity Coefficient at Infinite Dilution and Excess Molar Enthalpy at Infinite Dilution

Ref. [5]

Solute	γ_{12}^{∞} (gas–liquid chromatography)								$\Delta H_1^{E,\infty}$
	298.15 K	308.15 K	318.15 K	328.15 K	338.15 K	348.15 K	358.15 K	368.15 K	(kJ · mol^{-1})
Pent-1-yne	2.59	2.66	2.74	2.82	2.88	2.95	3.01	3.08	−2.3

Mixture 16: $CH \equiv CCH_2CH_2CH_2CH_3$ (1) + [BMPYR][CF_3SO_3] (2)

(1) Activity Coefficient at Infinite Dilution and Excess Molar Enthalpy at Infinite Dilution

Ref. [5]

Solute	γ_{12}^{∞} (gas–liquid chromatography)								$\Delta H_1^{E,\infty}$
	298.15 K	308.15 K	318.15 K	328.15 K	338.15 K	348.15 K	358.15 K	368.15 K	(kJ · mol^{-1})
Hex-1-yne	3.57	3.66	3.76	3.85	3.94	4.02	4.09	4.18	−2.0

Mixture 17: $CH \equiv CCH_2CH_2CH_2CH_2CH_3$ (1) + [BMPYR][CF_3SO_3] (2)

(1) Activity Coefficient at Infinite Dilution and Excess Molar Enthalpy at Infinite Dilution

Ref. [5]

Solute	γ_{12}^{∞} (gas–liquid chromatography)								$\Delta H_1^{E,\infty}$
	298.15 K	308.15 K	318.15 K	328.15 K	338.15 K	348.15 K	358.15 K	368.15 K	(kJ · mol^{-1})
Hept-1-yne	5.21	5.30	5.39	5.47	5.56	5.64	5.70	5.78	−1.4

Mixture 18: $CH \equiv CCH_2CH_2CH_2CH_2CH_2CH_3$ (1) + [BMPYR][CF_3SO_3] (2)

(1) Activity Coefficient at Infinite Dilution and Excess Molar Enthalpy at Infinite Dilution

Ref. [5]

Solute	γ_{12}^{∞} (gas–liquid chromatography)								$\Delta H_1^{E,\infty}$
	298.15 K	308.15 K	318.15 K	328.15 K	338.15 K	348.15 K	358.15 K	368.15 K	(kJ · mol^{-1})
Oct-1-yne	7.56	7.63	7.70	7.76	7.83	7.88	7.94	7.99	−0.7

Mixture 19: C_6H_6 (1) + [BMPYR][CF_3SO_3] (2)

(1) Activity Coefficient at Infinite Dilution and Excess Molar Enthalpy at Infinite Dilution

Ref. [5]

Solute	γ_{12}^{∞} (gas–liquid chromatography)								$\Delta H_1^{E,\infty}$
	298.15 K	308.15 K	318.15 K	328.15 K	338.15 K	348.15 K	358.15 K	368.15 K	(kJ · mol^{-1})
Benzene	1.41	1.44	1.46	1.48	1.50	1.53	1.54	1.56	−1.3

Mixture 20: $C_6H_5CH_3$ (1) + [BMPYR][CF_3SO_3] (2)

(1) Activity Coefficient at Infinite Dilution and Excess Molar Enthalpy at Infinite Dilution

Ref. [5]

Solute	γ_{12}^{∞} (gas–liquid chromatography)								$\Delta H_1^{E,\infty}$
	298.15 K	308.15 K	318.15 K	328.15 K	338.15 K	348.15 K	358.15 K	368.15 K	(kJ · mol^{-1})
Toluene	2.16	2.19	2.22	2.25	2.27	2.30	2.32	2.35	−1.1

Mixture 21: $C_6H_5C_2H_5$ (1) + [BMPYR][CF_3SO_3] (2)

(1) Activity Coefficient at Infinite Dilution and Excess Molar Enthalpy at Infinite Dilution

Ref. [5]

Solute	γ_{12}^{∞} (gas–liquid chromatography)								$\Delta H_1^{E,\infty}$
	298.15 K	308.15 K	318.15 K	328.15 K	338.15 K	348.15 K	358.15 K	368.15 K	(kJ · mol^{-1})
Ethylbenzene	3.43	3.45	3.46	3.48	3.50	3.51	3.53	3.54	−0.4

Mixture 22: *o*-$C_6H_4(CH_3)_2$ (1) + [BMPYR][CF_3SO_3] (2)

(1) Activity Coefficient at Infinite Dilution and Excess Molar Enthalpy at Infinite Dilution

Ref. [5]

Solute	γ_{12}^{∞} (gas–liquid chromatography)								$\Delta H_1^{E,\infty}$
	298.15 K	308.15 K	318.15 K	328.15 K	338.15 K	348.15 K	358.15 K	368.15 K	(kJ · mol^{-1})
o-Xylene		2.91	2.93	2.95	2.97	2.99	3.01	3.03	−0.6

Mixture 23: *m*-$C_6H_4(CH_3)_2$ (1) + [BMPYR][CF_3SO_3] (2)

(1) Activity Coefficient at Infinite Dilution and Excess Molar Enthalpy at Infinite Dilution

Ref. [5]

Solute	γ_{12}^{∞} (gas–liquid chromatography)								$\Delta H_1^{E,\infty}$
	298.15 K	308.15 K	318.15 K	328.15 K	338.15 K	348.15 K	358.15 K	368.15 K	(kJ · mol^{-1})
m-Xylene		3.44	3.45	3.47	3.48	3.50	3.51	3.53	−0.4

Mixture 24: *p*-$C_6H_4(CH_3)_2$ (1) + [BMPYR][CF_3SO_3] (2)

(1) Activity Coefficient at Infinite Dilution and Excess Molar Enthalpy at Infinite Dilution

Ref. [5]

Solute	γ_{12}^{∞} (gas–liquid chromatography)								$\Delta H_1^{E,\infty}$
	298.15 K	308.15 K	318.15 K	328.15 K	338.15 K	348.15 K	358.15 K	368.15 K	(kJ · mol^{-1})
p-Xylene		3.32	3.35	3.38	3.40	3.43	3.45	3.48	−0.7

Mixture 25: CH_3OH (1) + [BMPYR][CF_3SO_3] (2)

(1) Activity Coefficient at Infinite Dilution and Excess Molar Enthalpy at Infinite Dilution

Ref. [5]

Solute	γ_{12}^{∞} (gas–liquid chromatography)								$\Delta H_1^{E,\infty}$
	298.15 K	308.15 K	318.15 K	328.15 K	338.15 K	348.15 K	358.15 K	368.15 K	(kJ · mol^{-1})
Methanol	0.717	0.684	0.655	0.628	0.604	0.581	0.563	0.543	3.6

Mixture 26: C_2H_5OH (1) + [BMPYR][CF_3SO_3] (2)

(1) Activity Coefficient at Infinite Dilution and Excess Molar Enthalpy at Infinite Dilution

Ref. [5]

Solute	γ_{12}^{∞} (gas–liquid chromatography)								$\Delta H_1^{E,\infty}$
	298.15 K	308.15 K	318.15 K	328.15 K	338.15 K	348.15 K	358.15 K	368.15 K	(kJ · mol^{-1})
Ethanol	1.05	1.00	0.948	0.900	0.862	0.827	0.791	0.763	4.2

Ref. [8]

Solvent	$\gamma^{\infty}{}_1$					
	303.15 K	313.15 K	323.15 K	333.15 K	343.15 K	353.15 K
[MBPYR][OTF]	–	0.95	0.92	0.88	0.86	–

Mixture 27: 1-C_3H_7OH (1) + [BMPYR][CF_3SO_3] (2)

(1) Activity Coefficient at Infinite Dilution and Excess Molar Enthalpy at Infinite Dilution

Ref. [5]

Solute	γ_{12}^{∞} (gas–liquid chromatography)								$\Delta H_1^{E,\infty}$
	298.15 K	308.15 K	318.15 K	328.15 K	338.15 K	348.15 K	358.15 K	368.15 K	(kJ · mol^{-1})
Propan-1-ol		1.26	1.20	1.14	1.08	1.04	0.995	0.953	4.4

Mixture 28: 1-C_4H_9OH (1) + [BMPYR][CF_3SO_3] (2)

(1) Activity Coefficient at Infinite Dilution and Excess Molar Enthalpy at Infinite Dilution

Ref. [5]

Solute	γ_{12}^{∞} (gas–liquid chromatography)								$\Delta H_1^{E,\infty}$
	298.15 K	308.15 K	318.15 K	328.15 K	338.15 K	348.15 K	358.15 K	368.15 K	(kJ · mol^{-1})
Butan-1-ol				1.49	1.42	1.35	1.29	1.24	4.8

Mixture 29: H_2O (1) + [BMPYR][CF_3SO_3] (2)

(1) Activity Coefficient at Infinite Dilution and Excess Molar Enthalpy at Infinite Dilution

Ref. [5]

Solute	γ_{12}^{∞} (gas–liquid chromatography)								$\Delta H_1^{E,\infty}$
	298.15 K	308.15 K	318.15 K	328.15 K	338.15 K	348.15 K	358.15 K	368.15 K	(kJ · mol^{-1})
Water			0.865	0.830	0.794	0.765	0.739	0.712	3.8

Ref. [8]

Solvent	γ^{∞}_1				
	303.15 K	313.15 K	323.15 K	333.15 K	343.15 K
[MBPYR][OTF]	–	0.67	0.56	0.51	0.44

Mixture 30: $OCH_2CH_2CH_2CH_2$ (1) + [BMPYR][CF_3SO_3] (2)
(1) Activity Coefficient at Infinite Dilution and Excess Molar Enthalpy at Infinite Dilution

Ref. [5]

Solute	γ_{12}^{∞} (gas–liquid chromatography)								$\Delta H_1^{E,\infty}$
	298.15 K	308.15 K	318.15 K	328.15 K	338.15 K	348.15 K	358.15 K	368.15 K	(kJ · mol^{-1})
THF	1.38	1.39	1.41	1.42	1.43	1.44	1.45	1.46	−0.8

Mixture 31: SCH = CHCH = CH (thiophene) (1) + [BMPYR][CF_3SO_3] (2)
(1) Activity Coefficient at Infinite Dilution and Excess Molar Enthalpy at Infinite Dilution

Ref. [5]

Solute	γ_{12}^{∞} (gas–liquid chromatography)								$\Delta H_1^{E,\infty}$
	298.15 K	308.15 K	318.15 K	328.15 K	338.15 K	348.15 K	358.15 K	368.15 K	(kJ · mol^{-1})
Thiophene	1.02	1.04	1.06	1.08	1.10	1.11	1.13	1.14	−1.5

Mixture 32: $(CH_3)_3COCH_3$ (1) + [BMPYR][CF_3SO_3] (2)
(1) Activity Coefficient at Infinite Dilution and Excess Molar Enthalpy at Infinite Dilution

Ref. [5]

Solute	γ_{12}^{∞} (gas–liquid chromatography)								$\Delta H_1^{E,\infty}$
	298.15 K	308.15 K	318.15 K	328.15 K	338.15 K	348.15 K	358.15 K	368.15 K	(kJ · mol^{-1})
MTBE	5.67	5.65	5.64	5.63	5.62	5.60	5.59	5.58	0.2

4.3.2 Ternary Mixtures

Mixture 1: C_6H_{14} (1) + C_6H_6 (2) + [BMPYR][CF_3SO_3] (3)
(1) Selectivity at Infinite Dilution and Capacity at Infinite Dilution

Ref. [5]

Solvent	S^{∞}_{ij}	k_j^{∞}
$T = 298.15$ K	*n*-Hexane (1)/benzene (2)	Benzene
[BMPYR][CF_3SO_3]	39.6	0.71

Mixture 2: $CH_2(CH_2)_4CH_2$ (1) + C_6H_6 (2) + [BMPYR][CF_3SO_3] (3)
(1) Selectivity at Infinite Dilution and Capacity at Infinite Dilution

Ref. [5]

Solvent	S^{∞}_{ij}	k_j^{∞}
$T = 298.15$ K	Cyclohexane (1)/benzene (2)	Benzene
[BMPYR][CF_3SO_3]	18.2	0.71

Mixture 3: C_6H_{14} (1) + SCH = CHCH = CH (thiophene) (2) + [BMPYR][CF_3SO_3] (3)
(1) Selectivity at Infinite Dilution and Capacity at Infinite Dilution

Ref. [5]

Solvent	S^{∞}_{12}	k_2^{∞}
$T = 298.15$ K	*n*-Hexane (1)/thiophene (2)	
[BMPYR][CF_3SO_3]	70.1	0.98

Mixture 4: $CH_2(CH_2)_4CH_2$ (1) + SCH = CHCH = CH (thiophene) (2) + [BMPYR][CF_3SO_3] (3)
(1) Selectivity at Infinite Dilution and Capacity at Infinite Dilution

Ref. [5]

Solvent	S^{∞}_{12}	k_2^{∞}
$T = 298.15$ K	Cyclohexane (1)/thiophene (2)	
[BMPYR][CF_3SO_3]	25.2	0.98

Mixture 5: C_6H_6 (1) + SCH = CHCH = CH (thiophene) (2) + [BMPYR][CF_3SO_3] (3)
(1) Selectivity at Infinite Dilution and Capacity at Infinite Dilution

Ref. [5]

Solvent	S^{∞}_{12}	k_2^{∞}
$T = 298.15$ K	Benzene (1)/thiophene (2)	
[BMPYR][CF_3SO_3]	1.4	0.98

Mixture 6: C_2H_5OH (1) + H_2O (2) + [MBPYR][OTF] (3)
(1) Selectivity and capacity at infinite dilution γ^{∞}_1 for the separation of ethanol from water in ionic liquids

Ref. [8]

		[MBPYR][OTF]	
Separation problem	T (K)	S^{∞}_{12}	k^{∞}_2
	303.15	–	–
	313.15	1.41	1.48
Ethanol (1)/water (2)	323.15	1.66	1.80
	333.15	1.73	1.96
	343.15	1.94	2.26

4.4 08014-03001: 1-Methyl-1-octylpyrrodinium bis ((trifluoromethyl)sulfonyl)imide

Abbreviation:	[P1,8][NTf$_2$], [OMPYR][BTI]
Molecular formula:	$C_{15}H_{28}F_6N_2O_4S_2$
Molar mass:	478.51
Structure:	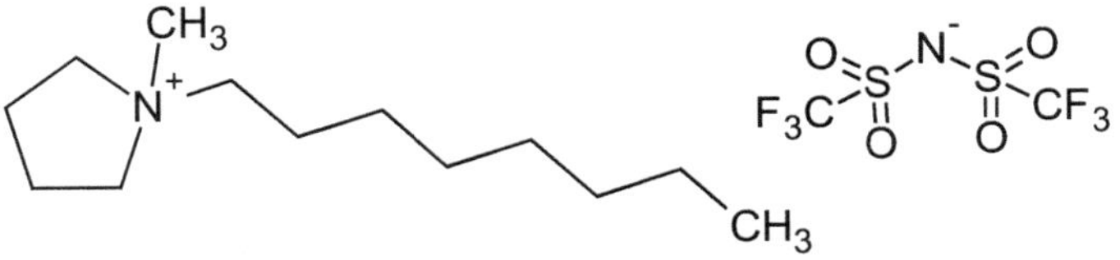

4.4.1 Binary Mixtures

Mixture 1: C_6H_{14} (1) + [OMPYR][BTI] (2)
(1) Activity Coefficient at Infinite Dilution

Ref. [6]

T/K	γ_1^∞	T/K	γ_1^∞	T/K	γ_1^∞	T/K	γ_1^∞
303.16	5.24	313.13	5.05	323.17	4.90	333.19	4.80

(2) Phase Equilibrium

Ref. [6]

x_1	P/kPa	x_1	P/kPa	x_1	P/kPa
363.15 K					
0.0104	8.652	0.0773	61.11	0.1168	88.93
0.0272	22.61	0.0874	68.52	0.1487	110.6
0.0428	35.09	0.0971	75.47	0.2029	142.8
0.0600	48.23	0.1060	81.64	0.2831	181.4

Mixture 2: C_7H_{16} (1) + [OMPYR][BTI] (2)
(1) Activity Coefficient at Infinite Dilution

Ref. [6]

T/K	γ_1^∞	T/K	γ_1^∞	T/K	γ_1^∞	T/K	γ_1^∞
303.10	6.90	313.15	6.73	323.09	6.54	333.13	6.36

Mixture 3: C_8H_{18} (1) + [OMPYR][BTI] (2)
(1) Activity Coefficient at Infinite Dilution

Ref. [6]

T/K	γ_1^∞	T/K	γ_1^∞	T/K	γ_1^∞	T/K	γ_1^∞
303.10	8.78	313.15	8.55	323.09	8.39	333.13	8.22

(2) Phase Equilibrium

Ref. [6]

x_1	P/kPa	x_1	P/kPa	x_1	P/kPa
363.15 K					
0.0100	2.262	0.0470	10.76	0.0872	18.91
0.0171	4.019	0.0571	12.89	0.1145	23.63
0.0273	6.425	0.0660	14.73	0.1667	30.85
0.0366	8.479	0.0743	16.43		

Mixture 4: 1-C_5H_{10} (1) + [OMPYR][BTI] (2)
(1) Activity Coefficient at Infinite Dilution

Ref. [6]

T/K	γ_1^∞	T/K	γ_1^∞	T/K	γ_1^∞	T/K	γ_1^∞
303.14	2.50	313.13	2.43	323.11	2.33	333.11	2.25

Mixture 5: 1-C_6H_{12} (1) + [OMPYR][BTI] (2)
(1) Activity Coefficient at Infinite Dilution

Ref. [6]

T/K	γ_1^∞	T/K	γ_1^∞	T/K	γ_1^∞	T/K	γ_1^∞
303.14	3.11	313.13	3.29	323.11	3.22	333.11	3.15

(2) Phase Equilibrium

Ref. [6]

x_1	P/kPa	x_1	P/kPa	x_1	P/kPa
363.15 K					
0.0114	7.770	0.0809	52.22	0.1270	79.86
0.0270	17.76	0.0930	59.54	0.1807	110.6
0.0450	29.00	0.1040	66.36	0.2453	145.5
0.0669	43.28	0.1144	72.45	0.3426	190.2

Mixture 6: 1-C_7H_{14} (1) + [OMPYR][BTI] (2)
(1) Activity Coefficient at Infinite Dilution

Ref. [6]

T/K	γ_1^∞	T/K	γ_1^∞	T/K	γ_1^∞	T/K	γ_1^∞
313.11	4.37	323.14	4.22	333.17	4.10	343.21	4.02

Mixture 7: 1-C_8H_{16} (1) + [OMPYR][BTI] (2)
(1) Activity Coefficient at Infinite Dilution

Ref. [6]

T/K	γ_1^{∞}	T/K	γ_1^{∞}	T/K	γ_1^{∞}	T/K	γ_1^{∞}
313.11	5.53	323.14	5.40	333.17	5.27	343.21	5.21

(2) Phase Equilibrium

Ref. [6]

x_1	P/kPa	x_1	P/kPa	x_1	P/kPa
363.15 K					
0.0094	1.889	0.0530	9.909	0.1013	17.82
0.0192	3.836	0.0643	11.93	0.1303	21.98
0.0312	6.060	0.0748	13.72	0.1838	28.81
0.0434	8.235	0.0845	15.21	0.2737	37.44

4.4.2 Ternary Mixtures

Mixture 1: 1-C_8H_{16} (1) + C_8H_{18} (2) + [OMPYR][BTI] (3)
(1) Phase Equilibrium

Ref. [6]

$x_{1,\mathrm{corr}}$	y_1	$1/\alpha_{12}$	$x_{1,\mathrm{corr}}$	y_1	$1/\alpha_{12}$	$x_{1,\mathrm{corr}}$	y_1	$1/\alpha_{12}$
342.45 K 90 mol.% (3)								
0.00460	0.0475	1.10	0.02677	0.2576	1.23	0.06362	0.6508	1.29
0.00907	0.0921	1.16	0.03545	0.3442	1.25	0.07260	0.7696	1.31
0.01453	0.1418	1.20	0.04531	0.4509	1.27	0.08134	0.8796	1.34
0.01816	0.1757	1.21	0.05274	0.5390	1.27			

4.5 08017-03009: Pyrrolidinium nitrate

Abbreviation:	[Pyrr][NO_3]
Molecular formula:	$C_4H_{10}N_2O_3$
Molar mass:	134.14
Structure:	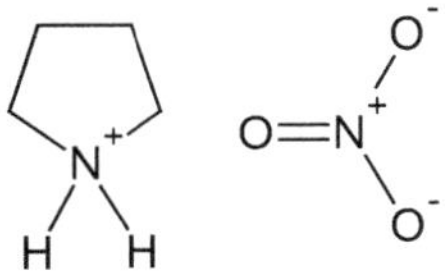

4.5.1 Binary Mixtures

Mixture 1: H_2O (1) + [Pyrr][NO_3] (2)
(1) Heat Capacity and Derived Properties

Ref. [9]

x_2	C_p/(J · mol^{-1} · K^{-1} ± 0.1 J · mol^{-1} · K^{-1})	$C^0_{p,m}$/(J · mol^{-1} · K^{-1})	C^E_p/(J · mol^{-1} · K^{-1})
298.15 K			
1.000	228.0	228.0	0
0.845	218.3	204.4	+14.0
0.722	198.8	185.6	+13.1
0.617	180.3	169.5	+10.9
0.349	121.4	128.6	−7.22
0.246	100.6	112.9	−12.3
0.171	87.51	101.4	−13.9
0.123	80.83	94.03	−13.2
0.091	77.44	89.20	−11.8
0.054	75.24	83.55	−8.31
0.025	73.92	79.10	−5.18
0.000	75.32	75.32	0

(2) Volumetric Properties and Refractive Index

Ref. [9]

x_2	298.15 K	303.15 K	308.15 K	313.15 K	318.15 K	323.15 K	328.15 K	333.15 K
ρ/g · cm^{-3} (±0.0002 g · cm^{-3})								
1.0000	1.1675	1.1644	1.1611	1.1570	1.1531	1.1495	1.1455	1.1412
0.9401	1.1649	1.1616	1.1582	1.1533	1.1504	1.1465	1.1423	1.1379
0.8302	1.1595	1.1554	1.1519	1.1462	1.1442	1.1401	1.1356	1.1308
0.7203	1.1527	1.1476	1.1438	1.1383	1.1359	1.1323	1.1271	1.1225
0.6503	1.1467	1.1414	1.1374	1.1326	1.1293	1.1261	1.1210	1.1160
0.4903	1.1286	1.1230	1.1190	1.1141	1.1095	1.1068	1.1006	1.0955
0.3403	1.0996	1.0944	1.0895	1.0834	1.0800	1.0765	1.0729	1.0681
0.2610	1.0731	1.0675	1.0620	1.0583	1.0550	1.0527	1.0507	1.0454
0.1702	1.0260	1.0234	1.0215	1.0188	1.0165	1.0144	1.0119	1.0073
0.1185	0.9971	0.9930	0.9917	0.9889	0.9874	0.9839	0.9809	0.9757
0.0941	0.9831	0.9805	0.9764	0.9731	0.9716	0.9683	0.9641	0.9613
0.0554	0.9623	0.9602	0.9564	0.9544	0.9538	0.9504	0.9465	0.9411
0.0306	0.9614	0.9572	0.9554	0.9535	0.9520	0.9489	0.9443	0.9414
0.0147	0.9698	0.9654	0.9640	0.9627	0.9618	0.9586	0.9555	0.9497
0.0000	0.9971	0.9957	0.9941	0.9926	0.9902	0.9880	0.9857	0.9831
V^E/cm^3 · mol^{-1} (±0.0005 cm^3 · mol^{-1})								
0.9401	0.0852	0.1076	0.1180	0.2000	0.1077	0.1332	0.1582	0.1730
0.8302	0.2285	0.3168	0.3467	0.4900	0.3374	0.3796	0.4377	0.4950

(continued)

x_2	298.15 K	303.15 K	308.15 K	313.15 K	318.15 K	323.15 K	328.15 K	333.15 K
0.7203	0.3810	0.5457	0.5956	0.7200	0.6185	0.6256	0.7376	0.7755
0.6503	0.5294	0.7049	0.7678	0.8400	0.8118	0.7960	0.8926	0.9644
0.4903	0.8691	1.0372	1.1002	1.1823	1.2461	1.2087	1.3695	1.4448
0.3403	1.3046	1.4311	1.5366	1.6751	1.6827	1.7031	1.7159	1.7780
0.2610	1.6921	1.8294	1.9564	1.9878	1.9986	1.9734	1.9440	2.0059
0.1702	2.2747	2.2948	2.2835	2.2895	2.2712	2.2508	2.2365	2.2941
0.1185	2.3263	2.3989	2.3759	2.3964	2.3595	2.3944	2.4053	2.4849
0.0941	2.2623	2.2874	2.3537	2.3917	2.3610	2.3909	2.4424	2.4470
0.0554	1.9744	1.9881	2.0442	2.0501	2.0024	2.0356	2.0761	2.1568
0.0306	1.4036	1.4667	1.4713	1.4777	1.4584	1.4834	1.5381	1.5496
0.0147	0.8467	0.9091	0.9045	0.9004	0.8695	0.8937	0.9104	0.9838
$V_{\Phi 2}/(\text{cm}^3 \cdot \text{mol}^{-1})$								
1.0000	114.87	115.18	115.51	115.92	116.31	116.68	117.08	117.52
0.9401	114.96	115.29	115.64	116.13	116.42	116.82	117.25	117.70
0.8302	115.15	115.56	115.93	116.51	116.72	117.14	117.61	118.12
0.7203	115.40	115.94	116.34	116.92	117.17	117.55	118.10	118.60
0.6503	115.68	116.26	116.69	117.21	117.56	117.90	118.45	119.00
0.4903	116.56	117.30	117.75	118.33	118.85	119.14	119.87	120.47
0.3403	118.70	119.39	120.03	120.84	121.25	121.68	122.12	122.74
0.2610	121.35	122.19	123.01	123.54	123.97	124.24	124.45	125.21
0.1702	128.24	128.66	128.93	129.37	129.66	129.90	130.22	131.00
0.1185	134.51	135.43	135.56	136.15	136.23	136.89	137.38	138.49
0.0941	138.91	139.48	140.52	141.33	141.40	142.08	143.03	143.52
0.0554	150.52	151.08	152.42	152.94	152.47	153.44	154.57	156.47
0.0306	160.80	163.17	163.66	164.27	164.03	165.22	167.41	168.22
0.0147	172.42	176.97	176.99	177.12	175.41	177.42	178.96	184.39
$V_{\Phi 1}/(\text{cm}^3 \cdot \text{mol}^{-1})$								
0.0599	19.48	19.88	20.09	21.48	19.99	20.45	20.91	21.21
0.1698	19.41	19.95	20.16	21.03	20.17	20.46	20.85	21.23
0.2797	19.42	20.04	20.25	20.72	20.40	20.46	20.91	21.09
0.3497	19.57	20.10	20.31	20.55	20.51	20.50	20.82	21.08
0.5097	19.77	20.12	20.28	20.46	20.63	20.60	20.96	21.15
0.6597	20.04	20.26	20.45	20.68	20.74	20.81	20.87	21.02
0.7390	20.35	20.56	20.76	20.83	20.89	20.90	20.87	21.03
0.8298	20.80	20.85	20.87	20.90	20.93	20.94	20.97	21.08
0.8815	20.70	20.81	20.81	20.86	20.86	20.94	21.00	21.14
0.9059	20.56	20.61	20.72	20.78	20.79	20.87	20.97	21.02
0.9446	20.15	20.19	20.28	20.31	20.31	20.38	20.47	20.60
0.9694	19.51	19.60	19.63	19.67	19.69	19.76	19.86	19.92
0.9853	18.92	19.01	19.03	19.06	19.07	19.14	19.20	19.32
1.0000	18.06	18.09	18.12	18.14	18.19	18.23	18.27	18.32

(*continued*)

x_2	298.15 K	303.15 K	308.15 K	313.15 K	318.15 K	323.15 K	328.15 K	333.15 K
$\overline{V}_2/(\mathrm{cm}^3 \cdot \mathrm{mol}^{-1})$								
1.0000	114.88	115.18	115.51	115.92	116.31	116.68	117.08	117.52
0.9401	114.89	115.31	115.54	116.03	116.34	116.74	117.17	117.62
0.8302	114.97	115.47	115.61	116.16	116.51	116.92	117.38	117.88
0.7203	114.63	115.16	115.29	115.83	116.30	116.68	117.19	117.67
0.6503	114.59	115.17	115.30	115.78	116.30	116.66	117.16	117.67
0.4903	114.82	115.82	115.86	116.35	116.67	117.00	117.56	118.08
0.3403	113.10	114.36	114.48	115.03	114.91	115.30	115.72	116.22
0.2610	111.50	112.69	113.03	113.47	113.24	113.58	113.93	114.46
0.1702	112.85	113.61	114.10	114.51	114.28	114.63	115.02	115.52
0.1185	118.11	118.71	119.29	119.72	119.58	119.98	120.40	120.91
0.0941	122.71	123.20	123.86	124.31	124.24	124.64	125.10	125.54
0.0554	134.08	134.56	135.06	135.47	135.54	135.94	136.38	136.90
0.0306	144.50	145.14	145.37	145.79	145.99	146.38	146.84	147.29
0.0147	152.75	153.51	153.54	153.95	154.20	154.60	155.02	155.53
$\overline{V}_1/(\mathrm{cm}^3 \cdot \mathrm{mol}^{-1})$								
0.0599	19.33	17.88	19.60	19.71	19.48	19.55	19.61	19.68
0.1698	18.96	18.57	19.70	19.87	19.20	19.28	19.38	19.49
0.2797	20.06	20.10	20.80	20.95	20.42	20.47	20.62	20.71
0.3497	20.11	20.13	20.70	20.80	20.52	20.55	20.69	20.81
0.5097	19.75	19.51	19.94	20.05	20.29	20.29	20.50	20.62
0.6597	20.96	20.68	20.98	21.14	21.46	21.52	21.58	21.69
0.7390	21.55	21.44	21.64	21.70	21.98	21.99	21.99	22.12
0.8298	21.22	21.18	21.16	21.19	21.34	21.36	21.39	21.49
0.8815	20.27	20.33	20.30	20.35	20.43	20.50	20.56	20.68
0.9059	19.75	19.78	19.85	19.91	19.97	20.04	20.14	20.19
0.9446	19.03	19.06	19.14	19.17	19.18	19.25	19.34	19.47
0.9694	18.58	18.66	18.69	18.73	18.76	18.82	18.92	18.98
0.9853	18.36	18.44	18.47	18.49	18.50	18.57	18.63	18.75
1.0000	18.06	18.09	18.12	18.14	18.19	18.23	18.27	18.32

Ref. [9]

x_2	n_D ±0.0002	$n^{\mathrm{id}}_{\mathrm{D}}$	$\Delta_\Phi n_D$
298.15 K			
1.000	1.3955	1.3955	0
0.845	1.3934	1.3937	−0.0003
0.722	1.3897	1.3919	−0.0022
0.550	1.3811	1.3884	−0.0073
0.349	1.3684	1.3813	−0.0129
0.246	1.3589	1.3751	−0.0162
0.171	1.3517	1.3684	−0.0167
0.123	1.3464	1.3623	−0.0159

(continued)

x_2	n_D ±0.0002	n^{id}_D	$\Delta_\Phi n_D$
0.091	1.3424	1.3572	−0.0148
0.054	1.3383	1.3495	−0.0112
0.025	1.3354	1.3415	−0.0061
0.000	1.3328	1.3328	0

Ref. [9]

T/K	$\overline{V}_2^{\infty}/(\mathrm{cm^3 \cdot mol^{-1}})$	$\overline{V}_1^{\infty}/(\mathrm{cm^3 \cdot mol^{-1}})$
298.15	180	19.33
308.15	182	19.36
318.15	185	19.38
328.15	188	19.42

4.6 08017-04021: Pyrrolidinium hydrogen sulfate

Abbreviation:	[Pyrr][HSO_4]
Molecular formula:	$C_4H_{11}NO_4S$
Molar mass:	169.21
Structure:	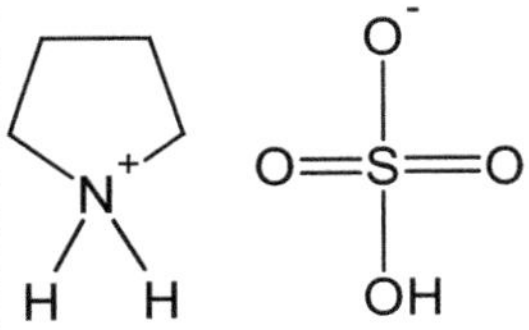

4.6.1 Binary Mixtures

Mixture 1: H_2O (1) + [Pyrr][HSO_4] (2)
(1) Volumetric Properties

Ref. [9]

x_2	ρ/g · cm^{-3}	$V^E/(\mathrm{cm^3 \cdot mol^{-1}})$	$V_{\Phi,2}/(\mathrm{cm^3 \cdot mol^{-1}})$	$V_{\Phi,1}/(\mathrm{cm^3 \cdot mol^{-1}})$
298.15 K				
1.0000	1.3421	0.0000	126.07	
0.4890	1.2905	0.3687	126.82	18.78
0.2980	1.2181	1.5221	131.18	20.23
0.1980	1.1391	2.6416	139.41	21.35
0.1370	1.0646	3.5136	151.72	22.13
0.0960	1.0060	3.8992	166.69	22.37

(continued)

x_2	ρ/g · cm^{-3}	V^E/(cm^3 · mol^{-1})	$V_{\Phi,2}$/(cm^3 · mol^{-1})	$V_{\Phi,1}$/(cm^3 · mol^{-1})
0.0850	0.9931	3.8343	171.18	22.25
0.0660	0.9711	3.6318	181.10	21.95
0.0250	0.9503	2.1669	212.75	20.28
0.0110	0.9645	1.1481	230.44	19.22
0.0000	0.9971	0.0000		18.06

Ref. [9]

T/K	$\overline{V}_2^{\infty}$/(cm^3 · mol^{-1})	$\overline{V}_1^{\infty}$/(cm^3 · mol^{-1})
298.15	240	15.3

4.7 08017-06005: Pyrrolidinium formate

Abbreviation:	[Pyrr][HCOO]
Molecular formula:	$C_5H_{11}NO_2$
Molar mass:	117.15
Structure:	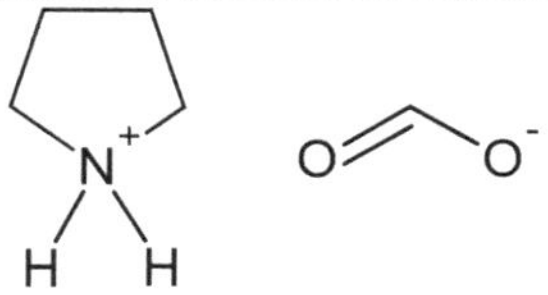

4.7.1 Binary Mixtures

Mixture 1: H_2O (1) + [Pyrr][HCOO] (2)
(1) Volumetric Properties

Ref. [9]

x_2	ρ/g · cm^{-3}	V^E/(cm^3 · mol^{-1})	$V_{\Phi,2}$/(cm^3 · mol^{-1})	$V_{\Phi,1}$/(cm^3 · mol^{-1})
298.15 K				
1.0000	1.1190	0.0000	104.73	
0.5799	1.1089	−0.2141	104.36	17.55
0.3784	1.0995	−0.3453	103.82	17.50
0.2626	1.0907	−0.4323	103.08	17.47
0.1892	1.0828	−0.4983	102.10	17.45
0.1336	1.0717	−0.4733	101.19	17.51
0.0967	1.0611	−0.4326	100.25	17.58
0.0616	1.0460	−0.3437	99.15	17.69

(continued)

x_2	$\rho/g \cdot cm^{-3}$	$V^E/(cm^3 \cdot mol^{-1})$	$V_{\Phi,2}/(cm^3 \cdot mol^{-1})$	$V_{\Phi,1}/(cm^3 \cdot mol^{-1})$
0.0372	1.0310	−0.2407	98.27	17.81
0.0173	1.0150	−0.1289	97.29	17.93
0.0000	0.9970	0.0000		18.06

Ref. [9]

T/K	$\overline{V}_2^{\infty}/(cm^3 \cdot mol^{-1})$	$\overline{V}_1^{\infty}/(cm^3 \cdot mol^{-1})$
298.15	94.0	17.7
298.15	148.5	13.2

References

1. Freire MG, Neves CMSS, Carvalho PJ, Gardas RL, Fernandes AM, Marrucho IM, Santos LMNBF, Coutinho JAP (2007) Mutual solubilities of water and hydrophobic ionic liquids. J Phys Chem B 111(45):13082–13089
2. Wang S, Jacquemin J, Husson P, Hardacre C, Costa Gomes MF (2009) Liquid–liquid miscibility and volumetric properties of aqueous solutions of ionic liquids as a function of temperature. J Chem Thermodyn 41(11):1206–1214
3. Comminges C, Barhdadi R, Laurent M, Troupel M (2006) Determination of viscosity, ionic conductivity, and diffusion coefficients in some binary systems: ionic liquids + molecular solvents. J Chem Eng Data 51(2):680–685
4. Kato R, Gmehling J (2005) Systems with ionic liquids: measurement of VLE and γ^{∞} data and prediction of their thermodynamic behavior using original UNIFAC, mod. UNIFAC(Do) and COSMO-RS(Ol). J Chem Thermodyn 37(6):603–619
5. Domańska U, Redhi GG, Marciniak A (2009) Activity coefficients at infinite dilution measurements for organic solutes and water in the ionic liquid 1-butyl-1-methylpyrrolidinium trifluoromethanesulfonate using GLC. Fluid Phase Equilib 278(1–2):97–102
6. Nebig S, Liebert V, Gmehling J (2009) Measurement and prediction of activity coefficients at infinite dilution (γ^{∞}), vapor–liquid equilibria (VLE) and excess enthalpies (H^E) of binary systems with 1,1-dialkyl-pyrrolidinium bis(trifluoromethylsulfonyl)imide using mod. UNIFAC (Dortmund). Fluid Phase Equilib 277(1):61–67
7. Lewandowski A, Świderska A (2004) New composite solid electrolytes based on a polymer and ionic liquids. Solid State Ion 169(1–4):21–24
8. Liebert V, Nebig S, Gmehling J (2008) Experimental and predicted phase equilibria and excess properties for systems with ionic liquids. Fluid Phase Equilib 268(1–2):14–20
9. Anouti M, Caillon-Caravanier M, Dridi Y, Jacquemin J, Hardacre C, Lemordant D (2009) Liquid densities, heat capacities, refractive index and excess quantities for {protic ionic liquids + water} binary system. J Chem Thermodyn 41(6):799–808

Mixtures Containing Piperidinium Ionic Liquids

5

5.1 09001-03001: *N*-methyl-*N*-propylpiperidinium bis((trifluoromethyl)sulfonyl)imide

Abbreviation:	[PP13][TFSI], [C_3mpip][Tf_2N]
Molecular formula:	$C_{11}H_{20}F_6N_2O_4S_2$
Molar mass:	422.41
Structure:	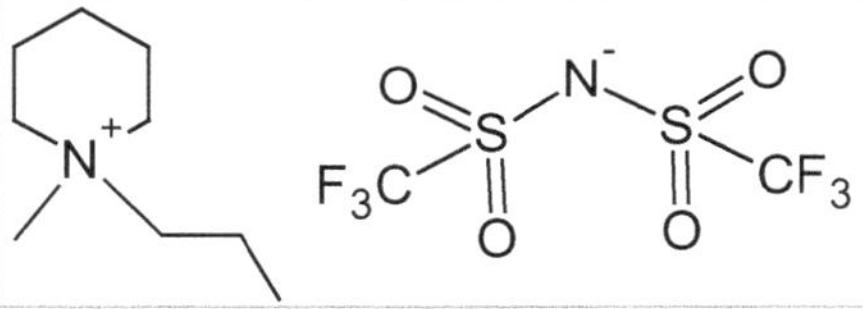

5.1.1 Binary Mixtures

Mixture 1: H_2O (1) + [C_3mpip][Tf_2N] (2)
(1) Solubility

Ref. [2]

T/K	$x_1 \pm \sigma^a$	T/K	$x_1 \pm \sigma^a$
288.15	0.188 ± 0.003	308.15	0.255 ± 0.005
293.15	0.199 ± 0.002	313.15	0.272 ± 0.002
298.15	0.212 ± 0.003	318.15	0.289 ± 0.005
303.15	0.231 ± 0.004		

[a]Standard deviation

S. Zhang et al., *Physicochemical Properties of Ionic Liquid Mixtures*,
DOI 10.1007/978-94-017-7573-1_5

5.2 09012–03001: 1-Methyl-1-octylpiperidinium bis((trifluoromethyl)sulfonyl)imide

Abbreviation:	[PP1,8][NTf$_2$], [MOPYR][BTI]
Molecular formula:	$C_{16}H_{30}F_6N_2O_4S_2$
Molar mass:	492.54
Structure:	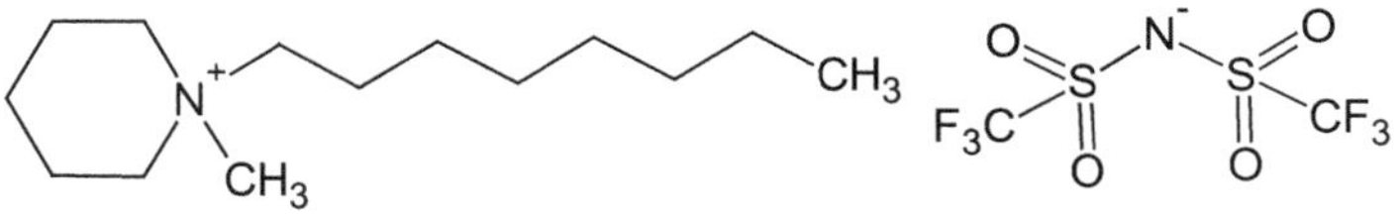

5.2.1 Binary Mixtures

Mixture 1: $C_6H_5CH_3$ (1) + [MOPYR][BTI] (2)
(1) Activity Coefficient at Infinite Dilution

Ref. [1]

	$\gamma^{\infty}{}_1$			
Solvent	303.15 K	313.15 K	323.15 K	333.15 K
[MOPYR]$^+$[BTI]$^-$	0.67	0.73	0.80	0.86

Mixture 2: $CH_3CH(CH_2)_4CH_2$ (1) + [MOPYR][BTI] (2)
(1) Activity Coefficient at Infinite Dilution

Ref. [1]

	$\gamma^{\infty}{}_1$			
Solvent	303.15 K	313.15 K	323.15 K	333.15 K
[MOPYR][BTI]	4.59	4.41	4.29	4.18

5.2.2 Ternary Mixtures

Mixture 1: $CH_3CH(CH_2)_4CH_2$ (1) + $C_6H_5CH_3$ (2) + [MOPYR][BTI] (3)
(1) Selectivity and capacity at infinite dilution $\gamma^{\infty}{}_1$ for the separation of methylcyclohexane from toluene for ionic liquids

Ref. [1]

Separation problem		[MOPYR][BTI]	
Methylcyclohexane (1)/toluene (2)	T (K)	$S^{\infty}{}_{12}$	$k^{\infty}{}_2$
	303.15	6.87	1.50
	313.15	6.09	1.38

(continued)

Separation problem		[MOPYR][BTI]	
Methylcyclohexane (1)/toluene (2)	T (K)	S^{∞}_{12}	k^{∞}_{2}
	323.15	5.34	1.24
	333.15	4.89	1.17

References

1. Liebert V, Nebig S, Gmehling J (2008) Experimental and predicted phase equilibria and excess properties for systems with ionic liquids. Fluid Phase Equilib 268(1–2):14–20
2. Freire MG, Neves CMSS, Carvalho PJ, Gardas RL, Fernandes AM, Marrucho IM, Santos LMNBF, Coutinho JAP (2007) Mutual solubilities of water and hydrophobic ionic liquids. J Phys Chem B 111(45):13082–13089

Mixtures Containing Pyridinium Ionic Liquids

6

6.1 11001-02001: *N*-butyl pyridinium tetrafluoroborate

Abbreviation:	$[Bpy][BF_4]$, $[BP][BF_4]$, $[BuPy][BF_4]$
Molecular formula:	$C_9H_{14}NBF_4$
Molar mass	223.02
Structure:	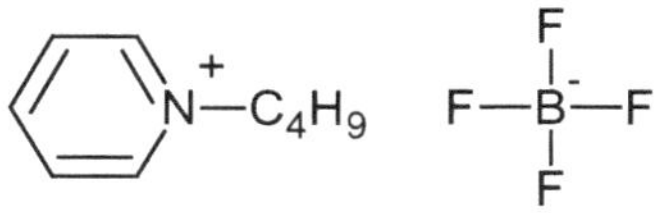

6.1.1 Binary Mixtures

Mixture 1: H_2O (1) + $[BP][BF_4]$ (2)
(1) Volumetric Properties and Transport Properties

Ref. [1]

Composition $V_{f,2}$	η/mPa · s	Composition $V_{f,2}$	η/mPa · s	Composition $V_{f,2}$	η/mPa · s
298.15 K					
0	0.93	0.80	7.29	0.96	41.95
0.20	1.33	0.85	10.80	0.98	75.35
0.40	2.03	0.90	13.90	1.00	145.23
0.60	3.35	0.93	25.50		

S. Zhang et al., *Physicochemical Properties of Ionic Liquid Mixtures*,
DOI 10.1007/978-94-017-7573-1_6

Ref. [2]

x_1	ρ/ (g · cm^{-3})	η/ (mPa · s)	V^E/ (cm^3 · mol^{-1})	$\Delta\eta$/ (mPa · s)	ρ/ (g · cm^{-3})	η/ (mPa · s)	V^E/ (cm^3 · mol^{-1})	$\Delta\eta$/ (mPa · s)
	T = 283.15 K				T = 288.15 K			
0.110	1.2196	210.64	0.218	−202.90	1.2160	153.77	0.231	−125.35
0.263	1.2131	84.04	0.374	−258.63	1.2095	60.23	0.386	−171.11
0.391	1.2062	44.91	0.428	−238.47	1.2027	35.82	0.444	−155.54
0.414	1.2047	41.27	0.435	−231.46	1.2013	32.98	0.449	−151.19
0.549	1.1944	22.77	0.423	−187.43	1.1908	18.49	0.461	−123.52
0.637	1.1849	15.85	0.387	−153.59	1.1813	13.01	0.430	−101.51
0.724	1.1717	10.78	0.331	−118.37	1.1680	10.72	0.379	−76.62
0.856	1.1358	5.53	0.192	−62.48	1.1324	4.58	0.232	−41.53
	T = 293.15 K				T = 298.15 K			
0.110	1.2127	115.56	0.212	−83.07	1.2092	88.10	0.214	−57.29
0.263	1.2060	47.41	0.398	−117.25	1.2025	38.11	0.421	−82.45
0.391	1.1991	29.05	0.476	−107.18	1.1955	23.97	0.511	−75.80
0.414	1.1976	26.81	0.483	−104.32	1.1941	22.21	0.519	−73.83
0.549	1.1871	15.37	0.496	−85.78	1.1835	12.99	0.544	−61.13
0.637	1.1776	10.82	0.471	−70.79	1.1739	9.16	0.523	−50.67
0.724	1.1642	7.61	0.427	−54.68	1.1606	6.55	0.478	−39.16
0.856	1.1289	3.88	0.270	−29.09	1.1254	3.36	0.322	−20.91
	T = 303.15 K				T = 308.15 K			
0.110	1.2058	67.11	0.227	−42.44	1.2024	53.43	0.238	−30.50
0.263	1.1990	31.12	0.448	−59.74	1.1955	25.87	0.470	−43.76
0.391	1.1920	20.01	0.546	−55.20	1.1884	16.93	0.577	−40.72
0.414	1.1905	18.63	0.554	−53.77	1.1869	15.81	0.587	−39.70
0.549	1.1799	11.10	0.587	−44.81	1.1763	9.61	0.626	−33.28
0.637	1.1703	8.00	0.571	−37.15	1.1666	7.01	0.613	−27.64
0.724	1.1568	5.65	0.536	−28.87	1.1531	4.92	0.577	−21.60
0.856	1.1219	2.94	0.366	−15.45	1.1183	2.60	0.405	−11.59
	T = 313.15 K				T = 318.15 K			
0.110	1.1989	43.19	0.247	−19.45	1.1955	35.47	0.256	−14.26
0.263	1.1920	21.69	0.490	−30.29	1.1885	18.48	0.509	−22.81
0.391	1.1850	14.51	0.595	−28.56	1.1814	12.53	0.622	−21.69
0.414	1.1834	13.58	0.617	−27.89	1.1799	11.77	0.644	−21.17
0.549	1.1727	8.32	0.656	−23.74	1.1692	7.41	0.684	−18.08
0.637	1.1629	6.13	0.654	−19.80	1.1593	5.43	0.692	−15.21
0.724	1.1496	4.34	0.614	−15.53	1.1459	3.85	0.653	−11.99
0.856	1.1147	2.32	0.442	−8.36	1.1110	2.09	0.479	−6.46
	T = 323.15 K				T = 328.15 K			
0.110	1.1921	29.49	0.262	−10.72	1.1887	24.88	0.269	−8.21
0.263	1.1852	15.87	0.514	−17.52	1.1816	13.77	0.541	−13.72
0.391	1.1779	10.93	0.646	−16.76	1.1743	9.58	0.678	−13.23
0.414	1.1764	10.22	0.669	−16.45	1.1729	9.06	0.693	−12.90
0.549	1.1654	6.57	0.734	−14.08	1.1618	5.86	0.765	−11.16

(continued)

x_1	ρ/ (g · cm^{-3})	η/ (mPa · s)	V^E/ (cm^3 · mol^{-1})	$\Delta\eta$/ (mPa · s)	ρ/ (g · cm^{-3})	η/ (mPa · s)	V^E/ (cm^3 · mol^{-1})	$\Delta\eta$/ (mPa · s)
0.637	1.1556	4.78	0.730	−11.94	1.1520	4.29	0.765	−9.50
0.724	1.1420	3.46	0.704	−9.39	1.1383	3.11	0.736	−7.50
0.856	1.1073	1.90	0.514	−5.07	1.1036	1.74	0.547	−4.04
	T = 333.15 K				T = 338.15 K			
0.110	1.1853	21.18	0.276	−6.34	1.1819	18.21	0.282	−4.98
0.263	1.1783	12.04	0.545	−10.83	1.1748	10.57	0.568	−8.71
0.391	1.1710	8.60	0.692	−10.38	1.1675	7.71	0.712	−8.30
0.414	1.1693	7.99	0.719	−10.29	1.1659	7.17	0.740	−8.25
0.549	1.1944	5.19	0.798	−8.99	1.1547	4.68	0.826	−7.29
0.637	1.1483	3.87	0.800	−7.63	1.1447	3.52	0.833	−6.20
0.724	1.1344	2.82	0.786	−6.03	1.1309	2.58	0.810	−4.91
0.856	1.0999	1.61	0.582	−3.24	1.0961	1.50	0.614	−2.62
	T = 343.15 K							
0.110	1.1785	15.89	0.286	−3.88				
0.263	1.1714	9.35	0.579	−7.10				
0.391	1.1641	6.91	0.727	−6.75				
0.414	1.1624	6.37	0.757	−6.79				
0.549	1.1512	4.24	0.850	−5.98				
0.637	1.1411	3.22	0.861	−5.09				
0.724	1.1271	2.37	0.847	−4.04				
0.856	1.0923	1.40	0.643	−2.14				

Mixture 2: CH_3OH (1) + [BP][BF_4] (2)
(1) Transport Properties

Ref. [1]

Composition $V_{f,2}$	η/mPa · s	Composition $V_{f,2}$	η/mPa · s	Composition $V_{f,2}$	η/mPa · s
298.15 K					
0	0.66	0.80	14.97	0.96	82.16
0.20	0.93	0.85	24.93	0.98	115.77
0.40	1.57	0.90	40.38	1.00	145.23
0.60	4.15	0.93	55.83		

Mixture 3: CH_2Cl_2 (1) + [BP][BF_4] (2)
(1) Transport Properties

Ref. [1]

Composition $V_{f,2}$	η/mPa · s	Composition $V_{f,2}$	η/mPa · s	Composition $V_{f,2}$	η/mPa · s
298.15 K					
0	0.40	0.80	32.70	0.96	113.33
0.20	1.25	0.85	63.56	0.98	124.00
0.40	–	0.90	73.94	1.00	145.23
0.60	9.33	0.93	84.46		

6.2 11003-02001: 1-Octyl pyridinium tetrafluoroborate

Abbreviation:	[OcPy][BF_4]
Molecular formula:	$C_{13}H_{22}NBF_4$
Molar mass:	279.12
Structure:	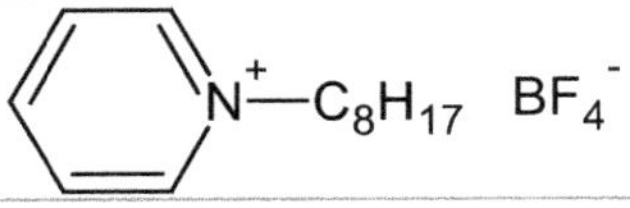

6.2.1 Binary Mixtures

Mixture 1: H_2O (1) + [OcPy][BF_4] (2)
(1) Volumetric Properties and Transport Properties

Ref. [2]

x_1	ρ/ $(g \cdot cm^{-3})$	η/ $(mPa \cdot s)$	V^E/ $(cm^3 \cdot mol^{-1})$	$\Delta\eta$/ $(mPa \cdot s)$	ρ/ $(g \cdot cm^{-3})$	η/ $(mPa \cdot s)$	V^E/ $(cm^3 \cdot mol^{-1})$	$\Delta\eta$/ $(mPa \cdot s)$
	T = 283.15 K				T = 288.15 K			
0.100	1.1222	403.80	0.045	−172.46	1.1186	288.86	0.046	−113.12
0.184	1.1205	261.38	0.157	−261.22	1.1170	192.63	0.167	−171.93
0.324	1.1172	161.38	0.314	−271.78	1.1136	121.31	0.336	−180.90
0.477	1.1131	116.44	0.328	−218.98	1.1095	89.03	0.359	−145.04
0.562	1.1103	63.83	0.272	−217.29	1.1068	49.84	0.299	−146.38
0.67	1.1055	65.55	0.169	−146.58	1.1021	50.65	0.193	−97.47
0.758	1.0991	51.80	0.091	−104.11	1.0959	37.80	0.107	−71.12
0.783	1.0966	54.27	0.070	−85.66	1.0936	38.63	0.080	−59.16
	T = 293.15 K				T = 298.15 K			
0.100	1.1151	213.16	0.052	−74.63	1.1115	159.12	0.059	−51.12
0.184	1.1134	145.61	0.174	−115.41	1.1099	112.29	0.173	−78.41
0.324	1.1101	94.11	0.344	−122.30	1.1067	71.96	0.337	−86.18
0.477	1.1060	69.94	0.378	−97.72	1.1025	55.07	0.391	−67.48
0.562	1.1031	39.60	0.340	−100.97	1.0996	31.77	0.373	−71.00
0.67	1.0984	40.42	0.237	−65.74	1.0946	32.69	0.299	−44.97
0.758	1.0924	28.74	0.144	−49.38	1.0885	23.24	0.212	−33.94
0.783	1.0899	29.21	0.126	−40.94	1.0860	22.73	0.191	−28.63
	T = 303.15 K				T = 308.15 K			
0.100	1.1079	122.48	0.072	−32.27	1.1044	96.16	0.086	−25.21
0.184	1.1063	86.00	0.195	−54.38	1.1029	68.34	0.195	−41.77
0.324	1.1032	57.88	0.360	−58.55	1.0998	47.09	0.369	−44.25
0.477	1.0990	44.37	0.415	−45.89	1.0955	36.11	0.444	−34.72
0.562	1.0962	26.47	0.380	−49.25	1.0924	22.22	0.444	−37.21
0.67	1.0913	26.82	0.309	−30.42	1.0876	22.27	0.362	−22.69

(continued)

x_1	ρ/ (g · cm^{-3})	η/ (mPa · s)	V^E/ (cm^3 · mol^{-1})	$\Delta\eta$/ (mPa · s)	ρ/ (g · cm^{-3})	η/ (mPa · s)	V^E/ (cm^3 · mol^{-1})	$\Delta\eta$/ (mPa · s)
0.758	1.0850	19.17	0.248	−23.03	1.0816	15.97	0.270	−17.19
0.783	1.0824	18.09	0.236	−19.83	1.0792	15.11	0.248	−14.69
	T = 313.15 K				T = 318.15 K			
0.100	1.1009	71.50	0.097	−23.74	1.0974	57.41	0.109	−16.06
0.184	1.0994	54.93	0.210	−31.47	1.0959	44.98	0.226	−21.68
0.324	1.0961	38.75	0.406	−32.94	1.0926	32.19	0.426	−23.14
0.477	1.0918	30.19	0.493	−25.42	1.0881	25.23	0.538	−17.71
0.562	1.0888	19.43	0.482	−27.25	1.0851	16.88	0.528	−19.18
0.67	1.0839	18.67	0.406	−16.67	1.0805	15.98	0.429	−11.33
0.758	1.0778	13.56	0.319	−12.53	1.0750	11.52	0.304	−8.67
0.783	1.0756	12.78	0.288	−10.67	1.0728	11.08	0.267	−7.09
	T = 323.15 K				T = 328.15 K			
0.100	1.0940	46.73	0.118	−11.73	1.0905	38.44	0.126	−9.77
0.184	1.0924	37.06	0.241	−15.99	1.0889	30.98	0.256	−12.79
0.324	1.0892	27.26	0.444	−16.78	1.0857	23.23	0.457	−13.11
0.477	1.0847	21.39	0.550	−12.80	1.0813	18.15	0.556	−10.08
0.562	1.0817	14.54	0.539	−14.19	1.0783	12.05	0.556	−11.67
0.67	1.0771	13.65	0.445	−8.13	1.0735	11.88	0.473	−6.12
0.758	1.0713	9.95	0.342	−6.17	1.0674	8.61	0.396	−4.73
0.783	1.0693	9.39	0.294	−5.12	1.0652	8.20	0.357	−3.81
	T = 333.15 K				T = 338.15 K			
0.100	1.0870	32.10	0.136	−8.70	1.0835	27.06	0.156	−7.27
0.184	1.0854	26.18	0.275	−10.85	1.0820	22.38	0.293	−8.79
0.324	1.0822	19.92	0.472	−10.84	1.0788	17.32	0.485	−8.57
0.477	1.0776	15.84	0.612	−8.07	1.0741	13.44	0.634	−6.69
0.562	1.0743	10.61	0.631	−9.49	1.0708	8.51	0.654	−8.43
0.67	1.0693	10.38	0.557	−4.88	1.0658	9.23	0.584	−3.64
0.758	1.0636	7.63	0.438	−3.68	1.0600	6.75	0.465	−2.79
0.783	1.0613	7.17	0.407	−3.03	1.0578	6.31	0.427	−2.30
	T = 343.15 K				T = 348.15 K			
0.100	1.0801	23.06	0.159	−6.13	1.0767	19.83	0.169	−5.21
0.184	1.0785	19.25	0.307	−7.25	1.0751	16.71	0.318	−6.03
0.324	1.0753	15.05	0.496	−6.97	1.0719	13.13	0.506	−5.77
0.477	1.0614	11.58	0.000	−5.55	1.0543	10.05		−4.65
0.562	1.0674	8.15	0.662	−6.26	1.0642	7.17	0.655	−5.20
0.67	1.0623	7.99	0.604	−2.97	1.0588	7.19	0.620	−2.22
0.758	1.0564	5.99	0.491	−2.15	1.0528	5.29	0.514	−1.72
0.783	1.0541	5.58	0.457	−1.76	1.0503	4.97	0.499	−1.35

6.3 11009-02001: 4-Methyl-*N*-butylpyridinium tetrafluoroborate

Abbreviation:	[4-MBpy][BF_4], [b4mpy][BF_4], [mebupy][BF_4]
Molecular formula:	$C_{10}H_{16}NBF_4$
Molar mass	237.05
Structure:	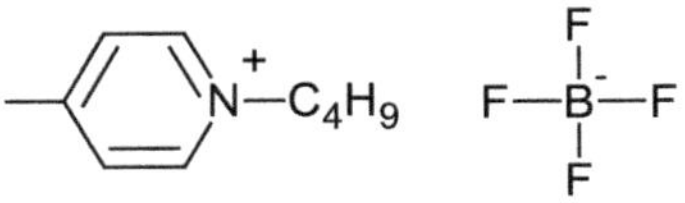

6.3.1 Binary Mixtures

Mixture 1: H_2O (1) + [b4mpy][BF_4] (2)
(1) Volumetric Properties and Transport Properties

Ref. [3]

x_2	ρ/(kg · m^{-3})	$10^9V^E_m$/ (m^3 · mol^{-1})	x_2	ρ/(kg · m^{-3})	$10^9V^E_m$/ (m^3 · mol^{-1})
298.15 K					
0.0502	1060.5	137	0.6014	1168.3	501
0.1042	1094.8	244	0.7228	1173.9	413
0.1512	1112.7	328	0.7493	1175.1	370
0.2006	1126.1	390	0.7830	1176.4	330
0.2543	1136.1	470	0.8449	1178.7	243
0.2948	1142.4	498	0.8731	1179.8	172
0.4000	1154.1	561	0.9504	1182.1	61
0.4913	1161.5	555			
318.15 K					
0.0327	1033.0	145	0.4841	1146.6	638
0.0994	1080.0	301	0.6083	1154.2	570
0.1505	1098.7	428	0.6715	1157.5	490
0.1989	1111.3	508	0.7125	1159.2	460
0.2360	1118.6	563	0.7299	1160.0	430
0.2921	1127.6	607	0.8124	1163.0	343
0.3237	1131.9	612	0.8565	1164.7	250
0.4169	1141.3	648	0.8888	1165.8	190

Ref. [1]

Composition $V_{f,2}$	η/mPa · s	Composition $V_{f,2}$	η/mPa · s	Composition $V_{f,2}$	η/mPa · s
298.15 K					
0	0.93	0.80	5.32	0.96	46.00
0.20	1.40	0.85	11.70	0.98	78.06

(*continued*)

Composition $V_{f,2}$	η/mPa · s	Composition $V_{f,2}$	η/mPa · s	Composition $V_{f,2}$	η/mPa · s
0.40	1.92	0.90	17.40	1.00	200.65
0.60	2.84	0.93	35.43		

Ref. [1]

	η/mPa · s					
Composition $V_{f,2}$	298.15 K	303.15 K	308.15 K	313.15 K	318.15 K	323.15 K
0	0.93	0.84	0.76	0.66	0.53	0.47
0.20	1.40	1.18	1.04	0.94	0.81	0.69
0.40	1.92	1.69	1.50	1.34	1.21	1.08
0.60	2.84	2.47	2.17	2.05	1.89	1.68
0.80	5.32	4.73	4.16	3.83	3.59	3.49
1.00	200.65	138.10	100.87	77.15	59.76	46.53

(2) Excess Molar Enthalpies

Ref. [3]

x_2	H^E_m/(J · mol^{-1})	x_2	H^E_m/(J · mol^{-1})	x_2	H^E_m/(J · mol^{-1})
298.15 K					
0.0072	104	0.0805	876	0.2972	1810
0.0145	198	0.0963	1015	0.3449	1812
0.0215	284	0.1144	1162	0.3974	1758
0.0303	380	0.1563	1460	0.4857	1586
0.0406	491	0.1863	1591	0.5954	1340
0.0526	613	0.2196	1701	0.7282	962
0.0659	740	0.2577	1772	0.8706	486
318.15 K					
0.0097	188	0.1187	1507	0.3830	2230
0.0226	392	0.1414	1681	0.4410	2162
0.0382	606	0.1695	1839	0.5071	2027
0.0565	825	0.2008	1972	0.5961	1758
0.0755	1041	0.2346	2088	0.6778	1421
0.0949	1228	0.2746	2191	0.7866	988
0.1132	1404	0.3253	2245	0.9058	428

Mixture 2: CH_3OH (1) + [b4mpy][BF_4] (2)

(1) Volumetric Properties and Transport Properties

Ref. [3]

x_2	ρ/(kg · m^{-3})	$10^9 V^E_m$/ (m^3 · mol^{-1})	x_2	ρ/(kg · m^{-3})	$10^9 V^E_m$/ (m^3 · mol^{-1})
298.15 K					
0.0539	880.6	−402	0.4548	1110.8	−545
0.0819	916.2	−510	0.5027	1120.9	−449
0.0950	931.5	−589	0.5533	1131.2	−451
0.1078	944.4	−613	0.6076	1140.4	−394
0.1549	984.7	−669	0.7174	1155.8	−280
0.2038	1017.3	−697	0.8053	1165.6	−151
0.3016	1064.3	−674	0.9129	1176.0	−56
0.3729	1088.5	−583			
318.15 K					
0.0489	855.9	−460	0.4888	1103.9	−633
0.0960	915.9	−720	0.5342	1112.7	−527
0.1439	961.3	−881	0.5959	1123.9	−474
0.1894	994.2	−943	0.6592	1133.5	−379
0.3003	1049.4	−885	0.7266	1142.3	−271
0.3551	1068.5	−790	0.8378	1154.8	−150
0.4349	1091.7	−737	0.9046	1161.4	−117

Ref. [1]

Composition $V_{f,2}$	η/mPa · s	Composition $V_{f,2}$	η/mPa · s	Composition $V_{f,2}$	η/mPa · s
298.15 K					
0	0.66	0.80	14.00	0.96	97.53
0.20	0.90	0.85	18.90	0.98	115.33
0.40	1.48	0.90	37.50	1.00	200.65
0.60	3.47	0.93	64.63		

(2) Excess Molar Enthalpies

Ref. [3]

x_2	H^E_m/(J · mol^{-1})	x_2	H^E_m/(J · mol^{-1})	x_2	H^E_m/(J · mol^{-1})
298.15 K					
0.0155	254	0.1711	1640	0.4270	1894
0.0359	485	0.1953	1735	0.5070	1718
0.0582	667	0.2214	1822	0.6029	1464
0.0832	907	0.2412	1895	0.7161	1137
0.1111	1131	0.2755	1963	0.8469	646
0.1397	1377	0.3109	2010		
0.1698	1578	0.3617	1986		
318.15 K					
0.0102	141	0.1969	2082	0.4968	2489
0.0238	277	0.2103	2214	0.5563	2385

(*continued*)

x_2	$H^E{}_m/(J \cdot mol^{-1})$	x_2	$H^E{}_m/(J \cdot mol^{-1})$	x_2	$H^E{}_m/(J \cdot mol^{-1})$
0.0405	410	0.2392	2337	0.6281	2230
0.0635	744	0.2699	2421	0.7109	1951
0.0898	1094	0.3021	2524	0.8042	1513
0.1174	1405	0.3442	2551	0.8793	998
0.1476	1687	0.3846	2590	0.9607	392
0.1778	1911	0.4397	2570		

Mixture 3: C_2H_5OH (1) + [b4mpy][BF_4] (2)
(1) Volumetric Properties

Ref. [3]

x_2	$\rho/(kg \cdot m^{-3})$	$10^9 V^E{}_m/(m^3 \cdot mol^{-1})$	x_2	$\rho/(kg \cdot m^{-3})$	$10^9 V^E{}_m/(m^3 \cdot mol^{-1})$
298.15 K					
0.0523	852.0	−281	0.5553	1110.8	−363
0.1023	901.7	−396	0.6045	1121.8	−295
0.1469	938.3	−477	0.6639	1134.2	−270
0.2023	976.0	−535	0.7127	1142.8	−180
0.3070	1030.5	−547	0.7447	1148.4	−162
0.3801	1059.5	−495	0.7964	1156.8	−138
0.4445	1080.8	−452	0.8673	1167.0	−69
0.5069	1098.6	−396	0.9171	1173.6	−35
318.15 K					
0.0464	827.9	−280	0.4947	1080.9	−597
0.0965	881.1	−574	0.5877	1103.6	−456
0.1445	921.6	−697	0.6821	1123.2	−383
0.2494	987.9	−760	0.7798	1139.8	−249
0.2968	1011.0	−739	0.8267	1146.8	−176
0.3855	1046.8	−696	0.8776	1153.8	−96

(2) Excess Molar Enthalpies

Ref. [3]

x_2	$H^E{}_m/(J \cdot mol^{-1})$	x_2	$H^E{}_m/(J \cdot mol^{-1})$	x_2	$H^E{}_m/(J \cdot mol^{-1})$
298.15 K					
0.0209	299	0.2460	2205	0.5522	1903
0.0461	623	0.2661	2318	0.6225	1629
0.0751	936	0.3000	2389	0.7120	1329
0.1068	1192	0.3402	2426	0.8071	931
0.1406	1498	0.3855	2386	0.9030	504
0.1751	1755	0.4296	2280		
0.2104	2014	0.4879	2115		
318.15 K					
0.0162	379	0.2364	2755	0.5506	2587

(*continued*)

x_2	$H^E_m/(J \cdot mol^{-1})$	x_2	$H^E_m/(J \cdot mol^{-1})$	x_2	$H^E_m/(J \cdot mol^{-1})$
0.0465	883	0.2498	2856	0.6196	2333
0.0750	1316	0.2824	2964	0.6923	2022
0.1080	1705	0.3237	3004	0.7625	1689
0.1427	2070	0.3722	2998	0.8348	1318
0.1780	2370	0.4286	2918	0.9014	892
0.2091	2605	0.4859	2777	0.9592	425

(3) Phase Equilibrium

Ref. [3]

x_2	T/K	x_2	T/K	x_2	T/K
0.060	275.69	0.091	276.39	0.133	276.18
0.066	276.00	0.103	276.37		
0.072	276.20	0.124	276.29		

Mixture 4: 1-C_3H_7OH (1) + [b4mpy][BF_4] (2)
(1) Volumetric Properties

Ref. [3]

x_2	$\rho/(kg \cdot m^{-3})$	$10^9 V^E_m/(m^3 \cdot mol^{-1})$	x_2	$\rho/(kg \cdot m^{-3})$	$10^9 V^E_m/(m^3 \cdot mol^{-1})$
298.15 K					
0.0154	815.6	−59	0.5972	1107.8	−239
0.0271	827.2	−95	0.6889	1129.1	−167
0.2988	1006.8	−340	0.6960	1130.6	−163
0.3987	1047.4	−305	0.7812	1147.9	−133
0.4965	1079.8	−262	0.8841	1165.9	−61
318.15 K					
0.0549	837.6	−284	0.4901	1064.0	−510
0.1008	875.1	−304	0.6039	1095.2	−379
0.1969	939.6	−485	0.6726	1111.2	−290
0.3103	997.6	−551	0.8248	1141.4	−143
0.3828	1027.5	−559	0.8967	1156.9	−35
0.4381	1047.4	−549	0.9191	1153.3	−22
0.4534	1052.4	−531			

(2) Excess Molar Enthalpies

Ref. [3]

x_2	$H^E_m/(J \cdot mol^{-1})$	x_2	$H^E_m/(J \cdot mol^{-1})$	x_2	$H^E_m/(J \cdot mol^{-1})$
298.15 K					
0.0120	132	0.3908	2421	0.6734	1650
0.0243	216	0.4344	2342	0.7520	1317
0.0308	301	0.4819	2261	0.8319	990
0.3198	2492	0.5374	2111	0.9181	579
0.3532	2462	0.6008	1912		

(continued)

x_2	H^E_m/(J · mol^{-1})	x_2	H^E_m/(J · mol^{-1})	x_2	H^E_m/(J · mol^{-1})
318.15 K					
0.0189	388	0.2581	2803	0.6964	2076
0.0540	960	0.2912	2895	0.7815	1693
0.0930	1467	0.3310	2963	0.8460	1292
0.1341	1899	0.3847	2945	0.9043	825
0.1775	2273	0.4488	2886	0.9568	436
0.2167	2546	0.5249	2720		
0.2312	2681	0.6069	2449		

(3) Phase Equilibrium

Ref. [3]

x_2	T/K	x_2	T/K	x_2	T/K
0.012	291.67	0.177	302.18	0.376	293.29
0.024	296.97	0.208	301.64	0.402	291.22
0.035	299.19	0.238	300.87	0.431	288.75
0.054	301.42	0.264	299.95	0.451	286.81
0.077	302.30	0.291	298.79	0.482	283.54
0.100	302.57	0.312	297.68	0.508	281.13
0.123	302.75	0.332	296.34	0.542	277.33
0.150	302.67	0.351	295.05		

(4) Interval of Immiscibility and Upper Critical Solution Temperature

Ref. [3]

Mixture	Inmiscibility interval at T		UCST	
	298.15 K	318.15 K	T/K	x_{IL}
[b4mpy][BF_4] + propanol	(0.028, 0.306)		302.8	0.123

Mixture 5: 1-C_4H_9OH (1) + [b4mpy][BF_4] (2)
(1) Volumetric Properties

Ref. [3]

x_2	ρ/(kg · m^{-3})	$10^9V^E_m$/(m^3 · mol^{-1})	x_2	ρ/(kg · m^{-3})	$10^9V^E_m$/(m^3 · mol^{-1})
298.15 K					
0.5948	1094.4	−97	0.7851	1141.8	−41
0.6920	1120.2	−87	0.8916	1163.9	−22
0.7390	1131.5	−71			
318.15 K					
0.0548	833.2	−85	0.5640	1071.4	−307
0.1169	875.7	−139	0.6758	1102.2	−280
0.2301	940.7	−218	0.7856	1128.1	−236
0.3406	992.1	−265	0.8427	1140.0	−169

(continued)

x_2	$\rho/(kg \cdot m^{-3})$	$10^9 V^E_m/(m^3 \cdot mol^{-1})$	x_2	$\rho/(kg \cdot m^{-3})$	$10^9 V^E_m/(m^3 \cdot mol^{-1})$
0.4471	1033.5	−315	0.8870	1148.6	−100

(2) Excess Molar Enthalpies

Ref. [3]

2	$H^E_m/(J \cdot mol^{-1})$	x_2	$H^E_m/(J \cdot mol^{-1})$	x_2	$H^E_m/(J \cdot mol^{-1})$
318.15 K					
0.0242	468	0.3344	2904	0.7232	1939
0.0613	944	0.3708	2946	0.7962	1509
0.1029	1419	0.4149	2958	0.8569	1162
0.1466	1795	0.4635	2886	0.9006	817
0.1938	2206	0.5162	2789	0.9435	538
0.2400	2515	0.5729	2608		
0.2864	2736	0.6488	2266		

(3) Phase Equilibrium

Ref. [3]

x_2	T/K	x_2	T/K	x_2	T/K
0.007	293.17	0.179	317.63	0.522	301.50
0.010	300.97	0.214	317.38	0.557	298.16
0.018	306.37	0.242	316.79	0.592	294.60
0.022	309.68	0.275	316.09	0.612	291.51
0.028	311.48	0.305	315.13	0.658	286.01
0.048	314.75	0.339	313.78	0.685	282.60
0.067	316.50	0.369	312.65	0.727	279.35
0.096	317.42	0.402	310.91	0.757	276.89
0.118	317.60	0.447	307.98		
0.145	317.67	0.483	305.79		

(4) Interval of Immiscibility and Upper Critical Solution Temperature

Ref. [3]

Mixture	Inmiscibility interval at T		UCST	
	298.15 K	318.15 K	T/K	x_{IL}
[b4mpy][BF_4] + butanol	(0.009, 0.557)		317.7	0.145

Mixture 6: 1-$C_5H_{11}OH$ (1) + [b4mpy][BF_4] (2)
(1) Volumetric Properties

Ref. [3]

x_2	$\rho/(kg \cdot m^{-3})$	$10^9 V^E_m/(m^3 \cdot mol^{-1})$	x_2	$\rho/(kg \cdot m^{-3})$	$10^9 V^E_m/(m^3 \cdot mol^{-1})$
318.15 K					
0.5651	1060.4	−259	0.8041	1126.1	−133
0.6035	1072.1	−231	0.8616	1139.6	−97

(continued)

x_2	$\rho/(kg \cdot m^{-3})$	$10^9 V^E{}_m/(m^3 \cdot mol^{-1})$	x_2	$\rho/(kg \cdot m^{-3})$	$10^9 V^E{}_m/(m^3 \cdot mol^{-1})$
0.6715	1091.9	−227	0.9120	1150.8	−64
0.7601	1115.3	−170			

(2) Phase Equilibrium

Ref. [3]

x_2	T/K	x_2	T/K	x_2	T/K
0.006	290.27	0.122	329.71	0.447	323.52
0.010	297.94	0.171	329.94	0.483	321.56
0.012	303.41	0.225	329.75	0.523	319.23
0.016	311.27	0.272	329.42	0.564	315.72
0.019	318.76	0.313	328.62	0.609	310.97
0.035	325.32	0.344	328.27	0.659	305.12
0.051	327.00	0.378	326.78	0.696	300.90
0.081	328.77	0.411	325.35	0.730	295.49

(3) Interval of Immiscibility and Upper Critical Solution Temperature

Ref. [3]

Mixture	Inmiscibility interval at T		UCST	
	298.15 K	318.15 K	T/K	x_{IL}
[b4mpy][BF_4] + pentanol	(0.009, 0.730)	(0.019, 0.564)	329.9	0.171

Mixture 7: CH_2Cl_2 (1) + [4-MBP][BF_4] (2)
(1) Transport Properties

Ref. [1]

Composition $V_{f,2}$	η/ mPa · s	Composition $V_{f,2}$	η/ mPa · s	Composition $V_{f,2}$	η/ mPa · s
298.15 K					
0	0.40	0.80	33.50	0.96	121.30
0.20	0.85	0.85	45.13	0.98	157.33
0.40	3.42	0.90	74.40	1.00	200.65
0.60	9.48	0.93	84.30		

6.3.2 Ternary Mixtures

Mixture 1: $C_6H_5CH_3$ (1) + C_7H_{16} (2) + [mebupy][BF_4] (3)
(1) Selectivity

Ref. [4]

n-heptane-rich phase			[mebupy][BF_4]-rich phase			
x_1	x_2	x_3	x_1	x_2	x_3	Experimental selectivity

(*continued*)

n-heptane-rich phase			[mebupy][BF_4]-rich phase			
x_1	x_2	x_3	x_1	x_2	x_3	Experimental selectivity
0.1 MPa						
313.2 K						
0.0000	1.0000	0.0000	0.0000	0.0095	0.9905	–
0.0367	0.9633	0.0000	0.0162	0.0080	0.9758	53.09
0.0645	0.9306	0.0049	0.0290	0.0079	0.9631	52.91
0.1021	0.8979	0.0000	0.0448	0.0079	0.9473	49.63
0.1388	0.8590	0.0023	0.0594	0.0077	0.9328	47.67
0.1558	0.8442	0.0000	0.0730	0.0088	0.9182	45.22
0.2319	0.7681	0.0000	0.1112	0.0091	0.8798	40.57
0.3183	0.6817	0.0000	0.1481	0.0085	0.8433	37.13
0.4162	0.5838	0.0000	0.1916	0.0084	0.8000	32.05
0.5653	0.4347	0.0000	0.2630	0.0082	0.7288	24.63
0.6638	0.3362	0.0000	0.2982	0.0070	0.6948	21.48
0.7759	0.2241	0.0000	0.3538	0.0062	0.6400	16.43
0.9011	0.0989	0.0000	0.4496	0.0041	0.5464	12.14
1.0000	0.0000	0.0000	0.5158	0.0000	0.4842	–
348.2 K						
0.0000	0.9951	0.0049	0.0000	0.0118	0.9882	–
0.0402	0.9598	0.0000	0.0166	0.0156	0.9678	26.15
0.0404	0.9531	0.0066	0.0182	0.0126	0.9692	34.26
0.0701	0.9299	0.0000	0.0334	0.0113	0.9553	39.32
0.1077	0.8923	0.0000	0.0483	0.0107	0.9410	37.43
0.1478	0.8522	0.0000	0.0669	0.0112	0.9218	34.33
0.2069	0.7804	0.0126	0.0929	0.0121	0.8950	28.98
0.2771	0.7229	0.0000	0.1008	0.0110	0.8882	23.87
0.3303	0.6600	0.0097	0.1455	0.0114	0.8431	25.59
0.4211	0.5719	0.0071	0.1836	0.0113	0.8050	22.04
0.5302	0.4668	0.0031	0.2307	0.0106	0.7587	19.15
0.5920	0.4080	0.0000	0.2420	0.0099	0.7481	16.85
1.0000	0.0000	0.0000	0.4911	0.0000	0.5089	–

(2) Solubility

Ref. [4]

Temperature					
Toluene in IL, maximum		*n*-Heptane in IL, average		IL in *n*-heptane, average	
313.2 K	348.2 K	313.2 K	348.2 K	313.2 K	348.2 K
0.52	0.49	0.0072	0.0107	0.0010	0.0045

Mixture 2: C_6H_6 (1) + C_6H_{14} (2) + [mebupy][BF_4] (3)
(1) Distribution Coefficients and Selectivity

Ref. [5]

N-hexane-rich phase			[mebupy]BF_4-rich phase			β		S
x_1	x_2	x_3	x_1	x_2	x_3	benzene	*N*-hexane	exp.
0.1 MPa								
T = 313.2 K								
0.000	0.990	0.010	0.000	0.020	0.980		0.020	
0.044	0.925	0.031	0.060	0.012	0.927	1.375	0.013	103.8
0.075	0.895	0.030	0.080	0.014	0.906	1.075	0.016	68.7
0.106	0.866	0.028	0.101	0.015	0.885	0.949	0.017	55.9
0.197	0.789	0.014	0.163	0.014	0.822	0.830	0.018	45.7
0.299	0.686	0.016	0.229	0.015	0.756	0.768	0.021	36.0
0.414	0.570	0.016	0.294	0.015	0.691	0.709	0.027	26.2
0.972	0.000	0.028	0.644	0.000	0.356	0.663		
T = 333.2 K								
0.000	0.978	0.022	0.000	0.017	0.983		0.017	
0.049	0.944	0.007	0.061	0.017	0.922	1.258	0.018	71.1
0.084	0.906	0.010	0.088	0.016	0.896	1.045	0.018	57.9
0.115	0.874	0.011	0.106	0.017	0.877	0.925	0.019	48.9
0.210	0.785	0.006	0.166	0.017	0.817	0.790	0.022	36.5
0.313	0.679	0.008	0.229	0.017	0.754	0.732	0.025	29.5
0.430	0.564	0.006	0.295	0.017	0.688	0.686	0.030	22.9
0.979	0.000	0.021	0.622	0.000	0.378	0.635		

Mixture 3: $C_6H_5C_2H_5$ (1) + C_8H_{18} (2) + [mebupy][BF_4] (3)
(1) Distribution Coefficients and Selectivity

Ref. [5]

N-octane-rich phase			[mebupy]BF_4-rich phase			β		S
x_1	x_2	x_3	x_1	x_2	x_3	ethylbenzene	*N*-octane	exp.
0.1 MPa								
T = 313.2 K								
0.000	0.992	0.008	0.000	0.009	0.991		0.009	
0.045	0.935	0.020	0.023	0.007	0.970	0.502	0.008	64.3
0.101	0.870	0.029	0.042	0.006	0.952	0.416	0.007	58.3
0.143	0.827	0.030	0.051	0.006	0.943	0.356	0.007	49.9
0.268	0.714	0.018	0.074	0.007	0.919	0.276	0.010	29.0
0.416	0.554	0.030	0.109	0.005	0.886	0.263	0.008	31.0
0.560	0.422	0.018	0.140	0.004	0.856	0.250	0.010	23.9
0.996	0.000	0.004	0.294	0.000	0.706	0.295		
T = 348.2 K								
0.000	0.982	0.018	0.000	0.008	0.992		0.008	
0.079	0.914	0.007	0.035	0.008	0.958	0.437	0.008	51.9

(*continued*)

N-octane-rich phase			[mebupy]BF_4-rich phase			β		*S*
x_1	x_2	x_3	x_1	x_2	x_3	ethylbenzene	*N*-octane	exp.
0.121	0.866	0.014	0.045	0.008	0.948	0.373	0.009	43.1
0.164	0.825	0.011	0.055	0.008	0.937	0.334	0.010	32.8
0.294	0.690	0.016	0.081	0.007	0.912	0.275	0.010	27.9
0.437	0.550	0.013	0.110	0.006	0.884	0.251	0.011	22.3
0.572	0.413	0.016	0.139	0.005	0.856	0.242	0.013	18.5
0.987	0.000	0.013	0.250	0.000	0.750	0.253		

Mixture 4: *m*-$C_6H_4(CH_3)_2$ (1) + C_8H_{18} (2) + [mebupy][BF_4] (3)
(1) Distribution Coefficients and Selectivity

Ref. [5]

N-octane -rich phase			[mebupy]BF_4-rich phase			β		
x_1	x_2	x_3	x_1	x_2	x_3	*m*-xylene	*N*-octane	*S* exp.
0.1 MPa								
T = 313.2 K								
0.000	0.992	0.008	0.000	0.009	0.991		0.009	
0.081	0.911	0.008	0.029	0.008	0.964	0.356	0.008	42.6
0.146	0.832	0.022	0.047	0.007	0.946	0.321	0.008	38.1
0.189	0.789	0.022	0.055	0.007	0.939	0.291	0.008	35.4
0.230	0.747	0.023	0.064	0.007	0.929	0.280	0.009	32.1
0.251	0.731	0.018	0.070	0.006	0.925	0.277	0.008	34.3
0.287	0.694	0.020	0.077	0.007	0.916	0.270	0.010	28.0
0.380	0.600	0.019	0.097	0.005	0.898	0.256	0.008	30.7
0.468	0.512	0.020	0.117	0.005	0.878	0.250	0.010	26.1
0.554	0.414	0.032	0.138	0.004	0.858	0.249	0.010	24.6
0.983	0.000	0.017	0.265	0.000	0.735	0.270		
T = 348.2 K								
0.000	0.982	0.018	0.000	0.008	0.992		0.008	
0.082	0.896	0.022	0.027	0.007	0.966	0.324	0.008	39.2
0.124	0.861	0.016	0.037	0.007	0.956	0.296	0.008	34.9
0.170	0.810	0.021	0.046	0.007	0.946	0.274	0.009	30.8
0.212	0.772	0.015	0.057	0.007	0.936	0.268	0.009	28.4
0.257	0.729	0.015	0.065	0.007	0.928	0.253	0.009	27.2
0.297	0.685	0.018	0.075	0.007	0.918	0.254	0.010	26.0
0.375	0.599	0.026	0.091	0.006	0.903	0.243	0.010	23.4
0.473	0.514	0.013	0.112	0.006	0.882	0.237	0.011	21.0
0.573	0.405	0.022	0.141	0.005	0.854	0.245	0.013	18.7
0.979	0.000	0.021	0.261	0.000	0.739	0.266		

6.4 11009-03001: *N* -Butyl-4-methylpyridinium bis [(trifluoromethyl)sulfonyl]imide

Abbreviation:	$[BM^4py][NTf_2]$
Molecular formula:	$C_{12}H_{16}F_6N_2O_4S_2$
Molar mass:	430.41
Structure:	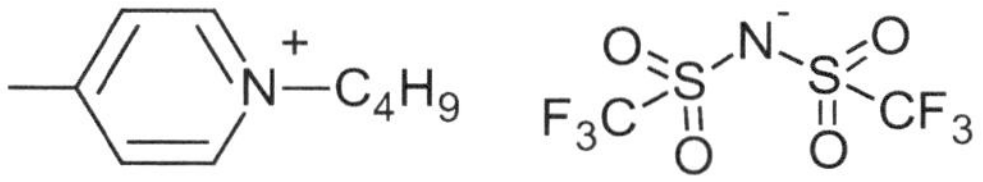

6.4.1 Binary Mixtures

Mixture 1: SCH = CHCH = CH (thiophene) (1) + $[BM^4py][NTf_2]$ (2)
(1) Phase Equilibrium

Ref. [6]

x_2	T^{LLE}/K	x_2	T^{LLE}/K
0.1427	269.0	0.0010	345.5[a]
0.1402	278.0	0.0009	333.3[a]
0.1375	287.6	0.0007	321.2[a]
0.1340	298.5	0.0007	320.4[a]
0.1295	308.9	0.0006	319.9[a]
0.1260	317.7	0.0006	314.0[a]
0.1227	323.6	0.0005	309.3[a]
0.1201	330.6	0.0004	299.2[a]
0.1177	337.1	0.0003	293.1[a]
0.1154	342.4	0.0003	290.2[a]
0.1122	348.3	0.0003	282.8[a]
0.1097	355.4	0.0002	277.3[a]
0.1068	364.1		
0.1029	367.6		

[a] Thiophene rich phase.

6.5 11009-04007: *N* -Butyl-4 -methylpyridinium tosylate

Abbreviation:	[BM⁴py][Tos], [BMPy][Tos]
Molecular formula:	$C_{17}H_{23}NO_3S$
Molar mass:	321.43
Structure:	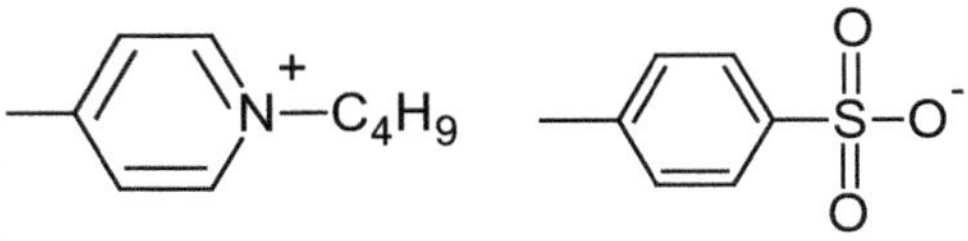

6.5.1 Binary Mixtures

Mixture 1: SCH = CHCH = CH (thiophene) (1) + [BM⁴py][Tos] (2)
(1) Phase Equilibrium
Ref. [6]

x_2	T^{SLE}/K	x_2	T^{SLE}/K
1.0000	324.8	0.6339	303.3
0.9980	323.5	0.5839	299.5
0.9693	322.3	0.5606	297.5
0.9428	320.9	0.5143	293.4
0.9061	318.9	0.4793	289.9
0.8584	315.9	0.4471	286.9
0.8384	314.7	0.4219	284.8
0.8046	313.5	0.3778	281.7
0.7834	312.2	0.3429	279.8
0.7652	311.2	0.3082	277.1
0.7298	309.3	0.2786	274.9
0.6976	307.5	0.2521	272.6
0.6645	305.2		

Mixture 2: C_2H_5OH (1) + [BMPy][Tos] (2)
(1) Phase Equilibrium

Ref. [7]

	Exptl		Wilson equation	
x_2	P/(kPa)	γ_1	P/(kPa)	γ_1
T = 373.15 K				
0.5412	49.2	0.494	47.2	0.474
0.5121	55.5	0.524	53.4	0.504
0.4894	60.3	0.543	58.7	0.529
0.4523	68.7	0.575	68.0	0.570
0.4321	74.4	0.600	73.5	0.593

(*continued*)

x_2	Exptl P/(kPa)	γ_1	Wilson equation P/(kPa)	γ_1
0.4043	81.6	0.626	81.4	0.625
0.3711	91.3	0.662	91.6	0.664
0.3436	95.9	0.666	100.6	0.697
0.3256	101.1	0.682	106.7	0.719
0.2941	107.8	0.694	117.9	0.757
0.0000	227.0	1.000	227.0	1.000

(2) Phase Equilibrium

Ref. [7]

x_2	Exptl T/K	γ_2	Wilson equation T/K	γ_2
1.0000	324.86	1.000	324.86	1.000
0.9035	321.67	1.050	321.06	1.039
0.8158	318.68	1.106	318.13	1.096
0.7420	315.01	1.142	314.92	1.140
0.6992	312.08	1.151	312.60	1.162
0.6598	310.13	1.178	310.10	1.177
0.6199	307.50	1.196	307.15	1.188
0.5898	304.40	1.187	304.63	1.192
0.5493	300.29	1.179	300.78	1.191
0.5052	295.61	1.171	295.95	1.179
0.4691	291.52	1.162	291.45	1.160
0.4224	285.02	1.128	284.85	1.123
0.3478	272.29	1.032	272.32	1.033
0.2984	262.06	0.940	262.52	0.951
0.2640	255.46	0.896	254.90	0.883

Mixture 3: 1-C_3H_7OH (1) + [BMPy][Tos] (2)
(1) Phase Equilibrium

Ref. [7]

x_2	Exptl P/(kPa)	γ_1	Wilson equation P/(kPa)	γ_1
T = 373.15 K				
0.3382	53.3	0.731	53.3	0.731
0.3010	58.5	0.758	59.7	0.773
0.2790	63.0	0.790	63.7	0.798
0.2499	70.3	0.845	69.0	0.830
0.2103	76.2	0.868	76.5	0.872
0.1929	80.4	0.895	79.9	0.889
0.1723	84.3	0.913	83.8	0.909
0.1502	89.7	0.945	88.1	0.928

(*continued*)

	Exptl		Wilson equation	
x_2	P/(kPa)	γ_1	P/(kPa)	γ_1
0.1268	93.5	0.957	92.5	0.947
0.1145	96.0	0.968	94.7	0.956
0.0000	112.7	1.000	112.7	1.000

(2) Phase Equilibrium

Ref. [7]

	Exptl		Wilson equation	
x_2	T/K	γ_2	T/K	γ_2
1.0000	324.86	1.000	324.86	1.000
0.9310	323.17	1.045	321.37	1.014
0.8721	320.64	1.069	318.97	1.040
0.8250	319.13	1.103	317.03	1.064
0.7770	316.61	1.121	314.90	1.088
0.7280	313.93	1.142	312.41	1.112
0.6631	309.88	1.167	308.51	1.138
0.6074	305.20	1.170	304.48	1.154
0.5623	301.46	1.178	300.66	1.160
0.5262	297.82	1.174	297.19	1.160
0.4910	294.31	1.174	293.43	1.154
0.4533	288.98	1.142	288.94	1.141
0.4080	283.33	1.126	282.86	1.115
0.3833	279.13	1.094	279.24	1.096
0.3493	274.13	1.073	273.81	1.065
0.3244	269.40	1.034	269.52	1.037
0.2811	259.70	0.940	261.34	0.980

Mixture 4: 1-C_4H_9OH (1) + [BMPy][Tos] (2)
(1) Phase Equilibrium

Ref. [7]

	Exptl		Wilson equation	
x_2	P/(kPa)	γ_1	P/(kPa)	γ_1
T = 373.15 K				
0.2622	35.7	0.939	36.8	0.967
0.2314	38.3	0.966	38.6	0.974
0.2064	40.1	0.978	40.1	0.979
0.1738	42.8	1.002	42.1	0.984
0.1500	44.6	1.013	43.5	0.988
0.1166	46.9	1.024	45.4	0.993

(*continued*)

	Exptl		Wilson equation	
x_2	P/(kPa)	γ_1	P/(kPa)	γ_1
0.0999	48.5	1.038	46.3	0.995
0.0775	49.8	1.040	47.7	0.997
0.0496	51.1	1.035	49.3	0.999
0.0000	52.0	1.000	52.0	1.000

Mixture 5: 1-$C_8H_{17}OH$ (1) + [BMPy][Tos] (2)
(1) Eutectic Temperature
Ref. [8]

$x_{2,e}$	T_e/K
0.184 ± 0.002	255.5 ± 0.5

Mixture 6: 1-$C_{10}H_{21}OH$ (1) + [BMPy][Tos] (2)
(1) Eutectic Temperature

Ref. [8]

$x_{2,e}$	T_e/K
0.209 ± 0.002	275.6 ± 0.5

Mixture 7: H_2O (1) + [BMPy][Tos] (2)
(1) Eutectic Temperature

Ref. [8]

$x_{2,e}$	T_e/K
0.205 ± 0.002	236.1 ± 0.5

6.6 11010-03001: 1-Dodecyl-3-methylpyridinium bis (trifluoromethylsulfonyl)imide

Abbreviation:	C_{12}MePyNTf $_2$
Molecular formula:	$C_{20}H_{32}N_2F_6O_4S_2$
Molar mass:	542.61
Structure:	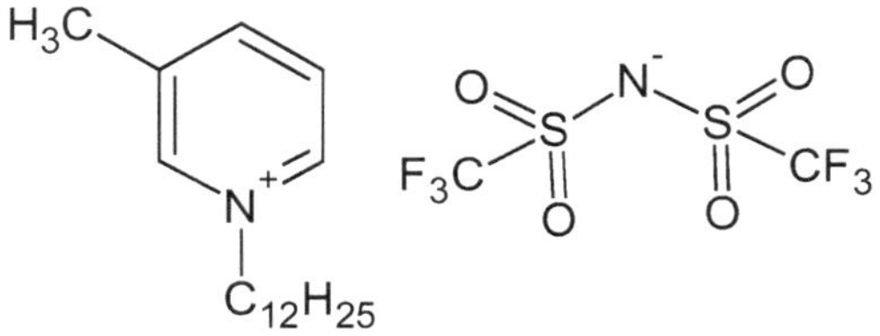

6.6.1 Binary Mixtures

Mixture 1: 1H,1H,2H,2H-$C_8F_{13}H_4OH$ (1) + $C_{12}MePyNTf_2$ (2)
(1) Cloud Point

Ref. [9]

x_2	T/K	x_2	T/K	x_2	T/K	x_2	T/K
0.050	267.7	0.163	281.0	0.269	283.0	0.638	273.6
0.083	275.4	0.200	281.9	0.331	283.1	0.560	278.0
0.120	278.5	0.227	282.7	0.418	282.5	0.508	281.0

6.7 11012-03001: 1-Tetradecyl-3-methylpyridinium bis (trifluoromethylsulfonyl)imide

Abbreviation:	$C_{14}MePyNTf_2$
Molecular formula:	$C_{22}H_{36}N_2F_6O_4S_2$
Molar mass:	570.66
Structure:	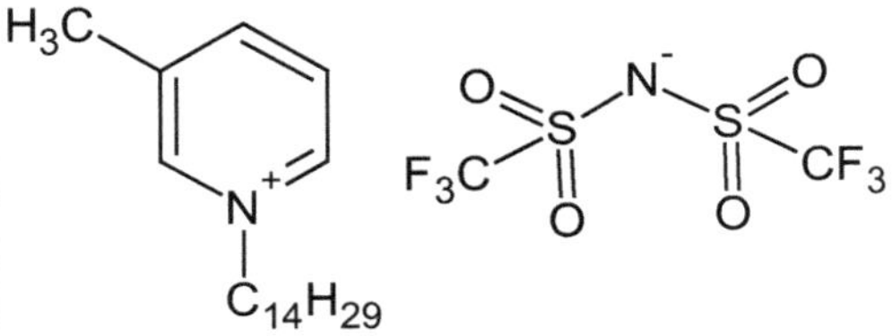

6.7.1 Binary Mixtures

Mixture 1: 1H,1H,2H,2H-$C_8F_{13}H_4OH$ (1) + $C_{14}MePyNTf_2$ (2)
(1) Cloud Point

Ref. [9]

x_2	T/K	x_2	T/K	x_2	T/K	x_2	T/K
0.020	267.2	0.153	292.6	0.647	283.6	0.426	292.7
0.034	277.5	0.191	293.1	0.610	286.2	0.378	293.2
0.045	281.2	0.222	293.2	0.569	288.4	0.340	293.5
0.066	286.7	0.820	263.5	0.528	290.0	0.314	293.5
0.082	288.5	0.768	271.1	0.493	291.2	0.270	293.5
0.106	290.7	0.705	278.1	0.460	292.1	0.239	293.4
0.126	291.7						

6.8 11025-03001: 1-Ethylpyridinium bis((trifluoromethyl) sulfonyl)amide

Abbreviation:	$[C_2py][NTf_2]$
Molecular formula:	$C_9H_{10}F_6N_2O_4S_2$
Molar mass:	388.31
Structure:	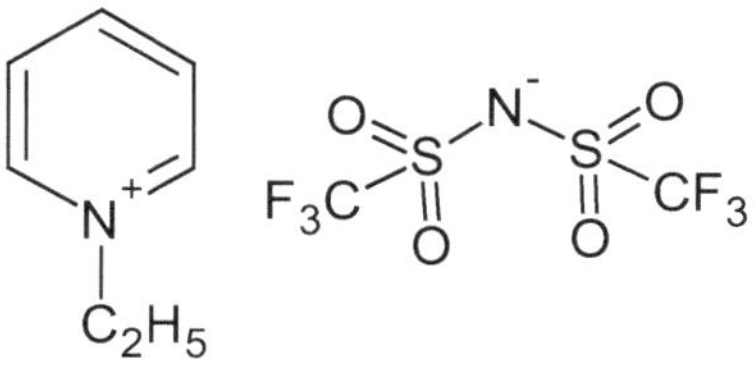

6.8.1 Binary Mixtures

Mixture 1: $[C_2py][NTf_2]$ (1) + $[P_{666,14}][NTf_2]$ (2)
(1) Phase Equilibrium

Ref. [10]

	Upper phase		Lower phase	
T/K	x_1	x_2	x_1	x_2
298.15	0.274	0.726	0.993	0.007
318.15	0.298	0.702	0.987	0.013
338.15	0.324	0.676	0.983	0.017
358.15	0.391	0.609	0.977	0.023
378.15	0.474	0.526	0.958	0.042
388.15	0.568	0.432	0.928	0.072
398.15	0.624	0.376	0.919	0.081
408.15	0.713	0.287	0.857	0.143

6.8.2 Ternary Mixtures

Mixture 1: C_6H_{14} (1) + C_6H_6 (2) + $[C_2py][NTf_2]$ (3)
(1) Distribution Ratio and Selectivity

Ref. [11]

Upper phase			Lower phase				
x_1	x_2	x_3	x_1	x_2	x_3	β	*S*
313.15 K							
1.000	0.000	0.000	0.048	0.000	0.952	–	–

(continued)

Upper phase			Lower phase				
x_1	x_2	x_3	x_1	x_2	x_3	β	S
0.937	0.063	0.000	0.039	0.078	0.883	1.24	29.7
0.857	0.143	0.000	0.037	0.171	0.792	1.20	27.7
0.762	0.238	0.000	0.035	0.271	0.694	1.14	24.8
0.646	0.354	0.000	0.037	0.371	0.592	1.05	18.3
0.493	0.507	0.000	0.036	0.483	0.481	0.95	13.0
0.331	0.669	0.000	0.032	0.614	0.354	0.92	9.5
0.164	0.836	0.000	0.024	0.688	0.288	0.82	5.6
0.085	0.915	0.000	0.015	0.716	0.269	0.78	4.4
0.000	1.000	0.000	0.000	0.773	0.227	0.77	–

6.9 11025-04018: Ethyl pyridinium ethylsulfate

Abbreviation:	$[C_2py][EtSO_4]$, $[Epy][ESO_4]$
Molecular formula:	$C_9H_{15}NO_4S$
Molar mass	233.29
Structure:	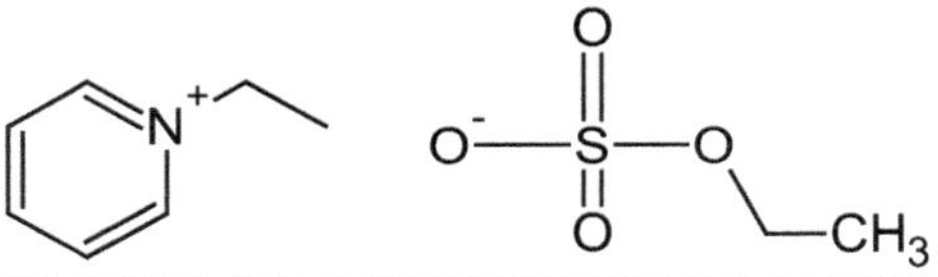

6.9.1 Binary Mixtures

Mixture 1: C_2H_5OH (1) + $[Epy][ESO_4]$ (2)
(1) Volumetric Properties, Transport Properties, Refractive index, Speed of Sound and Compressibility Properties

Ref. [12]

x_1	ρ/ (g · cm^{-3})	$10^3\eta$/ Pa · s	n_D	u/ m · s^{-1}	k_s/ T · Pa^{-1}	V^E/ (cm^3 · mol^{-1})	$10^3\Delta\eta$/ Pa · s	Δn_D	Δk_s/ T · Pa^{-1}
T = 298.15 K									
0.0000	1.2520	126.3	1.50252	1711	272.9	0.000	0.00	0.0000	0.0
0.0476	1.2453	103.3	1.50313	1700	277.7	−0.082	−17.04	0.0074	−28.2
0.1993	1.2208	58.4	1.49585	1661	297.0	−0.362	−42.98	0.0217	−114.6
0.2992	1.2012	40.3	1.48987	1630	313.5	−0.555	−48.51	0.0299	−167.6
0.5009	1.1472	17.27	1.47321	1546	364.9	−0.771	−46.30	0.0420	−256.6
0.6033	1.1092	10.79	1.46152	1491	405.4	−0.812	−39.96	0.0448	−287.3
0.7023	1.0620	6.63	1.44693	1428	462.0	−0.802	−31.73	0.0443	−299.6

(*continued*)

x_1	ρ/ (g · cm^{-3})	$10^3\eta$/ Pa · s	n_D	u/ m · s^{-1}	k_s/ T · Pa^{-1}	V^E/ (cm^3 · mol^{-1})	$10^3\Delta\eta$/ Pa · s	Δn_D	Δk_s/ T · Pa^{-1}
0.8016	0.9999	3.947	1.42734	1349	549.1	−0.759	−21.98	0.0389	−281.6
0.8990	0.9155	2.244	1.40078	1258	689.7	−0.571	−11.48	0.0262	−208.9
0.9476	0.8601	1.641	1.38342	1206	798.9	−0.374	−6.00	0.0157	−133.5
1.0000	0.7855	1.082	1.36023	1146	968.9	0.000	0.00	0.0000	0.0
T = 313.15 K									
0.0000	1.2421	57.6		1676	286.7	0.000	0.00		0.0
0.0476	1.2353	50.3		1665	292.0	−0.078	−4.64		−32.7
0.1993	1.2108	33.90		1625	312.7	−0.422	−12.39		−133.0
0.2992	1.1909	23.51		1593	330.8	−0.620	−17.12		−194.6
0.5009	1.1364	11.39		1508	386.8	−0.865	−17.78		−299.5
0.6033	1.0984	7.07		1453	431.4	−0.950	−16.28		−336.7
0.7023	1.0510	4.800		1388	493.8	−0.958	−12.93		−353.3
0.8016	0.9881	2.943		1308	591.6	−0.880	−9.15		−334.6
0.8990	0.9033	1.717		1214	751.5	−0.684	−4.84		−252.5
0.9476	0.8475	1.261		1160	877.5	−0.458	−2.54		−165.3
1.0000	0.7720	0.827		1093	1084.6	0.000	0.00		0.0
T = 328.15 K									
0.0000	1.2323	38.60		1641	301.1	0.000	0.00		0.0
0.0476	1.2253	34.14		1631	306.7	−0.065	−2.66		−37.8
0.1993	1.2007	22.38		1591	329.1	−0.464	−8.65		−153.5
0.2992	1.1807	15.20		1558	349.0	−0.694	−12.04		−224.5
0.5009	1.1259	8.015		1472	410.0	−0.994	−11.57		−347.1
0.6033	1.0877	5.198		1388	477.1	−1.101	−10.50		−373.2
0.7023	1.0399	3.642		1350	527.9	−1.111	−8.30		−412.5
0.8016	0.9766	2.287		1267	637.6	−1.032	−5.89		−393.2
0.8990	0.8901	1.344		1170	821.0	−0.719	−3.13		−298.5
0.9476	0.8341	0.987		1113	967.4	−0.479	−1.64		−196.3
1.0000	0.7586	0.641		1043	1211.4	0.000	0.00		0.0

Mixture 2: 1-C_3H_7OH (1) + [Epy][ESO_4] (2)

(1) Volumetric Properties, Transport Properties, Refractive index, Speed of Sound and Compressibility Properties

Ref. [12]

x_1	ρ/ (g · cm^{-3})	$10^3\eta$/ Pa · s	n_D	u/ m · s^{-1}	k_s/ T · Pa^{-1}	V^E/ (cm^3 · mol^{-1})	$10^3\Delta\eta$/ Pa · s	Δn_D	Δk_s/ T · Pa^{-1}
T = 298.15 K									
0.0000	1.2520	126.3	1.50525	1711	272.9	0.000	0.00	0.0000	0.0
0.0814	1.2372	95.2	1.50137	1689	283.3	−0.124	−20.99	0.0061	−37.2
0.1994	1.2127	62.7	1.49499	1649	303.2	−0.276	−38.82	0.0141	−86.4
0.3084	1.1861	43.6	1.48804	1608	326.0	−0.399	−44.40	0.0205	−127.5
0.4035	1.1590	30.03	1.48081	1565	352.1	−0.491	−46.12	0.0248	−157.1
0.5077	1.1238	20.12	1.47142	1516	387.2	−0.548	−43.08	0.0282	−183.0

(*continued*)

x_1	ρ/ $(g \cdot cm^{-3})$	$10^3\eta$/ Pa · s	n_D	u/ $m \cdot s^{-1}$	k_s/ $T \cdot Pa^{-1}$	V^E/ $(cm^3 \cdot mol^{-1})$	$10^3\Delta\eta$/ Pa · s	Δn_D	Δk_s/ $T \cdot Pa^{-1}$
0.6013	1.0858	13.68	1.46126	1466	428.6	−0.545	−37.89	0.0295	−196.4
0.7016	1.0367	8.840	1.44802	1409	486.1	−0.513	−30.27	0.0285	−197.6
0.8004	0.9769	5.597	1.43181	1347	564.3	−0.447	−21.22	0.0243	−177.2
0.9002	0.9006	3.483	1.41100	1279	679.0	−0.332	−10.94	0.0157	−121.0
0.9521	0.8524	2.676	1.39773	1244	758.6	−0.252	−5.29	0.0088	−71.8
1.0000	0.7996	2.017	1.38310	1207	858.4	0.000	0.00	0.0000	0.0
T = 313.15 K									
0.0000	1.2421	57.6		1676	286.7	0.000	0.00		0.0
0.0814	1.2274	48.0		1654	297.9	−0.161	−5.06		−42.9
0.1994	1.2027	34.23		1613	319.6	−0.327	−12.19		−99.7
0.3084	1.1761	25.21		1572	344.2	−0.490	−15.08		−147.7
0.4035	1.1485	18.31		1528	372.8	−0.552	−16.63		−182.4
0.5077	1.1130	12.91		1478	411.1	−0.611	−16.18		−213.4
0.6013	1.0748	9.161		1427	457.2	−0.616	−14.67		−229.6
0.7016	1.0252	6.131		1367	521.7	−0.561	−12.07		−231.7
0.8004	0.9653	3.988		1304	609.4	−0.509	−8.66		−209.8
0.9002	0.8889	2.507		1234	739.1	−0.397	−4.53		−146.5
0.9521	0.8403	1.921		1195	833.2	−0.276	−2.20		−87.0
1.0000	0.7874	1.430		1155	952.0	0.000	0.00		0.0
T = 328.15 K									
0.0000	1.2323	38.60		1641	301.1	0.000	0.00		0.0
0.0814	1.2176	32.49		1619	313.3	−0.192	−3.05		−49.4
0.1994	1.1927	23.31		1578	336.6	−0.381	−7.80		−115.2
0.3084	1.1659	17.27		1537	363.3	−0.565	−9.74		−171.0
0.4035	1.1383	12.21		1493	394.3	−0.666	−11.23		−211.9
0.5077	1.1026	8.898		1440	437.1	−0.743	−10.63		−247.8
0.6013	1.0642	6.523		1388	487.4	−0.760	−9.49		−268.3
0.7016	1.0143	4.494		1327	559.4	−0.704	−7.76		−272.0
0.8004	0.9541	2.977		1262	658.1	−0.649	−5.56		−248.1
0.9002	0.8771	1.874		1189	806.8	−0.492	−2.92		−174.8
0.9521	0.8281	1.411		1147	917.2	−0.330	−1.43		−103.6
1.0000	0.7748	1.041		1105	1057.0	0.000	0.00		0.0

6.10 11025-05001: 1-Ethylpyridinium hexafluorophosphate

Abbreviation:	[Epy][PF_6]
Molecular formula:	$C_7H_{10}NPF_6$
Molar mass:	253.12
Structure:	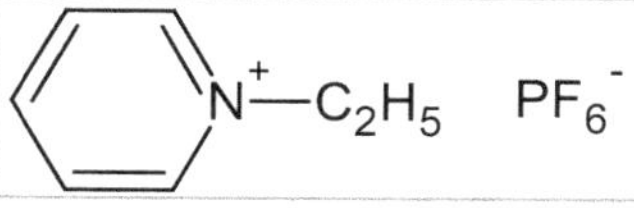

6.10.1 Ternary Mixtures

Mixture 1: H_2O (1) + C_2H_5OH (2) + [Epy][PF_6] (3)
(1) Solubility

Ref. [13]

T/K	10^2x_3	$10^2\Delta x_3$[a]	T/K	10^2x_3	$10^2\Delta x_3$[a]
$w_1 + w_2 = 1$					
$w_1 = 0.00$					
278.15	0.0084	0.00	313.15	0.0718	−13.79
283.15	0.0121	0.00	318.15	0.0976	−11.07
288.15	0.0171	0.00	323.15	0.1497	4.74
293.15	0.0224	−6.69	328.15	0.1891	1.64
298.15	0.0324	−1.85	333.15	0.2519	4.45
303.15	0.0561	19.61	338.15	0.3076	−0.49
308.15	0.0610	0.00	343.15	0.3894	−1.16
$w_1 = 0.10$					
278.15	0.1645	0.00	313.15	0.5109	−0.61
283.15	0.2001	0.00	318.15	0.5830	0.15
288.15	0.2404	0.00	323.15	0.6673	1.99
293.15	0.2883	0.94	328.15	0.7351	0.82
298.15	0.3348	0.24	333.15	0.8011	−0.75
303.15	0.3891	0.36	338.15	0.8891	0.19
308.15	0.4430	1.58	343.15	0.9633	−0.62
$w_1 = 0.20$					
278.15	0.2540	0.00	313.15	0.6811	−1.31
283.15	0.3025	0.00	318.15	0.7639	−0.38
288.15	0.3562	0.03	323.15	0.8488	0.39
293.15	0.4134	0.27	328.15	0.9343	0.94
298.15	0.4808	0.69	333.15	1.0020	−0.41
303.15	0.5409	0.72	338.15	1.0862	−0.03
308.15	0.6230	1.16	343.15	1.1647	−0.10
$w_1 = 0.30$					

(continued)

T/K	10^2x_3	$10^2\Delta x_3$[a]	T/K	10^2x_3	$10^2\Delta x_3$[a]
278.15	0.2414	0.04	313.15	0.7085	−0.49
283.15	0.2932	0.03	318.15	0.7946	0.08
288.15	0.3507	0.03	323.15	0.8831	0.67
293.15	0.4162	0.58	328.15	0.9705	1.03
298.15	0.4843	0.43	333.15	1.0473	0.43
303.15	0.5510	0.76	338.15	1.1210	−0.20
308.15	0.6252	1.09	343.15	1.1923	−0.70
$w_1 = 0.40$					
278.15	0.2352	0.00	313.15	0.7387	0.46
283.15	0.2901	0.03	318.15	0.8211	0.16
288.15	0.3516	0.03	323.15	0.8986	−0.58
293.15	0.4213	0.47	328.15	0.9846	−0.13
298.15	0.4913	0.24	333.15	1.0498	−1.42
303.15	0.5553	2.70	338.15	1.1523	1.15
308.15	0.6698	2.70	343.15	1.2077	−0.01
$w_1 = 0.50$					
278.15	0.2139	0.00	313.15	0.6007	−0.05
283.15	0.2568	0.00	318.15	0.6702	0.19
288.15	0.3043	0.03	323.15	0.7381	−0.01
293.15	0.3523	1.08	328.15	0.8032	−0.61
298.15	0.4152	0.72	333.15	0.8728	−0.61
303.15	0.4742	0.44	338.15	0.9374	−1.06
308.15	0.5381	0.54	343.15	1.0299	1.45
$w_1 = 0.60$					
278.15	0.1628	0.00	313.15	0.4673	−0.68
283.15	0.1964	0.00	318.15	0.5224	−0.54
288.15	0.2338	0.00	323.15	0.5804	−0.12
293.15	0.2736	0.48	328.15	0.6427	0.79
298.15	0.3211	0.50	333.15	0.6959	0.24
303.15	0.3705	0.89	338.15	0.7511	0.11
308.15	0.4162	0.36	343.15	0.8024	−0.36
$w_1 = 0.70$					
278.15	0.1204	0.00	313.15	0.3942	0.66
283.15	0.1477	0.00	318.15	0.4436	−0.18
288.15	0.1789	0.00	323.15	0.4947	−1.05
293.15	0.2109	1.42	328.15	0.5506	−1.31
298.15	0.2544	0.63	333.15	0.6141	−0.54
303.15	0.2991	1.20	338.15	0.6866	1.19
308.15	0.3435	0.49	343.15	0.7420	0.26
$w_1 = 0.80$					
278.15	0.0981	0.51	313.15	0.3245	1.05
283.15	0.1208	1.44	318.15	0.3741	1.60
288.15	0.1439	0.13	323.15	0.4298	2.50
293.15	0.1741	1.33	328.15	0.4857	2.44

(continued)

T/K	10^2x_3	$10^2\Delta x_3$[a]	T/K	10^2x_3	$10^2\Delta x_3$[a]
298.15	0.1932	5.30	333.15	0.5345	0.39
303.15	0.2340	2.05	338.15	0.5916	−0.52
308.15	0.2747	1.20	343.15	0.6434	−2.64
w_1 = 0.90					
278.15	0.0862	0.00	313.15	0.2737	−0.11
283.15	0.1035	0.00	318.15	0.3168	0.16
288.15	0.1235	0.08	323.15	0.3646	0.30
293.15	0.1453	0.83	328.15	0.4157	−0.02
298.15	0.1752	1.37	333.15	0.4735	−0.04
303.15	0.2036	0.54	338.15	0.5370	−0.09
308.15	0.2330	1.37	343.15	0.6079	0.07
w_1 = 1.00					
278.15	0.0474	0.00	313.15	0.1861	0.11
283.15	0.0581	0.00	318.15	0.2162	−3.33
288.15	0.0711	0.00	323.15	0.2603	−2.80
293.15	0.0857	1.17	328.15	0.3247	1.54
298.15	0.1051	0.29	333.15	0.3866	1.42
303.15	0.1301	1.84	338.15	0.4541	0.24
308.15	0.1588	2.83	343.15	0.5352	−0.39

[a] $\Delta x_3 = (x\text{-}x_{cal})/x$. where $x$cal. is the solubility value.

6.11 11038-04018: 1-Ethyl-3-methylpyridinium ethylsulfate

Abbreviation:	[empy][EtSO$_4$], EMpyESO$_4$, [EMpy][ESO$_4$], [empy][eSO$_4$]
Molecular formula:	$C_{10}H_{17}NO_4S$
Molar mass	247.31
Structure:	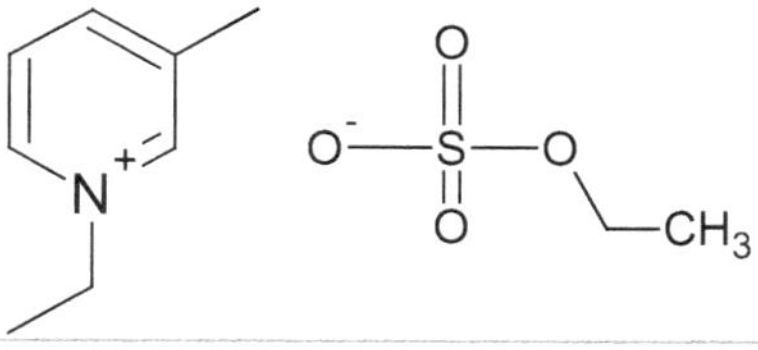

6.11.1 Binary Mixtures

Mixture 1: C_2H_5OH (1) + [empy][$EtSO_4$] (2)
(1) Volumetric Properties, Transport Properties and Refractive Index

Ref. [14]

x_1	ρ/g · cm^{-3}	n_D	$10^3\eta$/ Pa · s	V^E/cm^3 · mol^{-1}	Δn_D	$10^3\Delta\eta$/ Pa · s
298.15 K						
0.0000	1.22226	1.50666	161.4	0.000	0.0000	0.000
0.0609	1.21493	1.50392	116.7	−0.117	0.0062	−34.960
0.1108	1.20833	1.50181	96.9	−0.200	0.0114	−46.709
0.2066	1.19393	1.49730	68.0	−0.336	0.0209	−60.293
0.3083	1.17571	1.49126	45.35	−0.460	0.0297	−66.613
0.4004	1.15573	1.48469	31.11	−0.540	0.0367	−66.089
0.4998	1.12953	1.47600	20.04	−0.620	0.0425	−61.232
0.6034	1.09496	1.46448	12.170	−0.674	0.0462	−52.490
0.7034	1.05168	1.45003	7.452	−0.709	0.0464	−41.177
0.8452	0.96167	1.41977	3.322	−0.596	0.0369	−22.576
0.9389	0.87077	1.38864	1.812	−0.405	0.0195	−9.065
1.0000	0.78546	1.36023	1.082	0.000	0.0000	0.000
313.15 K						
0.0000	1.21254		77.1	0.000		0.000
0.0609	1.20503		55.2	−0.111		−17.208
0.1108	1.19838		48.57	−0.209		−20.082
0.2066	1.18386		36.25	−0.369		−25.091
0.3083	1.16553		25.91	−0.521		−27.676
0.4004	1.14551		18.88	−0.634		−27.680
0.4998	1.11907		12.922	−0.728		−26.057
0.6034	1.08430		8.317	−0.803		−22.760
0.7034	1.04072		5.332	−0.850		−18.117
0.8452	0.95009		2.502	−0.732		−10.131
0.9389	0.85852		1.388	−0.507		−4.099
1.0000	0.77200		0.827	0.000		0.000
328.15 K						
0.0000	1.20286		42.32	0.000		0.000
0.0609	1.19520		31.20	−0.114		−8.579
0.1108	1.18852		28.12	−0.228		−9.581
0.2066	1.17391		21.86	−0.416		−11.853
0.3083	1.15546		16.41	−0.597		−13.059
0.4004	1.13536		12.518	−0.740		−13.114
0.4998	1.10874		8.971	−0.856		−12.518
0.6034	1.07373		6.023	−0.950		−11.148
0.7034	1.02983		3.990	−1.006		−9.013

(continued)

x_1	ρ/g · cm^{-3}	n_D	$10^3\eta$/ Pa · s	V^E/cm^3 · mol^{-1}	Δn_D	$10^3\Delta\eta$/ Pa · s
0.8452	0.93848		1.943	−0.875		−5.150
0.9389	0.84610		1.086	−0.603		−2.102
1.0000	0.75855		0.641	0.000		0.000

(2) Osmotic Coefficient, Activity and Vapor Pressure

Ref. [15]

m_2/(mol · kg^{-1})	Φ_1	a_1	p/kPa
T = 323.15 K			
0.0302	0.845	0.998	29.406
0.0496	0.766	0.997	29.372
0.0797	0.740	0.995	29.314
0.1006	0.690	0.994	29.285
0.2002	0.615	0.989	29.138
0.3011	0.571	0.984	29.005
0.4010	0.567	0.979	28.855
0.5031	0.542	0.975	28.732
0.6850	0.512	0.968	28.523
0.9820	0.498	0.956	28.154
1.2494	0.487	0.946	27.842
1.6670	0.463	0.931	27.419
2.0605	0.452	0.918	27.008
2.4948	0.433	0.905	26.637
2.8906	0.426	0.893	26.258
3.2847	0.432	0.877	25.805
3.6865	0.429	0.864	25.416

Mixture 2: H_2O (1) + [empy][$EtSO_4$] (2)
(1) Volumetric Properties, Transport Properties and Refractive Index

Ref. [14]

x_1	ρ/g · cm^{-3}	n_D	$10^3\eta$/Pa · s	V^E/cm^3 · mol^{-1}	Δn_D	$10^3\Delta\eta$/Pa · s
298.15 K						
0.0000	1.22226	1.50666	161.4	0.000	0.0000	0.000
0.0526	1.22169	1.50583	127.6	−0.085	0.0083	−25.355
0.1412	1.22012	1.50427	102.1	−0.161	0.0222	−36.666
0.2445	1.21780	1.50193	77.9	−0.240	0.0379	−44.301
0.3233	1.21558	1.49986	62.3	−0.296	0.0495	−47.241
0.4174	1.21221	1.49685	46.87	−0.360	0.0629	−47.528
0.4972	1.20839	1.49329	36.11	−0.402	0.0732	−45.472
0.5880	1.20253	1.48806	26.83	−0.446	0.0838	−40.184
0.6975	1.19135	1.47838	17.72	−0.465	0.0932	−31.728
0.7928	1.17405	1.46398	11.177	−0.436	0.0954	−22.965
0.9005	1.12944	1.42959	4.927	−0.250	0.0798	−11.934

(continued)

x_1	ρ/g · cm^{-3}	n_D	$10^3\eta$/Pa · s	V^E/cm^3 · mol^{-1}	Δn_D	$10^3\Delta\eta$/Pa · s
0.9484	1.0862	1.39796	2.717	−0.095	0.0565	−6.451
1.0000	0.9972	1.33251	0.89	0.000	0.0000	0.000
313.15 K						
0.0000	1.21254		77.1	0.000		0.000
0.0526	1.21182		60.5	−0.058		−12.542
0.1412	1.21022		50.28	−0.126		−16.019
0.2445	1.20783		39.91	−0.192		−18.501
0.3233	1.20562		32.92	−0.246		−19.468
0.4174	1.20221		24.86	−0.302		−20.330
0.4972	1.19831		20.36	−0.335		−18.730
0.5880	1.19232		15.519	−0.367		−16.628
0.6975	1.18092		10.495	−0.371		−13.285
0.7928	1.16341		6.778	−0.334		−9.713
0.9005	1.11924		3.161	−0.164		−5.100
0.9484	1.0775		1.818	−0.044		−2.778
1.0000	0.9922		0.653	0.000		0.000
328.15 K						
0.0000	1.20286		42.32	0.000		0.000
0.0526	1.20202		34.93	−0.036		−5.190
0.1412	1.20035		29.83	−0.092		−6.587
0.2445	1.19802		23.49	−0.164		−8.604
0.3233	1.19572		19.76	−0.206		−9.043
0.4174	1.19223		15.511	−0.253		−9.357
0.4972	1.18824		12.779	−0.277		−8.749
0.5880	1.18214		9.894	−0.301		−7.837
0.6975	1.17042		6.847	−0.288		−6.307
0.7928	1.15261		4.508	−0.241		−4.659
0.9005	1.10866		2.197	−0.087		−2.468
0.9484	1.0680		1.315	0.004		−1.346
1.0000	0.9857		0.504	0.000		0.000

(2) Osmotic Coefficient, Activity and Vapor Pressure

Ref. [16]

m_2/(mol · kg^{-1})	Φ_1	α_1	p/kPa	m_2/(mol · kg^{-1})	Φ_1	α_1	p/kPa
T = 313.15 K				T = 333.15 K			
0.1419	0.956	0.9951	7.341	0.1419	0.948	0.9952	19.825
0.2108	0.927	0.9930	7.325	0.2108	0.925	0.9930	19.781
0.3168	0.897	0.9898	7.302	0.3168	0.887	0.9899	19.718
0.3829	0.873	0.9880	7.289	0.3829	0.864	0.9882	19.682
0.5036	0.846	0.9848	7.265	0.5036	0.836	0.9850	19.616
0.6837	0.828	0.9798	7.228	0.6837	0.811	0.9802	19.515
1.0336	0.813	0.9702	7.157	1.0336	0.784	0.9712	19.335
1.3260	0.808	0.9621	7.097	1.3260	0.782	0.9633	19.174

(*continued*)

m_2/(mol · kg^{-1})	Φ_1	α_1	p/kPa	m_2/(mol · kg^{-1})	Φ_1	α_1	p/kPa
T = 313.15 K				T = 333.15 K			
1.7255	0.807	0.9511	7.016	1.7255	0.782	0.9526	18.953
2.1099	0.813	0.9401	6.934	2.1099	0.792	0.9416	18.729
2.5006	0.826	0.9283	6.847	2.5006	0.802	0.9303	18.499
2.8922	0.843	0.9159	6.755	2.8922	0.809	0.9191	18.270
3.3025	0.859	0.9029	6.659	3.3025	0.817	0.9074	18.030

Mixture 3: 1-C_3H_7OH (1) + $EMpyESO_4$ (2)
(1) Osmotic Coefficient, Activity and Vapor Pressure

Ref. [15]

m/(mol · kg^{-1})	Φ_1	a_1	p/kPa
T = 323.15 K			
0.0483	0.831	0.995	12.043
0.1038	0.786	0.990	11.983
0.1979	0.704	0.983	11.899
0.2855	0.672	0.977	11.823
0.3882	0.660	0.970	11.731
0.4890	0.643	0.963	11.648
0.6946	0.628	0.949	11.477
0.9272	0.605	0.935	11.303
1.4193	0.577	0.906	10.955
1.6682	0.567	0.892	10.786
2.1077	0.555	0.869	10.497
2.5133	0.538	0.850	10.266
2.8833	0.536	0.831	10.030
3.2970	0.521	0.813	9.820
3.7124	0.515	0.795	9.593
4.0716	0.510	0.779	9.400

(2) Volumetric Properties, Transport Properties and Refractive Index

Ref. [17]

x_1	ρ/ (g · cm^{-3})	n_D	10^3 η/ (Pa · s)	V^E/ (cm^3 · mol^{-1})	Δn_D	10^3 $\Delta\eta$/ (Pa · s)
T = 298.15 K						
0.0000	1.22220	1.50666	161.4	0.000	0.0000	0.0
0.0863	1.20844	1.50223	126.4	−0.089	0.0062	−21.2
0.1279	1.20122	1.50020	112.4	−0.134	0.0093	−28.7
0.2201	1.18356	1.49524	85.83	−0.215	0.0156	−40.48
0.3243	1.16024	1.48866	62.26	−0.279	0.0219	−47.44
0.4177	1.13568	1.48167	48.86	−0.317	0.0264	−45.96
0.5126	1.10618	1.47321	34.16	−0.338	0.0296	−45.54
0.6093	1.06997	1.46270	24.62	−0.328	0.0310	−39.67

(continued)

x_1	ρ/ (g · cm^{-3})	n_D	$10^3\ \eta$/ (Pa · s)	V^E/ (cm^3 · mol^{-1})	Δn_D	$10^3\ \Delta\eta$/ (Pa · s)
0.7176	1.01990	1.44818	16.27	−0.324	0.0298	−30.75
0.8240	0.95678	1.42970	9.900	−0.304	0.0244	−20.162
0.8834	0.91304	1.41684	7.027	−0.265	0.0188	−13.580
0.9658	0.83778	1.39459	3.411	−0.144	0.0067	−4.053
1.0000	0.79950	1.38370	2.017	0.000	0.0000	0.000

Mixture 4: 2-CH_3CHCH_3OH (1) + $EMpyESO_4$ (2)
(1) Osmotic Coefficient, Activity and Vapor Pressure

Ref. [15]

m/(mol · kg^{-1})	Φ_1	a_1	p/kPa
T = 323.15 K			
0.0780	0.930	0.991	23.383
0.1057	0.888	0.989	23.322
0.1933	0.808	0.981	23.146
0.2645	0.759	0.976	23.020
0.3919	0.689	0.968	22.826
0.4970	0.666	0.961	22.657
0.7005	0.629	0.948	22.356
0.9707	0.594	0.933	21.990
1.2774	0.554	0.919	21.641
1.6817	0.519	0.900	21.208
2.0539	0.515	0.881	20.737
2.5589	0.504	0.856	20.161
2.9727	0.490	0.839	19.751
3.3132	0.487	0.824	19.379
3.7063	0.480	0.808	18.995

6.11.2 Ternary Mixtures

Mixture 1: C_6H_{14} (1) + C_6H_6 (2) + [EMpy][ESO$_4$] (3)
(1) Volumetric Properties, Speed of Sound and Refractive Index

Ref. [18]

x_1	x_2	ρ/(g · cm^{-3})	u/(m · s^{-1})	n_D
T = 313.15 K				
0.0133	0.0469	1.20255	1642.0	1.50170
0.0255	0.0470	1.20212	1641.1	1.50168
0.0129	0.0981	1.19597	1632.2	1.50207
0.0198	0.0976	1.19548	1631.4	1.50199
0.0144	0.1468	1.18851	1621.0	1.50238

(continued)

x_1	x_2	$\rho/(g \cdot cm^{-3})$	$u/(m \cdot s^{-1})$	n_D
0.0059	0.1983	1.18036	1609.0	1.50260
0.0074	0.2457	1.17216	1597.0	1.50287
0.0040	0.2452	1.17327	1598.6	1.50286
0.0018	0.3330	1.15728	1575.3	1.50331
0.0110	0.0224	1.20570	1646.4	1.50162
0.0224	0.0224	1.20504	1645.2	1.50147
0.0118	0.1189	1.19181	1625.9	1.50202
0.0224	0.1177	1.19020	1622.6	1.50162
0.0116	0.1658	1.18583	1617.1	1.50223
0.0224	0.1642	1.18405	1613.6	1.50198
0.0114	0.2176	1.17699	1603.9	1.50241
0.0224	0.2166	1.17675	1602.9	1.50212
0.0105	0.2705	1.16768	1590.1	1.50261
0.0105	0.3176	1.15902	1576.9	1.50288
0.0102	0.3677	1.14795	1559.9	1.50311

(2) Distribution Ratio and Selectivity

Ref. [18]

Alkane-rich layer		Ionic liquid-rich layer			
x^I_1	x^I_2	x^{II}_1	x^{II}_2	β	S
$T = 283.15$ K					
0.949	0.051	0.025	0.051	1.00	37.96
0.892	0.108	0.024	0.102	0.94	35.10
0.831	0.170	0.024	0.150	0.88	30.62
0.765	0.235	0.023	0.196	0.83	27.74
0.737	0.263	0.023	0.215	0.82	26.20
0.633	0.367	0.021	0.277	0.75	22.75
0.575	0.425	0.021	0.303	0.71	19.52
0.507	0.493	0.019	0.342	0.69	18.51
0.374	0.626	0.014	0.403	0.64	17.20
0.231	0.769	0.009	0.472	0.61	15.75
$T = 293.15$ K					
0.951	0.049	0.033	0.049	1.00	28.82
0.897	0.103	0.031	0.095	0.92	26.69
0.832	0.168	0.030	0.149	0.89	24.43
0.761	0.239	0.026	0.190	0.79	23.27
0.729	0.271	0.026	0.211	0.78	21.83
0.700	0.300	0.025	0.230	0.76	21.41
0.628	0.372	0.023	0.269	0.72	19.72
0.576	0.425	0.022	0.295	0.69	18.18
0.512	0.488	0.019	0.327	0.67	18.06
0.427	0.573	0.016	0.370	0.65	17.21

(continued)

Alkane-rich layer		Ionic liquid-rich layer			
x^{I}_1	x^{I}_2	x^{II}_1	x^{II}_2	β	S
0.359	0.641	0.014	0.400	0.62	15.99
0.220	0.780	0.009	0.466	0.60	14.57
0.154	0.846	0.007	0.499	0.59	14.01
$T = 298.15$ K					
0.950	0.050	0.033	0.047	0.94	27.06
0.896	0.104	0.031	0.094	0.90	26.12
0.832	0.168	0.029	0.140	0.83	23.91
0.760	0.240	0.028	0.188	0.78	21.26
0.728	0.272	0.027	0.209	0.77	20.72
0.643	0.357	0.023	0.257	0.72	20.13
0.573	0.427	0.020	0.295	0.69	19.79
0.515	0.485	0.018	0.326	0.67	19.23
0.374	0.626	0.012	0.391	0.62	19.47
0.196	0.804	0.006	0.475	0.59	19.30
$T = 303.15$ K					
0.947	0.053	0.030	0.052	0.98	30.97
0.896	0.104	0.027	0.095	0.91	30.31
0.826	0.174	0.025	0.145	0.83	27.53
0.762	0.238	0.024	0.187	0.79	24.95
0.720	0.280	0.023	0.207	0.74	23.14
0.636	0.364	0.021	0.254	0.70	21.13
0.569	0.431	0.020	0.293	0.68	19.34
0.503	0.497	0.019	0.324	0.65	17.26
0.348	0.652	0.013	0.394	0.60	16.18
0.246	0.754	0.009	0.446	0.59	16.17

Mixture 2: C_7H_{16} (1) + C_6H_6 (2) + [EMpy][ESO$_4$] (3)
(1) Volumetric Properties, Speed of Sound and Refractive Index

Ref. [18]

x_1	x_2	ρ/(g · cm^{-3})	u/(m · s^{-1})	n_D
$T = 313.15$ K				
0.0022	0.0467	1.20313	1642.8	1.50191
0.0041	0.0487	1.20178	1640.5	1.50169
0.0107	0.0481	1.20118	1639.1	1.50139
0.0114	0.0989	1.19510	1630.7	1.50202
0.0058	0.1002	1.19454	1629.8	1.50202
0.0087	0.1005	1.19459	1629.8	1.50204
0.0018	0.1519	1.18802	1620.7	1.50258
0.0069	0.1452	1.18800	1620.5	1.50220
0.0089	0.1493	1.18626	1617.3	1.50206
0.0049	0.1979	1.17904	1607.2	1.50248
0.0085	0.2005	1.17834	1605.6	1.50229

(*continued*)

x_1	x_2	$\rho/(g \cdot cm^{-3})$	$u/(m \cdot s^{-1})$	n_D
0.0035	0.2482	1.17171	1595.9	1.50287
0.0051	0.2465	1.17084	1594.7	1.50289
0.0032	0.2923	1.16322	1583.5	1.50315
0.0050	0.2991	1.16108	1579.9	1.50305
0.0018	0.3900	1.14349	1553.6	1.50368

(2) Distribution Ratio and Selectivity

Ref. [18]

Alkane-rich layer		Ionic liquid-rich layer			
x^I_1	x^I_2	x^{II}_1	x^{II}_2	β	S
$T = 283.15$ K					
0.948	0.052	0.008	0.045	0.87	103.17
0.837	0.163	0.008	0.128	0.79	82.42
0.786	0.214	0.009	0.164	0.77	67.06
0.729	0.271	0.010	0.199	0.74	53.65
0.676	0.324	0.010	0.231	0.71	48.29
0.558	0.442	0.011	0.298	0.67	34.22
0.433	0.567	0.009	0.364	0.64	30.84
0.303	0.697	0.007	0.431	0.62	26.81
0.159	0.841	0.000	0.509	0.61	
$T = 298.15$ K					
0.949	0.051	0.010	0.042	0.82	78.15
0.895	0.105	0.010	0.084	0.80	71.60
0.841	0.159	0.010	0.119	0.75	62.94
0.787	0.213	0.010	0.156	0.73	57.64
0.720	0.280	0.011	0.192	0.69	44.88
0.670	0.330	0.013	0.220	0.67	34.36
0.552	0.448	0.013	0.285	0.64	27.01
0.421	0.579	0.010	0.353	0.61	25.67
0.294	0.706	0.007	0.420	0.59	24.99
0.158	0.842	0.004	0.497	0.59	23.32
0.096	0.904	0.000	0.535	0.59	

Mixture 3: C_2H_5OH (1) + H_2O (2) + [empy][ESO_4] (3)
(1) Volumetric Properties, Transport Properties and Refractive Index

Ref. [17]

x_1	x_2	$\rho/(g \cdot cm^{-3})$	n_D	$10^3\,\eta/(Pa \cdot s)$	$V^E/(cm^3 \cdot mol^{-1})$	Δn_D	$10^3\,\Delta\eta/(Pa \cdot s)$
$T = 298.15$ K							
0.0559	0.9196	1.02817	1.37677	2.202	−0.335	0.0384	−2.643
0.1155	0.8615	1.00618	1.37705	2.554	−0.624	0.0373	−2.053
0.1918	0.7872	0.97826	1.37779	2.780	−0.879	0.0363	−1.524

(*continued*)

x_1	x_2	ρ/ ($g \cdot cm^{-3}$)	n_D	$10^3\ \eta$/ (Pa · s)	V^E/ ($cm^3 \cdot mol^{-1}$)	Δn_D	$10^3\ \Delta\eta$/ (Pa · s)
0.2634	0.7175	0.95224	1.37721	2.695	−0.994	0.0341	−1.323
0.3432	0.6397	0.92623	1.37591	2.635	−1.068	0.0309	−1.065
0.4352	0.5501	0.89992	1.37405	2.348	−1.107	0.0269	−0.985
0.6720	0.3194	0.84395	1.36843	1.748	−0.910	0.0158	−0.641
0.8189	0.1764	0.81607	1.36465	1.445	−0.619	0.0086	−0.359
0.0658	0.8853	1.06050	1.39835	3.138	−0.445	0.0555	−5.615
0.1401	0.8149	1.03049	1.39675	3.276	−0.734	0.0525	−4.867
0.2180	0.7411	0.99941	1.39389	3.206	−0.919	0.0482	−4.298
0.3049	0.6587	0.96677	1.39002	3.015	−1.033	0.0427	−3.775
0.3952	0.5731	0.93603	1.38603	2.731	−1.093	0.0370	−3.317
0.6019	0.3773	0.87548	1.37677	2.060	−1.011	0.0239	−2.291
0.0466	0.9420	1.00625	1.35735	1.692	−0.257	0.0216	−1.038
0.1018	0.8875	0.98746	1.36233	2.169	−0.546	0.0251	−0.466
0.1611	0.8289	0.96736	1.36564	2.479	−0.780	0.0269	−0.053
0.2316	0.7592	0.94382	1.36750	2.543	−0.954	0.0270	0.133
0.3119	0.6798	0.91842	1.36822	2.444	−1.044	0.0256	0.173
0.0849	0.8322	1.08841	1.42068	4.488	−0.582	0.0714	−9.725
0.1724	0.7526	1.05228	1.41499	4.225	−0.826	0.0646	−8.731
0.2651	0.6684	1.01510	1.40855	3.859	−0.980	0.0571	−7.768
0.3611	0.5810	0.97846	1.40186	3.363	−1.059	0.0493	−6.886
0.4610	0.4902	0.94290	1.39472	2.901	−1.077	0.0409	−5.914
0.5592	0.4009	0.91029	1.38805	2.499	−1.037	0.0331	−4.907
0.1191	0.7283	1.12032	1.44645	7.444	−0.724	0.0841	−17.965
0.3471	0.5398	1.03289	1.42552	4.852	−1.017	0.0637	−14.261
0.4518	0.4532	0.99275	1.41551	3.954	−1.053	0.0539	−12.267
0.5506	0.3715	0.95519	1.40581	3.239	−1.038	0.0445	−10.254
0.7437	0.2119	0.88211	1.38668	2.142	−0.816	0.0258	−6.018
0.8324	0.1386	0.84884	1.37740	1.723	−0.623	0.0168	−3.987
0.1016	0.5285	1.17767	1.48102	20.115	−0.612	0.0813	−40.178
0.2030	0.4688	1.15016	1.47277	15.065	−0.739	0.0775	−38.540
0.2995	0.4120	1.12126	1.46413	11.517	−0.827	0.0731	−35.729
0.4061	0.3493	1.08625	1.45364	8.503	−0.917	0.0673	−31.718
0.5036	0.2920	1.05000	1.44286	6.354	−0.922	0.0608	−27.439
0.6035	0.2332	1.00938	1.43053	4.707	−0.934	0.0529	−22.508
0.7012	0.1757	0.96443	1.41626	3.470	−0.868	0.0429	−17.303
0.0978	0.3563	1.19468	1.49186	36.488	−0.480	0.0616	−52.029
0.2306	0.3039	1.16567	1.48282	23.769	−0.631	0.0628	−51.883
0.3014	0.2759	1.14791	1.47724	18.706	−0.697	0.0628	−50.077
0.5002	0.1974	1.08679	1.45796	9.494	−0.801	0.0589	−40.025
0.6000	0.1580	1.04816	1.44605	6.633	−0.823	0.0548	−33.218
0.7009	0.1181	1.00135	1.43060	4.514	−0.794	0.0471	−25.554
0.8828	0.0463	0.88886	1.39397	2.076	−0.541	0.0246	−10.362
0.1956	0.1559	1.18684	1.49268	43.650	−0.415	0.0418	−61.364

(*continued*)

x_1	x_2	ρ/ ($g \cdot cm^{-3}$)	n_D	$10^3\ \eta$/ (Pa · s)	V^E/ ($cm^3 \cdot mol^{-1}$)	Δn_D	$10^3\ \Delta\eta$/ (Pa · s)
0.2915	0.1373	1.16747	1.48642	30.870	−0.532	0.0464	−61.755
0.3985	0.1166	1.14171	1.47807	20.634	−0.636	0.0501	−58.170
0.6000	0.0775	1.07627	1.45671	9.162	−0.762	0.0514	−43.598
0.7572	0.0471	0.99985	1.43149	4.453	−0.712	0.0439	−28.005
0.0443	0.9286	1.03153	1.37971	2.345	−0.303	0.0402	−2.226
0.0944	0.8799	1.00586	1.38277	2.731	−0.483	0.0410	−1.707
0.1526	0.8234	0.97712	1.38432	2.982	−0.567	0.0399	−1.300
0.2264	0.7517	0.94676	1.38536	3.112	−0.638	0.0376	−0.973
0.2957	0.6843	0.92275	1.38586	3.107	−0.664	0.0349	−0.792
0.3868	0.5958	0.89683	1.38617	2.967	−0.676	0.0311	−0.689
0.6264	0.3630	0.84824	1.38580	2.547	−0.607	0.0198	−0.468
0.7860	0.2079	0.82494	1.38501	2.273	−0.441	0.0117	−0.316
0.0557	0.8922	1.06016	1.40241	3.276	−0.394	0.0580	−4.655
0.1121	0.8389	1.02999	1.40185	3.534	−0.537	0.0551	−4.043
0.1795	0.7752	0.99697	1.40024	3.642	−0.616	0.0508	−3.513
0.2543	0.7045	0.96542	1.39840	3.582	−0.666	0.0459	−3.104
0.3400	0.6236	0.93466	1.39625	3.464	−0.687	0.0402	−2.686
0.5436	0.4313	0.87893	1.39189	2.966	−0.659	0.0275	−1.909
0.0391	0.9485	1.00726	1.36168	1.811	−0.247	0.0250	−0.783
0.0837	0.9045	0.98596	1.36754	2.276	−0.434	0.0287	−0.291
0.1346	0.8542	0.96171	1.37150	2.605	−0.529	0.0302	0.069
0.1958	0.7938	0.93717	1.37467	2.809	−0.608	0.0304	0.309
0.2665	0.7241	0.91321	1.37736	2.887	−0.645	0.0297	0.430
0.4603	0.5328	0.86717	1.38144	2.670	−0.678	0.0244	0.329
0.0704	0.8361	1.09166	1.42669	5.026	−0.497	0.0743	−8.462
0.1448	0.7692	1.05342	1.42208	4.867	−0.610	0.0673	−7.702
0.2252	0.6969	1.01625	1.41713	4.617	−0.679	0.0597	−6.960
0.3108	0.6199	0.98074	1.41215	4.345	−0.700	0.0519	−6.177
0.4031	0.5369	0.94709	1.40728	3.976	−0.701	0.0439	−5.407
0.0956	0.7479	1.11747	1.44815	7.846	−0.589	0.0836	−14.113
0.1915	0.6686	1.07334	1.43991	6.960	−0.656	0.0733	−12.884
0.2888	0.5881	1.03180	1.43191	6.104	−0.693	0.0634	−11.595
0.3927	0.5022	0.99060	1.42361	5.119	−0.700	0.0529	−10.288
0.4879	0.4235	0.95562	1.41657	4.547	−0.686	0.0440	−8.762
0.6912	0.2553	0.88815	1.40270	3.282	−0.568	0.0259	−5.544
0.1038	0.1883	1.19589	1.49678	58.404	−0.269	0.0357	−37.418
0.2167	0.1646	1.17083	1.48986	37.020	−0.349	0.0386	−46.979
0.3185	0.1432	1.14482	1.48261	27.653	−0.394	0.0402	−45.696
0.4164	0.1226	1.11618	1.47455	19.856	−0.430	0.0407	−43.241
0.5212	0.1006	1.08039	1.46443	14.398	−0.422	0.0397	−37.729
0.6176	0.0804	1.04203	1.45340	10.192	−0.412	0.0371	−31.845
0.7244	0.0579	0.99168	1.43903	6.781	−0.384	0.0320	−24.081
0.1017	0.5391	1.16720	1.47899	19.195	−0.537	0.0788	−29.921

(continued)

x_1	x_2	ρ/ $(g \cdot cm^{-3})$	n_D	$10^3\ \eta$/ $(Pa \cdot s)$	V^E/ $(cm^3 \cdot mol^{-1})$	Δn_D	$10^3\ \Delta\eta$/ $(Pa \cdot s)$
0.2017	0.4791	1.13385	1.47069	15.215	−0.601	0.0724	−28.656
0.4003	0.3599	1.06217	1.45248	9.400	−0.616	0.0580	−24.060
0.5051	0.2970	1.02173	1.44209	7.403	−0.599	0.0496	−20.564
0.5982	0.2412	0.98413	1.43226	5.986	−0.568	0.0415	−17.100
0.6988	0.1807	0.94177	1.42129	4.672	−0.527	0.0325	−13.135
0.1026	0.3681	1.18516	1.49008	32.143	−0.418	0.0602	−39.754
0.2960	0.2888	1.13083	1.47552	19.578	−0.503	0.0557	−37.257
0.3940	0.2486	1.09910	1.46692	14.858	−0.521	0.0522	−34.350
0.4947	0.2073	1.06299	1.45703	11.087	−0.525	0.0476	−30.280
0.5923	0.1672	1.02383	1.44628	7.939	−0.493	0.0419	−25.824
0.6936	0.1257	0.97879	1.43351	5.848	−0.477	0.0344	−20.026

6.12 11041-02001: 1-Butyl-3-methylpyridinium tetrafluoroborate

Abbreviation:	[bmpy][BF_4], [bmpyr][BF_4], [bpyr][BF_4], [b3mpy][BF_4], [3-MBP][BF_4]
Molecular formula:	$C_{10}H_{16}NBF_4$
Molar mass	237.05
Structure:	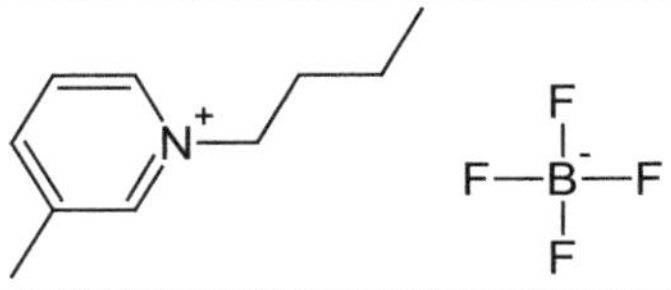

6.12.1 Binary Mixtures

Mixture 1: C_2H_5OH (1) + [bmpyr][BF_4] (2)
(1) Volumetric Properties, Heat Capacity at Constant Pressure and Excess Molar Enthalpies

Ref. [19]

x_2	293.15 K	298.15 K	303.15 K	308.15 K	313.15 K	318.15 K
	$\rho/g \cdot cm^{-3}$					
0	0.7896	0.7853	0.7810	0.7767	0.7723	0.7678
0.0073	0.8005	0.7963	0.7920	0.7876	0.7833	0.7789
0.0126	0.8078	0.8035	0.7993	0.7949	0.7906	0.7862
0.0231	0.8215	0.8173	0.8130	0.8088	0.8045	0.8003
0.0512	0.8551	0.8510	0.8468	0.8426	0.8384	0.8341
0.1003	0.9041	0.9001	0.8960	0.8919	0.8878	0.8837

(*continued*)

	293.15 K	298.15 K	303.15 K	308.15 K	313.15 K	318.15 K
x_2	ρ/g · cm^{-3}					
0.2569	1.0103	1.0064	1.0026	0.9987	0.9948	0.9909
0.3847	1.0654	1.0616	1.0579	1.0542	1.0504	1.0467
0.4841	1.0971	1.0934	1.0897	1.0861	1.0824	1.0787
0.5963	1.1249	1.1213	1.1177	1.1142	1.1106	1.1071
0.7107	1.1480	1.1445	1.1409	1.1374	1.1339	1.1304
0.7966	1.1622	1.1587	1.1552	1.1517	1.1483	1.1448
0.9514	1.1831	1.1796	1.1762	1.1728	1.1694	1.1660
1	1.1886	1.1852	1.1817	1.1783	1.1749	1.1716
x_2	C_p/J · mol^{-1} · K^{-1}					
0	111.3	113.3	115.3	117.5	119.7	122.1
0.0073	113.6	115.4	117.3	119.5	121.7	124.1
0.0126	115.2	117.1	119.1	121.2	123.3	125.6
0.0231	119.1	120.9	122.7	124.7	126.7	128.8
0.0512	129.4	131.0	132.6	134.3	136.0	137.9
0.1003	146.4	147.5	148.7	150.1	151.6	153.2
0.2569	190.3	191.3	192.4	193.6	194.9	196.2
0.3847	223.7	224.8	226.0	227.2	228.5	229.8
0.5031	252.6	254.2	255.7	257.3	258.9	260.4
0.5963	276.9	278.5	280.1	281.8	283.5	285.3
0.6919	301.6	303.4	305.3	307.3	309.4	311.5
0.7784	324.5	326.6	328.9	331.1	333.5	335.9
0.8711	350.0	352.1	354.4	356.9	359.5	362.2
0.8923	355.9	358.1	360.5	363.0	365.7	368.5
0.9514	372.8	375.1	377.4	379.7	382.0	384.4
1	385.8	387.7	390.1	392.8	395.9	399.3

Ref. [20]

x_2	ρ/ kg · m^{-3}	$10^9V_m^E$/ m^3 · mol^{-1}	x_2	ρ/ kg · m^{-3}	$10^9V_m^E$/ m^3 · mol^{-1}	x_2	ρ/ kg · m^{-3}	$10^9V_m^E$/ m^3 · mol^{-1}
T = 298.15 K								
0.0518	851.1	−268	0.2539	1006.2	−715	0.5895	1118.0	−391
0.0719	872.9	−369	0.3000	1028.2	−700	0.6640	1133.0	−271
0.1054	905.6	−526	0.4054	1068.6	−617	0.7275	1144.5	−212
0.1999	975.5	−668	0.5082	1098.9	−516	0.7682	1151.0	−149
T = 318.15 K								
0.0401	821.1	−317	0.2531	989.5	−776	0.5998	1106.3	−527
0.0671	851.0	−414	0.3062	1015.1	−772	0.6632	1118.8	−389
0.0992	882.7	−542	0.4041	1053.0	−715	0.7118	1127.7	−320
0.1497	924.5	−653	0.5002	1082.2	−645	0.7904	1141.0	−278

Ref. [21]

x_2	H^E(J · mol^{-1})	x_2	H^E(J · mol^{-1})	x_2	H^E(J · mol^{-1})
0.0306	554	0.3966	2599	0.8061	1258
0.1031	1356	0.5083	2495	0.8847	782
0.1976	2021	0.6001	2239		
0.3036	2460	0.692	1881		

Ref. [20]

x_2	H_m^E (J · mol^{-1})	x_2	H_m^E (J · mol^{-1})	x_2	H_m^E (J · mol^{-1})	x_2	H_m^E (J · mol^{-1})
T = 298.15 K							
0.0052	148	0.1263	1515	0.2305	2070	0.4689	2026
0.0159	414	0.1280	1539	0.2577	2155	0.5396	1777
0.0279	620	0.1450	1656	0.2878	2223	0.6277	1457
0.0457	849	0.1585	1693	0.3217	2275	0.7386	1074
0.0691	1088	0.1639	1748	0.3623	2284	0.8712	584
0.0972	1304	0.1838	1863	0.4085	2205	0.9405	304
0.1132	1413	0.2062	1968				
T = 318.15 K							
0.0123	262	0.1279	1721	0.2681	2531	0.5648	2295
0.0273	540	0.1612	1987	0.3063	2624	0.6576	1889
0.0472	842	0.1789	2127	0.3498	2694	0.7668	1364
0.0704	1125	0.2034	2278	0.4055	2699	0.8810	769
0.0980	1430	0.2335	2416	0.4787	2572		

(2) Phase Equilibrium

Ref. [20]

x_2	T/K	x_2	T/K	x_2	T/K
0.045	276.40	0.0924	278.21	0.1294	277.22
0.070	278.00	0.1195	277.57		

Mixture 2: H_2O (1) + [bpyr][BF_4] (2)
(1) Volumetric Properties, Heat Capacity at Constant Pressure, Excess Molar Enthalpies and Transport Properties

Ref. [19]

x_2	293.15 K	298.15 K	303.15 K	308.15 K	313.15 K	318.15 K
	ρ/g · cm^{-3}					
0	0.9982	0.9971	0.9957	0.9941	0.9922	0.9902
0.0494	1.0660	1.0633	1.0604	1.0575	1.0545	1.0513
0.0973	1.0959	1.0927	1.0894	1.0860	1.0825	1.0790
0.2312	1.1397	1.1360	1.1323	1.1287	1.1250	1.1212
0.3559	1.1573	1.1536	1.1500	1.1464	1.1427	1.1391
0.4035	1.1618	1.1582	1.1545	1.1509	1.1473	1.1437

(continued)

	293.15 K	298.15 K	303.15 K	308.15 K	313.15 K	318.15 K
x_2	ρ/g · cm^{-3}					
0.5195	1.1679	1.1643	1.1608	1.1572	1.1537	1.1502
0.7087	1.1795	1.1760	1.1725	1.1691	1.1656	1.1622
0.7512	1.1812	1.1778	1.1743	1.1708	1.1674	1.1639
0.8838	1.1832	1.1798	1.1763	1.1729	1.1695	1.1661
1	1.1886	1.1852	1.1817	1.1783	1.1749	1.1716
x_2	C_p/J · mol^{-1} · K^{-1}					
0	75.4	75.3	75.3	75.2	75.2	75.3
0.0015	76.7	76.7	76.8	76.9	77.0	77.1
0.0020	77.4	77.4	77.5	77.6	77.6	77.7
0.0027	77.7	77.7	77.7	77.8	77.9	78.0
0.0036	78.1	78.2	78.2	78.3	78.4	78.4
0.0102	81.0	81.1	81.2	81.3	81.5	81.7
0.0938	112.7	113.1	113.4	113.8	114.2	114.7
0.1885	142.4	143.1	143.8	144.5	145.2	146.0
0.2704	167.0	167.8	168.6	169.5	170.4	171.4
0.3620	193.8	195.0	196.1	197.3	198.4	199.5
0.4454	218.2	219.3	220.5	221.7	222.8	224.0
0.6432	275.2	276.9	278.6	280.3	282.1	283.8
0.8094	325.2	327.3	329.5	331.7	334.0	336.3
0.8791	346.0	348.3	350.8	353.3	355.8	358.5
0.9468	367.8	370.4	373.0	375.7	378.4	381.3
1	385.8	387.7	390.1	392.8	395.9	399.3

Ref. [20]

x_2	ρ/ kg · m^{-3}	$10^9 V_m^E$/ m^3 · mol^{-1}	x_2	ρ/ kg · m^{-3}	$10^9 V_m^E$/ m^3 · mol^{-1}	x_2	ρ/ kg · m^{-3}	$10^9 V_m^E$/ m^3 · mol^{-1}
T = 298.15 K								
0.0182	1027.1	33	0.2208	1131.4	315	0.6761	1171.8	332
0.0376	1050.3	66	0.2975	1143.6	384	0.8091	1176.6	246
0.0567	1067.1	106	0.3909	1154.0	417	0.8376	1177.6	209
0.0995	1093.7	176	0.4861	1161.5	419	0.8942	1179.4	127
0.1608	1116.8	262	0.5827	1167.2	403			
T = 318.15 K								
0.0466	1046.8	161	0.3038	1130.2	513	0.7551	1160.9	323
0.1004	1080.4	290	0.4108	1141.5	537	0.7904	1162.3	265
0.1523	1099.8	388	0.5062	1148.8	506	0.8123	1163.2	213
0.1991	1112.4	432	0.5822	1153.3	460	0.8988	1165.7	131
0.2510	1122.2	492	0.7042	1158.9	378	0.9180	1166.3	94

Ref. [20]

x_2	H_m^E $(J \cdot mol^{-1})$	x_2	H_m^E $(J \cdot mol^{-1})$	x_2	H_m^E $(J \cdot mol^{-1})$	x_2	H_m^E $(J \cdot mol^{-1})$
$T = 298.15$ K							
0.0060	78	0.0724	732	0.1713	1362	0.4051	1533
0.0137	167	0.0876	858	0.1993	1465	0.4511	1459
0.0226	266	0.1038	982	0.2307	1555	0.5139	1330
0.0328	374	0.1204	1103	0.2676	1619	0.6139	1093
0.0448	489	0.1234	1123	0.3091	1635	0.7899	631
0.0580	609	0.1415	1248	0.3573	1602		
$T = 318.15$ K							
0.0058	95	0.0774	927	0.1574	1545	0.3403	2148
0.0136	217	0.0961	1089	0.1845	1681	0.3889	2141
0.0231	352	0.1167	1250	0.2023	1819	0.4492	2064
0.0339	487	0.1225	1280	0.2358	1941	0.5301	1894
0.0459	623	0.1348	1377	0.2685	2041	0.6433	1532
0.0605	769	0.1378	1399	0.3026	2099	0.8206	841

Ref. [1]

Composition $V_{f,2}$	η/ mPa · s	Composition $V_{f,2}$	η/ mPa · s	Composition $V_{f,2}$	η/ mPa · s
298.15 K					
0	0.93	0.80	7.97	0.96	45.52
0.20	1.42	0.85	11.37	0.98	78.13
0.40	2.41	0.90	18.16	1.00	166.70
0.60	4.02	0.93	27.32		

Ref. [22]

x_2	$H^E/(J \cdot mol^{-1})$	x_2	$H^E/(J \cdot mol^{-1})$	x_2	$H^E/(J \cdot mol^{-1})$
303.15 K					
0.1023	1091	0.4891	2273	0.6659	1790
0.2013	1760	0.5761	2081	0.8154	1073
0.3036	2150				
0.4004	2319				

Mixture 3: CH_3NO_2 (1) + [bpyr][BF_4] (2)
(1) Volumetric Properties and Heat Capacity at Constant Pressure

Ref. [19]

	293.15 K	298.15 K	303.15 K	308.15 K	313.15 K	318.15 K
x_2	$\rho/g \cdot cm^{-3}$					
0	1.1402	1.1334	1.1266	1.1197	1.1129	1.1060
0.0517	1.1534	1.1474	1.1413	1.1353	1.1292	1.1231
0.1038	1.1615	1.1560	1.1505	1.1449	1.1394	1.1339
0.1952	1.1703	1.1654	1.1604	1.1555	1.1506	1.1456

(*continued*)

	293.15 K	298.15 K	303.15 K	308.15 K	313.15 K	318.15 K
x_2	$\rho/g \cdot cm^{-3}$					
0.3078	1.1767	1.1722	1.1678	1.1633	1.1588	1.1544
0.3563	1.1787	1.1743	1.1700	1.1657	1.1614	1.1571
0.3998	1.1801	1.1759	1.1716	1.1674	1.1632	1.1591
0.5121	1.1830	1.1790	1.1750	1.1710	1.1670	1.1631
0.5994	1.1846	1.1807	1.1769	1.1731	1.1692	1.1654
0.6856	1.1858	1.1821	1.1783	1.1746	1.1709	1.1672
0.7863	1.1869	1.1833	1.1797	1.1761	1.1725	1.1690
0.8842	1.1877	1.1842	1.1807	1.1772	1.1737	1.1703
0.9034	1.1879	1.1844	1.1809	1.1774	1.1740	1.1705
0.9410	1.1882	1.1848	1.1813	1.1779	1.1744	1.1710
1	1.1886	1.1852	1.1817	1.1783	1.1749	1.1716
x_2	$C_p/J \cdot mol^{-1} \cdot K^{-1}$					
0	106.6	106.9	107.3	107.7	108.2	108.6
0.0517	120.3	120.5	120.9	121.3	121.9	122.4
0.1038	134.2	134.7	135.2	135.7	136.4	137.1
0.1952	159.9	160.4	161.0	161.8	162.5	163.5
0.3563	205.2	206.0	206.9	208.0	209.2	210.6
0.3998	217.4	218.3	219.3	220.5	221.9	223.5
0.5121	249.2	250.2	251.4	252.8	254.4	256.3
0.5994	273.7	274.8	276.2	277.9	279.8	281.9
0.6856	297.7	299.0	300.6	302.5	304.6	307.0
0.7863	326.4	327.8	329.6	331.7	334.1	336.8
0.8842	353.6	355.3	357.3	359.7	362.4	365.5
0.9034	359.1	360.8	362.9	365.3	368.1	371.1
0.9410	369.8	371.6	373.7	376.2	379.0	382.3
1	385.8	387.7	390.1	392.8	395.9	399.3

(2) Excess Molar Enthalpies

Ref. [23]

x_2	$H^E/(J \cdot mol^{-1})$	x_2	$H^E/(J \cdot mol^{-1})$	x_2	$H^E/(J\ mol^{-1})$
303.15 K					
0.0992	−252	0.3978	−476	0.7038	−336
0.1547	−328	0.4973	−470	0.7950	−233
0.1998	−401	0.6053	−396	0.8525	−162
0.3029	−443				

Mixture 4: CH_3OH (1) + [b3mpy][BF_4] (2)

(1) Volumetric Properties, Excess Molar Enthalpies and Transport Properties

Ref. [20]

x_2	ρ/ kg · m^{-3}	$10^9 V_m^E$/ m^3 · mol^{-1}	x_2	ρ/ kg · m^{-3}	$10^9 V_m^E$/ m^3 · mol^{-1}	x_2	ρ/ kg · m^{-3}	$10^9 V_m^E$/ m^3 · mol^{-1}
T = 298.15 K								
0.0493	873.4	−366	0.2007	1014.7	−699	0.6009	1138.7	−485
0.0629	893.1	−482	0.2956	1061.6	−732	0.6913	1151.7	−377
0.0986	934.4	−580	0.4052	1098.2	−700	0.7611	1159.7	−241
0.1499	979.6	−629	0.5128	1122.8	−549			
T = 318.15 K								
0.0557	865.7	−515	0.3016	1048.4	−802	0.7088	1139.8	−354
0.1052	924.3	−697	0.3944	1079.5	−714	0.8550	1156.2	−202
0.1493	964.0	−814	0.4970	1104.7	−597	0.9465	1164.4	−91
0.2479	1025.1	−854	0.5868	1121.6	−476			

Ref. [20]

x_2	H_m^E (J · mol^{-1})	x_2	H_m^E (J · mol^{-1})	x_2	H_m^E (J · mol^{-1})	x_2	H_m^E (J · mol^{-1})
T = 298.15 K							
0.0057	109	0.1071	991	0.2672	1599	0.7167	1077
0.0149	228	0.1312	1133	0.3069	1617	0.7441	948
0.0279	386	0.1585	1287	0.3541	1604	0.8438	662
0.0440	546	0.1851	1391	0.4100	1578	0.9196	394
0.0635	699	0.2119	1502	0.4816	1512	0.9707	174
0.0842	846	0.2357	1560	0.5621	1379		
T = 318.15 K							
0.0159	246	0.1382	1406	0.2641	2142	0.4391	2253
0.0352	506	0.1674	1631	0.2984	2230	0.5504	1995
0.0581	723	0.1970	1799	0.3343	2276	0.6911	1460
0.0823	939	0.2147	1934	0.3787	2294	0.8526	801
0.1098	1176	0.2317	2043				

Ref. [1]

Composition $V_{f,2}$	η/mPa · s	Composition $V_{f,2}$	η/mPa · s	Composition $V_{f,2}$	η/mPa · s
298.15 K					
0	0.66	0.80	13.92	0.96	59.40
0.20	0.88	0.85	22.40	0.98	113.03
0.40	1.53	0.90	36.18	1.00	166.70
0.60	3.38	0.93	51.80		

Mixture 5: $1\text{-}C_3H_7OH$ (1) + [b3mpy][BF_4] (2)
(1) Volumetric Properties and Excess Molar Enthalpies

Ref. [20]

x_2	ρ/ kg · m^{-3}	$10^9 V_m^E$/ m^3 · mol^{-1}	x_2	ρ/ kg · m^{-3}	$10^9 V_m^E$/ m^3 · mol^{-1}	x_2	ρ/ kg · m^{-3}	$10^9 V_m^E$/ m^3 · mol^{-1}
T = 298.15 K								
0.0160	816.7	−111	0.5959	1106.9	−308	0.7993	1150.4	−169
0.4528	1066.5	−442	0.6934	1129.5	−250	0.8582	1160.4	−88
0.4940	1079.3	−408	0.7543	1141.9	−197	0.8962	1166.6	−54
0.5262	1088.6	−379						
T = 318.15 K								
0.0143	798.4	−67	0.2295	957.5	−496	0.6072	1095.8	−441
0.0249	808.8	−100	0.3112	997.8	−566	0.6405	1103.5	−359
0.0477	830.6	−212	0.4002	1033.6	−550	0.7096	1118.9	−326
0.1003	873.4	−239	0.4285	1043.6	−544	0.8001	1136.9	−272
0.1222	890.6	−375	0.4635	1055.3	−526	0.8522	1145.7	−169
0.1492	909.3	−417	0.5638	1084.7	−470	0.9028	1154.2	−157
0.2000	940.8	−455						

Ref. [20]

x_2	H_m^E (J · mol^{-1})	x_2	H_m^E (J · mol^{-1})	x_2	H_m^E (J · mol^{-1})	x_2	H_m^E (J · mol^{-1})
T = 298.15 K							
0.0139	163	0.3707	2345	0.4980	2117	0.7278	1246
0.0242	236	0.4045	2309	0.5458	1965	0.7939	959
0.0278	285	0.4278	2275	0.5983	1741	0.8653	685
0.3397	2341	0.4642	2222	0.6621	1521	0.9335	407
T = 318.15 K							
0.0232	417	0.2044	2311	0.3585	2773	0.6073	2248
0.0520	776	0.2311	2441	0.3982	2777	0.6740	1912
0.0852	1156	0.2591	2557	0.4404	2742	0.7504	1509
0.1233	1552	0.2896	2652	0.4891	2654	0.8323	1053
0.1608	1924	0.3223	2727	0.5441	2502	0.9131	598
0.2011	2235						

(2) Phase Equilibrium

Ref. [20]

x_2	T/K	x_2	T/K	x_2	T/K
0.0166	293.82	0.1713	304.18	0.3968	292.52
0.0287	299.20	0.2172	303.21	0.4215	290.13
0.0407	302.04	0.2402	302.48	0.4400	287.07
0.0762	303.83	0.2798	300.68	0.4633	284.65
0.1102	304.42	0.3206	298.41	0.4918	281.71
0.1386	304.55	0.3577	295.55		

Mixture 6: 1-C_4H_9OH (1) + [b3mpy][BF_4] (2)
(1) Volumetric Properties and Excess Molar Enthalpies

Ref. [20]

x_2	ρ/ $kg \cdot m^{-3}$	$10^9 V_m^E$/ $m^3 \cdot mol^{-1}$	x_2	ρ/ $kg \cdot m^{-3}$	$10^9 V_m^E$/ $m^3 \cdot mol^{-1}$	x_2	ρ/ $kg \cdot m^{-3}$	$10^9 V_m^E$/ $m^3 \cdot mol^{-1}$
T = 298.15 K								
0.0166	819.7	−46	0.8166	1147.7	−109	0.9040	1165.1	−67
0.7641	1136.6	−170	0.8558	1155.7	−76			
T = 318.15 K								
0.0191	805.5	−2	0.7777	1125.9	−265	0.9084	1152.1	−108
0.6961	1107.5	−411	0.8535	1141.6	−177	0.9479	1159.4	−75

Ref. [20]

x_2	H_m^E ($J \cdot mol^{-1}$)	x_2	H_m^E ($J \cdot mol^{-1}$)	x_2	H_m^E ($J \cdot mol^{-1}$)	x_2	H_m^E ($J \cdot mol^{-1}$)
T = 318.15 K							
0.0098	151	0.3603	2849	0.5101	2726	0.7667	1582
0.0250	286	0.3932	2860	0.5852	2473	0.8531	1104
0.0464	577	0.4450	2832	0.6705	2106	0.9346	521

(2) Phase Equilibrium

Ref. [20]

x_2	T/K	x_2	T/K	x_2	T/K
0.0072	293.72	0.2145	320.35	0.5067	300.56
0.0134	302.81	0.2421	319.51	0.5284	298.22
0.0232	311.12	0.2708	318.01	0.5527	295.25
0.0504	318.28	0.2942	317.17	0.5742	292.90
0.0930	320.78	0.3278	315.40	0.6033	289.48
0.1304	321.58	0.3733	312.80	0.6540	281.22
0.1678	321.34	0.4243	308.86		
0.1814	321.17	0.4633	305.27		

Mixture 7: CH_2Cl_2 (1) + [3-MBP][BF_4] (2)
(1) Transport Properties

Ref. [1]

Composition $V_{f,2}$	η/ mPa · s	Composition $V_{f,2}$	η/ mPa · s	Composition $V_{f,2}$	η/ mPa · s
298.15 K					
0	0.40	0.80	32.25	0.96	120.10
0.20	1.31	0.85	44.36	0.98	140.30
0.40	3.62	0.90	65.70	1.00	166.70
0.60	10.53	0.93	90.33		

6.13 11041-03001: 1-Butyl-3-methylpyridinium bis ((trifluoromethyl)sulfonyl)imide

Abbreviation:	[bmpy][TFSI], [bmpy][BMeI]
Molecular formula:	$C_{12}H_{16}F_6N_2O_4S_2$
Molar mass	430.39
Structure:	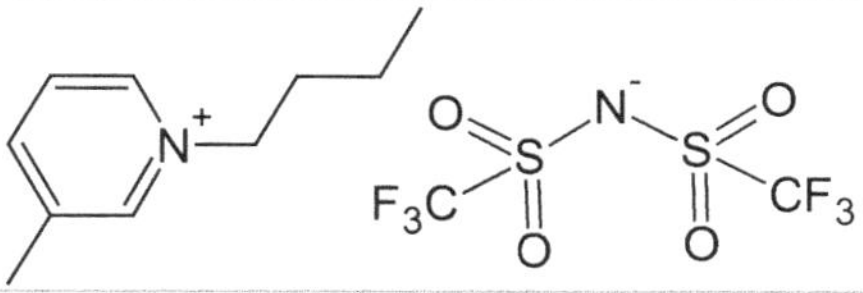

6.13.1 Binary Mixtures

Mixture 1: CH_2F_2 (1) + [bmpy][BMeI] (2)
(1) Solubility and Diffusivity

Ref. [24]

T/K	P/MPa	100 x_1	$10^{11}D$/ $m^2 \cdot s^{-1}$	T/K	P/MPa	100 x_1	$10^{11}D$/ $m^2 \cdot s^{-1}$
298.15	0.0096	1.0	6.8	298.15	0.5496	42.8	15
298.15	0.0995	10.0	7.8	298.15	0.7005	51.0	
298.15	0.2495	22.4	12	298.15	0.8506	58.6	
298.15	0.3995	33.1	15	298.15	1.0000	65.4	23

6.14 11041-04007: *N*-Butyl-3-methylpyridinium tosylate

Abbreviation:	[BM^{3}py][Tos]
Molecular formula:	$C_{17}H_{23}NO_3S$
Molar mass:	321.43
Structure:	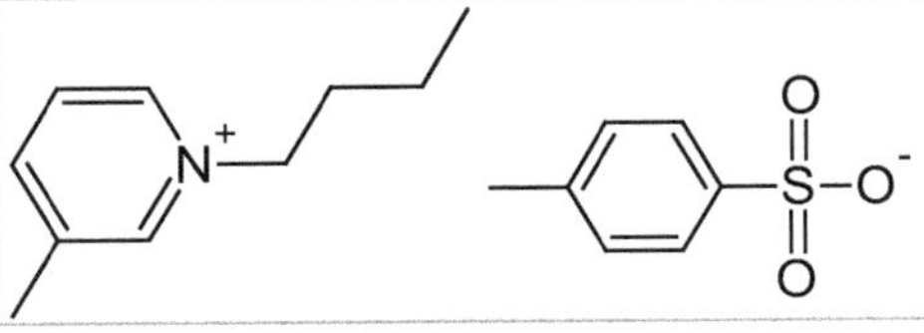

6.14.1 Binary Mixtures

Mixture 1: SCH = CHCH = CH (thiophene) (1) + [BM^{3}py][Tos] (2)
(1) Phase Equilibrium

Ref. [6]

x_2	T^{SLE}/K	T^{LLE}/K	x_2	T^{SLE}/K	T^{LLE}/K
1.0000	323.7		0.1705		296.8
0.9797	322.9		0.1687		306.0
0.9323	319.7		0.1670		313.0
0.8925	317.9		0.1654		317.3
0.8583	316.6		0.1637		322.6
0.8271	315.3		0.1622		326.0
0.7951	313.6		0.1607		330.2
0.7393	309.8		0.1595		334.5
0.6893	306.3		0.1580		338.4
0.6470	303.2		0.1567		341.5
0.6105	300.1		0.1555		345.0
0.5754	298.5		0.1543		348.9
0.5433	295.5		0.1531		351.7
0.4813	292.4		0.1520		355.5
0.1729		285.8			

6.15 11046-04007: *N*-Hexyl-3-methylpyridinium tosylate

Abbreviation:	[HM³py][Tos]
Molecular formula:	$C_{19}H_{27}NO_3S$
Molar mass:	349.49
Structure:	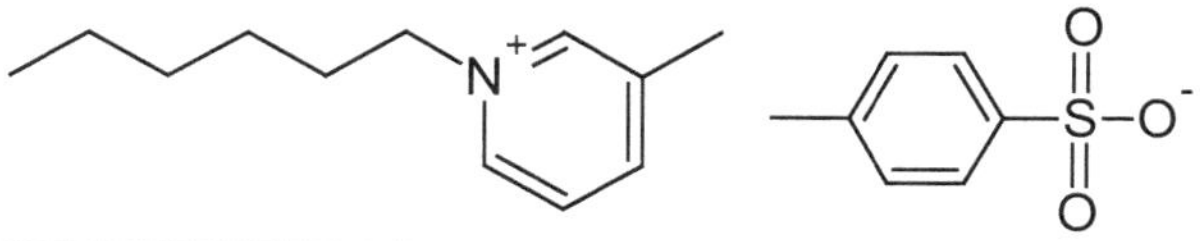

6.15.1 Binary Mixtures

Mixture 1: SCH = CHCH = CH (thiophene) (1) + [HM³py][Tos] (2)
(1) Phase Equilibrium

Ref. [6]

x_2	T^{SLE}/K	T^{LLE}/K	x_2	T^{SLE}/K	T^{LLE}/K
1.0000	329.3		0.2462	280.4	
0.9175	327.1		0.2177	273.6	
0.8534	325.3		0.1874	268.4	
0.8066	324.3		0.0431		294.6[a]
0.7578	322.5		0.0423		295.8[a]
0.7125	320.6		0.0404		297.0[a]
0.6759	318.8		0.0354		298.7[a]
0.6145	316.0		0.0302		300.2[a]
0.5608	312.4		0.0254		301.0[a]
0.5173	310.7		0.0218		301.3[a]
0.4787	309.0		0.0189		301.7[a]
0.4455	307.1		0.0167		301.5
0.4173	304.9		0.0150		300.6
0.3949	303.0		0.0132		299.4
0.3753	301.2		0.0114		297.6
0.3408	297.6		0.0100		295.2
0.3117	293.2		0.0088		293.5
0.2757	288.0				

[a] Thiophene rich phase.

6.16 11069-02001: 1-Butyl-2-methylpyridinium tetrafluoroborate

Abbreviation:	[b2mpy][BF_4]
Molecular formula:	$C_{10}H_{16}NBF_4$
Molar mass:	237.04
Structure:	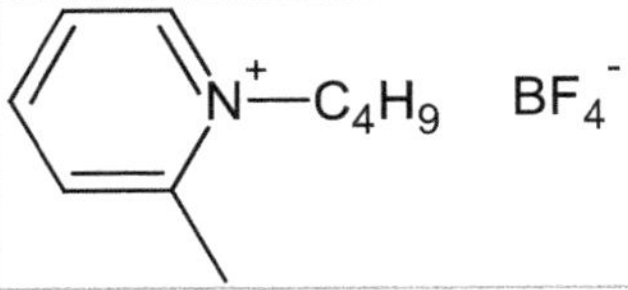

6.16.1 Binary Mixtures

Mixture 1: H_2O (1) + [b2mpy][BF_4] (2)
(1) Excess Molar Enthalpies

Ref. [25]

x_2	H^E_m/J · mol^{-1}	x_2	H^E_m/J · mol^{-1}	x_2	H^E_m/J · mol^{-1}
T = 298.15 K					
0.0764	768	0.1780	1273	0.4300	1457
0.0871	845	0.2071	1367	0.5129	1369
0.0988	921	0.2392	1416	0.6130	1190
0.1138	1010	0.2742	1464	0.7260	933
0.1315	1101	0.3137	1494	0.8524	558
0.1526	1201	0.3686	1494		
T = 318.15 K					
0.0801	1064	0.1755	1795	0.4925	1955
0.0923	1185	0.2078	1942	0.5883	1699
0.1072	1323	0.2558	2096	0.6399	1542
0.1255	1472	0.3296	2133	0.7806	1069
0.1482	1624	0.4389	2037	0.8577	746

(2) Volumetric Properties

Ref. [25]

x_2	ρ/kg · m^{-3}	10^9 V^E_m/m^3 · mol^{-1}	x_2	ρ/kg · m^{-3}	10^9 V^E_m/m^3 · mol^{-1}
T = 298.15 K					
0.0502	1066.29	149	0.5730	1184.33	499
0.1010	1102.15	261	0.6430	1188.35	455
0.1517	1124.07	350	0.7102	1191.74	391
0.1998	1138.48	414	0.7636	1193.96	354
0.2833	1155.54	487	0.8500	1197.32	251

(*continued*)

x_2	ρ/kg · m^{-3}	$10^9\ V^E{}_m$/m^3 · mol^{-1}	x_2	ρ/kg · m^{-3}	$10^9\ V^E{}_m$/m^3 · mol^{-1}
0.3386	1163.53	517	0.8992	1198.97	193
0.4100	1171.48	543	0.9259	1199.93	142
0.5112	1180.11	529			
T = 318.15 K					
0.0497	1053.69	228	0.5011	1165.60	609
0.1001	1087.79	382	0.5806	1170.96	569
0.1492	1108.60	496	0.7075	1177.86	441
0.2049	1125.13	572	0.6486	1175.12	480
0.2941	1142.77	640	0.7797	1181.05	349
0.3827	1154.47	661	0.8578	1184.06	235
0.4388	1160.16	651	0.9306	1186.72	88

Mixture 2: CH_3OH (1) + [b2mpy][BF_4] (2)
(1) Excess Molar Enthalpies

Ref. [25]

x_2	$H^E{}_m$/J · mol^{-1}	x_2	$H^E{}_m$/J · mol^{-1}	x_2	$H^E{}_m$/J · mol^{-1}
T = 298.15 K					
0.0655	751	0.3212	1735	0.6623	1275
0.1301	1195	0.3652	1766	0.7423	1047
0.1715	1379	0.4145	1752	0.8190	789
0.2223	1559	0.4658	1670	0.8928	473
0.2512	1637	0.5239	1582	0.9592	173
0.2832	1696	0.5900	1481		
T = 318.15 K					
0.0789	1124	0.2878	2241	0.5984	1842
0.1257	1516	0.3279	2246	0.6701	1648
0.1708	1868	0.3737	2244	0.7477	1423
0.1927	1989	0.4258	2162	0.8334	1096
0.2208	2094	0.4784	2074	0.9112	689
0.2576	2189	0.5373	1973	0.9778	199

(2) Volumetric Properties

Ref. [25]

x_2	ρ/kg · m^{-3}	$10^9\ V^E{}_m$/m^3 · mol^{-1}	x_2	ρ/kg · m^{-3}	$10^9\ V^E{}_m$/m^3 · mol^{-1}
T = 298.15 K					
0.0532	884.03	−492	0.5520	1146.32	−423
0.1033	945.97	−653	0.6388	1161.71	−360
0.1509	989.58	−716	0.7058	1171.44	−280
0.2096	1030.56	−751	0.7993	1182.91	−162
0.3049	1077.52	−698	0.8561	1189.09	−114
0.4071	1112.27	−588	0.9464	1197.68	−34
0.5097	1138.03	−504			

(*continued*)

x_2	ρ/kg · m^{-3}	10^9 V^E_m /m^3 · mol^{-1}	x_2	ρ/kg · m^{-3}	10^9 V^E_m/m^3 · mol^{-1}
T = 318.15 K					
0.0344	836.06	−415	0.4736	1117.21	−838
0.0806	902.54	−648	0.5358	1130.75	−753
0.1493	972.68	−885	0.6959	1156.94	−467
0.2235	1024.96	−1025	0.7856	1168.14	−314
0.3310	1074.96	−1008	0.8660	1176.85	−210
0.4150	1102.20	−925			

Mixture 3: C_2H_5OH (1) + [b2mpy][BF_4] (2)
(1) Phase Equilibrium

Ref. [25]

x_2	T/K	x_2	T/K	x_2	T/K
0.009	281.49	0.069	293.94	0.196	293.19
0.014	283.82	0.083	294.28	0.220	292.36
0.020	285.95	0.096	294.50	0.248	290.78
0.026	288.43	0.112	294.56	0.276	288.94
0.031	290.27	0.129	294.62	0.313	286.91
0.039	291.58	0.148	294.48	0.350	283.42
0.048	292.82	0.167	294.17	0.396	279.40
0.058	293.49	0.176	293.82		

(2) Excess Molar Enthalpies

Ref. [25]

x_2	H^E_m/J · mol^{-1}	x_2	H^E_m/J · mol^{-1}	x_2	H^E_m/J · mol^{-1}
T = 298.15 K					
0.0880	1100	0.2046	1871	0.4649	1976
0.1095	1260	0.2309	1984	0.5385	1850
0.1357	1466	0.2625	2058	0.6273	1622
0.1477	1556	0.3000	2095	0.7223	1330
0.1640	1652	0.3459	2090	0.8190	974
0.1832	1758	0.4016	2063	0.9158	499
T = 318.15 K					
0.0891	1330	0.2819	2530	0.6095	2161
0.1114	1560	0.3220	2601	0.6744	1888
0.1546	1917	0.3691	2600	0.7470	1617
0.1741	2048	0.4221	2568	0.7923	1415
0.1954	2178	0.4828	2454	0.8588	1067
0.2207	2314	0.5534	2304	0.9333	561

(3) Volumetric Properties

Ref. [25]

x_2	ρ/kg · m^{-3}	10^9 $V^E{}_m$/m^3 · mol^{-1}	x_2	ρ/kg · m^{-3}	10^9 $V^E{}_m$/m^3 · mol^{-1}
T = 298.15 K					
0.0360	853.45	−229	0.3860	1104.33	−472
0.0547	882.67	−338	0.5402	1143.01	−333
0.0948	932.65	−415	0.6246	1158.49	−264
0.1250	963.65	−488	0.7398	1175.45	−179
0.2676	1058.83	−557	0.8245	1185.45	−93
T = 318.15 K					
0.0527	838.96	−410	0.5814	1119.42	−720
0.1009	889.70	−604	0.6499	1134.27	−576
0.1518	933.02	−697	0.7104	1145.79	−441
0.2108	975.07	−836	0.7827	1158.72	−377
0.3129	1030.54	−925	0.8665	1171.06	−178
0.4397	1079.70	−857	0.9135	1177.48	−95

Mixture 4: 1-C_3H_7OH (1) + [b2mpy][BF_4] (2)

(1) Phase Equilibrium

Ref. [25]

x_2	T/K	x_2	T/K	x_2	T/K
0.006	293.76	0.072	319.81	0.323	314.04
0.009	301.67	0.094	320.03	0.357	311.55
0.013	306.49	0.118	320.24	0.392	309.17
0.017	309.13	0.144	320.26	0.426	307.04
0.021	311.22	0.172	320.19	0.461	304.42
0.027	313.84	0.202	320.04	0.500	299.79
0.033	315.68	0.235	318.89	0.540	296.91
0.041	317.10	0.263	317.68	0.581	294.01
0.056	318.82	0.292	316.36		

(2) Excess Molar Enthalpies

Ref. [25]

x_2	$H^E{}_m$/J · mol^{-1}	x_2	$H^E{}_m$/J · mol^{-1}	x_2	$H^E{}_m$/J · mol^{-1}
T = 298.15 K					
0.5405	2065	0.6444	1779	0.8128	1098
0.5779	1969	0.7131	1558	0.8980	686
T = 318.15 K					
0.0350	606	0.4475	2623	0.7660	1564
0.2685	2585	0.4956	2502	0.8214	1235
0.2961	2640	0.5464	2369	0.8691	941

(continued)

x_2	$H^E{}_m$/J · mol^{-1}	x_2	$H^E{}_m$/J · mol^{-1}	x_2	$H^E{}_m$/J · mol^{-1}
0.3263	2682	0.6020	2161	0.9183	652
0.3621	2707	0.6599	1991	0.9598	378
0.4028	2690	0.7138	1770		

(3) Volumetric Properties

Ref. [25]

x_2	ρ/kg · m^{-3}	10^9 $V^E{}_m$/m^3 · mol^{-1}	x_2	ρ/kg · m^{-3}	10^9 $V^E{}_m$/m^3 · mol^{-1}
T = 298.15 K					
0.5277	1102.19	−291	0.7807	1164.18	−125
0.6373	1132.22	−227	0.8554	1178.40	−77
0.7152	1150.44	−171			
T = 318.15 K					
0.0191	804.05	−103	0.6120	1113.67	−594
0.0362	821.60	−206	0.7555	1147.08	−470
0.2698	987.52	−609	0.7829	1152.29	−387
0.3386	1020.39	−641	0.8144	1157.94	−279
0.4200	1053.64	−669	0.8865	1170.83	−169
0.5385	1093.16	−638	0.9385	1179.17	−58

Mixture 5: 1-C_4H_9OH (1) + [b2mpy][BF_4] (2)
(1) Phase Equilibrium

Ref. [25]

x_2	T/K	x_2	T/K	x_2	T/K
0.007	315.38	0.088	336.51	0.365	330.89
0.010	319.93	0.111	336.86	0.411	328.30
0.013	323.18	0.138	336.80	0.452	325.31
0.017	325.16	0.174	336.83	0.491	322.19
0.021	327.24	0.202	336.57	0.532	318.08
0.028	330.14	0.222	335.94	0.574	315.12
0.039	332.82	0.250	335.27	0.616	312.89
0.052	334.58	0.284	334.47	0.660	310.62
0.070	335.96	0.323	332.96	0.704	309.95

(2) Excess Molar Enthalpies

Ref. [25]

x_2	$H^E{}_m$/J · mol^{-1}	x_2	$H^E{}_m$/J · mol^{-1}	x_2	$H^E{}_m$/J · mol^{-1}
T = 318.15 K					
0.5094	2599	0.7175	1946	0.9536	616
0.5661	2476	0.7995	1555		
0.6400	2249	0.8821	1122		

(3) Volumetric Properties

Ref. [25]

x_2	ρ/kg · m^{-3}	10^9 V^E_m/m^3 · mol^{-1}	x_2	ρ/kg · m^{-3}	10^9 V^E_m/m^3 · mol^{-1}
T = 318.15 K					
0.0038	794.43	−36	0.7145	1128.50	−433
0.5577	1084.46	−553	0.8334	1156.16	−288
0.6128	1101.18	−537	0.8943	1168.58	−167

Mixture 6: 1-$C_5H_{11}OH$ (1) + [b2mpy][BF_4] (2)
(1) Phase Equilibrium

Ref. [25]

x_2	T/K	x_2	T/K	x_2	T/K
0.005	316.68	0.090	351.12	0.394	347.94
0.009	324.85	0.122	351.39	0.406	346.46
0.012	330.97	0.156	352.26	0.432	344.76
0.015	335.34	0.194	351.90	0.452	343.35
0.019	338.50	0.232	351.61	0.472	342.44
0.024	341.38	0.280	350.69	0.487	341.39
0.029	343.75	0.314	350.06	0.512	340.25
0.038	346.38	0.337	349.44	0.578	337.56
0.050	348.68	0.352	349.09	0.623	337.26
0.057	350.17	0.373	347.97	0.394	347.94

6.17 11074-03001: 3-Methyl-1-propylpyridinium bis (trifluoromethylsulfonyl)imide

Abbreviation:	[C_3mpy][Tf_2N]
Molecular formula:	$C_{11}H_{14}N_2O_4S_2F_6$
Molar mass:	416.36
Structure:	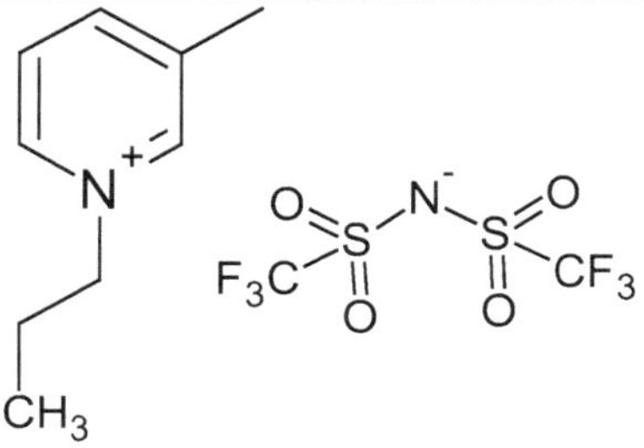

6.17.1 Binary Mixtures

Mixture 1: H_2O (1) + [C_3mpy][Tf_2N] (2)
(1) Solubility

Ref. [26]

T/K	$x_1 \pm \sigma^a$	$10^4\ (x_2 \pm \sigma^a)$
288.15	0.206 ± 0.001	3.51 ± 0.04
293.15	0.226 ± 0.004	3.66 ± 0.02
298.15	0.236 ± 0.004	3.75 ± 0.02
303.15	0.260 ± 0.006	3.90 ± 0.02
308.15	0.270 ± 0.003	4.27 ± 0.04
313.15	0.292 ± 0.001	4.42 ± 0.04
318.15	0.310 ± 0.005	4.76 ± 0.09

^{a} Standard deviation.

(2) Thermodynamic Properties

Ref. [26]

$(\Delta_{sol}H_m{}^0 \pm \sigma^a)$/kJ · mol^{-1}	$(\Delta_{sol}G_m{}^0 \pm \sigma^a)$/kJ · mol^{-1}	$(\Delta_{sol}S_m{}^0 \pm \sigma^a)$/kJ · mol^{-1}
298.15 K		
6.5 ± 1.5	19.56 ± 0.02	−43.9 ± 5.1

^{a} Standard deviation.

Mixture 2: CH_2F_2 (1) + [pmpy][BMeI] (2)
(1) Solubility and Diffusivity

Ref. [24]

T/K	P/MPa	100 x_1	$10^{11}D$/ m^2 · s^{-1}	T/K	P/MPa	100 x_1	$10^{11}D$/ m^2 · s^{-1}
283.15	0.0102	1.5	5.7	323.15	0.1003	5.8	19
283.15	0.1000	14.0	5.7	323.15	0.2498	13.8	20
283.15	0.2494	31.3	11	323.15	0.4000	20.8	
283.15	0.3996	45.5	13	323.15	0.5494	27.2	24
283.15	0.5493	57.3		323.15	0.6994	33.0	35
283.15	0.6996	68.3	20	323.15	0.8503	38.2	40
283.15	0.8495	78.2	27	323.15	0.9994	43.3	
298.05	0.0095	1.0	9.9	348.05	0.0103	0.3	39
298.05	0.1002	9.6	10	348.05	0.1002	3.7	39
298.05	0.2503	22.2	12	348.05	0.2502	9.0	40
298.05	0.3996	33.0	15	348.05	0.4002	13.9	40
298.05	0.5497	42.2	19	348.05	0.5493	18.4	
298.05	0.6992	50.7	22	348.05	0.7003	22.6	46
298.05	0.8496	58.0	24	348.05	0.8493	26.5	48
298.05	1.0004	64.5		348.05	1.0002	30.4	55
323.15	0.0103	0.7					

6.18 11075-04032: Pyridinium ethoxyethylsulfate

Abbreviation:	$[C_5H_5NH][C_2H_5OC_2H_4OSO_3]$
Molecular formula:	$C_9H_{15}NSO_5$
Molar mass:	249.29
Structure:	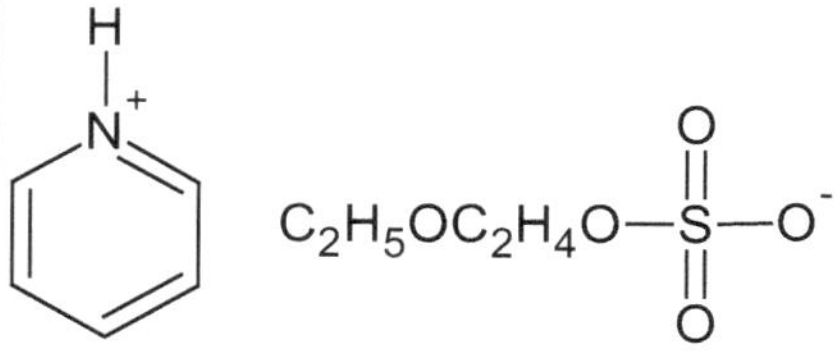

6.18.1 Binary Mixtures

Mixture 1: C_6H_6 (1) + $[C_5H_5NH][C_2H_5OC_2H_4OSO_3]$ (2)
(1) Phase Equilibrium

Ref. [27]

x_1	P (kPa)	x_1	P (kPa)	x_1	P (kPa)	x_1	P (kPa)
333.15 K							
0.0039	0.66	0.0372	6.09	0.2266	31.04	0.5589	51.70
0.0074	1.26	0.0553	8.83	0.2772	36.28	0.6071	51.70
0.0112	1.90	0.0769	11.95	0.3350	41.43	0.6512	51.71
0.0155	2.61	0.1083	16.37	0.3933	45.74	0.6906	51.71
0.0202	3.38	0.1414	20.84	0.4513	49.06	0.7259	51.72
0.0281	4.65	0.1810	25.79	0.5067	51.33		

Mixture 2: $C_6H_5CH_3$ (1) + $[C_5H_5NH][C_2H_5OC_2H_4OSO_3]$ (2)
(1) Phase Equilibrium

Ref. [27]

x_1	P (kPa)	x_1	P (kPa)	x_1	P (kPa)	x_1	P (kPa)
333.15 K							
0.0030	0.31	0.0323	3.14	0.1965	14.60	0.5137	18.38
0.0059	0.61	0.0461	4.35	0.2424	16.67	0.5627	18.38
0.0087	0.90	0.0643	5.87	0.2960	18.37	0.6082	18.38
0.0117	1.20	0.0899	7.89	0.3514	18.38	0.6495	18.38
0.0184	1.85	0.1194	10.03	0.4073	18.38	0.6872	18.38
0.0257	2.51	0.1553	12.32	0.4616	18.38		

6.19 11076-03001: 1-Decyl-3-methylpyridinium bis (trifluoromethylsulfonyl)imide

Abbreviation:	$C_{10}MePyNTf_2$
Molecular formula:	$C_{18}H_{28}N_2F_6O_4S_2$
Molar mass:	514.56
Structure:	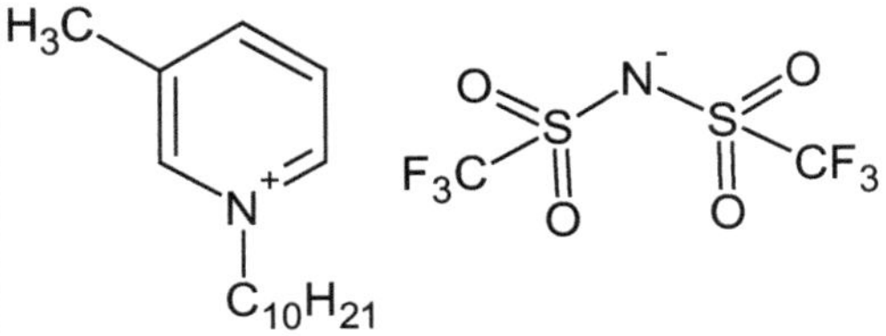

6.19.1 Binary Mixtures

Mixture 1: 1H,1H,2H,2H-$C_8F_{13}H_4OH$ (1) + $C_{10}MePyNTf_2$ (2)
(1) Cloud Point

Ref. [9]

x_2	T/K	x_2	T/K	x_2	T/K	x_2	T/K
0.111	266.8	0.313	274.6	0.483	273.0	0.694	263.9
0.146	271.0	0.349	274.0	0.529	272.2	0.736	263.0
0.183	273.0	0.386	273.9	0.575	269.4	0.066	259.4
0.220	273.7	0.424	273.0	0.616	269.0	0.097	265.7
0.278	274.5	0.438	273.6	0.655	266.8	0.118	267.8

6.20 11077-04007: 1,4-Dimethylpyridinium tosylate

Abbreviation:	$[M^{1,4}py][Tos]$
Molecular formula:	$C_{14}H_{17}NO_3S$
Molar mass:	279.36
Structure:	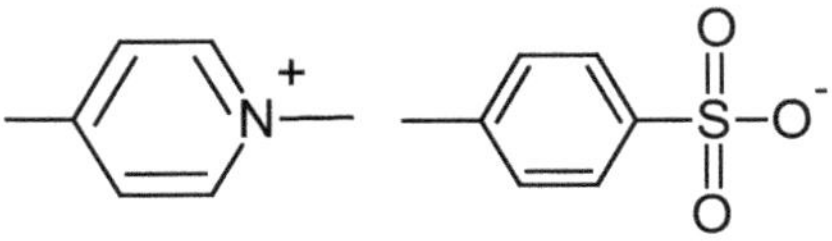

6.20.1 Binary Mixtures

Mixture 1: SCH = CHCH = CH (thiophene) (1) + [$M^{1,4}$py][Tos] (2)
(1) Phase Equilibrium

Ref. [6]

x_2	T^{SLE}/K	T^{LLE}/K	x_2	T^{SLE}/K	T^{LLE}/K
1.0000	424.8		0.2596	379.1	
0.6218	410.7		0.2472	378.4	
0.5639	407.8		0.2444	377.9	
0.5262	404.9		0.2419		389.5
0.4877	402.4		0.2393		400.6
0.4566	399.1		0.2364		412.3
0.4097	394.0		0.2341		421.0
0.3746	389.3		0.1940	377.8	
0.3374	385.6		0.1517	377.8	
0.2871	381.5		0.1192	377.8	

6.21 11078-04007: 2,4,6-Collidine tosylate

Abbreviation:	[$M^{2,4,6}$py][Tos]
Molecular formula:	$C_{15}H_{19}NO_3S$
Molar mass:	293.38
Structure:	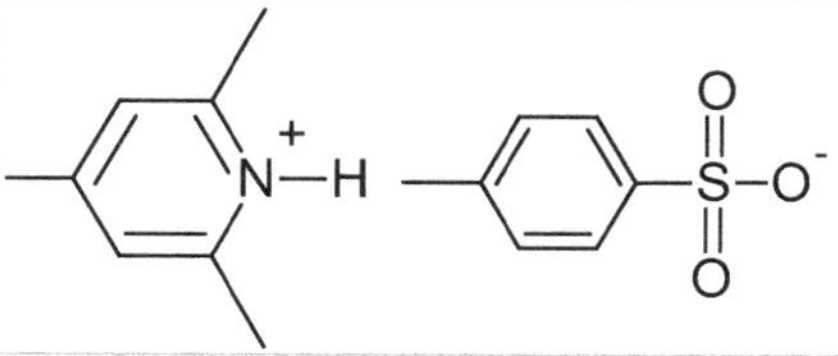

6.21.1 Binary Mixtures

Mixture 1: SCH = CHCH = CH (thiophene) (1) + [$M^{2,4,6}$py][Tos] (2)
(1) Phase Equilibrium

Ref. [6]

x_2	T^{SLE}/K	T^{LLE}/K	x_2	T^{SLE}/K	T^{LLE}/K
1.0000	403.7		0.0344	297.7	313.5
0.5056	347.3		0.0315	297.7	314.6
0.3545	317.2		0.0277	297.7	314.8
0.3010	310.0		0.0263	297.7	314.6

(*continued*)

x_2	T^{SLE}/K	T^{LLE}/K	x_2	T^{SLE}/K	T^{LLE}/K
0.2505	305.8		0.0246	297.7	314.4
0.1930	301.9		0.0341	297.7	314.2
0.1554	299.1		0.0229	297.7	314.0
0.1379	298.2		0.0212	297.7	313.9
0.1256	298.0		0.0201	297.7	313.4
0.1172	297.7		0.0196	297.7	313.4
0.1060	297.8		0.0168	297.7	312.4
0.0990	297.7		0.0145	297.7	311.7
0.0892	297.7		0.0140	297.7	311.3
0.0821	297.7		0.0090	297.7	307.1
0.0767	297.7	299.8	0.0054	297.7	
0.0640	297.7	303.5	0.0042	297.4	
0.0591	297.7	306.3	0.0029	295.1	
0.0520	297.7	308.1	0.0024	293.9	
0.0411	297.7	311.0	0.0021	293.4	
0.0386	297.7	312.1	0.0018	293.2	
0.0349	297.7	313.5			

6.22 11078-06005: Collidinium formate

Abbreviation:	[Col][HCOO]
Molecular formula:	$C_9H_{13}NO_2$
Molar mass:	167.20
Structure:	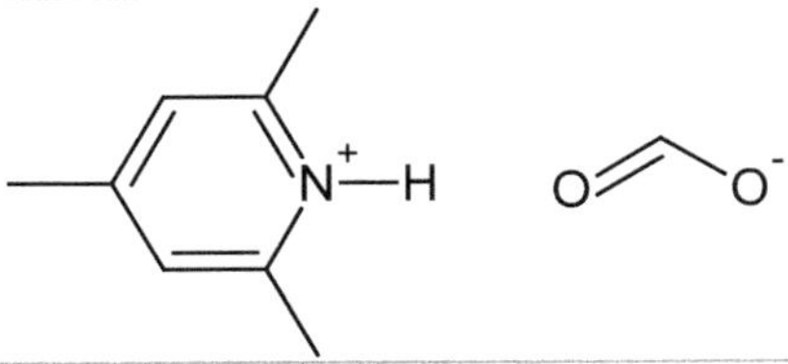

6.22.1 Binary Mixtures

Mixture 1: $H_2O(1)$ + [Col][HCOO] (2)
(1) Volumetric Properties

Ref. [28]

x_2	ρ/g · cm^{-3}	V^E/(cm^3 · mol^{-1})	$V_{\Phi,2}$/(cm^3 · mol^{-1})	$V_{\Phi,1}$/(cm^3 · mol^{-1})
298.15 K				
1.0000	1.0210	0.0000	163.76	

(*continued*)

x_2	ρ/g · cm^{-3}	V^E/(cm^3 · mol^{-1})	$V_{\Phi,2}$/(cm^3 · mol^{-1})	$V_{\Phi,1}$/(cm^3 · mol^{-1})
0.4872	1.0499	−2.6801	158.27	12.80
0.3022	1.0638	−2.8103	154.41	14.09
0.1996	1.0671	−2.3911	151.73	15.09
0.1380	1.0642	−1.9319	149.70	15.83
0.0973	1.0573	−1.5101	148.09	16.40
0.0665	1.0433	−1.0156	148.00	16.87
0.0438	1.0304	−0.6593	147.85	17.31
0.0263	1.0176	−0.3777	148.60	17.67
0.0119	1.0043	−0.1343		17.92
0.0000	0.9950	0.0000		18.06

References

1. Khupse ND, Kumar A (2009) Dramatic change in viscosities of pure ionic liquids upon addition of molecular solvents. J Solut Chem 38(5):589–600
2. Mokhtarani B, Sharifi A, Mortaheb HR, Mirzaei M, Mafi M, Sadeghian F (2009) Density and viscosity of pyridinium-based ionic liquids and their binary mixtures with water at several temperatures. J Chem Thermodyn 41(3):323–329
3. Ortega J, Vreekamp R, Penco E, Marrero E (2008) Mixing thermodynamic properties of 1-butyl-4-methylpyridinium tetrafluoroborate [$b_4$mpy][BF_4] with water and with an alkan-1ol (methanol to pentanol). J Chem Thermodyn 40(7):1087–1094
4. Meindersma GW, Podt AJG, de Haan AB (2006) Ternary liquid–liquid equilibria for mixtures of toluene + *n*-heptane + an ionic liquid. Fluid Phase Equilib 247(1–2):158–168
5. Meindersma GW, Podt A, de Haan AB (2006) Ternary liquid – liquid equilibria for mixtures of an aromatic + an aliphatic hydrocarbon + 4-methyl-N-butylpyridinium tetrafluoroborate. J Chem Eng Data 51(5):1814–1819
6. Domańska U, Królikowski M, Ślesińska K (2009) Phase equilibria study of the binary systems (ionic liquid + thiophene): desulphurization process. J Chem Thermodyn 41(11):1303–1311
7. Domańska U, Królikowski M, Paduszyński K (2009) Phase equilibria study of the binary systems (N-butyl-3-methylpyridinium tosylate ionic liquid + an alcohol). J Chem Thermodyn 41(8):932–938
8. Domańska U, Królikowski M, Pobudkowska A, Letcher TM (2009) Phase equilibria study of the binary systems (N-butyl-4-methylpyridinium tosylate ionic liquid + organic solvent, or water). J Chem Eng Data 54(5):1435–1441
9. Ferreira R, Blesic M, Trindade J, Marrucho I, Lopes JNC, Rebelo LPN (2008) Solubility of fluorinated compounds in a range of ionic liquids. Cloud-point temperature dependence on composition and pressure. Green Chem 10(9):918–928
10. Arce A, Earle MJ, Katdare SP, Rodríguez H, Seddon KR (2007) Phase equilibria of mixtures of mutually immiscible ionic liquids. Fluid Phase Equilib 261(1–2):427–433
11. Arce A, Earle MJ, Rodríguez H, Seddon KR, Soto A (2009) Bis{(trifluoromethyl)sulfonyl} amide ionic liquids as solvents for the extraction of aromatic hydrocarbons from their mixtures with alkanes: effect of the nature of the cation. Green Chem 11(3):365–372
12. González B, Calvar N, Gómez E, Domínguez I, Domínguez A (2009) Synthesis and physical properties of 1-ethylpyridinium ethylsulfate and its binary mixtures with ethanol and 1-propanol at several temperatures. J Chem Eng Data 54(4):1353–1358
13. Yang XZ, Wang J, Li GS, Zhang ZZ (2009) Solubilities of 1-ethylpyridinium hexafluorophosphate in ethanol + water from (278.15 to 345.15) K. J Chem Eng Data 54(1):75–77

14. González Ba, Calvar N, Gómez E, Macedo EnA, Domínguez An (2008) Synthesis and physical properties of 1-ethyl 3-methylpyridinium ethylsulfate and its binary mixtures with ethanol and water at several temperatures. J Chem Eng Data 53(8):1824–1828
15. Calvar N, González B, Domínguez Á, Macedo EA (2009) Vapour pressures and osmotic coefficients of binary mixtures of 1-ethyl-3-methylimidazolium ethylsulfate and 1-ethyl-3-methylpyridinium ethylsulfate with alcohols at T = 323.15 K. J Chem Thermodyn 41 (12):1439–1445
16. Calvar N, González B, Domínguez Á, Macedo EA (2009) Osmotic coefficients of binary mixtures of four ionic liquids with ethanol or water at T = (313.15 and 333.15) K. J Chem Thermodyn 41(1):11–16
17. Calvar N, Gómez E, González Ba, Domínguez An (2009) Experimental determination, correlation, and prediction of physical properties of the ternary mixtures ethanol and 1-propanol + water + 1-ethyl-3-methylpyridinium ethylsulfate at 298.15 K. J Chem Eng Data 54 (8):2229–2234
18. González EJ, Calvar N, González B, Domínguez Á (2009) (Liquid + liquid) equilibria for ternary mixtures of (alkane + benzene + [EMpy] [ESO_4]) at several temperatures and atmospheric pressure. J Chem Thermodyn 41(11):1215–1221
19. García-Miaja G, Troncoso J, Romaní L (2007) Density and heat capacity as a function of temperature for binary mixtures of 1-butyl-3-methylpyridinium tetrafluoroborate + water, + ethanol, and + nitromethane. J Chem Eng Data 52(6):2261–2265
20. Ortega J, Vreekamp R, Marrero E, Penco E (2007) Thermodynamic properties of 1-butyl-3-methylpyridinium tetrafluoroborate and its mixtures with water and alkanols. J Chem Eng Data 52(6):2269–2276
21. García-Miaja G, Troncoso J, Romaní L (2008) Excess properties for binary systems ionic liquid + ethanol: experimental results and theoretical description using the ERAS model. Fluid Phase Equilib 274(1–2):59–67
22. García-Miaja G, Troncoso J, Romaní L (2009) Excess enthalpy, density, and heat capacity for binary systems of alkylimidazolium-based ionic liquids + water. J Chem Thermodyn 41 (2):161–166
23. García-Miaja G, Troncoso J, Romaní L (2009) Excess molar properties for binary systems of alkylimidazolium-based ionic liquids + nitromethane. Experimental results and ERAS-model calculations. J Chem Thermodyn 41(3):334–341
24. Shiflett MB, Harmer MA, Junk CP, Yokozeki A (2006) Solubility and diffusivity of difluoromethane in room-temperature ionic liquids. J Chem Eng Data 51(2):483–495
25. Navas A, Ortega J, Vreekamp R, Marrero E, Palomar J (2009) Experimental thermodynamic properties of 1-butyl-2-methylpyridinium tetrafluoroborate [$b_2$mpy][BF_4] with water and with alkan-1-ol and their interpretation with the COSMO-RS methodology. Ind Eng Chem Res 48 (5):2678–2690
26. Freire MG, Neves CMSS, Carvalho PJ, Gardas RL, Fernandes AM, Marrucho IM, Santos LMNBF, Coutinho JAP (2007) Mutual solubilities of water and hydrophobic ionic liquids. J Phys Chem B 111(45):13082–13089
27. Kato R, Krummen M, Gmehling J (2004) Measurement and correlation of vapor–liquid equilibria and excess enthalpies of binary systems containing ionic liquids and hydrocarbons. Fluid Phase Equilib 224(1):47–54
28. Anouti M, Caillon-Caravanier M, Dridi Y, Jacquemin J, Hardacre C, Lemordant D (2009) Liquid densities, heat capacities, refractive index and excess quantities for {protic ionic liquids + water} binary system. J Chem Thermodyn 41(6):799–808

Mixtures Containing Isoquinolinium Ionic Liquids

7

7.1 12006-03003: *N*-tetradecyl-isoquinolinium bis((perfluoroethane)sulfonyl)imide

Abbreviation:	$[C_{14}isoq][BETI]$, $[C_{14}Iq][C_2C_2N]$
Molecular formula:	$C_{27}H_{36}F_{10}N_2O_4S_2$
Molar mass	706.71
Structure:	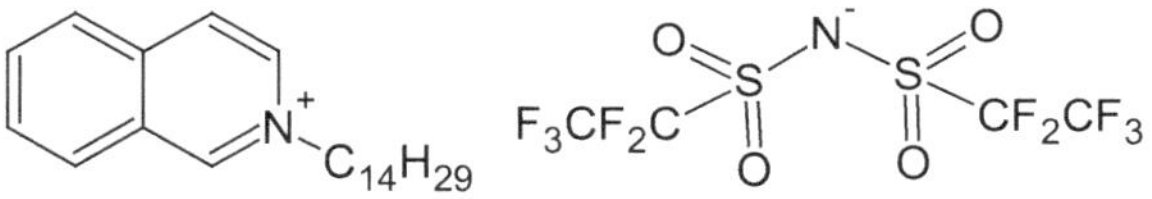

7.1.1 Ternary Mixtures

Mixture 1: H_2O (1) + $[C_{14}Iq][C_2C_2N]$ (2) + $[ThpA][C_2C_2N]$ (3)
(1) Volumetric Properties

Ref. [1]

x	$\rho/g{\cdot}cm^{-3}$	x	$\rho/g{\cdot}cm^{-3}$	x	$\rho/g{\cdot}cm^{-3}$	x	$\rho/g{\cdot}cm^{-3}$
298.15 K, x = the mole fraction of $C14Iq^+$							
0	1.153	0.03	1.157	0.2	1.175	0.8	1.263
0.005	1.153	0.04	1.159	0.3	1.189	0.9	1.278
0.01	1.155	0.05	1.160	0.5	1.219	0.95	1.290
0.02	1.156	0.1	1.165	0.7	1.248	1	1.297

S. Zhang et al., *Physicochemical Properties of Ionic Liquid Mixtures*,
DOI 10.1007/978-94-017-7573-1_7

Reference

1. Ishimatsu R, Kitazumi Y, Nishi N, Kakiuchi T (2009) Phase transition of a binary room-temperature ionic liquid composed of bis(pentafluoroethanesulfonyl)amide salts of tetraheptylammonium and N-tetradecylisoquinolinium and its surface properties at the ionic liquid|water interface. J Phys Chem B 113(27):9321–9325

8 Mixtures Containing Sulfonium Ionic Liquids

8.1 14002-03001: Triethylsulphonium bis[(trifluoromethyl) sulfonyl]imide

Abbreviation:	$[Et_3S][NTf_2]$
Molecular formula:	$C_8H_{15}F_6NO_4S_3$
Molar mass:	399.42
Structure:	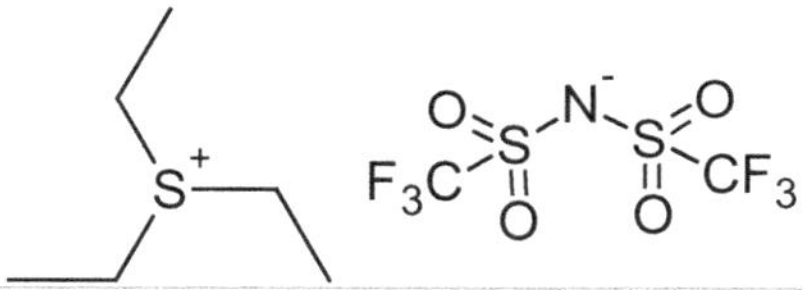

8.1.1 Binary Mixtures

Mixture 1: SCH=CHCH=CH (thiophene) (1) + $[Et_3S][NTf_2]$ (2)
(1) Phase Equilibrium

Ref. [1]

x_2	T^{LLE}/K	x_2	T^{LLE}/K
0.1999	251.8	0.1860	312.8
0.1959	273.7	0.1850	323.4
0.1921	284.8	0.1837	333.4
0.1910	291.4	0.1824	341.2
0.1901	297.2	0.1812	349.0
0.1880	305.7	0.1796	356.7
0.1863	313.8		

S. Zhang et al., *Physicochemical Properties of Ionic Liquid Mixtures*,
DOI 10.1007/978-94-017-7573-1_8

8.1.2 Ternary Mixtures

Mixture 1: C_6H_{14} (1) + SCH=CHCH=CH (thiophene) (2) + $[Et_3S][NTf_2]$ (3)
(1) Selectivity at Infinite Dilution and Capacity at Infinite Dilution

Ref. [2]

Solvent	S^{∞}_{12}	k_2^{∞}
T=298.15K	n-Hexane (1)/thiophene (2)	
$[Et_3S][NTf_2]$	30.9	1.08

Mixture 2: $CH_2(CH_2)_4CH_2$ (1) + SCH=CHCH=CH (thiophene) (2) + $[Et_3S][NTf_2]$ (3)
(1) Selectivity at Infinite Dilution and Capacity at Infinite Dilution

Ref. [2]

Solvent	S^{∞}_{12}	k_2^{∞}
T=298.15K	Cyclohexane (1)/thiophene (2)	
$[Et_3S][NTf_2]$	17.4	1.08

Mixture 3: C_6H_6 (1) + SCH=CHCH=CH (thiophene) (2) + $[Et_3S][NTf_2]$ (3)
(1) Selectivity at Infinite Dilution and Capacity at Infinite Dilution

Ref. [2]

Solvent	S^{∞}_{12}	k_2^{∞}
T=298.15K	Benzene (1)/thiophene (2)	
$[Et_3S][NTf_2]$	1.2	1.08

References

1. Domańska U, Królikowski M, Ślesińska K (2009) Phase equilibria study of the binary systems (ionic liquid + thiophene): desulphurization process. J Chem Thermodyn 41(11):1303–1311
2. Domańska U, Redhi GG, Marciniak A (2009) Activity coefficients at infinite dilution measurements for organic solutes and water in the ionic liquid 1-butyl-1-methylpyrrolidinium trifluoromethanesulfonate using GLC. Fluid Phase Equilib 278(1–2):97–102

9 Mixtures Containing Ammonium Ionic Liquids

9.1 15013-03001: Trimethyl-butylammonium bis((trifluoromethyl)sulfonyl)imide

Abbreviation:	$[N_{1114}][TFSI]$, $[Me_3BuN][NTf_2]$, $[N_{4111}][NTf_2]$
Molecular formula:	$C_9H_{18}F_6N_2O_4S_2$
Molar mass:	396.38
Structure:	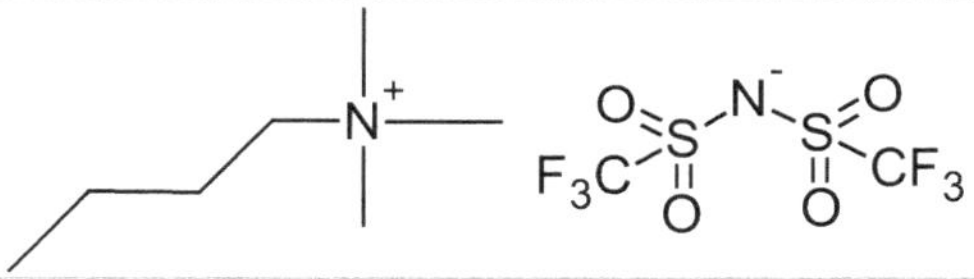

9.1.1 Binary Mixtures

Mixture 1: CH_3OH (1) + $[Me_3BuN][NTf_2]$ (2)
(1) Vapor Pressure and Activity Coefficient

Ref. [1]

	298.15 K		303.15 K		308.15 K		313.15 K	
x_1	p/Pa	γ_1	p/Pa	γ_1	p/Pa	γ_1	p/Pa	γ_1
0.0000	0	1.279	0	1.219	0	1.147	0	1.120
0.0850	1980	1.328	2411	1.274	2823	1.205	3419	1.171
0.1093	2502	1.339	3040	1.287	3658	1.219	4379	1.183
0.1308	2923	1.347	3608	1.297	4315	1.230	5171	1.192
0.2069	4799	1.367	5917	1.323	7202	1.261	8650	1.219
0.2166	5034	1.368	6230	1.325	7561	1.264	9102	1.222
0.2539	5963	1.371	7386	1.331	8945	1.273	10746	1.229
0.2973	7051	1.369	8729	1.333	10553	1.279	12748	1.234

(continued)

S. Zhang et al., *Physicochemical Properties of Ionic Liquid Mixtures*,
DOI 10.1007/978-94-017-7573-1_9

	298.15 K		303.15 K		308.15 K		313.15 K	
x_1	p/Pa	γ_1	p/Pa	γ_1	p/Pa	γ_1	p/Pa	γ_1
0.4400	10047	1.328	12605	1.304	15457	1.263	18703	1.222
0.5989	12676	1.232	16044	1.218	19893	1.194	24271	1.163
0.6450	13302	1.198	16902	1.186	20913	1.166	25631	1.140
0.7410	14489	1.125	18386	1.118	22840	1.106	28317	1.090
0.8624	15756	1.044	20028	1.041	25170	1.037	31522	1.031
0.9046	16118	1.023	20583	1.021	26007	1.019	32503	1.016
0.93664	16409	1.011	20997	1.010	26561	1.009	33321	1.008
0.96865	16750	1.003	21546	1.003	27400	1.002	34573	1.002
1.0000	16958	1.000	21880	1.000	27960	1.000	35450	1.000

(2) Activity Coefficient at Infinite Dilution and Partial Molar Excess Enthalpies at Infinite Dilution

Ref. [1]

Solute i	Temp range/K	γ_i^∞ (298 K)	$H_i^{E,\infty}$/kJ · mol^{-1}
Methanol	304.0–343.7	1.67	9.14

Ref. [1]

	γ_i^∞ (298 K)		$H_i^{E,\infty}$/kJ · mol^{-1} at 298 K	
Solute i	GC	Static	GC	Static
Methanol	1.67	1.28	8.7	7.1

Mixture 2: C_2H_5OH (1) + [Me_3BuN][NTf_2] (2)
(1) Vapor Pressure and Activity Coefficient

Ref. [1]

	298.15 K		303.15 K		308.15 K		313.15 K	
x_1	p/Pa	γ_1	p/Pa	γ_1	p/Pa	γ_1	p/Pa	γ_1
0.0000	0	2.378	0	2.199	0	1.983	0	1.815
0.0575	1097	2.225	1369	2.061	1625	1.864	1910	1.706
0.1524	2222	2.019	2678	1.874	3184	1.705	3872	1.562
0.1651	2420	1.995	2913	1.852	3461	1.687	4235	1.546
0.2540	3652	1.845	4555	1.720	5423	1.575	6280	1.447
0.3305	4665	1.737	5748	1.625	7000	1.497	7947	1.379
0.4141	5486	1.637	6839	1.539	8218	1.427	9774	1.320
0.5078	6226	1.538	7889	1.457	9677	1.361	11885	1.267
0.5919	6771	1.458	8652	1.391	10769	1.311	13488	1.229
0.6555	7219	1.399	9215	1.343	11596	1.275	14490	1.203
0.7041	7616	1.352	9736	1.307	12248	1.249	15212	1.185
0.7364	7731	1.320	9991	1.282	12607	1.230	15701	1.173
0.7969	7825	1.256	10227	1.230	13127	1.193	16569	1.149
0.8705	7837	1.165	10364	1.156	13501	1.138	17338	1.112
0.94893	7854	1.052	10430	1.053	13690	1.053	17790	1.051

(continued)

	298.15 K		303.15 K		308.15 K		313.15 K	
x_1	p/Pa	γ_1	p/Pa	γ_1	p/Pa	γ_1	p/Pa	γ_1
0.96847	7857	1.025	10439	1.027	13723	1.028	17856	1.029
1.0000	7876	1.000	10468	1.000	13768	1.000	17928	1.000

(2) Activity Coefficient at Infinite Dilution and Partial Molar Excess Enthalpies at Infinite Dilution

Ref. [1]

Solute i	Temp range/K	γ_i^∞ (298 K)	$H_i^{E,\infty}$/kJ · mol^{-1}
Ethanol	322.9–343.7	2.29	8.96

Ref. [1]

	γ_i^∞ (298 K)		$H_i^{E,\infty}$/kJ · mol^{-1} at 298 K	
Solute i	GC	Static	GC	Static
Ethanol	2.29	2.38	9.0	14.2

Mixture 3: 1-C_3H_7OH (1) + [Me_3BuN][NTf_2] (2)
(1) Vapor Pressure and Activity Coefficient

Ref. [1]

	298.15 K		303.15 K		308.15 K		313.15 K	
x_1	p/Pa	γ_1	p/Pa	γ_1	p/Pa	γ_1	p/Pa	γ_1
0.0000	0	3.888	0	3.625	0	3.248	0	3.112
0.0223	247	3.723	316	3.478	382	3.137	492	3.003
0.0336	332	3.644	422	3.408	514	3.083	644	2.951
0.0785	710	3.356	905	3.152	1125	2.883	1424	2.758
0.1456	1193	2.994	1535	2.830	1941	2.626	2504	2.512
0.1742	1355	2.860	1750	2.710	2223	2.529	2839	2.419
0.1841	1416	2.816	1824	2.670	2305	2.497	2951	2.389
0.2518	1772	2.543	2289	2.427	2917	2.295	3770	2.199
0.2937	1977	2.396	2567	2.295	3280	2.184	4200	2.096
0.3317	2134	2.274	2802	2.187	3615	2.091	4598	2.010
0.3857	2330	2.117	3082	2.046	4012	1.969	5122	1.898
0.4575	2519	1.931	3364	1.879	4410	1.823	5654	1.764
0.5025	2594	1.825	3478	1.783	4583	1.737	5955	1.687
0.5622	2665	1.694	3587	1.665	4766	1.630	6259	1.590
0.62471	2702	1.566	3658	1.549	4895	1.523	6471	1.494
0.65725	2715	1.502	3686	1.490	4940	1.469	6550	1.445
0.78237	2737	1.274	3737	1.275	5041	1.267	6723	1.260
0.87331	2747	1.125	3755	1.130	5076	1.127	6775	1.128
0.91542	2 750	1.066	3 755	1.070	5 076	1.069	6 782	1.071
1.0000	2786	1.000	3826	1.000	5199	1.000	6986	1.000

(2) Activity Coefficient at Infinite Dilution and Partial Molar Excess Enthalpies at Infinite Dilution

Ref. [1]

Solute i	Temp range/K	γ_i^∞ (298 K)	$H_i^{E,\infty}$/kJ · mol^{-1}
1-propanol	301.8–333.4	2.91	9.91

Ref. [1]

	γ_i^∞ (298 K)		$H_i^{E,\infty}$/kJ · mol^{-1} at 298 K	
Solute i	GC	Static	GC	Static
1-propanol	2.91	3.89	9.9	12.1

Mixture 4: C_5H_{12} (1) + [Me_3BuN][NTf_2] (2)
(1) Activity Coefficient at Infinite Dilution and Partial Molar Excess Enthalpies at Infinite Dilution

Ref. [1]

Solute i	Temp range/K	γ_i^∞ (298 K)	$H_i^{E,\infty}$/kJ · mol^{-1}
1-pentane	301.85–322.85	13.70	0.32

Mixture 5: C_6H_{14} (1) + [Me_3BuN][NTf_2] (2)
(1) Activity Coefficient at Infinite Dilution and Partial Molar Excess Enthalpies at Infinite Dilution

Ref. [1]

Solute i	Temp range/K	γ_i^∞ (298 K)	$H_i^{E,\infty}$/kJ · mol^{-1}
1-hexane	301.9–333.4	19.85	4.66

Mixture 6: C_7H_{16} (1) + [Me_3BuN][NTf_2] (2)
(1) Activity Coefficient at Infinite Dilution and Partial Molar Excess Enthalpies at Infinite Dilution

Ref. [1]

Solute i	Temp range/K	γ_i^∞ (298 K)	$H_i^{E,\infty}$/kJ · mol^{-1}
1-heptane	301.7–333.3	31.23	6.69

Mixture 7: C_8H_{18} (1) + [Me_3BuN][NTf_2] (2)
(1) Activity Coefficient at Infinite Dilution and Partial Molar Excess Enthalpies at Infinite Dilution

Ref. [1]

Solute i	Temp range/K	γ_i^∞ (298 K)	$H_i^{E,\infty}$/kJ · mol^{-1}
1-octane	301.7–333.3	46.98	7.54

Mixture 8: C_9H_{20} (1) + $[Me_3BuN][NTf_2]$ (2)
(1) Activity Coefficient at Infinite Dilution and Partial Molar Excess Enthalpies at Infinite Dilution

Ref. [1]

Solute i	Temp range/K	γ_i^∞ (298 K)	$H_i^{E,\infty}$/kJ · mol^{-1}
1-nonane	301.7–375.0	70.55	9.31

Mixture 9: $C_{10}H_{22}$ (1) + $[Me_3BuN][NTf_2]$ (2)
(1) Activity Coefficient at Infinite Dilution and Partial Molar Excess Enthalpies at Infinite Dilution

Ref. [1]

Solute i	Temp range/K	γ_i^∞ (298 K)	$H_i^{E,\infty}$/kJ · mol^{-1}
1-decane	302.2–354.1	104.86	10.43

Mixture 10: $C_{11}H_{24}$ (1) + $[Me_3BuN][NTf_2]$ (2)
(1) Activity Coefficient at Infinite Dilution and Partial Molar Excess Enthalpies at Infinite Dilution

Ref. [1]

Solute i	Temp range/K	γ_i^∞ (298 K)	$H_i^{E,\infty}$/kJ · mol^{-1}
1-undecane	322.8–364.6	154.67	11.58

Mixture 11: $C_{12}H_{26}$ (1) + $[Me_3BuN][NTf_2]$ (2)
(1) Activity Coefficient at Infinite Dilution and Partial Molar Excess Enthalpies at Infinite Dilution

Ref. [1]

Solute i	temp range/K	γ_i^∞ (298 K)	$H_i^{E,\infty}$/kJ · mol^{-1}
1-dodecane	333.3–354.1	248.09	15.44

Mixture 12: 1-C_5H_{10} (1) + [Me_3BuN][NTf_2] (2)
(1) Activity Coefficient at Infinite Dilution and Partial Molar Excess Enthalpies at Infinite Dilution

Ref. [1]

Solute *i*	Temp range/K	γ_i^∞ (298 K)	$H_i^{E,\infty}$/kJ · mol^{-1}
1-pentene	301.9–312.4	6.17	3.39

Mixture 13: 1-C_6H_{12} (1) + [Me_3BuN][NTf_2] (2)
(1) Activity Coefficient at Infinite Dilution and Partial Molar Excess Enthalpies at Infinite Dilution

Ref. [1]

Solute *i*	Temp range/K	γ_i^∞ (298 K)	$H_i^{E,\infty}$/kJ · mol^{-1}
1-hexene	305.9–364.55	10.60	3.91

Mixture 14: 1-C_7H_{14} (1) + [Me_3BuN][NTf_2] (2)
(1) Activity Coefficient at Infinite Dilution and Partial Molar Excess Enthalpies at Infinite Dilution

Ref. [1]

solute *i*	temp range/K	γ_i^∞ (298 K)	$H_i^{E,\infty}$/kJ · mol^{-1}
1-heptene	306.8–354.1	16.69	6.13

Mixture 15: 1-C_8H_{16} (1) + [Me_3BuN][NTf_2] (2)
(1) Activity Coefficient at Infinite Dilution and Partial Molar Excess Enthalpies at Infinite Dilution

Ref. [1]

Solute *i*	Temp range/K	γ_i^∞ (298 K)	$H_i^{E,\infty}$/kJ · mol^{-1}
1-octene	306.8–364.6	24.87	6.71

Mixture16: 1-C_9H_{18} (1) + [Me_3BuN][NTf_2] (2)
(1) Activity Coefficient at Infinite Dilution and Partial Molar Excess Enthalpies at Infinite Dilution

Ref. [1]

Solute *i*	Temp range/K	γ_i^∞ (298 K)	$H_i^{E,\infty}$/kJ · mol^{-1}
1-nonene	306.7–364.6	37.72	8.12

Mixture 17: 1-$C_{10}H_{20}$ (1) + [Me_3BuN][NTf_2] (2)
(1) Activity Coefficient at Infinite Dilution and Partial Molar Excess Enthalpies at Infinite Dilution

Ref. [1]

Solute *i*	Temp range/K	γ_i^∞ (298 K)	$H_i^{E,\infty}$/kJ · mol^{-1}
1-decene	306.2–354.1	55.53	8.51

Mixture 18: 1-$C_{11}H_{22}$ (1) + [Me_3BuN][NTf_2] (2)
(1) Activity Coefficient at Infinite Dilution and Partial Molar Excess Enthalpies at Infinite Dilution

Ref. [1]

Solute *i*	Temp range/K	γ_i^∞ (298 K)	$H_i^{E,\infty}$/kJ · mol^{-1}
1-undecene	322.6–375.0	87.88	9.55

Mixture 19: 1-$C_{12}H_{24}$ (1) + [Me_3BuN][NTf_2] (2)
(1) Activity Coefficient at Infinite Dilution and Partial Molar Excess Enthalpies at Infinite Dilution

Ref. [1]

Solute *i*	Temp range/K	γ_i^∞ (298 K)	$H_i^{E,\infty}$/kJ · mol^{-1}
1-dodecene	333.3–385.5	124.9	11.25

Mixture 20: $CH_2(CH_2)_4CH_2$ (1) + [Me_3BuN][NTf_2] (2)
(1) Activity Coefficient at Infinite Dilution and Partial Molar Excess Enthalpies at Infinite Dilution

Ref. [1]

Solute *i*	Temp range/K	γ_i^∞ (298 K)	$H_i^{E,\infty}$/kJ · mol^{-1}
Cyclohexane	301.0–333.3	5.76	5.33

Mixture 21: $CH = CHCH_2CH_2CH_2CH_2$ (1) + [Me_3BuN][NTf_2] (2)
(1) Activity Coefficient at Infinite Dilution and Partial Molar Excess Enthalpies at Infinite Dilution

Ref. [1]

Solute *i*	Temp range/K	γ_i^∞ (298 K)	$H_i^{E,\infty}$/kJ · mol^{-1}
Cyclohexene	312.1–354.2	7.65	5.90

Mixture 22: CH = CHCH = $CHCH_2CH_2$ (1) + $[Me_3BuN][NTf_2]$ (2)
(1) Activity Coefficient at Infinite Dilution and Partial Molar Excess Enthalpies at Infinite Dilution

Ref. [1]

Solute i	Temp range/K	γ_i^∞ (298 K)	$H_i^{E,\infty}$/kJ · mol^{-1}
1,3-cyclohexadiene	306.0–364.6	3.76	3.99

Mixture 23: C_6H_6 (1) + $[Me_3BuN][NTf_2]$ (2)
(1) Activity Coefficient at Infinite Dilution and Partial Molar Excess Enthalpies at Infinite Dilution

Ref. [1]

Solute i	Temp range/K	γ_i^∞ (298 K)	$H_i^{E,\infty}$/kJ · mol^{-1}
Benzene	322.8–364.5	1.37	1.74

Mixture 24: $C_6H_5CH_3$ (1) + $[Me_3BuN][NTf_2]$ (2)
(1) Activity Coefficient at Infinite Dilution and Partial Molar Excess Enthalpies at Infinite Dilution

Ref. [1]

Solute i	Temp range/K	γ_i^∞ (298 K)	$H_i^{E,\infty}$/kJ · mol^{-1}
Toluene	322.8–364.5	1.78	0.74

Mixture 25: $C_6H_5C_2H_5$ (1) + $[Me_3BuN][NTf_2]$ (2)
(1) Activity Coefficient at Infinite Dilution and Partial Molar Excess Enthalpies at Infinite Dilution

Ref. [1]

Solute i	Temp range/K	γ_i^∞ (298 K)	$H_i^{E,\infty}$/kJ · mol^{-1}
Ethyl benzene	322.9–354.3	3.01	2.33

Mixture 26: $C_6H_5C_3H_7$ (1) + $[Me_3BuN][NTf_2]$ (2)
(1) Activity Coefficient at Infinite Dilution and Partial Molar Excess Enthalpies at Infinite Dilution

Ref. [1]

Solute i	Temp range/K	γ_i^∞ (298 K)	$H_i^{E,\infty}$/kJ · mol^{-1}
Propyl benzene	333.3–364.5	4.33	2.15

Mixture 27: $C_6H_5C_4H_9$ (1) + $[Me_3BuN][NTf_2]$ (2)
(1) Activity Coefficient at Infinite Dilution and Partial Molar Excess Enthalpies at Infinite Dilution

Ref. [1]

Solute i	Temp range/K	γ_i^∞ (298 K)	$H_i^{E,\infty}$/kJ · mol^{-1}
Butyl benzene	333.3–375.0	6.5	2.79

Mixture 28: $C_6H_5C_5H_{11}$ (1) + $[Me_3BuN][NTf_2]$ (2)
(1) Activity Coefficient at Infinite Dilution and Partial Molar Excess Enthalpies at Infinite Dilution

Ref. [1]

Solute i	Temp range/K	γ_i^∞ (298 K)	$H_i^{E,\infty}$/kJ · mol^{-1}
Pentyl benzene	354.1–385.5	10.39	4.18

Mixture 29: 1-C_4H_9OH (1) + $[Me_3BuN][NTf_2]$ (2)
(1) Activity Coefficient at Infinite Dilution and Partial Molar Excess Enthalpies at Infinite Dilution

Ref. [1]

Solute i	Temp range/K	γ_i^∞ (298 K)	$H_i^{E,\infty}$/kJ · mol^{-1}
1-butanol	302.6–343.7	3.85	10.16

Mixture 30: 1-$C_5H_{11}OH$ (1) + $[Me_3BuN][NTf_2]$ (2)
(1) Activity Coefficient at Infinite Dilution and Partial Molar Excess Enthalpies at Infinite Dilution

Ref. [1]

Solute i	Temp range/K	γ_i^∞ (298 K)	$H_i^{E,\infty}$/kJ · mol^{-1}
1-pentanol	322.9–364.5	5.32	10.60

Mixture 31: 1-$C_6H_{13}OH$ (1) + $[Me_3BuN][NTf_2]$ (2)
(1) Activity Coefficient at Infinite Dilution and Partial Molar Excess Enthalpies at Infinite Dilution

Ref. [1]

Solute i	Temp range/K	γ_i^∞ (298 K)	$H_i^{E,\infty}$/kJ · mol^{-1}
1-hexanol	343.7–385.6	7.1	10.69

Mixture 32: 2-CH_3CHCH_3OH (1) + $[Me_3BuN][NTf_2]$ (2)
(1) Activity Coefficient at Infinite Dilution and Partial Molar Excess Enthalpies at Infinite Dilution

Ref. [1]

Solute *i*	Temp range/K	γ_i^{∞} (298 K)	$H_i^{E,\infty}$/kJ · mol^{-1}
2-propanol	302.6–354.1	2.54	9.70

Mixture 33: $(CH_3)_2CHCH_2OH$ (1) + $[Me_3BuN][NTf_2]$ (2)
(1) Activity Coefficient at Infinite Dilution and Partial Molar Excess Enthalpies at Infinite Dilution

Ref. [1]

Solute *i*	Temp range/K	γ_i^{∞} (298 K)	$H_i^{E,\infty}$/kJ · mol^{-1}
Iso-butanol	302.6–354.1	4.23	10.12

Mixture 34: $CH_3CH_2CH(OH)CH_3$ (1) + $[Me_3BuN][NTf_2]$ (2)
(1) Activity Coefficient at Infinite Dilution and Partial Molar Excess Enthalpies at Infinite Dilution

Ref. [1]

Solute *i*	Temp range/K	γ_i^{∞} (298 K)	$H_i^{E,\infty}$/kJ · mol^{-1}
sec-butanol	333.2–375.0	3.16	9.9

Mixture 35: $(CH_3)_3COH$ (1) + $[Me_3BuN][NTf_2]$ (2)
(1) Activity Coefficient at Infinite Dilution and Partial Molar Excess Enthalpies at Infinite Dilution

Ref. [1]

Solute *i*	Temp range/K	γ_i^{∞} (298 K)	$H_i^{E,\infty}$/kJ · mol^{-1}
tert-butyl alcohol	302.6–354.1	2.36	8.85

Mixture 36: $CH_3CH_2C(CH_3)_2OH$ (1) + $[Me_3BuN][NTf_2]$ (2)
(1) Activity Coefficient at Infinite Dilution and Partial Molar Excess Enthalpies at Infinite Dilution

Ref. [1]

Solute *i*	Temp range/K	γ_i^{∞} (298 K)	$H_i^{E,\infty}$/kJ · mol^{-1}
tert-pentanol	302.6–354.1	2.64	7.19

Mixture 37: $CH_2(CH_2)_4CHOH$ (1) + $[Me_3BuN][NTf_2]$ (2)
(1) Activity Coefficient at Infinite Dilution and Partial Molar Excess Enthalpies at Infinite Dilution

Ref. [1]

Solute *i*	Temp range/K	γ_i^{∞} (298 K)	$H_i^{E,\infty}$/kJ · mol^{-1}
Cyclohexanol	343.7–385.6	7.74	15.10

Mixture 38: CH_3CH_2CHO (1) + $[Me_3BuN][NTf_2]$ (2)
(1) Activity Coefficient at Infinite Dilution and Partial Molar Excess Enthalpies at Infinite Dilution

Ref. [1]

Solute *i*	Temp range/K	γ_i^{∞} (298 K)	$H_i^{E,\infty}$/kJ · mol^{-1}
Propanal	301.7–343.6	0.59	2.36

Mixture 39: $CH_3(CH_2)_2CHO$ (1) + $[Me_3BuN][NTf_2]$ (2)
(1) Activity Coefficient at Infinite Dilution and Partial Molar Excess Enthalpies at Infinite Dilution

Ref. [1]

Solute *i*	Temp range/K	γ_i^{∞} (298 K)	$H_i^{E,\infty}$/kJ · mol^{-1}
Butanal	301.8–354.2	0.81	1.1

Mixture 40: $CH_3(CH_2)_3CHO$ (1) + $[Me_3BuN][NTf_2]$ (2)
(1) Activity Coefficient at Infinite Dilution and Partial Molar Excess Enthalpies at Infinite Dilution

Ref. [1]

Solute *i*	Temp range/K	γ_i^{∞} (298 K)	$H_i^{E,\infty}$/kJ · mol^{-1}
Pentanal	301.8–354.2	1.74	2.67

Mixture 41: $CH_3(CH_2)_4CHO$ (1) + $[Me_3BuN][NTf_2]$ (2)
(1) Activity Coefficient at Infinite Dilution and Partial Molar Excess Enthalpies at Infinite Dilution

Ref. [1]

Solute *i*	Temp range/K	γ_i^{∞} (298 K)	$H_i^{E,\infty}$/kJ · mol^{-1}
Hexanal	322.8–364.7	1.99	1.08

Mixture 42: $CH_3(CH_2)_5CHO$ (1) + $[Me_3BuN][NTf_2]$ (2)
(1) Activity Coefficient at Infinite Dilution and Partial Molar Excess Enthalpies at Infinite Dilution

Ref. [1]

Solute *i*	Temp range/K	γ_i^{∞} (298 K)	$H_i^{E,\infty}$/kJ · mol^{-1}
Heptanal	354.2–396.0	3.17	8.39

Mixture 43: $CH_3(CH_2)_6CHO$ (1) + $[Me_3BuN][NTf_2]$ (2)
(1) Activity Coefficient at Infinite Dilution and Partial Molar Excess Enthalpies at Infinite Dilution

Ref. [1]

Solute *i*	Temp range/K	γ_i^{∞} (298 K)	$H_i^{E,\infty}$/kJ · mol^{-1}
Octanal	364.6–396.0	4.02	1.75

Mixture 44: $CH_3OOCCH_2CH_3$ (1) + $[Me_3BuN][NTf_2]$ (2)
(1) Activity Coefficient at Infinite Dilution and Partial Molar Excess Enthalpies at Infinite Dilution

Ref. [1]

Solute *i*	Temp range/K	γ_i^{∞} (298 K)	$H_i^{E,\infty}$/kJ · mol^{-1}
Methyl propanoate	311.7–343.8	0.96	1.09

Mixture 45: $CH_3OOCCH_2CH_2CH_3$ (1) + $[Me_3BuN][NTf_2]$ (2)
(1) Activity Coefficient at Infinite Dilution and Partial Molar Excess Enthalpies at Infinite Dilution

Ref. [1]

Solute *i*	Temp range/K	γ_i^{∞} (298 K)	$H_i^{E,\infty}$/kJ · mol^{-1}
Methyl butanoate	311.7–364.5	1.35	1.56

Mixture 46: $CH_3OOCCH_2CH_2CH_2CH_3$ (1) + $[Me_3BuN][NTf_2]$ (2)
(1) Activity Coefficient at Infinite Dilution and Partial Molar Excess Enthalpies at Infinite Dilution

Ref. [1]

Solute *i*	Temp range/K	γ_i^{∞} (298 K)	$H_i^{E,\infty}$/kJ · mol^{-1}
Methyl pentanoate	322.8–364.6	1.92	4.19

Mixture 47: $CH_3OOCCH_2CH_2CH_2CH_2CH_3$ (1) + $[Me_3BuN][NTf_2]$ (2)
(1) Activity Coefficient at Infinite Dilution and Partial Molar Excess Enthalpies at Infinite Dilution

Ref. [1]

Solute i	Temp range/K	γ_i^∞ (298 K)	$H_i^{E,\infty}$/kJ · mol^{-1}
Methyl hexanoate	343.7–375.0	3.24	5.32

Mixture 48: $CH_3OOCCH_2CH_2CH_2CH_2CH_2CH_3$ (1) + $[Me_3BuN][NTf_2]$ (2)
(1) Activity Coefficient at Infinite Dilution and Partial Molar Excess Enthalpies at Infinite Dilution

Ref. [1]

Solute i	Temp range/K	γ_i^∞ (298 K)	$H_i^{E,\infty}$/kJ · mol^{-1}
Methyl heptanoate	343.6–385.5	4.68	7.04

Mixture 49: CH_3COCH_3 (1) + $[Me_3BuN][NTf_2]$ (2)
(1) Activity Coefficient at Infinite Dilution and Partial Molar Excess Enthalpies at Infinite Dilution

Ref. [1]

Solute i	Temp range/K	γ_i^∞ (298 K)	$H_i^{E,\infty}$/kJ · mol^{-1}
Acetone	301.9–343.8	0.42	1.10

Mixture 50: CH_3CN (1) + $[Me_3BuN][NTf_2]$ (2)
(1) Activity Coefficient at Infinite Dilution and Partial Molar Excess Enthalpies at Infinite Dilution

Ref. [1]

Solute i	Temp range/K	γ_i^∞ (298 K)	$H_i^{E,\infty}$/kJ · mol^{-1}
Acetonitrile	301.8–343.8	0.52	2.49

Mixture 51: 1,4-$CH_2CH_2OCH_2CH_2O$ (1) + $[Me_3BuN][NTf_2]$ (2)
(1) Activity Coefficient at Infinite Dilution and Partial Molar Excess Enthalpies at Infinite Dilution

Ref. [1]

Solute i	Temp range/K	γ_i^∞ (298 K)	$H_i^{E,\infty}$/kJ · mol^{-1}
1,4-dioxane	343.8–354.3	0.74	0.11

Mixture 52: $CHCl_3$ (1) + [Me_3BuN][NTf_2] (2)
(1) Activity Coefficient at Infinite Dilution and Partial Molar Excess Enthalpies at Infinite Dilution

Ref. [1]

Solute *i*	Temp range/K	γ_i^∞ (298 K)	$H_i^{E,\infty}$/kJ · mol^{-1}
Trichloromethane	301.8–333.3	1.02	−0.56

Mixture 53: $CH_3NCH_2CH_2CH_2C=O$ (1) + [Me_3BuN][NTf_2] (2)
(1) Activity Coefficient at Infinite Dilution and Partial Molar Excess Enthalpies at Infinite Dilution

Ref. [1]

Solute *i*	Temp range/K	γ_i^∞ (298 K)	$H_i^{E,\infty}$/kJ · mol^{-1}
N-methyl-pyrrolidone	312.35–375.15	175.49	24.86

Mixture 54: $HCON(CH_3)_2$ (1) + [Me_3BuN][NTf_2] (2)
(1) Activity Coefficient at Infinite Dilution and Partial Molar Excess Enthalpies at Infinite Dilution

Ref. [1]

Solute *i*	Temp range/K	γ_i^∞ (298 K)	$H_i^{E,\infty}$/kJ · mol^{-1}
N,N-dimethyl-formamide	375.15–396.15	0.25	−1.73

Mixture 55: H_2O (1) + [N_{4111}][NTf_2] (2)
(1) Volumetric Properties

Ref. [2]

x_1	ρ/g · cm^{-3}										
	293.15 K	298.15 K	303.15 K	308.15 K	313.15 K	318.15 K	323.15 K	328.15 K	333.15 K	338.15 K	343.15 K
0	1.3995	1.3950	1.3905	1.3861	1.3816	1.3771	1.3726	1.3682	1.3637	1.3592	1.3548
0.0387	1.3984	1.3939	1.3894	1.3849	1.3805	1.3760	1.3715	1.3670	1.3624	1.3579	1.3534
0.0765	1.3970	1.3925	1.3880	1.3835	1.3790	1.3746	1.3700	1.3655	1.3609	1.3563	1.3518
0.1132	1.3953	1.3908	1.3863	1.3819	1.3774	1.3729	1.3684	1.3639	1.3593	1.3547	1.3502
0.1400	1.3941	1.3895	1.3850	1.3806	1.3761	1.3716	1.3671	1.3626	1.3580	1.3533	1.3487
0.1890	1.3924	1.3878	1.3832	1.3787	1.3741	1.3696	1.3651	1.3605	1.3558	1.3511	1.3464
0.2108	1.3918	1.3872	1.3825	1.3778	1.3732	1.3687	1.3641	1.3595	1.3548	1.3500	1.3451
0.2322	1.3912	1.3866	1.3818	1.3771	1.3724	1.3678	1.3632	1.3585	1.3537	1.3487	1.3438
	V^E_m/cm^3 · mol^{-1}										
0.0387	0.014	0.016	0.018	0.040	0.022	0.024	0.025	0.047	0.069	0.071	0.093
0.0765	0.074	0.078	0.083	0.106	0.110	0.094	0.117	0.140	0.163	0.186	0.210
0.1132	0.175	0.182	0.189	0.195	0.201	0.207	0.212	0.236	0.260	0.285	0.309
0.1400	0.204	0.231	0.240	0.247	0.255	0.262	0.269	0.294	0.319	0.363	0.406
0.1890	0.206	0.234	0.262	0.289	0.316	0.325	0.335	0.378	0.422	0.465	0.527
0.2108	0.161	0.189	0.234	0.295	0.323	0.333	0.360	0.404	0.448	0.508	0.605
0.2322	0.114	0.143	0.204	0.264	0.309	0.336	0.364	0.424	0.484	0.578	0.674

(2) Solubility

Ref. [2]

T/K	x_1	T/K	x_1	T/K	x_1	T/K	x_1
289.15	0.2043	323.45	0.2907	358.34	0.3876	332.15	0.9997
298.15	0.2322	339.75	0.3192	362.62	0.4108	318.15	0.9998
313.78	0.2564	345.82	0.3364	353.15	0.9996	298.15	0.9999

9.2 15032-01001: Tetrabutylammonium chloride

Abbreviation:	$[N_{4444}]Cl$
Molecular formula:	$C_{16}H_{36}NCl$
Molar mass:	277.92
Structure:	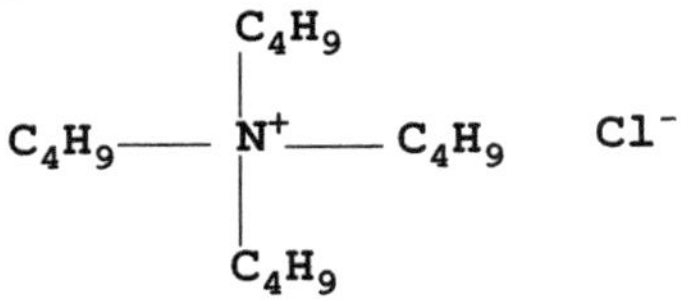

9.2.1 Binary Mixtures

Mixture 1: o-$C_6H_4(CH_3)_2$ (1) + $[N_{4444}]Cl$ (2)
(1) Solubility

Ref. [3]

T/K	x_2	T/K	x_2
324.65	0.1076	330.65	0.0817
334.15	0.0876	335.25	0.0701
339.35	0.0753	337.65	0.0611
342.85	0.0641	339.40	0.0529
345.15	0.0526	340.25	0.0439
346.35	0.0441	340.35	0.0419
346.65	0.0356	340.40	0.0376
346.45	0.0280	340.40	0.0366
345.85	0.0241	340.38	0.0341
344.55	0.0198	340.05	0.0280
335.95	0.0118	339.30	0.0237
		336.55	0.0176
		333.35	0.0144

Mixture 2: $C_6H_5C_2H_5$ (1) + $[N_{4444}]Cl$ (2)
(1) Solubility

Ref. [3]

T/K	x_2	T/K	x_2
312.35	0.1570	341.15	0.0446
328.85	0.1134	340.65	0.0404
336.75	0.0866	337.55	0.0319
340.55	0.0677	332.25	0.0236
341.85	0.0539	319.15	0.0155

9.3 15032-01002: Tetrabutylammonium bromide

Abbreviation:	[TBA]Br, [N4444]Br, Bu_4NBr
Molecular formula:	$C_{16}H_{36}NBr$
Molar mass:	322.37
Structure:	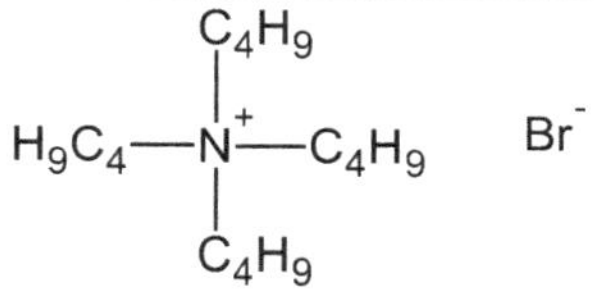

9.3.1 Binary Mixtures

Mixture1: IM (Imidazole) (1) + Bu_4NBr (2)
(1) Melting Point, Transport Properties and Electrical Conductivity

Ref. [4]

Molar ratio (Bu_4NBr: IM)	T_m(K)	η/(cp)											
		293.15 K	303.15.K	313.15 K	323.15.K	333.15.K	343.15.K	353.15.K	363.15.K	373.15.K	383.15.K	393.15.K	403.15.K
3: 7	294.15	810.0	314.0	144.8	75.0	44.0	27.8	18.2	12.5	10.4	8.6	7.2	6.2

Ref. [4]

Shear rate (s^{-1})	η/(cp)	Shear rate (s^{-1})	η/(cp)
296.15 K Molar ratio (Bu_4NBr:IM = 3:7)		343.15 K Molar ratio (Bu_4NBr:IM = 3:7)	
176	540	836	27.8
195	500	1006	27.3
251	480	1092	25.7
282	470	1403	25.5
344	460	2028	25.5
391	452		

Ref. [4]

Molar ratio (Bu_4NBr:IM)	conductivity ($S \cdot cm^{-1}$)											
	293.15 K	303.15 K	313.15 K	323.15 K	333.15 K	343.15 K	353.15 K	363.15 K	373.15 K	383.15 K	393.15 K	403.15 K
3: 7	0.24×10^{-3}	0.52×10^{-3}	1.0×10^{-3}	1.8×10^{-3}	3.0×10^{-3}	4.6×10^{-3}	6.5×10^{-3}	9.0×10^{-3}	1.2×10^{-2}	1.4×10^{-2}	1.7×10^{-2}	1.9×10^{-2}

9.3.2 Ternary Mixtures

Mixture 1: IM (1) + Bu_4NBr (2) + Solvent (3)
(1) Solubility in common solvents

Ref. [4]

Molar ratio (Bu_4NBr:IM)	Solubility (wt.%) (cm, complete miscible, im, immiscible)										
	Water	Acetonitrile	Ethanol	Dichloro-methane	Chloroform	Acetone	Ethyl acetate	Toluene	Benzene	Ethyl ether	Hexane
3:7	cm	cm	cm	cm	cm	cm	cm	im	im	im	im

Mixture 2: IM (1) + Bu_4NBr (2) + Compounds (3)
(1) Solubility of some compounds in Binary Mixtures

Ref. [4]

Molar ratio (Bu_4NBr:IM)	Solubility (wt.%) 343.15 K							
	CuO	Fe_2O_3	$MgCl_2$	LiCl	$ZnCl_2$	Glucose	Citric acid	Benzoic acid
3:7	0.07	0.01	1.62	7.13	18.42	18.83	25.96	39.20

9.4 15032-05001: Tetrabutylammonium hexafluorophosphate

Abbreviation:	[TBA][PF_6]
Molecular formula:	$C_{16}H_{36}NPF_6$
Molar mass:	387.43
Structure:	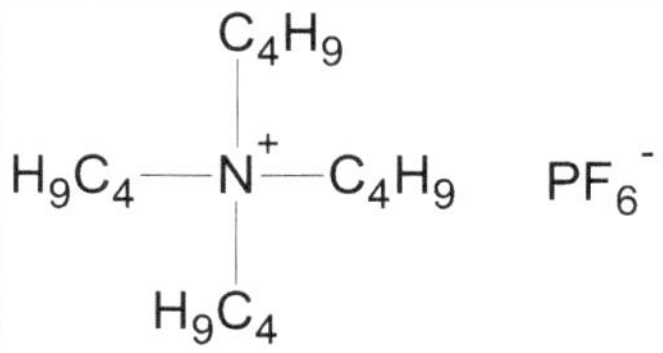

9.4.1 Binary Mixtures

Mixture 1: CH_3OH (1) + [TBA][PF_6] (2)

(1) Volumetric Properties, Speed of Sound and Compressibility Properties

Ref. [5]

m_2/ (mol · kg^{-1})	10^3 ρ/ (kg · m^{-3})	u/ (m · s^{-1})	10^6 V_φ/ (m^3 · mol^{-1})	10^{12} k_s/ Pa^{-1}	10^{14} · k_φ/ (m^3 · mol^{-1} · Pa^{-1})
T = 298.15 K					
0.0101	0.787566	1103.62	320.74	1042.49	−9.48
0.0131	0.787869	1103.85	322.48	1041.66	−7.60
0.0172	0.788270	1104.25	325.04	1040.38	−7.05
0.0227	0.788822	1104.73	325.92	1038.74	−6.25
0.0311	0.789663	1105.48	326.61	1036.23	−5.70
0.0386	0.790410	1106.14	327.01	1034.02	−5.34
0.0460	0.791134	1106.79	327.64	1031.86	−5.02
0.0523	0.791748	1107.31	328.04	1030.09	−4.66
0.0694	0.793383	1108.63	329.27	1025.52	−3.58
0.0963	0.795939	1110.61	330.10	1018.58	−2.45
0.1253	0.798621	1112.69	331.05	1011.37	−1.56
0.1587	0.801683	1115.06	331.49	1003.23	−0.93
0.1850	0.804020	1116.77	332.07	997.25	−0.32
0.2172	0.806921	1118.87	331.95	989.94	0.10
0.2431	0.809166	1120.43	332.27	984.45	0.57

Mixture 2: CH_3CN (1) + [TBA][PF_6] (2)
(1) Volumetric Properties, Speed of Sound and Compressibility Properties

Ref. [5]

m_2/ (mol · kg^{-1})	10^3 ρ/ (kg · m^{-3})	u/ (m · s^{-1})	10^6 V_φ/ (m^3 · mol^{-1})	10^{12} k_s/ Pa^{-1}	10^{14} · k_φ/ (m^3 · mol^{-1} · Pa^{-1})
T = 298.15 K					
0.0102	0.777663	1279.09	332.72	785.97	10.03
0.0129	0.777923	1279.14	333.18	785.65	10.12
0.0152	0.778152	1279.18	333.3	785.37	10.20
0.0221	0.778829	1279.29	333.89	784.55	10.46
0.0324	0.779824	1279.43	334.58	783.38	10.81
0.0379	0.780355	1279.50	334.82	782.76	10.94
0.0689	0.783317	1279.96	335.44	779.24	11.17
0.1039	0.786566	1280.45	336.25	775.42	11.42
0.1640	0.792030	1281.24	336.60	769.13	11.66
0.2112	0.796196	1281.83	336.89	764.40	11.82
0.2904	0.802895	1282.87	337.55	756.79	12.04
0.3790	0.810144	1284.09	337.74	748.59	12.15
0.5002	0.819480	1285.83	338.22	738.06	12.30

Mixture 3: CH_3SOCH_3 (1) + [TBA][PF_6] (2)
(1) Volumetric Properties, Speed of Sound and Compressibility Properties

Ref. [5]

m_2/ (mol · kg^{-1})	10^3 ρ/ (kg · m^{-3})	u/ (m · s^{-1})	10^6 V_φ/ (m^3 · mol^{-1})	10^{12} · k_s/ Pa^{-1}	10^{14} k_φ/ (m^3 · mol^{-1} · Pa^{-1})
T = 298.15 K					
0.0104	1.095371	1485.01	345.69	413.98	14.52
0.0185	1.095446	1484.85	345.80	414.04	14.73
0.0268	1.095520	1484.66	345.90	414.12	14.88
0.0313	1.095560	1484.54	345.95	414.17	14.95
0.0437	1.095667	1484.24	346.05	414.30	15.05
0.0529	1.095744	1483.99	346.12	414.41	15.12
0.0606	1.095808	1483.77	346.18	414.51	15.18
0.0698	1.095878	1483.55	346.29	414.60	15.20
0.0879	1.096013	1483.09	346.46	414.81	15.26
0.1105	1.096162	1482.61	346.73	415.02	15.27

Mixture 4: $CH_3CON(CH_3)_2$ (1) + [TBA][PF_6] (2)
(1) Volumetric Properties, Speed of Sound and Compressibility Properties

Ref. [5]

m_2/ (mol · kg^{-1})	10^3 ρ/ (kg · m^{-3})	u/ (m · s^{-1})	10^6 V_φ/ (m^3 · mol^{-1})	10^{12} k_s/ Pa^{-1}	10^{14} k_φ/ (m^3 · mol^{-1} · Pa^{-1})
T = 298.15 K					
0.0037	0.936522	1456.52	335.89	503.33	12.99
0.0069	0.936738	1456.51	336.11	503.22	13.12
0.0136	0.937186	1456.51	336.6	502.98	13.12
0.0194	0.937570	1456.49	336.87	502.78	13.21
0.0327	0.938442	1456.47	337.29	502.33	13.25
0.0397	0.938888	1456.45	337.63	502.11	13.31
0.0533	0.939757	1456.40	337.93	501.68	13.38
0.0668	0.940600	1456.36	338.35	501.25	13.43
0.078	0.941282	1456.35	338.68	500.9	13.45
0.097	0.942447	1456.28	338.94	500.33	13.51
0.115	0.943518	1456.26	339.3	499.77	13.53

Mixture 5: $OCH_2CH_2CH_2CH_2$ (1) + [TBA][PF_6] (2)
(1) Volumetric Properties, Speed of Sound and Compressibility Properties

Ref. [5]

m_2/ (mol · kg^{-1})	10^3 · ρ/ (kg · m^{-3})	u/ (m · s^{-1})	10^6 V_φ/ (m^3 · mol^{-1})	10^{12} k_s/ Pa^{-1}	10^{14} k_φ/ (m^3 · mol^{-1} · Pa^{-1})
T = 298.15 K					
0.0059	0.882641	1278.76	294.38	692.85	1.41
0.0080	0.882854	1278.81	298.05	692.63	3.50

(continued)

m_2/ (mol · kg^{-1})	$10^3 \cdot \rho$/ (kg · m^{-3})	u/ (m · s^{-1})	$10^6\ V_\varphi$/ (m^3 · mol^{-1})	$10^{12}\ k_s$/ Pa^{-1}	$10^{14}\ k_\varphi$/ (m^3 · mol^{-1} · Pa^{-1})
0.0116	0.883196	1278.92	303.77	692.24	5.42
0.0124	0.883265	1278.98	305.3	692.12	5.44
0.0145	0.883458	1279.05	307.46	691.89	6.08
0.0172	0.883617	1279.37	316.12	691.42	5.95
0.0223	0.884059	1279.28	318.50	691.17	8.48
0.0263	0.884409	1279.42	319.52	690.75	8.77
0.0396	0.885580	1279.65	321.00	689.59	9.97
0.0445	0.886001	1279.88	321.56	689.01	9.87
0.0619	0.887450	1280.05	323.61	687.71	11.02
0.0806	0.888991	1280.52	324.82	686.01	11.27
0.1172	0.891940	1281.32	326.31	682.89	11.69
0.1495	0.894415	1281.98	327.70	680.3	12.03
0.1822	0.896960	1282.53	327.99	677.78	12.24
0.1919	0.897693	1282.77	328.15	676.98	12.25
0.2248	0.900122	1283.37	328.75	674.52	12.43
0.2757	0.903903	1284.70	328.87	670.31	12.37
0.3755	0.910504	1286.72	330.91	663.36	12.75

9.5 15032-06042: Tetra-*n*-butylammonium pricrate

Abbreviation:	$N_{4444}Pic$
Molecular formula:	$C_{22}H_{38}N_4O_7$
Molar mass:	470.34
Structure:	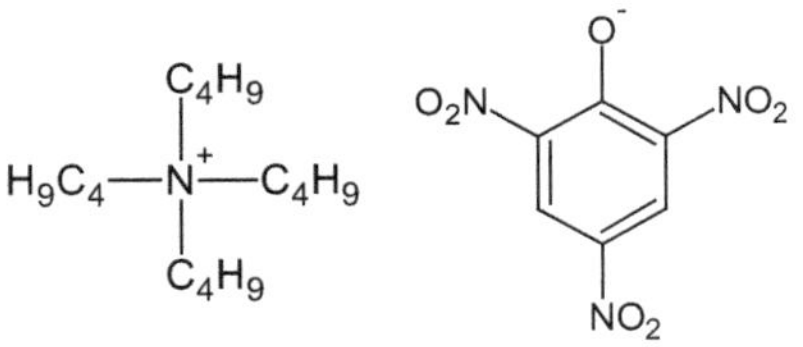

9.5.1 Binary Mixtures

Mixture 1: 2-CH_3CHCH_3OH (1) + $N_{4444}Pic$ (2)
(1) Dielectric Constant and Critical Temperature

Ref. [6]

$\epsilon(T_c)$	T_c/K	w_1
16.8	315.87	0.3500

(2) Refractive Index

Ref. [6]

T/K	n_D	T/K	n_D	T/K	n_D
327.90	1.41583	324.14	1.41748	319.13	1.41959
327.90	1.41581	322.89	1.41799	317.88	1.42011
326.65	1.41638	321.64	1.41852	316.62	1.42067
325.40	1.41694	320.38	1.41907	316.12	1.42085

Ref. [6]

T/K	n_u	n_l	T/K	n_u	n_l	T/K	n_u	n_l
315.85	1.41421	1.42904	315.53	1.40599	1.43828	314.12	1.39657	1.45130
315.83	1.41322	1.43029	315.46	1.40504	1.43943	313.87	1.39579	1.45289
315.81	1.41222	1.43136	315.42	1.40464	1.44005	312.93	1.39304	1.45759
315.79	1.41148	1.43213	315.37	1.40409	1.44076	312.86	1.39295	1.45796
315.77	1.41073	1.43290	315.27	1.40309	1.44181	311.86	1.39102	1.46221
315.75	1.41023	1.43348	315.22	1.40269	1.44239	310.86	1.38930	1.46583
315.72	1.40923	1.43439	315.17	1.40224	1.44310	309.86	1.38819	1.46911
315.68	1.40843	1.43540	315.02	1.40114	1.44457	308.85	1.38708	1.47223
315.64	1.40753	1.43636	314.87	1.40020	1.44600	306.85	1.38578	1.47761
315.61	1.40718	1.43675	314.62	1.39875	1.44789	304.84	1.38478	1.48248
315.55	1.40618	1.43804	314.37	1.39756	1.44978	302.84	1.38408	1.48694

n_u and n_l are the refractive indices of the upper and lower (salt rich) phase below the critical temperature.

Mixture 2: 2-$CH_3CH(OH)CH_2CH_3$ (1) + N_{4444}Pic (2)
(1) Dielectric Constant and Critical Temperature

Ref. [6]

ε (T_c)	T_c/K	w_1
13.5	318.32	0.316

Mixture 3: 2-$CH_3CH(OH)(CH_2)_2CH_3$ (1) + N_{4444}Pic (2)
(1) Dielectric Constant and Critical Temperature

Ref. [6]

ε (T_c)	T_c/K	w_1
10.0	328.38	0.292

Mixture 4: 1-$C_8H_{17}OH$ (1) + N_{4444}Pic (2)
(1) Dielectric Constant and Critical Temperature

Ref. [6]

ε (T_c)	T_c/K	w_1
9.5	298.55	0.336

Mixture 5: 1-$C_9H_{19}OH$ (1) + N_{4444}Pic (2)
(1) Dielectric Constant and Critical Temperature

Ref. [6]

ϵ (T_c)	T_c/K	w_1
7.9	308.68	0.325

Mixture 6: 1-$C_{10}H_{21}OH$ (1) + N_{4444}Pic (2)
(1) Dielectric Constant and Critical Temperature

Ref. [6]

ϵ (T_c)	T_c/K	w_1
6.4	318.29	0.3152

(2) Refractive Index

Ref. [6]

T/K	n_D	T/K	n_D	T/K	n_D
327.90	1.45738	323.89	1.45885	318.88	1.46068
327.90	1.45733	322.89	1.45920	318.88	1.46068
326.90	1.45769	321.89	1.45960	318.48	1.46084
325.90	1.45810	320.89	1.45995	318.38	1.46090
324.90	1.45841	319.88	1.46029		

Ref. [6]

T/K	n_u	n_l	T/K	n_u	n_l	T/K	n_u	n_l
318.28	1.45704	1.46520	317.91	1.45064	1.47262	315.68	1.44356	1.48539
318.26	1.45594	1.46636	317.91	1.45064	1.47262	315.03	1.44262	1.48768
318.24	1.45540	1.46699	317.79	1.44982	1.47396	314.23	1.44179	1.49014
318.22	1.45475	1.46762	317.70	1.44918	1.47476	313.21	1.44099	1.49329
318.20	1.45439	1.46816	317.58	1.44854	1.47556	311.93	1.44031	1.49659
318.17	1.45365	1.46878	317.48	1.44790	1.47627	310.34	1.43977	1.50032
318.14	1.45310	1.46941	317.38	1.44754	1.47716	308.34	1.43938	1.50468
318.10	1.45265	1.47012	317.23	1.44710	1.47822	305.27	1.43901	1.51092
318.06	1.45210	1.47075	316.96	1.44620	1.47964	303.26	1.43897	1.51443
318.01	1.45146	1.47155	316.62	1.44531	1.48132	298.25	1.43954	1.52260
317.96	1.45109	1.47209	316.20	1.44443	1.48327	293.25	1.44042	1.52959

n_u and n_l are the refractive indices of the upper and lower (salt rich) phase below the critical temperature.

Mixture 7: 1-$C_{11}H_{23}OH$ (1) + N_{4444}Pic (2)
(1) Dielectric Constant and Critical Temperature

Ref. [6]

ϵ (T_c)	T_c/K	w_1
5.4	326.98	0.303

Mixture 8: 1-$C_{12}H_{25}OH$ (1) + $N_{4444}Pic$ (2)
(1) Dielectric Constant and Critical Temperature

Ref. [6]

ϵ (T_c)	T_c/K	w_1
4.7	335.91	0.2951

(2) Refractive Index

Ref. [6]

T/K	n_D	T/K	n_D	T/K	n_D
347.96	1.45100	344.20	1.45242	339.18	1.45426
347.96	1.45098	342.94	1.45284	337.93	1.45472
346.70	1.45143	341.69	1.45334	336.68	1.45526
345.45	1.45192	340.44	1.45376		

Ref. [6]

T/K	n_u	n_l	T/K	n_u	n_l	T/K	n_u	n_l
335.90	1.45293	1.45912	335.51	1.44603	1.46726	331.91	1.43825	1.48446
335.88	1.45148	1.46069	335.44	1.44567	1.46780	331.14	1.43768	1.48681
335.87	1.45075	1.46141	335.34	1.44495	1.46887	330.23	1.43723	1.48921
335.86	1.45084	1.46145	335.23	1.44451	1.46950	329.13	1.43682	1.49212
335.85	1.45029	1.46190	335.10	1.44383	1.47066	327.82	1.43648	1.49535
335.82	1.45012	1.46231	334.94	1.44326	1.47182	326.27	1.43628	1.49888
335.80	1.44925	1.46315	334.75	1.44259	1.47280	324.40	1.43642	1.50317
335.78	1.44921	1.46329	334.53	1.44198	1.47410	322.20	1.43612	1.50707
335.76	1.44852	1.46387	334.26	1.44133	1.47535	319.55	1.43642	1.51183
335.74	1.44857	1.46405	333.94	1.44069	1.47682	316.40	1.43701	1.51718
335.68	1.44780	1.46494	333.56	1.44010	1.47860	312.65	1.43781	1.52310
335.63	1.44707	1.46583	333.11	1.43939	1.48038	308.17	1.43891	1.52933
335.58	1.44676	1.46628	332.57	1.43883	1.48216	302.84	1.44040	1.53597

n_u and n_l are the refractive indices of the upper and lower (salt rich) phase below the critical temperature.

Mixture 9: 1-$C_{13}H_{27}OH$ (1) + $N_{4444}Pic$ (2)
(1) Dielectric Constant and Critical Temperature

Ref. [6]

ϵ (T_c)	T_c/K	w_1
4.3	342.35	0.284

(2) Critical Temperatures of Phase Separation

Ref. [6]

w_1	T_p /K	w_1	T_p /K	w_1	T_p /K
0.0458	323.95 ± 0.20	0.2828	342.35 ± 0.05	0.4469	340.50 ± 0.05
0.0996	336.91 ± 0.05	0.2978	342.33 ± 0.05	0.4713	339.68 ± 0.05
0.1255	339.55 ± 0.05	0.2978	342.35 ± 0.05	0.5791	334.15 ± 0.20
0.1495	340.76 ± 0.05	0.3696	341.95 ± 0.05	0.6305	330.15 ± 0.20
0.2001	341.95 ± 0.05	0.4128	341.52 ± 0.05		

Mixture 10: 1-$C_{14}H_{29}OH$ (1) + N_{4444}Pic (2)

(1) Dielectric Constant and Critical Temperature

Ref. [6]

$\epsilon(T_c)$	T_c/K	w_1
3.6	351.09	0.2721

(2) Refractive Index

Ref. [6]

T/K	n_D	T/K	n_D	T/K	n_D
361.99	1.44791	357.48	1.44956	352.97	1.45132
361.99	1.44791	355.98	1.45014	352.47	1.45152
360.49	1.44840	354.47	1.45071	351.97	1.45163
358.98	1.44898	353.97	1.45114	351.47	1.45188

Ref. [6]

T/K	n_u	n_l	T/K	n_u	n_l	T/K	n_u	n_l
351.08	1.44851	1.45593	350.62	1.44244	1.46314	343.77	1.43444	1.48944
351.06	1.44760	1.45701	350.50	1.44163	1.46448	341.90	1.43397	1.49396
351.04	1.44705	1.45755	350.34	1.44083	1.46583	339.55	1.43388	1.49914
351.02	1.44642	1.45827	350.15	1.44031	1.46726	336.93	1.43430	1.50461
351.00	1.44605	1.45863	349.91	1.43953	1.46870	334.92	1.43479	1.50855
350.98	1.44615	1.45863	349.61	1.43866	1.46981	332.92	1.43508	1.51191
350.96	1.44551	1.45944	349.23	1.43799	1.47190	330.41	1.43535	1.51584
350.93	1.44487	1.46042	348.75	1.43734	1.47377	327.90	1.43600	1.51979
350.86	1.44406	1.46123	348.15	1.43644	1.47628	322.89	1.43744	1.52716
350.80	1.44361	1.46186	347.39	1.43594	1.47923	317.88	1.43912	1.53492
350.72	1.44288	1.46232	346.44	1.43520	1.48203			
350.68	1.44252	1.46286	345.26	1.43460	1.48552			

n_u and n_l are the refractive indices of the upper and lower (salt rich) phase below the critical temperature.

9.6 15039-03003:Tetraheptylammonium bis ((perfluoroethane)sulfonyl)imide

Abbreviation:	[ThpA][C_2C_2N]
Molecular formula:	$C_{32}H_{60}F_{10}N_2O_4S_2$
Molar mass:	790.95
Structure:	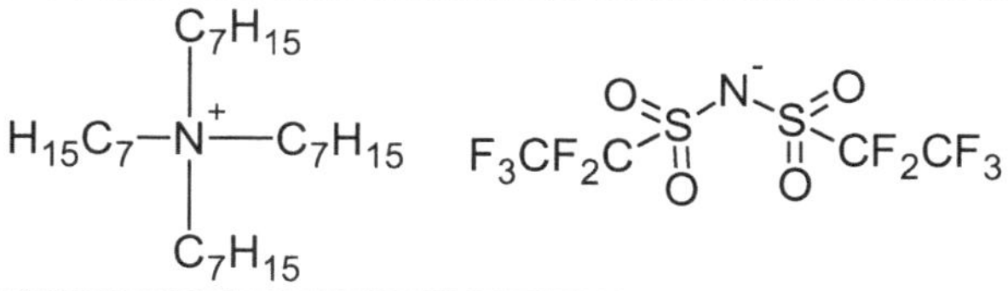

9.6.1 Ternary Mixtures

Mixture 1: H_2O (1) + [C_{14}Iq][C_2C_2N] (2) + [ThpA][C_2C_2N] (3)
(1) Volumetric Properties

Ref. [7]
These physicochemical properties data are listed in a mixture 1 of IL 12006-03003.

9.7 15067-03001: *N,N,N*-trioctyl-*N*-methylammonium bis ((trifluoromethyl)sulfonyl)imide

Abbreviation:	[N1888][NTf_2], [OMA][Tf_2N]
Molecular formula:	$C_{27}H_{54}F_6N_2O_4S_2$
Molar mass:	648.85
Structure:	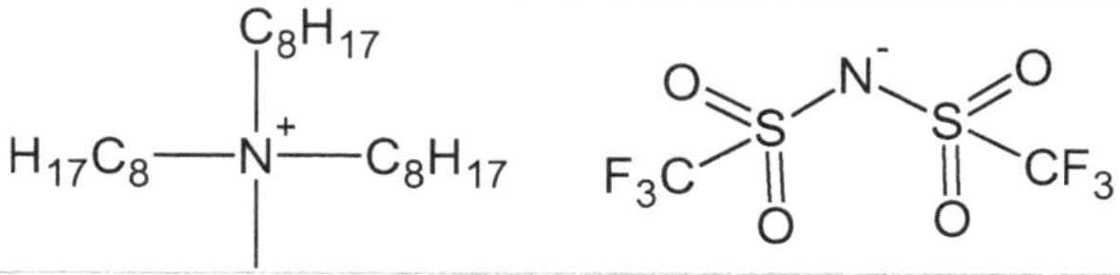

9.7.1 Binary Mixtures

Mixture1: CH_3OH (1) + [OMA][Tf_2N] (2)
(1) Volumetric Properties, Speed of Sound, Compressibility Properties and Transport Properties

Ref. [8]

x_2	$\rho/(g \cdot cm^{-3})$	$V^E_m/(cm^3 \cdot mol^{-1})$	x_2	$\rho/(g \cdot cm^{-3})$	$V^E_m/(cm^3 \cdot mol^{-1})$
298.15 K					
0.9345	1.1079	−0.224	0.4172	1.0770	0.910
0.8522	1.1057	−0.234	0.3500	1.0679	0.972
0.7084	1.0995	0.229	0.2125	1.0370	0.965
0.6529	1.0966	0.356	0.1291	0.9990	0.860
0.5362	1.0887	0.652	0.0781	0.9578	0.600
303.15 K					
0.9345	1.1037	−0.280	0.4172	1.0707	1.380
0.8522	1.1015	−0.292	0.3500	1.0615	1.396
0.7084	1.0950	0.288	0.2125	1.0311	1.164
0.6529	1.0915	0.626	0.1291	0.9941	0.877
0.5362	1.0826	1.180	0.0781	0.9535	0.546
313.15 K					
0.9345	1.0972	−0.344	0.4172	1.0640	1.339
0.8522	1.0947	−0.224	0.3500	1.0545	1.409
0.7084	1.0877	0.526	0.2125	1.0217	1.503
0.6529	1.0845	0.730	0.1291	0.9818	1.423
0.5362	1.0761	1.096	0.0781	0.9400	1.050

Ref. [8]

x_2	$\rho/(g \cdot cm^{-3})$	$u/(m \cdot s^{-1})$	$k_s/(T \cdot Pa^{-1})$	$\Delta k_s/(T \cdot Pa^{-1})$
298.15 K				
0.0336	0.8945	1097.8	1043.4	−1.8
0.1291	0.9990	1085.4	1034.6	−83.6
0.2125	1.0370	1074.5	1026.8	−138.7
0.4172	1.0770	1047.9	1008.0	−146.2
0.5362	1.0887	1032.4	997.0	−96.5
0.6611	1.0949	1016.1	985.4	−53.6
0.7084	1.0993	1010.0	981.1	−22.2
0.8522	1.1057	991.2	967.8	30.1

Ref. [9]

60 % [N_{8881}][NTf_2] + 40 % methanol	$\eta/mPa \cdot s^{-1}$			
	293.15 K	303.15 K	313.15 K	323.15 K
	10.98	10.8	9.16	8.4

Mixture 2: C_2H_5OH (1) + [OMA][Tf_2N] (2)
(1) Volumetric Properties, Speed of Sound and Compressibility Properties

Ref. [8]

x_2	$\rho/(g \cdot cm^{-3})$	$V^E_m/(cm^3 \cdot mol^{-1})$	x_2	$\rho/(g \cdot cm^{-3})$	$V^E_m/(cm^3 \cdot mol^{-1})$
298.15 K					
0.9393	1.1071	0.241	0.4286	1.0704	0.217
0.7774	1.0989	0.590	0.3353	1.0552	0.123
0.7338	1.0965	0.594	0.2728	1.0410	0.037
0.6307	1.0900	0.498	0.1731	1.0050	−0.042
0.5145	1.0802	0.360	0.0484	0.8955	−0.043
303.15 K					
0.9493	1.1027	0.342	0.4286	1.0625	1.194
0.7774	1.0922	1.663	0.3353	1.0495	0.440
0.7338	1.0896	1.697	0.2728	1.0361	0.153
0.6307	1.0818	1.940	0.1731	1.0034	−0.455
0.5145	1.0715	1.736	0.0484	0.8956	−0.467
313.15 K					
0.9493	1.0959	0.338	0.4286	1.0584	0.385
0.7774	1.0872	0.864	0.3353	1.0436	0.138
0.7338	1.0847	0.886	0.2728	1.0293	0.044
0.6307	1.0779	0.840	0.1731	0.9935	−0.113
0.5145	1.0680	0.649	0.0484	0.8833	−0.074

Ref. [8]

x_2	$\rho/(g \cdot cm^{-3})$	$u/(m \cdot s^{-1})$	$k_s/(T \cdot Pa^{-1})$	$\Delta k_s/(T \cdot Pa^{-1})$
298.15 K				
0.1731	1.0050	1116.1	965.2	−89.6
0.2728	1.0410	1098.8	963.9	−121.7
0.3353	1.0552	1087.9	963.0	−122.8
0.4286	1.0704	1071.6	961.8	−107.9
0.5145	1.0802	1056.6	960.6	−108.0
0.6307	1.0900	1036.4	959.1	−59.7
0.7338	1.0965	1018.4	957.7	−30.9
0.7774	1.0989	1010.8	957.1	−13.5
0.9493	1.1071	980.8	954.8	46.0

Mixture3: 1-C_3H_7OH (1) + [OMA][Tf_2N] (2)
(1) Volumetric Properties, Speed of Sound and Compressibility Properties

Ref. [8]

x_2	$\rho/(g \cdot cm^{-3})$	$V^E_m/(cm^3 \cdot mol^{-1})$	x_2	$\rho/(g \cdot cm^{-3})$	$V^E_m/(cm^3 \cdot mol^{-1})$
298.15 K					
0.9533	1.1071	0.138	0.4033	1.0594	0.183
0.8439	1.1015	0.296	0.3303	1.0452	0.101

(*continued*)

x_2	$\rho/(g \cdot cm^{-3})$	$V^E_m/(cm^3 \cdot mol^{-1})$	x_2	$\rho/(g \cdot cm^{-3})$	$V^E_m/(cm^3 \cdot mol^{-1})$
0.7833	1.0979	0.357	0.2152	1.0109	0.014
0.6882	1.0914	0.371	0.1146	0.9560	−0.025
0.5240	1.0762	0.285	0.0736	0.9191	−0.032
303.15 K					
0.9533	1.1030	0.084	0.4033	1.0548	0.253
0.8439	1.0973	0.288	0.3303	1.0406	0.153
0.7833	1.0937	0.346	0.2152	1.0063	0.040
0.6882	1.0871	0.393	0.1146	0.9515	−0.032
0.5240	1.0717	0.352	0.0736	0.9145	−0.032
313.15 K					
0.9533	1.0962	0.079	0.4033	1.0467	0.524
0.8439	1.0904	0.315	0.3303	1.0325	0.371
0.7833	1.0866	0.450	0.2152	0.9987	0.085
0.6882	1.0796	0.631	0.1146	0.9441	−0.059
0.5240	1.0637	0.678	0.0736	0.9072	−0.081

Ref. [8]

x_2	$\rho/(g \cdot cm^{-3})$	$u/(m \cdot s^{-1})$	$k_s/(T \cdot Pa^{-1})$	$\Delta k_s/(T \cdot Pa^{-1})$
298.15 K				
0.0736	0.9191	1187.7	868.4	21.1
0.1146	0.9560	1178.1	872.2	5.0
0.2152	1.0109	1154.7	881.5	−14.2
0.3303	1.0452	1127.9	892.2	−27.7
0.4033	1.0594	1110.9	898.9	−42.7
0.5240	1.0762	1082.8	910.1	−41.0
0.6682	1.0915	1049.2	923.4	−28.9
0.7833	1.0970	1022.4	934.1	−7.8
0.8439	1.1015	1008.3	939.7	9.5
0.9533	1.1071	982.9	949.8	42.6

9.8 15084-03009: Ethylammonium nitrate

Abbreviation:	[NHHH2][NO_3], EAN
Molecular formula:	$C_2H_8N_2O_3$
Molar mass:	108.10
Structure:	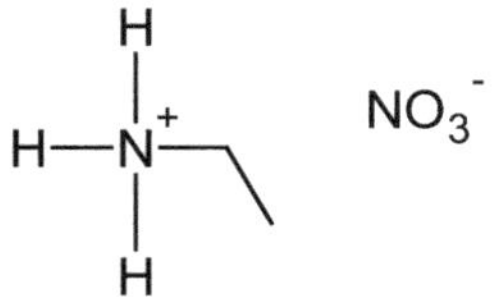

9.8.1 Binary Mixtures

Mixture1: H_2O (1) + EAN (2)
(1) Volumetric Properties

Ref. [10]

$C/(\text{mol} \cdot \text{L}^{-1})$	$\rho/(\text{kg} \cdot \text{dm}^{-3})$	$\varphi_{V,2}/(\text{cm}^3 \cdot \text{mol}^{-1})$	$\varphi_{V,2}$ -1.5 $C^{1/3/}$ $(\text{cm}^3 \cdot \text{mol}^{-1})$
298 K			
0.000	0.997047		
0.0674	0.998741	83.21	82.60
0.1035	0.999637	83.32	82.62
0.1338	1.000375	83.47	82.70
0.1684	1.001245	83.51	82.68
0.2115	1.002300	83.43	82.54
0.4433	1.007922	83.81	82.67

Ref. [10]

x_2	$C/(\text{mol} \cdot \text{L}^{-1})$	$V^E_{\text{mol}}/(\text{cm}^3 \cdot \text{mol}^{-1})$	$V'_2/(\text{cm}^3 \cdot \text{mol}^{-1})$	$V'_1/(\text{cm}^3 \cdot \text{mol}^{-1})$
298 K				
0.000	0.000	0.00	82.8	18.07
0.042	2.015	−0.22	85.6	18.0
0.100	4.027	−0.36	87.2	17.9
0.186	6.027	−0.44	88.1	17.8
0.336	8.111	−0.54	88.7	17.5
0.441	9.004	−0.50	89.0	17.4
0.625	10.048	−0.39	89.2	17.2
0.819	10.747	−0.21	89.3	16.9
1.000	11.198	0.00	89.3	16.7

9.9 15086-06003: 2-Hydroxyethylammonium acetate

Abbreviation:	$[\text{NHHH},(\text{CH}_2)_2\text{OH}][\text{OAc}]$, $[\text{HO}(\text{CH}_2)_2\text{NH}_3][\text{OAc}]$, [HMEA][Ac]
Molecular formula:	$C_4H_{11}NO_3$
Molar mass:	121.14
Structure:	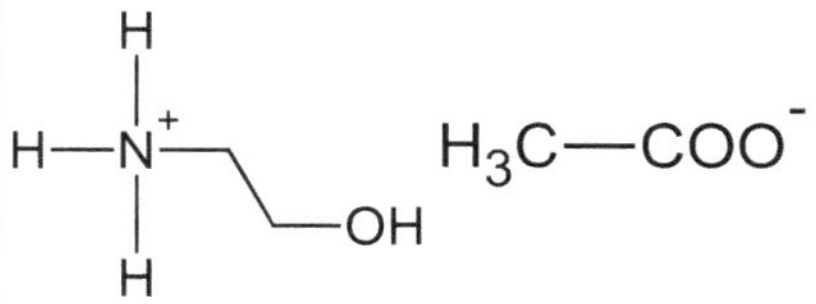

9.9.1 Binary Mixtures

Mixture 1: H_2O (1) + [HMEA][Ac] (2)
(1) Boiling Point

Ref. [11]

x_1	T/K	x_1	T/K	x_1	T/K
101.3 kPa					
0.9949	373.28	0.9575	374.58	0.9260	375.73
0.9908	373.43	0.9437	375.06	0.9005	376.73
0.9787	373.85				

Mixture 2: CH_3OH (1) + [HMEA][Ac] (2)
(1) Boiling Point

Ref. [11]

x_1	T/K	x_1	T/K	x_1	T/K
101.3 kPa					
0.9862	338.16	0.8981	341.02	0.8066	344.40
0.9730	338.57	0.8520	342.66	0.7523	346.63
0.9352	339.78				

Mixture 3: C_2H_5OH (1) + [HMEA][Ac] (2)
(1) Boiling Point

Ref. [11]

x_1	T/K	x_1	T/K	x_1	T/K
101.3 kPa					
0.9766	352.21	0.8478	356.30	0.7413	360.46
0.9416	353.31	0.7866	358.59	0.7121	361.80
0.9157	354.10				

9.9.2 Ternary Mixtures

Mixture 1: H_2O (1) + C_2H_5OH (2) + [HMEA][Ac] (3)
(1) Boiling Point

Ref. [11]

x_1	x_2	T/K	x_1	x_2	T/K
0.2144	0.6522	354.21	0.7802	0.1825	356.76
0.4036	0.4951	354.53	0.8298	0.1413	357.88
0.6002	0.3319	355.06	0.8690	0.1088	358.98
0.7036	0.2460	355.91			

Mixture 2: C_2H_5OH (1) + CH_3OH (2) + [HMEA][Ac] (3)
(1) Boiling Point

Ref. [11]

x_1	x_2	T/K	x_1	x_2	T/K
0.0842	0.8073	342.48	0.5279	0.4162	346.70
0.1831	0.7202	343.40	0.6093	0.3445	347.50
0.3088	0.6094	344.62	0.6783	0.2836	348.27
0.4014	0.5278	345.51			

9.10 15094-03001: (2-Hydroxyethyl)trimethylammonium bis [(trifluoromethyl)sulfonyl]amide

Abbreviation:	$[N_{111}(C_2OH)][NTf_2]$
Molecular formula:	$C_7H_{14}F_6N_2O_5S_2$
Molar mass:	384.32
Structure:	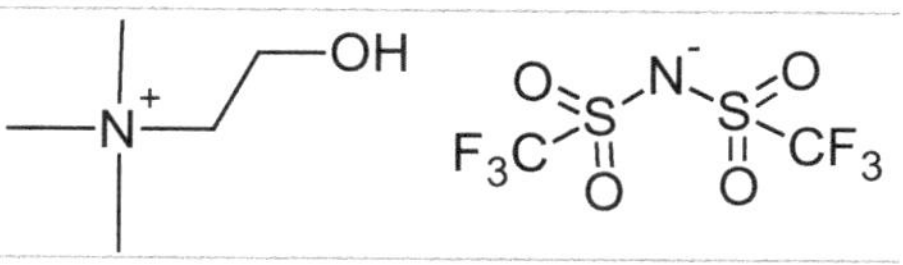

9.10.1 Ternary Mixtures

Mixture 1: C_6H_{14} (1) + C_6H_6 (2) + $[N_{111}(C_2OH)][NTf_2]$ (3)
(1) Distribution Ratio and Selectivity

Ref. [12]

Upper phase			Lower phase				
x_1	x_2	x_3	x_1	x_2	x_3	β	S
313.15 K							
1.000	0.000	0.000	0.015	0.000	0.985	—	—
0.946	0.054	0.000	0.014	0.037	0.949	0.69	46.3
0.880	0.120	0.000	0.014	0.080	0.906	0.67	41.9
0.754	0.246	0.000	0.015	0.135	0.850	0.55	27.6
0.608	0.392	0.000	0.013	0.206	0.781	0.53	24.6
0.421	0.579	0.000	0.018	0.290	0.691	0.50	11.7
0.298	0.702	0.000	0.016	0.341	0.642	0.49	9.0
0.207	0.792	0.000	0.015	0.395	0.590	0.50	6.9
0.109	0.891	0.000	0.008	0.422	0.570	0.47	6.5
0.000	1.000	0.000	0.000	0.491	0.509	0.49	—

9.11 15094-04002: Choline methanesulfonate (2-Hydroxyethyl-trimethylammonium methanesulfonate)

Abbreviation:	$[(CH_3)_3N(CH_2)_2OH][CH_3SO_2O]$
Molecular formula:	$C_6H_{17}NO_4S$
Molar mass:	199.27
Structure:	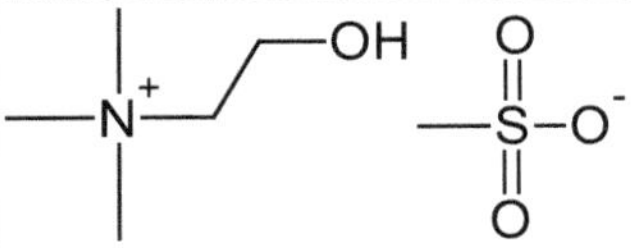

9.11.1 Binary Mixtures

Mixture 1: H_2O (1) + $[(CH_3)_3N(CH_2)_2OH][CH_3SO_2O]$ (2)
(1) Phase Equilibrium

Ref. [13]

x_1	p/mbar	x_1	p/mbar	x_1	p/mbar
303.15 K		313.15 K		323.15 K	
0.6603	9.3	0.6590	18.5	0.6561	30.0
0.6788	10.2	0.6782	20.0	0.6755	33.3
0.6957	11.0	0.6950	21.6	0.6932	35.6
0.7092	12.0	0.7087	22.7	0.7074	38.1
0.7247	13.1	0.7242	25.0	0.7232	40.4
0.7653	15.5	0.7650	29.7	0.7644	49.2
0.7969	18.2	0.7967	34.2	0.7964	57.2
0.8303	21.5	0.8301	40.1	0.7964	57.2
0.8727	26.9	0.8727	48.6	0.7964	57.2
0.9374	36.1	0.9374	63.8	0.8725	81.3
0.9716	40.8	0.9716	70.8	0.9374	106.4
1.0000	42.4	1.0000	73.8	0.9716	117.7
				1.0000	123.4

9.12 15094-06007: Choline Lactate (2-Hydroxyethyl-trimethylammonium 2-hydroxypropanoate)

Abbreviation:	$[(CH_3)_3N(CH_2)_2OH][CH_3CH(OH)COO]$
Molecular formula:	$C_8H_{19}NO_4$
Molar mass:	193.24
Structure:	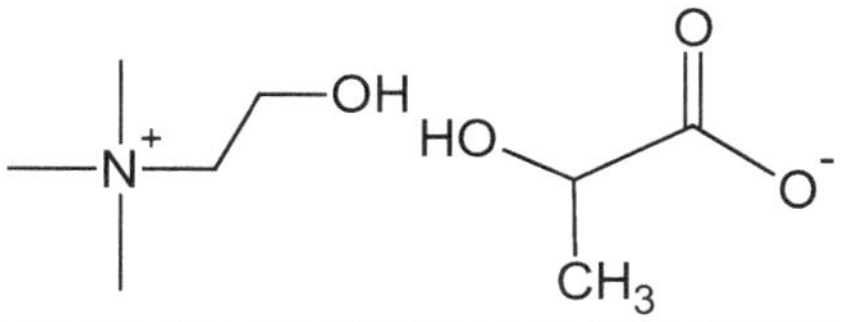

9.12.1 Binary Mixtures

Mixture 1: H_2O (1) + $[(CH_3)_3N(CH_2)_2OH][CH_3CH(OH)COO]$ (2)
(1) Transport Properties

Ref. [13]

x_1	η/cP	x_1	η/cP
293.15 K			
0.0272	5850	0.878	12.5
0.382	940	0.915	6.5
0.546	336	0.941	3.9
0.731	75		

(2) Phase Equilibrium

Ref. [13]

x_1	p/mbar	x_1	p/mbar	x_1	p/mbar
303.15 K		313.15 K		323.15 K	
0.1978	2.2	0.1974	3.0	0.1973	4.6
0.2776	3.7	0.2772	4.4	0.2771	7.4
0.3449	4.1	0.3439	6.7	0.3442	10.0
0.4196	5.3	0.4188	9.0	0.4185	13.5
0.4818	6.5	0.4807	11.4	0.4806	17.2
0.5720	9.6	0.5710	15.8	0.5709	24.3
0.6591	13.5	0.6583	21.8	0.6582	33.7
0.7309	17.4	0.7301	28.5	0.6582	33.7
0.8135	23.4	0.8130	39.4	0.6582	33.7
0.9175	34.4	0.9174	59.3	0.8129	63.3
0.9698	40.6	0.9698	70.0	0.9173	97.5
1.0000	42.4	1.0000	73.8	0.9697	116.7

(continued)

x_1	p/mbar	x_1	p/mbar	x_1	p/mbar
303.15 K		313.15 K		323.15 K	
				1.0000	123.4

(3) Excess Enthalpy

Ref. [13]

x_1	H^E/J · mol^{-1}	x_1	H^E/J · mol^{-1}	x_1	H^E/J · mol^{-1}
303.15 K		313.15 K		323.15 K	
0.1585	−3923	0.1687	−3023	0.1600	−1541
0.2269	−6323	0.2123	−4595	0.2662	−2716
0.3051	−8527	0.3065	−6560	0.3310	−3583
0.3528	−10059	0.3943	−8836	0.5059	−6469
0.4853	−14022	0.5019	−10885	0.6257	−8997
0.6710	−17243	0.6030	−12137	0.7785	−10864
0.8719	−18290	0.7380	−13791	0.8805	−8764
0.9078	−14847	0.8794	−14581	0.9555	−4596
0.9751	−7845	0.9535	−10331	0.9884	−1001
0.9878	−2529	0.9769	−6484		

9.13 15094-06034: Choline glycolate (2-Hydroxyethyl-trimethylammonium 2-hydroxyethanoate)

Abbreviation:	$[(CH_3)_3N(CH_2)_2OH][CH_2(OH)COO]$
Molecular formula:	$C_7H_{17}NO_4$
Molar mass:	179.21
Structure:	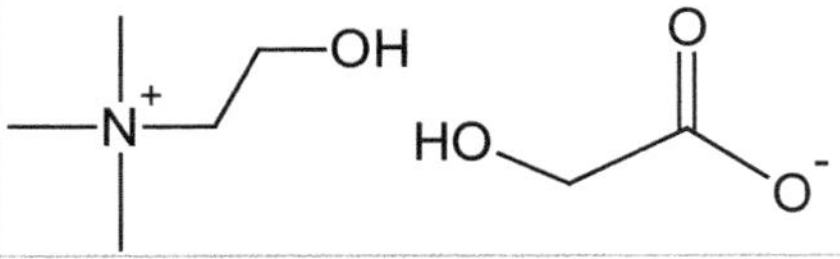

9.13.1 Binary Mixtures

Mixture 1: H_2O (1) + $[(CH_3)_3N(CH_2)_2OH][CH_2(OH)COO]$ (2)
(1) Transport Properties

Ref. [13]

x_1	η/cP	x_1	η/cP
293.15 K			
0.366	660	0.813	19.9

(*continued*)

x_1	η/cP	x_1	η/cP
0.536	209	0.871	9.2
0.718	53.2	0.937	3.3

(2) Phase Equilibrium

Ref. [13]

x_1	p/mbar	x_1	p/mbar	x_1	p/mbar
303.15 K		313.15 K		323.15 K	
0.3379	1.2	0.3369	2.1	0.3272	4.5
0.3987	1.6	0.3988	2.9	0.3917	5.3
0.4460	2.1	0.4458	3.7	0.4413	6.9
0.5026	2.4	0.5024	4.7	0.5000	9.1
0.5571	3.6	0.5569	5.9	0.5556	11.6
0.6300	4.9	0.6300	8.9	0.6293	16.8
0.6951	6.6	0.6951	12.6	0.6947	23.5
0.7580	9.8	0.7580	18.5	0.6947	23.5
0.8286	15.2	0.8286	28.1	0.6947	23.5
0.9216	29.1	0.9216	52.1	0.8286	50.2
0.9669	38.1	0.9669	67.0	0.9216	89.2
1.0000	42.4	1.0000	73.8	0.9669	112.8
				1.0000	123.4

(3) Excess Enthalpy

Ref. [13]

x_1	H^E/J · mol^{-1}	x_1	H^E/J · mol^{-1}	x_1	H^E/J · mol^{-1}
303.15 K		313.15 K		323.15 K	
0.3483	−3092	0.3400	−2018	0.3425	−993
0.4300	−3932	0.3792	−2268	0.3674	−1151
0.4573	−4087	0.4264	−2438	0.4192	−1294
0.5174	−4642	0.4739	−2574	0.5692	−1721
0.6020	−5375	0.5268	−2954	0.6811	−2005
0.6513	−5689	0.6185	−3247	0.7740	−2207
0.7001	−6245	0.7391	−3534	0.8726	2377
0.7959	−6319	0.8041	−3718	0.9543	−1992
0.8916	−5820	0.9032	−1500	0.9816	−1151
0.9543	−4195	0.9330	−1002		
		0.9748	−472		

9.14 15096-03009: Didecyldimethylammonium nitrate

Abbreviation:	[DDA][NO_3]
Molecular formula:	$C_{22}H_{48}N_2O_3$
Molar mass:	388.62
Structure:	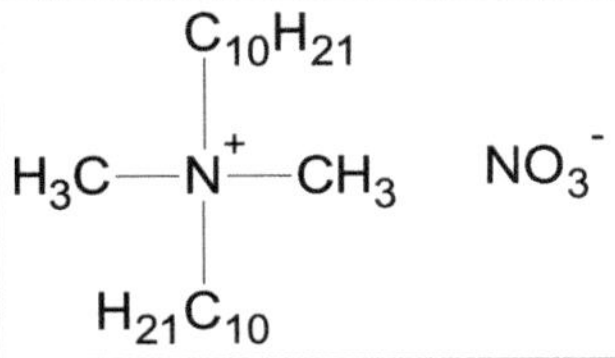

9.14.1 Binary Mixtures

Mixture 1: H_2O (1) + [DDA][NO_3] (2)
(1) Activity Coefficient

Ref. [14]

x_2	T^{SLE}_2/K	γ_2	x_2	γ^{SLE}_2/K	γ_2
0.0000	273.2		0.3845	284.5	1.97
0.0800	275.1	8.20	0.5573	289.6	1.46
0.1091	276.8	6.19	0.6722	294.4	1.30
0.1378	277.4	4.94	0.7402	298.3	1.24
0.1799	278.7	3.86	0.8554	300.5	1.11
0.2210	280.4	3.22	1.0000	304.8	1.00

Mixture 2: 1-C_3H_7OH (1) + [DDA][NO_3] (2)
(1) Activity Coefficient

Ref. [14]

x_2	T^{SLE}_2/K	γ_2	x_2	γ^{SLE}_2/K	γ_2
0.0000	146.7		0.3643	281.9	2.00
0.0042	223.5	58.45	0.4193	283.1	1.77
0.0084	230.9	34.24	0.4943	290.2	1.66
0.0247	240.9	14.42	0.5880	295.0	1.50
0.0622	248.7	6.68	0.7177	296.1	1.24
0.1214	258.1	4.07	0.8177	297.7	1.11
0.2140	269.2	2.79	1.0000	304.8	1.00
0.2870	275.3	2.30			

Mixture 3: 1-C_4H_9OH (1) + [DDA][NO_3] (2)
(1) Activity Coefficient

Ref. [14]

x_2	$T^{SLE}{}_2$/K	γ_2	x_2	$\gamma^{SLE}{}_2$/K	γ_2
0.0000	183.7		0.3747	283.9	2.00
0.0256	236.7	12.74	0.4865	291.9	1.73
0.0597	242.2	6.12	0.6159	299.9	1.52
0.1105	246.9	3.63	0.7831	302.0	1.23
0.2044	257.2	2.38	1.0000	304.8	1.00
0.3181	274.1	2.03			

Mixture 4: 1-$C_8H_{17}OH$ (1) + [DDA][NO_3] (2)
(1) Activity Coefficient

Ref. [14]

x_2	$T^{SLE}{}_2$/K	γ_2	x_2	$T^{SLE}{}_2$/K	γ_2
0.0000	258.2		0.4296	292.1	1.97
0.0474	267.2	12.21	0.5142	293.1	1.67
0.0885	269.0	6.73	0.5956	293.1	1.44
0.1387	275.0	4.73	0.7062	298.3	1.30
0.1940	279.5	3.63	0.8036	299.3	1.15
0.2500	282.8	2.96	0.9409	301.3	1.02
0.3384	288.4	2.37	1.0000	304.8	1.00

Mixture 5: 1-$C_{10}H_{21}OH$ (1) + [DDA][NO_3] (2)
(1) Activity Coefficient

Ref. [14]

x_2	$T^{SLE}{}_2$/K	$T^{LLE}{}_2$/K	γ_2	x_2	$T^{SLE}{}_2$/K	$T^{LLE}{}_2$/K	γ_2
0.0000	278.2			0.5130	304.7	325.3	1.95
0.0444	286.9		17.68	0.5572	304.7	331.9	1.79
0.0643	287.5		12.31	0.6354	304.0	340.5	1.56
0.0967	289.3		8.40	0.7027	303.8	349.7	1.41
0.1384	292.0		6.10	0.7890	303.8	355.8	1.25
0.1923	295.4		4.59	0.8270	304.3		1.20
0.2394	297.9		3.82	0.8893	304.7		1.12
0.3012	300.6	306.7	3.15	0.9325	303.8	363.1	1.06
0.3796	302.0	314.3	2.54	0.9774	304.0		1.01
0.4164	303.3	316.2	2.36	1.0000	304.8		1.00
0.4544	303.8	319.9	2.17				

Mixture 6: $C_6H_5CH_3$ (1) + [DDA][NO_3] (2)
(1) Activity Coefficient

Ref. [14]

x_2	$T^{SLE}{}_2$/K	γ_2	x_2	$T^{SLE}{}_2$/K	γ_2
0.0000	178.1		0.4671	282.8	1.58
0.0463	271.8	13.47	0.5651	283.5	1.32
0.0735	272.1	8.54	0.6820	290.4	1.21
0.1372	273.3	4.66	0.7594	294.9	1.16
0.2178	275.8	3.05	0.8632	300.0	1.09
0.2788	278.1	2.47	0.9501	304.7	1.05
0.3688	280.9	1.95	1.0000	304.8	1.00

Mixture 7: $C_6H_5C_3H_7$ (propylbenzene) (1) + [DDA][NO_3] (2)
(1) Activity Coefficient

Ref. [14]

x_2	$T^{SLE}{}_2$/K	γ_2	x_2	$T^{SLE}{}_2$/K	γ_2
0.0000	174.1		0.4825	291.7	1.75
0.0385	286.2	20.19	0.5769	294.8	1.52
0.0529	287.1	14.88	0.6823	299.3	1.37
0.0827	287.8	9.62	0.7759	302.5	1.25
0.1422	288.3	5.64	0.8604	303.4	1.16
0.2133	289.4	3.81	0.9189	304.4	1.08
0.2914	290.7	2.84	0.9676	303.4	0.97
0.3956	291.8	2.13	1.0000	304.8	1.00

Mixture 8: C_6H_{14} (1) + [DDA][NO_3] (2)
(1) Activity Coefficient

Ref. [14]

x_2	$T^{SLE}{}_2$/K	$T^{LLE}{}_2$/K	γ_2	x_2	$T^{SLE}{}_2$/K	$T^{LLE}{}_2$/K	γ_2
0.0000	178.8			0.5717	299.7		1.64
0.0512	298.0		17.88	0.6410	302.8		1.52
0.1245	296.5		7.21	0.6950	299.2		1.34
0.1934	298.3		4.75	0.7486	300.2		1.26
0.2770	298.0		3.30	0.8176	298.3		1.12
0.3445	298.0	323.3	3.10	0.8628	302.1		1.12
0.3756	297.9	315.0	2.40	0.9167	302.5		1.06
0.4463	298.3	309.4	2.06	0.9660	304.5		1.03
0.4868	298.0	303.7	1.95	1.0000	304.8		1.00
0.5182	300.9		1.83				

Mixture 9: $C_{16}H_{34}$ (1) + [DDA][NO_3] (2)
(1) Activity Coefficient

Ref. [14]

x_2	$T^{SLE}{}_2$/K	$T^{LLE}{}_2$/K	γ_2	x_2	$T^{SLE}{}_2$/K	$T^{LLE}{}_2$/K	γ_2
0.0000	291.1			0.3652	302.5	>433.1	2.66
0.0006		>433.1		0.4537		>433.1	
0.0047		>433.1		0.5748	301.7	>433.1	1.67
0.0405		>433.1		0.6807		>433.1	
0.0679		>433.1		0.7705	304.6	424.3	1.29
0.0943		>433.1		0.9159	304.0	411.9	1.08
0.1511		>433.1		0.9715	303.4	406.8	1.01
0.2151	301.8	>433.1	10.20	0.9918		393.7	
0.2860		>433.1		1.0000	304.8		1.00

9.15 15099-01002: Butyl(2-hydroxyethyl)dimethylammonium bromide

Abbreviation:	C_4Br
Molecular formula:	$C_8H_{20}NOBr$
Molar mass:	226.15
Structure:	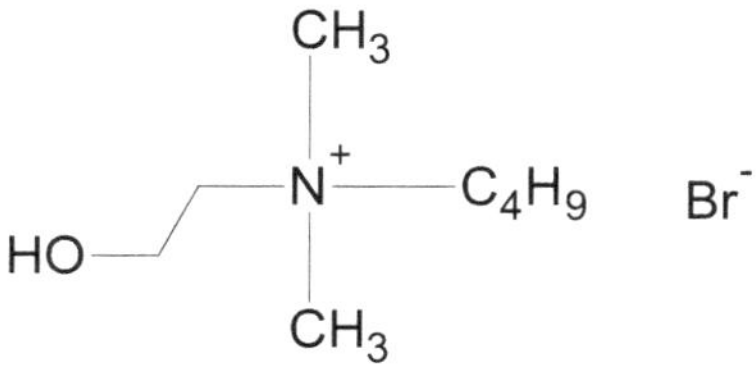

9.15.1 Binary Mixtures

Mixture 1: H_2O (1) + C_4Br (2)
(1) Volumetric Properties and Surface Tension

Ref. [15]

$C_2 \times 10^2$ (mol · dm^3)	ρ (g · cm^{-3})	σ (mN · m^{-1})
298.15 K		
44.98	1.0099	34.59
24.94	1.0032	40.01
9.91	0.9992	46.72

9.16 15099-02001: Butyl(2-hydroxyethyl)dimethylammonium tetrafluoroborate

Abbreviation:	C_4BF_4
Molecular formula:	$C_8H_{20}NOBF_4$
Molar mass:	233.06
Structure:	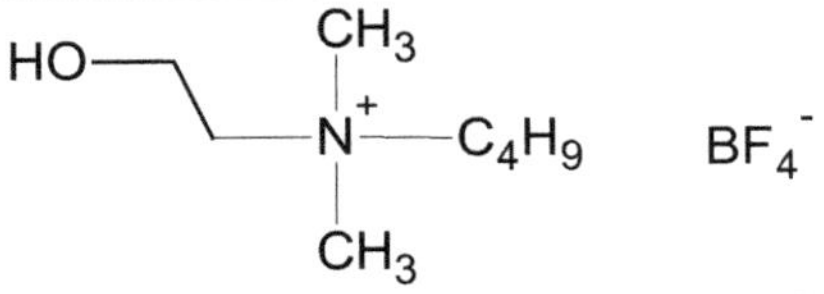

9.16.1 Ternary Mixtures

Mixture 1: C_6H_{14} (1) + *p*-$C_6H_4(CH_3)_2$ (2) + C_4BF_4 (3)
(1) Distribution Ratio and Selectivity

Ref. [16]

ionic liquid-rich layer		hexane-rich layer			
x_3^{II}	x_2^{II}	x_3^{I}	x_2^{I}	β	S
298.15 K					
1.00	0.00	0.00	0.00		
0.91	0.08	0.00	0.22	0.36	28.4
0.76	0.20	0.00	0.33	0.61	10.2
0.70	0.27	0.00	0.44	0.61	11.5
0.68	0.30	0.00	0.50	0.60	15.0
0.60	0.38	0.00	0.53	0.72	16.8
0.52	0.46	0.00	0.69	0.67	10.3
0.41	0.57	0.00	0.76	0.75	9.0
0.30	0.68	0.00	0.83	0.82	7.0

9.17 15132-01002: Propyl(2-hydroxyethyl)dimethylammonium bromide

Abbreviation:	C_3Br
Molecular formula:	$C_7H_{18}NOBr$
Molar mass:	212.12
Structure:	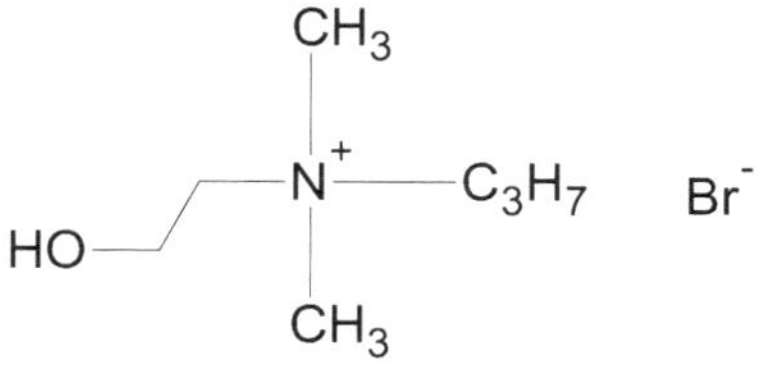

9.17.1 Binary Mixtures

Mixture 1: H_2O (1) + C_3Br (2)
(1) Volumetric Properties and Surface Tension

Ref. [15]

$C_2 \times 10^2$ (mol dm^3)	ρ (g · cm^{-3})	σ (mN · m^{-1})
298.15 K		
51.01	1.0190	47.72
25.13	1.0084	48.30
11.58	1.0079	51.99

9.18 15142-01001: Diethanolammonium chloride

Abbreviation:	[HDEA]Cl
Molecular formula:	$C_4H_{12}NO_2Cl$
Molar mass:	141.60
Structure:	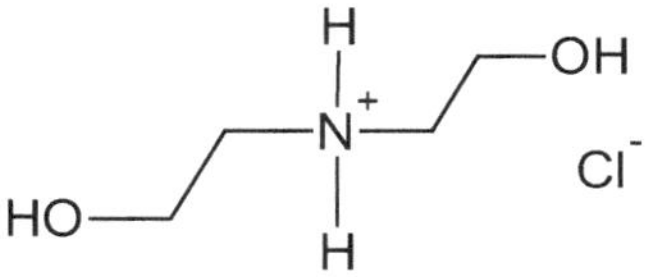

9.18.1 Binary Mixtures

Mixture 1: H_2O (1) + [HDEA]Cl (2)
(1) Boiling Point

Ref. [11]

x_1	T/K	x_1	T/K	x_1	T/K
101.3 kPa					
0.9933	373.50	0.9584	375.43	0.9264	377.36
0.9821	374.08	0.9424	376.35	0.9072	378.66
0.9684	374.83				

Mixture 2: CH_3OH (1) + [HDEA]Cl (2)
(1) Boiling Point

Ref. [11]

x_1	T/K	x_1	T/K	x_1	T/K
101.3 kPa					
0.9918	337.96	0.9283	339.67	0.8781	341.15
0.9727	338.47	0.9019	340.42	0.8318	342.72
0.9439	339.24				

Mixture 3: C_2H_5OH (1) + [HDEA]Cl (2)
(1) Boiling Point

Ref. [11]

x_1	T/K	x_1	T/K	x_1	T/K
101.3 kPa					
0.9801	351.95	0.8589	353.46	0.7729	354.42
0.9434	352.49	0.8079	354.03	0.7389	354.88
0.9125	352.88				

9.18.2 Ternary Mixtures

Mixture 1: H_2O (1) + C_2H_5OH (2) + [HDEA]Cl (3)
(1) Boiling Point

Ref. [11]

x_1	x_2	T/K	x_1	x_2	T/K
0.1298	0.7582	353.53	0.6847	0.2747	355.29
0.4146	0.5101	353.72	0.7586	0.2103	356.21
0.5898	0.3574	354.50	0.8205	0.1564	357.49

Mixture 2: C_2H_5OH (1) + CH_3OH (2) + [HDEA]Cl (3)
(1) Boiling Point

Ref. [11]

x_1	x_2	T/K	x_1	x_2	T/K
0.3470	0.5451	345.39	0.6108	0.3249	347.85
0.4616	0.4494	346.48	0.6972	0.2528	348.64
0.5245	0.3969	347.04	0.7644	0.1966	349.25

9.19 15142-06003: Diethanolammonium acetate

Abbreviation:	[HDEA][Ac]
Molecular formula:	$C_6H_{15}NO_4$
Molar mass:	165.19
Structure:	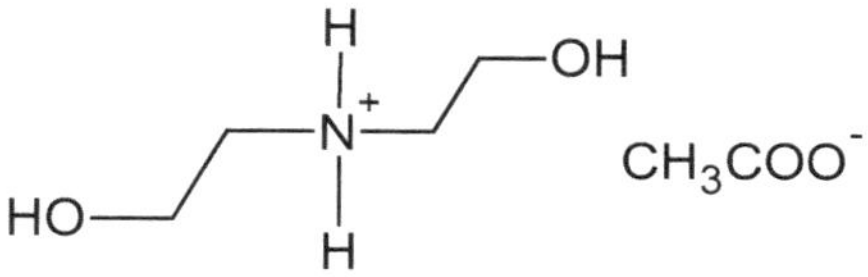

9.19.1 Binary Mixtures

Mixture 1: H_2O (1) + [HDEA][Ac] (2)
(1) Boiling Point

Ref. [11]

x_1	T/K	x_1	T/K	x_1	T/K
101.3 kPa					
0.9942	373.31	0.9671	374.13	0.9144	375.90
0.9936	373.33	0.9421	374.92	0.8935	376.75
0.9851	373.59				

Mixture 2: CH_3OH (1) + [HDEA][Ac] (2)
(1) Boiling Point

Ref. [11]

x_1	T/K	x_1	T/K	x_1	T/K
101.3 kPa					
0.9912	337.93	0.9266	339.62	0.8696	341.22
0.9834	338.13	0.8987	340.39	0.8395	342.12
0.9595	338.74				

Mixture 3: C_2H_5OH (1) + [HDEA][Ac] (2)
(1) Boiling Point

Ref. [11]

x_1	T/K	x_1	T/K	x_1	T/K
101.3 kPa					
0.9884	351.74	0.8857	354.09	0.8031	356.05
0.9672	352.29	0.8450	355.01	0.7778	356.74
0.9333	353.11				

9.19.2 Ternary Mixtures

Mixture 1: C_2H_5OH (1) + CH_3OH (2) + [HDEA][Ac] (3)
(1) Boiling Point

Ref. [11]

x_1	x_2	T/K	x_1	x_2	T/K
0.0959	0.8298	341.02	0.5132	0.4468	345.72
0.1876	0.7457	342.04	0.6050	0.3626	346.80
0.3125	0.6310	343.43	0.6760	0.2973	347.65
0.4156	0.5364	344.60			

9.20 15143-06003: Triethanolammonium acetate

Abbreviation:	[HTEA][Ac]
Molecular formula:	$C_8H_{19}NO_5$
Molar mass:	209.24
Structure:	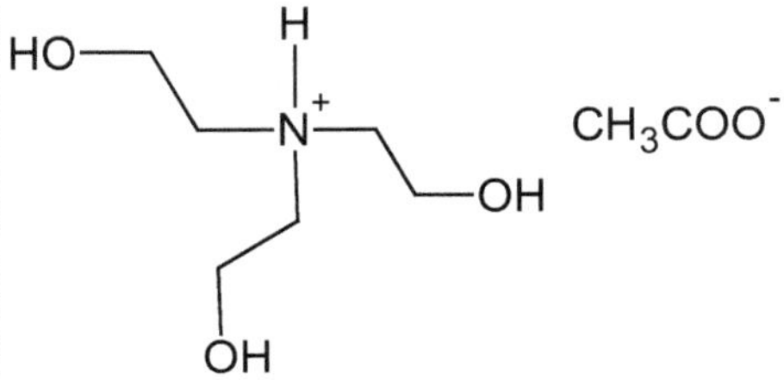

9.20.1 Binary Mixtures

Mixture 1: H_2O (1) + [HTEA][Ac] (2)
(1) Boiling Point

Ref. [11]

x_1	T/K	x_1	T/K	x_1	T/K
101.3 kPa					
0.9969	373.30	0.9779	374.30	0.9441	376.28
0.9949	373.42	0.9595	375.40	0.9177	377.81
0.9911	373.63				

Mixture 2: CH_3OH (1) + [HTEA][Ac] (2)
(1) Boiling Point

Ref. [11]

x_1	T/K	x_1	T/K	x_1	T/K
101.3 kPa					
0.9921	337.96	0.9583	339.09	0.9051	340.93
0.9883	338.09	0.9328	339.96	0.8907	341.46
0.9791	338.40				

Mixture 3: C_2H_5OH (1) + [HTEA][Ac] (2)
(1) Boiling Point

Ref. [11]

x_1	T/K	x_1	T/K	x_1	T/K
101.3 kPa					
0.9890	351.77	0.9063	353.86	0.8387	355.31
0.9673	352.40	0.8721	354.61	0.8150	355.80
0.9417	353.07				

9.20.2 Ternary Mixtures

Mixture 1: H_2O (1) + C_2H_5OH (2) + [HTEA][Ac] (3)
(1) Boiling Point

Ref. [11]

x_1	x_2	T/K	x_1	x_2	T/K
0.1896	0.7363	354.08	0.7646	0.2138	356.69
0.4474	0.5020	354.41	0.8245	0.1595	357.82
0.5784	0.3830	354.98	0.8628	0.1247	358.92
0.6945	0.2776	355.84			

Mixture 2: C_2H_5OH (1) + CH_3OH (2) + [HTEA][Ac] (3)
(1) Boiling Point

Ref. [11]

x_1	x_2	T/K	x_1	x_2	T/K
0.0519	0.8843	341.53	0.4448	0.5178	345.45
0.1255	0.8157	342.22	0.5238	0.4441	346.28
0.2225	0.7252	343.19	0.5923	0.3803	347.04
0.3426	0.6132	344.39			

9.21 15163-01002: Hexyl(2-hydroxyethyl)dimethylammonium bromide

Abbreviation:	C_6Br
Molecular formula:	$C_{10}H_{24}NOBr$
Molar mass:	254.20
Structure:	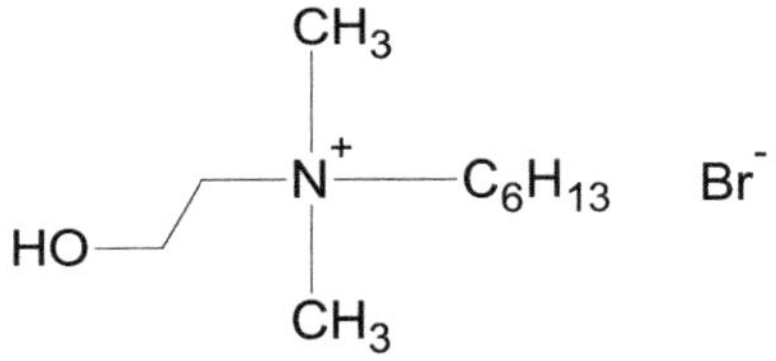

9.21.1 Binary Mixtures

Mixture 1: 1-$C_8H_{17}OH$ (1) + C_6Br (2)
(1) Volumetric Properties and Surface Tension

Ref. [15]

$x_2 \times 10^4$	ρ (g · cm^{-3})	σ (mN · m^{-1})
298.15 K		
7.6	0.8287	26.60
4.5	0.8268	26.57
1.8	0.8245	26.57
0.00	0.8226[a]	26.52[a]

[a]Volumetric Properties is 0.8216 g cm^{-3} at 298.15 K; surface tension is 26.52 mN · m^{-1} at 298.15 K

Mixture 2: H_2O (1) + C_6Br (2)
(1) Volumetric Properties and Surface Tension

Ref. [15]

$C_2 \times 10^2$ (mol · dm^3)	ρ (g · cm^{-3})	σ (mN · m^{-1})
298.15 K		
45.14	0.9628	35.35
25.05	0.9785	39.24
10.40	0.9852	48.07

9.22 15163-02001: Hexyl(2-hydroxyethyl)dimethylammonium tetrafluoroborate

Abbreviation:	C_6BF_4
Molecular formula:	$C_{10}H_{24}NOBF_4$
Molar mass:	261.10
Structure:	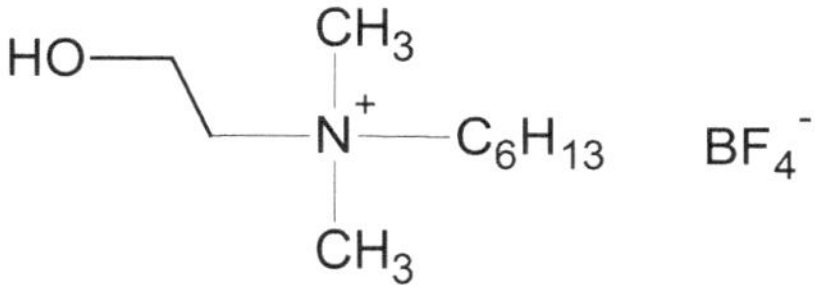

9.22.1 Ternary Mixtures

Mixture 1: C_6H_{14} (1) + p-$C_6H_4(CH_3)_2$ (2) + C_6BF_4 (3)
(1) Distribution Ratio and Selectivity

Ref. [16]

Ionic liquid-rich layer		Hexane-rich layer			
x_3^{II}	x_2^{II}	x_3^{I}	x_2^{I}	β	S
298.15 K					
0.99	0.00	0.00	0.00		
0.97	0.02	0.00	0.31	0.11	4.5
0.94	0.05	0.00	0.45	0.21	6.1
0.87	0.12	0.00	0.56	0.46	9.4
0.70	0.29	0.00	0.63	0.72	17.0
0.45	0.51	0.00	0.71	0.79	5.2
0.36	0.60	0.00	0.76	0.06	4.7

9.23 15164-01002: Ethyl(2-hydroxyethyl)dimethylammonium bromide

Abbreviation:	C_2Br
Molecular formula:	$C_6H_{16}NOBr$
Molar mass:	198.10
Structure:	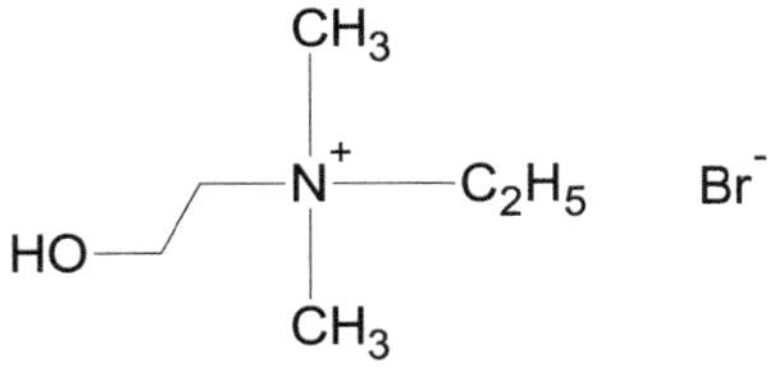

9.23.1 Binary Mixtures

Mixture 1: H_2O (1) + C_2Br (2)
(1) Volumetric Properties and Surface Tension
Ref. [15]

$C_2 \times 10^2$ (mol · dm^3)	ρ (g · cm^{-3})	σ (mN · m^{-1})
298.15 K		
44.48	1.0250	63.98
25.01	1.0122	65.23
10.14	1.0026	66.97
0.00	0.9974[a]	71.85[a]

[a] Volumetric Properties is 0.99692 g cm^{-3} at 299.09 K and surface tension is 71.94 mN · m^{-1} at 298.15 K.

9.24 15164-02001: Ethyl(2-hydroxyethyl)dimethylammonium tetrafluoroborate

Abbreviation:	C_2BF_4
Molecular formula:	$C_6H_{16}NOBF_4$
Molar mass:	205.00
Structure:	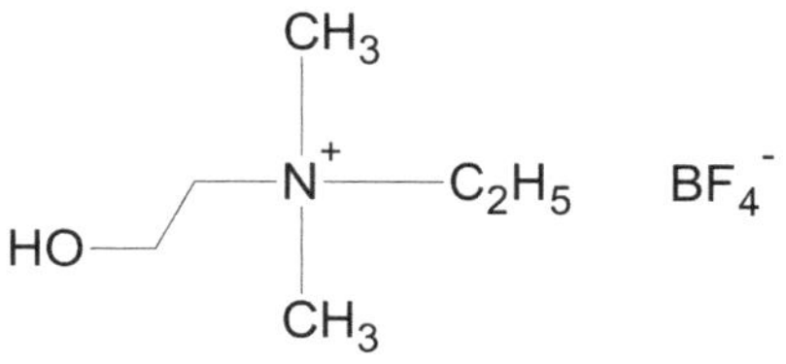

9.24.1 Binary Mixtures

Mixture 1: H_2O (1) + C_2BF_4 (2)
(1) Volumetric Properties and Surface Tension

Ref. [15]

$C_2 \times 10^2$ (mol · dm^3)	ρ (g · cm^{-3})	σ (mN · m^{-1})
298.15 K		
45.31	1.0127	49.00
24.87	1.0061	53.07
10.08	1.0030	61.79

9.25 15164-03001: Ethyl(2-hydroxyethyl)dimethylammonium bis(trifluomethylsulfonyl)imide

Abbreviation:	C_2NTf_2
Molecular formula:	$C_8H_{16}N_2O_5S_2F_6$
Molar mass:	398.35
Structure:	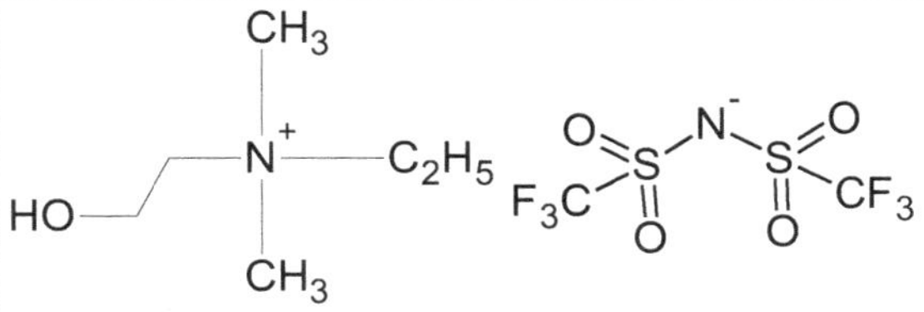

9.25.1 Binary Mixtures

Mixture 1: H_2O (1) + C_2NTf_2 (2)
(1) Phase Equilibrium

Ref. [17]

x_2	T_2^{SLE}/K	T_2^{LLE}/K	x_2	T_2^{SLE}/K	T_2^{LLE}/K
1.0000	279.16		0.4282	251.55	312.63
0.9510	277.36		0.3951	251.55	318.98
0.9294	275.39		0.3503	251.55	326.28
0.8492	268.06		0.3075	251.55	331.46
0.8021	263.50		0.2651	251.55	338.26
0.7529	259.37		0.2287	251.55	345.08
0.7108	256.18		0.1994	251.55	351.36
0.6665	252.73		0.1616	251.55	359.70
0.6253	251.50		0.1169	251.55	368.35
0.5865	251.48	278.50	0.0791	251.55	
0.5539	251.55	287.81	0.0114		359.52
0.5221	251.55	294.40	0.0053		331.46
0.4711	251.55	304.74	0.0000	273.15	

9.25.2 Ternary Mixtures

Mixture 1: C_6H_{14} (1) + C_6H_6 (2) + C_2NTf_2 (3)
(1) Distribution Ratio and Selectivity

Ref. [18]

Alkane—rich layer			Ionic liquid—rich layer				
x^I_1	x^I_2	x^I_3	x^{II}_1	x^{II}_2	x^{II}_3	β	S
298.15 K							
1.00	0.00	0.00	0.08	0.00	0.92		
0.83	0.17	0.00	0.05	0.21	0.74	1.2	20
0.65	0.35	0.00	0.03	0.33	0.64	0.9	20
0.59	0.41	0.00	0.02	0.36	0.62	0.9	26
0.51	0.49	0.00	0.02	0.37	0.61	0.8	19
0.43	0.57	0.00	0.02	0.37	0.61	0.6	14
0.36	0.64	0.00	0.02	0.39	0.59	0.6	11
0.30	0.70	0.00	0.02	0.41	0.57	0.6	9
0.20	0.80	0.00	0.03	0.55	0.42	0.7	5

Mixture 2: C_6H_{14} (1) + p-$C_6H_4(CH_3)_2$ (2) + C_2NTf_2 (3)
(1) Distribution Ratio and Selectivity

Ref. [18]

Alkane—rich layer			Ionic liquid—rich layer				
x^I_1	x^I_2	x^I_3	x^{II}_1	x^{II}_2	x^{II}_3	β	S
298.15 K							
1.00	0.00	0.00	0.02	0.00	0.98		
0.91	0.09	0.00	0.02	0.11	0.87	1.2	56
0.69	0.31	0.00	0.02	0.13	0.85	0.4	14
0.64	0.36	0.00	0.02	0.14	0.84	0.4	12
0.51	0.49	0.00	0.02	0.15	0.83	0.3	8
0.46	0.54	0.00	0.02	0.16	0.82	0.3	7
0.37	0.63	0.00	0.02	0.17	0.81	0.3	5
0.22	0.78	0.00	0.01	0.21	0.77	0.3	6

Mixture 3: C_6H_{14} (1) + m-$C_6H_4(CH_3)_2$ (2) + C_2NTf_2 (3)
(1) Distribution Ratio and Selectivity

Ref. [18]

Alkane—rich layer			Ionic liquid—rich layer				
x^I_1	x^I_2	x^I_3	x^{II}_1	x^{II}_2	x^{II}_3	β	S
298.15 K							
1.00	0.00	0.00	0.02	0.00	0.98		
0.80	0.20	0.00	0.02	0.05	0.93	0.2	10
0.77	0.23	0.00	0.02	0.07	0.91	0.3	12
0.57	0.43	0.00	0.02	0.13	0.85	0.3	9
0.50	0.50	0.00	0.02	0.14	0.84	0.3	7
0.48	0.52	0.00	0.02	0.15	0.83	0.3	7
0.44	0.56	0.00	0.02	0.17	0.81	0.3	7
0.35	0.65	0.00	0.02	0.18	0.80	0.3	5
0.26	0.74	0.00	0.02	0.26	0.72	0.3	5
0.14	0.86	0.00	0.02	0.29	0.69	0.3	2

Mixture 4: C_8H_{18} (1) + m-$C_6H_4(CH_3)_2$ (2) + C_2NTf_2 (3)
(1) Distribution Ratio and Selectivity

Ref. [18]

Alkane—rich layer			Ionic liquid—rich layer				
x^I_1	x^I_2	x^I_3	x^{II}_1	x^{II}_2	x^{II}_3	β	S
298.15 K							
1.00	0.00	0.00	0.01	0.00	0.99		
0.94	0.06	0.00	0.01	0.02	0.97	0.3	31
0.90	0.10	0.00	0.01	0.03	0.96	0.3	27
0.81	0.19	0.00	0.01	0.06	0.94	0.3	26
0.71	0.29	0.00	0.01	0.09	0.90	0.3	22

(*continued*)

Alkane—rich layer			Ionic liquid—rich layer				
x^I_1	x^I_2	x^I_3	x^{II}_1	x^{II}_2	x^{II}_3	β	S
0.63	0.37	0.00	0.00	0.10	0.90	0.3	
0.58	0.42	0.00	0.00	0.12	0.88	0.3	
0.43	0.57	0.00	0.00	0.18	0.82	0.3	
0.30	0.70	0.00	0.00	0.23	0.77	0.3	
0.18	0.82	0.00	0.00	0.26	0.74	0.3	
0.01	0.99	0.00	0.00	0.31	0.69	0.3	

9.26 15164-03004: Ethyl(2-hydroxyethyl)dimethylammonium dicyanamide

Abbreviation:	$C_2N(CN)_2$
Molecular formula:	$C_8H_{16}ON_4$
Molar mass:	184.25
Structure:	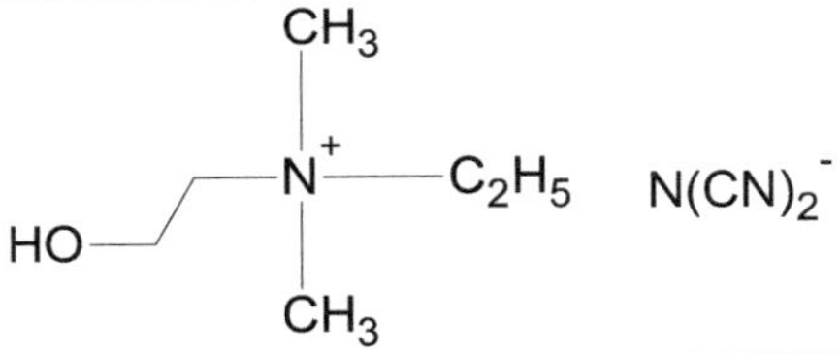

9.26.1 Binary Mixtures

Mixture 1: H_2O (1) + $C_2N(CN)_2$ (2)
(1) Volumetric Properties and Surface Tension

Ref. [15]

$C_2 \times 10^2$ (mol · dm^3)	ρ (g · cm^{-3})	σ (mN · m^{-1})
298.15 K		
44.87	1.0128	48.23
25.27	1.0079	56.80
10.01	1.0016	66.90

9.27 15164-05001: Ethyl(2-hydroxyethyl)dimethylammonium hexafluorophosphate

Abbreviation:	C_2PF_6
Molecular formula:	$C_6H_{16}ONPF_6$
Molar mass:	263.17
Structure:	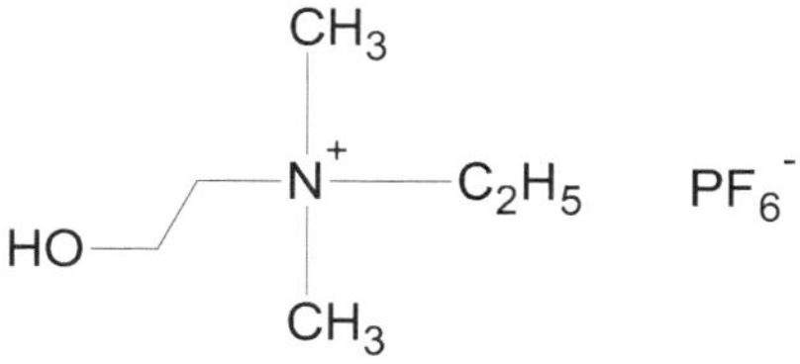

9.27.1 Binary Mixtures

Mixture 1: H_2O (1) + C_2PF_6 (2)
(1) Volumetric Properties and Surface Tension

Ref. [15]

$C_2 \times 10^2$ (mol · dm^3)	ρ (g · cm^{-3})	σ (mN · m^{-1})
298.15 K		
30.11	1.0293	45.30
14.55	1.0152	56.30
10.06	1.0076	61.80

9.28 15165-06005: Diisopropyl-ethylammonium formate

Abbreviation:	[DIPEA][HCOO]
Molecular formula:	$C_9H_{21}NO_2$
Molar mass:	175.27
Structure:	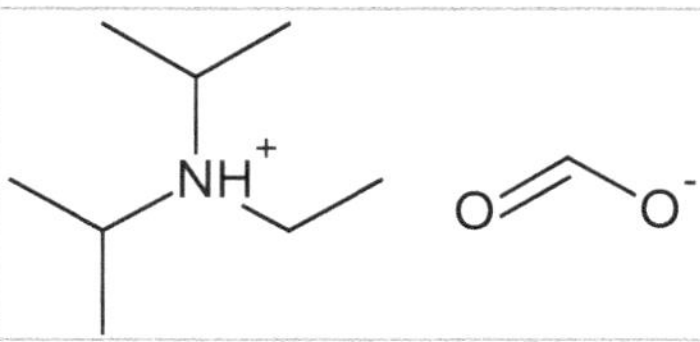

9.28.1 Binary Mixtures

Mixture 1: H_2O (1) + [DIPEA][HCOO] (2)
(1) Volumetric Properties

Ref. [19]

x_2	ρ/g · cm^{-3}	V^E/(cm^3 · mol^{-1})	$V_{\Phi,2}$/(cm^3 · mol^{-1})	$V_{\Phi,1}$/(cm^3 · mol^{-1})
298.15 K				
1.0000	1.0083	0.0000	173.82	
0.8343	1.0100	−0.2803	173.48	16.37
0.6958	1.0117	−0.4897	173.12	16.45
0.5848	1.0134	−0.6325	172.74	16.54
0.5383	1.0142	−0.6831	172.55	16.58
0.4777	1.0152	−0.7383	172.27	16.65
0.3680	1.0173	−0.7959	171.66	16.80
0.2887	1.0187	−0.7871	171.09	16.95
0.1522	1.0185	−0.5914	169.93	17.36
0.1069	1.0160	−0.4435	169.67	17.56
0.0776	1.0127	−0.3183	169.72	17.71
0.0461	1.0070	−0.1629	170.29	17.89
0.0250	1.0020	−0.0616	171.36	18.00
0.0043	0.9974	−0.0008	173.63	18.06
0.0000	0.9969	0.0000		18.06

Ref. [19]

T/K	$\overline{V}_2^{\infty}/(\mathrm{cm}^3 \cdot \mathrm{mol}^{-1})$	$\overline{V}_1^{\infty}/(\mathrm{cm}^3 \cdot \mathrm{mol}^{-1})$
298.15	173	16.30

9.29 15167-06003: *N,N*-dimethylethanolammonium acetate

Abbreviation:	[DMEA][Ac]
Molecular formula:	$C_6H_{15}NO_3$
Molar mass:	149.19
Structure:	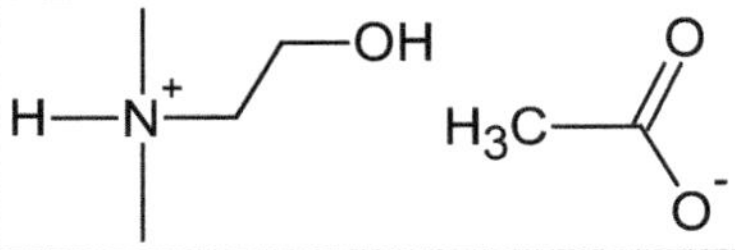

9.29.1 Binary Mixtures

Mixture 1: NH_3 (1) + [DMEA][Ac] (2)
(1) Solubility

Ref. [20]

T (K)	P (MPa)	$100x_1$ (mol%)	T (K)	P (MPa)	$100x_1$ (mol%)
283.2	0.136	47.7 ± 3.7	348.0	0.433	45.4 ± 4.7
283.2	0.198	62.0 ± 2.4	348.0	0.693	60.0 ± 3.1
283.2	0.288	71.6 ± 1.8	348.0	1.335	69.1 ± 2.0
283.2	0.316	76.8 ± 1.2	348.0	1.680	74.5 ± 1.3
283.2	0.415	81.9 ± 0.8	348.0	2.164	79.9 ± 1.0
283.2	0.491	86.5 ± 0.5	348.0	2.689	85.3 ± 0.6
298.1	0.163	47.5 ± 3.8	372.8	1.994	67.5 ± 2.2
298.1	0.278	61.6 ± 2.4	372.8	2.529	73.1 ± 1.3
298.1	0.431	71.3 ± 1.8	372.8	3.305	78.5 ± 0.7
298.1	0.500	76.5 ± 1.2	372.8	4.249	84.4 ± 0.5
298.1	0.641	81.6 ± 0.8	–	–	–
298.1	0.769	86.4 ± 0.5	–	–	–
322.7	0.277	46.6 ± 4.2	298.1	0.401	71.4 ± 2.0
322.7	0.463	60.9 ± 2.3	298.1	0.496	76.5 ± 1.2
322.7	0.786	70.4 ± 1.7	298.1	0.637	81.6 ± 0.8
322.7	0.980	75.7 ± 1.1	298.1	0.791	86.4 ± 0.5
322.7	1.250	80.9 ± 0.7	–	–	–
322.7	1.521	86.0 ± 0.5	–	–	–

9.30 15168-04017: Tris(2-hydroxyethyl)methylammonium methylsulfate

Abbreviation:	$NMe(CH_2CH_2OH)_3$, [MTEOA][MeOSO$_3$]
Molecular formula:	$C_8H_{21}SO_7N$
Molar mass:	275.33
Structure:	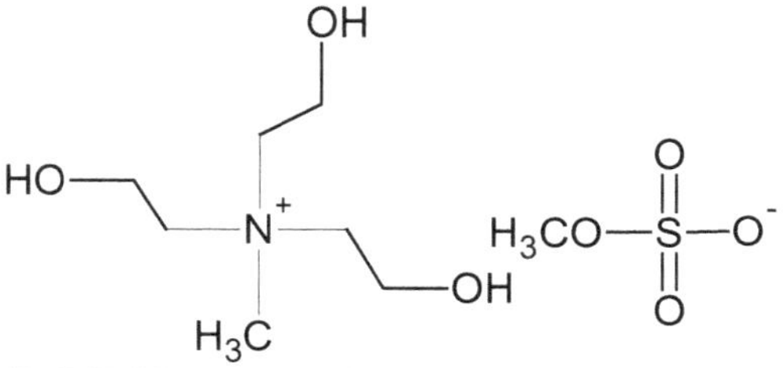

9.30.1 Ternary Mixtures

Mixture 1: H_2O (1) + C_2H_5OH (2) + [MTEOA][MeOSO$_3$] (3)
(1) Volumetric Properties, Transport Properties and Refractive Index

Ref. [21]

x_3	x_1	ρ/g · cm^{-3}	η/mPa · s	n_D	V^E/cm^3 · mol^{-1}	$\Delta\ln(\eta/\eta^o)$
T = 298.15 K and atmospheric pressure						
0.7530	0.1124	1.31685	480.2	1.47880	−0.490	0.816
0.7090	0.1642	1.31501	413.4	1.47809	−0.500	0.985
0.6702	0.2100	1.31328	371.5	1.47704	−0.518	1.160
0.5789	0.3175	1.30843	234.3	1.47504	−0.560	1.362
0.5015	0.4088	1.30291	161.3	1.47232	−0.576	1.551
0.4078	0.5193	1.29373	94.11	1.46736	−0.580	1.692
0.2580	0.6959	1.26775	32.00	1.45477	−0.532	1.701
0.1627	0.8081	1.23350	12.53	1.43874	−0.431	1.455
0.0803	0.9053	1.17125	4.199	1.41007	−0.283	0.960
0.0400	0.9529	1.11016	2.096	1.38250	−0.152	0.559
0.6164	0.1078	1.28615	251.3	1.47296	−1.052	1.130
0.5506	0.2030	1.28278	188.9	1.47091	−1.033	1.325
0.4806	0.3043	1.27823	134.7	1.46847	−1.003	1.498
0.4007	0.4199	1.27113	86.95	1.46487	−0.945	1.645
0.3373	0.5118	1.26381	59.16	1.46074	−0.916	1.724
0.2909	0.5790	1.25691	41.83	1.45521	−0.896	1.716
0.2047	0.7037	1.23668	20.43	1.44641	−0.784	1.630
0.1345	0.8054	1.20571	9.752	1.43142	−0.588	1.403
0.0665	0.9037	1.14685	3.654	1.40282	−0.340	0.918
0.0340	0.9508	1.09502	2.005	1.37802	−0.207	0.556
0.4869	0.1100	1.24775	122.8	1.46471	−1.465	1.326
0.4378	0.1998	1.24478	96.36	1.46278	−1.414	1.446
0.3753	0.3140	1.23994	70.39	1.45992	−1.339	1.593
0.3207	0.4139	1.23427	50.07	1.45663	−1.260	1.656
0.2721	0.5027	1.22749	36.19	1.45274	−1.173	1.690
0.2148	0.6073	1.21660	23.07	1.44684	−1.069	1.662
0.1628	0.7025	1.20104	14.20	1.43775	−0.922	1.561
0.1061	0.8061	1.17178	7.224	1.42210	−0.677	1.304
0.0545	0.9004	1.12238	3.245	1.39600	−0.407	0.884
0.0274	0.9499	1.07604	1.870	1.37214	−0.222	0.532
0.3622	0.1012	1.19167	52.72	1.45272	−1.549	1.358
0.3173	0.2125	1.18885	42.59	1.45114	−1.498	1.481
0.2711	0.3273	1.18501	31.69	1.44763	−1.435	1.532
0.2303	0.4284	1.18030	24.28	1.44363	−1.357	1.571
0.1960	0.5137	1.17506	18.95	1.43994	−1.284	1.581
0.1601	0.6028	1.16721	14.07	1.43423	−1.174	1.553
0.1144	0.7160	1.15106	8.655	1.42353	−0.962	1.409

(*continued*)

x_3	x_1	ρ/g · cm^{-3}	η/mPa · s	n_D	V^E/cm^3 · mol^{-1}	$\Delta\ln(\eta/\eta^o)$
0.0784	0.8056	1.13036	5.410	1.41028	−0.762	1.210
0.0384	0.9046	1.08721	2.687	1.38445	−0.424	0.809
0.0185	0.9541	1.04948	1.646	1.36260	−0.203	0.469
0.1775	0.1042	1.06098	11.83	1.42317	−1.372	1.166
0.1578	0.2037	1.06187	10.89	1.42127	−1.401	1.240
0.1357	0.3151	1.06269	9.374	1.41909	−1.409	1.266
0.1183	0.4028	1.06336	8.428	1.41691	−1.408	1.298
0.0974	0.5081	1.06310	7.158	1.41295	−1.346	1.301
0.0771	0.6108	1.06169	5.957	1.40779	−1.241	1.279
0.0604	0.6951	1.05884	4.907	1.40102	−1.108	1.218
0.0385	0.8054	1.05105	3.530	1.38821	−0.863	1.063
0.0193	0.9025	1.03357	2.221	1.36833	−0.490	0.753
0.0096	0.9515	1.01773	1.522	1.35307	−0.240	0.452
0.0000	0.0480	0.79106	1.158	1.36013	−0.194	0.082
0.0000	0.1116	0.79908	1.261	1.36100	−0.411	0.179
0.0000	0.1599	0.80532	1.326	1.36150	−0.549	0.238
0.0000	0.1761	0.80744	1.360	1.36168	−0.589	0.266
0.0000	0.2668	0.82006	1.496	1.36247	−0.793	0.378
0.0000	0.3010	0.82506	1.556	1.36266	−0.852	0.424
0.0000	0.3994	0.84069	1.726	1.36295	−0.987	0.545
0.0000	0.4970	0.85826	1.932	1.36280	−1.063	0.676
0.0000	0.6122	0.88242	2.161	1.36176	−1.073	0.809
0.0000	0.7116	0.90785	2.385	1.35974	−1.025	0.926
0.0000	0.7774	0.92661	2.353	1.35727	−0.927	0.924
0.0000	0.8145	0.93762	2.366	1.35540	−0.837	0.936
0.0000	0.8529	0.94942	2.193	1.35243	−0.720	0.867
0.0000	0.8967	0.96095	1.920	1.34783	−0.500	0.742
0.0000	0.9481	0.97677	1.432	1.34063	−0.242	0.458
0.0478	0.9522	1.12913	2.341	1.38887	−0.150	0.614
0.0967	0.9033	1.19506	5.054	1.41769	−0.233	1.030
0.1939	0.8061	1.25848	16.79	1.44581	−0.323	1.528
0.2421	0.7579	1.27598	27.00	1.45390	−0.346	1.655
0.2894	0.7106	1.28849	41.22	1.45949	−0.344	1.736
0.3872	0.6128	1.30602	86.07	1.46758	−0.305	1.765
0.4858	0.5142	1.31749	159.5	1.47335	−0.248	1.670
0.6064	0.3936	1.32714	290.1	1.47587	−0.178	1.396
0.6698	0.3302	1.33096	390.7	1.47878	−0.140	1.235
0.7822	0.2178	1.33636	569.0	1.48130	−0.067	0.798
0.8670	0.1330	1.33964	756.2	1.48270	−0.019	0.470
0.9398	0.0602	1.34207	960.7	1.48341	0.015	0.183
0.0482	0.0000	0.87674	2.260	1.38180	−0.572	0.402
0.0981	0.0000	0.95117	4.523	1.39899	−0.939	0.745
0.1509	0.0000	1.01410	8.351	1.41373	−1.190	0.986
0.1994	0.0000	1.06107	13.32	1.42476	−1.325	1.111

(continued)

x_3	x_1	ρ/g · cm^{-3}	η/mPa · s	n_D	V^E/cm^3 · mol^{-1}	$\Delta\ln(\eta/\eta^o)$
0.2420	0.0000	1.09716	19.62	1.43322	−1.494	1.198
0.2922	0.0000	1.13275	29.16	1.44083	−1.578	1.241
0.3882	0.0000	1.18780	58.72	1.45307	−1.707	1.264
0.4841	0.0000	1.22934	110.8	1.46222	−1.678	1.223
0.5923	0.0000	1.26446	204.3	1.46950	−1.426	1.073
0.6288	0.0000	1.27393	247.6	1.47153	−1.262	1.008
0.6881	0.0000	1.28867	333.6	1.47511	−1.086	0.888
0.7379	0.0000	1.29919	430.4	1.47666	−0.859	0.792
0.7878	0.0000	1.30946	518.1	1.47867	−0.713	0.626
0.8379	0.0000	1.31878	657.1	1.48018	−0.550	0.511
0.8738	0.0000	1.32439	775.7	1.48134	−0.346	0.423
0.9203	0.0000	1.33187	902.1	1.48278	−0.184	0.247

Mixture 2: H_2O (1) + CH_3OH (2) + [MTEOA][MeOSO$_3$] (3)
(1) Volumetric Properties, Transport Properties and Refractive Index

Ref. [22]

x_3	x_1	ρ/(g · cm^{-3})	η/(mPa · s)	n_D	V^E/(cm^3 · mol^{-1})	$\Delta\ln(\eta/\eta^o)$
298.15 K and atmospheric pressure						
0.0498	0.8079	1.09512	2.987	1.38928	−0.623	0.915
0.0985	0.7665	1.16497	5.692	1.41584	−0.681	1.204
0.1482	0.7242	1.20861	10.27	1.43326	−0.700	1.431
0.2003	0.6799	1.23906	17.76	1.44427	−0.703	1.599
0.2971	0.5976	1.27448	41.09	1.45798	−0.666	1.730
0.3873	0.5210	1.29501	80.45	1.46609	−0.624	1.743
0.5338	0.3964	1.31579	188.8	1.47396	−0.524	1.526
0.6079	0.3333	1.32293	282.8	1.47698	−0.458	1.389
0.6748	0.2765	1.32824	375.3	1.47887	−0.401	1.183
0.8000	0.1700	1.33609	645.1	1.48187	−0.295	0.811
0.8640	0.1156	1.33913	805.6	1.48293	−0.215	0.565
0.9452	0.0466	1.34246	1022	1.48383	−0.114	0.210
0.0430	0.6699	1.04645	2.825	1.38229	−0.965	0.981
0.1026	0.6282	1.13900	5.914	1.41404	−1.021	1.280
0.1880	0.5685	1.21014	14.81	1.43880	−1.017	1.568
0.2883	0.4982	1.25584	35.31	1.45444	−0.952	1.697
0.3859	0.4299	1.28283	72.49	1.46407	−0.855	1.696
0.4905	0.3567	1.30203	139.0	1.47061	−0.724	1.574
0.5814	0.2930	1.31406	228.7	1.47487	−0.604	1.402
0.7047	0.2067	1.32627	412.5	1.47904	−0.451	1.082
0.7950	0.1435	1.33299	613.2	1.48158	−0.319	0.812
0.8853	0.0803	1.33852	820.4	1.48326	−0.188	0.437
0.9427	0.0401	1.34158	1000	1.48388	−0.108	0.211
0.0487	0.5239	1.02383	2.813	1.38276	−1.161	1.007
0.0944	0.4987	1.09979	5.000	1.40679	−1.262	1.241

(continued)

x_3	x_1	$\rho/(g \cdot cm^{-3})$	$\eta/(mPa \cdot s)$	n_D	$V^E/(cm^3 \cdot mol^{-1})$	$\Delta \ln(\eta/\eta^o)$
0.2020	0.4394	1.19762	15.30	1.43848	−1.279	1.557
0.2959	0.3877	1.24327	34.52	1.45310	−1.202	1.671
0.3939	0.3338	1.27363	70.78	1.46287	−1.085	1.659
0.4905	0.2806	1.29399	130.8	1.46927	−0.926	1.553
0.5927	0.2243	1.30979	235.0	1.47448	−0.761	1.376
0.6855	0.1732	1.32060	361.8	1.47787	−0.592	1.117
0.7577	0.1334	1.32757	492.6	1.47997	−0.471	0.886
0.8252	0.0963	1.33309	652.9	1.48168	−0.351	0.665
0.8667	0.0734	1.33609	765.4	1.48263	−0.272	0.515
0.8953	0.0577	1.33792	829.4	1.48298	−0.204	0.382
0.0481	0.3763	0.98870	2.374	1.37737	−1.190	0.916
0.0978	0.3567	1.07356	4.484	1.40485	−1.348	1.178
0.1893	0.3205	1.16602	12.11	1.43325	−1.410	1.482
0.2662	0.2901	1.21310	24.17	1.44620	−1.395	1.593
0.3911	0.2407	1.26147	63.90	1.46093	−1.305	1.625
0.4855	0.2034	1.28492	119.9	1.46796	−1.155	1.544
0.5980	0.1589	1.30497	224.0	1.47354	−0.935	1.321
0.6740	0.1289	1.31545	345.9	1.47693	−0.781	1.182
0.7601	0.0949	1.32490	491.1	1.47947	−0.565	0.885
0.8531	0.0581	1.33350	712.7	1.48180	−0.356	0.556
0.8942	0.0418	1.33689	833.5	1.48317	−0.273	0.403
0.0418	0.1929	0.93735	1.666	1.37056	−1.074	0.703
0.0994	0.1813	1.04182	3.669	1.40379	−1.410	1.053
0.1424	0.1726	1.09590	6.073	1.41446	−1.547	1.229
0.1994	0.1612	1.14862	11.07	1.42947	−1.641	1.395
0.2859	0.1438	1.20324	24.38	1.44594	−1.629	1.524
0.3781	0.1252	1.24225	49.60	1.45652	−1.513	1.530
0.4837	0.1039	1.27368	107.6	1.46565	−1.351	1.499
0.5812	0.0843	1.29464	193.0	1.47157	−1.143	1.339
0.6722	0.0660	1.30985	306.8	1.47479	−0.940	1.108
0.7663	0.0471	1.32228	489.6	1.47939	−0.696	0.858
0.8986	0.0204	1.33643	852.9	1.48294	−0.390	0.403
0.9373	0.0126	1.33977	975.6	1.48361	−0.276	0.242
0.0000	0.0518	0.79640	0.618	1.32843	−0.235	0.106
0.0000	0.1058	0.80527	0.694	1.33005	−0.385	0.195
0.0000	0.1633	0.81558	0.777	1.33206	−0.551	0.278
0.0000	0.2185	0.82567	0.863	1.33358	−0.683	0.356
0.0000	0.2576	0.83313	0.928	1.33460	−0.770	0.409
0.0000	0.3118	0.84347	1.017	1.33600	−0.861	0.473
0.0000	0.3978	0.85994	1.160	1.33804	−0.944	0.561
0.0000	0.5067	0.88281	1.339	1.33995	−1.003	0.650
0.0000	0.6196	0.90743	1.504	1.34106	−0.956	0.709
0.0000	0.7050	0.92632	1.580	1.34097	−0.839	0.715
0.0000	0.7474	0.93580	1.600	1.34053	−0.756	0.707

(continued)

x_3	x_1	$\rho/(g \cdot cm^{-3})$	$\eta/(mPa \cdot s)$	n_D	$V^E/(cm^3 \cdot mol^{-1})$	$\Delta \ln(\eta/\eta^o)$
0.0000	0.7968	0.94677	1.569	1.33964	−0.635	0.662
0.0000	0.8519	0.95890	1.458	1.33817	−0.471	0.560
0.0000	0.9015	0.97019	1.308	1.33626	−0.308	0.427
0.0000	0.9466	0.98129	1.128	1.33450	−0.157	0.256
0.0516	0.0000	0.92209	1.373	1.36679	−0.839	0.532
0.0958	0.0000	1.00259	2.653	1.39035	−1.245	0.848
0.1459	0.0000	1.06925	5.171	1.40937	−1.511	1.128
0.2091	0.0000	1.13061	9.870	1.42698	−1.711	1.286
0.2545	0.0000	1.16435	15.60	1.43665	−1.800	1.393
0.2952	0.0000	1.18874	22.44	1.44345	−1.804	1.441
0.3943	0.0000	1.23382	50.97	1.45584	−1.719	1.495
0.4885	0.0000	1.26421	98.55	1.46397	−1.543	1.426
0.5295	0.0000	1.27509	128.2	1.46696	−1.469	1.373
0.5761	0.0000	1.28588	169.1	1.46985	−1.356	1.289
0.6788	0.0000	1.30556	307.4	1.47509	−1.092	1.092
0.7812	0.0000	1.32133	520.2	1.47959	−0.844	0.827
0.8385	0.0000	1.32813	654.1	1.48106	−0.613	0.613
0.9412	0.0000	1.33937	910.0	1.48396	−0.304	0.149

9.31 15170-03009: (Benzyl)dimethyldodecylammonium nitrate

Abbreviation:	$[N_{1,1,12,CH2Ph}][NO_3]$
Molecular formula:	$C_{21}H_{38}N_2O_3$
Molar mass:	366.53
Structure:	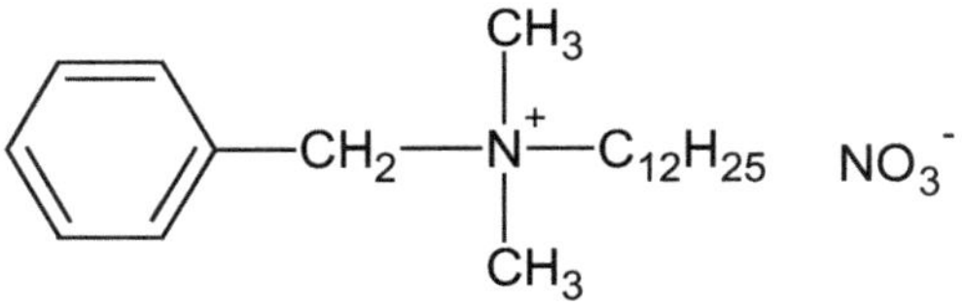

9.31.1 Ternary Mixtures

Mixture 1: H_2O (1) + $[N_{1,1,12,CH2Ph}][NO_3]$ (2) + $[N_{1,1,14,CH2Ph}][NO_3]$ (3)
(1) Phase Equilibrium

Ref. [23]

w_{2+3}	T_2^{SLE}/K	T_2^{LLE}/K	w_2	T_2^{SLE}/K	T_2^{LLE}/K
$[C_{21}H_{38}N][NO_3]$ (w = 0.4) + $[C_{23}H_{42}N][NO_3]$ (w = 0.6)					
0.0000	273.2		0.5657	281.8	

(continued)

w_{2+3}	T_2^{SLE}/K	T_2^{LLE}/K	w_2	T_2^{SLE}/K	T_2^{LLE}/K
0.0059	273.2	332.9	0.6373	289.9	
0.0547	273.2	351.1	0.6756	297.5	
0.1112	273.2	360.8	0.7170	302.1	
0.1659	273.2	357.2	0.7635	307.2	
0.2075	273.2	350.4	0.8284	310.6	
0.2543	273.2	340.0	0.8638	310.8	
0.2975	273.2	330.6	0.8938	311.0	
0.3449	273.2	321.5	0.9324	311.2	
0.3845	273.2	312.2	0.9579	312.2	
0.4264	273.2	302.8	0.9773	312.2	
0.4640	273.2	291.7	0.9779	312.1	
0.5140	273.2		1.0000	311.8	

Mixture 2: 1-C_3H_7OH (1) + $[N_{1,1,12,CH2Ph}][NO_3]$ (2) + $[N_{1,1,14,CH2Ph}][NO_3]$ (3)
(1) Phase Equilibrium

Ref. [23]

w_{2+3}	T_{2+3}^{SLE}/K	T_{2+3}^{LLE}/K	w_{2+3}	T_{2+3}^{SLE}/K	T_{2+3}^{LLE}/K
$[C_{21}H_{38}N][NO_3]$ (w = 0.4) + $[C_{23}H_{42}N][NO_3]$ (w = 0.6)					
0.0000	146.7		0.8782	301.5	
0.6412	256.3		0.9401	306.7	
0.6896	272.2		0.9966	311.4	
0.7744	286.1		1.0000	311.8	
0.8180	292.3				

Mixture 3: 1-C_4H_9OH (1) + $[N_{1,1,12,CH2Ph}][NO_3]$ (2) + $[N_{1,1,14,CH2Ph}][NO_3]$ (3)
(1) Phase Equilibrium

Ref. [23]

w_{2+3}	T_{2+3}^{SLE}/K	T_{2+3}^{LLE}/K	w_{2+3}	T_{2+3}^{SLE}/K	T_{2+3}^{LLE}/K
$[C_{21}H_{38}N][NO_3]$ (w = 0.4) + $[C_{23}H_{42}N][NO_3]$ (w = 0.6)					
0.0000	183.7		0.6721	250.0	
0.1738	224.0		0.7498	266.0	
0.2142	227.7		0.9023	295.0	
0.3649	228.8		0.9652	305.0	
0.5058	232.0		0.9843	309.4	
0.5482	234.6		1.0000	311.8	
0.6072	240.4				

Mixture 4: 1-$C_6H_{13}OH$ (1) + [$N_{1,1,12,CH2Ph}$][NO_3] (2) + [$N_{1,1,14,CH2Ph}$][NO_3] (3)
(1) Phase Equilibrium

Ref. [23]

w_{2+3}	T_{2+3}^{SLE}/K	T_{2+3}^{LLE}/K	w_{2+3}	T_{2+3}^{SLE}/K	T_{2+3}^{LLE}/K
[$C_{21}H_{38}N$][NO_3] (w = 0.4) + [$C_{23}H_{42}N$][NO_3] (w = 0.6)					
0.0000	228.2		0.8250	304.5	304.5
0.1318	243.2		0.8730	304.2	325.5
0.2488	247.8		0.8926	304.7	334.4
0.3867	254.9		0.9392	305.2	342.1
0.4830	259.8		0.9733	305.5	340.5
0.5974	266.0		0.9874	306.0	
0.6755	274.4		0.9880	306.0	
0.7576	286.0		1.0000	311.8	
0.8074	296.5				

Mixture 5: 1-$C_8H_{17}OH$ (1) + [$N_{1,1,12,CH2Ph}$][NO_3] (2) + [$N_{1,1,14,CH2Ph}$][NO_3] (3)
(1) Phase Equilibrium

Ref. [23]

w_{2+3}	T_{2+3}^{SLE}/K	T_{2+3}^{LLE}/K	w_{2+3}	T_{2+3}^{SLE}/K	T_{2+3}^{LLE}/K
[$C_{21}H_{38}N$][NO_3] (w = 0.4) + [$C_{23}H_{42}N$][NO_3] (w = 0.6)					
0.0000	258.2		0.7117	301.0	330.3
0.0495	272.3		0.7633	301.0	340.5
0.1279	276.1		0.8034	301.0	347.3
0.1842	277.7		0.8455	301.0	352.0
0.2844	277.8		0.9050	301.0	354.2
0.3870	279.8		0.9586	301.0	350.8
0.4721	284.8		0.9825	301.0	344.5
0.5525	290.1		0.9932	301.0	332.4
0.6089	301.0		0.9960	301.0	
0.6594	301.0	317.3	1.0000	311.8	

Mixture 6: 1-$C_{10}H_{21}OH$ (1) + [$N_{1,1,12,CH2Ph}$][NO_3] (2) + [$N_{1,1,14,CH2Ph}$][NO_3] (3)
(1) Phase Equilibrium

Ref. [23]

w_{2+3}	T_{2+3}^{SLE}/K	T_{2+3}^{LLE}/K	w_{2+3}	T_{2+3}^{SLE}/K	T_{2+3}^{LLE}/K
[$C_{21}H_{38}N$][NO_3] (w = 0.4) + [$C_{23}H_{42}N$][NO_3] (w = 0.6)					
0.0000	278.2		0.6699	294.6	323.8
0.0917	275.4		0.7308	294.6	330.4
0.1744	280.6		0.8036	294.6	338.5
0.2744	284.2		0.8393	294.6	340.6
0.3730	287.4		0.9062	294.6	342.2
0.4645	290.3		0.9655	294.6	341.6
0.5417	294.4		1.0000	311.8	

(*continued*)

w_{2+3}	T_{2+3}^{SLE}/K	T_{2+3}^{LLE}/K	w_{2+3}	T_{2+3}^{SLE}/K	T_{2+3}^{LLE}/K
0.6135	294.5	314.3			

Mixture 7: C_6H_6 (1) + $[N_{1,1,12,CH2Ph}][NO_3]$ (2) + $[N_{1,1,14,CH2Ph}][NO_3]$ (3)
(1) Phase Equilibrium

Ref. [23]

w_{2+3}	T_{2+3}^{SLE}/K	T_{2+3}^{LLE}/K	w_{2+3}	T_{2+3}^{SLE}/K	T_{2+3}^{LLE}/K
$[C_{21}H_{38}N][NO_3]$ (w = 0.4) + $[C_{23}H_{42}N][NO_3]$ (w = 0.6)					
0.0000	278.7		0.5637	269.0	
0.0435	278.7		0.6380	247.5	
0.0989	278.7		0.6750	266.2	
0.1676	278.7		0.7463	285.9	
0.2129	278.7		0.8040	303.5	
0.2759	278.7		0.8442		315.4
0.3266	278.7		0.8966		327.0
0.3806	278.7		0.9260		333.2
0.4399	278.7		0.9817		328.3
0.4934	277.5		1.0000	311.8	
0.5212	274.1				

$T_{2,e}$ =247.5 K; $w_{2,e}$ =0.638.

Mixture 8: $C_6H_5CH_3$ (1) + $[N_{1,1,12,CH2Ph}][NO_3]$ (2) + $[N_{1,1,14,CH2Ph}][NO_3]$ (3)
(1) Phase Equilibrium

Ref. [23]

w_{2+3}	T_{2+3}^{SLE}/K	T_{2+3}^{LLE}/K	w_{2+3}	T_{2+3}^{SLE}/K	T_{2+3}^{LLE}/K
$[C_{21}H_{38}N][NO_3]$ (w = 0.4) + $[C_{23}H_{42}N][NO_3]$ (w = 0.6)					
0.0000	178.2		0.7409	304.4	
0.0793	192.2		0.7660		311.8
0.1832	219.6		0.8025		320.4
0.3042	235.7		0.8569		340.5
0.4517	248.3		0.8944		346.7
0.5879	259.5		0.9335		355.3
0.6300	265.3		0.9826		348.5
0.6709	280.7		1.0000	311.8	

Mixture 9: C_6H_{14} (1) + $[N_{1,1,12,CH2Ph}][NO_3]$ (2) + $[N_{1,1,14,CH2Ph}][NO_3]$ (3)
(1) Phase Equilibrium

Ref. [23]

w_{2+3}	T_{2+3}^{SLE}/K	T_{2+3}^{LLE}/K	w_{2+3}	T_{2+3}^{SLE}/K	T_{2+3}^{LLE}/K
$[C_{21}H_{38}N][NO_3]$ (w = 0.4) + $[C_{23}H_{42}N][NO_3]$ (w = 0.6)					
0.0000	178.9		0.4542		
0.0014			0.5674		
0.0015			0.7461		338.2

(*continued*)

w_{2+3}	$T_{2+3}{}^{SLE}$/K	$T_{2+3}{}^{LLE}$/K	w_{2+3}	$T_{2+3}{}^{SLE}$/K	$T_{2+3}{}^{LLE}$/K
0.0037			0.8386		314.3
0.0059			0.8818	298.7	
0.0139			0.9340	302.1	
0.0828			0.9715	305.4	
0.1866			0.9906	307.0	
0.2725			1.0000	311.8	

Mixture 10: $C_{16}H_{34}$ (1) + $[N_{1,1,12,CH2Ph}][NO_3]$ (2) + $[N_{1,1,14,CH2Ph}][NO_3]$ (3)
(1) Phase Equilibrium

Ref. [23]

w_{2+3}	$T_{2+3}{}^{SLE}$/K	$T_{2+3}{}^{LLE}$/K	w_{2+3}	$T_{2+3}{}^{SLE}$/K	$T_{2+3}{}^{LLE}$/K
$[C_{21}H_{38}N][NO_3]$ (w = 0.4) + $[C_{23}H_{42}N][NO_3]$ (w = 0.6)					
0.0000	291.2		0.2125		
0.0009			0.2691		
0.0046			0.3557		
0.0176			0.5394		
0.0246			0.7254		433.2
0.0382			0.8500		367.6
0.0742			0.9217		326.3
0.0966			0.9743	305.6	
0.1363			0.9840	306.9	
0.1577			1.0000	311.8	
0.1869					

9.32 15171-03009: (Benzyl)dimethyltetradecylammonium nitrate

Abbreviation:	$[N_{1,1,14,CH2Ph}][NO_3]$
Molecular formula:	$C_{23}H_{42}N_2O_3$
Molar mass:	394.58
Structure:	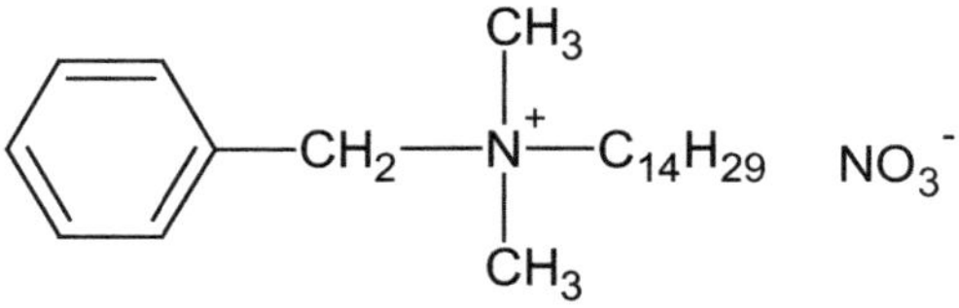

9.32.1 Ternary Mixtures

Mixture 1: H_2O (1) + $[N_{1,1,12,CH2Ph}][NO_3]$ (2) + $[N_{1,1,14,CH2Ph}][NO_3]$ (3)
(1) Phase Equilibrium
Ref. [23]
These physicochemical properties data are listed in a mixture 1 of IL 15170-03009.
Mixture 2: 1-C_3H_7OH (1) + $[N_{1,1,12,CH2Ph}][NO_3]$ (2) + $[N_{1,1,14,CH2Ph}][NO_3]$ (3)
(1) Phase Equilibrium
Ref. [23]
These physicochemical properties data are listed in a mixture 2 of IL 15170-03009.
Mixture 3: 1-C_4H_9OH (1) + $[N_{1,1,12,CH2Ph}][NO_3]$ (2) + $[N_{1,1,14,CH2Ph}][NO_3]$ (3)
(1) Phase Equilibrium
Ref. [23]
These physicochemical properties data are listed in a mixture 3 of IL 15170-03009.
Mixture 4: 1-$C_6H_{13}OH$ (1) + $[N_{1,1,12,CH2Ph}][NO_3]$ (2) + $[N_{1,1,14,CH2Ph}][NO_3]$ (3)
(1) Phase Equilibrium
Ref. [23]
These physicochemical properties data are listed in a mixture 4 of IL 15170-03009.
Mixture 5: 1-$C_8H_{17}OH$ (1) + $[N_{1,1,12,CH2Ph}][NO_3]$ (2) + $[N_{1,1,14,CH2Ph}][NO_3]$ (3)
(1) Phase Equilibrium
Ref. [23]
These physicochemical properties data are listed in a mixture 5 of IL 15170-03009.
Mixture 6: 1-$C_{10}H_{21}OH$ (1) + $[N_{1,1,12,CH2Ph}][NO_3]$ (2) + $[N_{1,1,14,CH2Ph}][NO_3]$ (3)
(1) Phase Equilibrium
Ref. [23]
These physicochemical properties data are listed in a mixture 6 of IL 15170-03009.
Mixture 7: C_6H_6 (1) + $[N_{1,1,12,CH2Ph}][NO_3]$ (2) + $[N_{1,1,14,CH2Ph}][NO_3]$ (3)
(1) Phase Equilibrium
Ref. [23]
These physicochemical properties data are listed in a mixture 7 of IL 15170-03009.
Mixture 8: $C_6H_5CH_3$ (1) + $[N_{1,1,12,CH2Ph}][NO_3]$ (2) + $[N_{1,1,14,CH2Ph}][NO_3]$ (3)
(1) Phase Equilibrium
Ref. [23]
These physicochemical properties data are listed in a mixture 8 of IL 15170-03009.
Mixture 9: C_6H_{14} (1) + $[N_{1,1,12,CH2Ph}][NO_3]$ (2) + $[N_{1,1,14,CH2Ph}][NO_3]$ (3)
(1) Phase Equilibrium
Ref. [23]
These physicochemical properties data are listed in a mixture 9 of IL 15170-03009.
Mixture 10: $C_{16}H_{34}$ (1) + $[N_{1,1,12,CH2Ph}][NO_3]$ (2) + $[N_{1,1,14,CH2Ph}][NO_3]$ (3)
(1) Phase Equilibrium
Ref. [23]
These physicochemical properties data are listed in a mixture 10 of IL 15170-03009.

9.33 15172-03001: Acetylcholinium bis (trifluoromethylsulfonyl)imide

Abbreviation:	$AcChNTf_2$
Molecular formula:	$C_9H_{16}N_2F_6O_6S_2$
Molar mass:	426.36
Structure:	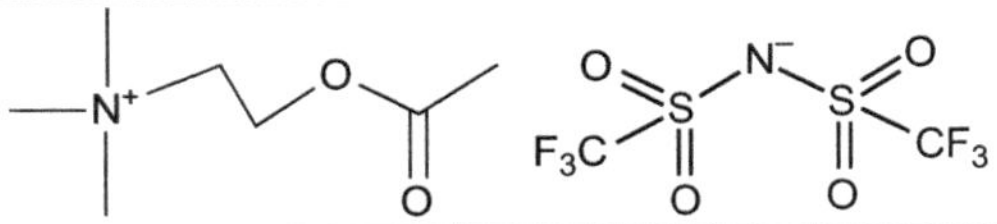

9.33.1 Binary Mixtures

Mixture 1: 1H,1H,2H,2H-$C_8F_{13}H_4OH$ (1) + $AcChNTf_2$ (2)
(1) Cloud Point

Ref. [24]

x_2	T/K	x_2	T/K	x_2	T/K	x_2	T/K
0.006	282.6	0.169	330.7	0.412	332.0	0.671	326.8
0.017	304.6	0.202	331.4	0.311	332.8	0.712	323.4
0.028	313.6	0.227	331.8	0.394	332.4	0.749	320.4
0.045	319.6	0.255	332.0	0.456	332.2	0.910	280.4
0.064	323.6	0.286	332.3	0.519	331.2	0.853	302.5
0.087	326.1	0.318	332.2	0.559	330.5	0.801	314.5
0.106	327.9	0.348	332.4	0.594	329.9	0.831	308.8
0.129	329.6	0.374	332.0	0.630	328.8	0.722	322.8

Mixture 2: 1-C_3H_7OH (1) + $AcChNTf_2$ (2)
(1) Cloud Point

Ref. [24]

x_2	T/K		T/K	x_2	T/K	x_2	T/K
0.008	308.6	0.063	333.5	0.173	333.4	0.629	275.7
0.017	319.8	0.076	334.1	0.190	332.9	0.553	292.3
0.023	324.7	0.093	334.3	0.181	333.2	0.422	311.5
0.031	328.2	0.108	334.4	0.212	331.8	0.340	321.4
0.041	331.0	0.125	334.3	0.248	329.5	0.242	329.8
0.051	332.5	0.153	333.9	0.289	326.4		

Mixture 3: C_2H_5OH (1) + $AcChNTf_2$ (2)
(1) Cloud Point

Ref. [24]

x_2	T/K	x_2	T/K	x_2	T/K	x_2	T/K
0.009	255.3	0.027	285.5	0.143	292.2	0.202	290.0
0.011	258.4	0.030	286.8	0.520	251.6	0.179	291.1
0.014	273.4	0.034	288.3	0.400	269.8	0.160	291.8
0.018	275.5	0.041	289.7	0.325	279.4	0.122	292.6
0.019	276.5	0.072	292.2	0.268	285.4	0.088	292.7
0.024	283.8	0.104	292.8	0.231	288.1		

References

1. Heintz A, Vasiltsova TV, Safarov J, Bich E, Verevkin SP (2006) Thermodynamic properties of mixtures containing ionic liquids. 9. Activity coefficients at infinite dilution of hydrocarbons, alcohols, esters, and aldehydes in trimethyl-butylammonium bis(trifluoromethylsulfonyl) imide using gas–liquid chromatography and static method. J Chem Eng Data 51(2):648–655
2. Wang S, Jacquemin J, Husson P, Hardacre C, Costa Gomes MF (2009) Liquid–liquid miscibility and volumetric properties of aqueous solutions of ionic liquids as a function of temperature. J Chem Thermodyn 41(11):1206–1214
3. Dittmar H, Butka A, Vale VR, Schröer W (2009) Liquid–liquid phase transition in the ionic solutions of tetra-n-butylammonium chloride in o-xylene and ethylbenzene: phase diagrams and corresponding state analysis. J Mol Liq 145(3):116–128
4. Hou Y, Gu Y, Zhang S, Yang F, Ding H, Shan Y (2008) Novel binary eutectic mixtures based on imidazole. J Mol Liq 143(2–3):154–159
5. Zafarani-Moattar MT, Shekaari H (2006) Volumetric and compressibility behaviour of ionic liquid, 1-*n*-butyl-3-methylimidazolium hexafluorophosphate and tetrabutylammonium hexafluorophosphate in organic solvents at T = 298.15 K. J Chem Thermodyn 38(5):624–633
6. Kleemeier M, Wiegand S, Schröer W, Weingärtner H (1999) The liquid–liquid phase transition in ionic solutions: coexistence curves of tetra-*n*-butylammonium pricrate in alkyl alcohols. J Chem Phys 110(6):3085–3099
7. Ishimatsu R, Kitazumi Y, Nishi N, Kakiuchi T (2009) Phase transition of a binary room-temperature ionic liquid composed of bis(pentafluoroethanesulfonyl)amide salts of tetraheptylammonium and N-tetradecylisoquinolinium and its surface properties at the ionic liquid|water interface. J Phys Chem B 113(27):9321–9325
8. Sibiya PN, Deenadayalu N (2008) Excess molar volumes and isentropic compressibility of binary systems {trioctylmethylammonium bis(trifluoromethysulfonyl)imide + methanol or ethanol or 1-propanol} at different temperatures. J Chem Thermodyn 40(7):1041–1045
9. Gan Q, Xue M, Rooney D (2006) A study of fluid properties and microfiltration characteristics of room temperature ionic liquids [C_{10}-min][NTf_2] and N_{8881}[NTf_2] and their polar solvent mixtures. Sep Purif Technol 51(2):185–192
10. Bou Malham I, Letellier P, Mayaffre A, Turmine M (2007) Part I: thermodynamic analysis of volumetric properties of concentrated aqueous solutions of 1-butyl-3-methylimidazolium tetrafluoroborate, 1-butyl-2,3-dimethylimidazolium tetrafluoroborate, and ethylammonium nitrate based on pseudo-lattice theory. J Chem Thermodyn 39(8):1132–1143
11. Wang JF, Li XM, Meng H, Li CX, Wang ZH (2009) Boiling temperature measurement for water, methanol, ethanol and their binary mixtures in the presence of a hydrochloric or acetic salt of mono-, di- or tri-ethanolamine at 101.3 kPa. J Chem Thermodyn 41(2):167–170

12. Arce A, Earle MJ, Rodríguez H, Seddon KR, Soto A (2009) Bis{(trifluoromethyl)sulfonyl} amide ionic liquids as solvents for the extraction of aromatic hydrocarbons from their mixtures with alkanes: effect of the nature of the cation. Green Chem 11(3):365–372
13. Constantinescu D, Schaber K, Agel F, Klingele MH, Schubert TJS (2007) Viscosities, vapor pressures, and excess enthalpies of choline lactate + water, choline glycolate + water, and choline methanesulfonate + water systems. J Chem Eng Data 52(4):1280–1285
14. Domańska U, Ługowska K, Pernak J (2007) Phase equilibria of didecyldimethylammonium nitrate ionic liquid with water and organic solvents. J Chem Thermodyn 39(5):729–736
15. Domańska U, Pobudkowska A, Rogalski M (2008) Surface tension of binary mixtures of imidazolium and ammonium based ionic liquids with alcohols, or water: cation, anion effect. J Colloid Interface Sci 322(1):342–350
16. Domańska U, Pobudkowska A, Żołek-Tryznowska Z (2007) Effect of an ionic liquid (IL) cation on the ternary system (IL + p-xylene + hexane) at T = 298.15 K. J Chem Eng Data 52 (6):2345–2349
17. Domańska U, Marciniak A, Królikowski M (2008) Phase equilibria and modeling of ammonium ionic liquid, C_2NTf_2, solutions. J Phys Chem B 112(4):1218–1225
18. Domańska U, Pobudkowska A, Królikowski M (2007) Separation of aromatic hydrocarbons from alkanes using ammonium ionic liquid C_2NTf_2 at T = 298.15 K. Fluid Phase Equilib 259 (2):173–179
19. Anouti M, Caillon-Caravanier M, Dridi Y, Jacquemin J, Hardacre C, Lemordant D (2009) Liquid densities, heat capacities, refractive index and excess quantities for {protic ionic liquids + water} binary system. J Chem Thermodyn 41(6):799–808
20. Yokozeki A, Shiflett MB (2007) Vapor–liquid equilibria of ammonia + ionic liquid mixtures. Appl Energy 84(12):1258–1273
21. Arce A, Soto A, Ortega J, Sabater G (2008) Viscosities and volumetric properties of binary and ternary mixtures of tris(2-hydroxyethyl) methylammonium methylsulfate + water + ethanol at 298.15 k. Chem Eng Data 53(3):770–775
22. Arce A, Soto A, Ortega J, Sabater G (2009) Mixing properties of tris(2-hydroxyethyl) methylamonium methylsulfate, water, and methanol at 298.15 K. Data treatment using several correlation equations. J Chem Thermodyn 41(2):235–242
23. Domańska U, Ba-kała I, Pernak J (2007) Phase equilibria of an ammonium ionic liquid with organic solvents and water. J Chem Eng Data 52(1):309–314
24. Ferreira R, Blesic M, Trindade J, Marrucho I, Lopes JNC, Rebelo LPN (2008) Solubility of fluorinated compounds in a range of ionic liquids. Cloud-point temperature dependence on composition and pressure. Green Chem 10(9):918–928

10 Mixtures Containing Phosphonium Ionic Liquids

10.1 17004-03001: Trihexyl-tetradecylphosphonium bis((trifluoromethyl)sulfonyl)imide

Abbreviation:	[P666,14][TFSI], [$P_{666,14}$][Ntf_2], [$P_{666,14}$][NTf_2]
Molecular formula:	$C_{34}H_{68}F_6NO_4PS_2$
Molar mass:	764.01
Structure:	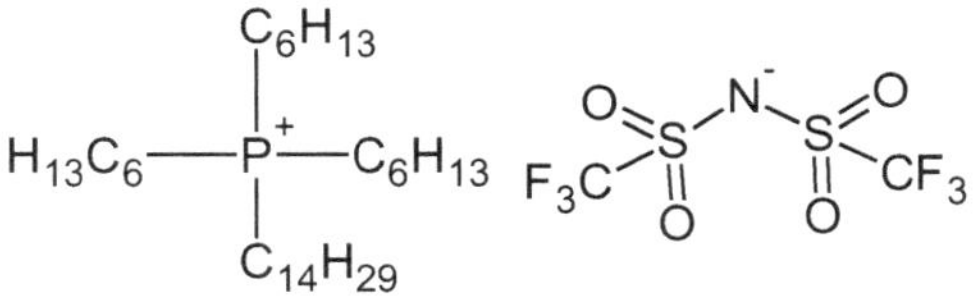

10.1.1 Binary Mixtures

Mixture 1: CF_4 (1) + [$P_{666,14}$][Ntf_2] (2)
(1) Equilibrium Pressure and Henry's Law Constant

Ref. [1]

T/K	$p/10^2$ Pa	$K_H/10^5$ Pa	$x_1/10^{-3}$	dev %
303.41	1053.56	374.25	2.67	−6.7
303.41	1095.30	341.13	2.93	+2.4
303.43	1045.89	337.23	2.97	+3.6
313.38	1131.44	370.43	2.70	+0.0
313.41	1088.30	388.70	2.57	−4.7
313.45	1080.41	347.77	2.88	+6.6
323.39	1167.77	407.90	2.45	+2.5
323.42	1123.13	409.76	2.44	+2.1

(*continued*)

S. Zhang et al., *Physicochemical Properties of Ionic Liquid Mixtures*,
DOI 10.1007/978-94-017-7573-1_10

T/K	$p/10^2$ Pa	$K_H/10^5$ Pa	$x_1/10^{-3}$	dev %
323.46	1115.28	431.75	2.32	−3.1
333.42	1204.43	524.25	1.91	−5.3
333.42	1149.74	488.34	2.05	+1.6
343.44	1150.29	612.93	1.63	+0.2
343.49	1241.01	605.45	1.65	+1.6

(2) Thermodynamic Properties

Ref. [1]

T/K	$\Delta_{solv}G^{\infty}$/kJ · mol^{-1}	$\Delta_{solv}H^{\infty}$/kJ · mol^{-1}	$\Delta_{solv}S^{\infty}$/J · mol^{-1} · K^{-1}
$p^0 = 0.1$ MPa			
303	14.74	−2.90	−58.2
313	15.39	−7.24	−72.3
323	16.19	−12.1	−87.7
333	17.15	−17.6	−104
343	18.28	−23.6	−122

Mixture 2: C_2F_6 (1) + [$P_{666,14}$][Ntf_2] (2)
(1) Equilibrium Pressure and Henry's Law Constant

Ref. [1]

T/K	$p/10^2$ Pa	$K_H/10^5$ Pa	$x_1/10^{-3}$	dev %
303.19	1156.68	183.62	5.45	+0.6
303.52	1055.40	184.57	5.42	+0.4
313.20	1195.90	210.35	4.75	−0.2
313.52	1091.18	207.85	4.81	−0.5
323.22	1234.99	234.31	4.27	+1.1
323.42	1126.58	235.09	4.25	+1.1
333.23	1274.28	280.25	3.57	−0.2
343.23	1313.48	338.18	2.96	−0.3

(2) Thermodynamic Properties

Ref. [1]

T/K	$\Delta_{solv}G^{\infty}$/kJ · mol^{-1}	$\Delta_{solv}H^{\infty}$/kJ · mol^{-1}	$\Delta_{solv}S^{\infty}$/J · mol^{-1} · K^{-1}
$p^0 = 0.1$ MPa			
303	13.14	−7.43	−67.9
313	13.86	−10.1	−76.5
323	14.67	−13.0	−85.8
333	15.58	−16.3	−95.8
343	16.59	−19.9	−106

Mixture 3: C_3F_8 (1) + $[P_{666,14}][Ntf_2]$ (2)
(1) Equilibrium Pressure and Henry's Law Constant

Ref. [1]

T/K	$p/10^2$ Pa	$K_H/10^5$ Pa	$x_1/10^{-3}$	dev %
303.43	1035.75	86.07	11.6	+2.0
303.43	1048.77	87.73	11.4	+0.1
303.64	696.38	88.58	11.3	−0.7
304.45	411.99	89.50	11.2	−1.1
313.44	1072.20	95.17	10.5	+1.1
313.44	1086.07	97.89	10.2	−1.7
323.47	1123.25	108.72	9.20	−0.7
333.53	1160.42	121.07	8.26	+2.1
343.50	1181.49	144.88	6.90	−1.0
343.70	1198.36	143.88	6.95	+0.0

(2) Thermodynamic Properties

Ref. [1]

T/K	$\Delta_{solv}G^{\infty}$/kJ · mol^{-1}	$\Delta_{solv}H^{\infty}$/kJ · mol^{-1}	$\Delta_{solv}S^{\infty}$/J · mol^{-1} · K^{-1}
$p^0 = 0.1$ MPa			
303	11.27	−6.19	−57.6
313	11.87	−8.27	−64.4
323	12.55	−10.6	−71.6
333	13.31	−13.1	−79.3
343	14.14	−15.9	−87.6

Mixture 4: $[C_2py][NTf_2]$ (1) + $[P_{666,14}][NTf_2]$ (2)
(1) Phase Equilibrium

Ref. [2]

These physicochemical properties data are listed in a mixture 1 of IL 11025-03001.

Mixture 5: $[C_2mim][NTf_2]$ (1) + $[P_{666,14}][NTf_2]$ (2)
(1) Phase Equilibrium

Ref. [2]

These physicochemical properties data are listed in a mixture 63 of IL 02003-03001.

Mixture 6: 1H,1H,2H,2H-$C_8F_{13}H_4OH$ (1) + $[P_{666,14}][NTf_2]$ (2)
(1) Cloud Point

Ref. [3]

x_2	T/K	x_2	T/K	x_2	T/K	x_2	T/K
0.015	283.8	0.180	306.1	0.686	278.2	0.163	306.0
0.021	290.3	0.234	306.1	0.013	280.9	0.215	306.0

(*continued*)

x_2	T/K	x_2	T/K	x_2	T/K	x_2	T/K
0.035	295.9	0.381	303.2	0.028	293.6	0.268	305.8
0.048	300.0	0.445	300.7	0.058	301.7	0.315	304.8
0.063	302.4	0.518	295.8	0.085	303.9	0.770	263.3
0.084	303.6	0.634	284.3	0.114	305.3	0.568	292.0
0.134	305.5						

10.1.2 Ternary Mixtures

Mixture 1: $[C_2mim][NTf_2]$ (1) + $[C_4mim][NTf_2]$ (2) + $[P_{666,14}][NTf_2]$ (3)
(1) Phase Equilibrium

Ref. [2]

These physicochemical properties data are listed in a mixture 13 of IL 02003-03001.

10.2 17004-04028: Tetradecyl(trihexyl)phosphonium 1,1,2-trifluoro-2-(perfluoroethoxy) ethanesulfonate

Abbreviation:	[6,6,6,14-P][TPES]
Molecular formula:	$C_{36}H_{69}F_8PSO_4$
Molar mass:	780.96
Structure:	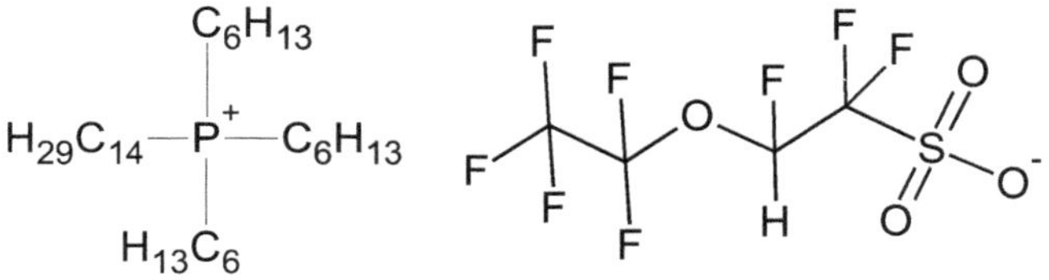

10.2.1 Binary Mixtures

Mixture 1: 1,1,1,2-$C_2H_2F_4$ (1) + [6,6,6,14-P][TPES] (2)
(1) Solubility and Diffusivity

Ref. [4]

T/K	P/MPa	100 x_1	10^{11} D/$m^2 \cdot s^{-1}$	T/K	P/MPa	100 x_1	10^{11} D/$m^2 \cdot s^{-1}$
283.10	0.0098	3.8	1.7	323.15	0.0101	1.1	–
282.90	0.0500	19.0	2.0	323.15	0.0501	6.8	9.5
283.05	0.1000	33.8	2.7	323.15	0.1001	13.4	9.9

(continued)

T/K	P/MPa	100 x_1	10^{11} D/m^2 · s^{-1}	T/K	P/MPa	100 x_1	10^{11} D/m^2 · s^{-1}
282.90	0.1499	45.2	3.3	323.15	0.1500	19.3	11
283.20	0.1994	55.1	5.6	323.15	0.2001	24.6	9.8
282.90	0.2496	63.4	10	323.15	0.2501	29.4	12
282.85	0.3000	71.8	12	323.15	0.3000	33.9	12
282.80	0.3495	79.9	–	323.20	0.3497	38.1	11
298.10	0.0100	1.8	2.8	348.05	0.0101	0.3	18
298.10	0.0500	12.1	4.2	348.10	0.0502	3.9	18
298.10	0.1000	22.8	4.6	348.05	0.1001	8.3	20
298.05	0.1499	32.0	5.5	348.10	0.1496	12.2	21
298.10	0.2000	39.7	6.6	348.10	0.2001	16.0	23
298.10	0.2503	46.2	8.9	348.10	0.2503	19.4	23
298.05	0.3004	52.0	9.6	348.10	0.3003	22.7	21
298.10	0.3504	57.4	11	348.10	0.3505	25.6	21

10.3 17005-04002: Tetrabutylphosphonium methanesulfonate

Abbreviation:	$[(CH_3CH_2CH_2CH_2)_4P][CH_3SO_3]$
Molecular formula:	$C_{17}H_{39}O_3SP$
Molar mass:	354.53
Structure:	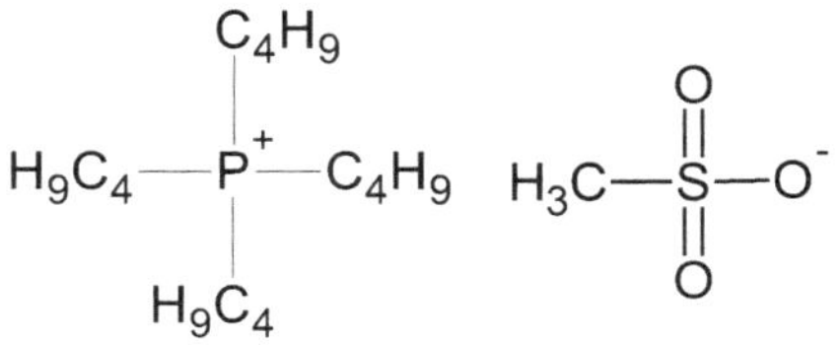

10.3.1 Binary Mixtures

Mixture 1: $1\text{-}C_8H_{17}OH$ (1) + $[(CH_3CH_2CH_2CH_2)_4P][CH_3SO_3]$ (2)
(1) Eutectic Temperature

Ref. [5]

$T_{2,e}$/K	$x_{2,e}$
247.64	0.2372

Mixture 2: 1-$C_{10}H_{21}OH$ (1) + [$(CH_3CH_2CH_2CH_2)_4P$][CH_3SO_3] (2)
(1) Eutectic Temperature

Ref. [5]

$T_{2,e}$/K	$x_{2,e}$
265.10	0.2929

Mixture 3: 1-$C_{12}H_{25}OH$ (1) + [$(CH_3CH_2CH_2CH_2)_4P$][CH_3SO_3] (2)
(1) Eutectic Temperature

Ref. [5]

$T_{2,e}$/K	$x_{2,e}$
279.83	0.3350

Mixture 4: C_6H_6 (1) + [$(CH_3CH_2CH_2CH_2)_4P$][CH_3SO_3] (2)
(1) Eutectic Temperature

Ref. [5]

$T_{2,e}$/K	$x_{2,e}$
271.87	0.1977

10.4 17005-04007: Tetra(*n*-butyl)phosphonium tosylate

Abbreviation:	[$(C_4H_9)_4P$][Tos]
Molecular formula:	$C_{23}H_{43}O_3SP$
Molar mass:	430.63
Structure:	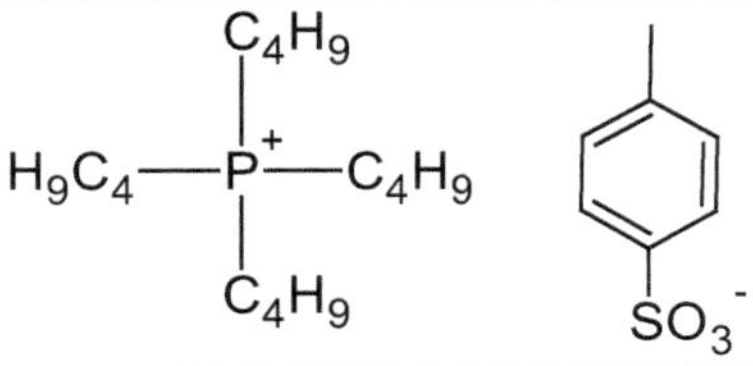

10.4.1 Binary Mixtures

Mixture 1: 1-$C_8H_{17}OH$ (1) + [$(C_4H_9)_4P$][Tos] (2)
(1) Eutectic Temperature

Ref. [6]

$x_{2,e}$	T_e/K
0.165	253.2

Mixture 2: 1-$C_{10}H_{21}OH$ (1) + $[(C_4H_9)_4P][Tos]$ (2)
(1) Eutectic Temperature

Ref. [6]

$x_{2,e}$	T_e/K
0.212	272.8

Mixture 3: 1-$C_{12}H_{25}OH$ (1) + $[(C_4H_9)_4P][Tos]$ (2)
(1) Eutectic Temperature

Ref. [6]

$x_{2,e}$	T_e/K
0.291	285.7

Mixture 4: C_6H_6 (1) + $[(C_4H_9)_4P][Tos]$ (2)
(1) Eutectic Temperature

Ref. [6]

$x_{2,e}$	T_e/K
0.248	268.0

10.5 17005-06010: Tetra(*n*-butyl)phosphonium alaninate

Abbreviation:	$[(C_4H_9)_4P][Ala]$, [TBP][Ala]
Molecular formula:	$C_{19}H_{42}NO_2P$
Molar mass:	347.51
Structure:	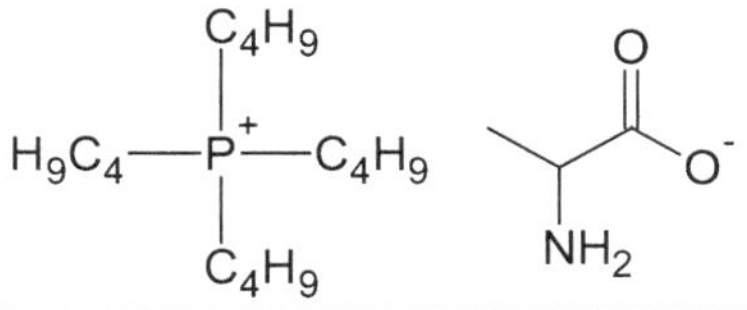

10.5.1 Binary Mixtures

Mixture 1: H_2O (1) + [TBP][Ala] (2)
(1) Transport Properties

Ref. [7]

Composition $V_{f,2}$	η/mPa · s	Composition $V_{f,2}$	η/mPa · s	Composition $V_{f,2}$	η/mPa · s
298.15 K					
0	0.93	0.80	33.18	0.96	150.45

(*continued*)

Composition $V_{f,2}$	η/mPa · s	Composition $V_{f,2}$	η/mPa · s	Composition $V_{f,2}$	η/mPa · s
0.20	2.23	0.85	52.90	0.98	210.85
0.40	6.02	0.90	88.43	1.00	338.50
0.60	13.63				

Mixture 2: CH_3OH (1) + [TBP][Ala] (2)
(1) Transport Properties

Ref. [7]

Composition $V_{f,2}$	η/mPa · s	Composition $V_{f,2}$	η/mPa · s	Composition $V_{f,2}$	η/mPa · s
298.15 K					
0	0.66	0.80	37.58	0.96	159.96
0.20	1.12	0.85	40.80	0.98	260.25
0.40	2.79	0.90	97.03	1.00	338.50
0.60	9.20				

Mixture 3: CH_2Cl_2 (1) + [TBP][Ala] (2)
(1) Transport Properties

Ref. [7]

Composition $V_{f,2}$	η/mPa · s	Composition $V_{f,2}$	η/mPa · s	Composition $V_{f,2}$	η/mPa · s
298.15 K					
0	0.40	0.80	33.20	0.96	204.80
0.20	0.82	0.85	49.00	0.98	244.80
0.40	3.17	0.90	102.20	1.00	338.50
0.60	10.35				

10.6 17005-06011: Tetrabutyl phosphonium valinate

Abbreviation:	$[(C_4H_9)_4P][Val]$, [TBP][Val]
Molecular formula:	$C_{21}H_{46}NO_2P$
Molar mass:	375.57
Structure:	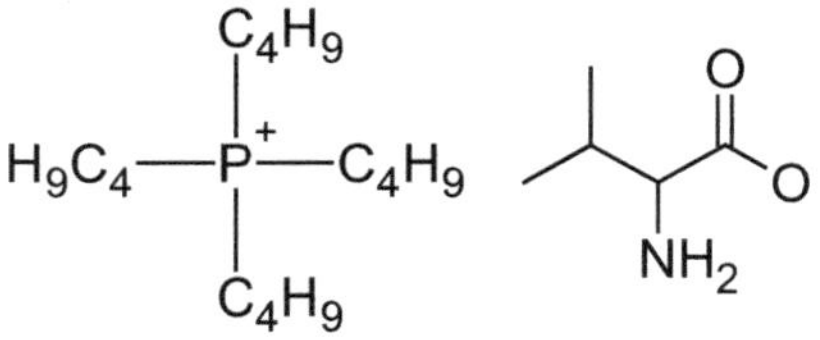

10.6.1 Binary Mixtures

Mixture 1: H_2O (1) + [TBP][Val] (2)
(1) Transport Properties

Ref. [7]

Composition $V_{f,2}$	η/mPa · s	Composition $V_{f,2}$	η/mPa · s	Composition $V_{f,2}$	η/mPa · s
298.15 K					
0	0.93	0.80	45.35	0.96	185.0
0.20	2.10	0.85	–	0.98	250.07
0.40	5.53	0.90	103.8	1.00	419.50
0.60	14.25				

Mixture 2: CH_3OH (1) + [TBP][Val] (2)
(1) Transport Properties

Ref. [7]

Composition $V_{f,2}$	η/mPa · s	Composition $V_{f,2}$	η/mPa · s	Composition $V_{f,2}$	η/mPa · s
298.15 K					
0	0.66	0.80	38.37	0.96	240.8
0.20	1.45	0.85	47.46	0.98	297.9
0.40	3.88	0.90	122.73	1.00	419.15
0.60	9.99				

Mixture 3: CH_2Cl_2 (1) + [TBP][Val] (2)
(1) Transport Properties

Ref. [7]

Composition $V_{f,2}$	η/mPa · s	Composition $V_{f,2}$	η/mPa · s	Composition $V_{f,2}$	η/mPa · s
298.15 K					
0	0.40	0.80	55.55	0.96	316.30
0.20	0.93	0.85	63.00	0.98	344.50
0.40	2.20	0.90	160.50	1.00	419.15
0.60	13.38				

10.7 17017-04029: Tributyl(tetradecyl)phosphonium 1,1,2,3,3,3-hexafluoropropanesulfonate

Abbreviation:	[4,4,4,14-P][HFPS]
Molecular formula:	$C_{29}H_{57}F_6PSO_3$
Molar mass:	630.79
Structure:	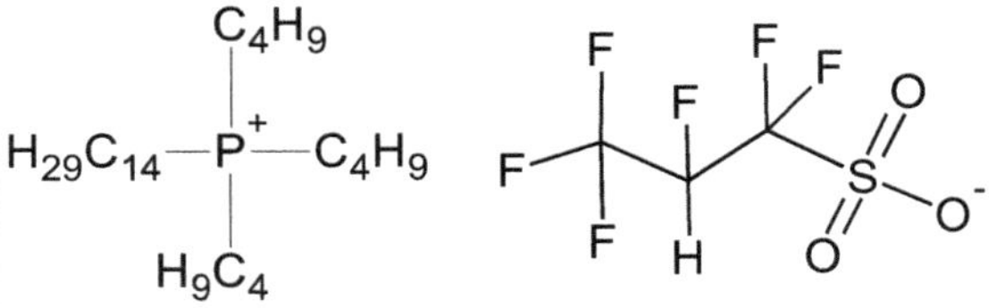

10.7.1 Binary Mixtures

Mixture 1: 1,1,1,2-$C_2H_2F_4$ (1) + [4,4,4,14-P][HFPS] (2)
(1) Solubility and Diffusivity

Ref. [4]

T/K	P/MPa	$100x_1$	$10^{11}\ D/m^2 \cdot s^{-1}$	T/K	P/MPa	$100x_1$	$10^{11}\ D/m^2 \cdot s^{-1}$
283.30	0.0099	3.2	–	323.15	0.0101	1.0	6.8
283.05	0.0500	15.2	1.3	323.15	0.0500	5.6	6.6
283.35	0.1000	28.9	1.7	323.15	0.0999	11.1	6.7
283.20	0.1500	40.3	2.7	323.15	0.1499	16.1	6.5
283.15	0.2000	49.4	4.8	323.15	0.2000	20.9	7.5
283.20	0.2503	58.7	5.4	323.15	0.2495	25.4	8.2
283.15	0.3002	67.2	14.9	323.15	0.2999	29.3	11.1
283.15	0.3493	76.3	20.2	323.15	0.3496	33.3	-
298.10	0.0100	1.9	1.7	348.10	0.0101	0.9	13.6
298.10	0.0500	10.4	2.6	348.10	0.0504	3.7	13.9
298.05	0.0999	19.4	3.2	348.10	0.1001	7.2	13.7
298.10	0.1499	27.3	3.7	348.05	0.1500	10.5	14.3
298.10	0.2002	35.0	3.6	348.10	0.2001	13.6	16.3
298.10	0.2500	41.8	4.8	348.05	0.2504	16.5	17.0
298.10	0.2999	47.7	7.4	348.10	0.3004	19.4	15.1
298.10	0.3502	53.0	10.1	348.05	0.3504	22.1	17.0

10.8 17040-04007: Tri-*iso*-butyl(methyl)phosphonium tosylate

Abbreviation:	[iBu$_3$MeP][Tos]
Molecular formula:	$C_{20}H_{37}O_3PS$
Molar mass:	388.54
Structure:	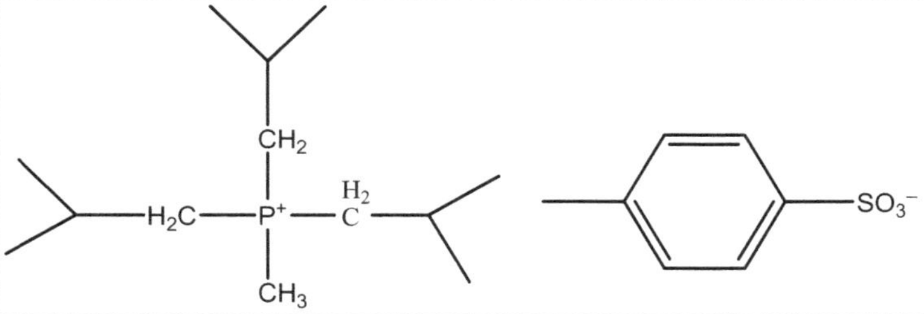

10.8.1 Binary Mixtures

Mixture 1: H_2O(1) + [iBu$_3$MeP][Tos](2)
(1) Transport Properties

Ref. [8]

Mass fraction IL ± 0.0001/ (w/w)	$(\eta \pm \sigma) \times 10^3$/ (Pa · s)	Mass fraction IL ± 0.0001/ (w/w)	$(\eta \pm \sigma) \times 10^3$/ (Pa · s)	Mass fraction IL ± 0.0001/ (w/w)	$(\eta \pm \sigma) \times 10^3$/ (Pa · s)
293.15 K					
0.0000	0.89	0.3750	4.0 ± 0.2	0.7500	23.0 ± 0.7
0.1250	1.65 ± 0.08	0.5000	6.9 ± 0.3	0.8720	68 ± 2
0.2500	2.6 ± 0.1	0.6250	11.6 ± 0.5	1.0000	1320 ± 13

Mixture 2: 1-$C_{10}H_{21}OH$ (1) + *i*-B$_3$MPTOS (2)
(1) Phase Equilibrium

Ref. [9]

x_2^{exptl}	T (K)	T^{calcd}(K)
0.2688	270.06	270.30
0.2368	271.92	271.97
0.1769	274.67	274.66
0.1491	276.04	275.73
0.1219	277.15	276.70
0.0989	277.72	277.44
0.0757	278.51	278.14
0.0489	279.06	278.89
0.0122	279.79	279.81
0.0000	280.09	280.09

Mixture 3: 1-$C_{12}H_{25}OH$ (1) + *i*-B_3MPTOS (2)
(1) Phase Equilibrium

Ref. [9]

x_2^{exptl}	T (K)	T^{calcd} (K)
0.6790	275.71	272.74
0.6452	276.39	274.67
0.6014	277.36	276.99
0.5558	278.78	279.19
0.5100	280.01	281.25
0.4678	281.8	283.01
0.4204	283.28	284.86
0.3809	284.77	286.31
0.3346	285.99	287.90
0.2955	287.43	289.18
0.2272	290.09	291.23
0.1944	291.54	292.15
0.1515	293.08	293.29
0.0912	294.75	294.75
0.0495	295.62	295.67
0.0000	296.67	296.67

Mixture 4: C_6H_{14} (1) + *i*-B_3MPTOS (2)
(1) Phase Equilibrium

Ref. [9]

x_2^{exptl}	T (K)	$(x_2^{calcd})'$ (calculated mole fractions of ionic liquid in IL-rich-phase)	$(x_2^{calcd})''$ (calculated mole fractions of ionic liquid in n-hexane-rich phase)
0.7556	308.40	0.7500	0.0005
0.7241	313.09	0.7254	0.0004
0.6593	324.31	0.7041	0.0004
0.6020	331.78	0.6581	0.0003
0.6991	316.86	0.6061	0.0003
0.5810	335.47	0.5781	0.0003
0.5578	338.12	0.5569	0.0003

Mixture 5: C_7H_{16} (1) + *i*-B_3MPTOS (2)
(1) Phase Equilibrium

Ref. [9]

x_2^{exptl}	T (K)	$(x_2^{calcd})'$ (calculated mole fractions of ionic liquid in IL-rich-phase)	$(x_2^{calcd})''$ (calculated mole fractions of ionic liquid in n-heptane-rich phase)
0.8195	15.86	0.8096	0.0010
0.8175	316.73	0.8056	0.0009

(*continued*)

x_2^{exptl}	T (K)	$(x_2^{calcd})'$ (calculated mole fractions of ionic liquid in IL-rich-phase)	$(x_2^{calcd})''$ (calculated mole fractions of ionic liquid in n-heptane-rich phase)
0.7917	320.28	0.7886	0.0008
0.7838	321.57	0.7822	0.0008
0.7638	324.67	0.7659	0.0007
0.6811	335.72	0.7205	0.0006
0.6345	343.18	0.6992	0.0005
0.6133	348.80	0.6453	0.0004
0.7108	332.45	0.5992	0.0004
0.5832	351.17	0.5782	0.0004
0.5536	353.78	0.5539	0.0004

Mixture 6: C_8H_{18} (1) + *i*-B_3MPTOS (2)
(1) Phase Equilibrium

Ref. [9]

x_2^{exptl}	T (K)	$(x_2^{calcd})'$ (calculated mole fractions of ionic liquid in IL-rich-phase)	$(x_2^{calcd})''$ (calculated mole fractions of ionic liquid in n-octane-rich phase)
0.8840	317.79	0.8666	0.0000
0.8429	322.31	0.8482	0.0000
0.8162	328.38	0.8198	0.0000
0.7705	337.41	0.7686	0.0000
0.7033	345.69	0.7100	0.0000
0.6811	348.51	0.6870	0.0000
0.6750	349.65	0.6771	0.0000
0.6413	353.78	0.6390	0.0000
0.6100	356.29	0.6137	0.0000
0.5801	360.10	0.5716	0.0000

Mixture 7: C_9H_{20} (1) + *i*-B_3MPTOS (2)
(1) Phase Equilibrium

Ref. [9]

x_2^{exptl}	T (K)	$(x_2^{calcd})'$ (calculated mole fractions of ionic liquid in IL-rich-phase)	$(x_2^{calcd})''$ (calculated mole fractions of ionic liquid in n-nonane-rich phase)
0.9605	320.04	0.9258	0.0000
0.9403	324.98	0.9108	0.0000
0.8654	332.22	0.8837	0.0000
0.8462	336.79	0.8630	0.0000
0.8395	337.66	0.8587	0.0000
0.7518	352.56	0.7613	0.0000
0.7322	355.78	0.7329	0.0000
0.7045	359.00	0.7011	0.0000
0.6836	361.51	0.6737	0.0000

Mixture 8: $C_{10}H_{22}$ (1) + *i*-B_3MPTOS (2)
(1) Phase Equilibrium

Ref. [9]

x_2^{exptl}	T (K)	$(x_2^{calcd})'$ (calculated mole fractions of ionic liquid in IL-rich-phase)	$(x_2^{calcd})''$ (calculated mole fractions of ionic liquid in n-decane-rich phase)
0.8641	347.02	0.9429	0.0000
0.8310	353.37	0.9067	0.0001
0.8033	357.30	0.8781	0.0002
0.7563	363.40	0.8380	0.0000
0.9663	330.56	0.8145	0.0000
0.9090	341.14	0.8069	0.0000
0.8108	356.40	0.7461	0.0000

Mixture 9: $C_6H_5CH_3$ (1) + *i*-B_3MPTOS (2)
(1) Phase Equilibrium

Ref. [9]

x_2^{exptl}	T (K)	$(x_2^{calcd})'$ (calculated mole fractions of ionic liquid in IL-rich-phase)	$(x_2^{calcd})''$ (calculated mole fractions of ionic liquid in Toluene-rich phase)
0.1287	320.93	0.1280	0.0072
0.1203	328.95	0.1270	0.0075
0.1144	332.52	0.1197	0.0094
0.1083	336.57	0.1156	0.0105
0.0972	345.13	0.1141	0.0109
0.0886	351.17	0.1104	0.0121
0.0830	354.28	0.1067	0.0133
0.0794	355.62	0.0980	0.0164
0.0755	357.48	0.0911	0.0192
0.0682	359.76	0.0874	0.0208
0.0615	360.9	0.0810	0.0239
0.0558	361.3	0.0779	0.0255
0.0427	361.62	0.0733	0.0282
0.0378	361.29	0.0702	0.0300
0.0314	360.32	0.0664	0.0325
0.0278	358.56	0.0644	0.0339
0.0155	339.35	0.0621	0.0355
0.0193	349.18	0.0604	0.0369
0.0097	322.01	0.0603	0.0369
0.0111	333.68	0.0466	0.0466

Mixture 10: $C_6H_5C_2H_5$ (1) + *i*-B_3MPTOS (2)
(1) Phase Equilibrium

Ref. [9]

x_2^{exptl}	T (K)	$(x_2^{calcd})'$ (calculated mole fractions of ionic liquid in IL-rich-phase)	$(x_2^{calcd})''$ (calculated mole fractions of ionic liquid in ethylbenzene-rich phase)
0.0064	320.38	0.2772	0.0050
0.0056	313.91	0.2719	0.0054
0.0105	361.65	0.2641	0.0060
0.0098	355.17	0.2589	0.0064
0.0092	340.32	0.2483	0.0073
0.0069	333.39	0.2448	0.0076
0.2769	309.47	0.2398	0.0081
0.2589	324.73	0.2287	0.0093
0.2447	336.27	0.2239	0.0098
0.2302	349.29	0.2214	0.0101
0.2242	353.17	0.2161	0.0107
0.2146	359.39	0.2132	0.0111

Mixture 11: $C_6H_5C_3H_7$ (1) + *i*-B_3MPTOS (2)
(1) Phase Equilibrium

Ref. [9]

x_2^{exptl}	T (K)	$(x_2^{calcd})'$ (calculated mole fractions of ionic liquid in IL-rich-phase)	$(x_2^{calcd})''$ (calculated mole fractions of ionic liquid in n-propylbenzene-rich phase)
0.4089	304.32	0.4080	0.0028
0.4006	308.88	0.4021	0.0030
0.3855	322.41	0.3910	0.0033
0.3708	334.51	0.3852	0.0035
0.3664	337.33	0.3712	0.0041
0.3512	349.69	0.3701	0.0041
0.3454	354.47	0.3666	0.0042
0.0038	317.75	0.3653	0.0043
0.0040	333.58	0.3555	0.0047
0.0042	338.38	0.3514	0.0049
0.0045	346.33	0.3458	0.0051
0.0050	354.22	0.3455	0.0051
0.0056	362.43	0.3358	0.0055

Mixture 12: $C_6H_5C_4H_9$ (1) + *i*-B_3MPTOS (2)
(1) Phase Equilibrium

Ref. [9]

x_2^{exptl}	T (K)	$(x_2^{calcd})'$ (calculated mole fractions of ionic liquid in IL-rich-phase)	$(x_2^{calcd})''$ (calculated mole fractions of ionic liquid in n-butylbenzene-rich phase)
0.5277	303.89	0.5256	0.0002
0.5134	317.84	0.5174	0.0004
0.5011	335.23	0.5019	0.0009
0.4935	346.29	0.4961	0.0011
0.4837	350.83	0.4893	0.0014
0.4625	364.36	0.4835	0.0016
0.0024	363.32	0.4659	0.0025
0.0016	340.52	0.4643	0.0026

10.8.2 Ternary Mixtures

Mixture 1: CO_2 (1) + H_2O (2) + [iBu_3MeP][Tos] (3)
(1) Solubility

Ref. [8]

T/K	x_{CO2} = 0.0303 p/MPa		x_{CO2} = 0.0501 p/MPa		x_{CO2} = 0.0801 p/MPa	
4 mol% of IL in water						
288.40	4.5		69.8			
289.05	4.5		66.0			
291.60	4.7		61.4			
295.15	5.4		53.5			
314.85	6.5		27.5			
324.45	6.7		21.5			
329.95	7.5		19.2			
349.45	8.5		17.3		93.80	
369.32	10.4		17.5		57.0	
T/K	x_{CO2} = 0.0316 p/MPa	x_{CO2} = 0.0500 p/MPa	x_{CO2} = 0.0806 p/MPa	x_{CO2} = 0.1028 p/MPa	x_{CO2} = 0.1231 p/MPa	
8 mol% of IL in water						
288.22	2.8	5.7	39.2			
295.45	3.6	5.9	27.8			
314.15	4.1	7.8	14.2	56.4		
332.63	5.2	10.0	14.2	33.5	76.1	
351.77	6.1	10.2	15.2	26.1	46.5	
369.49	6.4	11.2	15.2	24.1	36.2	
T/K	x_{CO2} = 0.0301 p/MPa	x_{CO2} = 0.0503 p/MPa	x_{CO2} = 0.0801 p/MPa	x_{CO2} = 0.1015 p/MPa	x_{CO2} = 0.1200 p/MPa	x_{CO2} = 0.1404 p/MPa
12 mol% of IL in water						

(continued)

276.85	0.8	4.0	8.2	35.1	99.7	
280.65	1.2	4.1	8.4	26.3	84.7	
282.65	1.3	4.1	8.5	22.7	77.5	
285.154	1.5	4.2	8.7	21.3	75.4	
287.70	1.7	4.3	8.8	18.0	59.1	
291.45	2.0	4.4	9.0	13.3	51.8	
293.60	2.1	4.8	9.1	12.1	45.1	
312.55	3.1	4.8	10.2	10.9	23.6	57.6
322.62	3.5	5.5	11.1	9.5	20.2	43.7
331.87	3.8	5.7	12.0	13.9	18.7	34.4
351.45	4.3	7.2	13.3	14.4	17.9	28.1
369.32	5.4	8.0	14.8	15.5	18.5	26.2

(2) Phase Equilibrium

Ref. [9]

T/K	*p*/MPa	*T*/K	*p*/MPa	*T*/K	*p*/MPa	*T*/K	*p*/MPa
4 mol% of IL in water and x_{CO2} = 0.0501							
278.85	3.9	288.05	4.9	298.15	6.1	307.35	6.8
284.05	4.5	293.25	5.3	306.15	6.5		
4 mol% of IL in water and x_{CO2} = 0.0801							
278.75	3.9	284.15	4.5	293.65	5.6	305.65	7.5
279.25	3.9	288.25	5.0	297.15	6.2		
8 mol% of IL in water and x_{CO2} = 0.0806							
280.65	4.1	286.35	4.7	290.25	5.0	295.25	5.2
281.65	4.3						
8 mol% of IL in water and x_{CO2} = 0:1028							
278.65	4.0	283.45	4.5	288.95	5.2	294.95	5.9
8 mol% of IL in water and x_{CO2} = 0.1231							
279.15	4.0	293.35	5.7	298.25	6.5	303.55	7.2
284.45	4.56						
12 mol% of IL in water and x_{CO2} = 0.1200							
282.35	4.5	286.15	5.2	290.75	5.5	298.35	6.6
283.65	4.5	287.85	5.2	293.65	6.0		
12 mol% of IL in water and x_{CO2} = 0.1404							
279.45	4.1	288.95	5.3	297.45	6.4	302.35	7.2
285.05	5.1	293.45	5.9				

References

1. Pison L, Canongia Lopes JN, Rebelo LPN, Padua AAH, Costa Gomes MF (2008) Interactions of fluorinated gases with ionic liquids: solubility of CF_4, C_2F_6, and C_3F_8 in trihexyltetradecylphosphonium bis(trifluoromethylsulfonyl)amide. J Phys Chem B 112(39):12394–12400
2. Arce A, Earle MJ, Katdare SP, Rodríguez H, Seddon KR (2007) Phase equilibria of mixtures of mutually immiscible ionic liquids. Fluid Phase Equilib 261(1–2):427–433
3. Ferreira R, Blesic M, Trindade J, Marrucho I, Lopes JNC, Rebelo LPN (2008) Solubility of fluorinated compounds in a range of ionic liquids. Cloud-point temperature dependence on composition and pressure. Green Chem 10(9):918–928

4. Shiflett MB, Harmer MA, Junk CP, Yokozeki A (2006) Solubility and diffusivity of 1,1,1,2-tetrafluoroethane in room-temperature ionic liquids. Fluid Phase Equilib 242(2):220–232
5. Domańska U, Casás LM (2007) Solubility of phosphonium ionic liquid in alcohols, benzene, and alkylbenzenes. J Phys Chem B 111(16):4109–4115
6. Domańska U, Paduszyński K (2008) Phase equilibria study in binary systems (tetra-*n*-butylphosphonium tosylate ionic liquid + 1-alcohol, or benzene, or *n*-alkylbenzene). J Phys Chem B 112 (35):11054–11059
7. Khupse ND, Kumar A (2009) Dramatic change in viscosities of pure ionic liquids upon addition of molecular solvents. J Solut Chem 38(5):589–600
8. Ventura SPM, Pauly J, Daridon JL, Lopes da Silva JA, Marrucho IM, Dias AMA, Coutinho JAP (2008) High pressure solubility data of carbon dioxide in (tri-iso-butyl(methyl)phosphonium tosylate + water) systems. J Chem Thermodyn 40(8):1187–1192
9. Domańska U, Paduszyński K (2009) (Solid + liquid) and (liquid + liquid) phase equilibria measurements and correlation of the binary systems {tri-iso-butyl(methyl)phosphonium tosylate + alcohol, or + hydrocarbon}. Fluid Phase Equilib 278(1–2):90–96

List of Compounds

Entry	Abbreviation	Name
1	H_2O	Water
2	C_5H_{12}	*n*-Pentane
3	C_6H_{14}	*n*-Hexane
4	C_7H_{16}	*n*-Heptane
5	C_8H_{18}	*n*-Octane
6	C_9H_{20}	1-Nonane
7	$C_{10}H_{22}$	1-Decane
8	$C_{11}H_{24}$	1-Undecane
9	$C_{12}H_{26}$	1-Dodecane
10	$C_{16}H_{34}$	Hexadecane
11	*trans*-3-C_6H_{12}	*trans*-3-Hexene
12	$CH_3C(CH_3)_2CH_2CH(CH_3)CH_3$	2,2,4-Trimethylpentane
13	1-C_5H_{10}	1-Pentene
14	1-C_6H_{12}	1-Hexene
15	1-C_7H_{14}	1-Heptene
16	1-C_8H_{16}	1-Octene
17	1-C_9H_{18}	1-Nonene
18	1-$C_{10}H_{20}$	1-Decene
19	1-$C_{11}H_{22}$	1-Undecene
20	1-$C_{12}H_{24}$	1-Dodecene
21	$CH{\equiv}CCH_2CH_2CH_3$	Pent-1-yne
22	$CH{\equiv}CCH_2CH_2CH_2CH_3$	Hex-1-yne
23	$CH{\equiv}CCH_2CH_2CH_2CH_2CH_3$	Hept-1-yne
24	$CH{\equiv}CCH_2CH_2CH_2CH_2CH_2CH_3$	Oct-1-yne
25	CH_3OH	Methanol
26	C_2H_5OH	Ethanol
27	1-C_3H_7OH	1-Propanol
28	1-C_4H_9OH	1-Butanol
29	1-$C_5H_{11}OH$	1-Pentanol
30	1-$C_6H_{13}OH$	1-Hexanol
31	1-$C_7H_{15}OH$	1-Heptanol

(*continued*)

S. Zhang et al., *Physicochemical Properties of Ionic Liquid Mixtures*,
DOI 10.1007/978-94-017-7573-1

Entry	Abbreviation	Name
32	1-$C_8H_{17}OH$	1-Octanol
33	1-$C_9H_{19}OH$	1-Nonanol
34	1-$C_{10}H_{21}OH$	1-Decanol
35	1-$C_{11}H_{23}OH$	1-Undecanol
36	1-$C_{12}H_{25}OH$	1-Dodecanol
37	1-$C_{13}H_{27}OH$	1-Tridecanol
38	1-$C_{14}H_{29}OH$	1-Tetradecanol
39	2-CH_3CHCH_3OH	2-Propanol
40	2-$CH_3CH(OH)CH_2CH_3$	2-Butanol or *sec*-Butanol
41	2-$CH_3CH(OH)(CH_2)_2CH_3$	2-Pentanol
42	$(CH_3)_2CHCH_2OH$	*iso*-Butanol
43	$(CH_3)_3COH$	*tert*-Butanol
44	$CH_3CH_2C(CH_3)_2OH$	*tert*-Pentanol
45	1-C_4H_9OD	Deuterated 1-Butanol
46	$(CH_3)_2CHCH_2OD$	Deuterated *iso*-Butanol
47	2-$CH_3CH(OD)CH_2CH_3$	Deuterated 2-Butanol
48	$(CH_3)_3COD$	Deuterated *tert*-Butanol
49	1,2-$HOCH_2CH_2OH$	1,2-Ethanediol
50	1,3-HOC_3H_6OH	1,3-Propanediol
51	1,4-$HOCH_2CH_2CH_2CH_2OH$	1,4-Butanediol
52	1,2-$C_6H_{12}(OH)_2$	1,2-Hexanediol
53	CH_3CH_2CHO	Propanal
54	$CH_3(CH_2)_2CHO$	Butanal
55	$CH_3(CH_2)_3CHO$	Pentanal
56	$CH_3(CH_2)_4CHO$	Hexanal
57	$CH_3(CH_2)_5CHO$	Heptanal
58	$CH_3(CH_2)_6CHO$	Octanal
59	$CH_3(CH_2)_7CHO$	Nonan-1-al
60	CH_3COCH_3	Acetone
61	2-$CH_3COC_2H_5$	2-Butanone
62	2-$CH_3COC_3H_7$	2-Pentanone
63	2-$CH_3COC_4H_9$	2-Hexanone
64	2-$CH_3COC_7H_{14}$	Nonan-2-one
65	3-$C_2H_5COC_2H_5$	Pentan-3-one
66	4-$C_3H_7COC_3H_7$	Heptan-4-one
67	$C_6H_5C_2H_4COCH_3$	4-Phenyl-butan-2-one
68	CH_3COOH	Acetic acid
69	CH_3CH_2COOH	Propionic acid
70	$HOOC(CH_2)_2COOH$	Succinic acid
71	$HOOC(CH_2)_3COOH$	Glutaric acid
72	$HOOC(CH_2)_4COOH$	Adipic acid
73	$HOC(CH_2COOH)_2COOH$	Citric acid
74	$HCOOCH_3$	Methylformate
75	$HCOOC_2H_5$	Ethylformate

(*continued*)

Entry	Abbreviation	Name
76	$CH_3OOCCH_2CH_3$	Methyl propanoate
77	$CH_3OOCCH_2CH_2CH_3$	Methyl butanoate
78	$CH_3OOCCH_2CH_2CH_2CH_3$	Methyl pentanoate
79	$CH_3OOCCH_2CH_2CH_2CH_2CH_3$	Methyl hexanoate
80	$CH_3OOCCH_2CH_2CH_2CH_2CH_2CH_3$	Methyl heptanoate
81	CH_3COOCH_3	Methyl acetate
82	$CH_3COOCH_2CH_3$	Ethyl acetate
83	$CH_3COOC_3H_7$	Propyl acetate
84	$CH_3COOC_4H_9$	Butyl acetate
85	$C_6H_5COOC_2H_5$	Ethyl benzoate
86	$CH_3OOC(CH_2)_4COOCH_3$	Dimethyl adipate
87	C_6H_6	Benzene
88	C_6H_5Cl	Chlorobenzene
89	C_6H_5Br	Bromobenzene
90	C_6H_5I	Iodobenzene
91	$C_6H_5CH_3$	Toluene
92	$C_6H_5C_2H_5$	Ethylbenzene
93	$C_6H_5C_3H_7$	Propylbenzene
94	$C_6H_5C_4H_9$	Butylbenzene
95	$C_6H_5C_5H_{11}$	Pentylbenzene
96	m-$C_6H_4(CH_3)_2$	*m*-Xylene
97	o-$C_6H_4(CH_3)_2$	*o*-Xylene
98	p-$C_6H_4(CH_3)_2$	*p*-Xylene
99	C_6H_5CHO	Benzaldehyde
100	$C_6H_5OCH_3$	Methoxy-benzene
101	$C_6H_5CH_2NH_2$	Benzylamine
102	C_6H_5OH	Phenol
103	C_6Cl_5OH	Pentachlorophenol
104	$HOC_6H_4C(CH_3)_2C_6H_4OH$	Bisphenol A
105	$CH_3C_6H_4OH$	(Hydroxymethyl)-benzene
106	$C_6H_5CH_2OH$	Benzyl alcohol
107	$C_6H_5CHOHCH_3$	1-Phenylethanol
108	4-$C_9H_{19}C_6H_4OH$	4-Nonylphenol
109	4-$C_8H_{17}C_6H_4OH$	4-Octylphenol
110	1-{4-$(CH_3)_3C$-C_6H_4}-CH_3CHOH	1-(4-Isobutyl phenyl)-ethanol
111	4-$CH_3C_6H_4CHO$	4-Methyl-benzaldehyde
112	$C_6H_5COCH_3$	Acetophenone
113	$(CH_3)_2CHCH_2C_6H_4COCH_3$	4-Isobutylacetophenone
114	C_6H_5COOH	Benzoic acid
115	CH_3CN	Acetonitrile
116	CH_3NO_2	Nitromethane
117	CH_2Cl_2	Dichloromethane
118	$CHCl_3$	Trichloromethane
119	1,2-$C_2H_4Cl_2$	1,2-Dichloroethane

(*continued*)

Entry	Abbreviation	Name
120	1,3-$ClCH_2CH_2CH_2Cl$	1,3-Dichloropropane
121	CH_3F	Fluoromethane
122	CH_2F_2	Difluoromethane
123	CHF_3	Trifluoromethane
124	CF_4	Perfluoromethane
125	C_2H_5F	Fluoroethane
126	C_2HF_5	Pentafluoroethane
127	1,1-$C_2H_4F_2$	1,1-Difluoroethane
128	1,1,1-$C_2H_3F_3$	1,1,1-Trifluoroethane
129	1,1,1,2-$C_2H_2F_4$	1,1,1,2-Tetrafluoroethane
130	CF_3CF_2H	Pentafluoroethane
131	C_2F_6	Perfluoroethane
132	C_3F_8	Perfluoropropane
133	C_6H_5F	Fluorobenzene
134	1,3-$C_6H_4F_2$	1,3-Difluorobenzene
135	1,4-$C_6H_4F_2$	1,4-Difluorobenzene
136	1,2,4-$C_6H_3F_3$	1,2,4-Trifluorobenzene
137	1,3,5-$C_6H_3F_3$	1,3,5-Trifluorobenzene
138	1,2,3,5-$C_6H_2F_4$	1,2,3,5-Tetrafluorobenzene
139	1,2,4,5-$C_6H_2F_4$	1,2,4,5-Tetrafluorobenzene
140	C_6HF_5	Pentafluorobenzene
141	C_6F_6	Hexafluorobenzene
142	CF_3OCF_2H	Difluoromethoxy trifluoromethane
143	CF_3OCH_3	Trifluoromethoxy methane
144	$CF_3CF_2CF_2OCH_3$	1-Methoxyheptafluoropropane
145	$CF_3CF_2CF_2CF_2OCH_3$	1-Methoxynonafluorobutane
146	$CF_3CF_2CF_2CF_2OCH_2CH_3$	1-Ethoxynonafluorobutane
147	$CF_3CF_2CF(OCH_3)CF(CF_3)_2$	1,1,1,2,2,3,4,5,5,5-Decafluoro-3-methoxy-4-(trifluoromethyl)-pentane
148	$CF_3CF_2CF_2CF(OCH_2CH_3)CF(CF_3)_2$	2-Trifluoromethyl-3-ethoxydodecafluorohexane
149	CF_3CH_2OH	Trifluoroethanol
150	1H,1H,2H,2H-$C_8F_{13}H_4OH$	1*H*,1*H*,2*H*,2*H*-Perfluorooctanol
151	1H,1H,2H,2H,3H,3H -$C_9F_{13}H_6OH$	1*H*,1*H*,2*H*,2*H*,3*H*,3*H*-Perfluorononanol
152	$CFCl_3$	Trichlorofluoromethane
153	$CDCl_3$	Deuterated chloroform
154	$CFCl_2CF_2Cl$	1,1,2-Trichloro-1,2,2-trifluoroethane
155	$CHCl_2CF_3$	2,2-Dichloro-1,1,1-trifluoroethane
156	$CF_3CHFCHFCF_2CF_3$ threo-H	Threo-2,3-dihydrodecafluoropentane (Threo-H)
157	$CF_3CHFCHFCF_2CF_3$ threo-D	Threo-2,3-dihydrodecafluoropentane (Threo-D)
158	$CF_3CHFCHFCF_2CF_3$ erythro-H	Erythro-2,3-dihydrodecafluoropentane (Erythro-H)
159	$CF_3CHFCHFCF_2CF_3$ erythro-D	Erythro-2,3-dihydrodecafluoropentane (Erythro-D)
160	$CH_3CH_2CH_2OCH_2CH_2CH_3$	Dipropyl ether

(*continued*)

Entry	Abbreviation	Name
161	$(CH_3)_2CHOCH(CH_3)_2$	Diisopropyl ether
162	$CH_3(CH_2)_3O(CH_2)_3CH_3$	Dibutyl ether
163	$(CH_3)_3COCH_3$	*tert*-Butyl-methyl ether
164	$(CH_3)_3COC_2H_5$	Ethyl *tert*-butyl ether
165	$C_2H_5OC(CH_3)_2CH_2CH_3$	*tert*-Amyl ethyl ether
166	$CH_3OC(CH_3)_2C_2H_5$	Methyl-1,1-dimethylpropyl ether
167	$HOCH_2CH_2OCH_2CH_2OC_2H_5$	Diethylene glycol monoethyl ether
168	$HOCH_2CH_2OCH_3$	Ethylene glycol monomethyl ether
169	$CH_3OCH_2CH_2OCH_3$	Ethylene glycol dimethyl ether
170	$CH_3(OCH_2CH_2)_2OCH_3$	Diethylene glycol dimethyl ether
171	$HO(CH_2)_2O(CH_2)_2OCH_3$	Diethylene glycol monomethyl ether
172	$HO(CH_2)_2O(CH_2)_2O(CH_2)_2OCH_3$	Triethylene glycol monomethyl ether
173	$CH_2(CH_2)_3CH_2$	Cyclopentane
174	$CH_2(CH_2)_4CH_2$	Cyclohexane
175	$CH_2(CH_2)_5CH_2$	Cycloheptane
176	$CH_2(CH_2)_6CH_2$	Cyclooctane
177	$(CH_2)_4CO$	Cyclopentanone
178	$CH_2(CH_2)_4CHOH$	Cyclohexanol
179	$CH_3CH(CH_2)_4CH_2$	Methylcyclohexane
180	$CH{=}CHCH_2CH_2CH_2$	Cyclopentene
181	$CH{=}CHCH_2CH_2CH_2CH_2$	Cyclohexene
182	$CH{=}CHCH{=}CHCH_2CH_2$	1,3-Cyclohexadiene
183	$CH_2{=}C(CH_3)COOCH_3$	Methyl methacrylate
184	$(CH_3O)_2C{=}O$	Dimethyl carbonate
185	$(CH_3CH_2O)_2C{=}O$	Diethyl carbonate
186	$CH_3CH_2COOCH{=}CH_2$	Vinyl propionate
187	$CH_3CH(Ph)COOCH_2CH_3$	*rac*-1-Phenylethyl propionate
188	$CH_3CH(Ph)OH$	*rac*-1-Phenylethanol
189	$C_4H_6O_3$	Propylene carbonate
190	$C_3H_4O_3$	Ethylene carbonates
191	$OCH_2CH_2CH_2CH_2$	Tetrahydrofuran
192	$CH_2CH_2CH_2CH_2SO_2$	Sulpholane
193	$1,4\text{-}CH_2CH_2OCH_2CH_2O$	1,4-Dioxane
194	$CH_3NCH_2CH_2CH_2C{=}O$	*N*-Methyl-pyrrolidone
195	$HCONH_2$	Formamide
196	$HO(CH_2)_2NH_2$	Ethanolamine
197	$HCON(CH_3)_2$	*N*,*N*-Dimethylformamide
198	$CH_3CON(CH_3)_2$	*N*,*N*-dimethylacetamide
199	$CH_3N(CH_2)_3CO$	N-Methyl-2-pyrrolidinone
200	$3\text{-}CH_3C_5H_4N$	3-Picoline
201	$(NH_2)_2CO$	Urea
202	$(C_5H_5)_2Fe$	Ferrocene
203	$1,4\text{-}CH_2CH_2OCH_2CH_2O$	1,4-Dioxane
204	$NH(CH_2)_5C{=}O$	Caprolactam

(continued)

Entry	Abbreviation	Name
205	$CH_2(CH_2)_3CH_2C{=}NOH$	Cyclohexanone oxime
206	$SCH{=}CHCH{=}CH$	Thiophene
207	CH_3SOCH_3	Dimethylsulfoxide
208	$p\text{-}H_2C{=}C(CH_3)C_6H_4CH_3$	Limonene
209	$(CH_3)_2C{=}CHCH_2CH_2C(OH)(CH_3)CH{=}CH_2$	Linalool
210	$[CH_2\text{-}CH_2\text{-}O]_n$	Poly(ethyleneoxide)
211	$[CH_2\text{-}CHOH]_n$	Poly(vinylalcohol)
212	$[CH_2\text{-}CHCN]_n$	Poly(acrylonitrile)
213	$HO[CH_2CH_2O]_nOH$(PEG-1500, $n{\approx}34$)	Polyethylene glycols
214	$HO[CH_2CH_2O]_nOH$(PEG-2000, $n{\approx}45$)	Polyethylene glycols
215	$HO[CH_2CH_2O]_nOH$(PEG-3400, $n{\approx}77$)	Polyethylene glycols
216	PEOEMA	Poly(2-ethoxyethyl methacrylate)
217	$C_{12}H_{22}O_{11}$	Sucrose
218	$C_5H_{10}O_5$	Xylose
219	$C_6H_{12}O_6$	Glucose
220	$C_6H_{12}O_6$	Fructose
221	$C_{10}H_8$	Naphthalene
222	$C_{10}H_{18}$	Decalin
223	IM	Imidazole
224	IM-C_6H_5	2-phenylimidazole
225	$C_{15}H_{12}N_2$	4,5-diphenylimidazole
226	$C_{21}H_{16}N_2$	2,4,5-triphenylimidazole
227	DODCI	3,3'-Di iethyloxadicarbocyanine iodide
228	I_2	Iodine
229	H_2	Hydrogen
230	O_2	Oxygen
231	CO_2	Carbon dioxide
232	SO_2	Sulfur dioxide
233	NH_3	Ammonia
234	$LiBF_4$	Lithium tetrafluoroborate
235	$LiPF_6$	Lithium hexafluorophosphate
236	$LiClO_4$	Lithium perchlorate
237	$AlCl_3$	Aluminium trichloride
238	NaCl	Sodium chloride
239	KCl	Potassium chloride
240	LiCl	Lithium chloride
241	RbCl	Rubidium chloride
242	CsCl	Caesium chloride
243	NaBr	Sodium bromide
244	NaI	Sodium iodide

(continued)

Entry	Abbreviation	Name
245	Na_2SO_4	Sodium sulfate
246	K_2HPO_4	Dipotassium hydrogen phosphate
247	K_3PO_4	Potassium phosphate
248	$C_6H_5K_3O_7{\cdot}H_2O$	Tri-potassium citrate
249	CuO	Copper oxide
250	Fe_2O_3	Ferric oxide
251	$MgCl_2$	Magnesium chloride
252	$ZnCl_2$	Zinc chloride

List of ILs

Name of ILs	ID
A	
Acetylcholinium bis(trifluoromethylsulfonyl)imide	15172-03001
n-Acryloyloxypropyl-*N*-methylimidazolium bromide	22042-01002
1-Allyl-3-methyl-imidazolium chloride	02199-01001
1-Amyl-3-methylimidazolium bis((trifluoromethyl)sulfonyl)imide	02016-03001
1-Amyl-3-methylimidazolium hexafluorophosphate	02016-05001
B	
(Benzyl)dimethyldodecylammonium nitrate	15170-03009
(Benzyl)dimethyltetradecylammonium nitrate	15171-03009
1-Butyl-2,3-dimethylimidazolium hexafluorophosphate	03010-05001
1-Butyl-2,3-dimethylimidazolium tetrafluoroborate	03010-02001
Butyl(2-hydroxyethyl)dimethylammonium bromide	15099-01002
Butyl(2-hydroxyethyl)dimethylammonium tetrafluoroborate	15099-02001
1-Butyl-3-methylimidazolium acetate	02011-06003
1-Butyl-3-methylimidazolium bis((trifluoromethyl)sulfonyl)imide	02011-03001
1-Butyl-3-methylimidazolium bromide	02011-01002
1-Butyl-3-methylimidazolium chloride	02011-01001
1-Butyl-3-methylimidazolium dibutylphosphate	02011-05013
1-Butyl-3-methylimidazolium dicyanamide	02011-03004
1-Butyl-3-methylimidazolium diethyleneglycolmonomethylethersulphate	02011-04019
1-Butyl-3-methylimidazolium ethylenglykolmonomethylethersulfate	02011-04036
1-Butyl-3-methylimidazolium hexafluorophosphate	02011-05001
1-Butyl-3-methylimidazolium 1,1,2,3,3,3-hexafluoropropanesulfonate	02011-04029
1-Butyl-3-methylimidazolium hydrogen sulfate	02011-04021
1-Butyl-3-methylimidazolium iodine	02011-01003
1-Butyl-3-methyl imidazolium methylsulfate	02011-04017
1-Butyl-3-methylimidazolium nitrate	02011-03009
1-Butyl-3-methylimidazolium octylsulfate	02011-04008
1-Butyl-3-methylimidazolium perchlorate	02011-01004
1-Butyl-3-methylimidazolium tetrafluoroborate	02011-02001
1-Butyl-3-methylimidazolium 1,1,2,2-tetrafluoroethanesulfonate	02011-04030

(*continued*)

S. Zhang et al., *Physicochemical Properties of Ionic Liquid Mixtures*,
DOI 10.1007/978-94-017-7573-1

Name of ILs	ID
1-Butyl-3-methylimidazolium 2-(1,2,2,2-tetrafluoroethoxy)-1,1,2,2-tetrafluoroethanesulfonate	02011-04031
1-Butyl-3-methylimidazolium thiocyanate	02011-15003
1-Butyl-3-methylimidazolium tosylate	02011-04007
1-Butyl-3-methylimidazolium tricyanomethane	02011-07002
1-Butyl-3-methylimidazolium trifluoromethanesulfonate	02011-04003
1-Butyl-3-methylimidazolium 1,1,2-trifluoro-2-(perfluoroethoxy) ethanesulfonate	02011-04028
1-Butyl-3-methylimidazolium 1,1,2-trifluoro-2-(trifluoromethoxy) ethanesulfonate	02011-04033
1-Butyl-3-methylpyridinium bis((trifluoromethyl)sulfonyl)imide	11041-03001
N-Butyl-4-methylpyridinium bis{(trifluoromethyl)sulfonyl}imide	11009-03001
1-Butyl-2-methylpyridinium tetrafluoroborate	11069-02001
1-Butyl-3-methylpyridinium tetrafluoroborate	11041-02001
N-Butyl-3-methylpyridinium tosylate	11041-04007
N-Butyl-4-methylpyridinium tosylate	11009-04007
1-Butyl-1-methyl-pyrrolidinium trifluoromethanesulfonate	08004-04003
N-butyl pyridinium tetrafluoroborate	11001-02001
C	
Choline glycolate(2-Hydroxyethyl-trimethylammonium 2-hydroxyethanoate)	15094-06034
Choline Lactate(2-Hydroxyethyl-trimethylammonium 2-hydroxypropanoate)	15094-06007
Choline methanesulfonate(2-Hydroxyethyl-trimethylammonium methanesulfonate)	15094-04002
2,4,6-Collidine tosylate	11078-04007
Collidinium formate	11078-06005
D	
1-Decyl-2,3-dimethylimidazolium bromide	03017-01002
1-Decyl-3-methylimidazolium bis((trifluoromethyl)sulfonyl)imide	02039-03001
1-Decyl-3-methylimidazolium bromide	02039-01002
1-Decyl-3-methylimidazolium chloride	02039-01001
1-Decyl-3-methylpyridinium bis(trifluoromethylsulfonyl)imide	11076-03001
Didecyldimethylammonium nitrate	15096-03009
Diethanolammonium acetate	15142-06003
Diethanolammonium chloride	15142-01001
1, 3-Dihexyloxymethylimidazolium bis((trifluoromethyl)sulfonyl)imide	02118-03001
1, 3-Dihexyloxymethylimidazolium tetrafluoroborate	02118-02001
Diisopropyl-ethylammonium formate	15165-06005
N,*N*-Dimethylethanolammonium acetate	15167-06003
1,3-Dimethylimidazolium bis((trifluoromethyl)sulfonyl)imide	02001-03001
1,3-Dimethylimidazolium chloride	02001-01001
1,3-Dimethylimidazolium dimethylphosphate	02001-05011
1, 3-Dimethylimidazolium methylsulfate	02001-04017
1, 2-Dimethyl-3-propylimidazolium bis((trifluoromethyl)sulfonyl)imide	03007-03001
1, 2-Dimethyl-3-propylimidazolium tri(trifluoromethylsulfonyl)methide	03007-07001

(*continued*)

Name of ILs	ID
1,4-Dimethylpyridinium tosylate	11077-04007
1-Dodecyl-3-methylimidazolium bis((trifluoromethyl)sulfonyl)imide	02046-03001
1-Dodecyl-3-methylimidazolium bromide	02046-01002
1-Dodecyl-3-methylimidazolium 1,1,2,2-tetrafluoroethanesulfonate	02046-04030
1-Dodecyl-3-methylpyridinium bis(trifluoromethylsulfonyl)imide	11010-03001
E	
Ethylammonium nitrate	15084-03009
1-Ethyl-3-butylbenzotriazolium hexafluorophosphate	07068-05001
1-Ethyl-3-ethylimidazolium diethylphosphate	02007-05012
Ethyl(2-hydroxyethyl)dimethylammonium bis(trifluomethylsulfonyl)imide	15164-03001
Ethyl(2-hydroxyethyl)dimethylammonium bromide	15164-01002
Ethyl(2-hydroxyethyl)dimethylammonium dicyanamide	15164-03004
Ethyl(2-hydroxyethyl)dimethylammonium hexafluorophosphate	15164-05001
Ethyl(2-hydroxyethyl)dimethylammonium tetrafluoroborate	15164-02001
1-Ethyl-3-methylimidazolium acetate	02003-06003
1-Ethyl-3-methylimidazolium bis((perfluoroethane)sulfonyl)imide	02003-03003
1-Ethyl-3-methylimidazolium bis((trifluoromethyl)sulfonyl)imide	02003-03001
1-Ethyl-3-methylimidazolium bromide	02003-01002
1-Ethyl-3-methylimidazolium chloride	02003-01001
1-Ethyl-3-methylimidazolium dicyanoamide	02003-03004
1-Ethyl-3-methylimidazolium diethyleneglycolmonomethylethersulphate	02003-04019
1-Ethyl-3-methylimidazolium diethylphosphate	02003-05012
1-Ethyl-3-methylimidazolium dimethylphosphate	02003-05011
1-Ethyl-3-methylimidazolium ethylsulfate	02003-04018
1-Ethyl-3-methylimidazolium hexafluorophosphate	02003-05001
1-Ethyl-3-methylimidazolium hydrogen sulfate	02003-04021
1-Ethyl-3-methylimidazolium tetrafluoroborate	02003-02001
1-Ethyl-3-methylimidazolium 1,1,2,2-tetrafluoroethanesulfonate	02003-04030
1-Ethyl-3-methylimidazolium thiocyanate	02003-15003
1-Ethyl-3-methylimidazolium tosilate	02003-04007
1-Ethyl-3-methylimidazolium trifluoroacetate	02003-06001
1-Ethyl-3-methylimidazolium trifluoromethanesulfonate	02003-04003
1-Ethyl-3-methylpyridinium ethylsulfate	11038-04018
1-Ethylpyridinium bis((trifluoromethyl)sulfonyl)amide	11025-03001
Ethyl pyridinium ethylsulfate	11025-04018
1-Ethylpyridinium hexafluorophosphate	11025-05001
H	
1-Heptyl-3-methylimidazolium bis((trifluoromethyl)sulfonyl)imide	02024-03001
1-Heptyl-3-methylimidazolium 1,1,2,2-tetrafluoroethanesulfonate	02024-04030
Hexyl(2-hydroxyethyl)dimethylammonium bromide	15163-01002
Hexyl(2-hydroxyethyl)dimethylammonium tetrafluoroborate	15163-02001
1-Hexyl-3-methylimidazolium acetate	02021-06003
1-Hexyl-3-methylimidazolium bis((trifluoromethyl)sulfonyl)imide	02021-03001

(continued)

Name of ILs	ID
1-Hexyl-3-methylimidazolium bromide	02021-01002
1-Hexyl-3-methylimidazolium chloride	02021-01001
1-Hexyl-3-methylimidazolium hexafluorophosphate	02021-05001
1-Hexyl-3-methylimidazolium hydrogen sulfate	02021-04021
1-Hexyl-3-methylimidazolium tetrafluoroborate	02021-02001
1-Hexyl-3-methylimidazolium thiocyanate	02021-15003
1-Hexyl-3-methylimidazolium trifluoromethanesulfonate	02021-04003
N-Hexyl-3-methylpyridinium tosylate	11046-04007
1-Hexyloxymethyl-3-methylimidazolium bis((trifluoromethyl)sulfonyl)imide	02237-03001
1-Hexyloxymethyl-3-methylimidazolium tetrafluoroborate	02237-02001
2-Hydroxyethylammonium acetate	15086-06003
1-Hydroxyethyl-3-methylimidazolium bis((trifluoromethyl)sulfonyl)imide	02125-03001
1-Hydroxyethyl-3-methylimidazolium hexafluorophosphate	02125-05001
1-Hydroxyethyl-3-methylimidazolium tetrafluoroborate	02125-02001
1-Hydroxyethyl-3-methylimidazolium triflate	02125-04003
(2-Hydroxyethyl)trimethylammonium bis{(trifluoromethyl)sulfonyl}amide	15094-03001
1-Hydroxy-1-propyl-3-methylimidazolium nitrate	02239-03009
I	
Imidazolium chloride	00001-01001
1-Isobutenyl-3-methylimidazolium tetrafluoroborate	02238-02001
M	
4-Methyl-*N*-butylpyridinium tetrafluoroborate	11009-02001
N-methyl-*N*-butyl-pyrrolidinium bis((trifluoromethyl)sulfonyl)imide	08004-03001
1-Methyl-1-butyl-pyrrolidinium trifluoromethanesulfonate	08004-04003
1-Methyl-3-ethylimidazolium chloride	02236-01001
N-methyl-*N*-hexylpyrrolidinium bis((trifluoromethyl)sulfonyl)imide	08005-03001
1-Methylimidazolium chloride	01001-01001
1-Methyl-3-methylimidazolium dimethylphosphate	02001-05011
1-Methyl-3-octylimidazolium diethyleneglycolmonomethylethersulphate	02029-04019
1-Methyl-1-octylpiperidinium bis((trifluoromethyl)sulfonyl)imide	09012-03001
1-Methyl-1-octylpyrrodinium bis((trifluoromethyl)sulfonyl)imide	08014-03001
3-Methyl-1-pentylimidazolium bromide	02016-01002
3-Methyl-1-pentylimidazolium trifluoroacetate	02016-06001
N-Methyl-*N*-propylpiperidinium bis((trifluoromethyl)sulfonyl)imide	09001-03001
3-Methyl-1-propylpyridinium bis(trifluoromethylsulfonyl)imide	11074-03001
N-Methyl-*N*-propyl-pyrrolidinium bis((trifluoromethyl)sulfonyl)imide	08003-03001
3-Methyl-1-vinylimidazolium methyl sulfate	02085-04017
O	
1-Octadecyl-3-methylimidazolium bis((trifluoromethyl)sulfonyl)imide	02059-03001
1-Octyl-3-methylimidazolium bis((trifluoromethyl)sulfonyl)imide	02029-03001
1-Octyl-3-methylimidazolium bromide	02029-01002
1-Octyl-3-methylimidazolium chloride	02029-01001

(*continued*)

Name of ILs	ID
1-Octyl-3-methylimidazolium hexafluorophosphate	02029-05001
1-Octyl-3-methylimidazolium tetrafluoroborate	02029-02001
1-Octyl pyridinium tetrafluoroborate	11003-02001
P	
1-Pentyl-3-methylimidazolium chloride	02016-01001
1-Propyl-2,3-dimethylimidazolium tetrafluoroborate	03009-02001
Propyl(2-hydroxyethyl)dimethylammonium bromide	15132-01002
1-Propyl-3-methylimidazolium bis((trifluoromethyl)sulfonyl)imide	02008-03001
1-Propyl-3-methylimidazolium bromide	02008-01002
1-Propyl-3-methylimidazolium iodine	02008-01003
Pyridinium ethoxyethylsulfate	11075-04032
Pyrrolidinium formate	08017-06005
Pyrrolidinium hydrogen sulfate	08017-04021
Pyrrolidinium nitrate	08017-03009
T	
Tetrabutylammonium bromide	15032-01002
Tetrabutylammonium chloride	15032-01001
Tetrabutylammonium hexafluorophosphate	15032-05001
Tetra-*n*-butylammonium pricrate	15032-06042
Tetrabutyl phosphonium alaninate	17005-06010
Tetrabutylphosphonium methanesulfonate	17005-04002
Tetra(*n*-butyl)phosphonium tosylate	17005-04007
Tetrabutyl phosphonium valinate	17005-06011
N-tetradecyl-isoquinolinium bis((perfluoroethane)sulfonyl)imide	12006-03003
1-Tetradecyl-3-methylpyridinium bis(trifluoromethylsulfonyl)imide	11012-03001
Tetradecyl(trihexyl)phosphonium 1,1,2-trifluoro-2-(perfluoroethoxy) ethanesulfonate	15042-04028
Tetraheptylammonium bis((perfluoroethane)sulfonyl)imide	15039-03003
Tributyl(tetradecyl)phosphonium 1,1,2,3,3,3-hexafluoropropanesulfonate	17017-04029
Triethanolammonium acetate	15143-06003
Triethylsulphonium bis((trifluoromethyl)sulfonyl)imide	14002-03001
Trihexyl-tetradecylphosphonium bis((trifluoromethyl)sulfonyl)imide	17004-03001
Tri-*iso*-butyl(methyl)phosphonium tosylate	17040-04007
Trimethyl-butylammonium bis((trifluoromethyl)sulfonyl)imide	15013-03001
Tris(2-hydroxyethyl)methylammonium methylsulfate	15168-04017
N,*N*,*N*-trioctyl-*N*-methylammonium bis((trifluoromethyl)sulfonyl)imide	15067-03001

List of Mixtures

A

Acetylcholinium bis(trifluoromethylsulfonyl)imide
Acetylcholinium bis(trifluoromethylsulfonyl)imide (1) + 1*H*,1*H*,2*H*,2*H*-Perfluorooctanol (2)
Acetylcholinium bis(trifluoromethylsulfonyl)imide (1) + 1-Propanol (2)
Acetylcholinium bis(trifluoromethylsulfonyl)imide (1) + Ethanol (2)
***n*-Acryloyloxypropyl-*N*-methylimidazolium bromide**
n-Acryloyloxypropyl-*N*-methylimidazolium bromide (1) + *n*-Hexane (2) + Thiophene (3)
n-Acryloyloxypropyl-N-methylimidazolium bromide (1) + Cyclohexane (2) + Thiophene (3)
n-Acryloyloxypropyl-N-methylimidazolium bromide (1) + Benzene (2) + Thiophene (3)
1-Allyl-3-methyl-imidazolium chloride
1-Allyl-3-methyl-imidazolium chloride (1) + Water (2)
1-Allyl-3-methyl-imidazolium chloride (1) + Ethanol (2)
1-Amyl-3-methylimidazolium bis((trifluoromethyl)sulfonyl)imide
1-Amyl-3-methylimidazolium bis((trifluoromethyl)sulfonyl)imide (1) + Water (2)
1-Amyl-3-methylimidazolium bis((trifluoromethyl)sulfonyl)imide (1) + Carbon dioxide (2)
1-Amyl-3-methylimidazolium hexafluorophosphate
1-Amyl-3-methylimidazolium hexafluorophosphate (1) + 3-Methyl-1-pentylimidazolium trifluoroacetate (2)

B

(Benzyl)dimethyldodecylammonium nitrate
(Benzyl)dimethyldodecylammonium nitrate (1) + (Benzyl)dimethyltetradecylammonium nitrate (2) + Water (3)

S. Zhang et al., *Physicochemical Properties of Ionic Liquid Mixtures*,
DOI 10.1007/978-94-017-7573-1

(Benzyl)dimethyldodecylammonium nitrate (1) + (Benzyl)dimethyltetradecylammonium nitrate (2) + 1-Propanol (3)
(Benzyl)dimethyldodecylammonium nitrate (1) + (Benzyl)dimethyltetradecylammonium nitrate (2) + 1-Butanol (3)
(Benzyl)dimethyldodecylammonium nitrate (1) + (Benzyl)dimethyltetradecylammonium nitrate (2) + 1-Hexanol (3)
(Benzyl)dimethyldodecylammonium nitrate (1) + (Benzyl)dimethyltetradecylammonium nitrate (2) + 1-Octanol (3)
(Benzyl)dimethyldodecylammonium nitrate (1) + (Benzyl)dimethyltetradecylammonium nitrate (2) + 1-Decanol (3)
(Benzyl)dimethyldodecylammonium nitrate (1) + (Benzyl)dimethyltetradecylammonium nitrate (2) + Benzene (3)
(Benzyl)dimethyldodecylammonium nitrate (1) + (Benzyl)dimethyltetradecylammonium nitrate (2) + Toluene (3)
(Benzyl)dimethyldodecylammonium nitrate (1) + (Benzyl)dimethyltetradecylammonium nitrate (2) + *n*-Hexane (3)
(Benzyl)dimethyldodecylammonium nitrate (1) + (Benzyl)dimethyltetradecylammonium nitrate (2) + Hexadecane (3)

(Benzyl)dimethyltetradecylammonium nitrate

(Benzyl)dimethyldodecylammonium nitrate (1) + (Benzyl)dimethyltetradecylammonium nitrate (2) + Water (3)
(Benzyl)dimethyldodecylammonium nitrate (1) + (Benzyl)dimethyltetradecylammonium nitrate (2) + 1-Propanol (3)
(Benzyl)dimethyldodecylammonium nitrate (1) + (Benzyl)dimethyltetradecylammonium nitrate (2) + 1-Butanol (3)
(Benzyl)dimethyldodecylammonium nitrate (1) + (Benzyl)dimethyltetradecylammonium nitrate (2) + 1-Hexanol (3)
(Benzyl)dimethyldodecylammonium nitrate (1) + (Benzyl)dimethyltetradecylammonium nitrate (2) + 1-Octanol (3)
(Benzyl)dimethyldodecylammonium nitrate (1) + (Benzyl)dimethyltetradecylammonium nitrate (2) + 1-Decanol (3)
(Benzyl)dimethyldodecylammonium nitrate (1) + (Benzyl)dimethyltetradecylammonium nitrate (2) + Benzene (3)
(Benzyl)dimethyldodecylammonium nitrate (1) + (Benzyl)dimethyltetradecylammonium nitrate (2) + Toluene (3)
(Benzyl)dimethyldodecylammonium nitrate (1) + (Benzyl)dimethyltetradecylammonium nitrate (2) + *n*-Hexane (3)
(Benzyl)dimethyldodecylammonium nitrate (1) + (Benzyl)dimethyltetradecylammonium nitrate (2) + Hexadecane (3)

1-Butyl-2,3-dimethylimidazolium hexafluorophosphate

1-Butyl-2,3-dimethylimidazolium hexafluorophosphate (1) + Water (2)

1-Butyl-2,3-dimethylimidazolium tetrafluoroborate

1-Butyl-2,3-dimethylimidazolium tetrafluoroborate (1) + Water (2)

Butyl(2-hydroxyethyl)dimethylammonium bromide

Butyl(2-hydroxyethyl)dimethylammonium bromide (1) + Water (2)

1-Butyl-3-methylimidazolium bis((trifluoromethyl)sulfonyl)imide (1) + Water (2) + 1,2-Hexanediol (3)
1-Butyl-3-methylimidazolium bis((trifluoromethyl)sulfonyl)imide (1) + Water (2) + Cyclohexanol (3)
1-Butyl-3-methylimidazolium bis((trifluoromethyl)sulfonyl)imide (1) + *n*-Hexane (2) + 1-Ethyl-3-methylimidazolium bis((trifluoromethyl)sulfonyl)imide (3)
1-Butyl-3-methylimidazolium bis((trifluoromethyl)sulfonyl)imide (1) + *n*-Heptane (2) + 1-Ethyl-3-methylimidazolium bis((trifluoromethyl)sulfonyl)imide (3)
1-Butyl-3-methylimidazolium bis((trifluoromethyl)sulfonyl)imide (1) + *n*-Octane (2) + 1-Ethyl-3-methylimidazolium bis((trifluoromethyl)sulfonyl)imide (3)
1-Butyl-3-methylimidazolium bis((trifluoromethyl)sulfonyl)imide (1) + 1-Pentene (2) + 1-Ethyl-3-methylimidazolium bis((trifluoromethyl)sulfonyl)imide (3)
1-Butyl-3-methylimidazolium bis((trifluoromethyl)sulfonyl)imide (1) + 2-Butanone (2) + 1-Ethyl-3-methylimidazolium bis((trifluoromethyl)sulfonyl)imide (3)
1-Butyl-3-methylimidazolium bis((trifluoromethyl)sulfonyl)imide (1) + 2-Pentanone (2) + 1-Ethyl-3-methylimidazolium bis((trifluoromethyl)sulfonyl)imide (3)
1-Butyl-3-methylimidazolium bis((trifluoromethyl)sulfonyl)imide (1) + Water (2) + 1-Ethyl-3-methylimidazolium bis((trifluoromethyl)sulfonyl)imide (3)
1-Butyl-3-methylimidazolium bis((trifluoromethyl)sulfonyl)imide (1) + Trihexyl-tetradecylphosphonium bis((trifluoromethyl)sulfonyl)imide (2) + 1-Ethyl-3-methylimidazolium bis((trifluoromethyl)sulfonyl)imide (3)
1-Butyl-3-methylimidazolium bis((trifluoromethyl)sulfonyl)imide (1) + *n*-Hexane (2) + Benzene (3)
1-Butyl-3-methylimidazolium bis((trifluoromethyl)sulfonyl)imide (1) + Water (2) + *iso*-Butanol (3)
1-Butyl-3-methylimidazolium bis((trifluoromethyl)sulfonyl)imide (1) + Cyclohexane (2) + Benzene (3)
1-Butyl-3-methylimidazolium bis((trifluoromethyl)sulfonyl)imide (1) + Water (2) + 1-Butanol (3) + Carbon dioxide (4)

1-Butyl-3-methylimidazolium bromide

1-Butyl-3-methylimidazolium bromide (1) + Water (2)
1-Butyl-3-methylimidazolium bromide (1) + Methanol (2)
1-Butyl-3-methylimidazolium bromide (1) + Ethanol (2)
1-Butyl-3-methylimidazolium bromide (1) + Trifluoroethanol (2)
1-Butyl-3-methylimidazolium bromide (1) + Water (2) + Dipotassium hydrogen phosphate (3)
1-Butyl-3-methylimidazolium bromide (1) + Water (2) + Potassium phosphate (3)
1-Butyl-3-methylimidazolium bromide (1) + Water (2) + Ethanol (3)
1-Butyl-3-methylimidazolium bromide (1) + Water (2) + Tri-potassium Citrate (3)

1-Butyl-3-methylimidazolium chloride

1-Butyl-3-methylimidazolium chloride (1) + Water (2)
1-Butyl-3-methylimidazolium chloride (1) + Ethanol (2)

1-Butyl-3-methylimidazolium hexafluorophosphate (1) + 2-Propanol (2) + Ethyl acetate (3)
1-Butyl-3-methylimidazolium hexafluorophosphate (1) + Carbon dioxide (2) + Acetone (3)
1-Butyl-3-methylimidazolium hexafluorophosphate (1) + Propylene carbonates (2) + Ethylene carbonates (3)
1-Butyl-3-methylimidazolium hexafluorophosphate (1) + Poly(2-ethoxyethyl methacrylate) (2) + Propylene carbonates (3)
1-Butyl-3-methylimidazolium hexafluorophosphate (1) + Poly(acrylonitrile) (2) + Sulpholane (3)
1-Butyl-3-methylimidazolium hexafluorophosphate (1) + Acetone (2) + Water (3)
1-Butyl-3-methylimidazolium hexafluorophosphate (1) + Methanol (2) + Carbon dioxide (3)
1-Butyl-3-methylimidazolium hexafluorophosphate (1) + Poly(2-ethoxyethyl methacrylate) (2) + Propylene carbonates (3) Ethylene carbonates (4)
1-Butyl-3-methylimidazolium hexafluorophosphate (1) + Poly(2-ethoxyethyl methacrylate) (2) + Propylene carbonates (3) + Lithium hexafluorophosphate (4)
1-Butyl-3-methylimidazolium hexafluorophosphate (1) + Poly(2-ethoxyethyl methacrylate) (2) + Propylene carbonates (3) + Lithium perchlorate (4)
1-Butyl-3-methylimidazolium hexafluorophosphate (1) + Poly(2-ethoxyethyl methacrylate) (2) + Propylene carbonates (3) + Ethylene carbonates (4) + Lithium hexafluorophosphate (5)
1-Butyl-3-methylimidazolium hexafluorophosphate (1) + Poly(2-ethoxyethyl methacrylate) (2) + Propylene carbonates (3) + Ethylene carbonates (4) + Lithium perchlorate (5)

1-Butyl-3-methylimidazolium 1,1,2,3,3,3-hexafluoropropanesulfonate
1-Butyl-3-methylimidazolium 1,1,2,3,3,3-hexafluoropropanesulfonate (1) + 1,1,1,2-Tetrafluoroethane (2)
1-Butyl-3-methylimidazolium 1,1,2,3,3,3-hexafluoropropanesulfonate (1) + Difluoromethane (2)

1-Butyl-3-methylimidazolium hydrogen sulfate
1-Butyl-3-methylimidazolium hydrogen sulfate (1) + Ethyl acetate (2) + Ethanol (3)
1-Butyl-3-methylimidazolium hydrogen sulfate (1) + Ethyl acetate (2) + Acetic acid (3)

1-Butyl-3-methylimidazolium iodine
1-Butyl-3-methylimidazolium iodine (1) + Water (2)

1-Butyl-3-methyl imidazolium methylsulfate
1-Butyl-3-methyl imidazolium methylsulfate (1) + Water (2)
1-Butyl-3-methyl imidazolium methylsulfate (1) + Ethanol (2)
1-Butyl-3-methyl imidazolium methylsulfate (1) + Nitromethane (2)
1-Butyl-3-methyl imidazolium methylsulfate (1) + 1,3-Dichloropropane (2)
1-Butyl-3-methyl imidazolium methylsulfate (1) + Diethylene glycol monoethyl ether (2)
1-Butyl-3-methyl imidazolium methylsulfate (1) + 1,2-Ethanediol (2)

1-Butyl-3-methyl imidazolium methylsulfate (1) + 1-Butyl-3-methylimidazolium tetrafluoroborate (2)
1-Butyl-3-methyl imidazolium methylsulfate (1) + Dipropyl ether (2)
1-Butyl-3-methyl imidazolium methylsulfate (1) + Dibutyl ether (2)
1-Butyl-3-methyl imidazolium methylsulfate (1) + Tert-butyl-methyl ether (2)
1-Butyl-3-methyl imidazolium methylsulfate (1) + Methyl-1,1-dimethylpropyl ether (2)
1-Butyl-3-methyl imidazolium methylsulfate (1) + 2-Pentanone (2)
1-Butyl-3-methyl imidazolium methylsulfate (1) + Pentan-3-one (2)
1-Butyl-3-methyl imidazolium methylsulfate (1) + 2-Hexanone (2)
1-Butyl-3-methyl imidazolium methylsulfate (1) + Heptan-4-one (2)
1-Butyl-3-methyl imidazolium methylsulfate (1) + *n*-Pentane (2)
1-Butyl-3-methyl imidazolium methylsulfate (1) + *n*-Hexane (2)
1-Butyl-3-methyl imidazolium methylsulfate (1) + *n*-Heptane (2)
1-Butyl-3-methyl imidazolium methylsulfate (1) + *n*-Octane (2)
1-Butyl-3-methyl imidazolium methylsulfate (1) + 1-Decane (2)
1-Butyl-3-methyl imidazolium methylsulfate (1) + Cyclohexane (2)
1-Butyl-3-methyl imidazolium methylsulfate (1) + Cycloheptane (2)
1-Butyl-3-methyl imidazolium methylsulfate (1) + Benzene (2)
1-Butyl-3-methyl imidazolium methylsulfate (1) + Toluene (2)
1-Butyl-3-methyl imidazolium methylsulfate (1) + Ethylbenzene (2)
1-Butyl-3-methyl imidazolium methylsulfate (1) + Propyl benzene (2)
1-Butyl-3-methyl imidazolium methylsulfate (1) + *o*-Xylene (2)
1-Butyl-3-methyl imidazolium methylsulfate (1) + *m*-Xylene (2)
1-Butyl-3-methyl imidazolium methylsulfate (1) + *p*-Xylene (2)
1-Butyl-3-methyl imidazolium methylsulfate (1) + Difluoromethane (2)
1-Butyl-3-methyl imidazolium methylsulfate (1) + 1-Propanol (2)
1-Butyl-3-methyl imidazolium methylsulfate (1) + 2-Propanol (2)
1-Butyl-3-methyl imidazolium methylsulfate (1) + Water (2) + Ethanol (3)
1-Butyl-3-methyl imidazolium methylsulfate (1) + Toluene (2) + *n*-Heptane (3)
1-Butyl-3-methyl imidazolium methylsulfate (1) + *n*-Heptane (2) + Ethanol (3)
1-Butyl-3-methyl imidazolium methylsulfate (1) + *n*-Hexane (2) + Ethanol (3)

1-Butyl-3-methylimidazolium nitrate

1-Butyl-3-methylimidazolium nitrate (1) + Water (2)
1-Butyl-3-methylimidazolium nitrate (1) + Carbon dioxide (2)
1-Butyl-3-methylimidazolium nitrate (1) + Ethanol (2)
1-Butyl-3-methylimidazolium nitrate (1) + 1-Propanol (2)
1-Butyl-3-methylimidazolium nitrate (1) + 1-Butanol (2)
1-Butyl-3-methylimidazolium nitrate (1) + Water (2) + Carbon dioxide (3)

1-Butyl-3-methylimidazolium octylsulfate

1-Butyl-3-methylimidazolium octylsulfate (1) + Water (2)
1-Butyl-3-methylimidazolium octylsulfate (1) + 1-Propanol (2)
1-Butyl-3-methylimidazolium octylsulfate (1) + Hydrogen (2)
1-Butyl-3-methylimidazolium octylsulfate (1) + Carbon dioxide (2)
1-Butyl-3-methylimidazolium octylsulfate (1) + 1,2-Ethanediol (2)

1-Butyl-3-methylimidazolium tetrafluoroborate (1) + *iso*-Butanol (2)
1-Butyl-3-methylimidazolium tetrafluoroborate (1) + 2-Butanol (2)
1-Butyl-3-methylimidazolium tetrafluoroborate (1) + *tert*-Butanol (2)
1-Butyl-3-methylimidazolium tetrafluoroborate (1) + Deuterated 1-Butanol (2)
1-Butyl-3-methylimidazolium tetrafluoroborate (1) + Deuterated *iso*-Butanol (2)
1-Butyl-3-methylimidazolium tetrafluoroborate (1) + Deuterated 2-Butanol (2)
1-Butyl-3-methylimidazolium tetrafluoroborate (1) + Deuterated *tert*-Butanol (2)
1-Butyl-3-methylimidazolium tetrafluoroborate (1) + Dimethylformamide (2)
1-Butyl-3-methylimidazolium tetrafluoroborate (1) + 3-Picoline (2)
1-Butyl-3-methylimidazolium tetrafluoroborate (1) + Carbon dioxide (2)
1-Butyl-3-methylimidazolium tetrafluoroborate (1) + 1-Butyl-3-methylimidazolium bis((trifluoromethyl)sulfonyl)imide (2)
1-Butyl-3-methylimidazolium tetrafluoroborate (1) + Trifluoroethanol (2)
1-Butyl-3-methylimidazolium tetrafluoroborate (1) + 1*H*,1*H*,2*H*,2*H*-Perfluorooctanol (2)
1-Butyl-3-methylimidazolium tetrafluoroborate (1) + Propylene carbonate (2)
1-Butyl-3-methylimidazolium tetrafluoroborate (1) + Thiophene (2)
1-Butyl-3-methylimidazolium tetrafluoroborate (1) + Benzaldehyde (2)
1-Butyl-3-methylimidazolium tetrafluoroborate (1) + Dichloromethane (2)
1-Butyl-3-methylimidazolium tetrafluoroborate (1) + *N*-methyl-2-pyrrolidinone (2)
1-Butyl-3-methylimidazolium tetrafluoroborate (1) + Threo-2,3-dihydrodecafluoropentane (87.12 ± 0.6% Threo-H) (2) + Erythro-2,3-dihydrodecafluoropentane (11.89 ± 0.6% Erythro-H) (3)
1-Butyl-3-methylimidazolium tetrafluoroborate (1) + Water (2) + Ethanol (3)
1-Butyl-3-methylimidazolium tetrafluoroborate (1) + Acetophenone (2) + Carbon dioxide (3)
1-Butyl-3-methylimidazolium tetrafluoroborate (1) + 1-Phenylethanol (2) + Carbon dioxide (3)
1-Butyl-3-methylimidazolium tetrafluoroborate (1) + 4-Isobutylacetophenone (2) + Carbon dioxide (3)
1-Butyl-3-methylimidazolium tetrafluoroborate (1) + Water (2) + 1-Propanol (3)
1-Butyl-3-methylimidazolium tetrafluoroborate (1) + Water (2) + Adipic acid (3)
1-Butyl-3-methylimidazolium tetrafluoroborate (1) + Water (2) + Glutaric acid (3)
1-Butyl-3-methylimidazolium tetrafluoroborate (1) + Water (2) + Succinic acid (3)
1-Butyl-3-methylimidazolium tetrafluoroborate (1) + Water (2) + 2-Propanol (3)
1-Butyl-3-methylimidazolium tetrafluoroborate (1) + Ferrocene (2) + Dimethylformamide (3)
1-Butyl-3-methylimidazolium tetrafluoroborate (1) + Water (2) + Sucrose (3)
1-Butyl-3-methylimidazolium tetrafluoroborate (1) + Carbon dioxide (2) + 1-(4-Isobutyl phenyl)-ethanol (3)
1-Butyl-3-methylimidazolium tetrafluoroborate (1) + Water (2) + Xylose (3)
1-Butyl-3-methylimidazolium tetrafluoroborate (1) + Water (2) + Glucose (3)
1-Butyl-3-methylimidazolium tetrafluoroborate (1) + Water (2) + Fructose (3)
1-Butyl-3-methylimidazolium tetrafluoroborate (1) + *n*-Hexane (2) + Propionic acid (3)

N-Butyl-3-methylpyridinium tosylate

N-Butyl-3-methylpyridinium tosylate (1) + Thiophene (2)

N-Butyl-4-methylpyridinium tosylate

N-Butyl-4-methylpyridinium tosylate (1) + Thiophene (2)
N-Butyl-4-methylpyridinium tosylate (1) + Ethanol (2)
N-Butyl-4-methylpyridinium tosylate (1) + 1-Propanol (2)
N-Butyl-4-methylpyridinium tosylate (1) + 1-Butanol (2)
N-Butyl-4-methylpyridinium tosylate (1) + 1-Octanol (2)
N-Butyl-4-methylpyridinium tosylate (1) + 1-Decanol (2)
N-Butyl-4-methylpyridinium tosylate (1) + Water (2)

1-Butyl-1-methyl-pyrrolidinium trifluoromethanesulfonate

1-Butyl-1-methyl-pyrrolidinium trifluoromethanesulfonate (1) + *n*-Pentane (2)
1-Butyl-1-methyl-pyrrolidinium trifluoromethanesulfonate (1) + *n*-Hexane (2)
1-Butyl-1-methyl-pyrrolidinium trifluoromethanesulfonate (1) + *n*-Heptane (2)
1-Butyl-1-methyl-pyrrolidinium trifluoromethanesulfonate (1) + *n*-Octane (2)
1-Butyl-1-methyl-pyrrolidinium trifluoromethanesulfonate (1) + 1-Nonane (2)
1-Butyl-1-methyl-pyrrolidinium trifluoromethanesulfonate (1) + 1-Decane (2)
1-Butyl-1-methyl-pyrrolidinium trifluoromethanesulfonate (1) + Cyclopentane (2)
1-Butyl-1-methyl-pyrrolidinium trifluoromethanesulfonate (1) + Cyclohexane (2)
1-Butyl-1-methyl-pyrrolidinium trifluoromethanesulfonate (1) + Cycloheptane (2)
1-Butyl-1-methyl-pyrrolidinium trifluoromethanesulfonate (1) + Cyclooctane (2)
1-Butyl-1-methyl-pyrrolidinium trifluoromethanesulfonate (1) + Pent-1-ene (2)
1-Butyl-1-methyl-pyrrolidinium trifluoromethanesulfonate (1) + Hex-1-ene (2)
1-Butyl-1-methyl-pyrrolidinium trifluoromethanesulfonate (1) + Hept-1-ene (2)
1-Butyl-1-methyl-pyrrolidinium trifluoromethanesulfonate (1) + Oct-1-ene (2)
1-Butyl-1-methyl-pyrrolidinium trifluoromethanesulfonate (1) + Pent-1-yne (2)
1-Butyl-1-methyl-pyrrolidinium trifluoromethanesulfonate (1) + Hex-1-yne (2)
1-Butyl-1-methyl-pyrrolidinium trifluoromethanesulfonate (1) + Hept-1-yne (2)
1-Butyl-1-methyl-pyrrolidinium trifluoromethanesulfonate (1) + Oct-1-yne (2)
1-Butyl-1-methyl-pyrrolidinium trifluoromethanesulfonate (1) + Benzene (2)
1-Butyl-1-methyl-pyrrolidinium trifluoromethanesulfonate (1) + Toluene (2)
1-Butyl-1-methyl-pyrrolidinium trifluoromethanesulfonate (1) + Ethylbenzene (2)
1-Butyl-1-methyl-pyrrolidinium trifluoromethanesulfonate (1) + *o*-Xylene (2)
1-Butyl-1-methyl-pyrrolidinium trifluoromethanesulfonate (1) + *m*-Xylene (2)
1-Butyl-1-methyl-pyrrolidinium trifluoromethanesulfonate (1) + *p*-Xylene (2)
1-Butyl-1-methyl-pyrrolidinium trifluoromethanesulfonate (1) + Methanol (2)
1-Butyl-1-methyl-pyrrolidinium trifluoromethanesulfonate (1) + Ethanol (2)
1-Butyl-1-methyl-pyrrolidinium trifluoromethanesulfonate (1) + Propan-1-ol (2)
1-Butyl-1-methyl-pyrrolidinium trifluoromethanesulfonate (1) + Butan-1-ol (2)
1-Butyl-1-methyl-pyrrolidinium trifluoromethanesulfonate (1) + Water (2)
1-Butyl-1-methyl-pyrrolidinium trifluoromethanesulfonate (1) + Tetrahydrofuran (2)
1-Butyl-1-methyl-pyrrolidinium trifluoromethanesulfonate (1) + Thiophene (2)
1-Butyl-1-methyl-pyrrolidinium trifluoromethanesulfonate (1) + *Tert*-butyl-methyl ether (2)

1-Butyl-1-methyl-pyrrolidinium trifluoromethanesulfonate (1) + *n*-Hexane (2) + Benzene (3)
1-Butyl-1-methyl-pyrrolidinium trifluoromethanesulfonate (1) + Cyclohexane (2) + Benzene (3)
1-Butyl-1-methyl-pyrrolidinium trifluoromethanesulfonate (1) + *n*-Hexane (2) + Thiophene (3)
1-Butyl-1-methyl-pyrrolidinium trifluoromethanesulfonate (1) + Cyclohexane (2) + Thiophene (3)
1-Butyl-1-methyl-pyrrolidinium trifluoromethanesulfonate (1) + Benzene (2) + Thiophene (3)

***N*-butyl pyridinium tetrafluoroborate**
N-butyl pyridinium tetrafluoroborate (1) + Water (2)
N-butyl pyridinium tetrafluoroborate (1) + Methanol (2)
N-butyl pyridinium tetrafluoroborate (1) + Dichloromethane (2)

C

Choline glycolate (2-Hydroxyethyl-trimethylammonium 2-hydroxyethanoate)
Choline glycolate (2-Hydroxyethyl-trimethylammonium 2-hydroxyethanoate) (1) + Water (2)

Choline Lactate (2-Hydroxyethyl-trimethylammonium 2-hydroxypropanoate)
Choline Lactate (2-Hydroxyethyl-trimethylammonium 2-hydroxypropanoate) (1) + Water (2)

Choline methanesulfonate(2-Hydroxyethyl-trimethylammonium methanesulfonate)
Choline methanesulfonate(2-Hydroxyethyl-trimethylammonium methanesulfonate) (1) + Water (2)

2,4,6-Collidine tosylate
2,4,6-Collidine tosylate (1) + Thiophene (2)

Collidinium formate
Collidinium formate (1) + Water (2)

D

1-Decyl-2,3-dimethylimidazolium bromide
1-Decyl-2,3-dimethylimidazolium bromide (1) + Water (2)

1-Decyl-3-methylimidazolium bis((trifluoromethyl)sulfonyl)imide
1-Decyl-3-methylimidazolium bis((trifluoromethyl)sulfonyl)imide (1) + Methanol (2)
1-Decyl-3-methylimidazolium bis((trifluoromethyl)sulfonyl)imide (1) + 1-Ethyl-3-methylimidazolium bis((trifluoromethyl)sulfonyl)imide (2)
1-Decyl-3-methylimidazolium bis((trifluoromethyl)sulfonyl)imide (1) + 1-Butyl-3-methylimidazolium bis((trifluoromethyl)sulfonyl)imide (2)

1-Decyl-3-methylimidazolium bis((trifluoromethyl)sulfonyl)imide (1) + 1-Hexyl-3-methylimidazolium bis((trifluoromethyl)sulfonyl)imide (2)
1-Decyl-3-methylimidazolium bis((trifluoromethyl)sulfonyl)imide (1) + 1-Octyl-3-methylimidazolium bis((trifluoromethyl)sulfonyl)imide (2)
1-Decyl-3-methylimidazolium bis((trifluoromethyl)sulfonyl)imide (1) + 1*H*,1*H*,2*H*,2*H*,3*H*,3*H*-Perfluorononanol (2)
1-Decyl-3-methylimidazolium bis((trifluoromethyl)sulfonyl)imide (1) + 1-Propanol (2)
1-Decyl-3-methylimidazolium bis((trifluoromethyl)sulfonyl)imide (1) + *n*-Hexane (2) + Benzene (3)

1-Decyl-3-methylimidazolium bromide
1-Decyl-3-methylimidazolium bromide (1) + Water (2)
1-Decyl-3-methylimidazolium bromide (1) + Dimethylsulfoxide (2)
1-Decyl-3-methylimidazolium bromide (1) + 1,2-Ethanediol (2)
1-Decyl-3-methylimidazolium bromide (1) + Dimethylformamide (2)

1-Decyl-3-methylimidazolium chloride
1-Decyl-3-methylimidazolium chloride (1) + Ethanol (2)
1-Decyl-3-methylimidazolium chloride (1) + 1-Butanol (2)
1-Decyl-3-methylimidazolium chloride (1) + 2-Butanol (2)
1-Decyl-3-methylimidazolium chloride (1) + *tert*-Butanol (2)
1-Decyl-3-methylimidazolium chloride (1) + 1-Hexanol (2)
1-Decyl-3-methylimidazolium chloride (1) + 1-Octanol (2)
1-Decyl-3-methylimidazolium chloride (1) + 1-Decanol (2)
1-Decyl-3-methylimidazolium chloride (1) + 1-Dodecanol (2)

1-Decyl-3-methylpyridinium bis(trifluoromethylsulfonyl)imide
1-Decyl-3-methylpyridinium bis(trifluoromethylsulfonyl)imide (1) + 1*H*,1*H*,2*H*,2*H*-Perfluorooctanol (2)

Didecyldimethylammonium nitrate
Didecyldimethylammonium nitrate (1) + Water (2)
Didecyldimethylammonium nitrate (1) + 1-Propanol (2)
Didecyldimethylammonium nitrate (1) + 1-Butanol (2)
Didecyldimethylammonium nitrate (1) + 1-Octanol (2)
Didecyldimethylammonium nitrate (1) + 1-Decanol (2)
Didecyldimethylammonium nitrate (1) + Toluene (2)
Didecyldimethylammonium nitrate (1) + Propylbenzene (2)
Didecyldimethylammonium nitrate (1) + *n*-Hexane (2)
Didecyldimethylammonium nitrate (1) + Hexadecane (2)

Diethanolammonium acetate
Diethanolammonium acetate (1) + Water (2)
Diethanolammonium acetate (1) + Methanol (2)
Diethanolammonium acetate (1) + Ethanol (2)
Diethanolammonium acetate (1) + Ethanol (2) + Methanol (3)

Diethanolammonium chloride
Diethanolammonium chloride (1) + Water (2)
Diethanolammonium chloride (1) + Methanol (2)

E

1-Ethyl-3-methylimidazolium bis((trifluoromethyl)sulfonyl)imide (1) + 1,2,4,5-Tetrafluorobenzene (2)
1-Ethyl-3-methylimidazolium bis((trifluoromethyl)sulfonyl)imide (1) + Pentafluorobenzene (2)
1-Ethyl-3-methylimidazolium bis((trifluoromethyl)sulfonyl)imide (1) + Benzene (2)
1-Ethyl-3-methylimidazolium bis((trifluoromethyl)sulfonyl)imide (1) + Hexafluorobenzene (2)
1-Ethyl-3-methylimidazolium bis((trifluoromethyl)sulfonyl)imide (1) + Pentafluoroethane (2)
1-Ethyl-3-methylimidazolium bis((trifluoromethyl)sulfonyl)imide (1) + Difluoromethoxy trifluoromethane (2)
1-Ethyl-3-methylimidazolium bis((trifluoromethyl)sulfonyl)imide (1) + 1,1,1-Trifluoroethane (2)
1-Ethyl-3-methylimidazolium bis((trifluoromethyl)sulfonyl)imide (1) + Trifluoromethoxy methane (2)
1-Ethyl-3-methylimidazolium bis((trifluoromethyl)sulfonyl)imide (1) + 1-Ethyl-3-methylimidazolium tetrafluoroborate (2)
1-Ethyl-3-methylimidazolium bis((trifluoromethyl)sulfonyl)imide (1) + Water (2)
1-Ethyl-3-methylimidazolium bis((trifluoromethyl)sulfonyl)imide (1) + 1-Methoxyheptafluoropropane (2)
1-Ethyl-3-methylimidazolium bis((trifluoromethyl)sulfonyl)imide (1) + 1-Methoxynonafluorobutane (2)
1-Ethyl-3-methylimidazolium bis((trifluoromethyl)sulfonyl)imide (1) + 1-Ethoxynonafluorobutane (2)
1-Ethyl-3-methylimidazolium bis((trifluoromethyl)sulfonyl)imide (1) + 1,1,1,2,2,3,4,5,5,5-Decafluoro-3-methoxy-4-(trifluoromethyl)-pentane (2)
1-Ethyl-3-methylimidazolium bis((trifluoromethyl)sulfonyl)imide (1) + 2-Trifluoromethyl-3-ethoxydodecafluorohexane (2)
1-Ethyl-3-methylimidazolium bis((trifluoromethyl)sulfonyl)imide (1) + Carbon dioxide (2)
1-Ethyl-3-methylimidazolium bis((trifluoromethyl)sulfonyl)imide (1) + *n*-Hexane (2)
1-Ethyl-3-methylimidazolium bis((trifluoromethyl)sulfonyl)imide (1) + Ethanol (2)
1-Ethyl-3-methylimidazolium bis((trifluoromethyl)sulfonyl)imide (1) + 1-Propanol (2)
1-Ethyl-3-methylimidazolium bis((trifluoromethyl)sulfonyl)imide (1) + Acetonitrile (2)
1-Ethyl-3-methylimidazolium bis((trifluoromethyl)sulfonyl)imide (1) + Difluoromethane (2)
1-Ethyl-3-methylimidazolium bis((trifluoromethyl)sulfonyl)imide (1) + Dimethyl adipate (2)
1-Ethyl-3-methylimidazolium bis((trifluoromethyl)sulfonyl)imide (1) + Ethyl Benzoate (2)
1-Ethyl-3-methylimidazolium bis((trifluoromethyl)sulfonyl)imide (1) + Benzylamine (2)

1-Ethyl-3-methylimidazolium bis((trifluoromethyl)sulfonyl)imide (1) + 1-Pentene (2)
1-Ethyl-3-methylimidazolium bis((trifluoromethyl)sulfonyl)imide (1) + 1-Hexene (2)
1-Ethyl-3-methylimidazolium bis((trifluoromethyl)sulfonyl)imide (1) + 1-Heptene (2)
1-Ethyl-3-methylimidazolium bis((trifluoromethyl)sulfonyl)imide (1) + Cyclopentane (2)
1-Ethyl-3-methylimidazolium bis((trifluoromethyl)sulfonyl)imide (1) + Cyclopentene (2)
1-Ethyl-3-methylimidazolium bis((trifluoromethyl)sulfonyl)imide (1) + Toluene (2)
1-Ethyl-3-methylimidazolium bis((trifluoromethyl)sulfonyl)imide (1) + 2-Butanone (2)
1-Ethyl-3-methylimidazolium bis((trifluoromethyl)sulfonyl)imide (1) + 2-Pentanone (2)
1-Ethyl-3-methylimidazolium bis((trifluoromethyl)sulfonyl)imide (1) + Methanol (2)
1-Ethyl-3-methylimidazolium bis((trifluoromethyl)sulfonyl)imide (1) + Trihexyl-tetradecylphosphonium bis((trifluoromethyl)sulfonyl)imide (2)
1-Ethyl-3-methylimidazolium bis((trifluoromethyl)sulfonyl)imide (1) + Chlorobenzene (2)
1-Ethyl-3-methylimidazolium bis((trifluoromethyl)sulfonyl)imide (1) + Bromobenzene (2)
1-Ethyl-3-methylimidazolium bis((trifluoromethyl)sulfonyl)imide (1) + Iodobenzene (2)
1-Ethyl-3-methylimidazolium bis((trifluoromethyl)sulfonyl)imide (1) + Acetonitrile (2) + Carbon dioxide (3)
1-Ethyl-3-methylimidazolium bis((trifluoromethyl)sulfonyl)imide (1) + Acetone (2) + 2-Propanol (3)
1-Ethyl-3-methylimidazolium bis((trifluoromethyl)sulfonyl)imide (1) + 2-Propanol (2) + Water (3)
1-Ethyl-3-methylimidazolium bis((trifluoromethyl)sulfonyl)imide (1) + *n*-Heptane (2) + Toluene (3)
1-Ethyl-3-methylimidazolium bis((trifluoromethyl)sulfonyl)imide (1) + *n*-Octane (2) + Ethylbenzene (3)
1-Ethyl-3-methylimidazolium bis((trifluoromethyl)sulfonyl)imide (1) + *n*-Hexane (2) + 1-Butyl-3-methylimidazolium bis((trifluoromethyl)sulfonyl)imide (3)
1-Ethyl-3-methylimidazolium bis((trifluoromethyl)sulfonyl)imide (1) + *n*-Heptane (2) + 1-Butyl-3-methylimidazolium bis((trifluoromethyl)sulfonyl)imide (3)
1-Ethyl-3-methylimidazolium bis((trifluoromethyl)sulfonyl)imide (1) + *n*-Octane (2) + 1-Butyl-3-methylimidazolium bis((trifluoromethyl)sulfonyl)imide (3)
1-Ethyl-3-methylimidazolium bis((trifluoromethyl)sulfonyl)imide (1) + 1-Pentene (2) + 1-Butyl-3-methylimidazolium bis((trifluoromethyl)sulfonyl)imide (3)

1-Ethyl-3-methylimidazolium ethylsulfate (1) + Water (2)
1-Ethyl-3-methylimidazolium ethylsulfate (1) + *n*-Hexane (2)
1-Ethyl-3-methylimidazolium ethylsulfate (1) + *n*-Octane (2)
1-Ethyl-3-methylimidazolium ethylsulfate (1) + Benzene (2)
1-Ethyl-3-methylimidazolium ethylsulfate (1) + Toluene (2)
1-Ethyl-3-methylimidazolium ethylsulfate (1) + Ethyl benzene (2)
1-Ethyl-3-methylimidazolium ethylsulfate (1) + 2-Pentanone (2)
1-Ethyl-3-methylimidazolium ethylsulfate (1) + Pentan-3-one (2)
1-Ethyl-3-methylimidazolium ethylsulfate (1) + 2-Hexanone (2)
1-Ethyl-3-methylimidazolium ethylsulfate (1) + Heptan-4-one (2)
1-Ethyl-3-methylimidazolium ethylsulfate (1) + Dibutyl ether (2)
1-Ethyl-3-methylimidazolium ethylsulfate (1) + *tert*-Butyl-methyl ether (2)
1-Ethyl-3-methylimidazolium ethylsulfate (1) + Ethyl tert-butyl ether (2)
1-Ethyl-3-methylimidazolium ethylsulfate (1) + Tetrahydrofuran (2)
1-Ethyl-3-methylimidazolium ethylsulfate (1) + Dimethylsulfoxide (2)
1-Ethyl-3-methylimidazolium ethylsulfate (1) + Ammonia (2)
1-Ethyl-3-methylimidazolium ethylsulfate (1) + Methanol (2)
1-Ethyl-3-methylimidazolium ethylsulfate (1) + 1-Propanol (2)
1-Ethyl-3-methylimidazolium ethylsulfate (1) + 2-Propanol (2)
1-Ethyl-3-methylimidazolium ethylsulfate (1) + Cyclohexane (2)
1-Ethyl-3-methylimidazolium ethylsulfate (1) + Hydrogen (2)
1-Ethyl-3-methylimidazolium ethylsulfate (1) + Carbon dioxide (2)
1-Ethyl-3-methylimidazolium ethylsulfate (1) + 1-Nonane (2)
1-Ethyl-3-methylimidazolium ethylsulfate (1) + 1-Decane (2)
1-Ethyl-3-methylimidazolium ethylsulfate (1) + 1-Undecane (2)
1-Ethyl-3-methylimidazolium ethylsulfate (1) + 1-Dodecane (2)
1-Ethyl-3-methylimidazolium ethylsulfate (1) + 1-Hexene (2)
1-Ethyl-3-methylimidazolium ethylsulfate (1) + 1-Heptene (2)
1-Ethyl-3-methylimidazolium ethylsulfate (1) + 1-Octene (2)
1-Ethyl-3-methylimidazolium ethylsulfate (1) + 1-Nonene (2)
1-Ethyl-3-methylimidazolium ethylsulfate (1) + 1-Decene (2)
1-Ethyl-3-methylimidazolium ethylsulfate (1) + 1-Undecene (2)
1-Ethyl-3-methylimidazolium ethylsulfate (1) + 1-Dodecene (2)
1-Ethyl-3-methylimidazolium ethylsulfate (1) + Cyclohexene (2)
1-Ethyl-3-methylimidazolium ethylsulfate (1) + Propyl benzene (2)
1-Ethyl-3-methylimidazolium ethylsulfate (1) + Butyl benzene (2)
1-Ethyl-3-methylimidazolium ethylsulfate (1) + Pentyl benzene (2)
1-Ethyl-3-methylimidazolium ethylsulfate (1) + 1-Butanol (2)
1-Ethyl-3-methylimidazolium ethylsulfate (1) + 1-Pentanol (2)
1-Ethyl-3-methylimidazolium ethylsulfate (1) + 1-Hexanol (2)
1-Ethyl-3-methylimidazolium ethylsulfate (1) + *iso*-Butanol (2)
1-Ethyl-3-methylimidazolium ethylsulfate (1) + *sec*-Butanol (2)
1-Ethyl-3-methylimidazolium ethylsulfate (1) + *tert*-Butanol (2)
1-Ethyl-3-methylimidazolium ethylsulfate (1) + *tert*-Pentanol (2)
1-Ethyl-3-methylimidazolium ethylsulfate (1) + Propanal (2)

1-Ethyl-3-methylimidazolium hydrogen sulfate

1-Ethyl-3-methylimidazolium hydrogen sulfate (1) + Ethyl acetate (2) + Ethanol (3)
1-Ethyl-3-methylimidazolium hydrogen sulfate (1) + Ethyl acetate (2) + Acetic acid (3)
1-Ethyl-3-methylimidazolium hydrogen sulfate (1) + 1-Propanol (2) + Propylacetate (3)
1-Ethyl-3-methylimidazolium hydrogen sulfate (1) + Acetic acid (2) + Propylacetate (3)
1-Ethyl-3-methylimidazolium hydrogen sulfate (1) + 1-Butanol (2) + Butylacetate (3)
1-Ethyl-3-methylimidazolium hydrogen sulfate (1) + Acetic acid (2) + Butylacetate (3)

1-Ethyl-3-methylimidazolium tetrafluoroborate

1-Ethyl-3-methylimidazolium tetrafluoroborate (1) + 1-Hexyl-3-methylimidazolium tetrafluoroborate (2)
1-Ethyl-3-methylimidazolium tetrafluoroborate (1) + 1-Ethyl-3-methylimidazolium bis((trifluoromethyl)sulfonyl)imide (2)
1-Ethyl-3-methylimidazolium tetrafluoroborate (1) + Threo-2,3-dihydrodecafluoropentane (Threo-H) (2)
1-Ethyl-3-methylimidazolium tetrafluoroborate (1) + Threo-2,3-dihydrodecafluoropentane (Threo-D) (2)
1-Ethyl-3-methylimidazolium tetrafluoroborate (1) + Erythro-2,3-dihydrodecafluoropentane (Erythro-H) (2)
1-Ethyl-3-methylimidazolium tetrafluoroborate (1) + Erythro-2,3-dihydrodecafluoropentane (Erythro-D) (2)
1-Ethyl-3-methylimidazolium tetrafluoroborate (1) + 1,3-Propanediol (2)
1-Ethyl-3-methylimidazolium tetrafluoroborate (1) + Water (2)
1-Ethyl-3-methylimidazolium tetrafluoroborate (1) + 2-Propanol (2)
1-Ethyl-3-methylimidazolium tetrafluoroborate (1) + 1-Propanol (2)
1-Ethyl-3-methylimidazolium tetrafluoroborate (1) + Lithium tetrafluoroborate (2)
1-Ethyl-3-methylimidazolium tetrafluoroborate (1) + Poly(ethyleneoxide) (2)
1-Ethyl-3-methylimidazolium tetrafluoroborate (1) + Poly(vinylalcohol) (2)
1-Ethyl-3-methylimidazolium tetrafluoroborate (1) + Poly(acrylonitrile) (2)
1-Ethyl-3-methylimidazolium tetrafluoroborate (1) + Ethanol (2)
1-Ethyl-3-methylimidazolium tetrafluoroborate (1) + Acetonitrile (2)
1-Ethyl-3-methylimidazolium tetrafluoroborate (1) + Threo-2,3-dihydrodecafluoropentane (87.12 ± 0.6% Threo-H) (2) + Erythro-2,3-dihydrodecafluoropentane (11.89 ± 0.6% Erythro-H) (3)
1-Ethyl-3-methylimidazolium tetrafluoroborate (1) + Water (2) + Ethanol (3)
1-Ethyl-3-methylimidazolium tetrafluoroborate (1) + Water (2) + 2-Propanol (3)
1-Ethyl-3-methylimidazolium tetrafluoroborate (1) + Water (2) + 1-Propanol (3)
1-Ethyl-3-methylimidazolium tetrafluoroborate (1) + Water (2) + Tetrahydrofuran (3)
1-Ethyl-3-methylimidazolium tetrafluoroborate (1) + Poly(acrylonitrile) (2) + Sulpholane (3)
1-Ethyl-3-methylimidazolium tetrafluoroborate (1) + Ethyl acetate (2) + Ethanol (3)

1-Ethyl-3-methylimidazolium tetrafluoroborate (1) + Iodine (2) + 1-Propyl-3-methylimidazolium iodine (3)

1-Ethyl-3-methylimidazolium 1,1,2,2-tetrafluoroethanesulfonate

1-Ethyl-3-methylimidazolium 1,1,2,2-tetrafluoroethanesulfonate (1) + Difluoromethane (2)

1-Ethyl-3-methylimidazolium thiocyanate

1-Ethyl-3-methylimidazolium thiocyanate (1) + Ammonia (2)
1-Ethyl-3-methylimidazolium thiocyanate (1) + Thiophene (2)

1-Ethyl-3-methylimidazolium tosilate

1-Ethyl-3-methylimidazolium tosilate (1) + *n*-Hexane (2) + Propionic acid (3)
1-Ethyl-3-methylimidazolium tosilate (1) + *n*-Hexane (2) + Vinyl propionate (3)
1-Ethyl-3-methylimidazolium tosilate (1) + *n*-Hexane (2) + *rac*-1-Phenylethyl propionate (3)
1-Ethyl-3-methylimidazolium tosilate (1) + *n*-Hexane (2) + *rac*-1-Phenylethanol (3)

1-Ethyl-3-methylimidazolium trifluoroacetate

1-Ethyl-3-methylimidazolium trifluoroacetate (1) + Water (2)
1-Ethyl-3-methylimidazolium trifluoroacetate (1) + 1-Ethyl-3-methylimidazolium acetate (2)
1-Ethyl-3-methylimidazolium trifluoroacetate (1) + Carbon dioxide (2)
1-Ethyl-3-methylimidazolium trifluoroacetate (1) + Carbon dioxide (2) + 1-Ethyl-3-methylimidazolium acetate (3)

1-Ethyl-3-methylimidazolium trifluoromethanesulfonate

1-Ethyl-3-methylimidazolium trifluoromethanesulfonate (1) + Ethanol (2)
1-Ethyl-3-methylimidazolium trifluoromethanesulfonate (1) + Trichloromethane (2)
1-Ethyl-3-methylimidazolium trifluoromethanesulfonate (1) + 1-Propanol (2)
1-Ethyl-3-methylimidazolium trifluoromethanesulfonate (1) + Water (2)
1-Ethyl-3-methylimidazolium trifluoromethanesulfonate (1) + Ethyl acetate (2)
1-Ethyl-3-methylimidazolium trifluoromethanesulfonate (1) + Methanol (2)
1-Ethyl-3-methylimidazolium trifluoromethanesulfonate (1) + Methyl acetate (2)
1-Ethyl-3-methylimidazolium trifluoromethanesulfonate (1) + Acetone (2)
1-Ethyl-3-methylimidazolium trifluoromethanesulfonate (1) + Nitromethane (2)
1-Ethyl-3-methylimidazolium trifluoromethanesulfonate (1) + Trichloromethane (2) + Ethanol (3)
1-Ethyl-3-methylimidazolium trifluoromethanesulfonate (1) + 1-Propanol (2) + Water (3)
1-Ethyl-3-methylimidazolium trifluoromethanesulfonate (1) + Ethyl acetate (2) + Ethanol (3)
1-Ethyl-3-methylimidazolium trifluoromethanesulfonate (1) + Ethyl acetate (2) + Methanol (3)
1-Ethyl-3-methylimidazolium trifluoromethanesulfonate (1) + Acetone (2) + Methanol (3)

1-Ethyl-3-methylpyridinium ethylsulfate

1-Ethyl-3-methylpyridinium ethylsulfate (1) + Ethanol (2)
1-Ethyl-3-methylpyridinium ethylsulfate (1) + Water (2)
1-Ethyl-3-methylpyridinium ethylsulfate (1) + 1-Propanol (2)

1-Ethyl-3-methylpyridinium ethylsulfate (1) + 2-Propanol (2)
1-Ethyl-3-methylpyridinium ethylsulfate (1) + *n*-Hexane (2) + Benzene (3)
1-Ethyl-3-methylpyridinium ethylsulfate (1) + *n*-Heptane (2) + Benzene (3)
1-Ethyl-3-methylpyridinium ethylsulfate (1) + Ethanol (2) + Water (3)

1-Ethylpyridinium bis((trifluoromethyl)sulfonyl)amide
1-Ethylpyridinium bis((trifluoromethyl)sulfonyl)amide (1) + Trihexyl-tetradecylphosphonium bis((trifluoromethyl)sulfonyl)imide (2)
1-Ethylpyridinium bis((trifluoromethyl)sulfonyl)amide (1) + *n*-Hexane (2) + Benzene (3)

Ethyl pyridinium ethylsulfate
Ethyl pyridinium ethylsulfate (1) + Ethanol (2)
Ethyl pyridinium ethylsulfate (1) + 1-Propanol (2)

1-Ethylpyridinium hexafluorophosphate
1-Ethylpyridinium hexafluorophosphate (1) + Water (2)+ Ethanol (3)

H

1-Heptyl-3-methylimidazolium bis((trifluoromethyl)sulfonyl)imide
1-Heptyl-3-methylimidazolium bis((trifluoromethyl)sulfonyl)imide (1) + Water (2)

1-Heptyl-3-methylimidazolium 1,1,2,2-tetrafluoroethanesulfonate
1-Heptyl-3-methylimidazolium 1,1,2,2-tetrafluoroethanesulfonate (1) + Difluoromethane (2)

Hexyl(2-hydroxyethyl)dimethylammonium bromide
Hexyl(2-hydroxyethyl)dimethylammonium bromide (1) + 1-Octanol (2)
Hexyl(2-hydroxyethyl)dimethylammonium bromide (1) + Water (2)

Hexyl(2-hydroxyethyl)dimethylammonium tetrafluoroborate
Hexyl(2-hydroxyethyl)dimethylammonium tetrafluoroborate (1) + *n*-Hexane (2) + *p*-Xylene (3)

1-Hexyl-3-methylimidazolium acetate
1-Hexyl-3-methylimidazolium acetate (1) + Water (2) + *tert*-Butanol (3)

1-Hexyl-3-methylimidazolium bis((trifluoromethyl)sulfonyl)imide
1-Hexyl-3-methylimidazolium bis((trifluoromethyl)sulfonyl)imide (1) + Perfluoromethane (2)
1-Hexyl-3-methylimidazolium bis((trifluoromethyl)sulfonyl)imide (1) + Methylcyclohexane (2)
1-Hexyl-3-methylimidazolium bis((trifluoromethyl)sulfonyl)imide (1) + *n*-Octane (2)
1-Hexyl-3-methylimidazolium bis((trifluoromethyl)sulfonyl)imide (1) + Toluene (2)
1-Hexyl-3-methylimidazolium bis((trifluoromethyl)sulfonyl)imide (1) + Water (2)
1-Hexyl-3-methylimidazolium bis((trifluoromethyl)sulfonyl)imide (1) + 1-Hexyl-3-methylimidazolium chloride (2)
1-Hexyl-3-methylimidazolium bis((trifluoromethyl)sulfonyl)imide (1) + *n*-Hexane (2)

1-Hexyl-3-methylimidazolium hexafluorophosphate (1) + Dimethyl carbonate (2)
1-Hexyl-3-methylimidazolium hexafluorophosphate (1) + Diethyl carbonate (2)
1-Hexyl-3-methylimidazolium hexafluorophosphate (1) + Acetone (2)
1-Hexyl-3-methylimidazolium hexafluorophosphate (1) + 2-Butanone (2)
1-Hexyl-3-methylimidazolium hexafluorophosphate (1) + 2-Pentanone (2)
1-Hexyl-3-methylimidazolium hexafluorophosphate (1) + Methyl acetate (2)
1-Hexyl-3-methylimidazolium hexafluorophosphate (1) + Butylacetate (2)
1-Hexyl-3-methylimidazolium hexafluorophosphate (1) + Carbon dioxide (2)
1-Hexyl-3-methylimidazolium hexafluorophosphate (1) + 1,1,1,2-Tetrafluoroethane (2)
1-Hexyl-3-methylimidazolium hexafluorophosphate (1) + Cyclohexane (2) + 2-Butanone (3)
1-Hexyl-3-methylimidazolium hexafluorophosphate (1) + Bisphenol A (2) + Water (3)
1-Hexyl-3-methylimidazolium hexafluorophosphate (1) + Phenol (2) + Water (3)
1-Hexyl-3-methylimidazolium hexafluorophosphate (1) + 4-Nonylphenol (2) + Water (3)
1-Hexyl-3-methylimidazolium hexafluorophosphate (1) + 4-Octylphenol (2) + Water (3)
1-Hexyl-3-methylimidazolium hexafluorophosphate (1) + Pentachlorophenol (2) + Water (3)
1-Hexyl-3-methylimidazolium hexafluorophosphate (1) + *n*-Hexane (2) + Ethyl acetate (3)
1-Hexyl-3-methylimidazolium hexafluorophosphate (1) + 2-Propanol (2) + Ethyl acetate (3)
1-Hexyl-3-methylimidazolium hexafluorophosphate (1) + *n*-Heptane (2) + Benzene (3)
1-Hexyl-3-methylimidazolium hexafluorophosphate (1) + 1-Dodecane (2) + Benzene (3)
1-Hexyl-3-methylimidazolium hexafluorophosphate (1) + Hexadecane (2)+ Benzene (3)
1-Hexyl-3-methylimidazolium hexafluorophosphate (1) + 1-Hexene (2) + Ethanol (3)
1-Hexyl-3-methylimidazolium hexafluorophosphate (1) + 1-Heptene (2) + Ethanol (3)

1-Hexyl-3-methylimidazolium hydrogen sulfate

1-Hexyl-3-methylimidazolium hydrogen sulfate (1) + Ethyl acetate (2) + Ethanol (3)
1-Hexyl-3-methylimidazolium hydrogen sulfate (1) + Ethyl acetate (2) + Acetic acid (3)

1-Hexyl-3-methylimidazolium tetrafluoroborate

1-Hexyl-3-methylimidazolium tetrafluoroborate (1) + Ethanol (2)
1-Hexyl-3-methylimidazolium tetrafluoroborate (1) + Methanol (2)
1-Hexyl-3-methylimidazolium tetrafluoroborate (1) + 1-Ethyl-3-methylimidazolium tetrafluoroborate (2)

1-Hexyl-3-methylimidazolium tetrafluoroborate (1) + 1-Butyl-3-methylimidazolium tetrafluoroborate (2)
1-Hexyl-3-methylimidazolium tetrafluoroborate (1) + Carbon dioxide (2)
1-Hexyl-3-methylimidazolium tetrafluoroborate (1) + 1-Pentanol (2)
1-Hexyl-3-methylimidazolium tetrafluoroborate (1) + Water (2)
1-Hexyl-3-methylimidazolium tetrafluoroborate (1) + 1-Propanol (2)
1-Hexyl-3-methylimidazolium tetrafluoroborate (1) + 1-Butanol (2)
1-Hexyl-3-methylimidazolium tetrafluoroborate (1) + 2-Butanol (2)
1-Hexyl-3-methylimidazolium tetrafluoroborate (1) + 2-Pentanol (2)
1-Hexyl-3-methylimidazolium tetrafluoroborate (1) + 1- Hexanol (2)
1-Hexyl-3-methylimidazolium tetrafluoroborate (1) + 1,1,1,2-Tetrafluoroethane (2)
1-Hexyl-3-methylimidazolium tetrafluoroborate (1) + 2-Butanone (2)
1-Hexyl-3-methylimidazolium tetrafluoroborate (1) + Ethyl acetate (2)
1-Hexyl-3-methylimidazolium tetrafluoroborate (1) + Acetonitrile (2)
1-Hexyl-3-methylimidazolium tetrafluoroborate (1) + Nitromethane (2)
1-Hexyl-3-methylimidazolium tetrafluoroborate (1) + Bisphenol A (2) + Water (3)
1-Hexyl-3-methylimidazolium tetrafluoroborate (1) + Phenol (2) + Water (3)
1-Hexyl-3-methylimidazolium tetrafluoroborate (1) + 4-Nonylphenol (2) + Water (3)
1-Hexyl-3-methylimidazolium tetrafluoroborate (1) + 4-Octylphenol (2) + Water (3)
1-Hexyl-3-methylimidazolium tetrafluoroborate (1) + Pentachlorophenol (2) + Water (3)
1-Hexyl-3-methylimidazolium tetrafluoroborate (1) + *n*-Heptane (2) + Benzene (3)
1-Hexyl-3-methylimidazolium tetrafluoroborate (1) + 1-Dodecane (2) + Benzene (3)
1-Hexyl-3-methylimidazolium tetrafluoroborate (1) + Hexadecane (2) + Benzene (3)
1-Hexyl-3-methylimidazolium tetrafluoroborate (1) + 1-Hexene (2) + Ethanol (3)
1-Hexyl-3-methylimidazolium tetrafluoroborate (1) + 1-Heptene (2) + Ethanol (3)

1-Hexyl-3-methylimidazolium thiocyanate

1-Hexyl-3-methylimidazolium thiocyanate (1) + Thiophene (2)

1-Hexyl-3-methylimidazolium trifluoromethanesulfonate

1-Hexyl-3-methylimidazolium trifluoromethanesulfonate (1) + Toluene (2)
1-Hexyl-3-methylimidazolium trifluoromethanesulfonate (1) + Methylcyclohexane (2)
1-Hexyl-3-methylimidazolium trifluoromethanesulfonate (1) + Ethanol (2)
1-Hexyl-3-methylimidazolium trifluoromethanesulfonate (1) + Water (2)
1-Hexyl-3-methylimidazolium trifluoromethanesulfonate (1) + Methylcyclohexane (2) + Toluene (3)
1-Hexyl-3-methylimidazolium trifluoromethanesulfonate (1) + Ethanol (2) + Water (3)
1-Hexyl-3-methylimidazolium trifluoromethanesulfonate (1) + *n*-Hexane (2) + Benzene (3)
1-Hexyl-3-methylimidazolium trifluoromethanesulfonate (1) + Methylcyclohexane (2) + Benzene (3)

***N*-Hexyl-3-methylpyridinium tosylate**

N-Hexyl-3-methylpyridinium tosylate (1) + Thiophene (2)

1-Hexyloxymethyl-3-methylimidazolium bis((trifluoromethyl)sulfonyl)imide

2-Hydroxyethylammonium acetate (1) + Ethanol (2) + Methanol (3)

1-Hydroxyethyl-3-methylimidazolium bis((trifluoromethyl)sulfonyl)imide

1-Hydroxyethyl-3-methylimidazolium bis((trifluoromethyl)sulfonyl)imide (1) + Dichloromethane (2)

1-Hydroxyethyl-3-methylimidazolium bis((trifluoromethyl)sulfonyl)imide (1) + 2-Propanol (2) + Dichloromethane (3)

1-Hydroxyethyl-3-methylimidazolium bis((trifluoromethyl)sulfonyl)imide (1) + 2-Propanol (2) + 1,2-Dichloroethane (3)

1-Hydroxyethyl-3-methylimidazolium hexafluorophosphate

1-Hydroxyethyl-3-methylimidazolium hexafluorophosphate (1) + Dichloromethane (2)

1-Hydroxyethyl-3-methylimidazolium hexafluorophosphate (1) + 2-Propanol (2) + Dichloromethane (3)

1-Hydroxyethyl-3-methylimidazolium tetrafluoroborate

1-Hydroxyethyl-3-methylimidazolium tetrafluoroborate (1) + Water (2)

1-Hydroxyethyl-3-methylimidazolium tetrafluoroborate (1) + Dichloromethane (2)

1-Hydroxyethyl-3-methylimidazolium tetrafluoroborate (1) + 2-Propanol (2) + Dichloromethane (3)

1-Hydroxyethyl-3-methylimidazolium triflate

1-Hydroxyethyl-3-methylimidazolium triflate (1) + Dichloromethane (2)

1-Hydroxyethyl-3-methylimidazolium triflate (1) + 2-Propanol (2) + Dichloromethane (3)

1-Hydroxyethyl-3-methylimidazolium triflate (1) + 2-Propanol (2) + 1,2-Dichloroethane (3)

(2-Hydroxyethyl)trimethylammonium bis{(trifluoromethyl)sulfonyl}amide

(2-Hydroxyethyl)trimethylammonium bis{(trifluoromethyl)sulfonyl}amide (1) + *n*-Hexane (2) + Benzene (3)

1-Hydroxy-1-propyl-3-methylimidazolium nitrate

1-Hydroxy-1-propyl-3-methylimidazolium nitrate (1) + Water (2)

1-Hydroxy-1-propyl-3-methylimidazolium nitrate (1) + Carbon dioxide (2)

1-Hydroxy-1-propyl-3-methylimidazolium nitrate (1) + Carbon dioxide (2) + Water (3)

I

Imidazolium chloride

Imidazolium chloride (1) + Water (2)

1-Isobutenyl-3-methylimidazolium tetrafluoroborate

1-Isobutenyl-3-methylimidazolium tetrafluoroborate (1) + 1-Hexene (2)

1-Isobutenyl-3-methylimidazolium tetrafluoroborate (1) + *trans*-3-Hexene (2)

1-Isobutenyl-3-methylimidazolium tetrafluoroborate (1) + 1-Hexene (2) + *trans*-3-Hexene (3)

M

4-Methyl-*N*-butylpyridinium tetrafluoroborate

***N*-methyl-*N*-butyl-pyrrolidinium bis((trifluoromethyl)sulfonyl)imide**

N-methyl-*N*-butyl-pyrrolidinium bis((trifluoromethyl)sulfonyl)imide (1) + 2-Pentanone (2)
N-methyl-*N*-butyl-pyrrolidinium bis((trifluoromethyl)sulfonyl)imide (1) + Ethanol (2)
N-methyl-*N*-butyl-pyrrolidinium bis((trifluoromethyl)sulfonyl)imide (1) + 2-Propanol (2)
N-methyl-*N*-butyl-pyrrolidinium bis((trifluoromethyl)sulfonyl)imide (1) + Diisopropyl ether (2)
N-methyl-*N*-butyl-pyrrolidinium bis((trifluoromethyl)sulfonyl)imide (1) + Poly (ethyleneoxide) (2)
N-methyl-*N*-butyl-pyrrolidinium bis((trifluoromethyl)sulfonyl)imide (1) + Ferrocene (2) + Dimethylformamide (3)
N-methyl-*N*-butyl-pyrrolidinium bis((trifluoromethyl)sulfonyl)imide (1) + *n*-Hexane (2) + Benzene (3)
N-methyl-*N*-butyl-pyrrolidinium bis((trifluoromethyl)sulfonyl)imide (1) + Cyclohexane (2) + Benzene (3)

1-Methyl-1-butyl-pyrrolidinium trifluoromethanesulfonate

1-Methyl-1-butyl-pyrrolidinium trifluoromethanesulfonate (1) + Ethanol (2)
1-Methyl-1-butyl-pyrrolidinium trifluoromethanesulfonate (1) + Water (2)
1-Methyl-1-butyl-pyrrolidinium trifluoromethanesulfonate (1) + Ethanol (2) + Water (3)

1-Methyl-3-ethylimidazolium chloride

1-Methyl-3-ethylimidazolium chloride (1) + Aluminium trichloride (2)
1-Methyl-3-ethylimidazolium chloride (1) + Aluminium trichloride (2) + Sodium chloride (3)
1-Methyl-3-ethylimidazolium chloride (1) + Aluminium trichloride (2) + Potassium chloride (3)
1-Methyl-3-ethylimidazolium chloride (1) + Aluminium trichloride (2) + Lithium chloride (3)
1-Methyl-3-ethylimidazolium chloride (1) + Aluminium trichloride (2) + Rubidium chloride (3)
1-Methyl-3-ethylimidazolium chloride (1) + Aluminium trichloride (2) + Caesium chloride (3)
1-Methyl-3-ethylimidazolium chloride (1) + Aluminium trichloride (2) + Sodium bromide (3)
1-Methyl-3-ethylimidazolium chloride (1) + Aluminium trichloride (2) + Sodium iodide (3)

***N*-methyl-*N*-hexylpyrrolidinium bis((trifluoromethyl)sulfonyl)imide**

N-methyl-*N*-hexylpyrrolidinium bis((trifluoromethyl)sulfonyl)imide (1) + Dimethylformamide (2)
N-methyl-*N*-hexylpyrrolidinium bis((trifluoromethyl)sulfonyl)imide (1) + *n*-Pentane (2)
N-methyl-*N*-hexylpyrrolidinium bis((trifluoromethyl)sulfonyl)imide (1) + *n*-Hexane (2)

O

1-Octyl-3-methylimidazolium bis((trifluoromethyl)sulfonyl)imide (1) + 2-Pentanone (2)
1-Octyl-3-methylimidazolium bis((trifluoromethyl)sulfonyl)imide (1) + Tetrahydrofuran (2)
1-Octyl-3-methylimidazolium bis((trifluoromethyl)sulfonyl)imide (1) + Diisopropyl ether (2)
1-Octyl-3-methylimidazolium bis((trifluoromethyl)sulfonyl)imide (1) + Ethyl *tert*-butyl ether (2)
1-Octyl-3-methylimidazolium bis((trifluoromethyl)sulfonyl)imide (1) + Methyl *tert*-amyl ether (2)
1-Octyl-3-methylimidazolium bis((trifluoromethyl)sulfonyl)imide (1) + Methyl acetate (2)
1-Octyl-3-methylimidazolium bis((trifluoromethyl)sulfonyl)imide (1) + Ethyl acetate (2)
1-Octyl-3-methylimidazolium bis((trifluoromethyl)sulfonyl)imide (1) + 1-Dodecane (2) + Thiophene (3)
1-Octyl-3-methylimidazolium bis((trifluoromethyl)sulfonyl)imide (1) + Cyclohexane (2) + Thiophene (3)
1-Octyl-3-methylimidazolium bis((trifluoromethyl)sulfonyl)imide (1) + *n*-Hexane (2) + Thiophene (3)
1-Octyl-3-methylimidazolium bis((trifluoromethyl)sulfonyl)imide (1) + *n*-Heptane (2) + Thiophene (3)
1-Octyl-3-methylimidazolium bis((trifluoromethyl)sulfonyl)imide (1) + Hexadecane (2) + Thiophene (3)
1-Octyl-3-methylimidazolium bis((trifluoromethyl)sulfonyl)imide (1) + *n*-Hexane (2) + Benzene (3)
1-Octyl-3-methylimidazolium bis((trifluoromethyl)sulfonyl)imide (1) + Methyl acetate (2) + Methanol (3)
1-Octyl-3-methylimidazolium bis((trifluoromethyl)sulfonyl)imide (1) + Ethyl acetate (2) + Ethanol (3)

1-Octyl-3-methylimidazolium bromide

1-Octyl-3-methylimidazolium bromide (1) + Dimethylsulfoxide (2)
1-Octyl-3-methylimidazolium bromide (1) + 1,2-Ethanediol (2)
1-Octyl-3-methylimidazolium bromide (1) + *N,N*-Dimethylformamide (2)

1-Octyl-3-methylimidazolium chloride

1-Octyl-3-methylimidazolium chloride (1) + Water (2)
1-Octyl-3-methylimidazolium chloride (1) + Methanol (2)
1-Octyl-3-methylimidazolium chloride (1) + Ethanol (2)
1-Octyl-3-methylimidazolium chloride (1) + 1-Propanol (2)
1-Octyl-3-methylimidazolium chloride (1) + *n*-Heptane (2)
1-Octyl-3-methylimidazolium chloride (1) + 1-Dodecane (2)
1-Octyl-3-methylimidazolium chloride (1) + Hexadecane (2)
1-Octyl-3-methylimidazolium chloride (1) + Ethanol (2) + Water (3)
1-Octyl-3-methylimidazolium chloride (1) + *n*-Heptane (2) + Methanol (3)
1-Octyl-3-methylimidazolium chloride (1) + Hexadecane (2) + Methanol (3)

1-Octyl-3-methylimidazolium tetrafluoroborate (1) + 2-Butanone (2)
1-Octyl-3-methylimidazolium tetrafluoroborate (1) + Methyl acetate (2)
1-Octyl-3-methylimidazolium tetrafluoroborate (1) + Ethyl acetate (2)
1-Octyl-3-methylimidazolium tetrafluoroborate (1) + Bisphenol A (2) + Water (3)
1-Octyl-3-methylimidazolium tetrafluoroborate (1) + Phenol (2) + Water (3)
1-Octyl-3-methylimidazolium tetrafluoroborate (1) + 4-Nonylphenol (2) + Water (3)
1-Octyl-3-methylimidazolium tetrafluoroborate (1) + 4-Octylphenol (2) + Water (3)
1-Octyl-3-methylimidazolium tetrafluoroborate (1) + Pentachlorophenol (2) + Water (3)
1-Octyl-3-methylimidazolium tetrafluoroborate (1) + *n*-Heptane (2) + Thiophene (3)
1-Octyl-3-methylimidazolium tetrafluoroborate (1) + 1-Dodecane (2) + Thiophene (3)
1-Octyl-3-methylimidazolium tetrafluoroborate (1) + Hexadecane (2) + Thiophene (3)
1-Octyl-3-methylimidazolium tetrafluoroborate (1) + *i*-Octane (2) + Thiophene (3)
1-Octyl-3-methylimidazolium tetrafluoroborate (1) + Toluene (2) + Thiophene (3)
1-Octyl-3-methylimidazolium tetrafluoroborate (1) + *n*-Hexane (2) + Cyclehexane (3)
1-Octyl-3-methylimidazolium tetrafluoroborate (1) + Water (2) + Tetrahydrofuran (3)

1-Octyl pyridinium tetrafluoroborate
1-Octyl pyridinium tetrafluoroborate (1) + Water (2)

P

1-Pentyl-3-methylimidazolium chloride
1-Pentyl-3-methylimidazolium chloride (1) + Water (2)

1-Propyl-2,3-dimethylimidazolium tetrafluoroborate
1-Propyl-2,3-dimethylimidazolium tetrafluoroborate (1) + Water (2)

Propyl(2-hydroxyethyl)dimethylammonium bromide
Propyl(2-hydroxyethyl)dimethylammonium bromide (1) + Water (2)

1-Propyl-3-methylimidazolium bis((trifluoromethyl)sulfonyl)imide
1-Propyl-3-methylimidazolium bis((trifluoromethyl)sulfonyl)imide (1) + Water (2)

1-Propyl-3-methylimidazolium bromide
1-Propyl-3-methylimidazolium bromide (1) + Water (2)
1-Propyl-3-methylimidazolium bromide (1) + Acetonitrile (2)
1-Propyl-3-methylimidazolium bromide (1) + Dimethylsulfoxide (2)
1-Propyl-3-methylimidazolium bromide (1) + Dimethylformamide (2)

1-Propyl-3-methylimidazolium iodine
1-Propyl-3-methylimidazolium iodine (1) + Iodine (2) + 1-Ethyl-3-methylimidazolium tetrafluoroborate (3)

Pyridinium ethoxyethylsulfate
Pyridinium ethoxyethylsulfate (1) + Benzene (2)
Pyridinium ethoxyethylsulfate (1) + Toluene (2)

T

Tetra-*n*-butylammonium pricrate (1) + 1-Nonanol (2)
Tetra-*n*-butylammonium pricrate (1) + 1-Decanol (2)
Tetra-*n*-butylammonium pricrate (1) + 1-Undecanol (2)
Tetra-*n*-butylammonium pricrate (1) + 1-Dodecanol (2)
Tetra-*n*-butylammonium pricrate (1) + 1-Tridecanol (2)
Tetra-*n*-butylammonium pricrate (1) + 1-Tetradecanol (2)

Tetrabutyl phosphonium alaninate
Tetrabutyl phosphonium alaninate (1) + Water (2)
Tetrabutyl phosphonium alaninate (1) + Methanol (2)
Tetrabutyl phosphonium alaninate (1) + Dichloromethane (2)

Tetrabutylphosphonium methanesulfonate
Tetrabutylphosphonium methanesulfonate (1) + 1-Octanol (2)
Tetrabutylphosphonium methanesulfonate (1) + 1-Decanol (2)
Tetrabutylphosphonium methanesulfonate (1) + 1-Dodecanol (2)
Tetrabutylphosphonium methanesulfonate (1) + Benzene (2)

Tetra(*n*-butyl)phosphonium tosylate
Tetra(n-butyl)phosphonium tosylate (1) + 1-Octanol (2)
Tetra(n-butyl)phosphonium tosylate (1) + 1-Decanol (2)
Tetra(n-butyl)phosphonium tosylate (1) + 1-Dodecanol (2)
Tetra(n-butyl)phosphonium tosylate (1) + Benzene (2)

Tetrabutyl phosphonium valinate
Tetrabutyl phosphonium valinate (1) + Water (2)
Tetrabutyl phosphonium valinate (1) + Methanol (2)
Tetrabutyl phosphonium valinate (1) + Dichloromethane (2)

***N*-tetradecyl-isoquinolinium bis((perfluoroethane)sulfonyl)imide**
N-tetradecyl-isoquinolinium bis((perfluoroethane)sulfonyl)imide (1) + Water (2) + Tetraheptylammonium bis((perfluoroethane)sulfonyl)imide (3)

1-Tetradecyl-3-methylpyridinium bis(trifluoromethylsulfonyl)imide
1-Tetradecyl-3-methylpyridinium bis(trifluoromethylsulfonyl)imide (1) + 1*H*,1*H*,2*H*,2*H*-Perfluorooctanol (2)

Tetradecyl(trihexyl)phosphonium 1,1,2-trifluoro-2-(perfluoroethoxy) ethanesulfonate
Tetradecyl(trihexyl)phosphonium 1,1,2-trifluoro-2-(perfluoroethoxy) ethanesulfonate (1) + 1,1,1,2-Tetrafluoroethane (2)

Tetraheptylammonium bis((perfluoroethane)sulfonyl)imide
Tetraheptylammonium bis((perfluoroethane)sulfonyl)imide (1) + Water (2) + *N*-tetradecyl-isoquinolinium bis((perfluoroethane)sulfonyl)imide (3)

Tributyl(tetradecyl)phosphonium 1,1,2,3,3,3-hexafluoropropanesulfonate
Tributyl(tetradecyl)phosphonium 1,1,2,3,3,3-hexafluoropropanesulfonate (1) + 1,1,1,2-Tetrafluoroethane (2)

Triethanolammonium acetate
Triethanolammonium acetate (1) + Water (2)
Triethanolammonium acetate (1) + Methanol (2)
Triethanolammonium acetate (1) + Ethanol (2)
Triethanolammonium acetate (1) + Water (2) + Ethanol (3)

List of Properties

Property category	Description
Activity properties	Activity coefficient at infinite dilution
	Activity coefficient
	Activity
	Molal activity coefficient
	The mean molal activity coefficient
Osmotic properties	Osmotic coefficient
Phase equilibrium	Liquid-liquid equilibrium
	Solid–liquid equilibrium
	Vapor–liquid equilibrium
	Henry's law constant
	Fugacity coefficients
	Eutectic composition
	The composition of points on the coexistence curve
	Compositions of the conjugate solutions
	Distribution ratio
	Distribution coefficients
	Capacities at infinite dilution/infinite dilution partition coefficients
	Separation factors
	Selectivity
	Selectivity at infinite dilution
	Liquid-liquid–vapor equilibrium
	Quadruple point temperature
	Eutectic temperature
	Liquid–vapor boundaries data
	Bubble point
	Binodal data
	Critical temperatures of phase separation
	Vapor pressure
	Boiling point

(*continued*)

S. Zhang et al., *Physicochemical Properties of Ionic Liquid Mixtures*,
DOI 10.1007/978-94-017-7573-1

Property category	Description
	Phase boundary pressure
Critical properties	Critical pressure
	Critical temperature
	Miscibility temperatures
	Interval of immiscibility
	Upper consolute temperature
Electrochemical properties	Electrical conductivity
	Molar conductivity
	Ionic association constant
	Limiting molar conductance
	Distance parameter
	Standard deviation of electrical conductivity
	Dielectric constant
	Relative permittivity
	Cathodic limit
	Anodic limit
	Theoretical Debye-HüCkel limiting slope
Thermal conductivities	Thermal conductivities
Enthalpy	Excess enthalpy
	Excess molar enthalpy
	Partial molar excess enthalpy at infinite dilution
	Excess molar enthalpy at infinite dilution
	Partial molar enthalpy at infinite dilution
	Excess molar partial enthalpy
Heat capacity and derived properties	Molar heat capacity
	Ideal heat capacity
	Excess molar heat capacity
	The excess partial molar heat capacity at infinite dilution
	Heat capacity at vapor saturation pressure
Surface tension	Interfacial tension
	Surface tension
	Surface tension deviation
Refractive index	Refractive index
	Molar refraction
	Deviations in molar refraction
	Deviations in the refractive index
	Ideal refractive index
	Deviation from ideal refractive index
Speed of sound	Speed of sound
	Excess speed of sound
Transport properties	Tracer diffusion coefficient of component
	Thermal diffusivity
	Diffusivity

(continued)

Property category	Description
	Diffusion coefficient
	Viscosity
	Absolute viscosity
	Viscosity deviation
	Dynamic viscosity
	Excess logarithm viscosity
	Kinematic viscosity
Volumetric properties	Density
	Excess volume
	Excess molar volume
	Apparent molar volume
	Partial molar volume
	Observed molar volume
	Excess partial molar volume
	Molar volume
	Coefficient of thermal expansion
	Isobaric thermal expansivity
	Apparent molar expansibility
	The infinite dilution apparent molar expansibility
	Isobaric expansivity
	The partial molar volume at infinite dilution
	The excess partial molar volume at infinite dilution
	The apparent molar volume at infinite dilution
	Perturbed volume
	Molar volume of the pure components
	The limiting apparent molar expansivity
	Limiting apparent molar volume
Compressibility properties	Isentropic compressibility
	Molar isentropic compressibility
	Deviations in isentropic compressibility
	Excess isentropic compressibility
	Excess molar isentropic compressibility
	Isothermal compressibility
	Apparent molar isentropic compressibility
	Apparent molar isothermal compressibility
	Excess isentropic compressibility
	Limiting apparent isentropic compressibility
	Limiting apparent isothermal compressibility
Internal pressure	Internal pressure for aqueous solutions
Thermodynamic properties of solution	Excess Gibbs free-energy
	The standard molar enthalpy of solution
	The standard molar enthalpy of solvation

(*continued*)

Property category	Description
	The standard molar Gibbs energy of solution
	The standard molar Gibbs energy of solvation
	The standard molar entropy of solution
	The standard molar entropy of solvation
	The standard enthalpy of solution
	The standard Gibbs free-energy of solution
	The standard entropy of solution
	The local molar Gibbs energy of solvation
	The local molar enthalpy of solvation
	The local molar entropy of solvation
	The enthalpy of solvation
	The entropy of solvation
	The Gibbs energy of solvation
	The relative apparent molar enthalpy
	The heat of mixing per mole of solution
	The excess partial molar entropy of the solvent
	The excess partial molar entropy of the solute
	The molar enthalpy of mixing for the solute
	The Gibbs free-energy of mixing per mole of solution
	The entropy of mixing per mole of solution
	Excess free energy of activation of viscous flow
	Activation energy
	Solubility
	Gordon parameter
Cloud point	Cloud point
Phase transition temperature	Glass transition temperature
	Crystalline temperature
	Melting point
Emission properties	The quantum yields
	The observed lifetime
	The radiative
	The non-radiative
	The rotational relaxation time

List of Symbols

m	Molality
$\tilde{m}$	Defined $\tilde{m} = n_{\text{solute}}/m_{\text{solution}}$
w	Mass fraction
c	Molarity
x	Mole fraction of liquid phase
y	Mole fraction of vapor phase
ρ	Density
$\Delta\rho$	The difference of the densities
V	Observed molar volume
V^o	Molar volume of the pure component
V^L	The liquid molar volume
V' or $\bar{V}$	Partial molar volume
V^{ex} or V^E or V_m^E or $\bar{V}^{exc}$	Excess molar volume
φ_V or V_Φ	Apparent molar volume
$u\ (V_\Phi)$	Standard uncertainty
$\bar{V}^E$	Excess partial molar volume
$\bar{V}_{\text{m,i}}^{E,\infty}$	The excess partial molar volume at infinite dilution
$\bar{V}_{\text{m,i}}^{\infty}$ or $\bar{V}^{\infty}$ or V^{∞}	The partial molar volume at infinite dilution
V_φ^0 or $V_{\varphi,i}^{\infty}$	The apparent molar volume at infinite dilution
Φ_V^0	Limiting apparent molar volume
P_{i}	Perturbed volume
β_S or k_S	Isentropic compressibility
β_T or k_T	Isothermal compressibility
k_S^E	Excess isentropic compressibility
$K_{S,m}$	Molar isentropic compressibility
$K_{S,m}^E$	Excess molar isentropic compressibility
k_φ or φ_{K_s}	Apparent molar isentropic compressibility
φ_{K_T}	Apparent molar isothermal compressibility
$\varphi_{K_S}^0$	Limiting apparent isentropic compressibility

(*continued*)

S. Zhang et al., *Physicochemical Properties of Ionic Liquid Mixtures*,
DOI 10.1007/978-94-017-7573-1

$\varphi_{K_T}^{0}$	Limiting apparent isothermal compressibility
Δk_s	Deviations in isentropic compressibility
α	Coefficient of thermal expansion
α_p	Isobaric thermal expansivity
$^{\Phi}E$	Apparent molar expansibility
E_{φ}^{0}	The infinite dilution apparent molar expansibility
Φ_E^0	The limiting apparent molar expansivity
P	Pressure
P_{i}	Internal pressure
$RD\ (p)$	The relative deviation of pressure
Δp	Experimental uncertainty
η	Viscosity
η^{o}	Defined $\eta^{\mathrm{o}} = 1$ mPa·s
$\Delta\eta$	Viscosity deviation
$(\ln\eta)^{\mathrm{E}}$	Excess logarithm viscosity
$\Delta\ln(\eta/\eta^{\mathrm{o}})$	The viscosity logarithm variations
α_{21}	The relative volatility
B_V	Deviation parameter
$\mathrm{S_E}$	The errors in the limiting properties
S_{K_S}	The isentropic experimental limiting slope
S_{K_T}	The isothermal experimental limiting slope
σ	Surface tension
δ_σ	Surface tension deviation
Φ_L	The relative apparent molar enthalpy
H^E	Excess enthalpy/excess molar enthalpy
$H_{1,2}^{E,\infty}$ or $H_i^{E,\infty}$	Partial molar excess enthalpy at infinite dilution
$\Delta H_1^{E,\infty}$	Excess molar enthalpy at infinite dilution
ΔG^{E}	Excess Gibbs free-energy
$\Delta \mathrm{G}^{*\mathrm{E}}$	Excess free energy of activation of viscous flow
$\Delta_{\mathrm{sol}}H_m^0$	The standard molar enthalpy of solution
$\Delta_{\mathrm{sol}}G_m^0$	The standard molar Gibbs energy of solution
$\Delta_{\mathrm{sol}}S_m^0$	The standard molar entropy of solution
$\Delta_{\mathrm{svt}}H_m^*$	The local molar enthalpy of solvation
$\Delta_{\mathrm{svt}}G_m^*$	The local molar Gibbs energy of solvation
$\Delta_{\mathrm{svt}}S_m^*$	The local molar entropy of solvation
$\Delta_{\mathrm{solv}}S^{\infty}$ or $\Delta_{\mathrm{solv}}S$	The entropy of solvation
$\Delta_{\mathrm{solv}}G^{\infty}$ or $\Delta_{\mathrm{solv}}G$	The Gibbs energy of solvation
$\Delta_{\mathrm{solv}}H^{\infty}$ or $\Delta_{\mathrm{solv}}H$	The enthalpy of solvation
$\Delta_{svt}H_m^0$	The standard molar enthalpy of solvation
$\Delta_{svt}G_m^0$	The standard molar Gibbs energy of solvation
$\Delta_{svt}S_m^0$	The standard molar entropy of solvation
ΔH_{mix}	The heat of mixing per mole of solution
ΔG_{mix}	The Gibbs free-energy of mixing per mole of solution
ΔS_{mix}	The entropy of mixing per mole of solution

(continued)

$\Delta_{sil}H^o$	The standard enthalpy of solution
$\Delta_{sil}G^o$	The standard Gibbs free-energy of solution
$\Delta_{sil}S^o$	The standard entropy of solution
$(S_1 - S_1^0)^E$	The excess partial molar entropy of the solvent
$(S_2 - S_2^0)^E$	The excess partial molar entropy of the solute
K_H	Henry's law constant
ΔK_H	Experimental uncertainty
S	Selectivity
S_{ij}^{∞}	Selectivity at infinite dilution
$\alpha_{2,1}$	Separation factors
λ	Thermal conductivity
n_D	Refractive index
$n_{\mathrm{D}}^{\mathrm{id}}$	Ideal refractive index
Δn_D	Deviations in the refractive index
$\Delta_{\Phi}n$	Deviation from ideal refractive index
R_{m}	Molar refraction
ΔR	Deviations in molar refraction
C_{P}	Molar heat capacity
$C_{\mathrm{p,m}}^0$	Ideal heat capacity
C_p^E	Excess molar heat capacity
$\bar{C}_{p,m,i}^{E,\infty}(T)$	The excess partial molar heat capacity at infinite dilution
f	Fugacity
$\Delta\ (f_1/m_1)$	Experimental uncertainty
φ_i	Fugacity coefficients
τ_f	The observed lifetime
k_{nr}	The non-radiative
K_{a}	Ionic association constant
R	Distance parameter
T_{g}	Glass transition temperature
T_{S}	Critical temperature of phase separation
T_{d}	Disappearance of the second liquid phase on heating and temperature
T_{P}	Critical temperature of phase separation
T_{c}	Crystalline temperature
T_{m} or T_{fus}	Melting temperature
T_{e}	Eutectic temperature
T_{a}	Appearance of the second liquid phase on cooling
b_{V}	Empirical parameter
E_a	Activation energy
u	Speed of sound
u^E	Excess speed of sound
k	Electrical conductivity
Λ	Molar conductivity
Λ_{o}	Limiting molar conductance
$\sigma\ (\Lambda)$	Standard deviation of electrical conductivity

(*continued*)

a	Activity
γ	Activity coefficient
$\gamma_{\pm}$	The mean molar activity coefficient
γ^{∞}	Activity coefficient at infinite dilution
Φ	Osmotic coefficient
D	Diffusivity
$s\ (D_{FC})$	Standard deviation for diffusion coefficient
β, D_2	Distribution ratio
K_i	Distribution coefficient
k_j^{∞} or k^{∞}	Capacity at infinite dilution
A	Slope of tie lines
Γ	Interfacial tensions
Φ_{f}	The quantum yields
k_{r}	The radiative
τ_{rot}	The rotational relaxation time
ϵ	Dielectric constant
G	Gordon parameter
S_V^{DH}	Theoretical Debye-HüCkel limiting slope
ϵ_r	Relative permittivity

GPSR Compliance
The European Union's (EU) General Product Safety Regulation (GPSR) is a set of rules that requires consumer products to be safe and our obligations to ensure this.

If you have any concerns about our products, you can contact us on

ProductSafety@springernature.com

In case Publisher is established outside the EU, the EU authorized representative is:

Springer Nature Customer Service Center GmbH
Europaplatz 3
69115 Heidelberg, Germany

www.ingramcontent.com/pod-product-compliance
Ingram Content Group UK Ltd.
Pitfield, Milton Keynes, MK11 3LW, UK
UKHW021557270926
475869UK00011B/214
* 9 7 8 9 4 0 1 7 7 5 7 1 7 *